MILES AIRCRAFT

THE EARLY YEARS

The Story of F G Miles
and his Aeroplanes
1925-1939

Peter Amos

An AIR-BRITAIN Publication

Copyright 2009 ©
Peter Amos
and Air-Britain (Historians) Ltd

Published in the United Kingdom by:
Air-Britain (Historians) Ltd

Website: www.air-britain.co.uk
for full details of publications,
membership and services.

Sales Dept:
41 Penshurst Road, Leigh,
Tonbridge, Kent TN11 8HL, England
Sales enquiries: sales@air-britain.co.uk

Correspondence regarding this book to:
Peter Amos
4 Castle Bungalows
Storrington, Pulborough
West Sussex RH20 4LB

And/or to the dedicated e-mail address:
milesaircraft@talktalk.net

All rights reserved. No part of the contents of this publication may be reproduced, stored in a retrieval system or transmitted in any form by any means, electronic, mechanical, photocopying, recording or otherwise without the prior permission of the author and Air-Britain (Historians) Ltd.

ISBN 978-0-85130-410-6

Edited by Malcolm Fillmore

Origination by Howard Marks, Winchelsea

Cover design by Dave Partington

Printed and bound in Poland:
www.polskabook.pl

Front Cover:
Miles M.2W Hawk Trainer Mk.II G-ADWT of 8 ERFTS overhead Reading Aerodrome, Woodley. Original painting by Les Vowles

Back Cover:
Original Miles Hawk transfer [Via The Miles Aircraft Collection]

Contents

Contents		3
Foreword		5
Background and Acknowledgements		6
Abbreviations		10
Preface and Introduction		11
Chapter 1:	The Early Days - F G Miles at Shoreham, Sussex - 1925 to 1931	15
Chapter 2:	The Hornet Baby and Southern Martlet	49
Chapter 3:	The Southern Metal Martlet	53
Chapter 4:	Test Flying for George Parnall & Co, Yate - 1929 to 1932	55
Chapter 5:	The Miles M.1 Satyr	57
Chapter 6:	Reading - 1932 to 1939	61
Chapter 7:	Miles M.2 Hawk (and variants)	97
Chapter 8:	Miles M.2E Hawk Speed Six (and variants)	111
Chapter 9:	Miles M.2F Hawk Major (and variants)	121
Chapter 10:	Miles M.2R Hawk Trainer (and variants)	133
Chapter 11:	Miles Pusher	139
Chapter 12:	Miles M.3 Falcon and M.3A Falcon Major	141
Chapter 13:	Miles M.3B Falcon Six (and variants)	149
Chapter 14:	Miles M.4 and M.4A Merlin	161
Chapter 15:	Miles M.5 and M.5A Sparrowhawk	167
Chapter 16:	Miles M.6 Hawcon	179
Chapter 17:	Miles M.7 and M.7A Nighthawk	183
Chapter 18:	Miles M.8 and M.8A Peregrine	189
Chapter 19:	Miles M.9 RR PV Trainer and M.9A Master	199
Chapter 20:	Miles M.9B Master Mk.1	211
Chapter 21:	Miles M.10 Queen Wasp	219
Chapter 22:	Miles M.11 Whitney Straight Special and M.11A Whitney Straight (and variants)	220
Chapter 23:	Miles X.2 Transport Project	235
Chapter 24:	Miles M.12 Mohawk	243
Chapter 25:	Miles Marathon Transport Project	255
Chapter 26:	Miles M.13 Hobby	256
Chapter 27:	Miles M.14 Magister (and variants) and M.14 Hawk Trainer Mk.III (and variants)	263
Chapter 28:	Miles M.15 Trainer	275
Chapter 29:	Miles M.16 Mentor	281
Chapter 30:	Miles M.17 Monarch	285
Chapter 31:	Miles M.18 Mks.I,II, III and IV	291
Appendix 1:	Joyrides - 5/- Flights	299
Appendix 2:	F G Miles' Letters of 1927	300
Appendix 3:	Fields used by F G Miles for joyriding – 1926 to 1931	303
Appendix 4:	The Shoreham Days - F G Miles Aeroplanes from 1925 to 1931	303
Appendix 5:	Southern Martlet and Metal Martlet Production	309
Appendix 6:	The Sevenoaks Days - Miles' flying from Penshurst – 1931 to 1932	313
Appendix 7:	Blossom Miles' flying – 1930 to 1935	314
Appendix 8:	Phillips & Powis - The Pre-Miles days	315
Appendix 9:	Miles M.2 Hawk (and variants) Production	317
Appendix 10:	Miles M.2E Hawk Gipsy VI, M.2L and M.2U Hawk Speed Six Production	324
Appendix 11:	Miles M.2F Hawk Major (and variants) Production	327
Appendix 12:	Miles M.2R Hawk Trainer Mk.I and M.2W to M.2Z, Hawk Trainer Mk.II Production	339
Appendix 13:	Miles M.3 Falcon and M.3A Falcon Major Production	343
Appendix 14:	Miles M.3B Falcon Six (and variants) Production	347
Appendix 15:	Miles M.4 and M.4A Merlin Production	352
Appendix 16:	Miles M.5 and M.5A Sparrowhawk Production	352
Appendix 17:	Miles M.7 and M.7A Nighthawk Production	355
Appendix 18:	Miles M.9 Rolls-Royce PV Trainer - Brochures	357
Appendix 19:	Miles M.11 Whitney Straight (and variants) Production	359
Appendix 20:	Miles M.14 Magister (and variants) and M.14 Hawk Trainer Mk.III (and variants) Production	367
Appendix 21:	Miles M.16 Mentor Production	407
Appendix 22:	Miles M.17 Monarch Production	409
Appendix 23:	Miles M.18 Production	411
Appendix 24:	The Drawing Office and Experimental Department in 1937	413
Appendix 25:	Under "B" Conditions - "B" marks used between 1934 and 1940	414
Appendix 26:	Miles M.14 Magisters in Egypt	416
Appendix 27:	Miles M.14A Magisters in Eire	418
Appendix 28:	The EMK Miles M.2 Hawk Proposal	419
Appendix 29:	Agreement between Phillips & Powis Aircraft Ltd and Rolls-Royce Ltd	420
Appendix 30:	No 8 Elementary & Reserve Flying Training School, Woodley	421
Appendix 31:	Summary of Constructors Numbers (Pre-War)	423
Appendix 32:	Registration, Type and C/n Cross Reference	424
Bibliography		428
Index		429

A pleasing study of F G Miles at his desk in 1942.

Foreword

Peter Amos has written what must be the seminal book on Miles Aircraft. It is more than just a catalogue of Miles aeroplanes and other achievements but is a history of a remarkable enterprise. Its historical completeness is illuminated by anecdotes and "side-stories" which provide context and humour, sometimes surprising.

I was born into that world and it is only now, with the perspective of hindsight, that I see just how remarkable my father and the business as a whole were. Things which I took completely for granted were, I now realise, quite out of the ordinary run.

Many people know of the Miles Magister and Master trainers on which so many of the pilots in the 1940s learned to fly and, since recent television coverage, of the Miles M.52 supersonic aircraft tragically cancelled shortly before it was due to fly. What may not be so well known is that Miles produced the first commercial photocopier, developed (although he did not invent) the ball-point pen, and developed chipboard as a way of using up the waste wood shavings from the manufacture of aeroplanes.

There have been two just previous books on Miles Aircraft. *The Book of Miles Aircraft* complied by A H Lukins, which was published by Harborough in 1946, with its classic foreword (reprinted in this book) by C G Grey, the editor of *The Aeroplane* weekly magazine; and Don Brown's *Miles Aircraft since 1925*, published by Putnam in 1970, in which Don, characteristically, wrote his own foreword!

The first of these books was one of three catalogue-style books on the aeroplanes of various manufacturers (the others being Westland and Bristol) whilst Don Brown's book fleshed out the bones with history and anecdotes - many of them personal.

I have fond memories of Don who had appointed himself as a sort of godfather to me. He told the story of giving me flying lessons - I must have been 12 or 13 at the time - and when he told me that I was climbing too steeply after take-off I apparently said that Miles always climbed like that! I also recall that he ended all letters to government departments "You remain, Sir, my humble and obedient servant"; it was some years before anybody noticed!

I have known Peter for some fifteen years, ever since he founded 'The Miles Aircraft Collection'. His interest in all things Miles - as he explains in the book - started long, long before that. His knowledge is encyclopaedic and his determination to track down and verify every fact is impressive. He constantly unearths stories and titbits about Miles activities from the obscurity in which they have lain for many years and which are a real source of delight to any reader. Not only has he brought the Miles story up to date but he has filled in fascinating elements which were not included in the previous books - correcting a few errors on the way - and which light and warm the whole story.

I hope that you will find Peter's book as good a read as I have - there is so much that I did not know, even having been born into the family - as well as a long-lasting source of reference on all things Miles.

Jeremy J Miles
October 2008

This book is dedicated

to the memory of

the early Shoreham 'Enthusiasts',

the Phillips & Powis 'Pioneers',

the Miles family and

Bert Clarke

Background and Acknowledgements

Many of the people named by Don Brown in the acknowledgements to his book, *Miles Aircraft since 1925*, were also known to myself and to these and the many other friends and colleagues around the world who have helped with the preparation of this book, I extend my thanks.

It is very difficult to mention all by name but before I start, I feel that Don should not have forgotten to mention E J Riding and A J Jackson in his acknowledgements as they contributed so much detailed history on the aircraft and provided many of the photographs used in his book. If I have omitted to mention any of my friends and colleagues, then I would ask them to please accept my most sincere apologies as, without their help over so many years, this history of Miles and their aircraft - the British aeroplane manufacturer with the most interesting history of all time - could never have been told.

Firstly, I would like to offer my most sincere thanks to the late Bert Clarke. Without his invaluable help over nearly 40 years I could never have made it. When struggling with a particular problem he would always come to my aid with a logical explanation from his meticulous researches. In my opinion and regardless of what he would say to the contrary (Bert was a very modest chap), he really was **the** historian of Woodley Aerodrome, Phillips & Powis and Miles Aircraft, certainly up to 1939.

Bert's father, Mr A Clarke, was, in September 1932, the 8th person to join Phillips & Powis Aircraft (Reading) Ltd and this qualified him to be one of the *Pioneer Employees*. It is of great interest to note that the next two to join the firm were none other than F G Miles and Mrs Blossom Miles, who became the 9th and 10th *Pioneer Employees* respectively!

Mr Clarke not only helped to set up the original Stores Department in 1932 but, together with his wife, ran the works canteen in a small wooden shed from 1933 until 1935, before returning to stores work. He retired from Western Manufacturing Estate Ltd (Miles Aircraft's successors) in 1952. Incidentally, the original canteen shed survived in Bert's back garden until his death and plans to move it to The Museum of Berkshire Aviation were only abandoned due to its rather fragile state.

Bert was born in *Eleven Elms* cottage in Tippings Lane shortly after Charles Powis acquired the *Hundred Acre Field* site to the west at the end of 1928. This site, plus some 30 acres which Powis later purchased, became Reading Aerodrome. The *Falcon Hotel*, which was built opposite *Eleven Elms* cottage in 1937, was deliberately badly damaged by fire in the late 1980s and demolished later to make way for a housing estate - but the cottage still stands.

As a child, Bert had the run of Woodley aerodrome and this started his lifelong interest in the activities of the firm, the personalities and of course, the aeroplanes. My correspondence with Bert started on 8th May 1967 and we corresponded regularly until his untimely death on 7th September 2004. It is to my eternal regret that he was denied the opportunity to see this book to which he had contributed so much. In his later years he checked every chapter as it was drafted - from the early days at Shoreham and Woodley to the history of each individual aeroplane produced by the firm up to the outbreak of war in September 1939.

In sharing his lifetime's study of the daily activities at Woodley Aerodrome and the correspondence he had with both F G Miles and Don Brown, we have been able to dispel many of the myths, mysteries and published 'misinformation' which has been written on the subject of Miles Aircraft over the years.

Bert also became closely acquainted with A H Lukins of the technical publications department of Phillips & Powis Aircraft Ltd and Miles Aircraft Ltd and who was the editor of the first pre-war *Miles Magazine*. In 1945, Lukins compiled the *Book of Miles Aircraft*, published by Harborough Publications and which contained excellent three-view drawings of Miles aircraft. Thanks to Bert, these drawings can now also be reproduced in this history, together with representative photographs of original Phillips & Powis three-view general arrangement drawings of some of the early Miles aircraft.

Words cannot adequately describe the most generous help, encouragement and continued support given to me by my very good friends Mary and Terry Kearney in the preparation of this work and, were it not for their help, I very much doubt if this book could ever have been written. I was using a typewriter in the early days of my writing, a nightmarish tool for such a task, so they not only very kindly provided me with a computer but also, very patiently, taught me how to get the best out of it!

My especial thanks go to F G's son Jeremy Miles for kindly offering to let me peruse his father's archives and for his help in adding to our knowledge of the Miles Group of Companies at Shoreham from the early 1950s onwards. Jeremy worked for F G Miles Ltd on the design of the Student etc and in 1963 became Managing Director of Miles Marine and Structural Plastics Ltd and then, in 1965, Commercial Director of Miles Electronics Ltd. I am most grateful to Jeremy for reviewing my draft history of that period and for kindly agreeing to write the foreword to this volume.

In the early 1980s, Rod Simpson, Editor of Air-Britain *Aviation World* (late Air-Britain *Digest*), was visiting Mr Clifton Lawrence Culling Nash, the Company Secretary of the Adwest Group plc, Woodley (in 1948, the name of Western Manufacturing Estate Ltd was changed to Adwest Group) on business. Mr Nash had joined Phillips & Powis in October 1933 as the 34th employee and inevitably the subject of Miles aircraft was raised. Rod discovered that, not only was Mr Nash interested in Miles aircraft but he had also kept records of the aeroplanes as they were built up to just after the outbreak of war. These he copied by hand and kindly sent me a copy of his exciting find. These records helped to fill many of the previously tantalising gaps in the early production records. Rod was also instrumental in saving many of the firm's photographs and negatives which were about to be destroyed during a works improvement.

The small hardback notebooks in which Mr Nash had kept his records had been professionally printed, with the columns being headed:

MACHINE NO. REGN. JOB NO. TYPE CUSTOMER COLOUR ENGINE DELIVERED

Although his records apparently started at Machine No.1, there were a few anomalies in his records prior to his joining the firm and, to complete his records, he obtained details of the earlier aeroplanes from other and somewhat doubtful sources. However, from October 1933, his accurate records have provided much information on the construction, customers and delivery dates of aircraft built prior to the outbreak of war. Shortly after that his records became less detailed as production intensified and the keeping of records of this nature undoubtedly became more difficult but these notebooks and the one surviving "Works Records" book which he saved have proved to be invaluable reference sources.

After his retirement in 1955, Mr Nash decided to write the history of Phillips & Powis/Miles Aircraft but died in 1985 before he could complete it and unfortunately the manuscript has never been found.

In 1985, Julian Temple was commissioned by the Adwest Group plc to write a book on Miles Aircraft and Adwest at Woodley and wrote articles in local newspapers to obtain memories and

Background and Acknowledgements

memorabilia from former employees and friends of the firm. The purchaser of Mr Nash's house fortuitously saw one of these articles and, having found Nash's records and the Works Record book under a seat in the summerhouse, met Julian and presented him with them and some other documents.

Shortly after this I met Julian and became deeply involved with his researches, the result of which was the book entitled *Wings over Woodley*, published by Aston Publications in 1987. As a result of our mutual interest in Miles Aircraft, Julian and I have become close friends and I cannot thank him enough for his help and encouragement with this epic tome.

Don Brown was not aware of Nash's records when he wrote his book - he did not join Phillips & Powis until 1942 having been employed before the war in the gas industry in Sussex. He somehow managed to 'escape' from his employment to join the Ministry of Aircraft Production at the A&AEE Boscombe Down, where he worked on the preparation of pilot's notes. This did not stop him from keeping in touch with Miles' activities at Woodley, but he was not privy to all that went on there during those hectic intervening years.

Don relied on his excellent memory for his writings but others for aircraft designations, registration details and what has since been found to be erroneous first flight dates. These were accepted at the time as 'gospel' being based on the information which was then available. However, thanks to many years of research, many of these 'errors and omissions' have now been corrected. This research benefited from studying many surviving pilot's log books, including the majority of those of F G Miles, three of George Miles and all of Blossom Miles, Don Brown, assistant test pilot Flt Lt Hugh Kennedy, pre-war instructor R E M B 'Bob' Milne, Wing Cdr John F Moir (Officer Commanding the Phillips & Powis operated Reserve Flying School at Woodley), Sqn Ldr Samuel E Esler and Sqn Ldr James C Nelson.

From these log books has come a veritable wealth of information including, from Hugh Kennedy, final and positive confirmation of the existence of the M.19 Master Mk IIs produced at Sheffield and Doncaster. Thanks to Peter Clegg for discovering the log books of Sqn Ldrs Esler and Nelson - these have provided so much useful information on postwar Miles production. My grateful thanks to Hugh Kendall who, in 1996, three years before he died, very kindly donated four of his log books appertaining to the time he was with Miles Aircraft and Handley Page (Reading) Ltd to *The Miles Aircraft Collection*. Hugh Kendall (not to be confused with Hugh Kennedy) was with Miles from September 1945 until the end and his log book was signed by the last chief test pilot, Ken Waller, on 31st March 1948. Hugh Kendall then became a Ministry-approved freelance test pilot and later returned to Woodley as a test pilot for Handley Page (Reading) Ltd.

Unfortunately, the log books of Tommy Rose were lost during a succession of house moves but those of Ken Waller's have since been found and await examination.

My most grateful thanks to Dennis Bancroft and to his wife Elizabeth for their help and encouragement in so many ways. Dennis was the chief aerodynamicist of Miles Aircraft Ltd and one of the main progenitors behind the design of the M.52, for his help in greatly increasing our knowledge of that aircraft and his unswerving faith in the undoubted success that it would have been, had it not been cancelled so near to completion. The M.52 was the first supersonic aircraft designed from the outset to have the all-moving, variable incidence tailplane and Dennis has since written the real story of its development. He also spent a considerable amount of time in unravelling the truth behind the cancellation of the project which would, if it had it only been completed and flown, undoubtedly have been successful in its objective of achieving supersonic flight. Dennis later joined F G Miles Ltd at Redhill in December 1948, taking a leading role in setting up the successful plastics division there and later at Lancing, and helped considerably with this and many other Miles projects.

To Capt Eric 'Winkle' Brown for his very interesting reminiscences of flying so many different types of Miles Aircraft and who, in 2008, volunteered to take on the unenviable task of writing the definitive story of the M.52.

To David Lockspeiser who was a student with the Miles Technical School at the time of the cancellation of the Miles M.52 and who, with Eric Brown, tried to persuade David's father, Ben Lockspeiser, (later Sir Ben) not to cancel the Miles M.52. I later came to know David very well while we were both working at Dunsfold - he as a test pilot with Hawker Aircraft and I with the Flight Development Department.

To Jim Pratt, George Miles' son-in-law, who has been a great help with providing valuable information from George's archives and for persevering with his research into the M.52 from the American angle.

To what was, until recently, the Adwest Group plc, Woodley for granting Julian Temple unlimited access to their archives while he was researching *Wings over Woodley*. To a protégé of Don Brown, who wishes to remain anonymous, for his invaluable help with many Miles matters and for allowing me to grant Don Brown's lifelong wish to publish the true facts behind the collapse of Miles Aircraft Ltd. To David Freeman, a lifelong friend of mine who emigrated to Australia many years ago, for his continued interest in Miles aircraft and for providing copies of the history cards for the Miles aircraft which were impressed into the Royal Australian Air Force during the war.

To Fred Lynn who I first met on the River Adur embankment at Shoreham Airport soon after the end of the war; Roy Brooks, renowned Sussex aviation historian; the late Sylvia Adams, avid researcher and historian into the life of the Pashley brothers with especial emphasis on Cecil L Pashley who taught her to fly; Tim Webb, joint author, with the late Dennis Bird, of the book *Shoreham Airport Sussex*, which was published in 1996, for their help with the background to the early activities of Miles at Shoreham. It was Tim who finally confirmed the precise boundaries to all the various airfields at Shoreham, from before the first war, to Easter's Field and on to the present day and my thanks to him for letting me to reproduce this map to clarify the locations from where Miles operated in the early days.

To my lifelong friend, the late Chris Dearden who learned to fly with the Redhill Flying Club in 1947 on 'Maggies' (Hawk Trainer Mk.III's really!) and held a genuine pilot's 'A' Licence enabling him to fly even four-engined aeroplanes! Chris loved all British light aeroplanes but naturally had a very soft spot for Miles aircraft. To his late widow Dorothy, my most sincere thanks for very kindly giving me the exclusive use of a large room in her bungalow in which to sort and collate some 4,000 plus photographs with which to illustrate this book.

To the local newspapers of Swindon for publishing my request for information on the Phillips & Powis factory at South Marston and to the ex-employees that responded. The first of these was Ron Maxfield, one of the pioneers sent there from Woodley to help set up the factory. Ron kindly provided much valuable information on the site and also introduced me to many others who had worked at South Marston and, with their help, the story of the first Phillips & Powis shadow factory at Swindon unfolded.

To the late Michael Stroud, who started his career with Miles Aircraft in their London Sales Office, for his help with the post-war period at Miles. To my many friends and specialists in Air-Britain, including John Havers for his most generous help with Miles aircraft in the Middle East and for his mutual love of the Miles Aerovan and also for managing to find photographic 'gems' of Miles aircraft from the early days at Shoreham. To Bernard Martin, Malcolm Fillmore, James Halley, Dave Partington, Ian Callier, Peter Green, Phil Jarrett, John Davis, Fransisco Halbritter, Luc Wittemans, Vic Smith, Ken Tilley and the late Jack Meaden - all for their invaluable help with Miles aircraft on the British and overseas civil registers; To Phil Butler for his patient and unstinting help in transcribing the RAF record cards and answering many queries. To Phil Jarrett for kindly lending me many hitherto unpublished photographs and to the late Ray Sturtivant for helping to unravel details of Air Ministry contracts and commenting on the detailed serial listings. To Terry Judge in Canada for his research in the National Archives of Canada and the wonderful album of Miles aircraft photographs which he kindly donated.

To the late Gordon Swanborough for lending me his superb collection of negatives of Miles aircraft which he had taken at Woodley on the occasion of the last Miles Open Day on 20th July 1947 and other post-war visits to Woodley. To Mike Hooks for access to his photographs. To the late Geoffrey Alington and his family for publishing his memoirs in the most fascinating book *A Sound in the Sky* from which I have taken the liberty of using extracts on his flying of Miles aircraft.

To the late Alex Henshaw for sharing his delightful memories of flying Miles aircraft and for kindly letting me use extracts from his book *Wings Across The Great Divide - Postwar flying in Africa in the 1940s*. To Martin Barraclough for kindly allowing me to reproduce his wonderful memories of flying Miles aircraft in Kenya and that truly memorable flight home to England in the Messenger VP-KJL.

My especial thanks to Josh Spoor, for his unbounding generosity and his perseverance in researching the records held at the National Archives (late Public Records Office) in such great depth to seek the truth behind the cancellation of the M.52. His knowledge of their catalogues seems to be second to none and I only have to ask him and nothing is ever too much trouble. To the late Matthew Nathan who worked for Power Jets Ltd in the early days and designed the combustion chambers for the W.2/700 jet engine which was to have powered the M.52. Matthew was also a keen photographer and were it not for him, the early records of the work undertaken at Power Jets would have been lost forever. He also took the only known colour film of the engine being run on the test bed at Ansty and assembled a montage of coloured photographs of the M.52 from the design to the mock-up, thus recording a valuable piece of aeronautical history for posterity. To the late Jim Hodge, designer of the W.2/700 engine, who introduced me to Jack Whiting who had helped to manufacture the W.2/700; special thanks to Jack for allowing me to copy the video of the original cine film of Power Jets activities that Matthew took.

To the late Richard Almond, Air Traffic Control Officer at Shoreham, for being instrumental in saving so many of George Miles' documents from destruction after he left in the 1960s; Bob Ruffle, another Air Traffic Control Officer at Shoreham, for his help with the Magisters sold to Latvia and Estonia; Greta F Felixson and Sigurjon "John" Valsson for their help with Miles aircraft in Iceland and to Phillip Treweek for his help in acquiring copies of the official records of all Miles aircraft registered in New Zealand.

To the late Grahame Gates who joined Phillips & Powis in 1942 as a weights engineer and stayed until the end in late 1947, for sharing his memories and records from those days and for his help with the Miles story at Shoreham, for whom he worked in the 50s and early 60s. Without his help, the very complex history of this era could not have been written but fortunately Grahame was there and kept detailed records of this and of his later involvement with Beagle-Miles from 1961 to 1964.

To Frank C Langmaid, manufacturing manager of F G Miles Engineering Ltd (later with Hunting Communication Technology Ltd), who took over where Grahame Gates left off and provided a very detailed insight into the companies sired by F G. Thanks to Frank and to Jeremy Miles also for their invaluable help with the recovery of the wind-tunnel model of the M.52, which they duly presented to The Miles Aircraft Collection. This last surviving 'piece' of the M.52 is now on display in The Museum of Berkshire Aviation at Woodley. My thanks also to Mrs Skinner, widow of Bill Skinner's son, for kindly donating the pre-war photograph album of chief test pilot Bill Skinner to The Miles Aircraft Collection.

To Terry Wilson for kindly lending me her collection of Phillips & Powis/Miles documents and drawings and for helping with details of costings for various Miles aircraft and projects; David Legge for the background history to Miles Aviation & Transport (Research & Development) Ltd at Ford Aerodrome, Sussex. To the late Graham H R Johnson, a lifelong Magister enthusiast and author of lovely little book *The Miles Magister*. He helped me considerably with its history and also restored the famous Maggie G-AFBS for the Skyfame Museum at Staverton (now statically displayed with the Imperial War Museum at Duxford). Graham and a group of friends then built an authentic Magister replica, to which he gave the serial *L6906* and, following negotiations between Graham and I, this magnificent specimen was presented to Julian Temple of The Berkshire Aviation Group at the Air-Britain Fly-in at Wroughton in June 1987 and can now be seen in The Museum of Berkshire Aviation at Woodley.

To Adrian Brook for completing the restoration of the Miles Magister currently flying as N3788 and for so kindly storing The Miles Aircraft Collection's airframes over the years. The late Ron Paine, for his love of Miles aeroplanes in general and 'his' G-ADGP in particular; to Monty Cook, who worked for Phillips & Powis from before the war and for Miles Aircraft later, who recalled so many valuable memories of his times at Woodley, as did Des Armour who, as a young man there, witnessed so many interesting happenings, not least of all the first of the batch of ex RAF Magisters being prepared for delivery to Turkey; Dave Welch for continually finding rare Miles photographs; Richard Riding (E J Riding's son) and to the late Roger Jackson (A J Jackson's son), for allowing me the use of their late fathers' photographs; Ken Hearn, Lloyd Robinson, Richard Payne, Tom Singfield, the Air Attachés of Chile and Sweden and Louis S Casey, who started the restoration of the Miles Mohawk.

To Peter Campbell for his help with Miles' activities at Shoreham in the 1950s and his delightful *Tails of the Fifties* trilogy, which recorded some fascinating stories from many Miles aircraft owners and pilots in the austere post war period. Thanks also to Peter and Terry Booker for organising the annual Great Vintage Flying Weekends (G-VFWE), which gave 'us Miles enthusiasts and lovers of old British light aeroplanes' something to look forward to every year.

To Phil Spencer of *The British Roundel* magazine for his meticulous research into early Air Ministry Contracts; the late Barry Abraham for kindly sharing his researches into South Marston and the MAP Civilian Repair Organisations; Philip Birtles for his help with the later noise-abatement trials with the Miles Student; Jeff Hargrave for first bringing the surviving Hawk Major in Montevideo to my attention. To Mike Stowe for his help with accidents to Miles Masters in service with the USAAF and USAF in this country during the war.

To Jean Fostekew, one of the last students of the Miles Technical School, for her help with the background details of the school and for her masterpiece *Blossom - A Biography of Mrs F G Miles*. I would also like to take this opportunity to extend thanks to her and her husband Ken for their tireless efforts, a true labour of love in helping to 'run' The Museum of Berkshire Aviation at Woodley and for their valiant efforts in helping to preserve (amongst all the other Berkshire aviation interests) the memory of Miles Aircraft and for sharing with me the 'gems' of information which are handed in to the museum.

John S Webb, founder of the Reading Sky Observer's Club at Woodley also deserves a mention for being one of the first to help preserve the spirit of Miles Aircraft and for being instrumental in the formation of Air-Britain.

To Peter Holloway for his infectious enthusiasm for Miles Aircraft, which culminated in him once actually owning **three** Miles aircraft at the same time, a Falcon Major, a Magister and a Messenger, and for undertaking a memorable pilgrimage to Sweden in 2002 in Falcon Major G-AEEG to celebrate its early life in Sweden. Thanks also to Peter for also helping in many other ways.

I am also indebted to *The Aeroplane*, its sister publication *The Aeroplane Spotter*, *Flight* and many other aviation magazines for the reproduction of contemporary news items, articles and advertisements on Phillips & Powis and Miles Aircraft. To Dave Robertson for making a study of aviation advertisements and sharing with me those he found on Miles aircraft.

To Tony Wingfield, a member of The Miles Aircraft Collection, who kindly volunteered to proof read the draft of this book, I extend my most grateful thanks. When I realised that this would be an unenviable task, I knew that it was not going to be easy to find someone, not only with a knowledge of the subject and who would also be keen enough even to want to undertake the job, but also someone who would be able to afford the time to undertake it. Tony, who was born and bred at Winnersh, within the circuit of Woodley, cultivated a love of Miles aircraft and got to know the pilots and others from the airfield who used his father's garage there. He also later learned to fly there with 8 RFS after the war, so he was well qualified for the job. As he was retired he thought that this would give him something to do during the long winter nights - it sure did! Tony recently reminded me that he started this mammoth task in January 2001 and was still involved in February 2006, following the many changes which have had to be made following the continual discovery of yet more material. In fact, I am sure that Tony will probably be the only person (apart from the long-suffering editor) to have read the whole book, appendices and all! Three changes of publishers have resulted in much effort in altering the presentation and, as a result of the ever increasing size of the content, this book can of necessity cover only the pre-war part of the story. It has been a very big undertaking and I would also like to thank his wife Pat who, with her knowledge of grammar, also helped to make the text that much easier to read.

To Les Vowles for very kindly painting the most evocative picture that adorns the front cover.

You will have noticed references to *The Miles Aircraft Collection* and it was Mary and Terry Kearney who, together with Julian Temple, Adrian Brook and myself, decided to form this club for Miles Aircraft enthusiasts on 29th March 1993, in celebration of the 60th Anniversary of the first flight of the first Miles M.2 Hawk. Were it not for Mary and Terry's encouragement, this long overdue 'Club' would never have been formed. Since then, *The Miles Aircraft Collection* has grown from strength to strength and, thanks to their continued help, will hopefully continue to grow and flourish.

Thanks also to all members of *The Miles Aircraft Collection* for their continuing support and patience while waiting for this book to finally appear.

Finally, to my long suffering wife Doris, who has had to put up with me and my 'fanaticism', as she calls it, for the past 40-odd years while I worked on this treatise in every available spare moment, I can but proffer a very special vote of thanks.

And last, but by no means least, a very special thanks to F G Miles, his wife Blossom, Charles O Powis, George H Miles and Don L Brown - the first Miles Aircraft enthusiasts.

Peter Amos
January 2009

King's Cup Air Race Competition, 5th September 1935: "A demonstration took place yesterday at Reading Aerodrome, Woodley, Berks., of the aircraft which is to take part in the King's Cup Air Race which takes place on Saturday. Mr. F G Miles, with his wife and Mr. C O Powis, Managing Director of the Phillips & Powis Aircraft Ltd, are seen with the Miles M.5 Sparrowhawk, which has been specially designed for the race by Mrs. Miles. [Keystone via P Jarrett]

Abbreviations

AACU	Anti-aircraft Co-operation Unit
A&AEE	Aeroplane & Armament Experimental Establishment
AAS	Air Armament School
AASF	Advanced Air Striking Force
AD	Aircraft Depot
AFDU	Air Fighting Development Unit
AFEE	Airborne Forces Experimental Establishment
AGS	Air Gunners School
AM	Air Ministry
AMDP	Air Member for Development & Production
AMRD	Air Member for Research & Development
ANS	Air Navigation School
AOC	Air Officer Commanding
AONS	Air Observer & Navigator School
AOS	Air Observers School
ASU	Aircraft Storage Unit
ATA	Air Transport Auxiliary
ATC	Air Training Corps
ATS	Armament Training Station
AUW	All-up Weight
B&GS	Bombing & Gunnery School
BLEU	Blind Landing Experimental Unit
CAG	Civil Air Guard
Cat	Category (see below)
CF	Communications Flight
CFE	Central Fighter Establishment
CFI	Chief Flying Instructor
CFS	Central Flying School
c.g.	Centre of Gravity
CinC	Commander-in-Chief
Cld	Cancelled
CLN	Clifton L Nash (P&P record keeper)
c/n	Constructor's Number
CofA	Certificate of Airworthiness
Comm	Communications
CofR	Certificate of Registration (registration number/date)
CRD	Controller of Research & Development (UK)
CRD	Civilian Repair Depot (Australia)
CRO	Civilian Repair Organisation
CRU	Civilian Repair Unit
DBF	Destroyed by fire
DBR	Damaged beyond repair
Dd	Delivered
DGRD	Director General of Research & Development
DTD	Director(ate) of Technical Development
EAAS	Empire Air Armament School
ECFS	Empire Central Flying School
EFTS	Elementary Flying Training School
ERFTS	Elementary & Reserve Flying Training School
ETPS	Empire Test Pilots School
E&WS	Electrical & Wireless School
FA	Flying accident
FIS	Flying Instructors School
Flt	Flight
Flt/Lt	Flight Lieutenant
F/O	Flying Officer
FPP	Ferry Pilots Pool
FTC	Flying Training Command
FTS	Flying Training School
FTU	Ferry Training Unit
G/C	Group Captain
GI	Ground Instruction
Grp	Group
GTS	Glider Training School
HAD	Home Aircraft Depot
HGCU	Heavy Glider Conversion Unit
IFTS	Initial Flying Training School
ITP	Instruction to Proceed
LRDU	Long Range Development Unit
MAP	Ministry of Aircraft Production
MCA	Ministry of Civil Aviation
ME	Middle East
MI	Major inspection
MoS	Ministry of Supply
MR	Major repair
MU	Maintenance Unit
NTU	Not Taken Up
(O)AFU	(Observers) Advanced Flying Unit
OTU	Operational Training Unit
PA	Peter Amos (Author)
(P)AFU	(Pilots) Advanced Flying Unit
P/O	Pilot Officer
P&P	Phillips & Powis
PRU	Photographic Reconnaissance Unit
PTS	Parachute Training School
RAAF	Royal Australian Air Force
RAE	Royal Aircraft Establishment
RCAF	Royal Canadian Air Force
Regd	Registered
Regn	Registration
RNAS	Royal Naval Air Station
RNZAF	Royal New Zealand Air Force
RS	Radio School
R&SU	Repair & Salvage Unit
Rts	Reduced to spares
SofAC	School of Army Co-operation
SAAF	South African Air Force
SAS	Servicing Aircraft Section
SFP	Service Ferry Pool
(S)FPP	(Service) Ferry Pilots Pool
SFTS	Service Flying Training School
SOC	Struck off charge
SofP	School of Photography
SofTT	School of Technical Training
Sqdn	Squadron
Sq/Ldr	Squadron Leader
SS	Signals School
Stn Flt	Station Flight
TEU	Tactical Exercise Unit
TFPP	Training Ferry Pilots Pool
TFU	Telecommunications Flying Unit
T/O	Third Officer
TOC	Taken on charge
TSCU	Transport Support Conversion Unit
TSTU	Transport Support Training Unit
TTC	Technical Training Command
TU&RP	Training Unit & Reserve Pool
u/c	Undercarriage
WFU	Withdrawn from use
W/C	Wing Commander

RAF – Damage Categories

A	Repairable on site
Ac	Repairable on site but not by unit
B	Beyond repair on site
E	Write-off
E1	For reduction to spares
E2	For scrap

Preface and Introduction

The book *Miles Aircraft since 1925* was written by Don L Brown OBE, MIMechE, AFRAeS, and published by Putnam in 1970. Don Brown became involved with F G Miles at Shoreham in the 1920's, when Miles first became interested in aviation. Indeed it was Don who actually helped to make the Miles story happen. Somewhat later, in 1942, Don became Personal Assistant to George Miles and Assistant Test Pilot of Miles Aircraft Ltd until the company's demise in late 1947. There was, therefore, hardly anyone more qualified than Don Brown to have written the first detailed story of Miles Aircraft and Don will always be remembered for his unbounded enthusiasm for Miles aircraft. In a letter to Bert Clarke responding to a request for details of the early days at Shoreham, Don said that he never had any 'official' position in either the club or the firm and was only a club member 'who arrived on the scene in 1928'. This rather understated his position. Don published many brief histories of individual Miles aeroplanes in the pre-war *Miles Magazine* and in *News - Miles Aircraft Works Magazine* and under the titles *Miles Milestones* and *Milestones* during the war; and in 1944 he wrote a little booklet, which he also called *Milestones*.

In the first of what were to become three volumes he gave his reason for writing thus:

I started to write 'Milestones' partly for the fun of it and partly so that there should be some record of the days when we had little knowledge, little money and lots of enthusiasm. Since then we have gained a little more knowledge, a little more money but, what is far more important, we still have just as much enthusiasm.

In his Foreword to *Miles Aircraft since 1925,* Don Brown wrote:

In 1968, after reading this story, F G Miles wrote, 'I am glad to endorse this book. Don Brown is an enthusiast - in Greek, enthousiastes - the god within, who made life for the lucky forever interesting.' The author would only add that, over forty years ago, he was privileged to be admitted to that small band of enthusiasts who, under the inspiring leadership of F G Miles, were operating an Avro 504K giving five-shilling joyrides, always with the ultimate object of learning to design and build aeroplanes. My job was to swing the propeller, take the money, hang on to the wingtip when taxi-ing and, between flights, to teach F G Miles mathematics.

Those who may think that the descriptions of frequent struggles with bureaucracy are exaggerated should read Sir Frank Whittle's book 'Jet', which describes the similar struggles in the early days of Power Jets Ltd. I would merely add that these strictures do not apply to civil servants in general but only to those who sat in offices in London and advised the Minister on projects submitted by the industry. I have nothing but admiration for the magnificent team at RAE Farnborough with whom we always enjoyed the happiest relations and whose work for the industry and indeed for the country was (and still is) invaluable. This book is written for the enthusiast and describes the work of the Miles brothers and their team of enthusiasts.

I was pleased to be present at the launch of *Miles Aircraft since 1925*, held appropriately in St Mary's Hall, Shoreham - where Don gave an entertaining talk on the Miles brothers and their aircraft. Thurstan James later reviewed Don Brown's book for *Aerospace* (it transpired that his views were very similar to mine, not least of all those aspects surrounding the debacle of the M.52), wherein he wrote:

No phenomenon of British aircraft construction deserves closer study than the outburst of the Miles efflorescence in the twenty years between 1928 and 1948. In that time 47 different types of Miles aeroplane were flown and a total of 5,644 were built between 1929 and 1946. The story is told in this book. It tells how a young man (F G Miles) without training or money but blessed with unique talents and energy, coupled with those of his wife (Blossom), the aid of a gifted brother (George) and certain enthusiastic adherents (among whom the author of this book was one) revolutionised the look of British light aviation, grew big enough to go into partnership with Rolls-Royce and became a fully fledged member of the SBAC.

The protagonists learnt to fly before they learnt to design. They test-flew their own aircraft. It was a long time before Miles aircraft were built by a firm bearing that name. In the beginning was the Gnat Aero Company. This grew into Southern Aircraft Ltd. For what was perhaps the firm's finest hour it was known as Phillips & Powis Aircraft of Reading - Charles Powis of that company played no small part in the Miles story, more than appears in this book. Out of ninety separate types dealt with, half are projects - but what projects!

Outstanding was the Supersonic Project literally built round a Whittle turbine. Designed during the closing stages of World War II, it had been ordered by the Government with the object of attaining the hitherto unbelievable speed of 1,000 mph. After the War ended, chicken-hearted Authority lost its nerve and cancelled the razor-winged projectile before completion so that the Americans, whom the same chicken Authority enabled to study the design, got there first. Subsequent tests with the air-launched rocket-propelled models showed that the straight-winged Miles design could have achieved its goal. Its success full-scale might have altered the whole pattern of Britain's post war aircraft progress.

The essential rightness of Miles designs is shown by the fact that though Miles Aircraft closed down in 1948, there were in 1969 still 59 Miles designs on the British Register. It seems designers who can build and test-fly their own designs have a certain something!

With regard to his last comment, where indeed could any other contemporary aircraft manufacturer have been found then (and none now), whose Chairman & Managing Director, Chief Designer and his assistant, could not only design and build a most remarkable range of very advanced, innovative and practical aircraft but who were also qualified to fly and to test them as well. In fact, I can do no better than repeat the words of A H Lukins who wrote in the introduction to *The Book of Miles Aircraft* in 1946:

To those who have noted the amazingly rapid growth of the Miles Aircraft organisation the pace achieved may seem phenomenal, but there are many hidden factors to be taken into account, not the least being that every Miles aeroplane is conceived and designed by people who are themselves pilots. For instance, the Company's Directorate and senior executives, in addition to the official testing staff and the OC of the Flying School, include the following experienced pilots: the Managing Director, the Technical Director and both their Personal Assistants; the Director of Training and Education, the Repair and Service Manager, the Transport Manager and the Chief Liaison Officer.

The Managing Director, the Technical Director, the latter's Personal Assistant and the Repair and Service Manager are all MAP approved test pilots, while a grand total of over 24,000 flying hours is shared by the Chief Test Pilot, the Managing Director, the Technical Director, the OC of the Flying School and the Repair and Service Manager. This total is the more remarkable in that it has been mainly compiled in comparatively short testing, instructional and demonstration flights. That this wealth of talent has been used to the best advantage is evidenced by the number of Miles successes, and that it will be employed to further future aeronautical progress is not in doubt.

Finally, a word on the general policy of Miles Aircraft. Mr Miles believes that the best and only way in which to manage an aircraft manufacturing business is to have complete trust and identity of interest throughout the whole of the personnel, from the youngest student to the Managing Director. To implement this contention he introduced a Joint Production Board in 1941 and set up subcommittees to deal with matters far beyond the usual scope of such bodies, matters which affected the social well-being of all Miles Aircraft employees. The successful results of this innovation, which are evident in the pages of this volume, must be a great source of satisfaction to all concerned, and should be a pattern for the future, not only of British Aviation, but of all British Industrial undertakings.

If only Miles Aircraft had been allowed to continue in the business of designing and manufacturing aircraft instead of being callously destroyed just two years later by a ruthless accountant - who knows what might have happened?

Thurstan also almost threw away a comment that - *Charles Powis of that company played no small part in the Miles story, more than appears in this book* - and with this I agree. Charles Powis was instrumental in creating the company in the first place and is deserving of considerably more credit. In my own small way, I have attempted to redress the balance and to ensure that his efforts are now more fully recognised.

Having lived through the Second World War as a boy, not unnaturally I became avidly interested in aeroplanes, eagerly awaiting *The Aeroplane Spotter* as it came out every fortnight. It was in this journal in 1944 that I saw a review of a book entitled *Milestones* by Don Brown and I immediately despatched the required 2/- postal order to Miles Aircraft Ltd at Woodley. I waited with impatience for it to arrive and when it finally came I found that Mrs F G 'Blossom' Miles had written the very simple introduction:

D L Brown has been associated with the building of Miles aeroplanes since the early days of the Gnat and the Martlet. He is now personal assistant to G H Miles, Design Director of Miles Aircraft Ltd, so a better chronicler would be hard to find.

Such was the demand for *Milestones* that it soon had to be reprinted. It was followed by a second and third volume which were also purchased as they were announced. I learned much from these little booklets but it was an announcement in *Diary Dates* in *The Aeroplane Spotter* for 12th July 1947 that was to be responsible for really arousing what was to become my lifelong interest in Miles Aircraft. It read:
Reading - The Annual 'At Home' of Miles Aircraft Ltd, to be held at Woodley Aerodrome on Sunday July 20th, promises to be even better than last year. About 20 different types of Miles Aircraft will be on show and it is hoped that RAF support will be forthcoming.

Sunday 20th July 1947 dawned bright and sunny so, together with two equally keen friends, David Freeman and Don Pearman, I cycled off in the general direction of Reading. This was to be a truly magical day - I will never forget that day as long as I live.

The Aeroplane Spotter for 26th July 1947 reported on the 'At Home' as follows:
On Sunday July 20th, Miles Aircraft Ltd of Woodley, near Reading, held its third annual 'At Home' and Air Display. Organised by the Miles Trust Fund Social Committee, with profit going to the Miles Emergency Trust Fund and the 'Battle of Britain' House Fund, the event was blessed with fine sunny weather, bringing well over 10,000 people to the aerodrome and factory. All the usual static events were available, from tours round the factory to cinema shows in the canteen. The late afternoon, when the maximum number of people were present, was given over to a Flying Display, the aircraft flown being those in the Static Park in front of the main factory buildings.

Although not all the aircraft present were demonstrated, the following gave fine performances: The Libellula, Aerovan, Messenger, Gemini, Sparrowhawk and the second prototype Marathon.

In the hangars could be seen some interesting Miles aircraft, such as an M.38 Mk.2A Messenger (G-AILI), experimentally powered by a Czech 150hp Praga E air-cooled, eight-cylinder, horizontally opposed motor, two Swiss-registered Miles aircraft, M.28 Mk.IV (HB-EED) and an M.65 Gemini Mk.1A (HB-EEC) (this was actually HB-EEA which was delivered on 23.7.47, or possibly the reporter meant the M.38 Messenger HB-EEC which was also present at the time), a South African Messenger (ZS-AVY) and three target-tug Martinets finished in the colours of the Aeronautica Militar (Portuguese Army Air Force), and the prototype Broburn Wanderlust.

In those far off days, when the roads were for cyclists, lorries and the occasional wealthy car owner, we three used to cycle many miles at weekends in search of airfields and aircraft, both 'active or inactive' - it mattered not as we were just interested in them all. On that particular Sunday however, we headed for Woodley, taking in Wisley, Brooklands, Fairoaks, Farnborough and Blackbushe on the way and as we lived in and around the Redhill area in Surrey in those days this meant a journey of some 150 miles - not uncommon for us in those days - we were much fitter, and younger, then!

We were not too sure just where Woodley Aerodrome was, apart from the fact that it was around 60 miles away but following a reasonable bit of map reading and a number of aeroplanes seen flying around, we rounded a bend in a country lane and suddenly we were there - Woodley Aerodrome - the home of Miles Aircraft Ltd. The 'At Home' and Air Display was in progress and we had made it - just in time.

That July Sunday was one of those typically gloriously sunny, summer days that we all remember from our youth and, to our surprise, we found that the gates were also open to the general public - we were in luck and what a sight met our eyes! In neat rows and on view to families and friends were 21 different Miles aircraft, ranging from the Southern Martlet - the first aeroplane to be built by Miles, at Shoreham in 1929, to the second prototype Miles M.60 Marathon I, the company's first all metal, four-engined airliner.

Hugh Kennedy carried out 4½ hours joyriding in his Gemini, G-AJWE during the day and conducted tours of the factory were in progress, and the spectacle of so many different types of aircraft from the same stable, displayed on the ground at the same time was truly astounding. Then a flypast of twelve all-different Miles aircraft types took place, in what Sqn Ldr T S 'Wimpy' Wade, *The Aeroplane* reporter covering the event, was to describe as meteorologically favourable circumstances (in other words - good weather!) And it really was!

While it is still not known exactly who flew each aircraft in this magnificent display, it is known that both Hugh Kendall and Sqn Ldr J C Nelson did not take part, being engaged elsewhere, but we do know from his logbook entries that Sqn Ldr S E Esler flew both the Master Mk II G-AHOB and an Aerovan. Some of these types flown were also very ably demonstrated and I still ponder the thought - how many other aircraft manufacturers of the day or at any other time before or since could have flown twelve completely different types of aircraft at one time? It is indeed a sobering thought that now, there is just one manufacturer of British (and I use the term loosely) aircraft left in this once great country of ours........

Unbeknown to us at the time however, storm clouds were gathering on the horizon and by the following November, after 15 years of steady growth and just when its future seemed assured, the company apparently ran into a 'financial crisis'. The reasons for this were never made clear at the time, or later for that matter but it was generally accepted by the aviation press that the company had gone into liquidation. This was not true and the real reasons why this once great company was forced into receivership (for this is what had actually happened to the firm) have never before been published.

Don Brown had written an insider account of the reasons behind the collapse of Miles Aircraft Ltd but he was unable to publish this in 1970, much as he had so dearly wanted to, as too many of the original 'players' (as Don called them) were still alive. Now, all of the 'players' have passed away; so not suffering the constraints under which Don found himself, I can, and will, carry out his wish and publish his account of the final days of Miles Aircraft Ltd. What was to unfold regrettably does some of the 'players' no credit at all and brings it all back to me with much sadness, because of 'what might have been'.

Now, finally, the record can be put straight. When I read Don's story of the 'financial collapse', I had to admit that, although it confirmed my original unproven thoughts on what had actually happened, the circumstances of the 'financial crisis' of November 1947 were far more sinister than I could have ever imagined. It will be shown (although not in this book) that this apparent 'financial crisis' was through no fault of any of the Board of Directors of Miles Aircraft Ltd.

On 18th January 1987, Don Brown died at the age of 81 years and Mike Hirst, a very close friend of his, wrote the obituary which appeared in *Aeroplane Monthly* for March 1987:

Donald Lambert Brown, born on 17th November 1905 at Hove, Sussex, followed a career in aviation that was active for almost half a century. He was determined to be a lawyer, but his father, who was chief engineer of the London, Brighton and South Coast Railway, insisted that he trained as an engineer - so an engineer he was, with a remarkable ability to use words with devastating efficiency. He started his career as an apprentice at the Portslade Gas works around 1919. Twenty years later he was the works engineering manager and a full member of three institutions, the Mechanical Engineers, Gas Engineers and Civil Engineers. Most notably, he was also an associate fellow of the Royal Aeronautical Society. His early life was exciting - he enjoyed flying, motorcycling and speed-skating and his aeronautical distinctions arose from his able engineering assistance to the Miles brothers. He learned to fly on an Avro 504K at Shoreham and was awarded an A Licence (No. 3761) in 1931.

Never one to be thwarted by red tape, he fought to be allowed to leave the gas industry during wartime. He was too old to fly for the RAF but he was appointed by the Air Ministry as an author of pilot's notes. This was not to be a long occupation and he was with the Miles team by early 1942.

He remained there for five years, holding the title of chief designer when the company folded. He loved flying the M.38 Messenger and M.65 Gemini, especially the latter, which had what he regarded as the most splendid handling characteristics of all aircraft. His involvement in the M.52 supersonic research aircraft project was a fine opportunity for a unique man of his period, as he had a thorough knowledge of gas dynamics, as well as of aircraft design. He always bitterly resented the British decision to let American engineers (who built the Bell X-1) see the project, and the subsequent decision to scrap it, thus leaving the distinction of being the first supersonic nation to America - with the X-1.

An often overlooked period of Brown's employment was with British South American Airways, where he spent almost two years as training and administration officer; the airline's disastrous reputation provided him with many a well-rehearsed tale for later life.

His engineering skills were in demand, however, and in 1949 he joined Power Jets Ltd as principal at the School of Gas Turbine Technology at Farnborough where, thanks to his congenial and infectious manner, he left an indelible impression on many visitors. When he set up the Power Jets school, he ordered 'The Aeroplane' and several other magazines for regular delivery from a newsagent in the town. A rather frumpish lady wrote the list of titles as he dictated them, and scolded him vividly for daring to ask for 'Nature'. In his inimitable style he did not enlighten the lady about the magazine's scientific content, but smirked to deepen her indignation.

Between 1959 and 1968 Brown was successively head of research and administration executive at Smiths Industries. His ability to bring together an engineering team of excellent spirit and capability was a cornerstone of the company's foundation as the world's automatic-landing pioneer. He flew regularly with the flight-test teams, and maintained his pilot's licence until 1970. His logbooks recorded over 120 types, ranging from the Avro 504K to the Hawker Hunter, and he was never more at home than on an airfield. His sole published book, Miles Aircraft Since 1925 (Putnam 1970), is a testimony to his enthusiasm, and his occasional articles in the aeronautical press, over many decades, never failed to be informative and readable.

Don Brown died on 18th January 1987, a few days after a fall outside his home. He leaves a legacy of humour, humility and ability that very few could ever challenge."

Don Brown's *Miles Aircraft since 1925* was never revised or even reprinted and over the years a wealth of new material has been discovered. Although I have drawn on his book as well as much of his unpublished material (as he 'was there' for most of the time and had such a good memory), this work is completely new and - dare I say it - probably the most 'definitive' book on Miles Aircraft that will ever be written.

I completed my National Service in 1950 and was fortunate in obtaining my first job in the aircraft industry as a junior draughtsman with de Havilland's, in their Regent Street, London drawing office. However, soon tiring of the early morning commuting, I moved to Tiltman Langley Laboratories Ltd at Redhill Aerodrome, where I was to spend the next eight very happy years engaged on sub-contract design work for the major players of the British aircraft industry. In total, I spent twenty-five years in the aircraft industry, somehow managing to survive the traumas of successive government's mishandling of this once great organisation but, by the mid 1970s, while working as a flight development engineer with Hawker Siddeley Aviation at Dunsfold, first with the introduction of the Gnat Trainer into RAF service and latterly on the P.1127 (forerunner of the RAF's Harrier), I became very disillusioned with the continual traumas of government versus the aviation industry and left to go my own way.

My one regret in life will always be that I never had the opportunity to work for Miles Aircraft Ltd at Woodley. My initial interest in Miles aircraft developed into a lifelong passion and this eventually led me to becoming the Specialist member on Miles Aircraft for Air-Britain, which I had joined on its formation in July 1948. Then, in 1985, I met Julian Temple who had just been commissioned to write a book on the history of Woodley Aerodrome by the Adwest Group plc, the ultimate successors to Miles Aircraft Ltd. It was not too long before I became involved, in an advisory capacity, and the result of his extensive research was published in 1987 under the title *Wings over Woodley*.

Julian, having amassed a considerable amount of research material for his book, very generously shared this with me and on the strength of this new information, plus what I had been collecting over the years personally and in collaboration with the late Bert Clarke, I decided that the whole story of Miles aircraft should now be written. This particular book begins the process and will tell the story of F G Miles, his businesses and aircraft between 1925 to 1939. In doing so, I will use the opportunity to dispel many of the myths about the firm and its products which have been published over the years.

It may seem a strange thing for an author to say but, for the benefit of future researchers and historians, beware, as it is not always advisable to believe the written word! Over the years, and following an incredible amount of research into Miles aircraft, many hitherto respected reference works, including Don Brown's own book, have been proved wanting in many respects. Quite often, this was not the fault of the author because, for example, Don was denied access to some of the considerable amount of material which my colleagues and I have since managed to unearth. Also, in a hand-written letter from Don to a reviewer of his book, he explained the great deal of trouble he had had with the publishers, who really only wanted a book on Miles aeroplanes - with no background, story or personnel details or indeed any detailed appendices. Don tried to persuade them that this would be impossible as it was the characters who made the story of Miles, and eventually, they allowed him to include a brief history on the background and personnel plus a few appendices. However, as this considerably increased the size of the book, without recourse to Don, the publishers made drastic cuts to his original manuscript, the original of which has not been found, which is a great pity as who knows what else Don might have recorded for posterity.

However, be all that as it may (and a lot of water has gone under the bridge since 1970) it must also be noted that Don relied principally on his incredible memory to write his book. He did not keep detailed notes during his time with the firm and leant very heavily on the late A J Jackson and the late E J Riding for production details of aircraft types and registrations etc. Unfortunately, some of the information which these two stalwarts of pre-war and early post-war British aviation history had carefully and painstakingly built up over the years, and which was regarded by all and sundry (me included!) as 'gospel', has had to be amended as a result of further research. As just one example, the exact type designations for the many different Miles aircraft have now been properly resolved.

Of the original band of 'enthusiasts' at Shoreham, all have now gone. F G Miles died on 15th August 1976 at the age of just 73 years, and his obituary, which appeared in *The Times* for 17th August 1976 read:

> Mr Frederick George Miles FRAeS, MSAE, whose brilliance in the design and production of light aircraft was unparalleled, died on August 15th....Such experience made him scornful of official timidity and misjudgement, he once compiled a bitter, but amusing, list of the engineering feats officially declared to be impossible but actually achieved.

Referring to the famous court case against F G Miles and Sir William Mount, after the 'financial collapse' of the firm (in November 1947), the obituary continued:

> The Old Bailey Jury stopped proceedings and threw out the case....he did not allow the flow of creative ideas to be cut off. He was Chairman of F G Miles Ltd to 1961; he widened his technical and business activities and plunged into electronics. His work was, as in his early days, at Shoreham. He was a Freeman of the Guild of Air Pilots and Navigators and a member of the Royal Aero Club."

Blossom Miles' increasing blindness became a concern to her and she eventually passed away on 6th April 1984. On 18th January 1987, Donald Lambert Brown died at the age of 81 years. Charles O Powis, who was still flying his own light aircraft until the early 1960s, died in retirement in Kenya in 1990. George Miles, the last of the original band of 'pioneers', who spent his final years in Edinburgh and was still actively engaged in the design of new aircraft of advanced concept to the end, died suddenly on 19th September 1999, at the remarkable age of 88 years.

It is of interest to note that George retained the complete set of drawings and some jigs and tools for the Miles M.100 Student and still, apparently, cherished a desire to see it placed in production - by someone - somewhere – sometime! It is understood that negotiations were actually proceeding with 'an interested party' shortly before his death with a view to him actually achieving his goal, although nothing ultimately came of this.

A H Lukins, who had worked in the technical publications department of Phillips & Powis Aircraft Ltd and Miles Aircraft Ltd and was the editor of the first *Miles Magazine*, in 1945, compiled the very first *Book of Miles Aircraft*, published by Harborough Publications. Now, while I may have grave reservations as to the sanity and leanings of the highly controversial C G Grey, one-time autocratic editor of *The Aeroplane*, I have to admit that the foreword he wrote to this book said it all and I can do no better than to reiterate his sentiments:

> *Miles and I first met at that quaint and dangerous little aerodrome at Shoreham-by-Sea, under the dignified shadow of Lancing College, where he and a sportsman named Bellairs, and Cecil Pashley, one of our earliest and best pilots, were running the Southern Aircraft Co Ltd. There he had learned to fly and had built a highly aerobatic single-seat biplane, which he called the Martlet. Also he had built a sort of control-tower at one end of the hut, which was the Club House - I am sure the Club House could have held as many as a dozen members, as well as the bar.*
>
> *Two storeys up in his tower, of solid timber not ivory, Miles had the most eclectic, esoteric and alluring library I have met. It ranged from Froude to Freud, from the Arabian Nights to Lawrence of Arabia, from Gibbon to Philip Gibbs, and one approached the building from either direction by a road of the sort usually marked "Not Adopted," which nobody would have thought of adopting,* with the added charm, on the Lancing side, of the probability of a pupil in the act of landing alighting on the top of one's car, because unless the lander scraped the road he finished in the ditch to the South-West.
>
> *But it was a joyous place, and I concluded that a man who could design, construct, fly, read, and drive as he did, must have a future, if he did not break his neck first. A year or so later he blossomed forth at the Woodley Aerodrome, Reading, of Phillips and Powis Ltd, with the Miles Satyr, another spectacularly aerobatic biplane, and with him was the charming lady who is not only his wife, and the mother of a rising family, but is now School Director, keenest critic, and tireless partner in all his enterprises. As the daughter of Sir Johnston Forbes-Robertson, one of our greatest actors, and of that very charming actress, Gertrude Elliott, she is entitled to have brains, but one would not expect the mathematical sort.*
>
> *Their first enterprise was the Miles Hawk. They had acquired a number of Cirrus engines at a knock-down price, and they designed a cheap-to-build but sturdy and good-looking two-seat monoplane to fit it. They sold the lot at about £400, and found such a good market that they went on to design and build more powerful and more highly finished machines, and so they progressed to King's Cup racers, and cabin tourers, and twin engines, and now to four-engined feeder-line types.*
>
> *But always they - F G and Mrs Miles work together so closely that though one writes of them in the plural, one thinks of them as one person - were seeking after some new thing. I always believe that his fantastic experiments with trailing-edge flaps, made out of flattened petrol tins, were the beginning of the adoption of flaps by designers in this country and of their approval by the experts of the R.A.F., and the highbrows of Farnborough.*
>
> *Brother George Miles came in quite early in the proceedings and proved to be as good a pilot as Fred. He also has progressive ideas - for example, the M.38, which can get out of or into a tennis court, given a few yards run out, and a clear approach, and his tandem monoplane, the Libellula, interest me still more.*
>
> *The constitution of the firm has had sundry changes. The Phillips and Powis interests were replaced by those of Rolls-Royce, leaving only the P & P name, and with the new interests came the Kestrel-engined advanced trainer. And more lately the Miles family interests acquired the whole outfit. This book tells of what the Miles family have designed and built up to the date of publication. Whatever may come after we may be sure that it will show enterprise, foresight, intelligence and originality worthy of those amusing early activities at Shoreham.*

Who will now ever know what might have been, if only Miles Aircraft Ltd had been allowed to continue in the business of aircraft manufacture after November 1947?

The 'Miles Aircraft Story' will never be complete but I would like to think that this effort is something of a tribute to the endeavours of that great band of enthusiasts and pioneers. The memory of Miles Aircraft Ltd *will* live on through the medium of The Miles Aircraft Collection and its members - some of whom own and regularly fly the last surviving Miles aeroplanes. I am pleased and proud to be the Honorary General Secretary of this organisation and also the editor of its twice yearly, *(new) Miles Magazine* and it is to be hoped that this book, so long overdue, will also help to keep alive the memory of the 'Miles family' - F G Miles, Blossom Miles, Charles O Powis, George H Miles, Don Brown, their dedicated team of enthusiasts, the 'Pioneers' and of course their remarkable aeroplanes.

Don Brown gave me with the inspiration to undertake this interesting, if somewhat daunting task and I would, therefore, like to offer him my most grateful thanks. 38 years have now passed since Don Brown's book appeared and I, with the intervening years, have been able to put 'some more meat on the bones' - I sincerely hope that Don (wherever he is) approves of my efforts.........

The Gnat in the workshop of the Star Laundry, Portslade. Left to right: Dennis Miles, F Wallis, F G Miles, Ru Hart and Cecil Pashley.
[Via A H Lukins]

Chapter 1: The Early Days - F G Miles at Shoreham, Sussex - 1925 to 1931

The aviation accomplishments of two brothers, Frederick George and George Herbert Miles, will be described as this story unfolds. Their story starts in 1925 at Shoreham, in the County of Sussex, England.

Frederick George Miles was the eldest of four sons of Frederick Gaston Miles, the proprietor of The Star Model Laundry at Portslade-by-Sea, a small town midway between Brighton and Shoreham-by-Sea in Sussex. The young F G Miles was known to his friends as 'Miles' and by others as 'F G' but rarely, if ever, as Fred, although the press often later referred to him thus. Much later he was known by his employees, more respectfully, as Mr Miles. However, as his great friend Don Brown called him 'Miles', I have taken the liberty to use this herein to save any confusion between him and his father, who shared the same initials.

In 1925, two of his three brothers, Reginald and Dennis, were employed in the family laundry business, while the youngest, George Herbert, who was born on 28th July 1911, was still to leave school. There is no doubt that Miles, although devoted to his parents, was the leading light in the family and had always displayed considerable independence, which had become apparent in 1916 when his father was in the Army in France.

In a letter to Bert Clarke, in the early 1960s, Miles gave auto-biographical details of his early life which, although similar to those related by Don Brown in his book, add more details and are reproduced here in their entirety, as they are 'from the horse's mouth':

Born Portslade, Sussex 22nd March 1903 of Frederick Gaston Miles and Ester Miles, both of Portslade.
My paternal grandfather was an architect in Portslade and was quite successful in a limited field. He was certainly not rich. My grandmother Miles came from an old local family of fishermen and farmers. She had a considerable share of French blood (hence perhaps my father's second name, Gaston). My father worked, as a boy, first in the Portslade gasworks and later in the Petersfield Laundry. From these he entered for and won an award at the big Industrial Exhibition at White City (about 1900?).

My maternal grandfather was a successful builder in Portslade (I still have interests in houses my grandmother left us), their name was Wicks. My grandmother was an extremely generous woman and was a major help to my early ventures. She herself ran a general shop which was very popular for her home-baked cakes etc. They were wonderful cakes; I suspect they were half brandy or rum.

After a brief session at a private infant's school in Worthing, I attended St Nicholas Church School in Portslade from about 7 years old until 11, when I became a 'train boy' at East Hove secondary school in Cromwell Road, Hove. I had a rather patchy education, doing well enough in things I liked and ignoring others as far as I was permitted but hovered around the three or four top places in class except when sent to the bottom for being a nuisance. At some point between 13 and 14, I left by failing to attend any more. The war made this possible I suppose. I worked at the Star Laundry for a while but there was much demand for what a boy brought up amid machinery could do and much demand for him because of the dreadful drain on men called to war. My father was one (of these) at about this time.

Fairly soon my uncle sent me to help an acquaintance who managed the local cinema. They were in trouble with an ancient and neglected gas engine which drove the electrical plant. These were in terribly bad condition and there were no skilled men available to patch them up and keep them running. I was rather flattered and joined the glamorous world of Chaplin and Fairbanks as a mechanic-projectionist. From there I was sent to a larger cinema in Hove where the manager was a wounded ex-serviceman who was faced with similar problems on a larger scale. By this time I was receiving an income which would have been considered good today and things then were a quarter of the present price.

From the Queen's in Hove, I progressed to a share in a cinema in Edward Street, Kemptown (then a slum) in partnership with an even younger youth whose father had provided most of the money. This new venture we called the Majestic, although our clientele referred to it more familiarly as the 'blood-hole'. This was rather unjustified although things got tough at times. The war ended, my father returned and, as well as working evenings in the Majestic, I drove one of the laundry vans for part of the day as an aid in the shortage of drivers. Shortly after, my uncle bought a redundant Ford Ambulance from one of the war surplus auctions. I converted it and hired it out with myself as driver. Soon I employed a second driver and by driving during the day and helping out at the cinema in the evenings, I piled up a royal income before the days of tax-collectors.

Don Brown recalled that the young Miles spent many weary but profitable months driving this van during the day and servicing it in the evenings ready for the next day's work. Miles was lucky in that he was blessed with good health and a strong physique and an example of his strength occurred one day when the van got a puncture in one of the front wheels. He had a spare wheel but no jack, so he removed the nuts securing the wheel in readiness, stopped a passer-by and, taking a firm hold of the chassis, braced himself for the effort and lifted the front of the van while the helper quickly removed the punctured wheel and replaced it with the spare.

The van finally expired and Miles, wearying of his role of transport contractor, was forced to consider what the next course of action would be in the development of his career.

Among other sidelines, a friend and I had bought and rebuilt several ancient motor-cycles. These we hired out to all-comers, irrespective of age or skill, at a few shillings an hour. The danger was small as the hirers spent most of their time pushing. We also started a team wiring houses and opened an electrical shop in Boundary Road, Hove.

The motor-cycle workshop and centre was part of my father's stables (horses were still in use because of fuel shortage) and it was here I first met a man, ex RAF, who eventually turned my thoughts towards flying. This was just after the war, perhaps in 1916 (in reality 1921 - PA). *My new friend's name was Wallis* (Frederick 'Wally' Wallis - PA) *and he became interested in an enterprise my friend and I had started as a logical follow up to the motor-cycle business. We had decided to design and build a sports three-wheeler car and had proceeded with the chassis made of ash and spruce. Wallis, who occasionally saw our struggles on the way home from his work as a garage engineer and lorry driver, became interested. I began about this time to lose interest in cinemas and spent much time on the car. Wallis came in evenings and Sundays and helped to an extent that nearly led to his divorce, for my cinema life had accustomed my parents and myself to late nights - or rather, early mornings.*

Wally told wonderful stories about aeroplanes, flying and pilots over the cocoa breaks and my growing interest was brought to the burning point (in 1922 - PA) *by a 5/- ride in an early joy-ride Avro 504K* (unfortunately, it is not known exactly when, or where, Miles took his joyride in the anonymous Avro – PA). *This led to plans to build an aeroplane and the car project was ruthlessly scrapped although it had reached the running stage. The Ash and Spruce was cut to longerons and ribs for its successor which we named the Gnat* (on account of its small size - PA).

At this point it should be mentioned that 'Wally' and one Charles Gates had been building a small biplane, resembling the Caudron G.3 pusher, while they were serving at Shoreham during the First World War. This was to have been powered by a three-cylinder Anzani engine and a photograph taken at Shoreham shows this actually installed in the partially completed 'podded' fuselage with some incomplete wings visible in the background. Charles Gates later emigrated to Australia and in 1984 recorded his version of the story of the days at Shoreham and how the aeroplane had come to be designed. According to Gates, the pair was considering disposing of their incomplete biplane when they met with Miles at Southwick in 1921 and it was eventually agreed that they would exchange the unfinished aeroplane for the partially-completed sports car. It is understood that no money changed hands as a result of the transaction.

However, if the Wallis/Gates aeroplane was used by Miles as the basis for the construction of the Gnat, then its 'podded' fuselage would have had to have been considerably modified and its wings appear to be much thinner than those used on the Gnat. Furthermore, the three-wheeler sports car, which Miles said had been ruthlessly scrapped for its ash members in order to use them in the construction of the Gnat, would have had to have still been in one piece at that time in order for it to have been swapped for the unfinished Wallis/Gates aeroplane.

The Charles Gates story, so much at variance with Miles' own recollections of how the Gnat came to be built, has been confirmed by Jeremy Miles (F G Miles' son), but this does not alter the fact that the two aeroplanes were so different in design. The introduction of this story of how the Gnat came to be built should in no way be allowed to detract from the fact that Miles fully intended to build and fly a small aeroplane - come what may - no mean achievement for someone of his age and very limited budget.

Don Brown recalled that Miles had acquired a small two-cylinder engine of 698cc, together with its propeller (which had probably previously been used in one of the aircraft flown in the light aeroplane trials at Lympne in 1923), which Miles hoped would provide sufficient power for the Gnat. A few drawings were produced and, with Miles' brother Dennis, two former school friends and Ruben Hart, a one-armed lorry driver, construction commenced.

The Gnat was virtually complete apart from its covering of fabric when Miles realised that he had not yet learned to fly. Unfortunately, it is not known what became of the Gnat but the photograph shows it to have been an attractive little biplane, but for reasons which we will now never know, it was never finished.

Wally Wallis knew a man named Cecil Lawrence Pashley, or *Pash* as he was more popularly known. Pashley was born in Great Yarmouth on 14th May 1891 and his brother Eric Clowes Pashley was born on 25th July 1892. They had both taught themselves to fly powered aeroplanes at Brooklands in November 1909 together with the other early pioneer enthusiasts. Pash did not bother to take his Royal Aero Club Certificate, No.106, until 18th July 1911 but he could have taken it much earlier had he felt the need to do so. Eric obtained his, No.139, on September 26th of the same year.

The two brothers then moved to Shoreham, where they opened a flying school and designed and built their own aeroplane, a small pusher biplane, for competition flying. This machine won its first race a few days after it had been completed on the occasion of the pylon race for the Shell Cup at the opening of the 1914 season. Work was just starting on a faster machine of the same type, fitted with a 100hp Gnome engine, when the first world war broke out. Shoreham Aerodrome was taken over by the War Office and their operations ceased.

During the war, Pashley became an approved Admiralty test pilot, while also acting as an instructor training a very large number of pilots for the RNAS and RFC. Eric joined the Royal Flying Corps in 1916 and served with 24 Squadron in France, flying DH.2s and was credited with 8 victories before he was killed in a flying accident in France on 17th March 1917.

Chapter 1: The Early Days - F G Miles at Shoreham, Sussex - 1925 to 1931

Cecil Pashley in his Farman Pusher biplane in 1912. *[Nonie Cellier]*

In 1920, Pash obtained his Pilot's 'B' Licence, when this was first introduced, and then he flew for the Central Aircraft Co, of Northolt. He owned his own Avro 504K, G-EATU, and by 1926 had amassed the astonishing total of 6,000 hours in the air. In a letter to Bert Clarke, dated 21st March 1968, Pashley pointed out, in no uncertain manner, that much of what had previously been written about him by Don Brown and others was in fact just not true! So, in order to dispel the many myths about Pashley's early life that have been circulated over the years, this letter is produced verbatim:

I was never employed in my father's office, having no interest in racing (he was a bookmaker); furthermore, I have never given up flying, the Avro you mention held a commercial C of A, was my property, and was regularly used on various jobs, it was not stored at Hendon but obtained its C of A there. After obtaining its C of A, it was based at Northolt and was regularly flown from that airfield. I had not at this time met F G Miles and there was very little flying going on of any kind. (The official record shows that Pash first registered G-EATU on 12th June 1920 and its initial C of A was issued on 25th February 1921, only to lapse at the end of its first year. It was not to be formally renewed until June 1926 – PA).

The Shoreham Aero Club or Brighton-Shoreham Aero Club to be more precise, was in existence in about 1912, but had no aeroplanes that would fly. It would be more correct to say that the club was restarted in the early 1920s, than that it was started then. It became a real flying club when my brother and I arrived and produced one or two good aeroplanes coupled with the ability to fly them, there was no other flying going on at the time, as far as I knew.

We bought up the bits and pieces from the Avro School that you mention and did manage to make one of the machines serviceable, this aeroplane was usually flown by a Mr Gear who later became our manager. One of the Avros, there were three altogether, was involved in a fatal accident before we made the purchase. I should also mention that I had not received any instruction, certainly not from Blondeau, I taught myself to fly. Hale became our partner when we were at Shoreham, he was not a separate organisation.

Mr Miles senior did not subsidise the Southern Aero Club, he was a director of the Company.

However, to return to the story - Miles, full of hope, hurried off to London anxious to meet this pioneer aviator and to try to coerce him into teaching him to fly. He also intended to try to persuade

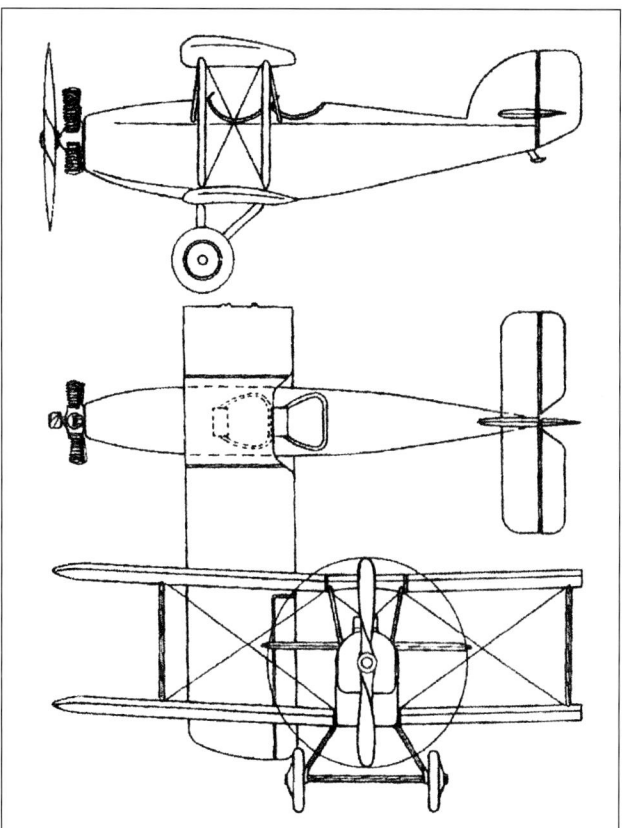

The Gnat.

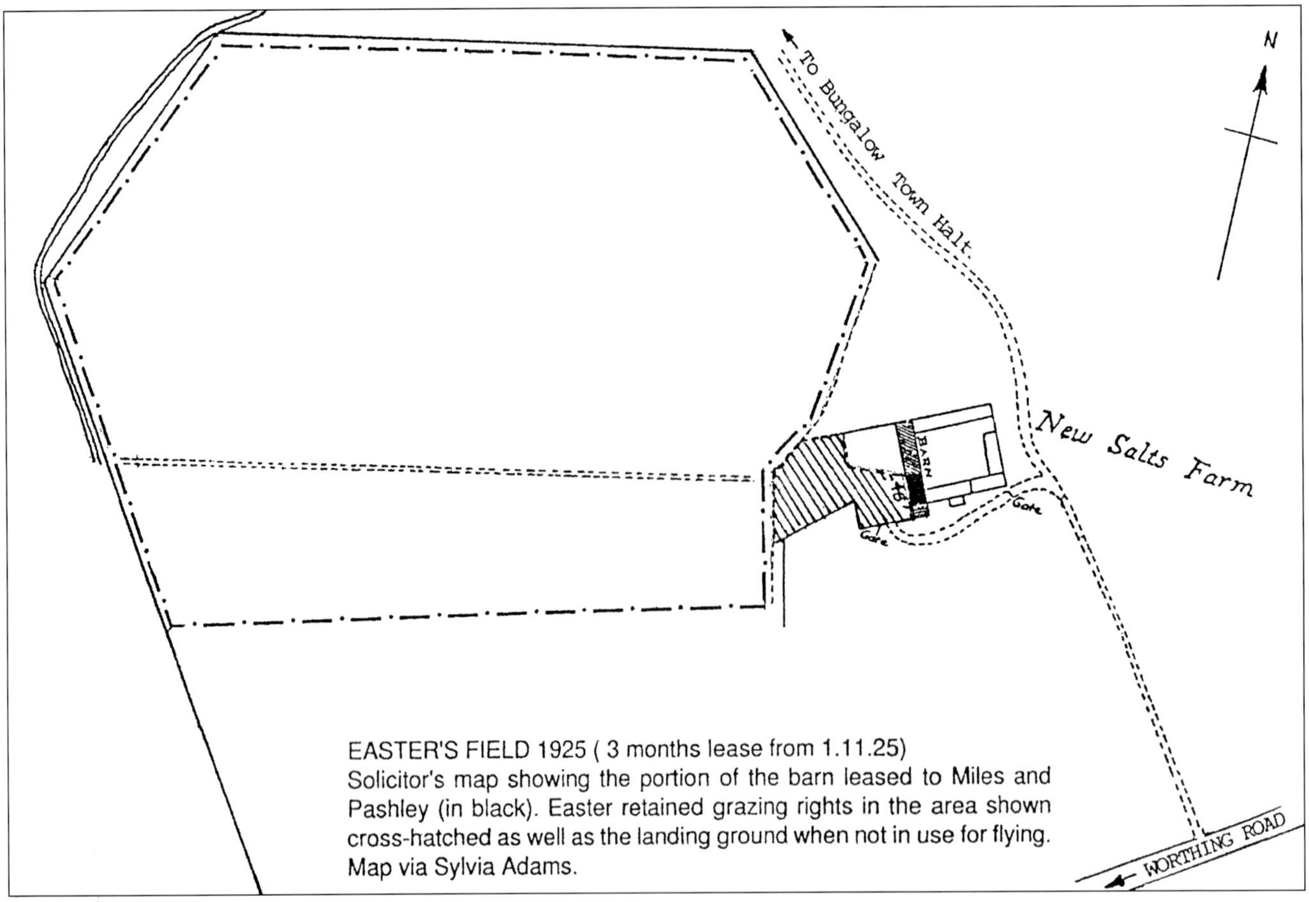
N
To Bungalow Town Halt
New Salts Farm
BARN
Gate
Gate
WORTHING ROAD
EASTER'S FIELD 1925 (3 months lease from 1.11.25)
Solicitor's map showing the portion of the barn leased to Miles and Pashley (in black). Easter retained grazing rights in the area shown cross-hatched as well as the landing ground when not in use for flying. Map via Sylvia Adams.

Solicitor's map of Easter's Field in 1925.

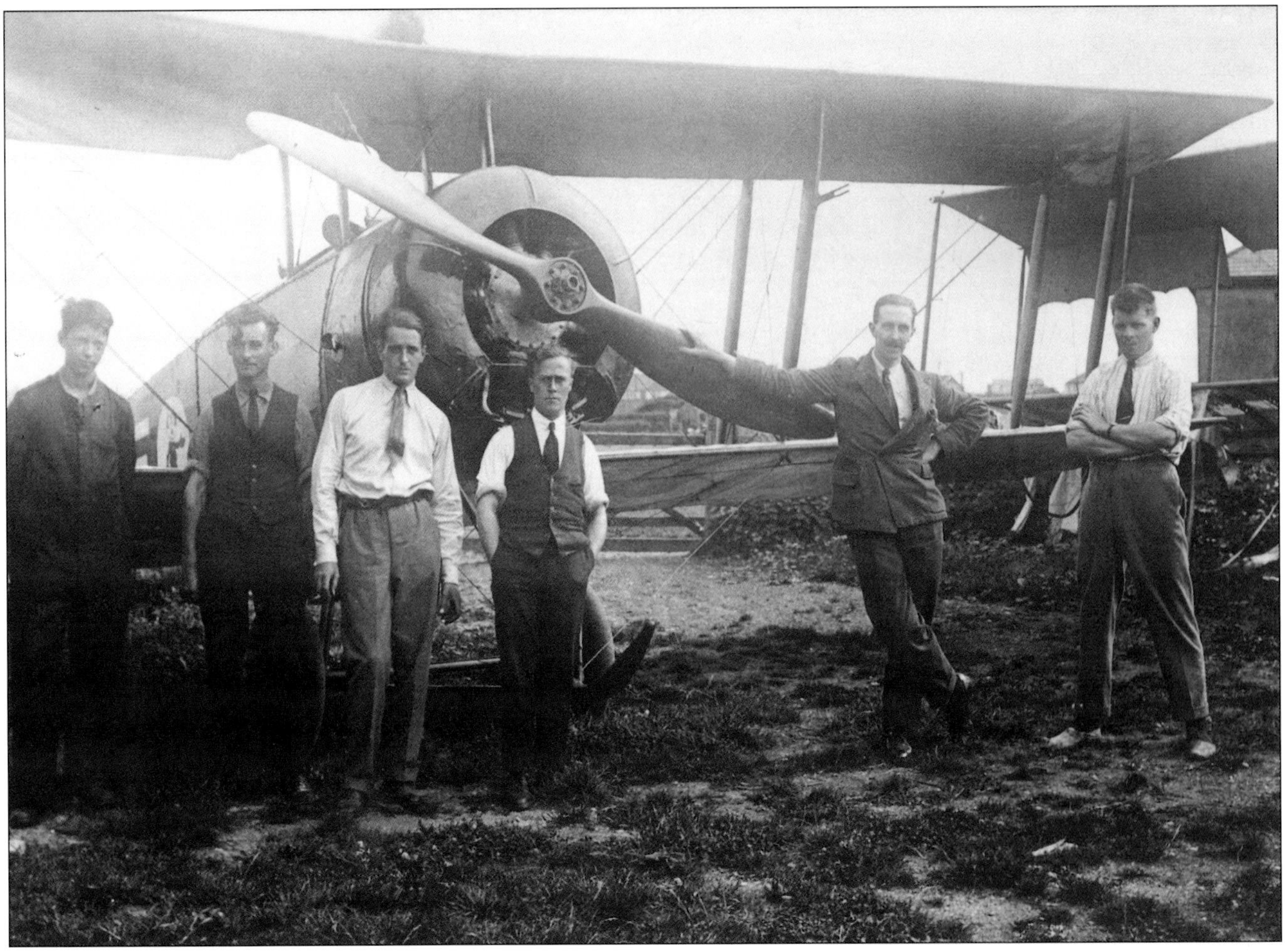

Easter's Field, 1926. Avro 504K G-EBJE with Cecil Pashley under the propeller and F G Miles on the right. Note the Grahame-White GW.15 on the right. *[Via B Clarke]*

CHAPTER 1: THE EARLY DAYS - F G MILES AT SHOREHAM, SUSSEX - 1925 TO 1931

Pashley to form a partnership to help him found a flying school and joyriding concern and he outlined his future plans for the eventual setting up of a company to design and build aeroplanes at Shoreham. Miles had probably formed a mental picture of what he expected Pash to look like but was somewhat surprised when he met a little man, not much over five feet tall, wearing a black jacket, pin-striped trousers and a bowler hat! Miles however persuaded Pash to agree to the proposed plans to form the partnership and with the necessary finance to be provided by Miles, or more accurately his father, the Gnat Aero & Motor Company was formed in 1925.

Miles and Pashley decided that Pashley's Avro should be re-assembled and flown to a hill north of Southwick, Sussex, where it could be dismantled and taken to Miles' workshop for overhaul. This workshop was in part of the stables of the Star Model Laundry at Portslade and there the Avro was stripped down preparatory to its reconditioning. This task took many weeks of hard work by Miles, his brothers and some friends and when the work was finished an AID Inspector by the name of Ashdown ripped open the fabric with his penknife in order to examine the interior. It was at this point that the floor of the workshop suddenly collapsed beneath him! As some of the work had been done with more enthusiasm than skill this timely co-operation on the part of the floor was probably not entirely unwelcome!

Their next problem was to find a suitable aerodrome and Miles knew of the barn, with *Sussex County Aero Club* painted in large white letters on its roof in a field at Shoreham, south of the railway line and to the west of New Salts Farm Road. This barn was in Easter's field and was all that remained of an attempt to form a flying club in 1919 by G Arthur Wingfield but which venture had proved premature and had gone into receivership in January 1921.

Miles arranged to lease the field and barn and then applied for a licence to operate it as an aerodrome. Refusal by the Air Ministry to grant Miles a licence, however, did not stop him from moving in, bringing the Avro 504K from Portslade and erecting it in the barn. They had to share the field with a herd of cows who, not content with the available grazing, launched regular attacks on the open fronted barn and its contents. These, including the precious Avro, were regularly severely mauled, even though a tarpaulin was securely lashed across the front of the barn after flying had finished at the end of the day. Heavy barbed wire entanglements were then constructed and these were dragged across the front of the barn in the evenings in an attempt to keep the cows out but the constant removal and replacement of these obstacles made for much inconvenience, especially when the rain turned the area surrounding the barn into a sea of mud. The inside of the barn was in no better shape either and often the floor was as muddy as the surrounding area.

Notwithstanding all these problems, the Gnat Aero & Motor Company, having from its inception possessed little more than the enthusiasm of its members, now boasted a real aeroplane, a qualified pilot, an aerodrome (albeit unlicensed) and a hangar (of sorts). The founders felt that progress was definitely being made.

Pashley put in a little flying practice before commencing his duties as flying instructor and joyride pilot and was soon ready to embark upon his new career. There were four pupils in addition to Miles, all intent on learning to fly and Miles finally started his flying tuition with Pashley, in G-EATU, on 26th November 1925. Progress was slow due to Pashley's caution, too slow in fact for the impetuous Miles, who soon considered he had reached the stage where he felt that he should go solo. Pash however insisted upon further instruction before entrusting their only aeroplane and means of sustaining their business to Miles' inexperienced hands. Miles' logbooks show that he was forced to carry out 15 hours 24 minutes dual instruction before finally being allowed to go solo. He made his first solo flight, after a 10-minute dual with Pash, on 19th May 1926 and the fact that he made this 10-minute 'refresher' flight before going solo implies that his log-book version of his first solo was correct. Or was it? – as there are two different versions of this 'historic' event!

The first of these appeared in an article in the *Miles Magazine* for February 1938 entitled: *Have you met Cecil Lawrence Pashley?* and this stated that…. *Pashley then met F G Miles and taught him to fly. Pashley insisted on fifteen hours dual and might have increased the hours had not his pupil got up very early one morning, 'stolen' a Grahame-White 15 Pusher and made his first solo.*

It is known that the Gnat Aero & Motor Company did, in fact, possess an unregistered Grahame-White GW.15 two-seat Boxkite which contemporary records state was purchased in February 1926. Miles makes no mention of ever having flown the Boxkite in his first log-book, but this omission could very well have been deliberate. If Miles had in fact made a first 'unofficial' solo in the Boxkite before his 'official' solo in the Avro 504K, then the flight would have been before 19th May 1926. The Boxkite was unregistered and without a current Certificate of Airworthiness and the flight would, doubtless, have been regarded by the authorities as somewhat irregular and questions would almost certainly have been asked of him!

It is of interest also to note that, on a page entitled 'Past experience' in a later log-book, Miles recorded *GW Type 15, 15-min* but, unsurprisingly, omitted to record the date upon which this flight took place! It seems to confirm that Miles did actually fly the Boxkite solo on probably just one occasion and suggests that this flight could have been his 'unofficial' first solo. The Boxkite only survived until 9th October 1926, when it was destroyed in a gale.

A second version of the story, recorded by Don Brown, recalled that Pashley's caution *did not suit Miles at all, so one morning he got up early, dragged the Avro out of the barn, started it up and went solo, and the next day calmly took over the instruction of two of the pupils.* While this version might be more plausible, with Miles having used a more likely aeroplane for his 'illicit' solo, it is unlikely that the second part of it was true, as Miles' records his next solo flight in G-EATU on 22nd May 1926. Miles' log book also records that he did not carry his first passenger until 19th June 1926 when he made his first 5-minute joyride in the recently acquired Centaur, G-EALL, after just six-hours solo. While the recorded 'facts' do not corroborate Don's version of the events, Don was probably there at the time and, allowing for his memory 'playing tricks', it was a good story anyway and should not be allowed to detract from the enterprise of the young Miles. Either way, on the front page of his Log Book, Miles recorded that he was then living at 42 St Aubyns, Hove, Sussex and that the 'No. of Licence' was 'Aeroclub 9003 A. 910' but he omitted to enter the 'Date of Expiration' (it was issued to him on 15th June 1926).

By this time Pashley was apparently being paid the princely sum of £3 per week for his labours (which represented quite a large wage in those days), while Miles only drew what he needed for current expenses. All running repairs on the aircraft were carried out by the licensed aircraft engineer, A H 'Jimmy' Hawes (GE 784), Miles and by his brother George in his spare time after school.

It is open to speculation as to what career George Miles would have chosen if Miles had not eventually decided to pursue a career in aviation but, after his 15th birthday in 1926, George started acting as unpaid assistant to his brother as a means of learning how to fly operate and repair aeroplanes. In between swinging the propeller (and with it the big rotary engine), George was largely responsible for the maintenance of the Avro 504K, which meant spending unlimited hours at Shoreham doing the dirtiest jobs imaginable. George was also a keen motorcyclist and motorist and he made himself responsible for the overhaul and tuning of the engines, soon becoming an expert at this job. He also acted as ground assistant to Miles and Pash when they were joyriding while he was endeavouring to get as much flying instruction as possible.

At the liquidation sale of Grahame-White's effects, which was held by auction at Hendon on 17th February 1926, Miles had purchased two aeroplanes, a number of engines and a quantity of tools and equipment. The aeroplanes were the Grahame-White GWE.6

Grahame-White GWE.6 Bantam G-EAFL acquired by Miles in February 1926. [P H T Green Collection]

Above and below: Two views of the Grahame-White GW.15 Boxkite at Easter's Field, Shoreham, in 1926. [R Almond Collection via A Brook]

Chapter 1: The Early Days - F G Miles at Shoreham, Sussex - 1925 to 1931

Avro 504Ks and 'gang' at Easter's Field in 1926. Note the 'new' Bessoneau hangar in the background. [R Almond Collection via A Brook]

Bantam G-EAFL and the unidentified Grahame-White GW.15 Boxkite, which figures in the story of Miles' first solo a few months' later.

The Bantam, which had been built in 1919, was a small single-seat biplane fitted with a Le Rhône engine of 80hp and it was rumoured that up to then no pilot had ever succeeded in landing it successfully. The Boxkite two-seat pusher training aeroplane was built during the First World War and was a fitted with an 80hp Gnome rotary engine. The true identity of this machine has never been established but a photograph shows it to have the number '24' inscribed in large numbers on the vertical tail surface, although the significance of this remains a mystery. Three Boxkites were actually registered to Grahame-White in 1919 but only one, G-EABD, actually took up its registration and it is therefore more than likely that this was '24' purchased by the Gnat Aero & Motor Company.

New pupils were beginning to arrive at the club and it soon became clear that more aeroplanes would be needed. In late May 1926 Pash heard that the Central Aircraft Company of Kilburn, for whom he had once worked, and which had ceased flying operations some years previously, still had some aeroplanes for sale and Miles was delighted at the possibility of adding to their fleet. He promptly borrowed some money from his father and went with Pash to inspect them. They turned out to be two Centaur IVs and the pair returned to Shoreham triumphant with G-EABI and G-EALL, which they had obtained for £30 each. The Centaur IV was a two-seat biplane of somewhat massive construction and G-EABI was fitted with a 70hp Renault engine while G-EALL had a 100hp Anzani radial engine. Miles flew G-EALL regularly but G-EABI is not recorded in his log book so it was probably cannibalised for spares. Neither ever aspired to official C of As.

The question of hangarage was by then becoming acute. The capacity of the barn was limited and the aeroplanes of that era, being constructed of wood and fabric, were not sufficiently weatherproof to be left in the open for long periods. By a lucky coincidence, the old film studios on Shoreham Beach had recently closed and amongst their possessions was a Bessoneaux hangar. Miles acquired this and it eased the situation for a while.

1926 was a busy year for the fledgling company as it attempted to become firmly established, albeit with a fleet of aged aeroplanes and equally aged hangars. It was also the year of the General Strike, which threatened to paralyse the country. Volunteers drove lorries and railway locomotives to keep communications open and even newspapers fell victim to the strike. However, the *Daily Mail* had a printing press in Paris for their continental readers and they arranged for Captain Dismore of Imperial Airways to collect loads of newspapers daily and fly them to various destinations in England for onward distribution. One such destination was Shoreham and Dismore would land the large single-engined de Havilland DH.34 there and leave a consignment, which Miles collected in an aged Calcott car and immediately delivered to Brighton. 1926 was also the year in which Pashley renewed his 'B' licence, a necessity when flying for hire or reward, and it was said that this was about the only licence the company held at the time!

Earlier in that year, Miles decided that they needed a more suitable site for their aerodrome and he located one about a quarter of a mile away to the north of the railway line, just to the west of the original first world war aerodrome. They leased this field for 30 shillings a week and moved in during June 1926. It should also be mentioned that they still made occasional use of the first world war airfield until at least 14th August 1930 'as the necessity arose'.

The new site was 400 yards long by 200 yards wide and was situated to the south-west of the centre of the present Shoreham Airport. They moved the Bessoneaux hangar there and the new field served as Shoreham Aerodrome for the next nine years until, in 1935, it became the Municipal Airport of the Brighton, Hove and Worthing District Councils.

Miles also formed the Southern Aero Club in June 1926 and Tamplins, the local brewery, was persuaded to subscribe £100 towards the cost of a clubhouse, on the strict understanding that its product would be exclusively retailed, of course! Recent research has indicated that this clubhouse might first have been erected on Easter's field. Two new hangars were later built, and the famous Miles Tower, in which Miles' office was located, was built in 1928.

An air-to-ground photograph of this airfield showing the Miles Tower, clubhouse, hangars and aeroplanes, taken much later in the mid 1930's, also shows the two new hangars, which had just been completed for the new Shoreham Airport, backing onto the new road running parallel with the railway embankment.

This excellent photograph shows the exact position of the Southern Aircraft site to the west of New Salts Farm Road, which at that time separated the site from the new municipal airport which was being constructed.

Miles then decided to hold a flying display at his new small, unlicensed field in order to gain publicity and to attract prospective members to the club. The display was due to take place on 19th June 1926 at 2 pm but it had hardly been advertised when the authorities began to take notice of the company's illicit activities and blatant disregard of the regulations. They issued an ultimatum to the effect that such operations must cease forthwith and this was quickly followed by a letter from no less a personage than the Director of Civil Aviation, Air Vice-Marshal Sir John William Sefton Brancker, announcing that he intended to visit them to ensure that his instructions had been obeyed.

Cecil Pashley recalled the occasion very well in an after-dinner speech on his early days at Shoreham which he gave at the Grand

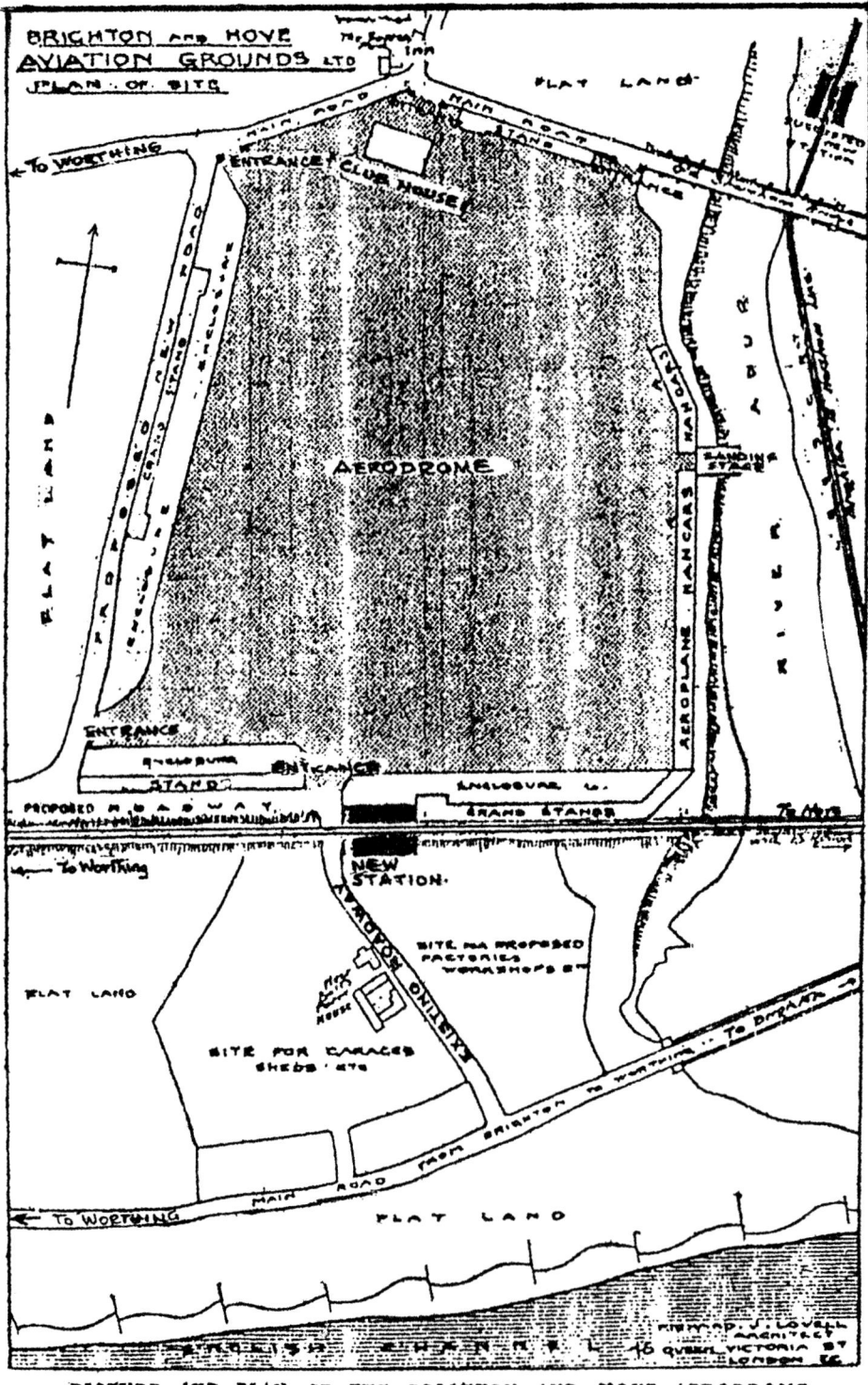

PICTURE AND PLAN OF THE BRIGHTON AND HOVE AERODROME.

Reproduced from an article in "The Daily Graphic" of March 1st 1911. This rather ambitious plan, which includes both a Grand Stand and some spectators' enclosures, also shows a row of hangars along the River Adur frontage. Note the new road replacing New Salts Farm Road on the western boundary. New stations were planned on both the adjacent railways, but only the Bungalow Town halt ever eventuated. Interestingly, at the top of the map there is the name "Brighton and Hove Aviation Grounds Ltd." However, it would seem that it had no connection with Wingfield. The London architect, J. Lovell, who prepared the map, could not have done much research on the site, as he placed the "Club House" almost in the notorious deep pond known as Honeyman's Hole!

The 'first' aerodrome at Shoreham.

CHAPTER 1: THE EARLY DAYS - F G MILES AT SHOREHAM, SUSSEX - 1925 TO 1931

Hotel, Brighton, at a ceremony in his honour, on 29th July 1939, to mark his having completed 10,000 flying hours - all gained giving flying instruction!

But I think that one of my most interesting and amusing experiences occurred when I returned to Shoreham in 1925 with Fred (Miles) and started the Southern Aero Club. We had a small but very enthusiastic following. I am happy to see that there are quite a few of them here tonight. With their help we reconditioned and erected our machines and built hangars. I, myself, was a pilot, ground engineer and labourer all rolled into one.

For some time we carried on successfully without any licences at all, till one day a very ominous letter arrived from the Air Ministry reminding us that if we wished to carry on we must obtain the necessary licences for our pilots and Ground Engineers, also C of As for the machines. This did not meet with our approval and we still carried on. Then we were informed that Sir Sefton Brancker intended coming to the club to put this matter right.

On the day of his arrival we were in a state of nervous prostration, fully expecting to be slung into prison and have our machines confiscated and, as he approached us we all felt like running away but in that friendly way of his for which he was so well known he said - I say you fellows, you must really must get these licences. I need hardly say that we got licences without further delay.

Sir Sefton was known to be a man of considerable force of character who had been appointed Director of Civil Aviation in May 1922. He was obviously well aware that the club was

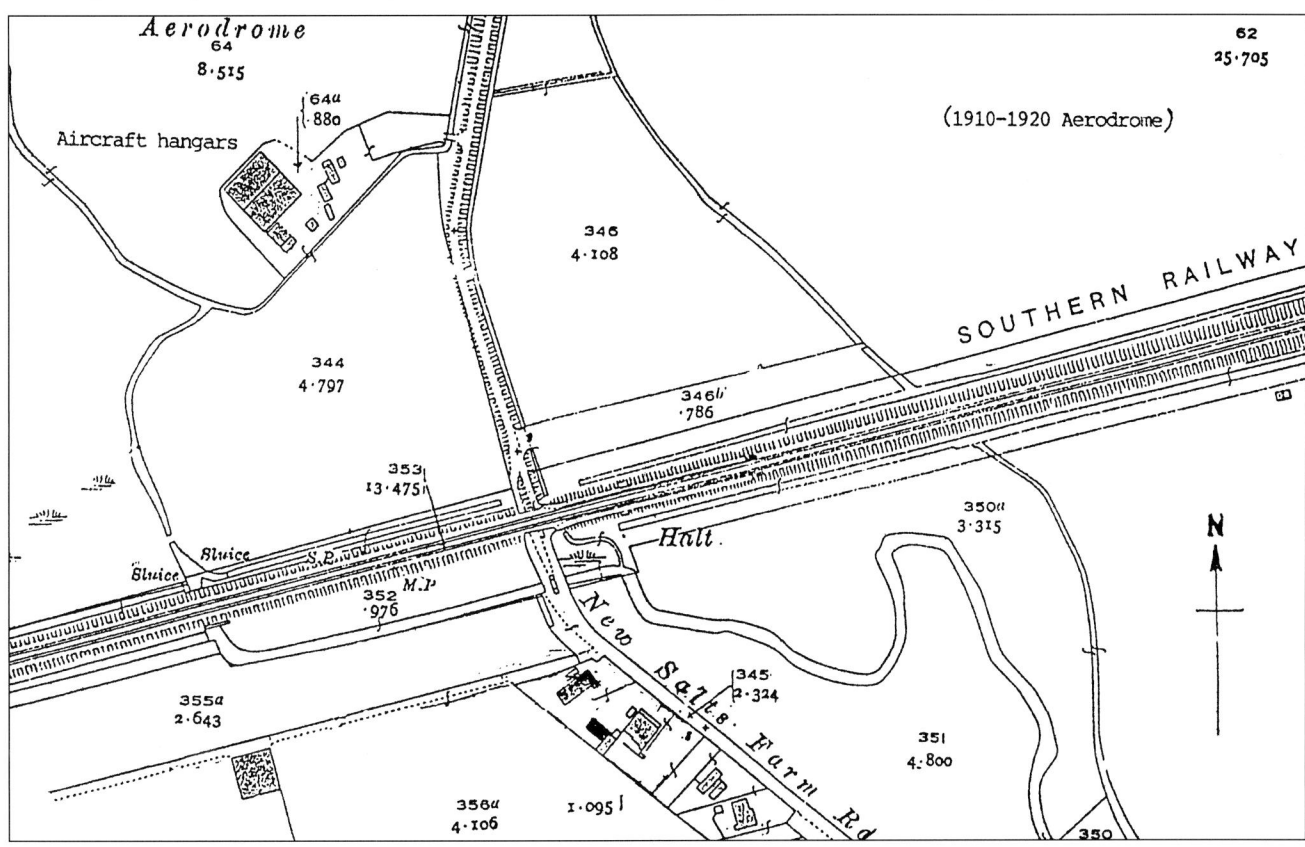

The new Southern Aircraft site is shown by 'Aircraft hangars' (top left).

General view of the Southern Aero Club, Shoreham, with Miles' tower under construction. [R Almond Collection via A Brook]

The new Southern Aircraft site is seen in the centre of this photograph from the NW, with the two new hangars for the municipal airport under construction at top left.

operating from an unlicensed field but before he visited Shoreham to see for himself he wisely decided to send one of his staff with the object of seeing 'how the land lay'. A few days before the advertised date of the display, a Bristol Fighter was seen circling the aerodrome evidently preparing to land. The Bristol weighed over a ton and had no brakes and, as it touched down on the small field, the inevitable happened. Luckily the field was bounded by ditches and not walls so no one was hurt in the ensuing crash, but quite obviously Miles felt that the game was up.

The day of the display duly arrived and, having heard nothing further from the Air Ministry, Miles decided to go ahead as he could not let the public down. He felt that if this was to be his 'swansong' then he should at least try to make a success of it in order to repay a little of the money which his father had lent him. One thing was for sure, he was certainly not going to close down until forced to do so. An hour or two before the display was due to commence, another Bristol Fighter was seen joining the circuit and instinctively Miles started to make his way to the place where he thought that it would go into the ditch.

To his surprise however, the aeroplane was brought in very slowly to make a perfect landing. Out stepped the dapper Sir Sefton, complete with monocle, and Miles, convinced that his fate was sealed, invited him into the clubhouse and offered him a beer, which Brancker accepted, remarking briskly *'So this is the Headquarters of the Independent Air Force!'* He took a brief look around the makeshift workshop and equipment and said, *'No licences, no Certificates of Airworthiness and no proper inspection, eh?'* Miles humbly explained the difficulties, of his determination to overcome them and of his hopes for the future - hopes which in those surroundings must have seemed pretty forlorn. The great man pondered for a moment and prepared to go. Miles awaited the final blow, which came in these few words, *'Really you chaps, you know this sort of thing mustn't go on indefinitely, although I admit that you are certainly not flying for much reward'.*

That remark was characteristic of Brancker who, until his most untimely death in the crash of the airship R.101 on 5th October 1930, had been friend and benefactor not only of Miles but of the British light aeroplane movement and civil aviation generally. Brancker had no patience with 'red tape' and regulations and was the champion of the light aeroplane movement in this country during those formative years. If anyone could have been called irreplaceable then it surely must have been him. To say that he was sadly missed would be an understatement and the light aeroplane movement never really recovered from his loss.

C G Grey, the editor of *The Aeroplane* magazine, wrote of Lt Col Francis Claude Shelmerdine, who succeeded Sir Sefton thus;
....succeeding Sefton Brancker was a difficult task. The happy bands of freebooters which made up British Civil Aviation in his day had become used to his throwing down barriers and bursting bonds. If anybody had tried to follow in his footsteps with his methods the imitator would have come a nasty cropper. Claude Shelmerdine's way was the "suaviter in modo" instead of the "fortier in re".' But nevertheless, Claude had a great ability in getting round obstacles, by strictly orthodox but unexpected ways to which nobody could object, instead of causing a fuss at home or abroad by trying to abolish them.

All this however, was to be in the future and meanwhile the display went ahead as planned. The programme, a copy of which has recently been discovered, is reproduced opposite:

The 'Events' commenced with Pash giving an aerobatic display, which consisted of just one loop from a great height, followed by a perfect landing - Pash was not one for dare-devil stunts. The flight by 'W F Miles' should have read as by 'F G' and from all accounts the display was an unqualified success.

Then, sometime between 29th July and 1st August 1926, the firm suffered its first major set-back with an accident to G-EATU, their one and only Avro 504K. The engine cut just as Pash was

```
PROGRAMME - Price 6d.


           The Southern Aero Club.

                 OPENING MEETING
                         at
                  Shoreham Aerodrome,

           SATURDAY, JUNE 19th, 1926, at 2 p.m.


                      OFFICIALS.

    Club Captain :   Flight-Lieut. Lee Roy L. Brown, D.F.C., R.A.F.
    Ground Engineers :  Messrs. Wallis & Miles.
            Aerodrome Officer :  Mr. B. Richardson.
    Hon. Medical Officer :  Surg.-Lieut. St. G. B. Delisle Gray, M.B., B.S.,
                    M.R.C.S., L.R.C.P., R.N.V.R.
    Traffic Officers :  Messrs. Parker & Greenfield.
    Information Bureau :  Messrs. A. D. S. Robertson & R. W. Hart.
            Member's Flights :   C. L. Pashley.
        Hon. Secretary :  Eng.-Captain F. J. Drover, R.N.

            BAND & DETACHMENT of the
            57th HOME COUNTIES' BRIGADE R.F.A.
              (by kind permission of Lieut.-Colonel J. H. Baines, M.C.)

            A FLIGHT of the 56th SQUADRON R.A.F.
              (by kind permission of Air-Vice-Marshall H.R.M. Brooke-Popham,
                        C.B., C.M.G., D.S.O., A.F.C.

            Air Officer Commanding Fighting Area R.A.F.
                will give a Special Display.
```

```
                       EVENTS

    1.  Exhibition of Flying - C. L. Pashley - Avro
                                    Chief Club Instructor

    2.  Flight by the first graduate pupil - W. F. Miles - Centaur

    3.  Aerobatics by Captain H. S. Broad - Moth
              (Demonstrator and Test Pilot - The De Havilland Co.)

         Display by the 56th Squadron R.A.F.
              (Squadron-Leader F. Vincent, D.F.C., R.A.F.)

    4.  Formation Flying and Flight Drill.

    5.  Ariel Combats.

    6.  Bombing Display.

    7.  Flying in the stone-age - The Hopping Bird

    8.  Aerobatics by Mr. Hinkler (the A. V. Roe Co.)

    9.  Demonstration by Pilot and Machine of the
              Gloucestershire Aircraft Co., Ltd

    10. Exhibition of Flying - Squadron-Leader F. Vincent,
              D.F.C., R.A.F. - Bristol Fighter


            NEW MEMBERS' FIGHTS as can be arranged.


    The above events will not necessarily take place in the order indicated,
         but the number of each event will be announced.

                    ----------o----------
```

becoming airborne after take-off from the South Downs and in the ensuing forced landing the Avro hit a ditch. Although badly damaged, the Avro was not written off as some reports claimed at the time but Pash sprained an ankle and Hawes, who was sitting behind Cecil Boucher (the Secretary of the Southern Aero Club) in the rear cockpit was thrown forward so violently that his teeth became firmly embedded in the back of the latter's neck. Apparently the most amusing part of this otherwise tragic episode was the sight of Boucher ruefully rubbing the back of his neck while Hawes feverishly danced around him trying to catch the broken dentures which were being scattered in all directions!

With the temporary loss of the Avro, the firm's only means of earning money was gone and Miles was forced once again to turn to his father for help. Again he very nobly came to the rescue with a cheque for £300, as Miles had heard that G-EBJE, another Avro 504K, was for sale at Brooklands. This belonged to John Cobb, the famous racing motorist, and on 2nd August 1926 Miles bought it, flew it back to Shoreham the same day and made a further five flights in it before nightfall! The original Avro, G-EATU, was rebuilt over the next couple of months and was flown again, by Miles, on 17th October 1926.

This episode emphasised the necessity of having at least two aeroplanes if the company was to avoid the risk of losing their only means of earning revenue, not only through misfortune but also during periods of overhaul.

Then, one evening, according to Don Brown, Miles heard that another Avro was for sale. This apparently belonged to Col G L P Henderson (who at the time ran the Henderson School of Flying at Brooklands) and it was supposedly lying dismantled in a hangar at Brooklands. Don recalled that Miles, happening to meet his old school friend Bert Hart in the street said; *Let's go and buy it - we can either use it for joyriding or sell it at a profit.* No sooner said than done, and with typical Miles' energy and drive they drove to Weybridge, arriving at 8.30 in the evening. They found digs for the night and Miles said *Let's just go along to the aerodrome and have a quick look at it.* Bert knew only too well what that meant but they found the Avro and as Bert had suspected, Miles suggested that they might as well make a start before going to bed. They arrived back at their digs at 2.30 in the morning but five hours later Miles leapt out of bed, roused Bert and without waiting for breakfast they rushed back to the hangar to complete the assembly.

By 9.30, they had the Avro roughly assembled but not rigged or checked and with nothing locked. *That's good enough*, said the exuberant Miles; *I'll fly it back to Shoreham and we will finish it off there. You take the car back there's a good chap* and without further ado Miles got in, Bert swung the propeller and half an hour later he landed back at Shoreham. Calling 'Jimmy' Hawes and Harry Hull he said - *here it is chaps, all it wants is a few split pins and some locking-wire and get a move on because I want to use it for joyriding after lunch!*

This lovely story is, unsurprisingly, not corroborated by the entries in Miles' log-book but he did in fact acquire the Avro 504K G-EAAY from Brooklands – not from Colonel Henderson though,

GNAT AERO CO., LTD.,

SHOREHAM AERODROM , SUSSEX,

are prepared to undertake the reconditioning of Avros during this winter at reasonable prices.

ALL 110 LE RHONE AND
AVRO 504K SPARES IN STOCK

MACHINES FOR HIRE OR SALE.

WRITE FOR PARTICULARS.

In May 1927, the firm changed its name to The Gnat Aero Co Ltd and was reformed as a limited company. This advertisement appeared in "The Aeroplane" for 31st August 1927.

but from Southern Counties Aviation Co and it was this machine which Miles flew from Brooklands to Shoreham on 8th October 1926. It was also on 8th October that one of the hangars at Shoreham collapsed in a gale and this destroyed the GW.15 Boxkite, although the three Avro 504Ks were extricated unscathed from the wreckage.

Miles did not record flying G-EAAY again until 16th October 1926, which implies that the ground engineers would have had a little more time in which to complete the necessary work. His last flight in G-EAAY was on 3rd April 1927, after which it was sold to Leslie Lewis, who subsequently formed another joy-riding concern, LJ Sky Trips Ltd with L A Jackson.

Their first accident, to Avro 504K G-EATU, in the summer of 1926.

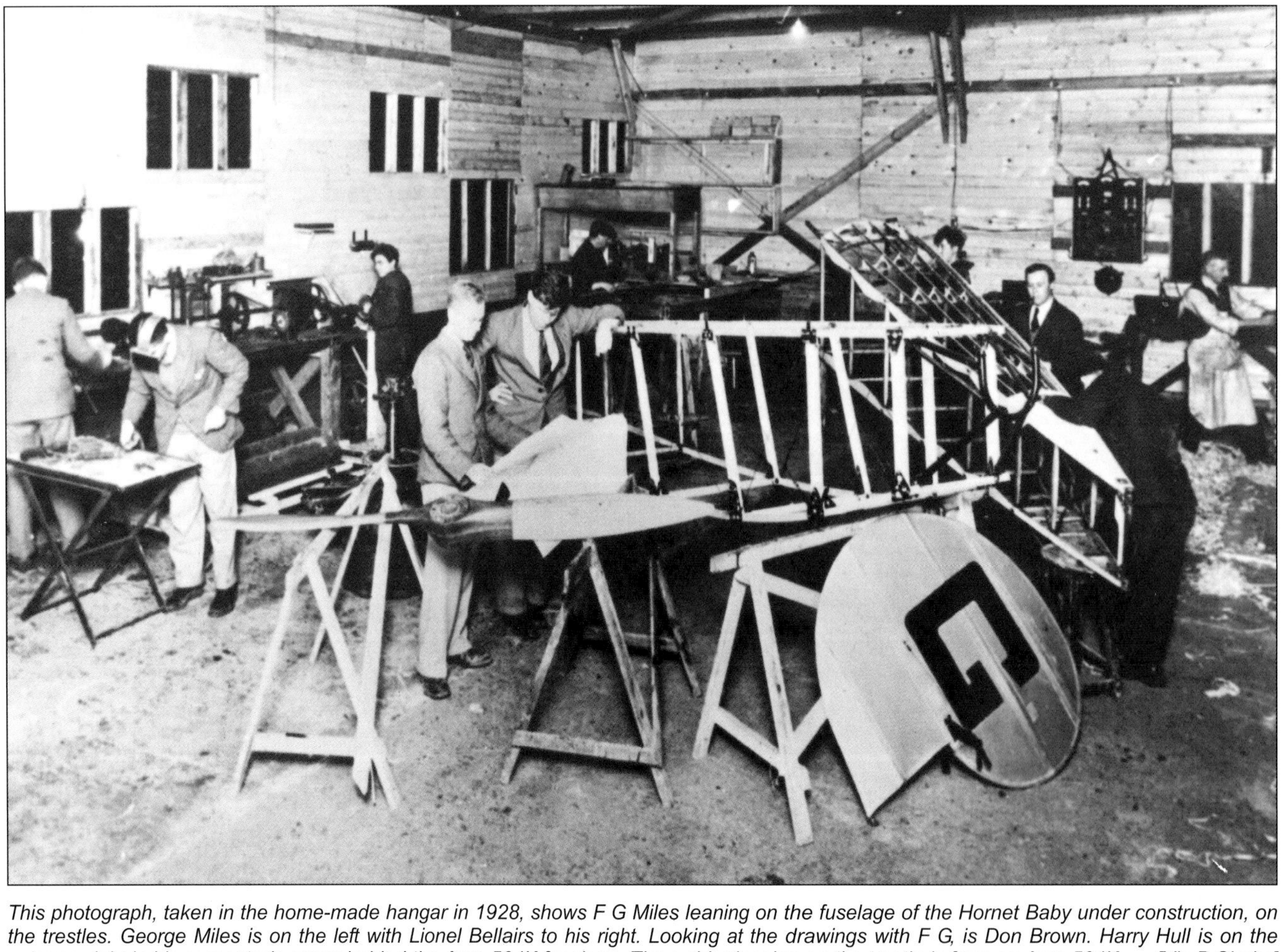

This photograph, taken in the home-made hangar in 1928, shows F G Miles leaning on the fuselage of the Hornet Baby under construction, on the trestles. George Miles is on the left with Lionel Bellairs to his right. Looking at the drawings with F G, is Don Brown. Harry Hull is on the extreme right in his carpenter's apron behind the Avro 504K fuselage. The rudder leaning on the trestle is from an Avro 504K. [Via B Clarke]

Chapter 1: The Early Days - F G Miles at Shoreham, Sussex - 1925 to 1931

Avro 504K G-EAJU at Shoreham in about May 1927. [Via B Clarke]

Details of the conditions under which the aircraft engineers worked at Shoreham in those early days were recalled by A E 'Ted' Hawes OBE, 'Jimmy' Hawes' son, who I happened to meet by chance at a meeting of the Shoreham Airport Society. He explained that Miles worked long hours and naturally expected all who worked for him to do likewise. On many occasions the engineers would work long into the night to keep the Avros flying and Pashley's wife Vera could often be seen sewing fabric 'into the small hours'. Ted Hawes related an amusing story concerning the primitive urinal facilities which existed in the hangar. Over the years the company had collected many spare parts for their rotary engines and among these were a large number of cylinder liners. Some enterprising spirit had set a row of these into the hangar wall at the appropriate height for the workmen and they had been in use as urinals for some considerable time when it was heard that the Russians were in desperate need for cylinder liners, for their intended purpose no less! Without more ado they were removed from the existing use and the men were set to work burnishing these for sale to that country!

Ted also told of how his father, Jimmy Hawes, had engaged in the perilous art of wing-walking, much to the fear and trepidation of his wife who was, not unnaturally, worried for the safety of the breadwinner of the family. She ultimately made him promise that he would never to do it again as she considered this sort of activity by a husband and father to be highly dangerous and irresponsible! Having giving her his solemn promise never to indulge in this pastime again, Mrs Hawes was somewhat aggrieved to see him hanging onto the wing of an Avro the very next time she visited the aerodrome!

In 1927, A V Roe & Co Ltd closed their factory at Hamble in order to concentrate their aviation activities in Manchester. From the sale of airframes and miscellaneous surplus stocks, Miles purchased a number of airframes and some half a dozen rotary engines at a total cost of less than £100. Although much of the timber in the wooden airframes had rotted while in store at Hamble, the metal fittings were serviceable and it was not long before a 'new' Avro 504K, G-EAJU was re-constructed from the assortment of spares. Don Brown also recalled that amongst the purchases was the fuselage of a Viper-engined Avro 504K, which had been modified into the Cierva C.8V Autogiro with the registration G-EBTX on it. This is somewhat curious but I recall seeing a photograph of this amongst the collection of Avro spares Miles acquired. G-EBTX was registered in September 1927 and crashed the following year, later to be rebuilt as the Avro 552A G-ABGO so possibly the accepted history of the Cierva having been based on the Viper Avro G-EAPR may be at fault.

About this time it was agreed that Miles should carry on with instructing and joyriding with G-EAJU at/from Shoreham, while Pash took G-EBJE further afield for joyriding only. The first place chosen for this latter activity was Wannock Glen near Eastbourne and with the additional revenue brought in by the second Avro, it was possible to give both Pash and Hawes a percentage of the takings in addition to their basic £3 per week. A 'Licence For Aerodrome', No.1011, dated 25th March 1927, 'Situated at Seaford, East Blatchington' (Sussex) was taken out by The Gnat Aero Company in order for it *to be used as a regular place of landing or departure by aircraft carrying passengers for hire or reward for a period of five days from the 14th day of April 1927 to the 18th day of April 1927 inclusive.* Its use was restricted to Avro (504K) and similar types of flying machines only and the licence was renewed for use between 9th July and 4th October 1927, so this must have been a good site chosen for their joyriding activities.

On days when business was brisk, Pash would continue joyriding until dusk, returning to Shoreham late in the evening, so late on some occasions that it was necessary to provide flares to enable him to land safely. The flarepath was produced by the simple expedient of having three people standing in a line across the aerodrome, each with a two-gallon can of petrol listening for the sound of the approaching Avro. Each person would then pour a little petrol on the grass and throw a lighted match on it! They would then feed the fires until Pash had landed. It was indeed testimony to Pashley's skill that, small as the field was and with only the three small fires to guide him in, he never once failed to land successfully at the first attempt. Don Brown also recalled that Pashley's skills were even more appreciated when flying as his passenger because, as he approached to land, the three fires appeared extremely small and gave little indication of the height above the ground.

Meanwhile, in accordance with the promise made to Sir Sefton Brancker that 'they would mend their ways as soon as possible' Miles decided to obtain a 'B' licence. With this he would be able

No. 1011.

Class. ---

1164

C.A. Form 7A.

AIR MINISTRY.

Air Navigation Act, 1920.

LICENCE FOR AERODROME.

1. Situation of Aerodrome	SEAFORD, East Blatchington. (see over)
2. Name of Licensee	The Gnat Aero Company.
3. Address of Licensee	Shoreham-by-Sea Aerodrome, Shoreham, Sussex.
4. Nationality of Licensee	British.

The above Aerodrome is hereby licensed by the Secretary of State under Article 7 of the Air Navigation (Consolidation) Order, 1923, made in pursuance of the provisions of the Air Navigation Act, 1920, as an Aerodrome to be used as a regular place of landing or departure by aircraft carrying passengers for hire or reward for a period of ....Five days........ from the ....14th........ day of ....April........ 1927 to the ....18th........ day of ....April........ 1927 inclusive, subject to the terms of the said Order or other Order under the said Act or of any Directions that may be issued by the Secretary of State, and subject also to the following conditions:—

1. Adequate first aid appliances must be kept at the Aerodrome.
2. "*Extract from Air Navigation (Consolidation) Order, 1923.*"

    ARTICLE 39.—"Nothing in this Order shall be construed as conferring any right to land in any place as against the owner of the land, or other persons interested therein, or as prejudicing the rights or remedies of any person in respect of any injury to persons or property caused by an aircraft."

3. This Aerodrome is licensed for use only by the licensee and by individuals specifically authorised by him.
4. Use restricted to Avro (504K) and similar types of flying machines only.

AIR MINISTRY, LONDON, W.C.2.

Date........25th March 1927........

Secretary of the Air Ministry.

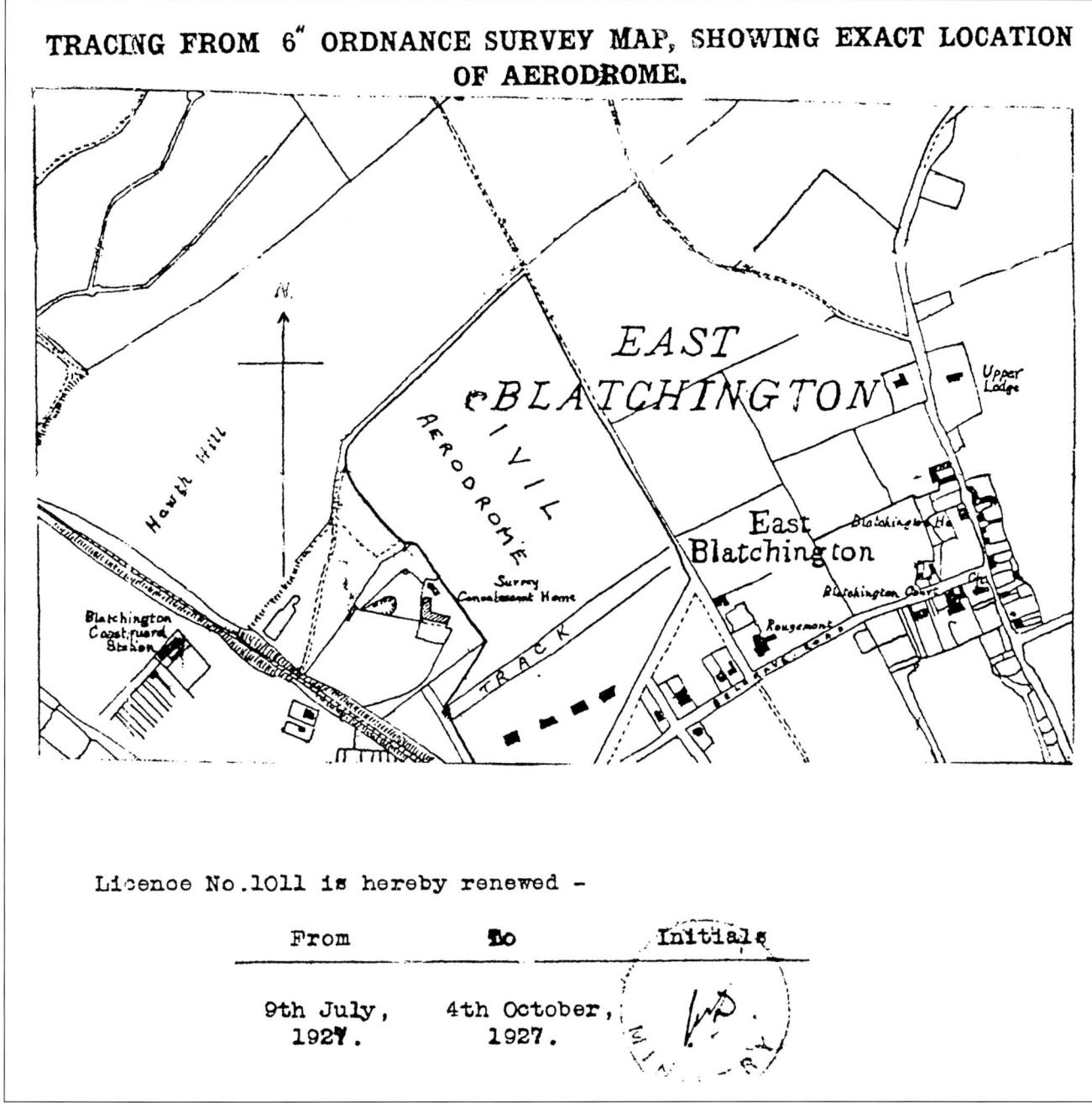

Left and above: The 'LICENCE FOR AERODROME' at Seaford, East Blatchington.

'legitimately' to carry fare-paying passengers and pupils. Having completed the prescribed 30 hours flying, Don Brown recalled that Miles then set out in the Avro to carry out the mandatory 200-mile cross-country flight which formed one of the tests. Unfortunately Don did not record the date upon which this epic event allegedly took place and it has not been possible to corroborate the story in Miles' log-book.

According to Don, the Avro's normal 110hp Le Rhône engine was being overhauled at the time so the Bantam's 80hp Le Rhône rotary engine was fitted in its place and with this lower-powered engine, the Avro's cruising speed was reduced to just 55 mph. On the day of the test, a strongish wind was blowing so Miles had to make several stops for refuelling and it soon became apparent to him that he would be unable to complete the course before nightfall. Nonetheless he carried on doggedly and on the return leg darkness fell, when close to Portsmouth, so he landed in a field and slept under a haystack, completing his flight early the next morning.

The nearest corroborative account to this from Miles' log book was a flight he made on 22nd October 1926. Miles had completed 82 hours 58 minutes flying when he left Shoreham in G-EBJE for Croydon, where he arrived one hour later. He then departed for Netheravon, which he reached in 1 hour 40 minutes and returned to Shoreham the same day in 1 hour 35 minutes. However, it was claimed, probably by Jimmy Jeffs, the control officer on duty at Croydon at the time, that *Miles started his night cross-country flight from Croydon, following the airway beacons to Lympne, where he landed before returning to Shoreham the next day.* However, Miles' log-books show that the first record of a flight to Lympne was not until 12th February 1928, when he left Shoreham in G-EBVL, returning the next day - and he made no comment to the effect that the flight was made at night.

With the passage of time memories fade and it is known that Don relied a lot on his memory when he wrote his book and it is possible that details of the two flights may have become confused, or Miles may simply have even omitted to record the details in his log book. Entries made by Miles in his log book were generally somewhat brief, if at all, to say the least!

Amongst the purchases Miles made from A V Roe at Hamble in 1927 was the single-seat Avro 534 Baby, G-EAUM, which he decided to 'modernise' by replacing the 35hp Green engine with a

The Avro Baby after Miles had fitted a 60hp ADC Cirrus I engine in 1927. [Via The Miles Aircraft Collection]

The re-engined Avro Baby at Heston on 5th July 1929 with its pilot, Flt Lt H H Leech (in flying kit) for the King's Cup Air Race. [Via B Clarke]

60hp ADC Cirrus I and installing two seats. Thus it became his first real attempt at 'design'. Unfortunately though, Miles had not yet learnt the meaning of the term 'centre of gravity' or for that matter the necessity of having it in the correct place and, in consequence, the engine was positioned 'by eye'! The resultant longitudinal trim, or rather the lack of it, would have hardly been acceptable even by contemporary standards but Miles first flew the modified Baby on 13th November 1927. The conversion was apparently successful, with the new engine weighing no more than the Green engine with its radiator. Don Brown recalled that when being flown without a passenger in the rear seat it was inclined to be nose-heavy and unless one was careful it could easily nose-over on landing, which it actually did on one occasion.

P/O H H Leech, an RAF friend of theirs who was serving with 43 Squadron at Tangmere on Armstrong Whitworth Siskins, flew the Baby in the Heston to Newcastle Race and subsequently bought it. Don flew as his passenger on one occasion when there was little or no wind and before opening the throttle Leech remarked dubiously *We may just make it!* The Baby's acceleration was poor and the lack of wind did not help but they did just make it.

Another aircraft acquired from A V Roe at Hamble was the Avro 547A Triplane G-EAUJ, which was fitted with three sets of Avro 504 wings and powered by a 160hp Beardmore engine. It had originally been designed for use as a taxi with four-seats in a cabin. Both Miles and Don toyed with the idea of modernising it but in the end they decided that the modification work necessary to make it into a reasonable proposition would involve more effort than to design a completely new aeroplane from scratch so they dropped the idea. Doubtless its three sets of 504 wings came in very useful for spares for their hard-worked fleet of Avro 504Ks.

In early 1927 the joy-riding side of the business, which was trading as the Gnat Aero Co, instigated another way of increasing their profits by the pilot signing postcards after a flight and selling them to passengers as souvenirs for 2d each! If business was very good, they sometimes gave them away! The Avro 504K G-EBJE had a Disney 'Felix' the cat character painted on the side of the fuselage under the rear cockpit and although this was considered to have been a crowd-puller and good for business, it is not known how this came to be used. Someone probably realised later that Walt Disney would have sued them had he known and so it was retained for only a fairly short time!

The Gnat Aero Co Ltd was formed on 16th May 1927, 'to acquire the unincorporated business from F G Miles and C L Pashley'. A

The Avro Baby G-EAUM, complete with exhaust system. [Via Phil Jarrett]

Avro 504K G-EBJE with the 'Felix' motif. [Via P Amos]

commentary on the firm's activities during those carefree days was given in a series of amusing letters written by F G Miles, which were published in *The Aeroplane* between April and September 1927. These are reproduced in Appendix 2, and a summary of the aeroplanes known to have been flown by Miles in the early days at Shoreham is shown in Appendix 4.

The following article was published in the Worthing Hall of Fame series in a supplement to the Shoreham Herald in January 2007:

F G Miles, the famous aircraft designer - Duck, it's that man Miles again!
At 8.50 on the evening of July 20th 1927, a startled police constable on duty just west of Worthing's Pier Pavilion ducked as a small biplane roared over his head, flying half over the Parade and a half over the beach, at a height be estimated as 'between 60 and 80 feet'. The plane turned towards the sea, flew back to the Pier and again swooped over the Pavilion. 'It proceeded to fly over the Pier at a height of about 15 feet and also over the Bandstand', reported the methodical constable, adding, 'This operation was repeated four or five times, after which the machine landed on the sands and ran along for some 250 yards. It turned and ran for another 250 yards before finally rising and flying off towards Shoreham'.

Constable Haizelden had plenty of time to get the plane's number, G-EAJU, and little difficulty finding the pilot. It was Frederick George Miles. In later years he became better known as F G Miles, famous aircraft designer and (in the 1930s) founder of the Miles Aircraft Company (company name incorrect! - PA).

Miles, who lived at Lancing, designed aircraft such as the Miles Magister, in which hundreds of fighter pilots trained prior to and during World War Two. So by a strange twist of fate the first man ever summoned for flying 'dangerously' low over Worthing later designed and built the planes that honed the skills of the fighter pilots who won the Battle of Britain and saved the nation.

In court back in 1927, F G Miles' lawyer contended: 'Mr Miles is a very safe pilot - so safe that I allowed him to take me up last evening!' (laughter in court). The defendant told the magistrate he was managing director and pilot-instructor of the Gnat Aeroplane Company, held three licences and was allowed to pass other pilots as efficient. He had taken nearly 700 people flying in the past three weeks and had no complaints from any of them. He had received one complaint from the police and that was for 'flying over a church on a Sunday, thereby disturbing the service'.

The magistrate convicted 'F G' of 'causing unnecessary danger to persons on land' but imposed a fine of only £5 - and that to include the witness's expenses.

Twenty years later, after world war two, 'F G' took me for a flight from Shoreham Airport in one of his two-seater Magister aircraft. By this time, local legend had been stretched to suggest that he had once flown UNDER Worthing Pier! 'Never, ever', he assured me and, recalling his court appearance in 1927, added that he never did fly over the Bandstand as alleged and at no time on that day flew lower than 80 feet, except when he came into land.

With a rueful grin he also admitted: 'I'm much safer flying a plane than I ever was driving a car!'

The famous test pilot, the late Wing Cdr Roland Prosper Beamont CBE DSO DFC DL FRAeS, had his first flight as a seven-year old boy in 1927 in G-EBJE with the 'Felix' motif, from a field at Summersdale Copse, Chichester, almost opposite where he lived at the time and about a mile north of St. Richard's Hospital. The pilot for this flight was probably Miles, as Roland Beamont stated in a letter to Sylvia Adams that; *'I don't remember much about the Avro pilot except that he seemed tall and smiled a lot'*. Pashley on the other hand was quite short but Roland also mentioned in the same letter that *'My flight was two smooth circuits over Summersdale looking down at our house in 'The Drive'* and Pashley was renowned for his 'smooth' flying. It could have been that even Pashley might have looked tall to a seven year old boy - we will probably never know who actually took him on his first flight.

The flight was almost certainly made between 28th April and Sunday 1st May 1927, as Miles records flights in G-EBJE from Chichester on every day during that period. Beamont recounts the event in his book *Flying to the Limit*:
Interest in flying was not for me a sudden urge or a desperate teenage mood to 'do something'! It began with a first flight in a "barnstorming" Avro 504 near my home at Chichester in 1927. My father seemed a little apprehensive at my demands to go and see 'the aeroplane in the field'; but when he had negotiated the fare for a flight, five shillings, he said: 'You'll be quite all right, I'll come too' (the Avro carried the pilot and two passengers). I said that would not be necessary at all, but he was determined and came anyway! The rush of wind in the open cockpit, and the sight of our village and home below in that short flight left an indelible impression.

Another contemporary who became involved in the proceedings at Shoreham was Anthony Graham Head, who lived in Hove. Graham, as he was generally known, noted in his diary; *'I commenced my flying career with Messrs Southern Aircraft Ltd at Roedean flying field at Easter 1928'*. Graham was taught to fly by Cecil Pashley and later undertook joyriding with him, mostly at Roedean. Graham proved to be a reliable and competent pilot and was later appointed honorary assistant instructor to look after the club work at Shoreham. In February 1930 he purchased the Avro 548A G-EBKN from Miles. Although Miles had moved to the new aerodrome north of the railway line in June 1926, Graham and Pashley (and probably Miles also) still occasionally flew from the old 1914-18 aerodrome in the South East corner (north of the railway and west of the railway bridge over the River Adur). Graham, made his last flight from there on 7th September 1935, with Don Brown, in the Simmonds Spartan G-AAGY, noting it as 'Shoreham No.2', a 3-minute flight to 'Brighton, Hove & Worthing Airport'. He noted in the remarks of his log book - 'End of old aerodrome'.

It should be mentioned here that, in a letter to Bert Clarke, Don Brown confirmed that he never had any official position in either the club or the firm and that he was only ever a member of the club *'who arrived on the scene in 1928'*

Meanwhile, Pashley and Graham Head had commenced joyriding in the Avro 504K G-EBJE from several small fields in the Brighton area, initially flying from a field between Roedean School, Brighton and the cliff top, from 29th July to 25th August 1927. They also used another field beside the road from Rottingdean to Ovingdean. A duplicate airfield licence issued on 6th May 1930 for the field at Roedean stated that the use of this field was restricted to aircraft of equivalent performance to Avro 504K type. They also occasionally used a field alongside Dyke Road, Hove, midway between the windmills at Patcham and West Blatchington. From Pashley's log books it has been found that he also flew from other sites, including Chichester (28.11.26 and 28.4.27/1.5.27), Seaford (15/18.4.27 and 26/28.8.28) and Bognor (19.7.27, 7/10.4.28, 21.7.28 and 1.8.28) to name but a few.

Miles and Don Brown initially conducted their joyriding from a field at Wilmington (which later became Wilmington Aerodrome) but in the early summer of 1929, Miles found what he thought would be an ideal site at Berwick Court, near Alfriston on the south side of the main Lewes to Eastbourne road. This was adjacent to Drusillas Tea Gardens (which still survives to this day, having been expanded to included a very popular zoo) and with the combination of the main road and Tea Gardens, Miles felt that this would draw the crowds. But first the field had to be licensed as an aerodrome so Don went to London, purchased a large-scale ordnance survey map of the district on which he marked the proposed site and presented it to the Air Ministry. A temporary licence was granted to F G Miles of Southern Aircraft Ltd and Don returned with it to Shoreham.

Chapter 1: The Early Days - F G Miles at Shoreham, Sussex - 1925 to 1931

Graham Head's Avro 548A, G-EBKN, at Shoreham in 1930. [Via B Clarke]

On the afternoon of the 17th July 1929, they flew over to Drusillas in the Avro 504K G-AACW but on arrival discovered to their dismay that, far from having no obstructions, the field was less than ideal, having telegraph wires at the north end and tall trees at the south! It was also so narrow that it could not be used in the other direction for landing, even with an Avro 504K, so operations had therefore to be carried out north and south according to the wind direction. Since by then they were committed, notice boards proclaiming '5/- Flights' were put out along the road.

The proceedings opened with a 10-minute aerobatic display which was calculated to attract the attention of passers-by and people in the locality, and soon a small crowd gathered. Don collected their money and handed out the five-shilling tickets as quickly as he could. The joyriding crew consisted of three people, the pilot, one to swing the propeller and hang on to the wingtips for taxiing and a third to help the customers in and out of the aeroplane, take the money and give out tickets. On occasions Don had to combine the duties of both members of the ground crew because George Miles had gone along with Pash to take the place of Graham Head, who was instructing at the club. If business was not too brisk they would stroll across to Drusillas, have tea and then resume joyriding. At about sunset they would pack-up, place the notice boards in the hedgerow ready for the next day and fly back to Shoreham, their pockets bulging with pound notes, eager to compare the amount they had taken with that of Pash and the rival team which had been operating from Roedean.

These pound notes were very precious because they were needed not only to pay the wages of the few paid members of the staff but also to buy the material for building the Martlet and for further extensions to their hangar. Don recalled that the days spent joyriding were amongst the most enjoyable they had ever experienced, the work was really hard and exhausting but great fun. In the intervals between joyrides they laid on the ground under the wing of the Avro while Don taught Miles basic mathematics, while at the same time swotting for an examination himself! Don later recalled that: *Miles never learned higher mathematics and that is all he knows to this day.*

The story of a typical day's joyriding, entitled '5/- Flights', appeared in the *Miles Magazine* for February 1938 and is reproduced in Appendix 1, with details of the fields known to have been used by Miles for his joyriding activities in Appendix 3.

By then, Miles had obtained both Pilot's 'A' and 'B' licences and four Ground Engineer licences in Categories A, B, C and D - no mean achievement. George, meanwhile, was undertaking all sorts of itinerant flying, buying and selling aeroplanes and engaging in photography etc.

Don Brown recalled that: *The reason why George Miles, Miss Nancy Birkett and myself all took three years to go solo was because F G was so fond of us three that he absolutely refused to allow any of us to go solo. We were told we could fly as much as we liked but not alone and it was only when he fell in love with Blossom much later and forgot all else that the three of us were at last able (rather surreptitiously) to go solo!*

A resume of how these early pioneers faired in their attempts to fly is as follows:

F G Miles received his first instruction from Pashley in the Avro 504K G-EATU on 26th November 1925 and went solo in G-EATU on 19th May 1926 after 15 hrs 53 mins;
Don Brown first flew, with Lionel Bellairs, in the Avian G-EBVA on 19th May 1928 and went solo on 17th March 1931 in the DH Moth G-EBZG after 15 hrs 35 mins;
George Miles received his first instruction from F G Miles in the Avian G-EBVA on 9th June 1928 and went solo in the Avian G-AADF on 13th September 1930 after 16 hrs 52 mins;
Graham Head received his first instruction from Pashley in the Avro 504K G-AACW on 27th September 1929 and went solo in the Avro 504K G-EBYB on 6th January 1930 after 9 hrs 5 mins.

George Miles entered, under 'Past experience', in his first log book: *About 150 hours as operating member of crew on the following types: Avro 504K; Avro 548; Avro Avian III and IV; Avro Baby, Gipsy Moth; Spartan Arrow and Desoutter.*

While George was learning to fly, he flew with Miles, Bellairs, Pashley and Head at various times, in six different aeroplanes of five different types, before eventually going solo while Miles was away!

Once solo, however, George followed Miles' example and soon started giving instruction and joyrides to unsuspecting passengers!

One day, a club member (probably Lionel Bellairs, who owned an Avro Avian), decided to fly over to Drusillas with his fiancée while Miles, Don and a friend flew over in an Avro 504K. The day's joyriding over, they prepared to return to Shoreham but the Avian refused to start. The trouble was water in the magnetos on account of several heavy showers which had fallen during the afternoon and, try as they may, it just would not start. Miles then suggested

Southern Aero Club.

FLYING MEETING,
SATURDAY, MAY 19th, 1928.

Programme — Price 3d.

POPE & BEESLEY, PRINTERS, SHOREHAM.

LIST OF CLUB OFFICIALS.

President:
Sir Cooper Rawson, R.N., V.R., M.P.

Vice-Presidents:
Captain Viscount Curzon, C.B.E., V.D., A.D.C., R.N.V.R., M.P.
Lord Leconfield, Lord Lieutenant of the County of Sussex, J.P.
Commander W. K. Stuart, V.D., M.R.C.V.S., R.N.V.R.

Secretary:
C. A. Boucher.

Club Pilots:
Messrs. C. L. Pashley, F. G. Miles.

Hon. Medical Officer:
Surg.-Lieut. St. George B. Delisle Gray, M.B., M.R.C.S., L.R.C.P., R.N.V.R.

THE SOUTHERN AERO CLUB.

Projected in 1923, and formed in its present guise in 1925, has provided cheap and efficient Instruction in the Art of Flying for pupils from all over the world. In arranging the present Entertainment the Organisers hope that the interest aroused will augment the membership of the Club, and thus assist materially in making and establishing England as the foremost Aviation country in the world.

Applications for membership should be made at the Club House.

Flying Members ... ... £2 : 2 : 0
Non-Flying Members ... ... £1 : 1 : 0
No Entrance Fee.

The programme for the Southern Aero Club Flying Meeting, which was held at Shoreham on Saturday 19th May 1928, is reproduced here.

EVENTS.

1. **Fly Past of Aircraft.**
   Machines will pass in following order.
   Avro Baby, Avro Avian, Avro 504K, Avro Gosport, Avro Lynx, D.H. Moth, Blackburn Bluebird, S.E.5a.

2. **Club Instructors Obstacle Race.**
   Race consists of pilots and crew running to machine, sorting out coats, helmets, overalls, etc., changing wheel on each machine, starting up and completing three circuits, and landing.
   Circuit—Norfolk Bridge, Sussex Pad, Lancing Railway Station and Aerodrome.
   Pilots, Messrs. C. L. Pashley, F. G. Miles.
   Prize. Pair of Flying Gloves and Box of Cigarettes.

3. **Demonstration by D.H. Moth.**

4. **Balloon Bursting Competition.**
   Balloons must be burst by lathe or tube attached to wing. Competitors are allowed three attempts.
   1st Prize, Silver Bowl, presented by F. P. Raynham, Esq.
   2nd „ Propeller Clock „ Southern Aero Club.
   3rd „ Fill up 20 Gallons Petrol Gnat Aero Co., Ltd.

5. **Stunting Exhibition.**
   Latest Avro Gosport.

6. **Wing Walking Demonstration.**
   Messrs. F. G. Miles and A. H. Hawes. Machine, Avro 504K. Weather permitting.

Prize for Smartest Machine.

EVENTS.

7. **Wing Folding Competition.**
   Competitors must unfold wings, start machine, make one circuit, land and fold wings.
   Circuit—Norfolk Bridge, Sussex Pad, Lancing Railway Station and Aerodrome.
   1st Prize, Silver Cigarette Box, presented by F. G. Miles, Esq.
   2nd Prize, Propeller Clock, „ Southern Aero Club.
   3rd „ Fill up 20 Gallons Petrol, presented by Gnat Aero Co., Ltd.

8. **Demonstration Blackburn Bluebird.**
   Pilot, Captain Blake.

9. **A Joy Riding Episode.**

10. **Taxi-ing Competition.**
    Competitors must taxi their machines between flags without touching.
    1st Prize, Silver Cup, presented by J. de Vere Naunton, Esq.
    2nd „ Propeller Clock, „ Southern Aero Club.
    3rd „ Fill up 20 gallons Petrol, presented by Gnat Aero Co., Ltd.
    Booby—Silver Cup.

11. **Height Competition.**
    Competitors must ascend to 2,000ft. with sealed barograph.
    1st Prize, Silver Cigarette Case, presented by Miss N. B. Birkett.
    2nd „ Propeller Clock, presented by Southern Aero Club.
    3rd „ Silver Match Box

12. **Bombing Native Citadel.**

> **44th (H.C.) Div. R.E. (T.) Band.**
>
> SATURDAY, 19th MAY, 1928.
>
> ## PROGRAMME.
>
> 1 March—" Father Rhine " ... ... Paul Lincke
> 2 Overture—" Light Cavalry " ... ... Suppé
> 3 Song Valse—" Janette " ... ... Horatio Nicholls
> 4 Selection—" On with the Show " ... ... Horatio Nicholls
> 5 Idyle—" The Glow Worm " ... ... Paul Lincke
> 6 Selection—" H.M.S. Pinafore " ... ... Sullivan
> 7 Ballad—" My Tumbledown Cottage of Dreams " ... Horatio Nicholls
> 8 Oriental Scene—" A Dervish Chorus " ... ... Sebek
> 9 Humoresque—" A Musical Switch " ... ... Alford
> 10 Intermezzo—" A Memory Garden " ... ... Farman
> 11 Fox Trot—" Persian Rose Bud " ... ... Horatio Nicholls
> 12 Regimental March ... ...
>   THE KING.
>   Bandmaster, H. W. BOND.
>
> IN THE FIELD ADJOINING
> **Aerial Rides can be had from 5/-**
> A marvellous opportunity you should not miss.
> Taste the thrills of Flying for only 5/-
> Safe or sensational flights at your choice.
>
> **Free Flights for Programme Lucky Numbers.**
>
> GNAT AERO CO., LTD., Shoreham Aerodrome.

a solution to the unfortunate owner and his fiancée *'Oh well, hop in with us and we'll all go back together'*! That meant five in the old Avro with an engine giving barely 110hp! The next question was, which way should they take-off? Miles was doubtful as to whether even he could take-off uphill, into wind and over the telegraph wires. The only other option was to take-off downwind towards the tall trees and this he elected to do. Opening the throttle, they roared off down the field, bumping over the rough ground, getting faster and faster with the trees at the end looming ominously close ahead of them. Don had almost given up hope of ever becoming airborne when Miles gave an almighty backward heave on the stick, the old Avro lurched into the air and staggered over the trees. After that the flight home was uneventful!

In August 1928, Miles finally disposed of the Avro 504K G-EAJU by selling it to John Colquhoun Don for his new barnstorming business. This was also based at Shoreham and Miles retained an involvement, recording a flight in 'JU to Swallowfield Park, Reading on 27th October 1928. Then, on 26th November 1928 Don, his wife Mrs E Don and Miles himself registered a new company named Dominion Aircraft Ltd of Shoreham and Gatwick. The new company was established with the aim, amongst other things *'To acquire and take-over as a going concern and carry on the business of Civil Aircraft Pilot now carried on by John Colquhoun Don etc, etc'*. Miles had 10 shares in Dominion Aircraft Ltd, Don had 750 and Mrs Don the remaining 740.

Dominion Aircraft Ltd lasted just over a year and the 'Return of the Final Winding-up Meeting of Dominion Aircraft Ltd', held on 31st January 1930, signed by the liquidator, noted that the property of the company had been disposed of. The Avro 504K G-EAJU had in fact been written-off in a crash near Brighton on 21st July 1929 but, on the formation of the company, two more 504Ks had been acquired. These were G-AACW and G-AACX, both almost certainly erected by Miles from his spares stocks. G-AACW was re-registered to Southern Aircraft Ltd in June 1929 but the fate of G-AACX is unknown, although it is suspected it was written off earlier in 1929. The company was reported to have had some operations in South Africa but nothing further is known. Miles' connection with it and his shareholding may simply have had something to do with the payment for the Avros!

By May 1929, with Miles about to enter the field of aircraft production, it was decided to change the company name to Southern Aircraft Ltd to give the operation more credibility, and this equally applied to their airfield. Shoreham aerodrome was not listed in the first edition of *The Air Pilot* but in the *UK Air Pilot*, (second edition) - 1929, the function of 'Shoreham-by-Sea (Lee's Barn) Aerodrome' it was fully described as shown.

They were obviously by then seeking to become recognised as a professional organisation.

At about this same time, a company called Phillips and Powis Aircraft (Reading) Ltd purchased the Avro 504K G-AAGG from Southern Aircraft Ltd for joy-riding purposes. Miles delivered it to them at Woodley on 4th May 1929, little realising that in three-years time events there would prove to be the turning point in his career.

Meanwhile, in the early part of 1929, Lionel Edward Richard Bellairs, a wealthy club member who had become a director of Southern Aircraft Ltd, wanted to have a single-seat aerobatic aeroplane of his own. Miles, encouraged by the success of his Avro Baby conversion, decided to embark upon a more ambitious modification of the type. This ended up amounting to almost a complete redesign, so Miles thought he should approach A V Roe & Co Ltd for permission to undertake the work. They readily agreed but on the strict understanding that the finished aeroplane would not in any way be associated with their name. Miles designed the new aeroplane to be fitted with an engine giving over double the power of the original Baby and this, in turn, necessitated the redesign of the empennage and undercarriage. In view of these major changes, Miles enlisted the services of a design draughtsman who was conversant with the airworthiness requirements and who possessed the necessary ability to undertake simple stressing.

Harry Hull, a very experienced aircraft carpenter, had joined The Gnat Aero Company Ltd in 1928 on his return from Canada the previous year and, as he had been in aircraft construction since 1913, his vast knowledge of types of wood and their uses made him a very valuable asset to the company. It had first been hoped to use as many parts as possible from the original Avro Baby but in practice this proved impossible although a few metal fittings were salvaged for re-use. Whilst the original intention undoubtedly gave birth to the myth that the new aeroplane was just 'a modernised Baby' and many aviation reference books claim this to have been the case, the new aeroplane, known originally as the Hornet Baby, was in practice a completely new design.

The single-seat Hornet Baby was first flown, by Miles, on 10th July 1929 having been registered G-AAII to Lionel Bellairs in June 1929. Its C of A was granted on 3rd October 1929 and it proved to be very popular with all who flew it but unfortunately this success was to prove a 'mixed blessing'. When accepting orders, the company had to insist on the customer paying cash 'up front' or making frequent progress payments, due to the company's limited capital. Many prospective customers were unwilling to accept these terms, especially when dealing with a small and virtually unknown company, but a few did and the first production aircraft, G-AAVD, by now named Southern Martlet (after the heraldic bird of the County of Sussex) was first flown, by Miles, on 24th March 1930.

It should be mentioned here that, in the context of the stories about Miles designing the Hornet Baby, Martlet and later the Metal Martlet, Don Brown recalled many years later that: *'Miles was the leader of our little gang, the driving force behind all our activities, the complete extrovert and, not unnaturally, he was credited with everything that we did!*

The completion of the first Martlet heralded a year of increasing activity for the company with four more Martlets being completed

SHOREHAM-BY-SEA (LEE'S BARN) AERODROME.

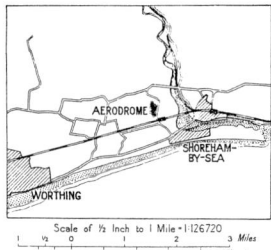

1. Introduction.
(a) *Function.*—Private licensed civil aerodrome. Not available for use, except in emergency, without permission being previously obtained from the controlling authority.

The use of this aerodrome is restricted to Avro 504K and similar types of aircraft.

(b) *Controlling authority.*—Southern Aircraft, Ltd., The Aerodrome, Shoreham-by-Sea, Sussex. The aerodrome is also used by the Southern Aero Club.

2. Location.
(a) *County.*—Sussex.
(b) *Latitude.*—50° 50′ N. *Longitude.*—0° 18′ W.
(c) *Magnetic variation* (1929).—13° W. Annual decrease, about 11′.
(d) *Local position.*—One mile W. of Shoreham-by-Sea, and 200 yards N. of the railway to Worthing.
(e) *Day landmarks.*—Configuration of the coast. The mouth of the River Adur. The town of Shoreham-by-Sea.
(f) *Night landmarks.*—Marine lights along the coast. Illumination of seaside towns.
(g) *Nature of surroundings.*—Flat and fairly open to W. and in the immediate vicinity of the aerodrome, with generally small, grass fields which are bordered by ditches and are liable to become soft in wet weather.

The South Downs approach to N. and E., and present many areas suitable for emergency landings.

3. Landing Area.
(a) *Dimensions.*—
N.—S.—450 yards.
N.E.—S.W.—300 yards.
E.—W.—250 yards.
S.E.—N.W. 300 yards.
(b) *Altitude above Mean Sea Level.*—10 feet (3 metres).
(c) *Surface conditions.*—Level, grass-covered surface. Hard, even in wet weather.

4. Obstructions.
(a) *North side.*—Ditch.
(b) *East side.*—Raised road, 4 feet high. Ditch.
(c) *South-east corner.*—Hangars, 15 feet high. Buildings, 10 feet high. Control tower, 24 feet high.
(d) *South side.*—Ditch. Railway embankment, 10 feet high, surmounted by telegraph wires, 20 feet high, 180 yards distant.
(e) *West side.*—Ditch.

5. Markings and Signals.
(a) *Day.*—
(i) *Markings.*—The boundary of the aerodrome is marked with red flags.
(ii) *Signals.*—Nil.
(iii) *Wind indication.*—A wind sleeve is flown near the hangars situated in the S.E. corner.
(b) *Night.*—
(i) *Markings.*—Nil.
(ii) *Signals.*—Nil.
(iii) *Wind indication.*—Nil.
(iv) *Floodlights, etc.*—Nil.
(c) *Fog.*—Nil.

6. Accommodation.
(a) *Hangars.*—

Number.	Structure.	Net Breadth.	Net Depth.	Door Height.	Door Width.
Two	Wood	ft. 65	ft. 26	ft. in. 12 0	ft. in. 42 0

(b) *Handling facilities.*—Landing party available.
(c) *Hotels, etc.*—Hotels at Shoreham-by-Sea.
(d) *Medical facilities.*—First-aid appliances at the aerodrome. Doctors at Shoreham-by-Sea.

7. Repair Facilities.
All repairs can be executed at the aerodrome. Ground engineers always in attendance.

8. Refuelling Arrangements.
Aviation fuel, oil and fresh water are available. Refuelling pumps are installed.

9. Communications.
(a) *Railway stations.*—Bungalow Town halt, 200 yards. Shoreham-by-Sea station, 1½ miles.
(b) *Local transport.*—Garages at Shoreham-by-Sea. Omnibus service to Shoreham-by-Sea passes the aerodrome.
(c) *Telegraph.*—At Shoreham-by-Sea. "Aerodrome Shoreham-by-Sea."
(d) *Telephone.*—At the aerodrome. Shoreham-by-Sea 168.
(e) *Radio.*—Nil.

10. Meteorological.
Reports may be obtained from the Air Ministry, London. (See page 41, § 27.)

11. Local Regulations.
(a) *Flying hours.*—Between sunrise and sunset.
(b) *Charges.*—By arrangement with the authority controlling the aerodrome. Normally, no landing fees are charged.
(c) *Special.*—

Reproduced from UK Air Pilot (Second Edition) - 1929.

G-AAII, the Hornet Baby, at Shoreham in 1929. [Via B Clarke]

The original caption to this photograph, which was taken from a contemporary society magazine, read: '**An Aviatrix of Famous Theatrical Parents: The Hon. Mrs. Freeman-Thomas.** The Hon. Mrs. [Blossom] Freeman-Thomas is the eldest daughter of the famous actor, Sir Johnston Forbes-Robertson and his equally famous wife, Gertrude Elliott, and was married in 1924 to the Hon. Inigo Freeman-Thomas, the only son of Lord Willingdon, the Viceroy-Designate of India, and Lady Willingdon. The raising of Lord Willingdon, who captained Eton and played in the Cambridge cricket team from 1886 to 1889, to an Earldom, gives Mrs. Freeman-Thomas the rank of a Viscount's wife. She is a keen flying woman and intends to pilot a Martlet single-seater in this year's King's Cup race, and in October to fly with her husband to India in a monoplane which is now being specially constructed for them. The aeroplane, to seat four, is to have its two engines mounted with a view to taking away fumes, noise and vibration from the passengers. Mrs. Freeman-Thomas is to pilot the machine part of the way.' Seen here in this superb photograph, Blossom is swinging the prop on a Martlet at Shoreham in about 1930.

and flown before the year was out. Lionel Bellairs kindly allowed the company to use his prototype for demonstration purposes while he ordered another one for himself and this was also used by the company. G-AAVD gained its C of A on 23rd June, just in time to be flown by Miles in the King's Cup Air Race of 1930. Flt Lt Richard Llewellyn Roger 'Batchy' Atcherley, a member of the victorious Schneider Trophy Team and who later rose to the rank of Air Vice-Marshal, ordered the second Martlet G-AAYX but, due to the shortage of company funds, this could not be completed in time for him, so this was taken over by Lionel Bellairs.

The next Martlet was G-AAYZ, built to the order of Capt The Rt Hon Frederick Edward Guest, MP and was fitted with a 120hp DH Gipsy II engine. This was flown in the 1930 King's Cup Air Race by his personal pilot Miss Winifred Spooner. The Marquess of Douglas and Clydesdale, who later made history by being the first man to fly over Mount Everest, also ordered a Martlet for the 1930 King's Cup, but it was not ready in time to be formally entered. This Martlet was G-ABBN, powered by an AS Genet II engine which had been specially tuned by Armstrong-Siddeley. It was later entered for the 1931 King's Cup Air Race but, unfortunately, on the evening before the race the engine suddenly packed-up and a standard one was installed in its place. This engine had not even been overhauled and inevitably it also packed-up and the Martlet had to retire at Shoreham.

The aerobatic displays given by Flt Lt H H Leech in the Martlet at many major air meetings in the early 1930s apparently created quite a sensation, as Leech, a serving RAF pilot, had the experience to show it off to its best advantage. The sixth and last Martlet, G-ABIF, was built to the order of The Hon Mrs Inigo Freeman-Thomas, Viscountess Ratendone (or 'Blossom' as she preferred to be known) and this made its first flight during late 1930/early 1931. Blossom will feature extensively later in our story.

Meanwhile, by November 1929, and probably following interest shown by Lionel Bellairs, Southern Aircraft Ltd had bought the Martinsyde F.4 G-EBMI (less engine) from ADC Aircraft Ltd. This machine had been designed as a single-seater fighter and was originally manufactured in 1918 for the RFC with serial D4295. It was later reconstructed by ADC Aircraft Ltd at Croydon and registered in their name as a civil aircraft in October 1925, gaining C of A No.966 (Category Aerobatic) on 15th April 1926. This certificate was renewed at the end of the first period of twelve months but, during those two years, the machine (according to the entries in its log book) was flown on only eight occasions and its total flying time amounted to rather less than 3½ hours. The C of A lapsed in May 1928 and the Martinsyde was left in a dismantled condition in the back of ADCs hangar at Croydon.

Miles decided that it should be safe enough to fly - provided it was restricted to straight and level flight! After its arrival at Shoreham, Miles found and fitted a 300hp Hispano-Suiza engine, making it the most powerful aeroplane that had been acquired by the company to date. It was apparently not put into commission with the new engine until March 1930. Miles' first flight in the Martinsyde was on 2nd May 1930 - and this was by way of an accident! Much later, Capt Eric Starling wrote to the editor of *Flight* concerning this incident and his letter was published on 13th May 1960. This prompted the editor to ask Miles for his comments. Miles replied with a copy of the original account, which had first appeared in the *Miles Magazine* for March 1938 - with the comment that *'this was a fairly watered-down version!'*

Miles' story was prefaced by a short pre-amble by the editor under the heading - *Unintentional flight:*

From time to time we hear of planes getting out of control on the ground and making pilot-less flights, usually with unhappy results. This story is on a similar theme, but with a happy ending. In the days when Southern Aircraft were buying up dismantled aeroplanes, re-assembling them and selling them to various customers, Mr Miles acquired a Martinsyde airframe, and a Hispano-Suiza engine to replace the missing Wolseley Viper. (This is an interesting comment as the Martinsyde F.4 was normally fitted with either a 275hp Rolls Royce Falcon III or a 300hp Hispano-Suiza, whereas the 200hp Wolseley Viper engine was supposedly only fitted to the Martinsyde F.6 but there may have been confusion over the Viper-engined SE.5As that Miles had also acquired – PA).

The men had hardly started on the work of re-assembly when a customer came along and said he would buy the machine if they could get a run-out of the engine, just to prove it was a good one. The aeroplane was immediately wheeled out on to the aerodrome, a large post was wedged under the wheels and the engineers began to try to start the motor. After they had been working fruitlessly for some time Mr Miles came out of the workshop in his shirt-sleeves and seeing they were in difficulties, came across to help. At this moment one of the GE's had the bright idea of doping the engine with ether, so Mr Miles swung himself up to the cockpit and, sitting on the edge with his legs hanging over the side, prepared to give contact and work the throttle.

The Avro 504K, G-EBJE, joyriding on the cliffs at Roedean School, Sussex, in 1927. [The John Havers Collection]

After two swings the motor started with a colossal roar. Before anyone could hang on to the wing, the aeroplane had jumped the makeshift blocks and started off down the field. Mr Miles immediately did two things - bunged the throttle forward on the chance that it was coupled up the wrong way; and switched off. He was right about the throttle, but the Viper only needed half the throttle movement of the Hispano, so that the full movement through the quadrant only dropped the engine revs very slightly. Switching off was a good idea too - only the switches were not wired up. By this time she was well away and he had to swing his legs over the side and drop quickly into the extremely hard, un-cushioned cockpit.

There he seized the controls and succeeded in steering the runaway safely through a line of parked Avros. By then she was doing 40 mph and rapidly nearing the ditch. He had to decide whether to turn over in the ditch at a slight risk to himself and certain bad damage to the machine, or to take her into the air and trust to luck and his skill to get her down in one piece.

As all pilots would he chose the air. Having once climbed to about 500 feet, he flew round and considered. Here was the position:- The motor was brand new and running at nearly full revs. He was not sure how much oil and petrol there was. This would not have been very important had the aeroplane been in flying trim, but the wheels had no tyres and were only held on by nails through the hubs. The wings were rigged by eye and none of the wires or cables was locked; the elevator and rudder were working, but of the ailerons only one of the four was coupled up - and that only by one operating cable to the bottom aileron, the return cables were not coupled at all. At high speeds it was possible to turn and bank on the rudder, but below 70 mph it was impossible to pick up the left wing if it dropped.

Having catalogued all these things in his mind, Mr Miles turned his attention to the engine. He found three different petrol cocks and, by experiment, discovered which one turned the petrol off. The carburettor of the Hispano held nearly a minute's worth of petrol, so very nice judgement was required to turn the petrol off at the right moment. He started his attempts to land by flying in straight from 500 feet at about 70 mph and then cutting the petrol a minute before he expected to touch down. Several times he thought he had misjudged and turned the tap on again just before the engine stopped. Once he had his wheels safely on the ground but the engine picked up again with a roar and off he went.

After this he suddenly had the idea of flying over to Tangmere, where he would have an enormous space to land on and the services of a fire-engine and an ambulance. But alas! as he got over Worthing one of the cylinders started missing and, thinking that others would soon follow suit, he turned back again. As he came back over the little 22-acre field that served Shoreham as an aerodrome, he saw that the ground staff had now organised a rescue party. He recognised the doctor's car parked on the edge of the field and saw the firm's car filled with men and fire extinguishers. Once more he got into position to do a straight approach, turned off the petrol cock and glided in at 70 mph. This time he had judged correctly. The Martinsyde sat down as sweetly and quietly as though she had been in perfect flying trim. Even the wheels stayed on.

This story has many morals attached to it, all too obvious to need repeating here.

As the result of this experience Miles made several good rules for himself on the subject of testing - some of which, one must admit, have since been 'slightly broken'.

Bellairs' Martinsyde remained at Shoreham (it was never certificated), was little used and was certainly not subjected to any very high speed manoeuvres. Its total flying time did not amount to more than 10 hours and on 5th August 1930 it was sold to F/O H H Leech, who was still with 43 Squadron at Tangmere. During the first week of August, the aeroplane was flown on several cross-country journeys, undoubtedly in contravention of the regulations, and Leech thought that he might use it instead of the Avro Baby G-EAUM for his occasional trips home to Newcastle. However, although hangarage would have been free of charge, Leech decided that it was going to be too expensive to run and so he either gave or sold it to Edwin Bigg of Bournemouth the same month. It was then flown to Woodley (although by whom is not recorded).

Its stay there was short lived for, on 24th August 1930, while engaged on a private pleasure flight (and probably also actually engaging in aerobatics), it crashed, killing the pilot, S W 'Pat' Giddy, then chief pilot and chief instructor of Phillips & Powis Aircraft (Reading) Ltd. Up to the time of the fatal accident, the machine had flown about 4 hours from Woodley. The accident report concluded that the accident was due to the pilot being deprived of control of the aeroplane owing to elevator or aileron flutter, or possibly both, which occurred during an exceptionally high-speed dive near the ground.

Meanwhile, in February 1930, Graham Head purchased the Avro 548A G-EBKN from Miles who had previously acquired it from the ADC at Croydon. This was really nothing more than the old faithful Avro 504K fitted with a more powerful 120hp Airdisco air-cooled eight-cylinder 'V' engine which drove an enormous airscrew through gearing and in consequence it had about the shortest take-off run ever, becoming airborne long before the throttle was fully open. It also had the steepest approach of any aeroplane they had flown up to then. These approaches could be made from about 200 feet altitude from 100 yards away and it would still touch down with ample room to spare, with a landing run of only a few yards. The engine had an external flywheel with no cowling over the top, so it was possible to lean over the front cockpit windscreen whilst in flight in order to make adjustments to the two carburettors as necessary!

In passing, Graham Head recorded in his diary that on Monday 24th February 1930, George Miles started work at Shoreham for 15/- per week!

A good indication of how the joy-riding team of Pashley and Head faired was recorded by Graham Head in his diary for 1930:

Saturday 19th April: *First day joy-riding at Roedean this year. Did not take a single passenger. Pitched the tent - the one with all the holes in it, and came home.*

Sunday 20th April: *Joy-riding at Roedean. Took about £11. Had to return a fiver owing to rain.*

Monday 21st April: *Bank Holiday. Took about £30 at Roedean. Lost one trade through a puncture since we had no spare wheel. Fine weather all day.*

Tuesday 22nd April: *Joy-riding at Roedean. Took about £5. I flew 'KN (G-EBKN) for 17 mins before we went.*

Wednesday 23rd April: *Wet.*

Friday 25th April: *Went to Roedean in afternoon but did not take a single passenger. Offered job as instructor at Shoreham Aerodrome.*

Saturday 26th April: *Flew 'KN for 30 min. Roedean in afternoon. Did not take more than £2. Graf Zeppelin flew over Brighton on a visit from Germany. Went up in 'JF (from Roedean but was actually G-EBJE) and had a look at her with Pashley.*

Sunday 27th April: *Took £20 at Roedean. Flew for 20 mins in evening.*

Sunday 25th May: Pashley and Head: *Flew to Peckham Farm, Chichester for joy-riding. Took about £5. Visibility bad.* This location was noted (later) as being: *South of Chichester near a convent south of the by-pass on the B2145 Selsey Road.*

Sunday 8th June: *Flew 'KN for 55 min. To Itford Hill where Herr Kronfeld was demonstrating soaring flight. Put 12 galls petrol in her. Went to Drusillas in afternoon with Pash. Took about £12. Came home via Itford where we landed.*

Sunday 22nd June: *Went to Drusillas. Took £14. Flew home round Ditchling and Hassocks as clouds were too low over the Hills.*

Monday 23rd June: *Attended Hove Police Court. Fred Miles got the case dismissed on a charge of low flying.* Graham added later: *this was Fred Miles' summons not mine as it might have read!*

So, joy-riding appears to have been quite a lucrative business on occasions but at other times it must have been very disappointing. A lovely story concerning a not too typical day in the life of a joy-riding aeroplane and Miles came in a letter from Stanley Payne in 1992 to the late Sylvia Adams:

I went with my mother and father on a Sunday in late July or early August 1930 (but see below – PA), *to Lancing. My father told me that a customer had given him a couple of horses to bet on and they had come up, so he was very happy to drive off on Sunday and spend money! He parked the car near the beach and the 4 children went for a swim. Father and mother walked over to a field and went for their 5/- flights. On their return he called us over for our trip.*

I cannot now remember where the exact site of the joy-riding field was but I remember seeing several 'railway carriage' dwellings along to the left. There was a deal kitchen table on which there were piles of postcards, which were being sold at 2d each and a book, which may have been the aircraft log or a book for passengers to sign disclaimers in. I asked a mechanic where the petrol was as I didn't see any cans and he said, 'We have plenty of petrol'. We 4 children went up together, Tom my eldest brother, 17, my elder sister Eileen, 15, 'Billy' (as I was known in the family), 12, and younger sister Josie aged 10 years all climbed aboard. We had an extra good flight. The mechanic said to my parents 'Miles loves children - he's giving them a good trip'. We went low over the sea to see some shipping then back over the field where Miles told them to wave at our parents. Suddenly there was a big bang (the magneto had failed) and Miles did a pancake landing in a nearby field but hit the bank of a drainage ditch and when the skid hit this it dug in and the plane turned over. At least one wing broke off near the root and folded back over the fuselage.

My elder brother helped to release Miles whose legs were partly caught. Miles got first aid from some Boy Scouts who were camping in the next field and they were helped by the mechanic. Miles helped to get the elder girl out and I was thrown out receiving very bad bruising on my back as I hit the grass. While I was still on the ground I saw some people running towards the plane with my father in the lead. People appeared with cameras with the mechanic shouting - warning them off and he nearly came to blows as it was bad for business!

The next week we went down again - joy-riding was still in progress and the mechanic offered a free flight for the children in compensation but my parents said NO! We were also offered a trip round the aerodrome instead and my parents agreed that we would pay a visit another weekend. We did but we couldn't find it!

Don Brown later recalled that Miles did *somehow manage to overturn G-EBVL while landing after a joyride at Southwick, 'near the railway works at Lancing'* (Don knew the area well but his memory must have let him down here as Southwick is to the east of Shoreham and Lancing is to the west! - PA), *fortunately without injury to either himself or his passengers* and this would seem to be the same incident referred to by Stanley Payne as being in August 1930.

But Miles recorded two flights in G-EBVL at Worthing (which is not far from Lancing) on 5th & 6th August 1928 although both recorded 'flights' were total times for joy-riding. The first totalled 2 hr 12 min and the second 2 hr 20 min. Miles made no comment about any accident in his log book but it is notable that there is no record of him flying G-EBVL again after this date and its C of A was not renewed when it expired on 8th January 1929. Indeed, it was not renewed until 2nd April 1931 (which would suggest a prolonged period of storage and subsequent repair) and on which date Graham Head noted in his diary that *G-EBVL crashed through engine failure while taking off, Pashley, who was flying her, was unhurt.* It was joy-riding at Roedean a couple of days later, so it couldn't have been badly damaged.

The Martinsyde F.4 G-EBMI.

[Via B Clarke]

In May 1930, everyone at Shoreham was delighted to hear that Sir Cooper Rawson, MP for Brighton, had been successful in obtaining a Government subsidy to enable the Southern Aero Club, of which he was the President, to reduce their charges for flying lessons. The local newspaper was moved to report that: *with a comfortable club house, the Southern Aircraft company to look after their machines, and Cecil Pashley as an instructor, the reduced charges must lead to an increase in Club membership.*

Part way through June, the Hon Freeman-Thomas and his wife Maxine Mary (Blossom) came to Shoreham to learn to fly. The Hon Freeman-Thomas was the sole surviving son of Lord Willingdon, one time MP for Hastings, who had since held a number of important administrative posts and was currently the Governor-General of Canada.

The Hon Inigo Brassey Freeman-Thomas held a number of directorships in the City, including one in Daimler Motors. His wife, Blossom, was a very attractive and talented person, a woodcut artist, engraver and a brilliant theatrical costume designer. Blossom took to flying 'like a duck to water' under the tuition of Miles and within a month had gone solo. In the meantime, her husband had become interested in the Southern Aircraft company and its potential, and particularly the Martlet aeroplane, and became a director of the company at the end of the year.

At the end of June, Brighton held its Annual Gala, which included a number of events, including a speedboat race. That year, as an added attraction, Miles and M H Volk organised an air show. Prior to the display, Miles made a number of flights over the Brighton beaches near the Aquarium to decide on the best air display area and to make sure that the spectators would get a good view of the aircraft, whilst ensuring their safety. These preliminary flights had been agreed with a senior police officer, but a lady, seeing what she thought was dangerous flying, complained to the police. They were unaware of the permission already given for the flights and issued a summons against Miles. He had to appear in Court before the Gala took place, but the magistrate, apprised of the circumstances, dismissed the case without a stain on his character, no mention being made of his previous conviction, however!

The display on Saturday was a great success: three German Klemm light monoplanes gave a demonstration of formation flying and Miles threw his new Martlet around the sky in gay abandon. Dudley Watts demonstrated his new DW2, which had a very low landing speed. Miles reappeared in an Avro 504K, with his mechanic H H Mason, for a display of wing-walking, and the air show closed with two Avro 504Ks, piloted by Miles and Pashley respectively, bombing the speedboat 'Roultia' with flour bombs. Miles was accompanied by his brother George and Graham Head, while Pashley had Nancy Birkett and Blossom in the rear cockpit. Verey lights were fired from both the motor boat and the aircraft, while the West Pier fired some blank shells from their starting cannon used for yacht races. It was all very exciting for the spectators; three hits were scored, two on the speedboat and one on a gentleman on the West Pier, who took it all in good part after being dusted off, leaving him almost as good as new!

In his diary for 25th July 1930, Graham noted *'Flew in 'YB (G-EBYB) to Roedean to escape AID'* and on August 14th wrote *'Flew 'YB into old aerodrome* (the site of the old 1914-18 Shoreham Aerodrome in the SE corner of Shoreham on the north side of the railway, by the railway bridge over the River Adur - PA) *to avoid AID inspectors who would certainly scrap it if they saw it. It is perfectly safe for the kind of flying that we do on it'*. On July 30th, Graham noted: *'Did my first instruction in 'YB'*, and on August 3rd, *'Took my first 12 passengers on 'YB, although I have only an 'A' licence'* - he added later *"first joyrides"*. Brancker was certainly correct in his earlier assumptions but despite their assurances, the rules still seem to have been flouted! Graham Head noted in his diary for August 11th, *Taxied downwind too fast and tore off a tyre and an aileron King post* and on 12th noted, *Had two consecutive forced landings in 'YB due to engine failure (petrol starvation). Was informed that tank was filled when it was not.*

On 22nd August 1930, he recorded, *My first crash in an aircraft of which I was in charge. Was giving Hoss instruction in 'engine offs' when we stalled in from about 200 ft due to loss of speed while turning away from the aerodrome too low. Completely wrote off 'YB, Hoss received a cut on cheek, I sustained slight wounds and a few fractured bones in my feet. Was taken home by George Miles* (who washed the blood off first to clean him up before he went home to his mother!). On August 26th, Graham wrote, *I should like to append the following note: That the accident was in no way my fault nor was there any possibility of my having averted it. It was caused solely through Hoss disobeying my orders deliberately thus surprising me by his action. Mr Yeatman corroborates my orders before take-off.* Interestingly enough, he also noted in his log book that Hoss had gone solo on August 22nd (before or after the crash was not recorded!). He also notes regarding Hoss: *'Engine off' right from 1200 ft!*. G-EBYB was made of sterner stuff though and its C of A was renewed on 3rd October 1930, despite its reported condition and the recent 'write off'.

In *The Aeroplane* for 12th February 1936, Don Brown was moved to write a few words on: *The Good Old Avro –*

Sir, I was pleased to see the article by the Earl of Cardigan. Naturally we at Shoreham have a soft spot in our hearts for the old Avro, because it was the machine with which we started and used exclusively throughout the first and most precarious years of our existence. Although in later years we had various light aeroplanes, we were never without at least one Avro - in fact at one time or another we built up fifteen 504's for ourselves and other people. These were fitted with all sorts of engines from the 80hp Le Rhône to the V-eight Airdisco and the Armstrong Siddeley Lynx.

All our original members learned to fly on the 504K, and, having since flown practically all the well-known types of light aeroplane, I must say that if one really wants to fly properly (which relatively few amateurs do) it is difficult to imagine a better machine than the old Avro. It had all the qualities required of a good training aeroplane and, although the technique of flying an Avro is rather different from that required in flying modern clean aeroplanes, it is still in my opinion the best machine that has been produced so far for training a pupil to fly safely and accurately.

The quick take-off due to the light wing-loading and slow-running rotary engine, together with the steep glide and low landing-speed were to my mind far more desirable qualities in any aeroplane than the "top speed at any price" ideas of today. In fact the tendency at the present day seems to me to rely upon the amazing reliability of the modern aero-engine as an excuse for producing thoroughly bad and dangerous aeroplanes, which are absurdly simple to fly in the air but which call for an unreasonable degree of skill if they have to be landed without the engine.

While no-one will deny that high performance is essential if an aeroplane is to fulfil its function as a useful vehicle, it would appear to be a retrograde step to sacrifice or at least risk all possible safety in order to attain a high performance before we have yet learned how to produce aeroplanes with a speed range of say 5-to-1. To return to the old Avro, how often did one catch fire in a crash? I only wish that we could say the same about many types of modern machine.

Again, surely no better proof could be given of the safety of the Avro than the fact that, although we were operating from what must have been one of the smallest aerodromes in the country for over nine years, we never had either a fatal or even a serious accident at Shoreham. In fact it was amazing sometimes what pupils could do in an Avro and still emerge unscathed.

There was one bright lad who went up to do the 'engine off' for his 'A' licence. He misjudged his landing and overshot, but, to our horror, instead of opening up again, he continued to glide down across the aerodrome finally arriving at the other side at a height of about ten feet. Being apparently of an optimistic nature, however, he did not abandon hope even then and although he had so far failed to touch down he simply thought that he would turn round and land back down wind! So, still 'engine off', he just put the Avro into a steep turn - at least it would have been steep if the wing hadn't touched the ground first. But he walked back to the hangars apparently quite unconcerned.

Another pupil who walked back undeservedly unscathed was one who suddenly decided, after having overshot, to do a steep climbing-turn down wind and 'engine off'. When he alighted (I won't say landed) the engine was still off – quite a long way.

You must not think that all our pupils used to do that sort of thing, as these were exceptional cases that merely serve to show what you can get away with in an Avro. You must forgive the length to which this letter has unintentionally grown. My only excuse is that it may perhaps be regarded as an epitaph to what I suggest is the most wonderful aeroplane that has even been produced.

Following this letter, the Editor C G Grey could not resist having the last word:

Most old-timers and many more recent recruits will agree with Lord Cardigan and Mr Brown. The extraordinary qualities of the good old Avro rather suggest that even now there might be a market for a large and lightly loaded machine of similar type. Certainly, if a club of young enthusiasts want to set about the co-operative production of a real aeroplane, they might do worse than build something on the lines of an Avro, and put a decent modern second-hand car motor into it, instead of trying to make soap-boxes fly with hotted-up motorcycle motors. (An obvious reference to the 'Flying Flea' craze which was then sweeping the country - PA)

Miles flew a considerable number of different types of aeroplanes from Shoreham in those early days and these are detailed in Appendix 4. However, one that deserves mention was the Clarke Cheetah G-AAJK. This was a very small single-seat, high-wing monoplane powered by a Blackburn Thrush 'V' twin engine and was so low powered that when one of its wheels met a slight obstruction whilst taxying, even the use of full throttle was insufficient to move it on and it needed a man to lift the wingtip in order to raise the wheel over the bump before it could proceed! Miles first flew the Cheetah on 6th April 1930, when it was owned by Lord Malcolm Douglas Hamilton but after a few minutes the engine stopped and he had to make a forced landing back at the aerodrome. No fault could be found so he took off again but the same fault manifested itself three or four times. Each time Miles just managed to get back to the aerodrome but on the last occasion he was just a little too far away and had to forced-land in a tiny field which was the only place available. The field was so small in fact that Miles had to come in very slowly with a high rate of sink and hope for the best. He made a skilful approach and finally dropped into the little field wondering if the undercarriage would stand up to the impact. This it did but unfortunately, the seat bearers did not and as he touched down there was a loud crack beneath him, the seat collapsed and Miles finished up with his feet sticking out through the fabric beneath the fuselage wondering if he would have to run to keep up with it!

The problem was ultimately found to be caused by vibration from the rigidly-mounted engine being transmitted to the aeroplane which, in turn, caused the petrol cock to turn itself off after a few minutes in the air. During the previous examinations it had been noticed that the fuel cock had been turned off but it was assumed that Miles had done this after landing. They sold the Cheetah later for £10! The reason for Miles' involvement with the Cheetah was not recorded but the aircraft was at the time being advertised for sale (at £75) alongside two SE.5A airframes and a Viper engine, offered at a package price of £25!

On Wednesday 21st May 1930, an advertisement appeared on the front page of the *Chichester Observer* newspaper announcing: *In conjunction with the presentation of the All-Talking Epic of the Air, 'Flight', The Southern Aircraft Company, Shoreham, will give a public exhibition of nerve-thrilling stunts over Chichester on Monday Afternoon, May 26th at 2 p.m. The public are invited to avail themselves of the opportunity of a 'Flip' at Wopple Field, Peckham's Farm, Chichester at the cost of 5s. per person.*

Admission Tickets to the Theatre (Exchange Theatre, West Street - PA) *will be on sale at the Flying Field, and anyone purchasing a Ticket will be entitled to half a "Flip" longer than those not purchasing an Admission Ticket. The Flying will go on all the week. SPECIAL NOTE. The film 'Flight' is scheduled to arrive by aeroplane at 12 noon on Monday May 26th at Wopple Field, Peckham's Farm, Chichester.*

This display is known to have included Miles flying the Martlet and for the record, the film *Flight* was about the Flying Devils Squadron in Nicaragua wiping out a bandit army and was made with the co-operation of the U.S. Marine Corps!

Later in 1930, a local newspaper reported on 'Distinguished Guests':

I think that everyone must admire the perpetual enterprise of the Southern Aircraft Ltd, Shoreham. They have always some new scheme in hand calculated to be of benefit not only to themselves, but to the district as a whole. Their fame is undoubtedly spreading widely over Europe and with it, of course, is always associated the name of Sussex. The purchase of land in that vicinity for a municipal aerodrome was a wise move.

On Sunday Viscount Willingdon (Viceroy of India) and Viscountess Willingdon visited the aerodrome with their son, Captain Freeman-Thomas, who is a director of Southern Aircraft Ltd. Mr Miles (managing director) told me that they are at present engaged in building a large machine which is to convey Mrs Freeman-Thomas, Captain Freeman-Thomas and himself to India next October where they will spend three months as the guests of the Viceroy.

Captain The Hon Inigo Brassey Freeman-Thomas (who became Viscount Ratendone in February 1931) had joined the board of Southern Aircraft Ltd in August 1930 and a press report stated that his wife was due to fly with him to India in February 1931 but the plans had been put back to October. The 'large machine' intended to have been used for the flight was the Falcon IV, a new twin-engined, four-seater designed by Basil Balfour Henderson, who had his own hangar and office alongside Southern Aircraft's. Henderson had previously designed the Hendy Hobo and 302, the wing for Bert Hinkler's Ibis amphibian and had also contributed to the design of the Percival Gull. He was a close friend of 'the Miles' gang' but quite independent business-wise. Contemporary reports stated that the Falcon IV, powered by two pusher Cirrus Hermes engines was being built by Southern Aircraft but it is not clear if this was, in fact, the case and indeed Don Brown thought otherwise.

The proposed expansion of Southern Aircraft Ltd made Miles and the other directors anxious that the Viscount should have an experienced pilot to fly him on his business trips for the company, and C W Bebb, who had recently left the RAF after ten years' service, ending up as an instructor on Siskin fighters, was approached. He joined the firm in early 1931, initially on a trial basis. One of his first 'official' flights was to fly Viscount Ratendone, who was a Captain in the Sussex Yeomanry, to his fortnightly camp on Salisbury Plain, on 13th March 1931, and at the end of the exercise he flew back to pick him up and take him to his home, just outside Hassocks. After lunch they were due to fly to Heston, but they apparently got a cool reception from Blossom on their arrival. She told them that she had no food in the house and that they would have to eat out. They found a nearby pub and on walking back to the aircraft, the Viscount turned to Bebb and suddenly said: "Blossom wants to leave me".

The Falcon IV plans were then unexpectedly shelved altogether, due to an unforeseen development as fate had intervened - Miles met Blossom, the wife of Inigo Freeman-Thomas and taught her to fly. She was given her first flying lesson by Miles on 2nd June 1930 in the Avro 504K G-EBYB and went solo in the same aeroplane on 25th July 1930, after 15 hours 18 minutes dual instruction. It was during the course of this instruction that Miles and Blossom fell in love..... things would never be the same again………

Shoreham Airfield, with Miles' famous 'tower' in the centre.   [R Almond Collection via A Brook]

A group snapped at the aerodrome showing the hangars, some of the club's machines, and a group of members. The names in the group, left to right, are: Mr C L Pashley (instructor), Miss N B Birkett (Secretary, Southern Aero Club), Mr F G Miles (Managing Director of Southern Aircraft Ltd), Lady Ratendone, Lord Ratendone, Mr Gordon, Mr Plant.   [Via P Amos]

Mrs Freeman-Thomas (as she then was) was born in 1902 and was christened Maxine Frances Mary Forbes-Robertson, although she was always known as Blossom, which she preferred. She was the eldest daughter of the famous stage actor, Sir Johnston Forbes-Robertson and his equally famous actress wife, Gertrude Elliott and had married The Hon Inigo Freeman-Thomas, the only son of the 1st Earl of Willingdon, in 1924.

In the book *Maxine* by Diana Forbes-Robertson (Blossom's younger sister) it has been possible to learn of her family background. Maxine Elliott (originally Jessie C Dermot), about whom the book was written, and her sister Gertrude Elliott (originally May Gertrude Dermot), were born of Thomas Dermot, who had arrived in Rockland, Maine as an Irish immigrant boy in about 1850. He had married Adelaide Hall, from one of the old pioneer families whose forebears had already been in the territory for about two hundred years.

Thomas was the son of Thomas MacDermot, who was born in County Galway on 18th December 1837, Tom's childhood memories could not have been particularly happy as he passed none down to his children. He had found his way to Liverpool where he became a 'dock-rat' and, after being involved in a fight, the wife of a New England sea captain took pity on him and persuaded her husband to take him on as a cabin boy on his return to America.

Captain David Ames, who wished to be entirely correct, inquired of the port authorities whether he could take the boy and the British official asked without much interest what Mrs Ames intended to do with him when she got to the other side. Captain Ames, usually poker-faced and forbidding on his quarter-deck, is supposed to have broken into a grin and replied, 'wash his face, send him to school and make a good Methodist of him.' Captain Ames stood bond for Tom on arrival, advising him to drop the Mac off his surname 'they don't like Micks over there,' he told Tom. New York and Boston docks were crammed with Macs and the very sound of the name caused resentment among other dock workers because the destitution of the Irish made them work for any wage offered.

Tom later became a militant Yankee, ready to take to fisticuffs in the shipyards of Rockland if anyone chaffed him as a Mick or a Limey. He did however preserve his clear good speech and passed this on to his children, refusing to allow them a Yankee twang, which assisted greatly the theatrical future of his two daughters.

The girl Tom Dermot married was easy to trace in Rockland as a local historian and genealogist had prepared a history of the region. There were eight generations of Halls, starting with George Hall who had come from Taunton, Devon in 1636 with his wife Mary and ending with Isaac Hall, who had married Sarah Hahn. Adelaide Hall was born to them in 1842 and was a girl of

Miles' first DH Moth, G-EBZG, acquired about May 1931. [Via B Clarke]

'exceptional delicacy of manner, high education, and beauty'. Tom, by now a master mariner, coming and going through the 1860s in a small coastal schooner, led Adelaide to the altar at the Methodist church on 8th March 1863 and within eight days of the wedding Tom was at sea again. They had six children.

The second of these, Jessie, was born on 5th February 1868 and became a famous actress, changing her name to Maxine Elliott. Her sister, May Gertrude was born on 14th December 1874 and also became a distinguished actress under the name of Gertrude Elliott. The name Elliott came about when Jessie became a student of Dion Boucicault, who at sixty-eight had behind him a career of fifty years as playwright, actor, and director. He advised that a stage name of importance was needed and starting with the first name, asked Jessie for the grandest name she knew. The grandest name I know, she said, was the name of a school friend's father, Maximilius. Boucicault felt it had the right ring but was too long so together they 'invented' Maxine. Then they went to work on the last name and after going through all her Rockland ancestors, she remembered Elliott, and thus Jessie Dermot became Maxine Elliott. May Gertrude, who looked up to her big sister and also decided on a stage career, duly changed her name to tour with Maxine as Gertrude Elliott.

Gertrude was always 'little sister' to Maxine, doing what she was told and trying desperately to please until on 22nd December 1900 and against Maxine's wishes she married Johnston 'Forbie' Forbes-Robertson, a 47-year old actor (and twenty years her senior) and described as 'a Victorian, a modern spirit, far in advance of his times'. They lived at 22 Bedford Square, London where Gertrude later became Gertrude Elsmere, Lady Forbes-Robertson. They were to have four daughters, Maxine, Jean, Chloe and Diana. The first daughter, Maxine Frances Mary Forbes-Robertson, (named after her famous aunt) was born on 21st September 1901 and as she grew up she longed to please her aunt Maxine and remained close to her.

When in the late 30s, Miles suggested to Aunt Maxine that the Rolls-Royce expansion into aircraft engines was making the company an excellent investment, she shook her head. 'I think war is coming. Assets in England will be frozen the moment it comes. I will keep my money in America..' Her money yes, but not herself and when war did come in September 1939, no amount of persuading would remove Maxine from the Château de l'Horizon on the coast of the French Riviera, near Cannes. She died at her home at the Château on 5th March 1940 and Blossom and Miles got special Home Office permission to fly out to the funeral. Maxine left an estate of a million dollars, divided among her family. How much of this was left to Blossom is not known but it undoubtedly helped to stand her in good stead in her later life.

Meanwhile, joyriding at Shoreham was on the increase and on 1st August 1930 Southern Aircraft Ltd purchased G-AATF, a Desoutter I three-seat, high-wing cabin monoplane, for taxi work. Given the company's usual parlous state, it is of interest to note that this and a second machine, G-AAWT purchased in March 1931, were fairly new machines and it is therefore suspected that although Miles ostensibly purchased the first one, it is possible that they were actually financed by Lionel Bellairs or Captain Freeman-Thomas, who as Directors of the company were considerably wealthier than the other members. Bellairs actually became the registered owner of G-AATF in March 1932.

The Desoutter was based on a design by Koolhoven in Holland but had been substantially modified by Marcel Desoutter and George Handasyde at Croydon in 1930. One day while Miles was flying one of the Desoutters, a cowling side panel came adrift in the air. This caused an anxious few minutes as there was a distinct possibility that it would come off and shatter the windscreen. Miles immediately applied opposite rudder so that the aeroplane flew along crabwise, thereby holding the cowling in position by air pressure, while Nancy Birkett and Don Brown, who were in the rear seat, moved and sat on the floor. Miles meanwhile held one arm across his face to protect his eyes in case the cowling came adrift but the return to Shoreham was made without further incident.

The aerodrome at that time was surrounded by ditches and one 'advantage' of this and the field's small size was that everybody had to learn to fly really accurately because, unless you approached at the correct speed, it was impossible to land in the space available. Pilots who had been trained at larger aerodromes often had the greatest difficulty in landing at Shoreham, because their instructors had allowed them to approach unnecessarily fast, rather than teaching them to fly accurately. After making three or four unsuccessful attempts to get down, some would give up and go away, some somehow managed to get down in one piece, while others just hoped for the best and sometimes finished up in the ditch! The latter option was chosen one day by a visiting pilot and that is how Miles acquired his first Moth!

Late one afternoon in 1930, a couple flew their DH.60X Moth, G-EBZG into Shoreham and after one or two unsuccessful attempts to land, opted for the policy of 'hoping for the best' and finished up in the ditch. Neither of the two occupants was hurt in the crash but the Moth was extensively damaged. The couple were taken to the clubhouse and given tea while they pondered over what to do about their aeroplane which was not insured. Somewhat optimistically Miles offered them £50 for the wreckage and, to his delight, they accepted. This acquisition was in the nature of a bonus as the company could not have afforded to purchase a new Moth at the time. Miles did at least have the decency to fly them home in the Desoutter free of charge, before hurrying back to Shoreham to examine the wreckage and to see how quickly it could be rebuilt.

The date of this incident has never been accurately established, but the Moth was flying again by October 1930 and was registered that month to the afore-mentioned Captain & Mrs Freeman-Thomas, who christened it *Jemimah*, by which name it was known for the rest of its time at Shoreham.

By January 1931, the structure of the club was revised, with its incorporation as Southern Aero Club Ltd. Nancy Birkett, who had been secretary both to the Southern Aero Club and to Southern Aircraft Ltd, continued as secretary of the Club as well as now being a director of the company. By April, the fleet had grown to three Avro 504K's, two Desoutters and one Moth and it was becoming evident that more staff were needed. On 30th March 1931, F/O Sydney Albert 'Bill' Thorn, who had previously been a test pilot at Farnborough, joined them from Croydon (where he had been Cirrus Aero Engine's test pilot), to act as instructor. Before Bill joined the club, methodical instruction was non-existent! The instructor merely sat in the other cockpit and let the pupil learn how to fly by making mistakes and endeavouring to correct them before it was too late! Little advice or instructions were given and the instructor did not even touch the controls unless it became necessary to avert disaster. With the arrival of Bill, things changed however, flying helmets and Gosport tubes made their appearance, together with an established system of instruction.

Don Brown, who had been receiving spasmodic instruction from Miles, Pashley and Bellairs, became Bill's first civilian pupil and he recalled how astounded Bill had been when he discovered how they had been carrying on. Bill Thorn stayed with the club until 1934 and later joined A V Roe at Manchester as a test pilot (and died in the crash of a prototype Tudor in August 1947). He was replaced as CFI by C W H Bebb, who had been instructing with the club since early 1931.

By early 1931, the Municipalities of Brighton, Hove and Worthing, in common with many other towns in the country, were beginning to think seriously of establishing an airport in the locality. They engaged the services of Sir Alan Cobham to survey the district to find the most suitable site for a municipal airport but apparently overlooked the fact that in August 1928, Miles and his father had already held preliminary discussions with the Mayor of Worthing regarding the possibility of establishing an aerodrome in the locality.

Of course, Miles had previously surveyed the district several times with a view to locating a suitable site. With considerable knowledge of the surrounding countryside, Miles was convinced that the most suitable site by far was that of the original Shoreham aerodrome, which had been situated to the east of their own field and which had been used as a flying school during the 1914-18 war. By 1920, it had been cleared and had reverted to grazing, but in 1929 Miles felt that the time had definitely come to acquire this site and he had turned once again to his father for help. In May 1929,

with the company's prospects looking considerably more secure and coupled with F G Miles senior's standing as a local businessman, Miles was able to borrow £7,000 to purchase the 150-acre site. The Municipalities of Brighton, Hove and Worthing, having tried unsuccessfully to find a suitable alternative to that of the old Shoreham aerodrome site, were thus obliged to enter into protracted negotiations with Southern Aircraft Ltd for the option to purchase the site.

The Hon Freeman-Thomas, who later became Viscount Ratendone, became an essential figure in the negotiations for the proposed joint Municipal Airport to be constructed at Shoreham Airport.

It had always been Miles' fervent hope that, if an agreement could have been reached with the three Municipalities on the choice of this site, then Southern Aircraft Ltd would be responsible for its establishment. Miles, Bellairs and Magnus Volk (the pioneer of the Brighton and Rottingdean Seashore Electric Railway), the Company's Directors, thought that they could undertake the construction of the new aerodrome in stages, or alternatively, obtain additional financial support elsewhere. However, they were unable to raise the extra money, so they turned to the most obvious 'customer', the three Municipalities.

The original intention that the company would manage the aerodrome for the Councils came to nought and the deal was not finally settled until 1933 when the company ended up selling the land to the Municipalities for £10,000. The deal was not as lucrative as they had hoped, as interest on the loan had accrued over the four years and High Court costs incurred in lifting the restriction of nuisance and obtaining permission to sell intoxicating liquor, had taken a fair share of the profits.

It was on 23rd March 1931 that Miles and Blossom made their first recorded joint visit in the Moth *Jemimah* to Reading Aerodrome, Woodley and subsequently Miles also made several solo flights there in the Martlet G-AAYX. However, before events were to take a sudden and totally unexpected turn in August 1931, the last occasion of note involving Miles at Shoreham was quite a spontaneous affair, which was reported on by F D Bradbrooke in *The Aeroplane* for 13th May 1931 as follows:

On May 9th, the Southern Aero Club gave the second luncheon within a week to much the same party of guests, but an important addition was the rash spirit who inspired Shoreham Aerodrome in the days when a customer from outside constituted an epoch. Mr G Arthur Wingfield was an honoured guest and regaled his fellow-guests with tales of a past which he insists was not misty, in that the Shoreham site was and is remarkably free from fog.

Sir Cooper Rawson, President of the Southern Aero Club, presided at luncheon and introduced Mr F G Miles. Mr Miles passed on the burden to Mr Wingfield with such promptitude that the guest of the day admitted he was scarcely prepared with the notes of his impromptu speech. He said he was very glad that the corporations of Brighton, Worthing and Hove had seen the virtue in Shoreham, which is on a direct line from Europe to central England, also from Berlin to New York. However, if they should change their minds, he felt he would like to try the same aerodrome again after 20 years. One thing which the members of both Parliamentary Houses there present might take to heart was the proximity of telegraph wires along the railway and power wires north of the aerodrome.

Viscount Gage agreed heartily about the wires. As a half-trained aviator he strongly objected to anything which reduced the useful area of the inadequate spot on which he was expected to land. Lord Gage omitted to mention the prospective increase in size of the "inadequate spot," but the matter of high wires does not lose anything in importance thereby.

After lunch there was a flying programme swiftly organised by the ruthless Mr J J Jeffs ('Jimmy' Jeffs, a Controller in the Control Office at Croydon and later Commandant of the new London Airport at Heathrow but in 1930 he was also a Royal Aero Club Observer - PA). His victims gave the onlookers some real emotion

The 'hair-raising' Martlet formation performing at Shoreham on 9th May 1931. [Via B Clarke]

CHAPTER 1: THE EARLY DAYS - F G MILES AT SHOREHAM, SUSSEX - 1925 TO 1931

to go on with. Mr F G Miles appeared from the Club side with the new metal Martlet (Hermes II) which he flung joyously about for a minute before landing it and pulling up short in a few yards with the wheel brakes. This is a most delightful mount for the purely sporting pilot. It is perfectly controllable at about 40 mph and seems very fast indeed when the Hermes II is let out.

Mr H H Leech RAF, then took it up to try its paces. We have seen this officer fly on several occasions, and his inclusion in the High Speed Flight recently was not needed to persuade us of his talents. All the same we deprecate the risks he takes with a neck which, however lightly he may esteem it, still represents some thousands of pounds' investment in his country. His vertical bank a foot or two off the ground at the first take-off was thrilling in the exact sense of the word, partly because such goings-on leave the pilot so utterly at the mercy of little circumstances over which he has no control. The Martlet was no such circumstance, however, as Mr Leech demonstrated in fifteen minutes of fine - sometimes too fine - flying.

No less hair-raising was the Martlet formation. Mr Miles flew the metal Martlet (Hermes II) in front, and at his wing tips were Mr Leech and Mr S A Thorn in Genet versions. In spite of the fine weather, or perhaps because of it, the air was none too smooth and it may be that old age is impairing our nerve but the final Prince of Wales' feathers came as rather a relief.

A flypast of the many representative types that were present included the Moth, Puss Moth, Avian, Avian Sports, Bluebird, Cutty Sark, Desoutter, Westland Widgeon, Redwing and Spartan. Later Flt Lt H M Schofield and Mrs Schofield arrived in the Autogiro and the former was immediately approached by Mr Jeffs with his best press-gang aspect and sent up for a demonstration. Afterwards he taxied his craft along the enclosure in reply to popular clamour.

Mr S A Thorn, who has been with Southern Aircraft Ltd at Shoreham for some time (actually only a month – PA), *then took the air with his old mount, the Avian Sports (Hermes II) (G-AAHJ – PA) which Mr Holman, of the Cirrus Hermes Engineering Co Ltd had brought down from Croydon. Their exhibition was quite up to the old standard, which is commendation enough.*

In all about 35 machines attended the meeting, Sqn Ldr the Hon F E Guest came in an Avro (Lynx), and among those who arrived by road were Flt Lt and Mrs Stainforth. Just as we were moving towards the clubhouse for tea, a Spartan Arrow arrived overhead and began a demonstration of non-stallage and general flying ability, from which we suspect that Lt Col L A Strange was the pilot thereof.

George Miles obtained his 'A' Licence on 9th May 1931, the day of the meeting, and Don Brown eventually made his first solo, in *Jemimah,* on the 17th. Shortly after this, someone suggested one evening, that it would be rather good fun to fly in the dark. Miles had already flown the Avro 504K G-EBJE at night, on 4th October 1928 and, as he was always ready for anything new, he promptly volunteered to be the pilot. On 28th May 1931, *Jemimah* was chosen to give some of the club members, including Don Brown and George Miles, their first experience of night flying. They waited until it got dark, wheeled her out, lit the usual three petrol flares and Miles took them up in turn.

It was an interesting and unusual experience and at 11.45 pm, when it came to Don's turn, Miles said to him *'Do your safety belt up tightly because we may finish up in the ditch'!* Miles did in fact overshoot on his first attempt and had to go round again but he made a perfect landing on the second attempt. Don remembered the disapproval with which Pash regarded these activities. Pash was older, wiser and more commercially minded than most of the members and felt strongly about anything they did which incurred any risk. He was all too aware of the precarious financial position of the company and made it very clear that he would not be a party to such activities. Nobody paid for night flying anyway and any damage which occurred as a result of it, would surely have incurred costly repairs and in his opinion it was just not worth the risk.

But then, in July 1931, Miles suddenly and somewhat surprisingly, announced his intention of emigrating and starting life afresh in South Africa. Possibly unbeknown to the others at the time, the fact was that he had fallen in love with Blossom, at the time the wife of Lord Ratendone, one of the wealthy directors of the firm. Realising that his love for Blossom would almost certainly cause a scandal at Shoreham when it was discovered, Miles made his last flight, from Shoreham to Croydon, in *Jemimah* on 4th August 1931 and then lost no time in selling his few possessions. With the money raised, he purchased the Simmonds Spartan, G-AAHA, and took this with him to Cape Town.

Shoreham without Miles seemed unthinkable and it must have taken the 'gang' some getting used to but the sudden loss of their leader was also a severe shock to the company. However, the situation had to be faced, so George Miles took over the management of the club, Nancy Birkett continued as the club secretary, Pashley carried on with the joyriding and Bill Thorn assumed responsibility for all instruction.

Graham Head recorded in his diary for Thursday 6th August 1931 that: *Miles & Bellairs went to Cape Town today.* However, until the recent discovery of this diary it had always been thought that Miles had gone there on his own, but Miles' son, Jeremy, believes that Lionel Bellairs probably accompanied Miles simply for company, although he may well also have paid his fare. When they arrived in Cape Town, Miles joined the local flying club and slowly started to assemble the Spartan. So slowly that the club members began to tease him, saying he was a lazy Englishman. But it was probable that Miles simply did not have the enthusiasm for the task, on realising that his spontaneous migration had been a terrible mistake especially after he had received a telegram from Blossom begging him to come home

However, he completed the assembly of the Spartan and in response to the teasing, was determined to 'show them' that he was not lazy. He took off, on his one and only flight from Cape Town, on 1st September 1931 and flew round the circuit then, aiming the aeroplane at the clubhouse, he ran the wheels across its corrugated iron roof – at right angles to the corrugations! The resultant noise undoubtedly awoke the members within to the fact that the Englishman had something to prove!

The club secretary was not amused, however, and reported the flight, which he considered to have been dangerous, to the local police chief. Very soon after, Miles received a telephone call from a friend of the racing driver Wolf Barnato. The friend, who worked in De Beers, warned him of the impending prosecution and said he should leave to go home with the utmost speed. To help him raise the fare, the friend said he would buy the Spartan from him. It was the weekend and unfortunately the friend could not arrange a banker's draft so he gave Miles £400 worth of sovereigns and a money belt in which to carry them, told him that he had booked his passage on a ship moored in the harbour and that Miles had better get a move on if he was to catch it!

The ship sailed before the police could serve the warrant and Miles, who was probably travelling steerage, happily found that the South African Rugby Team was aboard. They were travelling First Class and it is believed that Miles then had a very pleasant journey home, which he reached at the end of September. On his return, Miles found that Britain was in the process of abandoning the gold standard in the midst of the Depression, and that his £400 in gold was no longer going to be worth much. But, this problem was solved by Blossom and Miles taking a ferry to France, where the gold sovereigns were exchanged for about £1,200!

On their return from France, Miles and Blossom purchased a caravan and toured around Sussex and Kent before finally settling in Sevenoaks. Here, they parked their caravan in a farmer's field, where Blossom planted and nurtured a small garden. They also rented a small room over a shop for use as a drawing office and there they designed their first aeroplane together, which they called the Miles Satyr. Between December 1931 and March

1932 they often flew from nearby Penshurst Aerodrome and summaries of their flying from there are included in Appendices 6 and 7.

A report on the front page of the *Sussex Daily News* for 6th April 1932 read:

VISCOUNTESS DIVORCED.
IN LOVE WITH SHOREHAM FLYING INSTRUCTOR.
LORD RATENDONE'S PETITION.

Viscountess Ratendone, Inigo Brassey Freeman-Thomas, a Lloyd's underwriter, of Victoria Square, W, was granted a decree nisi for the dissolution of his marriage in the Divorce Court yesterday. He alleged adultery by his wife, formerly Miss Maxine Frances Mary Forbes-Robertson, with Mr Frederick George Miles, an instructor in aviation, with whom she became acquainted while learning to fly at Shoreham in 1930.

The case for Lord Ratendone, who married in October 1924, was that up to July last they were perfectly happy. Then his wife said she was in love with Miles. He tried unsuccessfully to persuade her to end the infatuation.

Miles and Blossom were married at the Holborn Registry Office on 6th August 1932, and Blossom was to be his inspiration and partner throughout the many vicissitudes which lay ahead.

Miles now had to get down to the serious business of earning a living as, up to then, he had enjoyed the comfort of having his father to fall back on, and also his wealthy backers in Southern Aircraft. His father, although far from being a wealthy man, had always shown complete confidence in Miles' future and had never hesitated to back his various ventures. His faith in his son had never wavered and Miles himself had never doubted that, in due course, he would be able to repay in full all that his father had so generously gambled on him. But Miles, then 29 years of age, had always said that until the age of thirty he proposed to devote the whole of his energies towards acquiring knowledge and experience and that he would not seriously attempt to make money until that period of training was complete.

Miles now had almost no possessions in the world apart from an innate confidence in himself and a devoted wife. So, together they started to design their first new aeroplane in that small Sevenoaks room, which was to see not just the birth of an aeroplane but also of the partnership which was to lead to so many great achievements.

On completion of the design of the Satyr, they were faced with the prospect of finding someone to make it. Miles, remembering his earlier association with Parnall (see Chapter 5), arranged for them to build it and, whilst it had always been assumed that Blossom provided at least some of the finance required for this venture, as she came from a wealthy family, it now seems likely that at least some of the finance came from Miles' fortuitous sale of the South African sovereigns!

While Miles was supervising the construction of the Satyr at Yate, he took the opportunity to undertake some local joyriding in the Simmonds Spartan G-AAGY which must have helped the finances. He first flew the Satyr G-ABVG from Yate on 31st July 1932 and found it to be a worthy successor to the Martlet, being economical, manoeuvrable and generally pleasant to fly. Further flights soon followed and on 24th August 1932 he flew it from Yate to Bristol (Whitchurch), where he made a local flight and another on the August 25th, before flying to Reading later that day. This visit was probably that which was to lead ultimately to the founding of a major aviation enterprise.

Miles and Blossom moved to Reading (with the Satyr) in October 1932, from where the story continues in Chapter 6.

Brother George meanwhile, remained at Shoreham, becoming acting secretary of the Southern Aero Club in 1933 and remaining closely associated with the club until the management was taken over by Brooklands Aviation Ltd in March 1935. They reformed it in September as the South Coast Flying Club, but George had, by then, rejoined his brother.

However, Miles had not quite finished with Shoreham, as a report in *Flight* for 14th June 1934 showed:

On May 26th, the Southern Club organised a very successful display, which was attended by several thousand people. Mr Cecil Pashley performed some really amazing crazy flying, and, by contrast, Mr F G Miles demonstrated the safety of flying by making a 'dead stick' landing on a given spot from 2,000 ft. In a race round a triangular course, Mr Jack Sale just won on a Fox Moth from Mr George Miles on the Hawk. As a grand finale, a motor car was driven over the aerodrome and bombed; much to everybody's delight, a direct hit by Mr George Miles sent it up in flames.

Unsurprisingly, F G Miles did not record any flights in his log book for 26th May 1934!

Southern Aircraft Ltd itself closed down on Friday 7th December 1934 and this sequence of events marks the end of the beginning - the first chapter in the story of the Miles brothers and their aeroplanes, at Shoreham, had ended.

Southern Aircraft's hangars, fleet and Miles' tower, in about March 1933. ['Flight']

The Hornet Baby takes off. [Via B Clarke)

Chapter 2:
The Hornet Baby and Southern Martlet

Three noteworthy events occurred at Shoreham during 1926 and 1927, which were ultimately to result in the appearance in 1929 of the Hornet Baby, a small high-performance single-seat biplane. The first was the arrival on the scene at Shoreham of the wealthy young gentleman Lionel Edward Richard Bellairs of Burwash, Sussex. The second was the purchase of a large quantity of airframes and miscellaneous items from A V Roe & Co Ltd at their sale of surplus stocks at Hamble in 1927, and the third was when Miles converted the Avro 534 Baby G-EAUM, which had been acquired from the said sale, to take a 60hp Cirrus I in place of its original 35hp water-cooled Green.

Bellairs had learned to fly at Shoreham and had then expressed a desire to own an aeroplane of his own. In late 1927 he acquired the Avro 504K G-EBVL, which Miles had built from the accumulation of airframes and spares bought from A V Roe. However, Lionel was also interested in aerobatics and he yearned for something smaller and livelier, so Miles modified the Avro Baby G-EAUM for him. This aircraft was registered to Bellairs and Miles jointly in November 1927. Having completed the Avro Baby modification, and with the numerous spare Avro parts remaining, Miles saw the opportunity of embarking upon a more elaborate conversion. He thus took his first steps towards becoming an aircraft constructor, which by then had become his firm ambition.

Miles had originally intended, for economy's sake, to use the wings and fuselage from the Avro Baby airframe he had bought and for which he had spares, and effect extensive modifications to produce a modern single-seat unequal-span, single-bay, staggered-wing biplane of appreciably higher performance.

He also chose a 75hp (max 82hp) ABC Hornet four-cylinder horizontally-opposed air-cooled engine with which to power the new machine, as this was approximately the same weight as the original Green water-cooled engine and radiator.

As the work amounted to almost a complete re-design of the Avro Baby, Miles approached A V Roe & Co for permission to undertake the work. They agreed but only on the strict understanding that the aeroplane would not in any way be associated with their name. In practice, it proved impossible to use many of the fittings and other parts from the original Baby as had first been hoped, but the story that they had been used gave rise to the myth that the Hornet Baby was simply a modernised Avro Baby.

During those hectic joyriding days of 1928, Miles somehow managed to find time to work on the new design but as ever the main obstacle was finance, or rather the lack of it! The wealthy Lionel Bellairs overcame this by making it clear that he was willing, either to put up the money in advance or to finance the project as it progressed.

As the new aeroplane was to be fitted with an engine giving over double the power of the Baby, this necessitated the redesign of the empennage and undercarriage and in view of the major changes, Miles enlisted the services of a design draughtsman named Horace Miles, who had previously been employed by A V Roe. Horace, a friend of Basil Henderson but no relation to F G Miles, was conversant with the current airworthiness requirements and also possessed the necessary ability to undertake simple stressing. Just five people, including to a lesser degree, Don Brown, worked on the design and construction of the first aircraft and these mostly in what spare time they had. And, in spite of the team's comparative ignorance of the official routine for designing and building aeroplanes, the Hornet Baby took form in a very short space of time.

Taking the basic concept of the Avro Baby, Miles manufactured a new welded steel-tube engine mounting and re-designed the undercarriage, with a new cross-axle 'V' type undercarriage with fairings over the oleo and coil spring compression legs (in place of the 'V' strut and bungee springing of the original Baby). He then designed a completely new tail unit with an unbalanced rudder replacing the original balanced one. The tapered ailerons of the Avro Baby were retained (although two Martlets were later completed with ailerons of parallel chord).

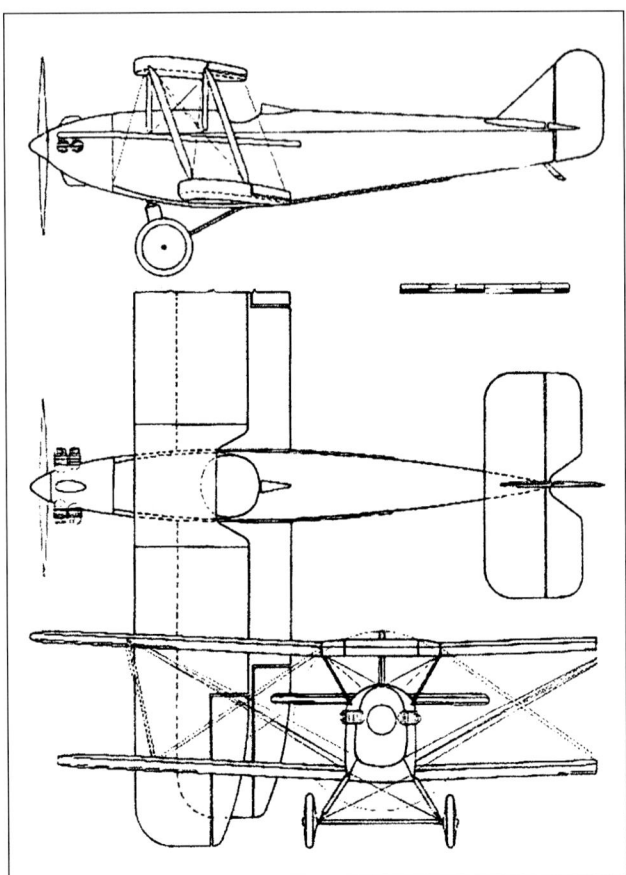

The Hornet Baby with the ABC Hornet engine.

In its final configuration, the wire-braced fuselage was constructed of rectangular section with ash longerons and spruce, while the wooden wings had box-spars with plywood-covered leading edges to the wings back to the front spar. A 15-gallon fuel tank was mounted in the centre section of the upper wing to give a straight gravity feed. A very small headrest was fitted and the whole airframe was fabric-covered.

Design work proceeded during 1928, concurrently with construction and there were occasions when there was some doubt as to which came first, as Miles was prone to go ahead with the manufacture of welded metal fittings, leaving the drawings and calculations to follow! As he was the only member of the team who could weld, it was difficult to stop him anyway!

The prototype Hornet Baby, as it was originally known due to its origins and engine, was registered G-AAII to Southern Aircraft Ltd in June 1929, but on 10th July 1929 it was renamed the Martlet. It was pushed out of its shed in an unfinished condition in order for Miles to undertake the eagerly-awaited taxiing trials and he climbed aboard and commenced to taxi to the far end of the field. However, before he reached there his enthusiasm got the better of him and he decided that it was ready for its first flight! He then turned into wind and took-off and the assembled onlookers were astounded to see the aircraft leave the ground after such an incredibly short take-off run. He entered the flight in his log book as 'July 10 Martlet G-AAII Shoreham -10 mins test'.

When the take-off run was measured during its official handling trials at the A&AEE Martlesham Heath in late August 1929 (Report No.542 dated 9/29), the distance to unstick was found to be only 87ft and a height of 66ft was reached after a take-off run of 795ft. The controls were found to be light, powerful and well co-ordinated, giving excellent manoeuvrability and it was also possible to do sideslip approaches at an indicated 45 mph.

On his return from Martlesham Heath on 31st August 1929, Miles called in at Hanworth Air Park, where he made a couple of local flights and stayed the night. The next day, September 1st, was the official opening of the aerodrome and Mr E S Milk, a local resident at the time, recalled his father telling him about the air display to mark the official opening thus: *A most unlikely looking very young pilot who, when he took off in a small single-seater biplane, electrified the crowd with his brilliant display.* This pilot was, of course, Miles but he made no comment about the display against the 45-minute flight he recorded in his log book, other than 'Hanworth local' - but it must have been quite some display!

No major alterations to the aeroplane were found to be necessary and its C of A was issued on 3rd October 1929. All the pilots who flew the Martlet were unanimous in their praise of the little machine. The Martlet in fact proved a winner from the start, with its main characteristics being short take-off, very steep angle of climb, extreme manoeuvrability, positive controls and remarkable docility. Don Brown commented that its handling characteristics were so superb as to be not easily forgotten.

The prototype was later fitted with an 82hp 5-cylinder Armstrong Siddeley Genet II engine. However, it was to suffer an unfortunate accident in the hands of an inexperienced pilot officer in the RAF, one Anthony Selway, newly posted to RAF Tangmere after training at RAF College Cranwell. He later rose to the rank of Air Marshal and among his reminiscences, extracted from *'Notes from my logbook'* by Air Marshal Sir Anthony Selway, KCB DFC and reprinted here, courtesy of the *Tangmere Logbook,* was: *'What I did to the Martlet'* -

In those days a number of us were very keen on getting in as much flying as possible in any aeroplane, at any time, anywhere. It was a thing we wanted to do more than anything else. Some of us were lucky in being able to borrow aeroplanes at that time. I was able to do this later on but was too new at the game to have the right contacts. One of the arch borrowers at Tangmere was Flying Officer Halliburton Leech, who was universally known as 'Girlie' Leech. There was nothing girlie about him except that when he got

G-AAII in about 1935, after it had been fitted with an AS Genet II engine. [Via B Clarke]

tight he used to giggle in a very high-pitched voice which made us all laugh. He was a very brilliant pilot and a great borrower of aeroplanes of any and every type. He took me over to Shoreham many times to see a friend of his called Miles. He was at that time busy modifying and rebuilding a very small biplane called an Avro Baby. Only one or two of them had been made by A V Roe & Co and then abandoned as a project. Miles had taken this machine and made into a single-seater and put a Genet engine into it and called it a Miles Martlet (At the time, the engine was still an ABC Hornet and it was never known as a Miles Martlet – PA).

Girlie Leech used to go over to Shoreham and test-fly this machine for Miles and he was allowed to take it away at times. On 9th March, 1930 he brought it into Tangmere and he let me have a go in it. It was a delightful aeroplane to fly and wonderful for aerobatics and I greatly enjoyed flying it. Later on we all went off to Hamble Aero Club and there Girlie Leech made a spectacular arrival in the Martlet; we of course having gone by car. We repaired to the clubhouse and met all the local members and after this Leech went up in the Martlet and gave a very good display of aerobatics as low as possible and perfectly executed. The local club members were very impressed, as usual, for their club was only equipped with De Havilland Moths and the scope for aerobatics was limited. When Girlie Leech had landed he said to me 'Would you like to have a go?' and I replied 'Certainly, I'll have a go!' And - this is the snag about competitive flying - I made a resolve to do even better. I did not care to recall at that moment that my total flying hours amounted to exactly 152 hours in the air on all types. Leech of course must have done at that time at least five times that amount. But pilots who have done anything up to 500 hours always believe they know everything there is to know and I was no exception to the rule. It is after 500 hours and a fright or two that you begin to take a little more care in what you do.

Well, up I went and put up much the same sort of show as Girlie had: loops, slow rolls, half-rolls-off-the top and so on and at the end of the show I thought of something that Girlie hadn't done and that was a 'bunt'. A bunt - for the uninitiated - is a reverse loop, that is, you fly level, slowing down a bit, and you then push the control column forward firmly and hold it there until you have dived over onto your back with your head on the outside of the loop rather than on the inside. This of course throws your whole weight into your shoulder straps and you get a little red in the face. Having got to the upside down position you roll out of it to your right side up position and away you go. You had to do the bunt with the engine throttled back as no engine of that type would run in the inverted position.

And so I followed this routine right over the centre of the airfield. All went well and I was congratulating myself on having successfully performed a bunt before the public, when I discovered that on opening the throttle to fly neatly away, nothing happened. The propeller became a motionless stick before my eyes and I had to do a little quick-thinking. I had never had a real forced landing before, only practice ones, and I seemed to have selected the very worst possible conditions in which to make my first one. Low down over a wood, down wind and no engine. But needs must and I came down in a slithering sliding turn trying to get into the wind before encountering Mother Earth. In front of me lay a very unattractive ploughed field (it would of course be ploughed!) into which I put the poor little Martlet. Up we went on to the nose, breaking the propeller, bang went the undercarriage and wheels came into my horrified gaze through the lower wings. And I sat there waiting for an irate Leech to appear, which he did, and as soon as the look of anxiety left his face when he found my unworthy self was undamaged, he asked me in words that I do not like to recall how I thought he was going to explain to the manufacturer how his borrowed aeroplane came to be in its present unflyable condition. All very awkward but I did meekly enquire why the Genet engine (it was still fitted with a Hornet – PA) had failed when it had gone perfectly well for him. This remained a mystery until it was found that when inverted the oil in the crankcase smothered the plugs and bridged the points, thus failing to produce spark. I could not but help wish it had been Girlie who had made this vital discovery. But I had learned something which I never forgot, which was never to trust an aeroplane not to let you down especially at the most critical moments.

I also learned something about human forbearance, for when we reached Shoreham by car late that night to explain to Miles that his one and only aeroplane, his personal invention and brainchild, had been wrecked by an inexperienced pilot officer who had no right to be flying it, we found him to be geniality itself and all he apparently wanted to know was 'whether the undercarriage radius rods had given way in the crash'. In point of fact it did not take long for him to make necessary repairs.

The original design also catered for a two-seat version, although no aircraft were to be built in this configuration but five more Martlets were built between 1929 and 1931 - registered G-AAVD, G-AAYX, G-AAYZ, G-ABBN and G-ABIF. These differed in a considerable number of respects, ranging from five different engine installations to redesigned ailerons, fins, and many other detail changes, which are shown in Southern Martlet production (Appendix 5).

Just one Martlet, G-AAYX, survives and after a very protracted restoration by the Shuttleworth Trust at Old Warden Aerodrome over some 14 years, on Monday 25th September 2000, it was flown for the first time in some 52 years by Andy Sephton, their chief pilot. It is fervently hoped that it will be maintained in flying condition, along with the rest of their collection, for many years to come.

Right: An advertisement for the Southern Martlet taken from a contemporary aviation magazine.

Below: G-AAVD, the second Southern Martlet. [Via B Clarke]

Southern Aircraft Ltd.

THE MARTLET

The Ideal high performance single-seater light aeroplane for the Flying Club and Private Owner.

Cruising Speed - 95 m.p.h.
Top Speed - 112 m.p.h.
Landing Speed - 40 m.p.h.

Price £550.

Write for particulars.
Demonstrations by Appointment.

Learn to Fly at Shoreham.
Private Owners Welcome.

SOUTHERN AIRCRAFT, Ltd.,
SHOREHAM-BY-SEA,
SUSSEX.

Telephone: Shoreham 168.

SPECIFICATION AND PERFORMANCE DATA

Dimensions:
- Span - upper wing 25ft 0in; lower wing 23ft 0in
- Length - Hornet engine 20ft 3in; Genet II approx 21ft 0in; Gipsy I and Gipsy II approx 20ft 3in, Genet Major approx 20ft 7in
- Height - 7ft 7in
- Wing area - untapered ailerons 180 sq ft; tapered ailerons approx 175 sq ft
- Wing section - RAF 15
- Wing loading - approx 5 lb/sq ft

Engine -	ABC Hornet	Genet II	DH Gipsy II	DH Gipsy I	Genet Major
Weight – empty (approx)	730 lb	705 lb	800 lb	790 lb	792 lb
AUW	1,040 lb	1,030 lb	1,105 lb	1,105 lb	1,105 lb
Performance					
max speed	112 mph	112.5 mph	130 mph	125 mph	
Cruising speed	95 mph	95.5 mph	95 mph		
Landing speed	40 mph				
Rate of climb	1,100 ft/min	1,100 ft/min	1,700 ft/min		
Still-air range	285 miles	280 miles			
Duration – approx 3 hours.					
Price – approx £600					

The Metal Martlet in about July 1931. Note the registration G-AAII of the Hornet Baby on top wing! [Via B Clarke]

Chapter 3: The Southern Metal Martlet

A number of well-known Service pilots who flew the Martlet expressed the opinion that it was the ideal machine for economical training in fighting and aerobatics, but at that time there was little hope of interesting the authorities in an all-wooden machine for that purpose, so in 1931 Miles started work on an improved metal version.

Aerodynamically, the Metal Martlet was very similar to its forerunner except that the pronounced stagger, which was a distinctive feature of the earlier machine, was eliminated to allow the wings to fold. Folding wings were fashionable at the time to economise on rented hangar space but the new design still retained the four ailerons, since they provided such sensitive and powerful lateral control. Several revolutionary features were incorporated in the new design and the new machine was one of the first in which welding was almost entirely avoided.

Although named the Metal Martlet, it had little in common with the original Martlet. The fuselage was built entirely of square section steel tubes, jointed with flitch plates and tubular rivets. No bracing wires were used in the structure and no rigging or adjustment was therefore necessary. Any member could be easily replaced in the event of damage without affecting the remainder of the structure. Detachable wooden fairings with doped fabric covering were used to give the fuselage an attractive and efficient external form. The wings and tail surfaces were of composite construction, covered with doped fabric, while the undercarriage was of the split axle type, with low pressure tyres and independently operated wheel-brakes.

It was financed, as with the Martlet, by Lionel Bellairs, with design work again carried out by Basil Henderson and Horace Miles. The latter, having both left A V Roe & Co in 1928, were now in partnership as The Hendy Aircraft Co, mainly designing aeroplanes to order.

The Metal Martlet, with c/n 31/1, was registered G-ABJW on 19th March 1931 to The Hon Mrs Inigo 'Blossom' Freeman-Thomas. However, for some inexplicable reason, the registration was initially mis-painted on the wings as G-AAII (the prototype Martlet built in 1929) and later mis-applied again, as G-AAJW, but this time the error went uncorrected. The Metal Martlet was first flown, by Miles, probably in March 1931, but the exact date is unknown due to his log book for that period having gone missing.

Powered by a 105hp Cirrus Hermes II, the Metal Martlet proved an immediate success and while retaining the short take-off and pleasant handling characteristics of the Martlet, it was appreciably faster due to its smaller size and larger engine. It was as manoeuvrable on the ground as in the air and the many RAF pilots who flew the Metal Martlet were of the opinion that it would have made an ideal aerobatic training aircraft for fighter pilots. It was also made to do 'a very funny dance to music' by Miles, which became a feature of the Shoreham Air Displays of the time.

An article written by FD Bradbrooke, appeared in *The Aeroplane* for 13th May 1931. In this he was moved to comment upon the demonstration following a Southern Aero Club luncheon at Shoreham of the Metal Martlet on 9th May 1931:

Mr F G Miles appeared from the Club side with the new metal Martlet (Hermes II) which he flung joyously about for a minute before landing it and pulling up short in a few yards with the wheel brakes. This is a most delightful mount for the purely sporting pilot. It is perfectly controllable at about 40 mph and seems very fast indeed when the Hermes II is let out.

Mr H H Leech RAF then took it up to try its paces. We have seen this officer fly on several occasions, and his inclusion in the High Speed Flight recently was not needed to persuade us of his talents. (Although this and other articles referred to Leech being a member of the High Speed Flight or Schneider Trophy Team, he was not and was an RAE and later A&AEE test pilot – PA). *All the same we deprecate the risks he takes with a neck which, however lightly he may esteem it, still represents some thousands of pounds' investment to his country. His vertical bank a foot or two off the ground at the first take-off was thrilling in the exact sense of the word, partly because such goings-on leave the pilot so utterly at the mercy of little circumstances over which he has no control. The Martlet was no such circumstance, however, as Mr Leech demonstrated in fifteen minutes of fine - sometimes too fine - flying.*

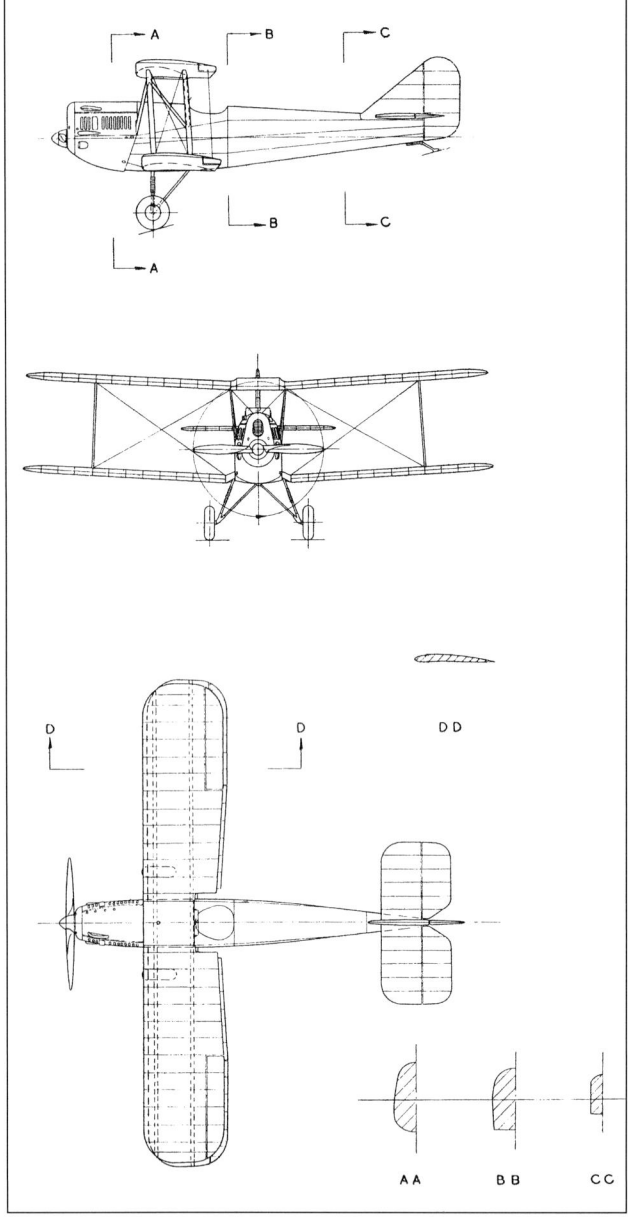

The Miles Metal Martlet.

A copy of a letter, dated 27th May 1931, from W R 'Roy' Maxwell, Director of the Ontario Provincial Air Service, Toronto to Sqn Ldr A T Cowley, Superintendent, Air Regulations, Department of National Defence, Ottawa, was recently found in the National Archives of Canada. Maxwell had a copy of a letter written by the Hon Mr Finlayson (Minister of Lands and Forests) to Southern Aircraft Ltd and it seems that the High Commissioner for Canada (The Hon Mr Ferguson) had written to this Department suggesting that they use one of the Metal Martlet aircraft. In acknowledgement, Maxwell pointed out that the Metal Martlet had not been adapted to floats and he wondered if they were in a position to supply him with any information which they might have on their files regarding the performance or otherwise of this particular machine. In his reply dated, 29th May 1931, Sqn Ldr Cowley regretted that, apart from the article in *The Aeroplane* they had no official information as to the performance of the Metal Martlet but, should he so desire they would make application to the Air Ministry for a confidential report. It is not known if this was taken any further, nor indeed what the initial intentions were.

A second Metal Martlet was laid down with c/n 31/2 and was registered G-ABMM to Walter Retlaw Westhead on 19th May 1931, but it was not completed and the registration was cancelled as 'uncomplete' in December 1932. Westhead later acquired the Martlet G-AAVD. The reason that the Metal Martlet was not proceeded with was almost certainly due to Miles' sudden departure to South Africa in August 1931, thereby ceasing his involvement with the company.

The prototype Metal Martlet was scrapped in November 1932, ending its days at Goody's scrapyard, Twyford, near Reading, Berks. It is possible that Miles or Blossom had taken it to Woodley when they moved there in August 1932.

SPECIFICATION AND PERFORMANCE DATA

Engine:	105hp Cirrus Hermes II
Dimensions:	Span 23ft 6in, span folded 9ft 4in; length 20ft 6in; height 8ft 3in; wing area 156sq ft
Weights:	disposable load 375lb
Performance:	max speed 130 mph; cruising speed 120 mph; take-off distance 150ft; initial rate of climb 1,400ft/min; landing speed 40 mph; landing distance 210ft; service ceiling 20,000ft; still air range 400 miles; duration 3½ hours

No less hair-raising was the Martlet formation. Mr Miles flew the metal Martlet (Hermes II) in front, and at his wing tips were Mr Leech and Mr S A Thorn in Genet versions. In spite of the fine weather, or perhaps because of it, the air was none too smooth and it may be that old age is impairing our nerve but the final Prince of Wales' feathers came as rather a relief.

Although the Metal Martlet was never issued with a C of A, Miles flew 'JW, as he recorded it, regularly from 24th April 1931 to 19th July 1931 but Blossom, despite being the registered owner, only seems to have flown it once – on 17th May 1931.

The Metal Martlet was registered G-ABJW, but flew as G-AAJW. The gentleman with the homburg hat obliterates the crucial letter!
[Via B Clarke]

F G flying the Metal Martlet in about July 1931. [Via B Clarke]

G-AAIN, the Parnall Elf II, which Miles test flew from Yate in June 1932. [Via P Amos]

Chapter 4:
Test Flying for George Parnall & Co, Yate - 1929 to 1932

In 1929 Miles carried out some test flying for aircraft manufacturer George Parnall & Co at Yate in Gloucestershire. This commenced on 20th March 1929, when he flew from Shoreham to Yate in his Avro Avian IV G-AADF and the visit was probably made in order for him to discuss the terms of his contract and to get the feel of the aerodrome, since he made a 30-minute 'local' flight there later in the day. Norman Hall-Warren, who worked for George Parnall & Co at the time, wrote of the company in *Air Pictorial* for February 1980 and gave details of Miles' activities at Yate:

Yate airfield was more suitable for agriculture than for flying! Not only was it small but it had a bumpy surface with a depression deep enough to hide from view a taxiing aircraft. For a final bad mark, at the longest run of about 500 yards and facing the prevailing wind, there were the works buildings and flight shed; to the rear of these obstacles and parallel to them was the railway line and a network of sidings. Beyond again was a housing estate, so it was vital not to have engine failure on take-off - yet this is what happened to F G Miles who was often employed to test Parnall prototypes!'

From Miles' early flying logbooks come brief details of his test flying for Parnall. His first test flight there was made on 8th April 1929, a 30-minute flight in the Parnall Imp G-EBTE, which had been modified to air-test the new 60hp Pobjoy P air-cooled radial engine. This engine had never been flown before but Miles made no comment as to the object of these flights in his log-book; the engine test flights included tests of endurance and were concluded on 15th May, after a total of 29 hr 50 min flying. Among these test flights were four of 3 hrs and two of 4 hrs duration.

Miles flew from Shoreham to Yate as and when required in G-AADF and on 26th June 1929 his objective was to carry out the first flight of a new two-seat light aeroplane, which Parnall had great hopes for. This was the Parnall Elf, G-AAFH, powered by a Cirrus Hermes I, but when Miles took it off on its maiden flight the next day, he little expected the problems he would face. Curiously, Miles made no mention of anything untoward in his log book during the course of what appeared to be a normal 20-minute test flight, but Norman Hall-Warren wrote of it:

Miles took the Elf into the air for its first flight and just after take-off, 200 feet above the works, the engine cut out. We watched the Elf turn and come back to land down-wind. We held our breath when the engine failed again on the second take-off and again for the third time and each time Miles turned back and brought our precious prototype safely to earth. But for the cool head and skill of Miles we could have lost our aeroplane, which was hoped would prove a turning point in the fortunes of the firm, for the Elf was intended as a competitor to the de Havilland DH.60 Moth. To have lost the prototype Elf would have ruined the project's chance of success, for there was no time to be lost as this aircraft was due to be exhibited at the International Aero Exhibition which opened at Olympia on 16th July 1929, only a fortnight after its first flight.

It would be interesting to speculate on Miles' feelings towards the Elf's somewhat unreliable behaviour after its 'first flight' but his earlier experiences at Shoreham and the many and various fields from which he had given joy-rides undoubtedly stood him in good stead and had prepared him for any eventuality. Norman implies that the three engine failures all occurred whilst Miles was trying to persuade the Elf to complete its first flight but, as Miles noted this as being of 20-minutes duration, the Elf must presumably have achieved full powered flight at some stage in the proceedings!

On 28th June 1929, Miles carried out an uneventful 30-minute flight in the Elf, followed by further flights, until 12th July when, following a 20-minute flight at Yate, he flew it to Shoreham (1hr 15min) and on to Martlesham Heath (1hr 35min) for its official trials later that day. A 20-minute flight at Martlesham on 18th July was followed by a flight to Heston on 19th, then to Farnborough on 20th before returning to Shoreham on July 21st. A number of cross-country flights were made from Shoreham, commencing on 21st July, for what appears to have been demonstration purposes; these included visits to Croydon, Brooklands and Heston, before returning to Croydon on 29th July 1929. Miles flew G-AAFH from Croydon to Yate on 1st August 1929 and made just one more flight in it, a local of 1hr 30min from Yate on 18th September 1929.

Miles was in South Africa when the first Elf II, G-AAIO, was tested - its C of A being issued on 2nd September 1931. He did not return to Yate until 11th March 1932, when he flew the Spartan Three-Seater G-ABJS from Croydon, probably to discuss the building of the Miles Satyr. On 20th March he flew an anonymous Elf from Yate to Bristol (a 15-minute flight), returning to Yate the same day and on 31st March 1932 he returned to Croydon in G-ABJS. His next flight at Yate, on 30th April 1932, was in the Elf II G-AAIO, against which he noted '1hr Bristol Airport - Gliding' and on 15th and 16th May 1932 Miles was again at Bristol, this time 'joy-riding' with the Simmonds Spartan G-AAGY, putting in 2hr 20min on the first day and 2 hours on the second.

Miles probably stayed in the area since, on 25th May 1932, he made a local flight from Yate in his Avian G-AADF and, on June 2nd, carried out a 35-minute 'test flight' in the second Elf II G-AAIN, which was in fact its first flight. Its C of A was issued on 15th June and it is of interest to note that G-AAIN survives with The Shuttleworth Collection at Old Warden in airworthy condition. Also on June 2nd, he flew the Bristol and Wessex Club's brand new DH.60M Moth G-ABWM from Bristol to Yate and on June 4th Miles was still at Bristol, joy-riding with the Spartan G-AAGY, during the course of which he flew 4hr 30min and carried 60 passengers. On June 8th, he made his last flight for Parnall, taking the Elf II G-AAIO from Yate to Farnborough and then on to Croydon (1hr 30min). Miles then flew the Spartan G-AAGY from Bristol to Capel Cynon, Wales on June 17th, apparently with the intention of earning some money as he carried out a total of 3 hours flying, including joy-riding.

It is of but passing interest to note that I saw G-AAGY just after the end of the second world war, close by a barn in the corner of the courtyard of a country club at Nutfield Marsh, near Redhill where it had been stored since early in the war. Still there in August 1947, with its wings folded and warped and its fabric hanging in tatters, G-AAGY was indeed a sorry sight and was scrapped soon after.

By 1932, Mr and Mrs Miles had settled in Sevenoaks, Kent, where they designed the Miles Satyr. As related in Chapter 5, this was built for them by George Parnall & Co and Miles' next flight at Yate was recorded as a 'Test flight' of the new aircraft on 31st July 1932. Two further test flights were made from Yate on August 10th and these were followed by a local flight on 11th. Then, on 21st August 1932, Miles made a return flight from Gatwick to Brooklands in an anonymous Robinson Redwing and on 24th August 1932, he flew the Satyr from Yate to Bristol (Whitchurch), making a local flight there the same day and another on the 25th before flying it to Reading later in the day.

Miles' association with Yate had ended and although he could not possibly have realised it at the time, his future was to be at Reading, which is where the next part of the story unfolds, in Chapter 6.

The Parnall Imp, G-EBTE, with the new 60hp Pobjoy P air-cooled radial engine.
[E Harlin via KE Wixey]

The Simmonds Spartan, G-AAGY, which Miles used for joy-riding while test flying for George Parnall & Co in May and June 1932.

Flt Lt John Pugh AFC gave the aerobatic displays in the Satyr with Luxury Air Tours in 1933. [Via B Clarke]

Chapter 5: The Miles M.1 Satyr

After Miles returned from South Africa in October 1931 to be reunited with Blossom, they acquired a caravan and toured around Sussex and Kent for a while. To have returned to Shoreham was obviously out of the question at that time, so they settled in Sevenoaks in Kent, where they lived in their caravan. It was there that they designed their first aeroplane together, from scratch, in a small rented room over a still unidentified shop. This was to be the last Miles biplane and it possibly owed much to the earlier Southern Martlet and possibly even something to the original Gnat. They called it the Satyr.

The Satyr was a very small, single-seat, high-performance aerobatic aircraft designed to be powered by a 75hp Pobjoy 'R' seven cylinder, air-cooled radial engine, geared to drive a large diameter propeller. The diminutive staggered biplane, its top wing of greater span than the lower, was only as high as the average man's shoulder, and it had streamlined interplane struts. The Satyr was constructed of wood with fabric covering and the control wires were located inside the wings and fuselage, with the gaps being sealed and the junctions faired. It had a straight-axle undercarriage and the fuel tank, originally mounted in the centre section of the top wing, was later moved to just behind the engine. The Satyr was originally intended purely for Miles' personal pleasure.

Between 20th March and 8th June 1932, Miles had test flown both the recently completed Elfs G-AAIN and G-AAIO for George Parnall & Co. They had little other work in hand and arrangements were made for him to build the Satyr in their factory at Yate under the supervision of Miles and Blossom. It was initially given the Parnall c/n J.7 (which followed on from Elf G-AAIN, c/n J.6) but this was later changed by Miles to c/n 1.

The Satyr was registered **G-ABVG** to Miles in mid-March 1932 and was first flown, by him, at Yate on 31st July 1932. The small aeroplane had a remarkable all-round flying performance, with all the delightful handling characteristics of the Martlet and, although the top speed was only 125 mph, the initial climb rate was later found to be an amazing 1,330ft/min. The Satyr had a virtually unobstructed view from the cockpit and in general appearance it bore a strong resemblance to the fighters of the day, although it was only a little more than half their size.

Miles and Blossom were married on 6th August 1932 and on the 24th Miles flew the Satyr to nearby Whitchurch, making a couple of local flights there. The next day, it is said, Miles flew to Shoreham to 'visit his parents', but as he called in at Reading 'on the way', it is almost certain that this was the occasion of Miles' visit to Reading especially to meet with Charles Powis of Phillips & Powis Aircraft (Reading) Ltd, to discuss his plans for a new monoplane light aeroplane. The recent discovery, in F G Miles' archives, of a document, written on Phillips & Powis Aircraft (Reading) Ltd headed notepaper, dated August 29th, 1932, tends to confirm that 25th August 1932 was, in fact, the date of the meeting. The document was entitled: 'Notes on the discussion concerning the proposed Agreement between Messrs Phillips & Powis Aircraft (Reading) Ltd, and Mr F. G. Miles', and this is reproduced in Chapter 6. His logbook shows that he never did get to Shoreham and likewise he never apparently returned to Bristol! On August 26th, he flew the Satyr to Farnborough and on August 30th he carried out a 50-minute 'test of climb' flight in which he recorded a 1,330ft/min climb rate at 55 mph. On September 4th, he climbed the Satyr to 10,000ft during a 40-minute flight.

On Saturday 24th September 1932, two tea-parties, or 'At Homes', were held at opposite ends of Woodley aerodrome to everyone's advantage - one by the Berks, Bucks and Oxon Flying Club and the other by the Reading Aero Club. *The Aeroplane* for 28th September 1932 reported on the event, and referred to the Satyr:

After tea Mr J F Lawn gave a demonstration of aerobatics in the new Miles Satyr, a single-seat braced biplane with a Pobjoy engine. It weighs 594 lbs empty, and has a loaded weight of 900 lbs. As demonstrated by Mr Lawn, it has a remarkable performance. Mr Miles' Martlets have been appreciated by connoisseurs for some time and the Satyr appears to have all the

The Satyr, with F G Miles, outside the clubhouse at Woodley on the 'At home' day, 24th September 1932. [Phillips & Powis via B Clarke]

Chapter 5: The Miles M.1 Satyr

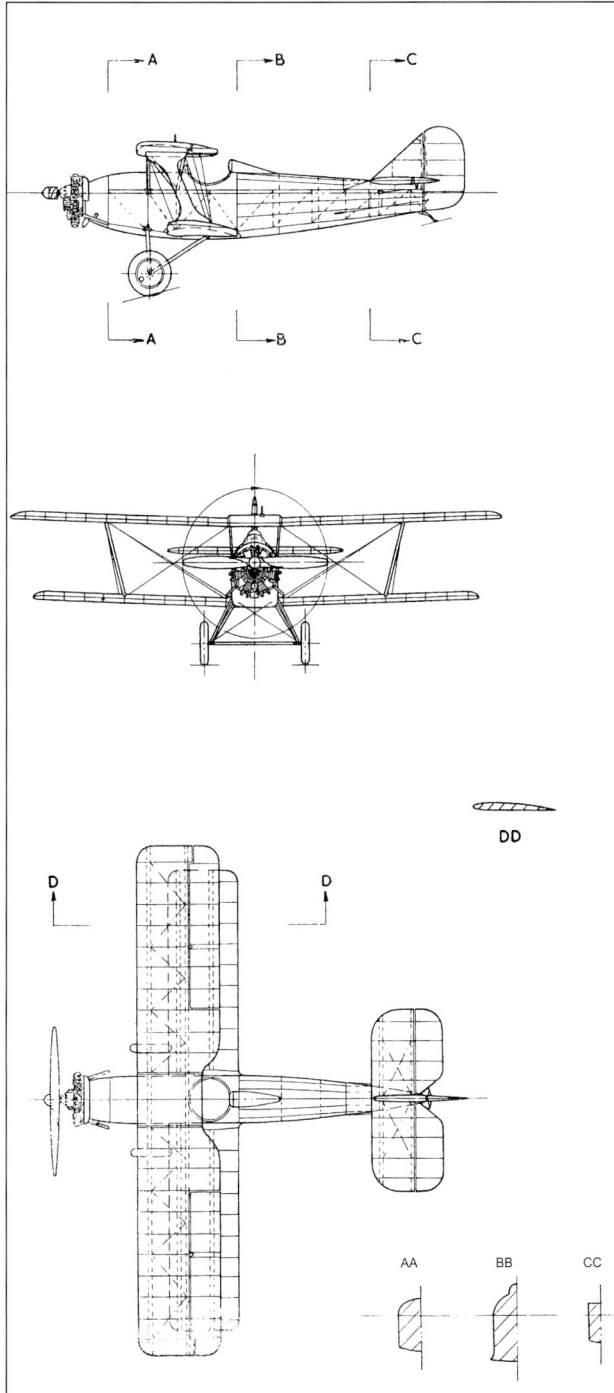

The Miles Satyr.

official trials, after which he returned to Reading via Hanworth on December 11th. The Satyr's C of A, No.3740 was issued to F G Miles on 1st February 1933.

On 29th March 1933, Miles flew his new M.2 Hawk monoplane for the first time, and after that he had little time for 'prolonged fun' in the Satyr, making his 34th and last flight in it on 5th April 1933. In total, Miles recorded 10hr 53min flying time in the Satyr.

The Hon Mrs Victor Bruce, with her company, Luxury Air Tours Ltd, had become involved with British Hospitals Air Pageants, an organisation giving flying displays and joyrides in aid of hospitals. She persuaded Miles to sell the Satyr to her and she took delivery of it just prior to the first display, arranged to take place at Hayes, Middlesex on 8th April 1933. Her personal pilot, Flt Lt John Pugh AFC was selected to give aerobatic displays in the Satyr, which were described as 'quite outstanding'. The Satyr, by now re-painted in a distinctive red and white chequer-board colour scheme, was formally registered to Luxury Air Tours Ltd, Croydon and Hanworth Park, on 21st April 1933. Although the Pageant's tour continued until October 8th, Mrs Bruce had flown the Satyr into telephone wires while landing at Stafford on 24th May 1933, causing slight damage to the wings and rear fuselage.

In the book *Nine Lives Plus* by The Hon Mrs Victor Bruce, she relates the story of how the Satyr came to grief:

Life with the circus was packed with incident, and for a while dogged by bad luck. Once, for instance, I had a five shilling bet with John Pugh that I could land the Satyr on the green outside Stafford in a shorter distance than he could in the Fairey Fox. I spotted the green with a large crowd waiting. It was bordered on two sides by small houses and just in front were telegraph poles, carrying a bank of about a hundred telephone wires. When you're flying, you do not see the wires, just the poles; one has to remember they are there. Unfortunately I forgot: the poles were hidden behind the houses. Thinking of my bet, I approached the green as slowly as I dared. Suddenly there was an almighty crash. I thought I had hit one of the houses, because the noise was terrific. Then I realised that I had flown into the telephone wires. The Satyr and I must have remained hanging in the wires for at least half a minute. Then the wires gave way and we fell fifteen feet to the ground with a bang. But I had won my bet. I was exactly four yards beyond the fence which bordered the green. I tumbled out of the plane with telegraph wires all round my neck. John Pugh said he did not think it worth while trying to beat me for five shillings. Except for a big bump on the head I wasn't hurt, but I was rushed off to the doctor who told me that I had slight concussion.

There was quite a crowd gathered to watch that exhibition. I heard one old prophet of doom mutter, 'Don't worry, this is only the beginning of it'. How right he was: an hour later C Evans (the parachutist) landed on the chimney of one of the houses I had missed; he had to be rescued by the fire brigade. The Satyr, of course, was badly damaged (and it was not insured). On top of the repair bill I got another from the telephone company: £80 for 80 broken wires.

As I felt sure the newspapers would be full of all this by the evening, I thought I had better telephone Victor to tell him I was all right. There was an unusual delay, so I asked to speak to the Supervisor. I told her I had an urgent message and must get through quickly. She replied: 'I'm sorry, but some silly fool has flown into the telegraph wires and we haven't a line to London, let alone Esher'.

While the Satyr was being repaired I hired a Gipsy Moth to take its place in the show, etc......'

The Satyr was repaired and was flying again at the Jersey Aerial Pageant on 31st August 1933. In October 1933, it was offered for sale by Air Dispatch Ltd, Croydon (another of Mrs Bruce's companies) but failed to attract any offers. It was again advertised for sale in *Flight* for 16th August 1934, but the Satyr still failed to attract any serious customers and it continued to appear at displays across the UK, often spiritedly flown by John Pugh. Its C of A was renewed, for the last time, on 2nd October 1935.

family charm. As it has yet to go to Martlesham its performance figures can only be alleged. They are maximum speed 122 mph, minimum speed 44 mph, and climb 1,400 ft/min.

It was reported that three Hawker Furies visited the aerodrome one day and the Satyr took-off in formation with them, *'like grown-ups going off to a party with a small child in tow'*! However, when it came to games, *'like a child, the Satyr could turn more somersaults and roll about quicker than the more serious grown-ups and it was only when the Furies started to race away that the little one was left behind'*! This incident is probably the one referred to by Miles in his logbook on 6th December 1932, where he recorded the brief comment 'Fight with Nimrod', against a 10-minute flight! The Hawker Nimrod was the Naval version of the Fury and it is possible that the reporter mistook them for Furies.

On 9th December 1932, Miles made a 10-minute 'Test for AID' in the Satyr before flying it to Martlesham Heath later that day for its

In conversation with Richard Riding in late 1977, Mrs Bruce said that the Satyr had then been 'half sold' to a '19-year old Japanese gentleman' who had taken it to Belgium where, during a flight in that country, this Oriental gentleman inadvertently entered a spin, from which he failed to recover. The pilot was killed and the aircraft was written-off. In fact, the Satyr's ultimate fate still remains a mystery, as this latter story has since been proved to have no foundation in fact.

Hitherto, the generally accepted account, recalled by Don Brown and others, was that the Satyr was written-off in September 1936 but in an incident identical to Mrs Bruce's crash. This would seem simply to have been an assumption made with reference to the aircraft's formal cancellation date in that month and is unlikely to be correct.

The Satyr's propeller was seen in the workshop of Mr Bristow in about 1960 by Bert Clarke but how it got there is not known. Mr Bristow was model maker to Phillip & Powis/Miles Aircraft and Vickers-Armstrong's and a few of his models survive today in the Museum of Berkshire Aviation at Woodley. Sadly this unique propeller now seems to have disappeared.

Following the Satyr's sale to the Hon Mrs Victor Bruce in early April 1933, it was repainted in red/white chequer-board. Note Hon Mrs Bruce's Fairey Fox, G-ACAS, in the background.
[Via Phil Jarrett]

The Satyr following its accident with the telephone wires at Stafford on 24th May 1933.
[Via B Clarke]

SPECIFICATION AND PERFORMANCE DATA

Engine:	75hp Pobjoy R
Dimensions:	span, upper wing 21ft 0in, lower wing 18ft 0in; chord, upper wing 3ft 6in, lower wing 3ft 0in; wing area 117 sq ft; length 17ft 8in; height 6ft 6in
Weights:	empty 594 lb; AUW 900 lb
Performance:	max speed 125 mph; cruising speed 110 mph; landing speed 40 mph; rate of climb 1,330ft/min; range, with normal fuel load of 12 gal, 200 miles; endurance 2 hours
Price:	£550

A DH Moth with some of the original ground staff and aviators outside the first hangar at Woodley in 1929. [Via B Clarke]

Chapter 6: Reading - 1932 to 1939

Charles Owen Powis was born in Canada on 29th September 1904, although he was in England by 1919, when he and E 'Jack' Phillips began a five-year apprenticeship with Herbert Engineering Ltd of Wolsey Road, Caversham, Reading, Berkshire. The company, which had repaired and overhauled aero-engines in the First World War, were then building the HE motor car. But in 1924, the company went into receivership and the two young motor engineers, who had not quite 'completed their time' were faced with finding alternative employment.

Thus, towards the end of 1924, Phillips and Powis decided to go into business in partnership, repairing and selling motor-cycles, under the name The Reading Motor Exchange. They commenced trading from 37A Erleigh Road, at the end of Donnington Road in East Reading and their business soon prospered. Rapidly outgrowing these premises, they purchased some land in the Oxford Road to the west of Reading town centre, where they had a purpose-built showroom, shop and workshop constructed for them. This was 470-478 Oxford Road, Reading and The Reading Motor Exchange had moved into these new premises by 1926. By 1928, the main premises were extended, other branches were opened and the business was now incorporated under the name Phillips & Powis Ltd.

In March 1928, further expansion of the business led to the opening of a cycle sales and repair shop at 34 West Street, Reading, where they also sold the 'P' bicycle. Then, in the summer of 1928, Charles and Pauline (whom he had married in 1927) flew to France for a holiday and this gave Charles the taste for aviation. Shortly after their return he decided to learn to fly and joined the Henderson School of Flying at Brooklands, where he was taught by his friend Oliver Edwin Simmonds, designer of the Simmonds Spartan. Gaining his pilot's 'A' Licence in October 1928, Powis purchased his first aircraft, the DH.60 Moth G-EBOT, from Miss Winifred Spooner, using funds from Phillips & Powis Ltd. Initially operating from Promenade Field, Caversham Bridge, Powis began to look for a site for a suitable permanent aerodrome near Reading and, in the meantime, he started demonstrating the Moth to potential customers both here and at Brooklands.

On 27th October 1928, the first Reading Air Pageant was held by the newly-formed Berks Bucks & Oxon Aeroplane Club at Sheep Bridge Farm, Swallowfield Park, Reading. Fifteen aircraft attended and Charles Powis appeared in the list of visitors and spectators. In November, Phillips & Powis Ltd purchased the Avro 504K G-EBWO, naming it 'Bluey' and this was used by Powis for flying demonstrations and joyriding in the locality. Then, on November 9th, Powis announced in the local press that he had found a site for the future Reading Aerodrome at Woodley, some three miles from the centre of Reading.

In early 1929, Phillips & Powis Ltd opened a new showroom and workshops at 10-24 South Street, Reading, from where they began selling three-wheeler motor cars for the first time. More significantly, on 19th March 1929, they formed a new company, Phillips & Powis Aircraft (Reading) Ltd, with a share capital of £5,000 and with Charles Powis being appointed Chairman and Managing Director. That Easter, the new company commenced operations from the south west side of their new aerodrome with the establishment of the Phillips & Powis School of Flying, which boasted one aeroplane (the Avro), a hangar (acquired with the help of Jack Coles, a haulage contractor from Shrewton, Witshire, who had dismantled the hangars at Stonehenge Aerodrome in the 1920s and transported one of the dismantled structures to Woodley) and a small two-room clubhouse.

Easter Monday, 1st April 1929, was a bad day for the new company when the Moth G-EBOT, which was used by the school for

Charles Powis with his wife, Pauline, in 1930. [Via B Clarke]

instruction, was damaged and the Avro 504K was written off in an accident at its former field at Caversham Bridge. A replacement was urgently required so, on May 4th, Powis purchased the Avro 504K G-AAGG, from Southern Aircraft Ltd, newly built-up from spares at Shoreham. The Avro was delivered to the Phillips & Powis School of Flying by F G Miles the same day and its Certificate of Airworthiness was issued to Phillips & Powis Aircraft (Reading) Ltd on 6th May. Although this was not believed to be the first meeting between Miles and Powis, it was probably the first to involve a direct business transaction between them.

By the end of 1929, the school fleet had increased to five machines and a repair shop and sales department had been established. Also, by this time, Powis had leased a small portion of the aerodrome to the embryonic Berks Bucks & Oxon Aeroplane Club (which, in September 1929, had been absorbed into National Flying Services and their organisation of light aeroplane clubs) and they erected their own clubhouse, hangar and workshops in the north east corner of the aerodrome.

Although in the summer of 1929, Charles and Pauline Powis had announced the formation of the Reading Aero Club, this was deferred and it was not until 13th September 1930 that a new subsidiary company, Reading Aero Club Ltd, was formed, initially as a social club for students and ex-pupils of the Phillips & Powis School of Flying. They set up a small clubhouse in a wooden hut on the aerodrome and, in January 1931, The Rt Hon Earl of Northesk was appointed president and Pauline Powis became the first secretary, although she relinquished this position within a few months.

On Saturday 16th May 1931, the new Reading Aero Club was officially opened and *The Reading Chronicle* for Friday 22nd May 1931 reported on the occasion thus:

The aerodrome on 9th June 1930. The original Phillips & Powis hangar is on the right, with 'Shell' painted on the roof. The original clubhouse is seen in the bottom right corner. The event in progress was the Reading Air Fete, organised by the Berks, Bucks and Oxon Aero Club.
['Flight' via B Clarke]

CHAPTER 6: READING - 1932 TO 1939

An Avro 504N, G-EBHD, being refuelled outside the original hangar, with the new clubhouse in the background, in about 1931.
[Eric Smith via B Clarke]

Reading's new aero club at Woodley was opened on Saturday by the Director of Civil Aviation (Lieut-Col FC Shelmerdine) in heavy rain. Lieut-Col Shelmerdine, who flew with Mrs Shelmerdine from Heston, Middlesex, in a Puss Moth, made one of his first official public appearances. A large clubhouse has been erected by Messrs Phillips and Powis, and already the club has a good membership. Lt-Col and Mrs Shelmerdine were met by the president of the club, Lord Northesk, who was accompanied by Lady Northesk and the Mayor and Mayoress of Reading, Councillor and Mrs F G Sainsbury. Lt-Col Shelmerdine and his wife were made life members of the club.

Lord Northesk, as president of the club and host, welcomed Lieut-Col and Mrs Shelmerdine, whose presence, he said, was very much appreciated. He had great pleasure in asking them to accept life membership cards of the Reading Aero Club and hoped that they would often be seen there. He hoped the aerodrome would be of very great use to them in their flying about the country.

The Mayor of Reading (Councillor F G Sainsbury) drew attention to the fact that Reading was one of the first towns to foster the spirit of civil aviation by the opening of an aerodrome. He, personally, was proud of the fact that such foresight was being evidenced by that ceremony, and asked Col Shelmerdine to accept the key with which he was to open the club-house.

Over 40 machines arrived, and following the opening there was a programme of aerobatics and stunt flying. Miss Amy Johnson, Miss Winifred Spooner and Miss Gower (daughter of Sir Robert Gower MP), were among those who entered for the ladies' handicap race for the president's cup. The race was flown in two heats, and the course was from the aerodrome to Wokingham church, on to Twyford station and back to the starting place. The competitors had to fly the course once in the heats and twice in the final. The results were Heat 1: Miss Amy Johnson, 88.5 mph 2, Miss Aitken, 87.5 mph; 3, Miss Crossley, 86 mph Heat 2: 1, Miss Burr, 78.5 mph; 2, Miss Gower, 93 mph; 3, Miss Spooner, 106 mph; 4, Miss Slade, 93 mph: 5, Mrs Young, 90 mph. Final: 1, Miss Aitken, 97.5 mph; 2, Miss Gower, 95.5 mph; 3, Miss Amy Johnson, 98.5 mph; 4, Miss Burr, 79 mph.

After Mrs Shelmerdine had presented the cup and prizes, Lord Northesk thanked her and Col Shelmerdine for coming that afternoon, and Mrs Shelmerdine for distributing the awards. They had had an excellent afternoon of flying in spite of the bad weather, and he wished to thank the private owners, and also aircraft manufacturers who had brought or sent machines. He, personally, was very proud of his position as president of the club, and he hoped the people in Reading and the surrounding district would support them and make it a thorough-going concern so that all over the country they would hear about it. Mrs Shelmerdine was presented with a bouquet by three-year old Miss Swann, the youngest member of the club (presumably the daughter of Douglas Swann, one of the founder directors of the club – PA). The Deputy Mayor and Mayoress were present.

The club-house served the club well for the next few years until it became increasingly hemmed in by the rapidly expanding Phillips & Powis aircraft factory and it was eventually replaced by the *Falcon* hotel on the aerodrome in October 1937.

A ceremony held on 18th April 1931 formally opened Bristol's Whitchurch Airport as Britain's third municipally-owned airport and this also marked the start of operations by Bristol Air Taxis, a subsidiary company of Phillips & Powis Aircraft (Reading) Ltd. With Stephen Cliff as pilot-manager, it operated the Desoutter Mk.II G-AAVO and was kept fairly busy with joy-rides and taxi work, including long distance charters to Naples in September 1932 and to Dusseldorf the following month. While Stephen Cliff established an excellent reputation, competition from resident Norman Edgar & Co's expanding fleet eventually proved too much and they took over Bristol Air Taxis and its Desoutter late in 1933, with Cliff returning to Woodley.

Powis's considerable business acumen was recognised on 1st March 1932 when Phillips & Powis Aircraft (Reading) Ltd was granted exclusive rights for one of the three main sales areas in the country for the de Havilland Aircraft Co Ltd. And, in *Flight* for 18th March 1932, it was announced: *Phillips & Powis Aircraft (Reading) Ltd - Increase of Capital. The nominal capital has been increased by the addition of £2,000 beyond the registered capital of £5,000. The additional capital is divided into 500 preference and 1,500 ordinary shares of £1 each, ranking pari passu with existing shares.*

In early April, F/O John Lawn joined the staff of the Phillips & Powis School as assistant instructor to the CFI, Lt Commander CW Croxford.

And so it was that, on 25th August 1932, Miles left Bristol (Whitchurch) Aerodrome in his newly built Satyr, en route to Shoreham, ostensibly with the intention of showing his new machine to his parents. However, his log-book shows that he was

BRITISH FLYING
READING
THE PHILLIPS AND

A pupil about to take off on an altitude test. Commander C. W. Croxford, better known as "Stiffey," the Chief Instructor, examining the barograph which records the height reached.

A conference on the tarmac. Left to right Mr. J. F. (Laddie) Lawn, Instructor; Commander Croxford, Chief Instructor; and Mr. C. W. Powis, Managing director and founder of the school.

The sales department can supply any make of new or second-hand aircraft. Mr. Powis examining the log of Jason III (in the background), which formally belonged to Amy Johnson.

Girl pupils moving a Moth on the aerodrome. Pauline Gower, one of the few lady pilots to hold a "B" licence, learned to fly at this school.

In the workshop. Overhauling a private owners machine. Note the grim humour of the pilot's device. Incidentally the skull and crossbones was the "coat-armour" of many famous war aces.

Above and right: Two pages on British flying schools – Reading Aerodrome – from a contemporary aviation magazine. [Via P Amos]

to get no further than Reading, where he had actually decided to have lunch and to meet Charles Powis.

It was there in the clubhouse that they discussed ideas for an 'ideal' aeroplane to replace the now ageing DH.60 Moth. The two-seat Moth, which had been sold in considerable quantities during the preceding six years, was the most widely used private owner's aeroplane at the time but the Moth and its nearest competitor, the Avro Avian, were braced biplanes of a bygone age.

Both Miles and Powis felt that the contemporary private owner would prefer the higher performance obtainable only with a clean, cantilever low-wing, two-seat monoplane but the only production aircraft of British design in this class and in small-scale production

Chapter 6: Reading - 1932 to 1939

SCHOOLS, No. 6
AERODROME
POWIS SCHOOL OF FLYING

"Live and Learn at Reading" is the slogan of the Phillips and Powis School. The clubhouse, seen from the aerodrome.

The business side. Mr. Powis (seated) in his office, with Mr. Lawn, going over the flying reports.

Tea time. The aerodrome is beautifully situated and is practically free from fogs or mist.

A corner of the Engine Shop. A mechanic at work on the valves of a de Haviland Gipsy engine.

A pupil opening the "umbrella" after "blind" flying.

in the early 1930's was the Percival Gull. Miles was confident that his proposed new aeroplane would sell in large quantities. In fact, so confident was Miles that it is now strongly suspected that he and Blossom had actually commenced work on the design for the new monoplane while they were still at Sevenoaks.

On 29th August 1932, a proposed agreement, written on Phillips & Powis Aircraft (Reading) Ltd notepaper was produced. Headed 'Notes on the discussion concerning the proposed Agreement between Messrs Phillips & Powis Aircraft (Reading) Ltd and Mr F G Miles' it stated:

That in consideration of Mr Miles taking up 1000 Preference Shares at par in the Company of Phillips & Powis Aircraft (Reading) Ltd, the Company agrees to place at Mr Miles' disposal a newly equipped Aircraft Workshop.

It is also agreed that Mr Miles has the right and is given the accommodation to build aeroplanes of his own design, and

The first Hawk flying over the new clubhouse in 1933. ['Flight' via B Clarke]

supervise the work thereof, and is also provided with the use of an office or drawing office. The nett cost of the work only to be charged to Mr Miles (suggest that the specified rate is agreed upon).

The Company agrees to place work in connection with the other branches of the Company's activities with Mr Miles from time to time payment for such service to be arranged (such as instructional flying, sales etc).

Mr Miles has the right to assist with or supervise the installation of new plant in the Workshop. It is understood that Mr Miles although not bound to any specified time will use his best endeavour to further the interests of the Company by any means in his power.

When present he will assist in the general supervision of the Workshop and renew his Inspectors' Licences.

In September 1932, as these arrangements with Miles were being finalised, Jack Phillips and Charles Powis decided to go their separate ways. Jack Phillips returned to motorcycle dealing and Charles Powis, who had decided that his future was to be in aviation, took over full control of Phillips & Powis Aircraft (Reading) Ltd (which rather surprisingly retained its full name).

Powis and Miles then finalised the agreement for their future relationship with the chief aim of forming an aircraft design and manufacturing company to build their monoplane. Any ensuing profits from the sale of this aeroplane were to be shared between

A busy scene inside the original hangar in 1934.

Chapter 6: Reading - 1932 to 1939

Miles and Phillips & Powis Aircraft (Reading) Ltd. Miles and Blossom moved to Reading about this time and all their subsequent flying was made from Woodley aerodrome. They set up a drawing office in part of the Phillips & Powis hangar and it was from there that Miles and Blossom traded initially under the name of the Miles Aeroplane Company, while the design of their new 'Type 2 Hawk' monoplane took shape. The prototype of the new aeroplane was to be built in the workshop and Blossom was to bear part of the cost but all selling rights belonged to Phillips & Powis Aircraft (Reading) Ltd.

These events were not to pass unnoticed and *The Aeroplane* for 12th October 1932 reported:

Mr F G Miles, formerly of Shoreham, has joined the staff of Phillips & Powis Aircraft (Reading) Ltd. He will with Mr Stisted (Henry Cockburn Gresley Heathcote Stisted, the racing motorist who was a Director and Business Manager of the firm - PA), *look after the technical activities of the firm, though the latter will concentrate on the sales and business aspects of the department. A new hangar is being built and the ambition of the directors is to have the finest repair-shop on the South-West side of London. They intend to develop intensively the repair and service side of their activities.*

Miles also formally became the Technical Manager of the Service Department of Phillips & Powis. He and Blossom spent the autumn and winter of 1932-33 working on the design of the new monoplane. The machine was to be constructed of wood and powered by the well tried, four-cylinder, in-line air-cooled 90hp Cirrus IIIA engine. This engine conformed exactly to the idea behind the aeroplane in that it was a thoroughly sound mechanism without the frills which add expense. The firm was very fortunate in being able to obtain a batch of fifty Cirrus IIIA engines which had previously been ordered by a company in Canada but cancelled prior to delivery and this 'windfall' enabled the first aeroplanes to be sold at the exceptionally low price of £395 each. The new aeroplane was also designed from the outset to have folding wings, which helped to keep down the cost of its operation.

Harry Hull, the carpenter who had built the Martlet and repaired the Avro 504Ks at Shoreham, joined Miles at Woodley in October 1932 to become works foreman and to build the prototype of their new aeroplane. Bert Clarke recalls that Harry regularly used to escape the pressures of work for a few minutes by cycling to the canteen, which Bert's mother and father ran, for a well-earned cup of tea whilst also using his 'ladies' bicycle to travel at great speed around the works so as to keep up with everything! Harry was still working in the factory in 1954, operating the large rubber die press, by which time he was well over 80 years of age and had been awarded the BEM for his services.

On 24th September 1932, at the same time as Miles was finalising his agreement with Charles Powis, two 'At Homes' took place at Woodley and *The Aeroplane* for 28th September 1932 reported thus:

A successful double was brought off at Reading on Saturday, September 24th. Two tea parties, or 'At Homes', were held at opposite ends of Woodley aerodrome to everyone's advantage. The weather was glorious with bright sunshine, a happy state of affairs. The Woodleyites tell you that the weather is always like that at Reading, especially if the air is murky elsewhere, but one suspects local patriotism.

I was in the happy state of being a guest at both parties, which dual role at least has the advantage that when some pedantic stickler for accuracy writes to point out an error, I can claim to have been at the other end of the aerodrome. The Reading Aero Club was 'At Home' to the Staff of Reading University, and some eighty visitors came to tea. Before eating, they were taken round the buildings, hangar, and repair shops, where Mr Pepper, Chief Ground Engineer, discoursed upon the mysteries of his profession. After tea, Mr J F Lawn gave a demonstration of aerobatics in the new Miles Satyr, a single-seat braced biplane with a Pobjoy engine. It weighs 594 lbs. empty, and has a loaded weight of 900 lbs. As demonstrated by Mr Lawn, it has a remarkable performance.

Mr Miles' Martlets have been appreciated by connoisseurs for some time and the Satyr appears to have all the family charm. As it has yet to go to Martlesham, its performance figures can only be alleged. They are; maximum speed 122 mph, minimum speed 44 mph, and climb 1,400ft/min. When Mr Lawn had at last reluctantly brought the Satyr down and had been greeted with the applause that his performance deserved, those of the visitors who had been lucky in the lottery were given joy-rides in Club machines. Other had to buy tickets. This visit is one of a number which have been arranged to bring home the activities of present-day civil flying to important local bodies and organisations. If forthcoming visits go off as well as the first, the evangelising should make a notable headway.

The 'At Home' of the Berks, Bucks and Oxon Aero Club was scheduled to begin at 15.00 hrs. Actually it began with the arrival of the visiting pilots. The earlier arrivals from London reported that conditions were not good there. But after getting through the low cloud and bad visibility around London they were agreeably surprised to find themselves flying into fine weather. Among the early arrivals were a contingent from Hanworth, including Mr McMullen in his own Autogiro. Mr McMullen had brought Mr Bradley with him as his first passenger.

Meanwhile two Desoutters were hard at work joy-riding. Mr Hayter looked in to get some petrol. He had gone to Gloucester for lunch in his Puss Moth but had been forced to land some three miles from the town by low cloud. Mr M H Findlay in a Gipsy Moth did aerobatics. Mr H Ward jumped with a parachute. Halfway down he opened a second, perhaps as an illustration that one is enough, for the second had little effect. Mr McMullen showed off his Autogiro.

The last event of the day was a Race of six machines to Wokingham Church and back, a total distance of 15 miles. Mr J J Jeffs was starter and Mr L De Loriol was judge. The entrants were: Mr H C Harley, Mr J E D Holder, and Miss J E Giles (Gipsy Moths); Mr A M Emmet, Mrs Battye, and Mr P A R Bremridge (Cirrus Moths). Mr Bremridge was first home, with Mrs Battye and Mr Emmet 2 and 3 seconds behind him. After this a great deal of fun was got out of the various events in the Motor Gymkhana, whose presiding genius was Mr Richard Ovey. Mr R E H Allen seemed to find it too easy, probably as a result of flying the Autogiro on point duty during the Derby.

While all this was going on Col The Master of Sempill arrived from Romford with some first-hand impressions of the meeting there. Lt-Cmdr Rodd also arrived in his Puss Moth G-AAYB. After dark the President of the Club, the Right Honourable the Earl of Cardigan, gave away the prizes. These were awarded as follows: - Concours d'Elegance (aeroplanes), Mr H C Harley; Race (aeroplanes), Mr Bremridge.

Both George Miles and Don Brown had kept in close touch with Miles' activities at Woodley and on 16th November 1932 they flew there in the Moth G-AAHI. This was to be the first of many visits to inspect progress on the new machine. During this busy period Miles was also listed as an instructor with the Phillips & Powis School of Flying, together with F/O John F Lawn, although he still managed to find the time to carry out test flights and the occasional joy-ride to help keep the money coming in.

Then, in the late afternoon of 29th March 1933, the new Miles monoplane emerged. It was still in primer paint and had no registration. The aeroplane was originally to have been named the Miles M.2 Ibex, but this was dropped as a result of possible confusion with the new Hinkler Ibis amphibian and as a result it was named Hawk instead. Thus was founded the Miles tradition for naming his aircraft after birds of prey.

The new aeroplane was started ready for Miles to undertake its first taxiing tests but it apparently behaved too well during the course of these tests for F G Miles' self control, so he took it straight off on its first flight! This momentous event was made exactly seven days after his thirtieth birthday - and appropriately in keeping with his long-avowed intention of making money from that day onwards. His ambition of becoming a manufacturer of aeroplanes was thereby taken a step nearer to realisation.

The Hawk proved to be an immediate success being simple and easy to fly, cheap to maintain and over 25 per cent faster than the Moth and the Avian and no modifications were needed. In fact, the Hawk was so successful that Miles and Blossom lost no time in inviting all and sundry to come and fly it. Miles gave Don Brown his first flight in the Hawk on 30th March 1933 and on May 5th Don made his first solo flight in it. De Havilland, the main manufacturers of light aeroplanes at the time, were even interested enough to send a contingent of their pilots to fly the Hawk including both Geoffrey de Havilland senior and junior. From that moment Miles' hopes were being realised and he never

This photograph was taken prior to February 1933 and shows that a second hangar has been added alongside the first. The caption in 'Air and Airways' for February 1933 read: 'WIDER STILL AND WIDER – Reading aerodrome from the air. At the left top end close to the road are the expanding works of Phillips & Powis (Aircraft) Ltd, and the Clubhouse of the Reading Aero Club. Along the same road in the middle are the buildings of the Berks Bucks and Oxon Club, which has now been taken over'. ['Air and Airways', February 1933]

The first Miles M.2 Hawk, in primer finish, March 1933. ['Popular Flying' via B. Clarke]

CHEAP AEROPLANES FOR EVERYBODY

Fifteen Pilots Test a £395 Two-Seater

The new £395 aeroplane at Woodley Aerodrome yesterday

Mr. and Mrs. F. G. Miles and Mr. Powis (tallest), managing director of the aerodrome.

SUCCESS IN FIRST FLIGHTS

An aeroplane which may revolutionise pleasure flying went through its first official tests at Woodley Aerodrome, near Reading, yesterday.

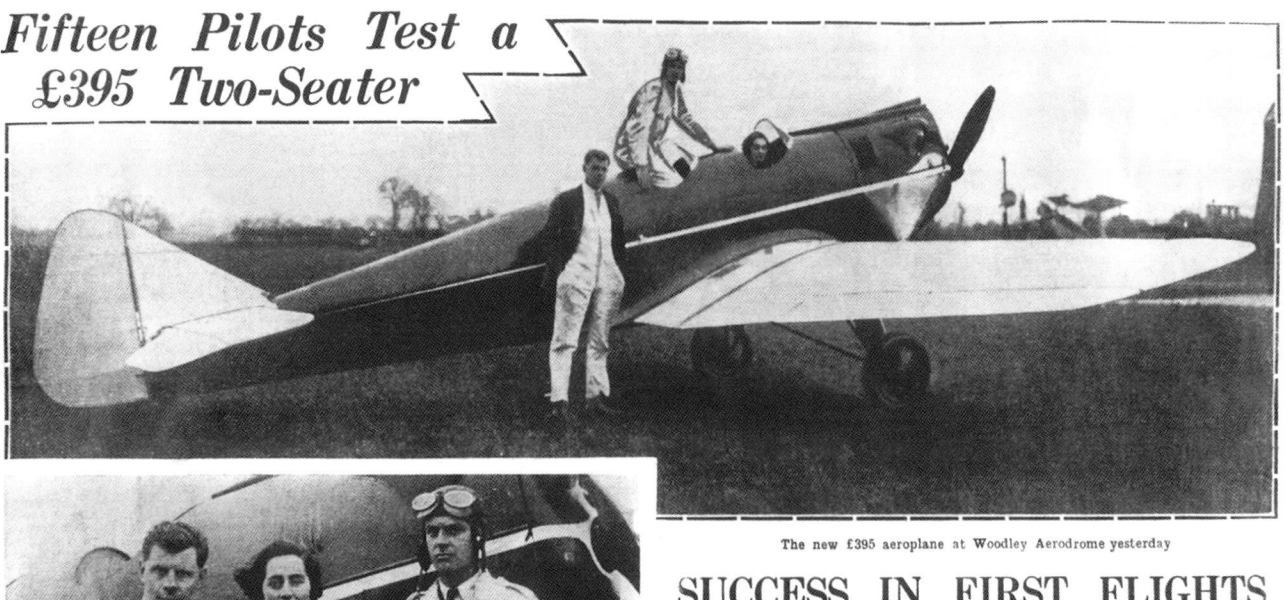

These photographs appeared in the 'Daily Mirror' for 3rd April 1933 – and show 'the new £395 aeroplane' at Woodley Aerodrome on the day before. The report continues by saying that the aircraft 'is one of the first low-wing, open-type monoplanes to be produced in England, and its price is £395 - almost half the lowest-priced two-seater 'plane at present'.

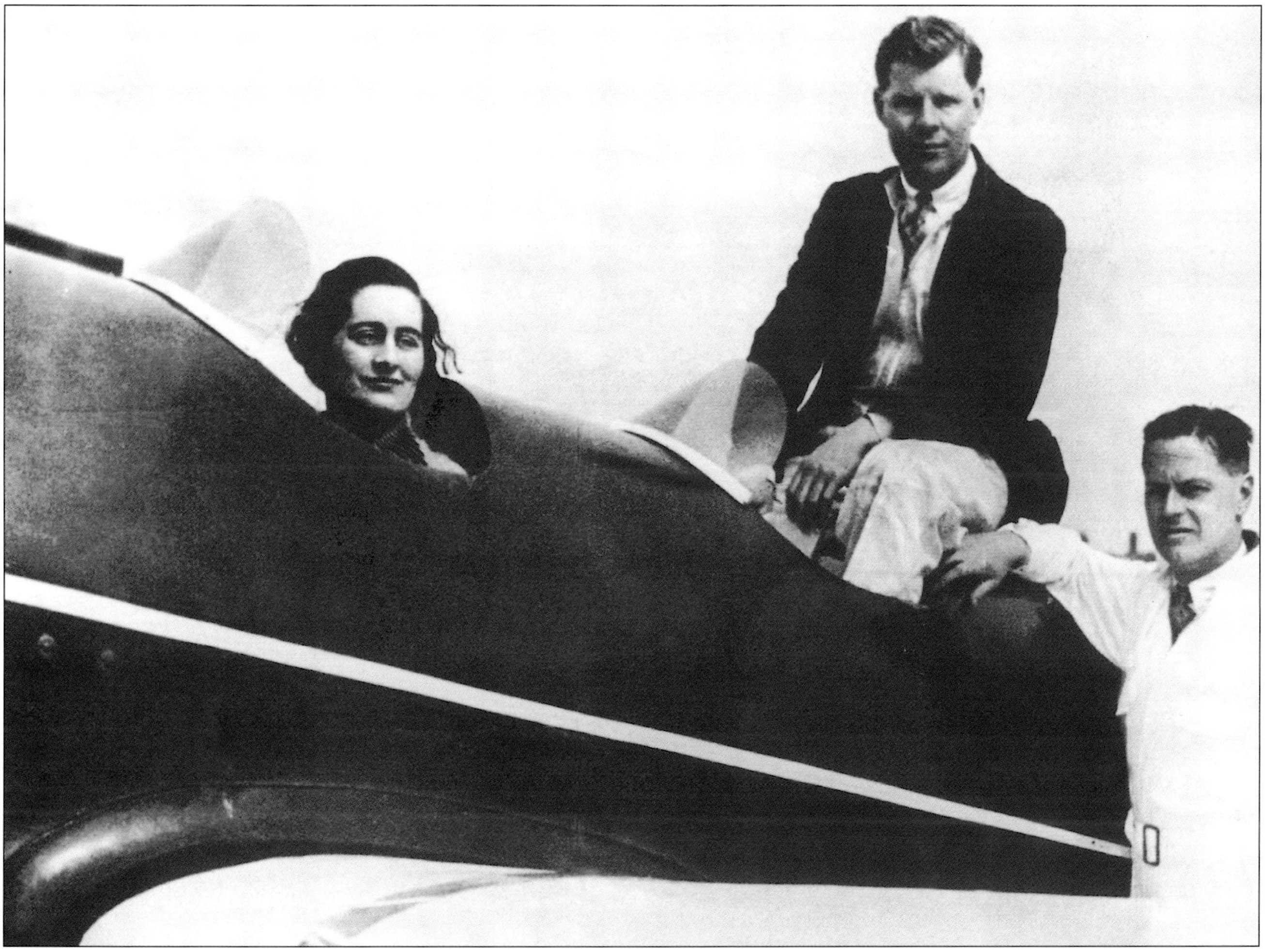

A successful designer and his latest production. Blossom Miles in the front cockpit of the Miles Hawk, with F G Miles in the rear cockpit. Charles Powis, in immaculate white flying overalls, leans nonchalantly on the fuselage. ['Sphere', 22nd July 1933]

looked back. However as orders began to pour in, the original supply of engines soon became exhausted and the price had to be raised from £395 to £450 in order to allow at least a small margin of profit.

An insight into the way Miles was thinking ahead even in those early days was given in *Flight* for 18th May 1933, under the headline 'Another International Air Race Design':

> Our readers will, by now, be well aware of the success achieved by Mr F G Miles as a designer of distinctive aeroplanes. His first machines, the Martlet and Satyr, showed that he possessed a true conception of what was required in a sporting single-seater, while his latest, the Hawk, proves his abilities in the general utility class.
>
> Of peculiar and topical interest, therefore, is the news that Mr Miles has completed the general layout for a small high-speed racing machine, using any in-line engine of about 6.5 litres capacity. The machine will have, he estimates, a performance in advance of anything which has been proposed so far, and should therefore be eminently suitable for entering in the International Air Races at Portsmouth in August. Mr Miles is confident that, not only can he get this machine out in time for the races if he is given an order at once, but also that the cost will be quite reasonable, and he is therefore anxious to hear from any Syndicate who wish to assist to uphold Britain's prestige on this occasion. Mr Miles is now in charge of all the design work for Phillips & Powis, and enquiries should be sent to him at Woodley Aerodrome, Reading.

Unfortunately it seems that Miles did not get any interest in his racer project at the time although by June the following year the M.2E Gipsy Six Hawk had been built for Sir Charles Rose and he raced it in the 1934 King's Cup Air Race.

An article appeared in the *Reading Standard* for 2nd June 1933, just two months after the first flight of the Miles Hawk, entitled *Reading's New Industry - Messrs Phillips & Powis to Manufacture 'Hawk' Aeroplanes. Next Week's Public Exhibition.*

> Reading - already possessing a name that is world famous by reason of its many important industries, seed growing, biscuit making, engineering, bridge building, etc. - has recently added to its list of trades the manufacturing of aeroplanes. Thus the town, which has always prided itself on its industrial progressiveness, will now own one of the most up-to-date businesses in the country, namely, the manufacturing by the firm of Phillips & Powis Aircraft (Reading) Ltd., of the 'Hawk' aeroplane, the cheapest and most reliable machine of its class in the world.
>
> Reading people will already have read in our columns a description of the new machine and details concerning its manufacture, and will welcome the opportunity which is being given to them by the firm next week of viewing the 'Hawk' in its constructional stages. The exhibition, which will be at the West Street shop of Phillips & Powis Motors Ltd, will demonstrate the engine and parts used in the construction of the aeroplane and also of the Satyr plane, both of which machines are under construction at the Woodley aerodrome. (The Satyr was, of course, a one-off and was not in production, having been built by Parnall at Yate prior to this – PA). With the exception of the engine (a Cirrus) the whole manufacture of the machine from the raw material is undertaken and carried out by this firm in its workshops, and it is gratifying to learn that the work progresses by leaps and bounds.
>
> Visitors to the exhibition will see a Cirrus engine, Pobjoy engine, propellers - wood and metal - spars, ribs, instrument boards, controls, and a variety of parts which are manufactured and used in the construction of the aeroplanes indicated.
>
> An added attraction being arranged by Messrs Phillips & Powis during the week is a competition in weight guessing. Anyone wishing to enter the competition may do so by entering the shop and placing on a card their name and address and the weight of the Cirrus engine as they estimate it to be as standing in the window. There will be no charge. The two nearest estimates to the correct weight of the engine (taken by the firm prior to it being placed in the window) will receive prizes in the form of free flights which will be given at the Aerodrome, Woodley. The demonstrations will undoubtedly be of great interest to the public generally. The exhibition will attract we feel more particularly by reason of the fact that this progressive and enterprising firm are at date doing much to advance the business interest of the town and provide work for its townspeople.

Flight for 22nd June 1933 announced an increase in capital for the firm - *Phillips & Powis Aircraft (Reading) Ltd. The nominal capital has been increased by the addition of £5,000 beyond the registered capital of £10,000. The additional capital is divided into 1,000 ordinary and 4,000 6 per cent preference shares of £1 each.* (The source of this new capital has not been identified – PA).

Then, in December 1933, Phillips & Powis bought the Berks Bucks & Oxon Aero Club from the Receivers of National Flying Services. Set up with a high profile in 1929 and the benefit of Government grants, NFS had an avowed intention of owning a network of

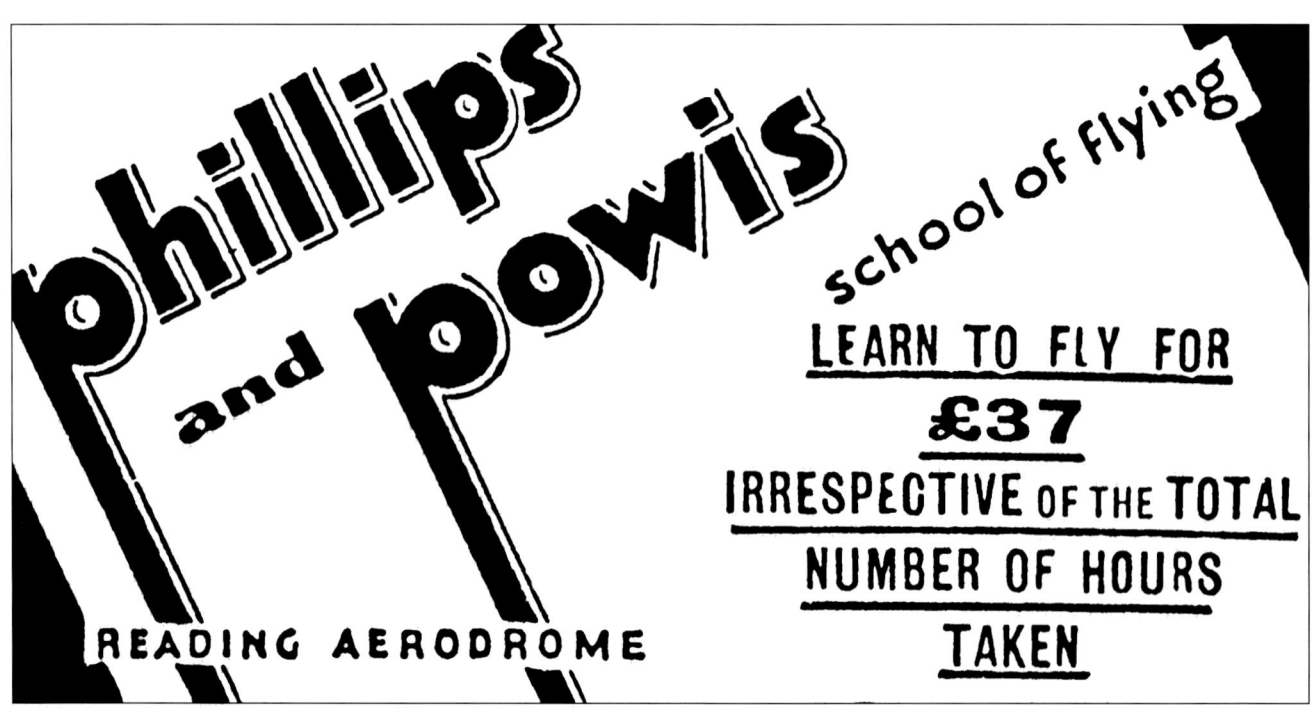

An advertisement from 'Flight' for 10th May 1933.

The expansion continues. A third hangar was erected in 1934. [Via B Clarke]

22 clubs and 100 landing grounds but had collapsed in ignominy, incurring substantial losses in the process earlier that year. Bob Milne, Berks Bucks' & Oxon's instructor, then joined the Phillips & Powis School of Flying.

The Aeroplane for 10th January 1934 reported 'Good Figures at Reading':

The annual Report from Woodley Aerodrome shows that flying time for members of the Reading Aero Club and the Phillips & Powis School of Flying (Woodley) for 1933, amounted to 1,862 hours, an increase of 209 hours on 1932. As this represents an average of just over five hours a day, the figure is very satisfactory. During 1933 the membership of the Club rose to 150, and the amalgamation of the Reading Aero Club and the Berks, Bucks and Oxon Aero Club brought another 60 members into the fold.

A Miles Martlet, which was added to the School Fleet towards the end of 1933, is a very popular acquisition. Pupils from foreign parts during the year came from Germany, France, Belgium, Norway, Sweden, Canada, India, Africa, Denmark, New Zealand, Ireland and Holland. The social activities of the Club flourished during the past year, the June 'At Home', the Autumn 'Ladies Only' Meeting, and the various Dawn Patrols, having been highly successful.

Miles Hawks are now in full production at Woodley, and this type has already been sold so far away as the Dutch East Indies, and Central Africa. The 1934 model is being produced at the rate of about 1½ per week. Six members of the Club have formed a syndicate and bought their own machine. The scheme has worked with complete success so far, and might well be adopted elsewhere.

The Old Year was wound up with a very successful Tramps Party, but the weird dress of some of the merry-makers is said to have impelled one lady who was visiting the Club for the first time to hurry away after one look, presumably thinking that she had inadvertently strayed into the local casualty ward.

For the record, the Reading Aero Club and Phillips & Powis Flying School charged Flying Members an entrance fee of £2 2s and £2 12s 6d annual subscription. Flying rates were £2 to £2 10s depending on the qualifications of the member with a special rate for casual instruction. These figures may sound ridiculously cheap by comparison with modern day rates of flying instruction averaging at about £145 per hour but everything is relative and one has to remember that the average weekly wage for the working man was then only in the order of £1 10s.

On 7th March 1934, Phillips & Powis Aircraft (Reading) Ltd entered into a contract with the Croydon-based Cirrus-Hermes Engineering Co Ltd to purchase the existing stocks of and rights to manufacture Cirrus Mks I, II and III aero-engines, spares and related assets. Just prior to this, Cirrus-Hermes had been taken over by Blackburn Aeroplane & Motor Co Ltd and they disposed of those activities they did not want, with all the Croydon staff being taken over by Rollasons. It is not known if any new engines were actually built by Phillips & Powis but meanwhile Blackburn went on to produce their new range of Cirrus engines.

The Berkshire Chronicle for Friday 16th March 1934 somewhat optimistically reported on *'Reading Aerodrome - Now an Important Aviation Centre'* thus:

A visit to Reading Aerodrome inspires the feeling that one has stepped into the future. At Woodley they have annihilated distance and mastered time, and are working to make Reading the centre of a vast network of air lines extending throughout the country and linking-up with the whole world. Their vision is not an idle dream, for much work has already been done, and the Aerodrome is now firmly established as one of the leading centres of civil aviation in the south of England, and is a growing aircraft factory already employing 160 men.

Although the idea of Reading becoming *...the centre of a vast network of air lines...* failed to materialise, it was clear that progress was indeed being made at Reading and it was during 1934

The two Miles M.2D Hawk Three-seaters in the centre, G-ACPD and G-ACPC. These were delivered on 8th and 5th April 1934 respectively.
[Via B Clarke]

that Phillips & Powis Aircraft (Reading) Ltd purchased the land which then comprised the whole of the Reading Aerodrome.

At the AGM of the Reading Aero Club, which took place in the clubhouse at Woodley on 11th March 1934, Cyril Nepean Bishop, the Club's Hon Secretary, reported that the year 1933 had been a good one and that the present year showed promise of being still better, as the hours for the first eight weeks of the year were 218 as against 99 for the same period of 1933. Club membership had risen from 100 to 135 and there had been a number of additional members from the Berks Bucks & Oxon Club. Mr Bishop expressed the gratitude of the Club to Mr and Mrs Powis for their untiring help and advice in the running of the Club and he coupled with this the names of Messrs Lawn and Milne, who "hovered around like guardian angels watching over the destinies of their fledglings."

It was also in March that a new, red-painted Hawk, (with the registration, G-ACOP, secured in honour of C O Powis) was allocated to the Phillips & Powis School of Flying and was christened *Ruddy Duck* by Mrs Blossom Miles.

An eye catching, half-page, advertisement appeared in *Flight* for 5th April 1934: *Phillips & Powis Aircraft (Reading) Ltd – Manufacturers of Hawk Aircraft and Cirrus Engines.*

'MILES FASTER ! MILES SLOWER ! MILES BETTER ! MILES HAWK !

Two seater £450 Complete. Three-seater (with Brakes) £550 Complete.'

The next advertisement for the Hawk in *Flight,* for 19th April 1934, incorporated the following commendation from Major Oliver Stewart, who wrote in *The Tatler* for 11 March 1934:

'The Miles Hawk' is more than an Aeroplane...... The Hawk is better finished - I am weighing my words - than any comparable British Aeroplane...... from every aspect the Hawk at £450, is an amazingly efficient production......

One year after the Hawk's first flight, the firm who had designed and built it were well and truly on the aviation map. An interesting article appeared in *The Garage and Motor Agent* for 24th March 1934, which gave an insight into how Charles Powis became involved with aeroplanes:

That motor trading and aircraft trading are kindred businesses in which existing motor agents should obtain an early footing, is a view often expressed by persons at present outside the aeronautical field. To test the strength of the contention by reference to one who has wide experience in both trades, a Garage and Motor Agent contributor recently journeyed to Woodley Aerodrome, near Reading, to interview Mr C O Powis.

Mr Powis started business, with Mr E J Phillips, by buying a used motorcycle, overhauling it and selling it at a profit. This initial effort led to the firm of Phillips & Powis becoming one of the leading motorcycle dealers in Berkshire. The firm then launched into the motor car sales field with conspicuous success, and Phillips & Powis Motors is known all over Berkshire, Buckinghamshire and Oxfordshire, particularly in connection with guaranteed used cars.

With commendable foresight, the Reading aerodrome at Woodley was in due course laid out by Messrs Phillips & Powis, who began a flying school, and during the last year sold some fifty aeroplanes, many of them on hire-purchase terms. Latterly, they have developed the constructional side of the aeroplane business.

The board of directors was strengthened some months ago by the inclusion of Mr F G Miles, a designer of aeroplanes and an engineer of repute. He is the designer of the new light Hawk aeroplane, which is being made at the Reading aerodrome. For this machine the firm has received orders from India, Africa, Surabaya, Penning (Penang – PA) *and the Continent. Among the well-known private owners in this country who have placed orders for the Hawk are Mrs Victor Bruce, Sir Alfred Beit, Mrs Patterson (the first woman pilot to get an instructor's licence) and Wing-Commander Probyn, who won the race for the Wakefield Cup at the International Air Meeting at Lympne last summer.*

Four questions were put to Mr Powis by our contributor, and these, with Mr Powis's replies are given below.

Q. 1 - What scope do you consider exists for members of the motor trade with good connections and premises to enter the aircraft business?
A. - The aircraft business is now in its infancy but has infinite possibilities. Those in it today are not making fortunes: the aircraft boom has not really started. When it does start, the trade will offer greater scope than any other. Therefore, to those looking for bread and butter, I say, you would do better to invest your money in a longer established trade which would be likely to show a more certain, immediate return. Nevertheless, I believe that those who want a mild gamble are on a good thing in the aircraft trade. The scope for its expansion is almost incalculable. The motor trade is

one of the most suitable from which to graduate into the aircraft business. Motor trade experience is invaluable in conducting aircraft sales and service. But nobody should think that aircraft business can be conducted from motor showrooms. The two businesses are quite distinct and different, and aeroplane manufacturers allow discounts only to those having an aerodrome or renting premises on an aerodrome.

Q. 2 - How should motor agents or garage proprietors who wish to enter the aircraft business make a beginning?
A. - They must start an aerodrome, or rent premises on an aerodrome. If they can start new aerodromes, so much the better for the aircraft trade in general. The more aerodromes the better for all concerned. The lack of them still hampers aircraft development. Then the trader must decide whether he will start a flying school or a flying club. If he decides on a school, then he will retain the full control of the organisation and take all the profit, if any. A school is generally considered to be the better-class training organisation, since it is a commercial concern and depends on its reputation and the results that it achieves. Our school has a steady flow of pupils because it has an international reputation for turning out good pilots. We draw pupils from all over the world, on account of our being so scrupulous about the standard of training given that our pupils pass the word on.

A flying club, on the other hand, is composed of a party of individuals who join with a view to sharing expenses and getting cheaper flying. The result often is that the general efficiency is lower than that of a school, although some clubs are very efficient as trainers of pilots. Such a club may be started by a trader in the first instance, but, if the true club spirit is to be retained, part control must be handed over at a later stage to a members' committee. Popular flying rates in a club must be low, with the result that there is little if any chance of any return for the promoters, unless the club is used to attract people to an aerodrome which has sales, garage, repair and maintenance services to offer.

Q. 3 - Can you give me any idea of the amount of capital required to launch into the aircraft business with a fair chance of success?
A. - Although a man can start in a comparatively small way, he cannot start on a scale similar to that of a small garage. It must be remembered that everything at an aerodrome is under Government supervision, which is very thorough. All responsible members of the staff - pilots, mechanics and so forth - must be fully qualified, and they are therefore relatively highly paid. In the aircraft business, there is no such thing as cheap labour. Moreover, machines must be licensed daily by a mechanic approved by the Air Ministry. That involves frequent overhaul, and a thorough overhaul of every machine is essential to obtain the annual renewal of the Ministry's certificate of airworthiness.

Clearly, the trader contemplating an entry into the aircraft trade requires considerable capital to finance these requirements, and to establish premises and his organisation on an aerodrome. Premises must be large enough to house his aircraft and stores. Overhead costs are therefore on a comparatively high level at the outset. If the trader intends to sell aircraft, he will find that prices average higher than those of motor cars. Most light aeroplanes cost from £700 to £1,400.

The machine now manufactured here - the Hawk - is less than £700; in fact £450. Machines varying in value from £400 to £900 will have to be taken in part exchange, also, which increases the amount of capital required. For these reasons, anyone who contemplates the conduct of aircraft sales and services from premises rented at an aerodrome requires at least £5,000.

Those who establish an aerodrome will require not less than £10,000. If only a school or club is launched, without a sales organisation, a smaller capital investment will be adequate. Two second-hand machines, costing £600 - £700, and a small clubhouse and hangar constitute the main items of outlay, so that £2,500 would probably be sufficient. I would emphasise the importance of having adequate reserves, however, because there is always the chance of being faced suddenly with a fairly heavy expenditure. The Air Ministry may call for modifications to be carried out on your machine, and in the unlikely event of accident the excess on the policy has to be met. On aircraft policies, the owner has always to bear a certain amount of each and every loss.

Q. 4 - Do you recommend persons embarking on aircraft sales to concentrate on selling a single type of machine?
A. - Yes. Concentration undoubtedly pays. The usual course is to build the reputation of one machine in the district. That is better than trying to sell everything.

In May 1934, Phillips & Powis School of Flying announced:
We are re-equipping our School of Flying with 'Miles Hawk Aircraft'. 'Hawk' instruction gives a 50% advantage to the pupil, over any other training. Time saved in learning to fly - confidence in flying - the safest and most practical aeroplane. Learn to fly in a 'Hawk' and save money.

Following the success of the first Hawk, the company began to receive enquiries for both the standard aeroplane and for a number of variants for sporting purposes. In early 1934, Don Brown also started badgering Miles to produce a de-luxe version for which he saw a market. This de luxe version finally appeared as the M.2F Hawk Major powered by a 110hp Gipsy III engine and was first flown, by Miles, on 23rd June 1934. It became an immediate success, being 50 per cent faster than the majority of private owners' aeroplanes available at that time, while still retaining the pleasant handling characteristics and low landing speed of the original Hawk.

Production models of the Hawk Major were powered by the 130hp Gipsy Major and Miles flew the first of these on 28th July 1934.

It was just prior to this that Miles had turned his attention to the subject of flaps in order to decrease the landing speed and he began experimenting with fixed 'bent aluminium' flaps screwed onto the underside of the trailing edge of the wing of a standard Hawk. The trials were so successful that the next stage in their development was to incorporate split trailing edge flaps into later production M.2F Hawk Majors, offered as an optional extra for an additional £50. Although Miles was not the inventor of the split flap, he was certainly the first in this country to recognise its value.

Flight for 16th August 1934 announced an increase in capital:
Phillips & Powis Aircraft (Reading) Ltd. The nominal capital has been increased by the addition of £10,000 beyond the registered capital of £15,000. The additional capital is divided into 5,000 6 per cent preference and 5,000 ordinary shares of £1 each. (Again, the source of this extra capital has not been identified – PA).

In the 1934 King's Cup Air Race, held on July 13/14th, the prototype Hawk Major, G-ACTD, flown by Tommy Rose the well-known sporting pilot, gained second place and in October 1934 another Hawk Major, ZK-ADJ, was entered for the great MacRobertson England to Australia Air Race. This succeeded in breaking the record for single-engined aeroplanes from England to Australia with a time of 5 days 15 hours 13 minutes and an average speed of 105 mph, to finish fifth in the handicap section. Had it not been for problems encountered on the last lap of the race across Australia, it would probably have won the handicap section of the race outright.

These successes were a triumph indeed for the new small company which was struggling to make a name for itself in the light aeroplane market. After sixteen M.2F Hawk Majors had been built, it was replaced on the assembly line by the M.2H Hawk Major, from November 1934. This became the first British aeroplane to go into production to be fitted with manually-operated split trailing-edge flaps as standard.

With both the Hawk and the Hawk Major selling well, Miles started on the design of the slightly larger M.3 Falcon, a four-seat cabin aeroplane to be powered by a 130hp DH Gipsy Major engine. The prototype, which only had three seats, was first flown, by Miles, on 23rd September 1934 and was purchased by Harold Leslie Brook for the 1934 MacRobertson England to Australia Air Race. This started on 20th October 1934 and *The Aeroplane* for October 24th reported on Brook's efforts thus: *H L Brook left Marseilles at 13.13 hours on Monday, with Miss Lay as passenger, where the Falcon had been detained with trouble of some kind but he continued to be dogged with bad luck on the outward journey.* However, he was more fortunate on the return flight in March 1935. Leaving Darwin on March 24th, he broke the light aeroplane record for the route and was accorded a tremendous reception upon his arrival at Croydon.

The production four-seat M.3A Falcon Major, fitted with the Gipsy Major engine, was first flown, by Miles, on 16th January 1935, and there was nothing to compare with its performance and accommodation at the time. This was followed by the M.3B Falcon Six, fitted with the 200hp DH Gipsy Six engine, which was first flown, by Miles, on 27th July 1935.

Don Brown, meanwhile, was still furthering his chosen career as a gas engineer at Portslade Gasworks, Sussex but had, throughout this time, remained in close touch with Miles at Woodley. He later recalled a number of amusing incidents which had occurred in the early days there, two of which are well worth relating:

In order to gain official recognition from the Air Ministry as an aircraft manufacturer, it was necessary for the company to first obtain design approval and on the strength of the Hawk's success, Miles felt that he could safely apply for this hallmark. An official from the Air Ministry duly arrived at Woodley to investigate and report on the company and he spent some time with Miles and Blossom in the tiny lean-to shed which they used as a drawing office. Anxious to make a favourable impression, they produced all the Hawk drawings and calculations and eagerly volunteered to be cross-examined on them. All was going well until the office door suddenly burst open and in marched an irate purchaser of a Hawk, dragging his eighteen-year-old daughter behind him. He seized his daughter's frock, pulled it up to the level of her shoulders, displayed the young lady's person soaked in dirty oil and exclaimed angrily, 'Look what your ******* aeroplane has done!'

Then there was the man who ordered and took delivery of a Hawk and was either unwilling or unable to pay for it. In those days, the firm could not afford to incur bad debts or to indulge in expensive litigation so clearly something had to be done about it. Unfortunately, the customer had possession of the aeroplane and refused to give it up. Desperate measures were called for, so one night Miles, Powis and one or two accomplices drove to the field where they knew the aeroplane was kept but to their dismay they found that the rascally 'owner', possibly suspecting a repossession had locked the Hawk in a shed which was surrounded by barbed wire.

This was an unexpected snag but to accept defeat was unthinkable and stumbling over one another in the darkness, amid whispered curses and exhortations to be quiet, the conspirators set to with a will. Some cleared the barbed wire away - not an enjoyable task in the dark while Miles tackled the lock. They felt sure they would be detected due to the noise they were making but their luck held and at last the Hawk was wheeled out, its wings spread and Miles climbed into the cockpit. The carburettor was flooded, the propeller swung and the engine started with a roar. Miles opened the throttle and took off straight ahead into the darkness while the others leapt into the car and made good their escape. Fortunately there was sufficient fuel to enable him to cruise around until dawn broke and he was able to land back at Woodley.

Despite much effort, it has not been possible to confirm which aeroplanes were involved in these incidents! By late 1934, the firm was in urgent need of more hangarage and factory space and on 2nd November 1934, they placed a contract for the erection of an 'Aeronautical School and hangar' at Woodley which in due course became the Experimental Department. Then, on December 18th, they placed a contract with Boulton & Paul Ltd for the construction of a large new assembly building in which it was planned to be able to increase production to six aeroplanes per week.

Flight magazine for 15th November 1934 reported that Flt Lt Thomas 'Tommy' Rose has been appointed Sales Manager to Phillips & Powis, as from January 1st 1935. This marked the start of Tommy Rose's long association with Miles, culminating with him being appointed Chief Test Pilot in November 1939 following the sudden and unexpected death of Chief Test Pilot Bill Skinner. Tommy Rose held this position until his retirement in January 1946. Born at Chilbolton on 27 January 1895, Tommy Rose had served in the RFC and later the RAF from 1917 until 1926. After leaving the service, he became an instructor at Midland Aero Club before, in 1929, joining Anglo-American Oil Co Ltd as their pilot. In 1931 he flew their Avro Avian G-ABIE *High Test* to South Africa and back and later, in September 1933, became chief instructor at the Northamptonshire Flying Club.

1935 brought a greater level of financial security to the company and at a Board Meeting on 15th February 1935, resolutions were passed to put in hand the voluntary winding-up of the existing company for the purpose of reconstruction with a declaration of solvency being signed and submitted to the Registrar of Companies. New Articles of Association were drawn up and at an EGM on 5th March 1935 the changeover was approved by shareholders. The 'new' company, Phillips & Powis Aircraft Ltd was formed on 11th March 1935 and converted to a public company on March 18th, with the sole aim of buying the existing business and assets of Phillips & Powis Aircraft (Reading) Ltd with effect from 1st November 1934 for £57,597. On March 18th, shares in the new company were offered for sale to the public ...*to help develop the business previously carried on by Phillips & Powis Aircraft (Reading) Ltd, designers and constructors of civil aircraft known as the Miles Hawk and the Miles Falcon.*

The public company had a share capital of £125,000 – 500,000 Shares of 5/- each. The public were offered 269,605 Shares at a price of six shillings a share. In the prospectus, the directors were identified as Charles Owen Powis (Aeronautical Engineer – Chairman and Managing Director, with a salary of £1,500 plus a percentage of profits); Frederick George Miles (Aircraft Designer – Technical Director) and Major George William Graham Allen, MC (Governing Director of John Allen & Sons (Oxford) Ltd). Miles, whose home address was now Land's End House, Sonning, Berks, agreed to remain with the company for a period of five years.

Major Allen, who lived at Iffley, near Clifton Hampden, Oxfordshire, was a private flyer of some renown, having his own private airstrip at his home. His first aircraft was the DH Gipsy Moth G-AAJJ, which he bought new in June 1929, replacing it with a Puss Moth, G-ABKD, in March 1931, which he kept throughout the thirties, donating it to the Secretary of State for Air on the outbreak of war.

With the additional funding secured, the 4-passenger M.4 Merlin was the next logical step. A development of the M.3B Falcon Six, it was built in early 1935 to the specification of Flt Lt George Birkett of the air taxi company Birkett Air Service Ltd, Heston. Birkett wanted an aeroplane which would carry five people, including the pilot, and cruise at 145 mph. The 200hp DH Gipsy VI engine was chosen to power the new machine and in order to get a good take-off performance it was proposed to use a variable-pitch propeller. The Merlin resembled a rather corpulent Falcon Six in appearance with the fuselage being over four feet wide, to enable the pilot and one passenger to sit in front, with three 'hefty' passengers to be seated comfortably on the wide seat behind. Radio was carried and the Merlin was designed to have dual control fitted if desired. It was first flown, by Miles, on 24th March 1935.

A Supplement to *The Reading Standard* of 10th May 1935 provided a good contemporary report on the company, including the following, which puts matters into perspective:

Chapter 6: Reading - 1932 to 1939

The advance in progress of aircraft manufacture, which is now taking its place as one of the important factors of the nation's trade, is recognised by all to have been remarkable. Nowhere has this progress been so marked as in Reading. Taking a long-sighted view of the possibilities [Mr Powis] became one of the pioneers of the aircraft industry. Extensive premises were leased at Woodley for the purpose of commencing flying school and business initially for the repair and sale of aircraft.

Throughout the years 1930, 1931 and 1932, the volume of business both as flying school and as a trading concern increased considerably and the firm took its place as one of the most widely-known schools of instruction in the South of England, pupils coming from all parts of the world. During this period the hangar accommodation had been considerably increased, and the staff supplemented to the extent of sixteen employees.

The progress of the firm is marked by the fact that Phillips & Powis Aircraft Limited are now represented in over twenty countries by agents, and Hawk and Falcon aircraft are being flown in all parts of the world, in countries so far apart as South America, Denmark, Greece, Switzerland, France, Spain, Portugal, Roumania, Iraq, India, Malaya, Australia, New Zealand and China; in fact the Hawk has 'girded the world'. The sale of this aircraft has already run into hundreds. It will be readily understood that this great output of machines and business involved therein have led to a great increase in the number of employees in the firm, which has risen from thirty-seven on 30th March, 1933, to some 350 today.

Between 1932 and 1934 two further extensive hangars were erected. In 1934 the firm purchased outright the property on which the aerodrome stands, and at the same time purchased from the National Flying Services Limited, hangars, club-house and buildings which had been erected by that company on a piece of land already belonging to Phillips & Powis Aircraft (Reading) Limited. As is common knowledge, the company has recently been reconstructed as a public company, with a capital of £125,000. Further extensive workshops and aeronautical accommodation have been provided during the past few months and these are in operation, the extent of the new workshops being 360 ft. by 75 ft.

The workshops are being fitted with additional plant of the most up-to-date type, and the firm which up to the date of reconstruction was putting out two machines per week is now budgeting to put out six machines per week. Here again further extensions and additional staff are being budgeted for. Without doubt Reading has taken its place on the map as the centre of aircraft enterprise and has already assumed the premier position as the home of the manufacture of light aircraft.

In April 1935, Miles realised that if he could win the 1935 King's Cup Air Race, it would have tremendous publicity value for the company. He therefore decided to enter as many Miles aircraft as possible for the race, including an entirely new single-seat racing aeroplane for himself. He informed Blossom of his decision about a month or so before the race and Don Brown recalled that when Blossom asked Miles how on earth he thought a new aeroplane could be produced in the time available he replied: *the design of it is your problem and you had better get on with it straightaway!*

Blossom realised that the only way of producing such a machine in the short time available would be to utilise a standard Hawk fuselage, delete the front cockpit, thereby shortening it by twelve inches, reduce the wingspan by five feet by attaching the wings directly to the fuselage and lower the undercarriage. The aircraft was to be powered by a high-compression DH Gipsy Major engine and the finished result was the Miles M.5 Sparrowhawk. Registered G-ADNL, this was first flown, by Miles, on 19th August 1935, less than three weeks before the race.

Also for the King's Cup, Phillips & Powis prepared eleven Hawk variants and two Falcons for themselves and other private owners, including the new M.3B Falcon Six, G-ADLC, which was to be flown by Tommy Rose whose perfect course-keeping made him an ideal racing pilot. The race was to be run over two days in two sections with the first consisting of a thousand-mile eliminating heat around a course from Hatfield to Scotland, across to Ireland and back to Hatfield on 6th September 1935. The Sparrowhawk was fitted with long-range tanks to enable Miles to cover the distance with only one stop for refuelling.

While many of the competitors nursed their engines for the second day's race Miles, with characteristic abandon, left the throttle wide open for the whole of the thousand miles and, thanks to the reliability of the Gipsy engine, he came in first at an average speed of over 170 mph.

The second day's race, on September 7th, was over a number of short laps and it soon became apparent that Tommy Rose in the Falcon Six would win the King's Cup, provided that his engine kept going. His course-keeping was as perfect as ever and an excited crowd saw him come round lap after lap, taking exactly the same time for each and maintaining a far higher speed than the handicappers had believed possible. Excitement increased when two M.2P Hawk Trainers were seen lying in second and third places and it was fervently hoped that the three Gipsy engines would stand the pace, since they were all flying at full throttle.

Tommy Rose's Falcon G-ADLC streaked across the finishing line to win at the astonishing speed of 176.28 mph, with the two Hawk Trainers, G-ADLN flown by F/O H R A Edwards and G-ADLB flown by Owen Cathcart Jones, who had been flying almost neck and neck throughout the race, coming in second and third at 157.84 and 157.53 mph respectively. Another Falcon, G-ADLS piloted by Samuel Harris and Laurence Lipton, came in fifth at about 147 mph to give Phillips & Powis their greatest success to date, taking four out of the first five places!

This was the first time that the first three aeroplanes to finish in a King's Cup Air Race had all been designed and built by the same manufacturer and this unique feat has never been repeated by any other aeroplane manufacturer since, or ever will be again.

During 1935, the large new assembly building was erected by Boulton & Paul in front of the original three smaller hangars and this allowed the construction of the Hawk, Hawk Major and Falcon, in all their variants, to be manufactured at the same time in less cramped conditions.

Another epoch-making event occurred in 1935 when the Royal Aircraft Establishment at Farnborough became interested in obtaining full-scale data relating to the effect on performance of wings of varying aspect ratio and thickness. Miles' reputation for turning out light aircraft speedily and cheaply led the Air Ministry to place an official contract with Phillips & Powis to build an aircraft capable of fulfilling the requirements of thick-wing research for the RAE. This led to the construction of the M.6 Hawcon – named to reflect its dual ancestry from the Hawk and Falcon. The Hawcon was first flown, by Miles, on 14th October 1935.

The Hawcon was built with four alternative, interchangeable sets of wings with different thickness/chord ratios. These wings were test flown in turn and much valuable information was gained. The Hawcon became instrumental in initiating the happy relationship between Miles and the RAE, which later resulted in a number of experimental light aeroplanes being built by Miles specifically for their research purposes.

While variants of the M.2 Hawk and Hawk Major were being used successfully by aeroplane clubs, on 25th November 1935 three of the new M.2W Hawk Trainer Mk.II's entered service with the Reserve Training School at Woodley, officially formed on that day.

With the expansion of the RAF from 1934, the five existing civilian-run Elementary & Reserve Flying Training Schools were not able to cope with the extra pilots needed and the Air Ministry decided to award contracts to other civilian companies already

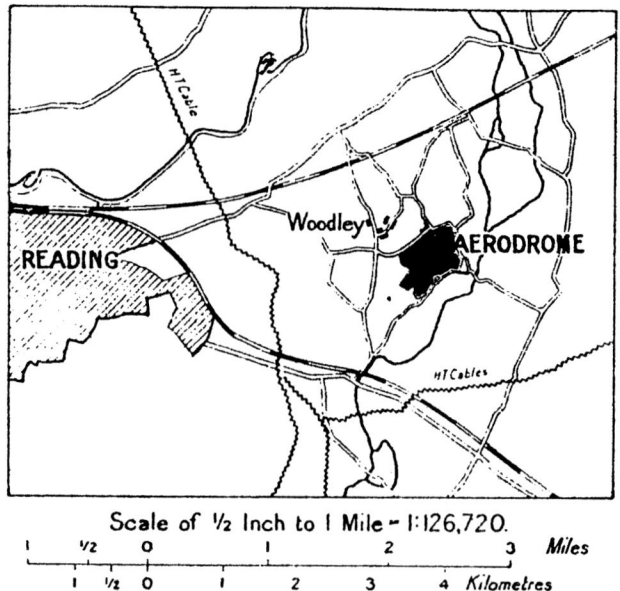

Telegrams:
"Sonning 114."

Telephone:
Sonning 114 or 115.

Lat. 51° 27′ N.   Long. 00° 53′ W.

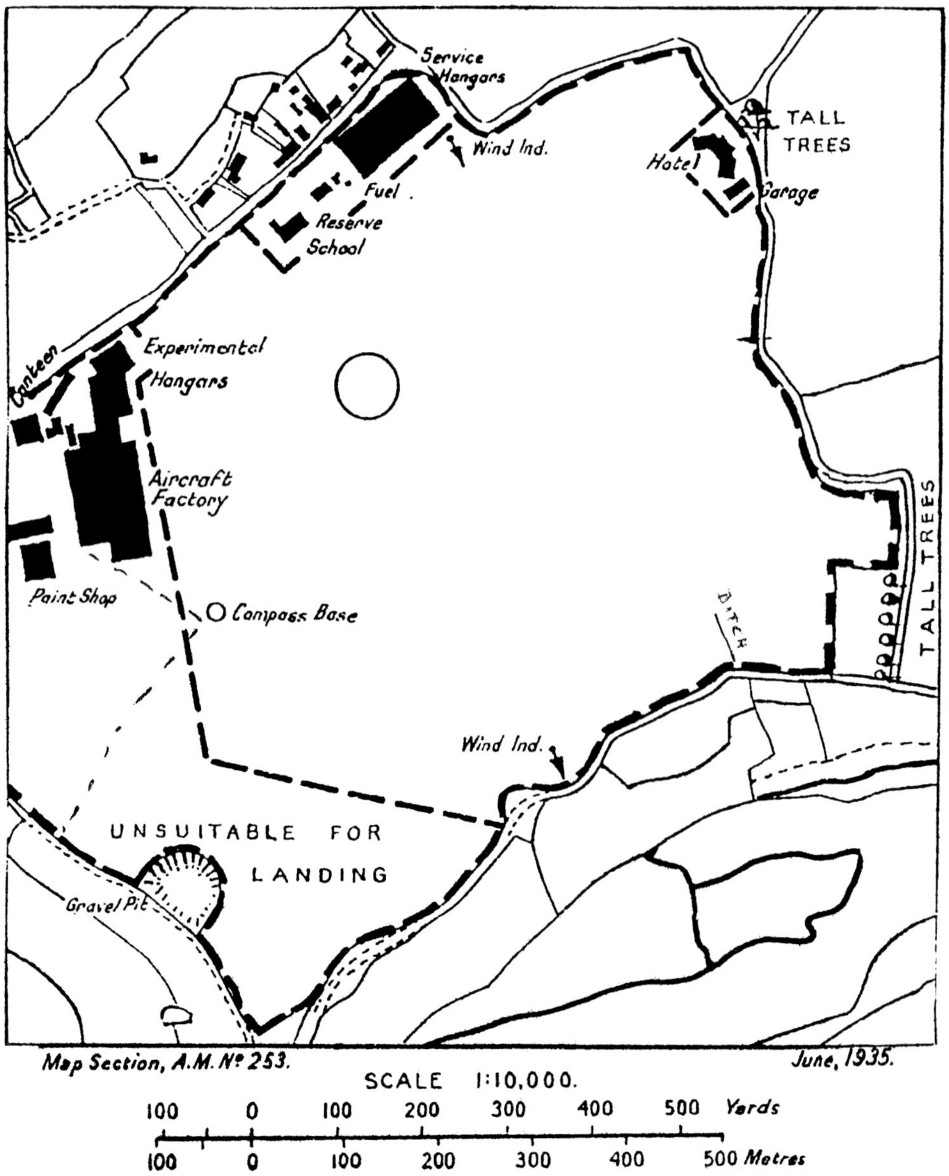

Seen at the top is a map showing Woodley Aerodrome, with below a large scale drawing giving details of the layout of the buildings in June 1935.

Chapter 6: Reading - 1932 to 1939

1. **Controlling Authority.**—Phillips and Powis Aircraft, Ltd., Reading drome, Woodley, Berks.

2. **Landing Area**
   (a) *Dimensions.*—
   N.—S. 850 yards.
   N.E.—S.W. 1,050 yards.
   E.—W. 750 yards.
   S.E.—N.W. 775 yards.

   (b) *Altitude above Mean Sea Level.*—150 feet (46 metres).

   (c) *Surface conditions.*—Grass covered.

3. **Obstructions Requiring Special Caution**
   (a) *West side.*—Aircraft factory, 32 feet high.

4. **Special Signals.**—Nil.

5. **Lighting**
   (1) *Obstruction lights.*—Red obstruction lights, mounted on the aircraft factory, aeronautical school and service hangar, are operated on request.
   (2) *Floodlight.*—A mobile floodlight is available on request.

6. **Facilities for Aircraft**
   (*Available daily from 0745 hours to dusk, and at other times on request.*)

   (1) *Refuelling.*—Aviation fuel, oil and fresh water available. Refuelling pumps are installed.
   (2) *Repairs.*—All normal repairs can be executed.
   (3) *Hangars.*—

Number.	Structure.	Net Breadth.	Net Depth.	Door Height.	Door Width.
		ft.	ft.	ft. in.	ft. in.
One	Steel and corrugated iron (*aircraft factory hangar*).	360	300	14 0	60 0
One	Steel and corrugated iron (*service hangar*).	70	180	15 0	40 0
Eight	Steel and corrugated iron (*lock-ups*).	12	30	15 0	12 0

7. **Facilities for Personnel**
   (1) *Transport.*—Railway station at Reading, 3½ miles. Omnibus services to Reading pass the aerodrome. Taxis available.
   (2) *Hotels, etc.*—Club house and public restaurant at the aerodrome.

8. **Local Regulations**
   (1) Low flying in the vicinity of the farm situated 1½ miles S.W., and of Sonning golf course, 1 mile N. of the aerodrome, should be avoided.

Details of Reading (Woodley) Aerodrome in 1935.

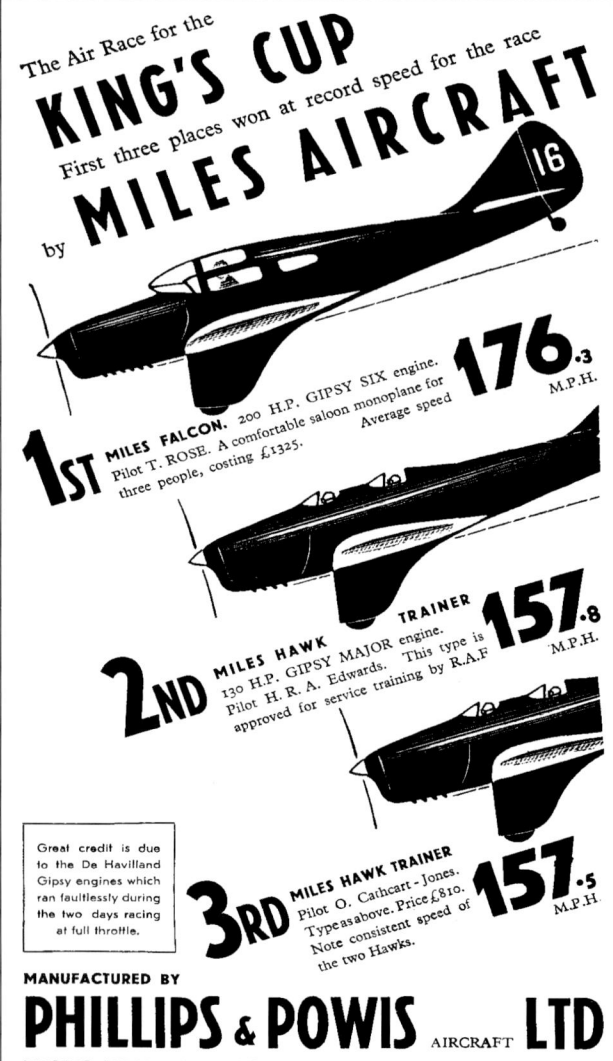

Above and left: Following their success in the King's Cup, it was inevitable that the firm would capitalise on it, as these contemporary advertisements show.

experienced in flying training. The formation of the Reserve Training School (which was officially known as No. 8 Elementary & Reserve Flying Training School) to be managed by Phillips & Powis Aircraft Ltd under Contract No.A113018 to the Air Ministry, was announced in June 1935. The School was opened on 25th November 1935, under the command of the New Zealand born Chief Flying Instructor Flt Lt James F Moir, who had previously been a qualified RAF test pilot at the A&AEE Martlesham Heath, with Capt E W Stewart as School manager. The fleet initially consisted of nine DH Tiger Moths (G-ADJB to G-ADJJ) and three Hawk Trainer Mk.IIs (G-ADWT to G-ADWV).

A new building was erected near to the repair and service department hangar to house the staff. Two adjacent houses, 'Westwinds' and 'Hawkhurst' were also used to accommodate the airmen and instructors. An emergency landing ground was set up at Sheffield Farm (also known as Theale Relief Landing Ground) and this was used by the School from soon after the commencement of operations.

By May 1937, the School's fleet had increased to 14 Hawk Trainer Mk.IIs, 4 RAF Magisters, 2 Hawker Hart Trainers, 2 Hawker Hart bombers and 2 Hawker Audax army co-operation aircraft, with the Tiger Moth fleet reducing to seven. By mid-1938 the fleet was further increased by the addition of 1 Hart Trainer, 4 Hart bombers and 2 Hawker Hinds and in 1939 it was again enlarged by the addition of 3 Fairey Battles and 2 Avro Anson Mk.I's.

In the period to 31st August 1939, 19 Ab Initio Courses were completed. These Courses varied in strength from 25 to 42 pupils and 454 pupils satisfactorily completed their courses and proceeded to Service Flying Training Schools. During the same period an average of 40 to 50 officers and airmen of the RAFO and RAFR carried out Annual Training each year. From May 1937 to August 1939 approximately 150 pilots of the RAFVR were trained of which 48 had reached Wing Standard prior to mobilisation.

CHAPTER 6: READING - 1932 TO 1939

A superb photograph of the aerodrome in 1935, showing the new Boulton & Paul assembly building linking the three hangars.

['Flight' via B Clarke]

Miles M.2R Hawk Trainer Mk.I, G-ADLN, outside the new assembly building.　　　[B&P Archives, Norwich, via B Buss & B Holmes]

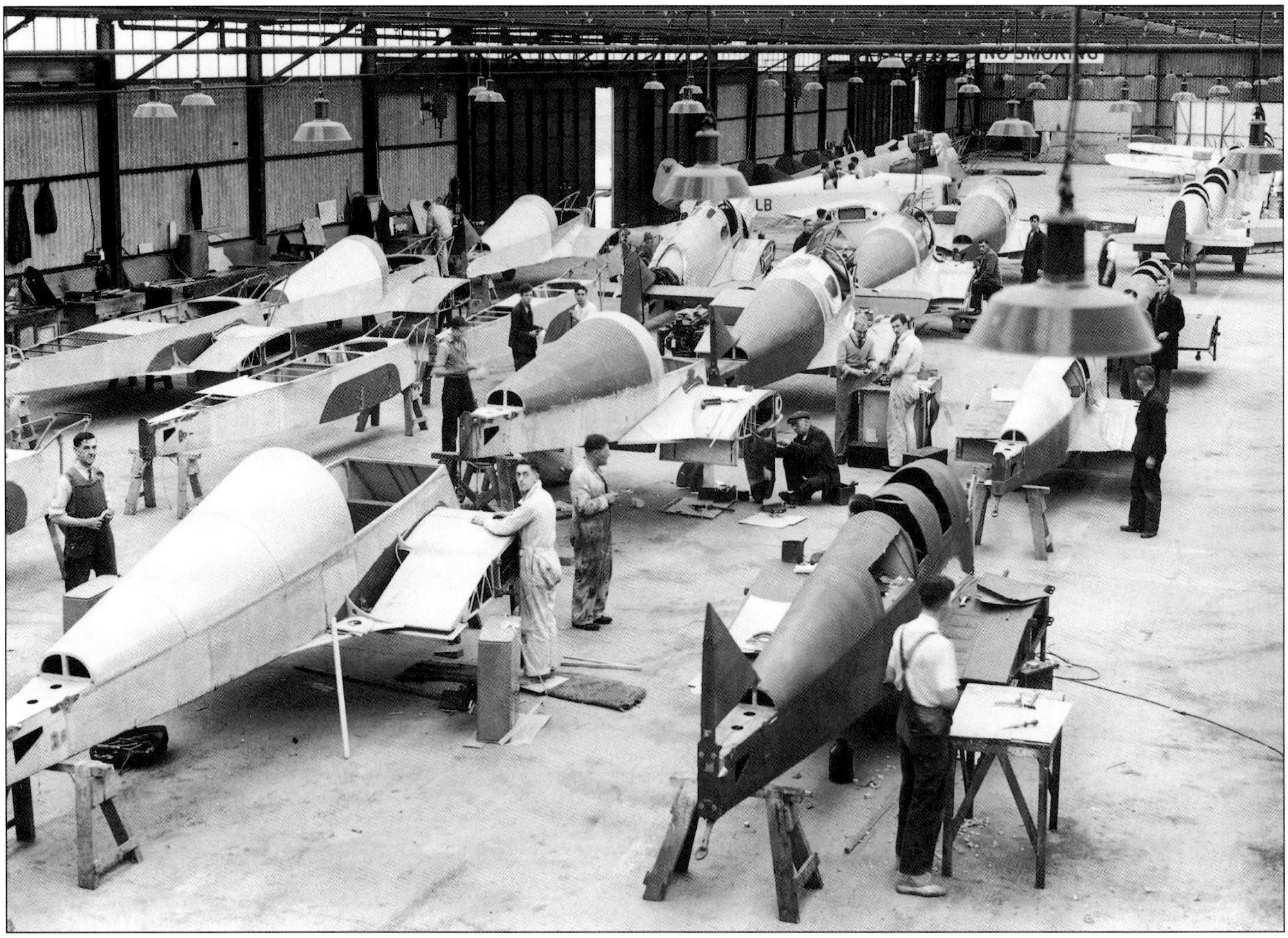

Miles Hawk and Falcon variants under construction in the new assembly shop in 1935. Note the Miles M.2R Hawk Trainer Mk.I, G-ADLB, near the door. [Via B Clarke]

Miles aeroplanes of various types under construction in the new assembly building. Note Tiger Moth G-ADJI of the Reserve Training School. [B&P Archives, Norwich, via B Buss & B Holmes]

CHAPTER 6: READING - 1932 TO 1939

At the time, the majority of civilian training schools were being supplied with Tiger Moths, but it was not surprising that Miles should seek permission from the Air Ministry to partially equip its fleet with his 'home-grown' product. Indeed, *The Aeroplane* for 4th March 1936 reported: *As the factory turns them (Hawk Trainers) out they (will) replace the Tigers and before the Summer (of 1936) the last Tiger Moth is scheduled to disappear from the Reading school for good.* (This was optimistic – the Tigers stayed with the School until 1939 and for a while thereafter, apart from G-ADJF which was sold off early in 1936 and G-ADJB which crashed later in the year – PA) The school aircraft were housed and serviced in a new extension to the Repair and Service Department hangar.

Leonard Hackett was initially responsible for the ground organisation of the Reserve Training School, having joined the Sales Department in January 1936. On the outbreak of war 'Reserve' was dropped from the title, which then become 8 Elementary Flying Training School and Flt Lt James Moir had, by then, been promoted to Squadron Leader.

The last new type to be built in 1935 was the M.7 Nighthawk and this was, in essence, a two-seat, dual-controlled trainer development of the M.3B Falcon Six. It was intended to be powered by either a 130hp DH Gipsy Major or a 200hp DH Gipsy Six engine. It had hydraulically-operated split trailing-edge flaps and, like most good aeroplanes of the day, started life as a private venture. The Nighthawk was first flown, by Miles, on 26th October 1935 and was awarded an Air Ministry development contract to include extensive spinning tests. These extended into early 1936 as officialdom had still not totally accepted low-wing monoplanes as being the way forward and needed to know much more about their spinning characteristics.

The next notable event took place in February 1936, when Tommy Rose took-off from Lympne in the M.3B Falcon Six G-ADLC on his record attempt for Cape Town, where he arrived 3 days 17 hours 37 minutes later, thus establishing a new record for the distance. Tommy also created a record for the return journey in March. It was also in early 1936 that Charles Powis and Tommy Rose made a 4,500 mile demonstration tour of non-European countries, flying the now-famous Falcon Six, *1935 King's Cup Air Race Winner,* G-ADLC and the M.5 Sparrowhawk G-ADNL.

By early 1936, Miles had decided in his own mind that it would be logical if all future RAF pilots were to proceed to a specially designed high-performance two-seat, advanced training monoplane after having completed their elementary flying training, in order better to prepare themselves for the new and advanced Hawker Hurricane and Supermarine Spitfire fighters which were about to enter service. Miles envisaged an advanced trainer of wooden construction powered by the Rolls-Royce 745hp Kestrel XVI engine. This engine was more representative of the Rolls-Royce Merlin engine then being fitted to the new fighters and although Miles had no idea how the project was to be financed, he commenced work on the design.

The problem of how to privately finance the construction of a prototype appeared at first to be insurmountable as the company could not afford it, so Miles decided to approach Rolls-Royce with his proposals. Fortunately, the directors of Rolls-Royce gave him a very favourable reception as they could see the force of his argument. The production of the Kestrel engine was nearing its end at the time, so they could also see the partnership as a means of keeping it in production. Although details were not announced until later, *The Aeroplane* for 25th March 1936 carried the following story:
  Stories have recently appeared in the financial pages of various daily papers coupling the names of Phillips & Powis Aircraft Ltd and Rolls-Royce Ltd. Such suggestions are not so improbable as they might seem. Mr F G Miles is one of our brightest and most enterprising designers. He has already acquired a great deal of experience in stressed-skin construction. Though such structures have been of plywood they might equally well have been of metal. Modern formulæ can be used as well for one as the other. And as various aircraft firms build aero-motors there seems to be no reason why an aero-motor firm should not build aeroplanes. Mr Sidgreaves and Mr Powis are both good business men. So why should they not do a deal?

The following week, *The Aeroplane* of 1st April 1936 reported: *Reliable informants asseverate that within the past week one of the principals concerned has been heard to say that the Rolls-Royce car people now hold a large proportion of the share capital of Phillips & Powis Ltd. According to rumour the idea is that the Rolls company is to induce Mr Miles to design a machine for a Merlin, to show what it can do. Perhaps the Heinkel 70 has persuaded Rolls-Royce Ltd that quite a lot is possible in a suitable civil machine which is out of the question in the modern heavily-equipped fighter. Or perhaps the company is considering the manufacture of a small motor to fit present and subsequent Miles designs. Or perhaps there is a really fast twin-motor transport in the offing. And for a long time there has been talk of an air-cooled high-powered Rolls.*

I cannot help but think that the date of that issue may have had something to do with the publication of some of these wild flights of fantasy by C G Grey as, on this occasion, his surmises were well wide of the mark! But, it is equally possible that a diversionary story was being put about whilst secret negotiations were in hand.

However, the reputation which Miles had already earned, together with the logic of his arguments, so impressed Rolls-Royce that they decided not only to back him financially to build a prototype of the advanced trainer but also to acquire a large financial interest in Phillips & Powis Aircraft Ltd. The Agreement was finally signed on 6th April 1936 (see Appendix 29) and provided for Rolls-Royce to inject £125,000 by subscribing for 500,000 3.5% 5/- convertible preference shares and which thereby gave Rolls-Royce the right to a 50% stake in the company. They also gained the right to appoint two directors and these were A F Sidgreaves OBE and A Wormald. The latter unfortunately died in December 1936 and was replaced by Lieut Col Maurice Ormonde Darby, OBE (who had, back in 1920, been the managing director of The Aircraft Disposal Co at Croydon and who had subsequently been involved in several aviation companies).

With the future of the company and the new aeroplane assured, design work on the high-performance M.9 Rolls-Royce Private Venture Trainer for the RAF commenced in earnest. In May 1936, at the first AGM of the new public company, a loss of £20,000 was announced and the shareholders agreed to the Rolls-Royce Ltd refinancing.

Meanwhile, also in May 1936, George Miles finally joined his brother at Woodley, having wound up Southern Aircraft Ltd and the Southern Aero Club at Shoreham. George had established that Cecil Pashley was assured of continuing employment with Brooklands Aviation Ltd, which had taken over the old club on 1st September 1935 and renamed it the South Coast Flying Club. Brooklands Aviation Ltd, as a result, now controlled four flying clubs, as, apart from the eponymous Brooklands Flying Club, they also ran Cinque Ports Flying Club at Lympne and Northamptonshire Aero Club at Sywell.

George had, in the interim bought, flown and sold a Westland Widgeon (G-EBRO) and replaced it with an Avro Avian IVM (G-AAWH). He had also gained further valuable experience in flying, maintenance and knowledge of aero engines and he started work at Woodley in charge of the new Menasco engine project.

In mid 1935, Charles Powis had commenced negotiations with the American aero-engine designer Al Menasco with a view to manufacturing a range of Menasco engines under licence at Reading. In March 1936, a licence agreement was signed by Phillips & Powis Aircraft Ltd to build four Menasco models, in normally aspirated and supercharged forms. These were all inverted air-cooled in-line engines and comprised the 125hp four-cylinder Pirate C4, the 150hp six cylinder supercharged Pirate C4S,

A photograph showing the Reserve Training School building with Tiger Moths and Hawk Trainer Mk.IIs lined up outside. Note also the extensions to the repair and service hangar.
[Phillips & Powis via Mrs Skinner & The Miles Aircraft Collection]

the 160hp six-cylinder Buccaneer B6 and the 200hp six-cylinder Buccaneer B6S. The engines were well known in America and in this country it was thought that they would have been of particular interest in bringing supercharging to the private owner.

Miles had already fitted a Menasco Pirate C4 engine in the Hawk Trainer G-ADLN and George's first job was to fly this aeroplane around the country for long periods at full throttle, in order to severely test the Menasco engine's reliability. During these flights, George was often accompanied by Don Brown.

It is not known if any Menasco engines were actually built at Reading but it is thought unlikely. However, one Menasco engine, probably from The Shuttleworth Collection's Desoutter G-AAPZ, survives in a wooden crate at Old Warden and this has a Phillips & Powis Aircraft Ltd brass plate attached to it, which could simply have been attached to an imported American engine. Another is installed in the Miles Mohawk, now restored and displayed in the RAF Museum, Hendon.

Miles' monoplanes meanwhile were gaining world-wide recognition and even the Soviet Union placed an order for one Hawk Trainer through Arcos Ltd, its British agent. Miles' growing reputation also brought him to the notice of Whitney Willard Straight, a wealthy young American. Whitney Straight, born in New York in 1912, was only five years old when his father, a banker, died of pneumonia leaving the family seriously wealthy. His mother later married Leonard Knight Elmhirst, the founder of Dartington School in Totnes, Devon and they moved to England in 1926. At the age of 19, Whitney Straight had bought his first aircraft, the DH.60M Moth G-AATB and replaced this shortly afterwards by Puss Moth G-ABNZ.

In 1932, Whitney Straight took up motor racing and formed his own team – Team Straight – but gave this up when he married Lady Diana Finch-Hatton in July 1935. He had probably first met Miles when he arranged for Basil Henderson to design the Hendy Heck for him in 1933 (and which was to be built by George Parnall at Yate). In April 1935 he formed the Straight Corporation Ltd with a view to owning and managing a number of flying clubs in different parts of the country and he planned to equip these clubs with what he considered to be the ideal aircraft for the private

George Miles leaning on the wing of a Miles M.14 Magister.
[Via B Clarke]

The fuselage production shop in January 1936. Note the cabin assembly for a Falcon Six in the background. [Via B Clarke]

owner. While most clubs were still operating the open-cockpit Moth, Whitney Straight felt that the time had come to have an aeroplane which would provide the comfort and convenience of the average car.

Miles and Whitney Straight collaborated in drawing up a specification for a two-seat, comfortable cabin aeroplane with side-by-side seating and folding wings, to be powered by a 130hp DH Gipsy Major engine. It had to be easy to fly, have a considerably improved performance over the old biplanes and be reasonably priced. The result of this collaboration was the M.11 Whitney Straight Special (G-AECT), which was first flown, by Miles, on 3rd May 1936.

Shortly after its first flight, the prototype made its public appearance at the Royal Aeronautical Society's Garden Party at Heathrow where it aroused much interest on account of its roominess and comfortable layout. The aeroplane went into production as the M.11A Whitney Straight and it had a large single-moulded Perspex windscreen which extended the full width of the cabin to provide an unobstructed view. Were it not for the outbreak of war in September 1939, considerably more of these machines could well have been produced.

In June 1936, it was reported that Flt Lt Tommy Rose had taken up an appointment with the old-established firm of distillers, Seager Evans and Co Ltd of London to fly their Miles Falcon. This moved C G Grey to observe of the company: *whose gin makes the World go round, but under complete control.* And, of Rose: *His interest in aviation will not suffer as he will continue to make short work of long distances in flying competitions generally.* However, Tommy Rose does not appear to have left Phillips & Powis to take up this appointment and equally no Falcons are known to have been registered to Seager Evans.

Miles M.11 Whitney Straight Special, with George Miles in the cabin and Lionel Tysoe standing by after having swung the propeller. [L Tysoe]

On 29th June 1936, three Miles types – Miles M.2W Hawk Trainer Mk.II, G-ADZD, Miles M.7A Nighthawk U6/10 and the Miles M.11 Whitney Straight Special, G-AECT – were lined up for inspection at the 5th SBAC Show, at Hatfield. This was the first time that Phillips & Powis had participated in the SBAC show. ['Flight' via B Clarke]

The bar of the Falcon Hotel with, from left to right: 'Blossom', Victor Burnett, 'Shell-Mex' and George Miles. [A H Lukins via B Clarke]

Meanwhile, the need had arisen for a new building to replace the old Reading Aero Club house, which was becoming hemmed in by the new factory buildings springing up around it. This resulted in the building of The Falcon Hotel and inevitably, C G Grey just had to say a few words on this, in his own inimitable style, in *The Aeroplane* for 3rd June 1936:

Long before the present association of interests between Woodley and Derby, Reading aerodrome showed signs of acquiring a Rolls-Royce atmosphere. For many years the car-park's glittering glow was a Berkshire landmark at a time when the Club's aerial fleet still consisted of sombre stick-and-string ships. Nowadays the sleek row of low-wing creations, chocked and couchant on the campus, shames even the shiniest speed-wagon in the paddock. As fame came to Woodley the cramping clutch of commercial production hemmed in the Club with hangars and a civil flying training school sprouted on the northern boundary to steal the week-day thunder of the flying amateurs. Now the old Club is to shift over to the far side of the aerodrome to make more room. The Reading Aero Club's new rendezvous will be, appropriately, the Falcon Hotel at the opposite corner of the landing area near the approach gaps used in the prevalent west winds of Woodley.

Just over a year later, *The Aeroplane* for 3rd November 1937 reported on the opening of this architecturally brilliant building (see page 86).

Meanwhile, another famous person to visit Miles in 1936 was none other than Colonel Charles Lindbergh, the young American aviator who had distinguished himself by making the first solo eastbound crossing of the Atlantic in 1927. Lindbergh was on an extended mission to Europe and having settled in Sevenoaks in Kent decided that he needed a fast, long-range, light aeroplane in which he and his wife could make business trips between the European capitals. He could not find what he wanted amongst the aircraft types then available, so he asked Miles if he would design a high performance, long range, touring aircraft to suit his requirements.

The M.12 Mohawk, as the new aeroplane was to be named, was the result and this was first flown, by Miles, on 22nd August 1936. It was no small compliment to Miles and the fledgling company that Charles Lindbergh should have placed the order to build the aeroplane with them. The Mohawk, registered G-AEKW, remained in Lindbergh's ownership until he returned it to Phillips & Powis for storage in April 1939 on his return to the States.

Don Brown had suggested to Miles earlier in 1936 that he should consider designing a twin-engined monoplane as there were very few aeroplanes of this type available at the time. Miles must have agreed as work soon commenced on the design of the eight-seat M.8 Peregrine. This new machine was fitted with a retractable undercarriage and was powered by two 200hp DH Gipsy Six engines. Just prior to its completion, however, Miles and Blossom left on a visit to the USA to study design methods and to continue discussions with Al Menasco over the manufacturing of his new engine at Reading. Therefore, it fell to Charles Powis to make the first flight of the Peregrine, on 12th September 1936. Powis had never before flown a twin-engined aeroplane but all went reasonably well.

Unfortunately, any plans they had for producing the Peregrine had to be abandoned when the company received an order for 90 of the new M.14 Magister elementary trainer. This decision was regrettable because no aircraft of comparable performance was available elsewhere and it was to be many years before Miles would again have the time to design and build a multi-engined aeroplane to undertake a similar role.

As a result of the experience gained with the Hawk Trainer in service with their Reserve Training School, together with Miles' continual representations, the Air Ministry had ultimately realised that the time had at last come to adopt monoplane training. On 28th October 1936, they issued a contract to Phillips & Powis to Specification 40/36 'for the Manufacture of Civil Hawk', a new elementary trainer, designed to incorporate the experience gained in operating the Hawk Trainer with the Reserve Training School. The Specification took the Reserve Training School's G-ADYZ as the sample and the first contract called for 90 aircraft. The new aeroplane was to be called the M.14 Magister, and the contract was the largest the company had received. In consequence it became necessary to considerably extend the factory yet again.

The Pilots' Pub

EVEN Mr. Eric Linklater's "Poets Pub" had nothing on the new pilots' pub at Woodley on the Reading aerodrome, so far as the amenities of civilisation are concerned. We only hope that its attractiveness will lure enough money-spending people to recompense the enterprising owner-brewers, H. G. Simonds Ltd., of Reading, not merely for their expenditure but for the intelligence that they have put into it.

The proper name of the pilots' pub is The Falcon Hotel and the sign-board, a very pretty piece of work, is a glorification of the well-known Miles Hawk trademark. Although in fact the pub has been at work for several weeks it was officially opened on Oct. 29 at a lunch given by the owners to the Press and various people in the Trade who are concerned with Reading, including Rolls-Royce Ltd., and Phillips and Powis Aircraft Ltd.

Mr. F. A. Simonds, who has a pretty wit, welcoming his guests and referring to the influence of Rolls-Royce at Reading, said that all our inclinations are in favour of Rolls-Royce if our incomes are not. He hoped that the hotel would be a haven of refuge for all connected with Reading aerodrome. The decorations had been done by artists (the Brothers Morton) who are already famous and will be more so in the future. He congratulated Mr. Guy Morgan of Morgan and Partners, the architects, on his work and said how pleased he was by a tour round the bed-rooms where he found double rooms complete with bathroom and a "built-in companion," —we discovered that this is in fact a peculiarly well-fitted wardrobe. He congratulated the Managing Director of Rolls-Royce Ltd., and hoped that he would be able to congratulate him more when there were more tangible results to show at Reading.

Lord Herbert Scott, Chairman of Rolls-Royce Ltd., denied being Managing Director. He said that when he came to the lunch he felt like Daniel in that if there was to be any after-lunch speaking it would not be for him. He thought that Mr. Simonds had misquoted the Rolls-Royce jest, which he understood was that there was no use in having Rolls-Royce ideas on a push-bike income. He said that Mr. Sidgreaves is in fact Managing Director of Rolls-Royce. He was delighted to see such a delightful hotel and rooms and he thanked goodness that there was no wireless in the rooms.

Mr. Miles thanked Mr. Simonds and said that the hotel is a revolution in spite of its being surrounded by most conservative people such as Rolls-Royce Ltd. At any rate two things about the aerodrome were ahead of everything else, namely Rolls-Royce motors and this new hotel, and yet they were both essentially English. Only the day before a well-known American motor maker had wished they had something like this hotel in the States.

Mrs. Miles was unfortunately absent, because of the grave illness of her father, Sir Johnston Forbes-Robertson.

Mr. Sidgreaves congratulated the Simonds firm on a very fine job. He said that they maintained a staff of surveyors and architects of their own and yet they had the courage in this special job to bring in a new man, Mr. Morgan.

Mr. Reginald Palmer, who makes, not takes, biscuits, returned thanks to Mr. Simonds and congratulated him on the wonderful building of which he hoped to see more.

Really this is a most remarkable pub. It is probably the first, at any rate in connection with aviation, which has had a whole foot of a column in *The Times* all to itself. The Reading aerodrome people and the Simonds firm ought to be grateful for such a valuable advertisement.

There is no space here in which to describe it in detail. We can only say that it is everything that an aerodrome hotel ought to be or can be within the limitations imposed by area and height. The only possible objection to it is that it ought to have been on the other side of the aerodrome, because where it is the sun will be in the eyes of the people sitting on the terrace all through the afternoon and evening. We were told that there were insuperable difficulties in the way of getting ground at the opposite side, but we cannot imagine why. After all the aerodrome belongs to the owners of the aerodrome and they could have had their own road on their own property to an hotel on that side.

Still we hope that the attraction of the pub as a pub will outweigh the distraction of sun headaches.

The idea of the pub originally was to replace the old Reading Aero Club which is now penned in by the vast factory buildings all round it. The pub now includes the Club, the members of which have a private lounge upstairs which is far too fine for any ordinary club members. Also there is a beautiful little club bar downstairs. The Morton Brothers did the furniture as well as the frescoes on the walls and they have done it extraordinarily well.

The management is in the hands of Captain and Mrs. Reginald Growden (he late R.A.F.), and if they can induce the chef to keep up the standard of the inaugural lunch the place will build up a reputation apart from its connection with flying.

The Reading Club is going to have a dance there on Nov. 12, Friday week, and we strongly recommend everybody who can do so to go to it.

The Falcon Hotel, which replaced the original clubhouse.   *[Reproduced from 'The Aeroplane' for 3rd November 1937]*

While the technical success of the company had never been in doubt, it was one thing to design a series of successful aeroplanes but quite another for the company to achieve financial success. Any new aircraft is always a doubtful business proposition until it is in quantity production and with the approach of 1937, the company's financial outlook was again looking somewhat gloomy. However, the worsening international situation with its associated repercussions on national policy was ultimately to benefit Miles and help him to establish the company as an aircraft manufacturer of considerable repute.

In March 1937, the Bristol Blenheim I, the first of the new breed of monoplanes for the RAF, entered service. This was followed in May by the Fairey Battle and in December by the Hawker Hurricane. Deliveries of the Supermarine Spitfire were not so well advanced though with the first of these not reaching the first squadron until August 1938. The technique of flying all these monoplanes, which were equipped with retractable undercarriages, flaps, variable-pitch propellers and other devices to enhance performance, was very different from that of the biplanes then in use. Although the new monoplanes were being steadily introduced, there was still no suitable aeroplane for training the pilots in the new techniques. While it should have been obvious to the Air Ministry that a suitable advanced trainer would be required prior to the introduction of the new breed of aircraft, they had, typically, not foreseen the need.

Miles had, in 1936, foreseen the necessity for an advanced trainer to fulfil this role and had commenced work on its design. The private venture, two-seat advanced trainer he envisaged would enable the trainee pilot to get the 'feel' of the new range of high-powered monoplane aircraft before being sent off solo on type. With the backing of Rolls-Royce, Miles completed the detailed project design study of the M.9 Advanced Trainer and submitted it to the Air Ministry with the suggestion that an order be placed forthwith, as he felt that the need for such an aircraft was already evident. Needless to say, his suggestion was initially rejected almost out of hand.

Rolls-Royce had designed the 745hp Rolls-Royce Kestrel XVI (VP) power-plant for the new advanced trainer as they hoped to obtain a contract to supply the whole package forward of the fireproof bulkhead. However, the M.9 Rolls-Royce PV Trainer was ultimately ordered into production as the M.9 Master with the 725hp Rolls-Royce Kestrel Mk. XXX engine. This was the ultimate variant of the Kestrel engine, with the compression ratio raised slightly to 6.2:1 but although all the engines which were used in the production of the M.9B and M.9C Master Mk.I and Mk.IA aircraft were Rolls-Royce Kestrels, they were in fact rebuilt and up-rated engines, which had previously been used in obsolete Hawker Hart variants and not the new engines as Rolls-Royce had originally hoped. While it was still good business for them, it was not quite what the company originally had had in mind!

In the early days of the Hawk, Miles and Blossom had, between them, coped single-handed with all the drawing, stressing and such calculations as were necessary but as the firm expanded they trained a few youngsters to help in the drawing office. One of these, Tommy Botting, was of considerable ability and later became a section leader in Miles' personal design team, until he left in 1936 as a result of his strong political convictions to go to Spain to fight in the civil war. With the advent of the M.9 Master, it now became necessary to establish a properly staffed design office, subdivided into sections and working in a more orthodox manner than had hitherto been the case.

By early 1937, the M.9 was taking shape in the drawing office and production of the M.14 Magister was getting underway in the

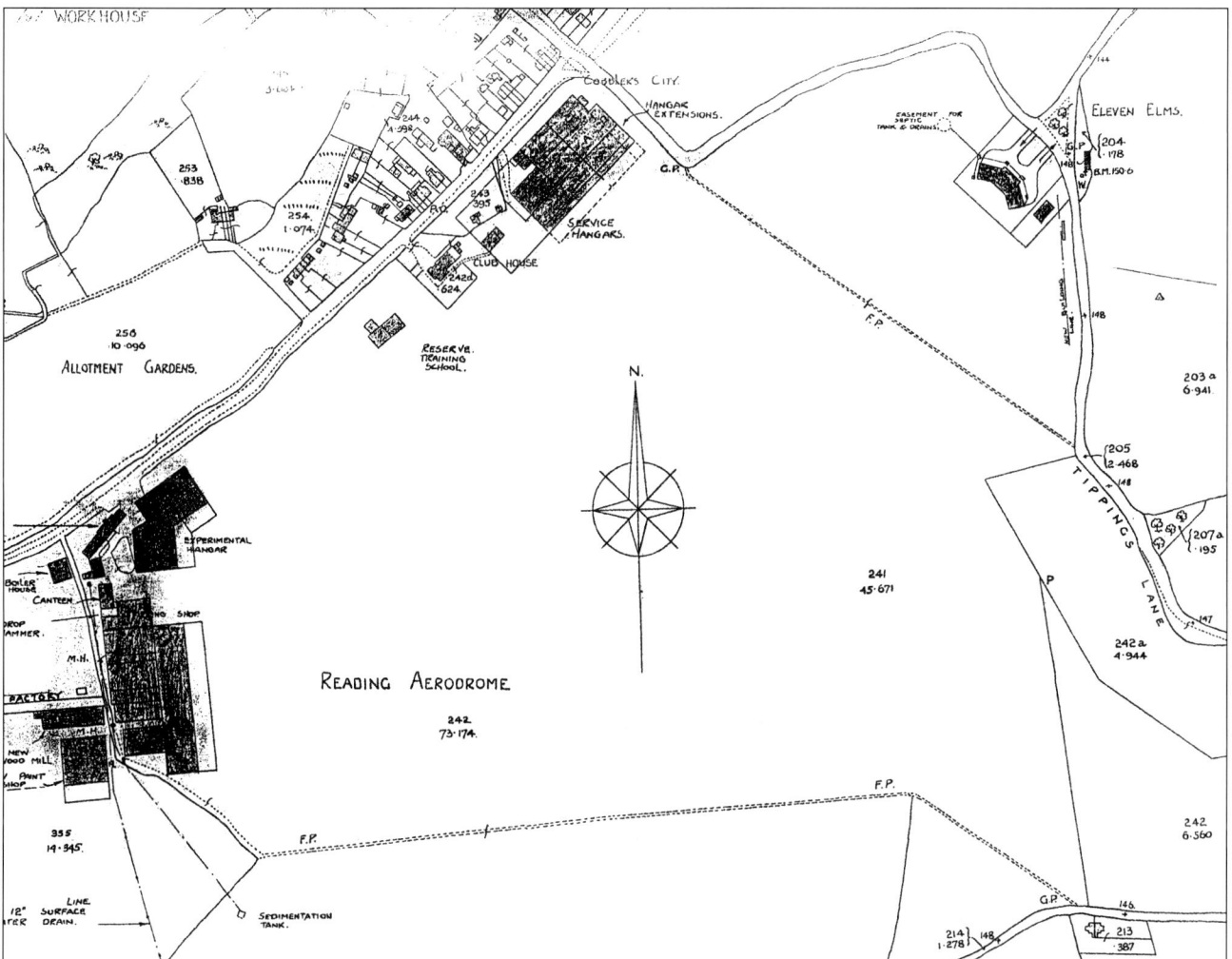

Plan of Woodley Aerodrome in 1937.

enlarged factory. On 20th March 1937, the first Magister was christened by Blossom in front of the assembled workers and press, after which Miles carried out its first flight. This milestone was celebrated by the workers and guests in time-honoured fashion, by the consumption of eighty gallons of the best local ale, which had been laid on by the company!

Meanwhile, the Air Ministry had been doing some hard thinking and, after discussion with the RAF, had come to the conclusion that perhaps a modern advanced trainer should be provided after all. On 12th June 1936, the Air Ministry issued Specification T.6/36 to Miles, Avro and de Havilland to design a general purpose single-engined training aircraft to replace the Hawker-class aircraft as existing stocks were exhausted. The new low-wing monoplane was to have *all the characteristics of the modern Service type aircraft*. In the interests of economy the aircraft was designed to have side-by-side dual control, for dual instruction or advanced training, with a third seat in a dorsal, rotatable gun turret behind the cockpit for gunnery practice. A large cabin was to be provided for radio training and other training equipment was to be installed. Avro submitted the Type 676 and 677 layouts and Miles the M.9 Kestrel but it was de Havilland with the DH.93 Don which was awarded the contract. Apparently de Havilland's took on the task with reluctance, remembering their experience of specification work a good many years previously but the German threat was serious and as the industry was at the disposal of the Government, they had little choice other than to comply.

The DH.93 Don was a wooden, low-wing monoplane, 'under'-powered by a 525hp DH Gipsy Twelve engine and was first flown on 18th June 1937. Continuous officially inspired 'improvements' were then made on such a scale that the structural weight increased to the point where the aircraft was literally no use as a flying machine. In an attempt to salvage something from this impossible situation, the turret and other heavy equipment was removed and the Don was converted to communications duties but even this proved unsuccessful and the order for 250 aircraft was subsequently cancelled. The 30 aircraft which had been delivered, were, wisely, relegated to ground instruction duties but the project had wasted a considerable amount of very valuable company resources and time.

Although Miles had little experience of official Air Ministry Specifications, he was still astonished to find how impractical they were. He had submitted the M.9 design to meet his view of the requirement of T.6/36, as he felt that the specification as written would not provide the type of trainer needed. Miles was probably not too surprised when his design was not accepted by the Air Ministry but this did not stop him taking the bold step of ignoring the official AM Specification and carrying on with the design of the M.9, which was already well advanced.

Miles remained confident that the M.9 Rolls-Royce Private Venture Trainer would provide the RAF with exactly what it required and with the abject failure of the DH.93 Don as an advanced trainer (albeit through no fault of de Havilland's who had only been trying to meet the impractical specification), the Air Ministry ultimately had to agree that Miles had been right all along in his appraisal of the RAF's actual requirement.

The M.9 Rolls-Royce PV Trainer was first flown, by Miles, on 3rd June 1937 (two weeks before the Don) and after a few inevitable teething problems had been ironed out, it became an immediate technical success. Its performance was in excess of the design specification and with a maximum speed of 296 mph at 15,000 ft, the rated height of the engine, it was only 14 mph slower than the contemporary Hurricane, which was powered by the Rolls Royce Merlin engine with 300 more horse-power. By the end of July 1937, the new aircraft had become known as the M.9 Kestrel, fortuitously in keeping with Miles' penchant of naming his aeroplanes after birds of prey. When Lord Swinton, the Secretary of State for Air, visited Reading Aerodrome in July 1937 (arriving in a DH Dragonfly), Miles accompanied him on a tour of inspection of the hangars and workshops, etc, before taking him for a flight in the new M.9 Kestrel. There is probably little doubt that Miles also took the opportunity of bending the Minister's ear!

On 6th June 1937, the first production Magister was delivered to the RAF and over the next three years 1,230 Magisters were to be ordered in 'dribs and drabs' for use by the RAF. Some of this relatively small number was to be diverted for use by the Irish Air Corps and the Royal Egyptian Air Force and the last Magister to be

The first Miles M.14 Magister at Woodley on 20th March 1937. The aircraft has just been christened, by 'Blossom' (note the damp patch on the ground under the propeller!), and F G Miles, with helmet and goggles, climbs aboard to make its first flight. Note the three Hawk Trainers and a Falcon in the background.
[Lionel Tysoe]

The Miles M.9 Private Venture Trainer in June 1937. ['Flight']

built was delivered to the RAF in January 1941. However, in spite of the RAFs first monoplane trainer becoming such an outstanding success, it was to be the biplane Tiger Moth elementary trainer that was kept in production throughout the war, with some 7,290 being delivered by 15th August 1945 from factories throughout the world. Why the Tiger Moth should have been chosen to fulfil the RAF's major elementary training role has always been a mystery to your author.

Ranald Porteous (later of Auster Aircraft fame), who had learned to fly in 1934 while with the De Havilland Technical College, wrote about his flying experiences around the world in *More Tails of the Fifties*. He also worked for Phillips & Powis for a time from 1938 and tells of this:

A chance arose to join the Miles organisation, starting as assistant to the famous, great and good (all three, in generous measure) Tommy Rose, who at that time ran the Reading Aero Club at Woodley. There was talk of occasional test-flying etc within the firm, so brandishing my recently-acquired instructor's licence, I bade fond farewell to Geoffrey (Alington) and Bunny (Spratt) (of the Gatwick-based firm Air Touring – PA) and to the recently recruited Tom Brooke-Smith, later to distinguish himself with Shorts and the GAPAN. It had been a fun time but Woodley was a more serious challenge.

Tommy Rose was beyond all praise. Bluff, jovial, kindly and extrovert, he was nevertheless shrewd in matters of human nature. His considerable fame and seniority rested lightly on his shoulders and to me, a relative whippersnapper, he combined the functions of benevolent boss, father-confessor and merry uncle. He had, to an exceptional degree, the rare gift of giving one a feeling of his real interest and genuine concern. He had lost none of this when I stayed with him many years later in Alderney.

The club aircraft, all very modern for their day, consisted of two Miles Hawk Majors, one Hawk (Trainer) fitted with a Menasco engine of power (130hp) and characteristics similar to the Gipsy Major, and one Whitney Straight, a most civilised and roomy side-by-side cabin two-seater. All these Miles aircraft flew beautifully, due to exceptionally efficient wing design, coupled with simple and reliable split flaps which gave just about the ideal degree of increased drag without unduly disturbing longitudinal trim. The Menasco engine, being American, rotated in the 'wrong' direction, which confused some novices no end as they experienced yaw and torque effects the reverse of what they had come to expect. In this regard it served as an excellent training tool.

I cannot look back on this very positive period, when I felt that I was learning so much (not all of it in the cockpit!), without remembering with horror one very nasty narrow squeak which may well have frightened me out of any drift towards over-confidence. I was nearing Woodley in one of the Hawk Majors with a charming, youngish woman as pupil, after a routine training flight. Her progress had been excellent so far and her aptitude markedly above the average. As we began our final approach she remarked to me through the Gosport tube that we were a trifle high and should she go round again or sideslip some of the height off? We had been practising sideslips during the previous week and she had become quite expert, so I agreed to the latter, whereupon, in a split second, she violently applied full rudder and opposite aileron with a considerable amount of up-elevator ... and we had executed a half-flick roll at what cannot have been more than about four hundred feet. Emitting what must have been a startled yelp, I managed to seize everything and nurse the Hawk downwards through its half-loop, steering at high speed between two rows of suburban houses on the edge of Sonning at below roof-top height. It was all very scary, not least to the lady in question, and I can only claim that, in thousands of hours instructing, nothing similar was allowed to happen again - thank God!

A favourite club member, with whom I greatly enjoyed flying, was Veronica Innes, lately 'Queen of Beauty' at the Runnymede Pageant. She had a charmingly insouciant way of tackling things. When war came she quickly graduated from the Civil Air Guard (which the club had served) into the Air Transport Auxiliary, but in her Civil Air Guard days at Woodley she was largely my responsibility - no hardship to me. One late afternoon she was authorised to do a triangular cross-country navigation exercise which, as far as I can remember, should have taken her to Suffolk, Cambridgeshire and home after about an hour and a half. When she was more than an hour overdue and a thickening haze warned of the approach of evening we were all decidedly anxious, the famous furrows on Tommy Rose's forehead being much in evidence. Just as we were on the point of initiating all sorts of emergency procedures by telephone, the Hawk's engine was heard and the aircraft appeared through the haze, clearly identifiable. My feeling of relief cannot be described. Apart from the responsibility, I was really fond of Veronica, who by now had landed and was taxiing in to a clamorous reception embracing 'Where the HELL have you been?', 'What went wrong?', 'Thank God you're back', 'Join us in the bar'.

Veronica's reply to all this was calm and clear. 'The visibility was very poor and I'm afraid I got lost for a while, but I found myself among some barrage balloons, so I knew it was Cardington and remembered a friend who farms near there, so landed in one of his fields for a chat'. ... Bless her!

At about this time I was, to my delight, entrusted with the prototype Miles Monarch, G-AFCR, for a demonstration to the French civil aviation authority at Le Bourget. The Monarch was a quiet and comfortable three-seater, somewhere between the four-seat Falcon and the two-seat Whitney Straight, which it resembled most closely. To my way of thinking, the controls and stabilities were perfectly harmonised and the machine was a pleasure to fly. My contacts at Le Bourget turned out to be a couple of gravely important Garlic-and-Gauloises gentlemen in blue pin-stripes and wearing Homburg hats. They spoke (or pretended to speak) no English so my halting schoolboy French was stretched to its limit. Having taken off into a smoggy autumn haze which, at two thousand feet, prevented one from seeing anything of the ground save vertically downwards, they proceeded to perform a series of lurching gyrations, timing various recoveries with a stopwatch. Sitting there helplessly in this grey horizonless void I suddenly felt an increasing panic lest I should be airsick for the first time in my life, thus disgracing myself, Miles Aircraft, King and Country .... in that order. Great was my relief when Mafia Boss No.1 gestured to me to take over and return to Le Bourget, which I found in the murk more or less by luck. In reply to my innocent query as to their opinion of the aircraft, their only comment was 'Ce n'est pas assez stable transversellement . . .'. This, I remain convinced, was 'une charretée de savetiers' – to coin a phrase. I returned to Woodley feeling that I should have done better but George Miles generously dispelled this and explained that the exercise had been something of a 'long shot' anyway.

Some months before the war I was transferred from the club to the Flying Training School (No.8 ERFTS). The training of pilots had by now become desperately urgent and we flew our Magisters, Hawks and even a Hart or two in all weathers for long hours. Our particular version of the Hawk Trainer had one noteworthy failing; its carburettor float tended to stick when subjected to negative g, as in a slow roll, thus cutting off the petrol from the jet. Sometimes it could be shaken loose by violent 'jinking', but more often one had to execute a model forced landing, with full 'patter' as matter of honour! Consequently each instructor had to choose his field or area, over which alone he could teach aerobatics. There were plenty of good fields within range of Woodley. My chosen plot was on a hillside between Henley and Wargrave. The surface was concave, the slope increasing markedly as one went uphill, rather like a ski-jump in reverse. In all but extreme winds, therefore, one landed uphill. The ensuing take-off downhill was always fun. Meanwhile life was restored to the Gipsy Major by opening its cowling and striking the carburettor sharply with a stone or, failing that, with one of easily removable control sticks. This resulted in an audible metallic 'clink' as the errant float dropped down into its proper position. This was scarcely high technology but I recall one training course during which no less than 91 dead-stick forced landings were recorded without any aircraft suffering so much as a scratch.

In the late spring of 1937, Pauline Powis separated from her husband Charles (they were to divorce the following year). At about the same time, claiming 'ill-health', Powis went on an extended trip or holiday in the USA but it was evident that his standing in the company had sunk dramatically. In a letter to Julian Temple in 1991, Menasco historian John Underwood recalled that Powis had told Al Menasco that business in his absence had taken a downturn and that Rolls-Royce were looking for a scapegoat. He believed that it was partly due to his 'domestic problems' and said that even Miles had turned against him. Powis failed to attend the company's AGM on 14th May 1937 and tendered his resignation as Chairman two days later. On July 10th, he resigned as a Director and Miles and Lt Col Ormonde Darby were appointed Joint Managing Directors in his place.

It transpired that Col Darby was behind the Rolls-Royce move to dispense with both Powis and Menasco. Darby did not want to continue with the licence agreement and was looking for a way out. Whilst the final outcome is not known, the Agreement was ended with Menasco taking a heavy loss. Powis is believed to have been talking to Blackburn Aircraft at the time with a view to them taking over the contract, because there was talk of an order for 500 engines from the Air Ministry in the offing, but nothing came of it.

Another (and perhaps even a major) reason behind Powis's decision to leave the company, although not officially recorded, could have been in connection with the untimely death on 16th January 1937, from bronchitis, of his young son Oliver - just 11 weeks old. It is understood that Charles Powis, in partnership with a Dutch boat-builder, subsequently became sole UK agent for the Menasco engine and it is of interest to note that, in December 1937, Powis was recorded as the new owner of the 'Rex' cinema in High Wycombe, Bucks as - *a newcomer to the business*. In 1940, Powis joined the RAF and ultimately reached the rank of Squadron Leader.

In July 1937, following Powis' departure, Rolls-Royce provided a loan of £30,000 to finance further factory extensions and new plant required for the Magister production. They also advanced £5,000 specifically for the private development of the Kestrel.

The M.9 Kestrel really owed its ultimate success to two salient features - firstly that Miles was a very experienced pilot and instructor and secondly that its handling characteristics virtually reproduced those of the new fighter aircraft with which the RAF was then being re-equipped. Finally recognising the urgent need for an advanced trainer, the Air Ministry had perforce to turn to Miles and on the 11th June 1938 they placed an order with Phillips & Powis for 500 of the re-named M.9 Master advanced training aircraft. This contract for the production version of the Kestrel was worth £2,000,000 and as such was the largest order by far which had ever been placed for a training aircraft at the time.

In order to cope with this huge contract, further extensions had to be made to the factory but despite this and the fact that the Kestrel had to be considerably modified to suit the requirements of the AM, the first production Master made its first flight within fourteen months of the receipt of the contract.

That the company had been experiencing troubled times is evidenced by the report prepared by Miles for the shareholders at the end of his first year as Managing Director and this was a pretty damning verdict on his predecessor:

The year 1937 has certainly been a critical one in the life of this company. Looking back, it seems obvious that the fundamental changes in organisation and management made at the beginning of the year have had a far-reaching effect. It is now accepted that the position at that time was serious. We held contracts for Whitney Straight aircraft and Magisters which were already a month behind, yet they showed no sign of reaching the production stage. The financial situation had become critical, partly owing to the delay in production, partly mis-spending and poor original estimates and partly to reluctance to face the situation. A definite change in policy was decided upon and put into action. Circumstances forced a slow start, but the changes were speeded up as all concerned became convinced of their necessity.

From the start of the new regime, strictest economy was enforced, one of the first rules being that no expense should be incurred without the written approval of the managing director. A number of wasteful non-productive jobs were reduced and the control of every form of capital expense was tightened to the last notch.

The board was reconstituted and a drive was started from the Rolls Royce end to bring the prices for Air Ministry aircraft in line with actual costs. This drive finally became more successful than was ever dreamed possible at the time. However, the negotiations in the early stages were nearly impossible owing to the unreliability of our cost figures.

In the factory, a complete reorganisation of manufacturing methods was undertaken against very considerable prejudice and opposition. All the expedients resorted to in order to get production going had to be simple since neither time nor money were available. Production of the Whitney Straight was brought up to two machines a week. This helped the financial position. A goal of six Magisters a week was set and achieved.

Much design and research work against the future has been done, especially in view of our small organisation. Relationships with the Air Ministry have been improved and various tenders have been successfully carried out. Now we have come to the end of the year and it can be said that the general position is improving enormously. The company has improved from a position of no confidence to one where we have ceased to lose money and can look forward with reasonable anticipation to profits.

The Air Ministry has been sufficiently impressed by our continued production to place further and larger orders with us in time to keep the works going without interruption next year. There is, I trust, a very much improved spirit among all concerned and I feel that we can look forward to the immediate future with the feeling that a good hard push to get new orders and better production will ensure our first good year in 1938.

A resumé of the Experimental Department's activities for 1937 has also survived in the form of an internal note and this makes for interesting reading:

Most of the work under the Research and Drawing Office was carried out in the Experimental Department. Types built or complete in 1937 include the Kestrel Trainer, the Master I, Queen Wasp (part) Research Twin (Miles M.8A Peregrine Mk.II, which first flew on 30th March 1938 - PA), M.17 Monarch.

Research or modification work was carried out on the following:-
a) *Hawcon (new wings)*
b) *Falcon (written down on last balance sheet)*
c) *Hobby (experimental work on wings)*
d) *Magister (all work on improvements to meet complaints, such as draughty cockpits)*
e) *Mohawk (various modifications)*
f) *T.1/37 (mock-up)*
g) *M.11c (modification and rectification)*
h) *Menasco Hawk (Experimental work on deck landing gear)*

In this department the metal working section is particularly good. Much work has been done on castings and a pattern making section developed. Various models and test rigs were built. All test work (including all techniques) were developed in this department.

A general view of Reading Aerodrome in the late 1930s, showing the extensions surrounding the original hangar, seen in the centre of the photograph.
[Phillips & Powis via Mrs Skinner]

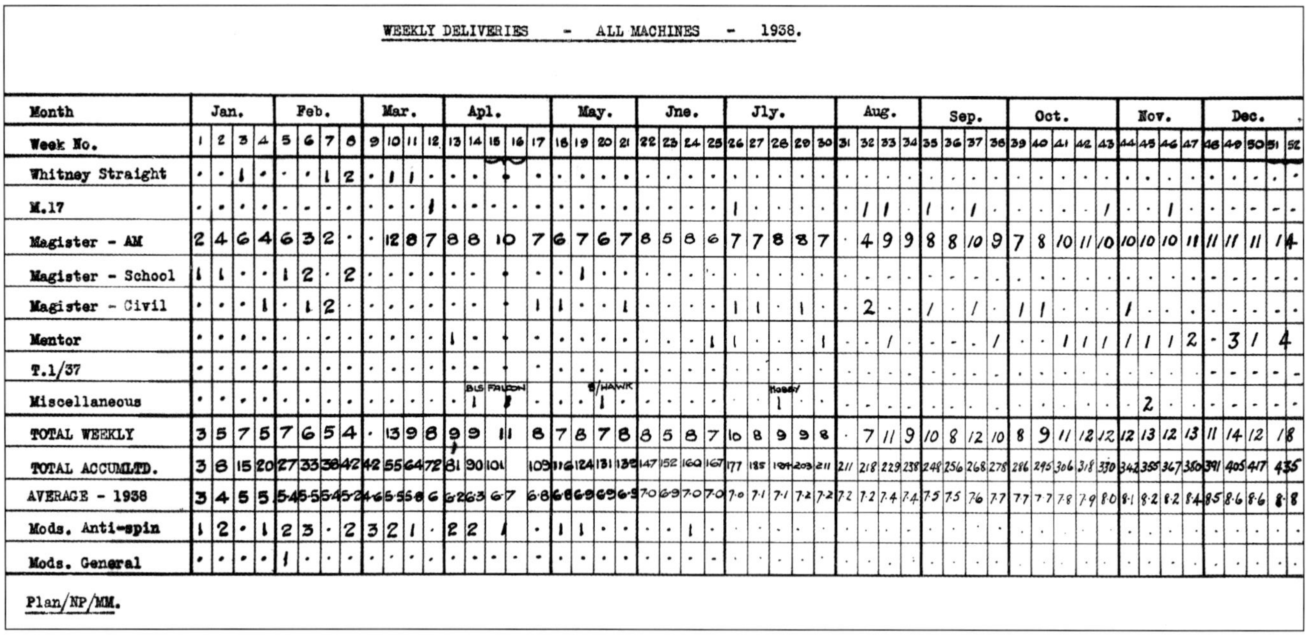

WEEKLY DELIVERIES - ALL MACHINES - 1938.

Month	Jan.				Feb.				Mar.				Apl.					May.				Jne.				Jly.
Week No.	1	2	3	4	5	6	7	8	9	10	11	12	13	14	15	16	17	18	19	20	21	22	23	24	25	26
Whitney Straight	·	·	1	·	·	·	1	2	·	1	1	·	·	·	·		·	·	·	·	·	·	·	·	·	·
M.17	·	·	·	·	·	·	·	·	·	·	·	1	·	·	·		·	·	·	·	·	·	·	·	·	1
Magister - AM	2	4	6	4	6	3	2	·	·	12	8	7	8	8	10		7	6	7	6	7	8	5	8	6	7
Magister - School	1	1	·	·	1	2	·	2	·	·	·	·	·	·	·		·	·	1	·	·	·	·	·	·	·
Magister - Civil	·	·	·	1	·	1	2	·	·	·	·	·	·	·	·		1	1	·	·	1	·	·	·	·	1
Mentor	·	·	·	·	·	·	·	·	·	·	·	·	1	·	·		·	·	·	·	·	·	·	·	1	1
T.1/37	·	·	·	·	·	·	·	·	·	·	·	·	·	·	·		·	·	·	·	·	·	·	·	·	·
Miscellaneous	·	·	·	·	·	·	·	·	·	·	·	·	·	BLS 1	FALCON 1		·	·	·	S/HAWK 1	·	·	·	·	·	·
TOTAL WEEKLY	3	5	7	5	7	6	5	4	·	13	9	8	9	9	11		8	7	8	7	8	8	5	8	7	10
TOTAL ACCUMLTD.	3	8	15	20	27	33	38	42	42	55	64	72	81	90	101		109	116	124	131	139	147	152	160	167	177
AVERAGE - 1938	3	4	5	5	5.4	5.5	5.4	5.2	4.6	5.5	5.8	6	6.2	6.3	6.7		6.8	6.8	6.9	6.9	6.9	7.0	6.9	7.0	7.0	7.0
Mods. Anti-spin	1	2	·	1	2	3	·	2	3	2	1	·	2	2	1		·	1	1	·	·	·	·	1	·	·
Mods. General	·	·	·	·	1	·	·	·	·	·	·	·	·	·	·	·	·	·	·	·	·	·	·	·	·	·

Month	Jly.				Aug.				Sep.				Oct.					Nov.				Dec.				
Week No.	27	28	29	30	31	32	33	34	35	36	37	38	39	40	41	42	43	44	45	46	47	48	49	50	51	52
Whitney Straight	·	·	·	·	·	·	·	·	·	·	·	·	·	·	·	·	·	·	·	·	·	·	·	·	·	·
M.17	·	·	·	·	·	1	1	·	1	·	1	·	·	·	·	·	1	·	·	1	·	·	·	·	·	·
Magister - AM	7	8	8	7	·	4	9	9	8	8	10	9	7	8	10	11	10	10	10	10	11	11	11	11	14	
Magister - School	·	·	·	·	·	·	·	·	·	·	·	·	·	·	·	·	·	·	·	·	·	·	·	·	·	·
Magister - Civil	1	·	1	·	·	2	·	·	1	·	1	·	1	1	·	·	·	1	·	·	·	·	·	·	·	·
Mentor	·	·	·	1	·	·	1	·	·	·	·	1	·	·	1	1	1	1	1	1	2	·	3	1	4	
T.1/37	·	·	·	·	·	·	·	·	·	·	·	·	·	·	·	·	·	·	·	·	·	·	·	·	·	·
Miscellaneous	·	HOBBY 1	·	·	·	·	·	·	·	·	·	·	·	·	·	·	·	·	2	·	·	·	·	·	·	·
TOTAL WEEKLY	8	9	9	8	·	7	11	9	10	8	12	10	8	9	11	12	12	12	13	12	13	11	14	12	18	
TOTAL ACCUMLTD.	185	194	203	211	211	218	229	238	248	256	268	278	286	295	306	318	330	342	355	367	380	391	405	417	435	
AVERAGE - 1938	7.1	7.1	7.2	7.2	7.2	7.2	7.4	7.4	7.5	7.5	7.6	7.7	7.7	7.7	7.8	7.9	8.0	8.1	8.2	8.2	8.4	8.5	8.6	8.6	8.8	
Mods. Anti-spin	·	·	·	·	·	·	·	·	·	·	·	·	·	·	·	·	·	·	·	·	·	·	·	·	·	·
Mods. General	·	·	·	·	·	·	·	·	·	·	·	·	·	·	·	·	·	·	·	·	·	·	·	·	·	·

Plan/NP/MM.

Chart showing weekly deliveries of new aircraft for 1938 and cumulative totals of aircraft built.

Whilst it is extremely difficult to calculate the actual number of aircraft produced on a yearly basis, the following table illustrates the company's rate of growth, based on known aircraft deliveries per year to the end of 1939. It does not include rebuilds of crashed aircraft, mock-ups, or unfinished airframes due to cancelled orders:-

	Total aircraft Built	Average No. dd per week
1933 (March to December – 39 weeks)	15	0.38
1934	54	1.04
1935	85	1.63
1936	58	1.12
1937	152	2.92
1938	409	7.87
1939	559	10.75
Total	1,332	

These figures were based on aircraft delivery dates from the production records kept by C L Nash and other firm's sources and particularly illustrates the problems experienced during 1936.

The photograph of a company chart entitled 'Weekly Deliveries – All Machines – 1938' (reproduced above) shows an accumulated total of 435 and an average of 8.8 aircraft delivered per working week for 1938. This figure included 6 Whitney Straights; 369 Magisters for the Air Ministry, 8 for the Reserve Training School and 17 for civilian use; 22 Mentors and 8 Monarchs. Also noted under 'Miscellaneous' were the BLS Peregrine, 1 Falcon and 1 Sparrowhawk. A similar chart for 1939 shows that a total of 574 aircraft were delivered, comprising 3 Monarchs; 461 Magisters for the Air Ministry, 5 civil Magisters; 24 Mentors; 79 Masters and 2 T.1/37.

While the company showed no profit for 1936 or 1937, by the end of 1938 the directors were pleased to report that a profit of £25,628 had been made and this rose to £100,000 in 1939 - the corner had been turned.

Although there was still little surplus money available, Miles was anxious that Phillips & Powis should be recognised as a firm which was also prepared to undertake full-scale research. One area in which he was particularly interested was that of boundary layer control, so he decided to undertake trials to determine whether the theoretical advantages claimed for control of the boundary layer to reduce skin friction could be achieved in flight. As far as he was aware, no one had ever attempted to install a means of controlling the boundary layer into a full sized aeroplane, so he built a very thick aerofoil section around a portion of the port wing of a Whitney Straight, covered its upper surface with a porous material and connected this by ducting to an enormous venturi mounted beneath its outer section. With this simple test, he was able to obtain some indication of the actual effects of boundary layer control, which enabled him to investigate the effect of drag on wings of varying thickness.

He concluded that, providing the increase in drag was not prohibitive at the modest speeds then prevailing, the structural advantages gained with thicker wings would be well worthwhile, especially in connection with the large airliners, which were also beginning to interest Miles at that time. It should also be noted that the M.9 Kestrel used a wing of greater thickness than those used previously by Miles in his earlier designs.

The results obtained by the boundary layer experiment were so encouraging that the RAE decided that this work merited further investigation and they accordingly placed a contract for a modified Peregrine. The wings of the second Peregrine were covered by a porous skin under which were a number of spanwise internal ducts connected in turn to a centrifugal blower. Unfortunately, with the accelerated expansion of the RAF and the rapidly worsening political situation in Europe, the RAE was reluctantly forced to abandon its programme of aerodynamic research, which had included the boundary layer programme, in favour of work which had more immediate application.

By 1938, with many Magisters in service, Miles felt that an even better trainer could be designed in the light of experience gained with this aircraft, the basic design of which was not only six years old but also based on his first attempt at producing a low-wing monoplane. Walter Gustav Kaepelli, who was born in Zurich, Switzerland, on 8 November 1913, had come to England in 1934 and in 1937 joined the staff of Phillips & Powis as Chief Draughtsman. Miles gave him the job of designing the Magister replacement for submission to the Ministry. This became the M.18 - the first aeroplane from the Phillips & Powis stable not to have been designed by Miles himself. (Walter was granted British nationality in June 1940 and changed his name by deed poll from Kaepelli to Capley).

The main considerations of the new aircraft, which was to bear little resemblance to its forebear, were to be ease of handling and serviceability. These characteristics of the Magister replacement were obtained, albeit at some sacrifice in performance, by the adoption of an entirely new wing of practically constant chord and thickness. The fuselage was 4 inches wider than that of the Magister in order to give considerably more 'elbow room' for the crew and the fin and rudder were placed at the extreme end of the fuselage. Single-piece wrap-round windscreens were fitted to both cockpits and the fixed undercarriage was spatted. Powered by a 130hp DH Gipsy Major I engine, the new machine was never named although it was sometimes referred to as the 'Magister Mk.II' and the prototype was first flown, by Miles, on 4th December 1938.

Although a contract was later raised for the production of 200 M.18 Trainers for the RAF, this was soon cancelled and apart from three further prototypes produced for different purposes later, no further M.18's were built.

The last civil aeroplane to be produced by the company before the outbreak of war was the M.17 Monarch. This was designed by George Miles, who had taken over as Manager of the Repair & Service Department from Capt E D Ayre in 1938. Although the majority of Monarchs were built in the Magister assembly shop, due to shortage of factory space, the last two were actually completed in the Repair & Service Department. The prototype Monarch G-AFCR was first flown, by Chief Test Pilot Bill Skinner, on 21st February 1938 and a report which appeared in a contemporary issue of *The Aeroplane* stated that the Monarch was *probably the most ownable aeroplane in the world*. Unfortunately, Government orders for the M.14 Magister and the M.9 Master meant that the Monarch could not go into full-scale production.

In June 1938, tragedy struck the Company when test pilot Wing Commander F W 'Freddie' Stent was killed in the Whitney Straight G-AEYI. His obituary appeared in the *Miles Magazine* for July 1938:

It is with the profoundest regret that we have to report the death of Wing Commander F W Stent MC in a flying accident on Tuesday

Wing Commander F W 'Freddie' Stent.

the 28th of June 1938. Wing Commander Stent joined this Company in 1936 on his retirement from the Royal Air Force where he had served for 23 years. He was widely known throughout the service and his duties with this Company brought him in contact with a great number of our owners and operators both at home and abroad all of whom will be deeply grieved at the loss of a charming friend. He played an important part in the progress of the Company and the many tasks entrusted to him he undertook with that determination which was so essential a part of his character. The Company, to whom he rendered great service, has lost a valued ally and every member thereof a true friend.

On 27th January 1939, the Air Minister, Sir Kingsley-Wood, officially opened the new very large factory extensions three weeks ahead of schedule. These had been built to accommodate the new Master assembly line and incorporated the first 'moving' aircraft assembly line in the country. He was also shown the mock-up of the new 'top secret' fighter in the Experimental Department. Initially known as the Munich Fighter, this was later given the designation Miles M.20/1.

The first production Master Mk.I, N7408, was first flown, by Miles, on 31st March 1939 and although only seven Masters had been delivered to the RAF by the outbreak of the war, the increase in production when the moving assembly line was in operation was dramatic, with a further 63 Masters being delivered to the RAF between 3rd September and the end of the year.

There is a reference in the National Archives to Contract B987363/39, for the loan of Hurricane and Spitfire aeroplanes to Phillips & Powis Aircraft Ltd, dated 9th May 1939 (and completed 1st January 1943), for flight tests to be carried out by Miles. Photographs taken at Woodley in 1940 show a Spitfire Mk.IA, P9444 (which was on the strength of the RAE Farnborough at the time - and which surprisingly still survives in the Science Museum, to whom it was presented in August 1944), and a Hurricane Mk.I, whose serial number was not visible (but possibly L1788 which

Walter Capley.

Sir Kingsley Wood (in centre) being shown the cockpit of the Miles M.20/1 'Munich Fighter' mock-up on 27th January 1939. Chief Test Pilot Bill Skinner is on the left and F G Miles on the right. [Phillips & Powis via Mrs. Skinner]

Blossom at her drawing board at 'Lands End', in August 1938. [Via P Jarrett]

Tommy Rose poses for the photographer in the cockpit of an early Miles Master. [Phillips & Powis via The Miles Aircraft Collection]

was also with the RAE at that time). Little is known of this contract but it is possible that these fighter aircraft were used in connection with the introduction of the Master into service with the flying training schools and also, later, in comparative trials between these two aircraft and the M.20 fighter.

In passing, it should be noted that in June 1939, *The Daily Mirror* reported that two mysterious fires had been started in the factory at Woodley and that sabotage was suspected, but nothing further was heard of this possible arson attack.

On Wednesday 15th November 1939, at the early age of 36 years, Harold William Chetwynd 'Bill' Skinner died from a cerebral haemorrhage at his home in Woodley while shaving. Born at Quetta, India, he was educated in India and Thetford, near Norwich. On leaving school he joined the RAF and from 1924 to 1929 served in Egypt. In 1932 he became an instructor at Cranwell and in November 1935 had joined the RAF Reserve School at Woodley as a Sergeant pilot/instructor. He was a remarkable pilot with great 'feel' and knowledge and on 1st April 1937, had been appointed Chief Test Pilot of Phillips & Powis Aircraft Ltd, a post which he held until his untimely death. This happened shortly after he had commenced test flying the Master Mk.II prototype but it was, however, the resolution of the spinning problems of the early Magisters and the valuable contribution to the development of the M.9 Master that should really have made Bill Skinner famous. These achievements were never fully recognised outside the firm.

Miles wrote a personal appreciation of Bill Skinner in *Flight* for 23rd November 1939:
In addition to the ordinary every day test work, he contributed a great deal of knowledge to the spinning of low-wing monoplanes. He was I believe, the first man in England to recover from an uncontrollable spin by means of a tail parachute. Certainly nobody knew for certain that it would work, and we did know the spin became uncontrollable. He had probably tested more purely experimental wings, gadgets and aeroplanes than any other pilot over a similar period and his reports were immortal.

Tommy Rose was appointed Chief Test Pilot in his place and it was he, together with George Miles, who completed the trials of the new Mercury-engined M.19 Master Mk.II prototype.

Bill Skinner, shortly after becoming Chief Test Pilot, poses by a Miles M.14 Magister. [Phillips & Powis via P Amos]

As the volume of flight testing increased, Tommy was joined by Flt Lt Hugh V Kennedy (RAFO), who had joined 8 E&RFTS as an instructor on 30th December 1938 and transferred to Phillips & Powis as a test pilot on 27th October 1939.

The story of Miles in the early days and his many and various activities at Reading has now reached September 1939. On the 3rd, the Second World War was declared and the balmy days of peace had gone forever.

The company had survived its various vicissitudes to emerge with a strong business partner in Rolls-Royce and finally, after a long haul, was beginning to show a profit, albeit almost entirely as a result of the military production in readiness for war. The story of The Wartime Years takes us beyond the scope of this particular book.

The newly-completed P&P Head Office at Woodley Aerodrome, built to the design of architects Guy Morgan & Ptnrs of London, photographed in early 1939.
[Phillips & Powis via The Miles Aircraft Collection]

Assistant Test Pilot Hugh Kennedy.
[Phillips & Powis via The Miles Aircraft Collection]

This photograph shows a typical day inside the Repair & Service Department Hangar and first appeared in the Miles Magazine for November 1938, on the opposite page to an open tin of sardines – for obvious reasons! The photo shows aircraft from the Reserve Training School, together with privately-owned aircraft. It was taken sometime in 1938.
[Phillips & Powis via P Amos]

The first Miles M.2 Hawk on an early flight in April 1933. ['Flight' via B Clarke]

Chapter 7:
Miles M.2 Hawk, M.2A Cabin Hawk, M.2B Hawk Single-seater, M.2C Hawk Sports Two-seater, M.2D Hawk Three-seater and M.2K Hawk variants

Miles apparently knew what he wanted before coming to Woodley and it is suspected that the idea of a clean low-wing monoplane stemmed from the Hendy 302. However, Miles wanted something cheaper.

The reason for the Hawk, as the new monoplane was to be known, was set out by Miles in an introduction to a special section on the aircraft, in *Flying* for 20th September 1933, where he explained the chief aims which he had in designing the machine:

Before I started on the Hawk I had always designed machines to suit my own individual taste - such as the Martlet and Satyr. During a chance visit to the Reading Aerodrome, (probably that on 25th August 1932 when Miles flew from Bristol to Reading in the Satyr - PA) Mr C O Powis suggested that a cheap practical aeroplane was badly needed, and when views were exchanged concerning types, the low-wing monoplane appeared to be the most suitable.

Although perhaps not the cheapest form of construction, the cantilever low-wing offers many advantages. Briefly our intentions in designing were to produce a machine that was sturdy and simple to handle and maintain. Mr Powis's company, Phillips & Powis Aircraft (Reading) Ltd, was able to make a contract with the Cirrus-Hermes Engineering Co to supply a batch of engines at a reasonable price and this gave us the opportunity of producing a machine at a lower figure than hitherto had been accomplished.

The design developed from the mental picture of a machine which should have a modern cruising speed with absolute simplicity and sturdiness. I believed that a low-wing type with a given wing loading would have a better take-off and landing performance than a high wing of the same loading. It avoids bracing wires to adjust or climb through, and by constructing everything of wood and covering with 3-ply, the result would be a carpenter's job throughout, with practically no maintenance for the owner to bother with. The plywood cover would add strength to the wing structure, be very durable and simplify repairs. The undercarriage was designed to be wide, for stability on the ground and to be out of the slipstream.

As we decided not to economise on the actual structure of the machine, it was necessary for us to save, apart from the saving already made on the engine, by carefully planned production methods. A great deal of thought was given to the designing of jigs and every component is made on a jig, which ensures interchangeability of parts.

One of the most difficult, but perhaps most interesting parts of the design was the folding wings. Unfortunately, the wing-folding apparatus increased the weight and, of course, the cost, but we felt that an unfoldable wing was not much use to the average private owner.

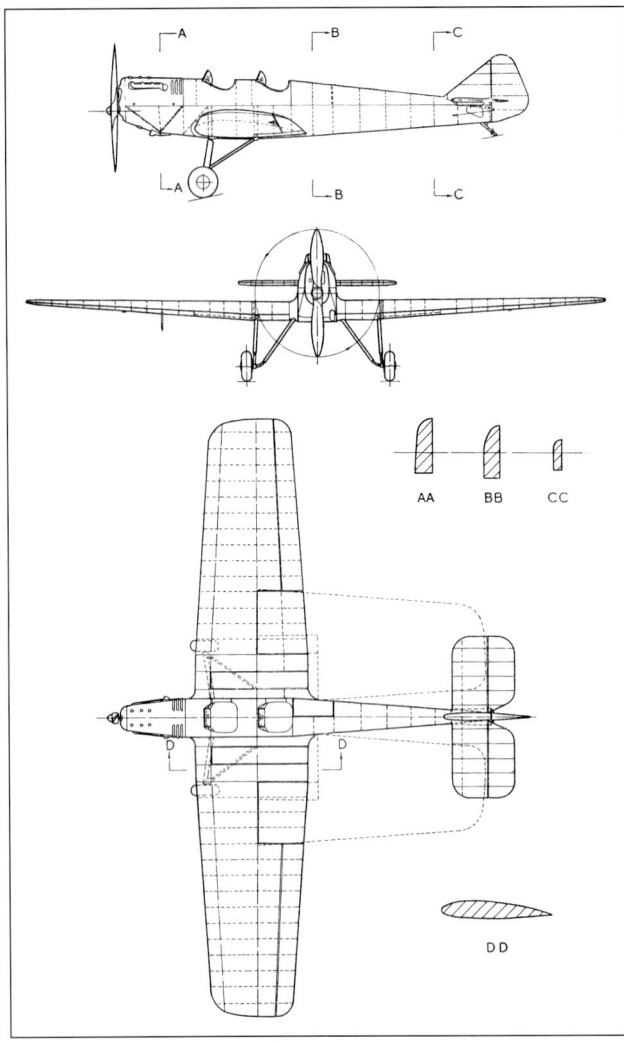

The Miles M.2 Hawk.

Although we designed for lightness as a matter of course, we were more concerned to have the machine simple and strong. The design actually conforms to the increased load factors which have just been made obligatory by the Air Ministry. Now that the machine has flown, and a number have been turned out, we feel we have been justified in choosing the low-wing as the performance all round is better than we expected.

When an aeroplane flies for the first time - not to mention subsequent occasions - it has often been so changed to confirm to new notions and fresh requirements that it has little in common with the machine which first took shape in the designer's dreams. The Hawk differs very little indeed from the first drawing which we made. The first machine was designed and built at the same time, and as a result the work on the construction and the design was simplified.

We only struck one snag during the work, when the figures for the wing began to look funny after a month's calculations. However, Mrs Miles took the blame for using a wrong factor in the formulae - and settled down to a month's arithmetic all over again.

The original design work was carried out by Miles and Blossom at Woodley under the name of the Miles Aeroplane Co and the new machine was built in a corner of the Phillips & Powis workshops by Harry Hull, the aircraft carpenter who had built the Martlets at Shoreham. The design was originally known as the Ibex, but was changed to Hawk to avoid confusion with the contemporary Hinkler Ibis amphibian.

Although Harry may have been helped by a fourteen-year-old lad, Tommy Botting (who later became a section leader in Miles' personal design team) and occasionally by Mr & Mrs Miles, he actually worked an average of 17 hours per day (119 hours for a 7-day week) for three weeks in his enthusiasm to complete the first Hawk. An interesting article appeared in *The Aeroplane* shortly after its first flight:

For some months Mr F G Miles has been at work in the new drawing office of Phillips & Powis Aircraft (Reading) Ltd, and the flying community has been looking forward to the appearance of something low in price. It emerged on March 29th, behaved too

The M.2 wing fold.

[Via B Clarke]

well in its taxying tests for Mr Miles' self-control and he flew it straight off. By dusk on March 30th it had flown six hours without suggesting to its proprietors any change except a widening of the outlet louvers of the engine cowling.

The Miles Hawk is a wooden low-wing two-seat cantilever monoplane of the greatest simplicity and sturdiness. The one departure from elemental necessity is the folding of the wings and as this naturally means more weight, Mr Miles obviously regrets doing it. He is quite right though to consider wing-folding a necessity in the modern lock-up shed. He has made an excellent job of it and has managed to put ample strength into a one-man folding system.

In the production machines the Cirrus III will be standard but the machine has been designed for in-line engines up to and including the Hermes IV and Gipsy Major.

The Hawk will be cheap partly because it has been designed for cheap production and partly because a batch of Cirrus engines came onto the market cheaply when a certain Canadian aeroplane firm stopped production. For these two reasons a number will be sold at a price estimated now at £395. When engines are bought at list prices the article will naturally cost more, still without getting out of the reasonable class. This explanation is to forestall the mistake of thinking that the Hawk is cheap in a nasty sense.

An article on building the Hawk, published in *Flight* for 24th August 1933, is of especial interest:

Few manufacturers will deny that there is often something to be learnt from the methods of those who come fresh into a business. In the aircraft trade in particular, each manufacturer regards his own methods as being, not perhaps better in a general sense, but certainly better suited to his own particular needs. Those who pride themselves on being willing to learn should take note of the construction of the new Miles Hawk at Reading. They will find much to interest them.

A case in point is the method which Phillips & Powis have evolved for attaching glued plywood to wing ribs, wing and fuselage covering and similar places. The old method of brass tacks was slow, and moreover the tacks were apt to split the struts or longerons, besides which they served only to hold the plywood to a very small extent. Mr Powis had a look round when he came into the game and found a wire staple machine just like that which many of us use for securing sheets of paper together, a sort of punch affair which you bang down with your fist. Now they use this at Reading for all places where plywood is glued on. In the case of the wing covering, the sheet of plywood is held in place by one man and another man goes bang, bang, bang, along with the punch. As soon as the glue has set the staples are withdrawn, leaving a clean surface upon which it is easy to get that high-class, motor-car-finish, which goes such a long way towards selling aeroplanes nowadays. The staples are left on inside the fuselage or on the wing ribs, not because the glue is not adequate but chiefly to save labour, though, of course, the staple in itself is better for holding the plywood on than an ordinary tack. We imagine that the saving of weight over tacks must also be quite considerable.

The factory has about double the floor space it started with, and orders are in hand which will keep them busy for a very long time. The jigs have been laid out for the fuselage, wing spars, wing assembly, fuselage sides, and all the small covered units, so that in point of fact production from them is ahead of final assembly.

The whole job is really rather impressive from a constructional point of view. Its ruggedness, combined with the natural apparent strength of the low-wing type, makes it look a very safe aeroplane, a factor which should assist its sales considerably, particularly as it is designed for those who are newcomers to aviation. The difficulty with aeroplane production is, as presumably in most other productions, that the designer, being enthusiastic about his work, goes on making improvements so that the wretched engineer finds it almost impossible to get down to economical methods.

The standard Hawk was a two-seat, low-wing, cantilever monoplane with open tandem cockpits. It was built entirely of wood and was designed to be very largely assembled by semi-skilled labour. As mentioned above, an example of unconventional yet effective production methods was the use of an ordinary office stapling machine for holding the plywood on to the wing spars during gluing. This process gave even tension and did not create the surface depressions normally associated with the use of brads, the staples being withdrawn and the holes filled when the glue had set.

The fuselage consisted of a plywood-covered 'box' structure, surmounted by a deep deck fairing extending from engine to tail which housed a large luggage locker behind the rear cockpit. The engine-bearers were made of wood, braced with steel tubes and the only other use of metal in the machine was for jointing, fittings, controls and tail skid etc. The centre section was constructed as a separate unit from the fuselage with two extremely strong box-spars which passed through slots beneath the seats and the lower longerons were joined beneath the spars by removable steel fittings. A 22½ gallon petrol tank was installed in the starboard wing stub but an additional tank could readily be fitted in the port stub to increase the range. The undercarriage was attached to the ends of the centre-section spar with each wheel being carried on an independent Dowty shock-absorber and half-axle hinged to the front spar, positioned by a radius rod to the rear spar. The Dowty undercarriage was not completed in time for the first Hawk, so a heavier, temporary undercarriage was borrowed from a Spartan Arrow for its first flights.

Each wing consisted of two tapering box spars connected by spruce ribs and covered with three-ply. The leading edge with its nose ribs and the trailing edge with its riblets were attached to this structure, thus minimising the result of accidental damage. A hinged flap at the inner end of the trailing edge permitted the wing to fold around the rear spar, this being locked back in its folded position by a hook from the tail plane. The tailplane, fin, rudder and elevators were of full cantilever design and were covered with fabric.

Full dual control was provided, although the front stick and the rod connecting the front and rear rudder bars could be removed. The rear cockpit was occupied by the pilot and contained the instrument panel surmounted by a small shelf. Instruments were not normally provided in the front cockpit, although a large shelf was available for maps and gloves.

The normal power plant for the standard M.2 Hawk was the 90hp Cirrus IIIA, a four-cylinder in-line, air-cooled engine, cowled-in with easily detachable aluminium panels. A Claudel-Hobson carburettor was fitted on the opposite side of the engine to the two BTH magnetos, the front one of which had an impulse starter. As the engine was not inverted, the oil was carried in the sump and circulated by a submerged pump to the main bearings. The petrol tank was below the carburettor level, so a dual AC engine-driven petrol pump was fitted which could be operated by hand to prime the system if it was drained. Short stub exhausts were fitted as standard.

Although a very early Miles drawing shows the Hawk fitted with a Pobjoy radial engine, this engine was never fitted to a production machine.

The first Hawk, in primer paint and with no registration marks, emerged in the late afternoon of 29th March 1933 and (as described above) 'behaved too well' in its taxiing tests for Miles' self-control! As the sun was setting, he took it straight off, recording 'Hawk Test' against a 5-minute flight in his log-book. Before darkness had fallen Miles had also taken Blossom for a 5-minute flight, and by dusk the next day, Miles had flown the Hawk for 1 hour 30 minutes, and its total flying time had reached 6 hours. The only change found necessary following these early flights was a widening of the outlet louvres of the engine cowling in order to improve the cooling.

On March 30th, Miles gave flights to Don Brown, Harry Hull, 'Tom', Parker and Colson, and on the 31st spinning and diving tests and load trials were carried out, probably by F/O Leech as Miles did not fly the Hawk again until April 2nd. The Hawk's behaviour was perfectly satisfactory.

On Sunday 2nd April 1933, the Hawk was demonstrated at Woodley in public for the first time and the following day it made headline news in the press. The *Daily Herald* for April 3rd reported: *The plane, which costs only £395, was put through its paces by an RAF test pilot, who pronounced it to be even better than it was said to be.*

On the same day, the *Daily Mirror* reported the story under the bold headline: *Cheap Aeroplanes for Everybody - Fifteen pilots test a £395 Two-Seater:*
  An aeroplane which may revolutionise pleasure flying went through its first official tests at Woodley Aerodrome, near Reading, yesterday. It is one of the first low-wing open-type monoplanes to be produced in England, and its price is £395 - almost half the lowest-priced two-seater 'plane at present. A new wing section enables the machine to take off in fifty yards. Everything in the 'plane is British, and arrangements are being made for the manufacture of the machine at Reading. The 'plane, which has one 90hp engine, has a maximum speed of 114 mph and carries sufficient fuel to keep it, with two people and their luggage, flying for four hours at 100 mph.

  The 'plane was tried out by fifteen pilots yesterday. At the close of this exhaustive test Mr Miles told the Daily Mirror that it had behaved perfectly and that he was very pleased with it. 'It has only been flown once before and that was on Wednesday when we had it up for a few minutes'. He added: 'It has been in the air solidly for six hours today. Five Royal Air Force pilots tried it and one of them has been stunting it quite hard. But it stood up to it'.

  'We believe that there is a market for this type of 'plane and are starting to build six tomorrow. We have already received three orders, and there is no doubt that people are interested judging by the number who have flown over to-day to see it'. Mrs Miles told the Daily Mirror yesterday that 'I have flown a number of different machines and there is only one I like better - and my husband designed that! But all his previous designs have been high-performance machines'.

  An ordinary open two-seater Moth aeroplane costs in the region of £750. The cheapest 'plane at present on the market is the single-seater Comper Swift at £550.

The *Aeroplane* for 5th April 1933 reported: *The great event of the week at Reading was the first appearance of the Miles Hawk (Cirrus III). By the evening of 2nd April it had been flown by some 15 pilots including Wing Cdr H M Probyn, Major H G Travers, F/L Bonham-Carter, F/L Pugh, F/L Clarkson, F/O Leech, Alan Muntz and others.* F D Bradbrooke wrote in the same issue:
  There could hardly be a better proof of the designers' satisfaction than by being allowed to fly the Hawk raw from the shop. As I was allowed to fly the Hawk during its first hours I am able to give some idea of what may be expected from the production model. Mr Miles had told me some weeks before that the minimum flying speed on the level would be 48 mph, and as he had never built a low-wing machine before he preferred not to count on any cushioning or downward component beforehand. Having seen beams of great strength and apparent weight built into the machine I was also prepared to find that the 90hp Cirrus would take it off rather sluggishly. I found Messrs Miles and Powis justifiably excited at the falsification of their estimates - on the right side.

  Naturally the machine had no load except myself and some petrol, but by comparison with equivalent machines in a similar state the Hawk definitely has nothing to fear. There must be something in this low-wing idea. The wing loading was probably about eight lbs per sq ft, but it took off at an indicated 43 mph after a short run and held a steep angle of climb at 55 mph.

  The cockpit is fairly large and seems to have more room than usual because the instrument board and map locker are well forward to give plenty of space to work. The turtle-decking comes well back with the windshield so that goggles are not necessary when cruising. The pilot's seat is on the rear spar and therefore gives an unusually upright position, as in the Klemm. This is certainly no disadvantage.

  The throttle control was bad because it was stubby and stiff, which can easily be rectified. There was some mechanical stiffness in the ailerons and no tail trimming gear had yet been rigged. These are the details of newness which affect present feel and must be allowed for in giving judgment. There could hardly be a better proof of the designers' satisfaction than my being allowed to fly the Hawk raw from the shop.

  The machine is as simple and clean in action as in appearance. There is nothing the least bit strange about its flying qualities unless it be in aerobatics or the spin, none of which I tried. There is a typical thick-wing feel about the ailerons, which are very slightly heavy at high speeds, but as they are effective down to 40 mph or less I should consider them hardly worth altering, particularly at the cost of balancing or other departure from extreme simplicity.

  Strangely enough the trim was almost perfect with my weight and the Hawk held a steady climb at full throttle or steady level cruising at about 1,850 rpm. The hands-off behaviour is very steady fore-and-aft, so with a tail-trimming gear the same should be true with all ordinary loads. The luggage will be aft, as usual, but the passenger is slightly forward of the CP. - (Which means that with the tail trimmed upwards there will be a higher degree of fore-and-aft stability. - Ed.)

  Top speed can never be judged accurately from a short trial with an uncalibrated airspeed indicator, but assuming that the instrument did not flatter the Hawk at both ends of the scale I expect to see an official top speed of at least 110 mph. I prefer not to say more because it seems unlikely. Side-slipping, landing, and taxiing are all too normal to call for comment. In its light state the Hawk touched down at an indicated 40 mph and did not run far. There are no wheel brakes but the ground angle is large.

  The engine was obviously in good form, and ran with a smooth alacrity which reminded one that the Cirrus III may be out-moded but not much out-classed. The significant thing is that two people with their luggage may expect to cruise at over 90 mph for four hours at about 23 miles per gallon in a machine which costs £395, is cheap to insure, and as cheap to maintain as a wholly wood and plywood structure and a notably simple engine can make it.

(Note - The machine did spinning and diving tests and load trials on March 3rd; its behaviour was perfectly satisfactory.)

Flight for 6th April 1933 reported:
  During a short trial recently, we found the Hawk particularly pleasant to fly. It did not seem to have any vices, and its controls were nicely harmonised. They appeared to be amply effective at the bottom end of the speed range, without being unduly heavy at full speed. The rudder is, perhaps, geared a little too low for our personal liking, but apart from that it was impossible to criticise the controls in general after a short trial of this nature. As a matter of fact, everything felt just 'right', and we found it one of those machines which immediately give the impression that it will never want to do anything it was not meant to do. Its landing properties are excellent. The glide is coarse at low speed, and when landing the machine sits down without showing any tendency to 'balloon off'. The take-off is more than admirable, and at a rough estimate we should say that 40 yards would easily see the machine in the air with two up on any ordinary day. It is, of course, at this stage impossible to give accurate performance figures, but with a calibrated ASI the top speed appeared to be in the region of 115 mph, while the machine 'touched down' just 'off the clock'.

Chapter 7: Miles M.2, M.2A, M.2B, M.2C, M.2D and M.2K Variants

The view forward is excellent for a single-engined tractor, as the cowling falls away sharply from the top and impedes the view either side as little as is possible with this layout. Torsionally the wing seems strong, it being impossible to flex it at all by hand. The Hawk also has the added advantage that the wings may be folded.

One likeable attribute of the Hawk is its ability to cruise comfortably at low engine rpm. For example, at 1,500 rpm a speed of 78 mph was shown on the ASI, and at this speed the engine was hardly audible. It should therefore make an economical and particularly pleasant machine for the private owner who has to consider the £sd of its flying very carefully. A good sized luggage locker is fitted in the top-decking behind the rear seat and will comfortably accommodate two ordinary size suit cases. We understand that Phillips & Powis will now be putting the machine into production and selling it at a very low figure, and we imagine that it should find a ready market amongst private owners who want a simple, easy and safe machine which, while being very economical, has an excellent turn of speed when needed.

At the time of going to press we hear that further trials have shown that this machine can only be spun by forcing it to do so with full controls and engine. It was difficult to keep it spinning for more that 1½ turns and it came out immediately when either the stick or rudder were centralised. Accidentally getting into a spin should, therefore, be a rather remote possibility.

The Aeroplane for April 12th reported: *The Miles Hawk has now done 30 hours flying with 53 different pilots. F/O Leech gave an aerobatic display in it on April 9th.*

The pilots included Miles' brother George and before the month was up, the number of pilots who had flown the Hawk had increased to 80 and its popularity was assured. *The Bystander* for April 26th noted that: *in the hands of the test pilot it has been dived to 200 mph and stunted so vigorously, so that no doubts as to its safety factor exist.*

It is of interest to note that Miles, with his customary contempt for authority, carried out all the early flights of the first Hawk with no registration marks and the machine still in its primer paint! The Hawk was painted by April 5th and Miles last flew it without registration on April 13th. It was not formally registered as G-ACGH until 19th April 1933, although Miles flew the Hawk as such for the first time on April 14th.

Due to its clean design, the Hawk's performance was 50% better than the contemporary biplanes of the same power and two people with their luggage could cruise in comfort, with a performance of over 90 mph for 4 hours, on about 23 miles per gallon. By nature of its construction, the Hawk was also cheap to maintain, simple to repair and cheap to insure. In view of its high performance, however, it was suggested that the control surfaces should be mass balanced in order to avoid the risk of flutter at high speeds. This phenomenon had already been experienced in the contemporary Comper Streak racing monoplane but, such was Miles' confidence in his calculations that, donning a parachute for the first time in his life, he climbed the Hawk to 10,000ft, stood it on its nose and dived it to its terminal velocity – and no sign of flutter occurred!

The technical capability of Phillips & Powis was judged from the fact that, although a comparative newcomer to the field of aeroplane production, it was soon put on the Air Ministry's list of firms approved for aircraft design.

Standard Hawks were initially sold at the remarkably low price of £395, partly because they had been designed for cheap production but also on account of the fact that Phillips & Powis were able to take advantage of a large batch of Cirrus IIIA engines which had fortuitously come on to the market at below list price, through the

This delightful posed photograph shows (from left to right) Charles Powis (in white flying overalls), F G Miles and Blossom Miles, with Master Carpenter Harry Hull, builder of the machine.
['Flight' via B Clarke]

An advertisement for the Hawk, which appeared in 'The Aeroplane' for 30th August 1933.

failure of a Canadian aircraft company which had ordered them. When the supply of cheap Cirrus engines was exhausted the selling price of the standard Hawk had to be raised to £450.

The first production Hawk, G-ACHJ, was tested at A&AEE Martlesham Heath, but their Report No 622, dated 7/33, was not quite so complimentary, stating that 'it was tiring to fly and had a long landing run' but this did not affect its popularity and orders for the Hawk began to pour in, with some 37 Hawk variants being delivered in the first year of manufacture alone. This heralded the success of the venture and thus began the start of a long line of successful aircraft to be designed and built at Woodley.

More workers had to be taken on in order to cope with the demand for the new machine and these included Mr L Williams, who joined the firm in early April 1933, as foreman in charge of fuselage production. Mr G Chapman, who joined about the same time, would be in charge of wing production.

An indication of how rapidly production gathered momentum can be seen from the following known first-flight data, which indicates that production was soon running at approximately three machines per month:

c/n 3	G-ACHJ	28th June 1933
c/n 4	G-ACHK	21st July 1933
c/n 5	G-ACHL	24th August 1933
c/n 6	G-ACHZ	pre 9th Sept 1933
c/n 7	G-ACIZ	19th October 1933
c/n 8	G-ACJC	26th Sept 1933
c/n 9	G-ACJD	7th October 1933
c/n 10	G-ACJY	6th Sept 1933
c/n 11	G-ACKI	16th January 1934
c/n 12	VT-AES	12th December 1933
c/n 13	PK-SAL	23rdth October 1933
c/n 14	G-ACLI	27 November 1933
c/n 15	G-ACLA	pre November 1933
c/n 16	G-ACLB	7th November 1933
c/n 17	G-ACMH	pre 24th November 1933
c/n 18	G-ACMM	pre 23rd December 1933
c/n 19	G-ACOB	12th March 1934
c/n 20	U1	4th February 1934

A contemporary Phillips & Powis brochure gave the following details:

The key note of the Hawk design is practicability. It was designed and produced by pilots after exhaustive discussions as to what constituted a perfect aeroplane for the private owner. The aeroplane had to be low in cost, robust in construction and much easier to maintain than the average light aeroplane.

At first sight, simple maintenance seemed inseparable from high first costs. As the design evolved, however, simplified construction led to simplified maintenance. In choice of materials, the designers were unanimous. The light aeroplane of metal, however attractive it sounds, is expensive to buy and very expensive to maintain. To keep weight down extremely light gauge tubes have to be used and the elaborate structures necessitated are difficult to inspect and replace.

In the Hawk, wooden construction is carried to its logical conclusion. Sitka Spruce and the finest Birch 3-ply are used throughout the aeroplane. The form of construction used in the wing gives the greatest strength for weight obtainable. Each surface consists of ply made in one piece, attached to the wing by a method which entirely eliminates tacks or screws. It can be given a smooth surface and finish comparable with that of a high class motor car. In strength and durability the special wing construction is greatly superior to the fabric covered type and it has also been proved cheaper to repair and re-cover.

CHAPTER 7: MILES M.2, M.2A, M.2B, M.2C, M.2D AND M.2K VARIANTS

Above and below: In August 1933, Hawk fuselages were being manufactured in one part of the hangar, and wings in the other. [Via B Clarke]

Fitting out and final assembly – Hawks in production in late 1933. [Via P Amos]

W/C H Probyn's 'Hawk', the winner of the Race for the Wakefield Trophy at the International Meeting at Lympne in 1933. Note the front cockpit has been faired over. [Via B Clarke]

G-ACHL was a standard M.2 Hawk and is seen here in about 1938. [Via B Clarke]

This advertisement appeared in 'Flight' for 31st May 1934.

The finish of the Hawk is absolutely weatherproof, and ensures long life to the outer surfaces and internal structure. The extreme aerodynamic 'cleanness' leads to practical cleanliness, because the absence of protruding fittings, struts and wires in the Hawk makes it very easy to wash and polish.

There is no other machine on the market of similar horse power that will carry two people plus 275 lbs of luggage at the cruising speed 100 mph at 4½ to 5 gals of petrol. This means that one can travel 100 miles at a fuel cost of 3/6 per person.

We are detailing below likely costs to the Private Owner; these figures are not estimated, but are taken from actual costs and experience. Examples are based on 100 hours flying a year.

Housing (a year)	30	0	0
Insurance (a year)	40	0	0
Renewal of Certificate of Airworthiness, etc.	30	0	0
Maintenance, etc.	5	0	0
Petrol (4¾ gph)	35	12	0
Oil (1 pint ph)	2	15	0
	£143	**7**	**0**
Cost per hour	£1	8	8

At average cruising speed of 100 mph, the cost will be 3½d per mile. This means that in a Hawk with full tanks, two people with 275 lbs of luggage can cruise at 1¾d a mile per person, or can fly from Croydon to Paris for 30/11d each

These figures show a saving of approximately 10/- an hour flying over other types of light aircraft of similar horse-power. The saving is obtained by a higher cruising speed with a given consumption of petrol and low maintenance costs.

The figures stated are practical figures and are if anything on the high side, for example, subject to the machine being looked after properly, the cost of renewal of Certificate of Airworthiness in all probability would be below £30, and a number of aerodromes will house a light aeroplane for £25 per year. It is easily realised how inexpensively a Hawk can be run when it is generally considered that a car of the light 10-12hp type, costs in the region of 5d a mile to run.

It is interesting to note that Miles flew the first Hawk, G-ACGH, for the last time on 18th July 1933 and then, on September 6th, he flew G-ACJY, the same Hawk rebuilt, for the first time. However, what caused the first Hawk to be rebuilt and re-registered, remains a mystery.

The *Eastern Evening News* for 10th April 1934 published a very interesting article regarding the Norfolk and Norwich Aero Club, which read: *During the weekend Messrs Phillips & Powis of Reading sent two Miles Hawk machines to the club for demonstration. These demonstrations were given by Mr A G Barrett, an old member of the club, Mr Stephen Cliff and Mr F G Miles, the designer. A very good display was given and many of the members were allowed to fly the machines. Their reports were very favourable indeed. The Miles Hawk is a low wing monoplane originally designed for the not-too-rich private owner, but it has many qualities that recommend it for club use. One of these is its low landing speed of only 40 mph. Another is that the position and width of its undercarriage and also its low centre of gravity allow it to stand up to continual hard use which must be undergone during training periods. Furthermore, it is a comparatively easy machine to fly and what is very important it only costs £450, which is considerably lower than any other light aircraft on the market. The Miles Hawk is fitted with a Cirrus IIIa engine and possesses a maximum speed of 116 mph with a cruising speed of 100 mph. The normal take-off is 80 yards.*

A 'de luxe' M.2 model, known as the Colonial Hawk was produced for operation overseas and although this was identical in appearance to the standard model, the structure was strengthened internally and the plywood of the mainplanes was fabric covered to ensure the longest possible life. Special attention was paid to the protective materials used throughout the aeroplane and more comprehensive equipment was provided, with a compass being fitted as standard. A special exhaust system, similar to that on the

M1, previously PK-SAL, the Miles M.2 'Colonial' Hawk single seat, seen here in the Dutch East Indies in mid-1936. Note the front cockpit is faired over.
[Via Nico Braas]

The Miles M.2 Hawk, G-ACUD, with the 'new Miles split flaps', seen here in August 1934. ['Flight' *via B Clarke*]

G-ADGR, a late production Miles M.2 Hawk, purchased from donations to the Gerald Royle Fund in 1935. [A J Jackson Collection]

later three-seat Hawk, could be fitted for a small extra cost to reduce the noise level in the cockpits without affecting the performance.

In July 1934, Miles began experimenting with fixed, bent aluminium flaps under the trailing edges of the wings on a standard Hawk, making his first recorded flight with these fitted to an M.2 Hawk, probably G-ACUD, on July 16th. The resultant steep angle of glide, increased coefficient of lift and consequently reduced landing speed, were so marked that manually operated split trailing edge flaps were fitted as optional extras to the later M.2F Hawk Major, the Gipsy Major development of the Hawk and as standard fit to the M.2H Hawk Major.

Whilst the Cirrus-powered Hawks were soon eclipsed by their Gipsy-powered development, they did enjoy considerable popularity. One such Hawk was registered to the locally-based Reading Aero Club, and its acquisition is a rather poignant story. It concerns Gerald Royle, who was known to many of the flying clubs in the country as he had been the youngest person ever to have gone solo (at Scarborough in 1933) but who was tragically killed near Scarborough in mid-1934. Reading Aero Club's Secretary, Cyril Albert Nepean Bishop, or *Bish* as he was universally known, naturally sent a message of sympathy to his parents on the Club's behalf.

A fund was set up by Club members in memory of Gerald Royle and Mr & Mrs George Royle responded with an offer to contribute £200. Bish later wrote: *This aircraft, bought with donations to the Gerald Royle fund, was nearing completion and after the original donation of £200 from the Royles I had members of the club making donations to the fund. I also had a little correspondence with the registration authorities for I had noticed that the registrations were then getting toward the G-ADG series, so I requested that G-ADGR be reserved for this new Hawk thereby perpetuating the initials of the Royle family who were all G - George, Georgina, Greig and the late Gerald. They would not promise but they did and by the end of May G-ADGR was ready and - thanks to the generosity of the late Lord Wakefield who, when told that we lacked £75 after all donations had been received, put up the amount himself.*

The Hawk was delivered to the Club on 2nd June 1935 and was christened by Lady Muskerry, a club member, in the presence of the Royles and a large number of club members and their friends. Bish himself took Greig Royle for a flight, which was also Bish's first in the aircraft. When, in November 1936, the club was being run down and Bish wanted to move to Brooklands, Mr Royle agreed that G-ADGR could be sold and, from the proceeds of sale, the members who had donated were repaid.

The M.2 Hawk was designed to take in-line engines up to and including the 120hp Cirrus-Hermes IV and 110hp DH Gipsy III, and in addition to the Colonial Hawk, four sub-variants of the standard model were ultimately produced as follows:-

M.2A Cabin Hawk This had the open cockpits replaced by an enclosed cabin (said somewhat cruelly by one of the employees at the time, to have been produced by a local manufacturer of greenhouses!). It had a 110hp Gipsy III engine and was built for Stephen Cliff, (pilot manager of Phillips & Powis' air taxi service at Bristol) who wanted to compete in the Egyptian Oases Rally in January 1934.

M.2B Hawk A single-seat, long-range version, this had a 120hp Cirrus Hermes IV engine and was fitted with a sliding canopy made by de Havillands and similar to that fitted to their Speed Fox Moth. This was built to the order of Man Mohan Singh for what proved to be an unsuccessful attempt on the England-Cape record in January 1934.

M.2C Sports Hawk This was similar to the standard Hawk but with a 110hp Gipsy III engine and was delivered in April 1934.

M.2D Hawk Three-seater This had an elongated rear cockpit to carry two passengers, with the centre section span increased by two feet and complete with wheel brakes. It sold for £550 and became very popular for joyriding and with travelling air circuses, which were then all the rage. It could be fitted with either the 90hp Cirrus IIIA or the 110hp Cirrus Hermes II engine and six were built between April and July 1934.

M.2K Hawk The sole specimen had a 110hp Cirrus-Hermes II engine and was delivered un-registered to The Royal Singapore Flying Club in January 1935.

Some 56 M.2 Hawks were built between March 1933 and July 1934, when it was replaced in production by the M.2F Hawk Major. Full production details of the M.2 Hawk are shown in Appendix 9.

Regrettably, none of the Cirrus Hawks built have survived, but, in 1986, John Evetts of EMK Aeroplane prepared a proposal for the construction of a replica and sent it to George Miles for his comments. A copy of this proposal, which unfortunately came to naught, is shown in Appendix 28.

CHAPTER 7: MILES M.2, M.2A, M.2B, M.2C, M.2D AND M.2K VARIANTS

Miles M.2B Hawk Single-Seater, VT-AES, for Man Mohan Singh, was first flown, by Miles, on 12th December 1933. ['Flight' via B Clarke]

G-ACLI, the sole Miles M.2A Cabin Hawk, outside the clubhouse in late 1933. ['Flight' via B Clarke]

UI, the first Miles M.2D Hawk Three-seater, with pilot and two passengers, in February 1934.
['Flight' via B Clarke]

Right: An advert for the Miles M.2D Hawk Three-seater, showing UI, which was first flown on 4th February 1934 and registered G-ACPC on 26th March 1934.
[Phillips & Powis via The Miles Aircraft Collection]

SPECIFICATION

The 1934 Brochure on the M.2 Standard Hawk gave the following specification and figures:

Construction:	Wooden fuselage and wings ply covered
Engine:	95hp Cirrus IIIA (or any other approved engine up to 120hp, to order)
Dimensions:	Wing area 169 sq ft; span overall 33ft 0in; span wings folded 13ft 10in; chord max 6ft 3in; length overall 24ft 0in; height 7ft 8in; wheel track 7ft 8in; aspect ratio 6.6 to 1 (not given in brochure)
Weights:	Empty, including all standard fixed equipment, 1,015lb; pilot 160lb; petrol 22½ gals 170lb; oil 20lb; payload (passenger and luggage or freight) 435lb; Loaded 1,800lb; disposable load 785lb (also given in advertisements as 800 lbs.); wing loading 10.6lb/sq ft (not given in brochure)
Performance:	Max speed at 1,000ft -116 mph; cruising speed at 1,000ft -100 mph; landing speed normal 40 mph; take-off normal 80 yds; landing run 100 yds; initial rate of climb normal 860 ft/min; ceiling - max 18,000ft, service 16,000ft; flight duration 4½ hrs; range (can be doubled) 450 miles; petrol consumption (cruising) 5 gals/hr (not given in brochure)
Standard Instruments:	Mounted on instrument board in rear cockpit: airspeed indicator, revolution indicator, aneroid, oil pressure gauge, inclinometer
Standard Equipment:	Palmer Airwheels 480 x 180; deep luggage locker; engine aircraft and journey log books; dual control
Standard Colour Scheme:	Planes and fuselage - Aluminium.
Price:	With Cirrus IIIA engine - landplane ex-works, ready for flight, £450.

Note:- Any other colour scheme (not more than two colour) involves an extra charge of £10.

The Miles M.2E Hawk 'Gipsy VI', which was flown by Sir Charles Rose in the King's Cup Air Race at Hatfield on 13th July 1934.
['Flight' via B Clarke]

Chapter 8: Miles M.2E Hawk Gipsy VI, M.2L Hawk Speed Six and M.2U Hawk Speed Six

Much misinformation has been written in the past concerning the background history of the 200hp DH Gipsy Six powered Hawk Gipsy VI/Hawk Speed Six, even by respected contemporary aviation historians. This was probably due to the fact that, although only three 'Speed Six's' were built, each was subjected to a considerable number of modifications throughout its life and only now, after many years of intensive research into their genesis, can the history and build state of each of these aeroplanes be finally confirmed. Before commencing a detailed study of these three aeroplanes, it is necessary to look at the way Miles was thinking ahead at Reading, in those early days.

An article appeared in *Flight* for 18th May 1933 under the heading 'Another International Air Race Design', and this gave the first insight into Miles' future plans:
> *Our readers will, by now, be well aware of the success achieved by Mr F G Miles as a designer of distinctive aeroplanes. His first machines, the Martlet and Satyr, showed that he possessed a true conception of what was required in a sporting single-seater, while his latest, the Hawk, proves his abilities in the general utility class. Of peculiar and topical interest, therefore, is the news that Mr Miles has completed the general layout for a small high-speed racing machine, using any in-line engine of about 6.5 litres capacity. The machine will have, he estimates, a performance in advance of anything which has been proposed so far, and should therefore be eminently suitable for entering in the International Air Races at Portsmouth in August. Mr Miles is confident that not only can he get this machine out in time for the races, if he is given an order at once, but also that the cost will be quite reasonable, and he is therefore anxious to hear from any syndicate who wish to assist to uphold Britain's prestige on this occasion. Mr Miles is now in charge of all the design work for Phillips & Powis and enquiries should be sent to him at Woodley Aerodrome, Reading.*

Unfortunately, Miles did not get any interest in his project at the time but early in 1934 Sir Charles Rose approached him with a view to building a sports-version of the Hawk which would be ideal for racing. Miles revised his earlier plans and based his new ideas on a modified Hawk airframe, with the front cockpit deleted to make way for a 200hp DH Gipsy Six engine mounted on metal bearers, a sliding canopy over the rear cockpit and neatly faired undercarriage legs. It should be mentioned that, contrary to reports, all three Hawk Speed Six variants were originally built with normal centre sections.

A report on the three versions of the 'Hawk' which were entered in the 1934 King's Cup Air Race appeared in *Flight* for 12th July 1934, with the comment on Sir Charles Rose's Hawk Gipsy VI: *It is a tribute to the robustness of the 'Hawk' that very few modifications in the way of strengthening have been found necessary for it to get its Certificate of Airworthiness.*

The first Hawk Gipsy VI, with an enclosed cabin and original shaped trouser fairings to the undercarriage, was given the designation M.2E and was first flown, by Miles, on 28th June 1934 on a 5-minute flight (although it had been registered G-ACTE in May 1934, the letters had probably not been painted on at this time). On July 11th, Miles demonstrated the machine, recording G-ACTE for the first time, and Sir Charles Rose took delivery of it in time to participate in the third heat of the 1934 King's Cup Air Race, held at Hatfield on Friday July 13th. Unfortunately, ignition trouble forced him to land at Northolt during this heat.

Sir Charles Rose then entered G-ACTE in the 'Round the Isle of Wight' and 'Portsmouth Trophy' races on July 21st, coming 9th in both races at 174 mph and 168 mph respectively. Then, on July 28th he entered it in the SBAC Trophy Race at Bristol, becoming the 'scratch' aircraft and, even though there were strong winds prevailing over the course, he achieved 2nd place, putting up the fastest time of the day at 164.75 mph. Sir Charles started from scratch in the Wakefield Cup race at the Cinque Ports International Meeting held at Lympne on September 1st and 2nd, and finished a very creditable 5th at 172.5 mph, to win the Yates Trophy for the fastest time. On 6th October 1934, Sir Charles could only manage 8th place from 13 starters in the London to Cardiff Race but he still put up the fastest time. However, the engine cut out at the end of the race, forcing him to make a half circuit of the aerodrome before overshooting past a dyke and into a hedge in the ensuing forced landing, during which the undercarriage was wiped off and the propeller and wingtip damaged (see photograph below).

During the subsequent repairs, the undercarriage trouser fairings were modified to 'fill-in' the trailing edges slightly but not quite to the same extent as those fitted to later model M.2H Hawk Majors. A short time after this, the cockpit canopy was removed and it was modified to have an open cockpit.

The aeroplane, by now known as the Hawk Speed Six, was then bought by William 'Bill' Humble, in time for him to take 4th place in the Grosvenor Trophy Race at Desford on 13th July 1935. He then entered it in the 1935 King's Cup Air Race held at Hatfield on September 6/7th, during which he averaged 177.79 mph to gain 13th place. On September 21st, Humble averaged the remarkably high speed of 182 mph in the London to Cardiff Race.

Two developments of the M.2E were also built in 1935 and these incorporated flaps, increased dihedral from 3.5 to 5.0 degrees and, possibly later, wider-track undercarriages, although there is still some doubt over this. The first was built for Luis Fontes, the racing car driver and operator of a speedboat firm in Torquay, Devon. Designated M.2L Hawk Speed Six, it was registered G-ADGP to him on 20th May 1935 and was first flown, by Miles (as an anonymous Hawk Gipsy VI), on June 2nd (prior to its first officially recorded 30-minute "test flight" by Tommy Rose on 6 June 1935 as recorded in its airframe log book). In its original form, the M.2L had a sliding canopy with a pointed windscreen, similar to the original one fitted to G-ACTE, but this was later changed to a raised one-piece, sideways-opening canopy with a one-piece curved windscreen.

Other modifications were also introduced during its career in an attempt to increase its speed still further and although G-ADGP was originally fitted with a Gipsy Six Srs I engine, which developed 185-190hp at max permissible rpm of 2,300, this was modified, with the help of Major Frank Halford, to become the DH Gipsy 1F, to produce 205hp. On July 13th, Luis flew G-ADGP into 2nd place, at 171mph, in the Grosvenor Race, two places ahead of Humble in G-ACTE. This was the fastest time and also a very creditable performance for its first race.

The third Hawk Speed Six was built for Luis Fontes' sister Ruth and this was designated the M.2U. This version had an open cockpit with a low, wrap-round one-piece windscreen and was fitted with a 220hp Gipsy Six 'R' racing engine, with higher compression ratio and high lift cams. It is believed that it was also fitted with a new type of undercarriage, which was quoted as being 'shorter and sleeker'. Registered G-ADOD, it was first flown, by Miles, on 17th August 1935.

Both Luis and Ruth Fontes entered and flew their Hawk Speed Sixes in the 1935 King's Cup Air Race alongside Humble in G-ACTE, but neither Luis nor Ruth flying G-ADOD under the pseudonym 'Miss R Slow' were placed.

Bill Humble raced G-ACTE in the London to Isle of Man Air Race on 30th May 1936 in bad weather, to gain 3rd place at 152 mph and to clock up the fastest time of the day. Then, on Whit Monday June 1st, he was scratch in the Manx Air Derby but gained 5th place with the fastest time of 179 mph.

The new Shoreham Airport was opened on 13th June 1936 and during the International Meeting which was held there the next day, Humble, as 'scratch' man again, came in 2nd in the South Coast Air

G-ACTE being recovered following the forced landing at Cardiff on 6th October 1934. [Via B Clarke]

Chapter 8: Miles M.2E, M.2L and M.2U

G-ACTE during rebuild at Woodley. Note later style undercarriage fairings similar to the M.2H Hawk Major. ['Flight' via B Clarke]

G-ACTE in September 1935 with later modification to open cockpit and race No 5 for the King's Cup Air Race. [Via B Clarke]

The Miles M.2U Hawk Speed Six G-ADOD seen here landing at Portsmouth for the start of the Schlesinger Air Race in September 1936.
['Flight' via B Clarke]

Miles M.2U Hawk Speed Six G-ADOD at Portsmouth preparing for the start of the Schlesinger Air Race on 24th September 1936.
[Via B Clarke]

G-ADGP with modified canopy for the 1937 King's Cup Air Race.
[Via P Amos]

Trophy Race with the fastest time of 174 mph but he was later awarded the Trophy when the first man home was disqualified. On July 10/11th, Humble entered the 1936 King's Cup Air Race held at Hatfield but could do no better than 5th place at 175.67 mph, whilst G-ADOD, entered by Viscountess Wakefield, was flown by Tommy Rose into 2nd place at an average speed of 184.20 mph. Bob Milne flew G-ADGP but was unplaced.

In the London to Newcastle Race on August 8th, Humble won with the fastest time of 183.75 mph, after the two Hawk Majors in 1st and 2nd places were disqualified. He also took part in other races during 1936 and during the course of the London to Cardiff Race on September 19th, he was scratch but still managed to gain 4th place and win the fastest time prize with a speed of 189.80 mph! It should be mentioned that, although he achieved 4th place, there were in fact only five starters - the handicappers were now becoming very hard on the Hawk Speed Six!

For the Schlesinger Portsmouth to Johannesburg Air Race in September 1936, G-ADOD was modified to have a new two-piece windscreen joined on the centre line, and a canopy with a sliding rear portion of increased depth, tapering into a raised headrest type rear fairing. The centre section was also removed and short spars fitted through the fuselage to reduce the wingspan by some 5 ft in a similar manner to that adopted on the M.5 Sparrowhawk. It was entered and flown by the New Zealander, Flt Lt Arthur Clouston. Clouston was at the beginning of his career as a long-distance pilot, which was to bring him much fame the following year when he, together with Betty Kirby-Green, brought the England - Cape record down to 45 hrs 5 mins in the DH.88 Comet G-ACSS.

There were two sections in the Schlesinger air race, a speed race and a handicap race, and the course set out for the race went via Belgrade and Athens to Cairo, then followed the route taken by Imperial Airways over Khartoum, Kisumu and Mt Pika, to Rand Airport, Germiston, a distance of 6,150 miles. Only nine aircraft were at Portsmouth for the start, with the departure time set for 6.15 am on Tuesday 29th September 1936. Clouston with 110 gallons of petrol on board and consuming 10 gallons per hour at a cruising speed of 170 mph, could theoretically, allowing for a reserve, cover 1,700 miles non-stop. He was second to arrive in Belgrade, the first stop, and in the same position at the next stop at Cairo, but was delayed in Khartoum with a seized engine.

An article in *Aeroplane Monthly* October 1982 gives details of the remainder of the flight:

Clouston arrived in Khartoum after a gruelling flight, where the RAF had to almost rebuild his engine overnight. Then, after various mechanical problems he reached Entebbe where he had to forced land in a swamp, having run out of fuel. Determined to reach Johannesburg, he enlisted the help of some natives, dragged the battered Hawk out of the swamp, refuelled with car fuel and made a very tricky take-off from a rough road. On October 1st, Clouston was in second place, four hundred miles from his goal when, in pitch darkness, the engine stopped due to oil pressure trouble. Too stupefied to be frightened, he thought this must be the end as he glided down towards unknown hazards. Suddenly, there were the sounds of disintegration; fragments flew through the air, then total silence. He thought he must be dead until he heard a woman's voice say 'Don't go near him he's dead'. He was able to inform her that he was not dead but had miraculously escaped injury. Only the engine of the Hawk was recognisable. He had landed at Gwelo, 130 miles south-west of Salisbury, in Southern Rhodesia but the M.2U had been written off in the process. (The engine was later returned to England where it is being restored to ground running condition and was recently seen by the author).

In the 1937 King's Cup held at Hatfield on September 10th, Tommy Rose flew G-ADGP into 10th place at an average speed of 185.00 mph. *The Aeroplane* for 22nd September 1937 also reported that Rose *gave a really good show with the black and white Speed Six at the Eastbourne Flying Club's Display at Wilmington aerodrome earlier in the month* - the colour quoted in the report confirming that this was G-ADGP.

While the Speed Six was achieving notoriety in the field of air racing, George Miles felt that the maximum speed of the M.2L G-ADGP could be still further increased, so, for the 1938 King's Cup Air Race he arranged for a number of major modifications to be carried out. The modifications included removing the centre section and fitting short spars through the fuselage to reduce the wingspan by some 5ft, as described earlier for G-ADOD in 1936 and for the M.5 Sparrowhawk.

A table of data on a Phillips & Powis drawing stamped 16th June 1938 gave the following details:

Span overall	28 ft 0.5 in
Length overall	24 ft
Wing area	146.3 sq ft
Area of Ailerons	17.4 sq ft
Area of Tailplane	17.2 sq ft
Area of Elevators	8.4 sq ft
Area of Fin	2.29 sq ft
Area of Rudder	7.24 sq ft
Undercarriage Track	8 ft 1¼ in

A plate on an undercarriage bracket on G-ADGP was recently discovered which stated: Serial No. R-PP 5495. Drg No L.51-L1. Mod. No. L-232. Inspector No. 3. Date 15.6.38 – so it all fits! And, while Ron Souch was carrying out the major overhaul of the airframe in 1987, cards with the date 17.6.38 were found in the wings in the vicinity of the repositioned undercarriage attachments. Other cards, dated 20.6.38, were found on the front fuselage bulkhead, which were probably in connection with the new 'V' shaped windscreen and extended top decking behind the modified canopy which was also fitted at that time.

These modifications to reduce the wing span undoubtedly accounted for the later, erroneous, reports to the effect that G-ADGP had been fitted with 'a Sparrowhawk centre section'. Other modifications included increasing the undercarriage track to 8ft 1¼in to place it outside the propeller slipstream; hinging the cabin coupé on the left side, and fitting a pointed two-piece 'V' shaped (in the plan view) windscreen, very similar to that fitted to G-ADOD as flown in the Schlesinger Air Race (although that fitted to G-ADOD had a sliding rear canopy), and modifying the rear fairing to have a new fairing tapering from the rear of the canopy to the fin.

The last King's Cup Air Race to be held before the outbreak of war was held at Hatfield on 2 July 1938. Luis Fontes in G-ADGP was 16th away and finished in 13th place at an average speed of 184.50 mph.

The Airframe Log Book records that G-ADGP was flown on 14.10.38, 20.10.38, 26/27/28.10.38 and 16.4.39. It was then grounded to enable further modifications that George Miles hoped would increase its speed for the next King's Cup Air Race to be carried out. The Log Book identifies these as including a modified cabin top and rear deck formers, together with a modified elevator con rod, aileron lever, con rod link plates and seat bearer. The airframe was then completely resprayed and the first flight after these modifications was made by George Miles on 16th May 1939 with further flights being made on May 18th, 23rd and 26th.

An advertisement appeared in the 'Aircraft for Sale' section of the *Miles Magazine* for May 1939, which showed a Hawk Speed Six with a black fuselage and white wings (while the registration was deleted in the photo, it could only have been G-ADGP) - *For sale - Over 190 mph; Gipsy Six Series I engine; airframe hours 150; engine hours 165; Coupé head; Sperry Horizon and Gyro mounted on anti-vibration panel; Electric Starter. Very easy to fly - Price £850.* But it was not sold.

On 27th May 1939, Luis Fontes flew G-ADGP to 4th place in the London to Isle of Man Air Race at 168.25 mph and on May 29th Tommy Rose took it to 2nd place in the Manx Air Derby at the very respectable high speed of 186.75 mph. The last two flights made by G-ADGP, before overhaul for renewal of its C of A at Woodley, were made on 30th May and 3rd June 1939.

Luis Fontes' Miles M.2L Hawk Speed Six G-ADGP at Hatfield for the King's Cup Air Race on 2nd July 1938. [Via P Jarrett]

G-ADGP after yet more extensive modification for the aborted 1939 King's Cup Air Race.
[Phillips & Powis via The Miles Aircraft Collection]

Its last pre-war C of A renewal was dated 12th August 1939 and photographs of G-ADGP taken in the late summer of 1939 show that it now had a very raked back, flatish curved one-piece deep windscreen with a sideways opening canopy, which formed a continuous straight line from the top of the windscreen arch into the raised top fuselage decking to fair into the leading edge of the fin.

All this additional modification work was in vain, however, as, with the outbreak of war, the 1939 King's Cup Air Race due to be held on September 2nd was cancelled. In any event, it would appear that, despite all the modifications carried out, the performance of G-ADGP was not significantly increased and none of the Hawk Speed Sixes built ever managed to beat the handicappers in the pre-war air races.

G-ADGP made its last flight before the ban on civil flying came into force, on 25th August 1939, a 30-minute local by Luis Fontes. The Airframe Log Book then records: *Aircraft dismantled and removed from Reading Aerodrome by owner for storage during hostilities. Signed April 1940.* It was taken to Fontes' Wimpole Street, London mews flat, where it was stored for the duration of the war. Luis Fontes joined the Air Transport Auxiliary only to lose his life while ferrying an RAF Wellington, R1156, at Llandow on 12th October 1940.

The Aeroplane for 3rd August 1945, carried an advertisement offering the Hawk Speed Six for sale by High Speed Motors Ltd at £750, stating it had been built for the late Luis Fontes for the King's Cup and was now based in 'London'. It was acquired by Miles Aircraft Ltd who restored it to airworthy condition and it made its first public appearance in the static display at the first post war RAeS Garden Party, organised by the Reading Branch, held at Woodley on 1st June 1946. The pre-war, 'slightly blown canopy' had been retained, but the aircraft was now painted cream overall with the registration yet to be applied. This was applied soon after (in red), probably before its first post-war flight on 23rd August 1946. G-ADGP was then used by the firm's test pilots, including Tommy Rose, Ken Waller and Hugh Kendall, with Tommy racing it with some success.

After Tommy Rose retired from test flying in January 1946, he was appointed general manager of Universal Flying Services Ltd at Fairoaks and while there he had G-ADGP on loan for use as his personal 'hack'. He regularly raced it around a 'home-made' local closed-circuit in the evenings to keep his hand in, using as 'markers' three conveniently spaced trees at Missen's Farm, near Chobham church, probably much to the annoyance of the local residents!

Hugh Kendall recorded flights in G-ADGP from Woodley on 9.7.46 'General'; 23.8.46 'Test'; 27.9.46 'Aerobatics'; 1.10.46 'Woodley-Beaulieu' and return (probably in connection with the Miles Monitors with the AFEE); 3.11.46 'Woodley-Lympne' and return; 8.11.46 'Woodley-Boscombe' and return (to fly the prototype Marathon G-AGPD on an u/c test); 18.1.47 Woodley-White Waltham & Aerobatics' and return; 16.4.47 'Woodley – Local - Rate of Roll & Aerobats'.

On 10th May 1947, G-ADGP was entered for its first post war air race at Portsmouth. Flown by Tommy Rose, as 'scratch' man of the eight entrants, he gained 2nd place in the three lap race for a cup presented by Edgar Percival. On Saturday May 24th, Tommy flew G-ADGP from Fairoaks to Ronaldsway in the Isle of Man, via Squires Gate, a 2-hour flight, to participate in the Tynwald Air Race of the Manx Air Derby on Bank Holiday Monday May 26th. Tommy won the Olley Challenge Trophy in convincing style during a flight of one hour duration, and *The Aeroplane* for May 30th reported: *Flt Lt Tommy Rose came over the edge of the perimeter track going like the proverbial bomb. He won the Manx Derby after a really magnificent race. So good had been the handicapping that it might have been anybody's race, and it was exciting to see a really good pilot show what could be done by good flying. His speed was 180 mph and that of Bruce Campbell (in the DH TK.2, G-ADNO) - second after an exciting dual with Rose - 179 mph.*

CHAPTER 8: MILES M.2E, M.2L AND M.2U

The *Miles Magazine* for July 1947 published a photograph of G-ADGP with the caption *'Old Warrior': First to cross the finishing line in the Manx Air Derby, Tommy Rose brought his Miles Hawk Speed 6 home to win after averaging a speed of 181 mph over the Isle of Man course. It is interesting to note that an aircraft designed and built in 1935, with a maximum speed of 185 mph, can in 1947 win a handicap race at a speed of 181 mph over a course of 156 miles.*

In a telegram sent to F G Miles after the race, Tommy Rose said: *Many thanks for the loan of the Old Warrior to make my somewhat throaty swan song. Combined ages of aircraft and driver believed worlds record* to which Miles replied: *Congratulations. Am certain that your positively last appearances will be as many as Melba's.* Hugh Kendall, who had participated in the same race with the Miles Gemini G-AISM, flew G-ADGP back from Ronaldsway to Woodley on May 27th, in 1 hour 45 min, and then made a return flight in it to Redhill the same day. He also flew G-ADGP to Welford and return on June 9th, but it was next seen by the public being flown very spiritedly by Ken Waller in the display and flypast of Miles aircraft at the last Miles Aircraft 'At Home' Day at Woodley, on Sunday July 20th. By then the canopy had been changed yet again, this time to a much larger 'blown bubble' type.

On 8th August 1947, Tommy Rose flew the Hawk Speed Six to Southend (Rochford) for the re-opening of Southend Municipal Airport celebrations, to be held the following day. In the Southend Cup Air Race, Rose came home in 2nd place, behind Ron Paine in the cleaned-up Hawk Trainer III G-AHNU who had won with a speed of 134 mph. Rose was, however, awarded the Southend Air Speed Cup and a prize of £50 for putting up the fastest speed of 178 mph. Tommy returned G-ADGP to Woodley on August 14th and it was test flown there by Hugh Kendall on September 19th. Hugh made his last flight, 'Test & Aerobatics', in the Hawk Speed Six on 30th September 1947 (recording it as 'G-AGPD', a transposition of the registration letters which coincided with the prototype Marathon which he had been regularly flying).

G-ADGP ready for the 1939 King's Cup Air Race, which was cancelled due to the outbreak of war in September of that year. [Via B Clarke]

G-ADGP at Woodley in June 1946 after refurbishment by Miles Aircraft Ltd following wartime storage. ['The Aeroplane' via P Jarrett]

The Hawk Speed Six was advertised for sale by W S Shackleton Ltd in *The Light Plane* for March 1948 at £575 but prior to the advert appearing, on 6th February 1948, it was purchased by Ron Paine, then Technical Director of Air Schools Ltd/Wolverhampton Aviation Ltd, from the Receiver & Manager of Miles Aircraft Ltd. G-ADGP had still flown less than 150 hours since new and Ron proudly flew it home to Derby on February 6th, but admitted that he now had to learn to fly it as well as Tommy Rose had done!

Ron Paine re-covered the fuselage and repainted it white with black wings, the reverse of its pre-war colour scheme. In an attempt to beat the handicappers, he decided to take some drastic action, and his first job was to approach de Havillands with a view to their agreeing to him up-rating the engine to Gipsy Six 1F standard. The carburettors were overhauled by Hobsons, the undershield was reduced from 3½ inches to 1½ inches and a finer-pitch propeller was fitted.

Ron's next job was to modify the undercarriage trousers as he had noticed that after take-off the wheels protruded 8 or 9 inches below their bottoms. He made a cable arrangement which held the legs in compression at all times and allowed the wheels to just protrude about 3½ inches. The fairings were then lowered and the whole area generally tidied up. He also cleaned up the wing root/fuselage fairing and the flap fairings, which were rather 'dirty', and Ron recalled that 'much balsa wood was consumed there before I was satisfied'. Noticing that there was a large gap between the rear spar and the leading edge of the ailerons, he re-profiled the ailerons and built up the spar to reduce the gap, at the same time taking the opportunity to remove the direct drive mechanism to the aileron rock shaft and fit Magister units to give differential ailerons. The final modification was to fair-in the rather troublesome aileron horn balances.

The 1949 King's Cup Air Race was held at Elmdon from Friday July 29th to Monday August 1st and that year there was a new system for picking finalists for the race by selecting them from heats or complete eliminating rounds spaced throughout the racing season. The King's Cup Race itself was held on July 30th and arranged in three heats so that the 13 finalists from the 36 entrants would be selected to race over three laps of the 20.315 mile quadrilateral course. It was restricted to British pilots, flying British-built aeroplanes (quite rightly), with a maximum speed at sea level of not less than 120 mph and a maximum sea level power of not more than 1,000hp. Ron performed well in his heat with G-ADGP, but in the final he had some trouble catching Tony Cole's Pobjoy Niagara-engined Comper Swift G-ABUS, which had been handicapped extremely well. After the last pylon, Ron steamed ahead and overtook Cole with about 200 yards to go to the finish but, just as he was feeling 'home and dry', he saw Nat Somers in the Gemini 3 G-AKDC ahead of him and already crossing the line! Ron gained 2nd place, but had increased the Hawk Speed Six's maximum speed by 5 mph to 184.00 mph. This he improved further on August 21st, when he attained an average speed of 188.75 mph in the main race at Thruxton.

The 1950 King's Cup Air Race was, much to Paine's delight, to be held at his home aerodrome at Wolverhampton and so he got to work once again to try to increase the maximum speed. One of the first modifications made was to fit 6:1 high-compression, oversized pistons in bored-out cylinders to the engine. The camshaft was then exchanged for one from a Gipsy Queen 2 so that the bottom end closely approximated to that of a Gipsy Six Srs II. These modifications necessitated larger carburettor choke tubes and Hobsons, who were situated just a few hundred yards away from the aerodrome, at Fordhouses, once again came to the rescue. With these modifications, the power output was raised to that of a 210hp Gipsy Six Srs II but they managed to keep the weight down by retaining the fixed-pitch airscrew. A ram air intake was fitted in the nose cowl, the flame trap was removed and the cowlings were tidied up by drawing them in an inch all round. A smaller tail-skid and shaft were fitted and faired-in, some warped ply skin was replaced and the whole aeroplane highly polished. Test flights proved that these modifications had been well worthwhile.

On 17th June 1950, the weather conditions were ideal and 36 entrants lined up for the start of the King's Cup. This was won by Edward Day in the Hawk Trainer III G-AKRV at 138.5 mph, but Paine, in G-ADGP managed to increase his lap speed to 192.83 mph to finish 5th. This was a new World 100km Closed Circuit Class C.1b record and the fastest that the Hawk Speed Six had flown up to that time. It was a record which it was to hold for some 25 years.

The first of what was to become an annual event, the Daily Express International Air Race, was held on 16th September 1950 along the South Coast from Hurn Aerodrome, near Bournemouth to Herne Bay in Kent. 75 aircraft were entered in the Battle of Britain Air Race (celebrating its 10th anniversary) and of the 67 aircraft that started, ranging from the diminutive Chilton DW.1A to the mighty Handley Page Halifax C.8, 61 finished but Paine could do no better than 15th place, which might indicate the handicappers had 'got wind' of his modifications.

The second race took place on 22nd September 1951 but this time it was over a course of 186 miles along the coast from Shoreham Airport to Whitstable in Kent and then back to finish at Brighton's West Pier. This time there were 62 entrants, of which 59 started. Paine found that by getting the aircraft down to about 50 feet above the water it would really motor! So, after a number of test flights, he fitted a boost gauge to the Hawk Speed Six which enabled him to fly accurately at just the right height to give maximum boost. With a following wind he managed, for the first time, to record a speed of 207 mph, winning his class and producing a nice surprise in the way of prize money!

The last Daily Express Race was held on 2nd August 1952 over a course from Shoreham Airport to Newhaven and thence to Reculver and back around the coast to Brighton West Pier. This time there were 46 entrants but Paine was unplaced.

There was no King's Cup Race in 1951, but on July 14th Paine won the Bristol Air Race at Whitchurch Aerodrome in very rough weather conditions over three laps of a 30-mile course at an average speed of 193 mph. In the 1952 King's Cup Air Race, he gained 7th place at 176.50 mph and, on 13th September 1952, won the Southern Aero Club Race. Another rough day was 16th May 1953 when Paine came home first to win the Goodyear Trophy at Wolverhampton and also gain 10th place in the Norton Griffiths Trophy Race at 176.50 mph. On June 20th, he took 6th place in the 1953 King's Cup Air Race at 187.00 mph and the following year, he won the Chiltern Hills Air Race at Denham on 8th August 1954. In 1955, Paine won the SBAC Trophy at Yeadon on May 30th and was placed 6th in the King's Cup Air Race at 187 mph on August 20th.

Ron Paine had been consistently in the top six finishers in the King's Cup and came in second at 189 mph in the 1959 event, but it was in 1962 that he had high hopes. However, in that year's King's Cup, on August 18th, he had to be content with a tie for second place with Dennis Hartas in the Tiger Moth G-ANZZ and in 1964, he came second (yet again) in the King's Cup at 187.5 mph.

In 1965, the SBAC made a magnificent gesture in presenting Ron Paine with the Trophy for life (this Trophy was awarded for the fastest speed in the King's Cup Race), for having logged the fastest time in King's Cup Air Races in 1958, 1959, 1962, 1963 and 1964, the last three years in succession. Unfortunately, by then being too well known to the handicappers, he was never to win the King's Cup.

In June 1965, having become very involved with the airline business, Paine could no longer afford to devote the time to racing most weekends throughout the season, so he reluctantly parted with his beloved G-ADGP to William Todd of Wolverhampton, who based it at Halfpenny Green. Following a slight accident to the canopy, it never put up such good performances as it had with Paine, being unplaced in the 1965 and 1966 King's Cup Races.

Chapter 8: Miles M.2E, M.2L and M.2U

Ron Paine, by the port undercarriage, with G-ADGP and friends at a race meeting in the 1950s. Note 'The Throttle Benders' insignia on the cowling. ['Flight' via P Amos]

After two seasons flying from Halfpenny Green, Todd sold it Tony Osborne, who was then setting up the British Historic Aircraft Museum at Southend Municipal Airport. G-ADGP was flown there on 23rd March 1967 and on May 20th Osborne flew it in the Manx Air Derby Challenge Trophy at Ronaldsway. G-ADGP had reportedly been purchased by Osborne with the intention of being raced 'to raise funds for the Museum' but after the Manx race he decided, for some reason, to strip it down and (as Ron Paine later recalled) 'to make a different animal of it altogether' (in other words, Ron did not like what he had done to it!). Indeed, having seen the speeds now being put up by it, he realised that it had regressed considerably.

During the course of these modifications the canopy was replaced yet again but this time by a much higher, straight-sided, framed one, which had to be faired into the top fuselage decking in the manner of a headrest, and this did absolutely nothing for the Hawk Speed Six's lovely lines. The wheels were also replaced by a pair of 'scooter type wheels', due to failure of the original wheels (these also, apparently, did absolutely nothing either for its braking capabilities!) The undercarriage fairings were also extended and the aircraft was given a new colour scheme, which did not suit it either! The C of A was renewed at Southend on 26th August 1967 and although G-ADGP was entered for the Goodyear Trophy Race at Halfpenny Green on August 28th it was unplaced.

G-ADGP was then put into storage, first at the museum at Southend and then at Stapleford Tawney (where it was seen in 1968 to have an open cockpit and to be painted white overall). On 26th September 1969 it was put up for sale and following a C of A renewal by Robinson Aircraft Ltd at Blackbushe, it was sold in May 1970 to the London Sports Car Centre Ltd, Shoreham. Later, they arranged for a pilot to fly it from Shoreham to Elstree but on arrival there it caught fire, owing to a fractured fuel line, and was badly damaged.

In February 1971, G-ADGP was acquired by David Hood who had been looking for a vintage aeroplane. David contacted Ron Paine and they went to Elstree together to inspect the damage. Ron decided that it was repairable and it was therefore dismantled and taken to his new base at Castle Donnington where, over the best part of the next year, he worked on it. Some 50% of the covering had to be replaced but it was finally restored in its beautiful cream and red finish once again. Ron could not resist the temptation to carry out some modifications which he had always wanted to do but which were not possible when the aircraft was all in one piece!

From an external point of view the most obvious of these was the new canopy. He had a cooper build a mould which fitted over him when seated in the cockpit, from which was made a much lower hood, somewhat similar to the mildly blown one fitted in 1939. The ailerons were modified and new trousers were made for the undercarriage and, following this restoration, G-ADGP was flown again on 13th February 1972. It then departed for Old Warden, where it had been agreed that it could then be kept.

On 15th July 1972, Ron Paine flew G-ADGP in the 50th Anniversary King's Cup Air Race at Booker and took 2nd place at 196.00 mph. He then flew it into 3rd place at Shobdon and 2nd place in the Goodyear Trophy at Halfpenny Green but this was to be the last time that G-ADGP competed in an air race. David Hood only flew the aircraft twice before selling it to Mike Stow in October 1978, but Paine continued to maintain it, renew its C of A and prepare it for the Shuttleworth Air Displays. Mike Stow was absolutely delighted with the aeroplane and used to 'put it through a spirited routine' at Old Warden. By 1980, it was still being taken to the occasional air show and, as with David Hood before him, Paine had an agreement with Stow that, should it ever be offered for sale again he would have first refusal.

However, for reasons unknown, Stow sold G-ADGP in 1984 to Roger Reeves, who still kept it at Old Warden but in February 1985, it was sold to Desmond McCarthy, who had it shipped to Florida, USA for re-sale. There was little interest there in vintage, wooden, British aeroplanes and G-ADGP would have languished at Tangerine, Florida had not restorer Ron Souch alerted American art dealer Tom Buffaloe, who subsequently rescued it. It was last flown in the States, from Lakeland, Florida on 26th March 1985 and was then dismantled and shipped to Ron's workshop at Sarisbury Green, near Southampton where, in October 1986 he commenced work on yet another restoration.

Ron Souch completely stripped the aircraft down and rebuilt it as near as possible to its original configuration. The post-war top decking was removed and the cabin replaced with something resembling the first one ever fitted to it. When the work was complete, Souch painted G-ADGP in a black and cream colour scheme to represent its original colours although, due to the centre section having been deleted before the war, and the undercarriage having been repositioned in 1938, the finished aircraft could not be truly representative of the original G-ADGP when it *was* finished in this colour scheme. In the opinion of the author, a self-confessed purist, this was, unfortunately, a retrograde step as G-ADGP looked

Roger Mills' Miles M.2L Hawk Speed Six at the Great Vintage Flying Weekend at Kemble in May 2002. [P Amos]

really beautiful in its immediate post-war configuration and in the 'house colours' of Miles Aircraft Ltd - cream with red trim.....

It was first flown, in its new 'hybrid' configuration, by Martin Barraclough, from RNAS Lee-on-Solent (to where it had been taken for re-assembly and flight test) on Tuesday 23rd May 1989. Tom Buffaloe and his wife watched this memorable event from the control tower. G-ADGP's tailskid was worn away on the runway and had to be replaced by a tailwheel, but this caused problems in taxi-ing. The Hawk Speed Six was then flown to Old Sarum's grass airfield and re-fitted with a tail-skid. By then, G-ADGP had accumulated a total of approximately 520 hours and, as it was Tom Buffaloe's intention to keep the aircraft in Britain (together with his DH.60 Moth G-AANV), he moved it back to Old Warden. G-ADGP still attended many events but, in the summer of 1999 Buffaloe decided that, as no buyer could be found in Europe, he would take it back to America.

G-ADGP made what was expected to be its last public appearance in England at the Great Vintage Flying Weekend at Kemble in May 1999 and was then dismantled for shipment from Cardiff docks to its owner's home in the USA. However, unbeknown to those of us who mourned its forthcoming departure, Capt Roger Mills, a British Airways Concorde pilot, was having very serious designs on it. As a boy, he had remembered seeing Ron Paine flying it from Wolverhampton, and, as he sat in his hotel room in New York one evening contemplating the somewhat exhorbitant price being asked for it, he finally made up his mind - he had to own the Hawk Speed Six, regardless of price. So he rang Buffaloe and a sale was concluded.

Meanwhile, the ship had sailed from Cardiff and had arrived in Rotterdam. Telephone calls were made and the Hawk Speed Six was off-loaded and placed on another ship to be returned to England - it had been a very close call indeed. Thanks to Roger's last minute decision to buy this unique Miles aircraft, a piece of British Aviation Heritage had been saved from a very uncertain future - English wooden aeroplanes do not fare well in the States - the Americans seem not to understand them and inevitably they rarely get the love and attention they deserve.

G-ADGP was registered to Roger Mills in July 1999 and was initially based at Fairoaks, where the runway necessitated the tailwheel to again be fitted. This was supplied by Ron Paine, who was equally excited by the Hawk Speed Six staying in this country. It was regularly flown to fly-ins and displays and following its appearance at the International Air Tattoo, Fairford, on 21st July 2003, Roger flew it to its new base at White Waltham where G-ADGP is now maintained in airworthy condition. It is also of interest to note that the original DH Gipsy Six engine (No.6174) fitted in G-ADGP has remained with the aircraft throughout its entire life and is still giving yeoman service.

For details of Miles M.2 Hawk Speed Six production see Appendix 10.

SPECIFICATION AND PERFORMANCE - *based on the M.2E except where noted otherwise*

Engine:	M.2E 200hp DH Gipsy Six; M.2L 205hp DH Gipsy Six Series 1F High Compression; M.2U 220hp DH Gipsy Six R
Dimensions:	M.2E Span 33ft 0in; length 24ft 0in; height 6ft 8in; wing area 169 sq ft; aspect ratio 6.6 to 1. The M.2L was modified in 1938 to have a reduced span from 36ft 0in to 28ft 1½in; wing area reduced from 176 sq ft to 143.30 sq ft
Weights:	Empty 1,355lb; pilot 160lb; petrol (33 gall) 250lb; oil 20lb; payload 115lb (if 16½ gall petrol only carried, the payload can be increased to 220lb); weight loaded 1,900lb; disposable load 545lb; wing loading 11.25lb/sq ft; power loading 9.50lb/hp
Performance:	Normal load: max speed at 1,000ft 185 mph; cruising speed at 1,000ft 160 mph, (both speeds dependent on type of engine fitted, a max speed of 195 mph and cruising speed of 170 mph was achieved); landing speed 45 mph; take-off run 80 yds; landing run with brakes 100 yds; initial rate of climb 1,500ft/min; ceiling 20,000ft; petrol consumption 13 gall/hr; flight duration (33 gall) 2½ hr; range (33 gall) 406 miles.

The 'Spirit of Manawatu' taxi-ing out from Mildenhall at the start of the MacRobertson England to Melbourne, Australia Air Race in October 1934.
[Via B Clarke]

Chapter 9: Miles M.2F Hawk Major and M.2G, M.2H, M.2J, M.2M, M.2N, M.2P, M.2S and M.2T variants

By the early summer of 1934, Miles and Blossom, with the success of the Cirrus Hawk behind them, turned their attention to the design of an improved replacement to take account of the lessons learned from their initial design and to fit a more powerful engine. The supply of cheap Cirrus IIIA engines was coming to an end and even though Phillips & Powis had acquired the rights to put the Cirrus engine into production, they decided in favour of the higher-powered DH Gipsy Major which, being inverted, would also improve the view for the pilot. The prototype of the new machine was fitted with the 110hp DH Gipsy III while the production version (to be known as the M.2F Hawk Major) was to have the 130hp Gipsy Major engine.

The airframe was basically the same as the Hawk with a few aerodynamic refinements. It retained the wing-fold but a metal engine-mounting was designed to take the new engine and a cantilever undercarriage was fitted, complete with streamlined trousers. Bendix brakes were standard fit and were efficient both for parking on the hand-brake and in differential braking action on the rudder-bar for taxi-ing. These overall improvements increased the AUW to 1,900lb but they gave the new fully aerobatic M.2F Hawk Major a maximum speed of 150 mph, which was considerably faster than contemporary light aircraft.

The prototype, ordered by Capt Geoffrey Shaw, Chairman of Sywell Aerodrome Ltd, was first flown, by Miles (who recorded it as a 'Hawk Gipsy III' in his log book), on 23rd June 1934. Registered G-ACTD, it was entered by Shaw in the 1934 King's Cup Air Race and although fitted with full dual-control, was flown as a single-seater by Flt Lt Tommy Rose, the instructor of the Northamptonshire Aero Club at Sywell. It came in second at a very creditable 147.78 mph, reaching a maximum speed of 156 mph during the race. This was to be the first of many racing honours to be achieved by Miles aeroplanes.

The Aeroplane for 11th July 1934 gave the first details of the new machine:

The Miles Hawk Major which has just been produced by Phillips & Powis Aircraft (Reading) Ltd is a cleaned-up version of the Miles Hawk with a 130hp Gipsy Major and a top speed of 150 mph. It is intended for those who are prepared to pay for a higher performance. All the characteristics of the Hawk have been retained and the makers claim that the Hawk Major alights at 42 mph and is easy to fly. Structurally the Hawk Major is like its prototype and has a wooden fuselage and wings covered with plywood. The most obvious alteration is the new undercarriage with Palmer wheels and Bendix brakes, which gives the latest Hawk its characteristic appearance. Tail-trimming gear and rudder bias control is provided and Moseley "Float-on-Air" cushions are standard.

The special Phillips & Powis finish obtained with Titanine materials and already described in The Aeroplane *is standard on the Hawk Major. It is rather better than that on most modern cars. Complete with standard equipment, compass and usual instruments the Hawk Major costs £750 ex works.*

"Pontius" wrote of a flight he made in the new Hawk Major in *The Aeroplane* for 1st August 1934, under the title *The Pilot's Point of View (New Series) – XVI:*

After the prototype Hawk Major had covered itself with distinction in the King's Cup Race it was put at my disposal for a trial. I am not used to aeroplanes which have been specially cleaned up for racing, and was a little chary of forming an opinion on a trial of an actually raced aircraft. But I was assured that the

Hawk Major, as it is to be sold to the public, is to be identical with the one tested, except that it will have the 130hp Gipsy Major instead of the 120hp Gipsy III, and so will be rather faster. Therefore, people will be able to buy a real racing aeroplane without having to pay for all those little extras which are normally prohibitive.

I can imagine no more sporting aeroplane than this. The speed range alone is enough evidence not only of its suitability for the comparatively inexperienced amateur, but also its practicability as a racing or fast cruising machine. The speed range, by the way, as confirmed by me in flight, is 42 mph to 156 mph.

On the ground the Hawk Major is a very clean and aesthetically pleasing job. The special Titanine satin finish gives an admirably clean effect to the plywood, which, as is well known, is the standard Miles covering. External fittings and projections are almost completely absent. The neatly and fully trousered undercarriage legs harmonise well and blend into the general design, so that they look like part of the aircraft instead of a nuisance to the designer. The engine cowling, although thoroughly utilitarian, is neat and obviously efficient.

The tanks, which are in the wing roots, are accessible, and I am glad to see that their filler caps are well protected. Tank gauges are visible from the pilot's cockpit. The cockpits are made reasonably accessible by small doors on the starboard side. The pilot's cockpit is well arranged, and neither is in any way cramped, as might perhaps have been expected from the sleek external lines. The instrument panel is full, and the quoted price for the machine includes the whole of the equipment with which this particular machine was raced in the King's Cup, which you may assume was the best that could be got.

The Hawk Major has mechanical brakes operated through hand and foot controls of Mr Miles' own invention, which deserve a word of comment. The control is so arranged that by an initial hand movement the steering effect of the brakes is brought into play, without interfering with the full useful scope of the rudder-bar; and the brakes of both wheels are applied more and more by further hand action. The result of this is that the rudder-bar is not impeded or obstructed by the application of the brakes, and any desired degree of differential effect can be got merely by suitably setting the hand-lever.

Incidentally, the hand brake lever is commendably convenient and is near the throttle control. And I should mention that the rudder-bar itself and the stick are well placed, and the controls move freely and evenly, which is not always a feature of aeroplanes which fold. The Hawk Major handles well when being driven on the ground, and the fact that it has a tail-skid instead of a wheel seems not to affect turns of small radius, even on bad surface. The tail-skid appears to be very substantial, and one does not get the impression that it is being racked unduly in turning.

Actual flying is extremely easy. The take-off is short and the acceleration very rapid. There is no tendency to swing when the tail comes up. Elevator control is at all times absolutely positive.

Shortly after the Hawk Major first flew, Owen Cathcart-Jones (who in October 1934 came fourth in the MacRobertson Air Race in a DH.88 Comet) was given the opportunity to fly it and he sent an undated but glowing letter to Charles Powis:

Thank you so much for the opportunity you gave me yesterday of flying the Hawk Major. I cannot yet get over the fact of what a really marvellous aeroplane it is, and I went back to London last night full of genuine enthusiasm and encouragement, with the knowledge that at last we had made some real progress in British Aviation. To my mind the Hawk Major is so far in advance of anything that has yet been produced, that it represents at least 10 years' progress straight off, without taking into consideration the remarkable value for the low price.

I am writing to tell you how much I appreciate having the pleasure of at last flying a <u>real</u> aeroplane. Whether I am pleased with it or not, makes not the slightest difference to anybody, but if my opinion is of the least value, I am only too pleased to say that I do consider that you have now got an aeroplane which embodies all the delights of flying without any of the disadvantages. Apart from the really remarkable performance and low landing speed, the thing which strikes me more than anything else is the positiveness of control at all speeds which makes it suitable for instruction even in the "ab initio" stage.

You have in the Hawk, an aeroplane which gives instant confidence and ease of control and I doubt very much if anybody could find an easier aeroplane to learn on, as it appears to be entirely devoid of any vices; and I gave it a most exacting trial yesterday. I have

Tommy Rose in the cockpit of G-ACTD, the first Miles M.2F Hawk, seen at Hatfield (with its front cockpit faired over) for the King's Cup Air Race on 13th July 1934. It gained second place in the race at an average speed of 147.78 mph. [Via T Singfield]

written to Marsinah's two young friends in Scotland and told them that they would be most foolish to consider any other aeroplane - as this is my honest opinion - as the Hawk Major can give them more than any other aircraft yet built. You must excuse this rather long winded letter, but after all these years of stagnation in British Aviation, it is such a breath of fresh air to see something developed which we can all be proud of. I shall try hard to think of some sufficiently reasonable excuse to come down to Reading again soon to have another flight if I may?

Again very many thanks, yours sincerely, Owen Cathcart-Jones.

Flight for 9th August 1934, carried an advertisement: *The Miles Hawk – Flies More Miles at Less Cost.*

The safest and easiest machine to fly - does not float in the glide - simple and robust construction - wide track undercarriage - you can land anywhere and taxi in the highest of winds. Ultra modern and clean, attractive lines give the Hawk an outstanding performance. Standard model cruises at 100 mph. Major model cruises at 135 mph. The good range and pay load make the Hawk an ideal private owner's aeroplane.

Cirrus III 95hp model, complete	£450
Hawk Major with Gipsy Major 130hp engine, complete with Wheel Brakes	£750

Much was being written about the new aeroplane and an article appeared in *Flight* for 16th August 1934 entitled: *Saving Time - A Fast Journey on a Hawk Major: Some Impressions of its Characteristics*

Flying is primarily useful because it is faster than other means of transport, and, within limits, the faster a machine the more likely it is to find favour with users. Not long ago we experienced at first hand just how much speed means when flying down to the Bristol Club's meeting in the new Hawk Major (Gipsy Major). The wind was blowing strongly from the west and most visitors were full of stories of the time they had taken over the journey. It took us just 38 minutes. The journey home again was even more spectacular. After the Bristol meeting we went on to Cardiff Airport, the home of the Cardiff Aero Club, and then, flying high to make the best of the wind, reached Reading in 34 minutes.

The Hawk Major, it will be remembered, has only a Gipsy Major engine and the cockpits have in no way been skimped as to size, yet this machine cruises at a very easy 135 mph, and perhaps even more important still, does not land any faster than the standard Hawk - 42 mph. Figures such as these make the idea of aeroplane ownership most attractive.

Although it is an open cockpit machine, the Hawk Major is particularly free from draughts, and there is no need to wear goggles except when visibility is poor and the pilot wishes to "crane out" rather more than usual. The clean design, largely consequent upon making the undercarriage a full cantilever one, gives the new Hawk a long flat glide if brought in fast, but, like its predecessor, the machine can be glided quite slowly with perfect safety and under perfect control so that landing in small fields has no terrors. Furthermore, the wheel brakes can, if necessary, be used hard.

In July 1934, Miles began experimenting with fixed aluminium flaps screwed to the underside of the trailing-edge of the wings of a Miles M.2D Hawk Three-seater and it would also appear from entries in Miles' log-book that fixed flaps were also fitted to other Hawks about this time for further trials. These flaps could easily be bent to the desired angle and it soon became apparent that the use of flaps resulted in a steep angle of glide, increased coefficient of lift and consequently a much reduced landing speed. Following the success of the trials it was decided to fit manually-operated split trailing edge flaps to production M.2F Hawk Majors as an optional extra.

C G Grey, editor of *The Aeroplane*, wrote of these flaps in the issue for 22nd August 1934:

Mr Miles' own particular version of the split trailing-edge flap is of special technical interest. It consists of a series of very narrow flaps (about 7-inch chord) below the wing and fuselage, which are hinged seven or eight inches in front of the trailing edge and are worked by a torque-tube which can be twisted from the cockpit. Usually when such flaps are pulled down the angle of the tail-plane has to be adjusted to counteract the movement of the centre of pressure, that is to prevent the machine from becoming nose-heavy.

The Miles flaps get over the difficulty by sheer good luck - according to their inventor. At any rate the various tests made at Reading prove that no tail-trimming is needed, for there is none on the machine, and I have seen Mr Miles flying circuits with his hands held up over his head. He says that if, as in other types, split-flaps do not extend under the fuselage, when they are depressed there is a flow between the inner ends underneath the fuselage and outwards towards the wing tips. The swirls from these inner ends, spreading outwards, actually pass outside the tail-plane. Thus the tail-plane is working in air which is flowing as it would were the flaps not down, and after a certain angle of flight the air passes away from the tail-plane on each side, and even the elevator fails.

Therefore, Mr Miles says, the flaps should extend right across the underside of the fuselage. Then there is no change of flow under the most important part of the wing, its centre. Also the tail-plane is working in air the flow of which has been strongly deflected downwards by the flaps. And, furthermore, the air flows inwards from the outer ends of the flaps towards the tail-end of the fuselage. The effect of this is that a down-load is automatically put onto the tail, which compensates for the backward movement of the centre of pressure caused by depressing the flaps. Thus the machine remains in balance, and the tail-plane and elevators are always working in a concentrated stream of air, which gives full control by elevator and rudder at angles of flight, which seem impossible.

The writer has watched Mr Miles flying circuits of the aerodrome at Woodley, taking off at about 35 mph and landing a couple of yards farther on, taking off and landing again, and doing landings in that way over and over again all around the aerodrome, - half a dozen times at least. He has also seen Mr Miles come in at 900 ft, put the nose of the machine down to an ordinary gliding angle, and sink under perfect control all the time, so that he landed short of the circle in the middle of the field.

The take-off and flying-speed and landing are about those of a pre-war kite, but the lateral and longitudinal controls are perfect. I agree with Mr Miles when he says that no sane person who exercises ordinary intelligence could possibly get into trouble with the arrangement of flaps. The first production machine is now going through. Its performance will cause a commotion wherever it is seen, and further development of the flaps should have far reaching effects.

In fact the split-flaps, which were offered as an optional extra on the M.2F Hawk Major, when fully opened, formed a continuous strip along the trailing edge. This projected almost vertically downwards from the bottom surface of the wing and therefore had a tremendous braking effect. It was also claimed that the piece beneath the centre section had an effect which was peculiar to itself, which was that it broke up the air flow on the tail surfaces causing turbulence and giving increased elevator and rudder control at speeds at and below the stall. The spilt flaps also increased the aileron control at and below the stall and had an effect of lowering the stalling speed so that the stall was not clearly defined. A Phillips & Powis advertisement for the Hawk Major appeared in *The Aeroplane* for 24th October 1934. This gave details of these flaps and is reproduced on the next page

The secret of the Hawk Major's popularity with private owners was that it was a light open two-seater, which combined a very high performance as regards speed, range and load carrying capacity with the exceptional low price – ex-works and ready for flight - £750. The Miles Split Trailing Edge Flaps were offered as an optional extra for an additional £50.

The MILES "HAWK MAJOR"
FLAPS FOR EFFICIENCY

Specially designed flaps on the trailing edge of the wing give the Hawk a sensational performance. The flaps reduce the landing speed and decrease the take-off and landing, and permit safe and practical operation from much smaller fields.
The flaps are lowered or raised in one second by a positive push-pull control inside the cockpit—no winding.

TAKE OFF NORMAL LOADING

	Time	Run
Without flaps	8·5 secs.	100 yds.
With flaps	7 „	64 „

PERCENTAGE IMPROVEMENT .. **36%**

LANDING RUN NORMAL LOADING

Without flaps or brakes.. .. 205 yds.
Without flaps but with brakes.. 120 „
With flaps and brakes .. .. 69 „
Percentage Improvement (without flaps but with brakes), 39%
Percentage Improvement (with flaps and brakes), 68%

PHILLIPS & POWIS AIRCRAFT (READING) LTD.
MANUFACTURERS OF "HAWK" AIRCRAFT & "CIRRUS" ENGINES.
READING AERODROME, WOODLEY, READING   Phone: SONNING 114-5.

Details of the new Miles split flaps appeared in an advertisement in 'The Aeroplane' for 24th October 1934.

The Hawk Major G-ACWV was delivered to Phillips & Powis on 1st September 1934 for use as demonstrator and this was almost certainly the machine used by Captain Neville Stack for his record breaking return flight to Denmark on Saturday September 8th. He flew from Woodley, via Heston, to Copenhagen, resting for just 45 minutes before returning via Heston the same day. He wrote to the firm expressing his thanks: *I am very impressed with the Miles Hawk Major and consider it to be the most outstanding light aeroplane I have ever flown. Its range of speed, controllability at all speeds down to the 'stall' and in bumpy weather, are remarkable, making it altogether the most pleasant and comfortable aeroplane to fly over long distances. It was particularly noticeable under exceptionally bad weather conditions, that though the machine was very badly thrown about there was no movement on the control column, but the machine responded the moment the stick was moved.*

For an open machine I found the cockpit most comfortable and when flying for nearly 10 hours at a stretch which included a certain amount of blind flying, I felt most comfortable and not at all fatigued. The machine was flying perfectly "hands off" which of course made blind flying much easier. The take off and landing speeds are remarkably low and I found no difficulty in landing with an absolutely full load, that is to say, fuel for 1500 miles, with a perfectly normal run, the same applied to the take off. The machine to me seems to be very robust and well made, and I have no hesitation in saying that I consider it to be the best light aeroplane I have ever flown. Yours very truly, T Neville Stack.

Another Hawk Major, ZK-ADJ, named *Spirit of Manawatu*, was entered by the Manawatu Aero Club in the 1934 MacRobertson England to Australia Air Race from Mildenhall to Melbourne, a distance of 11,333 miles. On 14th October 1934, Sqn Ldr Malcolm C MacGregor and Henry C Walker flew it to Mildenhall for the start of the race, which began at 6.30am on 20th October 1934. The Hawk Major arrived in Melbourne on October 27th, to be credited with the fastest time for a single-engined aircraft in the race, establishing a record time for a light aeroplane from England to Australia of 4 days 22 hrs 5 mins 46 secs (or 5 days 15 hrs 13 min depending upon which reference source is used). It came in 5th overall (in the handicap section) with an average speed of 105 mph and an elapsed time of 7 days 14 hrs 58 min.

Lady Blanche Scott Douglas (eldest daughter of the 9th Duke of Beaufort and widow of Capt George Scott Douglas) was F/O Cecil Victor Ogden's first pupil pilot after he joined the Bristol & Wessex Aeroplane Club as assistant instructor (in his spare time) in May 1934. Ogden was a test pilot with the Bristol Aeroplane Company, having joined them in early 1934. Lady Blanche soon gained her 'A' licence and in September 1934, after she had only been flying for about six to eight weeks, she bought a new Hawk Major, G-ACWY, in order to visit her friend, the Maharajah of Cooch Behar in Bengal, India! She invited F/O Ogden to accompany her as co-pilot on what was to become quite an adventurous flight.

The Bristol Evening Post for 16th August 1934 reported: *Lady Blanche told an Evening Post reporter today: 'I am making the flight simply for my own amusement. I have only been flying about six weeks or a couple of months. Flying Officer Ogden, who is an instructor of the Bristol & Wessex Aeroplane Club, and I will take turn and turn about at the controls. We shall take about three weeks over the flight, stopping at Cairo, Baghdad, and Karachi. My 'plane, which is now being built, will be a Miles Hawk Major, an open monoplane. We shall not carry wireless. We hope to start on November 20th, but have not yet decided whether it will be from Bristol'.*

F/O Cecil Victor Ogden (who hated his Christian name 'Cecil' and insisted on being called 'Micky'), later rejoined the RAF and flew Beaufighters in the North African Desert, finishing the war with the rank of Wing Commander. I was fortunate to meet him at Henstridge in April 1997 and talk to him about their epic flight. 'Micky' Ogden died on 21st October 2002 aged 99 years and in his obituary, which appeared in the *Daily Telegraph* for 13th December 2002, more details of their flight to India emerged:

The pair set off on November 20th 1934 (from Whitchurch) *in Lady Blanche's new Miles M.2F Hawk Major. They followed the route of the historic London to Australia air race, taking them through France, Italy, Greece, Palestine, Cairo and Baghdad. Continuing along the Persian Gulf by way of Bushire, the couple encountered a severe sandstorm, but managed to climb above it. Then, half way to Bandar Abbas, their plane developed engine trouble, and they had to make a forced landing in the desert.*

Now Lady Blanche and Ogden were marooned in the wastes of Persia. It was 10 days before they were rescued by local tribesmen, after which they made serviceable the engine of their plane and took off for Karachi. Further misfortune, however, awaited them at Bandar Abbas where, on landing, the aircraft ran into a hole and tipped on to its nose. Although neither Ogden nor Lady Blanche was injured, the propeller was badly damaged. Rather than incur the long delay from waiting for a replacement propeller, the couple took a considerable risk: they decided to fly on to Jask, where a new propeller was brought in from Karachi aboard a KLM airliner. The remainder of their journey to India was uneventful.

Lady Blanche Scott Douglas and F/O 'Micky' Ogden returned to Whitchurch on 16th February 1935.

Capt Neville Stack with the Miles M.2F Hawk Major G-ACWV, which he used for his famous record-breaking flight from Heston to Copenhagen and return on 8th September 1934. [Via B Clarke]

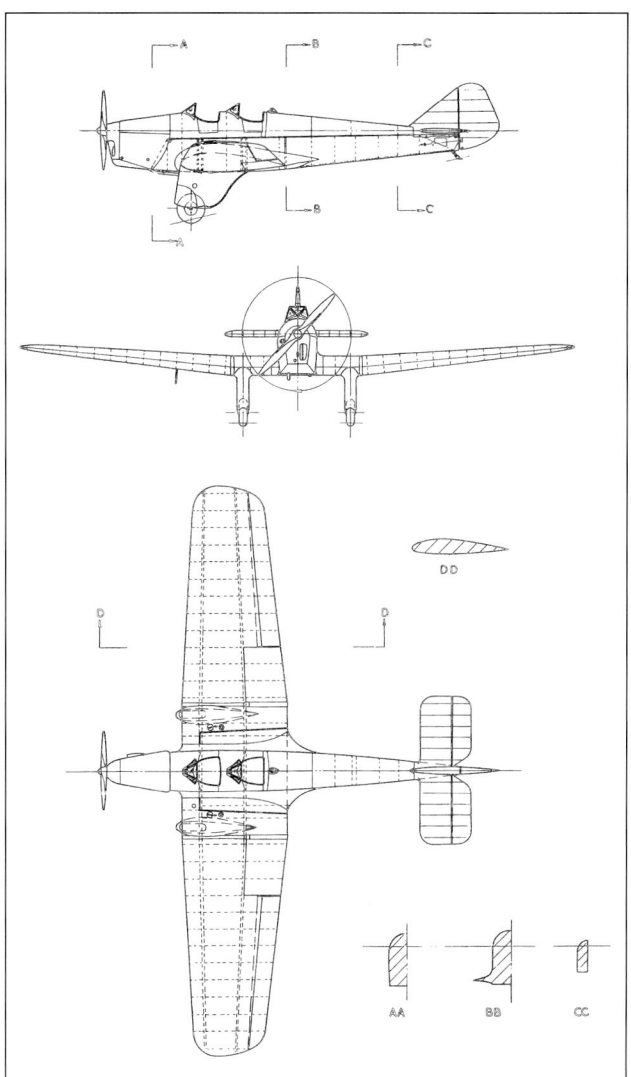

Miles M.2H Hawk Major.

M.2G

The one-off three-seat M.2G, G-ACYB, was in the nature of a hybrid aircraft. Ordered by the Club George Chazez in Switzerland in August 1934, it had an undercarriage similar to the M.2A Hawk (ie - strutted and with no trousers), a 130hp Gipsy Major engine mounted on metal engine bearers similar to the M.2F and a cabin over the rear two seats. It is not known if it had flaps or whether the wings were from an M.2 Hawk or the M.2F Hawk Major.

M.2H

After 18 M.2F Hawk Majors were built, they were replaced on the assembly line in November 1934 by the M.2H Hawk Major. The M.2H became the main production version of the Hawk Major and after the first few had been delivered, the design of the undercarriage fairings was modified (in two stages). The M.2H was fitted with manually-operated split trailing-edge flaps as standard (probably the first production aircraft of any type to be fitted with flaps as standard) and these were operated by a lever on the right-hand side of the cockpit, which moved over a notched quadrant of four positions, including full-down and closed.

'HAT' (believed to be H A Taylor) was given the opportunity of a demonstration of a 'flapped' Hawk Major and he wrote of his experiences in *Flight* for 13th December 1934, under the heading of 'Dual Personality': *The Flapped Hawk Major from the Amateur's Point of View - A Viceless Slow Approach Coupled with a High Cruising Speed.*

The almost incredible stories about the gentle approach of the Hawk Major with split flaps suggested the possibility that such performances were all in order for expert hands, but that the averagely competent amateur would not be encouraged to play similar games.

When, therefore, the opportunity of a demonstration occurred, through the good offices of Aircraft Distributors Ltd, I leapt at the chance while preparing myself for a serious disappointment and for the memorisation of a long list of 'don'ts'. As it happened, I discovered the only 'don't' for myself and the stories were found to be far from incredible.

By accident or otherwise, the Miles split flaps not only reduce the take-off run by a clearly perceptible distance, steepen the glide and reduce the landing speed, but they actually cause the lateral control to 'stay-put' in proportion to the lowered landing speed. And all that cannot sound nearly so much of a fairy story as the fact appears in practice. As far as I could gather, the rudder faded out first - an occurrence which will comfort those who look on the rudder as the father of all spins - and the fore and aft control remained to the bitter end, with the slipstream to help.

To one whose experience has been largely confined to conventional biplanes and high-wing monoplanes, the flapped Hawk Major requires a strange, but not difficult, technique of approach. Instead of a pronounced and obvious stall, there is an imperceptibly increasing sink at speeds below 55 mph, and the final approach speed must be gauged with this fact in mind. My 'don't' discovery was: 'Don't make full use of the characteristics if you value your undercarriage' - though a last-minute burst of throttle will save it. The latest models, incidentally, have air-wheels which will allow, perhaps, a little more latitude.

From the point of view of the comparatively inexperienced, the flaps allow the pilot (a) to come into a small aerodrome without the least chance of floating into the opposite hedge (the Hawk Major is a clean machine, as its 150 mph indicates); (b) immediately to vary the ratio of speed and sink as the boundary appears too near or too distant; and (c) to 'rumble', if necessary, on the flap control, which has only four positions, including 'full down' (which I did not use) and 'closed'. Normally, the control is put in the first notch for a quick take-off, in the same notch when the gliding speed has been pulled down to 80 mph, and in the second notch at 70 mph or thereabouts.

Apart from the fact that the lever is operated by one's stick hand, there are no difficulties. The nose drops pleasantly at each movement, and the machine takes up a 55 mph glide with the flaps down and without readjustment of the tail trimming lever. The outboard flaps drop down to 45 degrees in the second position, and the centre flap opens in the reverse direction - a system of balancing which accounts for the ease of operation.

The amount by which the actual glide is steepened varies, of course, with the speed, but it is out of all proportion to the size of

The Miles M.2G Hawk Major Three-seater, HB-OAS in September 1934. Note also the Hawk CH-380 in the background. [Via B Clarke]

The Miles M.2H Hawk Major G-ACYX at Broxbourne in August 1946. [Via B Clarke]

the flaps themselves. Some indication of this can be obtained from the fact that the loss of a thousand feet occupied a minute longer with the flaps up and that, when they were opened in level flight at 75 mph, the speed dropped to 60 mph, though it was difficult to make quite sure that the machine was not climbing slightly. The approach at 55 mph is a great deal steeper than that of a conventional training biplane and on my first two 'flapped' approaches, I undershot quite considerably.

Speaking generally, the outstanding impression was the difference, so to speak, between an aeroplane which was definitely travelling very fast and one which was floating down restfully and under full control. The flapped Hawk Major, in fact, possesses a dual personality.

William Fairweather, an agricultural and motor engineer, of Muzaffarpur in Bihar, India replaced his Cirrus-engined Hawk VT-AFD with a second-hand 1935-built M.2H, G-ADFC/VT-AKG, in November 1937 and had it shipped to India, completely erected, as personal baggage, in the hold of the ship he travelled in. On arrival in Bombay he flew it 1,200 miles to his home in just over 9 hours.

On 31st May 1938, he flew it the 320 miles to Calcutta in 2½ hours and also made several flights to his customers, landing on grounds barely 300 yards long. After the war, he wrote to Miles Aircraft Ltd about his wartime experiences in India and his story was published in the *Miles Magazine* for October 1946:

He flew the Hawk Major widely throughout India on business up to the outbreak of war, using it doing his 'Flying Boxwallah' round India. His age and eyesight prevented him from joining the RAF, so all he could do for the war effort was to give joy rides to raise funds for the Red Cross and drop leaflets on behalf of War Savings Weeks. In January 1942 the Japs were on the borders preparing to invade India and at that time he accepted a commission as a Squadron Leader to site airfields. When they saw that he was 48 years of age and wore spectacles he was told that the Service would purchase the Hawk and provide a pilot to fly him around! This was not acceptable to him so it was agreed that he would fly as a civilian at his own risk and maintain the aircraft himself. He was then given S/L's pay, plus for every flying hour for the use of the Hawk. He sited airfields around Calcutta and down the coast and throughout the South of India.

There were no existing airfields suitable for modern aircraft and in the stretch of 1,200 miles between Calcutta and Madras there were only three grass airfields, one 350 yards and the other two 750 yards. For two years he flew the Hawk 'here and there with no real servicing, except changing the oil.' It lay out in the rain, wind and sunshine; it landed on any old patch of ground or even roads. He camouflaged it with ordinary oil paint as nothing else was available. Odd minor damage was repaired on the spot by the local villagers from any material he could get. The Hawk became well known throughout India as it was 'the only one in the land.'

For three years and ten months he was on this job and flew 1,400 hrs in all weathers; not once was there a failure of either engine or airframe, which speaks highly for British manufacture. In May 1945 he thought he had lost the Hawk when the Service hangar in which he kept it at Bangalore collapsed during a cyclone and fell on top of it. The Hawk's complete top decking, windscreens and a top longeron were smashed; the weight of the hangar squashed the tyres practically flat and it looked like a write-off. However, within a fortnight, 'with the help of plywood and glue' she was in the air again and he flew her back to Muzaffarpur on the termination of his services. VT-AKG had flown over 1,800 hrs by this time, so it was then overhauled and re-painted in civilian colours (silver with red trim and named 'Robin' - PA) *as the heavy, rough camouflage paint had reduced the cruising speed to 120 mph.*

While in Bangalore it was examined by several structural experts sent out from England to enquire into the causes of wooden structure failure in other aircraft. It was found that inside was as good as new and the plywood covering to be in excellent condition. The engine was still giving 18 mpg and needing only minor replacement parts, even though at times, of necessity, any kind of lubricating oil had been used as had all grades of petrol from 67 up to 100 Octane.

Fairweather continued to operate VT-AKG right up until Indian independence in 1947, when he made his last flight in the Hawk Major, which was recorded for posterity in colour on a cine home movie, before returning home to England. He left it behind in airworthy condition but its subsequent fate is unfortunately not known.

Fourteen Hawk Majors were impressed into the RAF at the outbreak of war, and an HQ 41 Group Return (Ref: AIR 17/33) entitled 'Arrivals and Despatches for Week Ending and Stock in 41 Group ASUs at 0900 hours 6-12-45', gave the numbers of Hawk Majors surviving as follows: *5 Stock as at 0900 6-12-45; 3 Awaiting MAP disposal instructions. Non-Effective Aircraft. These non-effective aircraft require maintenance within 41 Group units.* (They probably comprised DG590 (G-ADMW), NF752 (G-ACYO) and NF750 (G-ADWT) - PA). *2 Awaiting/Undergoing Breakdown.*

Rather surprisingly, it has only been possible to account for 3 'Hawk Majors' being 'in stock' and qualifying for inclusion in this Return at this date. However, it is known that 4 'Hawk Majors' were still surviving at this time and these included: DG590 (G-ADMW), which was sold at the first sale of impressed light aircraft held at 5 MU Kemble in 12.45, NF752 (G-ACYO), which was sold at the same sale; and the Hawk Trainer II s/n NF750 (G-ADWT), also sold at the same event. One other Hawk Major, HL538 (G-AEGE), was at Watchfield between September 1943 and August 1947 (and was sold in 1948), so this couldn't have shown in this Return. It is not known which the other two 'Hawk Majors' that had survived to this date to be 'A/U Breakdown' were.

No Miles M.2F Hawk Majors survive and of the 43 M.2H Hawk Majors built, just three survive intact; one in store in Canada, one being restored to static display condition in Brazil and one at RAF Stafford with the RAF Museum Reserve Collection & Restoration Centre. This is subject to very low priority restoration to static display condition. However, one 'new' Miles M.2H Hawk Major has been built in Brazil from scratch using local indigenous timber, and the metal components from the original Uruguayan aircraft.

Apart from the production M.2F and M.2H series, a number of other variants are known to have been constructed. However, there is no record of an M.2I, and it is suspected that this wasn't used.

M.2J

A photograph, described by Don Brown as an 'unidentified two-seat cabin Hawk Major', appeared in *Miles Aircraft since 1925*, and this showed an otherwise anonymous Hawk Major with a large cabin and sliding hoods over both (or possibly three?) cockpits and a long streamlined top decking tapering aft to where the leading edge of the fin meets the fuselage. It appears to be painted in a dark colour with a natural metal cowling and has an M.2H type trousered-undercarriage (see photograph on page 129). Even after all these years of research, it has not been possible to positively identify this particular machine but it could be the one-off M.2J Hawk Major variant, only otherwise recorded by R A Saville-Sneath in his *British Aircraft Volume 1 (*although he stated that the M.2J was a three-seat open-cockpit version).

M.2M Hawk Major Three-seater

The M.2M was a three-seat version of the M.2H but with a cabin over the rear two cockpits. The first of two built was delivered in December 1934.

M.2N

We have to thank R A Saville-Sneath again for publishing the only reference to the M.2N Hawk Major, which is to be found in *British*

William Fairweather's Miles M.2H Hawk Major after the hangar collapse at Bangalore on 15th May 1945. [Via P H T Green Collection]

Aircraft Volume 1. He stated that the M.2N was, like the M.2H, 'fitted with Miles split trailing-edge flaps' and it is now known to have been the modified Hawk Major, G-ADGE.

M.2P Hawk Major de luxe

The next model to appear was the M.2P Hawk Major de luxe which was based on the M.2H Hawk Major and an advertisement in *The Aeroplane* for 31st July 1935, announced:

The de Luxe Miles Hawk Major (Gipsy Major) - Not only has new plant been installed, new methods of construction adopted, freeing the machine from all maintenance troubles, but the following alterations in design and detail improvements have been made:-

1. *Larger cockpits comfortably upholstered in real leather, fuselage 3 ins wider.*
2. *Increased wing span giving 6 sq ft extra wing area.*
3. *New design front bulkhead, reducing the engine vibration to a minimum.*
4. *Hydraulically operated flaps, in place of mechanically operated.*
5. *Specially finished instrument panel, including map tray and map pocket.*
6. *New type brake lever with quick release ratchet.*
7. *Headrest fairing.*
8. *Larger cockpit doors with new type fasteners.*
9. *Special finish on painted surfaces.*
10. *Improved type tail trim lever.*
11. *Improved position of compass.*
12. *New type control system, including longer stick. Controls beautifully balanced.*
13. *Larger luggage locker.*
14. *Improved and strengthened methods of attaching fairings etc.*

Price £810 – Flaps £60 extra

In fact the wingspan of the M.2P was increased to 34ft 0in and wing area to 174 sq ft, while the AUW was increased to 1,900lb. The first of five M.2P Hawk Major de luxe aircraft to be built was delivered in July 1935.

M.2R Hawk Trainer

The M.2P was followed in August 1935 by the M.2R which, although often also referred to as a Hawk Major de luxe, was really the first Hawk Trainer and was later referred to as the Hawk Trainer Mk.I. It was similar to the M.2P but with minor modifications. Four were built but as they were really the first of the Hawk Trainer series they are detailed separately in Chapter 10.

M.2S

The M.2S was a special single-seat, long-range version of the M.2H with a small sliding cabin hood and one was built for a Nigerian customer living in England, who based it at Broxbourne in Hertfordshire.

Its fate is described in the story of an ambitious but tragic attempt by one Mr Govind Nair, a 30-year old native of Travancore, India, which shows how not all hopelessly inexperienced, potentially world-travelling pilots got away with it. The very sorry tale, which appeared in *The Aeroplane* for 3rd November 1937, was probably written by C G Grey as it has his air of cynicism about it:

Mr Nair left Croydon on 28th October for an extended tour which was to include parts of America and at least two ocean-crossings. He crashed fatally near Rouen about two hours later. This end to the adventure was no real surprise to those who knew the circumstances. Mr Nair announced that he had 200 hours' solo experience, but in this country at least his flying seems to have been extremely local and intermittent. He took tuition at Brooklands Aero Club, with Phillips & Powis and the Herts & Essex Club, but there seems to be doubt whether his instructors at any of these had ventured to let him fly across country. He had probably flown as much as some who then did great flights, but if his abilities were exceptional they were so in a sense which boded ill for his plans.

By the help of compatriots he had bought a Miles Hawk Major and had extra tanks installed. Earlier and more ambitious plans came to nothing because he was refused permission to take off across the North Atlantic at this season with just enough petrol to reach Newfoundland in still air (a previous attempt to cross the Atlantic by the aeroplane's previous owner had also been thwarted by the authorities, probably for similar reasons - PA). As a compromise he proposed to do the Westward trip over the South Atlantic. He was uninsured for obvious reasons. Someone had remarked to one underwriter that a premium of about 95% would be needed. The underwriter replied that he wouldn't look at the business at any premium under 120%. After the event these pleasantries seem callous, but the late Mr Nair did nothing for aviation or its people which entitles him to even the usual allowances made post mortem to human fallibility.

Miles M.2H Hawk Major G-ADMW was one of the few to survive the war and be restored to the register. This photograph shows it at Croydon in the 1950s. [Via P Amos]

The mystery Hawk Major, possibly the Miles M.2J Hawk Major two/three-seat cabin, VR-RAH. [Via P Jarrett]

LN-BAH, one of the two Miles M.2M Hawk Majors built, seen at Woodley before it departed on delivery in December 1934. It tragically crashed at Kjeller, near Oslo, while on its delivery flight, on 10th December 1934. [OG Nordbø via B Clarke]

Left: G-ADGE, the sole Miles M.2N Hawk Major Deluxe 'Cook's Special', seen in 1935.
[A J Jackson Collection]

Right: This advertisement refers to the Miles M.2P Hawk Major Deluxe.

THE AEROPLANE　　Advertisements 18　　JULY 31, 1935

NEW!!

The DE LUXE MILES HAWK MAJOR
(GIPSY MAJOR ENGINE)

Not only has new plant been installed, new methods of construction adopted, freeing the machine from all maintenance troubles, but the following alterations in design and detail improvements have been made:

1. Larger cockpits comfortably upholstered in real leather, fuselage 3 inches wider.
2. Increased wing span giving 6 sq. ft. extra wing area.
3. New design front bulkhead, reducing the engine vibration to a minimum.
4. Hydraulically operated flaps, in place of mechanically operated.
5. Specially finished Instrument Panel, including map tray and map pocket.
6. New type Brake Lever with quick release ratchet.
7. Headrest fairing.
8. Larger cockpit doors with new type fasteners.
9. Special finish on painted surfaces.
10. Improved type Tail Trim Lever.
11. Improved position of Compass.
12. New type control system, including longer stick. Controls beautifully balanced.
13. Larger Luggage Locker.
14. Improved and strengthened methods of attaching fairings, etc.

Price £810

FLAPS £60 EXTRA

Manufacturers:
PHILLIPS & POWIS AIRCRAFT LTD.
READING AERODROME - - BERKS.
Telephone: Sonning 114 (4 lines).

His take-off at Croydon was seen by many besides those of his compatriots who went to wish him well and knew no better. A pilot asked, as the Hawk staggered round the sky almost fully stalled, how many hours the pilot had. When told 200, he asked; "And how many more do you think he will have?"

The estimate, in reply, "About two," was grimly accurate. The manner of his progress apparently did not change, because a pilot of an air liner is said to have reported by radio that a private aeroplane had been seen over the Channel obviously in trouble, and flying in a stalled attitude. The wreck of the machine and the death of the pilot apparently came about through stalling when trying to land in a field.

M.2T

Two Hawk Majors were fitted with Blackburn Cirrus Major R engines for the 1935 King's Cup Air Race and designated M.2T. Although designed for racing, both these engines let their pilots down and neither aircraft completed the course. The one flown by Alex Henshaw, who had been given his engine free by Blackburn in order to test it under racing conditions, was forced to ditch in the Irish Sea after crankshaft failure caused the propeller to part company from the engine, while the other, flown in the race by F D Bradbrooke had to force-land near Blackpool.

For details of M.2 Hawk Major production see Appendix 11.

SPECIFICATIONS

The following data was taken from an undated Specification and Performance leaflet but which depicted at its head a drawing of the prototype Miles M.2F Hawk Major.

CONSTRUCTION	Wooden Fuselage and Wings ply covered.
ENGINE	130hp Gipsy Major (actually a Gipsy Major I Series I but the prototype M.2F had a 110hp Gipsy III - PA).
STANDARD INSTRUMENTS	Compass.) Airspeed Indicator.) Revolution Indicator.) Mounted on Aneroid.) Instrument Board Oil Pressure Gauge.) in Rear Cockpit. Inclinometer.)
STANDARD EQUIPMENT.	Dual Control. Palmer Wheels and Tyres - 600 x 100, or Airwheels - 480 x 180; Bendix Wheel Brakes, with the 'P&P' special Interconnecting Rudder Bar Control and separate Hand Lever. New Type Single Strut Undercarriage of robust design; this undercarriage reduces the drag and improves the general efficiency of the machine considerably. Tail Trim and Rudder Bias Control. Deep Luggage Locker. Nickel-plated Control Levers. Map Shelf. 'Float-on-Air' Cushions. Cockpit specially finished. Cockpit Doors. Engine, Aircraft and Journey Log Books. Cockpit Covers
STANDARD COLOUR SCHEME.	Tail Plane, wings rudder and fin, Silver. Fuselage - Any single colour scheme, using the following colours, is offered without extra charge:- Blue, Green, Black, Grey, Dark Red and Silver.

NOTE: The Hawk Major is offered to the Private Owner who requires a machine of high performance and yet retains all the characteristics of a slower machine. This model lands at 42 mph and is exceptionally easy to fly; the high standard of finish through the machine, coupled with the really good performance, makes this model at the price it is offered, a very attractive proposition.

PRICE, with Gipsy Major Engine - Landplane ex Works, ready for flight . . . . **£750.**
        Miles Split Trailing Edge Flaps . . . . **£50** Extra.

TERMS OF PAYMENT: 10 per cent deposit when order is placed, remainder in cash on completion of the aircraft or against shipping documents in London. We reserve the right to alter the above from time to time. All sales are made subject only to the Company's Special Warranty.

Performance at:	**Normal Weight 1,650 lb**	**Maximum Weight 1,900 lb**
Max Speed at 1,000ft	150 mph	149.5 mph
Cruising Speed at 1,000ft	130-135 mph	129-134 mph
Petrol Consumption at Cruising	6 gall/hr	6 gall/hr
Range (32 gallons)	700 miles	695 miles
Landing Speed	42 mph	45 mph
Take-off Run (5 mph wind)	95 yd	160 yd
Take-off Run with Flaps	80 yd	140 yd
Landing Run with Brakes	95 yd	140 yd
Landing Run with Brakes & Flaps	70 yd	110 yd
Rate of Climb at Sea Level	1,300ft/min	1,000ft/min
Max Ceiling	20,000ft	18,500ft
Service Ceiling	18,000ft	16,500ft
Best Gliding Speed	60-65 mph	96-105 mph
Best Climbing Speed	65-75 mph	105-120 mph

NOTE: The performance figures are guaranteed to within 5 per cent., under normal conditions in this country, and the figures stated can be obtained by pilots of average ability; in each case these figures have been exceeded by our own Test Pilot. When machines are to be operated abroad, due allowance should be made for local climatic conditions and the altitude of the aerodrome used. The fuel consumption and the range are based on the engine manufacturers' claimed figures.

Weights:
Tare weight 1,150 lb; pilot 160 lb; petrol (32 gal.) 246 lb; oil (3 gal.) 28 lb; payload 316 lb.
 (If only 16 gal. petrol carried the payload can be increased to 441 lb)
Normal AUW - pilot, passenger and 20 gal. petrol 1,650 lb; payload 316 lb; disposable load 750 lb.
Loaded weight 1,900 lb

Overall Dimensions:
Length overall - 24ft; Height overall – 6ft 8in; Span - overall 33ft & with wings folded 13ft 10in.

General Data:
Incidence - 1½ deg. to thrust line. Dihedral - 5 deg. Wing section - Miles "P & P". Mean wing chord - 5ft 2in. Wheel track - 6ft. Locker capacity - 5½ cu ft. Wing area - 169sq ft. Area of ailerons - 17.4sq ft. Area of tailplane - 17.2sq ft. Area of elevators - 8.4sq ft. Area of fin - 2.24sq ft. Area of rudder - 7.24sq ft.

A later sales leaflet, showed the **M.2P Hawk Major de Luxe** G-ADLO, and this gave slightly different figures for the Specification and Performance of the Hawk Major. The most noticeable change concerned the wing span which was now stated to be 34ft and the wing area which is shown to have increased to 172sq ft. The 'normal' all-up weight had increased by 70 lb to 1,720 lb.

The construction and engine details were similar to the previous Hawk Major specification with the exception of standard instruments which were now mounted on leather-bound instrument board in the rear cockpit and engine switches and flap indicator were fitted as standard. The Standard colour scheme remained the same.

The Standard Equipment was as previous but with the following changes:
 Miles Hydraulic Split Trailing-Edge flaps.
 Extra large luggage locker.
 Cockpit comfortably upholstered in real leather.
 Improved type tail trim and rudder bias control.
 Leather-covered map tray and pocket.
 Large cockpit doors with newest fasteners.

Performance at:	Normal Weight 1,720 lb	Maximum Weight 1,900 lb
Max Speed at 1,000ft	150 mph	149.5 mph
Cruising Speed at 1,000ft	130-135 mph	129-134 mph
Petrol Consumption at Cruising	6½ gall/hr	6½ gall/hr
Range (32 gallons)	660 miles	655 miles
Landing Speed	42 mph	45 mph
Take-off Run (5 mph wind)	95 yd	160 yd
Take-off Run with Flaps	80 yd	140 yd
Landing Run with Brakes	95 yd	140 yd
Landing Run with Brakes & Flaps	70 yd	110 yd
Rate of Climb at Sea Level	1,300ft/min	1,000ft/min
Max Ceiling	20,000ft	18,500ft
Service Ceiling	18,000ft	16,500ft
Best Gliding Speed	60-65 mph	96-105 mph
Best Climbing Speed	65-75 mph	105-120 mph

Weights and Overall Dimensions:

A few other changes were apparent between the two leaflets. Tare weight had increased to 1,210 lb which reduced normal payload to 246 lb. The overall span increased by one foot to 34ft and the wing area to 172sq ft.

There is still some confusion over the wing span and wing area for the M.2F-M.2P Hawk Majors and later M.2W-M.2Z Hawk Trainers. The span and wing area of the M.2F Hawk Major prototype was quoted by Phillips & Powis as 33 ft 0 in and 169 sq ft respectively but by *The Aeroplane* at about the same time as span 33 ft 9 in and wing area 169 sq ft. A photograph of the M.2P was shown in the undated Phillips & Powis sales leaflet where the quoted span is given as 34 ft 0 in and wing area of 172 sq ft. The later M.2W is also recorded having a span of 34 ft 0 in and wing area of 174 sq ft, whereas the even later M.14 development had a span of 33 ft 10 in and a wing area of 176 sq ft. The folded span of 13 ft 10 in remained the same for all Hawk variants (but the wing did not fold on the M.14).

The Specification and Performance of the **M.2M Hawk Major Three-seater** varied from the standard and is quoted as follows:
Weights: empty (including all standard fixed equipment) 1,200 lb; pilot 160 lb; petrol (33 gal) 250 lb; oil (3 gal) 30 lb; payload 335 lb (if 16½ gal petrol carried the payload can be increased to 360 lb); loaded 1,975 lb; disposable load 775 lb
Performance: normal load: max speed at 1,000ft 150 mph; cruising speed 135 mph; landing speed 42 mph; take-off run 90 yd; landing run with brakes 80 yd; initial rate of climb 1,300 ft/min; ceiling 20,000ft; petrol consumption 6 gal/hr; flight duration (33 gal) 5½ hr; range (33 gal) 687 miles; oil consumption 1.75 pt/hr.

Two Miles M.2R Hawk Trainer Mk.Is taking off from Hatfield in the 1935 King's Cup Air Race. ['Flight' via B Clarke]

Chapter 10:
Miles M.2R Hawk Trainer Mk.I & M.2W, M.2X, M.2Y, M.2Z Hawk Trainer Mk.II

The M.2R of July 1935, of which four were ultimately built, was originally known as the Hawk Major de Luxe but was also referred to as the Hawk Trainer. Indeed, Miles referred to it as the Hawk Trainer in his log book following the first flight of G-ADLB, which he made on 31st July 1935. It was similar to the M.2P Hawk Major of June 1935 but with minor modifications, which included hydraulically-operated Miles split flaps in order to considerably quicken the ease and speed of their operation. The M.2R, like the M.2P, did not have a horn balanced rudder, although G-ADLN was later fitted with a mass balance weight to the rudder. However, the M.2R did not formally become known as the Hawk Trainer Mk.I until after the advent of the Hawk Trainer Mk.II.

Lt Owen Cathcart-Jones is recorded as having gone 'to South America in 1935 to demonstrate the Hawk Trainer' but little else is known about this. However, it possibly related to the ultimate purchase of the third M.2R, CC-FBB, by Franco Bianco of Magallanes, Chile in December 1935. This he later used for his epic Trans-Andean flight in June 1936, and a recently discovered letter from Bianco to Phillips & Powis dated 13th July 1936, gives an account of the flight, which is truly a testimonial to the Hawk Trainer and its engine:

Dear Sirs. Just back from my flight Magallanes-Puerto Montt-Santiago-Mendoza-Buenos Aires-Bahia Blanca-San Antonio-Trelew-Comodoro Rivadavia-Deseado-Rio Gallegos-Magallanes. As the extra tanks I wanted to have from you would have taken too long to arrive I decided to have these made here, and they were made of iron, and of course they weigh a lot. I put one tank in the passenger's seat with 110 litres (24 galls) and one in the locker with about 85 litres (18 galls).

I took off the 7th June at night time, two small fires having been lighted at the end of the landing ground so that I had to take off passing between them. Here I regretted very much that my instruments were not luminous and I helped myself with a small pocket light. Immediately after taking off I encountered the way full of clouds and had to take 6,000 feet to pass over them. I went like that for about 1½ hours until daylight started and at the same time the clouds finished.

Then I carried on flying normally, that is to say with good visibility and as low as the mountains permitted, sometimes 5,000 feet others 8,000 feet and the highest 12,000 feet, until Lago General Paz. So far I had followed the limit line between Chile and Argentine over the Andes. After this lake I found it impossible to carry on to the north owing to fog and bad snow storms, so I went for a while west trying to reach Canal Moraleda, but then the fog again made me go back south for about 150 kms (93 miles) when I found a small place between the fog to continue west and I arrived at the Estero Puyuhapi (Magdalena Island in Moraleda Canal).

The way to the north was very foggy again and I had to come right on the water, at say perhaps 10 metres (33 feet) and carry on like that during 3 entire hours with visibility practically nil; so I passed the Moraleda Canal and the Corcovado Gulf, in fog and rain and having at the same time a very strong west wind which made very dangerous the flight at so little altitude. Twenty minutes before arriving at Puerto Montt the visibility was a little better and I could climb about 200 metres (654 feet), always on water. I landed safely in Puerto Montt after 9.55 hours, flying without stops. I must remind you that during the 9.55 hours, the whole flight, not only are there no landing grounds but not even a free camp of 100 metres (109 yards) in which one could drop without killing himself and neither was there one single house nor anybody living there so everything depended upon the Trainer and its motor, and they both behaved beautifully.

The distance between Magallanes and Puerto Montt in a straight line is 1,600 km (994 miles). With the turn I was forced to make I travelled about 1,850 kms (1,150 miles). In Puerto Montt I was given a dinner by the Air Force Officers and Aero Club. The 9th June I left Puerto Montt and made a non-stop flight to Santiago, with again poor visibility and a north wind. It took me

6 hours to reach Santiago, the distance being about 600 kms (373 miles). In Santiago I was received by all the Aeronautical Authorities, Air Force Officers, Aero Club members and hundreds of people. I was given a Vermouth immediately in the same aerodrome, by the Aero Club, in the club house.

The flight was very much appreciated in Santiago and the quality of the aeroplane, because the difficulties of the trip Magallanes-Puerto Montt especially, are known by everybody there and the Air Force have always had many accidents and have lost several aeroplanes in that same route nearly every one of the few times that they came to Magallanes or tried to. And then they always had big multi-motor aeroplanes, flying boats or amphibians and with several people as crew. And then never the flight had been done non-stop, and also never a land machine had done it before.

I had any amount of receptions in my honour in Santiago from the Air Force, the Sub-Secretary of Aviation, the National Air Line, Aero Club, visited Ministers and even the President of the Republic. Everybody in Santiago admired the Trainer and spoke of its good qualities, but nearly all found that the price was too high as the exchange in Chile is very low, and nearly everybody (private owners) are used to cheap American aeroplanes which cost less than half the price of the Trainer. The Air Force Officers admired it very much. You can be assured that the Trainer has had in Chile the best publicity it could have.

I carried on the flight to Mendoza, crossing the Andes at about 15,000 feet and then Buenos Aires where after a few days came back to Magallanes, with the stops I mentioned in the beginning of this letter. During my way back from Buenos Aires I again found fog nearly all the route and again had to take the sea and travel over it at about 20 to 50 metres (65 to 96 feet) during at least 3/4 of the whole way, except the route Buenos Aires - Bahia Blanca in which I had to fly close on the railway line, which was the only way to obtain a little visibility.

The only trouble I had during my whole flight regarding the aeroplane or motor, was the carburettor getting dirty which made me go back to Buenos Aires the first day I tried to reach Bahia Blanca and that very nearly caused me a forced landing in very inadequate camps which were flooded owing to continuous rain. I am glad to express to you once more my satisfaction for the Trainer and I may add that I will probably go from Magallanes to Rio de Janeiro, to spend one month there, as soon as I get the permission from the Argentine, Uruguayan and Brazilian Governments. You may say, honestly, that the route Magallanes-Puerto Montt, is surely one of the most difficult in the world, and the Trainer did it. I remain, dear sirs, yours very faithfully, F Bianco*

In November 1935, the fourth M.2R Hawk Trainer was supplied, in component parts, by Phillips & Powis's Indian agents R K Dundas Ltd, to The Aeronautical Training Centre of India, New Delhi. Registered VT-AHF, this was the first aeroplane to be assembled in India (ignoring a couple of homebuild Heath Parasols) and was possibly nearer to the M.2W spec than the M.2R. A photograph taken at Delhi aerodrome on 27th January 1938, probably on the occasion of its first flight, was published in the *Miles Magazine* for March 1938. This showed VT-AHF: *successfully flight tested by Capt Pierce, Chief Instructor of the Centre and also showed an Indian Cadet, Capt Alan T Eadon, Governor and Principal of the ATCI, Mr MacWade, Chief Engineer with Capt Pierce seated in the rear cockpit.*

A report on The Aeronautical Training Centre of India, which appeared in *Indian Aviation* for 15th March 1936, (it should be mentioned that ATCI was established in 1935) stated:

... no assistance had been given by the Government to the Centre but when the Centre was fully established Government might consider the grant of bonuses assessed on the result of the advanced training given. The Centre was not debarred from undertaking ab initio training, but, under an arrangement with the Delhi Flying Club all ab initio training was given by the Club and that flying clubs which receive Government subsidies were, or should be, in a position to give ab initio training more cheaply than the Centre. The lease to the Centre of a building site on the Delhi aerodrome at commercial rates also contained the provision that half the number of students enrolled should be nominated by the Director of Civil Aviation, who would naturally do his best to secure the admissions of candidates recommended by the clubs. The Centre being an independent commercial organisation was liable only to such supervision as was provided in the Aircraft Act and by the terms of lease of land.

In connection with Miles' thinking of taking out a licence to manufacture the Menasco C.4 Pirate, the M.2R Hawk Trainer, G-ADLN, was first flown with the 125hp C.4S Pirate engine on 21st March 1936, and in April of that year, it was flown to A&AEE Martlesham Heath. Miles flew it back to Woodley on April 8th and from May 22nd onwards, George Miles commenced the famous

Miles M.2R Hawk Trainer Mk.I CC-FBB 'Saturno' with Franco Bianco, during his epic non-stop 900-mile flight across the Andes in June 1936.
[Via B Stainer]

endurance flying 'at full throttle' with a flight from Shoreham of 20 minutes, followed by a 4-hour flight the following day and a 2-hour flight the day after. Further flights followed nearly every day – on May 26th he flew Shoreham - Portsmouth - Southampton - Yapton - Shoreham -Wilmington - Shoreham and the next day Shoreham - Gatwick -Shoreham and so it went on with many hours being flown, mostly starting from Shoreham. He was often accompanied by Don Brown as passenger, but a couple of times it was Graham Head. On August 26th, he recorded two return flights to Heston from Woodley. Contrary to other reports, G-ADLN was still fitted with the Menasco engine when it was impressed in February 1941.

The lineage of the M.2W Hawk Trainer Mk.II can be traced back to the M.2P Hawk Major, itself an improved version of the M.2H Hawk Major. The M.2P had the wing span increased to 34ft 0in, which in turn increased the wing area to 174 sq ft. It also had dual control, wider cockpits to improve access for a crew wearing parachutes, hydraulically-operated Miles split-flaps, a headrest to the rear cockpit plus increased AUW and maximum speed. The production model of this improved new type became the M.2W Hawk Trainer Mk.II and it was originally designed for use by the Phillips & Powis operated Reserve Training School at Woodley, which had opened on 25th November 1935.

The M.2W Hawk Trainer Mk.II then became the first low-wing monoplane to receive Air Ministry approval for Service training. It retained the wing folding facility of the M.2 Hawk and Hawk Major, which had been a good selling feature of these machines, although this was probably rarely used, since the aircraft were kept in a hangar at night and were in constant use for the training of potential RAF pilots during the day.

The Phillips & Powis works records for G-ADVF state that it was 'Type Miles Hawk Trainer Mk.II M.2W subsequent to [M.2P] G-ADDK but [with] top decking to allow easy entry and exit with parachute. Windscreens altered. Dual control to flap operating pump. Dual control to tail trim gear. Parachute seats at back and front. Mass balance on rudder and ailerons. 'Acrobatic' (in the language of the day). All up weight 1,650lb. Control column shortened. Dual control brake system fitted. Dual control flap operating gear'. The cockpits actually had full Sutton harnesses fitted and were 3in wider, in order to accommodate pilots in full service kit. The seats were made of aluminium to accommodate the parachutes; the flaps were hydraulically operated and mass balances, in the form of projecting weights, were fitted to the rudder and ailerons. Full blind flying equipment was fitted to both cockpits. On 23rd December 1935, G-ADVF was formally designated the Type machine for the M.2W Hawk Trainer Mk.II.

After the first four M.2W's had been in service for a short time, the top of the rudder was modified on subsequent production aircraft to have an aerodynamically balanced 'horn'. The height of the fin was reduced in order to let the 'horn' project forward, and at the same time the chord of the rudder was increased at the level of the tailplane to increase its area and this gave the trailing edge of the rudder a rounded appearance. The first Miles aircraft to have the horn balanced rudder from scratch was G-ADZD, and this was given the designation M.2Y Hawk Trainer Mk.II. The surviving M.2W's were later fitted with the new horn balanced rudder, and the maximum AUW was increased to 1,720lb. Photographs confirm that all the M.2W aircraft used by the Reserve Training School at Woodley were modified to M.2Y standard and it was probably this rudder modification which has caused so much of the confusion over which actual type of Hawk Trainer Mk.II was in service with the Reserve Training School at the time.

Miles recorded a number of cross-country flights in anonymous 'Hawk Trainers' and 'Trainers' in late 1935 and it is possible that a flight he made in G-ADVF on 1st October 1935, to 'Andover and local', which took 30 minutes, could have been its first flight as it only normally took 15 minutes to get to Andover from Woodley. A 'Hawk Trainer' was demonstrated at the fifth SBAC Show, held at Hatfield on 29th June 1936, but it is not known which one this was.

Regardless of the many reports to the contrary, which have appeared over the years, the Phillips & Powis works records confirm that G-ADYZ was in fact the only M.2X Hawk Trainer Mk.II to have been built. This was subsequent to the 'Type M.2Y' G-ADZD, and was fitted with Redwing undercarriage struts and Theed flap vacuum operating gear. Regarding this new Theed flap vacuum operating gear, an article was published in *The Aeroplane* for 12th February 1936, entitled 'Vacuum Jacks for Flaps'. This explained the principle behind the system in great detail and although it omitted to mention to which Miles aeroplane it was in fact fitted, it must have been G-ADYZ:

Vacuum-operated jacks have been successfully applied to the operation of split flaps in this country. These vacuum jacks are a development of the Theed Vacuum Servo Equipment made by Hamilton Motors (London) Ltd, 466-490, Edgware Road, London, W2. Such equipment is extensively used in this and other countries to work brakes on commercial and private vehicles.

As one might expect, the first experimental work of applying these jacks to aeroplanes has been done by Mr F G Miles of Phillips & Powis Aircraft Ltd at Reading, which concern always shows commendable enterprise in trying out new ideas to make flying safer, cheaper and easier. The application of these vacuum jacks to aeroplanes is likely to appeal to the pilot or private owner because of the simplicity of operation. One merely turns a tap on or off. No pumping is required. The first time the jacks were tried out there was quite a crowd of curious spectators watching the flaps open and shut as if the machine were trying to flap its way off the ground.

Mr Theed has devised a most ingenious control, about the size of an ordinary throttle lever, by which the flaps can be set to any intermediate position the pilot may like to select. The system is quite simple. The suction to operate the jacks is obtained by making use of the depression in the induction-pipe. This is done by connecting a small tank to a suitable place in the induction system and incorporating a non-return valve in the outlet of the tank. Although the steady running conditions of an aero-motor do not provide the same high degree of suction as is obtained from even a small car engine, quite enough suction can be obtained by means of the tank to work the flaps several times after the motor had stopped. The cylinder is made double-acting so as to return the flaps.

The Theed system of flap operation was later to be used on the M.11A Whitney Straight, M.12 Mohawk, M.14 Magister and the M.17 Monarch.

The first M.2Y Hawk Trainer Mk.II, G-ADZD conformed to the M.2W G-ADVF but its 'acrobatic' category weight was increased to 1,750lb, and its rear tailplane spar was strengthened.

The M.2Z Hawk Trainer Mk.II conformed to the M.2Y G-ADZD with the exception of two 15-gallon aluminium fuel tanks, inverted flying system incorporating auxiliary aluminium fuel tank and reversed throttle and mixture controls. Ten of this model were supplied to the Rumanian Government in June and July 1936, for use by the Rumanian Air Force, and it was these aircraft which were referred to in a Phillips & Powis Aircraft advertisement for Hawk Trainers which appeared in *Aeropilot* in September 1936:

A number of Miles Hawk Trainers are being supplied to the Rumanian Air Force. This high performance Trainer is fully aerobatic and has a maximum speed 150 mph; cruising speed 130 mph; disposable load 690lb and is fitted with a 130hp DH Gipsy Major srs I engine. It is completely equipped with dual instruments and controls, parachute seats and hydraulically operated flaps.

On 15th December 1936, the M.2Y Hawk Trainer Mk.II G-AENT was delivered to the A&AEE Martlesham Heath for spinning trials (EFTS) but on December 19th, while being flown by Sgt 'Sammy' Wroath, it failed to recover from what was intended to have been a right hand spin of three turns, and crashed into a school playground near Woodbridge, Suffolk. Sgt 'Sammy' Wroath baled out

Miles M.2Y Hawk Trainer Mk.II G-AENT of the Reserve Training School, Woodley, seen at Brooklands on 11th October 1936. [A J Jackson Collection]

CHAPTER 10: MILES M.2R HAWK TRAINER MK.I & M.2W, M.2X, M.2Y, M.2Z HAWK TRAINER MK.II

successfully and luckily, being a Saturday afternoon, there were no casualties but, as the crash had occurred so early in the spinning programme, it was somewhat of a set-back. Sgt Wroath later became a commissioned officer and went on to command the Empire Test Pilot's School at Farnborough after the war.

The A&AEE, Martlesham Heath issued a report on the accident (Report No. M/708/F.1) dated February 1937, and this is reproduced below (*Ref: AVIA18/645*):

I.   <u>Description Of Accident</u>.

On December 19th 1936 the aircraft was undergoing preliminary spinning trials limited to 3 turns before starting prolonged spinning trials (8 turns). The pilot found some reluctance to recover in his first spin to the left, but the recovery from the second spin to the left was reasonable 1¼ turns. He was unable to recover from his first spin to the right and abandoned the aircraft by parachute, after 18-20 turns, at 1,800 feet. Eye-witnesses state that the aircraft did about 5 or 6 more turns and then went into a steep dive (about 45°). The aircraft hit soft ground and was completely wrecked: the wreckage (except for a few pieces of airscrew and one petrol tank filler neck) was not scattered. The pilot landed by parachute and was unhurt.

II. <u>General</u>.

The aircraft was a Hawk Trainer, delivered by the Contractor to this Establishment on December 15th 1936 for prolonged spinning trials (8 turns). Form 1090 and a Daily Certificate of Safety were supplied. As soon as the accident was reported the Form 700 was collected and found to be in order. This aircraft has a Certificate of Airworthiness in the Aerobatic Category (which covers 4-turn spins) but in order to familiarise the pilots with the particular type, preliminary 3-turn spins were being carried out with the centre of gravity 22 inches aft of the leading edge, (the permissible range being 19" to 24"). It had been flown by one pilot who found that it did not come out of a right hand spin as quickly as expected. He handed over to another pilot explaining the difficulty in order to get a second opinion.

III. <u>Examination After Crash</u>.

The aircraft was completely wrecked (a notable feature being the way in which the three-ply covering of the wings and front fuselage was shattered into small pieces) except the tail unit which, although damaged, was still operable. The rudder cables were intact. The elevator cable for forward movement of the control column was intact. The condition of the breaks in the other elevator cables, and those in the aileron cables, were consistent with their having taken place in the crash. The state of the remainder of the wreckage was such that it is not possible to diagnose any structural fault previous to the crash (c.f. para.VII(2)). The ballast was still lashed in the correct position.

IV. <u>Facts Established</u>.

This aircraft is similar to those in use at Messrs. Phillips & Powis Training School who report that no difficulty has been experienced in a spin. The only differences established were (a) this aircraft had about 4° increased rudder movement, and (b) a bag of ballast was lashed to the engine mounting inside the cowling (to adjust the centre of gravity to approximately the mid position). This particular aircraft was first registered in September 1936 (Certificate of Registration No.7348) and its Certificate of Airworthiness No.5684 was issued in October 1936. The flight reports by the two pilots concerned are attached as Appendices I and II.

V.   <u>Possibilities</u>.

A conference was called by the Director of Technical Development on 22nd December 1936, at which the accident was discussed with the designer and other representatives from the Firm. It was decided that model spinning tests should be put in hand, but in order to save time the Firm offered to carry out any tests whatever on a similar aircraft: the following possibilities were to be investigated:-

(a) That ballast on the engine mounting interfered with the air passing through the engine cowling and hence upset the flow over the tail.

(b) That the ballast on the engine mounting might cause an unfavourable inertia effect.

(c) That there was a difference of method of application of controls as between their pilots and ours. It was decided to ascertain the effect of deliberate mishandling of controls: if failure to recover could be established then the matter of degree could be found by trial and error.

At a later date it was noticed, on a similar aircraft, that application of the hand brake lever restricted the rudder movement and the Firm were invited to include a test covering the remote possibility of having entered the spin with full rudder movement and attempting recovery with restricted movement (i.e. an accidental application of the hand brake during the spin).

VI. <u>Results</u>.

In general the Firm are satisfied, as a result of these further tests, that this type of aircraft will come out of the spin under any circumstances tried, even with mishandled controls (i.e. control column moved first) as long as opposite controls are applied. The interference effect and inertia effects suggested at sub. paras (a) and (b) of Para.V are negligible. Even under the conditions of sub para (c) of Para. V.2, recovery was satisfactory. At the same time there has been at least one occasion when recovery has not been normal.

VII. <u>Conclusions</u>.

In view of this, and the fact that both Martlesham pilots found difficulty in recovery, there is a possibility that this type of aircraft can be made to develop an unusual type of spin from which recovery is not easy, but which does not normally develop, and which is difficult to reproduce deliberately. Conversely it would be peculiar that this difficult and rare form of spin should be obtained in successive flights by different pilots. Therefore, in spite of the inspections both before delivery to Martlesham and at this Establishment, and also of the fact that the tail controls functioned after the crash, the possibility of some structural failure (which had escaped notice during inspection) insufficient to affect the aircraft in normal flight but sufficient to upset its behaviour in a spin cannot be dismissed.

Appendix I. Report by First Pilot. Date: 19-12-36
Spinning Test. Aircraft: Miles Hawk. CG Position: 22" aft.

General remarks:-
First Spin. Aircraft entered spin quite easily. Spin gentle and no flick. Recovery by normal method of recovery. Recovered quite easily in ¾ to 1 turn.
Second Spin. Entered spin easily. Spin gentle and inclined to be flat - not very fast - no flick. On attempting recovery nothing appears to happen when controls are set for recovery. After a couple of turns the rate of spin "slows down" and the controls seemed to start to take effect and the aircraft gradually stopped rotating.

Appendix II. Report by Second Pilot.

(1) Spin to left 3 turns, apparent reluctance to recover after 2 turns. Engine used to recover in further half turn. Height lost in recovery 800 feet.
(2) Spin to left 3 turns, height in 5,200 ft, height commence recovery 4,700 ft. time 7 seconds. Time from commencement of recovery to complete recovery at 4,300 ft. 4 seconds. Recovery effected full opposite control 1¼ turns.

(3) Spin to right 3 turns. Controls set full opposite for recovery at 3½ turns, held for 4 turns. No apparent slowing down of spin or sign of recovery. Engine opened to full throttle for approximately 1½ turns no sign of recovery. Aircraft appeared to spin faster, throttle closed, controls brought to spin position and applied for recovery for further 2 turns. Removal of hands and feet from controls to prepare exit. Aircraft continued to spin four or five turns before final departure, which was slightly delayed owing to spin recorder becoming caught in dashboard. Spin at first appeared normal, rate of turn fast for type of aircraft. This appeared to increase to a high rate. Control column tending to be forced to the rear. Movement of controls had no effect on speed of rotation.

Height entered spin.	6,300 ft.	
Height after 3 turns.	5,700 ft.	Time 8 seconds.
Attempted recovery at	5,500 ft.	
Height left aircraft.	1,800 ft.	

After the Munich crisis of September 1938, the Hawk Trainer Mk.IIs which remained in service with 8 E&RFTS at Woodley were camouflaged on their top surfaces in line with RAF practice at the time. Nine such aircraft were still in service on the outbreak of war but were neither impressed into the RAF, nor pensioned off for ground training. With one exception (G-ADWT) their registrations were cancelled on 31st March 1941 and hitherto their fate has simply been described as reduced to spares.

What happened to these aircraft is explained by an article written by George Miles in the *Miles Magazine* for May 1942. The article was entitled: 'Wood or Metal? (for aircraft construction - PA) in which he refuted: *the contention that no wooden aeroplane has been constructed that can stand up to prolonged outdoor picketting.*

The statement that no wooden aeroplane can stand up to 12 months of outdoor picketting without becoming a serious strain on the resources of the ground staff is disproved in practice. Probably the best answer to this criticism is the case of several Hawk Trainer aircraft which were taken off our Elementary Flying Training School at the beginning of the war in the interests of standardisation. These aircraft were all over five years old and had done in the region of 2,000 hours (each - PA) hard school flying. For lack of space they were picketted for a year in the open without covers, and received no attention or maintenance whatsoever. It will be remembered that the first winter of the war was the worst we have experienced in this country for very many years and the following summer was one of the hottest.

Towards the end of 1940 we thought it would be interesting to examine one of the machines to find out just how much they had suffered from deterioration. Accordingly one of them was brought in and given a thorough inspection which included cutting out sections of the skin of the wing and fuselage to examine the condition of the glueing and plywood. To our surprise the aeroplane was found to be structurally in very good condition indeed. The paint work had suffered severely and in some cases had flaked off altogether (these aircraft were constructed before it became our regular practice to cover all wooden surfaces with madapollam prior to application of the finishing coats) but apart from the engine and certain metal fittings the work needed to put the machine in good airworthy condition was actually less than would have been expected on a normal C of A overhaul.

At about this time the writer needed a machine for experimental purposes, and, encouraged by the results of the above-mentioned inspection, brought over another of the Hawks from its dispersal point and used it immediately for several hours' flying without carrying out any work other than that necessary to ensure that the engine was airworthy. Although the type of flying carried out was of a fairly severe nature, no maintenance of any kind was necessary. It would be interesting to carry out similar tests with an orthodox metal aeroplane.

Unfortunately, it is not known for certain which two aircraft were involved in these tests, or whether either were subsequently impressed into the RAF. Just one M.2W Hawk Trainer Mk.II, G-ADWT, survives, and this is maintained in airworthy condition.

For details of Miles M.2 Hawk Trainer production see Appendix 12.

SPECIFICATION AND PERFORMANCE DATA

The Introduction to an undated Phillips & Powis Aircraft sales brochure entitled 'The Miles Hawk Trainer', referred to 'A new Training Aeroplane' and this is reproduced in full below. This document appears to have been written for the M.14 Hawk Trainer Mk.III and the three-view general arrangement drawing shows this machine. However, there were a number of anomalies; the max aerobatic AUW of the M.2W Hawk Trainer Mk.II was 1,650lb and the M.2Y Hawk Trainer Mk.II 1,750lb whereas the brochure showed an AUW of 1,900lb. The dimensions quoted on the drawing were: span 34ft 0in, length 21ft 9in and height 8ft 3in, whereas the later M.14 had a span of 33ft 10in, length 24ft 7½in and height 6ft 8in, so it must be assumed that the brochure was a preliminary preview of the M.14 Hawk Trainer Mk.III/Magister, but with dimensions based loosely on the existing M.2W/Z Hawk Trainer Mk.II. The span with the wings folded was quoted on the drawing as being 9ft (all Hawk variants had a folded span of 13ft 10in), whereas the wings were not designed to fold on the M.14. It is suspected, therefore, that the details referred to in this brochure were probably for the M.2W/Z Hawk Trainer Mk.II, although the drawing showed the features of the later M.14 Hawk Trainer Mk.III.

It should also be mentioned that there is still some confusion over the wingspan and wing area for the M.2F - M.2P Hawk Majors and M.2W - M.2Z Hawk Trainers. The span and wing area of the M.2F Hawk Major prototype was quoted by Phillips & Powis as 33ft 0in and 169 sq ft respectively, and by *The Aeroplane*, at about the same time, as span 33ft 9in and wing area 169 sq ft. A photograph of the M.2P was shown in an undated Phillips & Powis sales leaflet, but it is not known for which sub-variant the quoted span of 34ft 0in and wing area of 172 sq ft applied to. The M.2W is also recorded having a span of 34ft 0in and wing area of 174 sq ft respectively, whereas the M.14 had a span of 33ft 10in and a wing area of 176 sq ft. It is, however, of interest to note that the folded span of 13ft 10in remained the same for all Hawk variants, apart from the M.14 Hawk Trainer Mk.III/Magister, whose wings did not fold.

PRELIMINARY SPECIFICATION AND PERFORMANCE DATA

Engine:	130hp DH Gipsy Major
Dimensions:	span 34ft 0in; (M.14 span 33ft 10in) span wings folded 13ft 10in; length 24ft 0in (the drawing shows 21ft 9in, while the M.14 length was 24ft 7½in); height overall 7ft 0in (the drawing shows 8ft 3in); wing area 176 sq ft; wing section Miles 'P&P' 4; dihedral 5° (the drawing shows 6°, which is the same as the M.14 Hawk Trainer Mk.III/Magister)
Weights:	tare 1,240lb; pilot 200lb; passenger 200lb; petrol (20 gal) 154lb; oil (2½ gal) 22lb; pay useful load 84lb, aerobatic 9lb; AUW 1,900lb, aerobatic 1,825lb
Performance:	max speed 145 mph; cruising speed 125 mph; rate of climb at sea level 1,200 ft/min landing speed 45 mph; t/o run with flaps 521ft; landing run with flaps and brakes 300ft; max ceiling 22,000ft; service ceiling 18,000ft; range (with 22 gal) 400 miles.

The Miles Pusher. [Courtesy G N Wickner via A J Jackson Collection]

Chapter 11: Miles Pusher

In November 1919 Frederik 'Frits' Koolhoven, the chief designer for the short-lived British Aerial Transport Co Ltd, produced a little single-seat pusher aeroplane known as the BAT Crow. It soon faded from view but Don Brown later became convinced that this was the ideal layout for a club or private owner's aircraft, with its unobstructed view from the pilot's seat (which was only 2ft above the ground) providing the inexperienced pilot with ideal conditions for judging his landing. Don's views were confirmed when he inspected the two-seat Shackleton Murray SM.1 pusher, G-ACBP, of a similar layout, at Brooklands in 1933.

Miles was, at this time, fully absorbed in producing the Hawk but, in early 1934, Don Brown and George Miles both kept badgering him to design a pusher-type aeroplane of this layout. Don Brown recalled in a letter to Bert Clarke in 1967 that: *It represented our ideas of what a club aeroplane should be and I still think it was right.*

It was originally intended to use a very light engine (probably a Pobjoy) but during its construction, Miles changed his mind and fitted the heavier 95hp Cirrus III (which he had earlier modified to enable it to run inverted) and, in order to keep the centre of gravity in the right position, the new engine had to be moved further forward. As a result, an enormous portion of the trailing edge of the wing had to be cut out to clear the propeller. Thus the entire centre section virtually provided no lift and in Don's words: *It was hardly surprising that the machine never flew!*

In May 1934, Geoffrey Neville Wikner, an Australian aircraft designer and constructor (and cousin of Edgar Percival) arrived in England and, after various short-term jobs, joined Phillips & Powis in late 1934/early 1935. Although Don Brown never mentioned his name in connection with the Pusher project, nor indeed that he had ever been with Phillips & Powis, Wikner himself later wrote that:

I moved from Imperial Airways Ltd where I had been employed to assist in the design and construction of the first reclining passenger seats, which we fitted to the Handley Page HP.42, to Phillips & Powis Aircraft Ltd at Woodley as engineer in charge of the Experimental Workshop. I worked with Freddy Miles and helped with cockpit layouts and detail design on a number of types, particularly the Hawcon and Nighthawk in late 1935, but my most interesting project was the Miles Pusher project. F G had wanted to experiment with the idea of a tricycle undercarriage and, after discussions, he accepted my basic design and I was told to proceed. The design was intended for a Pobjoy motor with pusher-propeller and had side-by-side seating with dual controls. I built a 'backbone' at the front end to take the loads of a nosewheel and designed and installed a 'swing-over' control column with spectacle-type hand wheels, as was later used on both the Nighthawk and Peregrine.

In December 1934, Miles was finally persuaded by the urgent request of Don Brown and George Miles, but rather reluctantly and very much under protest, to build the Pusher. Its basic design, with Pobjoy engine, was completed in about July 1935, and construction then commenced, despite Miles' lack of enthusiasm for the project. Hawk wings, empennage and undercarriage - the nosewheel was to come later – were used in its construction to save money.

Wikner further wrote:

Miles had taken little interest in the project and as funds were very limited it had to be shelved. Rather than abandon the Pusher altogether I managed to get permission to fit a Cirrus engine which had been inverted some time in the past. This was a far heavier unit which meant it had to be installed further forward, necessitating a cut-out in the trailing edge, and had no cowlings to streamline it as had been the intention with the Pobjoy. To my knowledge Freddy Miles did fly it, I expect that without cowlings and with excessive drag the performance would have been very poor and it was then finally abandoned. If it had been given the backing I feel sure that the design could have been a winner, both in its tricycle form and as an amphibian flying boat. At this stage I was endeavouring to establish my own manufacturing business and Tommy Rose put me in touch with a racing motorist (Donald Marendaz – PA) *who was interested in building a light aeroplane.*

Wikner parted company with Phillips & Powis prior to starting work on the Marendaz Special at Maidenhead in February 1936. Meanwhile, on both October 3rd and December 30th 1935, Miles had flown the Shackleton Murray SM.1, G-ACBP, which was now jointly owned by Lord Apsley and Miss Dolly Miles (no relation) and based at Yate, probably in order to get the feel for flying this type of machine.

The Pusher was painted red and given c/n 230 but it was never allocated a Miles designation or even given a 'B' mark and no registration was applied for. Miles failed to record details of his first 'tentative hop' of the Pusher in his logbook, but it is believed to have taken place in early January 1936. He probably did not consider this event to be worthy of note, as the Pusher apparently showed little inclination to leave the ground, due to excessive drag. It was hardly surprising, therefore, that, with the very limited facilities at his disposal, Miles lost no time in dropping it. He considered that the Pusher was an unsuccessful project; he never had much enthusiasm for it, and saw it as yet another design to compete in the same market as the Hawk. From Wikner's account, the Pusher project must have been all over in a very short space of time!

The wings and empennage were probably re-used in the construction of a production Hawk and Don Brown later recalled that Miles elected to scrap the airframe rather than do the extensive modifications necessary to make it a success with the unsuitable engine. George and Don of course blamed Miles for the failure and remained convinced that there was still much to commend this layout.

Until 1980, no photograph of the Pusher had come to light and then one, presumably from Wikner, was published in the April-June 1980 issue of *Vintage Aircraft* magazine and this is reproduced here. Since then, another couple have also been discovered, but these are of very poor quality.

Don Brown wrote, much later, that there was a 'close resemblance' between the Miles Pusher and George Miles' M.100 Student basic jet trainer of 1957, but this, I feel, is stretching a point! Wikner also claimed in his letters that 'his' Pusher was the inspiration for the Student but once again, I think that this was just sour grapes on his part.

No Specification or Performance details have survived – if indeed any were made or recorded at the time!

Reading Aerodrome, Woodley, just after the completion of the new assembly shop, built by Boulton & Paul in 1935. Note 'ACME' - the peak of perfection - with 'Phillips & Powis Aircraft (Reading) Ltd', under, on one side of the roof of the original hangar. [Via P Amos]

Miles M.3 Falcon with 'B' mark U3 in September 1934. [Via P Amos]

Chapter 12:
Miles M.3 Falcon and M.3A Falcon Major

In 1934, Miles embarked upon the design of his first true cabin aircraft, a three-seater with the two passengers sitting side-by-side behind the centrally-placed pilot. It was to be powered by a 130hp DH Gipsy Major I engine and was to retain the tried and proven wooden construction used on the Hawk, comprising spruce members covered with birch three-ply. The wings were also arranged to fold in a similar manner to the Hawk and in the tradition of the new breed of monoplanes, was called the M.3 Falcon.

At about the same time as Miles was working on the design of the Falcon, Harold Leslie Brook of Harrogate became very interested in this aeroplane as he was considering entering the England to Australia MacRobertson Race. A contemporary issue of *The Aeroplane* gave a background history of Brook:

After serving for five years in France and India in the First World War, he remained a normal civilian until Yorkshire began to build and fly sailplanes and gliders. These occupations kept him mildly diverted until the approach of his 37th birthday. Then he began to yearn for horse-power. The York County Aviation Club at Sherburn-in-Elmet offered a likely fulfilment of this secret ambition. So, in August 1933, Brook placed himself in the hands of Instructor Cudemore, and after four hours' instruction became a soloist with serious designs on the MacRobertson Handicap, for which Phillips & Powis have built him the first of their Miles Falcons.

What happened between last autumn and this spring is now almost historic. Brook bought the Puss Moth Heart's Content *in which the Mollisons had crossed the Atlantic and, with a total of 43 hr in his logbook, pushed off solo from Lympne to survey the route to Melbourne. That was on March 28th 1934 at 5.20 am. By noon the incident had closed.*

Describing it a few days later, Brook said that while flying through very dirty weather over France, he was forced down from 12,000ft by ice formation on the wings and, before he knew how or why, the side of an unsuspected mountain was rushing up at him out of the murk. Guided by some uncanny sixth sense, he brought off a bloodless landing on the mountain proper. The scene of this epic of the air was Genolhac in the Cevennes. With some local help he salvaged the Gipsy Major, brought it back to England and has had it installed in Heart's Content II (there is no evidence this name was ever used – PA).

The prototype of the Falcon was thus 'manufactured for sale' and completed to Brook's order for the MacRobertson Race. It was fitted with the Gipsy Major I engine taken from the Puss Moth G-ABXY, and was registered G-ACTM to Brook on 11th June 1934. It differed from the later production machines in having a 'normal' windscreen with a glazed roof to the rear of the cabin.

Following an informal Tea Party held at the Reading Aero Club on Sunday 23rd September 1934, Miles took off at dusk on the successful maiden flight of the Falcon, which carried the 'B' mark U3. In October, the Falcon was submitted to A&AEE Martlesham Heath for C of A trials (Report No 660 dated 11/34) where it was found to have good handling characteristics but poor brakes and no emergency exit in the roof. However, the latter shortcomings were seemingly not considered to be as significant as the lack of a 'No Smoking' sign in the cabin!

The Falcon was handed over to Brook on October 12th. Although it was initially reported that 'if expectations are realised' Brook would be accompanied by two lady passengers, in the event, he left Mildenhall, *escorted by the beautiful Miss E M Lay*, on October 20th, at the start of what was to become an eventful race for him. This included a forced landing in Greece with propeller failure after which he became hopelessly lost in Afghanistan, probably due to his less than 50 hours flying experience.

He eventually completed the journey safely in 26 days 20 hours, arriving in Darwin on November 20th, long after the race had closed. Brook spent five months in Australia before fitting extra fuel tanks to the Falcon to give a 2,000 mile range for the return journey to England, with sponsorship apparently provided by Ovaltine. He set off on his own from Darwin on 23rd March 1935 on what was to become a record-breaking solo flight, during the course of which he made the 1,700 mile stage from Jodhpur to Basra non-stop. Brook duly arrived at Lympne in the record time from Australia to England of 7 days, 19 hours and 50 minutes.

In a letter to Phillips & Powis, dated 1st April 1935, Brook wrote:

I am sure you will be interested to know that I found the Miles Falcon to be a great deal faster than anticipated, and what is much more important on long-distance flights, it was exceedingly comfortable and easy to handle. I have tried it out under all sorts of climatic conditions out East and was impressed by the fact that in spite of the thin air it behaved just as well as it did on home aerodromes. I would like to mention that a number of well known pilots in Australia handled the machine and without exception they were highly delighted with its performance, both in its flying abilities and its comfort. Its appearance was also very favourably commented upon. The reliability of the aircraft has, of course, been proved up to the hilt.

I would like to conclude by saying that on my way back the Miles Falcon was flown very, very hard indeed and came through the test with flying colours.

Harold Brook also wrote an article on *Flying to Australia - and Back*, which was published in *AERO* magazine for May 1935, and extracts from this are reproduced below:

I suppose the most interesting thing about my recent record solo flight from Australia was the fact that I was unknown and comparatively inexperienced. Eyebrows were raised when I turned up at Mildenhall to enter the race to Australia. First of all, while most of the records have been set up or broken by young men in their twenties, I was 36 years old. I had not been associated with aviation, but had been in business. On the engineering side I had some knowledge and it was this which drew me into learning to fly. I found that a course of gliding was great help in learning balance. It helped me to go solo after only 4½ hours dual on powered machines.

When I made up my mind to enter for the England-Australia Air Race I wanted to see the route for myself and gather experience of flying abroad. The choice of a machine for a long-distance flight again gives the young pilot further experience of a kind of flying which the stay-at-home airman never gets. It so happened that Mollison's Puss Moth Heart's Content, *in which he had flown the North and South Atlantic Oceans, was in the market. I bought it; it had tanks for a range of 3,500 miles, but I altered this to 2,000 and had her a little more lightly loaded. Better vision was also obtained.*

When I left Lympne to fly to Australia in the Heart's Content *it was winter. Snow lay thick over Europe. This proved my undoing, for I got ice on the wings which forced me down on hills in the south of France. I returned a sadder and wiser man. The engine however, was salved and I had it fitted into a Miles Falcon monoplane which I chose for the Race. More flying practice in England brought my total up to 100 hours - just enough to be a competitor in the great air race.*

The first 1,000 miles in the Race presented no navigational difficulties. From Rome I crossed the Appenines and flew down the Adriatic coast to Brindisi. Then the sea crossing to Athens and after that the 800 mile stretch across the lower Cicilian mountains to Aleppo in the desert but as I had insufficient range I made for Cyprus. Then on to Beirut and Aleppo, after which I followed the Euphrates for 500 miles to Baghdad. For 1,500 miles I then flew with one wing over the shore and one wing over the water down to Karachi.

Flying eastwards across India presented some little difficulty until Calcutta was reached. But from then the rest of the route followed coast line, islands and archipelagos until Timor was reached. The dreaded 500 miles of open sea crossing had then to be faced, and that was an ordeal for the pilot of a single-engined machine. The crossing of Australia itself was more difficult as there were few landmarks, but in daylight and in good weather there was no reason why, after the experience of flying to Darwin, you should lose yourself in the Australian deserts.

Flight for 10th January 1935 carried an article entitled: *Improvements in the Falcon - some interesting alterations in a well-known Miles product.*

At Mildenhall, on the morning of the England-Australia race, the public saw the first four-seater to be produced by Phillips & Powis Ltd of Reading. This was the Miles Falcon, which had been previously described in Flight *for August 23rd 1934. Since that time considerable alterations and improvements have been made to this model, and the first details are published below.*

A point which will first be noticed when this latest Falcon comes out of the factory - as it will do very shortly - is the new windscreen. Instead of the sloping screen characteristic of British design for many years past, Mr Miles now employs one which actually slopes forward. He has already tried out this design in 'mock-up' form, with gratifying results and an increase of speed of approximately 4 mph. The small front panes are at a sharp angle and fine rain or snow should not collect upon them (It was also claimed that this sloping forward windscreen eliminated reflections from inside the cabin - PA). The top part of the screen, over these panes, is of 'Rhodoid' moulded to the correct shape. The general theory of this screen, as evolved in the United States on certain of the larger high-speed commercial air liners, is that the air in front of the

Miles M.3 Falcon G-ACTM with H L Brook taxi-ing out for a flight prior to the start of the 1934 England-Australia Air Race. The photograph was dated 18th October 1934.
['Flight' via M J Hooks]

screen is pushed forward and forms a cushion which tends to streamline all that part of the machine, so that the flow comes from the nose of the machine over the windscreen and flows aft, unbroken, behind it. With the sloping form of screen the air is shot steeply upwards over the screen, creating eddies above and behind it, thereby creating more drag than in the former case.

The internal arrangement of the Falcon has also been somewhat redesigned. It is now slightly wider and will provide extremely comfortable seating for four people. Dual-control has been arranged in a neat manner which obviates the necessity for having two control columns or a change-over wheel type of control. This has been achieved by jointing the stick well above the bottom pivot,

The welcoming party on the apron at Croydon Airport after Brook's record-breaking solo flight from Australia to England.
[Phillips & Powis via B Clarke]

Mr Brook with G-ACTM at Woodley on 1st April 1935, after his return solo flight.
[Via B Clarke]

F G Miles, Blossom, Harold Brook and Charles Powis at Woodley in April 1935. [Phillips & Powis via P Amos]

as is done in some military aircraft; reference to our sketches will show the details. An arm-rest will be arranged between the seats, thus allowing the pilot in the left-hand seat to fly with his right elbow on the rest, in which position the stick comes comfortably to his hand.

Very considerable interest has been aroused by this machine and although at the time of writing it has not yet flown, several orders have been placed. It is hoped that the cruising speed will be in the neighbourhood of 125-130 mph with the Gipsy Major 130hp engine. With the Gipsy Six, which can be fitted if a smaller payload can be accepted, the performance will be substantially higher. The balanced flaps, which have already achieved such success in the Hawk Major, will be used and for the present, at any rate, their operation will be very much the same, though the question of hydraulic operation is under consideration.

All the Hawk series are substantially the same as regards their construction, that is to say, the spars are boxed up with spruce booms and plywood webs and both the wings and fuselage are completely covered with plywood, which is secured to the appropriate members by glue. It will be remembered that in a description of the Hawk when it first came out, Flight *commented very favourably upon the method by which Mr Powis ensured perfect glueing while at the same time doing away with all pins and screws. This was by the use of a form of office stapling-machine with the turn-over mechanism removed, so that the small wire staples, of the kind normally used for clamping sheets of paper together, were forced straight into the plywood and through to the member underneath. This can be done exceedingly quickly and when the glue is dry the staples are removed.*

Mr Miles believes in an adequate margin of strength and his aircraft are, therefore, considerably stronger than the Air Ministry regulations require, so that for special circumstances a great deal more load can be carried. It is understood that this feature is being made use of in a subsequent type which will carry five people and for which orders have already been placed (the M.4 Merlin - PA).

The first production aircraft was designated the M.3A Falcon Major and was fitted with a 130hp DH Gipsy Major Srs I engine. It had the wider cabin with a modified windscreen and roof to accommodate four people, although one (G-AEEG), had seating for only three people with the pilot sitting centrally in front.

The M.3A Falcon Major G-ADBF was first flown, by Miles, on 16th January 1935. Dual control was provided as standard, as were the hydraulically-operated Miles split trailing-edge flaps. The clean trousered undercarriage of the Hawk Major was also incorporated in the design. Although the take-off, with four people aboard and full fuel tanks, was rather long by the standards of the day, the Falcon Major was easy and pleasant to fly, being only 5 mph slower than the Hawk Major. There was no other contemporary aeroplane of that power available at the time which provided such performance and accommodation.

The Falcon Major was produced simultaneously with the Hawk Major, and proved to be very popular overseas as well as on the home market. *Flight* for 21st February 1935, gave details of the flying characteristics of the Falcon Major under the title: *Flying Comfort - The Miles Falcon in Production:*

There should be a large demand for the Miles Falcon which has just gone into production, because it seems to have nearly

Chapter 12: Miles M.3 Falcon and M.3A Falcon Major

every desirable feature which private owners will require in an aeroplane of this type:

Strength - as with all Mr Miles' designs, the factors are more than the legal minimum;

Comfort - plenty of room for four people and ample space for luggage;

Speed - we found the ASI showing 122-125 mph at normal rpm;

Slow landing - with flaps down we touched down 'off the clock' and ran only a few yards;

Stability - hands or feet off, the Falcon flew safely.

Taking the flying characteristics in more detail, a prolonged trial showed that we were justified in thinking that the Falcon in its production form, would be even more pleasant to fly than the Hawk Major, although when we tried the latter machine it did not seem possible to ask for more. A slight increase in the dihedral of the wings has resulted in more lateral stability, so that the Falcon almost flies itself, but it has not, as is so often the case with a machine having this characteristic to a marked extent, lost any of its manoeuvrability. In addition, the Falcon is both directionally stable and has its fin surface so nicely proportioned that turns can be made, and the machine flown for all normal purposes, without touching the rudder bar.

The unusual shape of the windscreen enables the pilot to sit well forward, so that he has an excellent outlook when flying, landing or taxi-ing. The outlook when landing is also helped by the nose-down attitude in which the Falcon can be glided when the flaps are lowered. Probably there is no machine in which an approach is so easy as this. The hydraulically-operated flaps can be used in exactly the same way as the brakes on a car, and this ability to lengthen or steepen the glide at will proved of inestimable benefit when the engine cuts, as it did last Sunday, and we had to land 'from where we were' so to speak. Actually, the possession of these flaps robbed such a proceeding of any excitement, and the incident, which was due to a stuck fuel gauge showing a supply which was non-existent, served only to prove their great value.

In the air, it is delightful to throttle back, lower the flaps a few strokes of the hydraulic pump handle and then, with a 'waffle'" of engine, cruise along at less than 50 mph.

A number of pre-war testimonials relating to the flying qualities of Miles light aircraft have recently come to light, and one, dated Saturday, 1st June 1935, was from Sr Giorgio Parodi, Italy (who had acquired the M.3A Falcon Major, G-ADHC, in May 1935), who wrote from Albergo del Mehari, Tripoli. It read:

Dear Mr. Powis

I was yesterday very pleased to be able to write you of the Falcon's success. And also the Italian Authorities must be pleased because had it not been for the Falcon, the Rally would have been won by Capitaine Puget of the French Army piloting a beautiful (but frightfully uncomfortable) Caudron Rafale, coming straight from the Caudron Works with the declared intention to win.

Your machine behaved perfectly and gave more that I expected. The weight of the crew was (in Kilos) 73, 70, 60, 57 = 260 (585lb). I managed to cover something over 4,000 km during the 25 allowed hours and the average speed has been about 139 mph. I'm writing you immediately after the speed race which followed the Rally and I haven't yet the exact figures before me. I could have easily covered more km but some of the crew were not trained and I preferred to be reasonable not asking them too much (for them). I will send you the official figures as soon as possible.

The Falcon has won (the Raduno Sahariano – PA), as I expected, with a very big margin indeed. It was the only machine to have four people on board and although it was one of the lowest powered, it was the fastest of the lot (with the exclusion of course of the above mentioned Caudron Rafale). I went of course with the throttle wide open from start to finish and that amazing Gipsy engine which was giving 2,260 rpm at the beginning went up to 2,300 in the last stage over the torrid African desert. I must write to Messrs DH about it.

The handling of the machine, also if overloaded, does not require more than ordinary skill; sometimes I had to do hurried landings

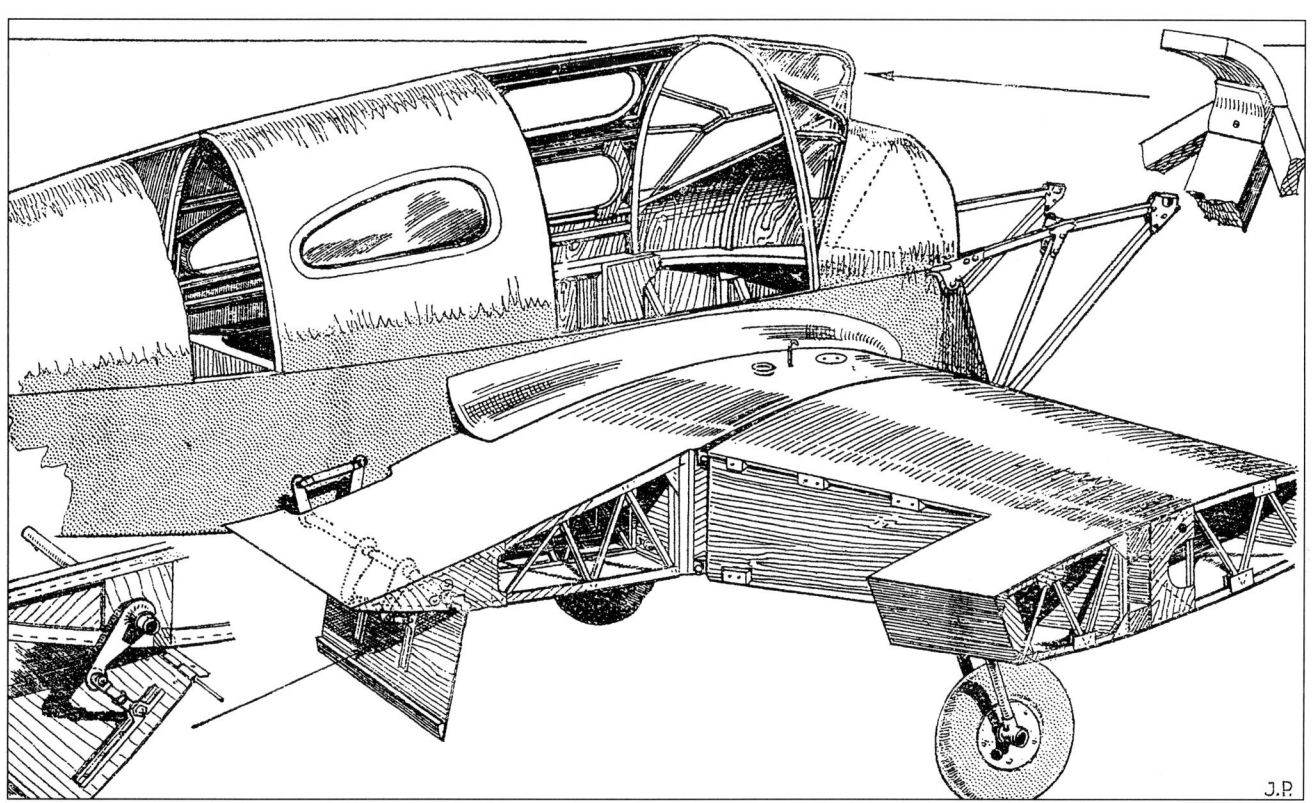

Structural details of the cabin of the M.3A, showing the formation of the windscreen. Part of the starboard wing flap is shown; the flaps are of the balanced type, with the centre balance portion operating forward, and thus in the opposite direction to the main flaps.

['Flight' 10th January 1935]

A delightful study of G-AEEG landing at Brooklands in the 1990s. [C.Knox]

and take-offs which were far from perfect, but the machine always stood the additional strain. Will you please tell Mr Miles, that I got with the Falcon the severest bumps of my life because of the wind over the African desert but the machine did not break, so he can please send me the C of A with something more than 2,200 lb. This certificate has not been asked to me but I prefer to have it in my pocket. Yours, Giorgio Parodi

The Falcon was widely used for touring and Geoffrey Alington who, with Edward 'Bunny' Spratt set up a small charter business called Air Touring, relates several stories in his book *A Sound in the Sky*. One concerns the time in August 1937 when the Falcon Major G-ADFH was chartered by someone who turned out to be one of Adolf Hitler's girlfriends. Although Geoffrey did not know it at the time, this was the (in)famous Unity Valkyrie Mitford, but he was briefed to go to Paris-Le Bourget to pick her up and fly her around Europe for two weeks, taking her wherever she wished to go. 'Bobo' (or 'Toto') as she called herself, also wanted to take Gaston Bergerie, a Deputy in the French Government and his wife, who was American. Geoffrey suggested that they went to the South of France but 'Bobo' wanted to go to Finland!

On 6th August 1937 we departed for Finland calling at Amsterdam and then on to Hamburg where, with Bobo's diplomatic passport, we were cleared through all entry procedures in double quick time. The following morning we flew on to Malmo and after refuelling we flew on to Stockholm where we spent two days. From Stockholm we flew to Helsinki crossing over 100 miles of water before reaching Finland. I thought that Helsinki was as far east as 'Bobo' wanted to go, but not a bit of it. After a night stop, she said she wanted to go on eastwards to Viipuri, a town very close to the Russian border. We remained two days at Viipuri and then my passengers said that they wanted to return to Helsinki and visit the Arctic Circle at Rovaniemi. I was left on my own while the 'Mixed Bag' went away to the Arctic Circle, or that is where they said they were going and I had no reason to doubt their word.

It was then decided that the places they would like to visit would be Danzig, Warsaw, Budapest, Venice and then back to Paris – quite a trip! 'Bobo' and Gaston Bergerie had, I thought, vanished from the screen of my life for ever; but it was not to be. Years later, when France capitulated, Gaston Bergerie was made Vichy Air Minister - but this was after Russia had attacked Finland and taken Viipuri. Had I flown him up there to spy out the land? I shall never know.

A large proportion of the 20 Falcon Majors built between January 1935 and May 1937 were exported - to Australia, France, Hong Kong, Italy, Palestine, Spain, Sweden and Switzerland. One even

The Miles M.3A Falcon Major in which Sr Giorgio Parodi flew to victory in the Raduno Sahariano in Tripoli, Libya, in May 1935.

[Via B Clarke]

A post-war photo of G-ADFH.

[Via B Clarke]

Miles M.3A Falcon Major VH-AAT at Redcliffe, Queensland, Australia, on 11th June 1975. [Air-Britain]

ended up in Malaya where, as VR-RAP with the Kuala Lumpur Flying Club. It was later impressed into service with the Malayan Volunteer Air Force, and a letter from Mr K G Hannett, published in the July 1946 edition of the *Miles Magazine*, gave details of its wartime service:

You probably do not know me, but I know you and know your aircraft even better, in fact were it not for the ruggedness of the Falcon in operations against the Japs some four years ago, I would not be writing this now. A member of the Kuala Lumpur Flying Club since 1935, I was Vice Captain and Hon Chief Instructor of the club from 1941 onwards.

As you probably know, all the Malayan Club aircraft were mobilised when the Japs started in December 1941, and they went into the war piloted by club members, unarmed, with no extra equipment (with the exception of the bomb racks which were even fitted to the Gipsy aircraft), and they did a pretty good job. You would be horrified at the tasks which were handed out to small aeroplanes, and to this day I am amazed at the way in which they stood up to the strain. Maintenance was almost nil and the territory over which we flew was either open sea or impenetrable swampy jungle with trees up to 150 feet high.

The worst experience we had was when my Danish navigator and I set off in the Falcon to fly from Pekanbaroe to Palembang, refuelling en route at Djambi, where we heard the news that Jap paratroops had landed at Palembang. By the next morning, Singapore had almost fallen, and we decided to head for Lahat, east of Sumatra, as it appeared to be the only possible landing ground. By this time the Falcon was in pretty bad shape, water having penetrated into fabric causing the structure to begin warping. At Lahat we found a first-class aerodrome, laid by the Dutch and the place was full of RAF who had evacuated Palembang. Everything at Lahat was pretty chaotic, and the only thing to do was to try to get to Java . . . a frightening proposition considering the type of country over which we would have to fly.

However, before dawn, we tanked up to maximum capacity - 45 gallons as far as I can remember, and the Dane and I (a total weight of 25 stone), plus all our gear and the remaining bottle of gin, packed ourselves into the aircraft. In the semi-darkness, with a ground mist and on wet grass, we started on the shakiest take-off I have ever attempted. It was more or less 'blind', with the added hazard of odd RAF planes taxiing over the aerodrome.

The Falcon got off twice, each time falling back, but at the third attempt it staggered into the air, half-stalled and finally lurched over some tree-tops, but we were airborne! To get on course meant a very gentle turn, the maximum height obtainable being under 1,000ft. By this time the flap gear had given way and continual pumping was necessary to keep them up at all - most of the time they were a third down and the empennage was so warped that it took a lot of stick and rudder to keep her on course. In this way we plodded on until tired but intact, we finally reached Batavia.

Just two M.3A Falcon Majors survive to this day. One, the three-seat G-AEEG, is maintained in airworthy condition by its proud owner, Peter Holloway, at Old Warden, and the other, VH-AAT, is also equally well maintained. This flies from Wangaratta in Northern Victoria, Australia, and is presently believed to be for sale.

For details of M.3 & M.3A Falcon Major production see Appendix 13.

SPECIFICATION AND PERFORMANCE DATA

Engine:	130hp DH Gipsy Major I
Dimensions:	span 35ft 0in (folded 15ft 10in); length 25ft 0in; height 6ft 6in; wing area 174.3 sq ft
Weights:	empty (M.3) 1,270lb; (M.3A) 1,300lb; AUW (M.3) 2,000lb; (M.3A) 2,200lb; wing loading 12.65lb/sq ft
Performance:	max speed 145 mph; cruising speed 125 mph; t/o run (5 mph wind) 510ft at 1,950lb and 675ft at 2,200lb; rate of climb 750ft/min; service ceiling 15,000ft; range on 32 gal, 615 to 667 miles depending on load.

Tommy Rose taxi-ing in after winning the 1935 King's Cup. Note the loss of the spinner. [Keystone via F G Miles Ltd]

Chapter 13: Miles M.3B, M.3C, M.3D, M.3E and M.3F Falcon Six Variants

In the spring of 1935, Miles decided to build a higher-powered three-seat version of the Falcon Major for racing purposes, to be fitted with the 200hp DH Gipsy Six engine and driving a Fairey metal airscrew. The additional 70hp over the Gipsy Major was expected to give the new machine a maximum speed of about 160 mph and a cruising speed of around 145 mph. The prototype was designated M.3B Falcon Six and the main noticeable difference between this and the Falcon Major was that the engine bulkhead was moved much closer to the wing leading edge to accommodate the larger engine. The pilot was seated centrally with two passenger seats behind. The M.3B, with an increased AUW of 2,350lb, had modified controls and a pointed top to the slightly larger rudder with the leading edge of the fin being faired smoothly into the top of the fuselage.

As the new Falcon Six was being readied for its first flight, Tommy Rose looked at it and remarked "it looks like Pregnant Percy", on account of its rather corpulent appearance brought about by the large cabin and the name stuck. The Falcon Six prototype was registered G-ADLC and was first flown, by Miles, on 20th July 1935 (not 27th July 1935 as often stated – even in Miles' own publications) and was promptly entered by Lady Wakefield of Hythe in that year's King's Cup Air Race, to be held on September 7th and flown by Tommy Rose.

In order to outwit the handicappers (who had handicapped the Falcon Six to do 157 mph), Tommy Rose decided to *fib just a little* over its performance! He could not blatantly lie as this would have prejudiced sales and put him in a difficult position, so he flew it all around the country and carried out many demonstrations, praising its handling and high cruising speed but complaining bitterly of its poor top speed. His first chance to 'bend the truth' came during the London to Newcastle race, in which he praised the Falcon Six's performance, saying that it had a high cruising speed but felt disappointed with its top speed. It was also about this time that it acquired the name 'Preggers.'

Thirteen Miles aeroplanes were entered for the 1935 King's Cup Air Race and, of these, seven had been specially built for the race. After qualifying 'at a gentlemanly speed' in the first day's

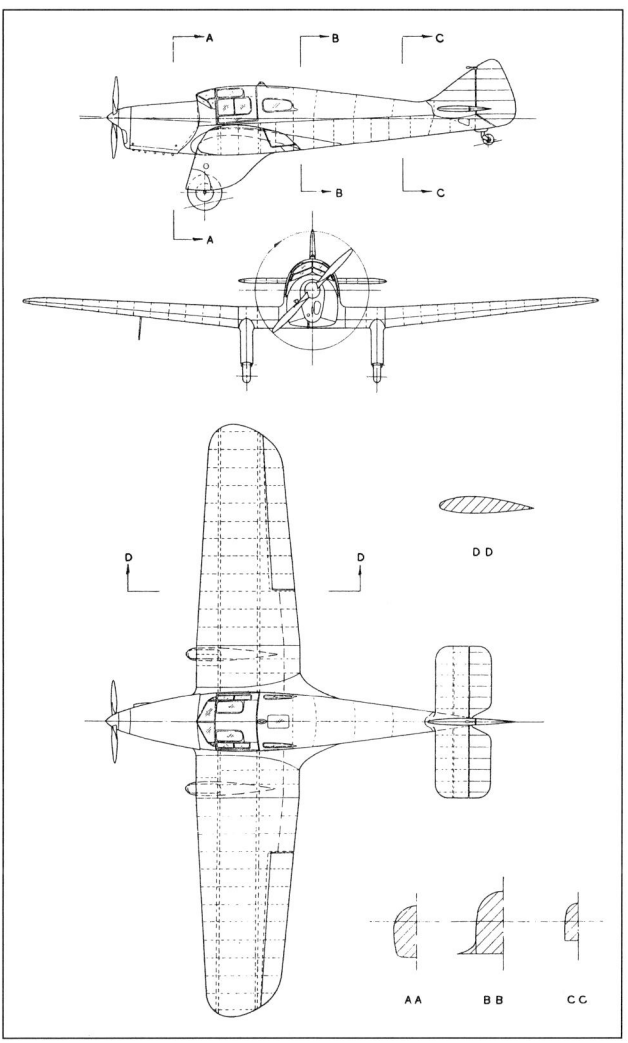

The Miles M.3B Falcon Six.

Tommy Rose sits by the first Miles M.3B Falcon Six, G-ADLC, at Hatfield, nonchalantly reading the 'News Chronicle', before the start of the 1935 King's Cup Air Race. ['Flight' via B Clarke]

Tommy Rose alighting after winning the King's Cup Air Race in 1935. ['Flight' via B Clarke]

Tommy Rose being presented with the King's Cup from Sir Phillip Cunliffe Lister. ['Flight' via B Clarke]

qualifying race round England, his complaints about the top speed of the Falcon Six proved to have been 'a tissue of lies', uttered in the hope that the handicappers would take notice, particularly after he had been forced into second place with the Hawk Major the previous year, by a run-away winner who had totally out-foxed the handicappers!

Thus it was that Tommy Rose won the King's Cup Air Race in the Falcon Six G-ADLC, beating its handicap by 19 mph, to romp home in first place at 176.28 mph. For the record, second and third places were gained by two Hawk Trainers and fifth place was gained by the sole M.3C Falcon Six at 167 mph.

After the race, Tommy Rose re-named the Falcon Six *'Preggers Primus'* (Primus being Latin for 'the first') and although it was duly christened thus, it was still usually known as *Preggers*. He then set out on a sales tour of Europe. Tommy Rose wrote of his experience in the 1935 King's Cup in the September 1935 issue of *Reading Review*:

For a number of years it has been my ambition to win the race for the Cup presented annually by His Majesty King George the Fifth, and this being the one year event, I was more anxious than ever to pull off this contest which is the Blue Riband of British Flying.....The first day of the contest consisted of a circuit of Britain, the turning points being Newcastle, Edinburgh, Renfrew, Port Patrick, Belfast, Dalbeattie, Blackpool, Manchester and Cardiff. Compulsory stops were to be made at Renfrew, Belfast, Manchester and Cardiff, and the race started and finished at Hatfield. The machines were divided into two groups, one with engines up to 150hp, and those with engines over that power.

We left Hatfield shortly after 8 am on the Friday and reached Newcastle at 10 am. The weather was slightly hazy at first, but the further north we went, the better it became, until in Scotland we ran into brilliant sunshine, which with the exception of a shower or two in Ireland, remained, with us for the rest of the circuit. We arrived at Renfrew at 11 am and landed at Newtownards, Belfast at 11.30 am. Here we refuelled and continued to Manchester. I have flown over Manchester hundreds of times, but never before have I seen the WHOLE of the town from the air in one piece! Cardiff was reached at 1.15 pm and we arrived back at Hatfield at 1.50 pm. As the first ten machines in each class were eligible to compete in the final heat, and we knew the approximate speeds of our rivals, there was no need to go 'all out' during the circuit; but we nevertheless made an average over the course of 153 mph, including stops, which was not at all bad going. Actually we started tenth and finished ninth.

The next day was quite a different affair, consisting of seven laps of a fifty-mile circuit lying round Broxbourne, Henlow and Hatfield. The weather was again ideal, with a slight breeze blowing from the north. Being one of the fastest machines in the race, we had the doubtful pleasure of seeing twelve other competitors get away before it came to our turn. However, we did not have to start thirteenth, as we were scheduled to go at the same moment as the other Falcon entered, and it got away just in front of us!

An advert from 'The Aeroplane' of 5th February 1936 for the Miles Falcon.

CHAPTER 13: MILES M.3B, M.3C, M.3D, M.3E AND M.3F FALCON SIX VARIANTS

At the end of the first lap we had picked up two places, and were lying eleventh, and by the end of the third circuit we had advanced to ninth. Fourth time round and we were third. At the end of this lap we had a brief moment of excitement, as there was a sudden noise and jar from the front of the machine. I thought, 'Well, that's that', but nothing else happened and we carried on as before. It afterwards transpired that the airscrew spinner, presumably having been hit by a bird, had come off, luckily without striking the machine.

By the end of the fifth lap, we were in the lead and it was then just a matter of keeping going at full throttle and holding one's course. The Gipsy Six and the Falcon stood up to their respective reputations, and we eventually passed the post, some minutes before the second man home, who incidentally was also flying a Miles machine.

The most dangerous part of the two days was then to come, for on landing I was seized and hauled out of the machine on the shoulders of the men who had made it. They had come over from Reading in seven motor coaches to see their handiwork win. Lady Wakefield was most excited at seeing her machine come in first. She had not been in good health recently, but said that my win had been as good as a tonic to her. I am very grateful indeed that she allowed me to be her doctor.

In The Aeroplane's own flight-test shortly after the race:
 . . . . the winning pilot told us that he flew round the entire eliminating course with his feet off the rudder bar and for most of the way with his hands off the stick. And he said that in the final he flew feet off most of the time. When we heard this we were rather unconvinced, and put the statement down to the modesty of a great pilot. But now we have taken the air in it ourselves, we agree that with a first-class navigator as passenger, the pilot might have crossed his legs and read the newspaper.

Without minimising in any way the considerable personal skill of the very popular winner of the greatest British race of the year we can safely say that the performance and propensities of his Miles Falcon put it in a class by itself.

With two aboard, tanks full, and the Gipsy Six awoken with one prod of the electric starter, we were airborne in less than ten seconds in the flat calm of last Friday. We demanded full speed ahead, and without increased din sundry needles went round till the airspeed stood at 180 mph. With the stick alone perfect turns were made both ways and then we went back to cruising speed. At 150-160 mph turns up to 60 degrees of bank were easy to begin, maintain or stop without the rudder. At what must have been about 40 mph the elevator and rudder-bias levers on the floor under the dash-board could be set to keep the machine flying level and hands-off at about half throttle, without any feeling of instability or insecurity. The lateral control was still quite sensitive enough to do reasonable turns either way and bumps left the trim unscathed.

Side-slipping is unnecessary. Putting the nose down quite sharply made the descent still steeper without greatly affecting the airspeed. Landing was so easy that we felt that there must be a catch in it. There wasn't. The only non-standard feature about the machine was the pilot, and the fact that such pilots cannot be drawn from stock every day of the week shows the foresight and good fortune of the designer-builders and entrant of the winning machine.

Tommy Rose then made an attempt on the England-Capetown record. A 75-gallon fuel tank was installed in G-ADLC and he duly set off in January 1936, only to make a crash landing at Abbeville in France shortly after. The machine was repaired and he departed from Lympne a second time on 6th February 1936. *The Aeroplane* for 12th February 1936 reported on the flight as follows:
 Flight Lieut Tom Rose DFC, who won the race for the King's Cup on 7th September 1935 at 176 mph in the Miles Falcon (Gipsy 200hp) G-ADLC, started from Lympne on February 6th in the same machine, twenty-five minutes after midnight (GMT). He landed at Malta (1,260 miles) at 09.20 hrs (GMT) and left at 10.40 hrs. At 17.15 hrs (GMT) he landed at Almaza, Cairo (1,106 miles) and left at 22.10 hrs. He arrived at Khartoum (1,037 miles) at 12.00 hrs (GMT) on February 7th and left at 12.25 hrs.

On February 8th he reached Kisumu (1,088 miles) at 05.30 hrs (GMT) and left again at 06.55 hrs. He landed at dusk 80 miles south of Salisbury (Southern Rhodesia) after missing the town in a bad storm. He damaged the wheel fairings, which were repaired at Salisbury when he returned there in the morning. He left Salisbury (1,260 miles) at 06.00 hrs on February 9th (GMT) and landed at Capetown (1,383 miles) at 18.03 hrs - (total 7,134 miles).

His route followed that of Imperial Airways and his time was 3 days, 17 hours and 37 minutes. He beat the previous best time made by Mrs Mollison, who flew the shorter Western route (6,300 miles) (in a Puss Moth in November 1932 - PA) *by 13 hrs 19 mins* (her actual time being 4 days, 6 hours, 56 minutes - PA). *He was fresh when he arrived, though he had had but two hours' sleep and had not used the special tablets to prevent sleepiness, which he had carried with him.*

One particular highlight in the trip was the very short time (14 hours) which he took to reach Cairo (2,366 miles). He had two mishaps. A petrol leak was caused by carelessly packing a suitcase against an ordinary metal petrol-pipe. He was lucky it did not happen over the Mediterranean. This forced him back to Wadi Halfa for repairs. The second mishap was that he had to return to Salisbury to refuel after passing it in failing light during a bad patch of weather. Without these troubles he would certainly have clipped one whole day from the previous best flight to the Cape.

Flt Lt Tom Rose had a distinguished war record in the RFC and RAF. He brought down more than a dozen enemy machines and earned the DFC. For some years he was chief instructor to the Midland Aero Club at Castle Bromwich. Then he went to the Anglo-American Oil Co Ltd (and while working for them in 1931, he had a go at the record to the Cape in the Avro Avian G-ABIE High Test but failed to reach it on the outward flight and then failed to break the record on the return flight - PA). *Thereafter he was chief instructor and manager at Sywell. And always and everywhere he was immensely popular. In 1934 he went as sales manager to Phillips & Powis Aircraft Ltd of Woodley Reading. Besides all that he earned four caps for RAF rugger against the Army and Navy in 1924-7 and he is one of the best all-round sportsmen and pilots whom we have the pleasure of knowing.*

G-ADLC naturally had a Gipsy Six 200hp motor with a Claudel Hobson carburetter, two BTH magnetos, KLG plugs, Auto-Klean filters, Superflexit pipe-lines, a Fairey metal airscrew, a Weyburn camshaft, Geo Salter and Co valve-springs, and a Peto and Radford battery. The engine bearers were of Reynolds tube. The motor was supplied with Shell and Castrol from tanks made by DJ Hawkins of Early Road, Reading, who also made the cowling and wheel fairings of British Aluminium.

Palmer wheels and tyres with Bendix brakes were fitted to an undercarriage of Reynolds tube sprung by Geo Salter and Co springs. The airframe of the Falcon was built of spruce supplied by Bambergers and of Saro laminated plywood, covered with fabric supplied by the Bessbrook Spinning Company of Aldersgate Street, EC, and doped with Titanine.

The cockpit held a Short and Mason 'Sestrel' compass, Sperry directional gyro and artificial horizon, Smith's instruments and an Ever-Ready electric torch. Rumbold did the upholstering and soundproofing, the Harley landing light and navigation lights were supplied and fitted by Mechanisms Ltd of Croydon. Brown Brothers and Rubery Owen made and supplied every nut and bolt and screw in its construction and Lancegaye Safety glass, Rhodoid Plastilume moulded windows, by May and Baker, were fitted to the cabin.

The pilot wore an Irvin Air Chute and the pleased expression which goes with it. The sleep-preventers, the mere threat of which kept

him awake, as they were unused, were prepared to the prescription of Dr MacLachlan, medical officer in charge at Woodley Aerodrome and Reserve School, and are reputed to be capable of making a dormouse win a non-stop waltz competition. The hot water bottle was by Boots Ltd.

After the record breaking flight to the Cape, Tommy Rose sent a telegram to the firm which read *'Preggers definitely Primus'*! On 3rd March 1936 he left Capetown on his return trip and *The Aeroplane* for 11th March 1936 reported thus:

Flight Lt Tom Rose DFC left Capetown on his return trip to London on March 3rd at 04.08 hrs (GMT) in the same Miles Falcon (Gipsy Six) G-ADLC which won the King's Cup in 1935, and in which he made his recent record flight to the Cape in 3 days 17 hrs 38 mins. The previous best recorded time for the return journey was made in 1935 by David Llewellyn and Mrs Jill Wyndham, who flew the Hendy Heck home in 6 days 12 hrs 3 mins, but were denied an official record for the flight because they failed to average more than 150 km per hr over the Great Circle course between the two places. Rose had waited two days for a break in the general bad weather.

He arrived in Bulawayo at 12.30 hrs and at Salisbury at 16.40 hrs. On March 4th, after being twice forced back by low cloud and heavy rain, he left Salisbury at 10.45 hrs and reached Mbeya for the night. On March 5th he got to Kisumu, Lake Victoria, at 09.40 hrs, and took off for Juba at 11.30 hrs. On March 6th he arrived at Khartoum at 09.30 hrs and reached Cairo the same day at 20.07 hrs, thus fracturing all records between the Cape and Cairo.

Going out he had landed at Malta but he was forbidden to land there on the way back - perhaps the Admiralty feared lest the superior performance of the Falcon might dishearten the inhabitants. So he was obliged to go by way of Tunis. On March 7th at 04.00 hrs he left Cairo and arrived at Benghazi, Tripoli, at 10.40 hrs. His chances of beating the record by more than 24 hrs began to look promising, but he was arrested by the Italians for flying over a prohibited area. He was detained during the rest of March 7th while the authorities conferred. A telegram to his wife, a telephone from her to Mr Harold Perrin, c/o Mr Lindsay Everard at Ratcliffe, and a cable from him to Marshal Balbo had immediate effect. The Falcon was de-gaoled to start next day.

On March 8th he left early, took off from Tunis at 14.26 hrs, and landed at Cannes at 16.30 hrs. On March 9th at daybreak he left Cannes and arrived at Croydon at 11.05 hrs with five hours in hand. The Lord Mayor, Sir Percy Vincent, and the Lady Mayoress received him at the Mansion House at 13.00 hrs.

The total distance by the route he took is 7,863 miles. And considering that he lost a whole day in Tripoli, his time - 6 days 7 hrs, - was very good going. The Great Circle Course is just on 6,000 miles, so his speed is not enough to be homologated as a record.

And, in Tommy Rose's own words, once again taken from the *Reading Review*, this time of April 1936:

On the Mediterranean crossing, that is between Malta and the coast of Africa, I was not so lucky. The crossing was one of 700 miles and it was raining hard all the time. I had to fly at 50 feet most of the way, and as I have often done before (and probably shall again), I said to myself 'If I can only get to land I'll never do this again'. Some time after leaving Cairo, I suddenly noticed a strong smell of petrol in the cabin, and on investigation found that a petrol pipe had been fractured owing to a suit-case vibrating against it, so a forced landing was an immediate necessity. It was just dawn and by sheer good luck I was able to pull off a landing in very wild country, 150 miles from the nearest living person, and 100 miles from water. I managed to repair the leak and took off again, arriving at Khartoum at noon, where the Falcon was refuelled. The next lap was to Kisumu, which I reached at 5.30 am on the morning of the 8th, and after an hour's rest I carried on to Salisbury. During this leg of the course after flying across miles and miles of bush country, I suddenly came across a little farmhouse set in a small clearing. In front of this farm there was a flagstaff, and on top of this fluttered a small Union Jack. This sight aroused my emotions very intensely, and I felt very proud to think that I was flying under the same flag.

Just south of Salisbury, which I completely missed owing to bad visibility, I ran into a torrential cloudburst and was literally forced to the ground. This may seem strange to those used to flying in England, and if I had been told of it beforehand I should not have believed it myself. However, the good old machine which had come through so much could not stand this immense weight of water, and was forced lower and lower. At the critical moment a clearing appeared ahead and she landed at full throttle and without damage. In the morning I got off again and returned to Salisbury, where the tanks were refilled, and took off on the last lap to Cape Town, which was reached at about 6 pm just as dark was falling.

The next three weeks were spent in making business flights around the Union, and on the morning of March 3rd I regretfully bade farewell to South Africa, and took of at 4.08 am for England. The weather did not behave very kindly to me on the way back, as I had to contend with headwinds nearly all the way. At times these reached 60 mph and brought my cruising speed from 150 down to 90 mph. The flight back was done along the same route as the outgoing journey, with the exception that I had to cut out Malta owing to the aerodrome being prohibited for landplanes at the moment. I decided to land at Benghazi aerodrome in Tripoli. As I reached the 'drome the Italians fired every kind of rocket they could into the sky in order to bring me in to land. Down I came and was nearly arrested for flying over an Italian prohibited area! I said that I did not know it, but this did not help. However, when the authorities learned that I was trying to beat the record they were very helpful, and I was enabled to cable to England, where Commander Perrin of the Royal Aero Club at once sent a telegram to General Balbo, who is governor of Italian Tripoli, and on Sunday morning I was 'released'. I reached Cannes on Sunday night and left for Croydon early on Monday the 9th, arriving there at four minutes past eleven.

With its success assured, the Falcon Six was to be produced in several different models to suit individual tastes. The sole M.3C G-ADLS differed from the M.3B in having dual control and four seats, and this aircraft came in fifth in the 1935 King's Cup Air Race. The later M.3D Falcon Six was similar to the M.3B but was strengthened in certain respects and had different equipment installed, while other versions had modified undercarriages.

The Falcon Six was an extremely pleasant aeroplane to fly and thanks no doubt to its 5 degrees of dihedral, as compared with 3½ degrees on the Hawk, its stability was quite exceptional. Owing to this high degree of stability however, the controls were perhaps a little heavy judged by modern standards, but on a cross-country flight, one had only to point the nose in the required direction and the heading never varied, thereby reducing pilot fatigue on long flights.

The story of Falcon Six ZK-AEI was told in the Christchurch Press between 29th December 1935 and 2nd March 1936, from which extracts were later published in the *Aviation Historical Society of New Zealand Journal*:

Towards the end of 1934 it was announced that the Union Steam Ship Company would apply for a licence to operate a Palmerston North to Dunedin trunk airline service. As a result of this application Union Airways of New Zealand Ltd was formed early in 1935. Squadron Leader Malcolm MacGregor MC, who had been appointed service manager to the company, left for England at the end of May 1935 to purchase suitable aircraft for the route and chose three DH 86's ZK-AEF, ZK-AEG and ZK-AEH and additionally a Miles M.3B Falcon Six ZK-AEI for taxi and charter work. The Falcon arrived in Wellington about the middle of December of that year. MacGregor went there from Christchurch on the evening of 18th December to see and test fly it for himself.

By 31st December all was set to make its first cross-country flights; the first took place between Wellington and Christchurch when ZK-AEI, declared by the newspapers to be the fastest aircraft in

CHAPTER 13: MILES M.3B, M.3C, M.3D, M.3E AND M.3F FALCON SIX VARIANTS

New Zealand, took only 82 minutes on the job. The pilot was Mr MacGregor with passengers Mr N S Falla, managing director of the Union Steam Ship Co Ltd, and Mr F M Clarke, technical adviser to Union Airways. The party then transferred to DH.86 ZK-AEG to carry on to Dunedin, where they had lunch. In the meantime Mr G R White returned the Falcon Six to Blenheim in 72 minutes against a head wind, from where Mr MacGregor, who returned to Blenheim from Dunedin, flew ZK-AEI back to Wellington. On the return trip to Christchurch, which was reached at 7 pm, head winds were again encountered. However, the time of 74 minutes was the quickest time in which the two centres had ever been linked up to that date.

Newspaper reports of that time also stated that the Falcon Six had been flown between Wellington and Palmerston North, the company's northern terminal, in 32 minutes average time for a mileage of 85. The average time to fly across Cook Strait, including a climb to 5,000 feet, was 22 minutes during early January 1936. The newspaper also quoted a maximum speed of nearly 190 mph with the 200hp Gipsy Six motor, but mentioned that it would be slightly less with all seats occupied. During January faster speeds were attained between centres with the Falcon Six and scarcely a day went by without a new record being made.

However, only a month later the Falcon Six was completely wrecked. On the morning of 18th February 1936 a young sheep farmer from Irishman Creek station, Mr C L F Hamilton (later known for making jet-boats, and also in the gliding world), arrived at Wellington on the steamer from Christchurch, and chartered the Falcon Six to fly to Hamilton on urgent business the same day. Piloted by A V Jury, Mr Hamilton completed his business and they flew back to Palmerston North the following day. Then, shortly after 2 pm Mr Hamilton left Palmerston North for Rongotai, this time piloted by Mr MacGregor.

For the greater part of the flight both outward and inward the weather could not have been worse and on arrival at Palmerston North there were very heavy rain storms and poor visibility. Mr Hamilton later wrote: we left Palmerston North in the rain and flew out a few miles to sea, following the coast down to Rongotai. On arrival there, I think cloud base would have been about 500 feet and it was raining extremely heavily, wind at gale force and visibility a few hundred yards. Water was lying over the centre of the aerodrome but there was a fairly clear patch in front of the hangars. I remember Mac opening the port window. There were two buildings with the anemometer mast in between. I do not remember any words being spoken; it was a particularly noisy machine, and in fact, we spoke very little all the way down. We approached from over the sea, took a half circuit round the aerodrome coming close over the houses on the hill of the town side from the north. MacGregor obviously intended to avoid the water and land in the dry patch. He might have forgotten for a moment the anemometer mast, or more possibly the turbulent conditions prevailing at Rongotai were the cause of his hitting it.

As we came over the edge of the aerodrome the anemometer mast cut the starboard wing off a few feet from the fuselage. When this occurred I remember feeling that something had happened and almost simultaneously a crash. I heard nothing from MacGregor and since there would not have been half a second in between the loss of the wing and the crash, I do not think he would have had time to realise what had happened. The first thing I knew we were upside down on the ground. I had only a few inches clear for my head and could not see MacGregor properly. My first thought was of fire and felt I could move around and kick a hole in the fuselage. But when I moved I found it painful and as there was no sign of fire and I heard voices alongside, almost immediately, I lay still while the plane was lifted and someone pulled me out.

I was not knocked out but I was dazed; I remember asking after MacGregor but getting only rather vague replies. I gather he did not regain consciousness and died shortly afterwards in hospital. I remember being examined by a doctor and being surprised when he told me that no bones were broken.

C G Grey wrote an obituary of Squadron-Leader MacGregor in The Aeroplane for 26th February 1936:

All who have come into contact with him will learn with the deepest regret of the death of Squadron-Leader Malcolm C MacGregor, late RFC and RAF, who has for years been regarded in his own country as the leading aviator in New Zealand. We in this country remember him and his partner Walker as having put up a marvellous performance in a Miles Hawk in the MacRoberston Race. They beat all other light aeroplanes in the race and reached Australia in 5 days 15 hrs 13 min, - well ahead of record time.

After a distinguished career in the War 1914-18, MacGregor went home to New Zealand, and as soon as aviation began to develop in that Dominion he took a part in it. Most people in this country heard of him first as the pilot who flew an air-mail service, all on his own, from the South to the North of New Zealand to send the Christmas mails home in 1933. After that he came to this country for the great MacRobertson Race. He and his partner Walker were two of the most popular aviators at Mildenhall and their splendid performance in the Race gave great satisfaction.

He was killed in just one of those absurd accidents which take from us so many of our best men. Surely the supreme irony of Fate is that a man who flew gallantly and dangerously in the War, and so skilfully from England to Australia, a man who had already done such fine flights in New Zealand, should be killed merely by hitting a post with his wing-tip. Not only was MacGregor the most famous aviator in New Zealand, but he was also one of the best beloved. Everybody had a good word to say for 'Old Mac', whether as a pilot or as an individual. The last time we saw him was at the King's Cup Race in September last at Hatfield. And wherever he went one always saw him surrounded by friends.

In late 1935, an M.3B Falcon Six was ordered by the Air Ministry and given the serial K5924. It was used for research by the RAE into wings with leading-edge slats and modified flaps and was delivered to Farnborough in April 1936. In late 1937, the RAE decided that they would also like to investigate the stalling characteristics of various wings with high taper, because certain types of aircraft with highly tapered wings were found to have vicious wing-dropping tendencies at the stall. K5924 was selected for these tests and duly delivered to the RAE in April 1938, with three sets of high taper outer wings, with a taper ratio of 4.45 to 1. All these had the same root section (Modified Clark YH), with a root chord of 88in and the root thickness of 0.19 chord, which necessitated the construction of a new centre section. The aspect ratio was 6.75. The tip sections at Rib M, 136in from where the outer wing joined the centre section, with a tip chord of 27.7in, were Clark YH, NACA 4415 and Gottingen 387 and the high taper wing span was 154in from the centre section to the tip. The wing span remained at 35ft.

The wing with the Clark YH tip produced bad stalling characteristics and at the stall the wing dropped sharply with the aircraft tending to spin. The wing with the NACA tip was good with the stall being quite mild, first one wing and then the other dropped 30 degrees. The third wing, with the Gottingen tip, was not so good, first one wing dropped 45 degrees then the other dropped 70 degrees followed by the nose but, provided that the control column was held central, there was no tendency to spin. This latter was generally comparable with the standard Falcon wing (which had a Clark YH tip but a taper ratio of only 1.6 to 1.0) and produced a reasonable stall although if the control column was held back after the stall, a wing would drop sharply and a spin would start. The RAE later issued Report No.2166 entitled *Tests on Falcon with high taper wing* (1937).

Concurrent with these tests, Phillips & Powis also built three low-taper outer wings for the RAE for comparative purposes and these were fitted to the M.3E Falcon Six L9705 during 1938. All these wings had the same root section (Modified Clark YH) with a root chord of 75in. The aspect ratio was 6.75. The tip sections at Rib M, 136in from where the outer wing joined the centre section, with a tip chord of 50.87in, were NACA 23009, NACA 4415 and RAF 28 and the low taper wing span was 154in from the centre

K5924, the Miles M.3B Falcon Six Special at Woodley, before application of the serial, in March 1936. The aircraft was delivered to RAE Farnborough in April 1936. [L. Tysoe]

Chapter 13: Miles M.3B, M.3C, M.3D, M.3E and M.3F Falcon Six Variants

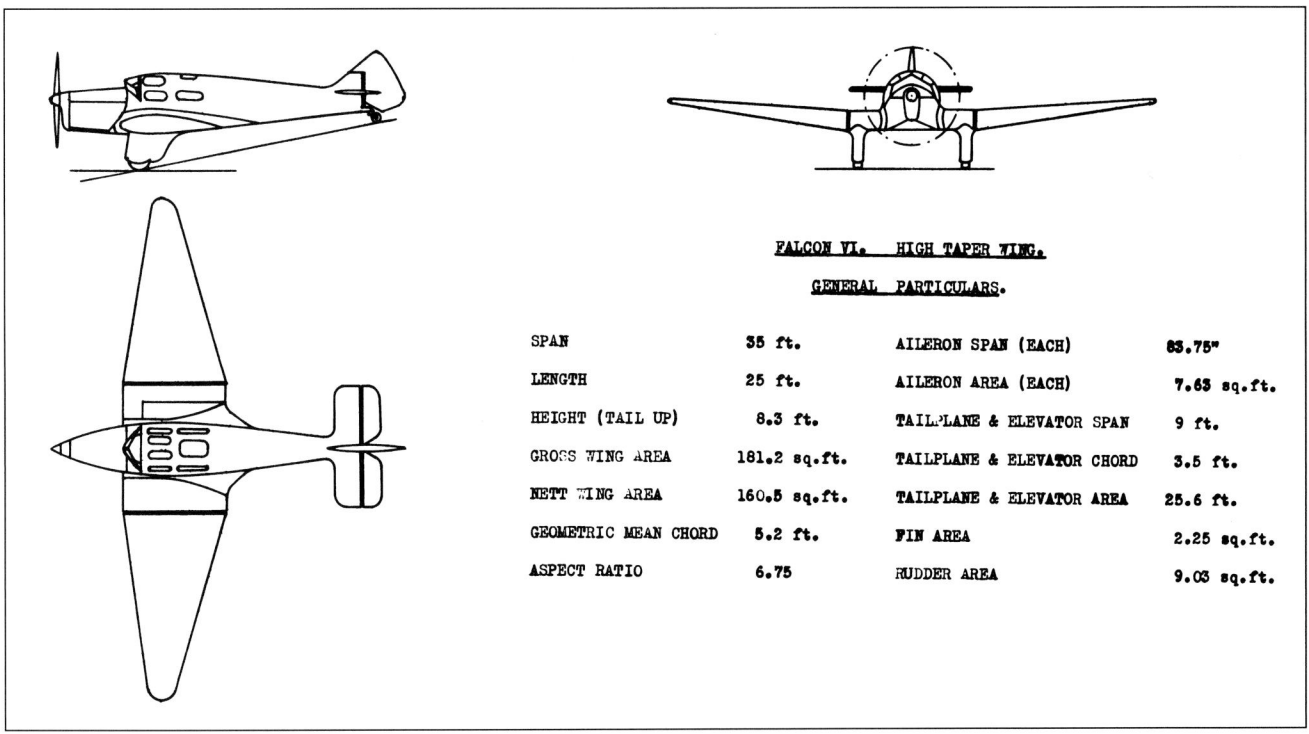

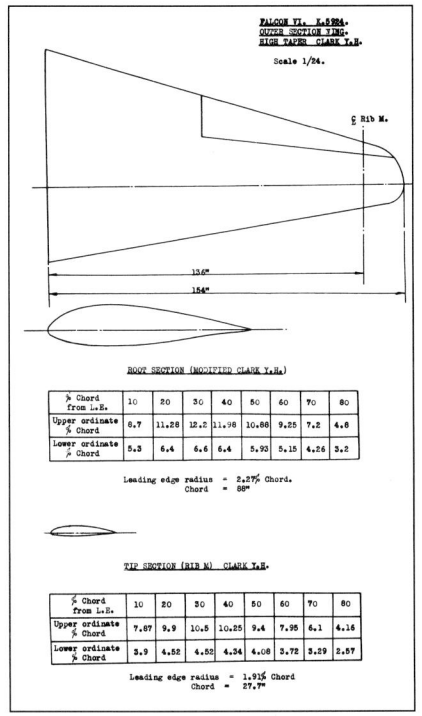
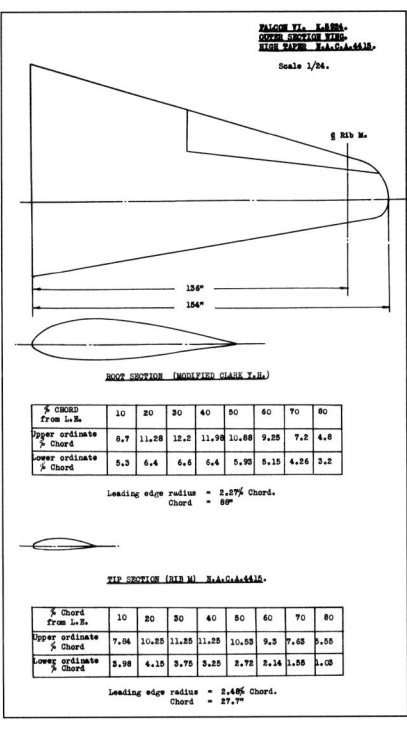
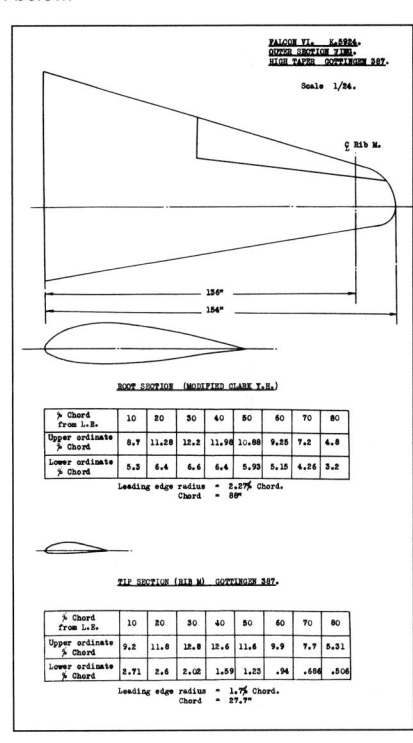

Above and below: G.A. of Falcon VI with the three different outer-section high-taper wings shown below.

section to the tip. The taper ratio in each case was 1.6 to 1.0. Both the NACA tips gave a good stall with no tendency to spin and the RAF 28 tip had generally good characteristics, except that with the CG aft and flaps up there was a tendency to spin. Again the wing span remained at 35ft.

An RAE Report, entitled *Flight Tests of a Falcon aircraft fitted with an Irving Flap*, dated 1938 (ref AVIA 6/2243) referred to experiments undertaken on K5924, although no further details of this flap are known.

In early 1939, the RAE wanted to obtain full-scale flight data on the Piercy laminar-flow wing sections and Phillips & Powis built three sets of alternative wings for fitting to K5924 for comparative tests. Each wing was approximately 19% thick at the root with the maximum thickness being at 0.4 chord, 0.5 chord and 0.6 chord respectively. The tests showed that as the maximum thickness moved aft, the drag approached the calculated value for laminar flow until, at 1.5 chord, this value was obtained. Moving the maximum thickness further after gained little value. When fitted with these wings, the aircraft became known as the Falcon Six Special.

In 1944, L9705 was modified to flight-test the biconvex laminar flow section wing (in the low speed range), which had been chosen for the M.52 high-speed research aircraft. A wooden biconvex wing of the same shape and section as that proposed for the M.52, but with straight leading and trailing edges, was fitted to the Falcon Six, which also had a narrow-tracked undercarriage attached directly to the fuselage in order to keep the wing free of all excrescences. The razor-sharp leading and trailing edges of the wing led to L9705 being known as the 'Gillette' Falcon. Given the

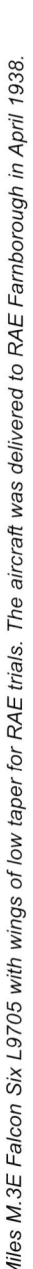

Miles M.3E Falcon Six L9705 with wings of low taper for RAE trials. The aircraft was delivered to RAE Farnborough in April 1938. ['Flight' via B Clarke]

CHAPTER 13: MILES M.3B, M.3C, M.3D, M.3E AND M.3F FALCON SIX VARIANTS

RAF Experimental Aircraft No.237 for identification purposes, the 'M.52 Falcon Six fitted with biconvex wings', as it was known by the firm, was first flown, by Hugh Kennedy, on 11th August 1944. The original tailplane was retained for the early flight trials but was changed later to one of the same shape and section as the M.52, but still with elevators. This was later changed for an 'all-moving' variable incidence tailplane (with no elevators) - the first of its type to be so fitted to an aeroplane (discounting those used on aeroplanes of the early pioneers) - in order to be more representative of that envisaged for the M.52 high-speed research aircraft.

In August 1939, another Falcon Six was sold to RAE for tests with spoilers. This became the sole M.3F with serial R4071. The later M.7 Nighthawk and the projected M.10 Queen Wasp radio-controlled target both owed their origins to the Falcon Six.

Few stories about the use of impressed aircraft during the war have ever appeared in the aviation press but Lt Alan Peter Goodfellow RNVR told me of a flight he made in the Falcon Six AV973 (previously G-AFBF) in December 1944, which could have had an unhappy ending. The flight was made shortly after he had completed his test pilot's course (No.2) with the ETPS at Boscombe Down, together with Ken Waller (later Chief Test Pilot at Miles Aircraft), Jimmy Orrell (Avro), Dick Northway (Bristol), Zurakowski (Gloster Aircraft) and others. Peter had been posted to Westland Aircraft at Yeovil to gain experience in production-testing newly-built Seafire IIIs. The RTO there had left some luggage at Hendon and asked Peter if he would collect it in the firm's Falcon. So off goes Peter, via Boscombe Down (to see his friends there) and Woodley, where he decided to call in to see Ken Waller, Hugh Kennedy and Tommy Rose.

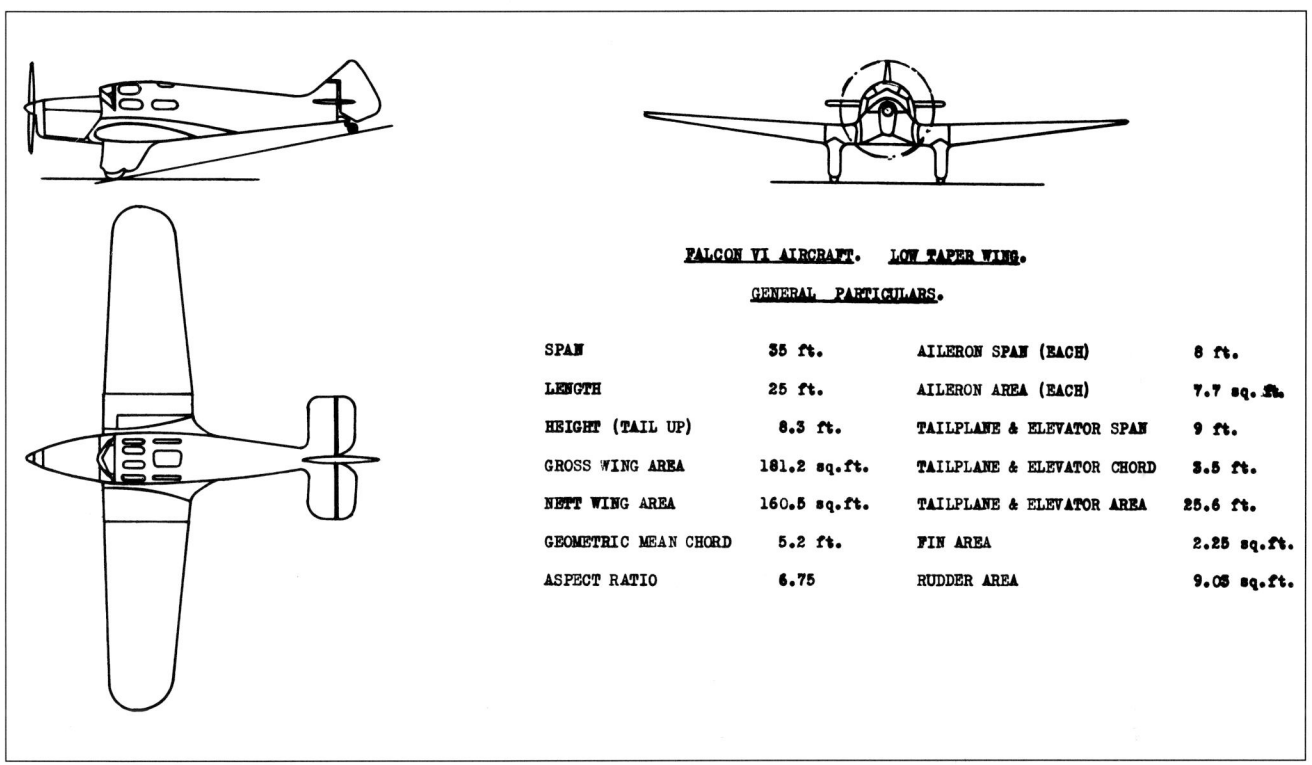

Above and below: G.A. of the Falcon VI with the three different outer-section low-taper wings shown below.

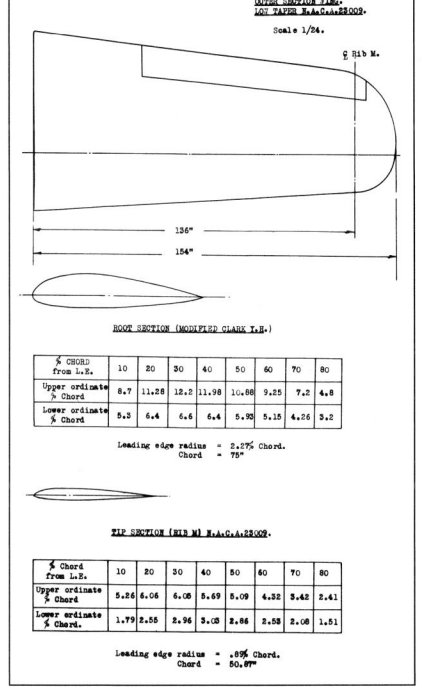 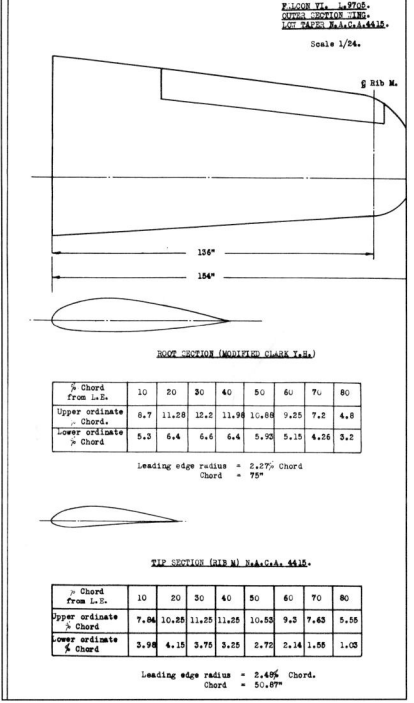 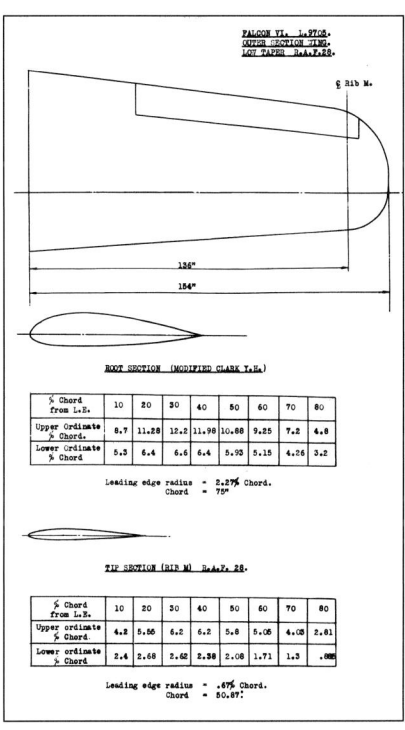

After the usual pleasantries they adjourned to the White Hart at Sonning - the Miles test pilot's local - for the odd pint, or two! However, things must have got somewhat out of hand because he recalled not getting to bed until five in the morning - somewhat the worse for wear and probably not remembering the main purpose of his flight that morning!

After he finally awoke and breakfasted, he repaired to Woodley and, probably against his better judgment (if he had had any, still being somewhat 'under the weather'!), he took off and headed for Hendon. The next thing he remembered was landing there at in excess of 120 knots, the cruising speed he had used to get there! Now, the aerodrome at Hendon, not being over large, was soon used up and it was only just before he went through the hedge at the end of the runway, did he realise that he had landed somewhat fast! He managed to stop, by dint of a smart ground loop, fortunately without any damage to the aircraft or himself! Needless to say, this close shave had an instant sobering-up effect on him and he immediately decided against calling in at Woodley on the return flight to Yeovil, choosing to land at Odiham instead!

For details of M.3 Falcon Six production see Appendix 14.

SPECIFICATION AND PERFORMANCE DATA

An undated Phillips & Powis Aircraft Ltd sales leaflet entitled *An important advance in Light Cabin Aircraft Construction - New and Improved Miles Falcon Six* gave details and specifications:

Miles Falcons, since their introduction last year, have proved their worth over thousands of miles of private ownership. They have proved to give advantages that no other design can approach. The Falcons are beautifully finished, and with their great speed range and attractive price are outstanding value in the aircraft world.

In accordance with their policy of introducing improvements in their aircraft as and when desirable, Phillips & Powis Aircraft Ltd have improved the Falcons in appearance, comfort and in many engineering refinements. Not only has new plant been installed and new methods of construction adopted freeing the machines from maintenance troubles, but alterations in design and detail improvements have been incorporated.

Construction:	Wooden. Fuselage and wings ply-covered.
Engine:	200hp Gipsy Six
Standard Instruments:	Compass; airspeed indicator; revolution counter, aneroid; oil pressure gauge; inclinometer; 8-day clock.
Cabin Furnishings:	A new standard of comfort has been established by the cabin furnishing and upholstery of the Falcon. Plenty of leg and head room has been provided for the largest passengers, and seats are fitted with safety belts. The upholstery is of grey antique leather with head lining and carpets of material to match. The pilot is placed centrally in front in a comfortable seat with arm rests and the two passengers behind in side-by-side seats of ample dimensions.
Standard Equipment:	Miles' split trailing edge flaps, hydraulically operated; cabin controlled cold air ventilation; luggage locker and rack for light articles behind passenger seat; fire extinguisher; emergency exit; rudder bar compensator; Bendix wheel brakes; flap control indicator; wooden airscrew (Standard model); metal airscrew (Speed model).
Standard Loose Equipment:	Tool roll and grease gun; engine, aircraft and journey log books; engine and aircraft instruction books.
Wheels:	Dunlop wheels and tyres, 400 x 180; Speed Model - 600 x 100.
Standard Colour Scheme:	Tailplane, wings, rudder and fin - silver. Fuselage - any single colour scheme, using the following without extra charge: blue, green, black, grey, dark red and silver.
Optional Equipment:	Electric self-starter and battery; landing lights; navigation lights.

Performance - With maximum permissible all-up weight of 2,500 lb.

SPEED SIX	Maximum speed 180 mph; cruising speed 160 mph; landing speed 40-45 mph. Take-off run with flaps 140 yards; landing run with brakes 130 yards. Standard range 560 miles; maximum range 880 miles. Initial rate of climb 1,125 ft/min; climb to 5,000 ft 4 min 55 sec; climb to 10,000 ft 11 min 34 sec. Service ceiling 20,000 feet.
STANDARD MODEL	The performance figures are as above with the exception that the top speed is reduced by 5 mph.
Weights:	Tare weight 1,550 lb; pilot 170 lb; petrol 260 lb; oil 25 lb. Two passengers 320 lb; luggage 175 lb. AUW 2,500 lb.

The dimensions were not given in the leaflet but those for both the Speed Six and Standard Model were as follows:

Dimensions:	Span 35ft 0in (folded 15ft 10in); length 25ft 0in; height 6ft 6in; wing area generally quoted as 174.3 sq ft but pre-war Phillips & Powis brochures and Milestones for May 1939 both quote 181.50 sq ft (Note: the span on both the high and low taper outer wings tested on K5294 and L9705 was 35ft as on the standard model and both had gross wings areas of 181.20 sq ft with aspect ratios of 6.75 - PA)
Weights:	AUW M.3B 2,350lb, M.3D 2,500lb; wing loading: 13.5 to 15.2lb/sq ft

Miles M.3E Falcon Six:

Engine:	205hp DH Gipsy Six II
Dimensions:	Span 35ft 0in, (with biconvex wings as M.52 Falcon Six - span 29ft 0in; length 25ft 0in)
Weights:	AUW 2,500 lb
Performance:	Max speed 164 mph.

The prototype Miles M.4 Merlin on 1st April 1935 before the application of 'B' mark U8. ['Flight' via B Clarke]

Chapter 14:
Miles M.4 and M.4A Merlin

The M.4 Merlin was a development of the M.3B Falcon Six and was built in early 1935 to the specification of Flt Lt George Birkett, of the air taxi company Birkett Air Service Ltd, Heston. Birkett wanted an aeroplane which would carry five people, including the pilot, and cruise at 145 mph. The 200hp DH Gipsy VI engine was chosen to power the new machine, and in order to get a good take-off performance it was proposed to use a variable-pitch propeller. The propeller, it was hoped, would have been in production by the time the aircraft was built but, in the event, the aeroplane was ready before the propeller and a fixed-pitch wooden propeller had to be fitted instead.

The Merlin resembled a rather corpulent Falcon Six in appearance with the fuselage being over four feet wide to enable the pilot and one passenger to sit in front with three 'hefty' passengers to be seated comfortably on the wide seat behind. Radio was carried and the Merlin was designed to have dual control if desired. The prototype Merlin, which was actually 'manufactured for sale', was of standard Miles wooden construction, with spruce members covered with three-ply birch, and featured hydraulically operated balanced split flaps and folding wings. The firm's brochure described it thus:

> *The Miles Merlin is designed as an economical private charter or feeder line aeroplane. It can be used as either a mail carrier or a comfortable miniature air liner for five persons (pilot and four passengers), or alternatively it can be converted easily into an aerial ambulance. Considering the low horse-power of this machine, it has an outstanding performance, and we claim that it can be operated as cheaply as the average 3-seater light aircraft of today. The cabin of the Merlin is large and roomy and luxuriously furnished. This, coupled with its high performance and well balanced controls, makes it a delightful aeroplane to fly.*

It is agreed that the Miles Patented Split Trailing-Edge Flaps are one of the greatest advances made in civil aviation in recent years, and on the Merlin they not only reduce the landing speed, but make the approach and landing a great deal easier than that of the

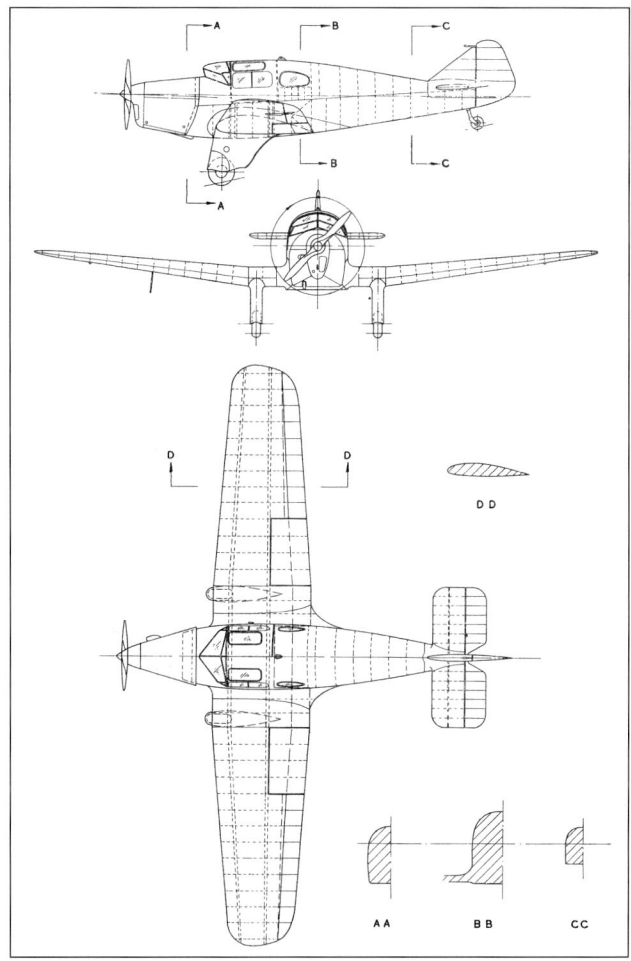

The Miles M.4 Merlin.

normal aeroplane. Because of its easy flying qualities we feel that the Miles Merlin will also appeal to the private owner who requires more room and carrying capacity than offered by the average light aircraft.

The prototype Merlin, with 'B' mark U8, was first flown, by Miles, on 24th March 1935 *(not the 11th June 1935 as stated in many reference works, including 'A pictorial précis of some of the aircraft designed and built by Miles Aircraft Ltd' and published by them!)*, and its take-off capability with five people and their luggage was later demonstrated to be quite remarkable. With a genuine cruising speed of 140 mph and a landing speed, with flaps, of just 50 mph, the Merlin was really quite a remarkable aircraft.

The Merlin was flown to the A&AEE, Martlesham Heath in April 1935 for CofA trials and their Report No.671, dated June 1935, criticised it for poor brakes and poor undercarriage damping but still recommended it for a CofA. Although registered G-ADFE on 5th April 1935, it continued to fly as U8 for some time after and a photograph taken at the A&AEE dated 12th April 1935 shows it as such. It was eventually delivered to Captain Birkett on 11th June 1935.

Flight for 4th April 1935 gave a very good description of the Merlin and the reason for its creation under the following sub-heading: *New five-seater with Gipsy Six engine and very complete standard equipment: high performance with economical running indicate fitness for taxi work.*

Not many months pass without rumours of yet another new machine about to be produced by Phillips & Powis at Reading. This firm has been in existence as aircraft constructors for only a little over a year, but already it has produced the Hawk, Hawk Major, Hawk Racer, Falcon, and now the Merlin. Mr Miles, the firm's designer, is, incidentally, probably unique in the aviation industry, as he has, in his drawing office, the help of his wife. Mrs Miles is, with him, equally responsible for the originality and success of the machines they have designed and built.

The Merlin is the outcome of collaboration between the constructors, Flt Lt Birkett (Birkett Air Service Ltd) and the Tata interests, who operate extensive airlines in India. The first model is being delivered to Flt Lt Birkett for his taxi services. It is particularly interesting as, with the exception of racing machines, it is the first standard British civil aeroplane to be designed to use a controllable-pitch airscrew, and it is also being marketed with an unusually full equipment - a commendable feature which we hope to see become general practice. In view of these features, let it not be thought that the Merlin is suitable only for the commercial operator. That is far from the case; it should prove equally attractive to the private owner, who wishes to buy a machine with a high performance and one which is ready to go anywhere at any time without the necessity for extra equipment being purchased.

In general, the Merlin is a development of the Falcon - a machine very much in the news at the present time owing to Mr Brook's record flight in the prototype from Australia. The seating accommodation has been increased so that five persons can now be carried in perfect comfort; one beside the pilot in front and the other three on a sofa seat behind. Furthermore the luggage-locker, which lies immediately behind and slightly above this latter seat, is deep enough for a stretcher case to be carried, and a patient can be transported without removal from the stretcher.

Structurally, the Merlin is a low-wing wooden-built monoplane with box plywood and spruce spars, plywood covered wings and fuselage, and cantilever undercarriage; it differs from the earlier Miles machines in small details only. The span of wing has been increased and this has resulted in an increase of the taper and aspect ratio. A further departure from the Falcon is an increase of the dihedral angle of the wings to 7 degrees, which has had the effect of making the Merlin very stable and comfortable to fly.

The general design of the cabin is very like that of the Falcon; the forward-sloping windscreen has been retained, as has the neatly hinged cabin door, while large windows on each side of and above the seats make the cabin particularly light and give a pleasantly airy impression. The engine, a DH Gipsy Six, is carried on a welded steel-tube mounting and the neat cowling which surrounds it is obviously responsible to no small extent for the exceptional performance of the machine. Like all modern P&P machines, the Merlin is fitted with the Miles wing-flap gear, which is hydraulically operated by a small quick-acting pump near the pilot's left hand.

A short flying trial soon showed that the Merlin is a very considerable advance over anything which Mr Miles has yet designed. In spite of the fact that the new Ratier controllable-pitch

Miles M.4 Merlin at the A&AEE Martlesham Heath in April 1935. [Air Ministry Crown Copyright via P Jarrett]

Chapter 14: Miles M.4 and M.4A Merlin

Miles M.4 Merlin G-ADFE of Birkett Air Service Ltd at Heston in September 1935. [J Edelsten via Croydon Airport Society]

airscrew was not then available and that a standard wooden airscrew had to be used, the performance was quite outstanding. With five people and full tanks the take-off was below the 200-yard mark, and with flaps down the landing speed was, by ASI, about 46 mph. In the air it is at once noticeable that the increased dihedral of the wings gives that high degree of positive lateral stability which, we consider, is a desirable feature of commercial aeroplanes. The Merlin can be flown 'hands-off' for long periods with perfect safety, even without the feet on the rudder bar; in other words, directional as well as lateral stability is ample and definite.

Furthermore, we are assured by the designer that bumps are corrected without any undue lurching; so that the machine may be considered as one which can be safely left to fly itself while maps are examined or lunch is eaten. This is a very desirable feature, particularly when it is combined with a high degree of manoeuvrability, and when, as in the Merlin, the fin areas are so proportioned that turns can be made without touching the rudder. The occasion of our flight was one of fine drizzling rain, so it was possible to prove the practicability of the unusual windscreen design; it was extremely pleasing to find that there was no difficulty at all in sitting forward so as to get very near the windscreen - a great help in conditions of this kind.

The ratio of gross weight to tare weight is unusually good for this class of aeroplane, thus there is yet another reason why the machine should prove admirable for taxi operators. That is really just what would be expected when it is known how the Merlin came to be conceived. Even with the high percentage payload which can be carried in the present model the figure is about 800 lb - the range is excellent, the tankage for 44 gallons of fuel permitting journeys of about 800 miles without refuelling, and when the controllable-pitch airscrew is available both this and the cruising speed will probably be improved.

The last truly commercial aeroplane - that is, one which enabled operators to make profits - to be built as the direct outcome of a designer giving an operator what he asked for was, perhaps, the DH Dragon. This was designed as a result of collaboration with the late Mr E Hillman and proved to be one of the most successful machines of recent times. It is interesting to note that it was built of wood just after a phase when most machines were built in metal. Now we have the Merlin, also built in similar circumstances and also of wood. If it finds as much favour among operators abroad as has the other, it will certainly appear that the old argument that aeroplanes for abroad had to be of metal has ceased to carry any truth.

In conclusion, the completeness of the equipment is again worth mentioning. The controllable-pitch airscrew will, as soon as it is available, enhance the already outstanding performance; the electric starter, navigation lights, and landing lights built into the leading edge are what we hope to see as standard on all aeroplanes before long.

Birkett's Merlin was to undertake many long distance charter trips, including one to Addis Ababa with a newspaper man covering the Italian invasion of Abyssinia in 1935/6. In fact, most of the photographs of the war in Abyssinia used by the newspapers were flown to Cairo in the Merlin. On 10th December 1935, Birkett sent a letter to Phillips & Powis giving details of that experience:

Having successfully completed a fifteen thousand mile charter, flying between Europe and Africa in connection with the Italian Abyssinian disputes, I wish to congratulate you on the performance of my machine - the Miles Merlin. The ease of handling this machine both in the air and on the ground made the flying of eleven hours a day for consecutive days, with stops, a comfortable and tireless proposition. The range of one thousand four hundred miles was very comforting in view of the long hops that I was often compelled to make, whilst the average speed of 140 mph on a light throttle enabled me to cover fifteen hundred miles per day. This was, of course, with the extra tanks fitted and carrying one passenger.

My passengers were agreeably impressed with the comfort and roominess of the cabin, also the steadiness of the machine in bumpy conditions. This machine is a worthy addition to the range of your productions. Yours faithfully, Birkett Air Service Ltd.

In 1950, George Miles commenting on this flight wrote: *The charter flight of 15,000 miles at 1,500 miles a day on a 200hp five-seater could not be surpassed today.*

G-ADFE roamed all across Europe until the outbreak of war in September 1939 when, unfortunately, all trace of it was lost. It was possibly destroyed by the German bombing somewhere in France or the Low Countries in early 1940.

Miles M.4A Merlin VH-UXN of Roberts Airways in 1936. Note the previous registration VH-UBN still on the wings. [E Coates Collection]

Chapter 14: Miles M.4 and M.4A Merlin

Tata Sons Miles M.4A Merlin seen with Chief Engineer McWade and Peter Menezes in mid-1935. [Via P Amos]

The second Merlin, designated M.4A, registered VT-AGP, was built to the order of Mr J R D Tata of Tata Sons Ltd, Bombay, and was handed over to the customer on 12th June 1935. Two further M.4A Merlins were built in 1936 and in the light of experience gained with the first two aircraft, the plywood on the fuselage sides in the area of the centre section was increased in thickness to 2.5 mm and the flap gear was modified. One of these was delivered to the Victorian and Interstate Airways Ltd, Melbourne while the other was a second machine for Tata Sons Ltd.

The sole Australian example, initially painted as VH-UBN (later changed to VH-UXN), went into passenger service between Melbourne-Deniliquin-Hay in August 1936, and was soon averaging a weekly utilisation of 25 hours, covering 3,233 miles and carrying 22 passengers. The service was only delayed on very rare occasions when the weather was extremely bad, and this regularity was responsible for the service also being well patronised for mail and freight in addition to passengers. 'Aerotopics' in the *Miles Magazine* for March 1938 published:

This extract from a letter received from the owners of Miles Merlin VH-UBN which is being operated on an Australian air service, is proof of the sterling qualities of Miles Aircraft. "We think we should tell you at this stage, how satisfied we are with the operation of the Merlin on our service. As you know, we had difficulty in handling this machine off the ground with anything like a load when we first got it, but as a result of the experience we have had, we have now developed a technique which enables us to get the machine off the ground with quite a short run with any load we like to put in her.

For instance, recently the writer, who weighs 14 stone, had four 14 stone passengers, a full load of luggage weighing approximately another 150 to 200 lbs, and tanks almost full, and the machine came off in comparatively calm air in something like 250 to 300 yards. The method we use is to head the machine into wind then, disregarding the stick, push the throttle wide open, and clamp it there. Then, as the machine develops a run, gradually get the tail up, but not too high, and when 55 to 60 mph are showing on the Airspeed Indicator, start pumping on the flaps, up to three double pumps (it takes nine double pumps to get our flaps fully down). As the flaps are pumped in this way, the machine becomes slightly nose heavy, and if pressure is borne back on the stick, it appears that she is just pumped off the ground. The flaps are then pumped up slowly after she has obtained 300 to 400 ft in height.

It should be noted that the Merlin was more than 300 lbs overweight in this trial! The remarkable performance of a 200hp five-seater aircraft which could be used from the smallest aerodromes and yet cruise at 145 mph will be readily appreciated.

Tata Airlines commenced operations, for both passengers and mail, with their first Merlin, VT-AGP, on 26th November 1935 and a report on *The Tata Mail Services* appeared in *The Aeroplane* for 8th January 1936:

The Karachi - Madras service is primarily for mails, but many passengers have been carried. The demand for seats between Bombay and Karachi is increasing, but the mail loads on that particular section have become heavy and often the company has

A rare photograph showing both Tata Sons Miles M.4A Merlins together at Bombay in 1936. [Via B Clarke]

not been able to carry passengers. On 16th October we recorded that Tata engineers had recently assembled and flown a Miles Merlin at Yervada aerodrome, Poona. This is cruising at 135-140 mph. The scheme is to fly the whole Karachi - Madras trip in the day. (In fact, Tata were ultimately to claim a record time for the flight to Hyderabad - PA). *At present the machine leaves Karachi on Mondays and Fridays at 06.30, calls at Ahmedabad at 10.20 and Bombay at 13.40 and lands at Hyderabad for the night at 18.10. It goes on at 06.30 next morning and gets to Madras at 09.55. The Northbound service leaves Madras at 14.00 hrs on Mondays and Fridays, stops the night at Hyderabad and goes on to Karachi by 18.10 the following afternoon.*

In the monsoon, from June to September, the service goes from Ahmedabad by way of Poona to Hyderabad, cutting out Bombay because of bad weather between Bombay Island and Ghats, 30 miles East of Bombay, which are the edge of the Deccan and rise to about 2,400 ft. From October this year passengers are being carried all along the line, and if the Merlin proves satisfactory it may become standard.

The Merlin obviously proved satisfactory as Tata Airlines ordered a second machine which was delivered in March 1936. Soon after this, Tata very nearly lost one of their Merlins and the story of this was told by Keki R Gazder AFRAeS of Air-India in 2001 in his book *Gypsies to Jets*. Barring the odd error in the story, it is still an object lesson for all pilots:

One afternoon I was called out to do a test flight on one of our Miles Merlin aircraft. The Merlin was a low-wing monoplane powered by a Gipsy Major engine (see below - PA). *It carried three passengers and a pilot* (actually four passengers and a pilot - PA). *It was underpowered and consequently notoriously earthbound. This characteristic, coupled with the short take-off run available at Poona, made it extremely difficult and the pilots hated flying it. On this occasion the Merlin had had an engine-change, hence the test flight.* (One can only hope that the Indians had not actually changed the engine to a Gipsy Major, which would almost certainly have given the take-off problems experienced, as the Merlin was definitely not under-powered with its Gipsy Six engine – PA).

I reported to the engineer in charge, collected the form, checked the engine logbook and carried out the usual mandatory formalities before the test flight. I then got into the aircraft, started the engine, warmed it up and operated the controls to ensure that they were free. I operated the flaps to ensure that they were working satisfactorily, then I held the stick right back to keep the tail on the ground and opened the throttle wide. I checked the full-power RPM, the oil-pressure and tested the magnetos. I was satisfied that everything was just right.

I raised my right arm over my head and waved it left, right, left, right, the signal to remove the chocks. When the mechanic had given me the all-clear, I opened the throttle and taxied out for take-off. From the corner of my eye I saw the Chief Engineer rush out of the hangar frantically waving his arms, I looked, he was signalling to me with his palms turned up and was waving his arms up and down from the elbow. I interpreted this signal as 'Hurry up and get off the ground'. I made a rude signal with my two fingers and carried on.

When I reached the extension, I turned the aircraft round, opened the throttle and started my take-off. The aircraft moved forward gathering speed and yet I felt that something was wrong. The acceleration was sluggish. I had used up practically three-quarters of the available run and yet I had not reached the lift-off speed. 'God, these bloody Merlins!'

The short fence at the end of the field was rushing at me at tremendous speed; still there was no flying speed. To cut the engine and try to stop the aircraft was out of the question. I reached the irrevocable decision. I held the aeroplane down until the very last second, then heaved back on the stick and got airborne, missing the fence by inches. I barely gained 50 feet when the aircraft started to shudder. I knew I was approaching the stall, I eased the stick forward just a fraction, the shuddering continued. I had gained a little more height, so the stick was eased forward a shade more. After an eternity the shuddering stopped but the air-speed indicator was hovering dangerously near the stalling speed. I felt trapped. There was not enough speed to climb and the trees ahead were getting taller.

A couple of seconds later I was brushing past the treetops, I was told later that there was no more than six inches between my wheels and the tree-tops. I kept flying straight and tried to find out the cause of the lack of flying speed. I checked the engine RPM again and again, but there was nothing wrong. Suddenly my eye caught the flaps indicator. To my disgust it showed flaps FULL DOWN. There was the culprit; the drag created by 'full flaps' was slowing me down. I had been hanging on my propeller all this time but luckily I managed to fly across the belt of trees and was now flying over level barren ground.

My carelessness in not checking the flaps again before take-off had nearly cost me my life. Then, in my agitation I made another mistake which, by rights, should have killed me. As soon as I saw that I had 'full flaps', I pulled the lever to the 'Up' position instead of gradually milking the flaps up. The action ended in near disaster, the aeroplane just sank down. I lost all my precious altitude and I hit the ground with my wheels. Most fortunately the patch I hit was fairly level and very hard. The aircraft bounced in the air and I was up to 60 feet again. I was so shaken that I do not think I did anything consciously to fly the aeroplane but somehow it kept flying and gained speed. Then it suddenly dawned on me what the Engineer meant by the frantic waving of his arms. It was very clear now that he was telling me to get my flaps up before I taxied out.

Although the Phillips & Powis brochure mentioned that the Merlin was capable of conversion to an air ambulance, there is no record of it ever being used in that capacity.

Some sources might claim that the Merlin was a failure, as only four were produced but this was not due to aircraft's performance but due to the fact that perhaps there just wasn't the right market for it. With all the DH aircraft, Fox Moths and Dragons etc rolling off the assembly line, perhaps it was either just a little too expensive for the operators, or ahead of its time.

For details of Miles M.4 Merlin production see Appendix 15.

SPECIFICATION AND PERFORMANCE DATA

Engine: 200hp DH Gipsy VI
Dimensions: span 37ft 0ins; length 25ft 10ins; wing area 196 sq ft
Weights: tare 1,700lb; load, made up as follows: pilot 160lb, four passengers 640lb, petrol 397lb, oil 35lb, luggage 118lb; disposable load 1,350lb; AUW 3,050lb, (except for c/n 274 where the AUW was increased to 3,150lb); wing loading 15.5 lb/sq ft
Performance: max speed 155 mph (with wooden airscrew); cruising speed 140 mph; rate of climb 900 ft/min; landing speed 50 mph; range 560 miles
Equipment: VP airscrew (new type Ratier), self-starter, landing lights (Harley's), navigation lights, wireless (standard but the price varied according to the set used), turn and bank indicator (ditto).
Basic price: £1,630 ex works.

Miles M.5 Sparrowhawk G-ADNL in its original form with three-piece windscreen in August 1935.   ['Flight' via B Clarke]

Chapter 15: Miles M.5 and M.5A Sparrowhawk

In early July 1935, just eight weeks before the King's Cup Air Race, Miles announced his intention of flying in the race. There was no time to design and build an entirely new aeroplane, so Blossom solved the problem by suggesting that standard Hawk Major components should be used wherever possible in its construction. The shops were full with 12 other Miles aircraft, some works' entries and others belonging to private owners, but all being prepared for the race. The design of the new aircraft, to be named the Sparrowhawk, became the sole responsibility of Blossom, who also undertook to do the chasing - no easy job at such a time.

A standard Hawk Major fuselage was taken and shortened from the front by 12 inches, the rear top decking was lowered as far as practicable; the pilot was to sit practically on the floor with his legs passing over the front spar and the centre section was reduced in span by approximately 2 ft 6 in on each side to the width of the fuselage. The outer wings, fitted with Miles split trailing-edge flaps, were attached directly to the fuselage, thereby reducing the span by 5 ft to 28 ft 1½ in. A high-compression 140hp DH Gipsy Major engine was fitted and the shortened, new type single-strut undercarriage 'of robust design' was moved outboard to clear the slipstream from the propeller. This undercarriage was intended to reduce the drag and improve the general efficiency of the machine considerably. Three fuel tanks were provided; one in the fuselage behind the engine, and one in each wing to make it possible to complete the first day's course of 953 miles with only one stop for refuelling.

The prototype M.5 Sparrowhawk was registered G-ADNL to Miles on 12th August 1935 and was first flown, by him, on August 19th with its CofA being issued two days later. From all accounts, the Sparrowhawk's handling characteristics proved to be excellent, with a speed range of over 4 to 1 and, with a landing speed of 42 mph, the Sparrowhawk was exceptionally easy to fly.

As originally flown, the Sparrowhawk had a three-piece windscreen like the Hawk Major, but this was soon changed to a

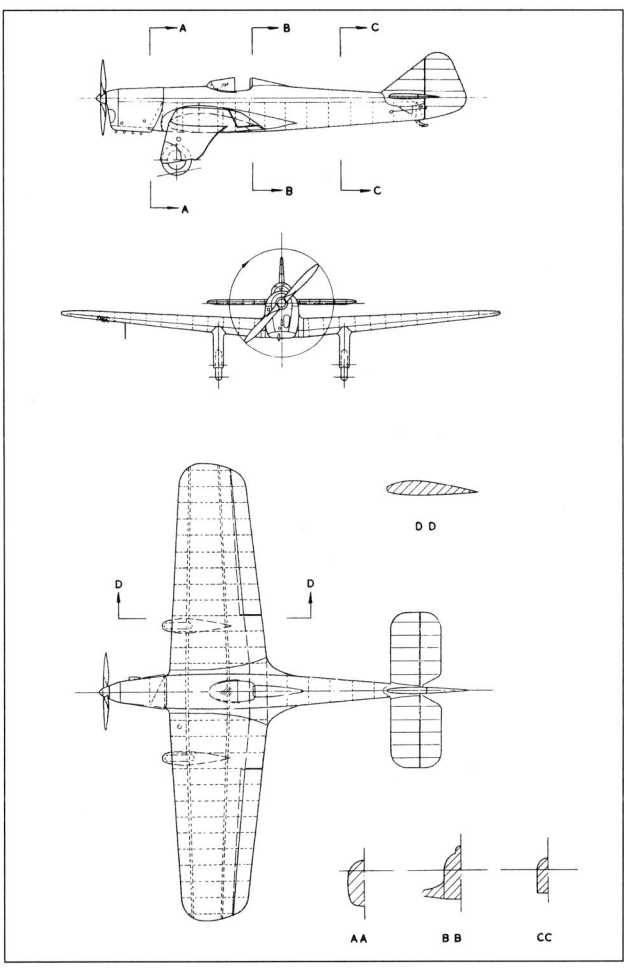

The Miles M.5 Sparrowhawk.

lower and longer one-piece wrap-round screen, and it is of interest to note that two different photographs of G-ADNL taken at the time (both with Race No.9), show the two different windscreens, so it must have been changed just before the 1935 King's Cup Air Race.

On August 25th, Miles carried out a 2 hour 40 min 'King's Cup practice' flight before flying to Hatfield and round the course of the race on the 27th, noting that on the last leg of the course, from Woodford to Cardiff, he 'lost Ruth Fontes'. Then, on August 28th, he made a return flight to Cardiff etc before flying the second day's course on the 31st.

On September 5th he flew to Hatfield for the start of the race and flew part of the course. The King's Cup Race, which was to be flown over two courses, was held at Hatfield on 6/7th September 1935. The first day's race was an un-handicapped circuit of Britain starting at Hatfield, then to Newcastle, Edinburgh, Glasgow, Belfast, Dalbeattie, Blackpool, Cardiff and back to Hatfield, with compulsory landings at Glasgow (Renfrew), Belfast (Newtownards) and Cardiff. The second day's race was seven laps of a 50-mile course from Hatfield to Broxbourne and Henlow before returning to Hatfield.

On September 6th, Miles was the fifth competitor away and dead-heated for second place at the Glasgow control. He still held second place when he reached Belfast which, thanks to his long-range tanks was his only refuelling point. A fierce duel resulted between the two designer-pilots, Miles and Edgar Percival (in his Mew Gull G-ACND) and, at Cardiff, the last control, Miles touched down just as Percival was taking-off on the last stage to Hatfield. Miles left Cardiff fractionally under three minutes behind the Mew Gull and passed Percival who, in his effort to increase his lead, had made a slight navigational error of judgement, while Miles, flying a straight course, deservedly won the race, having averaged 163.84 mph for the 953 miles circuit.

On the second day, Miles completed the seven laps and came in 11th at 172.38 mph owing to severe handicapping. He was later heard to remark that his only regret was at not arriving in time to see Tommy Rose come in first in the Falcon Six G-ADLC at 176.28 mph!

Nonetheless, Phillips & Powis (with 11 Hawks of varying types and two Falcon Sixes entered) succeeded in gaining an outstanding victory in the 1935 King's Cup Air Race, by taking first, second, third and fifth places, a very creditable performance and one never to be repeated by any manufacturer.

On 12th October 1935, G-ADNL was flown twice by George Miles and Flt Lt J F Moir, a test pilot with the A&AEE Martlesham Heath. Moir had flown to Woodley in the Hawker Fury K2876 to carry out a 'Handling New Type' test flight, of 20 minutes duration, in the 'Miles SSR-5', which must have been G-ADNL (with the entry perhaps translated as 'single-seat racer [type M] 5'). Shortly after, on 25th November 1935, James Moir joined Phillips & Powis on the formation of their Reserve Training School at Woodley, as chief flying instructor, although he made his first flight there on 21st November 1935 in the Hawk Trainer Mk.II G-ADVF.

Designer and Pilot: "Entrant for 1935 King's Cup Air Race, Mr F G Miles, and his wife. Mrs Miles designed the Miles M.5 Sparrowhawk which Mr Miles will fly".
[London News Agency Photos Ltd via P Jarrett]

Miles flew G-ADNL occasionally after the 1935 King's Cup and one notable flight, which he made on 30th January 1936, was from Reading to Paris, via Heston, in 1 hr 45 minutes. CofA tests of the Sparrowhawk G-ADNL 'at increased all up weight' were undertaken by the RAE in 1936 and it may have been in connection with these tests that G-ADNL was flown six times (for a total time

Miles M.5 Sparrowhawk G-ADNL with later moulded windscreen for the King's Cup Air Race in September 1935.   ['Flight' via P Amos]

Miles M.5 Sparrowhawk, flown by Pat Maxwell, refuelling at Shoreham on the first lap of the 1936 King's Cup Air Race on 10th July 1936.
[Fox Photos via P Jarrett]

of 1hr 55 min) by James Moir at Woodley between 7th and 10th May 1936 and thrice on 25th May 1936 (55 mins total) (the RAE Report, dated 1936, ref AVIA 6/4636).

G-ADNL was flown by Pat Maxwell in the 1936 King's Cup Air Race from Hatfield on July 10/11th, round a course of 1,380 miles, at an average speed of 165.74 mph, to gain 9th place and win the speed prize. In the 1937 King's Cup, held at Hatfield on September 10/11th, G-ADNL was flown by Wing Cdr Freddie Stent, who averaged 172.50 mph to gain 7th place.

F G Miles in the Sparrowhawk G-ADNL just prior to the start of the 1935 King's Cup Air Race. ['Flight' via B Clarke]

The second Sparrowhawk, registered G-ADWW, also had a modified Hawk Major fuselage and was entered by its owner, Gordon McArthur, in the 1936 King's Cup Air Race, but was disqualified in the eliminating race. This machine was originally intended to have been a two-seater but this was changed to single-seat before completion. On 3rd September 1936, it was sold to James Hopkins Smith Jnr (a pilot with Pan American Airways) and was despatched to the USA in two crates aboard the *SS Georgic,* arriving at Long Island, New York on 27th September 1936. It was the first British-built aeroplane to be imported into the USA under the new reciprocal agreement (Executive Series No.69) and was also the first Miles aircraft in the USA.

In the US Department of Commerce Application for registration NC191M, details of repairs and alterations were described thus: 'Dural strip - 14 inches long - full width of bottom of fuselage - placed over fabric aft of exhaust stacks. Two hinges and one hold-down pin added to both sections of wings supporting wing-flaps (see affidavit attached)'. The affidavit, signed by Smith, stated that: *'The undersigned on September 3rd 1936 paid to Phillips & Powis Aircraft Company, Ltd, the full purchase price for the Miles Sparrowhawk Aircraft G-ADWW on the condition that it be delivered to him ex-steamer New York City free from all liens of any nature. Payment was made through the Guaranty Trust Company, Lombard Street, London, and the ship was delivered in New York on September 27th. The vendor neglected to forward a bill of sale but has been requested by cable to do so and on receipt of this document it will be handed over to the Department of Commerce.*

A Letter of Approval Group 2-524 was issued 15th October 1936, stating that 2.3 hours total flying time had been logged in England. With reference to the repairs and alterations mentioned above, it was, in fact, Aero Trades Inc at Roosevelt Field, Long Island who had placed the 14 inch long Dural strip over the full width of the bottom of the fuselage over the fabric covering aft of the exhaust stacks and added the two hinges and hold down pins to both sections of the wings supporting the wing flaps. Test flights showed that *'the speed is not quite as high as expected but the general flying characteristics and manoeuvrability at low speeds are excellent'*. Smith was at first interested in a Miles' proposal to fit the Sparrowhawk with a Menasco C4S engine and even flew the Sparrowhawk to the Pacific coast in November 1936 to discuss the possibility with Mr Menasco but, in a letter to L A Hackett (Phillips & Powis sales manager), Smith concluded that the *'performance of*

Miles M.5 Sparrowhawk NC191M at Burlington, WV, in original single-seat configuration. [Via P Amos]

Miles M.5 Sparrowhawk after modification to two-seat configuration at Burlington, WV. [Via P Amos]

the ship with the Gipsy has been so satisfactory, however, that I do not believe the change will be made' - it never was. During the trip, the Sparrowhawk *'caused much interest at each airport visited'* and these included Philadelphia (Pennsylvania), Washington (DC), Dayton (Ohio), Indianapolis (Indiana), Fort Lordsburg (New Mexico), Tucson and Phoenix (Arizona), San Diego and Los Angeles (California), and Boston (Massachusetts).

In December 1936, the US licence was revoked pending further details from Great Britain, but the Sparrowhawk operated on a temporary licence until, in March 1937, Phillips & Powis Aircraft Ltd confirmed that it met Aeronautics Bulletin 7A. In June 1937, Max Karant of Chicago requested ownership details but, if this was a pending sale, it never occurred and NC191M was placed in store at Bridgehampton, New York for 8 months. On 4th June 1938, it was flown under a temporary authority to Roosevelt Field for inspection (flying time to date 202 hours) and on 1st May 1939, Smith (by now married with a family and having acquired a Fairchild 24) sold the Sparrowhawk to Perry Boswell Jnr of University Park, Maryland. NC191M was repainted from its sky blue to white with red trim on the undercarriage trousers, fitted with a steerable tailwheel and named *Electron*.

Boswell flew the Sparrowhawk extensively for trips including California and New Mexico and all over the eastern USA. It competed in small races and was demonstrated at air shows in the mid-Atlantic states in 1939 and 1940, including the 1939 Langley Day Air Races at College Park, Maryland, where it won its displacement class. A letter from Aero Trades Inc dated 16th January 1940 to L V Kerber, Chief Airworthiness Section, Civil Aviation Authority, Washington DC, stated that:

We have been requested by one of our customers to convert a Sparrowhawk into a two-seater (but making no mention of the fact that the customer also wanted to make it a closed two-seater - PA). *I believe this is the only ship of its kind in the country at present. It has a cover over the front cockpit, and we would like to know if*

Chapter 15: Miles M.5 and M.5A Sparrowhawk

it is permissible to make the ship into a two-seater, and install a sliding pyralin hatch over the two cockpits.

Their reply was dated 1st February 1940:

This refers to your letter of January 16, concerning the proposed conversion of this airplane from single place to two place. As you no doubt know, this airplane was manufactured under the terms of a Great Britain Airworthiness Certificate. We regret that we cannot approve alterations of this nature without information from Great Britain as to the structural integrity of the airplane with the proposed alterations and flight testing by an engineering inspector of the Authority. As an alternate method, complete compliance with CAR 04 may be proven.

In 1940, the Langley Day Air Races were held at Hybla Valley Airport, Washington, and the Sparrowhawk not only won its displacement class again, but also took second place in the unlimited class for standard licensed aircraft, losing a potential first place to a 450hp Beech 17 in the last half lap of the four-pylon course. Of this race, Perry Boswell later wrote: *Due to the agility of the Sparrowhawk, I was able to lead until the last half of the last lap by staying low and tight.*

Perry Boswell joined the Royal Canadian Air Force in December 1940 and became F/Lt P Boswell, O/C Test Flight, No.9 Repair Depot RCAF, St. Johns, Quebec, obtaining permission to operate the Sparrowhawk as his personal transport there, still under its American registration. On 30th June 1941, L Scholfield of Hyattsville reported to the CAA that he had made major repairs to landing gear and both wings (the repair was actually to the nose ribs, right landing gear struts, right wing and spar) and the following day, it was inspected and licensed. The Sparrowhawk was flown from Burlington to Canada in January 1942 and in order to make operation tolerable in the cold Canadian weather it was later fitted with a cockpit canopy and raised rear decking. In this configuration it was timed over a measured course at 184 mph, which compared with its original maximum speed of about 170 mph, was quite an improvement. Tyres became a problem until some tractor front tyres of the appropriate size were found (after the war Fiat 500 tyres were used to good effect). In May 1942 Perry Boswell joined the USAAC and on 19th February 1943, having attained the rank of Captain, he wrote to the Department of Customs, Washington, DC:

Forwarded herein are the facts regarding storage and use of aircraft NC191M which is registered and licensed in the name of the undersigned. The license has expired and it is at the present hangared in dead storage at No 9 Repair Depot, (RCAF), St John, Quebec, Canada under a 24 hour guard. The reason for the above being that the year and a half previous to May 1942 I was in the RCAF and upon leaving that service entered the US Army Air Forces. Before departure from Canada permission was obtained from the Station Commanding Officer to hangar my aircraft there for the duration. This action was deemed advisable as the aircraft would be guarded and hangared and not subjected to lack of consideration or acts of vandalism as could be expected from a commercial airport. This aircraft has been in Canada since its departure from Burlington, VT in January of 1942 and was flown there on local flights at the above station until the expiration of its current license. Your authority is requested to continue the storage of this aircraft in its present status and in addition the forwarding of any existing regulations or requirements which have not been complied with as yet. The CAA Branch Office at La Guardia Field, NYC was notified of the location of NC191M but no letter of approval has been received to date. It is hoped that the above meets with the approval of your demand.

The Sparrowhawk was later moved to storage at the de Havilland Canada factory at Toronto, where it remained until January 1945 when it returned to Buffalo, New York. In order to renew its expired CofA, the canopy and rear decking, which had never been an approved modification, had to be removed. A photograph taken at Burlington in 1946 shows NC191M with two open cockpits, but the FAA Dossier does not record this modification, or when and by whom it was carried out. Other photographs also taken at Burlington in 1946/7 show it to have a single cockpit, so it was probably modified there in about 1946. Boswell applied for a new licence on 11th July 1946, which was approved on 15th November 1946 through to 1st August 1947, but as he was not using the Sparrowhawk much by then, on 25th September 1946 he sold it to Carl L Conrad, Hyattsville, Maryland, West Virginia (although it initially remained at Burlington Airport), to whom it was registered on 22nd November 1946. On 18th July 1947, Conrad applied for licence, which was approved on 8th August 1947 (although no flying time had been logged since 13th July 1947), for authorisation for a ferry flight from Burlington Airport to Cumberland Airport for inspection. On 2nd August 1948, Conrad applied for licence, which was approved on 21st September 1948, although details were deleted on the form.

Some six years later, Perry Boswell, by now living at Delray Beach, Florida and tiring of his replacement aircraft, re-located the Sparrowhawk, still in West Virginia but unlicensed and in dead storage, and with less than 600 hours total time. On 6th August 1954, he once more became its legal owner, having paid $690 plus the $110 hangarage fee incurred by its previous owner and obtained a temporary permit to ferry it to Miami. Although engine compression was very poor, apparently due to stuck piston rings (as the valves etc, appeared satisfactory), the oil tank was filled with heavy duty oil plus other additives in the hope that the rings would free themselves en route. He set off, planning only one stop on the

Miles M.5 Sparrowhawk NC191M fitted with a cabin in mid-1942 for operation in Canada. [R Almond Collection via A Brook]

1,000 mile journey, since the aircraft had long-range tanks giving a total capacity of 60 gallons and a range of 1,000 miles. However, the rings did not free themselves and the resulting high oil consumption limited the range to about 200 miles, by which time the oil pressure was down to 10 psi and still falling, so the delivery flight took 1½ days! Boswell flew the aircraft to his cattle ranch at Pompano Beach (between West Palm Beach and Miami), Florida, where it was dismantled and taken by cattle truck to his home at Delray Beach for rebuild.

It was then painstakingly restored by four Canadian residents (UK emigrants, one of whom, Jimmy Jenkins, had worked for Miles Aircraft Ltd immediately after the war), in exchange for free accommodation during the working holiday. During the overhaul, consideration was given to increasing its performance by reducing the wingspan by 4 ft, installing a higher-powered Gipsy Major and fitting a constant speed propeller, or alternatively by substituting a 190hp Lycoming engine. The latter change was not favoured though, because 'it would lose the graceful English period appearance and become just another US special'.

In the event, none of the proposed changes were carried out and it was finally re-assembled in its original configuration and repainted at the local airport; the CofA being issued on 4th December 1956. The same month, Boswell sold the Sparrowhawk to a friend, George Roberts, an ex USAAF P-51 pilot, who kept it at Lantana Airport, just south of West Palm Beach, Florida, using it around Florida and for occasional trips. On 27th October 1957, it was sold to Sunrise Sales Co, Palm Beach although it is thought that George Roberts was probably involved with this business. At an air show at Lantana on 5th April 1959, George Roberts was invited to fly a friend's P-51, and it is recorded that 'he allowed a friend who did most of the maintenance on the Sparrowhawk' to fly it in a 'fly-by', which also implies that he still had an interest in the Sparrowhawk.

The friend was an ex Air Force Sergeant (non-flying) with limited civil flying experience and during the fly-by he pulled up nearly vertically, stalled, recovered and then stalled twice more. On the third stall it fell into a spin at low altitude; recovery action was apparently taken and the rotation stopped, but at too low an altitude to complete the recovery. The aircraft hit the runway almost vertically and disintegrating, 'virtually exploding as if it were blown apart by dynamite' to quote its former owner, who was a witness. The pilot was killed instantly. George Roberts was very upset and a few days later piled the remains of the Sparrowhawk and burned them, making certain that everything was destroyed completely. Someone, however, still has the engine data plate, which was apparently salvaged, and this still survives as a memento, which, together with a photograph of NC191M are co-mounted on a wall plaque.

There is a very strange twist to this sad tale however, as John Underwood, a Miles aircraft enthusiast living in the United States, claims that NC191M was offered for sale in '*Trade-a-Plane*' sometime after 1980 but, as its remains were supposedly 'totally destroyed by burning' by its last recorded owner, the reason behind the advert is unclear.

Three further Sparrowhawks, with newly designed fuselages, were designated M.5A, and these were laid down in late 1935 but, of these only the second was completed and sold at the time. This Sparrowhawk was built for the private owner who required a machine of high performance while still retaining all the characteristics of a slower aeroplane and was offered for sale at £825. A contemporary sales leaflet, prepared by Phillips & Powis, ended:

We look upon the Sparrowhawk as one of Mr F G Miles' best designs; it is definitely a thoroughbred, possessing beautifully harmonised controls and outstanding general abilities. For the private owner who appreciates performance (and after all the main excuse for flying is to get from "A" to "B" quickly) the Sparrowhawk is unique. It gives a performance on a mere 130hp, equal to a machine with more than double the power.

Those extra miles an hour are available at no extra cost to the operator, pure efficient design has given these unequalled speeds. The fact that this high performance has been obtained without affecting the ease of handling and low speed landing qualities makes the Sparrowhawk all the more remarkable. It is an aeroplane that will give the most hardened aerobatic pilot a 'thrill' and at the same time be an entirely suitable aeroplane for a pupil to carry out his first solo flight.

Apart from the fact that the design responsibility for the Sparrowhawk rested with Blossom - a fact conveniently overlooked in the leaflet - *that* said it all!

The new M.5A was registered G-AELT and was purchased in July 1936 by the young and well-known South African airman, Victor Smith who, in 1932 and 1933, had made three gallant but failed attempts to break the England - Cape record (then held by Amy

Victor Smith with his Miles M.5A Sparrowhawk G-AELT in September 1936. ['Flight' via B Clarke]

CHAPTER 15: MILES M.5 AND M.5A SPARROWHAWK

Miles M.5A Sparrowhawk G-AELT, Miles M.2U Hawk Speed Six G-ADOD and Mew Gull ZS-AHM at Portsmouth on 29th September 1936, ready for the start of the Schlesinger Air Race to Johannesburg. ['Flight' via B Clarke]

Johnson) in the Comper Swift G-ABRE. Smith had now entered in the Portsmouth to Johannesburg Schlesinger Air Race, which was to start on Tuesday, 29th September 1936.

South African millionaire entrepreneur Isidor Schlesinger had decided to offer £10,000 for an air race from London to Johannesburg on the occasion of the opening of the great Empire Exhibition in the latter city on 15th September 1936. As in the Melbourne race of 1934, the underlying motive was to promote air transport with the mother country and the handicap formula was devised to give best chances to the aircraft with the best characteristics as an airliner. There was one big difference with the Melbourne race however; participation in the Schlesinger race was limited to British subjects with aircraft (and engines) of British origin only. This probably accounted for both the limited number of entries and the relative lack of interest in the race and what happened during its course, outside of the British Empire.

There were two sections in the race, a speed race and a handicap race. The first prize, apart from a large cup, was to be £4,000, and the course set out for the race went via Belgrade and Athens to Cairo, to further follow the route taken by Imperial Airways, over Khartoum, Kisumu and Mt Pika, to finally reach Rand Airport, Germiston. The distance to be covered was 6,150 miles.

Of the 14 entries for the race, which started from Portsmouth on 29th September 1936, only nine aircraft were present for the start and, with the departure time set for 6.15 am, Victor Smith's Sparrowhawk was delayed as his propeller swung abortively when the flag fell, probably because the mixture was too rich for a hot engine. He finally took off after the others had gone. The Sparrowhawk had been handicapped to do 149.30 mph and it carried 100 gallons of petrol, sufficient for 11 hours effective range. With a cruising speed of 160 mph and a maximum speed of 180 mph, he was hoping to cover between 1,600 and 1,800 mile stages at a time. This would have involved him in just one stop for fuel, at Belgrade, where he arrived fourth. It is reported that Smith *looked so cold in his open cockpit without an overcoat*, that a kind Yugoslav mechanic insisted that he should borrow his leather coat, asking that Smith should return the coat, or its equivalent weight in gold, when he got to Johannesburg. Apparently, when Smith took off at 2.25 pm, he looked more comfortable.

Smith suffered serious trouble with the oil supply to his engine and had to force-land at Skopje in Yugoslavia. It transpired that the return valve in the oil tank was choked and this in turn had caused the engine to become practically filled with oil. Having sorted that problem out, Smith continued on to Khartoum, where he abandoned the race; it is not recorded if the Yugoslav mechanic ever saw his coat again! The Sparrowhawk (and Victor Smith) eventually arrived in South Africa, where Smith converted the aircraft into a two-seater cabin machine to his own design, which he then submitted to Phillips & Powis for approval. It was later recorded that: *'These drawings would have been a credit to any qualified aeronautical designer, and it is interesting to note that his machine has been entered by Lord Wakefield of Hythe for the Governor-General's Air Race, to be held at Durban in 7.38'*, in which Phillips & Powis Aircraft Ltd wished him every success.

The first and third uncompleted M.5A airframes were placed in store against future orders, and the first was later completed for Lionel Bellairs, who had earlier sponsored the Southern Martlet at Shoreham. Although registered to Bellairs as G-AFGA on 16th March 1938, he decided to purchase a Whitney Straight instead and so it was sold on to Bill Humble of Doncaster. It was delivered to Humble during the third week of May 1938, in time for him to compete in the London to Isle of Man Air Race in June, but he only managed 11th place.

C G Grey wrote of the race in *The Aeroplane* for 15th June 1938: *The brightest remark of the morning at the start of the race to the Isle of Man came from Bill Humble who extracted a collar from inside the high neck of his polo jersey and said cheerfully,- 'I always swim best without a collar',- a tragi-comic commentary on*

Miles M.5A Sparrowhawk G-AFGA at Hatfield prior to the start of the Hatfield to Ronaldsway, IoM, race on 4th June 1938, in which it was flown by Bill Humble. [Via B Clarke]

an organisation which tempts people in single-motor aeroplanes to fly over 40 or 50 miles of water, far out of gliding distance of land on the off chance of winning a cup and a small money prize. Humble also flew in the 1938 King's Cup Air Race held at Hatfield on July 2nd where he gained 7th place at an average speed of 169.50 mph. G-AFGA subsequently returned to Woodley where it was dismantled on expiry of its CofA in May 1939.

Both G-AFGA and the uncompleted third M.5A Sparrowhawk were later modified for high-lift flap research for the RAE Farnborough. G-AFGA, with 'B' mark U5, was used as a means of investigating cheaply, quickly and positively, a new type of flap, which another firm was contemplating for use on a projected shipborne fighter then being designed by them. While it has not been possible to positively identify this 'other firm', an interesting article on the Hillson Bi-Mono aeroplane, which was designed to explore the possibilities of a 'slip-wing' arrangement, appeared in *The Aeroplane* for 7th July 1944. In this it was stated that, in 1938, J Helmut Stieger, the originator of the monospar wing and subsequently Chief Designer of Blackburn Aircraft Ltd, had drawn up designs for a large trans-Atlantic flying-boat with an extremely high wing-loading, and that: *A special Miles Sparrowhawk, built to an order from Blackburn's, was used to test 'slip-wings' and high-lift devices.*

Whilst this statement implies that U5 was built to an order from Blackburn, no details of a 'slip-wing' being fitted to it or any other Sparrowhawk have ever come to light and Don Brown, who would almost certainly have been aware of it had it been tested and flown, never mentioned it. It is therefore considered highly unlikely that the idea was pursued. Although the idea to fit a Sparrowhawk with high-lift devices may have stemmed from this suggestion in the first instance, it has not been possible to confirm who actually placed the order, but it is more likely to have been completed to a contract from the RAE Farnborough. In the event, the specially designed Hillson Bi-mono was built by F Hills & Sons Ltd to test the slip-wing theory and the first release of the upper wing took place from this machine on 16th July 1941.

U5 was completed with large-chord trailing edge flaps, which could be extended rearward and downward along curved guides. These flaps were also coupled to large drooping ailerons in order to allow a high wing-loading, while keeping the stalling speed low enough to permit operation from an aircraft carrier. Much positive data was obtained from the tests as regards the lift obtainable and the handling characteristics and, in a modified form, the idea was later incorporated in a proposed fighter.

In order to test another idea, the third M.5A Sparrowhawk, with 'B' mark U3, was completed by the Experimental Department at Woodley in early 1940. This was fitted with large retractable double-slotted flaps extending for 45% of the span. The forward flap was 0.26 in chord and the rear flap was 0.13 chord. A reduction in stalling speed of 33% was obtained as compared with normal spilt flaps, and Don Brown recalled that the flaps were of such large chord that one had to remember to raise them before landing and Denis Bancroft recalls that they extended to one foot below ground level when fully down! In addition, the flap operation lever was so far forward that one had also to release one's harness in order to reach it. Finally, the stick had to be kept back during take-off because, if the tail came up, the propeller was liable to hit the ground because of the very short undercarriage.

Nevertheless, this Sparrowhawk was apparently delightful to fly with exceptionally crisp lateral control. This machine was later converted to standard configuration and given 'B' mark U-0223, before being registered G-AGDL, and was used by the firm as a hack. It was flown by Group Capt 'Bush' Bandidt to the Isle of Man, via Squires Gate, on Saturday, 24th May 1947, where he had entered it in The Manx Air Derby on Bank Holiday Monday, May 26th but, unfortunately, he could manage no better than 8th place in the race.

The late Ranald Porteous recalled his experiences in flying G-AGDL in *More Tails of the Fifties* as follows:

The Sparrowhawk, which I was still enabled to use from time to time for aerobatic displays, was a joy. It was immensely clean, with crisp controls and stabilities perfectly harmonised, cutting through turbulent air as straight as an arrow, without the slightest trace of 'jinking', yaw or Dutch-roll. It seemed a obey one's thoughts, always with effortless accuracy, and was by far the best and most enjoyable light aircraft I have ever known in the context of low-level aerobatic display flying. With most light aircraft one had to employ a certain amount of skill and downright cunning to achieve continuity while maintaining height during such displays, but in the Sparrowhawk the problem was to avoid accumulating more and more height as one's programme progressed!

In the Sparrowhawk I devised a pattern which suited its merits and which I called a 'four-leaf clover'. This required a display crowd arranged in 'L' form, which was quite usual. The Sparrowhawk would be flown very fast and low just inside one arm of the 'L', parallel to the spectators, then brought up into an absolutely vertical climb, aileron-rolled through precisely 90 degrees, the loop then being completed to bring the aircraft

Miles M.5A Sparrowhawk U5 with large-chord trailing edge flaps retracted and drooping ailerons on a re-designed wing of reduced span in 1939. *[Phillips & Powis via The Miles Aircraft Collection]*

U5 shown with flaps and ailerons fully down in 1939. *[Phillips & Powis via The Miles Aircraft Collection]*

down just inside the second arm of the 'L'. . . and so on, four times. Simple stuff, but easy for the Sparrowhawk and I was told that it 'looked artistic'! Sadly, this gem of an aircraft ended its days when a Rolls-Royce test pilot took off with the fuel tap wrongly set (the pilot was F Kirk, who was seriously injured in the crash, at Tollerton on 19th June 1948 – PA).

Meanwhile, on 17th March 1939, Geoffrey Alington, who at the time ran a small air taxi and charter firm at Gatwick, had flown to Reading to 'try out' G-ADNL, which was for sale. As he later wrote in *A Sound in the Sky – Reminiscences of Geoffrey Alington*: *I flew up to Reading to try it out. I fell in love with it immediately - it was a very fast young lady having a maximum speed of about 180 mph, cruised at an easy 150 mph and had an extremely ladylike stall which was gentle and occurred at about 42 mph. All this on an engine of 130hp. Having flown it, I immediately wanted to purchase it and set about completing the deal. So, on 17th April, less than three weeks after I first flew her, I had managed to pay for her and collect her from Reading. On that first flight from Reading to Gatwick, she sped me home in only 12 minutes averaging 190mph. I was 'pedalling' fast, but we also had a 30 mph tail wind to help us. I was well satisfied with her performance. She would ably fulfil what I had in mind for her. I named her 'Angela III'.*

Miles M.5 Sparrowhawk G-ADNL in the post-war racing colours of Geoffrey Alington. [G Swanborough Collection]

CHAPTER 15: MILES M.5 AND M.5A SPARROWHAWK

Miles M.5A Sparrowhawk U3 fitted with double-slotted, high-lift flaps for RAE trials in February 1940.
[Phillips & Powis via The Miles Aircraft Collection]

U3 shown with flaps retracted in February 1940.
[Phillips & Powis via The Miles Aircraft Collection]

U3 with flaps fully down in February 1940.
[Phillips & Powis via The Miles Aircraft Collection]

Miles M.5A Sparrowhawk G-AGDL at Woodley while being used as a hack by the firm during the war. [Via B Clarke]

His firm, Air Touring, had acquired a lucrative Air Ministry contract for Army Co-operation Flying and as the Sparrowhawk was capable of flying at over 150 mph, the rate paid was higher than for slower single-engined aeroplanes. On April 19th, Geoffrey flew up to Catterick to do two days flying for Army Co-operation with the guns and his reminiscences continue: *By now I felt very much at one with the Sparrowhawk and realised that she had, like a good trained horse, no vices and would answer to my slightest touch, so I had no qualms in landing her in a field at Swinhope, our family home, on the way back to Gatwick.*

This intensive Co-op flying continued on through July. Some days I was flying eight hours a day in my Sparrowhawk over Coventry, Leicester, Nottingham and Leamington. When in the Coventry area I often used to land at Major Bonniksen's small aerodrome at Tachbrook near Warwick, where he had a small flying club, The Leamington and District Flying Club. At the outbreak of war Air Touring came to an abrupt end and the Air Ministry eventually purchased all our aircraft, apart from the Sparrowhawk, as they felt a single-seater was no use to them. Being a bachelor it was my most treasured possession, apart from my bull terrier. I spent many hours worrying about its eventual fate. I finally came to the conclusion that it should be broken up as its hangar space might be required or a German bomb would blow her to pieces.

Luckily, unknown to me at the time, fate was going to be very kind to me and we should fly happily together throughout the war over bomb-scarred England until the sirens stopped their mournful wailing and the church bells proclaimed their peals of peace. This lucky break had come about as a result of Geoffrey becoming Chief Test Pilot for Austin Motors, first at Longbridge and then Elmdon and Geoffrey recalled, sometime after the first flight of Austin's first Stirling on 22nd February 1941: *Somehow I wangled my Sparrowhawk G-ADNL, had her camouflaged and then used her as a personal hack. She was fast and the Squadron pilots loved her where ever she landed. She was always much in demand for 'a whip round the circuit'.*

Post war, G-ADNL was flown by Alington in a number of air races and in the King's Cup Air Race at Wolverhampton on 17th June 1950, but while lying in 26th place, the magneto drive sheared, causing him to force-land at the disused Perton airfield. Although it was repaired, Geoffrey sold the Sparrowhawk soon after to Fred Dunkerley and, in December 1950, Dunkerley arranged to have it converted by F G Miles Ltd at Redhill into the M.77 Sparrowjet. It eventually emerged as such at Shoreham in December 1953, and its subsequent history will be documented at another time.

The M.77 was first flown, by George Miles, from Shoreham, on 14th December 1953. In the 1957 King's Cup Air Race, held at Baginton on July 14th and at the ripe old age of 22 years, the 'new' jet-propelled machine, flown by Fred Dunkerley, won at a speed of 228 mph, being the first jet aeroplane to do so. Following its success, the Sparrowjet returned to its base at Barton, where it was seen under dust covers in August 1959. The following year, it was moved to the BEAC engineering base at Heathrow (where it shared a hangar with the Nash Collection) and then to RAF Upavon, where the airframe was destroyed in a hangar fire in July 1964.

Rather surprisingly, the registration G-ADNL was restored to Kathleen Dunkerley in November 1991 as an M.5 Sparrowhawk. This was for a rebuild based on a few metal parts from the original G-ADNL, which had been left over and stored from the Sparrowjet conversion, plus some which were salvaged from the fire, together with the wings and some structural parts from the Hawk Trainer III G-ANWO. These are being used by Tim Cox at Bristol in the construction of a replica of the original prototype Sparrowhawk for Fred Dunkerley's son, Alan. The finished product will, it is understood, be flown under a Light Aircraft Association Permit as a 'Miles Magister derivative'.

SPECIFICATION AND PERFORMANCE

Engine:	130hp DH Gipsy Major Standard or 140hp DH Gipsy Major High compression
Dimensions:	span 28ft 1½in; length 23ft 6in; height 5ft 7in; wing area 147sq ft; chord root 75.00"; chord tip 50.87"; wing section modified Clark YH
Weights:	empty 1,080lb; pilot 160lb; fuel (22 gal) 170lb; oil (3 gal) 28lb; payload 312lb; AUW 1,750lb; max AUW 2,200lb; wing loading 11.1-15.0lb/sq ft
Performance:	with HC engine: max speed 180 mph at 1,000ft; cruising speed 155-160 mph at 1,000ft; fuel consumption at cruising speed 8½ gal/hour; range (22 gal fuel) 415 miles; landing speed 42 mph; with Standard engine the max and cruising speeds were reduced by 5-6 mph and the range was increased to 504 miles.

Miles M.6 Hawcon with 25% Wing 'A'.

Chapter 16: Miles M.6 Hawcon

In the early 1930s, the transition from braced biplanes to cantilever monoplanes necessitated wings of far thicker section than previously thought feasible. This was because of the fear of excessive profile drag. Profile drag is the drag incurred from frictional resistance of the blades passing through the air. It does not change significantly with angle of attack of the airfoil section but increases moderately as airspeed increases (an explanation of the various types of drag is given on page 181).

In view of the paucity of actual flight data on the effect of increasing wing thickness, in 1935, Miles considered that a research aircraft should be built to test several interchangeable wings of various thicknesses to show exactly what difference the thickness of a wing made to the speed of an aeroplane. The Royal Aircraft Establishment at Farnborough welcomed the opportunity of supplementing their wind-tunnel results with some full-scale measurements and accordingly arranged for a contract to be placed with Phillips & Powis for a research aircraft to be designed and built for this purpose. Thus began a long and friendly partnership between Miles and the RAE for full-scale research, the results of which were to yield much valuable data to their mutual benefit and which was also made available to the rest of the aircraft industry.

The aeroplane was named the Hawcon by combining the names Hawk and Falcon and was a low wing, two-seat monoplane, with a 200hp DH Gipsy Six engine. The fuselage was a modified form of Falcon Six but with two seats in tandem and a normal sloping windscreen. It had a fixed spatted single-strut cantilever undercarriage and was largely constructed from standard Hawk parts.

Three interchangeable sets of wings were ordered and built at the same time. One set had approximately the same thickness-to-chord ratio as the standard Hawk and Falcon, one set was considerably thicker, and one set considerably thinner. Whilst it was generally

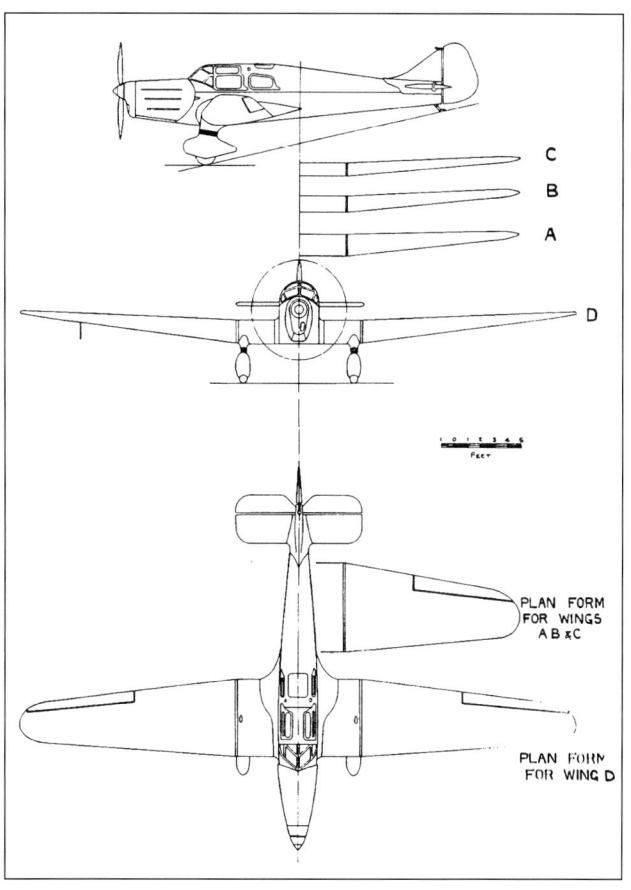

The Miles M.6 Hawcon, showing all wings tested. [Via P Amos]

accepted that a thin wing was theoretically faster, a thick wing had obvious advantages by way of strength and the possibilities of housing the undercarriages or engines within its structure. The Hawcon was designed to discover whether the loss in speed occasioned by the use of a thick wing would be balanced by the gain in strength and usefulness.

The first three interchangeable sets of wings all had the same plan form and section and a span of 33ft 0in and an aspect ratio of 6.76, but with the following thickness-to-chord ratios; wing 'A' 0.25 (Root) and 0.125 (Tip), wing 'B' 0.20 (Root) and 0.10 (Tip) and wing 'C' 0.15 (Root) and 0.10 (Tip). The surface of the wing had to be as smooth as practically possible as the section chosen was a very fast one. The planform followed that used in the Hawk and Falcon, but with a chord taper of 0.5 instead of 0.6.

From the design point of view, a certain amount of difficulty was found in making the thinnest section wing 'C' strong enough and, in consequence, the spar was nearly solid, which made the wing very heavy. However, wing 'A', the thickest of the original three sets of wings, was well over strength since, had it been designed strictly to the desired strength factor, the spar flanges would not have been deep enough to have given adequate gluing surfaces for the webs. This wing was therefore considerably lighter and more rigid than the other two.

A mock-up of the M.6 Hawcon was built with c/n **187**, and one actual aircraft **K5925** (c/n **238**) 'plus two extra wings' was built to Contract No. 414249/35 for 'drag research by the RAE Farnborough', with an Instruction to Proceed (ITP) dated 31st July 1935 and a projected cost of £1,630 plus £469. The Hawcon was painted cream, and carried the serial in black, but no RAF roundels.

The Hawcon was first flown, by Miles, on 14th October 1935 (not 29th November 1935 as stated in many reference works) and he made a further flight on October 15th, before delivering it to Farnborough on 28th November 1935 (although C L Nash recorded that it was delivered on 29th).

It would appear that the initial tests were carried out on just the 'A' and 'B' wings, with four flights being made in November, five in December, four in January 1936, seven in February and six in March, but unfortunately, it is not known which wings were fitted for which tests.

On 17th March 1936, the Hawcon was returned to Woodley, probably to have the 'C' wing fitted, and, following just one flight by Miles, on April 16th, it was returned to Farnborough on April 21st, where it carried out six test flights with wing 'C' in April, before being returned to Woodley on April 30th.

Miles M.6 Hawcon in November 1935. ['Flight' via B Clarke]

Miles M.6 Hawcon with 25% Wing 'A' in November 1935. [Via B Clarke]

The Hawcon was next flown, by Miles, on 3rd June 1936, before being returned (by pilot Dalrymple) to Farnborough on June 5th (RAE log has June 4th), for 'drag and flap research'. The Hawcon made four flights during June but was returned to Woodley later the same month, probably by road. The reason for this return was probably due to a flight made by F/O Arthur Clouston, a test pilot at the RAE: *'following a very bumpy flight I landed at Farnborough and upon opening the throttle to taxi back to the hangar was appalled to see that the port wing was dragging on the ground. A few more minutes in the rough air would undoubtedly have removed the wing'.*

Clouston, with an observer, had been en route to the test area on the south coast near Tangmere Aerodrome to carry out speed runs, when he had run into some bumpy weather with cloud covering the South Downs. He decided to abandon the tests and return to Farnborough, which proved to be a wise decision as the main spar had been broken in the severe turbulence! RAE issued a report in 1936 entitled '*Spar failure on Hawcon aircraft K5925*' (National Archives ref AVIA 6/4635).

The Hawcon's next recorded flight was its delivery back to Farnborough on 6th July 1936, following which, it carried out three further test flights in that month. C L Nash recorded that the Hawcon was fitted with special wings and centre section in July 1936, which, although giving the impression that the new 'D' wing was fitted at that time, probably just meant that it had been repaired after the structural failure (the 'D' wing was not, in fact, fitted until late 1937/early 1938). K5925 was flown twice in August 1936, six times in September and four times each in October, November and December.

In late 1937, the fourth wing, 'D' was built This differed considerably from the first three in having a high aspect ratio of 9.65 and the thickness-to-chord ratio of 30% at the root, tapering to 15% at the tip. The wing was designed and constructed by the Experimental Department in just two months. This illustrated the value of using cheap light aeroplanes for full-scale research. The 'D' wing had an increased span centre section and the wing span was increased from 33 ft 0 in to 39 ft 5 in overall. On this 'thick' wing, the profile drag was increased a further 12 per cent, together with a rather low value of lift coefficient (C_L) max - 1.06 without flaps and 1.60 with flaps.

The lift coefficient C is the number to relate the lift generated by an aerofoil, the dynamic pressure of the fluid flow around the aerofoil and the planform area of the aerofoil. Lift coefficient may be used to relate the total lift generated by an aircraft to the total area of its wing and in this application, it is called the aircraft lift coefficient C_L.

Due to bad weather, the Hawcon was not flown in January or February 1937 but it made seven flights in March and flew every month until September 1937. On 13th October 1937, Fraser flew it back to Woodley, where, it is believed, it was finally fitted with the 'D' wing. It remained at Woodley until delivered back to Farnborough on 1st March 1938, following which it carried out 20 test flights during March and 25 'drag trials' in April 1938.

The overall results of the tests gave a fairly clear picture of the variations of profile drag with wing thickness and enabled a designer to select the best compromise of wing thickness between the ever-conflicting requirements of aerodynamic and structural considerations. For aeroplanes designed to operate up to speeds of 200 mph, it was concluded that the wing thickness should not exceed 20 - 25 per cent but later, with the advent of laminar flow aerofoils, it was found that for aircraft operating at speeds up to 400 mph, the wing thickness should not exceed 12 - 15%.

In the flying tests, the difference in speed between the first three sets of wing sections was small and the top speeds, which are shown in the specification, show that the extra drag of the thickest wing ('A') only made a difference of 5 mph. This slight

AN EXPLANATION OF VARIOUS FORMS OF DRAG

Drag is simply force that opposes the motion of an aircraft through the air. However it does have separate components that comprise it.

Total Drag produced by an aircraft is the sum of the ***Profile*** drag, ***Induced*** drag, and ***Parasite*** drag. Total drag is primarily a function of airspeed. The airspeed that produces the lowest total drag normally determines the aircraft best-rate-of-climb speed, minimum rate-of-descent speed for autorotation, and maximum endurance speed.

Profile Drag is the drag incurred from frictional resistance of the blades passing through the air. It does not change significantly with angle of attack of the airfoil section, but increases moderately as airspeed increases.

Induced Drag is the drag incurred as a result of production of lift. Higher angles of attack which produce more lift also produce increased induced drag. In rotary-wing aircraft, induced drag decreases with increased aircraft airspeed. The induced drag is the portion of the ***Total Aerodynamic Force*** which is oriented in the direction opposing the movement of the airfoil.

Parasite Drag is the drag incurred from the non lifting portions of the aircraft. It includes the form drag and skin friction associated with the fuselage, cockpit, engine cowlings, rotor hub, landing gear, and tail boom to mention a few. Parasite drag increases with airspeed.

The graphic illustrates the different forms of drag versus airspeed:

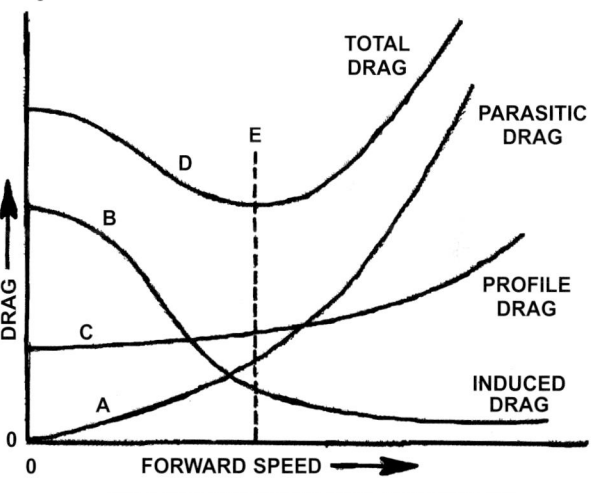

Curve "A" shows that parasite drag is very low at slow airspeeds and increases with higher airspeeds. Parasite drag goes up at an increasing rate at airspeeds above the midrange.

Curve "B" shows how induced drag decreases as aircraft airspeed increases. At a hover, or at lower airspeeds, induced drag is highest. It decreases as airspeed increases and the helicopter moves into undisturbed air.

Curve "C" shows the profile drag curve. Profile drag remains relatively constant throughout the speed range with some increase at the higher airspeeds.

Curve "D" shows total drag and represents the sum of the other three curves. It identifies the airspeed range, line "E", at which total drag is lowest. That airspeed is the best airspeed for maximum endurance, best rate of climb, and minimum rate of descent in autorotation.

Miles M.6 Hawcon with 15% Wing 'C' in November 1935. [Via B Clarke]

Miles M.6 Hawcon with 30% Wing 'D'. [Via B Clarke]

disadvantage may well have been outweighed, except perhaps in the case of a racing machine, by the advantage of strength, rigidity and useful space inside the wing structure.

The Hawcon's last flight was made on 20th May 1939 and movements control at Farnborough recorded that *K5925 undershot and landed on rough ground at Farnborough 'while testing flaps'*. It is unlikely that the Hawcon was repaired and it was struck off charge in October 1939.

In November 1939, the RAE issued a report entitled 'Flight and Model Tests on the Hawcon aircraft with wings 30% thick at the root' (ref AVIA 6/2324) and this showed that at a Reynolds No. of 9,000,000, the coefficient of drag was found to vary from 0.0065 for the 15% wing, to 0.0080 for the 25% wing. This proved that, over the speed range tested, the increase in drag due to thickness was not nearly so serious as had been anticipated. The report summarised the results as follows:

Flight-testing began in April 1936 with a specially built Miles Hawcon K5925 with a thickness of 15% at the wing root (wing 'C' - PA). This was later increased to 20%, 25% and finally 30%. These increases showed that there was no serious effect on the aerodynamic qualities of the aircraft. The ailerons became less effective but the elevators remained unhampered.

The report was issued in November 1939, after the demise of the Hawcon, much later than the tests on the original 'A' and 'B' wings, which had been carried out between November 1935 and March 1936. The 'C' wing was probably tested between April 1936 and October 1937, and the 'D' wing from March 1938. The order in which the wings were tested was not, however, confirmed in the report.

SPECIFICATION AND PERFORMANCE DATA

Engine: 200hp DH Gipsy Six

Dimensions: span, wings 'A' 'B' & 'C', 33ft 0in; wing 'D' 39ft 5in; length 25ft 0in; height 6ft 6in; wing area, wings 'A' 'B' & 'C' 161sq ft

Weights: empty, applicable to wings 'A' 'B' & 'C' 1,550lb; AUW 2,400lb; wing loading, wings 'A' 'B' & 'C' 14.9lb/sq ft; mean wt during RAE trials wing 'D' 2,350lb.

Performance: the difference in speed between the initial three sets of interchangeable wings is shown below, together with their respective thickness/chord ratios:

wing	span	t/c root	t/c tip	aspect ratio	speed mph
'A'	33ft 0in	0.250	0.125	6.76	176
'B'	33ft 0in	0.200	0.100	6.76	181
'C'	33ft 0in	0.150	0.100	6.76	178

Wing 'D' was designed and built to the following specification:

'D'	39ft 5in	0.300	0.150	9.65	180

Miles M.7 Nighthawk first prototype, G-ADXA (previously U1). Note cutaway bottom rudder.
[Real Photographs Co via A Madden and D Sykes]

Chapter 17:
Miles M.7 and M.7A Nighthawk

The M.7 Nighthawk was, in essence, a two-seat, dual-control trainer development of the M.3B Falcon Six and was originally intended to be powered by either a 130hp DH Gipsy Major or a 200hp DH Gipsy Six II engine. It had hydraulically-operated split trailing-edge flaps, a horn balanced rudder, and two retractable Harley landing lamps were fitted in the port wing for night flying.

Like most good aeroplanes of the day, the Nighthawk started life as a private venture, but the firm could ill afford even such luxuries as the blind flying instruments, which are now taken for granted these days. These were so expensive that it was not easy to find enough money for two sets of them for use in the Nighthawk. However, a Certificate of Airworthiness was applied for on 24th September 1935 and the prototype Nighthawk was first flown, by Miles, on 26th October 1935.

The Aeroplane for 2nd October 1935 published an article by Miles in which he expressed his views on 'Modern Flying Instruction':

This flying instruction business is becoming complicated. Training is no longer a matter of taking an inexperienced person and teaching him to make a few simple circuits of the aerodrome in a slowish machine, although it was for a long time considered enough. Ten or twelve hours' dual instruction, a few simple questions on how many lights a trawler should carry in the Bay of Biscay, and a new air pilot was launched in the world of romance. A lot of them stayed there and did not actually go up in the air any more, once having obtained their licences. A goodly number went on and by a process of trial and error turned themselves into first-class flying people in the course of a few months or years.

It is getting a much bigger job now. Compare the old 'B' Licence qualifications with the new. To have accomplished thirty solo hours in the air and to have been able to draw a triangle on a map or to rotate a calculator without in the least knowing why one did it used to be sufficient. Nowadays the minimum flying experience required is over 100 hours and experience has to be gained in instrument flying and night flying. Also the whole thing has been stiffened up and the tests are now done under much stricter supervision than erstwhile. And that's the way things are going.

But there's more than that to it. It is really, in fact, becoming more necessary to acquire a much wider knowledge of the art of flying before one can be said to be even moderately experienced. Certainly transport companies in civil life, as well as the powers that be in the military world are asking much more to-day than they did yesterday. For instance, to do even a lot of one's early flying on a lightly loaded biplane and thereafter, with one's probationary period over, to be launched on the more modern type of aeroplane isn't enough.

I think troubles have already arisen through differences in the characteristics of the machines of a few years ago and the machines of to-day. Quite a different technique is required with the more heavily loaded thick-wing section of the present-day aeroplane. They take-off differently, one can pull them up in the air a good deal more than one dare pull up some types of biplanes. There are flaps and things to learn to use. And then one is flying inside in the dry instead of outside in the draught as one did during instruction.

In fact, one may reasonably suppose that a lot of time is lost through not learning to fly in the type of machine that one will fly thereafter, for under present circumstances there are so many things left to learn when one has gained the accolade of the Air Ministry 'B' Licence, the Blind Flying Certificate and a pat on the shoulder from the Guild of Air Pilots, giving one permission to instruct. Only the fringe of navigation has been touched, one has still to learn the use of the instruments while one is tucked away in a cabin instead of being blown away in the open.

Learning is easier in a great many ways, but different. It is easier because one has the room to move freely and is not battered by all the winds of heaven. It is different because one has to accustom oneself to manage without that very useful information hitherto imparted by the said winds. And the outlook is different with windows all round instead of the plein-air. To sum up, the feel is different; one flies more by instrument and less by eye; more by observation and less by feel. The senses are aided and extended by more complicated instruments whose messages have to be translated into terms of knowledge, and there is wireless navigation and landing at night, and all sorts of things which should now be considered as elementary instead of advanced instruction.

To put on paper exactly what one feels about this position is difficult, but we have tried to put it in wings and wheels by making the Nighthawk. We have tried to arrange an aeroplane which is a type and an example of modern, fast, low-winged, flapped, head-lighted, wirelessed and instrumented flying. It is a dual-control machine, with the instructor and instructed sitting side-by-side in comfort. It has full dual controls and instruments, and is arranged to represent as nearly as we can the special and actual feel of the sort of machine the modern pilot flies.

We have fitted alternative types of control for instance. The central stick, the normal stick and wheel control. They are easily inter-

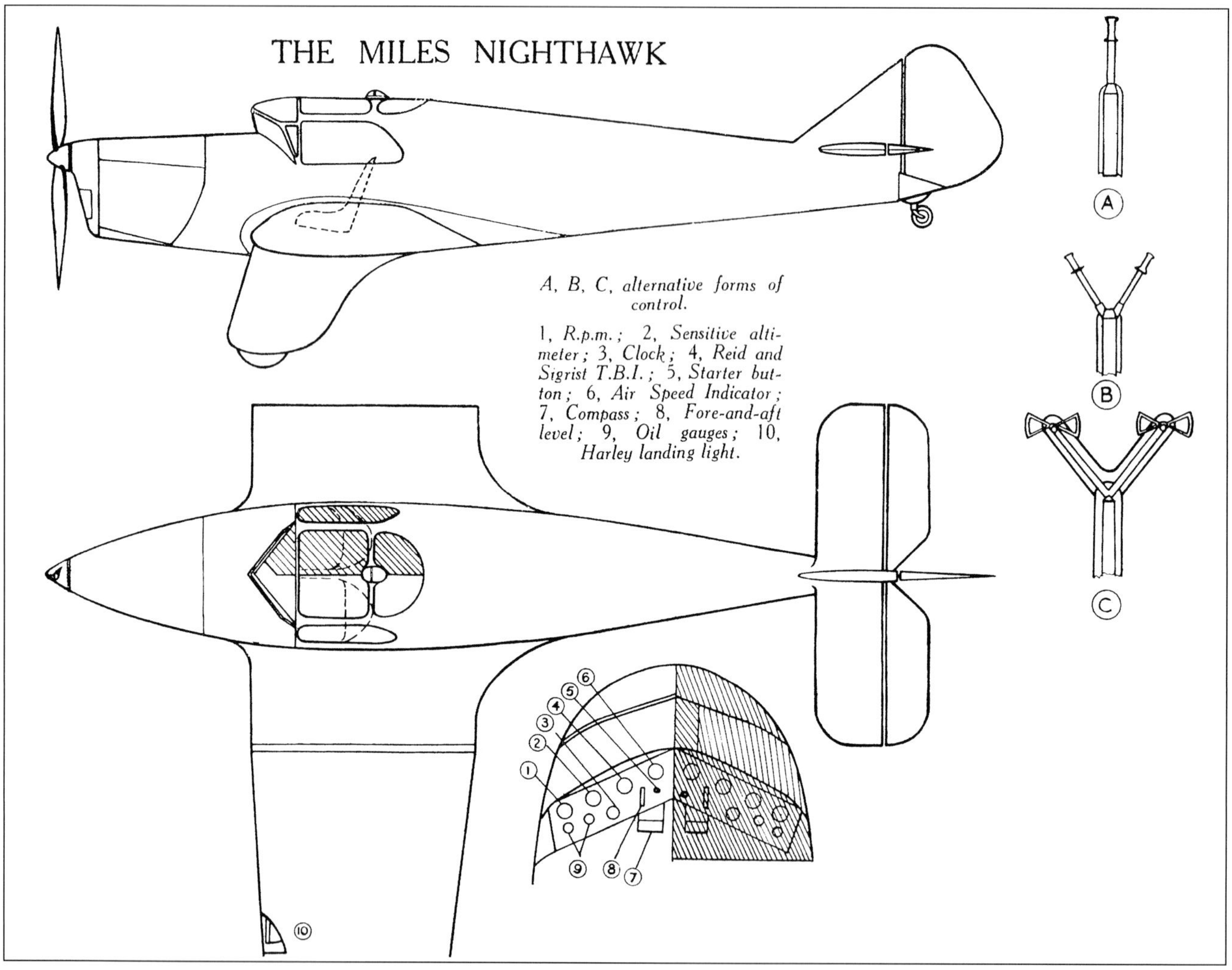

THE MILES NIGHTHAWK

A, B, C, alternative forms of control.

1, R.p.m.; 2, Sensitive altimeter; 3, Clock; 4, Reid and Sigrist T.B.I.; 5, Starter button; 6, Air Speed Indicator; 7, Compass; 8, Fore-and-aft level; 9, Oil gauges; 10, Harley landing light.

The Miles M.7 Nighthawk.

The prototype Miles M.7 Nighthawk cabin showing wheel control in January 1936. [Via B Clarke]

The prototype Miles M.7 Nighthawk showing the curtain in position for blind flying. [Via B Clarke]

F G Miles donning his parachute before setting off on another spinning flight in the Nighthawk. ['Flight' via B Clarke]

changeable so that experience can be gained on each kind. The machine is trimmed by tabs, because the new and ultra-fast machines are trimmed by tabs. The machine's speed and speed range are also comparable to the world's latest and fastest productions. Not equal to, but comparable with.

One hasn't to make the leap from 100 mph to 250 mph. We have kept the difference in speed as small as possible taking into account the fact that the Nighthawk has been built at a very reasonable price, so as to extend its uses as a training machine. It is fitted with flaps so that the pupil can be taught to use them properly under all conditions instead of just guessing their uses. It is fitted with the latest type of landing lights and night landing instructions can be given on it without external assistance such as flares, thus duplicating conditions that may arise in practice as nearly as possible.. One cabin can be curtained off for blind flying and there is an extra seat behind the instructor where a third pupil can observe and discuss with the instructor the mistakes made by a pupil under the hood.

It is unnecessary to say that we have made it as strong as a training machine should be.

We have to thank Blossom for providing the evidence that the prototype M.7 Nighthawk made its early flights with the 'B' mark U1, as there is no photographic or other written evidence to show what marks it carried. Much confusion has been caused over the years due to many reputable reference works mistakenly recording that the first prototype carried the 'B' mark U5, which was in fact the mark carried by the **second** prototype. Blossom recorded U1 when she made two flights in it with Miles on 4th December 1935 and when she next flew in it, again with Miles, on 11th December 1935, she recorded the registration G-ADXA. Following its initial trials the prototype was fitted with a horn balanced rudder and by that time the registration G-ADXA had been applied.

The new aeroplane aroused so much interest in official circles that Phillips & Powis were awarded a development contract to AM Specification 24/36, with the possibility of the machine later going into production. This development contract required a considerable amount of test flying, including spinning tests with a wide range of c.g.. This was quite a novelty in those days as people had not yet become accustomed to low-wing monoplanes, which were regarded as requiring a 'new technique' in flying and were certainly not the sort of machines considered to be spun light-heartedly.

However, Miles, who was the chief test pilot as well as the chief designer, had no hesitation in undertaking any tests required, and time after time he took the Nighthawk up and spun it, first to the right, then to the left, then he landed and had it re-loaded to a new c.g. position, to be followed by yet more spins. Don Brown remembered walking into Miles' office one day and seeing him sitting at his desk with his head buried in his arms, apparently weeping bitterly. Actually he was only trying not to be sick, because few people could stand up to a spin in an enclosed cabin, let alone half a dozen consecutive spins.

So successful were the trials of the Nighthawk that the Ministry decided to extend the c.g. range still further aft, thus reducing both the longitudinal stability and the likelihood of spin recovery. The Ministry were very interested to find out just how far one could go before 'something happened' - especially if someone else was doing the 'finding out how far one could go!'

By now, Miles was sharing the test flying with Wing Cdr Frederick William Stent MC. Freddy Stent was born in Cyprus in 1891 and was a good friend of Miles, having joined the Sales department of the firm in January 1936. He was a very experienced pilot who had flown with the RFC in the 1914-18 war and had recently retired from the RAF. Flt Lt James Moir, a former RAF test pilot at A&AEE Martlesham Heath, who had joined the Reserve Training School as Chief Flying Instructor, also helped with test flying as his duties permitted.

The Nighthawk had already been spun with tanks both full and nearly empty. It had proved reluctant, but recovery had been satisfactory at all c.g. positions, within the design limit, which was in fact 21.8in aft of the normal position. On 21st January 1936, James Moir had spun it with full tanks for twelve turns and found recovery difficult. It was later reported that 'immediately' after Moir had landed, Stent flew the Nighthawk and reported the spin controllable. Then the tanks were topped up and Stent did a further test. After eleven turns, from some 10,000ft altitude, the right hand spin began to flatten out and the nose rose - the limit had been reached. However, the use of the word 'immediately' gave rise to the story that the last flight of the Nighthawk was made on the January 21st but this was not the case. This next flight, with the topped-up tanks, was, in fact, made on Wednesday January 22nd and during the spin test, Freddy Stent found that neither the use of full control movements nor full throttle would effect recovery so, at the twenty-third turn, he did the only thing left to him and baled out.

This was not an easy task for a fourteen-stone man who had never had occasion to practice baling out, even under much more favourable circumstances than in a spin in an enclosed cabin, and with little time to spare. However, in successfully escaping from the aircraft, Stent qualified for membership to the Honourable Company of Caterpillars, as he had saved his life by the use of an Irving Air Chute. The Nighthawk, left to its own devices, retained a sense of humour right up to the end, because the spot on which she chose to crash was a piece of land belonging to a local landowner who positively loathed the sight of aeroplanes!

A report in *The Aeroplane* for 22nd January 1936 (the same day the Nighthawk crashed) under the title *An Ornithological Metamorphosis,* makes for interesting reading:

Phillips & Powis Aircraft Ltd, Reading Aerodrome, Woodley, will shortly offer the Miles Nighthawk, at present a military training machine, in civilian guise as a four-seater of particular refinement. Control will be by wheel, and ball bearings will be used throughout the control system. The machine will include all kinds of other refinements and will be known as the Miles Falcon Series IV. The top speed will be better than 170 mph.

In the event the Nighthawk was rejected by the Air Ministry as a trainer and was never produced as the 'Falcon Series IV' either, but it was later considered suitable for communications duties and a production order dated 29th December 1937 was placed with Phillips & Powis for the manufacture of 45 aircraft (a small order by later contract standards but a large enough one for them at the time) for a modified version to AM Specification 38/37. This was to be known as the M.16 Mentor and in practice it was to bear little resemblance to the Nighthawk, since it incorporated a considerable number of modifications in order to meet the specification. The Mentor is described in Chapter 29.

Following the crash of the prototype, Harry Hull (who was reckoned to be 'a bit of a slave driver' - and remained so until he finally retired many years after normal retiring age!) was given permission to commandeer all the men he wanted from anywhere in the factory to build a second machine in as short a time as possible. This machine was built in the Experimental Department in the incredibly short time of 240 hours, and a plaque was placed inside the aircraft to record this incredible feat, bearing the inscription:

At a time when it was of the utmost importance to us to have a Nighthawk flying, Harry Hull and the men with him of the Experimental Department built this machine from beginning to end in 10 days and made a good job of it. This is to thank them.

> *Here is an example of Harry Hull's powers,*
> *He built this aircraft in two forty hours.*
> *And though he could easily do it again,*
> *From making it necessary, please refrain.*

Needless to say, when anything went wrong with that machine everyone immediately said: 'What do you ****** well expect!'

CHAPTER 17: MILES M.7 AND M.7A NIGHTHAWK

Once thought to depict the first prototype Miles M.7 Nighthawk, this photograph actually shows the second aircraft. Note the straight rudder stern post and repositioned tailwheel. ['Flight' *via B Clarke*]

The second prototype Miles M.7 Nighthawk, U5, with F G Miles making adjustments, on 5th February 1936. ['Flight' *via B Clarke*]

The Miles M.7A Nighthawk Hybrid, G-AGWT, at Woodley in 1946. Note the Falcon Hotel still in wartime camouflage in the background.
[Miles Aircraft via The Miles Aircraft Collection]

This second Nighthawk, G-AEBP, was built from scratch with a horn-balanced rudder which extended straight to the bottom of the sternpost (unlike the first prototype on which the bottom of the fuselage extended aft under the tailplane to form an angular bottom edge to the rudder) and it also had a curved leading edge to the fin to give it greater area. This machine, which carried the 'B' mark U5 for its early tests, was first flown, by Miles, on 5th February 1936, with just the word 'spins' recorded against the flight in his log book. This suggests that the spinning trials must have recommenced immediately.

In connection with the tests, a report in *Flight* for 13th February 1936 stated that: *F G Miles has evolved a scheme for stopping a flat spin with the release of a small parachute anchored to the tail of the machine* and, as a result of this, presumably no further problems were experienced. Much valuable knowledge was gained as a result of these tests, both by the firm and by the authorities. G-AEBP was delivered to the A&AEE Martlesham Heath on 9th February 1936, still as U5, for tests but, although it was deemed acceptable for C of A, it was said to have failed to meet military standards. A&AEE Report No.691, dated 6/36, stated that: *the plywood construction was not considered robust enough, there was only one stick and the view was poor*.

Two further Nighthawks were built and because of their differences to the prototypes, they were designated M.7A. They had four seats and reversed throttle and mixture controls and were supplied to the Roumanian Government in August 1936.

Then, in 1940, a four-seat hybrid version of the M.7A Nighthawk was completed, and this utilised the wings from the second, uncompleted Mohawk. This had a single-piece moulded Perspex windscreen and was fitted with a 205hp DH Gipsy Six engine, which drove a two-position variable-pitch propeller. In this guise, it was flown by the firm as a research and communications aircraft throughout the war, first under 'B' mark U5 and later U-0225. Registered G-AGWT to Miles Aircraft Ltd in December 1945, this last remaining Nighthawk was later purchased from the firm by Hugh Kennedy, who, after retirement from test flying, formed Raceways Ltd at Woodley. This was an air taxi company whose main objective was to ferry jockeys and racing personnel around the country, although joy-riding and ad hoc charters were also carried out as the opportunity arose.

The Nighthawk was considered to be an extremely nice aircraft by all who flew her, but Don Brown recalled a rather disconcerting incident in this aircraft when the variable-pitch propeller developed an oil leak. At first this merely resulted in the windscreen becoming rather bespattered with oil but, in a matter of minutes, as the leak rapidly became worse, the entire windscreen was soon completely opaque. The aircraft was at full load and the side window was too small to stick his head out for landing, so Don resolved the matter by making a long sideslip approach, then at the last moment, kicking off drift and flaring completely blind to make a perfect landing.

G-AGWT was later sold in East Africa, where later, due to corrosion from continually landing on beaches, its undercarriage had to be replaced by one from a Percival Proctor! It eventually returned to the UK but, on 1st March 1963, flown by Brian McAllister, the Nighthawk departed Southend for Kenya (although it has also been reported as being en route to Singapore). It got no further than Marseilles (Marignane) Airport, in Southern France, however, where it was seized by the authorities for reasons unknown. G-AGWT was later sold for scrap at Lignane, near Aix-en-Provence, France, where it was last reported on 29th June 1965 in an advanced state of decomposition. A sad end for such an interesting aeroplane.

For details of Miles M.7 Nighthawk production see Appendix 17.

SPECIFICATION AND PERFORMANCE DATA

Engine: 200hp DH Gipsy Six II
Dimensions: span 35ft 0in; length 25ft 0in; wing area 181sq ft
Weights: empty 1,650lb; AUW 2,400lb; (M.7 U5, with DH Gipsy Six II and metal vp propeller 2,650lb); wing loading 13.8lb/sq ft
Performance: DH Gipsy Six II engine: max speed at 1,000ft 150 mph; cruising speed at 1,000ft 130-135 mph; landing speed 40-45 mph.

"Good Luck!" This photograph of the Miles M.8 Peregrine, U9, was possibly taken just prior to the aircraft's first flight. Len Hackett, the Sales Manager, is seen on the right. The gentleman with the parachute could be Charles Powis, but this is not confirmed as I have never seen him wearing a trilby hat!.
[Phillips & Powis via The Miles Aircraft Collection]

Chapter 18:
Miles M.8 Peregrine Mk.I and M.8A Peregrine Mk.II

As there was no really efficient twin-engined aeroplane in the 6-8 passenger category available in 1936, Miles felt that there could be an abundant market for an aircraft of that size, so he set about the design of what was to become the M.8 Peregrine. This was to be the first twin-engined aircraft of Miles design and he originally planned to fit it with a fixed trousered undercarriage but this was later changed to retractable undercarriage, making it also the first Miles design to incorporate this feature.

The Peregrine was originally to have been powered by either two 205hp DH Gipsy Six Series II or two 160hp Napier Javelin engines and it was designed to maintain height in the event of one of the engines ceasing to function. Electric starters were to be provided and the engines were to drive two-position DH variable-pitch airscrews. Both the undercarriage and the split trailing-edge wing-flaps were to be operated by compressed air rams. Controls on the dashboard were for either a one or two-man crew.

The Peregrine was, to quote from the preliminary specification:

A completely modern design and the effective use of Miles split trailing-edge wing-flaps have characterised all Miles products up to the present time, and now in the Miles twin-engined Peregrine we have produced an aeroplane not only endowed with the beneficial features of its smaller brethren, but brought into line with the most ultra-modern practice by the inclusion of variable-pitch airscrews and a retracting undercarriage. No attempt has been made to depart from the advanced and well-tried practice that has distinguished Miles aircraft, but provision for inspection facilities has been extended to the utmost possible limits in order to ensure low maintenance to the air operator. Commercially, the aeroplane is a very fine proposition, carrying 1,320lb of payload for 562 miles at a speed of 160 mph on 400hp.

Miles' typically simple and strong wooden design and construction was used to allow for ease of maintenance and repair. The wings were of the cantilever type, with two-spars and plywood covering to take the drag and torsional stresses. Saro plywood was specially made for the job, and the method of attachment enabled a highly polished finish to be obtained, thereby enhancing not only its appearance but also its performance, due to the reduction in skin friction. The main cabin was 12ft 8in long, 4ft 6in wide and 5ft high, and was unobstructed by shear members or bulkheads. The 'passenger saloon' was designed to accommodate six people in lounge seats providing the greatest degree of comfort. Ample arm and leg room was provided, and warm or fresh air could be introduced at will by the pilot, who was enabled to maintain an equable temperature by means of a thermometer situated alongside each seat. Light luggage racks were situated above each row of seats.

The Peregrine was scheduled to be completed by the end of August 1936, before Miles and Blossom left on a visit to the USA in early September but, unfortunately, and despite the best efforts of the workers, it was not ready for its first flight until the evening before their departure. Although it was only partially completed, Miles decided to attempt the first flight but, by the time the engines had been started and warmed up, the aerodrome was already in semi-darkness. Miles refused to be daunted and taxied out for the first fast runs and initial hops, but during the course of these the brakes seized and by the time they had been rectified, it was quite dark. In consequence, the Peregrine was the first Miles aeroplane (and for a long time afterwards, the last) on which Miles did not personally make the first flight.

Some days later, with the Peregrine nearer completion, Charles Powis, the Managing Director of the firm, elected to undertake its initial flight, despite the fact that although a very able and experienced pilot on single-engined light aeroplanes, he had never before flown a twin-engined machine. There was also some uncertainty as to the exact position of the c.g. so, as an added safeguard, Ray Bournon, a young engineer who had joined the firm in the mid-1930's and who had been involved in the Peregrine's design, very pluckily volunteered to go along as 'mobile ballast', prepared, in answer to signals from the pilot, to dash down into the tail end or to crawl under the dashboard if required! Bournon was later to become a senior designer under George Miles at the Experimental Department in Liverpool Road, Reading.

On 12th September 1936, the great moment for take-off duly arrived and, as the Peregrine, with 'B' mark U9, taxied out, George Miles unobtrusively took his seat at the wheel of the 'ambulance' (a Ford van containing a pillow and a blanket!) while Don Brown did likewise at the wheel of the 'fire engine' (another Ford van containing a small fire extinguisher!). Both started up their engines - the usual precaution when a new design makes its first flight. After an incredibly short run, the Peregrine left the ground and proceeded to climb at a dangerously steep angle. The onlookers little realised that Powis was pushing forward with all his weight on the control wheel while Bournon was huddled under the dashboard, wondering how wise he had been to volunteer as ballast! When the aircraft had climbed to about 30ft, one engine momentarily spluttered but fortunately picked up again, and after making a wide circuit, Powis came in to make a perfect landing. The faulty trim, which had caused the problem, was soon rectified.

An article appeared in a contemporary issue of *The Aeroplane* under the heading: *Latest from Woodley. Advent of the Peregrine, the first twin-engined Miles type: A Light Transport to the modern formula: Two Gipsy Six IIs or Menascos*

Generally advanced design and the use of the effective Miles split trailing-edge flaps have conferred on the family of single-engined monoplanes produced by Phillips & Powis Aircraft Ltd, some highly desirable features making for extraordinary performance, pleasant handling characteristics over the wide speed range, and economy of operation. The new Peregrine light transport monoplane has been endowed not only with the more beneficial features of its forebears, but is brought into line with the most modern practice by the inclusion of variable-pitch airscrews and a retractable undercarriage. For the commercial operator its virtues may be summarised as 2,200 lb disposable load carried for 562 miles at 160 mph on about 400hp.

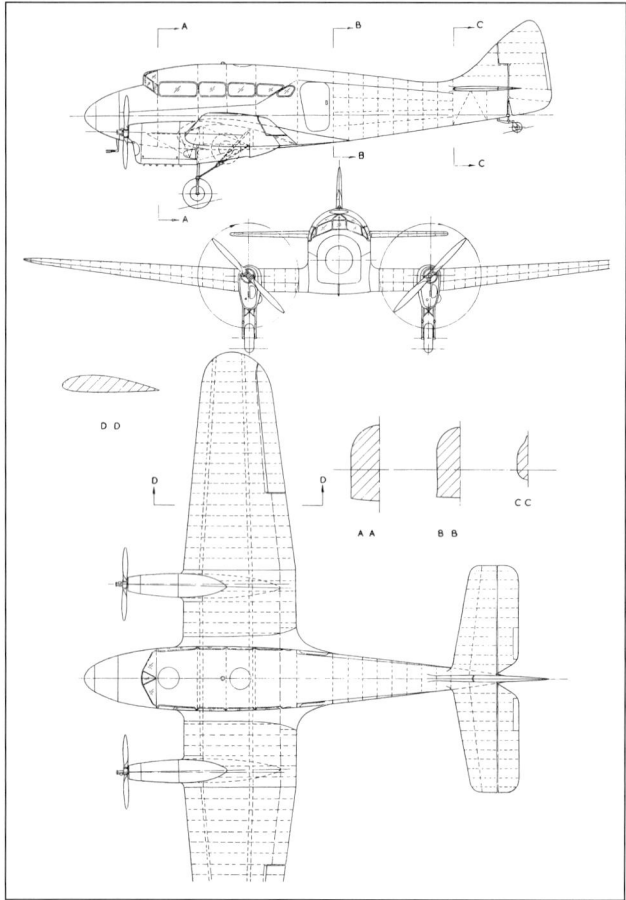

The Miles M.8 Peregrine.

Although conceived primarily as a civil type for commercial or private employment, the Peregrine is readily adaptable to serve as a twin-engined training machine or, with armament, as a military general-purpose model of no little merit.

No attempt has been made to introduce radical departures from standard Miles structural practice. The low cantilever wing has two spars with the Saro plywood covering taking drag and torsional stresses. Mass-balanced, differentially operated Frise ailerons and Miles three-piece split trailing-edge flaps operated by compressed air are specified. The rams for the flaps are accessible through inspection doors in the centre section which, as usual, is parallel in chord and is given no dihedral. The main portion of the fuselage, which embodies the cabin, is of semi-monocoque construction, but aft of the luggage compartment, located immediately behind the cabin, the structure becomes a true monocoque with laminated frames and formers.

Production machines will be a foot longer than the prototype, the extra length resulting from a re-designed barrel-shaped nose portion forward of the front spar.

A particularly commendable feature is the provision of ample inspection facilities. The nose portion of the fuselage hinges on its starboard side and swings aside, revealing the controls, cable conduits and brake operating gear, and there are two downward-folding panels in the floor, followed by a series of small servicing doors in the fuselage bottom for the inspection of all moving parts. The hinges of these folding portions are concealed, leaving a perfectly smooth exterior. Of orthodox layout and construction, the

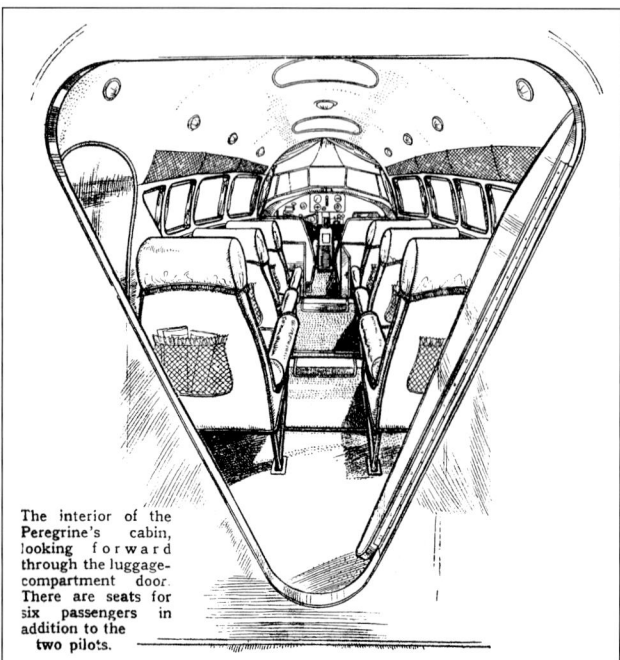

Illustration showing the interior of the Peregrine's cabin.

Chapter 18: Miles M.8 Peregrine Mk.I and M.8A Peregrine Mk.II

Above: The first prototype Miles M.8 Peregrine at Woodley in October 1936.
   ['Flight' via B Clarke]

Right: Miles M.8 Peregrine – the 'office' – in November 1936.
   ['Flight' via B Clarke]

Below: U9 taxi-ing out at Woodley, flaps fully extended.
   ['Flight' via B Clarke]

tail unit has a wooden, plywood-covered fin and tailplane and metal rudder and elevator, the latter surfaces incorporating trimming tabs.

The undercarriage is in two completely retractable portions, one beneath each engine nacelle. The retracting motion is a rearward and upward one and is effected by compressed air from a Dunlop system, which also supplies the wheel brakes. Lockheed Airdraulic struts are used not only for the main undercarriage, but for the tail wheel. A means of covering the gap admitting the wheels into each nacelle is being developed. The wheels are low-pressure, ball-bearing mounted Dunlops.

Power is derived from two of the new De Havilland Series II Gipsy Six in-line inverted air-cooled engines, delivering a maximum of 205hp each to De Havilland two-bladed, two-position variable-pitch airscrews of 7ft 6in diameter. These units are mounted in streamlined nacelles forward of the leading edge of the centre section. Their cowlings, incidentally, differ somewhat from those common on Series I engines, having, in place of the large outlet slot formed by the rear of the cowling, two comparatively small outlets and a pair of smaller louvres. Spinners for the airscrew hubs are available. The oil tanks form the tail fairings of the engine nacelles and fuel is carried, as usual, in wing tanks. Fuel consumption is given as 31 gall/hr, permitting the range of 562 miles.

The cabin is entered from the port side and is normally arranged to seat six passengers, the rear seats, nearest the door, being arranged to slide on rails to facilitate entry. The first and second pairs are mounted respectively on front and rear spars. Soundproofing and general upholstery are by Rumbold and great attention has been paid to the comfort and convenience of passengers. Light luggage racks are fitted above each row of seats and to each seat there is a lamp, ashtray and container for small articles. Immediately behind the cabin comes the luggage compartment, which is accessible from the cabin through a bulkhead door and from the exterior of the machine through a second door in the fuselage side. The door-locking arrangements are particularly ingenious, permitting a flush surface on the exterior.

A pair of Harley retractable landing lights is mounted in the wings, and the equipment includes standard Demec navigation lights and Rotax lights in the cabin and cockpit.

There are a number of usefully original features in the layout of the control cabin of the Miles Peregrine, even though allowance has usually to be made for necessary 'jury-rigging' in the case of any prototype. In its present form the Peregrine is not fitted with dual control but this will be optional on production models, which will also be arranged with a partition between the control cabin and the passengers' compartment. Both flying and auxiliary controls are of the Simmonds-Corsey type.

Most interesting of these control features is the way in which the secondary controls are arranged in a bank below the dash and between the two forward seats. At the base of the dashboard, in the centre, is the throttle and mixture control quadrant, while below, at the side of the centre column, are the elevator tab, rudder tab and controls for the Amal flame traps. On the face of this column are two easily read indicators, one for the flap and the other for the undercarriage position. In addition, a warning buzzer will be installed for the latter.

Both the undercarriage and the flaps are pneumatically operated from the same system as that used for the brakes; their respective controls are on the dash at the pilot's left. The airscrew pitch-changing controls are of the push-pull variety on the dash to the right of the central bank. On the right, too, in the case of the prototype, is the latest type of Radio Transmission Equipment receiver and homing set. The present instrument layout includes the normal instruments of Smith's manufacture, with a Hughes' rate-of-climb indicator, as well as a Reid and Sigrist Gyorizon and liquid fore-and-aft level.

As in the case of the Miles Whitney Straight, the shaped Rhodoid screen is arranged with sliding sections in its main body so that both first and second pilots can obtain an 'open-air' view in rain or bad visibility. The apertures are so placed that there is no direct airflow through them, and only a slight draught on either side of the direct line of vision. This draught would be very useful as an indirect means of ventilation in hot weather.

The major controls are of the normal type, though the spectacle control has no visible column in the ordinary sense of the term, the horizontal shaft sliding through the dashboard rather after the manner of that in the American Stinson. There is, consequently, no restriction to leg movement.

While at Woodley recently we had the chance of flying with the designer in the machine, both lightly and fully loaded. The acceleration provided by the DH vp airscrews is quite phenomenal and the Peregrine was off the ground with a full load in something like eight seconds after a 200 yard run. Although the stalling speed in this state is, with the flaps up, probably very little below 60 mph, Mr Miles climbed her at this speed, while the rate-of-climb indicator showed something in the region of 1,000 ft/min. At the time, the undercarriage was not retracted, and it is probable that this rate of climb could be considerably improved upon. For the same reason, the indicated maximum and cruising speed were of no interest, though the former was about 160 mph.

As a matter of interest to those who are not accustomed to flame-trap controls, these are open at the take-off and closed at cruising rpm, or below, in coarse pitch, unless the induction temperature rises above 40 deg C. In its present form the elevator tab trimmer will not quite hold the machine on full throttle and fine pitch, but certain minor modifications will be made to the tail unit of production machines.

It is possible to fly comfortably and without loss of height on full load, with either throttle cut back, and the rudder bias will hold the turn in one direction. Some slight rudder pressure is necessary when turning around the running engine. At the stall, flaps down, the machine shows no sign of dropping a wing - at least when the stall is made fairly gently, some aileron control remains even after rhythmic shudderings indicate that the airflow is breaking away.

With both the flaps and the undercarriage down, the approach is extremely steep, and the former are so efficient that, if the machine is brought in slowly, a little motor is advisable in order to extend the period of hold-off. On neither of the two landings did the machine come to rest much more than 150 yards from the boundary markers.

A version with Menasco Buccaneer B6S (200hp Buccaneer Supercharged Model B6S-4 - PA) engines will eventually be available, and will have the following performance (the speed figures in brackets relate to sea-level operation): Maximum speed, 194 mph (182 mph); cruising speed, 172 mph (162 mph); climb to 5,000 ft, 4.76 min; and climb to 10,000 ft 10.53 min.

On 14th October 1936, after his return from the United States, Miles flew the Peregrine for the first time and with 410hp to play with 'his exuberance knew no bounds'! On October 18th he flew it both to Brooklands and return and Gatwick and return; on October 20th he took his son Jeremy for a flight in it and on 24th he flew it to Brooklands and return. This was followed on the 25th by a flight to Shoreham and return via Gatwick. October 29th saw Miles posing the Peregrine for air-to-air photographs for *The Aeroplane* photographer and the next day he repeated the performance for *Flight*, with George Miles also on board. Then, on November 1st, Miles took Blossom for a flight and retracted the undercarriage for the first time and on November 5th he took George Miles for a flight. It should be noted that all these flights were made with 'B' mark U9. Miles' log book, in which he recorded these flights, finished on 1st November 1936 and, unfortunately, his next one is missing, so we have no further details of Miles flying the Peregrine.

CHAPTER 18: MILES M.8 PEREGRINE MK.I AND M.8A PEREGRINE MK.II

Miles M.8 Peregrine in flight with the undercarriage down (above) and with the undercarriage retracted (below), in November 1936.
[Both 'Flight' via B Clarke]

Flight for 14th November 1936 published an article on the Peregrine under the title: *A Twin from Reading. The new Miles Peregrine twin-engined monoplane: civil and military versions: 188 mph with two Gipsy Sixes.*

'Bigger and better' seems to be the keynote of the Miles development programme. Since the original Hawk design met with such widespread success there have been produced the Hawk Major, with its training and de luxe variations, the Falcon and the Merlin. The latest and largest member of the family is the Peregrine twin-engined cabin machine, a low-wing monoplane particularly suited to feeder-line work, for use as a luxury private-owner type, for training in flying multi-engined aircraft and where a fast economical machine is required for military purposes.

During the following year the Peregrine, which had been registered G-AEDE on 21st February 1936 but retained its 'B' marks U9 throughout its life, carried out much demonstration flying, usually piloted by Miles, who at that time (and sometimes even later!) used to specialise in extremely steep climbs straight off the ground. One veteran pilot, who was treated to such a demonstration in the Peregrine by Miles, remarked afterwards that you did not sit in your seat, you spent most of the time lying on the back of it!

However, the inherent stability of the machine was such that, after Miles had levelled off at the top of the climb, he would leave the cockpit and make his way to the cabin, where he would sit and chat affably with his passengers, not appearing to notice that their manner was somewhat tense and their smiles a little forced!

An amusing feature of the prototype Peregrine, and one which would have had to be rectified if it had gone into production, was the operation of the retractable undercarriage. The undercarriage was designed to be pneumatically operated but unfortunately, there was no provision on the engine for driving a compressor. The reservoir quickly became exhausted and the only way of replenishing it was by means of a bicycle pump - an operation which caused some surprise, and no little amusement, at the expense of the perspiring member of the crew to whom the task had been allotted.

The Peregrine was entered in the Schlesinger Air Race from Portsmouth to Johannesburg, which was scheduled to start on 29th September 1936. There were only 14 entries for the race, which included the Peregrine, G-AEDE, entered by Bateman Scott of Newbury, Berkshire (with Race No.11). It was to have been flown by Flt Lt Hugh Edwards (who had gained second place in the 1935 King's Cup Air Race in the Hawk Trainer G-ADLN) and Sqn Ldr Brian Thynne (an Airwork pilot whose home and private landing ground was in Findon, Sussex and who was also the Officer Commanding 601 (County of London) (Fighter) Squadron, AAF).

The Peregrine, which had been fitted with a homing wireless device in order to aid navigation in bad weather and was now estimated at being able to cruise at 160 mph, unfortunately could not be readied in time and so had to be withdrawn. Bateman Scott also entered the Peregrine, with the same crew, for the Paris-Saigon-Paris Race, which was due to take place in October 1936, but presumably for the same reasons, it did not participate in this race either. Indeed, Bateman Scott also had the Peregrine down for the July 1936 King's Cup air race, well before its actual flight. However, just who Bateman Scott was remains a mystery, despite much research.

On the whole, the Peregrine was an extremely good and efficient aeroplane, being some 30 mph faster than its contemporaries. The finest proof of this was that it would still have been so nine years later, in 1946. However, just when the company should have started putting it into production, they received their first big contract from the Air Ministry for 90 of the new M.14 Magister elementary trainer, and, as this completely absorbed the whole of the existing production capacity, everything else had to be dropped.

Both the Government of New Zealand and Union Airways had intended to place orders for the aircraft, but Miles' inability to put it into production led to the New Zealand Government buying the Airspeed Oxford, and Union Airways, Lockheeds, instead.

Miles flew the Peregrine to and from Farnborough on 17th April 1937 and chief test pilot Bill Skinner flew it on 9th June 1937 with George Miles. It was flown again by Miles, with George, on June 16th and 23rd. This latter flight was the Peregrine's last recorded one and it had been withdrawn from use and dismantled by December 1937. In not proceeding with the Peregrine, Miles lost the opportunity of producing what could well have been one of the most efficient aeroplanes in its class ever to be built.

While test-flying the Peregrine, Miles was also engaged upon research into the possibility of controlling the wing boundary layer by suction, thereby reducing wing drag, delaying the stall, increasing the rate of climb and generally improving the overall aerodynamic efficiency. The initial experiments were carried out on the prototype M.11 Whitney Straight, on which a section of hollow perforated wing was wrapped around the existing port wing with suction being provided by a large venturi mounted below the wing. The most promising results obtained from these experiments, led to a Government grant of about £10,000 being given to the firm towards full-scale experiments to be carried out on a new Miles Peregrine.

Miles M.8A Peregrine Mk.II

A second Peregrine, designated M.8A Peregrine Mk.II, was built for the RAE Farnborough, and although there has been some speculation that this machine was converted from the first prototype, it was a completely new aircraft built from scratch specifically for the purpose. Photographs of the Peregrine Mk.II under construction in the Experimental Department show that it had a different-shaped fuselage with a circular nose section, as opposed to the 'slab' sided nose on the first machine. The cabin windows were of a different shape and were also positioned higher up the fuselage than on the first prototype machine and about 600 design man-weeks were expended on the project.

The Peregrine Mk.II was fitted with two 260hp Menasco Super Buccaneer Model C6S-4 engines and George Miles recalled later that: *In order to concentrate our limited experimental resources, we also decided to use this aircraft for the trial installation of two Menasco CS.6 engines - this decision turned out to be most unfortunate since insufficient development had been done on this basically excellent engine and our boundary layer programme was much curtailed by power plant failures.* It should be mentioned that George noted the engine's designation incorrectly.

The M.8A Peregrine had a metal-skinned empennage, the first time such a covering had been used on a Miles aeroplane, and a full Magister-type horn-balanced rudder. The upper surfaces of the

Miles M.8A Peregrine II s/n L6346 under construction in the Experimental Department in early 1938.
[Phillips & Powis via The Miles Aircraft Collection]

Chapter 18: Miles M.8 Peregrine Mk.I and M.8A Peregrine Mk.II

The Miles M.8A Peregrine L6346 nearing completion. [Phillips & Powis via The Miles Aircraft Collection]

wings were covered with perforated skins over a series of full-span channels, which were, in turn, connected to the vacuum pump. George Miles also recalled that: *The nature of the suction equipment was dictated by finance and availability and consisted of a Ford '10' automobile engine, installed in the cabin, driving a centrifugal blower unit taken from a commercial air conditioning unit. Naturally, it was both heavy and inefficient but it served its purpose for the initial trials.*

The M.8A Peregrine was given serial L6346 and was first flown on 30th March 1938. George Miles went on to say that *the limited amount of flying carried out on this aircraft indicated that a striking reduction in drag and stalling speed could be obtained at a very small potential cost in power and weight. At this stage it became evident that to continue the work effectively it would be necessary to re-engine the Peregrine and carry out some improvements to the blower system. Finance for this work could not be obtained from official sources and the firm was unable to allocate any further funds for purely private venture research. With great regret, therefore, the project was abandoned at the moment when the path to success seemed clearly marked.*

Miles wrote an article in *Flight* for 26th January 1939, on the subject of: *Sucking away the Boundary Layer - Results of some Full-scale Experiments*, extracts from which follow:

Naturally, we felt that the results of these preliminary experiments were sufficiently encouraging for us to make a more ambitious and careful investigation. With the assistance of the Air Ministry we were able to convert the Peregrine design into a sort of flying laboratory. The entire upper surface of the wings was covered with a perforated sheet similar to that used in the first experiments, and channels within the wings led to an exhauster in the cabin. A Ford car engine was installed to drive the exhauster, and a comprehensive instrument panel was fitted so that we could measure the air sucked away, and the actual power required to do it.

It is very unfortunate that, owing to more urgent work on rearmament, we have been unable to complete the test programme which we originally laid down, and that for the time being at least, the matter has had to be shelved. Even so, during last summer we were able to carry out a number of preliminary experiments amounting in all to 40 hours flying.

Soon after the aeroplane had first flown we discovered that the suction had a very noticeable effect on the performance and instead of measuring the profile drag and the lift, as in the previous experiments, we made careful measurements of the change in the rate of climb and gliding angle, and later of top speed, though this was not so noticeably affected. In measuring the rate of climb a series of tests were made with the engines throttled to 1,800 rpm in order to prevent the overloading of the engines in a prolonged climb. During each flight the engines were being run under climbing conditions for practically an hour and a half and, in consequence, it will be appreciated that there was a tendency for the engines to overheat on full throttle.

On the throttled climbs, averaged over about 20 flights, there was a 29 per cent improvement for an extra expenditure of 8hp in sucking away the boundary layer, at a blower efficiency of 60 per cent. To have obtained this extra increase in rate of climb by means of the aircraft engines themselves an extra 17hp would have been required, so that you will see we got quite a sensible advantage by sucking away the boundary layer. Next we made a number of glides with the engines throttled back at the minimum rate of descent. Here the reduction in minimum gliding angle averaged 19 per cent, or in other words, the Cl/Cp increased by approximately the same amount. This improvement was obtained for a suction hp of 6½ at the same exhauster efficiency.

The Miles M.8A Peregrine Mk.II, L6346, in 1938. F G Miles has his back to the propeller, and Bill Skinner is leaning on the engine nacelle. [Phillips & Powis via Mrs Skinner]

Chapter 18: Miles M.8 Peregrine Mk.I and M.8A Peregrine Mk.II

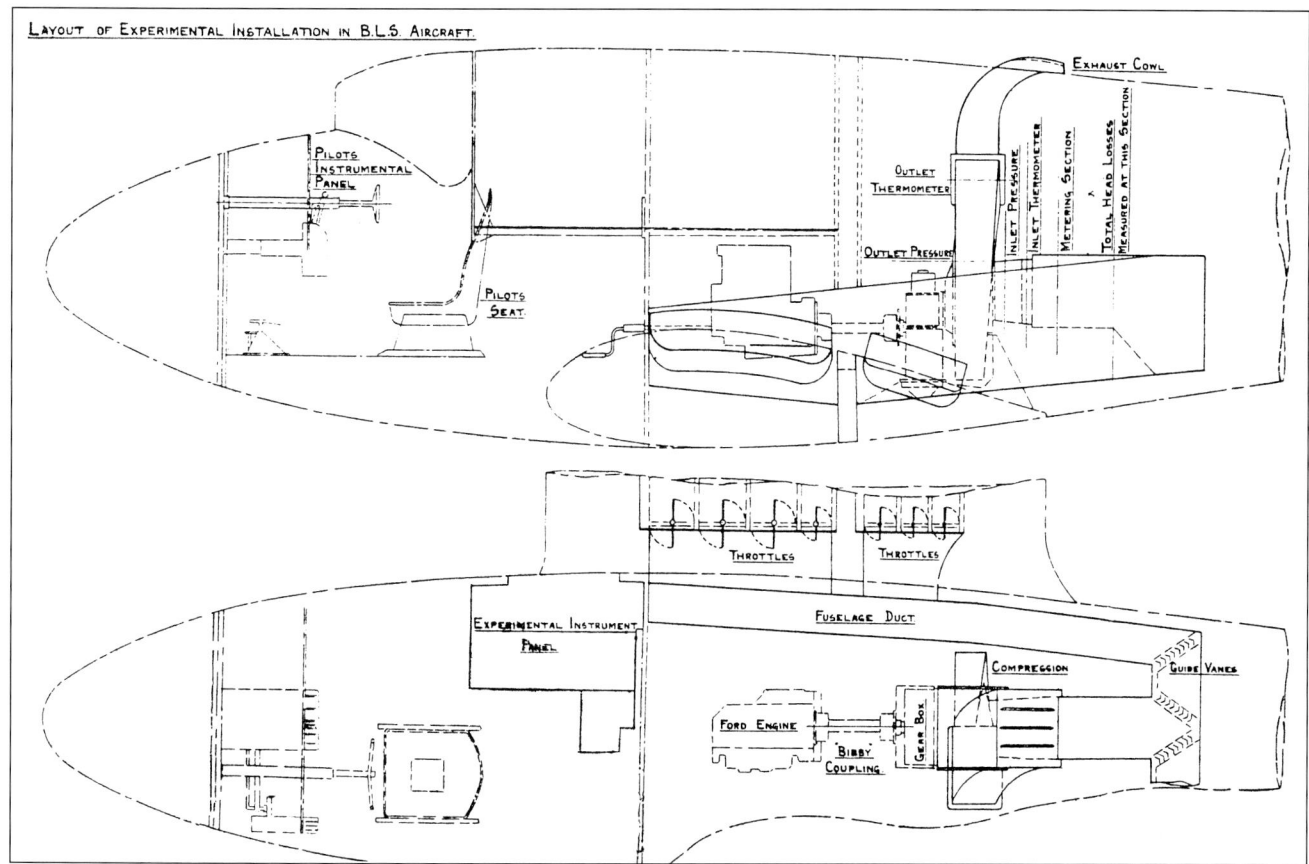

Above: The layout of experimental installation in the Boundary Layer Suction (B.L.S.) aircraft.

Seen at right and below are photographs showing details of the B.L.S. system installed in the Miles M.8A Peregrine.
[Phillips & Powis via The Miles Aircraft Collection]

Lastly, we made a few runs at top speed and found a 4-5 per cent increase in speed for an additional power expenditure of 13hp, compared with 27hp which would have had to be developed by the engines in order to get the same improvement. Analysis has shown that at top speed the effect of removing the boundary layer was to reduce the total drag of the aeroplane by 8 per cent, or assuming that all the improvement was due to reducing the wing wake, this was equivalent to 22 per cent decrease in the profile drag of the wings. Needless to say, the experiments are not yet sufficiently advanced to draw any general conclusions, but we do think that there are possibilities of a practical nature in this device.

Perhaps the idea of sucking over large areas will not be a thing of the near future, if it should ever be proved worth while, but very probably it will be used shortly to eliminate unavoidable drag caused by interference such as at the junction of the fuselage and wings, or between the fuselage and tail unit. Moreover, in spite of the somewhat discouraging results obtained in the wind tunnel on quite small models, I am of the opinion that the quantity of air to be withdrawn for a reasonable aerodynamic improvement is very much dependent on the Reynolds number.

This is not merely a guess on my part but has been substantiated by some theoretical work which we have done, which shows that if a means can be found by which the boundary layer can be removed in such a way that it never becomes turbulent, or in other words, the effect of the suction is to produce everywhere an extremely thin laminar layer such as one finds normally on the leading edge of a wing, then the quantity of air to be removed is an inverse function of the Reynolds number.

Consequently, whereas on the model scale the quantity is discouragingly large, at the higher Reynolds numbers encountered in flight the quantity falls to very much lower values. Although the theory cannot be checked at the higher Reynolds numbers, it is in reasonable agreement with Schrenk's model results. This probability of a marked scale effect is encouraging, more especially as I do not think that with the small amount of suction used in our experiments we could have got the improvements we found unless such a scale effect exists.

It remains now to find out whether it is possible to suck away the boundary layer without affecting it adversely. Is the solution a relatively weak suction spread over the entire surface and carefully graduated to conform with the pressure distribution? Or does it consist of a number of concentrated suction points placed on the surface where the flow is critical? Or is it a combination of the two? Or will some other technique be found, at present unsuspected?

The success of a surgical operation depends very largely upon the skill and highly developed technique of the surgeon. So also in the particular problem we have been discussing crude and clumsy methods may be the bar to success.

An intensive research over the range of Reynolds numbers from model to full-scale is needed, I think, to find out whether in reality it is possible to operate on the laminar boundary layer present initially in such a way that its natural instability does not become sufficient, even in the presence of an adverse pressure gradient, to cause it to become turbulent.

If we can find a suitable process which will enable us at any rate to approximate satisfactorily to the mathematicians' no doubt idealistic conceptions, we shall then be able to say quite justifiably that we have "covered a comfortable bit of ground."

The total expenditure on the M.8A, including the firm's contribution, was approximately £20,000 - by contemporary standards a ridiculously cheap experiment but, with the threat of war on the horizon, and the panic expansion of the RAF getting underway, all official work not directly connected with the war effort was dropped. In consequence, the RAE experiments were brought to a premature close, the engines and outer wings of the M.8A were removed and the aircraft dumped behind the Repair and Service hangars at Woodley.

Some 25 years later, Handley Page was awarded a full-scale development contract to continue the research, which was carried out first on a modified Vampire and later with a Hawker Siddeley 125 executive jet.

Production details:

300 M.8 Peregrine prototype; painted silver. Fitted with 2 x 205hp DH Gipsy Six srs II engines with DH vp propellers. Regd **G-AEDE** (CofR 6762) 21.2.36 to Phillips & Powis Aircraft Ltd. Regn never used. First flown as **U9** by Charles Powis 12.9.36; flown by Miles from 14.10.36. No C of A issued. Flown to and from Farnborough 17.4.37; flown at Woodley by an RAE pilot 11.5.37; still flying up to 23.6.37. Wfu at Woodley soon after and regn cld 12.37.

485 M.8A Peregrine Mk.II. Fitted with 2 x 260hp Menasco Super Buccaneer C6S-4 engines. ITP dated 30.9.36 for the manufacture of one modified M.8 Peregrine for wing boundary layer suction control research by the RAE Farnborough, with a projected cost of £10,000 under Contract No.567581/36 dated 9.12.36. Given serial **L6346**, it was first flown 30.3.38. Initial flight tests commenced during the second week of April 1938. Dd 4.39 to RAE Farnborough. SOC 9.8.39. Scrapped, Woodley.

SPECIFICATION AND PERFORMANCE DATA

M.8 Peregrine Mk.I

Engine: 2 x 205hp DH Gipsy Six Series II
Dimensions: span 45ft 0in; length 33ft 0in; height 9ft 10in; wing area 300 sq ft; cabin volume 218 cu ft; luggage compartment volume 25 cu ft
Weights: tare 3,200lb; disposable load 2,200-2,300lb; empty 3,850lb; fuel (80 gal) 600lb; oil (6 gal) 60lb; two crew 320lb; passengers and luggage 1,320lb; wing loading 18.33lb/sq ft; AUW 5,500lb, which could be increased by another 200lb if required
Performance: max speed at sea level 180 mph; cruising speed at sea level 160 mph; climb to 5,000ft 6.07 min, climb to 10,000ft 14.25 min, climb to 16,000ft 21.5 min; ceiling 23,000ft; single-engined ceiling 5,000ft; run to unstick 600ft; landing speed (with flaps) 52 mph; range at cruising speed 562 miles

M.8A Peregrine Mk.II

Engine: 2 x 260hp Menasco Super Buccaneer C6S-4
Dimensions: span 45ft 0in; length 33ft 8½in; height 9ft 10in; wing area 300 sq ft
Weights: tare 3,200lb; disposable load 2,300lb; fuel (80 gal) 600lb; oil (6 gal) 60lb; wing loading 18.33lb/sq ft; AUW 5,500lb
Performance: max speed 182 mph; cruising speed 177mph; rate of climb 1,050ft/min

The Miles M.9, showing the interior of the cockpit to advantage. ['Flight' via B Clarke]

Chapter 19:
Miles M.9 Rolls-Royce Private Venture Trainer/ Kestrel and M.9A Master Prototype

In order to set the scene for the events which were to follow, it is first necessary to go back to 1934 when the RAF was equipped with biplane fighters such as the Hawker Fury, with a top speed of 207 mph, the Hawker Demon, capable of 182 mph, and the Bristol Bulldog with a maximum speed of 174 mph. The first production Gloster Gauntlet, the RAFs penultimate biplane fighter, made its first flight on 17th December 1934 and had a maximum speed of 230 mph.

In early 1934, Lord Rothermere, the proprietor of the *Daily Mail* newspaper, had expressed a desire to obtain a fast and spacious aeroplane for his personal use. His aviation-minded organisation, appreciating the potential of what is now commonly known as the business or corporate aircraft, found that the Bristol Aeroplane Company had already drawn up the outline of the Bristol 142, a light transport aeroplane in this category and Rothermere arranged to purchase this machine. The Bristol 142, named *Britain First*, was first flown, with 'B' mark R12, from Filton on 12th April 1935, and to the surprise of all concerned was found to have a maximum speed of 307 mph, some 77 mph *faster* than the Gloster Gauntlet!

This machine raised serious thoughts in the minds of the RAF as to the state of the service's equipment. Lord Rothermere, however, was never to use the aircraft as he presented it to the Air Council, who promptly ordered a military version, the Bristol 142M, (later to become the Blenheim), and this made its first flight on 25th June 1936, with initial deliveries to the RAF commencing March 1937.

On 6th November 1935, the prototype of the new Rolls-Royce PV.12 (later RR Merlin) powered Hawker monoplane fighter made its first flight from Brooklands. This machine, with a calculated maximum speed of approximately 315 mph and later named the Hurricane, became the first fighter to be ordered for the RAF to have a maximum speed of over 300 mph. Unfortunately, due to the decision to fit the new 1,030 hp Mk.II version of the Merlin engine in the production aircraft, a revised nose profile had to be designed to accommodate its angled rocker-box flanges and this delayed the first flight of the production aircraft until 12th October 1937, with deliveries to the RAF finally commencing in December 1937.

In December 1935, Miles began to formulate his ideas for a high-performance fighter trainer. On 5th March 1936, the need for such a trainer really became apparent, when the prototype Supermarine Spitfire, powered by a 990hp Rolls-Royce PV.12 Merlin 'C' engine, made its first flight from Eastleigh. Just five days later, the prototype single-engined monoplane light bomber, the Fairey Battle, powered by the 1,030hp Rolls-Royce Merlin I engine, made its first flight from the Great West Aerodrome, with deliveries to the RAF squadrons commencing in May 1937. On 14th May 1938, the first production Spitfire I, fitted with the 1,030hp Rolls-Royce Merlin II engine, giving it a maximum speed of 364 mph, made its first flight, with deliveries to the RAF commencing in August 1938. But the RAF were still without an advanced fighter trainer.

At Woodley, on 20 March 1937, the new Miles Magister ab initio trainer made its first flight and this was the first monoplane trainer to be delivered to the RAF. Miles had for some time been very seriously concerned by the fact that, while speeds were increasing at an unprecedented rate, there were still no thoughts in official circles for a high-performance advanced fighter trainer in which to train and familiarise future pilots in line with these advances. With the RAF finally becoming equipped with high-performance monoplane fighters, pilots were being faced not only with flying a new type of aeroplane, but also having to cope with all the new devices which had been incorporated, such as retractable undercarriages and flaps, etc. This would result, somewhat inevitably, in a number of accidents while pilots were converting from obsolescent biplanes to the modern monoplane fighters.

Although Miles realised that to design and build such an advanced trainer at that time was far beyond the financial resources of Phillips & Powis Aircraft Ltd, that did not stop him from concentrating his thoughts on the preliminary design for a high-performance aeroplane in the hope that some means of financing the project would be found. He felt that the advanced fighter trainer must be of a high-performance and to resemble as closely as possible the new fighters but with two seats in tandem and a clear-view cabin with good all-round vision for both the pupil and the instructor.

It has not been possible to precisely date the commencement of the new advanced trainer project, soon to be designated the M.9, but it is thought to have probably been in December 1935, as the M.10 and M.11 designs were started in about December 1935/January 1936.

In early 1936, Miles and Charles Powis approached Rolls-Royce Ltd with a view to hearing their views on the possible use of their Kestrel engine in the projected advanced trainer. The Rolls-Royce Kestrel was probably the most suitable engine in its class and was powering many successful military aircraft at the time. Miles thought that it would suit his advanced trainer admirably but his initial approach was perhaps without any thoughts at that time of any financial involvement in the firm by the engine manufacturer. Rolls-Royce, obviously thinking of their future Kestrel engine position, agreed with Miles about the urgent need for an advanced trainer and probably also realised the problem facing Miles with regard to financing the new trainer. So, probably soon after the first meeting, Rolls-Royce must have decided to consider backing the project financially.

This did not mean that Rolls-Royce had any real interest in the firm's other products, they were simply looking after their own interests with regard to the future of the Kestrel engine, which they hoped to keep in production should the new trainer be adopted for the RAF. However, the successful outcome of the negotiations meant that Miles could now make a start on the detailed design work for the new M.9, secure in the knowledge that it had a real chance of coming to fruition. In the event, Rolls-Royce's hopes of keeping the Kestrel in production came to nought, as the Air Ministry chose to use Kestrel engines taken from the obsolescent Hawker Hart series of aircraft as they were struck off charge. However, Rolls-Royce did at least receive the contract to recondition the engines but this was obviously not as lucrative as building new ones. Another by-product of Rolls-Royce's financial interest in Phillips & Powis was that the new trainer then assumed two names - one, for obvious reasons, being the M.9 Rolls-Royce Trainer and the other being the M.9 Private Venture Trainer.

On 25th March 1936, just after the first flight of the Spitfire, a press report appeared which was to surprise a considerable number of people:

Stories have recently appeared in the financial pages of various daily papers coupling the names of Phillips & Powis Aircraft Ltd and Rolls-Royce Ltd. Such suggestions are not so improbable as they might seem. Mr F G Miles is one of our brightest and most enterprising designers. As various aircraft firms build aero-motors there seems to be no reason why an aero-motor firm should not build aeroplanes! Mr Sidgreaves and Mr Powis are both good business men. So why should they not do a deal?

In mid April 1936, the local Reading press reported that it was understood that Rolls-Royce Ltd would subscribe a substantial amount to the new capital of Phillips & Powis Aircraft Ltd and gave details of amounts involved, shares, etc, which raised new optimism despite the prior financial loss which the firm had experienced. Then, at the AGM of the company, held on 1st May 1936, the story in the press was confirmed. The local press then reported:

The announced relationship between Rolls-Royce Ltd and Phillips & Powis Aircraft Ltd indicates the former is going to finance the development of new types of aircraft………Mr Miles had already gone far towards creating a revolution in British aviation on a very limited expenditure. Thus they looked towards great steps forward now that money is available, etc, etc.

It was also about this time that Miles tentatively approached the Air Ministry with his ideas for a new advanced trainer but they advised him that his ideas were too futuristic!

The new Hurricanes and Spitfires, with speeds in excess of 300 mph and with retractable undercarriages and flaps, would have to be flown by pilots who were still being trained on the outmoded dual-control version of the obsolescent Hawker Hart operational biplane. The Hart, although one of the most successful aeroplanes of its day, had none of the handling characteristics of the new breed of fighters and none of the new devices mentioned above. These new developments, while advantageous, would greatly increase the workload of the service pilot and, in addition, the handling characteristics of a clean high-performance heavily-loaded monoplane fighter were very different from those of a light low-performance biplane light bomber. The Hart Trainer was, therefore, manifestly unsuitable for training pilots to fly the new monoplane fighters.

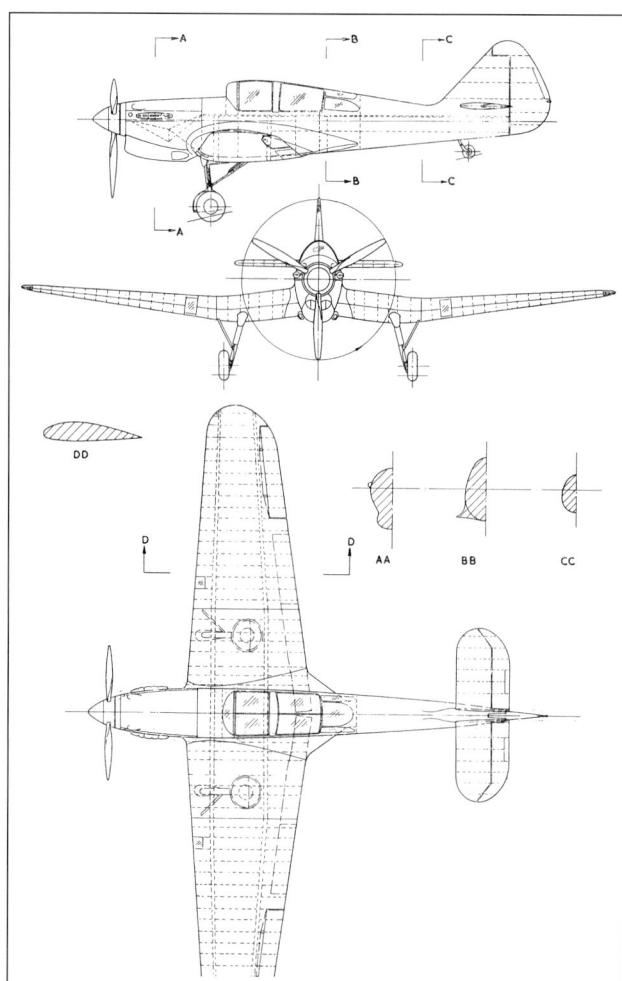

The Miles M.9 P.V. Advanced Trainer.

Even though these new operational monoplanes had been ordered, still no action had been taken to provide a suitable advanced training aeroplane on which to teach Service pilots the handling characteristics of the new types. Then, on 12th June 1936, finally realising the urgent need for such an aircraft, the Air Ministry issued Specification T.6/36, for a 'multi-role' advanced training aircraft.

De Havilland Aircraft Co Ltd tendered the DH.93 Don in a valiant attempt to meet a specification which even DH thought to be impractical. This machine turned out to be ill-conceived and considerably under-powered with its 525hp DH Gipsy Twelve engine and, following trials at the A&AEE Martlesham Heath in September 1937, it was found that the Don exhibited other faults beside its poor performance, which would make it unable to comply with Specification T.6/36 for a *multi-role* advanced trainer for the RAF.

However, by then the need for such an aircraft had become so desperate that 250 Dons were ordered straight off the drawing board - an unprecedented step in those days. It is of interest to note that in the event only 50 were built. Deliveries to the RAF were made between April 1937 and March 1939 but, of these, only 11 were used by Flying Training Schools, with the remainder being used by station flights for communications. The final 200 were cancelled and by March 1939 all the surviving Dons had either been relegated to ground instruction or struck off charge, with many of them not having being used in any role at all!

Miles had naturally studied the complex Specification T.6/36 at the time of its issue but had decided that, although his M.9 would not be able to match it and still have the high-performance that he deemed so necessary for an advanced trainer, neither would any other aircraft manufacturer be able to come up with a suitable design to meet the specification either! Miles maintained that, having foreseen the need back in December 1935 for an aircraft but with a *single* purpose - to train future RAF pilots to fly the new breed of high-performance aeroplanes - it was absolutely essential that the main purpose of such an aeroplane should not be lost by attempting to make it into a *multi-role* trainer, with the additional weight of all the equipment necessary to train air gunners, wireless operators, bomb-aimers, etc.

Something was needed, and needed fast, to replace the ageing and outmoded biplanes, and Miles therefore persevered with his design, which would offer a much better performance with its 745hp Rolls-Royce Kestrel XVI engine, of much greater power than the Don's 525hp DH Gipsy Twelve. The Kestrel XVI was still in production at the time and Rolls-Royce, having taken a large financial interest in Phillips & Powis Aircraft Ltd, agreed to back this new trainer venture, even to the extent of designing and supplying the complete power-plant for the prototype at their own expense. The fact that the RAF was also very familiar with this engine, which was still available in large numbers, would obviously be in its favour, even though it was soon to become obsolescent.

The M.9 'Rolls-Royce Advanced Trainer' was Miles' answer to what the Service really required - a two-seat, very advanced training aircraft for pilots. This was to be of standard Miles wooden construction, fitted with a thick wing (23.85% thickness/chord at the root tapering to 6% at the tip) based on an NACA 230 aerofoil, have a retractable undercarriage, split flaps and a variable pitch propeller. The wing loading was designed to be similar to that of the new operational fighters and thus, in every respect, the M.9 was representative of these aircraft. By careful attention to aerodynamic cleanliness, the maximum design speed was calculated to be 295 mph at 16,500ft, i.e. much faster than the Fairey Battle and only about 15 mph slower than the Hurricane, despite the fact that the latter was a single-seater with a 1,030hp Merlin II engine, giving 285hp more than the Kestrel XVI.

Although it is generally accepted that Miles then submitted his completed design to the Air Ministry, it is now known that he did not actually submit it until much later. When he did finally submit his proposal for an advanced trainer with but one aim, the Air Ministry responded that, although they agreed it had a fine performance and great potential, they considered it to be premature and turned it down, saying that it failed to comply with their recently issued but highly impractical Specification T.6/36 for an advanced trainer.

In vain, Miles tried to convince the Ministry that the necessity for such a machine was obvious, especially to anyone who was a pilot (as he was), but the civil servants with whom Miles was arguing were not pilots and were therefore quite unable to appreciate the arguments he put forward. Miles was so convinced that his arguments were right and that an advanced trainer similar to his M.9 Advanced Trainer would have to be provided for the RAF, *and soon*, that, with the financial backing of Rolls-Royce Ltd, he made the bold decision to go ahead with the aeroplane and build it as a Private Venture. This was no mean undertaking, as to design and build a new aeroplane of this magnitude was an extremely costly procedure, and one which the firm could ill afford, especially as the finished machine would not be a saleable product in the civil market in the event of the Air Ministry refusing to buy it.

However, nothing really worthwhile in life is ever acquired without effort, risk and bold decisions, and here was a gamble, which, if successful, would raise Miles above the status of a designer of light aeroplanes to which he had hitherto restricted himself. In this decision, Miles was fully supported by his co-director Charles Powis and by Rolls-Royce Ltd, with the latter undertaking to design and supply the power plant.

With the basic design of the M.9 completed and detailed drawings being finalised, construction of the new aeroplane commenced in the Experimental Department in about July/August 1936 and as it began to take shape, it could be seen to be one of the most beautiful looking single-engined aeroplanes that had ever been produced. Although no more than eight men were employed on the project at any one time, the 'Rolls-Royce Private Venture Trainer' flew within a year of commencing work. Just 400 man weeks were expended on the project, which is estimated to have cost the company just £2,500. The Miles 'Rolls-Royce Private Venture Trainer', with c/n **330**, was registered **G-AEOC** on 6th October 1936 (CofR 7382) to Phillips & Powis Aircraft Ltd but this registration was not taken up and no C of A was ever issued. The registration was cancelled in May 1938.

Miles had never before flown a high-performance aeroplane but on 3rd June 1937 he climbed aboard the superbly cream finished but unmarked M.9 (although it had been allocated the 'B' mark **U5**, this was not applied for its first flight) and took it off on its maiden flight. The onlookers' excitement knew no bounds as the beautiful looking machine streaked away into the distance. The new aeroplane was more than 100 mph faster than anything that Phillips & Powis had built up to then but it proved to be a delight to fly and, moreover, it possessed just those handling characteristics which were common to the Hurricane and Spitfire.

Not unnaturally, as is inevitable with a new design, a number of minor problems were encountered, including one with the undercarriage which developed an annoying habit of suddenly lowering itself after being retracted. But, as Don Brown later recalled, it was better than remaining retracted when it was required to be down as some undercarriages had been known to do! It was also found that the M.9 had a tendency to yaw and hunt at cruising speed but this was soon rectified by altering the section used in the tailplane, elevator and rudder.

On 5th June 1937, Miles flew a 20 minute 'Test' with George Miles and on June 22nd George Miles recorded a 30 minute flight, 'Test of larger fin in the MRRT', U5, with F G Miles. On 23rd June 1937, the M.9 was flown to and from Farnborough and the following day it was flown to Hendon for the eighteenth and last RAF Display, where Miles was pleased and doubtless very proud to *'show'* the Air Ministry *'the fastest trainer in the World'* - at its first public appearance. Demonstrated there also on June 24th, 25th and 26th, it caused quite a sensation - its maximum speed was, after all, quoted by the firm as being 296 mph.

The M.9, with 'B' mark U5 replaced by new type No.'2' for the show (the number '2' was carried on the starboard engine cowling and on the port side of the rear fuselage), surprised everyone with its remarkable performance, while the DH Don I, which was also exhibited at Hendon, looked impressive but failed to impress performance-wise. The M.9 was flown to RAE Farnborough on June 26th and two days later it was demonstrated at the sixth SBAC Display held at Hatfield on 28/29th June 1937. By this time, the local press were becoming ecstatic in their praise for the new aircraft, reporting that:

The new Miles Rolls Trainer was the first practical result of the fusion of Phillips & Powis Ltd and Rolls-Royce Ltd.

Lord Swinton, the Secretary of State for Air, visited Reading Aerodrome in July 1937 (arriving in a DH Dragonfly from Northolt) and was welcomed by Miles, who accompanied him on a tour of inspection of the hangars and workshops, etc. No doubt Charles Powis would also have wished to have shared the honours of showing Lord Swinton around, were it not for the fact that he had tendered his resignation earlier in the month, after complications in his business and private life. Following the tour of inspection, Miles took Lord Swinton for a flight in the M.9.

All Miles aeroplanes had, up to then, been named after birds of prey, and so, in mid-July 1937, and with the name *Kestrel* already

The Miles M.9 apparently being prepared for flight.   ['Flight' via B Clarke]

The Miles M.9 with F G Miles in the cockpit, probably just prior to its first flight, which was made on 3rd June 1937.
[Phillips & Powis via The Miles Aircraft Collection]

Chapter 19: Miles M.9 Rolls-Royce Private Venture Trainer/Kestrel and M.9A Master Prototype

The Miles M.9 Rolls-Royce P.V. Trainer No.2 of the new types at the 18th and last RAF Display at Hendon on 26th June 1937.
['The Aeroplane' via Real Photographs]

The M.9 P.V. Advanced Trainer. Note the prototype Miles M.11 'Straight Special' G-AECT in the background.
[Phillips & Powis via The Miles Aircraft Collection]

being used informally by the workers in the factory for the M.9, it was decided to formally name the prototype 'Kestrel', appropriately in keeping with Miles' penchant.

The Kestrel was the first tangible result of the link-up between Phillips & Powis Aircraft Ltd and Rolls-Royce Ltd, a combination which was to have far reaching effects in the future. Without Rolls-Royce's involvement, it is open to speculation as to whether either the prototype Kestrel or its production successor, the Master would ever have been built. The Kestrel engine itself was provided free of charge by Rolls-Royce, who also financed its installation, as it was obviously in their interest to see the project succeed with a view to keeping the Kestrel engine in production. Furthermore it should be noted that the M.9's airframe was also stressed to take the 1,030hp Rolls-Royce Merlin II engine and with that motor, it would then have had the phenomenal performance, for a two-seat training aeroplane, of somewhere in the region of 325 mph in level flight!

It is also of interest to observe that, although the de Havilland tender for the Specification T.6/36 had, by then, been accepted by the Air Ministry, Miles had decided that, despite his misgivings that the specification was 'impossible' to meet, he would submit a design to the Air Ministry which would try to comply with it. Undated, but contemporary, Phillips & Powis brochures described this as 'The Miles Kestrel high speed Trainer', and drawings

Chief Test Pilot Bill Skinner climbing into the Miles M.9 'Kestrel' at Woodley. [Phillips & Powis via The Miles Aircraft Collection]

contained therein show this to have been similar in appearance to the M.9 advanced trainer. It was projected in six different versions:

    I.   Dual Instruction
    II. Blind Flying
    III. Reconnaissance
    IV. Rear Gun Turret
    V. Rear Turret and Wireless (although this version did not show the turret!)
    VI. Bomb Aiming

These brochures show the M.9 in flight with the number "2" painted on the fuselage (as applied for the RAF Display on 26th June 1937) and also refer to the high-speed trainer as the Kestrel, so the brochures must have been issued after **mid** July 1937, when that name was first used officially. The introduction to the brochures stated:

*The aircraft described herein is designed to give a high performance for dual instruction and advanced training **to the original** (the bold text is mine - PA) specification T.6/36. It may also be considered as a two-seater fighter or a general purpose aircraft with a speed range of 5½ to 1.*

After careful consideration of the various operational requirements the two-seater tandem arrangement was decided upon. This, it was considered, would give the best performance, whilst the somewhat restricted view of the instructor in the rear seat would be offset by his familiarity with the aircraft. It was also felt that flying conditions for the pupil would more closely resemble those of the aircraft he would subsequently be flying.

The cooling duct for the engine has been developed to give minimum drag and throughout the design, aerodynamic cleanliness has been most carefully studied to produce an aircraft of exceptional performance without the extravagant use of Power. Messrs Rolls-Royce Limited have developed the "Kestrel" Series Engines for many years and in its present form it is probably the most reliable power unit that exists today.

Interestingly, one of these brochures had a grey cover with a logo at the top consisting of the word "HOLLINDA" in a stylised bird, and was entitled 'Miles Kestrel high speed Trainer'. This brochure differed from others in also having metric comparisons for the dimensions, weights and performance figures (some of which also differed from their English counterpart) but it bore no reference to Phillips & Powis, either on the cover or within the contents. Instead the cover bore at the bottom the inscription "Technische Handelmij. Hollinda NV, Lange V IJ Verberg 13 - Den Haag", and it is suspected that Phillips & Powis were trying to sell the multi-role concept aircraft to the Dutch, but for some reason chose not to put their name on it. Literally translated, the company 'Technische Handelmij' means 'technical firm' in The Netherlands. The content of these brochures is shown in Appendix 18.

Following completion of the manufacturer's trials at the end of July 1937, the Kestrel was delivered to Rolls-Royce at Hucknall, where their chief test pilot, Capt Ronald Shepherd was to fly it in connection with the engine installation and cooling development. It is suspected that the radiator may have been slightly deepened at this time. Ronnie Shepherd had previously briefly acted as a flying instructor with Phillips & Powis and elsewhere before becoming a freelance test pilot in 1931, but he was also, apparently, a gentleman with a very short 'fuse'! This was to manifest itself one day when the Kestrel's engine failed at about 100 feet on take-off!

He did the only thing possible in the circumstances and belly landed it in a field, sliding on until the machine came to rest right opposite a public house. He promptly demanded a pint of beer and sat in the cockpit drinking it! The police arrived within minutes and, not unnaturally, demanded to know: 'What was goin' on ere?' Apparently Shepherd was livid, saying; *I can't even have a pint without the b***** police poking their noses in!*

It is also of interest to note that the Heinkel He.70 (G-ADZF), which had been delivered to Rolls-Royce on 27th March 1936 for use as a RR Kestrel test-bed, was at one stage fitted with an engine installation identical to that of the Miles Kestrel.

CHAPTER 19: MILES M.9 ROLLS-ROYCE PRIVATE VENTURE TRAINER/KESTREL AND M.9A MASTER PROTOTYPE

By October 1937, the Ministry were now apparently showing some interest in the new aircraft and on October 30th it was delivered to RAE Farnborough where it 'underwent tests' before being flown back to Woodley by test pilot Bill Skinner on November 9th. The Ministry were then faced with a difficult problem; it was a year since the requirement for an advanced trainer had been officially recognised but there was still no sign of it and it was now needed with the utmost urgency *and* in quantity.

Miles had been proved to be correct in his assessment of the requirement, as he so often had been, and as there was by then only one solution to the problem, the Air Ministry had to swallow its pride and order the M.9 Kestrel. The fact that they had earlier condemned it as being premature, despite its impressive performance, and that it failed to comply with the original specification had somehow to be forgotten. The question was then how quickly could it be got into production? The whole episode became a triumph, not only for Miles but also for private enterprise, to which many of the best aeroplanes of the time owed their existence.

In about November 1937, when it is believed that the Air Ministry interest was finally aroused, it was requested that the Kestrel be sent to the A&AEE at Martlesham Heath for trials. The M.9 was modified to have i) a deeper chin-mounted radiator and ii) the underside of the rear fuselage deepened from just forward of the trailing edge of the wing. This gave the impression that the tailplane had been raised, which it had not at that time. The empennage of the prototype was retained in its original position and the original canopy was also retained, with hinged sideways opening (to starboard), front and rear cockpit hoods. The rear cockpit was also fitted experimentally with a tilting windscreen in order to give the instructor a better view for landing. However, the initial canopy modification, which raised the roof line to just above the front canopy, left the forward top of the rear canopy blanked out.

The modified M.9, retaining its original cream colour scheme and 'B' mark U5, visited RAE Farnborough on 30th November 1937, before being delivered to A&AEE Martlesham Heath in December. Photographs of the machine, taken at Woodley and Martlesham Heath in December 1937, clearly show these modifications.

The first report on its behaviour, dated 4th February 1938, concluded that, in its present state, although the handling qualities were generally satisfactory (apart from the severe directional oscillation at high speeds and unsatisfactory stalling characteristics, particularly at aft c.g.) *the Kestrel was found to have many serious shortcomings; the view was bad, the stall was without natural warning, the rudder with dashpot damping made taxiing difficult, and uncontrollable directional oscillations occurred at speeds above 250 mph. An aileron mass-balance broke off and wedged the control surface and the rear cockpit cover failed, the take-off distance to a height of 50 feet was 540 yards, i.e. some 40 yards longer than usually specified and the landing distance was 110 yards longer.*

The next report issued by the A&AEE, dated 26th March 1938, stated that three experienced test pilots had been testing the M.9 Kestrel since December 1937 and that during the course of the trials it had undergone some modifications. No spinning was attempted at this time, as the aircraft had not been fitted with an anti-spin parachute.

The Kestrel was returned to Woodley for further modifications to be carried out and these included: i) the fitment of a widened windscreen, also increased in depth by nearly two inches, to improve the pilot's view; ii) the fitment of a new framed front canopy of increased height to accommodate the new windscreen, which also brought it into line with the rear canopy; iii) the front pilot's seat raised by two inches; iv) the rear canopy was redesigned and the roof fitted with a glazed front top and a tilting windscreen, which enabled the instructor to raise his seat by some nine inches in order to be able to look over the pupil to give him a better view for landing; v) the rear top decking to the fuselage was raised and 'flared out' to where the fin joined the fuselage *(a photograph, taken at Woodley, shows that the leading edge of the fin had lost the smooth flare to the top of the fuselage, thereby confirming that the top of the rear fuselage was now 'wrapped' round the leading edge of the fin)*; vi) the underside of the rear fuselage was increased in depth, which also helped to give the impression that the tailplane had been raised but this was definitely not the case, and vii) the directional control was modified.

The Kestrel was then re-delivered to the A&AEE on 12th April 1938 for further testing and it is thought that these modifications

The Miles M.9 'Kestrel' Second Phase with F G Miles in the instructor's position. [Phillips & Powis via The Miles Aircraft Collection]

led to the M.9 prototype becoming known as the 'M.9 Kestrel Second Phase'. Between April 12th and 22nd, the Kestrel was flown by four test pilots and, following these tests, a further report was issued by the A&AEE, dated 20th May 1938. This report suggested that although the additional keel surface aft had improved directional stability, continuing undamped oscillations required further modifications.

The Air Ministry were by then becoming very anxious over the situation regarding an advanced trainer and were seriously considering ordering aircraft 'from abroad' (the North American Harvard from America in fact) and this despite the fact that Miles had been trying to convince them of the most urgent need for such an advanced trainer since early 1936. So, in spite of the early criticisms of the M.9 Kestrel, it was agreed that it should be ordered for production, albeit in a modified form.

Nobody was really surprised, and everybody was pleased at the factory at Woodley when, on 11th June 1938, the Air Ministry officially announced that an order had been given to Phillips & Powis Aircraft Ltd for the production of M.9 Kestrel advanced trainers. The order, to Contract No.779602/38, dated 11th June 1938, was for the manufacture of 500 M.9B Master Mk.I aircraft to AM Specification 16/38. The specification, dated 15th August 1938, was issued to Phillips & Powis on 5th September 1938.

C G Grey, the incorrigible editor of *The Aeroplane*, wrote in the 15th June 1938 issue, under the heading *A National Disgrace*:

Either the Air Ministry has gone crazy or there are some super-political or higher financial influences at work of which we ordinary taxpayers know nothing. On June 10 the following official notice appeared in the British newspapers (relevant extracts from which appear below):

'The Government have given consideration to the report of the Air Mission which recently visited the United States of America and Canada. The mission were instructed to inquire primarily into the possibility of purchases for early delivery of aircraft having certain special characteristics. A considerable number of proposals were made, not all of which fulfilled the conditions above referred to, and eventually the mission recommended that negotiations should be entered into with two USA firms for a total of 400 aircraft.

The Government accepted this recommendation and contracts have now been negotiated for (1) 200 aircraft suitable for general reconnaissance duties (the militarised version of the twin-motor Lockheed 14, later to become the Hudson), and (2) 200 aircraft suitable for advanced training duties (the North American BT-9B two-seat trainer, later to become the Harvard).

To order American aeroplanes of any sort is an imbecility. To order any foreign aeroplanes in these categories is a national disgrace etc, etc. There is still less excuse for ordering the North American machines. These are very ordinary low-wing fighter-trainers with fixed undercarts. The type has a 400hp Whirlwind motor, quite a good motor, and has a top speed of 171 mph. That is much less than the speed of the Miles Kestrel trainer - which could easily be built in vast quantities and would use up hundreds of Kestrel motors which are obsolete because they have been superseded by the Merlin.

A rough estimate puts the total of the order at about £4,000,000. That such a sum should go to America when we have nearly 2,000,000 unemployed in this country and tens of thousands of competent engineers and wood-workers out of work, and acres of factory floors unused is a crime against British Capital and Labour. Before this disgraceful deal, ramp, racket, or whatnot was made public some preliminary warnings were let out by British newspapers.

The *Telegraph* said:- *'It is understood that Mr A H Self, Deputy Under-Secretary to the Air Ministry, and leader of the British mission to the United States, has reported favourably to Sir Kingsley Wood, the Air Minister, on the possibilities of obtaining from them a supply of training 'planes within the necessary period. Each company, I understand, could supply about 150 machines within a few months. Additional orders by the Air Ministry may be placed elsewhere'.*

A few months for deliveries sounds optimistic. What about the time needed to get prototypes over here for test at Martlesham? And what about arranging for AID Inspectors to go to America? The US Aircraft Industry is not going to get away dumping dud trainers on us today as it did in 1914-15-16. If the pro-American expert advisers of the Air Ministry try to plant on the RAF Flying Training Schools or on the Elementary Schools a bunch of American 'airplanes' which have not passed our Stress Department, and have not been flight-tested at Martlesham, and have not been inspected by the AID from the raw material right up to the assembly shop, exactly as all the English machines are tested, there is going to be a bigger row in the House than Air Affairs have yet produced.

*We **will** not have the lives of British pupil-pilots risked on machines which have not passed AID, - just to save the political skins of a panic-stricken Cabinet etc, etc.*

POST SCRIPTUM. - Since the foregoing was written, the news - hitherto a secret known to everybody - has been allowed to leak out that big orders have been given to Phillips & Powis Aircraft Ltd of Reading for their Kestrel Trainers, which is in the same class as the North American etc, etc.

......But we know that the British orders were placed in effect, though not officially, months before the Mission even went to the United States. And we know, as soon as the Mission left, that its instructions were to interest itself in trainers and not in first line aircraft...... There are also big orders in hand for the Miles Magister primary trainer and for the Tiger Moth. But there is still plenty of spare factory space in England, and still plenty of unemployed skilled engineers, - so spending £4,000,000 in the States is still a national disgrace.

To say that C G Grey, as ever, felt strongly about such matters, would not be putting too fine a point on it! However, it did need to be said by somebody - and who better?

It had taken one year from the first flight of the M.9 Kestrel for the Air Ministry to place the order with Phillips & Powis for the production version. The company had in fact already been given the Instruction to Proceed (ITP) dated 20th May 1938, but the production aircraft were to be powered by the de-rated, 715hp Rolls-Royce Kestrel XXX engine. The reasons for fitting this de-rated engine are explained in Chapter 20 on the M.9B Master Mk.I.

The order for 500 Masters was to make history, being the largest contract ever placed by the Air Ministry for a training aircraft for the RAF, and it was valued at over £2,000,000. The contract also demanded that the aircraft were to be delivered in the shortest possible time. The sheer size of the order necessitated considerable extensions to the factory and a new assembly building had to be built. This incorporated an ingenious track system, on which it was planned that the whole production line would move forward as a complete unit every four hours, the assembly having been carefully divided into a number of stages, with each operation entailing four hours work. This assembly line, as applied to aircraft manufacture, was believed to have been the first of its kind, although it was apparently followed almost immediately by Junkers in Germany.

In keeping with the demand that the new aircraft should be delivered in the shortest possible time, the first 500 aircraft were delivered to the RAF between 4th August 1939 and 27th September 1940, at an average of 10 aircraft per week.

The production aircraft were to be a development of the M.9 Rolls-Royce PV Trainer prototype and, in line with the RAF policy of naming trainers after the learned profession (as befitting their status, e.g. Tutor, Magister, Oxford, Don etc), the new aircraft was

CHAPTER 19: MILES M.9 ROLLS-ROYCE PRIVATE VENTURE TRAINER/KESTREL AND M.9A MASTER PROTOTYPE

The Miles M.9A Master prototype N3300 after rebuild from the M.9 'Kestrel' (Second Phase). Note the modified cabin.
[Phillips & Powis via The Miles Aircraft Collection]

named the Master. The production aircraft were designated M.9B Master Mk.I in about January 1939. These would, thankfully, also be without all the original 'essential' additional requirements called for in Specification T.6/36, which had included a rear gun turret, wireless, bomb aiming and photographic reconnaissance equipment! In other words, the RAF was finally to get an advanced trainer aircraft specifically designed for the purpose of training pilots to fly high-performance aircraft.

The Air Ministry had however, by this time, decided that, in the event of the new Miles Masters not being available in time, they should, after all, place an order with North American Aviation for 200 of their new trainers. These were to become the Harvard Mk.I (N7000-N7199) in the RAF and they were delivered between December 1938 and October 1939. The first of the 500 Miles M.9B Master Mk.I's (N7408-N8081, N9003-N9017) was delivered to the RAF on 4th August 1939. It should have been and could quite easily have been delivered much earlier, if only the Air Ministry had committed themselves.....

Miles M.9A Master prototype

Following the trials of the M.9 Kestrel at the A&AEE Martlesham Heath, the Air Ministry issued Contract No.757691/38, dated 13th June 1938, with a projected cost of £16,000, to cover conversion of the prototype M.9 to the M.9A Master. The Kestrel was taken into the Experimental Department, where it was stripped down and completely rebuilt. The fuselage was cut off at the rear cockpit and rebuilt with a six-inch extension; the tailplane, elevator and trim tabs were enlarged to production standard, and fitted in the original, lower, position. A production type fin and larger rudder was fitted. The front and rear cabin hoods were fitted with jettison gear and the rear cabin hood was further modified. The tailwheel leg was lengthened, but the deepened chin mounted radiator was retained. The finished result was painted silver overall and given the serial **N3300**. Although it is not known for certain when this serial was issued, it was probably in May 1938.

The Master prototype, while looking more purposeful than the Kestrel, had lost the beautifully sleek lines of its forebear. The personal photograph album of Bill Skinner, very kindly donated to

The Miles Aircraft Collection by the widow of his son, shows three undated photographs of N3300 taken at Woodley. These were probably taken soon after its first flight, as there is no anti-spin parachute fitted but they show that, at some stage, the tailplane had been moved to a position above the intersection of the top fuselage with the fin, roughly mid-way between the rudder hinges. They also show that wool tufts had been strategically placed in the area around the under-surface of the intersection of the leading edge of the tailplane and the fin. Likewise, a large number of wool tufts were placed in the upper, triangular, portion of the centre section of the wing, from the rear of the front cockpit to the intersection with the trailing edge of the outboard wing, and also including the wing root fairing.

Unfortunately, the Air Ministry individual record card for N3300 does not record much detail of its movements, just: 'Director General Research & Development, 5th October 1938'. However, the official ledger (now kept by the Air Historical Branch) records that it was TOC by DGRD on 5th October 1938, and it is this entry which tends to imply that it was delivered to the A&AEE at Martlesham Heath on that date. However, it is now suspected that this was only a 'paper' date relating to its transfer to DGRD and that this was when the aircraft was taken into the Experimental Department to be converted to meet the Ministry requirements. No photographs of N3300 were to appear in the aviation press until March 1939, which implies that the extensive modifications had not been completed until about this time. Its actual first flight date in the new configuration is not known but was probably early in January 1939, as a report in a local newspaper dated 14th January 1939 stated that: *the first Master has appeared and gone to Martlesham for tests.*

However, it should be noted that this report is at variance with the contemporary report on the official opening of the new factory extension on 27th January 1939, which stated that; *it was regretted that the Master was unable to perform in the air, but the prototype is undergoing modifications to the tail*. It is suspected that this somewhat ambiguous report meant that the first production Master Mk.I was unable to perform in the air because it was not yet complete and the prototype Master N3300 was having its tailplane position moved, as the three photographs in Bill Skinner's album show, and these must have been taken at this time.

The M.9A Master prototype showing the deeper chin radiator and machine gun in starboard wing.
[Phillips & Powis via The Miles Aircraft Collection]

In fact, at the time of the opening of the new factory extension, N3300 had already been returned to Woodley and was back in the Experimental Department for further modifications. Following... *a great deal of redesign work,* which included *modifications to the tail and anti-spin parachute frames fitted,* N3300 was delivered back to A&AEE Martlesham Heath in June 1939 under Contract No.B996095/39, dated 14th June 1939, for further tests, still fitted with its 745hp Kestrel XVI engine.

For the record, it should also be noted that, on 24th March 1939, Miles commented, with reference to building an aeroplane comparable with any other machine (i.e. the new Master): *6 months ago we started to re-design down to every single nut, bolt, screw - and detail. If the plane is flying next month we shall have taken 7 months to do it.* This statement actually referred to the first production Master Mk.I, which was first flown, by Miles, later the same month, on 31st March 1939.

A photograph of N3300 flying alongside N7001, the second Harvard for the RAF, was taken on 24th April 1939, appearing in *Flight* for 18th May 1939. The Master prototype had been loaned to Phillips & Powis under Contract No.B996095/39 dated 9th May 1939 and a Harvard (presumably this one) was loaned to Phillips & Powis Aircraft Ltd under an undated Contract No.B7599/39.

A local newspaper, reporting on the *Empire Air Day* held at Woodley on 20th May 1939, stated that *The new Master was even more masterful as it swept by like a silver streak!* Then, on 23rd May 1939, a 'parade of aeroplanes' was laid on at Northolt for inspection by Members of the Houses of Lords and Commons, and N3300 was amongst those present.

The next A&AEE report, issued on 3rd August 1939, stated that:
The Master had been modified in several respects to make it more representative of the production aeroplane. Stability was generally satisfactory and the directional 'hunting' had finally been eliminated by the modified fin and rudder. The feel of the controls was the same as before except for the rudder which was very heavy and became almost immovable in high-speed dives. However, after preliminary diving tests had been made, they had to be curtailed following the failure of the starboard wheel fairing in a dive at 310 mph, causing the aeroplane to be returned to the makers for repair.

The pilot's direct vision panel also blew in but before that, it had been established that, when trimmed level at full throttle, a stick push force of 20lb was required to hold the aeroplane in a dive. It was, however, known that a further modification to the production Master was intended and that this would entail moving the radiator away from its chin position to a point below the front spar (see Chapter 20 on the M.9B Master Mk.I for full details - PA). As this would have affected the balance of the aircraft, dives and handling at limiting positions of the c.g. would be carried out on a production aircraft.

The report finally concluded that the M.9A Master prototype was suitable for the service, but *it had a 'heavy' rudder, a lack of stall warning and that it exhibited an ease of pulling into a stall on spin recovery.*

Meanwhile, an illustration of the spirit which in those pre-war years characterised the aircraft industry in general, and Phillips & Powis Aircraft Ltd in particular, was provided by an incident which occurred after the modified M.9 prototype had been purchased by the Air Ministry.

The design of the Kestrel provided for a maximum permissible diving speed of 330 mph indicated, which represented a true speed of 470 mph at about 20,000ft altitude, but this speed could only be reached in a steep dive. Miles, always a great lover of publicity, and even more so if associated with the unusual or dramatic, decided that they should see what could really be achieved in the right conditions. So one day between Sunday 5th November and Saturday 11th November 1939, when there was a howling northerly wind blowing, Bill Skinner took-off in N3300 and climbed steadily away to the north in the teeth of the gale, arriving over Oxford near the aircraft's service ceiling of between 25,000 and 30,000ft. Turning south, he then opened the throttle wide and put the nose down into a steep dive, arriving over Farnborough some minutes later, having reached a true speed of 504 mph!

Miles promptly sent a telegram to the Air Minister which read: *Kestrel has today achieved 504 mph* and equally promptly came the reply: *Congratulations on meteoric flight* - all very childish but it did tend to confirm that this speed had in fact been reached. Surely both the Air Ministry and the RAE would not have wanted to be involved in such a story if it had not been true. It is also interesting to note that Victor Burnett, the Air Correspondent of the *Sunday Express* newspaper at the time, was given a flight in the Kestrel by Miles, during the course of which they reached an indicated airspeed of 400 mph.

Then, tragically, on Wednesday 15th November 1939, only a few days after this momentous flight, Bill Skinner collapsed in his bathroom while shaving and died of a cerebral haemorrhage.

The Miles M.9A Master with raised tailplane and wool tufts on wing root. [Phillips & Powis via Mrs Skinner and P Amos]

The Miles M.9A Master with tailplane in final position. ['Flight' via B Clarke]

Leading edge slats, with wool tufts, were fitted experimentally to the Master prototype to investigate wing drop at the stall.
[Phillips & Powis via The Miles Aircraft Collection]

N3300 with leading edge slats and in RAF camouflage.
[Phillips & Powis via The Miles Aircraft Collection]

Although it will never be able to be proven, one theory for Bill Skinner's tragic death was later put forward by George Miles and this related to Bill's passion for inverted flying. George recalled that he was once flying back from a trip to Hatfield in loose formation with Bill Skinner, who was flying another Miles aircraft, and that shortly after take-off Bill rolled his aircraft, and then proceeded to fly back to Woodley inverted! The fact that most machines would not fly inverted for very long, suggests that he must have been flying one of the new M.2Z Hawk Trainer Mk.II's which had been modified for inverted flying for a Roumanian Government order.

Early in the war, both Miles and the Air Ministry were anxious to find some means of improving the behaviour at the stall of the new Hurricanes and Spitfires. These aircraft dropped a wing very sharply at the stall and this was regarded as indicative of a tendency to spin. To investigate the problem, Miles fitted a pair of leading edge slats along the outer portion of the wing of N3300 and with these in the open position, the stall was found to be very mild, and 30 degree gliding turns could be made at 1.1 Vs. The photograph above shows that they were fitted experimentally with wool tufts placed at regular intervals along them and although the date of this experiment is not known, the aircraft was camouflaged, so it could have been after September 1939. The only disadvantage with this solution to the problem was that means would have had to be provided for opening and closing the slats, and their fitment would have caused serious delays in delivery of both the Hurricane and the Master, both of which were by then in full production and urgently required.

Another experiment carried out on N3300 involved five pairs of large fixed slots through the centre of each wing aft of the front spar. This also gave a satisfactory stall with no tendency to spin, and again 30 degree gliding turns could be made at 1.1 Vs. However, it would have been necessary to provide means of closing the slots during flight, since their drag at high speeds was appreciable, resulting in a marked loss of performance and that, in turn, would have disrupted production, something which could not be tolerated in wartime. In the event, neither slats nor slots were fitted to production aircraft.

It was probably in connection with one or the other of these remedial devices that Hugh Kennedy, assistant test pilot, made his first flight in N3300 on 22nd February 1940, recording a 20 minute: 'Test Flight - NACA Wings'. Hugh also test flew N3300 on a number of occasions between 10th October 1940 and 15th March 1941, carrying out; *'handling, stability trials and test flights in accordance with ADM.293'*.

On 20th August 1941, N3300 was received at 6 MU Brize Norton where it was officially 'Struck off Charge' on that day, but it was almost certainly returned to Woodley later, as the contract under which it was loaned to the firm was not completed until 11th April 1942. It was then dismantled and its original Rolls-Royce Kestrel XVI engine was installed in the new Miles wind tunnel on the Davis Farm site, where it was used for driving the fan.

A sad end for such a once beautiful machine.

SPECIFICATION & PERFORMANCE

Miles M.9 Rolls-Royce Private Venture Trainer/Kestrel

Engine:	745hp Rolls-Royce Kestrel XVI (VP)
Dimensions:	span 39ft 0in; length 30ft 0in; min height (tail down) 10ft 0in; wing area 235 sq ft; wing section NACA 230
Weights:	empty 4,160lb; AUW 5,340lb (brochure gives AUW 4,699lb); wing loading 22.72lb/sq ft (brochure gives 20.0lb/sq ft)
Performance:	max speed generally quoted to be 296 mph at 14,500ft but the firm's brochure quotes variously 295/296 mph at 16,500ft; cruising speed 253 mph at 16,500ft; landing speed 60 mph

Miles M.9A Master Prototype

Engine:	745hp Rolls Royce Kestrel XVI (VP)
Dimensions:	span 39ft 0in; length 30ft 6in; min height (tail down) 10ft 0in; wing area 235 sq ft; wing section NACA 230
Weights:	not known
Performance:	max speed 226 mph at 14,500ft; landing speed 60 mph

For comparison the sizes and weights of the early Hurricane and Spitfire were:-

	Hurricane	Spitfire
Span:	40ft 0in	36ft 10in
Wing area	258 sq ft	242 sq ft
AUW	5,800lb	6,000lb
Wing loading	23.9lb/sq ft	23.2lb/sq ft

N7408, the first production Miles M.9B Master Mk.I prior to take-off on its first flight. Note the nose-mounted radiator and blanked out rear cockpit.
[Via Phillips & Powis and The Miles Aircraft Collection]

Chapter 20:
Miles M.9B Master Mk.I

Contract No.779602/38 was placed by the Air Ministry with Phillips & Powis Aircraft Ltd on 11th June 1938 for 500 M.9 Master Mk.I aircraft to Specification 16/38, dated 15th August 1938, for the 'Production of the Phillips & Powis Private Venture Trainer'. File No.788121/38/RDA3, which was issued to Phillips & Powis Aircraft Ltd on 5th September 1938, decreed that it was:

To meet Operational Requirement OR.58, the aircraft shall be constructed in strict accordance with the drawings and schedules covering the construction of the 'Miles Kestrel Trainer' aircraft N3300 in the form which that aircraft is accepted by DTD, as the prototype of the production aircraft, excepting for such amendments as may be introduced to meet the requirements of this specification and to make detail alterations to facilitate production.

Engine: Kestrel engines supplied on embodiment loan are to be installed. The installation is to comply, as far as is practical, with the requirements of DTD.1028 paragraph 15 to 20 inclusive. The engines shall be suitably cooled for operation under tropical conditions. A constant speed airscrew of a type to be agreed by DTD shall be fitted and its strength shall be adequate for the conditions of the diving tests prescribed in ADM.292.

Consideration shall be given to improvements in respect of the Stability at the Stall, Rudder oscillation, and Control in a dive, with a view to reducing to a minimum the adverse features mentioned in the reports of A&AEE on the prototype aircraft. Consideration shall be given to the following maintenance features: Engine mounting, Tail wheel (to be detachable), Fuel tank (improved shape), Elevators (to be detachable without removing tailplane), Wing tips (to be detachable).

Crew: 2 (Pilot and pupil).
Armament: One Browning gun with 300 rounds of ammunition.
Performance:
    Maximum speed at 14,400 ft: 226 mph
        at sea level: 193 mph
    Range at economic cruising: 484 miles
    Service ceiling: 27,000 ft.

The 745hp Rolls-Royce Kestrel XVI engine of the M.9 Rolls-Royce PV Trainer/Kestrel was to be replaced in the Master by the 715hp Rolls-Royce Kestrel XXX, which gave 585hp at 2,750 rpm at 15,000ft (or 12,000ft as given by other sources). This output compared unfavourably with the 745hp at 3,000 rpm at 14,500ft of the Kestrel XVI engine fitted to the prototype. The maximum speed was also reduced from 295 mph at 16,500ft (although other sources gave the maximum speed as 296 mph at 14,500ft) to 250 mph at 15,000ft; however the Master was still to be the world's fastest trainer. The reason given by the Air Ministry at the time for replacing the Kestrel XVI with the de-rated Kestrel XXX was that they suspected (wrongly, as it subsequently transpired), that pupils would be *'incapable of handling such high power after the much lower powered Magisters and Tiger Moths of the EFTS's'.*

Jack Angell, the Rolls-Royce representative at Woodley at the time, recalled that the real reason the engine was changed was because the Rolls-Royce Kestrel XXX was *'more rugged'* and could therefore, *'take the extra strain imposed under training conditions with the great numbers of take-offs and landings'.* It should be noted that the Kestrel XXX engines with which the Masters were to be powered, were not new, having been rebuilt and uprated by Rolls-Royce from existing Kestrel engines. Rolls-Royce were disappointed not to have been able to provide the brand new engine and installation, but they did at least obtain the contract to rebuild the existing Kestrel engines, mostly Mk.Vs, which had been taken from obsolete RAF Hawker Harts and Hinds.

During 1938, the Air Ministry asked Miles to consider substituting the 810hp Bristol Mercury radial air-cooled engine for the liquid-cooled Kestrel XXX engine, of which stocks were claimed to be running low. The nine cylinder Mercury engine, still in full scale production for use in the Bristol Blenheim bomber, was readily available. The necessary design modifications were put in hand immediately and the designation M.19 allotted to the re-engined aircraft, but the design was shelved when it was discovered that there were still stocks of Kestrel XXXs available after all.

On 27th January 1939, three weeks ahead of schedule, Sir Kingsley-Wood, the Air Minister, officially opened the new very large factory extension at Woodley which had been built to accommodate the Master assembly line. This extension incorporated the first 'moving' aircraft track assembly line in the country, on which completed units moved forward every four hours as each operation had been given four hours to complete. The new 'Track Assembly Line' for Miles Master production at Woodley hit the newspaper headlines in August 1940 and was also fully described by Miles in *Flight* for 22nd August 1940, under the heading 'Track Assembly-Line System in use at the Works of Phillips & Powis.'

Although only seven Masters had been delivered to the RAF by the outbreak of war, the increase in production when the moving assembly line came into operation was dramatic, with a further 63 Masters being delivered to the RAF between September 3rd and the end of 1939.

In 1939, when people were beginning to associate mass production with all-metal stressed skin structures, the study of the construction of a comparatively large aeroplane in wood in large numbers, and at a high rate, was considered by *The Aeroplane* to be both refreshing and instructive. It was also considered that, for quickness of getting into production, the wooden aeroplane had advantages over its metal counterpart. In original design and construction, the wooden prototype was also cheaper and quicker to build, being designed and built in less than a year.

Miles had written earlier in *The Aeroplane* for 9th November 1938 on the advantages of using wood for aircraft construction:
It is easy to design in wood - provided one remembers one is designing in wood. I prophesy that, if we are ever allowed really to adapt a modern high-speed design to wood, we will promise to cut down production time and put up the production rate to something really exciting.

Miles also laid down four points in favour of wooden construction:
1. Aeroplanes can be built as fast, strong, stiff, and as durable in wood as in metal.
2. They can be built more quickly, either in small numbers or in mass quantities, than similar metal aircraft, if the designer, works manager and the purchaser recognise certain fundamental simplifications inherent in the material, and take advantage of them.
3. There are unlimited supplies of suitable labour and suitable material. For metal there are neither.
4. If an emergency (polite term for war) makes us build really large quantities of fighters or light bombers, sooner or later we shall place most of our faith in wood.

He qualified these remarks by saying that they applied to military or civil aircraft of fairly light types. He did not argue about really large civil transport aeroplanes - but he did believe in wood for defence in war.

The first production M.9B Master Mk.I, **N7408**, was first flown on 31st March 1939 and photographs taken at the time show that it had a deep chin-mounted radiator. The rear cockpit was faired over and the fuselage appeared to be in primer with no markings, although the wings were fully painted with the serial clearly visible on their undersides. Monty Cook, a Philips & Powis employee, who witnessed the first flight, recalled that it could be plainly seen that the pilot was experiencing great difficulty in maintaining level flight on the circuit and the flight very nearly ended in disaster.

The problem was attributed to a design error. In those days the company did not have a weights office and as a result the c.g. was wrongly calculated using a 'dry' engine, i.e. without oil and coolant. When the aircraft took-off, the pilot found that the c.g. was too far forward causing excessive nose-down pitch and making the aircraft virtually uncontrollable. The only solution at that late stage in the proceedings was simply to reposition the radiator under the centre section. This gave the aircraft a somewhat ungainly appearance and did nothing to improve its performance, which was already lower than that which Miles had originally envisaged for his high-speed trainer.

Then, on 19th May 1939, N7408 crashed at Rolls-Royce's airfield at Hucknall (the reason for its non-appearance at the Empire Air

Miles M.9B Master Mk.I fuselages under construction. [Phillips & Powis via The Miles Aircraft Collection]

By the number of bodies around N7408, the first production M.9B Master Mk. I, it is suspected that this photograph was taken soon after its near-fatal first flight.
[Phillips & Powis via The Miles Aircraft Collection]

The inquest! N7408 with 'Tich' Edwards in whiteh overalls; 'Johnny' Johnson, the Superintendent of Final Assembly and Flight Shed, is immediately to his right.
[Phillips & Powis via The Miles Aircraft Collection]

Miles M.9B Master Mk.Is on the final assembly line in the original Assembly Shop. Note the wingtips and surround to the roundels which were yellow in colour. This photograph was 'passed for publication' by the Press Section of the Air Ministry on 24th July 1939. ['Flight']

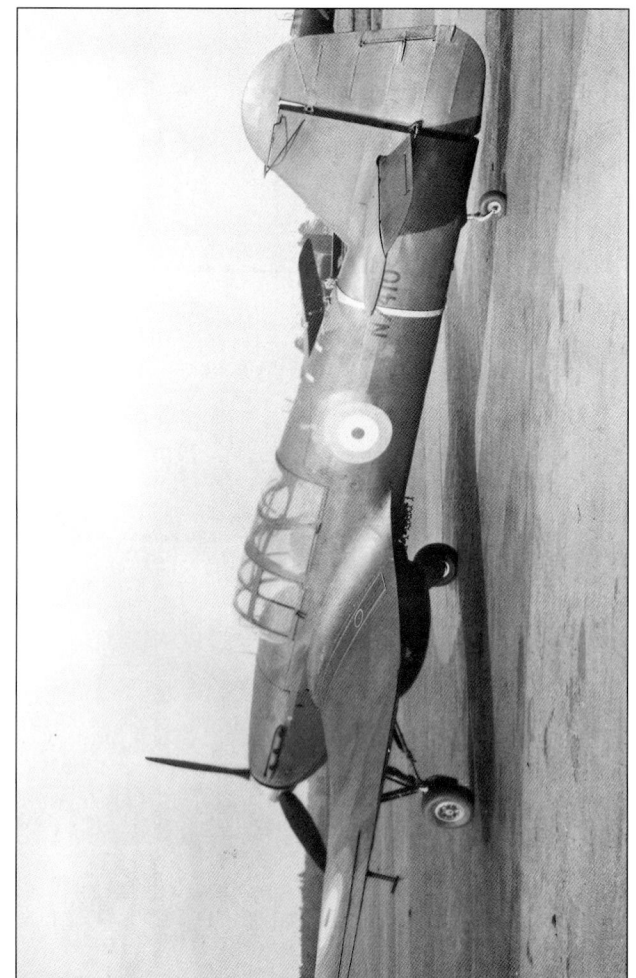

Photographs showing M.9B Master Mk.I N7410 at the A&AEE. The photographs, from the original format, were dated 31st August 1939.

[Air Ministry via The Miles Aircraft Collection]

Chapter 20: Miles M.9B Master Mk.I

The fighter-like front cockpit of the Miles M.9B Master Mk.I, showing the famous Miles control box on the left and the gun button on the control column.
[Phillips & Powis via The Miles Aircraft Collection]

Day at Woodley on May 20th). The reason for the crash and the extent of the damage is not recorded but it was returned to Woodley, rebuilt by the Repair & Service Department and ultimately delivered to RAF Sealand on 24th June 1940.

The Aeroplane for 26th July 1939 reported (correctly): *The most noticeable difference between the prototype and the production model is that the radiator has been moved back from the nose to a ventral position. The final position of the radiator was governed, not so much by cooling or aerodynamic considerations, as by considerations of weight distribution.*

However, by the time *The Aeroplane's* sister journal *The Aeroplane Spotter* published details of the radiator's re-positioning in the issue for 25th January 1945, they had completely overlooked what had been correctly published nearly six years before in *The Aeroplane* and stated: *Cooling difficulties which arose in the Master prototype, N3300 (with the deep chin radiator) resulted in the radiator being moved back to a position under the wings in the Master Mk.I.* In fact, 'cooling difficulties' were not experienced with the Master prototype, and the latter report shows how 'misinformation' can so easily happen, the moral being - don't always believe all that you read - although I expect you to believe what I write!

The second and third production aircraft, **N7409** and **N7410**, were despatched to the A&AEE Martlesham Heath but on the declaration of war on 3rd September 1939, the A&AEE moved to Boscombe Down on the edge of the Salisbury Plain, and the two Masters went with them on the same day.

The production M.9B Master Mk.I differed in a number of respects from the prototype M.9A; the aft portion of the fuselage was made deeper, the tailplane slightly raised, the areas of the fin and rudder increased slightly, the cockpits made more roomy and the cockpit canopy further improved so that the instructor in the rear seat could get an even better view when landing.

The wingspan of the production M.9B Master Mk.I was the same as the prototype M.9 Kestrel, but the length was increased to 30ft 8in and the height to 10ft 0in. The modifications however had reduced the maximum speed to 226 mph at 14,000ft, and the cruising speed to 196 mph, but had increased the landing speed to 85 mph, and, with an AUW of 5,356lb, the climb to 10,000ft took 5.9 minutes. The Master Mk.I was fitted with a Rotol three-bladed variable-pitch constant-speed propeller, of right-hand tractor, and the canopy was hinged on the starboard side. A 'GA of Armament' drawing, dated 31.12.38 (and later amended to also include the Master II and III), showed two flares carried internally in the port wing, eight bombs carried under the wings, one Browning .303 machine gun installed in the starboard wing and a G.42 cine camera mounted in the leading edge of the port wing for practice purposes. Additionally, there was a ring and bead sight for the rear cockpit (mounted internally on the starboard side), a reflector gunsight mounted on the coaming for the front cockpit and twin landing lamps installed in the port wing.

Even with all these changes, the performance of the Master Mk.I was still far higher than any other training aircraft then available in the world. The handling characteristics remained so similar to those of the Hurricane and Spitfire that the thousands of pilots who were trained on the Master felt perfectly at home on their first flight in either of these two modern fighters.

By 1939, stocks of reconditioned Kestrel engines were really running low and the Ministry asked Miles to fit the Bristol Mercury

These photographs show the Miles M.9B Master Mk.I with the instructor in normal flight position (above) and in landing position (below).
[Phillips & Powis via Mrs Skinner and P Amos]

Chapter 20: Miles M.9B Master Mk.I

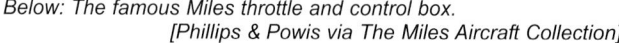

Above: Chief Test Pilot Bill Skinner on the wing with Blossom and F G Miles in discussion by an early production Master Mk.I.
[Phillips & Powis via The Miles Aircraft Collection]

Right: The Miles M.9C Master Mk.IA, development details of which continue beyond the remit of this particular publication.

Below: The famous Miles throttle and control box.
[Phillips & Powis via The Miles Aircraft Collection]

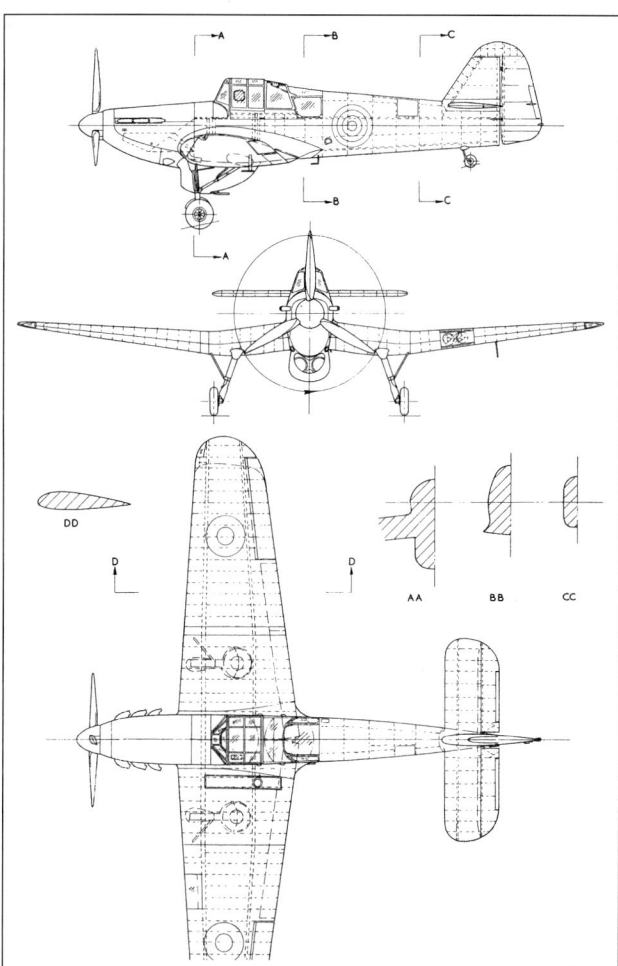

engine, of which it was claimed there were 'unlimited stocks' available. The earlier design was resurrected and the M.9B Master Mk.I N7422 was taken from the assembly line in June 1939 to be fitted with a 840hp Bristol Mercury VIII engine (as fitted to the Bristol Blenheim Mk.I). With the increased span tailplane of the M.9C it became the prototype M.19 Master Mk.II and was first flown, by chief test pilot Bill Skinner, during October 1939. However, following the crash of this prototype during its official trials with A&AEE at Boscombe Down, on 10th January 1940, the M.9B Master Mk.I N7447 was hastily converted to become the second prototype Mk.II in order for the trials to be completed and this was first flown, by Flt Lt Kennedy, on 1st April 1940.

The development, mass production and operational history of the Master Mk.I and Mk.II continued throughout the Second World War and extend the Master story well beyond the timescale and scope of this particular publication.

The first production Miles M.9B Master Mk.I, N7408, after the radiator had been repositioned to under the centre section.
[Phillips & Powis via The Miles Aircraft Collection]

SPECIFICATION AND PERFORMANCE DATA

Engine: * Rolls-Royce Kestrel XXX. The engine, engine mounting and engine cowling of the Master are designed and supplied by Rolls Royce Ltd. The engine is fitted with a three bladed Rotol constant speed airscrew.
Max HP for take-off 725 hp at 2,750 rpm; Max HP at 15,000 ft 585 hp at 2,750 rpm; cruising HP at 14,750 ft 495 hp at 2,400 rpm.

Dimensions: span 39ft 0in (later reduced to 35ft 9in); length, tail up 30ft 8in; height, tail down 12ft 8in; wing area (gross) 235 sq ft; wing area (nett) 210 sq ft; aspect ratio 6.5; wing section, root NACA 23024, tip NACA 23006

Weights: * tare 4,156; fuel (70 gal) 525lb; oil (7 gal) 63lb; coolant (10 gal) 110lb; pilots (2) 400lb; equipment (removable) 101lb; AUW 5,352lb, max normal full load; wing loading 25.4lb/sq ft, based on nett wing area; engine loading 9.2 lb/hp.

Armament: one Browning .303 machine-gun in starboard wing; one G.42 camera inboard of the two flares which were carried internally between the spars in the port wing; eight 25lb practice bombs carried on mountings attached to the rear spar under the wings.

Performance: max speed at sea level 196 mph, max speed at 15,000ft 250 mph*; cruising speed at 14,750ft 228 mph*; max economical cruising speed at 10,000ft 160 mph; max design diving speed 330 mph IAS; stalling speed, flaps up 76 mph IAS, flaps down 59 mph IAS; distance to unstick 720ft; distance to 50ft 1,425ft; landing run 1,125ft; landing run from 50ft 2,325ft; gliding speed flaps up 110 mph ASI*; gliding speed flaps down 80 mph ASI*; touch down speed 62 mph ASI*; range at economical cruising speed 484 miles; duration at economical cruising speed 3 hr; initial rate of climb 2,100ft/min; time to 10,000ft 7 min 40 sec*; time to 15,000ft 13 min; time to 20,000ft 21 min; service ceiling 26,800ft; absolute ceiling 28,200ft

*Engine details, weights and performance figures marked * are taken from the original Miles Master I Pilots' Notes issued by Phillips & Powis Aircraft Ltd in 1939.*

No photographs or drawings of the Miles M.10 Queen Wasp exist, but the Miles M.3D Falcon Six, on which it was based, is shown in this view of G-AEDL. [The A J Jackson Collection via B Clarke]

Chapter 21:
Miles M.10 Queen Wasp

In late 1935, the Air Ministry issued Specification Q.32/35, which called for a radio-controlled pilotless target aircraft to replace the DH Queen Bee (the pilotless target version of the DH Tiger Moth, although based on the DH Moth Major). Three firms submitted project designs; Miles with the M.10, a radio-controlled pilotless development of the M.3D Falcon Six fitted with a 205hp DH Gipsy Queen I engine; Percival Aircraft with their Type H, and Airspeed, who ultimately won the contract with their AS.30.

The M.10 was submitted as a landplane or twin-float seaplane, the latter being Miles' second proposal for a seaplane, the first being one for a proposed 'Hawk Two Seat Low Wing Seaplane', which was not proceeded with. This project awakened Miles' interest in radio-controlled pilotless aircraft however and this later manifested itself in the Hoopla Flying Bomb and the M.47, M.47A and M.49 pilotless radio-controlled target aircraft projects of 1942.

Contract No.489568/36 was issued to Phillips & Powis Aircraft Ltd and a mock-up of the M.10 Queen Wasp was built in 1936. C L Nash recorded that c/n 279 was a 'Miles Queen Wasp - Mock-up only' but unfortunately, he did not record a date, although this would probably have been in early 1936. He also noted that c/n 433 and c/n 482 were to have been Miles M.10 Queen Wasps.

It cannot be confirmed whether Phillips & Powis Aircraft Ltd ever actually received an instruction to go ahead with the manufacture of the two M.10 Queen Wasps, although the Air Ministry issued serials K8889 and K8890 for them on 17th April 1936. In the event, construction of two prototype aircraft was commenced in 1937 and Don Brown recalled that a total of about 120 design man-weeks were expended on the M.10 project, but they were not completed.

Miles M.10 Queen Wasp 'production'

279 P&P works records originally stated that c/n 279 was a type M.3B Falcon Six but this was crossed out and amended to read 'M.3D Falcon Six (Queen Bee)'. Under 'Subsequent & Modifications' the notes read 'Subsequent to G-AEAG No.266 but fitted Onions (Ribbesford Co) undercarriage. Land type. Also to be suitable for seaplane – floats'. The customer was given as the Air Ministry and it was noted that it was a 'Queen Bee aircraft for Farnborough'. However, C L Nash recorded that c/n 279 was a 'Miles Queen Wasp - Mock-up only', to Job No.H5720.

433 & 482 C L Nash recorded that these were for two M.10 Queen Wasps, both to Job No. 5007. They were 'Part constructed - Order cancelled' but he did not record any dates. In an attempt to date the order for these aircraft, c/n 433 was sandwiched between the end of the first order for M.14 Magisters for the RAF (90 aircraft with Job No.5003, ITP dated 4.11.36, with the last, c/n 432 being delivered on 19.2.38), and the start of the order for one prototype and 44 production M.16 Mentors (Job No.5002 for the prototype, and Job No.5021 and 5023 for the production aircraft, ITP dated 24.8.36, with the first, c/n 434, being delivered on 28.3.38). C/n 482 followed one Magister and two Hawk Trainers, with the last of these, c/n 480, being delivered on 4.2.37. However, it is likely that the c/n's for the production batches of Magisters and Mentors were allocated at the time of the placement of the orders and that would account for the apparent discrepancy in the dates.

It should be mentioned that some sources claim that the serials K8889 and K8890 were issued to Percival Aircraft Ltd for their Type H but this is incorrect as they were definitely issued to Phillips & Powis Aircraft Ltd.

SPECIFICATION AND PERFORMANCE

Engine: 205hp DH Gipsy Queen I
Dimensions: span 35ft 0in; length 25ft 0in; height 6ft 6in; wing area 174sq ft
Weights: not known
Performance: not known

Above: G-AECT coming in to land with flaps fully down, in September 1936. ['Flight' via B Clarke]

Chapter 22:
Miles M.11 Whitney Straight Special and M.11A, M.11B and M.11C Whitney Straight Variants

Mr Whitney Willard Straight was the eldest son of Mr Willard Straight (a wealthy United States diplomat, businessman and financier) who was born in New York in 1912. His father had died of pneumonia in 1918 and in 1925 his mother married Leonard Knight Elmhirst, the owner of Dartington Hall in Devon. Whitney Straight, who had been educated and had lived for ten years in England, had learned to fly at Haldon, Devon at the age of 16 years, and by his 17th birthday he had gained his 'A' Licence, having already flown more than 50 hours solo.

Rich, cheerful, handsome, physically strong and always enthusiastic, he went up to Cambridge in 1931 to read moral sciences and, as undergraduates were not allowed to keep cars during their first year, he kept his DH Puss Moth G-ABNZ at Marshall's Fen Ditton aerodrome! From there he would fly to Brooklands to take part in motor racing, and it was this which led to his decision to leave Cambridge and to devote himself to the sport, forming Team Straight in 1933. In July 1935, he married Lady Daphne Finch-Hatton, daughter of the Earl of Winchelsea and, as a result, gave up motor racing to concentrate on aviation.

He had already, in the summer of 1933, commissioned an aircraft to his required specification – this was the Hendy Heck G-ACTC, which was designed by Basil Henderson and built by Westlands at Yeovil. Unfortunately, soon after its first flight in July 1934, it was damaged in a collision with a cow during a forced landing and missed its primary purpose as Whitney Straight's entry in the King's Cup Air Race. It is thought that Whitney Straight never took delivery of the Heck, although it was registered to his mother until September 1935 (as a US citizen he was not allowed to have a British-registered aircraft).

On 17th April 1935, The Straight Corporation Limited was formed by Whitney Straight and Frederick Gwatkin, with the stated intention of operating 'airports and flying clubs'. In *The Aeroplane* for 12th February 1936 it was reported that Whitney Straight had applied to the Home Secretary for British naturalisation as: *Now that all his business interests are this side of the Atlantic he wishes to discard his American citizenship, which has debarred him in the past from competing in such events as the Air Race for the King's Cup. If his application is granted he may fly in this race in the future.*

Whitney Straight wrote in the editorial of the *Straightaway Review* for 1938: *We realised at the outset that an investment in civil aviation was one from which no dividends could be expected for a considerable time. We felt, however, and still feel that such an investment will, over a period of years, prove justified.* He went on to say that, although they had no wish to be aerodrome owners, their first stage of development was, in partnership with various public bodies, to undertake as the specialists in airport management and operation, the management of not more than fifteen aerodromes. By the end of 1938 The Straight Corporation were operating aerodromes and aero clubs at Ramsgate, Ipswich, Plymouth, Clacton, Exeter, Haldon, Inverness, Weston-super-Mare and Swansea. He had also taken a major shareholding in Western Airways Ltd at Weston-super-Mare.

Whitney Straight became convinced of the urgent need for a really efficient general-purpose light aeroplane. What he wanted was a two-seat, dual-control machine, which combined the high-speed cruising comfort of the long-distance tourer with the slow-landing, viceless performance of the club trainer. With this in mind and having witnessed the recent success of Miles aeroplanes in the King's Cup and the England to Melbourne Air Races, in early 1936, Whitney Straight approached Miles, who enthusiastically applied his acknowledged designing genius to the difficult task of building an aeroplane to this exacting specification.

The pilot and passenger were to sit side by side in a roomy and comfortable cabin instead of in draughty open cockpits, making communication much easier than by having to shout at each other through Gosport speaking tubes, which was the normal way of things at the time. The aircraft was also to be reasonably fast, easy and safe to fly and economical to operate and maintain.

In construction, the new aeroplane followed the usual Miles practice, being made of wood throughout, as this proven form of construction had the advantage of combining great strength and durability with economic maintenance and repair. The fuselage was of box form with spruce longerons and formers with plywood covering, while the two-spar full cantilever-type wings were covered in plywood and were designed to fold, being hinged on the rear spar with a section of the inner wing built as a folding flap. A feature of the Whitney Straight Special was the use of Port Orford plywood. This material was lighter than the more usual birch plywood and, in consequence, it was possible to use a thicker gauge for the same weight, to give a more robust structure. Only

The prototype Miles M.11 Whitney Straight Special under construction in April 1936. Note the entrance door on the starboard side. [Via P Amos]

The completed Miles M.11 Whitney Straight Special G-AECT. ['Flight' via B Clarke]

the tailplane and control surfaces were fabric covered, and the whole structure was protected both internally and externally against the effects of weather and extreme climatic conditions. The aircraft was jig-built throughout in order to achieve a high standard of precision in manufacture.

Access to the roomy cabin was by a large door on the starboard side, hinged longitudinally just past the centre line of the machine. The cabin was especially soundproofed and the two extra-wide seats were mounted on a single frame with a disappearing armrest. The windscreen of the prototype was constructed with separate panels but this was changed on the production aircraft to a one-piece moulded windscreen. Full dual control, with two independent vertical control columns, two throttles and twin parallel-motion adjustable length rudder-bars were fitted and a highly efficient system of Miles split-flaps was incorporated. These flaps consisted of aerodynamically-balanced double surface flaps in three sections extending across the fuselage. The outboard wing flaps were hinged at the forward edge and commenced to operate before the fuselage section, which in turn was hinged along the rear edge and so designed that when it came into action the air pressure forced it down and balanced out the air resistance acting on the other two sections. The flaps were operated by a Theed vacuum-ram, manufactured by Sir George Godfrey & Partners. This was operated from a vacuum tank exhausted from the induction system of the engine which was first tested on the M.2X Hawk Trainer Mk.II.

The prototype M.11 was originally known as the Whitney Straight Special and was powered by a 130hp DH Gipsy Major Series I engine. This gave it a maximum speed of 145 mph, a cruising speed of 130 mph and a landing speed of 38 mph. It is of interest to note that the M.11 was the last Miles aeroplane to incorporate wing fold (the Mohawk had it incorporated but it was probably never used). A Phillips & Powis photograph of a three-view drawing of the first machine entitled 'GA. of Miles Whitney Straight Special', with the indistinct date of 7.9.36, gave the following details:

Overall length 25'; Overall height 6' 6";
Wing span 35' 8"; Centre section span 9' 5 7/8";
Span with wings folded 16' 6".

The Whitney Straight Special was first flown, without any registration or 'B' mark, by Miles on 3rd May 1936 (not 14th May 1936 as many reference works state, including Miles' own publication 'A pictorial précis of some of the aircraft designed and built by Miles Aircraft Ltd'!) and although he made no comment against the 30-minute flight in his log book, it apparently came well up to his expectations, being both comfortable and easy to fly.

Enthusiasm for the new aeroplane grew daily and all who tried it were unanimous in their praise. Charles Lindbergh also flew it, although not on Wednesday May 6th, when he first visited Woodley. On that occasion, Lindbergh requested that no details of his visit be given to the press, but one newspaper reported the visit saying that Lindbergh had: *tried several (Miles) aeroplanes.* This may have been an over-statement, although it is known however that while at Woodley he flew a Nighthawk, probably U-6 (subsequently G-AEBP).

Miles next flew the Whitney Straight Special on May 9th, the same day that it was registered G-AECT, and on May 16th, when Lindbergh again visited Woodley, Miles gave him a 5-minute flight in the aeroplane before he made a solo flight in it.

The Aeroplane for 20th May 1936 reported that Lindbergh had again visited Woodley: *...to see the Whitney Straight Special masterpiece. He flew it alone and expressed satisfaction at its behaviour.* Another report stated that: 'Charles Lindbergh expressed the view that he considered that the aeroplane was the best machine of its type that he had ever flown'. *The Aeroplane* also commented that: *...its performance at both ends of the speed scale puts it in a class by itself among light aircraft, whilst its phenomenal stability is an enormous credit to its designer.*

On May 22nd, Don Brown flew from Shoreham to Woodley with George Miles in the Avian G-AAWH to collect the Hawk Trainer G-ADLN and while they were there, Miles gave Don a demonstration of the Whitney Straight Special's handling capabilities. This turned out to be one demonstration which Don would not easily forget, as he later recalled:

As I had expected (and feared) the take-off was made in the usual F G style by hitting the tailwheel smartly on the ground and taking the air in an almost perpendicular attitude. But worse was to follow. Having climbed practically vertically to 800ft with the throttle fully open and the stick fully back, F G promptly closed the throttle, applied full rudder, and cheerfully let go of the controls altogether. As I expected, one wing flicked down and I thought 'Here is where we spin - probably for the last time'. To my astonishment, however, the Whitney Straight merely did a short sharp dive and came out on an even keel - which is precisely what F G had intended to demonstrate. I have never forgotten, however, how awfully close the ground looked during those terrible few seconds when I was expecting a spin to develop. On that occasion, as can be seen, the laugh was on me, but a few days later it was on F G.

It was the practice among civil pilots in those days always to leave the petrol turned on, in order to safeguard against risk of air-locking. When starting out, it was easy to overlook checking that the petrol was on and instead simply note the contents of the fuel tanks. In the RAF however, it was the convention to turn the petrol off after landing.

Shortly after Don's demonstration 'fright', a prospective customer turned up and asked for a demonstration of the Whitney Straight Special. Full of enthusiasm and 'bubbling over' with sales talk, Miles hurried him into the prototype and treated him to a take-off similar to that experienced by Don. Unfortunately, the machine had just been flown by an RAF pilot who, in accordance with his training, had turned the petrol cock off after landing. The aeroplane leapt into the air with Miles fully intending to cut the throttle as before at 800ft, but imagine his feelings (and those of the unfortunate passenger) when the engine cut itself at 50ft! With the aeroplane 'hanging on the propeller' and in the hands of a less skilled pilot, that would have been the end of the story, but Miles' quickness of thought and action were as well known as his startling demonstrations. Below them lay a tiny patch of rough ground about 50 yards square and into this Miles sank the Whitney Straight.

The machine was designed with a fixed undercarriage but after this landing one could be excused for thinking it had a retractable one! The two occupants' heads hit the cabin roof pretty hard but they were both able to walk out of the machine relatively unscathed, which many hold to be the definition of a good landing - especially under such circumstances! Needless to say, Miles made no comment against any such demonstration flight in his log-book and the machine was quickly repaired. Miles had been flying the Whitney Straight throughout May and until 22nd July 1936, when he recorded a 5-minute flight, so this, therefore, could have been the occasion on which the fuel cock had been turned off before he took off.

On 29th June 1936, Miles and Blossom flew the Whitney Straight Special to Hatfield, where Miles ably demonstrated it at the Society of British Aircraft Constructors Display, before returning to Woodley later in the day.

While Lindbergh's Mohawk was being built, Miles lent him G-AECT, which he used for a business trip to Germany, probably in July 1936, when he landed at Staaken Airport, Berlin. On 16th August 1936, Lindbergh flew it from Lympne to the Continent, where he landed at Hanover, and from there went on a tour which included calls at Warsaw, Moscow, Kharkov, Rostov, Kiev, Odessa, Cracow, Olmuty, Prague, Stuttgart, Paris, Morlaix, Rotterdam and Berlin.

The Whitney Straight marked an enormous advance, both in comfort and performance, over any corresponding type of aeroplane at that time. It handled exceptionally well from the start and it was soon realised that the original performance figures, already about 50 per cent faster than the contemporary biplanes, were likely to be further improved upon. The effect of the large-area, finger-tip control, vacuum-operated 90-degree flaps, was perhaps the most striking feature of the machine's capabilities. Thanks to these two-position flaps, the Whitney Straight could be pulled off the ground at 50 mph and climbed away steeply and, with the flaps fully down, a really steep angle of glide was obtainable, making it easily possible to approach over high obstacles and still land in a restricted space. If there was anything like a wind blowing, and the necessity arose, the machine could be made to descend in a truly autogiro-like manner under complete control. The intermediate flap position was introduced for take-off and this could also be used on the glide if the pilot wished. The machine, therefore, had a variety of approach angles and this was immediately appreciated as a great safety factor, particularly in the event of an emergency landing.

The controls were smooth and well balanced at all speeds and exceptional liberties could be taken with the machine even near the stall. Apart from the intensive test flying which followed, great attention was paid to detail, finish and general layout. The cabin was upholstered with special soundproofing materials until the best possible results were obtained and at all times the opinions of pilots and private owners were sought, in order that criticism and advice could emanate from sound practical experience.

While the prototype machine was undergoing its trials, Phillips & Powis were making preparations for its large scale production. The production aircraft, which were designated M.11A Whitney Straight (reference to the 'Special' having been dropped about this time), differed from the prototype in having the door on the port side, a slightly modified undercarriage and it also became one of the first aeroplanes to incorporate a one-piece moulded Perspex windscreen, entirely free from structural members. This gave a very fine and completely unobstructed view from the cabin and became a very popular feature with all who were to fly it. The Whitney Straight had a range of 570 miles, a ceiling of 20,000ft and was originally sold for £985.

Although primarily produced as a club and private owner's aeroplane, the Whitney Straight became an ideal machine for carrying out research work and the prototype was later used for a number of very interesting experiments. In 1936, E F Relf, then superintendent of the aerodynamics department of the National Physical Laboratory (and after the war Principal of the College of Aeronautics at Cranfield), delivered the James Forrest Lecture entitled, 'Modern Developments in the Design of Aeroplanes' before the Institution of Civil Engineers. During the course of the lecture he said: *'... the revolutionary discovery we need as regards surface friction is to find a means to make the boundary layer flow remain laminar over a much greater proportion of the surface, or - in other words - to prevent turbulence from developing in the boundary layer itself."* This remark seemed to justify a continuation of the research work which Miles had already done on the drag of wings and, since thick wings offered such advantages structurally, Miles decided to find out whether any means could be discovered for reducing their drag.

Miles recorded 'boundary layer' against a 20-minute flight in G-AECT on 13th July 1936, with the same comment against a one-hour flight on July 20th and these two flights referred to the experiments in which Miles was very keenly interested - the idea of preventing, or at least delaying, the onset of turbulence.

An article by Miles appeared in *Flight* for 26th January 1939, entitled *Sucking Away the Boundary Layer - Results of Some Full-scale Experiments*. Leading in to the article, the Editor summed it up thus:

Two distinct methods of controlling the boundary layer are possible: re-energising the flow which has been slowed-down, and sucking it away. As a result of experiments carried out by Phillips & Powis at Reading, Mr F G Miles comes to the conclusion that the latter method is most promising. In this article he describes in his usual interesting style some of the experiments carried out and the results obtained. The outstanding conclusion is that scale-effect is large, and that therefore at high speeds the quantity of air to be dealt with is smaller than at low speeds. This fact seems to promise possibilities of making boundary layer control worth while. Further research is urgently needed.

Miles, in his article, said: *The subject of boundary layer control is, of course, a very large one, and in the general sense of the term it obviously includes any mechanical devices such as a slat, or diruptor, for manipulating the boundary layer flow in a favourable way, as well as the idea of re-energising it with compressed-air jets or by sucking it away. It is with this latter aspect that we have been concerned during the last three years. In spite of a fair amount of work in the laboratory both with jets to re-energise the retarded boundary layer flow and other experiments in which the boundary layer has been sucked away, it is still doubtful whether we can hope to gain a sensible increase of performance by such methods when applied to the actual aeroplane.*

On the whole, it would appear to be more satisfactory to suck away the boundary layer than to re-energise it by means of compressed air, for although one can get a greater aerodynamic advantage in the latter case as compared with the former, the first method does not require so much power to obtain a given improvement, the saving in power being well worth the slight loss in aerodynamic gain.

Before we began our experiments no attempts had ever been made to try the idea of boundary-layer suction in the air. It is interesting to note that one authority, well known for his scientific work on the subject, considered that it would be too dangerous to make such experiments. However, we decided to make a few tentative flight experiments.

At the time the first Miles Whitney Straight was being used for experimental work, and we considered that it would be suitable for the job. A short length of very thick wing, having a thickness/chord ratio of 30 per cent, was strapped on to the outside of the normal wing. The upper surface of the subsidiary section had a perforated metal sheet laid on and the boundary layer over this part of the surface was sucked away by a very large venturi slung beneath the experimental part of the wing. (Note: the normal Whitney Straight's wing section was 18% thickness/chord Clark 'Y' and the subsidiary section' thickness/chord ratio was later increased to 40 per cent. - PA).

The Test Rig
The large plates which are shown in the first two photographs were put there to prevent the spilling of air over the sides. With this rig we made a number of flights, and although from the pictures one may get the impression that the aeroplane was difficult to handle, I found that the control was scarcely affected, and even in landing there was nothing tricky. By fixing a 'pitot comb' similar to the Farnborough pattern in the wake of the experimental wing and connecting it to a manometer in the cabin we measured the profile drag with and without suction. To prevent suction we simply blocked up both ends of the venturi and covered the perforated sheet with doped fabric.

With the whole of the suction surface open we obtained negative results, but by restricting the suction to the rear third, we found that there was a very marked reduction of the profile drag, amounting to 26 per cent. Naturally, we tried repeating the experiment several times in order to establish that this improvement was, in fact, taking place. Although we did not take any definite measurements of the change in lift such as by surface pressure measurements, there was, undoubtedly, an increase in lift with suction, for we found from the wake measurements that with suction we got a marked increase in downward inclination of the wake.

In addition, we made some experiments with wool-tufts stuck on to the upper surface, as will be seen from the second photograph. I am not exaggerating when I say that, without suction, all the

Chapter 22: Miles M.11 Whitney Straight Special and M.11A, M.11B and M.11C Whitney Straight Variants

The Whitney Straight Special G-AECT, showing the modified port wing with large venturi, for boundary layer suction experiments, in July 1936. 'The Aeroplane' caption asked: What is a true streamline? A Miles Whitney Straight which is being used for experimental research. The thickness of the blown-out section of the wing is 30% of the chord. A thickness of 40% of the chord has actually been successfully flown. The loud-speaker horn seems to be some sort of a venturi and suggests suction. The pipes from the Pitot gear to the cabin can be seen on the side of the fuselage. This line of research is interesting, and if it be pursued diligently we may in twenty years' time know something about air-flow and streamlines.

Another view of the modified wing with wool tufts on the upper surface. [Phillips & Powis via The Miles Aircraft Collection]

wool-tufts, right up to the maximum ordinate, showed complete reversal of flow even at 100 mph, the speed at which we made our experiments. When the suction action was allowed to take place we saw a very different state of affairs. All the wool-tufts pointed downstream, and only those near the trailing edge showed indications of turbulence.

Naturally, we felt that the results of these preliminary experiments were sufficiently encouraging for us to make a more ambitious and careful investigation. With the assistance of the Air Ministry we were able to convert the Peregrine design into a sort of flying laboratory. (For details of these further experiments see Chapter 18 – PA).

F G Miles talking to Whitney Straight (in cabin) of a Miles M.11A Whitney Straight, during a demonstration at Heston on 22nd January 1937. Note the entrance door on the port side on production aircraft. [Associated Press via P Jarrett]

Miles M.11A Whitney Straight with F G Miles standing behind the fuselage and Whitney Straight alighting at Woodley in January 1937. ['Flight' via B Clarke]

However, to return to the fortunes of the Whitney Straight, the Associated Press photograph shows Mr Whitney Straight in the cabin and F G Miles standing on the wing of a new Whitney Straight, which was demonstrated by Whitney Straight and Miles at Heston Airport on 22nd January 1937. A small number of aviation experts had gathered to inspect and fly the first three production Whitney Straights, G-AENH, G-AERS and G-AERC (even though the latter did not receive its CofA until 5th February 1937) and praise was unstinted. Major C C Turner, aviation correspondent of the *Daily Telegraph* reported that; *'he had not flown a more comfortable light aeroplane, nor one so well finished in every respect. The control was exceptionally easy and the craft was so beautifully balanced that it 'flew itself', almost as if fitted with an automatic pilot'.*

An article by *Low Wing* which appeared in *Log Leaves*, in *Contact* magazine for March 1937, makes for very nostalgic, but interesting, reading:

I do not often get the chance to visit the Airport of Heston, but on a recent fine Saturday I went there and was struck by the number of cabin machines which grace that aerodrome. I suppose that as this is the Society aerodrome this is a natural phenomenon, and the place was full of Leopard, Hornet and Puss Moths. There were a few old Gipsy I Moths, a Miles Falcon or two, a Miles Merlin and one or two Percival Gulls, but, apart from the Moths and the Airwork School Avro Cadets, all the machines were of the cabin type. Jersey Airways have their headquarters at Heston and, during the afternoon, one of their pilots appeared to be getting a DH.86 on his 'ticket' and was putting down some very nice landings. (Ah - the sweet nostalgia of it all! - PA).

The machine that was attracting all the attention during the afternoon was the latest in trainers, the Miles Whitney Straight Special. By courtesy of Mr Gordon Marshall of Messrs Whitney Straight, I was enabled to try out this aircraft, and found it all that it was claimed to be. All Miles types are nice to fly, and this is by far the nicest yet. With its flat out speed of 145 mph and low landing speed of 40 mph, it makes a definite step forward in training aircraft.

With the flaps down (the lowering of which only necessitates the movement of a small lever on the dash) the machine definitely will not spin. With the air speed wavering around the 40 mph mark, and the nose well up, I tried kicking the rudder both ways, with no other result than a slight change of direction. There seemed hardly any tendency for the nose to drop either, the machine appearing to sink downwards, at all times under control. Another definite step forward is the sound-proofing of the cabin; this has been brought to a fine art, and conversation was easy without raising the voice at all.

Steady production and delivery against long-standing orders followed and by the early spring of 1937 the Whitney Straight had taken its place amongst the leaders in the civil aviation market. George Miles toured extensively across the continent giving demonstrations of the Whitney Straight in a number of countries, during the course of which he put in an incredible number of hours of really hard flying, both in demonstrating the machine itself and having to sit in it while all and sundry flew it. On one occasion, he flew over the Alps at a height of 17,000ft which, with no oxygen and the small unsupercharged engine, was no mean feat.

George took no ground staff on the tour and so, in addition to doing all the flying, he also undertook all the maintenance himself. He had no radio and with the weather forecasts as unreliable then as they are now, he was faced with further difficulties on this score when, on one occasion, the weather closed in on him entirely while he was flying up a valley between two ranges of hills in the middle of France. No suitable landing ground was available and yet George managed to put the machine down safely in a tiny field on the side of a hill.

Orders were obtained in France and in Switzerland, where the greatest interest was shown in the machine. The aerodromes in

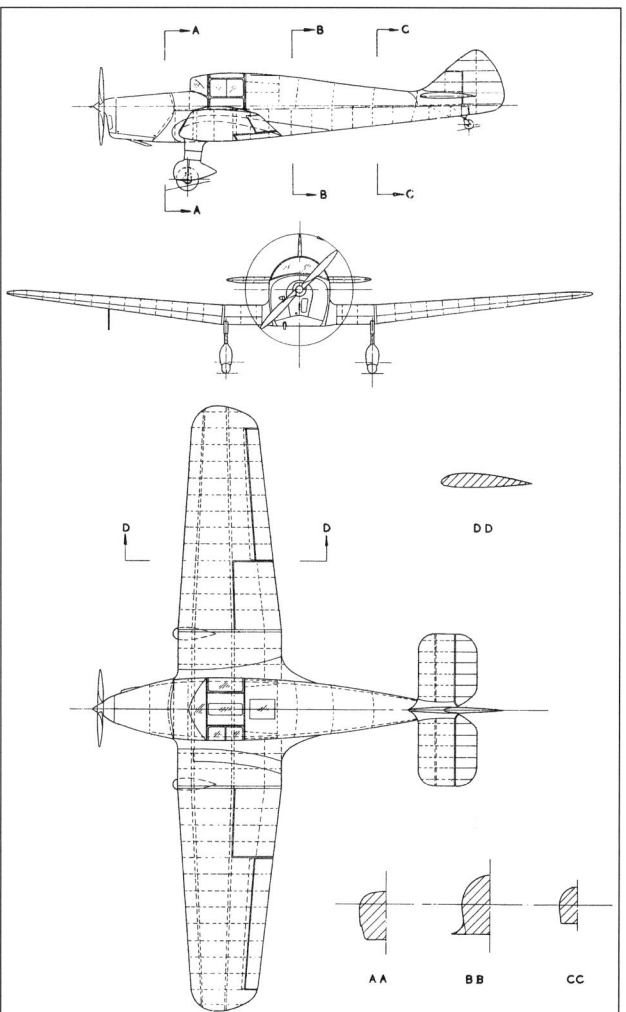

The Miles M.11A Whitney Straight.

those countries were often necessarily small and occasionally in Switzerland at some altitude and there, the good take-off and landing characteristics of the Whitney Straight were naturally greatly appreciated. The Whitney Straight was looked upon as one of the best available light aeroplanes able to operate under such conditions in those countries.

Among Whitney Straight's team was one Charles Amherst Villiers, a brilliant engineer who had long been engaged in experimental work on both car and aero motors. Villiers was well known for the supercharging of Bentley cars and by early 1937, his company, Villiers Hay Development Ltd, had developed his original 125hp Villiers Maya I engine of 1936 to give a conservative 135hp. Together with his wife, Mrs Maya Villiers, Charles purchased the third production Whitney Straight, G-AERC, to use as a test-bed for the 135hp Villiers Maya lightweight four-cylinder in-line, air-cooled inverted engine. The modified Whitney Straight was designated M.11B and a description of this engine and flight trials appeared in *The Aeroplane* for 14th April 1937. The Maya engine was exceptionally light, weighing only 280lb, which gave a weight/horsepower of 2.07lb/hp, the best ratio seen for a motor of that class at the time.

In March 1937, Miles decided to fit a 150hp Menasco C.4S (supercharged) engine into a Whitney Straight. C L Nash recorded that a 125hp Menasco C.4 Pirate engine was installed in the prototype Whitney Straight Special, noting against this entry, the date 1st April 1937. Don Brown however, noted in his log-book, against a flight on April 10th, that G-AECT was fitted with a Menasco 'B4S' engine. As there is no record of a 'B4S' engine, it is likely that this designation was simply in error but, for the record, the Menasco B.4 produced only 95hp and would not have been used.

Two views of a Miles M.11A Whitney Straight under construction in January 1937. ['Flight' via B Clarke]

Chapter 22: Miles M.11 Whitney Straight Special and M.11A, M.11B and M.11C Whitney Straight Variants

The Miles M.11B Whitney Straight G-AERC with the Villiers Maya engine, in January 1937. ['Flight' via B Clarke]

Miles M.11 Whitney Straight Special G-AECT fitted with Menasco C.4S engine. George Miles and Menasco engineer Andy Anderson are seen in attendance. [L Tysoe]

At about the same time as the Menasco engine was fitted, G-AECT was also modified to have two external flaps in tandem. Although the split-flap, which Miles had pioneered in this country was then in almost universal use, he felt that it was still possible to achieve an even more efficient type of flap, so for this experiment he mounted two flaps of aerofoil section, in tandem, aft of the trailing edge of the wing. These were initially fixed in position but if the trials turned out to be successful, they could subsequently be made retractable.

The results achieved were, in fact, most encouraging, with a very marked increase in lift coefficient and corresponding large decrease in stalling speed and take-off distance being obtained, with, moreover, a very much lower drag than that associated with split-flaps. The results obtained were so remarkable, that the machine was then entered for a competition, held at Odiham, for army co-operation aircraft of both British and foreign manufacture. A large number of firms entered aircraft, most of which were fitted with some sort of high-lift device to give the short take-off and

G-AECT fitted with two external flaps in tandem, showing wool tufts on flaps and also over the adjacent wing and aileron. ['Flight' via B Clarke]

Another view of the slatted flaps showing linkage and attachment to original flap. ['Flight' via B Clarke]

steep climb required by the army, and, in order to obtain an accurate record of the take-off achieved by each competitor, a camera was installed by the runway.

When George Miles made his first take-off in the modified Whitney Straight, he not only left the ground long before reaching the camera but also climbed so steeply that the camera could not be tilted to a sufficiently steep angle to photograph him! The camera was then moved considerably nearer the starting point but the same thing happened again and once again the camera had to be moved to a position, hopefully better suited to be able record the take-off and climb of the Whitney Straight. However, with the camera in this position it was found impossible to record the climb of the other competitors, who were still on the ground after passing the camera and at the point where the Whitney Straight was about 50ft in the air!

Subsequent tests showed that almost equally good results were obtainable with one auxiliary aerofoil flap having a chord equal to the combined chord of the two flaps originally fitted. George later fitted a fully retractable version of this type of flap as standard to the M.28, which he designed later and produced in 1941. This flap proved to be so successful that it was incorporated on all subsequent aircraft designed by George Miles while he was with Miles Aircraft Ltd, although these were not always retractable.

From March 1937, G-AERS was used to transport its owner, Second Lieutenant Rex King-Clark, of the 1st Manchester Regiment, some 55,000 miles, embracing 23 countries and 98 different airfields, from York in the west to Bali in the east. Shortly after taking delivery, Rex was posted to Ismailia, Egypt and was granted permission to fly there in the Whitney Straight, being asked by the War Office to observe and report on Italian military activity in the North African province of Cyrenaica en route. Leaving York on the morning of 4th March 1937, the journey took 10 days but he didn't think that his report on the Italian activity had a great effect on the outcome of the war. In Palestine, in July 1938, G-AERS was fitted with a Mauser machine pistol with a 20-round magazine attached to one of the undercarriage legs and this was operated by *'a bit of string pulled from the cabin...all you'll have to do to fire is to point the aircraft at the target and pull the string'*. His assistant, Cpl Howbrook suggested that 36 grenades be carried and dropped through the floor but that idea was frowned upon by Rex as it would have meant cutting a hole in the floor of the cabin *'I don't think old ERS would be too keen on that - we would also have to ruin the priceless rug which covers the floor!'* In the end Cpl Howbrook sat with the grenades on his lap and a successful test firing/bombing was carried out at low level in the Sea of Galilee *'Cpl Howbrook pulled back the window on his side of the cabin, took out a grenade, pulled out the pin and thrust his arm up to the shoulder out of the window. It was easy to see when he'd let go by the look on his face and at once I pulled ERS up and around into a steep climbing turn - there appeared on the blue rippled water far below a little splash of white...'*. This then constituted Capt Orde Wingate's air force in Palestine, although it is doubted that he ever knew about it!

It has been recorded that the RAF had a standard Whitney Straight on trial for five months, during which time it was used for full duties by 13 Army Co-operation Squadron (which happened to be based at RAF Odiham at the time), where it apparently 'gave great satisfaction', but this report should more correctly have read that it was the modified prototype and not a production aircraft.

In August 1937, it was announced that the price of the Whitney Straight had to be increased by £100. *The Aeroplane* 11th August 1937 reported: *From now onwards, the selling price of the machine in its standard form will be £1,100, ex Reading Aerodrome. Whitney Straight Ltd have tried hard to avoid this increase, but the rapidly rising costs of aircraft materials and labour have made it finally necessary. In spite of this rise, a new method of rating aeroplanes, now being tried out by us, shows the Whitney Straight monoplane still to be very good value for money compared with other types.*

Three Whitney Straights were entered for the King's Cup Air Race held at Hatfield on 10/11th September 1937 and Brig Gen Arthur Lewin, at the age of 63 years, flew G-AEZO into second place at an average speed of 144.50 mph, finishing less that 2.5 minutes behind the winner. Fourth place went to Whitney Straight G-AEVH, flown by Sqdn Ldr Arthur Vere Harvey at an average speed of 142.40 mph.

Unfortunately, on 10th October 1937, Brig Gen Lewin ran out of fuel while flying his Whitney Straight south over the Sudan en route to Kenya, and it was wrecked in the subsequent forced-landing in the Nile swamps south of Malakal. Reports on the accident appeared in *The Aeroplane* for 20th and 27th October 1937:

Brig-Gen AC Lewin and Mrs Lewin, on their way back home to Kenya in their Miles Whitney Straight, had a forced landing on 10th October after leaving Malakal. After some anxiety, they were located by the Empire flying-boat Cassiopeia *on 12th October as it passed over on the regular run. They were in one of the vast swamps of the Anglo-Egyptian Sudan, 150 miles South of Malakal. Their machine was overturned but they were, as appeared later, unhurt. Captain Caspareuthus, of the* Cassiopeia*, did what he could, which was to drop supplies and advise Khartoum by wireless of their position.*

A service aeroplane left Malakal to make further contact, but was unable to find General and Mrs Lewin till next day. Then further food and medical supplies were dropped. There was no possiblity of making a successful landing so a ground rescue party started from Kongor, 15 miles to the South. This party, which consists mainly of Dinkas, a primitive native people, made surprising progress across the swamps for more than 12 miles, in fact to within three miles of the marooned travellers, by 16th October. They were organised and largely guided by means of aeroplanes. They had to turn back some distance and find another route on 17th October because of the depth of swamp. They reached the Lewins on 18th October.

On 17th October a Service machine dropped message apparatus and later picked up a note written by Mrs Lewin. She expressed great concern at all the trouble caused, and hoped that none of the searchers had suffered anything untoward. They said they would not be able to walk through the swamps, and asked for hammocks. From the air they seemed well and active, considering the hardships they must be enduring.

General Lewin has said that he was forced down because he overshot his mark (the General overshot Malakal due largely to a compass error, combined with the breaking of his time-of-flight clock - PA). *His time-of-flight clock was broken and - as he found out to his cost - that instrument is almost as important as a compass when flying over a country of very sparse landmarks. Eventually his motor dried up (for lack of fuel) and he came down in the nicest-looking bit of swamp he could see. The machine nosed over in a few inches of water and bumped General Lewin and his wife about the heads and shoulders.*

Dinkas were organised at the nearest village to go and fetch the Lewins. Dinkas are primitive natives of the district, and are probably the best example in the world of the adaptation of the human frame to special conditions. Thousands of years of swamp-dwelling have produced an enormously tall and thin race, with legs rather like storks. (The story that Dinkas roost on one leg is said to be inaccurate, by the way). A clearing was made at Kongor, the nearest village, and the Lewins were evacuated from there by air to Malakal on 20th October, after the Dinkas had more or less carried them to the village. General and Mrs Lewin paid a tribute to Mr MacPhail, Mr Marwood (District Commissioner at Bor), No.47 Squadron, RAF, and the kindness of the Dinkas for the difficult and sometimes dangerous work which they had all done so efficiently.

Proof of the safe and easy handling characteristics of the Whitney Straight was that the General and his wife were relatively uninjured in this difficult forced-landing, even though the wheels entangled in the grass, the aircraft overturned and crashed on its back. In fact, such was his satisfaction with the Whitney Straight that he later purchased a replacement.

An excellent shot of Brig Gen Arthur Lewin's G-AEZO at Hatfield for the 1937 King's Cup Air Race. ['Flight' via B Clarke]

Chapter 22: Miles M.11 Whitney Straight Special and M.11A, M.11B and M.11C Whitney Straight Variants

Miles M.11A Whitney Straight HB-URO at Geneva in 1953. [P Amos]

A Whitney Straight was the only British aeroplane chosen to be shown at the second Milan Aeronautical Exhibition held from 2nd-17th October 1937, where it attracted a great deal of favourable interest and, immediately after the exhibition, it was bought by Gaspare Bona of Carignano, Turin.

Before production of the Whitney Straight ceased in 1938, they had been sold in Australia, Belgium, Chile, France, Germany, Italy, New Zealand, Rumania, Singapore and Switzerland and some 30 had been bought for private and flying school use in the UK, including by Air Service Training, Hamble, who bought one for general instruction. The sale of a Whitney Straight in Singapore produced favourable comment in the Royal Singapore Flying Club's report for February 1938 and this was published in the *Miles Magazine* for July 1938:

Miles Whitney Straight VR-SBB. This machine arrived towards the end of the month and most pilots qualified to fly it have tried it out. Those who should know say it is the best light aircraft they have ever flown. Criticism there has been but on items that it is possible to put right. The chief one is that it is very hot in the cabin. This will be improved by fitting an overhead blind, and possibly another window capable of being opened. The noise in the cabin is not excessive. Its take-off is excellent and with flaps down it can be landed almost anywhere.

The editor of *The Miles Magazine* was moved to comment that: *Even these minor criticisms have been overcome in the Monarch (see Chapter 30).*

After the first twenty-five Whitney Straights had been built, two further batches of ten and fifteen respectively were constructed and with the exception of the fiftieth, which C L Nash recorded was a 'fuselage only', all were delivered between January 1937 and September 1938. Unfortunately, due to the large contracts then being received for the production of the Magister and Master trainers for the RAF, no further production of the Whitney Straight could be undertaken.

On 5th April 1938, Don Brown flew U4, a production aircraft previously registered G-AEYI, which he noted was fitted with a 130hp Gipsy Major engine (although this was probably the 140hp DH Gipsy Major Series II engine) with a variable-pitch propeller, which had been installed in the aircraft at about that time. This combination, known as the M.11C, gave a really astonishing take-off and climb performance - especially in the hands of Miles! Unfortunately, this aircraft was to crash at Harefield, Berkshire on 28th June 1938, killing its pilot, W/C Freddie Stent.

With the advent of war, a number of Whitney Straights were impressed by the RAF for communications duties, both in this country and abroad and some were used by the ATA as taxi aircraft. Owen Maclaren of Andrews Maclaren Ltd, working for Airwork Ltd at Heston, modified 'his' (as he described Miss Rosemary Rees') Whitney Straight G-AFGK by fitting it with his own design of cross-wind landing undercarriage. He demonstrated this at Langley to an audience of test pilots which included Phillip Lucas, Hawker Aircraft's Chief Test Pilot. He made three successful take-offs with the wheels set at 30 degrees to the heading of the aircraft and at right angles to one of the highest winds ever experienced in England. The Met Office gave the wind as gusting 75 mph at Kew (10 miles from Langley) and 108 mph at Pembroke Dock.

On the whole it could be said that the Whitney Straight was one of the most popular and successful aeroplanes ever produced by the firm but it should be left to H A Taylor of *Flight* to have the last word:

Nothing in this world is perfect - and certainly no vehicle, whether on land, on sea, or in the air, can ever be foolproof. But within the limits imposed by present day knowledge of aircraft design the Miles Whitney Straight appears to be the last word in viceless compromise. It is fast for its power and accommodation, yet it lands at something rather less than 40 mph and can be flown in any reasonable attitude at its lowest speed without evincing any noticeable tendency to turn and bite.

An HQ 41 Group Return (Ref: AIR 17/33) entitled 'Arrivals and Despatches for Week Ending and Stock in 41 Group ASUs as at 0900 hours 6-12-45', gave the numbers of Whitney Straights surviving as follows: 4 Stock as at 0900 6-12-45; 1 Ready for issue; 3 Awaiting MAP disposal instructions - Non-Effective Aircraft. These non-effective aircraft require maintenance within 41 Group Units. (The four Whitney Straights were probably DR611 (DR617 and G-AETS), DJ713 (G-AEUX), DR612 (G-AEVA) and DP855 (G-AEVL), which were all sold at the first disposal of impressed light aircraft held at No.5 MU Kemble in December 1945).

For the record, there were actually 9 Whitney Straights still serving in the RAF on this date; the other 5 being EM999 (G-AERV) possibly at 13 MU Henlow, DP854 (G-AEVG) with FC Comm Sqdn, DJ714 (G-AEWA) with Northolt Stn Flt, BD183 (G-AFJX) possibly with RAF Station Grimsetter, and NF747 (G-AFZY) with 50 Grp Comm Flt, Woodley.

After the war, the surviving Whitney Straights continued to give sterling service for many years and the story of the escapade of one of them, G-AERV, was told in the Irish newspaper *The Tyrone Constitution* for Friday 24th May 1963, under the heading: *Pilot Escapes with Scratch as Plane Crash Lands at Seskinore - Forestry men rush to his aid on hilltop.*

A 37 year-old Ulster pilot crash landed his plane on a flat hilltop about a mile on the Omagh side of Seskinore on Wednesday afternoon, and stepped out of it slightly shaken and suffering only from a scratch on his nose. Two local men, who were at work in the Ministry of Agriculture's forest nearby, saw the plane come down and rushed to the flyer's aid, but he did not require any help. The pilot was Robert T Boyes, freelance photographer, of

The only photograph to show VR-SBB, in Singapore in 1938.
[Via B Clarke]

62 Victoria Avenue, Newtownards. He was one-and-a-half hours out from St Angelo, Enniskillen, in the privately-owned aircraft and flying at 800 feet when the engine began to fail.

He said afterwards 'At that height it is very difficult to pick your landing spot, so I had to make a quick decision. I was coming in for a perfect three-point landing, but the wheels stuck in the soggy ground and the plane pitched forward. I was thrown against the dashboard and cut my nose'. Boyes, who has been flying since 1955, has never had a mishap before. He is at present working at St Angelo which is being prepared for re-opening as a civil airport. The plane, a one-seater, single-engine Miles Whitney Straight, was not badly damaged. The propeller was shattered as it nosed into the grass covered field and there was damage to both wheels. Boyes hopes to have it removed to Newtownards within the next few days.

First man on the scene was forestry employee Samuel A Carson, of Seskinore, who was working about 200 yards away. 'I saw the plane come down and ran as fast as I could towards it. But when I got there the pilot was already climbing out and there was nothing very much for me to do'. A short distance behind Mr Carson was Gerald McAtee, Augharonan, who was working with him. Omagh Fire Brigade and police from Fintona were quickly on the scene also. The police put a guard on the aircraft, but the firemen, after standing by for a period, returned to the station. The forced landing took place near the Pond crossroads.

The aircraft touched down, ran for a about five yards and then bogged, coming to rest about 30 feet from one of several small piles of stones scattered over the field. The port u/c leg was broken off in the forced landing and a temporary wheel and strut assembly, made from 3" steel tubing and two wheelbarrow wheels (borrowed from a hardware shop in Gortin) by Gary Pentland, was fitted to the port u/c attachment point in order that the aircraft could be moved from the site.

On May 23rd, the Whitney Straight was towed, tail first behind Gary's father's Standard 10 van, while he and another man walked alongside in case of emergency. It was towed to Newtownards on May 24/25th and repaired. G-AERV's CofA lapsed in April 1966 and, sometime after this, Boyes lent it to the Belfast Transport Museum. The Irish daily newspaper, *The Newsletter* for 21st August 2002 carried a photograph of G-AERV sitting outside the old Belfast Transport Museum on its undercarriage having been towed there from Newtownards, and the article read: *You can't park that here, mate.*

The plane in the street is a Miles Whitney Straight which was built in the late Thirties and finally owned by R T Boyes, a flying instructor at Newtownards who decided to lend it to the Belfast Transport Museum in Witham Street, off the Newtownards Road. On the chilly but dry Sunday morning of January 30th 1966, Pat Bott (a local photographer - PA) *and Graham Skillen took off the wings and towed the rest to Belfast behind Pat's Morris Minor Traveller. The photograph was taken in McTier Street just before the Miles was towed into the museum's store.*

With regard to the museum, while there can be no doubt about the place to which it was taken in the first instance, the Folk & Transport Museum at Caltre, Co Down, for some reason the family towed it back to Newtownards later. G-AERV was later acquired by Mr Sherry, the owner of a chicken farm at Upper Ballinderry, near Crumlin, Co Antrim, who stored it in a dilapidated shed with a leaking roof with the wings stacked alongside the walls, where, over the years, they warped out of shape. In due course the cabin collapsed and although all the controls and instruments remained intact, the fabric covering on the tailplane rotted away and the oleo legs, which were slightly bent, began to rust.

By then G-AERV was deemed 'a basket case' by many who saw it with a view to its purchase (and many genuine efforts to purchase it over the years were made by members of The Miles Aircraft Collection). Such attempts were all thwarted by an unreasonable owner who was asking around £40,000 for it. So it was very pleasing (and somewhat surprising) to hear that, after the farm was sold in early 2002, G-AERV had been collected by Ron Souch of Aero Antiques for rebuild to flying condition on behalf of a purchaser, Richard Allen Seeley. By early 2008, Ron had completed the work with brand new materials and undercarriage (using the originals as patterns) and with a new windscreen manufactured by a firm in Switzerland. By September 2008, the aircraft was complete and in absolutely pristine condition, having been painted in its pre-war colours of silver and Cambridge blue with Oxford blue outlining. On 17th September, Ron Souch took it off on its maiden flight from Durley Airstrip, near Southampton.

There are only two other Whitney Straights surviving in the world, in anything like complete condition. One of these is stored in a museum in Canada (CF-FGK), but this now has a broken windscreen. The other (G-AEUJ) is at Sleap, awaiting the completion of a rebuild which was begun in about 1979. It was taken by road to a Warwickshire farm on 13th February 1982, where it was *'hoped that it would be flying by the end of the year'*. However, nothing further was done and it was later taken to Cosford. In early 2008, it was moved to Sleap, where it is to be hoped that restoration will, eventually, be completed.

For details of M.11 Whitney Straight production see Appendix 19.

SPECIFICATION AND PERFORMANCE

Engine:	130hp DH Gipsy Major Series I
Dimensions:	span 35ft 8in, (wings folded 17ft 2in; note -. this is at variance with the original drawing which shows 16ft 6in); length 25ft 0in; height 6ft 6in; wing area 187 sq ft; aspect ratio 6.8
Weights:	empty 1,250lb; pilot 170lb; pass 170lb; petrol (30 gal) 231lb; oil (2 gal) 19lb; luggage and/or extra equipment 160lb; AUW 2,000lb; wing loading 10.7lb/sq ft
Performance:	max speed at sea level 145 mph; cruising speed at 1,000ft 130 mph; rate of climb at sea level 850ft/min; stalling speed with flaps 38 mph; recommended landing speed 45 mph; ultimate range on 30 gal fuel 570 miles; duration 4.4 hrs; with the 135hp Villiers Maya engine fitted the max speed at 3,000ft 152 mph; cruising speed 136 mph; rate of climb at sea level 1,000ft in less than 1 min and 2,000ft in less than 2 min
Price:	£985.

A model of the Miles X.2 Transport Aeroplane. ['The Aeroplane']

Chapter 23:
Miles X.2 Transport Aeroplane Project

Shortly after the 1914-18 war, the Air Ministry became interested in the theories of a Russian inventor, Woyevodsky, who rightly suggested that the high drag of the braced biplanes of that date could be greatly reduced and the performance correspondingly increased, by having an aerodynamically clean cantilever monoplane fitted with a retractable undercarriage. Woyevodsky further suggested that both the fuselage and the wings should be of aerofoil section, and later wind tunnel tests of his models appeared to confirm his claims.

The Air Ministry asked Westland Aircraft Ltd to build a twin-engined aircraft, as suggested by Woyevodsky, but in the event the aircraft was built with a single-engine and with a fixed undercarriage. The Westland Dreadnought J6986 was completed in May 1924 and on May 9th, after some fast taxi-ing and short hops, it took-off from Yeovil. Shortly after take-off, Capt Stuart Keep, Westland's test pilot, was unable to retain control and the aircraft stalled and crashed from a height of about 100 ft. The project was then abandoned.

Twelve years later, Miles was anxious to build a large four-engined civil transport and he could see no fundamental reason why the Dreadnought should have crashed. He considered that the basic ideas of Woyevodsky were right and capable of development so, early in 1936, he began the design of a large transport aeroplane to this principle.

This design study became known as the Miles X.2 (there is no record of an X.1), and this large transport was intended to have the maximum degree of aerodynamic cleanness attainable, with a wide shallow aerofoil section fuselage merging almost imperceptibly into the wings with very large root fillets. The four engines were buried in the wings and drove airscrews through extension shafts.

Miles wrote the introduction to the Phillips & Powis brochure on the X.2 high-speed long-range transport aeroplane and submitted the brochure to the Air Ministry in early 1938. Although the brochure was undated, all the drawings were dated 24th January 1938 (with the exception of one dated 23rd January 1938), while the Advance Design and Production Programme chart was dated 31st January 1938. The brochure also included two 'diagrams of arcs of fire' for the armament of 4x27mm cannon and 14 Browning machine guns installed in five turrets. positioned at the side, top and bottom of the aeroplane, but it gave no details of bomb load for the proposed Miles X.2 Bomber, which was obviously considered to be a very heavily armed bomber version of the Miles X.2. Strangely, the brochure made no mention of this project.

This introduction, slightly amended, also appeared in *The Aeroplane* for 27th April 1938 and the Editor commented on the Miles X.2 thus:

Here we have the latest ideas of Mr Fred Miles of Phillips & Powis Aircraft Ltd for a high-speed long-range transport aeroplane. The photograph above, realistic though it appears, was taken of a metal model lent to us by Mr Miles. A drawing of the design appears below. In the article which follows Mr Miles discusses his plans for this machine. His selection of such a low aspect-ratio is interesting. It makes for stiffness, a reduction of weight and decreased skin-friction drag.

The Miles 'X.2' Transport Aeroplane –
Introduction by F G Miles

The design for the Miles 'X.2' which we describe briefly herein, came into being substantially as it now appears, over two years ago. Specifications were sent out to various interested parties but perhaps at that period our project was a little premature.

Now it is claimed that British transport aircraft are not up to the standard set in other parts of the world and efforts should be made to put British design ahead as soon as possible. I submit that efforts to do so will fail if we allow circumstances to force us into starting designs on either too orthodox or too unorthodox lines. Either of these alternatives may be adopted by the designer goaded into action by the critic of to-day. The Miles 'X.2' is a striking advance in general arrangement - but its design does not bring in fresh problems of control and stability. As far as the relative positions and areas of lifting and control surfaces are concerned, all existing design criteria apply.

We have now worked on the Miles 'X.2' for two years and during that time we have found no reason to modify the basic design. A careful analysis has been maintained of tendency in international design and the curve of progress would certainly seem to extrapolate in the general direction of the Miles 'X.2'. During those two years, we strained our organisation, which was at that time small, to the utmost to obtain research and experimental data and although, of course, expense has had to be limited, I think we can say that we have done much useful work on such things as the use of wings of large thickness/chord ratios; boundary layer control; high-lift devices; structural design etc.

There is still much to be done, but as I have indicated in my later remarks and by the diagram on page 4 our present research can be extended and completed concurrently with the new work and the building of the Miles 'X.2'. A lot of work and time have gone into the preparation of plans to build it and I believe we have not left much out of account. By this careful planning we hope to achieve several things:-

To finish the whole undertaking in a reasonable time.
To design and build the Miles 'X.2' in stages so as to allow for revision and decision by the other parties to the venture.

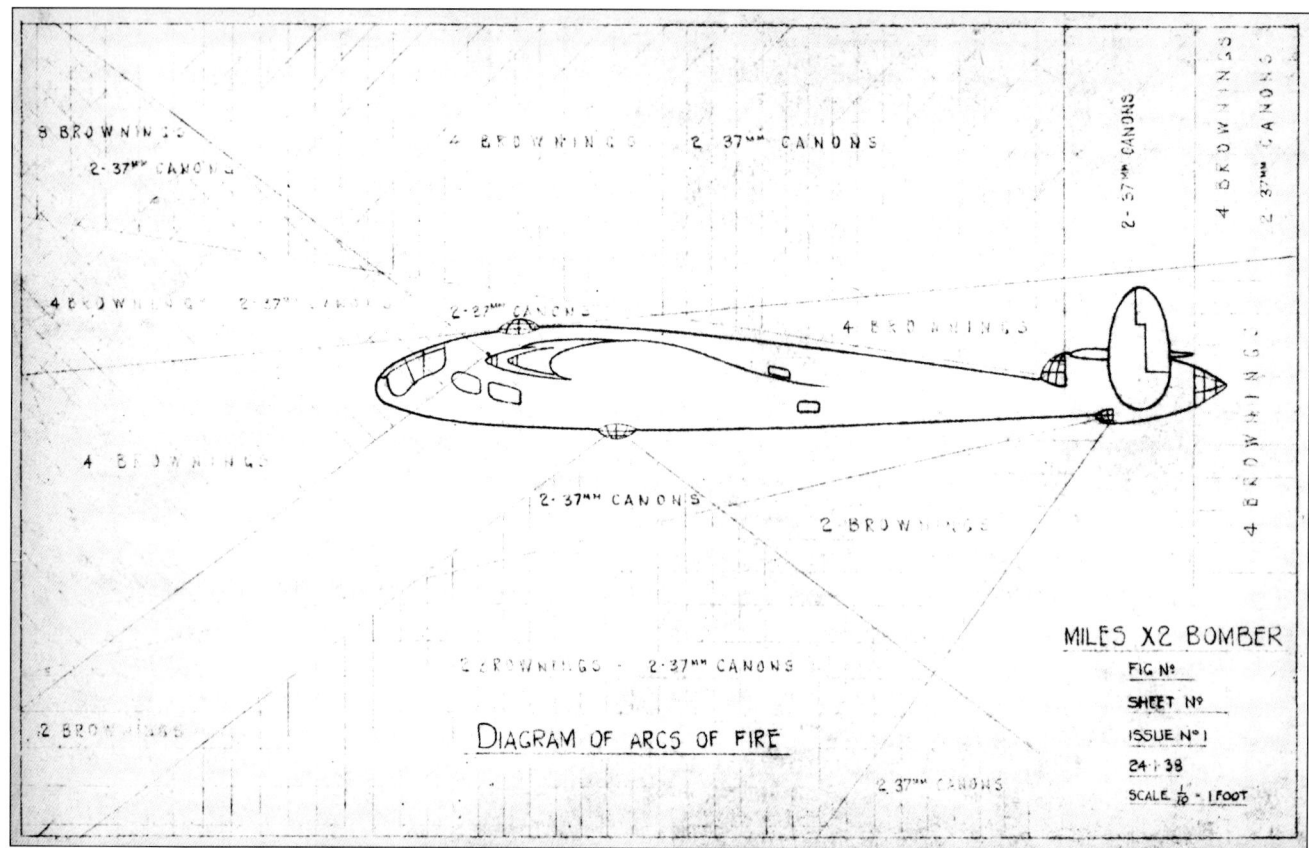

Above and below: Two original drawings of the Miles X.2 Bomber.

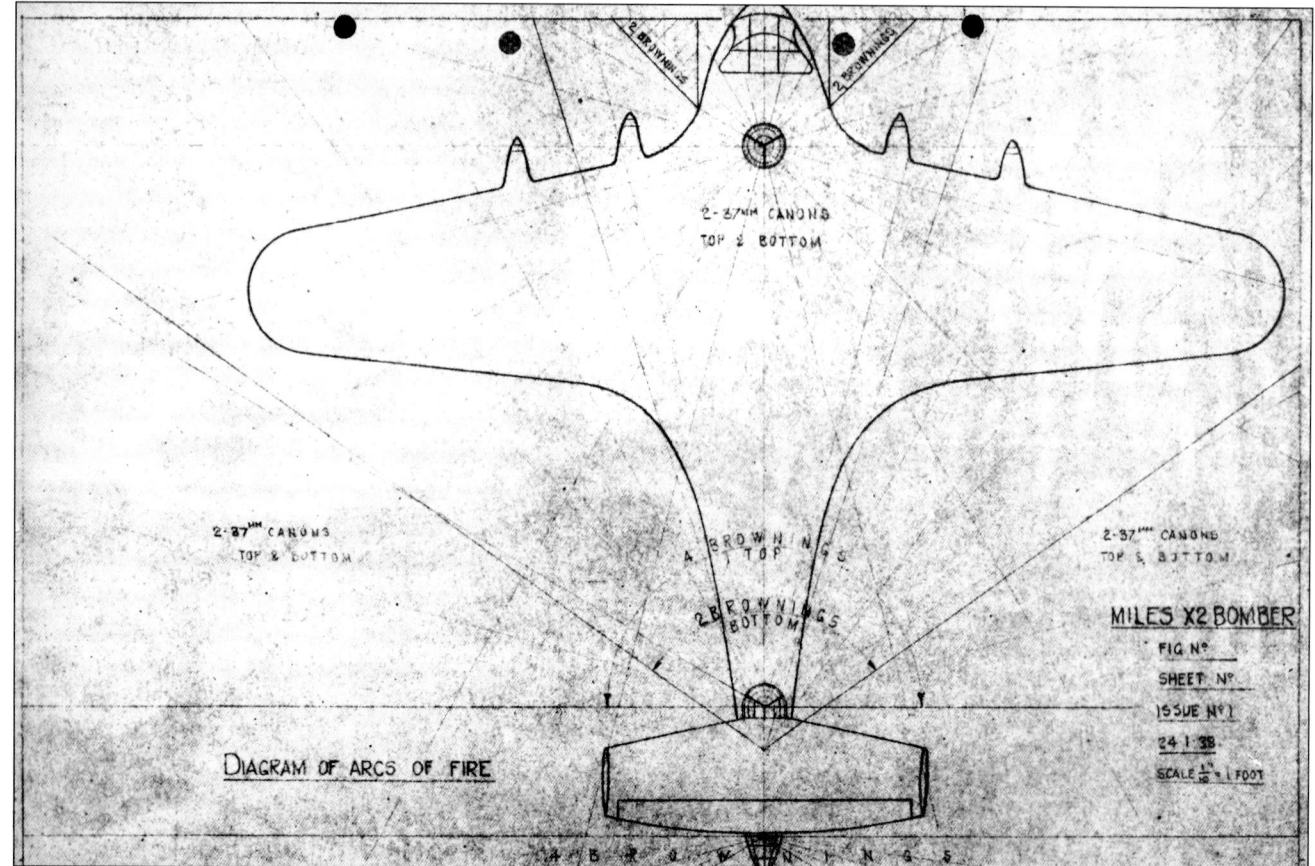

To allow us to go forward in such a way as to take advantage of any new discovery or knowledge that may become available.

On the following pages I propose to give a short summary of our plans. It is obvious that the building of, not only the prototype, but perhaps up to twenty large transport aircraft, is a big undertaking in which many interests may share. Therefore, financial and other arrangements need to be laid out in such a way as to avoid unnecessary risk without slowing up production. We shall require, as a Company, to know that our work in each of the necessary phases will be guaranteed as regards expense. It is also reasonable for us to know that the work we put into the design will, in the end, be of benefit to this company.

I therefore suggest that the actual prototype be ordered on the basis of a definite minimum sum to begin with, and that progress payments be made covering expenses throughout the time occupied in building the aircraft. The end of each stage would provide an opportunity to consider the progress made and if we fail to live up to our promises or in the early stages the reduced scale prototype should prove the design unsatisfactory in any respect, then the losses can be cut and kept very small.

We have no fear of failure ourselves; we recognise the greatness of the undertaking but perhaps I may be pardoned for mentioning that we have successfully organised a very large proportionate output of our present types and have, at the same time, probably carried out more full-scale research than organisations ten times our size. With our present backing I am confident we can succeed.

Summary of Design and Constructional Plans

The basic principle of the scheme is to divide the work into well-defined stages such that all concerned will be able to know the exact state of affairs at the end of or during each stage. In this way the existence of undue delay or waste will be quickly detected and should it be found desirable at any time to abandon the project a minimum of loss will be ensued.

First Stage. This stage is, in a way, the most important, since the degree of advance in the design will depend very largely on the experimental work. It would include:-
a) Building of a reduced scale flying prototype;
b) Construction of a very complete mock-up of the cabin and part of the outer wing;
c) Completion of research work now in hand;
d) Further research on pressure cabins, tricycle chassis, etc.
e) Planning of manufacturing stages.
f) Preparation of tentative plans for enlargement of plant and buildings at a later stage.

Second Stage. As soon as the first stage had been approved and the main features of the programme dealt with, the building of the prototype would begin. The layout of the equipment in the mock-up would be worked out in detail and the flight trials on the scale flying model performed. At this point the initial stages of production would also be put into effect, ie - the extending of the shops and plant, and enlisting of further specialised personnel.

Third Stage. This stage would be mainly concerned with assembly, flying trials and any necessary modifications of the prototype, and production would, by now, be well under way.

If the contract for the prototype is placed with us, I am sure we can successfully fill any reasonable demand by working on the plans roughly outlined above. It would receive the whole of my energy and attention as I believe there is a very promising future for large transport aircraft.

(Author's note: Miles was indeed a man of vision and years ahead of his time - witness the countless numbers of 'large transport aircraft' flying today - of which, unfortunately, none are now of wholly British manufacture. However, with regard to the First Stage a), it should be mentioned here, to save any later confusion, the projected reduced scale flying prototype of the Miles X.2 differed from the later M.30 'X' Minor in a number of respects. The 1938 reduced-scale aircraft had two 125hp Menasco C.4 Pirate engines, each one lying on its side buried in each wing, driving airscrews through extension shafts and although it had a wing area of 200 sq ft (the same area as the original wing on the later M.30 'X' Minor), its loaded weight was only 2,000lb, thereby giving a wing loading of just 10lb/sq ft. It had twin fins and rudders with a central dorsal fin and the retractable tricycle undercarriage was mounted on the fuselage. For the record, the M.30 'X' Minor later had its wing area increased to 223.24 sq ft but its wing loading was 21.2lb/sq ft and an AUW of 4,240lb. - PA).

General Description of the X.2

The body which merges into the wing and in which the crew and passengers are seated is divided conveniently into three main compartments by the spars. The shape developed in the 'X' layout is particularly suitable for transport and the volume available for passengers and freight is, it is claimed, larger than on other aircraft of similar size. The total cabin volume is 3,068 cubic feet. Of this 350 cubic feet are allotted to baggage in the rear fuselage, with an additional 200 cubic feet in the wing-root.

There is room for 2,500 gallons of petrol in the wing. When the full load of petrol is not required, however, some of this space, amounting roughly to a total of 400 cubic feet, may be used for baggage. With a crew of two pilots and a steward, and allowing for two lavatories the passenger volume is 2,631 cubic feet. When carrying a crew of two pilots, navigator, wireless operator, engineer and steward, the passenger volume is 2,118 cubic feet, after deducting the space required for two lavatories and dressing rooms.

The engines are housed within the wing and are completely accessible while the aircraft is in flight. The depth of the wing is sufficient to allow an engineer to attend to the engines in flight without difficulty. The sumps and cocks etc of the petrol and oil tanks will also be capable of easy inspection by the engineer when the aeroplane is in the air.

The Miles X.2 is designed to the following weights: 26,983 lb. tare; 61,128 lb. gross; 34,145 lb. disposable = 55.9% gross.

The Structure

General.
The Miles X.2 is to be of all-metal construction with flush-riveting wherever it is desirable on surfaces exposed to the air flow.

Wings.
The wing structure is to be of the stressed skin type. It will consist of two sheet metal webs to take the shear, the skin being reinforced either with stringers or corrugated sheet, so that it can take not only shear due to torsion and drag but also the end load due to bending. This arrangement gives a lighter structure than that of the more usual spar construction.

Fuselage.
For the fuselage, the semi-monocoque type of construction with former rings and stringers is proposed. This type of construction gives a very good strength for weight ratio and since there are no diagonal members it allows absolute freedom to the passengers. The design of the centre section spars is also worthy of note. These virtually form two bulkheads in the form of hoops having the same shape as that of the fuselage in their vicinity. On either side and integral with the hoops are stub spars connecting to the outer wing.

Empennage.
The tail plane, elevators, fin and rudder (in fact this was actually twin fins and rudders - PA) will all be of proved construction and call for no further comment at this stage.

Undercarriage and Flaps.
These will be operated either hydraulically or electrically according to requirements. It is perhaps appropriate to remark here that we are engaged in testing both a split flap design and a variable area wing to facilitate take-off and landing and that we shall bear in mind these developments in producing the Miles X.2.

Passenger Accommodation

General.
The adaptability of interior arrangement is a feature of the Miles X.2. To illustrate this we are including drawings of possible passenger accommodation ranging from 38 to 8 passengers. This variable capacity and the maximum comfort to the passengers under all conditions is made possible by the exceptionally large volume available in the cabin and by the convenient arrangement of the windows. Sky lights are arranged to give ample natural light to the cabin under all conditions. The gangway has a minimum width of 19 inches, though it is mainly much wider than this. The height of the cabin is in most parts 7 feet and nowhere is it less than 6 feet 6 inches.

Lighting.
It is proposed that each passenger shall have a reading lamp if required. These lamps will be so fitted that they are convenient to use when the aeroplane is fitted as a sleeper. In addition, each compartment receives an even illumination from central lights running the length of the cabin.

Air Conditioning.
In accordance with the latest practice the Miles X.2 will be fitted with a complete air conditioning plant. One method proposed is that air is taken in through slot type entries in the leading edge of the wing. It then passes through a rain separator and a filter and is finally led through branch pipes to the cabin. In one branch an exhaust heated muff is fitted. If this muff over the exhaust is not approved by the operators, modifications to our scheme can easily be introduced, such as a hot water heating system. This would, however, be slightly heavier.

The hot air is introduced to the cabin through a duct running along its length. The flow of air is controlled by the pilot, a thermometer being provided to indicate the temperature of the air in the duct. The outlet for the air is through two ducts running along the floor. The cold air supply may be controlled by the passenger. This may be done through the normal 'Punka' louvre, or alternatively by means of a flexible tube running along the arm of the passenger's seat. When he wishes to direct a flow of air on to his face the passenger lifts the tube, which remains in any position in which it is placed. At the same time the flow of air is automatically released.

It is proposed to introduce 17 cubic feet of conditioned air per minute per person. The baggage compartments have controllable air conditioning to protect any perishable goods the aircraft may be carrying.

Sound Proofing.
Every precaution will be taken to ensure that satisfactory methods of sound proofing are employed. Our Experimental Department has the problem of sound proofing as an important item in its programme, and it is hoped that it will be possible to employ a method which will serve the double purpose of reducing heat losses in the cabin as well as reducing noise.

Typical arrangements of Cabin.
In order to illustrate typical arrangements of the Miles X.2, we have taken three cases in detail, - firstly, the aircraft carrying 38 passengers and 700 gallons of petrol; secondly, the aircraft carrying 32 passengers and 1,400 gallons of petrol; and thirdly, an arrangement with 18 passengers and 2,500 gallons of petrol. These are only three cases of a large number of possible arrangements.

Two further examples of cabin arrangements for 20 passengers and 8 passengers respectively are also proposed.

Case One – Cabin Arrangement for 38 Passengers

This layout we consider is suitable for the shorter European trips such as that from London to Paris. The petrol provided is 700 gallons, which is more than sufficient for the flight to Paris and back to London. There is accommodation for a crew consisting of first pilot, second pilot and one or two stewards. Also, full flight and engine instruments are to be fitted, together with radio and Lorenze system.

The chairs to be fitted are what we consider to be the most comfortable and suitable for air transport.

The Weight Analysis is as follows:
Empty weight including petrol 5,428 lbs (700 gallons); oil 404 lbs (43 gallons) - 37,335 lbs.
4 crew @ 180 lbs each (First pilot; Assistant pilot (Navigation & Wireless duties) and Two stewards) - 720 lbs.
38 Passengers @ 170 lbs each - 6,460 lbs.
Passenger's baggage @ 80 lbs per passenger - 3,040 lbs (Payload = 9,500 lbs).
Total weight - 47,555 lbs. This gives a wing loading of approximately 27 lbs. per sq. ft.

Case Two – Cabin Arrangement for 32 Passengers

A suitable arrangement for longer European flights such as one from London to Moscow is, we suggest, the 32 passenger layout. The passengers have been arranged to sit in groups of four, each group having a separate compartment. Every individual has a reading lamp and cold air supply. Suitable central lights are provided in the cabin. It is not intended to fit angle-poise lamps, the type we have in mind being a special universal lamp. This gives a beam of light the width of which can be varied, and is arranged to shine over the shoulder of the passenger. It is controlled by the movement of a small knob. The amount of fuel to be carried is 1,400 gallons.

The Weight Analysis is as follows:
Empty weight including petrol 10,850 lbs (1,400 gallons); oil 808 lbs (86 gallons) – 44,606 lbs.
5 crew @ 180 lbs each (First pilot; Second pilot, Wireless operator, Flight Engineer and one steward) - 900 lbs.
32 Passengers @ 170 lbs each – 5,440 lbs.
Passenger's baggage @ 80 lbs per passenger – 2,560 lbs (Payload = 8,000 lbs).
Total weight – 53,506 lbs. This gives a wing loading of approximately 30 lbs. per sq. ft.

Case Three – Cabin Arrangement for 18 Passengers

The arrangement which we have taken as being the most suitable for trans-Atlantic flights is the 18 passenger layout. In this layout the passengers are seated in a front and rear compartment. The small front compartment seats 4 people and the rear 14 people. The middle bay is made into a comfortable lounge with lavatories and dressing rooms.

The Weight Analysis is as follows:
Empty weight including petrol 19,375 lbs (2,500 gallons); oil 1,433 lbs (153.6 gallons) – 54,648 lbs.
7 crew @ 180 lbs each (First pilot; Second pilot, Navigator, Flight Engineer, Relief Engineer and two stewards) – 1,260 lbs.
18 Passengers @ 170 lbs each – 3,060 lbs.
Passenger's baggage @ 120 lbs per passenger – 2,160 lbs (Payload = 5,220 lbs).
Total weight – 61,128 lbs. This gives a wing loading of approximately 35 lbs. per sq. ft.

Other Possibilities of Cabin Layout

Two other passenger combinations comprise: i) seating for 20 persons with the layout being very similar to that of Case Three. Here, however, the seats in the rear cabin are arranged in groups of four, as in Case Two, and tables are provided; and ii) for

8 passengers with facilities for sleeping. Each passenger has his own compartment with chair and table. When required the chair and table may be converted into a bed and every endeavour will be made to ensure that the passengers are comfortable. For this reason economy of space has been made of secondary importance and it is understood that such an arrangement is considered to be a satisfactory proposition.

Accommodation for the Crew

The pilot and second pilot are seated side-by-side in the nose, with the wireless operator behind and close to the second pilot. On long flights accommodation has been made for a navigator and a flight engineer. These arrangements are by no means final and if necessary could be altered to suit operators' requirements. Every effort has been made to ensure a good view for the pilots in all directions.

All supporting members for the windscreen have been reduced to minimum dimensions, so that interference with the pilot's vision is negligible. Opening panels in the windscreen for bad weather flying will be fitted with deflectors to prevent rain, etc. from entering the cabin. Work has been carried out by our Experimental Section on deflectors and clear vision openings and a type is being developed which shows promising results. A special opening is provided in the roof of the machine for the navigator to take celestial readings. Whilst doing so he is shielded from the wind.

In the layout of the cabin special care has been taken to allow ample room for the crew, a separate chair being fitted for members of the crew during periods of relaxation. A door is fitted in the underside of the fuselage to obviate the necessity of the crew passing through the passengers' quarters when the aircraft is on the ground.

Engine Installation

A new type of engine has been selected and is housed completely in the wing with extension shafts to the airscrews. Cooling is obtained by leading edge slots and after passing round the engine the air is discharged at the trailing edge of the wing. Adjustable flaps control the volume of air. The whole arrangement is extremely clean aerodynamically.

A streamline tunnel in the wing by which the engineer has access to the engines during flight has been designed. The engines are easily removable and a block-and-tackle system is to be installed in the wing by means of which the engines may be lowered through openings in the lower surface.

Engine Performance Data

The following engine powers are based on test bench conditions of still air entry to the carburettor:-
    International rating - 900 BHP at 3,600 RPM at 10,000 ft.
    Maximum power rating - 930 BHP at 3,800 RPM at 11,000 ft.
    Maximum take-off power - 1,000 BHP at 4,000 RPM at sea level.
    Fuel consumption:
        a) At max. continuous cruising conditions - 47 galls/hr.
        b) At max. climbing conditions - 60 galls/hr.
        c) At max. level flight at max. power altitude - 70 galls/hr.
    Oil consumption:
        a) At max. cruising conditions - 15 pts/hr.
        b) At max. climbing conditions - 20 pts/hr.
        c) At max. all-out level conditions - 30 pts/hr.

(The problems of engine installation in such conditions had been, and were being, investigated by Rolls-Royce Ltd, in co-operation with the Design Department of Phillips & Powis Aircraft Ltd. - PA).

Tank Layout

The fuel tanks which are to be of welded construction are arranged in the wings and have a total capacity of 2,500 gallons. All petrol and oil cocks will be easily accessible to the engineer during flight. The oil tanks are fitted in the leading edge of the wing and have a generous surface area, whilst arrangements are to be made for oil

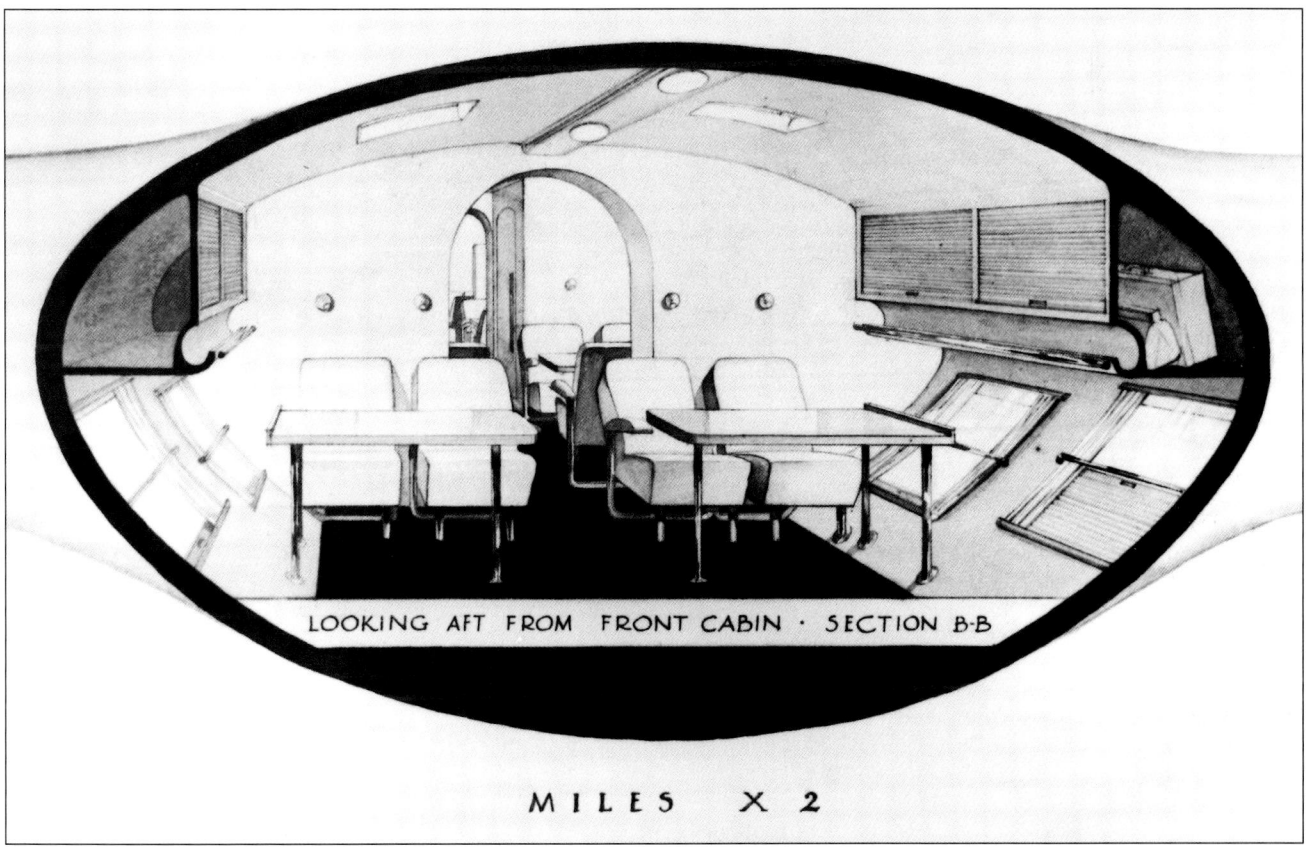

Artist's impression of the interior of the Miles X.2, looking aft from the front cabin.   *[Phillips & Powis via The Miles Aircraft Collection]*

coolers to be fitted inside the air cooling ducts. Separate petrol tanks are provided for extra high octane fuel during take-off and initial climb. Dump valves will also be fitted to all tanks.

Instruments and Technical Equipment

In specifying the technical equipment we have been guided by the experience of air transport operators who have helped us with their valuable co-operation. Bearing in mind the fact that there is a demand for an increasingly comprehensive instrument layout, we have taken great care to make the equipment conform to the ideal laid down by the operator and to allow for additional equipment which may be considered desirable in the future.

The primary flying instruments are as follows: Sperry Gyroscope; Sperry Artificial Horizon; Sensitive Altimeter; Coarse Altimeter; Rate of Climb Indicator; Airspeed Indicator; Drift Indicator; Compass; Chronometer; Watch with seconds hand; Full automatic pilot installation.

The navigational equipment is as follows: Sextant; Bygrave Slide Rule; Norris Tables; Nautical Almanac; Parallel rule and dividers; Compass Deviation Card; Airspeed Indicator; Altitude Computor; Radio, Directional Finder and Lorenze System. These three latter instruments are described under separate headings.

On longer flights when an engineer is carried it is essential that he shall be able to check the running of the engines independently of the pilot. For this reason it is proposed to duplicate the engine instruments, one set being mounted on the main instrument panel for the use of the pilot, and the other set to be installed on a separate panel for the convenience of the engineer.

The instruments to be included are:

Tachometers; Oil pressure gauges; Fuel pressure gauges; Fuel pressure warning lights; Oil thermometers at inlet and outlet; Carburettor induction thermometer; Cylinder Head Thermo Couples arranged to measure maximum and minimum temperatures; Boost pressure gauges; Fuel contents gauges; Flow meters; Vacuum gauges; Engine Synchroniser; Auxiliary hand pump; Indicators and warning devices for flaps and undercarriage.

Electrical Equipment

The electrical equipment consists of: Navigation Lights; Landing lamps; Cockpit lighting and instrument panel floodlighting; Cabin illumination; Electrical starting of power units; Autosyn remote indicating engine instruments; Wireless Accumulators.

The main electrical power supply will come from a 4KW 110-volt 800-cycle alternator driven by a small petrol engine, the alternator being excited from a 24-volt accumulator charged by an engine driven generator. Current for the starters will come from the 24-volt system.

Alternatively, the electrical power supply may be provided by a 24-volt DC Installation of the same power as that of the proposed 110-volt system, though this will, of course, result in an increase of weight due to heavier cables and heavier generator.

Radio

The proposed installation for long range work is a Marconi Aircraft Equipment, Type AD 67A/6872B/5062C/ 626B. This combination was used with great success on the Transatlantic crossings by the Empire Flying Boats. Direction finding and visual course indication is catered for by an additional receiver, a rotatable loop and a visual indicator. With a satisfactory installation and an efficient ground station, the maximum range on medium waves using a trailing aerial will be: Air to ground Telephony - 250–300 miles; Air to ground MCW Telegraph – 400-500 miles; Air to ground CW Telegraphy - 600– 750 miles.

Shortwave transmission range will be dependent upon the factors governing the propagation of these waves, i.e. wave length selected, time of day or night, season of the year etc. Provided the correct conditions are selected, ranges varying from several hundred to several thousands of miles may be obtained. Emergency working can be arranged for by a light 1 hp petrol engine arranged to drive the rotary transformer.

Blind Landing System

A complete Lorenz blind landing system is provided. Used in conjunction with gyro compass this enables complete blind approaches and landings to be effected with safety and precaution.

De-Icing Equipment

In this design de-icing equipment is to be provided. The Goodrich pulsating type has been chosen since it has been found to be the most effective and reliable in service. If, however, later experience should prove other equipment to be more successful, the improved type would be substituted. Provision is to be made for installations on the leading edge of the wing, fin and tail plane and de-icing slingers are to be fitted to the airscrews.

General Aerodynamic Description

The main objective in the aerodynamic design of the Miles X.2 has been to obtain an aeroplane of the utmost cleanness possible with existing knowledge and experience. At the same time consideration of its general external form shows that it is quite orthodox in conception and is simply a logical advancement on existing designs. Very careful performance analyses have been made, revised and checked during the initial design stages and it is considered that the figures quoted are a reliable indication of the aeroplane's capabilities.

In order to indicate the degree of aerodynamic cleanness achieved, the following table has been appended to show the magnitude of each component of drag. Drag Analysis of the Miles X.2 at 250 mph at an AUW of 55,000 lbs:

Skin friction drag	54%	Interference drag	9%
Induced drag	11%	Cooling drag	5%
Drag due to the wake	21%	Total	100%

This gives an aerodynamic efficiency, defined as the ratio of the skin friction and the induced drags to the total drag of 65%. Thus an appreciable improvement on present day standards is to be expected. Moreover, this advance has not been obtained at the expense of stability or control. The empennage has been designed to conform to existing requirements and the lateral control is quite normal.

Dimensions and Aerodynamic Data

> Max. Allowable AUW - 61,128 lbs.
> Gross wing area - 1,762 sq ft.
> Span - 99 ft. (the reason for the span being 99 ft, and not larger, was that the maximum span of a standard hangar then available was only 100 ft. - PA)
> Aspect ratio - 6
> Max. wing loading - 35 lbs/sq ft.
> Power loading @ 61,128 lbs. all-up weight:-
>   a) all-out level flight @ 3,800 rpm @ 11,000 ft - 16.5 lbs/BHP.
>   b) all-out level flight @ 3,800 rpm @ sea level - 18.5 lbs/BHP.
>   c) at take-off with max power - 4,000 rpm - 15.3 lbs/BHP.
> Overall length of aircraft - 78.8 ft.
> Max. cross sectional area of fuselage - A - 115 sq ft
> Wetted area of fuselage - E - 1,750 sq ft; E/A = 15.3
> Fin and rudder area - 113 sq.ft.
> Tail plane and elevator area - 285 sq.ft.

CHAPTER 23: MILES X.2 TRANSPORT AEROPLANE PROJECT

X.2 - Estimated Performances

	Case 1	Case 2	Case 3

1. Speed (mph) at all-out level power (allowable for 5 minutes only) @ 3,800 rpm.

Altitude feet	W - 47,555 lbs w - 27 lb/sq ft	W - 53,506 lbs w - 30 lb/sq ft	W - 61,128 lbs w - 35 lb/sq ft
0	262	259	255
5,000	280	278	274
11,000	301	298	295
15,000	298	294	290
20,000	295	290	280
25,000	280	269	242

2. Cruising speeds

   a) Maximum continuous cruising @ 3,400 rpm

Altitude	Case 1	Case 2	Case 3
0	248	246	243
5,000	263	260	258
11,000	280	276	273
15,000	276	272	265
20,000	260	255	243

   b) Cruising at 60% all-out level power @ 3,300 rpm

Altitude	Case 1	Case 2	Case 3
0	209	206	200
5,000	223	222	221
11,000	241	235	227
15,000	234	224	203
20,000	210	193	164

   c Cruising at 45% all-out level power @ 3,300 rpm

Altitude	Case 1	Case 2	Case 3
0	183	177	161
5,000	191	181	163
11,000	208	197	170
15,000	197	-	-
20,000	180	-	-

3. Stalling speed - at Sea Level – Without/with split flaps.

	Case 1	Case 2	Case 3
Without	76	81	86
With	65	70	75

4. Climb

	Case 1			Case 2			Case 3		
Altitude (ft)	Rate (fpm)	Time (min)	Climb (ft)	Rate (fpm)	Time (min)	Climb (ft)	Rate (fpm)	Time (min)	Climb (mph)
0	1,140	0	142	940	0	145	735	0	155
5,000	1,145	4.3	156	930	5.3	159	720	6.8	166
11,000	1,045	9.6	174	840	11.6	179	650	15.2	185
15,000	835	13.7	180	660	17.0	184	460	22.4	188
20,000	560	20.7	182	420	26.3	189	260	37.0	192

Initial rate of climb at sea level with take-off power (fpm).

Case 1	Case 2	Case 3
1,540	1,300	1.040

5. Ceiling (ft).

	Case 1	Case 2	Case 3
Absolute	30,000	28,500	25,500
Service	28,000	26,500	23,500

6. Range (miles).

   i. Cruising at 45% all-out level power - 3,300 rpm – a = still; b = 40 mph head wind

Altitude (ft)	a	b	a	b	a	b
0	1,150	900	2,420	1,890	4,350	3,360
5,000	1,190	950	2,430	1,920	4,300	3,340
11,000	1,200	970	2,450	1,970	4,300	3,340

   ii. Cruising at 60% all-out level power – 3,300 rpm

Altitude (ft)	a	b	a	b	a	b
0	1,050	850	2,240	1,810	4,120	3,300
5,000	1,060	880	2,260	1,860	4,130	3,380
11,000	1,080	910	2,300	1,930	4,200	3,500
15,000	1,275	995	2,510	2,080	4,520	3,730
20,000	1,275	1,030	2,640	2,130	4,650	3,730
25,000	1,180	1,000	2,560	2,080	4,600	3,680

Note. In estimating the range, allowance has been made for fuel consumption during 15 minutes running up on the ground. The reduction in weight of the aircraft due to consumption of fuel during flight has also been taken into account.

7. Take-off run at Sea Level (yards).

	Case 1	Case 2	Case 3
With 4 engines	465	597	778
With 3 engines	670	856	1,120

8. Performance with one outboard engine 'dead'.

   a Speed at all-out level power (allowable for 5 minutes only).

Altitude			
0	227	225	221
5,000	242	240	235
11,000	260	257	253

   b. Speed at maximum continuous cruising

0	213	211	203
5,000	225	222	215
11,000	241	235	227

   c. Service ceiling maintainable (ft)

21,000	18,500	15,000

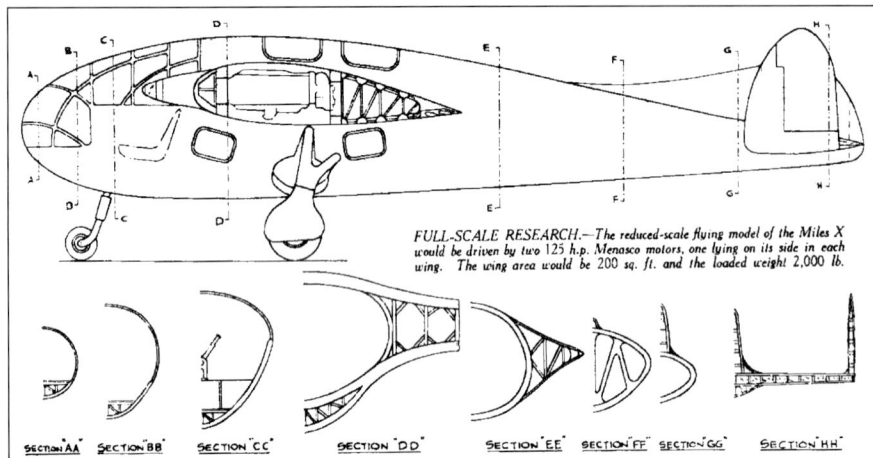

Left: The proposed Miles 'X' Minor.

The 'X' Minor

The reduced scale prototype for the X.2 is approximately one-third the full size of the 'X', and contains the basic principles to be introduced in the larger aircraft. It is to be built of wood and will carry a pilot and observer, and suitable research apparatus. Provided we obtained exemption from Air Ministry inspection, it is considered that we could build this machine in about six months.

The engines proposed are Menasco C.4s, placed inside the wing and on their sides. The following figures indicate the size of this aeroplane:
   Tare weight – 2,000 lbs. Gross weight – 2,840 lbs.
   Wing area – 200 sq ft. Wing loading – 14.2 lbs per sq ft.
   Power loading at 125hp per engine – 14.5 lbs per hp.

It is of interest to note that Specification 14/38 (Specification held in abeyance) was issued to Short Bros for a Long Range Civil Land Plane as follows:

Requirements - It is required to develop a type of all metal, long range landplane suitable in all respects for operation on scheduled air services. Two versions of the aeroplane are required, the first with an operating height of 10,000 ft and the second with an operating height of between 20,000 ft and 25,000 ft.

With 18 passengers, a payload of 7,500 lb is required. The accommodation should be capable of housing in comfort 18-20 day or night passengers, and a small promenade should also be provided. The mean cruising speed is to be not less than 250 mph at 10,000 ft or not less than 275 mph at 20,000-25,000 ft when using the engine maker's recommended maximum economic cruising power. The range is not to be less than 3,000 miles against a 30 mph headwind when carrying the specified crew and payload at the above speed. The passengers' and pilots' compartments should be supercharged to simulate conditions at 4,000 ft when flying at 20,000-25,000 ft. Stability and controllability, particularly at 'low' speeds, will be regarded as important features of the design.

The aircraft must be fitted with an Automatic Pilot and an approved form of anti-icing or de-icing equipment should also be provided. Crew: 1 Commander, 1 First Officer, 1 Navigator, 2 W/T Operators, 2 Stewards. Radio: Medium and short wave receiver and transmitter, Independent medium wave D/F receiver with rotatable loop, VHF approach receiving equipment.

The aeroplane, fully loaded, is to comply with the requirements specified for normal category certificate of airworthiness. When fully loaded it must clear a 66 ft screen in not more than 1,200 yards. It must comply with the normal category requirements of AP.1208, Design Leaflet F1, paragraphs 2, 3 and 4, when carrying full complement of crew, payload and equipment and fuel for 25% of the specified range. The details of the aircraft should be designed for easy maintenance and the ability to change an engine in one hour is considered desirable.

Contract No.762587/38 was awarded to Short Bros to cover the construction of three prototype S.32s to Spec 14/38, and they had actually built three fuselages by March 1939, but with the outbreak of war, the specification was cancelled on 13th November 1939. Folland Aircraft were also developing projects to this specification.

However, by this time the Air Ministry had lost interest in civil aviation and research work (other than that directly concerned with the urgent expansion of the RAF) and Miles was only offered a small development contract valued at £25,000. The value of this contract was totally inadequate for the construction of; i) an aircraft the size of the Miles X.2, ii) a reduced-scale twin-engined flying model of the Miles X.2 to be built of wood for research purposes to AM Specification 42/37E, and iii) a mock-up of the cabin, so Miles was forced to decline the offer. The project was then abandoned and Spec. 42/37E was not proceeded with.

Miles however was not a man to be put off by the lack of official interest shown in the project by the Air Ministry and he continued development of the basic design as and when project staff and time could be spared. The later Miles X.3 to X.15 projects are evidence of his further commitment.

Miles and Blossom with Col Charles Lindbergh and the Mohawk, probably on the occasion of its official handover on 1st February 1937 – it certainly looks cold enough!
[Phillips & Powis via J D Galloway]

Chapter 24:
Miles M.12 Mohawk

On 21st May 1927, Captain Charles Lindbergh, a young American aviator, distinguished himself by flying non-stop from America to Paris in 33½ hours, thereby completing the first solo eastbound crossing of the Atlantic. However, on 1st March 1932 Lindbergh's first son was kidnapped from their home in Hopewell and after two and a half months' suspense his body was found. It was to be over four years before the murderer was caught and brought to justice and by this time Lindbergh had had more than enough of publicity. On 21st December 1935, he departed secretly by freighter for England.

Charles and Mrs Anne (Morrow) Lindbergh rented Long Barn, Weald, near Sevenoaks, Kent (a house once owned by Vita Sackville-West and her husband Harold Nicholson) for their stay in England and soon after their arrival Lindbergh decided that he needed a fast, reliable, long-range light aeroplane in which he and his wife could make business trips between the capitals of Europe.

The fact that Miles had been in negotiation with Al Menasco, the American manufacturer of the Menasco in-line engines, since 1935, with a view to building a range of his engines under licence at Reading, may have influenced Lindbergh when it came to finding somebody to produce a specialised aeroplane to suit his needs, as it is known that Lindbergh 'rather fancied an American motor' for his machine.

It is recorded in the book *The Menasco Story* by Ralph J Schmidt that: *It was during this period of public interest over the arrival of the famous American flyer Charles Lindbergh and family from America, and in Lindbergh's supervised design and purchase of the Menasco-powered English Miles Mohawk monoplane, which he flew throughout Europe, that great interest was generated in the American Menasco inverted 'Buccaneer' engine, by British airplane designers and flying enthusiasts. Consequently, in March 1936, a royalty agreement was signed with Owen Powis and Fred Miles of Phillips & Powis Aircraft Ltd of Reading England, for a license to manufacture Menasco engines in that country. Menasco engines had now truly become of age both nationally and internationally, with engines in America, Canada, Germany, England, Spain and South America.*

Although it is recorded that Lindbergh's first visit to Phillips & Powis at Woodley was on 6th May 1936, (when the local press reported that he had visited the firm and had tried several Miles aeroplanes), C L Nash recorded that the Sparrowhawk G-ADNL was flown by Colonel Lindbergh on May 5th. Lindbergh also flew the second Nighthawk, as U6, on May 6th and it must have been during the course of these early visits that the subject of Miles building him a fast touring aeroplane for his planned trips around Europe and beyond was discussed. Reports appearing in the aviation press about July/August 1936 stated that Lindbergh was spending a good deal of time at Woodley *investigating the various Phillips & Powis Aircraft Ltd products - so we expect some-day to see something produced at Reading which will fulfil Colonel Lindbergh's own specification.*

Lindbergh obviously liked the Miles aeroplanes he had flown and it was no small compliment to Miles and his fledgling company that he decided to place the order to design and build this new aircraft with them. Miles later wrote *Lindbergh was the perfect person to work for. I had a good deal of time with him and he knew what he wanted and just how possible it was to meet his wishes. He knew as well as I did the sort of compromise that one has to accept, when designing and making an aeroplane.*

The result of the discussions between Lindbergh and Miles was the M.12 Mohawk. Lindbergh had written to Al Menasco regarding the suitability of the 6-cylinder, supercharged, 200hp Menasco B6S engine for the new machine and Al's reply was apparently so satisfactory that he asked Miles to arrange for this engine to be fitted. Al himself came to Reading in early June 1936 to discuss plans for the commencement of manufacture of his engines under licence, the first of which was to be the 125hp Menasco C-4 Pirate.

The Mohawk was designed to give an impressive cruising speed of 170 mph, have a range of 1,400 miles and be fitted with Miles' split trailing edge flaps, which were operated by means of a Theed vacuum jack, as in the M.11A Whitney Straight.

The Phillips & Powis Aircraft Ltd 'G.A. of Mohawk' drawing (below) was dated 17th June 1936 and by late June/early July the detailed design of the new aeroplane was nearly complete. Al Menasco sent his representative, Byron Anderson, over from the USA to liaise with George Miles with regard to the plan to produce Menasco engines at Woodley and also jointly to supervise the installation of the B6S engine into the Mohawk. Apparently the Air Registration Board (ARB) were not too keen on the Menasco engine but George Miles persevered with trying to get the 125hp Pirate C-4 engine certified.

It had been reported that the Mohawk was built to incorporate wing fold, as on the previous Miles types, although, as it was not intended to be used, this was fixed in place and the joints taped over. Wing fold was also shown on the 1/72nd scale drawing in *The Book of Miles Aircraft*, compiled by A H Lukins (who worked in Technical Publications at Miles Aircraft Ltd). However, wing fold was not shown on the original P&P G.A. drawing, shown below, which shows the outer wings attached to the centresection in a similar manner to those of the later M.14 Magister, with a strap covering the joint between the outer wing and the centre section.

The Mohawk was constructed under Job No.15235 but Phillips & Powis Aircraft Ltd applied for Certificates of Registration for two Mohawks (the second one to Job No.15289) on 6th July 1936. They were building the second machine for another customer, James McArthur, but this was to have a different section fuselage. Some years ago I acquired a pre-war postcard at a collector's fair which depicted a blue-painted aeroplane with white trim but no registration and which was not unlike the Mohawk. The painting was entitled 'Mills Majestic' but unfortunately both the printer and the artist chose to remain anonymous. Apart from the obvious error in the spelling of Miles, it does beg the question - if the second Miles M.12 had been constructed, would it have been called the Miles Majestic?

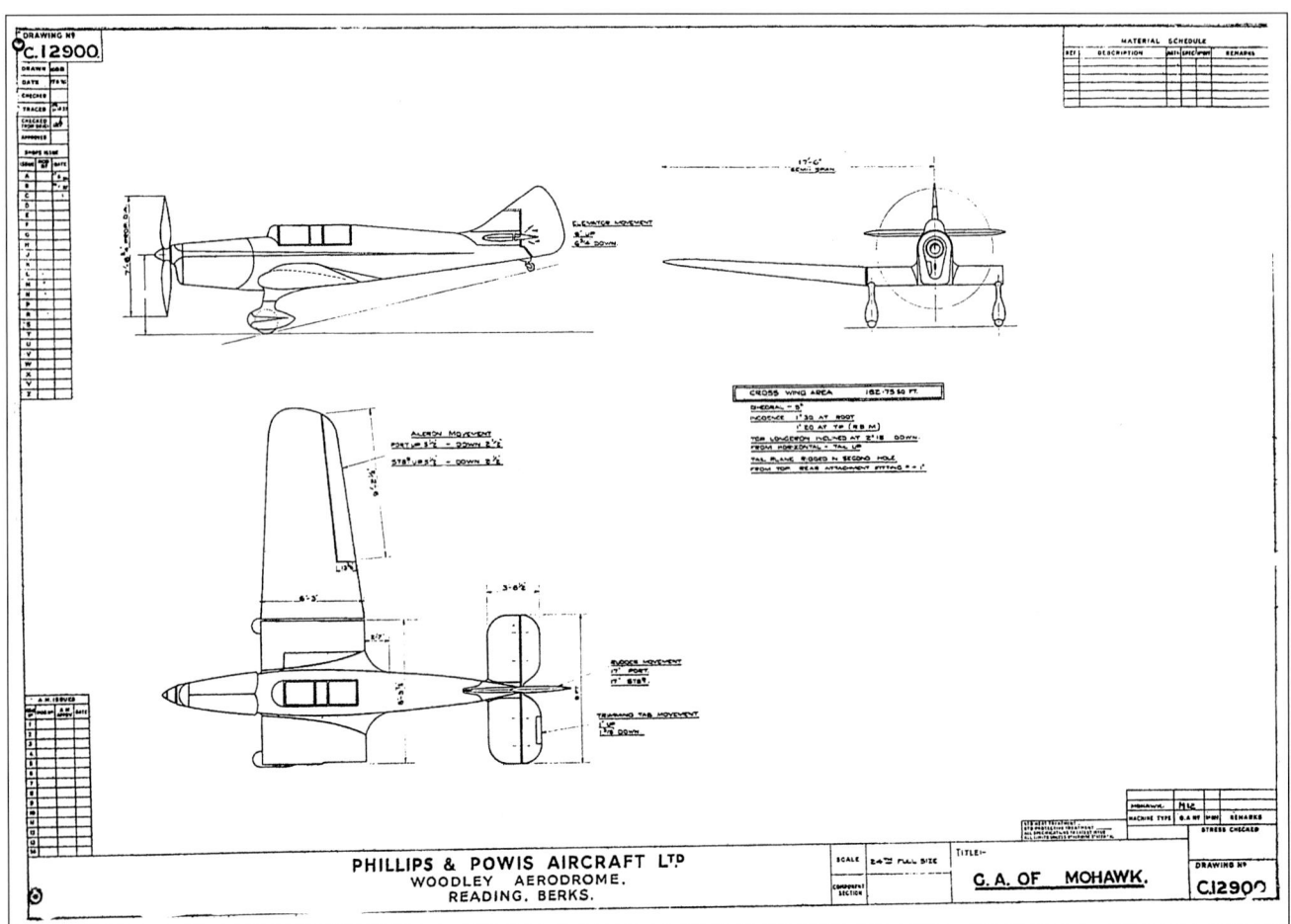

The Miles M.12 Mohawk original general arrangement drawing.

Chapter 24: Miles M.12 Mohawk

Miles made the first flight of the Mohawk, of 10 minutes duration, on 22nd August 1936, with the 'B' mark U8 chalked on the fuselage. A recently discovered photograph shows that after the chalk possibly 'wore off', or the aircraft was painted, 'U8' was replaced by a somewhat angular U8 in masking tape on the fuselage!

Lindbergh also made a test flight of 15 minutes on the August 22nd and flew it again, with George Miles as passenger (who did not record a registration), on 24th September 1936. Therefore, it must be confirmed, categorically, that the first flight date of 28th January 1937 given in so many reference works (and even in Miles Aircraft's own publications) is incorrect. Once again, you cannot always believe what you read, as this was, in fact, the date its C of A was issued!

On 17th September 1936, Lindbergh wrote to Mr A S Menasco:
  Thank you for sending me a copy of the new B6S handbook. We are still making tests on the plane and there are a number of details to be finished before accurate performance figures can be obtained. However, I am very impressed with what I have seen of your engine and its operation up to the present time. The need for a controllable pitch propeller is quite apparent on take-off. I hope that it will be possible to obtain one in the near future.

Full details of the M.12 Mohawk were given in an article which appeared in *Flight* for 5th November 1936, under the title *Lindbergh Buys British - His New Miles Mohawk Described. A High-performance 'Custom-built' Tourer*:
  When Colonel Lindbergh settled in this country he surveyed our range of touring machines, and, although unstinting in his praise, was unable to find exactly what he wanted. So, some months ago, he 'went into a huddle' with Mr F G Miles with a view to placing an order for a machine designed expressly to meet his requirements. The fruit of this collaboration was the Miles Mohawk, a two-seater, long range, high-speed monoplane with a supercharged Menasco Buccaneer engine. Phillips & Powis are quite prepared to build replicas, but do not regard the machine as a production type.

  To say that the lines of the Mohawk are the most attractive yet conceived on the P and P drawing boards is praise enough of its appearance. It is registered G-AEKN (and this registration was actually painted on the fuselage of the aircraft in error for G-AEKW but this was corrected later - PA) *and finished in orange and black at Col. Lindbergh's express wish, because this combination shows up most readily under all weather conditions.*

  It is essentially a typical Miles two-seater, but differs from its forebears in its cockpits, undercarriage, tail unit and its supercharged power plant. Structurally it is of wood and follows standard Miles practice; in fact, the outer panels of the wings are similar to those of the Hawk Major, with two spars and ply covering taking drag and torsional stresses, but with strengthened spars to take care of the greater all-up weight and a few minor modifications to permit the installation of extra fuel tanks. Miles split trailing-edge flaps are incorporated. In addition to the outboard tanks there are two in the centre section. The basic design of the wooden stressed-skin fuselage and fabric-covered tail unit does not depart from previous Miles technique, although certain members have been specially strengthened.

  English eyes will soon settle on the Menasco Buccaneer B6S six-cylinder, in line, inverted air-cooled engine supercharged by a centrifugal blower to give 200hp at 2,250 rpm at 4,500 ft. A detailed description of this power plant was given in Flight *of September 10th this year. Incidentally, it might be useful to remember that, externally, a Menasco installation may be distinguished from a Gipsy in that the cooling air chute is on the starboard instead of on the port side. The particular engine in the Mohawk has been imported from America, but the licence for the B6-S and other Menasco models was lately secured by Phillips & Powis. The engine is mounted on blocks of Ferodo brake lining and small springs, allowing a 1/8in. compression. For the time being the airscrew is a fixed-pitch metal Fairey, but provision has been made for a variable-pitch type. Hamiltons, incidentally, are not yet marketing a VP screw suitable for Menasco, although De Havillands, their English licensees, have developed successful models for engines of similar ratings.*

  Naturally the supercharger (driven at 8.75 times engine speed) results in a somewhat longer power unit than is normally seen in light aeroplanes, but happily Col. Lindbergh asked for rudder pedals instead of a bar so the final installation of engine and controls was made in less space than might have been imagined. An Eclipse generator and electric starter are fitted and the induction pressure is read off on a Smith's boost gauge.

  The tandem seats are mounted on front and rear spars and are sheltered by as neat a transparent roof as we have seen. This roof, with its long tail fairing tapering away to the fin, is of an inverted U section merging smoothly into the maximum cross section of the fuselage and resulting in a hollow along the fuselage. Col Lindbergh will normally fly from the front seat, with Mrs Lindbergh behind, and the reach of the adjustable rudder pedals has been designed accordingly.

  Between the seats is a fixed portion of the transparent enclosure, but flexible panels immediately above each occupant meet on the centre line of the fuselage, being pushed down into the side walls for entry, exit and "open cockpit" flying. The concave side of the enclosure permit one eye to be put outside in bad weather, leaving the greater portion of the face sheltered. A capacious luggage locker with a recess for suitcases is located behind the rear seat, and is accessible both from the cockpits and the starboard side of the fuselage. Behind this is stowage for a tent, a 10lb. collapsible dinghy and similar articles which might prove of value on journeys in inhospitable regions. The backs of the seats can be removed and laid flat to form with the seat bottoms a make-shift bed.

  The dual flying controls are conventional, but incorporate a patent Miles tail trimming device. At low speeds the elevator tabs become comparatively ineffective, so the operating gear is planned to adjust the tabs in the normal way at high speeds and at low speeds, for landing, to load the elevator control. The undercarriage is a single-strut type by Lockheed, incorporating brakes and Airdraulic struts. In view of the fact that this type has also been adopted on the new version of the Hawk Trainer, it may be deduced that future Miles designs using fixed undercarriages will be equipped with something similar. A Short float undercarriage has been designed, and is particularly noteworthy in that there are no cross struts. Large fairings cover the groups of attachment struts.

  Among the items of equipment introduced at the express wish of Col Lindbergh are three parachute flares mounted horizontally in the rear end of the fuselage. These flares are made by the International Flare Signal Co of Tippecanoe City, Ohio, USA, and are fired electrically from the sides of the machine, travelling about 40ft before blossoming out. The great advantage of this type of installation is that the flares do not lose altitude before giving their light and do not tend to vibrate out of the machine.

  Complete performance figures have not yet been taken, but the maximum speed is believed to be in the immediate neighbourhood of 200 mph. With the ample wing area and the Miles split flaps the landing speed should be quite low.

A small three-view drawing of the Mohawk which accompanied the article also showed floats apparently mounted on the undercarriage attachments (but not wing fold). A photograph of the prototype M.38 Messenger under construction later shows a float of a similar type hanging on the wall in the background and it is possible that this may have been acquired for the M.10 or possibly the Mohawk.

The article also showed photographs of the completed M.12 Mohawk including one of Lindbergh flying it '*on test, showing the front cockpit open*'. As the article appeared in November 1936, it

'Col Charles Lindbergh, the famous American flyer, has purchased a new 'plane, which is here at Reading Aerodrome, where he took over his new machine'. Note the 'B' mark U8 in masking tape on the fuselage. [Graphic Photo Union, London]

Charles Lindbergh with a thoughtful George Miles in front of the Mohawk. [Phillips & Powis via The Miles Aircraft Collection]

is all the more surprising that many normally respected sources of historical information should have given the date of its first flight as 28th January 1937.

Lindbergh wrote to Louis S Casey, Curator of the Smithsonian National Air & Space Museum, on 18th November 1973:

In regard to the Mohawk, while I worked closely with Miles in laying out specifications and watching the construction, I do not recall having drawings in my personal files. If there are such drawings in my files they should be at the Sterling Memorial Library in files to which I have restricted access without written authorization. So far as I know, the Mohawk was the only plane of its kind built by Phillips & Powis. I laid down the basic specifications. F G Miles was the engineer. The first Menasco engine installed in the plane was replaced by a Menasco of more advanced design.

A year of two ago, I received a letter from Miles to the effect, as I recall, that the Mohawk had been sold after World War II for a higher price than I had paid for it. This was a surprise to me because I had previously received only vague rumours about the plane after I gave it to the British government soon after the start of the war. Miles said nothing about the plane having crashed (according to my memory of his letter), so probably the report you received to the effect that it "survived till 1946 in Spain" should be checked for accuracy.

CHAPTER 24: MILES M.12 MOHAWK

It is of interest to note that George Miles later recalled:...*the delay in handing over the Mohawk to Lindbergh was mainly due to unsatisfactory manufacturing standards on the sub-contracted canopy. This had to be re-manufactured and Lindbergh's movements determined the eventual delivery date.* John Underwood also later recalled that: *They were doing a lot of testing, also the propellers were not satisfactory and he was most anxious to have controllable pitch.*

In November 1936, Lindbergh flew the Mohawk to Baldonnel Aerodrome, Dublin (still carrying the incorrect registration G-AEKN and with no C of A!), in connection with a proposal for a joint Imperial Airways/Pan American Airways base on the Shannon and on 21st he gave the Irish President, Eamon de Valera his very first flight - in the Mohawk,. George Miles flew the Mohawk, solo, but with no registration recorded, on local flights from Woodley on January 13th, 14th and 19th 1937 and it was also flown by Flt Lt John Moir on a 15-minute 'u/c test' on the 13th. Its C of A was eventually issued on Thursday 28th January 1937.

Although C L Nash recorded that the Mohawk was delivered to the customer on 29th January 1937, it was, in fact, officially handed over to Lindbergh at a ceremony at Woodley on Monday 1st February 1937. After the handover ceremony, Lindbergh and his wife Anne

The M.12 Mohawk in October 1936 with the starboard rear cabin window retracted into the cabin side wall. ['Flight' via B Clarke]

Charles Lindbergh at the controls of the Mohawk soon after its first flight. [Phillips & Powis via The Miles Aircraft Collection]

(who was, by then, expecting a baby) immediately left Woodley for Lympne to clear Customs on the first leg of a flight to India. Lindbergh had not told the press where he was going, but the local evening newspaper for February 2nd carried the headline *Lindbergh Safe* and stated that he had landed near Pisa, Italy before going on to Rome the following day, a 170 miles journey which apparently took them 2 hours 5 minutes to complete, due to having encountered strong headwinds en route. This dispelled the rumours that he had crashed on the Hungarian border! For reasons probably due to their past bitter experience with the press, the Lindberghs left Rome under the same veil of secrecy as that in which they had arrived, en route to India.

On February 4th they met Mussolini at his Venice Palace in the afternoon and then left Rome, refusing to reveal their destination, under a veil of secrecy. They may then have flown to Athens, although the reports in various American newspapers do not make this clear. Their next stop was Tunis followed by Tripoli in Libya, where they spent the day and met Air Marshal Italo Balbo, Governor General of Libya and famous trans-Atlantic airman. They left Libya on February 12th to fly along the North African coast to the British base of Mersa Matruh and then on to Cairo. *Nevada State Journal* for February 25th reported that: *Lindbergh and his wife motored to Poona tonight to inspect the airdrome which Indian air mail planes use during monsoons. They arrived in their Miles Mohawk plane today from Jodhpur after being unreported for almost two days on their flight from London. Instead of flying eastward to Allahabad as was at first reported they turned southward to Bombay.* Little is known of their other movements in India, other than that they arrived at Karachi on February 20th, nor which route they took on their return.

Having reached India safely in easy stages, and conducted their business, they arrived back at Gatwick Aerodrome on 9th April 1937.

After his return, Lindbergh told Miles that he was more than satisfied with the Mohawk's excellent performance in all respects - in fact it was exactly what he wanted. The Lindberghs subsequently made many fast non-stop flights around Europe, where the Mohawk with its distinctive colour scheme soon became well known at all the principal airports.

A Phillips & Powis report, dated 20th April 1937 concerning the 'Menasco saga', commented: '*The Mohawk has flown about 200 hours - mainly under tropical conditions. In a letter to F G Miles, Colonel Lindbergh stated: the engine has functioned well except the magneto shaft breakage and the push rod trouble. At the present time I know of no reason to blame the former on the engine designs. It seems more likely due to a flaw in the shaft. This engine has now been removed and will be replaced with another engine of the same type with provision for a hydraulically controlled variable-pitch propeller. The original engine will be stripped down for a thorough examination - although on superficial examination it appears in excellent condition and was running perfectly when Colonel Lindbergh returned*'.

George Miles made two 20-minute flights in the Mohawk on May 27th and 28th, and on June 11th he carried out a 45-minute flight. On 15 July 1937, he carried out a 35-minute 'Test of Hamilton Airscrew' and that was the last recorded flight by the firm of the Mohawk at Woodley. On 27th September 1937, Lindbergh wrote to Byron Anderson, Menasco representative and engineer at Woodley: *Many thanks for your September 23rd letter advising me of the crack which you located in the mounting ledge of my engine. I am glad that you will be able to have the old engine installed in time for our trip to the Continent. Please let me know at once if you foresee any delay which will keep the plane from being ready on October 5th.*

Have you formed an opinion as to the cause of the crack? Is there any sign of a flaw in the casting, or do you feel that this might be due to the type of engine mounting in the Mohawk? If you send the engine back to California for inspection and repair can you give me a rough estimate of when it will be back here? The old engine and propeller will be entirely satisfactory for our trip to the Continent next month. With best regards, Charles A Lindbergh.

There was at that time a rumpus in the newspapers over Lindbergh's 'business trips between the capitals of Europe' but to clarify the matter CG Grey wrote in *The Aeroplane* for 19th October 1938:

'How does he come to be Europe's busybody?' - asks the Sunday Express of Oct 16th about Col. Charles Lindbergh. The answer is so simple. Charles Lindbergh is technical adviser to Pan-American Airways. Pan-American Airways naturally want to know what every other country is doing. Charles Lindbergh is, presumably, paid a handsome wage to find out. He is a most amiable person when he gets away from absurd newspaper stories and offensive newspaper people (nothing changes - PA). He has a charming wife who has written one of the best flying books ever published ('North of the Orient') and has just produced another which, according to those who have read it before publication, is a better book than the other.*

Can you imagine a nicer job than having plenty of money to roam about the World to see what aviation is doing in all countries and just to write home about it when you think it is worth while? Naturally, having rightly established for himself a reputation as one of the World's most skilful pilots and air navigators, his opinion is asked by all sorts of people, - from the officers commanding the World's Air Forces, and the managers of the great air lines, down to mere Society folk who are interested in hearing his views of people whom he has met in other countries.

Some people seem to resent his friendship with the more responsible people in Germany. Seeing that Charles Lindbergh is practically pure Scandinavian, one would naturally expect a certain sympathy between him and the Germanic people. Equally some people have resented his friendship with the Russians. And the absurd stories published in this country about him and the Russians seem to show that he is not particularly popular there.

The article went on about the relative merits of Russian and German Air Forces but concluded: *They call him 'a lackey of German Fascism and its aristocratic British patron' and a 'political speculator'. Also they add: 'Nobody asked him to come here (Russia) and he was only allowed to do so because the United States authorities asked for that permission'. They also charge him with having said that he was offered the post of Chief of Civil Aviation in the Soviet. We judge from what he told us in Berlin that the most surprised and amused person over the whole affair is Lindbergh himself.*

On Easter Monday 18th April 1938, Lindbergh and his wife visited Woodley and stayed for a few hours. Then, in June 1938, the Lindberghs moved to the island of Illiec on the north coast of Brittany, France. Lindbergh continued to travel throughout Europe in the Mohawk (some say at the behest of the American government) where he was given access to the military aviation facilities of Britain, France, Czechoslovakia, Germany and even Russia, where a news film showed the Mohawk taxi-ing at Moscow airport. Col Lindbergh *'called in at the aerodrome (Woodley) earlier of this first week of August 1938'* and this was probably to collect the Mohawk after a service.

During the course of one notable tour which commenced at Lympne on 16th August 1938, the Lindberghs visited Hanover, Warsaw, Moscow, Kharkov, Rostov, Kiev, Odessa, Cracow, Olmuty, Prague, Stuttgart, Paris, Morlaix, Rotterdam and Berlin, before returning to England. The *Sunday Express* said of Lindbergh's visit to Russia: *As an honoured guest he was given unusual facilities to see Russia's Air Force and, together with Stalin, they watched a display by the Soviet Air Force.*

On 21st October 1938, Lindbergh flew the Messerschmitt Bf109 but shortly after this he was forced by 'such severe weather' to leave the Mohawk in Berlin and continue his journey by surface transport. The Mohawk remained in Berlin until 18th January 1939 when Lindbergh collected it and flew back to London. On 4th April 1939, Lindbergh landed at Lympne from Morane's Puteaux aerodrome in France to clear customs before flying on to Woodley

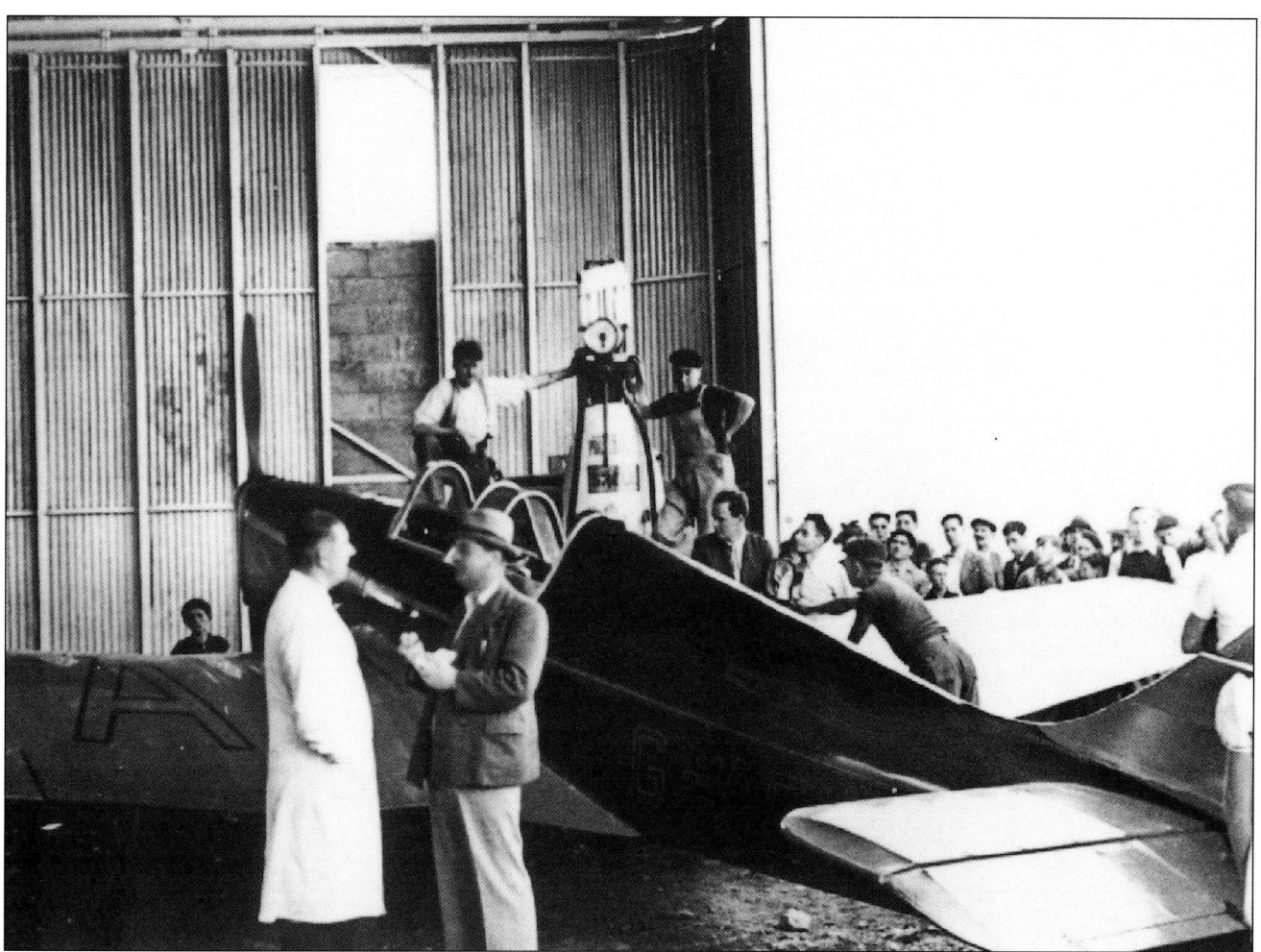
The Miles M.12 Mohawk at Morlaix, soon after Lindbergh had landed for refuelling in August 1938. [Via J Gregory]

later in the day. It is understood that this was probably the last flight he made in the Mohawk, as shortly after this he began preparations to move back to the USA. The Mohawk was then placed in storage at Woodley pending a decision as to its future.

The Lindbergh files on the Miles Mohawk, amounting to some 600 pages of information, are kept in the Sterling Library of Yale University and in March 1995 limited access was obtained by Robert B Price of New York. In a letter to Julian Temple in April 1995, he confirmed that the files contained documents ranging from receipts for petrol purchased in Bahrain, to detailed listings of the emergency equipment in the Mohawk. There was also a Phillips & Powis drawing of a proposed roll-over reinforcement for the centre section of the cockpit.

The Mohawk was surveyed by the Air Ministry at Woodley at the end of August 1939 with a view to its impressment should hostilities commence. It was dismantled for storage at Woodley soon after the outbreak of war but was impressed into the RAF on 31st October 1941 and delivered to RAF Turnhouse on 8th November 1941 as HM503. In the aforementioned letter of 18th November 1973 to Lew Casey, Lindbergh wrote: *My gift of the Mohawk to the British Government was outright. I did not ask for any payment. (I received a letter from the British Government, about 1940, saying the plane was wanted in the war effort and offering compensation).*

The rumour prevailing at Woodley following the Mohawk's impressment was that it was intended for use by the Duke of Hamilton - hence the delivery to Turnhouse.

An interesting letter from 'An RAF Corporal', was published in *The Aeroplane Spotter* for 27th July 1944:

With reference to your drawing of the Miles Mohawk in the Spotter's ABC LXXX in your issue of June 1st 1944, you may be interested to know that I worked on this aeroplane at an RAF aerodrome some time ago and that it was then camouflaged with RAF markings. The original power plant, a Menasco Buccaneer, gave a great deal of trouble through overheating and the aeroplane could not be flown for more than half an hour without the cylinders and oil temperatures rising to dangerous levels, and ground running also had to be kept to a minimum for the same reason.

The only other Menasco Buccaneer in this country was later installed, but the overheating still continued and a Hamilton constant-speed airscrew was fitted in place of the original Fairey-Reed type. This greatly reduced the overheating on ground runs, but it was so heavy that the Mohawk never flew with this airscrew fitted. The Fairey-Reed airscrew was then returned to the manufacturers where the pitch was considerably coarsened and the trouble was somewhat reduced. There were four petrol tanks in the mainplanes, but the two outboard tanks had been sealed off and only the inner two were used.

The airfield which could not be named at the time for security reasons was probably RAF Turnhouse. It does seem very strange that the RAF should have experienced problems with overheating as there is no record of Lindbergh having experienced any problems during the course of his many flights, most of which were of long duration. The Mohawk was officially recorded as having been returned to Woodley on 21st May 1943 for the fitment of a Fairey-Reed propeller of considerably coarsened pitch, but Hugh Kennedy made a 15-minute 'handling' flight in it at Woodley on 17th May 1943.

The Mohawk was delivered to Maintenance Command Communications Squadron, RAF Andover on 14th September 1943 and John A Painter, who was a fitter there, had a letter published in *Aeroplane Monthly* for October 1987 which throws some new light on the Mohawk at Andover:

Miles M.12 Mohawk at Redhill Aerodrome on 29th August 1947, following restoration by Southern Aircraft (Gatwick) Ltd, being prepared for the Lympne International Air Races, held between 30th-31st August 1947. [P Wood via P Jarrett]

G-AEKW at Lympne for the Folkestone Trophy Race in August 1947. [The A J Jackson Collection]

Mention of the Miles Mohawk in M J Hardy's concluding 'Redhill Recollections' (May Aeroplane) brought back a flood of memories of even earlier years. I was an ex-apprentice from Halton, stationed with No.15 (P)AFU at Andover. The station commander, a Grp Capt Lowe, as I recall, occasionally made sorties to White Waltham to return with odd aeroplanes which were placed in my tender care. One was a dainty side-by-side Tipsy; another was the Miles Mohawk referred to by Mr Hardy. The year was 1943.

The logbook showed it to have been flown extensively by Charles Lindbergh, often with his wife, if I recall correctly. I seem to remember that we were intrigued to find references to extensive aerial survey flights in the Soviet Union. The aircraft had been designed for that purpose, apparently, which accounts for the impressive 1,400 miles range mentioned by Mr Hardy. Soon after the Mohawk's arrival we received a packing case containing a brand-new Menasco Buccaneer engine. Bert Lilley, a crackerjack fitter sergeant, and I installed the new engine, and I repainted the aircraft. Eventually we started it, only to have oil thrown over the whole machine. The problem, we discovered, was that some of the pistons had rings missing!

I cannot remember how we found new rings but in the interim the good group captain decided he wanted a Hamilton Standard constant-speed propeller (no doubt the one referred to by Mr Hardy) fitted in place of the Fairey-Reed prop. Where it came from I do not know, but I fitted it to the Mohawk. Here I have to disagree with Mr Hardy's statement (and the afore-mentioned RAF Corporal's' - PA) that the aircraft never flew with this heavy propeller. It did: after I was 'invited' into the back seat by Grp Capt Lowe! I clearly remember climbing on board with great trepidation. I was quite unhappy with the prop and, without any weight and balance data, I had no idea how far the c.g. had moved.

In addition, each cockpit was enclosed by flexible sheets of clear plastic which ran in curved channels as they were lifted on either side of the seat to meet overhead - and, when closed, they had a habit of jamming! We did get off the ground after a very long run on Andover's grass surface. We made a simple, wide circuit of the airfield, the aircraft vibrating quite badly, and landed somewhat shakily in one piece. I believe the group captain shared my relief, because I do not remember the Mohawk being flown again. Shortly afterwards, the unit moved to a newly-opened airfield called Babdown Farm. The Mohawk stayed at Andover.

On 1st February 1944, shortly after this episode, the Mohawk was delivered to 5 MU Kemble for storage and it does not appear to have been flown by the RAF after this. In December 1945, over 50 civil light aeroplanes, which had been impressed for service in the RAF, were offered for public sale by the Ministry of Aircraft Production and these were lined up for inspection in one of the "underground" storage hangars at Kemble. Tender forms were obtainable from the MAP at Millbank, London.

Although Lindbergh's Mohawk was also at Kemble at that time, a report simply stated that it 'may be sold in the near future'. It was later purchased by Southern Aircraft (Gatwick) Ltd and restored to the register by them on 28th May 1946. Following refurbishment, it was advertised for sale in the July 1947 issue of *The Light Plane* as a: 'Supercharged two-seater with CofA. Finished in high-gloss, 1,000 miles range, cruising speed 170 mph'. Although no price was given, it was being offered at £2,500. Its C of A was renewed on 7th August 1947 and during the course of its restoration the Mohawk was repainted in high-gloss maroon overall with white trim, being seen (and photographed) in these colours by the author at Redhill Aerodrome soon after.

The Mohawk was flown by Wing Cdr Earle (with Race No.3) in the Lympne International Air Races held on 30th/31st August 1947, gaining second place in Heat 1 in the Folkestone Trophy Air Race at a speed of 138.5 mph. A later owner was Bruno Pini (a member of the Herts & Essex Aero Club, Broxbourne) who removed the sliding canopy sides in 1949 and flew it as an open-cockpit two-seater. The Mohawk's C of A was renewed for the last time on 2nd March 1949 and a photograph taken at about that time shows Neville Browning either offering up, or removing, a variable pitch propeller from the propeller shaft. Unfortunately no details of this propeller are known.

In October 1949, Bruno Pini and Neville Browning left England in the Mohawk to participate in the Oran International Rally in North Africa but on 1st January 1950, during the return journey, engine trouble forced them down in Spain. The pair returned to England leaving the Mohawk where it had landed, apparently with little or no visible damage. In February 1950, the Mohawk was apparently purchased by the Granada Aero Club, who considered 'rebuilding it' but nothing came of this although it was heard of in 1958 as 'awaiting restoration'.

In February and March 1976, *Aeroplane Monthly* published an article by John Grierson on Lindbergh and in the July 1976 issue, a letter from L S Casey, Washington, DC was published:

As a postscript to John Grierson's excellent article on Charles Lindbergh, I can confirm the rumour that Lindbergh's Miles Mohawk is alive (though not well) and restorable. When I talked with Mr Grierson it was rumoured that the aircraft was last recorded in Spain. Since that time, with the help of a number of very dedicated persons, the Mohawk has been saved and is now in custody of the Aeroflex Museum Foundation, Kenly, North Carolina.

This was followed by a letter from Jay Miller of Texas: *I was fortunate enough to have played a key role in the recovery of the Miles Mohawk. In December 1973 my good friend Connie Edwards (of Battle of Britain movie fame) was visiting associates at Tablada Air Base outside Seville, Spain. While there, Connie spotted an unusual aircraft in a junkyard nearby and took the time to photograph it. All was forgotten for a while, but upon returning and having the film processed, he forwarded the slides on to a mutual friend, who, in turn, forwarded them on to me.*

I had already concluded that the aeroplane in the photographs was a Miles product - exactly which, I did not know. Pulling Don Brown's outstanding 'Miles Aircraft since 1925' from the shelf, I started my research. It was not long before I knew what Connie had found. I immediately contacted Connie, who in turn called a friend of his at the Smithsonian. Shortly afterwards, my 'phone rang and the original slides were soon on their way to Louis Casey, Curator of the National Air & Space Museum.

Lou quickly came to the same conclusion - the aeroplane in the Spanish junkyard was the long-lost Mohawk. He proceeded to contact associates in Spain and here in America, and finally, in November 1975, the aircraft was brought here. The latter effort had been sponsored by Lou's good friend Dolph Overton - a renowned antique and vintage aeroplane collector in his own right. Today, the Mohawk is in storage with other Dolph Overton aircraft and artefacts in Kenly, N Carolina.

Restoration plans are presently being discussed, and it is assumed that the aeroplane will eventually be reassembled and rebuilt for display purposes. Its condition is poor. Much of the woodwork is rotted, and the fabric is beyond any hope, and the cockpit is almost non-existent. The engine, said by Lou not to be the original Menasco (though, it is a Menasco) is also in sad shape. The propeller, however, is immaculate. It seems that a local air base lounge operator removed it shortly after Connie's visit in 1973. It was later discovered by the Mohawk recoverers, hanging over the bar!

An undated and untitled page detailing aircraft stored and owned by the Museum at Orlando gives further interesting (and partially conflicting) information on what happened next:

Louis (Lew) S Casey, the curator of the Smithsonian National Air & Space Museum, Washington, DC started to search for the Mohawk in 1969 and after six years he eventually traced it to a scrapyard near Seville, Spain, where it had been sitting for some years. He purchased the remains, which were fairly complete, including the engine, the prop however was missing. He then had a stroke of luck, while arranging shipment to the USA, he struck up a conversation with one of the local flying fraternity and naturally mentioned his acquisition. To his surprise the man asked if he would like the propeller from the aircraft - and promptly had it removed from its resting place on the wall of the local flying club bar and presented it to him!

Lew had the aircraft shipped back to South Carolina where it was then stored at Santee with the Wheels and Wings collection and moved down to Orlando with them in 1978. The airframe is currently stored in the main museum building at Orlando (not on display) and is in a rather sorry condition. The fuselage is now in two pieces and the ply is damaged and de-laminating in places, - however, it is remarkably complete, even down to the cowlings and windscreen, so should pose no major problems in rebuilding. The fuselage carries no marks and the airframe is painted matt white on the upper surfaces (probably due to years of exposure in the sun - PA) and pale blue on the undersides. The engine has been overhauled by the Embry-Riddle Aeronautical University and is displayed in the museum. Some work has already been done on the tail-plane, but full rebuilding is likely to be done in Washington soon.

Lew Casey moved the airframe to his home in Fork Union, Virginia in 1976 with a view to continuing the restoration to static display condition but first he had to dig a basement under his house in order that he could commence work! Lew wrote to *Vintage Aircraft* in early 1977:

One of my colleagues brought your October-December '76 issue to my attention. The article on the Mohawk was of particular interest to me since I had a part in its recovery. What was not mentioned were the years spent in tracing the aircraft with the active interest of Gen Lindbergh. To my regret he died just before we were able to obtain it for restoration.

I am particularly interested in obtaining a photo of the instrument panel. I know what instruments were fitted and their manufacturer but not their locations on the panel. The second item of prime interest is a drawing or drawings of the wing structure, particularly

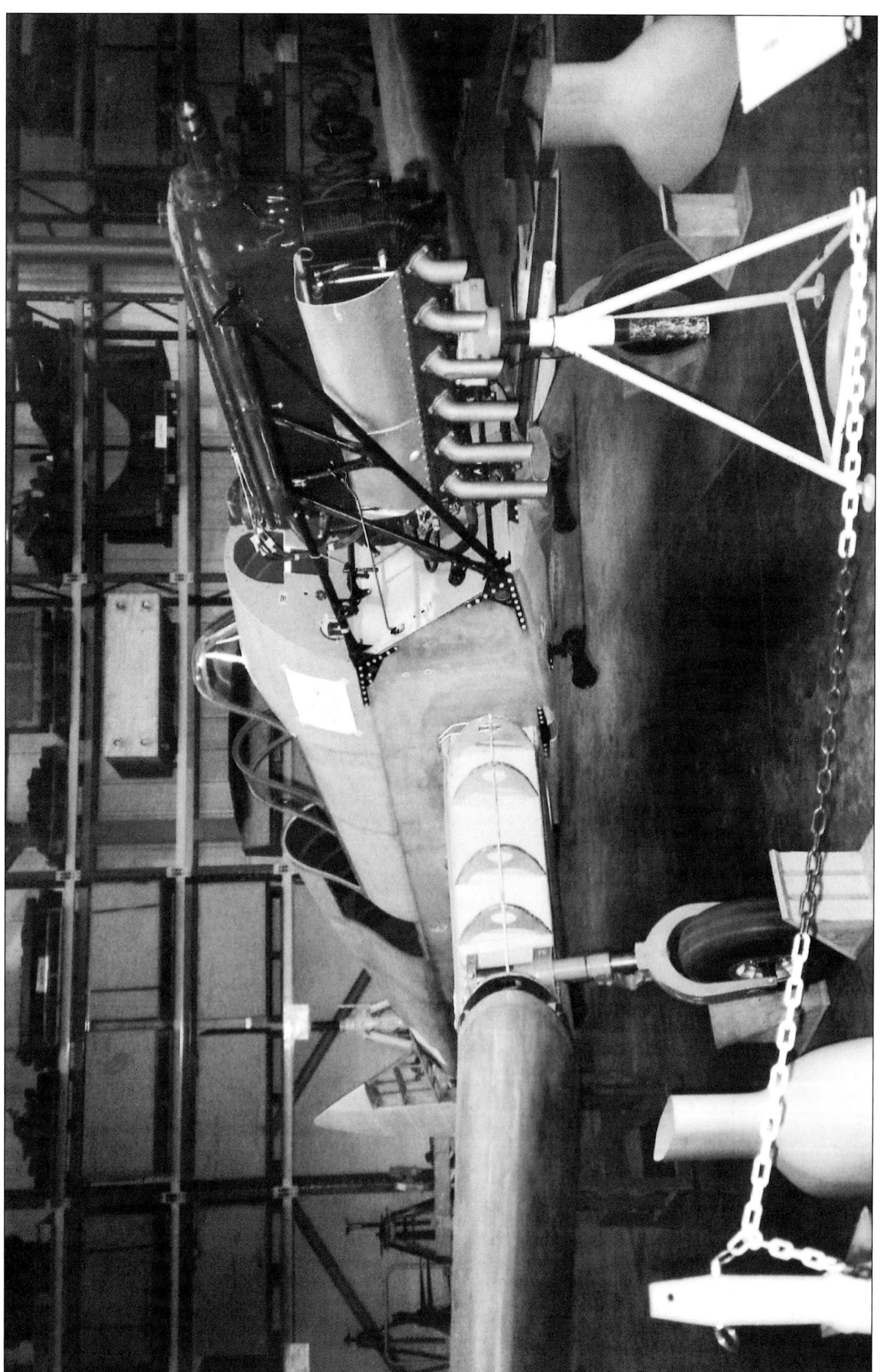

Miles M.12 Mohawk at Cosford Restoration Centre on 2nd March 2008. [T Bamford]

the spar construction and ribs. Since it was a one-off aircraft it used a wing of Miles Falcon configuration according to the CofA but Doug Bianchi feels that the Whitney Straight and Mohawk wings were identical. In answer to the engine question, the present engine is indeed the second engine. The first one, which gave so much trouble on the Lindbergh's Indian trip, is in the NASM collection. Etc, etc.

By 1984, Lew was slowly progressing with the rear fuselage but was still hampered by lack of drawings. However, The Miles Aircraft Collection, of which he later became a member, was able to help him in a small way with the positioning of the flap torque tubes through the fuselage.

He was also trying to locate a pair of Whitney Straight wheel spats or drawings for same for the main undercarriage, which were being sought to replace the originals, which had gone missing over the years.

On 5th March 1993, Lew wrote to Julian Temple with a progress report:

The Mohawk is progressing though slowly at this time of year because I am using epoxy adhesive which does not cure at low temperatures that prevail at this time of year. I have the fuselage joined together having removed structure in the rear cockpit area that was dry rotted. Side skin panels are in place to give rigidity to structure but top and bottom are open to allow a draftsman to gain access to make drawings.

Which brings to mind the second reason for writing this letter. From your experience cataloguing the Miles drawings do you think it possible that Falcon drawings exist that would give me a clue as to the flap installation. That area of the fuselage was powder but the metal fittings seem to be all there if I can locate the drawings of similar installation that would allow me to make the installation reasonably accurate. I believe I sent you a copy of the drawing list that your Civil Aviation Authority sent me with the Mohawk file.

Further on the plane, the right wing is coming along, most of nose ribs were missing but with the two or three still intact I have had a draftsman friend project the others. As a result that wing is nearing completion of the internal structure which required loosening almost all gussets, steel brushing old glue and re-gluing with epoxy. Very time consuming to say the least, much worse than building new but that's not the name of the game. In this case I am preserving the aircraft, not restoring for flight as some numb nut would, I'm sure, try to fly it some time in the future after it has left my personal control. To this end I am deliberately leaving out the bonding strips which will make it unairworthy without in any way changing the structure. I do have photos all of which are camera dated and form a historical sequence.

Rather surprisingly, it was always Lew's wish that when complete the Mohawk should return to England but following his decision to move to a smaller house, he decided that he had had enough. He had been working on the Mohawk in his spare time for many years but was not getting any younger and the project was still only about 2/5th's complete. So, in July 2000 he decided that he would like to donate it to the RAF Museum. They duly despatched Darren Hammond, one of their researchers, to Virginia to meet with Lew and to discuss the arrangements for the uncompleted project to be packaged and returned to England.

The Mohawk was shipped to England, where it arrived at Felixstowe Docks on 16th October 2000. It was then taken to the RAF Museum Reserve Collection and Restoration Centre at RAF Wyton, arriving on October 19th. On 30th March 2001, the fuselage and centre section was taken for restoration to Skysport Engineering Ltd at Rotary Farm, Beds, for fitting out the cockpit and attaching the undercarriage. It was next taken to the RAF Museum at Cosford on 3rd May 2002 for further restoration work but instead was put on static display in an uncompleted state due to shortage of funds. It returned to Skysport on 26th September 2004 where work on the fuselage was completed, the outer wings were restored and the ailerons recovered.

The 'Mills Majestic'! Could this have been the name given to the proposed production version of the Miles Mohawk? [P Amos]

Miles M.12 Mohawk with incorrect registration G-AEKN. ['Flight' via B Clarke]

On 16th February 2006 it was returned to Cosford for completion, assembly and repainting in its original colours and finally on 18th August 2008 it was taken by road to the RAF Museum at Hendon for display in the Milestones of Flight building.

Although the Mohawk was originally intended to have been a 'one-off' aircraft, the construction of a second airframe was actually commenced by Phillips & Powis for James 'Gordon' McArthur, who owned Sparrowhawk G-ADWW. However, construction was abandoned, possibly when McArthur decided to join the RAF full-time.

Production details

298 M.12 Mohawk (P&P works records state 'Conforms to M.7A Nighthawk, G-AEBP but fitted with 'Hawcon' type fuselage, modified top decking and cabin. Onions single strut u/c. Menasco B6S engine. Modified 'Hawk' type flying controls, wing tankage increased now two 28 gals in centre section and two 20 gals. in outer wings; Theed flap operation gear; AUW 2,700 lb. Constructed to Job No.15235 to the order of Col Charles A. Lindbergh, Long Barn, Weald, Sevenoaks, Kent'). Regd **G-AEKW** (CofR 7078) 14.7.36 to Phillips & Powis Aircraft Ltd (as nominee for Lindbergh, who could not, as a US citizen, have it regd in his own name). First flown, by Miles, as **U8**, 22.8.36. The regn was initially incorrectly applied as G-AEKN. CofA No.5775 issued 28.1.37; dd to Lindbergh 1.2.37. CofA lapsed 27.2.40. Regn cld 8.11.41 as sold. Impressed as **HM503** 31.10.41 and dd RAF Turnhouse 8.11.41. To 5 MU Kemble 2.4.42. To Miles Aircraft Ltd, Woodley 21.5.43 for fitment with Fairey-Reed propeller. To Maintenance Command Comm Sqdn, Andover 14.9.43. To 5 MU Kemble 1.2.44. Sold .46 to Southern Aircraft (Gatwick) Ltd. Regd **G-AEKW** 28.5.46 to Southern Aircraft (Gatwick) Ltd, Gatwick. CofA renewed 7.8.47. Regd 20.2.48 to Ernest G F Lyder, Bexley, Kent. Regd 28.5.48 to Bruno P Pini, Broxbourne. CofA lapsed 6.8.48. Sidescreens removed to make it open cockpit and CofA renewed 2.3.49. Forced landed in Spain 1.1.50 apparently with little damage. Regn cld 1.1.50 as destroyed. Sold 2.50 to Granada Aero Club. Not repaired or regd locally and stored at Tablada Air Base, Seville. Remains found in December 1973 and purchased by Louis S Casey, Washington, DC. Shipped 11.75 to Aeroflex Museum Foundation premises, Kenly, North Carolina; subsequently to Wheels & Wings collection, Santee, South Carolina. To Casey's home at Fork Union, Virginia in 1976 and under slow rebuild to static condition. Later donated to the RAF Museum.

301 M.12 Mohawk Regd **G-AEKX** (CofR 7079) 14.7.36 to James Henry Gordon McArthur, Bournemouth. Not completed. (P&P works records state 'conformed G-AEKW with the exception of the fuselage which was to be standard Hawcon'). Its wings, which had been completed, were subsequently fitted to the hybrid Miles M.7A Nighthawk c/n 286. The AUW was, like G-AEKW, to be 2,700 lb but it was also to be fitted with 75-gallon aluminium fuel tanks in the cabin which would have taken it to 3,200 lbs. Regn cld 12.37.

The only pre-war photograph to show the correct registration of G-AEKW in full on the Mohawk. [Via P Amos]

SPECIFICATION AND PERFORMANCE DATA

Engine:	200hp Menasco Buccaneer B6S-4 six cylinder, inverted, air-cooled, centrifugally supercharged, (200hp at 2,250 rpm at 4,500ft alt).
	The first Menasco engine was replaced by a Menasco of more advanced design, but exact model unknown
Dimensions:	span 35ft 0in; length 26ft 3in; height 6ft 8in; dihedral 5°; wing area 182.75 sq ft; incidence 1 deg 15 min; aspect ratio 6.7; wing loading 14.3lb/sq ft
Weights:	tare 1,605lb; AUW 2,700lb
Performance:	max speed initially quoted as being 'about 200 mph', later amended to 185 mph; cruising speed 170 mph; landing speed 44 mph; range 1,400 miles

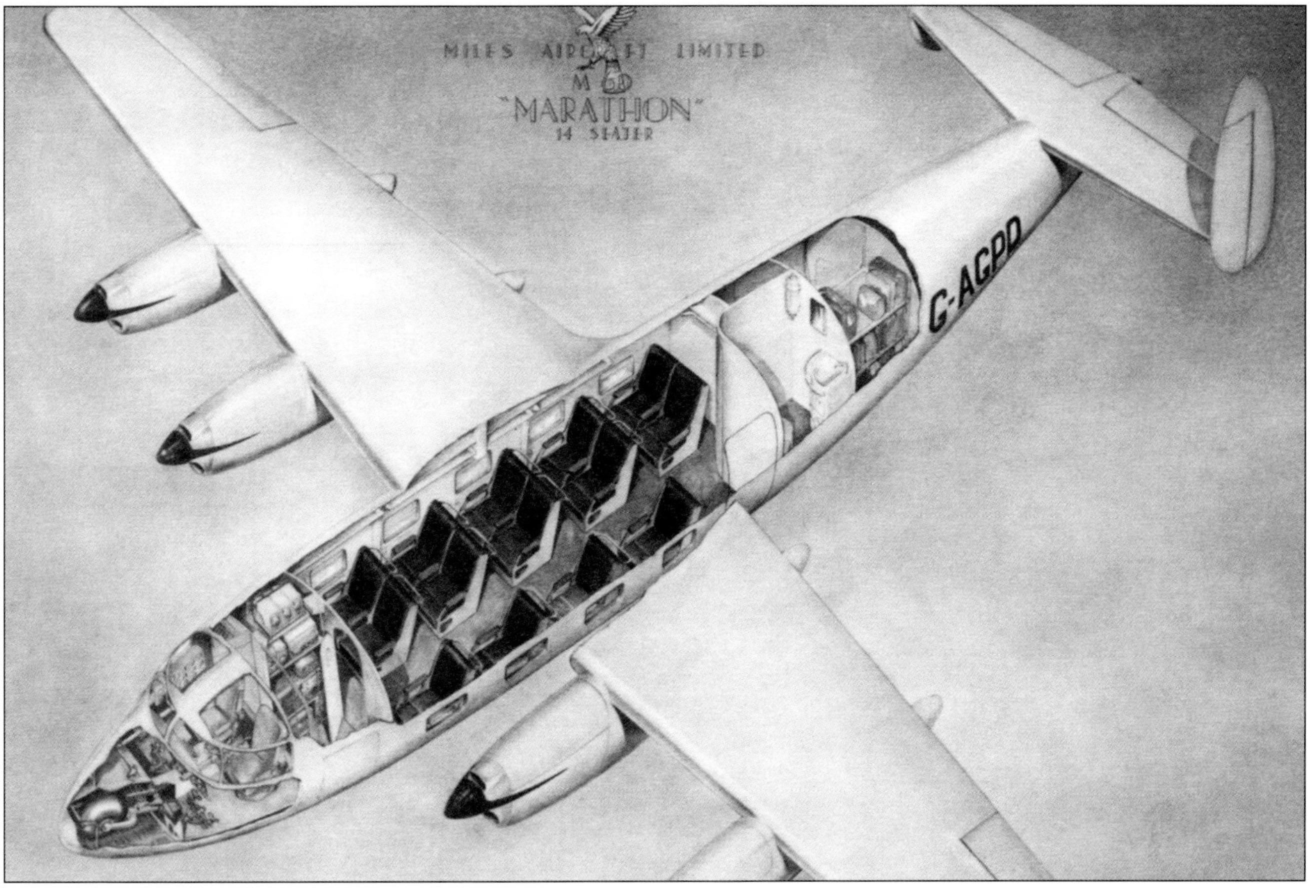

An artist's impression of the Miles M.60 Marathon of 1944, in 14-seat configuration. [Miles Aircraft Ltd via The Miles Aircraft Collection]

Chapter 25:
Miles Marathon Transport Project of 1936

Miles always prided himself in anticipating official requirements and towards the end of 1936, following the very successful trials of the Peregrine, he decided to submit a design to Imperial Airways Ltd for a feeder-line transport aeroplane, which he felt would soon be needed.

The projected design study for a four-engined all-metal monoplane to carry 12 passengers was to have a maximum cruising speed of 184 mph at 10,000 ft and a still air range of 700 miles.

The design project, which was never given a Miles type number, was named the Marathon and was Miles' first design for an all-metal aeroplane. It was intended as a replacement for the DH.86 biplane airliner, which carried ten passengers at a speed of 140 mph.

Although this, the first Miles Marathon, was not proceeded with, eight years later, in the summer of 1944, Miles Aircraft Ltd were invited to tender for the official Brabazon 5a Specification.

This called for a four-engined all-metal monoplane to carry 14 passengers, with a specified cruising speed of 175 mph at 10,000 ft and a still air range of 750 miles. Hardly surprisingly, Miles named his Brabazon offering the M.60 Marathon.

Unfortunately, no specification or anticipated performance details of the first Miles Marathon are known.

Right: A Phillips & Powis poster showing an artist's impression of the 'original' Miles Marathon.

A view of the Hobby being run-up, showing the wide track undercarriage. ['Flight' via B Clarke]

The Miles M.13 Hobby under construction. ['Flight' via B Clarke]

Chapter 26: Miles M.13 Hobby

Miles and Blossom decided to embark upon the design of a racing aircraft for Miles to fly in the 1937 King's Cup. This machine was to be powered by the new 140hp DH Gipsy Major Series II engine, fitted with a DH controllable-pitch propeller and was designated the M.13. It was designed to have a maximum speed of over 200 mph but, to achieve this performance with only 140hp, meant that it would have to be the smallest Miles aeroplane ever built. With a wing area of just 78 sq ft - less than half that of the Hawk - it certainly was! It was also the first Miles light aeroplane to be fitted with a retractable undercarriage.

So, Miles and Blossom, assisted by the young Chief Draughtsman Tommy Botting, lost no time in preparing the design as the race was due to be held just eight month's later. A name had also to be found for the aircraft and as all previous Miles aeroplanes had been named after birds of the Hawk species, it was decided that this aircraft should also be so named.

Miles had used practically all of the obvious Hawks so, one evening after dinner, a solemn conference was held and reference books were consulted. Suddenly came a roar of laughter from Miles who was looking at a book on Falconry. He had just read of a species of Falcon called the Hobby, which was so small that it could only eat butterflies and this limitation in its diet caused it to fly into such tempers that it choked itself! It was thought that Hobby would be just the name for the new machine.

The wings and fuselage were of spruce construction covered by a birch plywood skin and small split flaps were fitted, the whole to be covered by doped fabric with a high gloss finish to reduce skin friction. The construction of the aircraft was entrusted to Brian Swann, who did the metalwork, with Jack Sullivan, a young Irishman, being responsible for the woodwork.

Don Brown (who was not working for the firm at the time and regrettably made assumptions, some of which have, unfortunately, not always proved to be correct) subsequently wrote: *In those happy days before the war when flying was a sport and not a state controlled business, the enthusiasts who entered for the King's Cup Air Race almost invariably left it until the last minute to prepare their machines for the race. Why this should have been so is not clear but nevertheless almost everyone was like that and it was no uncommon sight to see a number of competitors feverishly finishing off their machines on the starting line. If all went well one used to put in the last split pin, hop into the cockpit and start the engine just as the starter's flag fell.*

In the case of the Hobby however, Miles had a spot of bad luck which was just sufficient to prevent the aircraft from flying in the race. The Hobby was the first light aeroplane which he had built with a retractable undercarriage and partly due to the usual shortness of time and partly due to the fact that the design of a retractable undercarriage is to some extent a specialist job, the design and construction of the undercarriage was 'farmed out' to a specialist firm. Such an arrangement naturally involves close liaison work between the two design staffs, in that the inwards retracting undercarriage, when finished must obviously retract into the holes in the wing provided for it.

However, about two days before the race, which was due to take place on 10th September 1937, the Hobby was ready for assembly and then the ghastly truth became evident - the undercarriage would not retract into the holes provided for it. It was too late for recriminations or for arguing about whose fault it was. Only one thing was possible - and that was virtually impossible - to redesign and rebuild the entire wing within 48 hours. The problem had been caused by virtue of the fact that due allowance had not been made

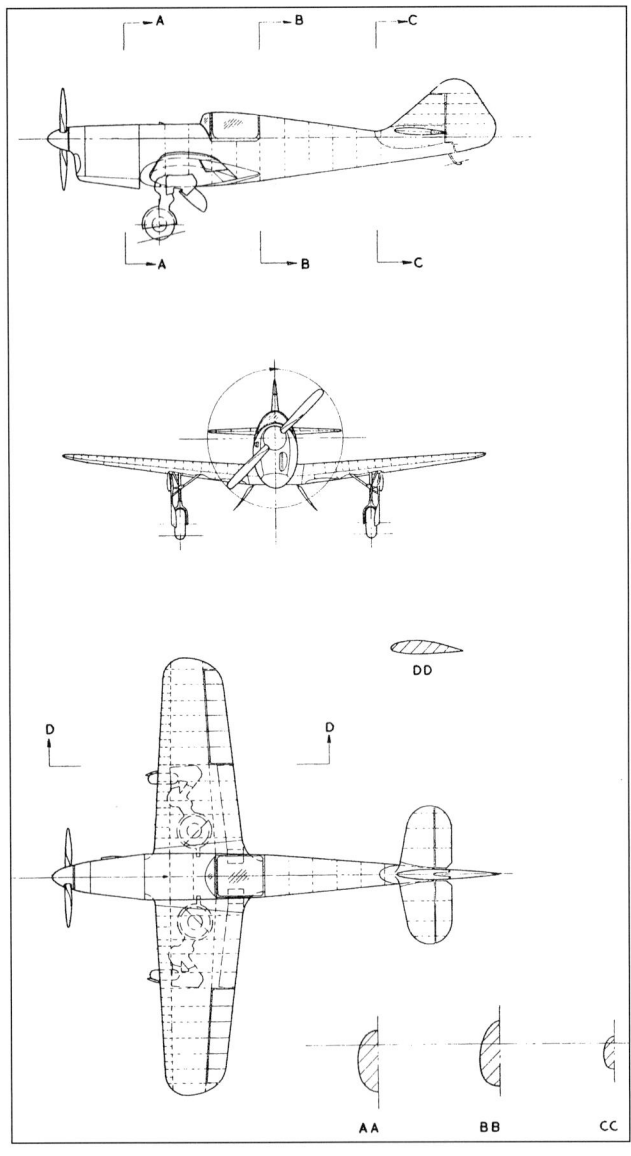

The Miles M.13 Hobby.

for the undercarriage extension once the weight had come off and the aircraft had become airborne. This had in turn led to the wheel wells being designed and built in the wrong position.

The team worked like Trojans non-stop for those two days and nights, but they just failed to make it in time for the race.

In fact, a photograph taken at the time (seen below) shows no less than 17 people working on the aeroplane *at one time*!

Miles carried out the engine runs and taxi-ing trials in the evening on 3rd September, with the aid of car headlamps, as the aerodrome was by then in darkness, and on 4th September 1937, Miles took the Hobby off on its first flight.

The Hobby was given the 'B' mark **U2**, but when Miles took it off on its first flight, it was still in primer with no identity marks. But Don Brown's analysis of the undercarriage issue was not, apparently, correct, as KP Godfrey in a letter in the July 1978 issue of *Aeroplane Monthly*, recalled:

Having read the article on the Miles M.13 Hobby in your February 1978 issue, may I take this opportunity to correct a few points? The 'young draughtsman named Tommy Botting', young though he may have been, was Chief Draughtsman at the time, having joined the Miles team as early as 1934. The disaster with the undercarriage was in the basic design, not in its non-retraction into the hole in the wing. The retraction tests were carried out in the hangar in the usual way, prior to roll-out. The problem was discovered on first taxiing trials, carried out at midnight using car headlights to mark out available aerodrome length.

The late F G Miles was at the controls, and after the first run came back complaining about the harshness of the telescopic shock absorber struts. The problem was clearly seen on the piston tubes, where pressure marks in the fully extended area indicated that all was not well. The rake forward angle of the strut was too great. The overnight cure to this was to simply reduce the rake angle. This done, the aircraft was again put through taxiing trials and flown - with wheels locked down, of course.

The hard work then came my way, as I was one of the young enthusiastic draughtsmen who had to alter the wing to suit the modified undercarriage geometry. I consider myself very fortunate to have worked as a draughtsman for such a great pioneer as the late F G Miles.

Seventeen 'workers' trying to complete the Hobby in time for the 1937 King's Cup Air Race! [Phillips & Powis via The Miles Aircraft Collection]

Chapter 26: Miles M.13 Hobby

Miles running up the Hobby. ['Flight' via B Clarke]

Miles takes the Hobby off on probably its first flight, with the newly-completed Falcon Hotel in the background. ['Flight' via B Clarke]

Although the Hobby proved to be fast and comfortable, there was just not enough time to take it to the A&AEE Martlesham Heath to obtain a Certificate of Airworthiness. Nevertheless, painted red, it was flown to Hatfield for the start of the race but having missed the official deadline for participants was unable to compete, much to the disappointment of Miles and his team.

C L Nash recorded that the Hobby was given the c/n 1-Y, although this has been interpreted by others as 1.7 or 1-7. The writing in the ARB ledger is open to interpretation but the reason for this 'one-off' c/n has never been established. The original application for a C of A had been made on 22nd January 1937 and the Hobby was registered **G-AFAW** (CofR 8027) to Frederick George Miles on 26th July 1937. It was shown at an air display held at Woodley in the spring of 1938 but the registration was never used and was cancelled in April 1938, with the application for C of A being cancelled on 15th May 1938.

An extract from a Phillips & Powis Research Report, believed to have been issued in December 1937, gives some interesting information on the use of the wing selected for use on the Hobby:

In the USA very extensive use has been made of the NACA 230 wing. This wing is used on the Miles Kestrel Trainer and on the Miles Hobby. The latter is being used at the moment as an experimental machine. Although from every aerodynamic point of view the section is desirable, it was found that the stall of the 230 wing was rather sharp, and it was necessary for us to find out the effect of wing tip wash-out. It is from this point of view that experiments are being carried out.

The Hobby wing had a wash-out of approximately 2 degrees at the tip. The wing was very sharp on the stall. This was considered undesirable and the wing at present is being fitted with a tip which gives no wash-out. The machine has not yet been flown with this wing. We are making completely new wings of normal Hawk section, in order to get a comparative check on handling characteristics and performance.

There was a very sharp stall on the Hobby wing and the knowledge we obtained of how to check and improve this may allow us to use a wing of the NACA 230 section on the T.1/37, which is a Trainer. It may be possible to make further application of this wing section later when we have found out all that it is necessary to know from the Hobby experiments.

In the Miles Kestrel Trainer it was found that the aircraft tended to yaw and hunt slightly when cruising. According to an American report improvements could be made by using a special aerofoil

The Hobby in the RAE wind tunnel. [The A J Jackson Collection]

A photograph taken in 1938 showing the Hobby in the Experimental Department with the wing covering stripped off. Note the retracted undercarriage . . . the cause of the earlier problem. [Via P Amos]

The Hobby in its red finish with 'B' mark U2 and no undercarriage fairings. [Via T R Judge]

section in the tail unit. A similar section was used on the Hobby, which had a very thick tailplane and rudder. The 'weaving' effect was eliminated and the new section is being used on the T.1/37.

In May 1938, 'One Hobbyhawk' was purchased from Phillips & Powis Aircraft Ltd by the Air Ministry for use by RAE Farnborough against Contract No.746818/38, with a projected cost of £2,250, and with the Instruction to Proceed being dated 11th April 1938. The RAE were eager to obtain the Hobby on account of its very small size, since this would enable them to erect it bodily inside their largest 24ft wind tunnel, and give them the unique opportunity of comparing the information gained from wind tunnel tests with the actual known performance of the same machine in flight. C L Nash recorded that the Hobby was delivered to Farnborough on 6th July 1938, still in its red finish and as U2. The Farnborough log, however, records that it was flown there on 13th July 1938 by Flt Lt Burke as **L9706**, painted silver.

The Miles M.13 Hobby, seen at Farnborough in full RAF regalia, after the wind tunnel tests. Note the patch covering 'U2' beneath serial L9706 and the Hawcon parked behind. [Via P Jarrett]

It was later flown at Farnborough by Burke (30 min) and Pebody (15 min) on July 18th; by Clouston (30 min) on August 8th; by Pebody (25 min) on 20th May 1939; by Wilson (40 min) on May 25th and by Pebody again on May 23rd and 24th (25 min and 50 min).

Much valuable information was obtained from the tests, which were also aimed at analysing the drag, thereby enabling it to be reduced by design improvements. The tests comprised positioning the Hobby at the incidence of minimum drag and progressively sealing all the leaks, measuring the drag at each step. In addition, engine cooling drag and cooling flow were measured as well as the drag at the junction of the wing root and fuselage. Tests were also made with the aeroplane completely sealed and the tail unit removed in order to estimate the drag of the empennage. While the aeroplane as flown produced 32.2lb drag at 100ft/sec, 2.7lb was saved with all the leaks sealed. However, 1.9lb of this arose from the engine cowling, which could hardly be effectively sealed in practice owing to the need for easy removal to facilitate engine access. It was suggested that the engine's internal cooling drag might be reduced by fitting a louvre on the exit to control the flow. In the process of fairing and sealing the aeroplane, no major source of drag was discovered and flow over the wing root junction was found to be smooth with apparently little room for improvement.

It was, however, recommended that further tests be carried out, including flight tests at maximum speed to check overall minimum drag, and to look into ways of 'seeing if the tail could be cleaned up'. These were probably in connection with the following flights, which were conducted mainly by Pebody – 6th June 1939 (1 hr 40 min), June 7th (30 min), June 8th (30 min – 'speed course') and June 10th (1 hr 10 min). Heycock also flew it for 5 mins on June 6th. Although the maximum design speed was 207 mph, the RAE found that the maximum speed obtainable was only 196 mph but no reason for this deterioration was given. From a handling point of view, the Hobby was not one of the best Miles aeroplanes, as it had a tendency to drop a wing sharply at the stall and the landing speed was inclined to be high, due to its relatively high wing loading and small flaps.

The Hobby was SOC on 24th July 1939 and was acquired from Farnborough by Hugh Kennedy early in the war. Taken to Woodley by road, it was stored in one of the two Robin hangars on the Davis Farm site used by the Miles Technical School, where it was seen by Grahame Gates in about January 1943. Grahame was one of a group of trainees who were clearing out these hangars and at that time it was minus its engine and was standing on its firewall with the tail tied to a roof support beam. Grahame recalled being told that it had travelled that way on a Queen Mary aircraft transporter from Farnborough by night, blacking-out a large area of Berkshire on the way when its tail hit an overhead power line! Why Hugh Kennedy acquired the Hobby is not clear, but Don Brown recalled that he had bought it for its engine. It is just possible that he wanted to restore it after the war, but it is not known if it was actually sold with its engine or even if its documentation was in fact still in existence. It is believed that the Hobby was later taken to the Repair & Service Department where, along with other various prototypes, it was subsequently broken up.

SPECIFICATION AND PERFORMANCE DATA

Engine:	140hp DH Gipsy Major srs II, fitted with a DH controllable pitch propeller
Dimensions:	span 21ft 5in; length 22ft 8in; wing area 78 sq ft; aspect ratio 5.78; u/c track 9ft 6in; dihedral 6°; wing section NACA 230
Weights:	empty 1,140lb; AUW 1,527lb; wing loading 19.58lb/sq ft
Performance:	max design speed 207 mph

The hammer falls as Blossom christens the first Miles M.14 Magister on 20th March 1937. ['Flight' via B Clarke]

Chapter 27: Miles M.14 Magister, M.14A Magister Mk.I, M.14B Magister Mk.II, M.14 Hawk Trainer Mk.III, M.14A Hawk Trainer Mk.III and M.14B Hawk Trainer Mk.II

In 1936, as a result of the highly satisfactory results being obtained with the Hawk Trainers being used by the Reserve Flying Training School, Miles decided to design an elementary trainer development of it for the RAF. This decision was to be of considerable significance as the new aeroplane was ultimately to become i) the first low-wing monoplane in the history of the Service to be adopted as a trainer ii) the first service aircraft to be fitted with split trailing-edge flaps iii) a radical departure from the policy which had been laid down by the Ministry in that only metal aircraft would henceforth be accepted for Service use and iv) the first service aeroplane in which magnesium alloy castings were to be used for stressed parts.

The AM Specification T.40/36, dated 28.10.36, called for an aircraft *'to meet Operational Requirement OR.44 and to be constructed in accordance with the drawings and schedules covering the design and construction of the standard civil Hawk aeroplane G-ADYZ, except as modified by other requirements in this specification'.* The specification also called for the aircraft *'to be manufactured upon jigs which will ensure their interchangeability and that spares will fit without adjustment. A set of master parts will be held by the Air Ministry as standards for interchangeability'.*

The special requirements called for included:-
i) The rudder bar to be raised by 1.25 inches
ii) The tail skid to be replaced by a tail wheel
iii) A strong point to be provided in the fuselage to protect the occupants in the event of the aeroplane overturning on the ground
iv) The single leg design of undercarriage to be fitted
v) Duplicate controls to be fitted for trailing edge flaps, elevator trimmer flap and mixture control
vi) The Bendix floor-mounted unit for differential control of brakes to be fitted
vii) A fixed trimming device to be fitted to the rudder
viii) The elevator chord to be increased by one inch
ix) Sutton Harness to be fitted in each cockpit
x) Dashboard instruments to be repositioned as in the mock-up and a clock provided
xi) The standard grey-green finish to be used for the cockpits; a yellow external finish is required
xii) Exhaust manifolds to be fitted

The specification also demanded that *'provision was to be made for two crew with parachutes and that the aeroplane, fully loaded,*

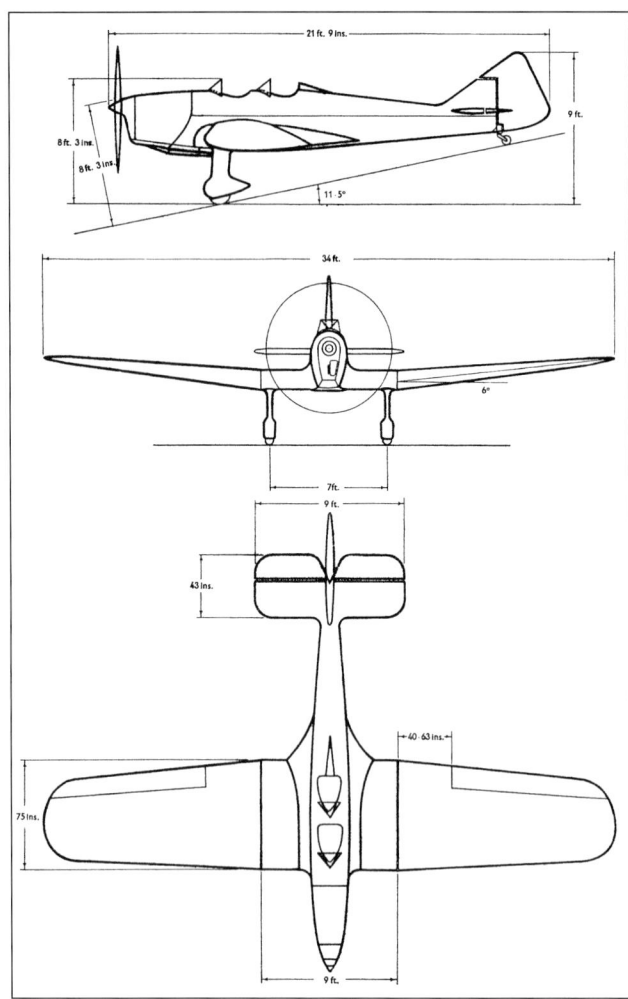

A three-view drawing of the Miles M.14 'Hawk' Trainer taken from a Phillips & Powis brochure.

shall be able to withstand an impact with the ground at a vertical velocity of 11 ft/sec. The torsional and flexural stiffness of the wings shall be satisfactory under the worst conditions likely to be encountered. Stiffness measurements will be made'.

The new machine differed externally from the M.2W Hawk Trainer Mk.II principally in having the wingspan reduced by 2 inches, a spatted undercarriage, a tailwheel in place of a skid and a slightly different profiled rudder. Although not mentioned in the special requirements listed above, provision was also made for full blind-flying equipment. The metal engine cowling was originally hinged along the centreline on the top of the fuselage, but this was somewhat unwieldy and prone to damage in strong winds, so it was later modified to have two separate removable side panels. The original solid-tyred tailwheel was mounted on the rear of the rudder stern post, but this was later replaced by a pneumatic tyre and also moved forward of the rear fuselage frame.

A Magister 'used for mock-up only' was built in 1936, with c/n 329, and the first production batch of Magister Mk.I's for the RAF was ordered to AM Spec. 40/36P (later to Spec. 37/37P, which was even later superseded by Spec. Magister I/P2). The Instruction to Proceed, dated 4th November 1936, to Contract No.568682/36, was issued to Phillips & Powis Aircraft Ltd, 'for the manufacture of 90 civil Hawk aircraft', with a projected cost of £72,000.

If only the Air Ministry had then had the courage to have ordered the Magister in substantial quantities, instead of in 'dribs and drabs' as was to be the case, the production costs could have been greatly reduced and larger numbers produced in a shorter time. As it turned out, Magister production for the RAF was then ordered as follows: Contract No.602402/37 (26 aircraft) undated, Contract No.706823/37 (214 aircraft) dated 23.12.37; Contract No.778435/38 (254 aircraft), ITP dated 25.5.38; Contract No.778435/38 (100 aircraft), ITP dated 12.11.38; Contract No.778435/38 (100 aircraft), ITP dated 10.2.39; Contract No.778435/38 (145 aircraft), ITP dated 8.5.39; Contract No.B53337/39 (300 aircraft), ITP dated 9.2.40, making a total of 1,229 aircraft. This total however, inevitably, does not reflect the actual number of Magisters delivered to the RAF as some were diverted to Eire and Egypt.

In fact, I find it ironic that the last *monoplane* Magister elementary trainer was delivered to the RAF in January 1941, while the obsolescent *biplane* DH Tiger Moth remained in production until almost the end of the war.

Unfortunately, the original contract, the largest ever received by the firm, taxed the firm's manufacturing resources to the full, necessitating some considerable additions to the factory and causing any thought of putting the twin-engined Miles Peregrine into production to be abandoned.

Concurrent with the Magister on the assembly line was the civil Hawk Trainer Mk.III. This was also planned to be jig built and included a number of features not listed in the specification, but which were also fitted to the Magister. These included a blind flying hood to the rear cockpit, a more robust engine cowling, increased rudder movement 'ensuring a more rapid response and facilitating immediate recovery from spinning', increased aerobatic AUW to permit advanced aerobatics with the aircraft fully loaded, and the top-decking aft of the pilot especially strengthened to protect the pilot in the event of the aeroplane overturning on landing.

The first M.14 Hawk Trainer Mk.III (c/n 331) was registered G-AETJ, although not taken up, on 15th February 1937 painted in red primer, with 'B' mark U2 and RAF roundels (the latter probably for publicity purposes) and was formally christened by Blossom Miles on 20th March 1937, following which it was first flown, by Miles, before some 800 employees. Photographs of U2 taken at the christening ceremony show it to have the following minor differences from the later RAF Magister: i) longer u/c spats, ii) pitot under the wing and iii) a 'flat' intake on the rear lower side of the starboard engine cowling.

The Aeroplane for 24th March 1937 reported on the occasion under the heading: *All Keyed-up*

Mrs F G Miles formally inaugurated the first of the new production Miles Hawk Trainers, to be known in future as the Miles Magister, on March 20th, at the Reading works of Phillips & Powis Aircraft Ltd, by exploding a bottle of the best over its nose cowling in the heartiest style. Mr Miles took it up immediately afterwards to prove that it would fly, about which there seemed to be no doubt whatever; and a flight of three trainers, flown by Flt Lt Moir and Messrs Skinner and Roxburgh, showed that its immediate predecessors were also fast and handy.

The occasion was essentially a family party, and some 800 employees dealt faithfully with eight barrels of beer in celebration of an exacting job well launched. The tooling-up of the Magister has been particularly thoroughly and carefully done at great cost of time, trouble and money, but the reward in output and accuracy of standardisation justifies the spirit of artistic joy which permeated the main hangar and found vent in song.

And today, 70 years on, it is pleasing to note that there are still three airworthy Maggies in this country giving equal joy to the pilots who fly them and to the people who see them perform regularly.

Aeropilot for April 1937 saw it slightly differently and their article: *Magister Makes its Bow* also makes for interesting reading:

Free beer to aircraft builders! And then they say there's a shortage of them. Anyway, that is what happened at the initiation of a Miles Hawk Mark III trainer monoplane at Reading aerodrome, where Mr F G Miles, the designer, demonstrated the first of the large batch of these machines ordered by the Air Ministry for training RAF pilots. 'Magister' is the type name given to the machine.

A thousand of the workers employed on building these machines and visitors who were present at the demonstration were all given refreshment in the traditional British style. The Magister, a shapely low-wing monoplane with open cockpits for two, was planned to meet the requirements of a Service in which the monoplane is rapidly displacing the biplane. Elimination of bracing struts and wires, and a general 'cleaning up' of design, with adoption of cantilever monoplanes, has caused important changes in flying characteristics, which in turn have demanded further scientific aids so that pilots may be able to handle their craft with the ease and safety of the older biplanes machines.

For example, the better streamlining of the cantilever monoplane means that drag, or head resistance, is less. Speed is thereby increased but at the same time the lesser drag produces a more "critical" gliding angle. Movable flaps on the trailing edges of the wings, and other aerodynamical devices, provide an antidote, increasing the drag at the will of the pilot and giving added 'lift' which permits landings to be made at speeds much lower than would otherwise be possible.

The machine is built almost entirely of wood, but is extremely robust, with ample safety factors to meet the stresses of aerobatic flying. The wings have a covering of plywood, and their safety factor is such that the covering can be practically destroyed without the risk of failure of the wing structure as a whole. The two cockpits are wide, with comfortable parachute seats to which aerobatic harness is attached. Both cockpits are equipped with three-piece windshields of safety-glass, and power is derived from a de Havilland Gipsy Major 130hp engine.

The new aeroplane was officially named the Miles Magister in April 1937 but its Certificate of Airworthiness was only issued in July 1937 and, unexpectedly, to the Royal New Zealand Air Force. U2 had in fact been sold to John Gamble of Wellington, Phillips & Powis's agent in New Zealand as ZK-AEX. Delivered on 10th June 1937, it was registered to the Auckland Aero Club, but, unfortunately, was to crash on take-off on Great Barrier Island in March 1939.

While the flight testing of the Hurricane and Spitfire was being undertaken at the A&AEE Martlesham Heath, Sgt 'Sammy' Wroath, one of the three 'A' Flight test pilots there, was given the task of testing what the contemporary report chose to call 'the prototype for the new Miles Magister' (but this was, in fact, G-AENT, the M.2Y Hawk Trainer Mk.II).

Up to this time, neither the Hawk nor the Hawk Major had experienced any problems with recovery from spins but the M.2W, M.2X and M.2Y Hawk Trainer Mk.IIs differed from their predecessors by having larger cockpits similar to the later Magister, in order to enable the crew to gain easier access while wearing parachutes. But no trouble had been experienced with spin recovery with the Hawk Trainers used by the Phillips & Powis Reserve School, whose crews also wore parachutes.

Therefore, when Sgt Wroath took off on his first spinning sortie on 19th December 1936, with the aeroplane loaded to the maximum all-up weight and with the c.g. 10% aft of basic, he envisaged no undue problems with the tests. At 8,000ft he put it into a left hand spin and recovered after the three turns, usual for the initial test. But from a right hand spin the aeroplane jibbed a bit before recovering, so Sammy climbed up and repeated it, this time being meticulous to apply full right rudder and to get the stick right back. After three turns recovery action produced no effect; after more turns Sammy was getting nowhere - the spin just went on and on. In the end, the green fields started to look awfully large. He undid his straps and

The 'Daily Sketch' of 22nd March 1937 had as its caption to a similar photo: 'The Flying Golliwog – Mr Miles, the aeroplane designer, fixing his mascot to a new Miles-Hawk monoplane after his wife had performed the naming ceremony at Reading Aerodrome'.
['Flight' via B Clarke]

'Flight' for 25th March 1937 used the following caption for this photograph: 'Magister Artium: Mr F G Miles, past-master of the art of aircraft demonstration, reveals the qualities of the Miles Magister Trainer (DH Gipsy Major) at a "family gathering" of Phillips & Powis employees last week, when the prototype was christened with due ceremony by Mrs F G Miles. The Magister is basically similar to the Hawk Trainer, but has been revised in detail. Note, for example, the new undercarriage and the tailwheel.'
['Flight' via B Clarke]

L5912, the first true Miles M.14 Magister for the RAF. ['Flight' via B Clarke]

had got his right hand over the high cockpit side; however, the rather bulky 'A&AEE Spin Recorder' strapped on to his left leg got caught beneath the dashboard. So he had to get back into the cockpit, slide the spin recorder down his left leg, and climb out again. He did not have long in his parachute - time to avoid landing in the River Debden, and barely time to avoid a substantial wood before landing very heavily on his back - fortunately in a ploughed field - while the pilotless aeroplane went on to crash into a school playground.

At the 'wash-up' in the Air Ministry, Sammy felt himself a lonely figure at the end of the table, with Miles reporting that they had carried out many spins without difficulty. It was decided that another prototype was to be prepared and flown to Martlesham. There, Flt Lt E W Simonds, one of the other 'A' Flight pilots, repeated the right hand spin in the first production RAF aircraft L5912 on 22nd July 1937. This too failed to recover and he also had to bale out, but sadly he came down in the water and drowned, while the Magister crashed into the River Debden near Felixstowe.

The second enquiry revealed doubts about the exact c.g. at which Miles' spinning tests had been carried out, and they were directed to prepare another aeroplane and demonstrate the spin at Martlesham. Five days later a company test pilot flying the M.14 Hawk Trainer Mk.III, G-AEZS, loaded to represent L5912 and fitted with a primitive spin-recovery tail parachute, put the aeroplane into a right hand spin, but he also failed to recover after 30 turns. This time the aircraft was saved by the tail-parachute, but the firm was very mystified by these accidents, because the Magister and its predecessors had been repeatedly spun, under, as they thought, every conceivable condition and no difficulty had ever been experienced in recovering from the spin.

A thorough programme of investigation was at once initiated, both with flight trials and wind-tunnel tests and a significant feature revealed by the flight trials was that, if the engine happened to stop either during the spin or on entry, recovery became impossible. The wind-tunnel tests conducted at Farnborough also indicated that the spinning characteristics would be greatly improved if the tailplane were to be raised by 6in, flattening the top of the rear fuselage and fitting 'fillets' (anti-spin strakes) ahead of the tailplane on top of the fuselage. These modifications were incorporated on the 22nd production aircraft, L5933, in October 1937 (which was also fitted with a tail parachute at the base of the fin post) and 46 sets of spins were then successfully carried out by A&AEE Martlesham Heath.

Miles M.14A Magister, L5933, was the first to have the deepened rear fuselage and anti-spin strakes in October 1937.
[Phillips & Powis via The Miles Aircraft Collection]

This modification resulted in the change of type designation to M.14A. During the course of these trials a new standard spin recovery technique was also drawn up and this involved the use of full rudder to oppose the direction of the spin (previously centring the rudder was recommended) and then centring the stick. A standard, fully modified aircraft, L6905 was then flown with satisfactory results.

By the time that L5945 emerged, the rear windscreen had been changed from a three-piece, similar to that fitted to the front cockpit, to a single wrap-round piece of Perspex, in order to obviate excessive draughts being experienced by the rear occupant. However, it is not known on which aircraft this modification was first introduced. Among the 'Special Requirements' to be incorporated in 'Specification No. Magister I/P.3 for the Manufacture of Magister Aircraft, dated 25th July 1938, were 16 modifications, including the more visible Mod. No.33 'to provide a larger rear windscreen' and Mod. No.43 'to fit a pneumatic tail wheel'.

The late Air Commodore A W Mermagen and I entered into correspondence in the late 1990s with regard to his experiences with the early Magisters while he was on the staff of the CFS at Upavon in 1937 and he recalled:

I took delivery of a Miles Magister, L5913 on 14th June 1937 for the purpose of assessing its suitability for the inverted flying formation taking place at the RAF Pageant at RAF Hendon.

I had been given the job of leading a three aircraft formation (inverted) and had started practising with Avro Tutors. At the time it had been suggested that the new Miles Magister might be suitable for this rather difficult task. I flew the Magister and gave it a thorough testing - it aerobatted easily, loops, half roll off loops and rolls but we found it difficult to formate with because of lack of drag (being a monoplane in the inverted position). The Tutor, a biplane was easy to handle inverted and to keep close formation. The carburettor fitted to the Magister, I remember as being fully efficient but not so reliable as the Tutors', which had been subjected to a lot of 'g' testing. We therefore, were forced to abandon the Magister in favour of the Tutor and the formation was very successfully displayed on the day - RAF Pageant, 26th June 1937. The three of us who participated, self, F/L G Stephenson and Sgt Colin Scragg (later to reach Air Vice Marshall rank) were a little sad that we couldn't use the Magister.

These Magisters were painted in the red and white 'sunburst' stripes similar to the Avro Tutors of the RAF Display Team and had been modified for inverted flying in readiness for the display but were presumably repainted later in standard RAF training colours.

After the Magister had been in service for a while, spinning problems again manifested themselves and, following a spate of accidents, Vernon Brown, the Chief Inspector of Accidents, Air Ministry, issued a Report, dated 22nd June 1938 entitled: Summary of Accidents to Magister Aircraft (Ref: AVIA 5/42 95481).

Since 24th January 1938, there have been eight serious accidents involving Magister aircraft of the modified type. To these must be added an accident, with fatal consequences, involving a Magister of the unmodified type which was being tested at the Aeroplane and Armament Experimental Establishment on 22nd July 1937 (L5912 referred to above - PA). All but one of these accidents occurred during 'ab initio' training at Elementary and Reserve Flying Training Schools; four involved candidates for Royal Air Force service, three involved members of the Royal Air Force Volunteer Reserve whilst the remaining accident caused the death of two young officers of the regular Royal Air Force.

In only two of the accidents did the occupants of the aircraft escape by parachute. The accidents were as follows:-

(1) Magister L6901 (from Hamble) off Calshot on 24th January 1938.

(2) Magister L6895 (from Hamble) at Sarisbury on 25th January 1938.

(3) Magister L6906 (from Hamble) at Bursledon on 16th March 1938.

(4) Magister L6916 (from Reading) at Sonning Common on 28th March 1938.

(5) Magister L5972 (from Hamble) at Locksheath on 7th May 1938.

(6) Magister L5948 (from Royal Air Force, Wyton) at Wyton on 10th May 1938.

(7) Magister L6894 (from Hamble) at Bursledon on 13th June 1938.

(8) Magister L6910 (from Brough) at Brough on 15th June 1938.

A brief description of each of these accidents follows:-

(1) The pilot was undergoing Volunteer Reserve 'ab initio' flying training and had completed 27¼ hours dual and 24 hours 10 mins solo flying, all in Magisters. After 40 mins dual instruction in slow rolls to the right he was sent up in the same aircraft to practise this manoeuvre and to carry out spins to right and left. Some 20 minutes after taking off the aeroplane was seen to dive at high speed from about 4,000 feet to 1,000 feet when it was almost pulled out of the dive. Instead of recovering, however, it continued in a steep diving turn to the left until it struck the water. The pilot was killed.

(2) The pilot was undergoing an 'ab initio' course for the Royal Air Force and had completed 25 hours dual and about the same amount solo flying, all except 1 hour 50 minutes being carried out in Magisters. The pilot was sent up to practise acrobatic manoeuvres. After 25 minutes flight he executed a voluntary right hand spin from a height of about 4,000 feet. When the spin had continued for about 7 turns the pupil attempted to recover but was unsuccessful and he resorted to his parachute at a low altitude and escaped uninjured.

(3) The pilot was a Volunteer Reserve 'ab initio' pupil who had completed 2 hours 20 mins dual in a [Avro] Cadet and 9¼ hours dual and 1¼ hours solo in Magisters. After 30 minutes dual instruction the pupil was ordered to carry out one circuit solo and land. The pilot allowed his aeroplane to enter a thick cloud in which he lost control and resorted to his parachute. He landed safely.

(4) The pilot was a regular airman (Air Gunner), with 200 hours air experience, under 'ab initio' training as an airman pilot. At the time of the accident he had completed 12 hours 50 mins dual and 1 hour 35 mins solo all in Magister aeroplanes. The pupil was detailed to practise medium turns and sideslipping at 2,000 feet for 40 minutes. About 40 minutes after taking off the aeroplane was seen to descend in a steep left hand spiral dive from a height of about 1,500 feet. This continued to the ground. The pilot was instantly killed.

(5) The pilot was undergoing a Volunteer Reserve 'ab initio' course of flying training and had completed 11 hours dual and 7 hours 20 mins solo flying, all in Magisters. After 20 minutes dual instruction in steep turns the pupil was sent off in the same aeroplane to practise steep turns, and climbing and gliding turns for one hour. Thirty minutes later the aeroplane was seen to fall into a spin at a height from which the pilot was unable to recover in time to avoid a crash. The pilot was killed.

(6) The occupants of this aircraft were Pilot Officers, Royal Air Force. One of the pilots, an ex Cranwell Cadet, had completed 302 hours solo flying but had no previous experience of Magister aeroplanes, his co-pilot had flown solo for 361¼ hours and had flown a Magister on one previous occasion for a total of 30 minutes. Some 15 minutes after the aeroplane had taken off it returned to the aerodrome, landed and took off again as though the pilot was executing a practise landing. About 5 minutes later it was seen about 4½ miles south of the aerodrome descending in a spiral dive which continued to the ground. Both occupants were killed.

(7) The pilot, a Royal Air Force 'ab initio' pupil, had completed approximately 14 hours dual and 9½ hours solo flying, all in Magisters. After 30 minutes dual instruction in spinning and

acrobatics the pupil was detailed to practise steep turns with and without engine, acrobatics, spinning and air navigation. About 40 minutes after the flight commenced the aircraft was seen to descend in a steep left hand spiral. The pilot abandoned the aircraft at a low altitude and was killed.

(8) The pilot, a candidate for Royal Air Force commission, had completed 7 hours dual and 16 hours solo in Moth aircraft in New Zealand. Since his arrival in this country he had received a further 15 hours 20 minutes dual and 10¾ hours solo in Magisters. While manoeuvring to land and when at a height of about 600 feet the aircraft was seen in a flat spin to the right with the control column held over to the left. The spin became steeper and the aeroplane fell into the Humber. The pilot was drowned.

From evidence collected during the investigation of these accidents it appears that Magister aircraft have the following characteristics:-

(a) Climbing. During the climb the aeroplane yaws to the right and requires left rudder to keep it straight.

(b) Gliding. During the glide the aeroplane yaws to the left. During a left hand gliding turn right rudder is necessary and for a right hand gliding turn an excessive amount of right rudder is required.

(c) Diving. During a dive the aeroplane tends to yaw to the left; under this condition a backward movement of the control column

Miles M.14 Magister L5916 painted in the colours of the CFS Aerobatic Team for the 1937 Hendon Display. Note the fairing over the front cockpit.
['Flight' via B Clarke]

L5916 inverted.
['Flight' via B Clarke]

has the effect only of steepening the dive, for unless right rudder is applied to correct the yaw, the elevators are inoperative. When the yaw is corrected, recovery from the dive is normal.

(d) Spinning. Contrary to accepted principles, the aircraft will spin off a straight stall without any yawing couple being applied. There is, moreover, a tendency to fall into a spin unexpectedly during an inaccurately executed manoeuvre. The tendency of the aircraft to remain in a dive unless it is on a straight course further complicates the recovery from a spin, for, if on recovery from a right hand spin the aircraft is allowed to swing to the left, it immediately assumes a steep dive as in (c) above.

Opinion

Until the results of further tests and research are known, it is not possible to give a definite opinion as to the cause of these accidents. On the evidence available, however, it appears that accidents Nos. (1), (4), (6) and (7) were due to the peculiar diving characteristics and Nos. (5) and (8) to the tendency of this type of aeroplane to fall into a spin. Nos. (2) and (3) were almost certainly due to the pilot's lack of flying experience.

As a result of these investigations full scale trials and wind tunnel experiments have been suggested to verify the aerodynamic qualities mentioned in paragraph 3(c). It is understood that these trials are already in hand and that additional research has also been initiated. Certain flying restrictions have, in the meantime, been imposed.
Vernon Brown, Chief Inspector of Accidents

Nothing more seems to have been heard about the Magister's spinning habits.

At this point, it must be emphasised that the re-designed, increased aspect ratio (ie taller) rudder, which was fitted to the Magister later, was not in any way (contrary to much popular misconception over the years) connected with the cure for these early spinning problems. This new rudder was found to be necessary as a result of a further problem, which became evident after the Magister had been in service for some time. This characteristic was duly reported thus: *It has been found that several accidents to Magister aeroplanes have been due to insufficient elevator control in the sideslip.*

However, an RAE Report entitled: 'Tests on the Magister in sideslip' (No. BA.1541, dated June 1939), noted that other changes which had been made in the evolution of the Magister from its civilian prototype, the Hawk Trainer Mk.II, and these included:-

(a) Fuselage increased in depth and made flat on top.
(b) Larger wing root fillets.
(c) Elevator chord increased by 1 in.
(d) Undercarriage fairing altered.
(e) Engine moved forward 3½ in.
(f) Larger windscreens.

The RAE carried out flight and wind tunnel tests on a Magister, the results of which were published as Reports No. BA.1541 and BA.1519 'Tests on the Magister in the sideslip'. 'Part I. Flight Tests' June 1939 and 'Part II. Wind tunnel tests' March 1939. The AM reference was a DTD letter dated 14th June 1938 and the Summary of the Report stated that:

Reasons for enquiry

In consequence of accidents which appeared to have been caused by large nose-down pitching moments in diving flight at high speeds it was required to examine the behaviour in dives of a Magister and to determine the conditions under which such pitching moments would be produced by involuntary sideslip.

Range of investigation

Diving tests have been made on Magister L5965. The effect of slipstream on rudder trim has been investigated, and rudder, elevator and sideslip angles have been measured in dives. Other types of low-wing monoplane have been flown for comparison (Hawk Trainer G-AEAX, Falcon L9705 and Mentor L4392 - PA).

Conclusions

In order to balance the effect of slipstream, the original rudder of low aspect ratio was provided with a fixed aerodynamic bias, adjusted to give trim at cruising speed. In high-speed dives with engine throttled, this bias held the rudder over to port, unless counteracted by pressure on the rudder bar, and produced a large angle of sideslip. This in turn caused a nose-down pitching moment, so that the effect of releasing the rudder at speeds over 150 mph ASI was to put the aeroplane into a swinging dive, recovery from which required an excessive pull on the stick unless the rudder was first centralised. The trouble has now been overcome by fitting a rudder of higher aspect ratio.

The report went on to conclude that the origin of the relation between sideslip and pitching moment remained obscure. Marked nose-heaviness in sideslip is not peculiar to the Magister. Model tests (a 1/7th scale model of the Magister was used in the wind tunnel) suggested that in all monoplanes it is associated with high tail plane position relative to the wing wake.

The increased aspect ratio rudder was fitted to Magister L8168, which was delivered to the A&AEE Martlesham Heath for trials on 13th June 1938 (it is of interest to note that the drawing of this new rudder, No.C.8796, was dated 19th July 1938). This new rudder was found to give a general improvement in handling and no further problems with the Magister were experienced.

This was the last major modification to be carried out on the Magister and it was incorporated in new build aircraft, probably with effect from L8168 (c/n 658) and retrospectively fitted to all surviving Magisters and Hawk Trainer Mk.III's from sometime after 3rd September 1938. A Phillips & Powis hand-written note on the subject of the increased aspect ratio rudder found by Terry Wilson (whose father built all the rudders for the M.16 Mentor!), dated 3rd September 1938, was entitled: <u>Magister I</u> - <u>Incorporation of modified rudder of increased aspect ratio.</u> <u>Modification No. Magister I/53.</u> <u>Class I.</u> *This modification provides for the fitting of a rudder of increased aspect ratio.* The note went on to detail each operation required to remove the original rudder and fit the new one, and it was signed by the Inspector-in-Charge AID.

The Air Ministry specification for the manufacture of the M.14 Magister was issued by the DTD on 25th July 1938 - Specification No. Magister I/P.3. Under it: *The aircraft shall be constructed in strict accordance with the drawings and schedules covered by the Magister I, DIS issue No.16 except as modified by this specification. Specification No. DTD 1000 shall be deemed to form part of this specification and shall be read in conjunction with it. The marking of aeroplanes and aeroplane parts shall be in accordance with the requirements of Specification No. DTD 1003 and any corrigenda thereto. Equipment shall be fitted or provided for in accordance with Appendix A No. 839.*

The Special Requirements of the manufacturing specification provided: *the following modifications shall be incorporated subject to their being technically approved by the Resident Technical Officer:*

<u>Mod. No.19</u>	Provision of Class II protective treatment with exception of the pilot's and passengers' seats.
<u>Mod. No.25</u>	To reposition vent holes in the outer tube of the oleo legs.
<u>Mod. No.27</u>	To strengthen bottom of rear seat.
<u>Mod. No.28</u>	To strengthen engine cowling.
<u>Mod. No.29</u>	To strengthen oil tank.
<u>Mod. No.30</u>	To strengthen oil tank mounting.
<u>Mod. No.31</u>	Improved arrangements for retaining the blind flying hood in the stowed position.
<u>Mod. No.32</u>	To strengthen welding at the end seams of the fuel tanks.
<u>Mod. No.33</u>	To provide a larger rear windscreen.
<u>Mod. No.34</u>	To provide stowage for First Aid pack.
<u>Mod. No.35</u>	To improve front throttle rod to prevent wear at the fire proof bulkhead.

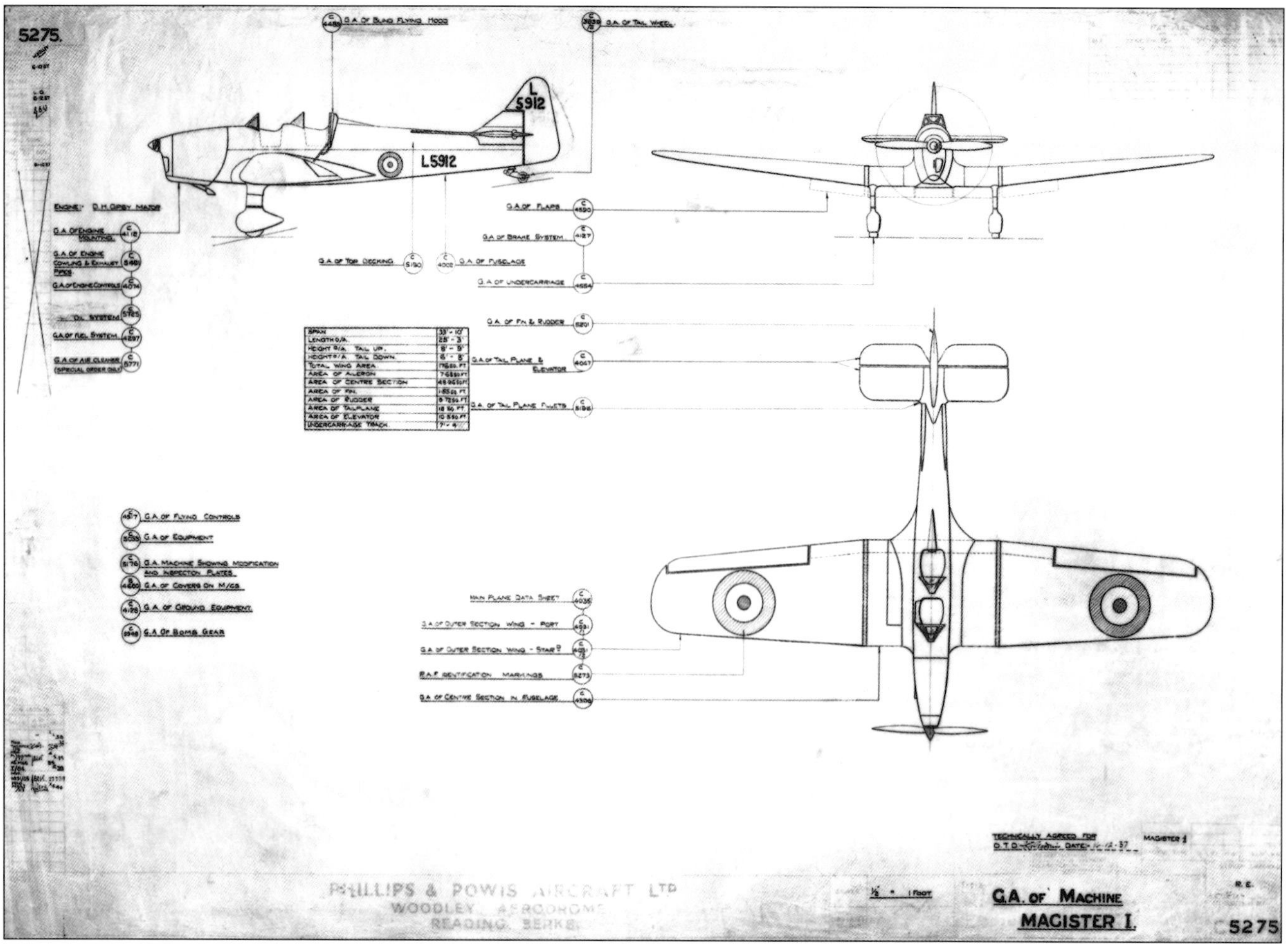

This drawing shows the final version of the Miles M.14A Magister but with original centre line opening cowling. Note the outline of the original rudder still visible.
[Phillips & Powis via The Miles Aircraft Collection]

A photograph of L8338 which shows the taller rudder and new cowling to advantage.
['Flight' via B Clarke]

Mod. No.36	To improve the attachment of the main undercarriage wheel spats.
Mod. No.38	Stowage for course height indicator.
Mod. No.40	To provide increased adjustment of pilot's safety harness in both cockpits.
Mod. No.41	General amendments.
Mod. No.43	To fit a pneumatic tail wheel.

And the provision for P.6 and Huson compasses as alternative installations.

Although the Magister had initially gained an unenviable reputation for being unable to recover from spins, it was to become the only low-wing cantilever monoplane with full Air Ministry approval for aerobatics, including protracted spins. The author can also vouch for its positive entry and recovery from spins, as he carried out a number of these manoeuvres in an ex RAF 'Maggie' while learning to fly with Jean Bird in the early fifties at Redhill. Entry into the spin was, in fact, very sudden, with the nose disappearing under itself as rudder was applied at the stall and, even though you were tightly strapped in the cockpit by the Sutton harness, you felt that you were going to be ejected from the cockpit very smartly in a downwards direction! The recovery, however, was quite straightforward and executed in the normal manner.

David Ogilvie, who was with the Shuttleworth Collection at the time, had accumulated a total 1,600 hours instructing on Magisters when he wrote in *Flight International* for 4th July 1981 on the subject of 'Why the Maggie needed strakes'

Sir. - The Miles Magister needed anti-spin strakes (Flight, May 16 page 1498) because without them the recovery results were wholly unpredictable. Even with strakes fitted, the Maggie would respond favourably only if given the traditional treatment of full opposite rudder, followed by a suitably positive pause before the stick was moved well forward. Any variation in recovery technique resulted in a spin that would be stopped only by terra firma. However, as far as I know, the proper procedure always produced the desired response. (Agreed - PA).

The Tiger Moth was different. Mark I Tigers, built for civil use, showed no deficiencies in spin recovery, but some specimens of the RAF's beefed-up Mark II revealed occasional reluctance to recover. By late 1940, 22 spin recovery incidents had been reported (including one by an instructor with more than 2,000hr on the type) and extensive tests proved that strakes provided the total cure. The story that only Tiger Moths with bomb racks failed to recover was untrue; the racks worsened the situation, but the basic fault was there without them. Today, despite some Tiger Moths boasting pre-war Mark I registrations, all of those flying are structurally Mark II's, so they should be treated as such.

An interesting Magister feature (see Flight for December 28th 1972), was that the elevators could be blanketed by the rudder. In straight and level flight a progressive application of rudder, countered by opposite aileron to keep the wings level, would result in the nose going down. With more rudder, this could be extended until the stick could be held almost fully back and the nose would continue to drop. Sudden centralising of the rudder would raise the nose so sharply that there might have been a risk of structural failure.

In the Shuttleworth Collection we have flyable specimens of both the Magister (one of only two in the world) and the Tiger Moth; however, in 1956 the Maggie was the subject of over-reaction about wooden aeroplanes by the (then) Air Registration Board, so it is in the non-aerobatic category and lives a right-way-up existence. Its spinning days belong to the history book. (Apart from the Shuttleworth example, the other two airworthy examples in the UK are still fully aerobatic - PA)

Over-reaction by the then ARB sums it up nicely and I wholly concur with David's statement based on his substantial experience. Just what irreparable damage was done to the British light wooden aeroplane by the hasty, and ill-conceived, action of the ARB, can be seen by the fact that there are so very few wooden aeroplanes left in this country. The attitude of the ARB, which was sparked off by the Civil Aviation Authority in Australia, was to produce a very jaundiced view of the strength and condition of the glue in wooden light aeroplanes manufactured before and just after the war, which culminated in the, quite often, unnecessary destruction of so many of our lovely British light aeroplanes. Memo Aero No.58, entitled 'Structural Testing of Wooden Aircraft', issued by the Department of Aircraft Design, The College of Aeronautics, Cranfield in December 1964, completely vindicated the strength of glued joints in elderly British wooden aircraft, and this will be reproduced in a future publication detailing the post-war M.65 Gemini.

The civil version of the Magister, known as the M.14A Hawk Trainer Mk.III, was also produced in limited numbers before the war and a few, which were similar machines, were fitted with the 135hp Blackburn Cirrus Major II engine. These became known as the M.14B Hawk Trainer Mk.II. The Mk.III was used by the Phillips & Powis Reserve Training School at Woodley and by the flying clubs operated by the Straight Corporation at Ramsgate, Exeter, Ipswich and Weston-super-Mare. The few civil and military Mk.IIs were supplied to Blackburn Aircraft Ltd for their Reserve Training Schools at Brough and Hanworth.

The Hawk Trainer Mk.III was also supplied before the war to the air forces of Eire and Egypt, and single examples were sold to Iraq, Russia and the air forces of Estonia, Latvia and Australia. In 1937, the Australian Air Board was looking for a new elementary trainer for use by 1 FTS and by their Citizen Air Force and had asked Laurence Wackett of the Commonwealth Aircraft Corporation to consider the availability and suitability of types for local manufacture.

Wackett looked at both the Magister and the Percival Gull and considered the latter inferior because of its fabric-covered wings and lack of aerobatic certification. He strongly recommended the Magister, which he also considered could be fitted with Sperry instruments and an enclosed cockpit.

The Air Board, however, did not agree and considered the Magister unsuitable for local manufacture or general adoption as the RAAF's standard elementary trainer. They were also concerned as to the cost of acquiring a licence for local manufacture. Their deliberations led, in due course, to the emergence of the CAC Wackett. However, one Magister was authorised to be purchased 'to allow RAAF pilots to become familiar with the type and its modern refinements such as flaps'. The purchase was also to allow for the testing of items prior to their incorporation in the indigenous product and also to trial the Gipsy Major Srs II engine against the Srs I.

Contract No.687916/37 was signed for one Magister, c/n 547, and this was dispatched from Woodley on 12th January 1938 arriving in Australia on February 18th, where it became A15-1. The engine, which had been removed due to a misunderstanding, arrived later. Whilst it had been intended to test the Magister at No.1 Air Depot in March 1938, prior to it being issued to the Training Depot for testing by any RAAF officer, this decision was reversed in order to avoid it being put into the hands of a bunch of unskilled pilots! In the event, it was 1 FTS at Point Cook which carried out the flight testing.

In July/August 1938, A15-1 was lent to CAC (presumably for them to see what use they could make of its design and manufacture) before being returned to 1 FTS, where it suffered some damage in a heavy landing in September. It was probably repaired but in April 1940, it was handed over to the RAAF Engineering School as an instructional airframe. An article on this machine appeared in *Air Enthusiast* for January/February 1997.

With a maximum speed of 140 mph and a landing speed of 45 mph, the Magister had a far better overall performance than any other elementary trainer then in use and, with its low-wing monoplane characteristics and mechanically operated spilt trailing edge flaps, it reproduced in a safe manner the handling techniques associated with the new operational aircraft, which were then beginning to enter service with the RAF. The flaps were later modified to be operated by vacuum jacks in a similar manner to the M.2X Hawk Trainer Mk.II.

Above: The only photograph to show Miles M.14A Magister s/n 159 of the Estonian Air Force, seen in the centre.
[Copyright E Reissar via P Branke and R Ruffle]

Left: The Miles M.14A Magister A.15-1 for the Royal Australian Air Force trials. [Via P Jarrett]

Below: Miles M.14A Magisters L5990 to L5996 inclusive in final assembly awaiting delivery in February 1938.
[Phillips & Powis via The Miles Aircraft Collection]

Above: Miles M.14A Magisters of No.29 E&RFTS Luton lined up in 1938.
['The Aeroplane' via P Jarrett]

Right: The famous publicity photograph showing eleven Magisters lined up at Woodley prior to delivery. What is never seen, because the image is so small, is that whilst the RAF Magisters L8334 to L8343 can be easily distinguished, the Royal Egyptian one on the end is always missed. By dating this, the Egyptian machine could be either L217, L218 or L219. Note L8338 in the middle of the line.
[Phillips & Powis via
The Miles Aircraft Collection]

The ease of handling and safety of the Magister used to be well demonstrated by George Miles, who frequently landed it with both hands held above his head, and also by Bill Skinner, the chief test pilot, who used to formate on other aircraft joining the Woodley circuit while flying the Magister inverted!

In August 1939, the Hawk Trainer Mk.III, U6, (c/n 538) was fitted with an M.18 wing for trials, the purpose of which has never been established but could have been in connection with the proposed Magister Mk.II.

A total of 838 M.14/A and M.14B Hawk Trainers Mk.III & II and Magisters had been delivered to a number of home and overseas customers by the outbreak of war on 3rd September 1939 and of these, 784 were Magisters delivered to the RAF.

All known contracts awarded to Phillips & Powis Aircraft Ltd by the Air Ministry with regard to the Magister, prior to 3rd September 1939 are listed below:

Date	Contract	Description
4.11.36	568682/36	The manufacture of 90 'Civil Hawk' aircraft
Undated	602402/37	Production of 26 aircraft
7.4.37	625317/37	Interchangeablity gauges
27.10.37	694309/37	ITP - incorporation of alterations
11.11.37	687916/37	ITP - production
9.12.37	707948/37	ITP - mock up and testing of alterations
20.12.37	714406/37	ITP - spares
23.12.37	706823/37	ITP - production of 214 aircraft
24.2.38	739244/38	ITP - incorporation of mods at CFT Schools
28.4.38	763808/38	ITP - spares
4.5.38	768124/38	ITP - spare parts schedule
20.5.38	776850/38	ITP - spares
24.5.38	777475/38	ITP - spares
25.5.38	774080/38	ITP - mock up of mods
25.5.38	778435/38	ITP - production of 254 aircraft
14.7.38	801171/38	ITP - spares
30.7.38	808653/38	ITP - spares
15.9.38	824716/38	ITP - parts for incorporation of mods
15.9.38	827642/38	ITP - experimental work
18.10.38	957847/38	ITP - spares
12.11.38	778435/38	ITP - production of 100 aircraft
1.12.38	976444/38	ITP - L5933 incorporation of mods
11.1.39	974065/38	ITP - mod spares
19.1.39	974064/38	ITP – spares
24.1.39	976294/39	ITP - re-designing exhaust system
10.2.39	778435/38	Production of 100 aircraft

Date	Ref	Description
3.4.39	984808/39	ITP - ancillary equipment
4.4.39	990017/39	ITP - spares
26.4.39	994091/39	ITP - engine exhaust manifolds
3.5.39	995340/39	ITP - Gipsy Major engine spares
8.5.39	778435/38	Production of 145 aircraft
15.5.39	996872/39	ITP - sets of parts for Mod.68
25.5.39	992388/39	ITP - mock up and testing of mods
1.7.39	748304/39	ITP - supply of laminated wood spares
17.7.39	10718/39	ITP - L8168 installation of tail parachute
19.7.39	B5907/39	Loan of Magister - no further details
20.7.39	11314/39	ITP - N3891 mod to take BLS tail wheel
19.8.39	16801/39	ITP - spares
2.9.39	19218/39	ITP - engine spares

For details of Magister and Hawk Trainer production to 3rd September 1939, see Appendix 20.

Above and below: The Miles M.14A Hawk Trainer Mk.III 'U6' (G-AEZS) fitted with the M.18 wing, in August 1939.
[Air Ministry via The Miles Aircraft Collection]

SPECIFICATION AND PERFORMANCE DATA

Engine: M.14, M.14A 130hp DH Gipsy Major I; M.14B 135hp Blackburn Cirrus Major II

Dimensions: span 33 ft 10 in; length 24 ft 7½ in; height overall, airscrew horizontal, tail down 6 ft 8 in; gross wing area 176 sq ft; net wing area 168 sq ft; aspect ratio 6.5; dihedral 6°; wing section modified Clark YM; thickness/chord ratio, root 19.2%, tip 9.0%

Weights: empty 1,286 lb; fuel (20 gal) 150 lb; max fuel capacity 21 gal; oil (2½ gal) 22 lb; pilot 200 lb; passenger/pupil 200 lb; useful load 42 lb; AUW 1,900 lb, aerobatic 1,845 lb; wing loading 10.7 lb/sq ft

Performance: max speed at sea level 140 mph, 130 mph at 5,000 ft, 125 mph at 10,000 ft; cruising speed 122 mph; stalling speed 52 mph IAS with flaps up, 43 mph IAS with flaps down; distance to unstick 630 ft with a 5 mph wind; distance to 50 ft 1,200 ft with 5 mph wind; landing run 420 ft with 5 mph wind; landing run from 50 ft 975 ft with 5 mph wind; initial rate of climb 750 ft/min; time to 5,000 ft 7.6 min; time to 10,000 ft 18.8 min; service ceiling 16,500 ft; absolute ceiling 19,000 ft; range 367 miles; duration 3 hrs

This photograph is believed to show the M.15 mock-up. Note the cockpit doors on the port side, whereas most photographs of the M.15 show them on the starboard side. Note also 'mock-up' on the engine. [Phillips & Powis via The Miles Aircraft Collection]

Chapter 28:
Miles M.15 Trainer Mk.I and Mk.II to Specification T.1/37

Specification T.1/37 (for the preliminary Operational Requirement OR.46) for an 'Ab initio Training Aeroplane', dated 22nd February 1937, was issued because *The Director of Training requires a replacement for the Tutor*. It was to be powered by a 200hp DH Gipsy Six Series II engine, and of the eight companies who put forward designs to meet this requirement - Airspeed (AS.36), De Havilland (DH.96), Fairey Aviation, General Aircraft (GAL.32a/b), Heston Aircraft, Parnall Aircraft, Percival Aircraft and Phillips & Powis Aircraft, three were ultimately contracted to build prototypes to Specification T.1/37E. These were Heston Aircraft, Parnall Aircraft and Phillips & Powis Aircraft.

Miles' experience of designing training aeroplanes suggested that the desired performance would be difficult to obtain within the limits set by the specification but nonetheless, he carried out some preliminary calculations and model experiments. The results of these, combined with his experience in the design of training aircraft, also confirmed his belief. In the end, Miles decided, rather against his better judgement, to try to meet the specified requirement, although he felt that the result would not be a good training aeroplane.

As this specification has been the subject of much debate over the years, it is reproduced here in its entirety:

Reference No. 609723/37/R.D.A.3.

Air Ministry Directorate of Technical Development - Confidential Specification No. T.1/37 - Ab Initio Training Aeroplane

Specification of Particular requirements to accompany the Contract Agreement.

This Specification is to be regarded for contract purposes as forming part of the Contract Agreement and being subject to the same general conditions.

Approved by: R H Verney (Air Commodore), Director of Technical Development.
Date: 22nd February 1937.

I. GENERAL
1. This specification and the Appendix B attached thereto detail particular requirements. Relevant general requirements are given in Specification DTD.1028 and it is essential that these shall be as completely fulfilled as the particular requirements of this specification, except where they are obviously inapplicable to an aeroplane of this type.

II. OPERATIONAL REQUIREMENTS
2. The main operational requirements are specified in Appendix B to this Specification circulated under cover of contracts letter, reference 610567/37/C4.A dated 13th February 1937.

Pilot's Cockpits
3. The comfort and convenience of the two cockpits shall be considered as of importance; in particular they shall be as draught free as possible. The seats should be adjustable in flight through a 4" vertical range and the rudder bars are to be adjustable in flight through a 6" range longitudinally.
4. Identical operational controls shall be fitted in both cockpits; instrument equipment shall also be duplicated.
5. Each seat is to be provided with the standard fighting harness.
6. Reference Specification DTD.1028 para.8(a). Cockpit heating is not required.

The first prototype Miles M.15 Trainer. [J Schrier (Den Haag) via B Clarke]

III. DESIGN REQUIREMENTS

7. The mock-up must be ready for examination within three months of the date of receipt of the order.

8. In order that a model or models of the complete aeroplane may be constructed for test of its spinning characteristics, the contractor is to supply to the Chief Superintendent, Royal Aircraft Establishment (BA Dept), drawings (in duplicate) showing:
   (1) its general arrangement in front elevation, side elevation and plan,
   (2) its wing sections,
   (3) the distribution of its mass relative to each of the three principal axes,
   (4) the position of its centre of gravity.

The drawings shall be forwarded sufficiently early to enable the model to be made and tested at the Royal Aircraft Establishment before the construction of the aeroplane is so far advanced that alteration to the fuselage and tail unit would be difficult.

The Contractor shall forward with the above-mentioned drawings a completed Aerodynamic Data Sheet, a pro-forma of which will be supplied to him by the DTD (RDA.3) at or about the time when the contract is placed.

Construction (Materials).
9. Reference Specification DTD.1028 para.12(a) and (b). Wood or composite construction should be substituted for metal.

Engine Installation
10. Reference Specification DTD.1028 para.15. The object of this clause is to ensure ease and rapidity in the changing of engines operating under Service conditions. Whilst such conditions will not apply to the aeroplane under consideration, in the main, the general principles outlined in this paragraph should be embodied in the design.

Engine Starting
11. The aeroplane is to be equipped with electric starting and a reserve hand starting gear. The electric starter shall be suitable for 24 volts. Provision shall be made for a ground starting plug so that the engine can be started under cold conditions from an outside supply. Operation of the press switch for starting shall be duplicated making control possible from each of the cockpits. It will be necessary to provide a starting accumulator in the aeroplane, and a generator on the engine to keep it charged.

Engine Cooling
12. Provision is to be made for satisfactory cooling of the engine under tropical conditions or the cooling arrangements shall be easily adaptable to meet such conditions.

Fuel System
13. The total fuel capacity shall be sufficient for the endurance called for in the Appendix B para.3(ii) after allowance has been made for ¼ hr. ground running.

Oil System
14. The net capacity of the oil tank(s) shall be sufficient for an endurance calculated at the maximum rates of consumption, 2 hours more than that for which fuel is provided after reckoning the ¼ hr. allowance for ground running.

Fire Extinguisher
15. A fire extinguisher shall be fitted in each cockpit and from each cockpit it shall be possible to extinguish a fire inside the engine cowling.

Maintenance
16. The component parts of the aeroplane are to be suitable for conveyance by Royal Air Force MT vehicles. For this purpose the maximum sizes of the various components shall not exceed those for the undermentioned categories as given in ADM.340:
   Category 'B' main portion of fuselage and/or centre portion of planes.
   Category 'B' other sections of the fuselage and sections of planes other than wing tips.
   Category 'A' All other components.

17. The ground clearance at the tail unit shall be such that there is a minimum of 9" clearance when the tail wheel tyre is flat and the elevators are in their down position.

18. The documents described in contracts circular No.22 shall be supplied.

Load for strength requirements & Flight trials
19. (a) Complete equipment in accordance with the tender Appendix A.
   (b) Fuel and oil for the specified endurance plus the necessary allowances.
   (c) Crew of two with parachutes 200 lbs. each.

Structural Strength
20. The strength of the aeroplane when flying in the fully-loaded condition shall not be less than is defined by the ultimate factors stated hereunder:-
   (1) Factor throughout the structure with the centre of pressure in its most forward position in normal flight. 10.0
   (2) Factor throughout the structure with the centre of pressure in its most backward position in horizontal (normal) flight, the flaps being down 6.3
   (Note - in the copy of the specification in my possession 'the flaps being down' and '6.3' was crossed out and '7.5' was pencilled above the latter figure – PA)
   (3) Factor throughout the structure in a dive
   (a) to an assumed terminal velocity of 450 mph IAS with the aeroplane in the attitude corresponding to maximum tail load (see AP.970 (May 1935) Chapter 2 para.4.) 2.0
   (Note - *in the copy of the specification in my possession the last line in brackets was crossed out and 'ADM341' was pencilled alongside – PA*) or
   (b) to the terminal velocity assuming the airscrew drag to be zero, if this speed is less than that quoted in (a) the aeroplane to be assumed in the attitude corresponding to maximum tail load 2.2
   (4) Factor in a dive as in (3) above, assuming any one wire to be cut. 1.0
   (5) Factor for the wing structure at the angle of incidence corresponding to an inverted stall with the centre of pressure at 0.36 of the chord. 5.0
   (Note - *in the copy of the specification in my possession 'with the centre of pressure at 0.36 of the chord' was crossed out – PA).*
   (6) Factor under 'up' and 'down' gusts (normal to the flight path) of 25 ft/sec (indicated) in an accelerated dive at 1.5 times maximum level speed with the aeroplane in the true terminal velocity attitude 1.5
   (7) Factor when diving with flaps down along a flight path at 70° to the horizontal at the steady gliding speed appropriate to this condition.
   (i) In a steady glide 2.2
   (ii) Under 'up' and 'down' gusts of 25 feet/second (indicated) normal to the flight path 1.5

(Note - *in the copy of the specification in my possession para '(7)' was crossed out in its entirety – PA).*

In items (2) and (6) above and in the Tail Plane Strength Requirements (ADM.341) the maximum and normal top speeds as defined in ADM.345 are to be used, where relevant, at an altitude of 3000 ft.

21. The aeroplane, at 1.1. times its weight, fully-loaded shall be able to withstand an impact with the ground at a vertical velocity of 10 feet per second, and at this velocity the impact load on the undercarriage is not to exceed 3.3 times the weight of the aeroplane, fully-loaded. Compliance with this requirement shall be demonstrated by means of approved dropping tests at the Royal Aircraft Establishment. The ultimate factor for the undercarriage when subject to this impact load shall not be less than 1.33 and for the remainder of the structure not less than 1.5 (this difference between factors is always to be maintained whatever modifications may be made eventually in the design of the aeroplane). The ultimate factors at 1.1 x weight, fully-loaded, when the aeroplane is at rest on the ground shall not be less than the following:-
   For the undercarriage 4.0
   For the remainder of the aeroplane 4.5

The ultimate factor throughout the structure under a side load at the axle equal to the weight of the aeroplane, fully-loaded, shall be not

The first prototype Miles M.15 in primer, probably after its first flight. Note the cockpit doors on the port side of the fuselage.

[Phillips & Powis via The Miles Aircraft Collection]

less than 1.1. For the side load to be assumed in stressing cantilever undercarriages reference shall be made to the Airworthiness Department, RAE.

22. When considering the undercarriage stressing cases in sub-paras. I(iii), (iv) (v) (vi) (vii) of Chapter III of AP.970 (May 1935), 1.1 x W should be substituted throughout for W.

23. In calculating the strength of the structure under fin and rudder loads, the load specified in AP.970, Chapter III, para.10, and Chapter V, Section II, para.2, is to be increased by trebling the ordinates of the present load distribution curve over the region from the fin leading edge to 0.3 of the fin chord aft of the leading edge.

24. The ultimate factor for control surfaces and systems under tail-to-wind loads in accordance with AP.970 (May 1935 edition), Chapter III, para.24 shall be 2.

25. The detail requirements (where appropriate) of Air Publication No.970 (edition of May 1935) are to be satisfied.

Acceptance Tests

26. Prior to delivery of the first aeroplane it shall have been certified to the DTD by the Contractor that the aeroplane has been subjected by a contractor's pilot to the following tests:-
 (1) General flying tests in accordance with Aircraft Design Memorandum No.291.
 (2) Diving tests in accordance with Aircraft Design Memorandum No.292.
 (3) Tests of lateral stability at the stall in accordance with Aircraft Design Memorandum No.293.
 (4) Spinning tests in accordance with Aircraft Design Memorandum No.294.
 (5) Acrobatic flying tests in accordance with Aircraft Design Memorandum No.295.

Centre of Gravity

27. The calculated co-ordinate normal to the datum line may be accepted, provided that the error in the calculated value of the co-ordinate parallel to the datum line is not greater than 2% of the mean chord when compared with its measured value.

CONFIDENTIAL - SPECIFICATION T.1/37. APPENDIX B.
Training Requirements for Replacement of the
Ab-Initio Training Aircraft.
The Director of Training requires a replacement for the Tutor.

2. General.
The aircraft should be a single-engined low or mid wing open two-seater monoplane with the general characteristics of the modern monoplane, although refinements for speed such as the retracting undercarriage and variable pitch airscrew are not considered necessary. It should be a tandem two-seater with dual control and full range of instruments (including instrument flying panel) in both cockpits.

3. Performance.
(i) Speed.
 A maximum of speed of not less than 150 mph at maximum rpm at sea level with normal full load.

(ii) Endurance
 An endurance of not less than 4½ hours at 2/3 maximum power at maximum cruising rpm at sea level.

(iii) Climb.
 A climb to 10,000 ft in not more than 10 minutes with normal full load (the climb must be good to ensure that time is not wasted in climbing up after aerobatics).

(iv) Landing and Take-off.
 When landing the aircraft without the use of flaps, it must come to rest in 500 yards after coming in over a 50 ft obstacle. Landing run not to exceed 200 yards. When taking-off with normal full load, the aircraft must clear a 50 ft obstacle in 500 yards.

4. Engine.
 The aeroplane is required to be fitted with the Gipsy VI, Series II, engine:-
 (i) With electric starting, to include a generator for charging accumulators.
 (ii) A fixed pitch wooden airscrew with Schwarz covering to be fitted.
 (iii) The engine is to be well silenced.

5. Equipment.
 (i) Brakes and tail wheel.
 (ii) Inter-communication by Gosport tube.
 (iii) Night flying equipment, including landing lamps and two parachute flares.
 (iv) Instrument flying hood over pupil's front seat.

6. Controls.
 (i) A good but not too sensitive fore and aft trimming device of similar design to that on the advanced training type.
 (ii) Flaps to be fitted, of easy operation.
 (iii) It is important that there should not be an excessive change of trim with the engine's 'on' or 'off' and with the flaps raised or lowered; with the flaps in the fully lowered position the pilot should still be able to take-off if he makes a mistake in landing, and the elevator should be capable of overcoming the flaps, in case he should omit to raise them.

7. Special Considerations.
 (i) Both pilots' seats must have a very good all round view to facilitate dual instruction, and adequate room. The seats must be arranged so that the pupil sits in front of the instructor, and the range of fore and aft control should be such that the aeroplane can be flown from the front seat quite as satisfactorily with no instructor in the rear seat, and without the addition of ballast.
 (ii) Easy maintenance and repair is essential.
 (iii) The aircraft must have a high degree of stability.
 (iv) Both cockpits must have comfortable seating and be free from draught.
 (v) Exit in emergency must be easy.
 (vi) There must be a strong point above the fuselage to take the shock in case of overturning on the ground.
 (vii) The addition of crash-proof tanks is desirable but not essential.
 (viii) The aeroplane must be capable of full acrobatics including good recovery from long spins.

8. Construction.
It is essential that the aeroplane should be designed for rapid production, ease of maintenance and good durability under storage conditions. It should be of wood or composite construction.

9. Load Factors (Added by DTD)
(a) Factor throughout the structure with the centre of pressure in its most forward position
 in normal flight.   10.0
(b) Load factor throughout the structure with the centre of pressure in its most backward position in horizontal normal flight.   7.5
(c) Factor in a terminal velocity dive.   2.0
(d) Other factors as normally required in Service Specifications and in AP.970.

During 1938/39, the various entries for the Specification T.1/37 were subjected to searching official service trials, but none of them came up to the standards demanded by the Service for an elementary trainer and the type was not ordered into production. Parnall Aircraft Ltd produced the private venture J1 (Parnall 382 Heck III) and Heston Aircraft the JA.3 and JA.6, while Percival Aircraft also built a mock-up of their project with the design letter 'S' (later designated P.20) in 1938. The latter bore a strong resemblance to the 'Open Gull', but the project failed to attract support and was dropped.

Phillips & Powis submitted the M.15, which was accepted and the Instruction to Proceed (ITP) with the manufacture of two

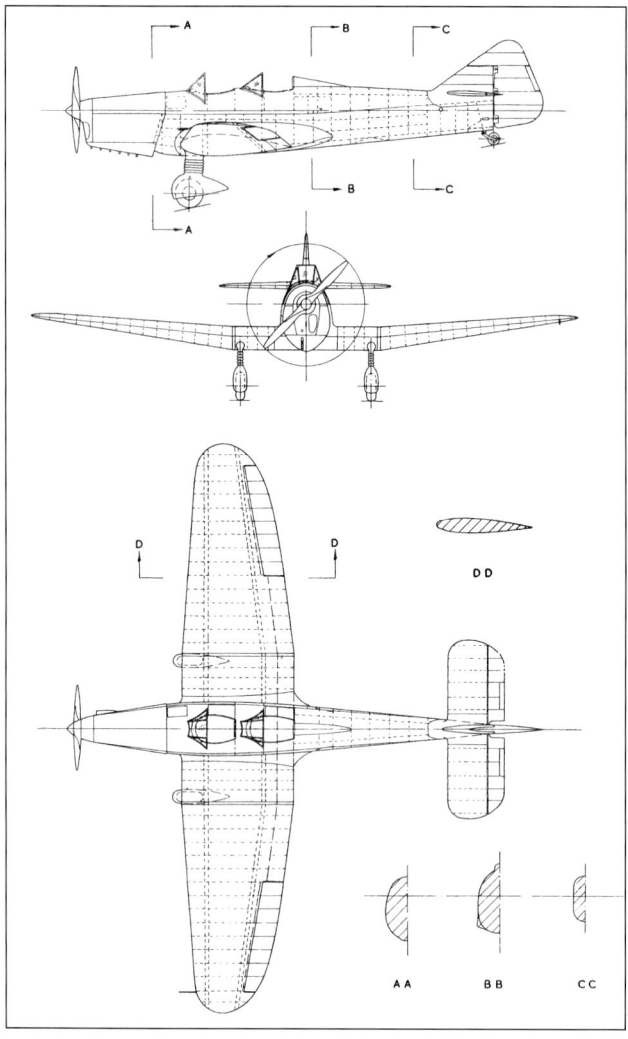

The Miles M.15 to T.1/37.

'Elementary Trainers for the RAF with DH Gipsy Six engines', was issued on 8th October 1937 to Contract No 678259/37 (with a projected cost of £500). This cost appears to have been in error for £5,000.

As the specification mentioned a mock-up, it is surprising that C L Nash recorded no c/n for it, as c/ns were often allotted for mock-ups of projected new types. It is, however, possible that the 'mystery' photograph, showing an unidentified machine similar to the M.15, which was probably taken in the Experimental Department, shows the mock-up of the M.15. The fuselage, with centre section, standing on its undercarriage, has a smaller, three-piece windscreen and headrest to the rear seat than with the final design, and has a DH Gipsy Six/Queen engine marked 'MOCK-UP' installed.

The photograph also shows the cockpit doors and the wing walkway on the port side, whereas both the M.15 Mk.I and Mk.II had the cockpit doors on the starboard side. The c/n 796 is blank in the works records and its timing would fit with one possibly reserved for the M.15 mock-up.

The first prototype M.15 Trainer (c/n **641**) was built during 1938 and was first flown, as **U1**, by Bill Skinner, probably on 22nd September 1938. It was also flown on that day by Flt Lt J F Moir, OC of the Reserve Training School at Woodley, and he, with his previous experience of testing RAF aeroplanes at the A&AEE, usually flew new prototype aircraft on the same day as their maiden flight. The prototype, later given serial **L7714,** was delivered to the A&AEE Martlesham Heath for service trials on 4th February 1939. Here it was found to be 'overweight, with a resulting lengthy take-off run and poor rate of climb. Aerobatics were impossible, due to the engine overspeeding', but on 3rd April 1939, it was delivered to the AMDP at RAE Farnborough, although it was possibly returned to Woodley by 3rd September 1939. On 5th December 1939, it was issued to 6 MU Brize Norton (possibly for storage). L7714 was returned to Phillips & Powis on 21st November 1942, but its ultimate fate was not recorded.

The various aircraft submitted were subjected to comprehensive official trials in 1938/39, and the results confirmed Miles' misgivings, in that none of the entries were deemed suitable for adoption as a Service trainer, although the Parnall Heck was said to be pleasant to fly, stall, spin and aerobat.

Following the trials, Miles stopped work on the second prototype, the Mk.I (c/n **1076**), which had been allocated serial **L7717.** CL Nash recorded that it 'proved obsolete before completion, salvage used on (c/n) 1077. AM instruction'.

In an effort to explore every possibility offered by the Specification and at the same time to produce a good aeroplane, in 1939, Miles built the M.15 Mk.II, against the same contract. A Phillips & Powis 3-view general arrangement drawing of the 'T.1/37 Mk.II', dated (unreadable) in 1939, showed that the Mk.II differed from the first prototype in a number of respects. These included squared wingtips, squarer tips to the tailplane, a small three-piece frameless type windscreen to the front cockpit and a much larger three-piece windscreen with stronger frame to the rear cockpit, which had no headrest.

The M.15 Mk.II, **P6326** (c/n **1077**) was built in 1939 from parts salvaged from the uncompleted second prototype and was probably powered by a 210hp DH Gipsy Queen II engine (although C L Nash records that it was fitted with a 200hp DH Gipsy Six engine). It was painted yellow overall but its first flight date is unknown. Flown to RAE Farnborough by Tommy Rose on 23rd May 1939, it was returned to Woodley by Bill Skinner on 26th May 1939. Delivered to 'A' Performance Squadron, A&AEE Martlesham Heath in July 1939 (not recorded on its record card), it had departed A&AEE by 17th November 1939. P6326 was taken on charge by AMDP (no date recorded), possibly at Woodley, and was delivered to RAF Kinloss on 1st July 1940, although no reason for this movement has been found. It was later returned to Phillips & Powis but there is only one flight recorded after this. The flight was made by Hugh Kennedy, who noted it with 'B' mark **U-0234**, on a 40-minute 'handling flight' on 16th March 1942. The reason for this flight, so long after the requirement had passed, is not known. The M.15 Mk.II was then placed in storage in a Robin hangar on the Davis Farm site, where it was seen early in 1943 by Grahame Gates. It was later moved to the Repair & Service Department, where it was broken up.

The hoped for improvement in performance had failed to materialise and in the end both the official candidates for AM Spec T.1/37, the M.15 and the Heston Aircraft projects were rejected and the private venture Parnall J1 also failed to secure an order.

SPECIFICATION (Performance Data unknown)

Miles M.15 Mk.I Engine: 200hp DH Gipsy Six Series II
Dimensions: span: 33ft 5in; length 29ft 6in; height 10ft 5in; wing area 200 sq ft; aspect ratio 5.6
Weights: empty 1,830lb; AUW 2,530lb; wing loading 12.6 lb/sq ft

Miles M.15 Mk.II Engine: 210hp DH Gipsy Queen I or 200hp DH Gipsy Six Series II
Dimensions: span: 33ft 6in; length 29ft 0in; height (tail down) 10ft 5in; wing area 200.25 sq ft; aspect ratio 5.59
Weights: not known

L4393, the second Miles M.16 Mentor Mk.I, delivered to the A&AEE Martlesham Heath on 28th March 1938. Note the deeper rear windows.
[Phillips & Powis via The Miles Aircraft Collection]

Chapter 29: Miles M.16 Mentor

During 1937 the Air Ministry invited tenders, to AM Specification 38/37, for a new three-seat cabin aeroplane, to be powered by a 200hp DH Gipsy Six engine, to fulfil a requirement for a variety of duties, including night flying, navigational training, instrument flying, radio training and communications. The requirements of this quite unreasonable Specification included the provision of full dual-control, with doors on both sides of the cabin, the provision of blind-flying instruments and landing lamps. The aircraft was also to be capable of carrying a considerable Service load, having due regard to the size and power of the machine.

Miles submitted a tender of a development of the Nighthawk, designated the M.16. This was of standard Miles wooden construction, but it also incorporated the one-piece moulded Perspex windscreen pioneered on the earlier Whitney Straight. The M.16 also incorporated split trailing-edge flaps, which were operated from a Theed vacuum double-acting jack, which was in turn operated via a reservoir exhausted from the induction manifold of the engine. The first Automotive Lockheed oleo undercarriage was also fitted to the M.16.

A mock-up 'Nighthawk/Mentor', with c/n 287, was built in 1936-37, and it is on record that the drawing office expended a total of 630 man-weeks on the design.

The M.16, by then named the Mentor, was deemed satisfactory 'for communications duties', and 45 were ordered under Contract No.534848/36. The Instruction to Proceed was dated 24th August 1936 and the projected cost of the contract was £40,500 (however, the final cost was to increase to £66,100 + £32,327). The prototype Mentor L4392, c/n 434, was first flown, by Bill Skinner, on 5th January 1938 and its performance was, as expected due to its multi-role configuration, much lower than that of the Nighthawk as the modifications had completely altered the handling characteristics for the worse. It was also much heavier and more sluggish on the controls. The prototype was submitted to the A&AEE Martlesham Heath for its official trials in 1938, in its original form with the fin and rudder shape similar to that of the early M.14 Magister. As expected, the Mentor was found to be unsuitable for the role for which it was intended and was considered 'not to have been an unqualified success'.

The A&AEE Report No 725, dated 9/38 (ref: AVIA 18/655), showed that the Mentor had a long take-off run of 450 yards, thereby making it unsuitable for small airfields, a poor rate of climb (the climb was actually claimed to be impossible with the flaps down!), a very uncomfortable cabin on account of engine noise, numerous leaks which let in rain, an effective blind flying hood but one which made communication between pilots very difficult at normal engine speeds, etc, etc! The Mentor was returned to Phillips & Powis Aircraft Ltd for modifications.

It is of passing interest to note that Contract No.984808/39, dated 3rd April 1939, was placed for 'ancillary equipment' for the Mentor but no further details of what this comprised are known.

Although probably having no bearing on the poor take-off and climb performance, a redesigned, more triangular rudder of greater area was fitted, probably after May 1939. The fitting of this new rudder resulted in very large changes of directional trim with engine power, to such an extent that, on the approach, there was a tendency to yaw and roll to the left. The clear-view panel

The first prototype Miles M.16 Mentor L4392 under construction in the Experimental Department in 1937. Note that the rear windows are not so deep as on the second and subsequent production aircraft. The M.12 Mohawk is just visible in the right background. [Phillips & Powis via The Miles Aircraft Collection]

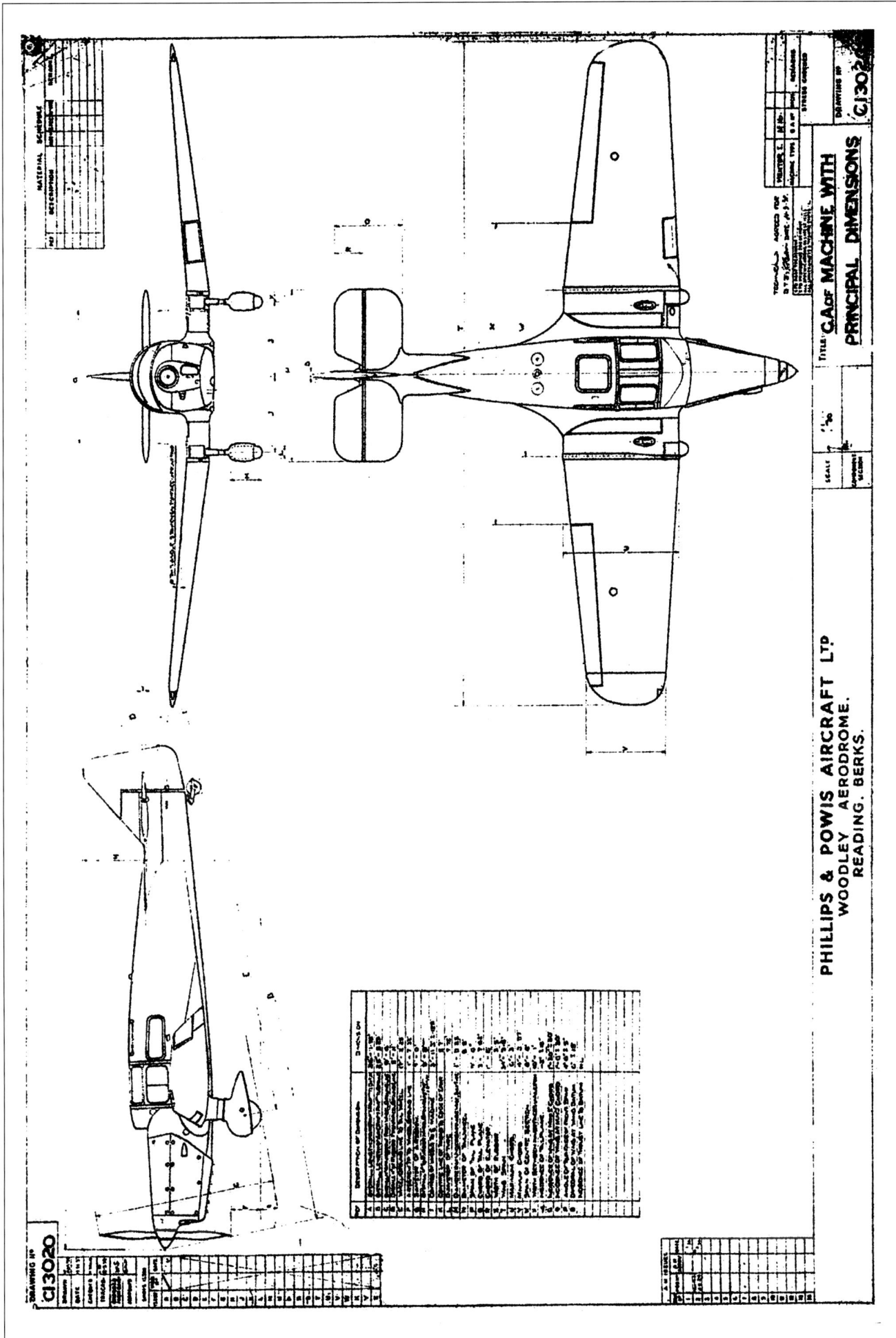

The Miles M.16 Mentor original G.A. drawing.

The Miles M.16 Mentor L4420 at Croydon, early in the war. This aircraft later became G-AHKM. The Magister to the rear, P6353, stalled on approach near Laurencekirk, Kincardine on 8th December 1942. [Via P Jarrett]

was repositioned and provided with a better seal, new undercarriage legs were fitted and the pilot's seat was modified. The radio and wireless operator's seat were removed and together these modifications resulted in a reduction of AUW from 2,884lb to 2,667lb, although both these figures are at variance with the firm's figure of 2,710lb AUW for the original aircraft.

The prototype Mentor was later returned to A&AEE Martlesham Heath, who found that the rudder control was 'lighter and better harmonised with the other controls', the exhaust noise had been reduced and the repositioned clear-view panel was satisfactory, although it still leaked. The A&AEE were still not happy with the entry to the cabin with pilots wearing full flying kit and that the seating was tiring on long flights due to the position of the main spar under the pilot's knees which caused his weight to be thrown forward onto his spine. Controls and communication was now improved but the fact that the pilot had to let go of the control column to apply the brakes was criticised. However, it was judged not to be suitable for communications because of its poor take-off and rate of climb.

In mid 1939, the second Mentor, L4393, was fitted with a two-pitch propeller, which reduced the take-off run by some 15 yards and marginally improved the rate of climb. However, the rudder bias, which had been recommended following the earlier tests, had apparently still not been fitted at this time.

In spite of its shortcomings, the order for 45 Mentors was upheld (originally to Specification 38/37P, but this specification was later cancelled) and they were all delivered to the RAF between March 1938 and June 1939. As delivered, they were probably still fitted with the early rudder but, by the end of 1939, all the surviving Mentors had been retrospectively fitted with the larger and more triangular shaped rudder.

The Mentor had an undistinguished career as a two-seat communications aircraft and little of note about it is recorded. An unidentified Mentor made a forced landing in Phoenix Park, Dublin, Ireland on 8th December 1942 'during the emergency' while being flown by an American pilot, who had become 'lost on a training mission'. The aircraft was moved to Baldonnel and flown out later.

An HQ No.41 Group Return (Ref: AIR 17/33) entitled 'Arrivals and Despatches for Week Ending and Stock in 41 Group ASUs as at 0900 hours 6.12.45', gave the numbers of Mentors surviving as follows: 12 Stock as at 0900 6-12-45; 1 Awaiting MAP disposal instructions - Non-Effective Aircraft. The non-effective aircraft require maintenance within 41 Group units; 11 Awaiting/Undergoing Breakdown.

It has only been possible to identify the one Mentor 'Awaiting MAP disposal instructions'; this was s/n L4420, which was sold in 1946. The other 11 probably comprised some of the 22 which were SOC in 1944.

Only one Mentor survived to be sold on the civil market; L4420, which was registered G-AHKM on 18th April 1946 and although its post war career was destined to be short, it did participate in a number of air races before it was written off in a crash on 1st April 1950.

For details of Miles M.16 Mentor production see Appendix 21.

SPECIFICATION AND PERFORMANCE DATA

Engine: 200hp DH Gipsy Six Srs I
Dimensions: span 34ft 9½in; length 26ft 1¾in; height 9ft 8in; wing area 181sq ft; aspect ratio 6.7
Weights: empty 1,978lb; AUW 2,710lb; wing loading 15lb/sq ft
Performance: max speed 156 mph; rate of climb 780ft/min; service ceiling 13,800ft.

The Miles M.17 Monarch being demonstrated outside the Falcon Hotel at Woodley in 1938. ['Flight' via B Clarke]

Chapter 30:
Miles M.17 Monarch

The popularity of the Whitney Straight, both in this country and abroad, gave George Miles the idea that he could produce an even better version to incorporate all its most desirable features. By 1937, George had become the Technical Director and Chief Designer of the firm and had also carried out most of the sales tours and demonstration flying on the Whitney Straight. He took on the task of designing and producing an improved version of the Whitney Straight, despite the fact that the works were fully occupied with the production of the Magister for the RAF. Miles, however, confirmed that orders would be taken for the new aircraft, to be called the M.17 Monarch and which was the first aeroplane to be wholly designed by George Miles.

The Monarch was superficially similar to the Whitney Straight, being a two-seater with full dual control, but with the option of an additional third seat in place of the luggage space on the starboard side behind the normal passenger's seat and with a window of its own. In detail however, it was radically re-designed. The centre section of the wing, which had a span of 9ft 7½in, was unique to the Monarch and although it did not have wing-fold, the outer wings were interchangeable with those of the Magister. Numerous other standard parts were incorporated, making the Monarch attractive from the point of view of spares and serviceability. The Monarch also had a special streamlined one-piece moulded Perspex windscreen, of the type pioneered by the company, which was deeper than that of the Whitney Straight and gave a considerably improved view. The soundproofing was even more carefully done and the door was moved to the starboard side, enabling passengers to enter and leave while the pilot remained in his seat.

Aerodynamically, it had revolutionary features, including the new Miles 'Glide Control' - a simple adaptation of the normal Miles vacuum-operated flaps. The Miles split trailing-edge flaps on the Monarch were in five portions which extended all the way from aileron to aileron. They were worked from a Theed vacuum double-acting jack, manufactured by Sir George Godfrey & Partners and operated in turn via a reservoir exhausted from the induction manifold of the engine. The control to the jack was perfectly positive, thereby enabling the flaps to be stopped in any position. The jack operated a torque tube to the outer ends of the centre section and from there the movement was taken by links to the flaps under the outer wings.

The 'Glide Control' was obtained by a simple and foolproof series of mechanical devices to link the throttle with the flap operation. When flaps are used to control the glide on a normal aeroplane, there is no safeguard to prevent that sudden and sometimes dangerous sink which occurs when the flaps are raised. With the 'Glide Control' however, all that was necessary when it became evident that the machine was undershooting on a landing was to flick the lever to the 'up' position. The flaps would then immediately close to the maximum lift position of approximately 15 degrees, but no further. As a result of this action, there would be an instantaneous reduction in drag but no dangerous loss of lift. When maximum drag was again needed the control was returned to the 'down' position and the gliding angle again becomes very coarse, with no intervening period of additional lift which is always inclined to upset the pilot's judgement. By suitable operation of the lever, any intermediate position of the flaps could be obtained.

Another advantage of the system was that if, for any reason, the landing was badly misjudged and it became necessary to use full throttle to go round again, the pilot was not suddenly faced with the necessity of climbing at a speed very near the stall with the flaps right down in order to gain any height. Miles noticed that this characteristic of a 'flapped' aeroplane was sometimes very disconcerting to the inexperienced pilot. With the 'Glide Control' however, all that was necessary was to open the throttle when the flaps would automatically return to the position of maximum lift.

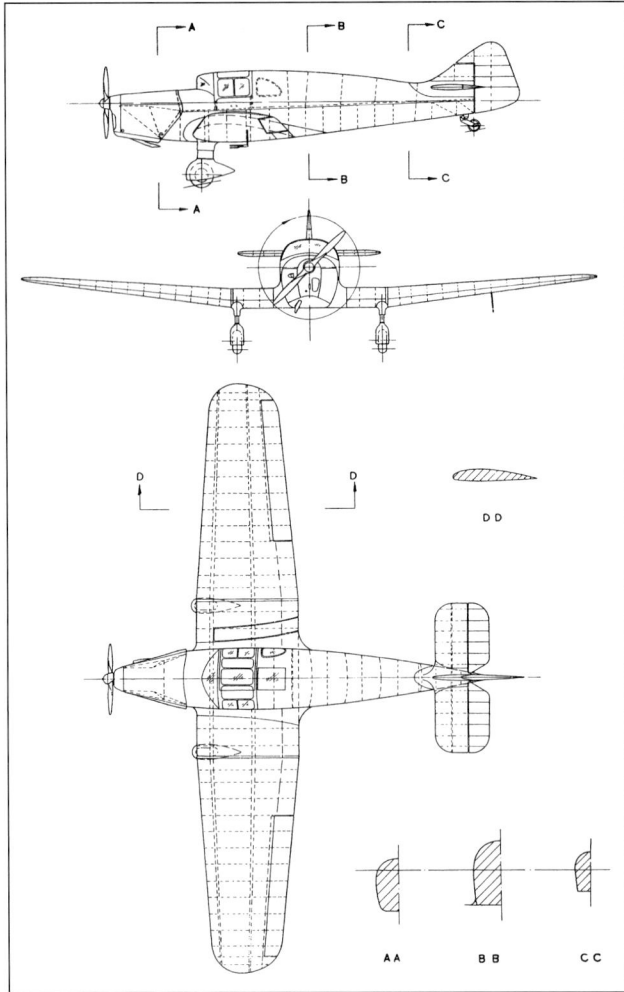

The Miles M.17 Monarch.

Considerable use was made of high-strength, but low weight, magnesium alloy castings in the controls throughout the Monarch. The standard of finish on the Monarch was claimed to be higher than was attainable as standard on any other aircraft in the country, if not the world, and the high polish not only gave the machine a very handsome appearance, but also increased the performance considerably.

Standard tankage was provided for a still-air range of over 600 miles, while larger tanks, giving duration of over seven hours and over 900 miles, were a standard extra. The undercarriage was designed to take a vertical velocity of 15ft/sec and the wheels could be removed by the extraction of a single pin.

The prototype M.17 Monarch, registered G-AFCR, was built in the Experimental Department during late 1937 and was first flown, by Chief Test Pilot Bill Skinner, on 21st February 1938, still in primer finish and with the 'B' mark U1. From early March, George Miles demonstrated the Monarch, both in this country and abroad, and a

The possibility of inadvertently taking-off with flaps down was also ruled out, as the flaps would move to the take-off position as soon as the engine was opened up. These advantages were obtained without the use of any new or untried components.

A patent for the new 'Glide Control', under the heading of 'Wing Flap Control' was first applied for on 3rd May 1938 by Phillips & Powis Aircraft Ltd and F G Miles of Reading and on 10th January 1940, this was accepted: *The invention is a control system which normally provides interlocking between the throttle and the flap control levers and relieves both these levers of the necessary effort to keep the flaps locked in position.*

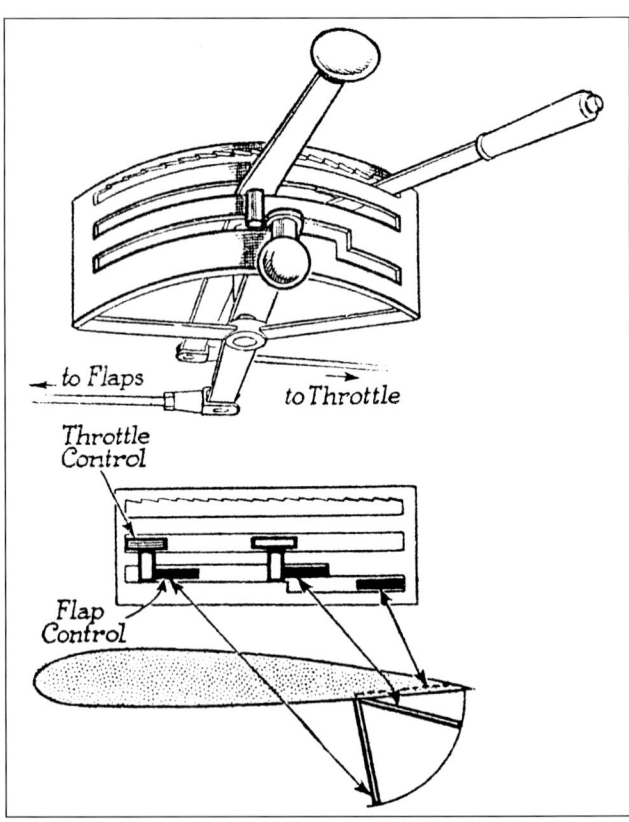

Diagrammatic and perspective drawings which show the simple operation of the interconnection and gate arrangement of the throttle and flap controls on the Monarch.

The prototype Miles M.17 Monarch in primer prior to its first flight, which was made by Bill Skinner on 21st February 1938. [Via P Amos]

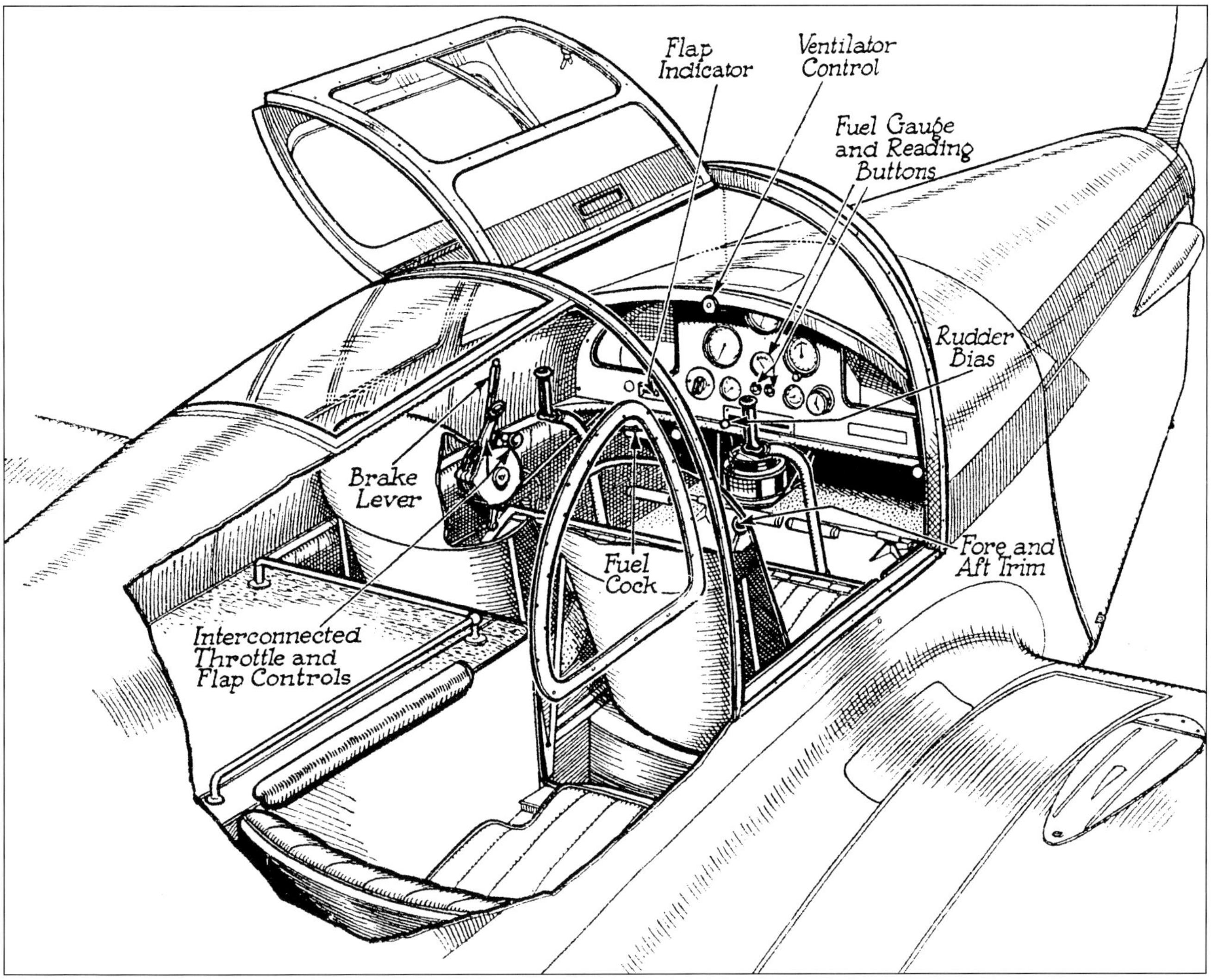

Chapter 30: Miles M.17 Monarch

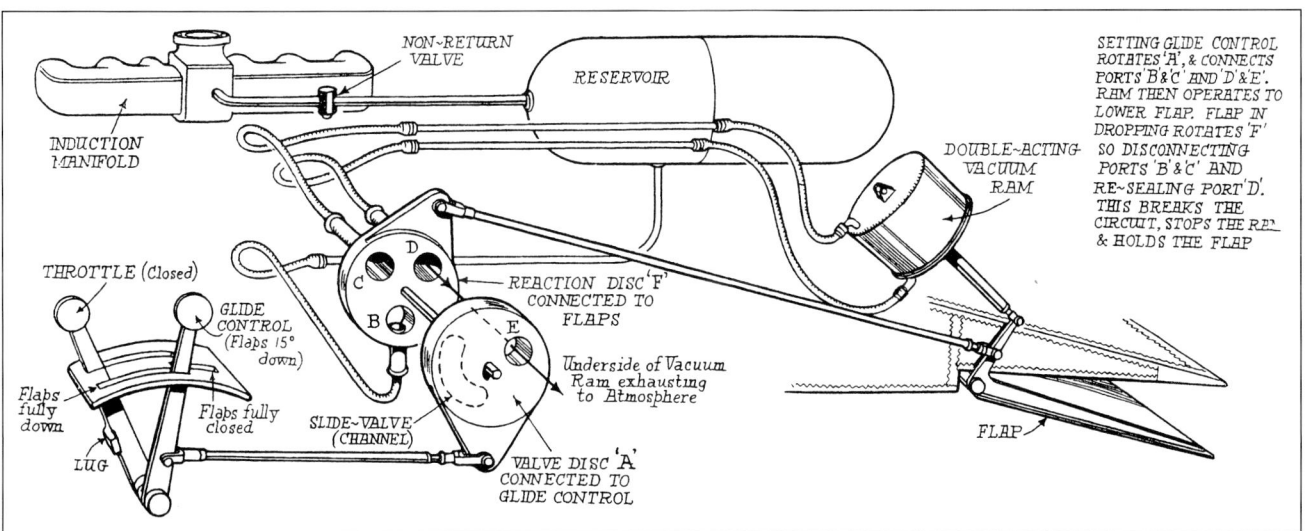

The layout of the controls, the third seat and the luggage compartment, are clearly shown in this cutaway sketch of the interior of the Monarch. The flap control is shown in the 25-degree position with the throttle half open. [Reprinted from 'Flight' 23rd June 1938]

The Miles Glide Control. [Reprinted from 'The Aeroplane' 22nd June 1938]

feature of these demonstrations was his 'hands and feet off' landing technique which he had first demonstrated on the Magister.

The introduction to the 1938 Specification to the Monarch read:
From the most successful aeroplane of 1937 we have developed the new Miles Monarch - and it concluded thus: *Many would have been satisfied with the success and acclaim with which the Miles Whitney Straight has been greeted, quite reasonably putting their trust in the fact that last year's aeroplane is likely to prove the world's best in its class for years to come. But we have re-designed, re-tooled and re-tested. The result is the Miles Monarch.*

The Monarch was luxuriously finished, both internally and externally and was pleasant and easy to fly, and at £1,250, proved to be one of the most popular private owner's aeroplanes. It could also be fitted with a Rotax Electric self-starter as an optional extra.

The Aeroplane for 22nd June 1938 reported that the Monarch was:*...the most ownable aeroplane in the world...* and, had it not been for the war, it would undoubtedly have been produced in greater numbers. Unfortunately, the main factory was fully committed to the production of the Magister and the Master, so it was not possible to produce the Monarch in the quantity necessary to meet the demand.

Hitherto, many reports have stated that all production Monarchs had been built 'on a small assembly line laid down by George Miles in the Repair & Service Department.' Indeed, Reg Kent, a foreman in that Department, recalled just that in discussion with Bert Clarke many years later, noting that George Miles had been appointed manager of the Repair & Service Department in about November 1938. However, it is now known that the first five Monarchs were actually built in 'Final Assembly' in the main factory. The 'Air Correspondent' of one local newspaper reported in early October 1938: *Construction of the Monarch has now been transferred to the Repair & Service Department.* This was probably in late September or very early in October 1938 and confirmation of this was made with the chance discovery of a photograph taken in late 1938 by P/O Gordon Owen Arscott, a pupil of 8 E&RFTS Woodley. This depicted two Monarch fuselages (with c/n's 792 & 793 in chalk on their fins), complete with centre-sections, undercarriage and tail units but with no windscreen, cabins or outer wings, being pushed from the Final Assembly, where their construction had obviously reached that stage, to the Repair & Service Department for completion.

The first production Monarch, G-AFGL, was sold to Airwork Ltd at Heston as a demonstrator. It is likely that this was also the aircraft which was fitted later with the R Sensand de Lavaud automatically variable-pitch propeller for test purposes, as the trials were carried out by Airwork Ltd. An account of a flight in the modified machine was given in *The Aeroplane* for 26th October 1938: *By courtesy of Airwork Ltd, we tried the airscrew fitted to their demonstration Monarch. It gives about 1,950 rpm etc, etc. Flight* for 17th November 1938 reported that the Monarch fitted with the de Lavaud propeller was *to be shown at the coming Aeronautical Exhibition at Paris* (held from 25th November to 11th December 1938).

Group Captain L G S Payne, the Air Correspondent of the *Daily Telegraph*, wrote a glowing account of flying the new Miles Monarch in the issue for Thursday, 11th August 1938, under the heading: *New Aeroplane that is almost fool-proof in landing. Descent in small area helped by Glide Control*

Airwork Ltd of Heston, gave me an opportunity to try out the new Miles Monarch, the latest of a long line of successful civil aircraft built by Phillips & Powis to the design of Mr F G Miles. I flew the machine from Heston to Reading aerodrome, where I found Mrs Miles engaged in the hangars. The wife of the designer is herself a pilot and a designer and has co-operated with her husband in drawing up the plans for some of his most successful aircraft.

In external appearance the Miles Monarch closely resembles its predecessor, the well-known Miles Whitney Straight. It is powered with same engine, the 130hp inverted de Havilland Gipsy Major, which is unquestionably one of the most reliable civil aero-engines in the world. In reality, the Monarch is a greatly improved machine embodying many new features and improvements in manufacturing technique. The aeroplane is now available either as a two-seater at £1,250 or as a three-seater at £1,325. These prices seem reasonable considering the up-to-date specification and the excellence of the finish.

The machine I flew was a three-seater, and I was able to try out its take-off with one, two and three people. On the flight to Reading I was accompanied by Mr Lacayo, one of the staff pilots of Airwork Ltd. Neither he nor I are lightweights, but the machine took off easily in about 150 yards. On the return journey to Heston I was alone. I was interested to note that the take-off with two people seemed almost as good as with one. After returning to Heston I took the Monarch off again with two passengers on board. A slightly longer run was required, but the machine came off easily and gave no impression of being overloaded.

The outstanding feature of the Monarch is the glide control by flaps which can be actuated with the throttle. This is a new feature. The Monarch is virtually fool-proof when it comes to landing. Even when there is no wind - and there was very little when I flew it - it can be flown in over the edge of the aerodrome at height of about 200 feet. Speed is reduced, the flaps are lowered, and the machine sinks to the ground in a steady glide of from 60-65 mph. Instead of landing far up the aerodrome, as might have been expected, one finds that the machine has stopped near the boundary of the aerodrome over which it passed. The difficulty of judging the glide-in to land, always a bugbear to inexperienced pilots is thus almost eliminated.

On any average-sized aerodrome one can be inside the area of the aerodrome before starting to land. Moreover, the pilot has the comforting feeling that, if he has made an error in his approach, he can immediately raise the flaps, either by opening the throttle or by advancing the flap lever, without any risk of causing the machine to sink. This feature of the Monarch, which is known as the Miles glide-control flap, must be tried to be appreciated. It certainly

C/n 792 and probably c/n 793 being pushed past the Reserve Training School, from the Magister Final Assembly Shop to the Repair & Service Department, for completion in October 1938. Strangely enough, both these aircraft still survive. [G Arscott via P Amos]

makes the aeroplane easier to land - and, if necessary, to land in a confined space - than any that I have previously flown.

Landing is also assisted by an exceptionally wide undercarriage. Bendix brakes and Lockheed air-draulic struts, which are designed to withstand a vertical drop of 15 feet a second. The cabin is roomy and luxuriously appointed. The field of view obtainable through the one-piece wind-screen is excellent. The machine is easy manoeuvrable, and the speed is all that the makers claim. I found that the machine, with the engine throttled down to 2,050 revolutions a minute would easily maintain its cruising speed of 125 mph. This, though not fast according to modern standards, should be ample for most private owners unless they belong to that limited class of civil pilots that is always trying to set up records.

In short, the Miles Monarch is a most pleasant and easy machine to fly and should be one of the most popular machines among civil pilots for some years to come.

An interesting letter from Airwork Ltd appeared in *The Aeroplane* for 7th September 1938 under the heading *The First Monarch*:

We note that in the current issue of your paper there is a statement that Messrs W S Shackleton have sold the first Miles Monarch (G-AFJU - PA) to Sir Victor Warrender. This is not correct. As you are probably aware we have been appointed the main selling agents for Great Britain for this type of aircraft and we have taken an order previous to that which we received from Messrs W S Shackleton. We have also received further orders since their order, and we have every reason to believe that quite a number of these machines will be sold during the current year.

The Miles M.17 Monarch G-AFGL with the Risensand de Lavaud automatically variable pitch propeller fitted by Airwork for trials, at Heston.   [Phillips & Powis via The Miles Aircraft Collection]

The official statement from W S Shackleton Ltd, which caused the letter to be written, announced that the concern had sold the 'first Miles Monarch for delivery to a private owner in England'. The ambiguous nature of this statement caused the misunderstanding. Another Monarch, OO-UMK, had just been sold to a member of the Belgian Government, Monsieur Camille Gutt (who later became Finance Minister to the exiled Belgian Government in Great Britain). When Belgium was invaded in May 1940, OO-UMK was flown to Woodley, where it was used by Phillips & Powis for communications duties, first with the 'B' mark U-0226, and later as G-AGFW, before being returned to its owner with RAF marks later in the war.

Sir Victor Warrender was another prominent Monarch owner, having taken delivery of G-AFJU in September 1938. He was MP for Grantham and Financial Secretary to the War Office and an article on him appeared in the *Miles Magazine*: *Although until 1938 he had never flown an ordinary aeroplane (having previously owned and flown an autogyro) he was able to fly away in his Miles Monarch after only half-an-hour's dual. This is certainly no mean achievement on which we congratulate the pilot, who, after calling the Monarch the ideal private owner's aeroplane, says: 'I find her very easy to fly, in fact she really flies herself. Landings and approaches are as simple as falling off a log and require the minimum of human judgement'.*

At the opening of the large new factory extensions on 27th January 1939, Miles gave the Air Minister, Sir Kingsley Wood, a flight in the prototype Monarch, and by May 1939, eleven Monarchs had been completed.   Were it not for the outbreak of war in September 1939, the Monarch would undoubtedly have been built in quantity and C L Nash recorded that c/ns 799 to 810 were reserved for the next batch of twelve Monarchs.

A Miles M.17 Monarch undergoing tests in the Repair & Service Dept. The Magister s/n N3806 was delivered on 12th November 1938, so the M.17 could possibly have been c/n 792, which became G-AFLW.
[Phillips & Powis via The Miles Aircraft Collection]

An HQ 41 Group Return (Ref: AIR 17/33), entitled 'Arrivals and Despatches for Week Ending and Stock in 41 Group ASUs as at 0900 hours 6-12-45', gave the numbers of Monarchs surviving as follows: 2 Stock as at 0900 6-12-45; 2 Awaiting MAP disposal instructions - Non-Effective Aircraft. These non-effective aircraft require maintenance within 41 Group units.

Two of these Monarchs were probably W6461 (G-AFCR) and X9306 (G-AFJU), which were both sold at the first sale of impressed light aircraft, held at 5 MU Kemble in December 1945. For the record, there were actually 4 Monarchs 'in stock' in the RAF at these dates, the other 2 being W6463 (G-AFRZ), which was sold at 5 MU Kemble in 1946, and W6462 (G-AFJZ), which was SOC at 60 MU Salvage Centre, York, on 16th December 1948.

Like the earlier Miles' types, the Monarch was not without its racing successes and in 1957, G-AIDE was flown by Walter Bowles in the National Air Races at Baginton held on 12/14th July 1957; winning both the Goodyear Air Challenge Trophy and the Goodyear Cup at 131.5 mph and gaining 3rd place in the King's Cup Air Race at a speed of 137.75 mph (an event where Miles aircraft gained a hat-trick, with Fred Dunkerley in the Sparrowjet G-ADNL and Sqdn Ldr James Rush in Falcon Six G-AECC coming in first and second). At the National Air Races at Baginton the following year, on 9/12th July, Bowles flew G-AIDE to 13th place in the 1958 King's Cup at a speed of 139.0 mph and won the Norton Griffiths Challenge Trophy at 136.16 mph and second place in the Osram Cup.

G-AFLW was flown to third place in the King's Cup in 1983, and to fifth place in the King's Cup in 1984; the pilot, Dr Ian Dalziel thereby gaining the British Air Racing Championship.

Only three Miles Monarchs survive to this day; G-AFJU, now being rebuilt to flying condition by Aero Classique in France for Peter Bishop, having previously been statically displayed in The Museum of Flight, East Fortune Airfield, Scotland; G-AFLW, which was flown until about 1995 and is stored at White Waltham; and G-AFRZ (flown after the war as G-AIDE), which was stored for many years at RAF Cosford and has recently been moved to Sleap, where it is hoped it will, in due course, be restored to airworthy condition.

For details of Miles M.17 Monarch production see Appendix 22.

SPECIFICATION AND PERFORMANCE DATA

Engine:	130hp DH Gipsy Major Srs I, with Fairey Reed metal airscrew
Dimensions:	span 35ft 7in; length 25ft 11¾in (note the CofA for G-AFJU gives the length as 25ft 7¼in); height 8ft 9¼in; wing area 180 sq ft; aspect ratio 7.04 to 1; wheel track 9ft 0in; width of cabin 3ft 7in
Weights:	three-seater: empty 1,390lb; petrol (30 gal) 225lb; oil (2½ gal) 23lb; pilot 170lb; 2 pass 320lb; luggage 22lb; payload 342lb; disposable load 800lb; max permissible AUW 2,150lb; wing loading 11.92lb/sq ft
Performance:	at 2,000lb: max speed at sea level 145 mph; max speed at 5,000ft 140 mph; max speed at 10,000ft 134 mph; cruising speed at sea level (73% power/weight ratio - "pwr") 130 mph; cruising speed at sea level (70% pwr) 125 mph; stalling speed at sea level 45 mph; stalling speed with flaps 40 mph; take-off run to unstick (5 mph wind) 435ft; landing run (5 mph wind) 300ft; initial rate of climb at sea level 850ft/min; climb to 10,000ft 16.40 mins; service ceiling 17,400ft; absolute ceiling 20,000ft; normal range (73% pwr) (30 gal) 600 miles, duration 4.6 hrs; normal range (70% pwr) (44 gal) 620 miles, duration 5.0 hrs; max range (70% pwr) 910 miles, duration 7.3 hrs
Price:	£1,250 for two-seater and £1,375 for the three-seater.

The first Miles M.18, 'U2', later known as the M.18 Mk.I, shortly after its first flight on 4th December 1938.
[Phillips & Powis via The Miles Aircraft Collection]

Chapter 31:
Miles M.18 Trainer Mk.I, Mk.II, Mk.III and Mk.IV HL

By 1938, there were a considerable number of Magisters in service but Miles felt that an even better trainer could be designed in the light of experience gained. The basic design of the Magister was not only six years old but it owed its origins to his first attempt at producing a low-wing monoplane.

In 1936, Walter Gustav Kaepelli joined the team at Woodley as chief stressman. Kaepelli, who was born in Zurich, Switzerland, on 8th November 1913, had come to England in 1934. He was taught to fly by British Air Transport Ltd at Gatwick, in DH.60 Moths, and qualified for his 'A' licence in June 1934. In June 1940, he was granted British nationality, and he then changed his name to Capley. Soon after his arrival, however, Miles gave him the job of designing a replacement for the Magister, and he was to become very well respected by Miles in later years.

The main requirements of the new design, which was designated the M.18, were to be ease of handling and serviceability. Although constructed of wood, as in previous Miles aircraft, the M.18 was to bear little resemblance to its forebear and the required characteristics were obtained, albeit at some sacrifice in performance, by the adoption of an entirely new wing of practically constant chord and thickness. The root section was a modified Clark YH with 75in chord, and with a thickness of 0.18c, while the tip section was NACA 4415, with 66.3in chord and a thickness 0.15c. The fuselage was 4in wider than that of the Magister in order to give considerably more 'elbow room' for the crew, and the fin and rudder was placed at the extreme end of the fuselage. Split flaps were incorporated, single piece wrap-round windscreens were fitted to both cockpits and it had a fixed undercarriage with spats.

Powered by a 130hp DH Gipsy Major I engine, the new machine was never named, although it was sometimes referred to as the 'Magister Mk.II'. The prototype was first flown, by Miles, as U2, on 4th December 1938 and he noted that the M.18's controls were light, well-harmonised and effective, right down to the stall, which was itself innocuous. Don Brown recalled that at first there was a tendency for the tailplane and elevators to blanket the fin and rudder in a spin and although this did not cause a serious problem, the fin and rudder were moved forward 22 inches on the second and subsequent M.18s to improve spin recovery characteristics. The first prototype, however, retained the fin and rudder in the original position throughout all its subsequent modifications.

Following makers trials, the M.18 was flown to the A&AEE Martlesham Heath for CofA tests and preliminary service assessment, to Contract No.B983811/39, dated 7th March 1939. In their subsequent Report No.M/738, issued in April 1939, they noted that *'Construction was wholly of wood, the wing loading was 10 lb/sq ft and it was fitted with split flaps. It was noted that the type was not designed to be flown solo from the front seat and that the aircraft was of very light construction. It was thought not to be sufficiently robust for service training from rough airfields'*. For the trials, the M.18 was flown at 1,153 lbs, plus 200 lb of fuel (26 gallons), 22 lb of oil (2½ gallons), pilot and passenger 400 lb, ballast 25 lb, giving a total AUW of 1,800 lb'.

Entry and exit was by doors on the starboard side and was considered to be satisfactory, although they thought that hand grips should be provided. The cockpits were comfortable for medium-sized pilots but too small for larger ones. The seats were not adjustable for height or the rudder bars for length. The view from both cockpits was satisfactory but the windscreens were flimsy and could become misshapen. Communication between cockpits was by speaking tubes, which worked well.

Taxi-ing was good in winds up to 30 mph and the brakes were adequate but the differential braking was poor. Take-off was easy and a tendency to swing could be easily checked. Initial rate of climb was very good. The ailerons were effective but had excessive friction in the control cables. Harmonisation of the controls was not regarded as good but the aircraft appeared stable in all axes. The stalling speed was 50 mph and was straight and gentle and could be easily controlled. Aerobatics were easy and pleasant.

The aircraft was dived up to 210 mph where it remained steady and had no undue vibration. Recovery was normal but the windscreen in the rear cockpit vibrated in a dive. Spinning was tested and found satisfactory. (In view of Don Brown's comments above regarding spinning, this seems surprising - PA). The M.18 would not spin with flaps down. Approach and landing was simple but opening up for an overshoot put the flaps from fully down to maximum lift position.

In their report, the A&AEE gave the following recommendations:

i) Fore and aft trim had separate controls as had the elevator tab and springs. They should be replaced by a single lever.
ii) The flaps should be intercoupled with the elevator to avoid a change of trim when flaps were lowered.
iii) The windscreens should be modified to reduce draughts. They were also considered to have been too flimsy.
iv) Friction in the aileron circuit should be eliminated.
v) Seat and rudder bar in rear cockpit to be made adjustable.
vi) Rear instrument panel to be replaced by a standard service panel.
vii) Mixture control should be incorporated in the throttle box.
viii) There was no provision for a blind flying hood.
ix) There were no control locking devices fitted.
x) There was no strong-point to take the weight of the aircraft if it overturned on landing.

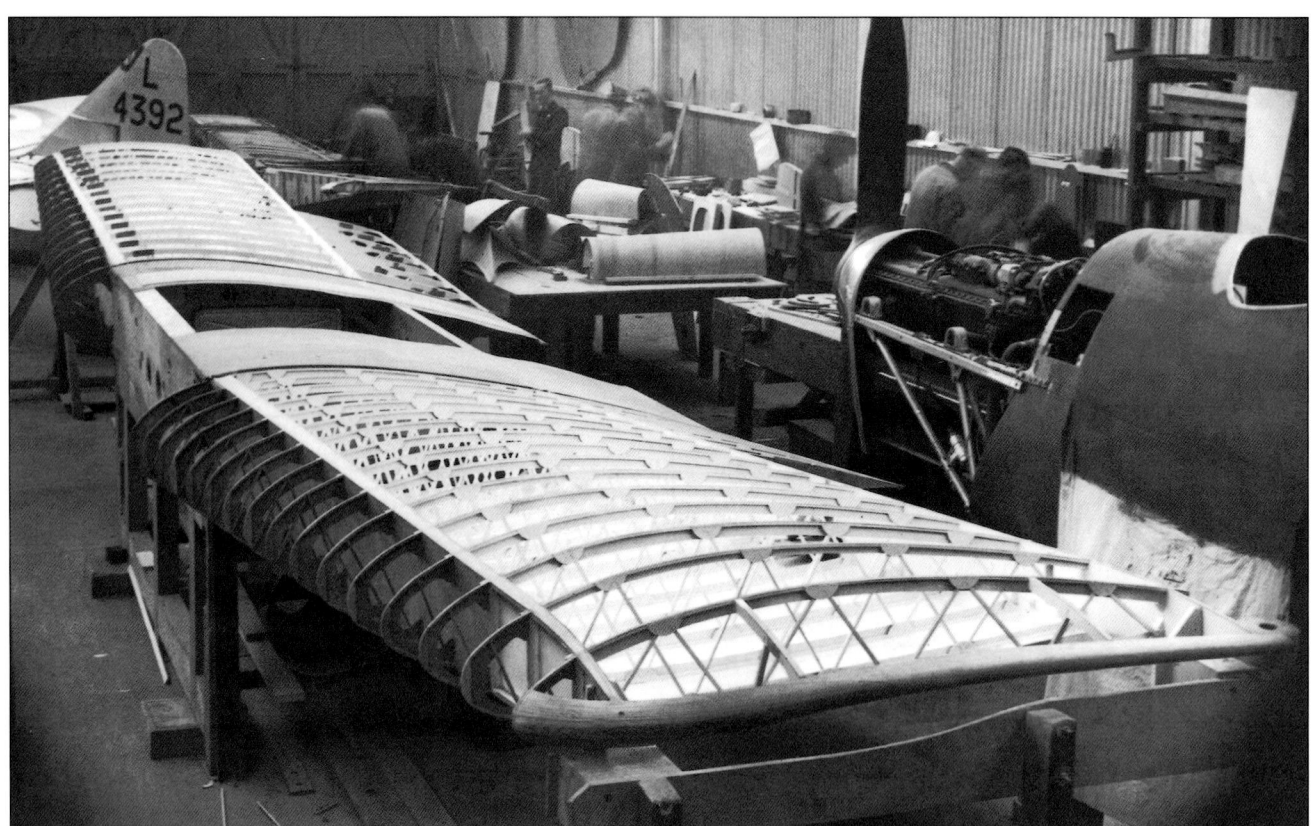

The wing of the prototype Miles M.18 under construction in the Experimental Department. The first prototype Miles M.16 Mentor, L4392, seen in the background, was delivered in April 1938. [Phillips & Powis via The Miles Aircraft Collection]

The Miles M.18 Mk.I, shown after conversion to tricycle undercarriage configuration. Note the small windscreen, long stroke undercarriage, and retained tailwheel. [Phillips & Powis via The Miles Aircraft Collection]

Chapter 31: Miles M.18 Trainer Mk.I, Mk.II, Mk.III and Mk.IV HL

Miles M.18 Mk.I prototype with 'P', short stroke undercarriage, fairings over protruding undercarriage on top of wing, and large three-piece windscreen.
[Via P Amos]

They concluded that if these deficiencies could be removed, the M.18 could be a viable *ab initio* trainer - but it was not robust enough.

The same contract number was also used for the 'Loan of PV Trainer No. M.18', dated 11th May 1939, and this presumably also covered its return to the A&AEE Martlesham Heath for further trials following the incorporation of the recommended modifications.

On 12th June 1939, chief test pilot Bill Skinner carried out spinning tests on the M.18 fitted with an early form of anti-spin strakes. In his report, dated 13th June 1939, he described the modifications as 'Fins along top side of fuselage immediately in front of tailplane'. Tested with an AUW of 1,583.5 lbs and with the CG 25.39" aft of leading edge wing root chord, he reported:

'Spin No.1 - Entry from a slow turn employing full aileron drag. Spin commences by being fairly steep and fast, after about 8 turns it went very flat. Strong tendency for the ailerons to snatch when applied in the spin. The rudder was not effective. Full opposite rudder was applied at the same time the control column was put fully forward and into the spin. After approximately 4 turns, the left wing went down, followed slowly by the nose. After a further two turns, the aircraft stopped spinning. No further spinning was attempted as this spin had shown that the modification had not cured the trouble. I cannot help feeling that the solution to the problem lies in eliminating aileron drag'

In August 1939, an M.18 wing was fitted to the Hawk Trainer Mk.III, U6 (G-AEZS), probably for comparison trials but no details have emerged.

Although this publication is intended to detail Miles aeroplanes produced up to September 1939, it is expedient, owing to the 'one-off' nature of the M.18, to cover all 4 prototypes here for completeness.

The prototype M.18 was later submitted for service assessment, where it evoked favourable comment from all who flew it and, as a result, the type was ordered into immediate full-scale production. Contract Acft/1272/C.23(c) was raised in about July 1941 for the manufacture of 200 single-engined Trainers (M.18) (allotted serials DW515-562; DW578-614; DW635-684; DW712-756 and DW777-796). However, hardly had this decision been taken, than the entry in the ledger was crossed out in red ink, with the note 'contract not placed'. There were undoubtedly good reasons for this decision but they have yet to be found, and the obsolescent Tiger Moth biplane remained in production as the standard RAF elementary trainer until July 1944.

In 1941, probably as a consequence of the arrival of American aircraft with tricycle undercarriages for the RAF, Miles modified the prototype M.18 to have a tricycle undercarriage, to give pilots a feel for the different take-off and landing techniques. In order to effect the modification, the nosewheel had to project into the front

The same aircraft now fitted with a four-bladed propeller.
[Phillips & Powis via The Miles Aircraft Collection]

cockpit, so this was faired over and the aircraft was flown from the rear cockpit, which was fitted with a very small three-piece windscreen. The tailwheel was retained, as the cg was so close to the line of the main wheels that the aircraft could be landed as a tricycle or tailwheel aircraft! The first recorded flight of the M.18 (by then with the wartime 'B' mark U-0222) with the tricycle undercarriage, was made by Hugh Kennedy on 11 August 1941 and it is believed that it was this version which was given the RAF Experimental Aircraft Number 173 for identification purposes.

Early photographs show this modification, but with no prototype 'P' in a circle on the fuselage but subsequent photographs show a later modification to U-0222 with a much larger three-piece windscreen and shortened nose and main undercarriage legs (the latter protruding above the wings into small fairings), a 2-bladed propeller and with a prototype 'P' on the fuselage.

U-0222 (still with a tricycle undercarriage) was later fitted with a small diameter 'four' bladed (two-bladed on two-bladed) propeller and a photograph, dated September 1941, showing this modification, noted also that it was fitted with a DH Gipsy Major I engine. However, engines on the M.18 were often changed, from Gipsy Major to Cirrus Major, and vice versa, etc, so it is interesting to note which engine was fitted at this time. The four exhaust stubs were also replaced by a single exhaust pipe at some time.

The long suffering U-0222 was also converted at some time into a glider but it is not known when, or if, it was ever flown in this configuration. It was also given the RAF Experimental Aircraft Number 204 for identification purposes, but to which of the many configurations of this machine this applied is not known.

An interesting experiment, about which little is known, involved an early use of plastic in aircraft structures and this was carried out in 1942 by Aeroplastics Ltd, c/o The Fairey Aviation Co Ltd, Hayes. Mr S Hart-Still, a director of the company had developed a new material of construction, known as *Aeroplastic*, which comprised a resin-bonded fibre-base material of the thermo-setting type. This had been specially developed for structural applications, and one of the test pieces manufactured was a moulded tailplane for the M.18. Photographs, stamped The Fairey Aviation Co Ltd and dated

The Miles M.18 Mk.I Tricycle, but now fitted with four-bladed (2 + 2) propeller. [Via R W Simpson]

The long-suffering Miles M.18 Mk.I with 'clipped' wings, in October 1942. [Phillips & Powis via The Miles Aircraft Collection]

19th March 1943, show a complete M.18 tailplane (less elevator) with a very smooth finish, but gave no indication as to whether this was ever actually fitted to an aircraft. In 1942, the RAE issued a Report entitled 'Moulded tailplane for M.18 Trainer aircraft', (ref AVIA 6/13017; also AVIA 6/13035).

In late 1942, the wingspan of the first prototype, still with tricycle undercarriage, was reduced from 31ft 0 ins to 22 ft 1.5 ins, and the aspect ratio from 5.2 to 3.7, in order to conduct tests on the effect of wings with very low aspect ratio. In this form it had a two-bladed propeller and possibly carried 'B' mark U-0239, as Hugh Kennedy recorded many flights in it with this mark about that time.

A general arrangement drawing entitled 'GA Clipped Wings' was dated 1.10.42 and a plan view drawing titled 'T.18 with clipped wings', possibly dated 1.9.42 (indistinct on photograph) gave the specification as follows:-

	Full span	Clipped
Span	31.00ft	22.12ft
Gross wing area	183.2sq ft	134.4sq ft
Aspect ratio	5.24	3.63
Aileron area	17.10sq ft	8.25sq ft
Tare weight	1247.50lb	1193.00lb
Engine HP	130	130

Hugh Kennedy recorded his first flight in U-0239 on 29th October 1942 and this may have been the first flight of the 'T.18 with clipped wings'. However, with the increase in wing loading from 6.81lb/sq ft to 8.88lb/sq ft, the take-off performance suffered considerably and as a result it took nearly the whole length of the airfield to unstick. This problem caused the trials to be swiftly abandoned and the outer wing sections were replaced soon afterwards!

In early 1943, a series of tests was initiated with the Miles Messenger to ascertain the feasibility of landing a light aeroplane onto a small deck mounted on the stern of merchant ships, with the aid of a net as a safeguard in the event of failure to engage the arrester gear.

It was proposed to use the prototype Messenger, renamed the M.38A Mariner for the trials but, before these took place, George Miles initially undertook high-speed taxiing tests into a net with U-0222, and although he actually later flew the M.38A into a net at Woodley, the M.18 was never flown into the net.

There were some attempts at this time to generate overseas interest in the M.18. In a Ministry of Aircraft Production report on 'Applications during the past twelve months for the export or sale of licences for manufacture abroad', dated 3rd December 1943, it was stated, under Chile, that:

An application by Miles Aircraft Ltd to sell manufacturing rights of the Magister, M.18 and M.38 is under consideration. One difficulty in the way of the proposal, is that of finding a suitable local timber in substitution of spruce. Arrangements have been made for the Chilean Ambassador and Air Attaché and certain Chilean industrialists to visit the firm.

In the same report, under Brazil, it was stated that:

Miles Aircraft Ltd were authorised in February 1943 to sell the manufacturing licence for the M.18.

Nothing came of either of these proposals, however, neither did anything come of a later note in the same report which stated, under Australia and New Zealand, that:

Miles Aircraft Ltd have been authorised to furnish details of Magister and M.18 but they have not yet applied to sell the manufacturing rights of these aircraft.

In June 1943, U-0222, still with its tricycle undercarriage but with its original full-span wings, was temporarily fitted with the new two-cylinder, horizontally-opposed, two-stroke Jameson engine, built by J L Jameson Ltd of Ewell, Surrey. Although this engine was intended to have been supercharged (when it would have delivered 90hp), the one fitted was normally-aspirated and although ground engine-runs were carried out, these were considered unsatisfactory and the M.18 never flew with this engine, neither were cowlings fitted. With the engine's extremely light weight, it was necessary to extend the nose of the M.18 in order to keep the c.g. in the right position.

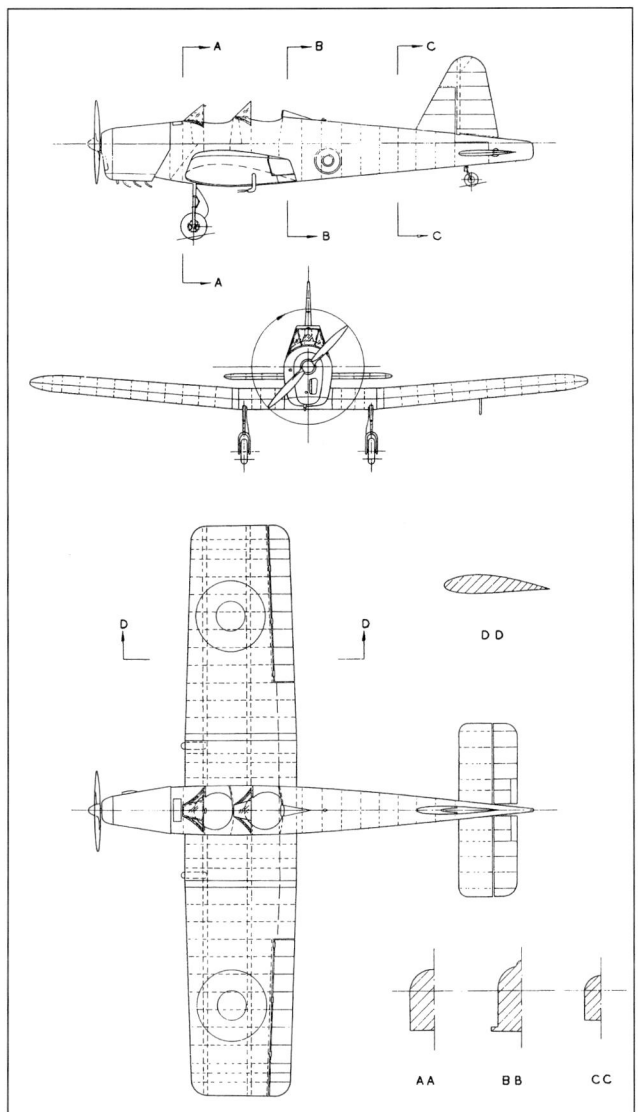

The Miles M.18 Mk.II.

A confidential Miles Aircraft 'Second Report – Aircraft', dated 30th May 1945 (which detailed the wartime production of Miles aircraft), stated under 'Unofficial development work initiated by the firm without any known official recognition' (this should not really have surprised 'the Men from the Ministry' as Miles could not stand bureaucracy or 'Red Tape' - PA), that 'Fitment and trials of a Jameson 2-stroke engine have been proceeding'.

In 1945, Miles Aircraft Ltd (probably George Miles, given his interest in engines) suggested to Jameson that they should design and build a prototype flat four-cylinder engine of between 110 and 120hp. This was probably about the time that experiments with the earlier two-stroke twin engine were finally abandoned. The result of this enquiry was for Jamesons to produce the 110hp Jameson FF normally-aspirated, horizontally-opposed, four-cylinder, four-stroke, air-cooled engine. This engine was delivered to Woodley for installation in the first prototype M.18, still with its tricycle undercarriage but otherwise in its original form.

The engine was designed to give a maximum output of 110hp at 3,300 rpm, driving a two-blade airscrew through a 0.619:1 reduction gear. A very low fuel consumption was claimed for this engine as a result of the design of the induction system. (It has been suggested that it was the fitment of *this* engine which required the nose of the M.18 to be extended to keep the c.g. in the right position, but this was actually only necessary for the trials of the two-stroke Jameson engine undertaken in June 1943).

One or two hops with the new Jameson engine were carried out towards the end of 1945 and George Miles actually flew the M.18 round the circuit at Woodley, but always within gliding distance of the aerodrome, with the engine completely uncowled.

Comparative figures for the 90hp Jameson and the 90hp Blackburn Cirrus Minor, an inverted air-cooled engine of similar power, were as follows:-

	Jameson FF	Cirrus Minor II
Cruising output	90hp @ 3,000 rpm	90hp @ 2,300 rpm
Fuel consumption	4.8 gal/hr	6.6 gal/hr
Weight	280lb	250lb

In Miles Aircraft correspondence, dated 12th February 1944, a forward swept wing was planned, or suggested, for trials on the M.18 Mk.I but nothing further is known of this project.

The second M.18, the Mk.II, was fitted with the more powerful 150hp Blackburn Cirrus Major III engine and had the fin and rudder moved forward 22 inches. It was first flown, as U8, but by whom is not known, in November 1939. Following trials by the A&AEE, an appendix to the initial report was issued on 27th May 1941, but the same complaints, plus a few more, emerged:

The seats and rudder bars were not adjustable and the rear seat was too high but the cockpits now had standard blind flying panels. The fuel contents gauge on top of the tanks in the wings was not easy to read from the rear cockpit. During night flying, the three in one navigation light on top of the fuselage behind the pilot was unsatisfactory because of reflections in the windscreen. Cockpit lighting was good and all instruments were now luminous but the fuel gauges could not be read. It was easy to fly at night and the view of the flarepath was good from both seats.

Further tests were undertaken by the A&AEE, after modifications had been carried out, at an AUW of 1,944lb and these were covered in a report dated 2nd July 1941. There was little change in performance from the first prototype. However, the report gave the following performance figures:

Max speed at sea level 130 mph; service ceiling 12,500 ft take-off run 260 yards and landing distance 495 yards and it should be noted that, while the take-off distance was similar to that obtained during firm's tests, 240 yards, the landing distance differs considerably from that given in the works trials, which was 177 yards.

The M.18 Mk.II, by then with 'B' mark U-0224, was delivered to 759 (Fleet Requirements) Squadron at Yeovilton in January 1941, on 'Short loan, possibly for Service Trials', before going to the CFS in February 1941, where it was flown by Flt Lt H de C A Woodhouse on February 5th. An undated Contract No.Acft/1865 was placed sometime in 1942 for the 'supply of one M.18 aircraft' but, unfortunately, no further details of this are known. However, the timing of this would seem to coincide with the M.18 Mk.II being given the serial HM545 for service assessment with the Empire Central Flying School at Hullavington in late 1942. The service trials were apparently successful and Don Brown recalled that: *In 1942 it was again ordered for immediate full-scale production but again the Ministry reversed their decision.*

Although there is no record of a second contract being placed for the production of the M.18, a note in the confidential Miles Aircraft report of 30th May 1945 (mentioned above), stated that: 'An order was received in 1942 for M.18's - subsequently cancelled before put into production'. This time, no serial numbers appear to have been allotted. In the main body of the aforementioned report, under the 'History of Miles M.18 and M.28', it was stated that:

At various times during 1943 these designs were pressed by Miles Aircraft Ltd as a new elementary trainer to replace the Tiger Moth. In April 1943 the Chief Executive planned to replace Tiger Moth by Miles M.18 with full production at Morris Motors. A meeting was held by DGT (Air) and both these designs were considered and turned down. It was decided to draft an entirely new specification, and put this out to tender by a number of firms. Production of the M.18 is again under discussion at present for elementary training due to the termination of Tiger Moth production at Morris Motors Ltd, in July 1944.

It really was a wonder how we ever got anything done during the war, with supposedly responsible people 'in high places' being offered two tried and tested, modern, monoplane trainers - the M.18 and M.28 - and then turning both down in preference for a completely new specification 'to be put out to tender'. This 'logic' could only have been inspired by the Ministry's attitude towards Miles and their disapproval of his way of doing things - i.e. not always bothering to wait for the wheels of the Ministry to slowly turn but just getting on with the job. On reflection, it was thinking such as that outlined above which was to result in such 'trainers' (and I use the term loosely) as the ill-conceived Percival Prentice.

Miles M.18 Mk.II was allotted serial HM545 on 11th December 1942 for the first series of service trials.

[Phillips & Powis via The Miles Aircraft Collection]

The Miles M.18 Mk.II reverted to 'B' mark U-0224 in 1943, following the service trials.

[Via B Clarke]

Chapter 31: Miles M.18 Trainer Mk.I, Mk.II, Mk.III and Mk.IV HL

Miles M.18 M.III U-0238. [Via B Clarke]

The Miles M.18 Mk.IV H.L. JN703, in its original form.
[Phillips & Powis via R W Simpson]

In the event, only two further M.18's were built.

The M.18 Mk.III was based on the second prototype but was fitted with a sliding canopy over both cockpits in a style similar to that used on the Canadian DH.82C Tiger Moth. Fitted with a 150hp Blackburn Cirrus III engine, the M.18 Mk.III was first flown, as U-0238, by Don Brown (accompanied by George Miles) on 17 October 1942 and was given the RAF Experimental Aircraft Number 194 for identification purposes.

Other reports of the first three M.18's being fitted with engines differing from those referred to above could well be correct, as the engines were almost certainly changed at various times in each of the first three prototypes.

The fourth M.18 resulted from an RAE proposal in late 1941 for Miles to build a small, wooden aircraft (a Magister was in fact suggested) in which would be incorporated a special high-lift wing, the aerodynamic details for which would be supplied by the RAE. This was required in order to obtain full-scale data on the elaborate arrangement of high-lift devices (comprising controllable leading-edge slats over the whole span, together with full-span slotted flaps of 0.40c with inset ailerons), before approval was given for the projected shipborne Supermarine S.12/40 (later Seagull) to be built.

To counteract the anticipated pitching moment with the high-lift devices, the area of the tailplane and elevator was increased by over 20% from those of the standard M.18, which had normal split trailing edge flaps. The Instruction to Proceed (ITP) with the construction of a prototype was dated 24th January 1942, at an estimated cost of £5,000. F H Robertson AFRAeS was in charge of the Experimental Design Office at Phillips & Powis at that time and he was given the difficult task of modifying an M.18 to meet the RAE requirements.

Robertson later wrote an article on his design, the M.18 Mk.IV High Lift, entitled: 'A Guinea Pig for the Seagull' and this appeared in *The Aeroplane* for 30th April 1948.

Designated M.18 Mk.IV HL, this machine was fitted with a 150hp Blackburn Cirrus Major III engine and was first flown, as U-0236, by Tommy Rose in October 1942. It was given the RAF Experimental Aircraft Number 147 for identification purposes, and was later allotted the serial JN703. Delivered to the RAE in January 1943, it achieved, during tests, a C_L max (lift coefficient) of 3.2 (measured in flight), with slats fully open and flaps at 30 degrees. However, in this configuration the aircraft was laterally unstable, and aileron control was poor at low speeds, which bore out almost exactly the RAE wind-tunnel forecast.

The M.18 Mk.IV HL was probably the first aeroplane to fly in this country, possibly even in the whole World, at a lift coefficient in excess of 3 and the only serious criticism levelled at the aircraft's performance was the lack of adequate lateral control near the stall. Robertson said he thought that the ailerons were not big enough, but he believed that the RAE later fitted spoilers, in addition to the ailerons, with quite good results.

In July 1943, in a further effort to overcome these shortcomings, the aircraft was fitted with upturned wingtips and the fin area was increased at about the same time, by the simple expedient of fitting a large triangular 'glove' over the leading edge of the original fin. This was secured by two wing nuts in order to be able to change the fin shape at will. The aileron area was increased in December 1943, but the handling characteristics were still considered to be unsatisfactory. It is of interest to note that a photograph of JN703 with the upturned wingtips, shows that it still retained the early RAF roundels.

Cdr Eric Brown first flew JN703 at the RAE Farnborough on 18th May 1944, where an investigation into the stalling characteristics of the wing was being carried out. He found that:

In flight the control deficiencies became immediately apparent. The ailerons were fairly light but their feel and effectiveness was somewhat lost by the large amount of friction throughout their range of movement. The effect of the slats in the open position (achieved by a mechanical handle) was to cause a slight lag in response, which developed into sluggishness when the flaps were lowered (also by a mechanical handle). The elevators were light and effective in their range, but inadequate to control the aircraft with slats open and flaps down under power-off conditions. However, the inadequacy of the rudder control was the limiting factor in carrying out normal manoeuvres such as sideslips and steep turns, and the rudder was almost completely useless as an aid to taxi-ing.

I found that the aircraft could not be stalled with the slats open, irrespective of whether the flaps were up or down, even with the stick right back and elevator trim wound fully aft. With the slats closed, the flaps-up stall occurred sharply at 55 mph after some buffeting warning, while with flaps down the right wing dropped very sharply and without warning, putting the aircraft into a steep

Miles M.18 Mk.IV H.L. s/n JN703 in its final configuration with upturned wingtips. Note the censored background.
[F G S via P H T Green Collection]

right-hand dive. As a result of these tests the RAE recommended modifications to the aircraft which were presumably incorporated by Miles Aircraft.

The use of an existing light aircraft for the tests of a new wing of unknown handling qualities, modified at a moderate cost, avoided the ordering and construction of a new and expensive aircraft to test a theory which was ultimately to prove unsuccessful. As a result of the tests, the projected shipborne aircraft was not proceeded with.

During the war, George Miles had been experimenting with a new engine, while Miles had been developing a new lightweight electric autopilot, and shortly after the end of the war, they both decided to extend the range of the firm's activities still further by producing their own electric propellers and electric actuators.

A number of fixed-pitch wooden propellers had already been built and tested during the war, with the ultimate objective of actually producing an electrically-controlled (in the pitch-change sense, by an electric switch) variable-pitch fully-feathering propeller.

There is some dispute as to whether the prototype was manually or electrically operated, but it was designed, built, and fitted to an M.18 (probably the first prototype). Owing to its fully feathering capability, dead stick landings became not infrequent, as the tests were rarely conducted at an altitude which would permit the rather slow process of unfeathering (electrically or manually) to be completed. Ron Dack was involved in the design of a 'production' version but, owing to the financial crisis in November 1947, this was never completed or run. It should be noted that neither of these were constant-speed propellers.

Of these four variants of the M.18 built, at least ten recognisably different versions were to ultimately appear (there could even have been others), and these were flown under thirteen different identities. It is rather surprising, considering all the modifications and trials which were carried out on these four aircraft during the war, that one, the M.18 Mk.II, G-AHKY, should have survived to the present day. Unfortunately, although still in relatively good condition - it was still flying up to 1989 - this may now only be seen on static display in the Museum of Flight at East Fortune.

For details of M.18 production see Appendix 23.

SPECIFICATION AND PERFORMANCE DATA

M.18 Mk.I

Engine:	130hp DH Gipsy Major or 150hp Blackburn Cirrus Major
Dimensions:	span 31ft 0in; length 24ft 10in; height 9ft 4in; wing area 183 sq ft; aspect ratio 5.2
Weights:	empty 1,300lb; fuel (24 gal) 180lb; oil (2.5 gal) 23lb; pilot 200lb; pass 200lb; AUW 1,903lb; wing loading 10.6lb/sq ft
Performance:	max speed 130 mph or 135 mph with the Blackburn Cirrus III; cruising speed 117 mph; stalling speed 56 mph IAS with flaps up, 50 mph with flaps down; run to unstick 720ft; distance to 50ft, 1,260ft; landing run 531ft; landing over 50ft, 840ft; rate of climb 780ft/min; time to 5,000ft, 8 min; time to 10,000ft, 22 min; service ceiling 12,400ft; absolute ceiling 14,100ft; range 376 miles; duration 3.2 hr

It should be noted that dimensions quoted for the span and length of the first prototype, shown in other sources, quote differing figures from those shown above as follows - span 32ft 2in; length 24ft 8in; and span 30ft 11in; length 24ft 11in.

M.18 Mk.IV HL

Engine:	150hp Blackburn Cirrus Major
Dimensions:	span 30ft 0in; length 24ft 10in; height 9ft 4in; wing area 147sq ft; aspect ratio 6.12; wing section, root NACA 23018, tip NACA 23010
Weights:	empty 1,420lb; AUW 2,000lb; wing loading 13.6lb/sq ft.

Appendix 1: Joyrides – 5/- Flights

The story of a typical day's joyriding, entitled '5/- Flights' appeared in the *Miles Magazine* for February 1938. Its female author was anonymous but may have been Nancy Birkett and it is reproduced here to give a feel for those carefree days:

5/- Flights

Pilots 1 and 2 were F G Miles and C L Pashley. The years ago were many.........

August Bank Holiday used to be a good day for joy-riding and we had hoped to make a clean-up with our two Avros, operating from two fields about ten miles apart. Our arrangements were all made and some advance publicity had been done, but alas! a previously undiscovered iron stake in the long grass of one of our joy-ride 'aerodromes' bent one Avro's undercarriage the night before, and the wretched machine had to return home on a lorry. We were pretty good at quick repairs in those days, but an undercarriage is bound to take a certain amount of time to mend, and we knew there was no chance of getting it through by Monday morning, though we had faint hopes for the evening.

Bank Holiday dawned bright and hot - the ideal day for raking in the five-bobs. We decided to try to operate from both fields with one aeroplane, and to hope that by the time the evening rush started our second machine would be ready. We sent a car off to one field with a book of tickets and instructions to our tout to try and gather a small crowd for our arrival timed for two hours later. We set off in the one remaining aeroplane, loaded with pilot, ground engineer and myself as female tout. The pilot took nothing. The GE took a complete kit of tools, a tin of petrol and a syringe to help start the old Le Rhone Gnome, ground screws and lunch in a handkerchief. I took books of tickets, a bag of silver change (very heavy), a camera (never used), and lunch in a paper parcel. At the last minute we added three bags of cherries and some lemonade, but of course forgot the opener.

We arrived at the field, which was of the minimum possible size and, having unloaded, parked the aeroplane near the gate to the main road. I then got out my tickets and my sweetest smile and prepared to force holiday-makers into the air. The first to appear, from an apparently deserted countryside, were dozens of small boys.

'Coo - an aeroplane'.
'W'a's it doing?'
'It's broken'.
'No it ain't'.
(Me): 'Don't go near the prop sonny'. (Smile.)
'Don't touch the ailerons!' (No smile.)
'Stop waggling that rudder!' (Threatening gesture.)

Same dialogue with variations from Pilot and GE

Then more hopeful clients began to arrive. A party of young people looked shyly over the hedge. I advanced and opened the gate for them. 'Wouldn't you like to have a joy-ride for five shillings? The country looks beautiful from the air today'.

Giggles and nudgings and 'I will if you will' from two of the girls. Before they had time to repent I tore off two tickets and thrust them into their hands, they were then too embarrassed to refuse. By this time there were a few more people peeping at us and some hardy spirits had come in at the gate. Now it was up to me to get two more passengers before the aeroplane had finished its two-minute circuit. In spite of much experience I advanced upon two of the lads.

'I expect you've been up before (flattery), but perhaps you'd like another trip?'
'No, never been up'.
'Oh! in that case you must have a shot'. (Appreciative laughter from lookers-on.)
'It's as safe as the ground and you can't let the girls have it all their own way'.
'No. I wouldn't mind going up if I could keep one foot on the ground!' Despairing laughter from me at this original remark. (Originally made by the original man originally asked to go for a joy-ride.)

Half a minute to go before the Avro was ready for the next load the situation was saved by two old ladies anxious to try a new sensation.

'Shall we need our sunshades?'
'No, Madam, I think it would be better if I looked after them for you'.
And off they went happily.

Soon the moment came to fly to the other field. I was left behind to try and keep some of our little crowd together for an hour. Most of them wandered off, but some stayed and some new ones came. By the time the old Avro reappeared I had quite a respectable number waiting to go up. There was a certain amount of discussion as to who should have first go, but a little firm handling soon put that right.

We had a goodly crowd, but round about lunchtime we found a moment to eat our sandwiches. The pilot sat in the cockpit between flights and ate his, richly seasoned with castor oil. The lemonade took a bit of handling but we managed it in the end, only breaking one bottle. The cherries had somehow disappeared. I remembered having had one or two, and the pilot and the GE said they had dipped into the bags once or twice. Short weight, I suppose. We got some more cherries down the road and I was left to organise the customers, while the Avro went to deal with the No. 2 field again.

The afternoon was busier and people were inclined to clamour. The Avro did shorter and shorter trips and dashed feverishly back and forth between the two fields.

I had a disgruntled passenger. He had already had a trip, he told me, with one of the big circuses, so I persuaded him to try again. He came down looking scornful and said 'Very ordinary that was. Why when I went up with X now, we landed through fence and smashed bottom of aeroplane. Shook up proper, we was!' I suppose one can't cater for everyone.

Things were getting pretty hectic by this time. No leisure for tea. We decided to have a few cherries. Unfortunately the bags had become empty again. Small boy got us some more. Larger and larger crowds gathering. Oh! for our second aeroplane. Money was rolling in. What a clean-up!

I had been deserted for nearly three-quarters-of-an-hour and any minute I was going to have to start giving money back. Agony! Agony!! Suddenly, blissful moment, I heard far away the distinctive note of our second aeroplane. The sun was low but there was a good two hours of light yet. No more worry, smiling faces, more crowd pouring in. Oh! well done chaps!! NOW for a bumper taking.

Pilot No.2 started whizzing round the field. People were boosted in and out of the cockpit almost before they knew they had had a flight.

'Put your right foot here, put your left here, in you go!'
'Left foot here, right foot here, give me your hand, out you jump!'

Thirty glorious minutes and then a pretty girl, who had been hesitating for some time, suddenly decided she would like a trip. She was hoisted in with much help from my many swains and some shrill cries from her. Round she and her boy-friend went and down they came. She started to get out. I saw her swing her leg over the side of the cockpit. I shouted as loudly as I could: 'Don't step on the wing! Don't put your foot there!'

But it was too late. With a quick grace she jumped. Her high-heeled shoes landed one on each side of the spar, there was a ripping noise, and two legs shot right through the wing.

There she was (and the dramatic moment is seen above left in cartoon form!).

I gave back the money to the disappointed joy-riders while Pilot No.2, GE No.2 and the swains prised her out. It took some time. She was unhurt and quite unembarrassed. By the time we had patched and doped the wing it was nearly dark. We threw away the empty cherry bags and flew home.

Appendix 2: F G Miles' Letters of 1927

On 18 May 1927, the firm owned by Miles and Pashley was re-organised as a limited company and registered as The Gnat Aero Company Ltd. A commentary on the firm's activities during those early carefree days was given in a series of amusing letters written by F G Miles, which were published in *The Aeroplane* magazine between April and September 1927.

Background: Southern Activities and 'Gnatural' History

The Southern Light Plane Club was formed at Shoreham in August 1925. In 1927 its President was Cdr Sir A Cooper Rawson MP, RNVR, its Secretary was Cecil A Boucher and the two flying instructors were Cecil L Pashley (Aviator's Certificate 106, B Licence 529) and Frederick G Miles (B Licence 810). The Membership was 46, with 21 flying, and among the aircraft used for instruction and joyriding between 1925 and 1928 were the Avro 504Ks G-EAAY, G-EAJU, G-EATU, G-EBJE, G-EBVL and G-EBYB.

April 27 1927 - Southern Activities
The Gnat Aero Co of Portslade have been very active of late. Unfortunately their dual instruction machine G-EATU was badly damaged on the 9th of this month by a pupil who had been trying an engine-off landing from 6,000 feet and, owing to the engine getting cold on the way down, had to land with a dead engine. He was unable to reach the aerodrome, so landed in another field where he met a wire fence and damaged the undercarriage and lower planes in about two feet of water. Dismantling the remains meant a day's paddling.

The firm recently sold G-EAAY to Mr Lewis, lately of the Southern Counties Company. During the Easter Holidays they did some joy-riding at Seaford on G-EBJE. Sundry members of the club hope to journey to the Hampshire Pageant by air on May 15th.

June 8 1927 - The Gnat Aero Co. at work
An interesting and at the same time amusing letter from Mr Miles of the Gnat Aero Co, of Portslade and Shoreham, shows that the South Coast is having its share of flying. The letter read as follows:

We have completed another week's joy-riding, at Chichester this time, not very profitable financially, but no doubt invaluable as regards experience. The most interesting thing that happened was that the Chief Constable of the County stood at the gate and took tuppences for us while we showed his friends over the machine. We could not persuade him to go up, however.

The rebuilding of 'TU proceeds apace, the only part we can use of the original machine appears to be the centre section, which is actually undamaged. We hope to complete it in a few weeks now.

The only other thing I have to say is: Have you any publication dealing with the right way to answer joy-ride crowd questions and jocular remarks? If so I should be very grateful for it, having exhausted all my rhetoric on our two attempts this year. The remarks we have to cope with are very much as follows - 'One foot on the ground', 30%; 'I'm certain to be sick', 33%; 'Don't want to die yet', 20%; 'Safer down here', 10%; and more shortly, 'Not me!', 6.5%. This leaves 0.5% who go up at all.

Thus I suppose it is a good job we are getting some sunshine now and the 'Constant Nimbus' seems not so much in evidence. The above meteorological pun is the work of a pupil. It seemed a pity to waste it as I was writing at the time. The aerodrome is in fine condition now, we have removed the floats from the machines and re-fitted wheels.

June 15 1927 - Gnatural History
The whole firm (Pashley, our secretary, ground engineer, boy and self) have all been travelling every-whither in search of pieces of Avro (not 'pieces of eight') with which to replace various and sundry components which were lacking in 'ATU owing to our pupil's contretemps with the fence, as faithfully reported recently.

Things are brightening now at Shoreham, the vanguard of the visitors is keeping us moderately busy with flips, the Bantam is flying regularly and well, Pashley is getting quite skittish with it - See large bills - 'Daring exhibitions on special machines, by well-known pilots' - also the dual machine comes back into commission today - AID and weather permitting.

Sunday, May 23rd, after hearing of Lindbergh's success, we held an 'Air-Mindedness day', cutting prices from 5s. a flight to 3s. a flip,- a distinction with a definite difference. It was a most successful day in every way and everybody was satisfied except the accountant. Among our aerial visitors we numbered the Australian aspirant, Mr Rooke, on his Moth (G-EBQJ – PA); he came and went as prophet unhonoured, I fear, although we showed almost indecent interest in the various green baize bags and things which he has tacked onto the fuselage.

As proof of our progressiveness we are installing a telephone and a petrol pump for the convenience of visitors. The former has now arrived and I have to spend most of my spare time watching it, as everybody is so fascinated that they ring up everybody else on the least pretext, the £'s worth of calls which the canny Post Office insists on in advance being nearly used already.

July 6 1927 - Flying at Shoreham
We have been joy-riding at Selsey during the last stormy days and are still doing so. Passengers want rather a lot of persuading to fly when the whole ground staff are holding the machines down through squalls, they mostly think that aeroplanes are such terribly flimsy things.

One voluble visitor was heard pacifying a timid friend with the following allegory - the said friend having awaited her more venturesome companion's return from a trip in the blue in fear and trembling, meanwhile finding ominous auguries in whole hosts of natural phenomena, such as the sun disappearing behind a cloud as they took off. (This sentence seems to have got entirely out of control, I'll have another try). As I was saying, this voluble and venturesome visitor explained to her friend.-'It's just like going to a doctor for a hoperation. You puts yourself in 'is 'ands, and if it's Gawd's will yer pulls through'.

G-EAJU is making good time. The pupil who piled 'TU likes this one lots better. He lectured us recently on all the various combinations of misfortunes which caused him, an entirely innocent and blameless party, to suppose that he could get over a fence and into a field, instead of flying over a field and into a fence, which he did. However, he is flying 'JU now very well, what time he is not explaining to nervous neophytes how very dangerous a thing is an aeroplane.

By the way, the petrol pump is installed and we are now entirely at the service of aerial visitors in the way of fuel and oil; also, our ground engineer, who holds A B and C licences for all types of aircraft and/or seacraft, up to three engines, is at the disposal of any private owner who cares to avail himself of his assistance or advice. We can undertake the reconditioning of any type of machine.

I hear that several clubs are having to turn pupils away owing to all the flying time being booked up. Well! It might possibly interest the said pupils to know that we are not in that fortunate position, but we have plenty of machines and pilots here for the members of the Southern Aero Club. We have the assistance of Flt Lt G Birkett, lately of the RAF, in which he has had ten years as instructor on all types. He is keeping his hand in here. As you know the Club offers Avro 504K tuition at 50s an hour.

The more childish among us derived a great deal of innocent pleasure from the fact that the Air Ministry has mentioned Shoreham in their new blue 'Notices to Airmen', saying that civil aircraft on certain routes must pass over us in foggy weather. I do prefer the new summer shade of blue to the 'greenery yallery' one it supersedes, don't you? PS - Since looking at the notice again I find that it refers to Shoreham in Kent, but can't be bothered to re-write.

July 20 1927 - More Gnatural History
We have been to Wannock Glen, near Eastbourne this week, joy-riding, but through landing on bad ground to avoid cows, Pashley had an undercarriage go and was unable to carry on. We took the two-seater over to his help complete with Ground Engineer, spare undercarriage, one wheel, diagonal struts, etc., and did the job on the field. By special request of the crowd, undaunted by the dismal spectacle of 'JE lurching on one wheel, I took up a number of passengers on the two-seater.

The crowd commentary for this week, I append. (Here I think I must interpolate again; my letters seem mostly such. This time I just wanted to point out that I use 'crowd' poetically, purely in view of its alliterativeness. Actually the crowd usually consists of the Oldest Inhabitant, his son, his son's wife and subsequent generations. Even so, circumstantial evidence indicates a dearth of anti-family-limitational literature in out otherwise well-read Sussex villages). That's the end of the brackets for a moment.

To resume, the third or fourth generation of our audience remarked to fifty per cent of the second or third-apropos of Pashley's latest landing – 'Did it just go up?' Pashley heard, and being ever a stickler for strict fact, he said 'No, it's just come down'. Whereupon the hitherto disinterested mother, with pleasurable anticipation writ large upon her face, inquired eagerly: 'Was anybody hurt? Where are they'. The lady ought to be an 'Airplane' correspondent to the daily press.

And now, I suppose, the Weather - with a capital W. - We are having Weather now at Shoreham and when we have had some Weather....we have some more WEATHER. And when I say WEATHER I mean Weather ad infinitum and to surfeition. 'For the rain continueth; and the rain continueth; broad and deep continueth; great beyond our knowing'. More rain than we want in fact. Damply yours, - F G M.

September 7 1927 – More Gnatural History
The following effusion from the head of the firm indicates that the Gnat Aero Co, of Shoreham, is still as active as ever:

First I must get off my chest the story of the lady and the Telescopic Tubes. In this case it was I who provoked the naive remark that follows: Primarily the cast - A daring husband, an admiring, if somewhat apprehensive, wife and myself. The husband paid his five shillings and took his ticket and his seat and his flight. Wife watched the machine round and, regaining her sang froid as 'bus approached terra firma once more, showed her air-mindedness by evincing an almost intelligent interest in conversation with the poor defenceless ground engineer.

After making several wild guesses as to the purpose of various parts, she capped all by pointing out the shock-absorber covers and inquiring, 'What are those black things under the engine? Oh! I see! Of course they're the pilot's feet!!!!!!' - Mind you, I have never been proud of my feet. But....

I suppose you notice the lack of news in these narratives. It is awfully difficult to find a subject to discuss other than pseudo-humorous ones. Certainly, I might get serious and say that on Sunday we took up 117 passengers and one got out making 115, and then perhaps on Monday we took up 36 of the same kind only fatter. These statistical observations are all very well and show a proper appreciation of mathematics no doubt. Nevertheless, they are easier to print with a cash register than a typewriter. And I can't see why Samuel Smith, who flew for 23 minutes on Thursday, the 24th, should be allowed to inflict his troubles on either you or myself.

One thing did happen a week or so ago that is worth recording. In fact it is almost worth a treatise pointing out the ideal spirit in which the British Public might approach aviation (and aviators). The instance in question was a gentleman, a very old gentleman, who insisted on paying one pound for a five shilling flight and who said when I ventured to remonstrate (a little sotto voce perhaps), 'Make it two pounds if you like', - this in a slightly aggrieved tone. I argued

'A Shocking Affair' – A spectator's idea of the function of the Avro undercarriage shock-absorbers.

no further. After taking him for a two-minute flip - he would not hear of a longer one - I brought him down. Then, imploring me not to be offended, he gave me a pound tip.

Thereafter, greatly cheered by the fact that I was not offended, he gave a pupil, who had assisted in starting the machine, another pound. The pupil, who by the way has just ordered a specially fitted Avro from us, paid it into the office fund, so we also benefited by that. Now this was very nice. Twice more he returned and repeated the whole performance except the tips, not only going up himself but paying for his friends as well at pre-war rate. We have not seen him since though tender memories linger still. - F G M.

September 14 1927 - Gnatural History Again
A surprise visit to the scene of Pashley's labours the other day brought to my astonished vision a most peculiar sight. It was Pashley sitting on the grass with his trouser-legs rolled up, and our advance agent, Boucher, spraying the legs thus exposed with a dope syringe. This was being done most solemnly and seriously. Then, when Boucher had finished his task, Pashley returned the compliment.

Earnest inquiry elicited the solution to the mystery. Apparently the harvest spiders which inhabit that part of the globe are of a particularly virulent nature, and spraying with petrol is, they tell me, very good, if somewhat odorous, preventative. (For the benefit of others similarly afflicted one may say that paraffin is less volatile, more adhesive, and equally effective – Ed.)

Personally I have been spending my time and money at Seaford. At least owing to apathy and lack of air-sense on the part of the inhabitants, I have not been doing much flying. In fact I passed a good deal of the time lying on the grass with my face in my hands, preserving that schoolgirl complexion and waiting for customers. Which gave a small girl occasion to remark to her nurse: 'Look, Nanny, the pilot is crying'. He was nearly. From ennui.

Incidentally, while I write of joy-riding, I must tell you of some of the objections and objectors that one encounters. At Wannock an irate and choleric gentleman, with some sort of a military prefix to his name, informed Pashley that he (Pashley) was flying directly, or very nearly so, over his (the military gentleman's) thatched cottage. And moreover - he continued - Pashley was endangering the said thatched cottage to an unwarranted degree, since, said he, if one of the passengers (who, he remarked in passing, were fools to go up anyway) should throw out a lighted cigarette or a match, it would be all UP with his beautiful but inflammable thatched cottage.

After Pashley had reassured him on this point, he came back with another grievance. In effect he said (and repeated later by letter), 'I have been given to understand that certain persons have been asked to stand by while the aeroplane takes off, and this while they have been using a public footpath that runs across the flying ground.

Unless I am given an assurance that this will not happen again I shall take the necessary steps to see that any person shall have the full, free and unfettered rights to the aforesaid footpath'.

Of course I spoke to Pashley severely about fettering people on a public footpath, but I made little impression, I fear. Ah, well! The joy-rider's life is a compound of trouble. And there is no joy in him.

EPILOGUE

A letter from Don Brown appeared in *The Aeroplane* for 12th February 1936:

The Good Old Avro

Sir, - I was so pleased to see the article by the Earl of Cardigan. Naturally we at Shoreham have a soft spot in our hearts for the old Avro, because it was the machine with which we started and used exclusively throughout the first and most precarious years of our existence. Although in later years we had various light aeroplanes, we were never without at least one Avro - in fact at one time or another we built up fifteen 504s for ourselves and other people. These were fitted with all sorts of engines from the 80 hp Le Rhône to the V-eight Airdisco and the Armstrong Siddeley Lynx.

All our original members learned to fly on the 504K, and, having since flown practically all the well-known types of light aeroplane, I must say that if one really wants to fly properly (which relatively few amateurs do) it is difficult to imagine a better machine than the old Avro. It had all the qualities required of a good training aeroplane and, although the technique of flying an Avro is rather different from that required in flying modern clean aeroplanes, it is still in my opinion the best machine that has been produced so far for training a pupil to fly safely and accurately.

The quick take-off due to the light wing-loading and slow-running rotary engine, together with the steep glide and low landing-speed were to my mind far more desirable qualities in any aeroplane that the 'top speed at any price' ideas of today. In fact the tendency at the present day seems to me to be to rely upon the amazing reliability of the modern aero-engine as an excuse for producing thoroughly bad and dangerous aeroplanes, which are absurdly simple to fly in the air but which call for an unreasonable degree of skill if they have to be landed without the engine.

While no one will deny that high performance is essential if an aeroplane is to fulfil its function as a useful vehicle, it would appear to be a retrograde step to sacrifice, or at least risk, all possible safety in order to attain a high performance before we have yet learned how to produce aeroplanes with a speed range of say 5-to-1.

To return to the Avro, how often did one catch fire in a crash? I only wish that we could say the same about many types of modern machine. Again, surely no better proof could be given of the safety of the Avro that the fact that, although we were operating from what must have been one of the smallest aerodromes in the country for over nine years, we never had either a fatal or even a serious accident at Shoreham. In fact it was amazing sometimes what pupils could do in an Avro and still emerge unscathed.

There was one bright lad who went up to do the 'engine off' for his 'A' licence. He misjudged his landing and overshot, but, to our horror, instead of opening up again, he continued to glide down across the aerodrome finally arriving at the other side at a height of about ten feet. Being apparently of an optimistic nature, however, he did not abandon hope even then and although he had so far failed to touch down he simply thought that he would turn round and land back down wind! So, still 'engine off', he just put the Avro into a steep turn - at least it would have been steep if the wing hadn't touched the ground first. But he walked back to the hangars apparently quite unconcerned.

Another pupil who walked back undeservedly unscathed was one who suddenly decided, after having overshot, to do a steep climbing-turn down wind and 'engine off'. When he alighted (I won't say landed) the engine was still off - quite a long way.

You must not think that all our pupils used to do that sort of thing, as these were exceptional cases that merely serve to show what you can get away with in an Avro.

You must forgive the length to which this letter has unintentionally grown. My only excuse is that it may perhaps be regarded as an epitaph to what I suggest is the most wonderful aeroplane that has ever been produced. (Signed) Donald Brown.

(Most old-timers and many more recent recruits will agree with Lord Cardigan and Mr Brown. The extraordinary qualities of the good old Avro rather suggest that even now there might be a market for a large and lightly loaded machine of a similar type. Certainly, if a club of young enthusiasts want to set about the co-operative production of a real aeroplane, they might do worse than build something on the lines of an Avro, and put a decent modern second-hand car motor into it, instead of trying to make soap-boxes fly with hotted-up motorcycle motors. – Editor)

Trust C G Grey, the editor, to have the last word on the subject!

Appendix 3:
Fields Used by F G Miles for joyriding and/or visiting – 1926 to 1931

It was commonplace in the 1920s and 30s to operate from unlicensed fields and, given the performance, particularly of such types as the Avro 504K, this was not in any way hazardous. There follows a listing, compiled from F G Miles' log books, of the locations of his joy-riding activities and/or casual visits.

Torr Hill, Sussex - 4.2.26 - noted just once while learning to fly but not identified (the only known Torr Hill being near Plymouth).
Southwick, Sussex – 26.6.26 - in Centaur G-EALL - first field outside Shoreham
Worthing, Sussex - 26.6.26 - in Avro 504K G-EATU
Portslade, Sussex - 26.6.26 - in Avro 504K G-EATU
Arundel (also Arundel Castle), Sussex
Blackrock, near Brighton, Sussex
Palace Pier, Brighton, Sussex - surely not literally!
Partridge Green, near Horsham, Sussex
Westbourne, near Thorney Island, Sussex
Eastbourne, Sussex
Rottingdean, near Brighton, Sussex
Chichester, Sussex. One of the many fields used by Miles and Pashley in the Chichester area has been identified by W/C Roland P Beamont, CBE, DSO, DFC, DL, FRAeS as being almost opposite The Drive where he lived at the time. This field was just North of Summersdale Copse, between Fordwater Road and the River Lavant (approximately 1 mile to the North of St Richard's Hospital) and now mostly built over. Other fields used by Pashley and Miles in the Chichester area included Wopple Field, Peckham's Farm, to the East of the Convent,
Crawley, Sussex
Seaford, Sussex
Hassocks, near Brighton, Sussex
Shotter Mill, near Haslemere, Surrey
Wannock, Sussex - possibly Wannock Glen, near Polegate, Sussex
Selsey, Sussex
Bognor, Sussex - possibly a field on the right of and about 2/3rds along Chalcroft Lane, North Bersted, going towards Pagham.
Lewes, Sussex
Burwash, near Hawkhurst, Sussex
Roedean, near Brighton, Sussex
Newick, near Uckfield, Sussex
Littlehampton, Sussex
Wadhurst, Sussex, near Tunbridge Wells, Kent
S'over? - probably an abbreviation for Southover, near Lewes
Findon, near Worthing, Sussex. Miles flew there on 22.6.28, 26.1.29 and 3.2.29 but does not appear to have carried out joyriding from there.
Cooksbridge, near Lewes, Sussex
Bosham, near Chichester, Sussex
Emsworth, near Thorney Island, Hampshire
Basingstoke, Hampshire
Berwick, near Alfriston, Sussex. Berwick Court, near Drusillas Tea Garden
Newhaven, Sussex
Chailey, near Lewes, Sussex
Warblington, near Emsworth, Hampshire
St. Margaret's Bay, near Dover, Kent

Apart from Bognor and Berwick and the Chichester fields, the exact location of the many other fields used are not identified.

Appendix 4:
The Shoreham Days – F G Miles' Aeroplanes from 1925 to 1931

This is the history of the 27 various aeroplanes acquired by F G Miles (or his various firms), at Shoreham prior to August 1931, when he left to go to South Africa. It has been compiled with the help of his log-books and those of George Miles and Don Brown and is presented in approximate chronological order of acquisition. Details of other aeroplanes flown by Miles are also given.

1. Avro 504K G-EATU (ex E3045); fitted with 80hp Le Rhone. Regd (CofR 542) 12.6.20 to Cecil Lawrence Pashley and operated by Central Aircraft Co, Northolt. CofA No.460 issued 25.2.21. CofA lapsed 24.2.22 and stored. To Shoreham 4.25 on formation of Gnat Aero & Motor Co. Miles received first flying lesson on this machine 26.11.25 and went solo in it 19.5.26 but CofA not formally renewed until 2.6.26. Operated by Southern Aero Club. Suffered engine failure on take-off from the South Downs on or shortly after 29.7.26 and damaged; repaired (prior to 17.10.26). Badly damaged 9.4.27 by a pupil trying an engineless landing at Shoreham from 6,000 ft and who missed the aerodrome, hit a wire fence, damaging the undercarriage and putting the lower wings in two feet of water! In forced landing in water; rebuilt by 6.27. Regd (CofR 1472) 1.9.27 to The Gnat Aero Co Ltd, Shoreham. Crashed Shoreham 5.9.27; probably not repaired. An undated photograph shows the fuselage to have been broken aft of the rear cockpit and this may have been of this incident. Regn cld 1.29 as wrecked.

2. Grahame-White GWE.6A Bantam G-EAFL (c/n E.6A); fitted with 80hp Le Rhone. Regd (CofR 142) 20.6.19 (as K-153) to The Grahame-White Co Ltd, Hendon. No CofA issued. Stored Hendon and regn cld 10.1.23. Sold at auction 17.2.26 to Gnat Aero Co (but never regd to them). Re-assembled, as time permitted, it was first flown by Cecil Pashley. First flown by Miles for 5 minutes 10.3.27; second (and last) flight by Miles of 10 minutes duration on 17.3.27. On one of these flights, Miles recorded that the engine failed on a test flight although he did not note that he also 'overturned' it on landing! It is possible that the Bantam never flew again, being relegated to the bonfire on the Guy Fawkes celebrations on 5.11.28.

3. Grahame-White GW.15 Boxkite two-seat pusher G-EABD (c/n 403). Regd (CofR 30) 15.5.19 to The Grahame-White Co Ltd, Hendon. No CofA issued. Stored Hendon and regn cld 10.1.23. Sold at auction 17.2.26 to Gnat Aero Co (but never regd to them). Miles possibly made his first 'unofficial' solo in this aircraft (prior to his 'official' first solo on 19.5.26). No Boxkite flights were recorded by Miles in his flying log-book, but on a page headed 'Past Experience' in a later log-book, he recorded against 'Hours flown on each type', 15-min against a 'GW. Type 15'. A photograph survives which shows the Boxkite outside a hangar at Shoreham, with No.24 painted in a white square on the vertical tail surface, but the significance of this number remains a mystery. The Boxkite was destroyed when the hangar in which it was kept collapsed in a gale on 8.10.26.

4. Central Centaur Mk.4A three-seater G-EALL (c/n 203 but officially CF4/201); fitted with 100hp Anzani engine. Regd (CofR 300) 8.19 to Central Aircraft Co, Kingsbury (later Northolt). CofA No.308 issued 5.3.20. CofA lapsed 4.3.22 and stored. Purchased from Central Aircraft Co and regd (CofR 1261) 5.26 to Gnat Aero & Motor Co. CofA not renewed but Miles flew it (for the first time) on 24.5.26 and gave his first joyride, of 5-min duration, in this aircraft on 19.6.26 to a passenger by the name of 'Poole'. Miles started dual tuition in it the following day and made a total of 77 flights in G-EALL, with the last being on 2.7.26 and believed wfu soon after. Regd (CofR 1345) 3.3.27 to The Gnat Aero Co Ltd. Regn cld 4.30.

5. Central Centaur Mk.4 three-seater G-EABI (c/n 201); fitted with a 70hp Renault engine. Regd (CofR 35) 7.5.19 (as K-108) to Central Aircraft Co, Kingsbury (later Northolt). No CofA issued. Purchased from the Central Aircraft Co and regd (CofR 1262) 5.26 to Gnat Aero & Motor Co. No record of any flying by Miles, implying that it was purchased for spares. Regd (CofR 1346) 3.3.27 to The Gnat Aero Co Ltd. Regn cld 4.30.

6. Avro 504K 3-seater G-EBJE (no c/n; built from spares by Southern Counties Aviation Co, Brooklands; fitted with 110hp Le Rhone. Regd (CofR 1088) 7.24 to Geoffrey Victor Peck, Dulwich (joint proprietor of Southern Counties Aviation Co). No CofA issued and crashed Belper, Derby 11.10.24; repaired. Regd (CofR 1229) 1.26 to John Rhodes Cobb, (the famous racing driver), Esher (based at Brooklands). CofA No.927 issued 21.1.26. Regd (CofR 1297) 8.26 to FG Miles who flew it to Shoreham on 2.8.26 as replacement for G-EATU. Regd (CofR 1699) 16.7.28 to Southern Aircraft Ltd, Shoreham. 'Worked the South Coast' giving joy-rides and an undated photograph shows G-EBJE with the u/c damaged, following what must have been a heavy landing; repaired. Miles made his first night flights in G-EBJE from 4.10.28 until 4.11.28. Last flown by Miles 23.9.29. On 16.4.30, G-EBJE was flown for the first time in 1930 preparatory to joy-riding but forced landed at Shoreham through petrol starvation; CofA renewed 17.4.30. CofA lapsed 29.9.34 and regn cld 12.34. Sold (in mid/late 1930s) to Richard Grainger Jeune Nash, t/a the International Horseless Carriage Corporation, Brooklands and stored. Nash's collection was purchased by the Royal Aeronautical Society in 12.53 and was later stored at Hendon and subsequently Heathrow. Rebuilt to static condition at RAF Abingdon in 1963 with the wings from G-EBKN (ex E449); and subsequently (in 1968) transferred to the RAF Museum, Hendon, where it is now on permanent static display in the RAF Museum, Hendon; painted as E449.

7. Avro 504K G-EAAY (ex C724). Regd 6.5.19 (as **C724**) to The Grahame-White Co Ltd, t/a London Flying Club, Hendon. CofA No.15 issued 9.5.19. Regd **G-EAAY** (CofR 25) 31.7.19 to same owner. CofA lapsed 25.8.22. Regd (CofR 1116) 10.9.24 to Lady Utica Beecham, Hendon. CofA renewed 8.9.24. Regd (CofR 1157) 12.5.25 to Liverpool Aviation Co, Liverpool. CofA lapsed 7.9.25. Regd (CofR 1255) 23.4.26 to Southern Counties Aviation Co, Brooklands; named "Jake". CofA renewed 16.5.26. Collected by Miles and flown to Shoreham 8.10.26; regd (CofR 1313) 22.10.26 to Frederick G Miles, operated by Gnat Aero & Motor Co, Shoreham; also reported as owned by Nancy B Birkett & Lt Rupert L Preston. Flown by Miles 6.3.27 on return flight to Crawley and on 3.4.27, he made two 3-minute local flights. Regd (CofR 1363) 11.4.27 to Leslie Allin Lewis, Shoreham. Crashed on landing Homefield landing ground, Wokingham 16.4.27 and broke in two; pilot LA Lewis; repaired. Regd (CofR 1448) 20.7.27 to LJ Sky Trips Ltd, Shoreham (later Bekesbourne). CofA lapsed 6.1.30. Regn cld 4.30. Regd (CofR 2731) 14.7.30 to Launcelot Richard Gladwin-Errington, Bekesbourne. CofA renewed 25.7.30. Crashed prior to 24.10.30. Regn cld 12.32.

In early 1927, Miles purchased a number of airframes and engines from the miscellaneous surplus stocks of A V Roe & Co Ltd, Hamble before they moved to Manchester. It has only been possible to positively identify three complete aircraft from this collection but others were built up from the surplus stocks as required. The three complete aircraft were:-

8. Avro 504K G-EAJU (ex H2592) Regd (CofR 255) 8.19 to AV Roe & Co Ltd. CofA No.136 issued 20.8.19. CofA lapsed 19.8.20. Regd (CofR 1389) 5.27 to The Gnat Aero Co Ltd, Shoreham; also operated by Southern Aero Club. First flown, by Miles 4.6.27; CofA renewed 4.6.27. G-EAJU 'worked the South Coast' giving joy-rides for about a year; an undated photograph shows it with the word 'Sunols' painted on the fuselage below the wings, sitting on its belly, with no u/c, in a water filled ditch! Regd (CofR 1713) 7.8.28 to John Colquhoun Don, Shoreham; for barnstorming; flown by Miles to the last Air Pageant to be held at Swallowfield Park, Reading on 27.10.28, where he probably met Charles O. Powis for the first time. Later operated by Dominion Aircraft Ltd (which was formed by Mr & Mrs Don and FG Miles on 26.11.28 'to acquire and take over as a going concern and carry on the business of civil aircraft pilot now carried on by John Colquhoun Don)'. Crashed Brighton 21.7.29. Regn cld 1.30.

9. Avro 543 Baby G-EAUM (c/n 543/1, later 5062); fitted with 35hp Green engine. Regd (CofR 561) 12.7.20 to AV Roe & Co Ltd. CofA No.669 issued 3.8.23. CofA lapsed 2.8.24. Regd (CofR 1516) 11.27 to Lionel Edward Richard Bellairs & FG Miles, Shoreham. Fitted with 60hp ADC Cirrus I air-cooled engine by The Gnat Aero Co Ltd and first flown in this configuration, by Miles, on a 5 min 'Test' on 13.11.27. CofA renewed 13.7.28. Regd (CofR 1702) 17.7.28 to Rudyard A Whitehead, Ealing (based Shoreham). flown (and crashed) by him in the July 1928 Kings Cup Air Race. Regd (CofR 1749) 6.9.28 to P/O Haliburton Hume Leech, RAF Tangmere (but not dd to him until 24.12.28, presumably after repairs completed). Flown by Leech in 1929 Kings Cup Air Race. Purchased from Phillips & Powis Aircraft and regd (CofR 2252A) 9.29 to P/O Hugh Robert Arthur Edwards, Heston (later RAF Grantham). Last flown by Miles 22.12.29. Edwards won the Handicap Race at the Teignmouth Flying Meeting on 5.9.31 in G-EAUM over a 25-mile circuit. Regd (CofR 3505) 11.31 to Roper Brown, Southend-on-Sea. Purchased in 1931 for £47 by to Geoffrey Young & Douglas O'Hanlon, Brooklands and rebuilt by Redwing Aircraft Co Ltd at Croydon for £160; regd (CofR 3952) 10.9.32 to Geoffrey Lawrence Young (only) (but still owned jointly). CofA lapsed 8.11.34. Remains burnt at Yeovil 18.1.35.

10. Avro 547A Triplane G-EAUJ (c/n 547A/1) Regd (CofR 558) 9.7.20 to AV Roe & Co Ltd; this was a large four-seat cabin aircraft only ever used for demonstration purposes. CofA No.413 issued 22.8.20 after trials at Martlesham Heath 3.8.20-14.8.20. CofA lapsed 21.8.22 but probably dismantled at Hamble earlier, as no buyer could be found for it. Purchased by Miles in 1927 and taken to Shoreham; however, Miles and Don Brown ultimately decided that to modernise it would involve more work than to design a new aircraft, so it was later scrapped. The wings, being standard Avro 504K type, were probably used on G-AAED and other 504K's which were built up from spares.

On 18.11.27, Miles made one flight from Shoreham in **SE.5A G-EBTO** owned by Will Thomson Hay, the music hall artiste, who kept it at Stag Lane; it is possible that Will Hay was appearing in Brighton or Worthing at the time.

11. Avro 504K G-EBVL (built up from ex AV Roe & Co spares) First flown, by Miles 24.12.27. Regd (CofR 1536) 12.27 to Lionel Edward Richard Bellairs, Shoreham and Burwash. CofA No.1270 issued 9.1.28. Also operated by Southern Aero Club (by 3.28) and named 'Bovril'. Regd (CofR 1701) 7.28 to Southern Aircraft Ltd, Shoreham. Used regularly by Miles for joy-riding, with 2 hr 12 min 'local' flying being recorded at Worthing on 5.8.28 and 2 hr 20 min 'local' at Worthing on 6.8.28. CofA expired 8.1.29 and it seems likely that Miles overturned G-EBVL while landing after a joyride at Southwick,

APPENDIX 4

near the railway works at Lancing on 6.8.28 (see story section). Presumed stored but later rebuilt and CofA renewed 2.4.31 but on same day Pashley crashed with engine failure while taking-off; he was unhurt and the aircraft was barely damaged; it was joy-riding at Roedean on 4.4.31. Regd (CofR 3364) 18.8.31 to Francis Colbourne Fisher, Christchurch. CofA lapsed 13.3.37 and regn cld at 31.12.38 census. The skeletal remains were still in open storage at Christchurch in 4.39.

At Shoreham, on 5.3.28 (and again on 9.3.28 and 28/29.4.28) Miles flew the **Boulton & Paul P.9 G-EBEQ** owned by Lt Hugh Kennedy (regd to him 13.12.27 and kept at Stag Lane).

12. Avro 504K G-EBYB (built up from spares); fitted with 80hp Le Rhone engine. Regd (CofR 1635) 5.28 to The Gnat Aero Co Ltd. First flown, by Miles, 26.5.28. CofA No.1413 issued 26.5.28. Operated by Southern Aero Club and 'worked the South Coast' giving joy-rides. *Flight* for 27.12.28 reported that G-EBYB was 'back in service with the top-plane intact, after damage a fortnight ago due to gale'. Engine replaced by a 110hp Le Rhone and flown as such by George Miles on 27.2.29. Regd (CofR 2627) 22.5.30 to The Southern Aero Club. It was in G-EBYB on 2.6.30 that Miles gave Blossom Freeman-Thomas her first flying lesson. CofA lapsed 2.10.31 and wfu Shoreham. Regn cld 2.33.

13. Avro 594 Avian III G-EBVA (c/n R3/CN/104) Regd (CofR 1517) 11.27 to AV Roe & Co Ltd, Hamble. CofA No.1297 issued 27.1.28. Purchased by Lionel Bellairs and flown from Hamble to Shoreham by Miles on 1.5.28; regd (CofR 1700) 7.28 to Lionel Edward Richard Bellairs and used by Southern Aero Club for instruction. It was in this aircraft that Miles made his first overseas flight, from Croydon to Le Bourget, on 9.5.28, returning via Lympne and Croydon on 12.5.28. On 9.6.28, George Miles made his first flight in 'VA as 'Pupil under Instruction' with FG Miles, a 16 min flight. On 10.9.28 Miles delivered G-EBVA to Croydon; regd (CofR 1798) 26.10.28 to Donald James Hamilton-Lister, Croydon. Regd (CofR 2307) 10.29 to Surrey Flying Services, Croydon. Regd (CofR 3491) 11.31 to Percy Cruttenden, Bexhill-on-Sea (but based Woodley). Regd (CofR 3809) 6.32 to F/O Geoffrey S Shaw, of 608 Sqdn, RAF Thornaby-on-Tees. Regd (CofR 5541) 11.1.35 to Richard Carver Pick & Walter Mason, based Newton House, Leeming. Regd (CofR 7405) 22.10.36 to Walter Mason (only). Destroyed by fire at Brooklands 24.10.36. Regn cld 11.36.

On 28.7.28, Miles flew **DH.60X Moth G-EBZF** (c/n 644) on a 10-minute local flight at Shoreham; it was en route to Spain where it was being delivered to Dr Antonio Habsburgo, Barcelona (it became M-CAAK).

On 2.8.28 at Shoreham, Miles flew the **Blackburn Bluebird III G-EBWE** (c/n 629/15) which had been delivered 15.7.28 by Blackburns to Mrs Ruth Knowles of The Friend Ship Fellowship Organisation of Ramshurst Manor, Tonbridge. It was being used by her on a tour of schools to promote the Organisation and was officially named 'The Friend Ship' by Lady Heath in August 1928.

14. Avro 594 Avian IV G-AADF (c/n R3/CN/204) Regd (CofR 1831) 12.28 to Southern Aero Club Ltd. CofA No.1725 issued 14.12.28; collected from Manchester by Miles and flown to Shoreham on 16.12.28 (a 3hr 45min flight). On 24.2.29, Miles achieved his 500th flying hour in this aircraft. Regd (CofR 2362) 12.29 to Stephen George Stevens, Brighton (based Shoreham). Regd (CofR 5415) 26.10.34 to Harold Leslie Alfred Powell, Westcliff-on-Sea (based Pitsea). Seized at Penshurst and sold by Sheriff of Kent by auction 23.4.35. Regd (CofR 5866) 10.5.35 to Frederick Wiliiam Green, Cambridge. Regn cld 31.3.40. Impressed as instructional airframe **2074M** and to 5 SofTT Locking.

On 5.1.29, Miles flew **DH.60X Moth G-EBYJ** (c/n 647) on a 15-minute local at Shoreham. It was owned at time by Harrington 'Tony' Law and normally based at Hendon.

On 10.2.29, Miles flew the prototype **Simmonds Spartan G-EBYU** (c/n 1) from Shoreham. Owned by the manufacturer, it was based at Hamble and was destroyed in a crash nr Bury St Edmunds 10.3.29.

15. SE.5A G-EBPA (ex F7016 – but regd as E7016) Regd (CofR 1293) 30.7.26 to Mrs Sophie Mary Eliott-Lynn (who became Lady Heath 11.10.27), Stag Lane. CofA No.1103 issued 12.4.27. Wfu and CofA lapsed 11.4.28. Lady Heath left England for the USA 11.28 (not returning until 1933). Sold (without engine) and regd (CofR 1879) 12.2.29 to FG Miles. Miles collected it from Hamble on 16.2.29 and during the course of the ferry flight back to Shoreham he recorded an altitude of 7,000 ft, the highest flown by any member of the club at that time. CofA not renewed, although Miles flew it again on 17.2.29, 25.2.29, 3.3.29, 10.3.29 and 17.3.29. In 4.30, Southern Aircraft Ltd were selling 2 SE.5A airframes and a spare Viper engine for £25. There is no evidence of a sale and regn cld 3.32.

On 2.3.29, Miles flew **Avro 504K G-EBWO** (c/n nil) owned by Phillips & Powis Ltd from Brooklands to Shoreham. It had been modified by Phillips & Powis in 1928 to have a 100hp Anzani radial engine and wings from an Avro 504N with tapered ailerons and tanks under the upper wings but retaining the Avro 504K undercarriage with skid. It may have gone to Shoreham for an engine change as, when Miles flew it again, on a local flight on 17.3.29, it was apparently fitted with a 110hp Le Rhone. G-EBWO crashed on landing Caversham Bridge 1.4.29.

16. Avro 504K G-AAED (built by Southern Aircraft Ltd from ex AV Roe spares) Regd (CofR 1861) 28.1.29 to John Sale and John Anthony Barnett, Hove (based Shoreham). First flown, by Miles, 3.3.29 (and again on 7.3.29 and 12.3.29). CofA No.1928 issued 12.3.29. Spun in on landing Stag Lane 4.7.29; Barnett & passenger killed. Regn cld 1.30.

On 24.3.29, Miles flew **Avro 504K G-AAFT** (ex J8379) from Brooklands to Shoreham and then back to Brooklands 25.3.29. On 27.3.29, he flew it back to Shoreham and the following day, flew it to Maylands Farm, Romford. The Avro had been regd 25.3.29 to Art Forsyth, t/a Inland Flying Services, Maylands and its CofA was issued 26.3.29 – so the flights were probably in connection with some pre-delivery work being done.

Whilst at Brooklands on 26.3.29, Miles flew **DH.60X Moth G-AAAG** (c/n 697), then owned by the Dutch motorcycle speedway rider, Lucien Niewenhuizen

17. Avro 504K G-AAGG (built by Southern Aircraft Ltd from ex AV Roe spares) Regd (CofR 1945) 4.29 to Phillips and Powis Aircraft (Reading) Ltd who wanted it for joy-riding; dd 4.5.29 to Woodley via Croydon by Miles on what would appear to have been its first flight. CofA No.2023 issued 6.5.29. CofA lapsed 7.5.30. Regd (CofR 3065) to Rollason Aviation Co, Croydon; CofA renewed 2.4.31. Regd (CofR 3443) 29.9.31 to Scottish Eastern Aircraft Services Ltd, Earlston. CofA lapsed 4.5.33. Regn cld 12.35.

18. Avro 504K G-AACW (built from spares, probably by Southern Aircraft Ltd). Regd (CofR 1814) 19.11.28 to Dominion Aircraft Ltd, Shoreham (a company of which Miles was a shareholder along with John and Mrs E Don). Not delivered. Regd (CofR 2012) 6.29 to Southern Aircraft Ltd. First flown by Miles 4.6.29. CofA No.2129 issued 27.7.29. G-AACW 'worked the South Coast', giving joy-rides and on 18.7.30, Graham Head recorded that he: w*ent up with Miles in 'CW for the purpose of taking part in the production of 'The Yellow Mask' by British International Talking Pictures Ltd. Doubled for hero (Lupino Lane) in a spin. Filmed two more spins with automatic camera strapped to tail etc.* On 19.7.30 he noted: *More film work for BIP. Let bombs off from 'CW etc. Flew for approx 25 min all told.* Regd (CofR 2879) 3.11.30 to Home Counties Aircraft Services Ltd, Gatwick. Spun in and crashed Horley 25.1.31; killing Lawrence Irving-Bell & 2 passengers. Regn cld 7.31.

On 10.7.29, Miles first flew the **Hornet Baby G-AAII** – see Appendix 5.

On 23.7.29 and 27.7.29, Miles flew **Westland Widgeon IIIA G-AAJF** (c/n WA.1776) at Heston. It was probably Westland's demonstration aircraft at the time.

Whilst at Heston on 26.7.29, Miles flew Miss Eleanor 'Susan' Slade's **DH.60X Moth G-EBSA** (c/n 414) from Heston to Stag Lane and return. Miss Slade was the company secretary of Airwork Ltd at the time. On the same day at Heston, Miles also flew the **Henderson-Glenny HFS.II Gadfly G-AAEY** (c/n 1).

On 4.8.29 and later on 5 & 6.9.29, Miles flew Arthur Lionel Finch Hill's newly-acquired **DH.60 Moth G-EBOT** (c/n 272), which was based at Shoreham. The latter flights were probably in connection with the renewal of its CofA (effected 5.9.29).

On 8.9.29, Miles flew **DH.60G Moth VP-KAC** (c/n 1004) on a 30-minute 'local' from Shoreham. The Moth, owned by Wilson Airways, had been flown back to England from Kenya by Tom Campbell-Black and would return to Kenya on 21.9.29, this time flown by Flt Lt FA Swoffer.

On 15.11.29, Miles made just one flight in the locally-built **Hendy Hobo G-AAIG** (c/n 1) which had been first flown the previous month.

19. Avro 548A G-EBKN (ex E449). Regd (CofR 1130) 12.24 to Aircraft Disposal Co Ltd, Croydon; fitted with 120hp Airdisco V8 engine, driving an enormous geared propeller. CofA No.849 issued 17.4.25. Change of name to ADC Aircraft Ltd 30.7.25. CofA lapsed 29.6.28. Bought by Miles from ADC in 1929. Sold & regd (CofR 2419) 6.2.30 to Anthony Graham Head, Shoreham, for £100 for his personal use. First flown, by Pashley and Graham Head 9.2.30, then a solo fight by Miles and then, at dusk, with Pashley and George Miles. Graham Head's diary notes *'highly satisfactory take off and climb. Wind E 30 mph. Revs 1,950. Slightly left wing heavy and considerably tail heavy. She was flown altogether for about 15 mins'*. On 16.2.30, Graham Head *'flew for 5 mins as a passenger with FG Miles. Engine warmed for 12 mins. Revs not known as FGM took her up without running her up (I make this note as a record of his infallible carelessness)* - whoops! Graham Head made his first solo flight in G-EBKN on 23.2.30. The CofA was never renewed and on 17.7.30, with the engine running very badly, Graham Head semi-forced landed it in a field to the SW of Shoreham aerodrome. He got her back and continued to look for the cause of the trouble but failed. On 6.11.30, Graham *asked Miles to have 'KN unrigged till she was needed.* It never was needed; regn cld 12.32. The wings of G-EBKN were later used in the restoration of the Avro 504K G-EBJE by RAF Abingdon in 1969.

20. Martinsyde F.4 G-EBMI (ex D4295). Regd (CofR 1202) 10.25 to ADC Aircraft Ltd, Croydon. CofA No.966 issued15.4.26. CofA lapsed 6.5.28 and aircraft dismantled and the engine removed. It laid in ADC's hangar until bought by Southern Aircraft Ltd 11.29 and taken to Shoreham. Here, it was 'put into commission again with a new engine' and regd (CofR 2482) 31.3.30 to Lionel Edward Richard. Bellairs, Shoreham (to become the only privately-owned F.4 in the country). The CofA was never renewed and Miles made just one 'involuntary' flight, of 10-min duration, on 2.5.30 (see full story in the main text). G-EBMI remained at Shoreham until 5.8.30, it was used 'very little' and was not subjected to any high-speed manoeuvres; total time-in-air was no more than 10 hours. During the first week of August, it was flown on several cross-country journeys (a contravention of regulations) and finally arrived at Reading where it was sold (but not regd) to Edwin Denis A Bigg of Bournemouth. Don Brown recalled that during its re-assembly at Shoreham the wooden airframe had 'markedly deteriorated' and that the new owner was told not to indulge in aerobatics because of this. The advice went unheeded and on the 24.8.30, after about 4 hours flying at Woodley, Pat Giddy, chief pilot and instructor with Phillips & Powis executed a number of sharp turning manoeuvres and two loops at heights of not less than 2,000ft and then proceeded to carry out low flying at high speeds. On levelling out from its fourth and final dive across the aerodrome at Woodley the aeroplane developed a violent 'flutter', which to all witnesses on the ground appeared to involve the ailerons and wingtips rather than the tail unit. A moment later, the machine turned nose downwards, still fluttering violently at a speed of up to 200 mph. It remained intact until it struck the ground, killing the pilot. In the opinion of the Inspector of Accidents *'this type of aeroplane as originally designed and with "separate" elevators (i.e. not positively coupled together - PA) should not, in the light of present-day experience, be given a Certificate of Airworthiness covering aerobatics.* Regn cld 10.30.

On 8.3.30, Miles flew **DH.60G Moth G-AAAL** (c/n 809) on a local flight and, on 25.3.30, flew it from Shoreham to Heston, via Farnborough and Brooklands. At the time, it was regd to Lt John Reginald Bryans, based at Fort Grange Aerodrome, Gosport.

On 24.3.30, Miles first flew the **Southern Martlet G-AAVD** (c/n 201) – see Appendix 5.

21. Clarke Cheetah G-AAJK (c/n CC.1) A biplane built by F/O John Clarke at Brough from various parts and powered by a Blackburne Thrush engine, this was regd to him (CofR 2056) 6.29 and with CofA No.2432 issued 17.9.29. Clarke was killed in a Siskin accident at Brough 11.10.29 and the aircraft was put up for disposal 'by RAF Station ballot' in 12.29. It was either won by or otherwise acquired by Lord Malcolm Douglas-Hamilton and was duly regd to him (CofR 2459) 3.30. He may have asked Miles to sell it, since Miles first flew it, at Shoreham, on 6.4.30 and again on 8.4.30. The Cheetah was anonymously advertised for sale in The Aeroplane the same month for £75, alongside 2 SE.5A airframes and a Wolseley Viper engine (for £25). Last flown, by Miles, for 3 min, on 2.5.30; Miles had been having trouble with fuel starvation and had pulled off a few successful forced landings until his luck ran out on this flight when it 'disintegrated' on the last landing! Its remains were reportedly then sold for £10! Regd (CofR 2830) 9.30 to Alfred Cecil Thomas, reportedly on behalf of 4 NCOs from 600 (City of London) Squadron, repaired and based at RAF Hendon. The CofA finally lapsed 3.7.34 and it then passed to the local aircraft 'scrap dealer' Capt Christopher PB Ogilvie at nearby Willesden Green. It was then sold and regd (CofR 6980) 12.5.36 to Richard Adrian Hopkinson of Hemel Hempstead and he appointed Luton Aircraft Co to rebuild it. They brought in Harold Best-Devereux, who used it as the basis for a new aircraft, financed by Richard Hopkinson's uncle, Martin Hopkinson and which emerged from Luton's Gerrards Cross works in 10.37 as the Martin Monoplane G-AEYY.

By 4th May 1930, Miles had amassed 785 hr 43 min flying time and the last flight he recorded in his third log book was made in the Avro 504K G-AACW on that day. Unfortunately, his next log book is missing, and the fifth log-book starts 376 hr 19 min later, on 15.4.31. Known events of interest that took place during that period are shown below.

22. Avro 616 Avian IVM G-AAVM (c/n 416) Regd (CofR 2464) 3.30 to Henlys (1928) Ltd, Heston. CofA No.2442 issued 24.3.30. Graham Head on 10.4.30 records: 'Hermes Avian G-AAVM down at Shoreham', which implies that it was dd to Shoreham on that date. It is possible that it was purchased by Southern Aircraft Ltd at that time, although it is recorded as still being operated by Henlys in 5.30, so the actual sale may have been later. On 19.1.31, Graham flew with Bellairs in G-AAVM (noting it as a Sports Avian) *and did a couple of fairish landings.* On 29.4.31, Graham noted *Sale crashed G-AAVM through engine failure when taking off; not his fault.* Whilst Graham also stated 'write off', this accident is not recorded by the Air Ministry and it must have been repaired as it was used by the Southern Aero Club and flown by Miles regularly from 28.6.31 until 4.8.31 when he severed his connection with Shoreham. Regd (CofR 5037) 5.34 to unknown party (possibly Southern Aero Club?). On 25.7.34, it crashed near Shoreham Aerodrome - Graham Head noting in his diary for that day *'Visited Aerodrome; discovered that 'VM had crashed. Apparently Pash was giving a trial lesson when the pupil froze on the controls, causing crash into ploughed field. Pash - cuts and bruises; pupil unconscious'.* On 7.12.34, Air Travel Ltd offered £15 for the wreckage and on 9.12.34, Graham Head *'wrote to Air Travel quoting £17 10s for 'VM wreckage'.* Regn cld 12.34. It was rebuilt and regd (CofR 6461) 6.11.35 to Air Travel Ltd, Penshurst. Dd 1.36 & regd (CofR 7103) 29.5.36 to South Staffs Aero Co Ltd, Walsall. CofA lapsed 3.2.38; renewed 27.2.39. Regd 20.3.39 to The Bedford School of Flying Ltd, Barton-in-the-Clay. Sold at auction by bailiff 19.12.40. Regn cld at census 1.12.46.

Prior to 18.6.30, the **Southern Martlet G-AAYZ** (c/n 203) was first flown, probably by Miles – see Appendix 5.

23. Desoutter I G-AATF (c/n D.19) Regd (CofR 2361) 12.29 to Walter Lawrence Hope, operated by Air Taxis Ltd, Stag Lane. CofA No.2388 issued 28.1.30. Purchased 1.8.30 by Miles for taxi work and regd (CofR 2778) 8.30 to Southern Aircraft Ltd, Shoreham. Flown by Bellairs, with Graham Head as passenger, on 1.8.30 and by Miles (with Graham Head) to Burgess Hill and back on 5.8.30. On 16.4.31 Miles left Lympne on a tour of the Continent, visiting Brussels, Cologne, Paris, Tours, Borest and Amsterdam, before returning via Lympne on 24.4.31. Regd (CofR 3666) 3.32 to Lionel Edward Richard Bellairs, Shoreham. Regd (CofR 4162) 7.2.33 to The Rollason Aviation Co Ltd, Yapton. Regd (CofR 4641) 22.8.33 to Lawrence Dundas, The Earl of Ronaldshay, Portsmouth. Crashed into fence and house Hanworth 9.5.34. Regn cld 8.34.

In 8.30, **Southern Martlet G-ABBN** (c/n 204) was flown, probably by Miles – see Appendix 5.

On 21.9.30, Miles first flew the **Southern Martlet G-AAYX** (c/n 202) – see Appendix 5

24. DH.60X Moth G-EBZG (c/n 676) Regd (CofR 1679) 6.7.28 to John Scott Oliver, Stag Lane. CofA No.1514 issued 16.7.28. Damaged on landing Shoreham (unknown date); remains sold to Southern Aircraft Ltd for £50. Regd (CofR 2869) 10.30 to Mrs Maxine FM Freeman-Thomas (Viscountess Ratendone), Shoreham; named 'Jemimah'. Repaired by Southern Aircraft; first record of post-rebuild flight was by George Miles and Pashley on 8.3.31. Flown by Miles 15.4.31 (but almost certainly earlier in the period of the missing log-book). Operated by Southern Aero Club. On 23.6.31, Miles and Blossom made their first 'joint' visit to Reading, in

G-EBZG, returning to Shoreham via Heston and Croydon the same day. Last flown by Miles, from Shoreham to Croydon on 4.8.31, which was also to be his last flight in England before he departed to South Africa later that month. Don Brown made his first solo, a 10 min flight, in G-EBZG on 17.5.31. Sold 4.32 and regd (CofR 4123) 11.1.33 to Edward Anthony James Lytton, Viscount Knebworth, Hendon. *(Knebworth was killed 1.5.33 in the crash of 601 Sqdn Hart K2974 at Hendon).* Regd (CofR 4639) 21.8.33 to Lt Richard Hatt Noble Graham, High Wycombe (based Heston). Regd (CofR 5078) 16.5.34 to F/Lt Harry Melville Arbuthnot Day, of 4 FTS Abu Sueir, Egypt. CofA lapsed 28.2.37. Regn cld 24.5.38 as wfu.

Prior to 6.2.31, Miles first flew the **Southern Martlet G-ABIF** (c/n 205) – see Appendix 5.

Probably in March 1931 (date not known) Miles first flew the **Southern Metal Martlet G-ABJW** (c/n 31/1) – see Chapter 3.

25. Austin Whippet G-EAPF (c/n AU.2) Regd (CofR 399) 11.19 to The Austin Motor Co Ltd, Cofton Hackett. Sold 1920/21 but only regd (CofR 1002) 7.23 to (aircraft scrap dealer) Christopher PB Ogilvie, Hendon. Regd (CofR 1060) 3.24 to Flt Lt Frank Ormond Soden, Brighton (based RAF Gosport, later RAF Northolt). First CofA No.830 issued 17.7.24. Dd 15.7.26 and regd (CofR 1341) 7.2.27 to The Midland Aero Club Ltd, Castle Bromwich. Regd (CofR 1627) 2.5.28 to ER King, Brackley (based Sywell). Regd (CofR 1958) 23.4.29 to Herbert MacDonald Pearson, RAF Calshot. CofA lapsed 27.4.29 and later test-flown from Hamble but u/c collapsed. Sent to Miles for what became a protracted overhaul later in 1929 but when the costs became excessive, it was sold to Miles for £15 in 1931. Graham Head noted 3.7.31 - *towed Austin Whippet to United Services Garage behind Austin 7* and on 4.7.31, *delivered Austin Whippet on Regent roof with Yettie and George and Regent staff.* This was in aid of a promotion for the showing of the classic film *Hell's Angels* on 5.7.31. On 13.8.31, Graham noted that *Mr F. towed the Austin Whippet to the New Empire Cinema and we installed it in the Assembly Rooms.* On 28.8.31, Graham wrote, *dismantled Austin Whippet and moved it back to aerodrome.* It is believed that it was scrapped sometime thereafter although the regn was cld 11.32 as sold.

26. Desoutter I G-AAWT (c/n D.23) Regd (CofR 2517) 4.30 to Cirrus Aero Engines Ltd, Croydon and flown by Maj John V Holman. CofA No.2478 issued 7.5.30. Crashed Croydon 28.5.30; repaired. Sold 3.31 and regd (CofR 3107) 4.31 to Southern Aircraft Ltd, Shoreham. Graham Head noted 15.3.31 *Went up in G-AAWT (inverted Hermes Desoutter) with Pash and Mrs Pash 5 min.* Miles made a return flight to Stag Lane in G-AAWT on 28.4.31 and flew to Woodley on 14.5.31, returning to Shoreham on 16.5.31. Regd (CofR 3439) 9.31 to Herts & Essex Aero Club Ltd, Broxbourne. Regd (CofR 3926) 3.8.32 to Phillips & Powis Aircraft [Reading] Ltd, Woodley. Regn cld 10.33 as sold (with CofA renewed 13.9.33). Arr Surabaya 11.33 & assembled at Morokrembangan Naval Base. Regd in Dutch East Indies as **PK-SAN** 12.33.

On 8.6.31, Miles flew **DH.80A Puss Moth G-AAZW** (c/n 2090) at Stag Lane. At the time, it was owned by Geoffrey Garnett of Bradford and was based at Sherburn. On 24.6.31, Miles flew the **DH.60G Coupe Moth G-AAHI** (c/n 1082) on a 30-minute local from Shoreham. It was owned by John Dalgety of Romsey and was based at Hamble. Graham Head flew the same aircraft on 1.12.31 and occasionally to 7.12.32.

On 28.6.31, Miles flew the prototype **Civilian Coupe G-AAIL** (c/n 1) at Hanworth. This was still owned by the manufacturer which was based at Hedon, near Hull. Oddly, the CofA had lapsed on 4.6.31 and was not renewed until March 1933.

On 19.7.31, Miles took the opportunity of flying (for 5 minutes only) Lord Halsbury's **Potez 36 F-ALJC** (c/n 2359). Although regd G-ABNB it actually retained its French regn whilst operated by CD Barnard's Air Tours from June 1931, before returning to France on 23.11.31.

27. Simmonds Spartan G-AAHA (c/n 23) Regd (CofR 1973) 6.5.29 to Charles Coombes, t/a Shanklin Flying Services, Apse, Shanklin. CofA No.2077 issued 27.6.29. Regd (CofR 3194) 5.31 to Spartan Aircraft Ltd, Somerton. Regd (CofR 3209) 5.31 to Flt Lt Frank George Gibbons, Plymouth [based Heston]. Collected by Miles from Woodley 29.7.31 & regd (CofR 3346) 4.8.31 to Frederick G Miles, Shoreham and taken by him to Cape Town. On 1.9.31 he made his only flight in it in South Africa before returning to England within the month, having sold the Spartan locally. Regn cld 3.32 as sold. Regd **ZS-ADC** [CofR 82] 3-4Q.32 to WD Mackay.

George Herbert Miles

The early flying log books of George Miles have survived and the following details all his first flights in any particular aircraft, made while he was at Shoreham. They include some from Woodley in the period when he was still working at Shoreham and before he started full-time work at Woodley in about May 1936.

Date	Regn	Type	Notes
9.6.28	G-EBVA	Avian III	first flight – 16 min local with F G M 'pupil under instruction'
25.12.28	G-AADF	Avian IV	first solo 13.9.30 – of 8 mins.
23.7.29	G-AAFH	Parnall Elf	demo flt - Shoreham-Croydon with F G M
26.9.29	G-AACW	Avro 504K	with Pashley
6.1.30	G-EBYB	Avro 504K	with Pashley from Shoreham
9.7.30	G-AABF	Bluebird III	with F G M and Graham Head on local flights
3.1.31	G-AAVM	Avian IVM	with Pashley; also on 16.9.32 1st recorded flt with Don Brown – Shoreham-Ford-Tangmere-Shoreham
8.3.31	G-EBZG	DH.60X Moth	with Pashley; on 28.5.31 went night-flying with F G M
9.4.31	G-AAWT	Desoutter I	Southern Aircraft Ltd; with Pashley
14.4.31	G-ABKI	Avian Sport	
31.7.31	G-AAII	Martlet	
22.11.31	-	BAC IV glider	with CH Lowe-Wylde – 1 min flt
6.12.31	G-AAHI	DH.60G	with Jack Sale
31.1.32	G-ABJS	Spartan 3-str	local flt at Penshurst
31.5.32	G-AAGI	DH.60G	
7.8.32	G-ABIF	Martlet	
6.9.32	G-ABVN	Monospar ST.4	PS&IoW Avn; Shoreham with Charles Eckersley-Maslin
3.10.32	G-EBZC	DH.60X	Woodley-Yate with F G M
6.3.33	G-ABVG	Miles Satyr	Woodley
30.3.33	(H.101)	Miles Hawk	Woodley with F G M; soloed on 4.4.33- no regn recorded
15.4.33	G-ABUR	Gull Four	Shoreham with Flt Lt Cyril Colman
21.4.33	G-ACCE	DH.84 Dragon	Brian Lewis & Co, Heston with Flt Lt Christopher Clarkson
22.4.33	G-AAGR	Avian IV	
30.4.33	G-ABZF	Avro Cadet	Shoreham with Pashley
2.7.33	G-ABVB	Wessex	PS&IoW Avn; solo at Shoreham (identity not recorded)
25.7.33	G-EBRO	Widgeon III	purchased - see full details page 308

26.8.33	G-EBQH	DH.60X	Woodley-Shoreham-Woodley 'Sports Moth'	
19.11.33	G-ABLS	DH Puss Moth	Shoreham	
21.1.34	G-AAYX	Martlet	Woodley	
13.2.34	G-AALT	DH.60G	Woodley	
4.3.34	G-ACGB	DH Fox Moth	Shoreham with Jack Sale	
11.3.34	G-ACMH	Miles Hawk	Shoreham	
18.4.34	G-ACNW	Miles Hawk	Woodley	
8.5.34	G-AAHE	Avian IV	Shoreham	
16.5.34	-	Miles Hawk	with F G M	
26.5.34	'G-ACNV'	Miles Hawk	In Southern Aero Club display, Shoreham (probably G-ACNW)	
26.5.34	G-ACSY	AS Courier	In Southern Aero Club display, with Cyril Colman	
2.6.34	G-ACHF	Spartan Arrow	Lympne-Shoreham	
20.7.34	G-AAVP	Avian IVM	Shoreham; later acquired – see below	
8.8.34	G-ACSA	DH Moth Major	Shoreham	
15.8.34	G-AAWH	Avian IVM	Southern Aero Club; later purchased – full details below	
11.11.34	G-ACWW	Hawk Major	Shoreham	
4.5.35	-	Miles Merlin	Woodley with F G M	
5.5.35	G-AAAC	DH.60G	Shoreham	
9.5.35	G-ADBI	Falcon Major	Shoreham-Woodley-Shoreham; with Tommy Rose	
26.5.35	G-AAIV	DH.60G	Wilmington-Shoreham	
29.5.35	G-ABIW	Avian IVM	Shoreham	
14.7.35	G-ACRE	Avro 504N	Shoreham-Penshurst	
3.8.35	-	Monospar	Shoreham; with Kenneth Seth-Smith (no regn shown)	
25.8.35	G-ADGB	Avro 504N	Shoreham	
28.8.35	G-AAGY	Spartan	Woodley-Shoreham	
25.9.35	G-ADLN	Hawk Trainer	Woodley (identity unconfirmed)	
12.10.35	G-ADNL	Sparrowhawk	Woodley	
12.10.35	G-ABWL	DH.60G	South Coast Aero Club, Shoreham	
20.11.35	G-AADA	DH.60G	South Coast Aero Club, Shoreham	
26.1.36	G-AAIN	Parnall Elf	Woodley-Shoreham	
7.2.36	G-AAJJ	DH.60G	Shoreham	
23.2.36	G-ADDB	BK Swallow II	Shoreham	
18.4.36	-	AS Envoy	Shoreham; with Cyril Colman, Pashley & Lord Amherst	
19.4.36	G-ADWH	ST.25 Jubilee	Shoreham; Kenneth Seth-Smith & Don Brown	
10.5.36	G-AECT	Whitney Straight	Woodley (identity unconfirmed)	
22.5.36	G-ADLN	Hawk Trainer	Menasco C4S engine - Woodley/Shoreham endurance flying	
28.5.36	G-ABJT	DH.60G	Shoreham-Wilmington-Shoreham	
18.6.36	G-ABAO	DH.60G	Shoreham-Hawkhurst-Shoreham	
20.9.36	-	Grunau Baby	glider – 4 min flight at Brighton	
16.5.37	-	Heston Phoenix	Shoreham; with A Amping	

Apart from the endurance flights in G-ADLN, most of George Miles' subsequent flying was then carried out from Reading, with occasional visits to Shoreham.

George Miles purchased three aircraft in his own right during this time. Their histories are as follows:

Westland Widgeon III G-EBRO (c/n WA.1682) Regd (CofR 1393) 8.27 to Westland Acft Works. CofA No.1195 issued 27.8.27. Operated by Harald Penrose [5.29] as Westland demonstrator. Regd (CofR 2522) 4.30 to Sqn Ldr The Hon Ralph Alexander Cochrane, RAF Andover. Sold 6.31 and regd (CofR 3306) 14.7.31 to John Glaholme Ormston, Broxbourne. Regd (CofR 4538) 7.6.33 to Maurice Carey Wilks, Farnham (based Castle Bromwich). Sold 25.7.33 for £175 to George Herbert Miles, Shoreham; dd same day. Flown by him regularly around the country, on 29.7.33 he flew Shoreham-Reading-Brooklands-Shoreham and on 30.7.33 he flew Shoreham-Ford-Shoreham. On 30.8.33, George recorded a 40-min flight to 8,000 ft and on the same day Graham Head flew it for the first time with George from Shoreham. On 31.3.34 Graham Head flew it to Woodley with George Miles as passenger and on the return, with George piloting, he had to land on the South Downs to the SE of Bramber on account of fog. While returning from Reading to Shoreham on 5.4.34, George was forced to land at Steyning due to fog and had to complete his journey the next day. Sold 2.5.34 for £155 and regd (CofR 5233) 12.7.34 to Samuel Frank 'Sammy' Youles, Shoreham. Regd (CofR 5854) 4.5.35 to Stephen George Stevens & Anthony Graham Head, Shoreham. Regd (CofR 6482) 7.11.35 to Ivor Percival Tidman, Harry Samuel Griffiths & Samuel Bert Yardley, Walsall. Regd (CofR 7701) 22.2.37 to Walter Partington, op by Bournemouth Flying Club, Bournemouth. Regd (CofR 8144) 18.10.37 to Anthony Ord Humble-Smith, Bournemouth. CofA lapsed 23.3.38. Regn cld 1.12.46 at census.

Avro 616 Avian IVM G-AAVP (c/n 417) Regd (CofR 2467) 3.30 to Neville Hart Player, Woodley. CofA No.2445 issued 29.3.30. Regd (CofR 3020) 2.31 to Airwork Ltd, Heston; operated by Airwork School of Flying. Regd (CofR 3729) 4.32 to Henlys Ltd, Heston. Sold 1.5.32 to Mr Bancroft but probably not dd. Regd (CofR 4163) 7.2.33 to Philip Harvard Johnson, Hedon. Regd (CofR 4598) 13.7.33 to Henlys Ltd, Heston. Operated 8.33 with KB Radio titles at Haldon display to demonstrate ground to air radio. Sold 4.34 to Victor N Buchan, a member of the Southern Aero Club, Shoreham. Flown by George Miles 20.7.34 and by Don Brown, with Victor Buchan, on 19.8.34. Flown Penshurst-Shoreham by George Miles 14.7.35. Its C of A expired 23.6.36 and it was wfu and pushed into the back of the hangar at Shoreham. Its existence was later remembered by George Miles when he was established at Woodley and he bought it with the intention of reconditioning and selling it. Ferried from Shoreham to Woodley by Don Brown, on a special one day Permit to Fly issued on 6.2.37. CofA renewed 23.4.37. Regd (CofR 7857) 3.5.37 to Leslie Vernon Farrington of Wood Norton, nr Evesham. CofA lapsed 22.4.38. Advertised for sale 8.38 'with new CofA' but not sold (or re-certified). Regn cld 15.8.45 in post-war census. Actually remained in store in Evesham until at least 1953.

Avro 616 Avian IVM G-AAWH (c/n 418) Regd (CofR 2501) 4.30 to Claude Pierrepont 'Philip' Hunter, Rhos Ucha Aerodrome, Wrexham. CofA No.2468 issued 18.4.30. Regd (CofR 3574) 1.32 to Home Counties Aircraft Services Ltd, Gatwick. Regd (CofR 3673) 3.32 to Evelyn Francis Ness, Gatwick. Regd (CofR 4331) 11.4.33 to Kenlyn Ltd, Eastbourne. Flown by George Miles at Brooklands 15.8.34 and dd to Shoreham (where again flown by George Miles 25.8.34). Regd (CofR 5371) 25.9.34 to Southern Aero Club Ltd, Shoreham. Regd (CofR 6435) 25.10.35 to George Herbert Miles, Shoreham (later Woodley). Regd (CofR 7337) 10.9.36 to Ralph Harold Henderson, Hanworth (later Gravesend). Advertised for sale 11.37 but probably not sold. Regd 22.4.39 to The Isle of Wight Flying Club Ltd, Lea. CofA lapsed 20.6.39. Sold 9.39 to Essex Aero Ltd, Gravesend. Impressed 31.3.40 as **2083M** and allocated to RAF Eastchurch. Transferred to 3 SofTT Blackpool 5.5.44. SOC Cat E2 as scrap 25.7.45.

Appendix 5:
Southern Martlet and Metal Martlet Production

2 SH (also 1 & 200)   Prototype built to the order of Lionel Edward Richard Bellairs, Shoreham (originally known as Hornet Baby); fitted with 75hp (max 82hp) ABC Hornet. Regd **G-AAII** (CofR 2026) 6.29 to Southern Aircraft Ltd; painted silver overall with black registration. First flown, by Miles, 10.7.29 - a 10-minute 'test'. (It had been entered by Bellairs in the 1929 King's Cup Air Race at Heston 5.7.29 (Race No.42) but was not ready). Much flying then carried out by Miles, culminating in a 10-minute 'local' from Shoreham 19.8.29. To A&AEE Martlesham Heath for official CofA trials (probably shortly after 19.8.29 since Miles seemingly flew Avian G-AADF back and forth to Martlesham 3 times on 24.8.29) and next flown by Miles on a 1-hour flight from Yarmouth (Isle of Wight) to Shoreham later on 24.8.29 (the Martlesham trials report is believed to have been dated 5.9.29). Miles flew G-AAII to Hanworth (a 1-hour flight) 31.8.29, where it made its first public appearance on the official opening of the Air Park; returning to Shoreham 1.9.29. CofA No.2255 issued 3.10.29 as the Southern Martlet (now with c/n 1 and which was later changed to c/n 200). On 4.10.29, Miles recorded a 10-minute flight at Newcastle; on 17/18.1.30 he made two local flights at Le Bourget before returning to Shoreham via Berck and Lympne 19.1.30. Damaged in forced landing nr Hamble 9.3.30 (see story); repaired. Regd (CofR 3255) 6.31 to Miss Edith Nancy Beynon Birkett, secretary of Southern Aircraft Ltd, who made her first flight in G-AAII on 29.7.31. Her nephew, Chris Greenhow, recalled in January 2003, *'being a very good woodworker Nancy had a hands-on approach to the building of Martlet G-AAII and was not satisfied with the performance when it was fitted with an ABC Hornet engine, so she bought an 82hp 5-cylinder Armstrong Siddeley Genet II for £40, which had previously been fitted to a Cierva Autogiro. Nancy transported it in the 'dicky' seat of her Brooklands Riley and then fitted it to G-AAII herself!* This was in fact the lightest of all the different engines to be fitted to the Martlet and resulted in G-AAII then having a longer nose, it was also fitted with a longer fin. Nancy must have enjoyed flying the Martlet because she recorded a total of 93 flights in her log-book - most local flights at Shoreham but a few cross-country flights were made as follows:- 1.8.31 Croydon; 6.10.32 Portsmouth; 24.11.32 Hamble; 26.11.32 Hailsham; 6.4.33 Ford; 8.4.33 Hailsham; 10.4.33 and 16.4.33 Storrington (these latter must have been to the original private airstrip at Parham, nr Storrington). After her 93rd flight in G-AAII, on 29.4.33, she noted -'II sold'. Regd (CofR 4491) 8.5.33 to Arthur Robert 'Buster' Frogley, the speedway rider who had formed Herts & Essex Aero Club at Broxbourne with his brother Roger Frogley. Regd (CofR 5349) 10.9.34 to Robin Baillie Nuthall, Hanworth. Regd (CofR 5434) 9.11.34 to Angus Hunter Tweddle, Hanworth. Regn cld 12.35 as sold. Regd in Ireland as **EI-ABG** 19.6.36 to G R S Pennefather of Kilworth, Co Cork (based Fermoy Aerodrome). Fate unknown although put up for sale by owner 8.36. Regn cld 9.11.48 as destroyed but this probably long post-dated its actual demise.

Blossom, Viscountess Ratendone, flew G-AAII just once, on 27.7.31. George Miles made his first flight in G-AAII on 31.7.31. Graham Head flew it for 10 mins on 25.9.31, noting in his diary, 'this is first time I have flown a single-seater machine and in his log book 'First flight in type (and last!)', the latter comment crossed out.

201   Martlet 101 - 2nd built but first built entirely from scratch; painted silver overall. Fitted with 82hp (max 88hp) AS Genet II. Built to the order of Lionel Bellairs, a director of Southern Aircraft Ltd (but never formally regd to him). Regd **G-AAVD** (CofR 2451) 3.30 to Southern Aircraft Ltd; (reportedly not 'officially' registered until 7.5.30). First flown, by Miles, 24.3.30 *(Graham Head took a 5-minute 35mm film of G-AAVD in the air on 4.4.30).* Flown by Lionel Bellairs 10.4.30 to 16,500 ft; flown by Flt Lt Atcherley 13.4.30. CofA No.2571 issued 23.6.30. Graham Head noted in his diary for 3.7.30 that he assisted Miles to change the Genet Minor engine in 'VD prior to King's Cup race; flown by Miles in the 1930 King's Cup Air Race (Race No.98 below the word *'Martlet'* on the rudder); the race started from Hanworth 5.7.30 but Miles retired with engine trouble at Woodford. Regd (CofR 3559) 1.32 to Walter Retlaw Westhead, Shoreham (who must have had a prior interest as he was the formal entrant in the 1930 Kings Cup Air Race and indeed was shown as owner in *Flight* 18.7.30 and in the 1931 *Register of Civilian Aircraft*). Regd (CofR 3836) 7.32 to James Frederick Alexander, Winterbourne Stoke, Salisbury (based High Post). For sale 3.33. Regd (CofR 5301) 8.8.34 to Alexander Robert Ramsay, Canterbury (based Bekesbourne). For sale at Woodley 3.35. Regd (CofR 6612) 31.12.35 to Henry MacDonald Goodwin, Bewdley (based Walsall and later Castle Bromwich) (probably purchased earlier in year, since advertised for sale 10.35 by Goodwin as 'just rebuilt'). Advertised for sale during 1936 (including by Laughton Goodwin & Co, Kidderminster 5.36) and CofA lapsed 28.10.36. Regd (CofR 7872) 13.5.37 to F/O Lionel Woodland Saben (of RAF Cranwell, later RAF Scampton and RAF Eastchurch); CofA renewed 22.6.37. For sale 6.38; regd (CofR 8898) 23.11.38 to Horace Whitaker, Banstead, Surrey (based Redhill). CofA lapsed 11.8.39 and later sold from RNAS Worthy Down by Lt Horace "Chunky" Whitaker RNVR to William H C Blake, Kings Worthy, Winchester for £35 (Blake was the builder of the 1930 Blake Blue Tit, which survived to be donated to the Shuttleworth Trust in 1968). Sold c.1942/43 to Lt McLauchlan, an Air Traffic Control Officer at RAF Turnhouse for £25. Last noted 8.44 in derelict condition at Turnhouse aerodrome. Regn cld in 1.12.46 census.

G-AAVD was flown just once by Blossom on 15.1.31.

202   3rd ordered, but 4th to be completed. Originally ordered by Flt Lt Richard Llewellyn Roger Atcherley (then C in C of the Staff College, Cranwell), who specified that it should be fitted with a special auxiliary fuel tank for inverted flight. Unfortunately, due to the limited funds available (they did not ask a new customer for money upfront), its building took so long that Atcherley was unable to await delivery and finally cancelled the order. The unfinished aircraft was then taken over by Lionel Bellairs, who paid for the aircraft to be completed. Fitted with the heavier 5-cylinder 103hp (max 110hp) AS Genet Major I engine, which entailed shortening the top longerons to give a sloping firewall; it was also fitted with parallel-chord ailerons. Regd **G-AAYX** (CofR 2588) 5.30 to Southern Aircraft Ltd (possibly formally regd 14.7.30); painted polychrome blue overall with black registration. *(The registration date of 1.6.30 now quoted by the CAA is a 'made-up' date when the files were computerised, the original information being lost).* First recorded flight was, by Miles, 21.9.30, but this may not have been its first flight (as there is a page missing from his log-book for that period). CofA No.3121 issued (belatedly – for reasons not known) 8.7.31. Flown by Miles to Reading 27.7.31, then on to Croydon, before returning to Reading and back to Shoreham same day; flown to Reading again 29.7.31 and appears to have been left there (Miles departed to South Africa shortly afterwards). Miles flew G-AAYX just after his return from South Africa, on 30.12.31 and probably from Penshurst and he took it with him to Woodley. By 1932, the fuselage had been painted black with a white skull and crossbones motif forward of the cockpit and with the inscription 'Old Glory' below; the wings, cowlings and u/c struts were painted silver; the black parts being later over-painted bright red, apart from the fin and rudder which were probably silver with red leading edges and tapered stripes. Regd (CofR 3860) 7.32 to Phillips & Powis Aircraft (Reading) Ltd (and used initially as personal aircraft by F G Miles but by end of 1933 by Reading Aero Club 'for use by members considered competent by the CFI to fly it'). CofA renewed 11.5.33; Miles recorded two test flights on 11.5.33 and 12.5.33 and last flew it, on a demonstration, on 12.7.33. Suffered a bad crash sometime in 1934/35, although place and date remains unknown (but after 8.9.34, when it was flown by CFI Bob Milne and prior to its CofA lapse 5.7.35). Rebuilt, probably at Woodley (or reportedly by someone at Gatwick) and CofA renewed 1.3.37; flown

G-AAVD in a new colour scheme in about 1936/37.
[Via J Havers Collection]

An early photograph of G-AAYX, the third Martlet. [Via B Clarke]

by George Miles from Woodley to Shoreham on 10.4.37 and on 11.4.37, when he had two forced landings at Shoreham; flown again by George Miles from Shoreham 16.4.37. Sold 5.37 and regd (CofR 7991) 13.7.37 to Guy Kempston Lawrence, who kept it at West Malling. Regd (CofR 8288) 3.1.38 to William Kenneth Vinson, Orpington (initially remained based at West Malling, but later to Ramsgate Airport). CofA renewed 31.3.38, at which time it was painted silver with red registration. Regd (CofR 8885) 17.11.38 to Michel Noel Mavrogordato (Lord Nuffield's personal pilot, and also a member of the Oxford Flying Club) based Witney. Dismantled on the outbreak of war and carefully stored in the loft of a 'dry barn somewhere near Oxford' for the duration. CofA lapsed 23.7.40. Advertised for sale in *The Aeroplane* 1.2.46 and *Flight* 7.2.46 by Mavrogordato (now of Westerham, Kent) for £200, quoting 228 total hours flown. He advertised it again in *The Aeroplane* 2.8.46, now with a price of £150 and giving an address in Downton, Wiltshire. Regd 16.10.46 to Butlin's Ltd, London W1, and handed over to Miles Aircraft at Woodley for a complete overhaul. The overhaul found few parts needing replacement, although it was fitted with Magister wheels with low pressure tyres. First flown after overhaul without marks by Don Brown 4.6.47 (and whose only observation was that it was rigged tail heavy). Immediately afterwards, it was flown by F G Miles himself and, after the trim was adjusted, flown again by Don Brown and finally by George Miles. On 10 & 13.6.47, it was flown by Hugh Kendall and its CofA was finally renewed 20.6.47. Repainted in Cambridge blue with cream trim and propeller, it was handed over to Butlins and performed aerobatic displays at various Butlin's Holiday Camps, normally based at Broom Hall, Pwllheli. CofA lapsed 12.7.49. Cld 7.8.49 and regd 16.1.50 to Ronald Wright Clegg, Maurice Oliver Tiffany & Louis Emile Martin, t/a The Ultra Light Aircraft Association, London SW1. Noted dismantled on a trailer in a scrapyard at Staverton 2.50, although only "passing through" en route to F G Miles Ltd at Redhill Aerodrome for overhaul. F G Miles Ltd moved its operations to Shoreham in 1952/53 and the Martlet less its Genet engine was probably amongst the last items to be transported from Redhill in 7.53. The Genet Major had been taken on 20.5.50 in an ex-London taxi by Tommy (T H) Marshall and a 'gang' of members from South Hampshire Ultra-Light Aeroplane Club at Christchurch (including the author) to Air Service Training Ltd, Hamble, where someone was going to overhaul it but it was never seen again! The airframe remained stored in a blister hangar at Shoreham on the southern perimeter road, standing on its wheels but without wings and was still there in 7.57. Its registration was finally cancelled 11.12.59 as pwfu. By 3.59, it had been acquired by The Shuttleworth Trust at Old Warden where it remained in store until restoration to airworthy condition commenced in about 6.86 with a new Genet Major engine having being found from somewhere! Regd 22.4.87 to The Shuttleworth Collection, Old Warden and re-regd 8.4.93 to The Shuttleworth Trust. Its first post-rebuild flight was by chief pilot Andy Sephton on 25 September 2000, having been repainted in the red and silver colour scheme it last wore at Woodley before the war. Its permit to fly was issued 8.5.01 and it is maintained in airworthy condition at Old Warden, flying two or three hours a year. Nominally re-regd 12.12.03 to The Richard Shuttleworth Trustees.

G-AAYX after it had been repainted red and silver. [Via B Clarke]

Having 'lost' its engine while with F G Miles Ltd at Redhill in 1949(!), G-AAYX later moved to Shoreham, where it is seen languishing in an old blister hangar in July 1955. [C Holland via B Clarke]

G-AAYX basking in the sunshine at Woodley, while awaiting delivery to Butlins Ltd, in June 1947. [Miles Aircraft Ltd via F Langmaid]

The beautifully refurbished G-AAYX at Old Warden in 2002.
[Via P Amos]

G-AAYZ, the fourth Martlet, with original engine. *[Via B Clarke]*

203 3rd to be completed and fitted with 110hp (max 120hp) DH Gipsy II. This was the most powerful Martlet to be built; the fuselage was reduced in length by removing a whole bay from the front fuselage. Initially, the top of the engine protruded above the line of the top of the fuselage, being somewhat crudely faired in to the rear, just in front of the cockpit and with the four exhaust pipes protruding upwards at an angle to exhaust over the top wing. Regd **G-AAYZ** (CofR 2592) 5.30 to Southern Aircraft Ltd. Completed to the order of Captain The Hon Frederick E Guest MP; first flown (probably by Miles) 18.6.30. CofA No.2614 issued 27.6.30. Regd (CofR 2870) 10.30 to Guest and based Hanworth. On 5.7.30, flown by Guest's personal pilot, Miss Winifred Spooner, into 14th place in the King's Cup Air Race (Race No.65), which started from Hanworth, at an average speed of 125.5 mph. Flown by Miles 20.6.31, 6.7.31 and 19.7.31. Regd (CofR 3455) 10.31 to F/O Edward Cecil Theodore Edwards, based Hendon and flown by his brother F/O Hugh Robert Arthur Edwards in the July 1932 King's Cup Air Race (Race No.18), during which he forced landed at Runcorn due to a split fuel tank and ran into a hedge. CofA renewed 24.7.33 and put up for sale in 10.33 (but not sold). Engine replaced in 1934 (or 1932?) by a 90hp (max 98hp) DH Gipsy I and visibility improved by lowering the engine to make the rocker boxes roughly in line with the top of the fuselage, a longer fin also fitted, the cockpit widened, a headrest added, fairings added to the undercarriage struts and axle, and the aeroplane painted in a darker colour. In the July 1934 King's Cup Air Race (Race No.12), it was flown by Hugh Edwards into 6th place in the second heat of Round 1 on Friday 13.7.34 at 119.8 mph. Flown just once more by Miles (although recorded as G-AAXZ in his log-book) on 13.9.34 (also Miles' last recorded flight in a Southern Martlet before the war); painted red overall. Final CofA renewal 24.1.35 and at some time thereabouts flown by Geoffrey Wikner from Skegness to Woodley, where it was photographed outside the new hangar. Regd (CofR 6373) 7.10.35 to Marius Maxwell, London SW1 (based Hanworth) but who was badly injured (and possibly died) in the crash of Percival Gull G-ABUV near Nice on 2.11.36. Fate unknown and regn cld in the 12.37 census.

204 5th built; fitted with 82hp (max 88hp) AS Genet II. Regd **G-ABBN** (CofR 2682) 6.30 to Southern Aircraft Ltd and built to the order of the Flt Lt The Marquess of Douglas & Clydesdale. First flown (probably by Miles) in 8.30. CofA No.2685 issued 8.8.30. Delivered to new owner 8.30 and regd (CofR 3111) 4.31 to the Marquess and based at Dungavel in Lanarkshire or Heston. Regd (CofR 3206) 5.31 to Sir John Lawrence Baird, Lord Stonehaven, based Hanworth. Flown on his behalf by Mogens Bramson in the July 1931 King's Cup Air Race (Race No.28) but retired at Shoreham due to engine failure. Regd (CofR 3590) 1.32 to (Sir Alan Cobham's) National Aviation Day Ltd and, painted in two shades of green, it was used for aerobatic displays in the season's flying circus tour commencing 4.32. No longer in use by 9.32 and regn cld 2.33. Regd (CofR 4296) 30.3.33 to aircraft dealers Henlys Ltd, Heston and CofA renewed 31.3.33. No sale known and CofA lapsed 30.3.34. A photo in *Aeroplane Monthly* August 2002 shows it at Woodford on 19.9.34. Fate unknown and regn cld 8.35.

G-AAYZ after repainting and now with 90hp Gipsy I engine.

G-ABBN, the fifth Martlet, at Hanworth Park in July 1931.
[Via B Clarke]

Southern Martlet G-ABBN in new colour scheme at Woodford, while touring with National Aviation Day Displays Ltd during the 1932 season.
[E J Riding via B Clarke]

Above: The Martlets G-ABBN and G-ABIF waiting for the starter's flag in the King's Cup Air Race at Hanworth Park on 25th July 1931.

Left: G-ABIF flying in 1933.
[Via B Clarke]

205 6th and last Martlet built; fitted with 82hp (max 88hp) AS Genet II and with parallel chord ailerons. Regd **G-ABIF** (CofR 2983) 1.31 to Viscountess Ratendone (Mrs Maxine 'Blossom' Freeman-Thomas), Shoreham. First flight date unknown (but Graham Head noted in his diary for 4.2.31 that Miles had taxied Mrs F-T's Martlet into the fence, 'thereby not improving same'). Flown just once by Blossom 6.2.31; Miles recorded five flights between 9 and 29.5.31. Regd (CofR 3217) 30.5.31 to Miss Jean Forbes-Robertson (Blossom's sister) and CofA No.3051 issued 30.5.31. Miles made a return flight to Brooklands 3.6.31, flew to Hamble and back 4.6.31 and made a return flight to Farnborough 23.6.31. Temporarily fitted with an enlarged centre section fuel tank and flown by F/O H H Leech (Race No.29) in the July 1931 King's Cup Air Race but retired at Sherburn-in-Elmet. On Miles' return to England, his first flight was in G-ABIF from Penshurst 9.12.31. Flown 3.1.32 by Miles to Woodley and regd (CofR 3598) 1.32 to Phillips & Powis Aircraft (Reading) Ltd. Regd (CofR 3823) 6.32 to Theodore Cecil Sanders, Woodley (and by 11.32 it was being operated on loan by Reading Aero Club). Flown by Nancy Birkett 20.8.32 and Miles carried out a test flight 28.10.32. Fitted with a Fairey Reed airscrew and flown by Sanders in the July 1933 King's Cup Air Race (Race No.7) but was forced to retire. Last flown by Miles 23.1.34. Flown by Graham Head from Shoreham 15.6.34, noting in his logbook 'Tried [out] machine' and in his diary for the same day '*Next flight Tranum* (probably the well-known parachutist and pilot John Tranum) *undershot and crashed in ditch - wrote off front end and cut on forehead*'. CofA renewed 7.9.35 and regd (CofR 6311) 18.9.35 to Joseph Ayers Keep, Shoreham. Regd (CofR 7230) 31.7.36 to Hubert Charles Wardle, Ealing (but based Gatwick) and CofA lapsed 6.9.36. An unconfirmed report suggests it flew with the CWA Scott's Flying Display tour in 1936, but possibly only briefly. Regd (CofR 8106) 10.9.37 to Air Travel Ltd, Gatwick. Regd (CofR 8202) 10.11.37 to Airports Ltd, Gatwick (the company which took over Air Travel Ltd in 9.37). CofA renewed 18.3.39 and regd 30.3.39 to Guy Duncan Tucker, Mill Hill (and based Hatfield). CofA lapsed following the outbreak of war and the Martlet was given to a local ATC Squadron for ground instructional purposes (possibly No.2203 Squadron, Hatfield). Regn cld 1.12.45.

METAL MARTLET

31/1 Regd **G-ABJW** (CofR 3067) 19.3.31 to The Hon Mrs Maxine Francis Mary, Viscountess Rathendone, Hurstpierpoint (based Shoreham); fitted with 105hp Cirrus Hermes II. Initially mispainted as G-AAII, then corrected erroneously to G-AAJW. First flown, by Miles, c3.31. No CofA issued. Flown at least up to July 1931. Regn shown as cld 11.30 (but probably error for 11.31). Scrapped and to Goody's scrapyard, Twyford, nr Reading.

31/2 Regd **G-ABMM** (CofR 3182) 19.5.31 to Walter Retlaw Westhead, London EC2. Not completed and regn cld 12.32.

Appendix 6:
The Sevenoaks Days – Miles' flying from Penshurst, Kent – October 1931 to March 1932

On 4th August 1931, Miles made his last flight from Shoreham, in the DH.60X Moth G-EBZG, before leaving for South Africa. He returned to England in October and set up home in a caravan near Sevenoaks in Kent. He flew from nearby Penshurst Aerodrome – the first in an anonymous Comper Swift (but probably Lionel Bellairs' G-AAZD) before buying the Spartan Three-seater G-ABJS, the only aircraft he owned in this brief pre-Woodley era.

Spartan Three-seater G-ABJS (C/n 56) Fitted with Hermes II engine. Regd (CofR 3060) 3.31 to Spartan Aircraft Ltd, Somerton. CofA No.2953 issued 31.3.31. Operated (6.31-8.31) on C D Barnard Air Tours. Regd 12.31 to Phillips & Powis Aircraft (Reading)] Ltd. Delivered late 12.31 and regd (CofR 3667) 3.32 to Frederick G Miles. Regd (CofR 3724) 4.32 to unknown party; possibly to Air Trips, a joy-riding business formed by Dorothy Spicer and Pauline Gower, to replace G-ABKK, which was damaged in a crash 26.3.32. Regd (CofR 5419) 30.10.34 to Major Edward George Clerk, Witney, who then took a "flying circus" to Australia. CofA lapsed 16.4.35. Regn cld 12.35 as sold. Regd **VH-UUU** (CofR 537) 20.7.35 to Edward George Clerk, t/a Austral Air Services Pty Ltd, Toowoomba, Qld. Destroyed in gales Tennant Creek, NT 17.1.36. Regn lapsed 19.7.36.

Miles' log books give details of his flying from Penshurst:

1931:
? October	?	Comper Swift	20 min local (probably G-AAZD)
9 December	G-ABIF	Southern Martlet	30 min local
30 December	G-AAYX	Southern Martlet	10 min local
30 December	G-ABJS	Spartan 3-seater	30 min local
31 December	ditto		30 min local

1932:
2 January	G-ABJS	Spartan 3-seater	Penshurst-Woodley and return – 1 hr 30 min
3 January	G-ABIF	Southern Martlet	Penshurst-Woodley – 1 hr (Blossom flew there in G-ABJS to collect him)
3 January	G-ABJS	Spartan 3-seater	Woodley-Penshurst – 30 min
12 January	ditto		Penshurst-Blue Barns, Colchester and Ipswich – 55 min
13 January	ditto		Ipswich-Croydon – 1hr 20 min
15 January	ditto		Croydon-Penshurst – 20 min
17 January	ditto		Penshurst-Stoke-on-Trent and Hooton Park – 2hr 30min
18 January	G-ABTE	Klemm L.25B.XI	15 min local at Hooton Park
19 January	G-ABJS	Spartan 3-seater	Hooton-Penshurst – 2hr 30min
23 January	ditto		30 min local
31 January	ditto		25 min local
6 February	ditto		Penshurst-Brighton and return – 45 min
9 February	ditto		Penshurst-Croydon – 20 min
11 February	ditto		Croydon-Cowes – 40 min
12 February	ditto		Cowes-Bristol – 1hr 45min
13 February	ditto		Bristol-Penshurst – 1hr 18min
16 February	ditto		12 min local
18 February	ditto		Penshurst-Croydon – 20 min
11 March	ditto		Croydon-Yate – 1hr 20min
20 March	unknown	Parnall Elf	Yate-Bristol-Yate
31 March	G-ABJS	Spartan 3-seater	Yate-Croydon

Appendix 7:
Blossom Miles' flying – 1930 to 1935

Mrs Maxine Freeman-Thomas or 'Blossom', as she preferred to be called, was taught to fly by F G Miles at Shoreham in 1930. She made her first flight on 2nd June 1930 and most of her flying training was carried out in G-EBYB with Miles. Blossom flew many other light aircraft and these are shown, as recorded in her surviving flying log book, with the date of her first flight in each:-

1930 Shoreham

2 June	G-EBYB	Avro 504K	first flight, with F G Miles
17 June	G-AACW	Avro 504K	
27 June	G-EBYB	Avro 504K	first flight with Cecil L Pashley
11 July	G-AABF	Blackburn Bluebird III	
25 July	G-EBYB	Avro 504K	first solo
27 July	G-ABAF	DH.60G Moth	Fielden - Ford (Aerodrome - PA)
31 July	G-AADF	Avro Avian IV	
13 October	G-EBZG	DH.60X Moth	
31 October	G-EBYP	Avro Avian IIIA	
30 November	G-EBZG	DH.60X Moth	LERB (Lionel Bellairs - PA)

1931 Shoreham

5 January	G-AAVM	Avro Avian IVM	
15 January	G-AAVD	Southern Martlet	5 min flight
6 February	G-ABIF	Southern Martlet	
16 March	G-ABDO	Robinson Redwing 2	
8 April	G-AAWT	Desoutter I	Shoreham - Le Bourget, with F G Miles
22 April	G-AATF	Desoutter I	Poix - Ostende
29 April	G-ABKI	Avro Sports Avian	
1 May	G-AADE	Westland Widgeon III	
17 May	G-ABJW	Southern Metal Martlet	

By May 27th, Blossom had completed 100 hr 34 min flying

18 June	G-ABLB	DH.80A Puss Moth	
27 July	G-AAII	Southern Martlet	
16 September	G-EBZG	DH.60X Moth	Her last flight made from Shoreham.

F G Miles departed to South Africa in August 1931, to return soon after. The following year, they married and settled in Sevenoaks, using the local aerodrome at Penshurst for their flying.

1932 Penshurst - every flight made from there is recorded below:

3 January	G-ABJS	Spartan 3-Seater	to Reading
15 January	G-ABJS	Spartan 3-Seater	
31 January	G-ABJS	Spartan 3-Seater	
16 February	G-ABJS	Spartan 3-Seater	
20 March	G-ABJS	Spartan 3-Seater	
16 May	G-AAGY	Simmonds Spartan	to Bristol and '?' (unreadable - PA), but 40 min return
25 May	G-AADF	Avro Avian IV	to Yate

1933 Woodley - every flight made from there is recorded below:

On 13th February 1933, Blossom flew the DH.60X Moth, G-EBUS, to Reading (Woodley), with F G Miles, from an unrecorded place, and thereafter, all flights were made from Woodley.

30 March	-	Miles Hawk	5 min solo
3 April	-	Miles Hawk	5 min, with F G Miles
3 April	-	Miles Hawk	10 min solo
5 April	G-ACAS	Fairey Fox I	with F G Miles
11 April	-	Miles Hawk	2 flights
15 April	G-ACDI	D.H.82 Tiger Moth	
16 April	G-ACGH	Miles Hawk	15 min, with F G Miles
12 September	G-ACJY	Miles Hawk	10 min, at Cannes
7 November	G-ACLB	Miles Hawk	

1934 Woodley - every flight made from there is recorded below:

21 March	-	Miles Hawk 'Gipsy'	
21 March	G-ACNW	Miles Hawk	
22 March	G-ACNW	Miles Hawk	
17 July	-	Miles Hawk	5 min - experimental flaps
2 August	G-ACVM	Miles Hawk 'Gipsy Major'	

1935 Woodley - every flight made from Woodley is recorded below:

4 November	G-ADLC	M.3B Falcon Six	Farnborough - Reading
24 November	G-ADWT	M.2W Hawk Trainer	with F G Miles
24 November	G-ADWV	M.2W Hawk Trainer	with F G Miles
24 November	G-ADWU	M.2W Hawk Trainer	with F G Miles
24 November	G-ADVF	M.2W Hawk Trainer	with F G Miles
4th December	U1	M.7 Nighthawk	two flights, with F G Miles

(Note: This represents the only record of the markings U1 carried for the early flights of the first prototype Nighthawk - PA)

11 December	G-ADXA	M.7 Nighthawk	with F G Miles

This was the last recorded flight in the surviving log books of Blossom Miles.

Appendix 8: Phillips & Powis Aircraft (Reading) Ltd – The pre-Miles Days

While Phillips & Powis is, these days, synonymous with Miles, they had been in the aeroplane business for four years prior to his arrival in October 1932. As well as a thriving aircraft dealing business, they operated an aero club, a flying school and briefly an air taxi business (the latter at Bristol Whitchurch).

The original aviation business was formed by Jack Phillips and Charles Powis in 1928 and initially operated from Caversham Bridge Promenade Field, Reading before moving to Reading Aerodrome (Woodley) in April 1929. This coincided with the formation of Phillips & Powis Aircraft (Reading) Limited on 19th March 1929.

The directors of the company were initially just Messrs Phillips and Powis but by March 1932, Major George William Graham Allen of The Elms, Iffley, Oxford had become a director. He owned DH.60G Moth G-AAJJ from June 1929, replacing it with DH.80A Puss Moth G-ABKD in March 1931 and also had a private airfield at his home.

Also by March 1932, Henry Cockburn Gresley Heathcote Stisted of Hawley Lodge, Blackwater, Hants had become a director. A racing motorist and a business manager, he bought his first DH.60X Moth, G-EBSP from Phillips & Powis in October 1929, replacing it with DH.60G Moth G-AADC in January 1930. This he wrote-off at Chipping Warden on 14th April 1930 and his third Moth was DH.60G G-AAKI registered in January 1931 (and sold in June 1931).

By June 1932, Henry Rudolph Trost had also been appointed a director. He had previously been the UK representative for Junkers (as Trost Bros Ltd) and then a director of Robinson Aircraft Ltd in December 1930, becoming Technical and Managing Director of the successor Redwing Aircraft Co Ltd in December 1931. At the same time John Kenworthy, the founder and chief designer of Robinson Aircraft/Redwing Aircraft Co, was also appointed a director of P&P. It is not known how long they stayed as directors and the circumstances surrounding their departure are unknown.

Phillips & Powis School of Flying

Operations commenced in April 1929, with Chief Instructor F/O Ronald Thomas 'Shep' Shepherd and Assistant Instructor F/O J H Caulfield. Shepherd moved on to Nottingham Flying Club in August 1929 and Caulfield joined the National Flying Services later the same year. They were replaced by F/O Herbert Basil Gordon Michelmore and S W 'Pat' Giddy. The latter became CFI in April 1930 but was killed in the crash of Martinsyde G-EBMI on 24th August 1930.

In February 1930, Michelmore was apparently seconded to Ferranti Ltd to operate DH.60G Moth G-AAKN as a radio research aircraft and on narrow beam navigation trials. The Moth, which was owned by P&P, was damaged on landing Barton on 17th May 1930, but was probably repaired as it remained in operation until the following year.

In April 1930, John Henry Arthur Wells was appointed an instructor and in October 1930, Flt Lt Reginald Lee Bateman (whose flying log books are now with The Miles Aircraft Collection) was appointed Chief Instructor and was joined at the same time by Flt Lt RA Seaton.

G-EBVD DH.60 Cirrus Moth at Woodley in the late 1920s.
[G Vlasto via B Clarke]

G-AAKN DH.60G Moth (operated for Ferranti in 1930-1931) at Woodley, proudly announcing the fact that Phillips & Powis were 'authorised "Moth" agents'. [Via B Clarke]

Above: Three DH.60 Moths and a Spartan, all with wings folded, outswide the Repair & Service hangar. [Via B Clarke]

G-EBVC DH.60X Moth of the Phillips & Powis School of Flying. [Via B Clarke]

DH.60X Moth, G-EBZI or G-EBZL, ouside the Repair & Service Department, Woodley, in about May 1935. [Via B Clarke]

J F Lawn joined in January 1931 and Cdr Charles Wilfred 'Stiffey' Croxford was appointed Chief Instructor in September 1931 replacing Bateman, who made his last flight (in G-EBUS, a 55-minute test) on 23rd December 1931, leaving to join The Shamba Coffee Company.

Aircraft known to have been operated by the School include:

DH.60 Moth	G-EBOT	April 1929 to August 1929
Avro 504K	G-EBWO	April 1929; crashed Caversham Bridge 1.4.29
Avro 504K	G-AAGG	May 1929 to (probably) May 1930 (when CofA lapsed)
DH.60 Moth	G-EBNY	June 1929 to (probably) November 1929 (when CofA lapsed)
DH.60X Moth	G-EBVC	September 1929; crashed Woodley 14.6.30
DH.60X Moth	G-EBSP	January 1930; crashed Woodley 3.8.33
DH.60X Moth	G-EBUS	April 1930 to at least December 1931
DH.60X Moth	G-EBWL	June 1930; crashed Hanworth 16.7.31
DH.60X Moth	G-EBZL	July 1930; still operated July 1932
DH.60X Moth	G-EBZC	March 1931 to (not known)
DH.60G Moth	G-AALT	February 1932 to (possibly) June 1934 – used for blind/night flying training

Aircraft Sales

A considerable number of aircraft (mainly DH Moths) were bought and sold in the early days. Also, in about 1932, Phillips & Powis were appointed sole concessionaires of de Havilland aircraft for Bucks, Berks and Oxon, Midlands, West of England and Wales. The following aircraft are known to have gone through their hands between 1929 and 1933 (after which their activities concentrated almost exclusively on Miles products):

Avro Avian	G-EBXE; G-EBZM; G-AATL; G-ACBV
Desoutter	G-AAWT
DH.60X Moth	G-EBRT; G-EBRX (twice); G-EBSP; G-EBWI; G-EBZI; G-EBZZ (twice); G-AAAC
DH.60G Moth	G-EBQH; G-EBZR; G-AAAO; G-AABJ; G-AADP; G-AAEA; G-AAEU; G-AAFX; G-AAGT; G-AAHR; G-AAJS; G-AAJW; G-AAKN; G-AAKU; G-AALD; G-AARM; G-ABDV; G-ABFD; G-ABPT; G-ABUB
DH.60M Moth	G-AAGE; G-AAIF; G-AARB; G-AAZJ
DH.80A Puss Moth	G-AAVA; G-AAZV; G-ACDU
DH.83 Fox Moth	G-ABVI
Klemm L.26	G-ABBU
Southern Martlet	G-ABIF
Spartan	G-AAGY
Spartan 3-seater	G-AAHV; G-ABJS; G-ABKK
Westland Woodpigeon	(advertised for sale November 1929)

Bristol Air Taxis

This was formed at Bristol Whitchurch and was opened for business by pilot/manager Stephen Cliff on 18th April 1931 using Desoutter G-AAVO. The business was sold to Norman Edgar & Co in late 1933. The Desoutter had previously been operated by P&P on joy-riding services on Swansea Sands in June 1930.

Reading Flying Club

This was formed in June 1929 and became a limited company on 13th September 1930. It was initially formed as the social club for students etc of the School of Flying. It took over the training activities in December 1933 when the locally-based Berks, Bucks and Oxon Aero Club, which had been part of the failed National Flying Services, was acquired.

It is possible that the Club operated DH.60 Moth G-EBOT (owned by DE Swann) and DH.60G Moth G-ABOV (owned by Charles Powis) on an informal basis in 1931 and, by the end of 1933, the Martlet G-AAYX had been added to the fleet 'for use by members considered competent by the CFI to fly it'!

Appendix 9: Miles M.2 Hawk, M.2A Cabin Hawk, M.2B Hawk Single-seater, M.2C Hawk Sports Two-seater, M.2D Hawk Three-seater and M.2K Hawk Production

G-ACGH – the first Miles M.2 Hawk, with F G Miles in white flying overalls, seen at Woodley in 1933. [A J Jackson Collection]

H.101 M.2 Hawk. Powered by 90hp Cirrus IIIA. First flown (in primer and with no registration markings), by Miles, who recorded 'Hawk Test 5 minutes' in the early evening of 29.3.33 (the Dowty u/c was not ready at the time of its completion so it was fitted with an u/c borrowed from a Simmonds Spartan). Flown by George Miles, with FG, for 15 minutes on 30.3.33 and on 4.4.33 George flew to Reading (in Martlet G-AAII) to make a 10-minute solo local flight. Painted red/silver (but still with no regn) by 5.4.33. Regd **G-ACGH** 19.4.33 (CofR 4337) to Frederick George Miles, Reading Aerodrome. Last flown without regn by Miles on 13.4.33 but recorded by him as G-ACGH on 14.4.33; flown to Brooklands 15.4.33, to Farnborough 1.5.33 and to the Dawn Patrol at Woodley on 7.5.33. Last recorded flight was by Miles on 18.7.33. No CofA issued and none apparently applied for. Regn cld 12.33 and fate unconfirmed but a note made by 'Kit' Keatings at Woodley at the time says that it crashed at Woodley, probably on the aerodrome (hence no record elsewhere). CLN noted that G-ACJY (c/n 10) was *'Rebuild No.1 G-ACGH delivered to Phillips & Powis Aircraft Ltd 6.9.33'*.

Note: Whilst there has been much speculation over the years as to the true c/n for the prototype Hawk, FG Miles confirmed in a letter to Bert Clarke many years later that H.101 was correct, although, in theory, it should perhaps have been the otherwise unused c/n 2 (to follow the first true Miles aeroplane, the Satyr with c/n 1).

2   Not used, for reasons outlined above.

3   M.2 Hawk; painted Portland (grey). First production aircraft. Regd **G-ACHJ** 25.5.33 (CofR 4383) to Wing Cdr Harold Melsome 'Daddy' Probyn, Farnborough. First flown, by Miles, 28.6.33. CofA No.3971 issued 4.7.33; dd 6.7.33. Entered by Probyn (who started as one of the strongest favourites) in the 1933 King's Cup Air Race at Hatfield on 8.7.33 but, when doing extremely well, the engine lost a push-rod and he had to forced land at East Harling, fortunately without damage. Probyn won the Wakefield Trophy race at Lympne 22.7.33 (Race No.14), with the front cockpit faired over. Probyn then had a coupe cabin fitted over both cockpits and a long exhaust pipe to clear the gases and smoke from the cockpits. Fitted with a Gipsy II engine 1.34 and flown by Probyn to Cairo (where he had been posted as Senior Staff Officer, RAF Middle East) with his wife in 2.34. For sale 4.35 but not sold. In 12.36, Probyn flew it to Capetown and then returned to England in 4.37 (to take up posting as Wing Commander 12 Fighter Group (for this Probyn claimed the longest UK-Cape flight - 3 years there and back!). Regd 10.6.37 (CofR 7933) to John Charles Neilan, Ards Airport, Co Down. CofA lapsed 25.11.37. Regd 2.6.38 (CofR 8537) to Fitzmaurice John Commerell Bloomfield, Lionafillon House, Gracehill, County Antrim (based at a private strip at Lionafillon House and at Newtownards); CofA renewed 17.6.38. Written off in crash at Clyst St. George, Devon 20.7.38. Regn cld 1.1.39 following the 1938 census.

4   M.2 Hawk; painted silver. Regd **G-ACHK** 25.5.33 (CofR 4384) to James Christian Victor Kiero Watson (Aviation Manager and Nominee for Germ Lubricants Ltd), Stag Lane. First flown, by Miles, 21.7.33 and dd same day. CofA No.3984 issued 22.7.33. Flown by Watson in the Wakefield Trophy Race at Lympne 22/23.7.33. Regd 24.10.33 (CofR 4688) to Edwin Denis A Bigg, Woodley (an ex RAF

A Miles M.2 Hawk in the final assembly hangar in August 1933. [Via B Clarke]

Officer who departed Woodley 30.11.33 for Nairobi, Kenya to form a flying school and air taxi business). On 16.12.33, while taking-off after a forced landing in the Cherangani Hills, near Kitale, Kenya, the aircraft hit trees and crashed; Bigg and his passenger M Soudan were both slightly injured. Regn cld wef 16.12.33.

5   M.2 Hawk; painted silver. Regd **G-ACHL** 25.5.33 (CofR 4385) to Francis Delaforce Bradbrooke (then staff member of *Aeroplane* magazine and, in July 1933, the editor of the new *Flying* magazine), based Heston. First flown, by Miles, 24.8.33. CofA No.4012 issued 25.8.33; dd same day. Flown by Geoffrey Cecil Harrison Last (the managing director of British Air Transport Ltd) from Gatwick 'local' on 15.8.34. Regd 26.6.35 (CofR 5984) to The Hon Anthony Bingham Mildmay, Shoreham Place, Sevenoaks (based Gravesend). Regd 2.12.35 (CofR 6554) to Ralph Harold Henderson, Hanworth. Regd 23.5.36 (CofR 7094) to Hubert Newton Peake, based at Cherry's Field, Sulgrave, nr Banbury, Oxon. Regd 11.12.36 (CofR 7528) to John Hedley Lewis, Whitchurch. Regd 30.4.38 (CofR 8476) to Southern Motors & Aircraft Co (John Coxon), Hamsey Green. Sold 26.11.38 and regd 13.1.39 to George William Alexander, Gatwick (but probably intended for his newly-formed Horton Kirby Flying Club). Sale presumed aborted and regd 1.2.39 back to Southern Motors & Aircraft Co (now Gatwick). Regd 28.7.39 to George de Freitas-Socchilary, Sutton (based Gatwick). He was a British Army officer serving in Palestine and flew the Hawk to Lydda, Palestine in 9.39. By 12.39, G-ACHL was being sold to Pensah Steinberg & Abraham Schechterman of Tel Aviv, representing the Etzel Organisation, Lydda but the sale reportedly fell through after lengthy delays before settlement was reached; however, the Hawk was formally regd to them 26.11.40. CofA lapsed 25.7.40. The report that it was impressed by the RAF as AX838 at Lydda 26.11.40 is unlikely, since this serial is not recorded in the ledger of serials allotted. Regn cld 1.11.43 as wfu. Fate unknown. *Reports that it was purchased or used by Lord Apsley or The Arab Legion in Transjordan are not true.*

6   M.2 Hawk; painted silver. Regd **G-ACHZ** 6.33 (CofR 4398) to Lt Col William E Duncan, (to be based Heston) and regd 4.7.33 (CofR 4583) instead to his wife Mrs Magda Duncan. *Lt Col Duncan was posted to India in September/October 1933 and it seems likely the Hawk was re-allocated prior to delivery.* First flight date unknown (*Miles was in France and Switzerland between 6/24.9.33*). CofA No.4017 issued 9.9.33. Sold 9.33 to Dr D Mintzman and Dr U (Henry) Whitby, Hatfield. Forced landed in field at Upshire, nr Waltham Abbey, Essex while being flown by 'the Dr and his passenger' 15.11.33 (or 14.11.33) causing damage to port wing and u/c. Flown at Woodley by Bob Milne 13.7.34 and by Miles 17.7.34, who noted 'forced landing' (probably at Woodley) against the

20-minute flight. Regd 20.8.34 (CofR 5314) to Eric Ian Woods, notionally of Reading Aero Club, Woodley who bought it for a flight to Northern Rhodesia but which plans were abandoned 10.34 due to bad weather; it is possible that the regn **VP-RAC** was reserved for this aircraft (but see also G-ACKI, c/n 11). Regd 2.11.34 (CofR 5426) to Robert Christian Ramsay, t/a Kent Flying Club, Bekesbourne. Regd 13.1.36 (CofR 6633) to Airsales & Service Ltd, Bekesbourne (which had been formed by Ramsay 4.4.35 to take over Kent Flying Club). Crashed near Bekesbourne 25.6.39. Regn cld 25.7.39.

7 M.2 Hawk; painted silver. Regd **G-ACIZ** 8.33 (CofR 4423) to Dr D Mintzman and Dr U (Henry) Whitby, Hatfield (but not delivered since replaced by G-ACHZ above). First flown, by Miles 19.10.33. CofA No.4027 issued 19.10.33. Regd 10.33 to Mrs Gabrielle Ruth Millicent Patterson, Lympne (later based Tollerton) and collected by her early in the morning of 21.10.33 ready for the Ladies Meeting held at Woodley that day and during which she won the 'Punctuality Prize'. *(Mrs Patterson [nee Burr] was the first woman pilot to get an instructors licence (by 24.3.34), and became the CFI of the South Staffs Aero Club in May 1936).* Possibly crashed in early 1934 (although no confirmation of this can be found) other than a note by CLN against c/n 101, 'Fuselage No for Rebuild Hawk No 7'; implying a rebuild in about April 1934. Aircraft fitted with manifold exhaust stubs and flown by her (the only woman competitor) in the 6th Heat of the 1934 King's Cup Air Race on 13.7.34 (Race No.41). In September 1934, Gabrielle raced G-ACIZ in the Wakefield Trophy at Lympne. Regd 9.11.35 (CofR 6483) to Wilfred James Martin, Grantham (based Tollerton). Offered for sale 12.35 but presumed not sold. Crashed Tollerton Aerodrome 30.6.36. Regn cld 12.36.

8 M.2 Hawk; painted silver. Regd **G-ACJC** 29.7.33 (CofR 4426) to Mrs Beatrice Shedden MacDonald, Gatwick. First flown, by Miles 26.9.33. CofA No.4028 issued 30.9.33. Beatrice was the wife of Sqn Ldr Alan MacDonald of the Iraq Intelligence Staff and G-ACJC was dd to Baghdad for her by F/O HRA Edwards in 10.33 (the press reported '... although bad weather and some bad aerodromes were encountered during the 47 hours' flying, the engine and airframe behaved perfectly. 'Neither gave me the slightest qualm', he added.' The average ground speed for the flight was 95 mph and the petrol consumption was 5 gal/hr). Returned to Woodley 8.34 and then probably part-exchanged for Hawk Major G-ADAB (c/n 137). Regn cld 12.34 on sale abroad. *(Recorded by CLN as becoming OY-DEY but this regn was never issued and he may possibly have confused it with c/n 107, regd OY-DEI in January 1935).* Regd **OY-DAK** 7.3.35 (CofR 62) to Ljtn. Flemming Lonborg, Hellerup (based Kastrup). Crashed off Gilleleje 20.4.35, owner and passenger killed. Regn cld 4.35.

9 M.2 Hawk; painted silver. Regd **G-ACJD** 29.7.33 (CofR 4427) to Sir Alfred Lane Beit Bt. MP, Heston. First flown, by Miles, 7.10.33. CofA No.4034 issued 14.10.33 and dd same day. Crashed Crichel Park Golf Course, nr Wimborne, Dorset 9.8.34 (and replaced by Hawk Major G-ACYZ). Regn cld 12.34.

10 M.2 Hawk; painted high gloss red. Regd **G-ACJY** 1.9.33 (CofR 4447) to Frederick G. Miles, Woodley. CLN records 'Rebuild of No 1 G-ACGH' and since no fate has ever been discovered for the first Hawk, its rebuild to production standard seems quite possible. The first recorded flight of G-ACJY was by Miles 6.9.33. CofA No.4013 issued 5.9.33. Miles then departed for Lympne with Blossom on the first stage of a flying business/holiday trip to the South of France and Switzerland. Their journey took them to Le Bourget, Lyons, Cannes, Lyons, Geneva (where they took part in an International Aviation Exhibition arranged by Phillips & Powis's Swiss Agent), Le Bourget and Lympne before returning to Reading on 24.9.33 after 20 flying hours without any problems. On the following day, 25.9.33, Brian Augustus Davy (an instructor with Airwork School of Flying) took off from Woodley on a trial flight, but at 6.20pm, on too low a landing approach, the starboard wing struck a telegraph pole and separated from the aircraft, which turned over and fell to ground on its back; the pilot sustained slight injuries. Regn cld 12.33.

11 M.2 Hawk; painted silver. First Hawk to be fitted with wheel brakes. Regd **G-ACKI** 13.10.33 (CofR 4457) to George Joseph Armstrong-Evans, Wolverhampton. First flown, by Miles 16.1.34. CofA No.4149 issued 20.1.34; dd 22.1.34. The owner intended to fly to Northern Rhodesia in October 1934 (and the regn **VP-RAC** may have been reserved for this flight – but see G-ACHZ, c/n 6). Crashed at Morston, Norfolk 25.8.34, killing the passenger P/O Peter Hilary Pomery Simonds and fracturing both thighs of the unnamed pilot. Regn cld 12.34.

12 M.2B Hawk Single-seater; painted silver. Fitted with 120hp Cirrus-Hermes IV and sliding transparent canopy (made by DH and similar to that fitted to the Fox Moth Special) to give range of 2,000 miles, cruising speed 140 mph, max speed 160 mph, climb to 10,000ft in 9 min 25 sec; (with a full load the M.2B 'Hermes Hawk' carried more than its own weight). Regd **G-ACKW** 10.33 (CofR 4470), probably to Phillips & Powis Aircraft (Reading) Ltd, Woodley. First flown, by Miles, as **VT-AES** 12.12.33 (which regn was allocated to Man Mohan Singh, then chief pilot to The Maharajah of Patiala, but never officially issued). CofA No.4122 issued 21.12.33 to Man Mohan Singh; the Hawk was handed over 22.12.33. Regd **G-ACKW** 11.1.34 (CofR 4822) to Man Mohan Singh (but probably remained painted as VT-AES). Departed Croydon for Cape Town record attempt 20.1.34 but crashed same day in France. *Flight 25.1.34 reported Man Mohan Singh started in his Miles "Hawk" (Hermes IV) from Croydon in the early hours of January 20th. The heavily loaded machine swerved somewhat on taking-off, but the pilot managed to make a very good correction and got away quite safely. Le Bourget was reached and flown over, and the journey continued past Paris. After a while oil began to pour on to the windscreen, and the oil pressure dropped to nothing. The engine was still running well, and Man Mohan Singh hoped to continue until daylight, but after a while he was compelled to land near Carcomb (Vaucluse). Seeing a wood under him, he decided to try a 'pancake' landing in the tree tops, and all would have been well had not a tree taller than the rest caught his wing and swung him around, causing him to crash rather heavily.* Accident Report CA.49 stated that the owner/pilot Man Mohan Singh attempted a forced landing nr Montargis but misjudged his location in the early hours (6.45am) and struck the top of a tree crashing into a wood at Griselles. Singh was seriously injured, fracturing his leg. Regn cld 12.36.

13 M.2 Colonial Hawk Single-seater; painted silver. 1934 model externally similar to the standard Hawk but with the front cockpit faired over and a special streamlined headrest to the rear cockpit, designed for minimum maintenance and featured internal strengthening and extra cooling louvres on the cowling. Regd **G-ACKX** 18.10.33 (CofR 4471) to Phillips & Powis Aircraft (Reading) Ltd. First flown, by Miles 23.10.33 with Dutch East Indies regn PK-SAL applied. UK regn cld 10.33 on sale in Dutch East Indies and 'delivered' 24.10.33 but remained initially based at Woodley. CofA No. 4068 issued (as PK-SAL) 27.10.33. Regd **PK-SAL** 22.12.33 to Jhl. HL Krayenhoff (Manager of Shell Petroleum's local aviation depot), Sourabaya and put into service with the Sourabaya Aero Club, 'where it gave great satisfaction'. On 20.2.34, Krayenhoff flew non-stop from Sourabaya to Bandoeng, a distance of about 345 miles, in 5 hours, flying against a headwind of some 23 mph. On 22nd, after a flight of 72 miles, he reached Batavia in 53 minutes and, after completing business there and demonstrating the Hawk to

PK-SAL, the Miles M.2 'Colonial' Hawk at Sourabaya, in the Dutch East Indies, in 1934. Note the front cockpit opened up.

[Via B Clarke]

G-ACLI, the one-off Miles M.2A Cabin Hawk for Stephen Cliff. First flown, by Miles, on 27th November 1933. *['Flight' via B Clarke]*

members of the aero club, he returned to Sourabaya on 24th, the distance of 400 miles being covered in 4 hours 10 minutes. The Cirrus IIIA engine ran faultlessly on AeroShell oil and Shell aviation gasoline and the average consumption's were 3½ to 5¼ pints and 5 gallons per hour respectively. Fitted with a 98hp DH Gipsy I engine (from Pander E PK-SAJ) in May/June 1936 and dd to Kreeg, later de Para-Militaire (Auxiliary Squadron). Temporarily marked as **M1** between 5.36 and 5.7.36. Reverted to **PK-SAL** later in 1936 to Vliegclub, Sourabaya. Crashed Darmo 19.11.39 and regn cld same date.

14 M.2A Cabin Hawk; painted red/silver. Fitted with 110hp DH Gipsy III engine. Regd **G-ACLI** 18.10.33 (CofR 4683) to Stephen Bartram Cliff (a director of Phillips & Powis), Woodley (obtaining a slightly 'personalised' regn!). First flown, by Miles, 27.11.33. CofA No.4101 issued 1.12.33 and dd to owner the same day. Flown to Egypt by Cliff on 4.12.33 for the International Circuit of the Oasis meeting to be held between 22nd and 26th February 1934 at Cairo (Race No 23) but was disqualified for being 40lb overload. The press reported '*this was sheer absence of mind on Mr. Cliff's part, as he had petrol for 739 miles and did not need so much. He averaged 115 mph round the circuit and only 117.4 mph in the race when he failed to get his engine to take full throttle*'. CofA lapsed 20.12.35; renewed 19.5.36. Burnt out in a hangar fire at Brooklands 24.10.36. Regn cld 12.36.

15 M.2 Hawk; colour not recorded. Regd **G-ACLA** 24.10.33 (CofR 4474) to Sub Lt Philip Alexander Roby Bremridge RN, Woodley. No CofA issued. Miles did not record flying G-ACLA, although he made flights in an unidentified Hawk on 13.1.34, 15.1.34 and 30.1.34. Regn cld 12.33 and fate unknown but it seems likely the sale was cancelled (since Bremridge bought Percival Gull G-ACIS instead) and airframe re-allocated with a new c/n; (significantly CLN has nothing against c/n 15).

16 M.2 Hawk; painted silver. Regd **G-ACLB** 24.10.33 (CofR 4475) to Edward Devereaux Spratt, Hamble. First flown, by Miles 7.11.33. CofA No.4073 issued 8.11.33; dd 13.11.33. CofA lapsed 7.11.34 and regn cld 12.35. Fate unknown and owner bought a new Hawk G-ACPW in April 1934 so its demise was probably prior to then. Possibly sold abroad with one prospect being the unidentified M.2 Hawk regd CC-FBN to F Bermudez of Punta Arenas, later regd (by 9.47) CC-PBL (CofR 0054).

17 M.2 Hawk; painted silver. Regd **G-ACMH** 11.33 (CofR 4737) to Phillips & Powis Aircraft (Reading) Ltd. First flight not made by Miles. CofA No.4092 issued 24.11.33. Flown by Miles 25.11.33 and retained by company for use as demonstrator. Regd 9.2.34 (CofR 4856) to Mrs Helen Marcelle Barnes, Caterham, Surrey (but based Shoreham) and dd 11.3.34. Flown by Graham Head on 13.3.34 on four dual circuits with Pashley and then solo same day, noting in his diary 'very nice aeroplane'. He flew it again, from the old (1914-18) Shoreham Aerodrome on 20.3.34 'as our field waterlogged' and from then on regularly from Shoreham, making his last flight on 28.8.34. Entered by Mrs Barnes in July 1934 in the Portsmouth Trophy Race, flown by Victor N Buchan of Brighton. Regn cld 9.34 as sold (and Mrs Barnes then acquired the new Hawk Major G-ACWW). In September 1934, it was delivered by Phillips & Powis to Denmark for LJ Abild, Aabenraa; departing Heston at 8.30 am and flown up the Franco-Belgian coast where heavy rain was encountered at Ostend before reaching Rotterdam in 2 hr 10 min; on to Hamburg (311 miles) in 2 hr 30 min and to Copenhagen in a further 1 hr 40 min (700 miles at average speed of 112 mph in 6 hrs 15 mins). Crashed Aabenraa 22.9.34 (while still in British marks) and with no Danish regn having been allotted.

18 M.2 Hawk; painted red. Regd **G-ACMM** 12.33 (CofR 4742) to George Lawrence Harrison, Sale, Cheshire (based Barton). CofA No.4113 issued 23.12.33. First flight not made by Miles; CLN records dd 30.12.33 (but confirmed in use by owner 23.12.33). Harrison (a committee member of the Reading Aero Club) was flown to the Vincennes, France, Whitsuntide Aviation Meeting by Stephen Cliff in 6.34. Regd 20.5.35 (CofR 5888) to Edward Ernest Marsh, Stourbridge (based Castle Bromwich). Regd 6.5.38 (CofR 8491) to Richard Arthur Walley, Wolverhampton. Sold 12.4.39 and regd 21.7.39 to Eric William Brookhouse of City Garage, Stoke-on-Trent (based Meir). Transferred 5.6.40 to owner's premises at City Garage, 3-7 Lonsdale Street, Stoke-on-Trent. Fate unknown, but presumed scrapped during war. Regn cld 20.2.46 as wfu.

19 M.2C Hawk Two-seat Sports; painted silver. Fitted with 110hp Gipsy III. Regd **G-ACOB** 2.3.34 (CofR 4887) to Phillips &

UI, the first Miles M.2D Hawk Three-seater. [Via B Clarke]

Miles M.2 Hawk G-ACMX after it appears to have come to grief in Dublin. Presumably repaired, it was re-registered EI-ABQ on 17th December 1938. [Via J Masterton]

Powis Aircraft (Reading) Ltd. First flown, by Miles 12.3.34 who recorded it as a 'Hawk (Gipsy)'. CofA No.4232 issued 29.3.34. Regn cld 5.34 as sold. Dd 10.4.34 to Mme Madeleine Charnaux, Neuilly-sur-Seine, France (who traded in her unidentified Moth, which had been collected from France by Charles O Powis in 9.33). Regd **F-AMZW** 2.7.34 (CofR 3816) to Mme Madeleine Charnaux. Regd 6.35 to Count Raymond Holt, Marseille. Probably destroyed during war.

20 M.2D Hawk Three-seater; painted yellow. Fitted with 110hp Cirrus II. First flown, by Miles 4.2.34 as **U1**, one day before the new 'B' marks officially came into use. Regd **G-ACPC** 26.3.34 (CofR 4931) to Air Pageants Ltd, Croydon. CofA No.4237 issued 3.4.34; dd 5.4.34. Took part in Air Pageants' managed Sky Devils Air Circus tour from April to October 1934, which visited 170 towns. Air Pageants Ltd went into receivership 3.10.34; CofA lapsed 2.4.35 and fate of aircraft unknown. Regn cld 5.37.

21 M.2 Hawk No identity confirmed but CofA No.4145 issued 15.1.34 to Man Mohan Singh (which implies it was completed and flown in early January 1934). Probably intended for The Maharajah of Patiala, but never regd in India (and presumably not delivered due to Singh's injuries in the crash of VT-AES/G-ACKW on 20.1.34). The suggestion that this was a replacement for c/n 12 can be discounted as its CofA date predates the accident by 5 days. CLN states that c/n 21 was a '*Standard Hawk reregistered as No.103, G-ACTN*' which thus suggests that the order was cancelled and the aircraft returned to stock for re-sale.

22 M.2 Hawk; painted silver. Not flown by Miles. Shipped to India with CofA No.4172 issued 2.2.34 to Tozer Kemsley & Milbourn (an international trading company). Regd **VT-AFD** 14.3.34 (CofR 212) to JW Ross, c/o Walford Transport Ltd, Calcutta (based at Dum Dum Airport); (CLN says dd 31.4.34 sic). Regd 16.12.35 to William Fairweather, Muzaffarpur, Bihar; flown for 400 hrs until he replaced it with Hawk Major G-ADFC. Sold 9.37 and regd 29.11.37 to Babu (the Hindi equivalent of Mister or Esquire) Chanoo Prased Singh, CPS Estate, Bharpura, Dharbanga. Regn cld 4.47 in census.

23 M.2 Hawk; painted red. Regd **G-ACMX** 8.1.34 (CofR 4752) to Frederick Robert Hill, Dublin (based Baldonnel). First flown, by Miles 23.1.34. CofA No.4144 issued 25.1.34; dd to owner in Dublin 28.1.34. Regn **EI-AAV** reported as allocated in 1934 but ntu and the aircraft remained as **G-ACMX**. Regn cld 6.7.38 following CofA lapse on 27.5.38. A photograph taken in Ireland shows G-ACMX in a

Miles M.2 Hawk G-ACNX at Meir in 1936. [Via B Clarke]

damaged condition on the back of an Irish lorry but accident details unknown. Regd **EI-ABQ** 17.12.38 to same owner, presumably following rebuild. Flown to England on 11.8.39 and Irish regn cld 8.2.40. Reported as sold in England and restored as G-ACMX in 2.40 (but no such entry in the official register although CofA shown as renewed 2.3.40). Unconfirmed report as crashed 27.1.44 (but this seems unlikely).

24 M.2 Hawk; painted blue. Regd **G-ACNX** 2.34 to Lady Anne Cathleen Elizabeth Nelson (the owner of the P&P agency Everson Flying Services), Kildonan Aerodrome, Finglas, Co Dublin. CofA No.4182 issued 20.2.34. Miles did not fly G-ACNX until 22.2.34, so its first flight must have been made by someone else previously. Departed from Woodley 24.2.34 for Dublin, flown by Everson's chief pilot Capt Hamilton and ground engineer Joe Currie. Regd **EI-AAX** 26.3.34 to same owner. Regn cld 12.34 on sale in England; returning to England 18.12.34. Regd **G-ACNX** 25.1.35 (CofR 5564) to Henlys Ltd, Heston; CofA renewed 31.1.35. Probably sold to British Air Transport Ltd, Redhill (flown by their managing director Geoffrey Cecil Harrison Last on local flights in February/March 1935). Severely damaged when it hit a tree while landing at Malmesbury, Wilts 12.4.35 while being flown by Flt Lt Maynard Dudding, (an instructor for Surrey Flying Services Ltd, Croydon) who was killed in the crash. Considering the extent of the damage sustained in the crash it is surprising to record that it was in fact rebuilt. Purchased by CR Pitt, Hamsey Green 11.35 and CofA renewed 5.11.35. Regd 5.12.35 (CofR 6573) to Edgar William Brent Grotian, Gottingham, Yorks (based Meir). Regd 3.3.36 (CofR 6792) to Staffordshire Airplanes Ltd, t/a North Staffordshire Aero Club, Meir. Regn cld 7.1.41 by Secretary of State and impressed into RAF as **DG578**. Dd to 24 MU Ternhill. Relegated to GI airframe **2617M** 24.6.41 and issued 1.4.42 to No.1211 ATC Sqdn (Derby Wing), Swadlincote, Derbyshire.

25 M.2 Hawk; painted red. Regd **G-ACOC** 3.34 to Cdr Charles Wilfred Croxford RN, (previously a CFI at Woodley), t/a Yorkshire Air Services, Newton House, Catterick. First flown, by Miles 7.3.34. CofA No.4206 issued 10.3.34; dd 5.4.34. Sold 2.37 and regd 30.6.37 (CofR 7965) to Philip Penrose Bradley, Tollerton. Regn cld 1.38 as sold (probably abroad).

26 M.2 Hawk, painted red. Regd **G-ACNW** 19.2.34 (CofR 4871) to Phillips & Powis Aircraft (Reading) Ltd. First flown, by Miles 14.3.34. CofA No.4197 issued 17.3.34 and dd to the Phillips & Powis School of Flying. On 15.10.34 Miles recorded two flights in G-ACNW on 'burble creator' tests (photographs show that these trials were carried out on M.2D G-ACVR but it is possible that these were also tested on G-ACNW). Crashed Woodley 21.9.35. Regn cld 12.35.

27 M.2 Hawk; painted silver. Regd **G-ACRB** 4.34 to Edward A Bayley & Charles Bilson (who in May 1934 formed The Walsall Aircraft & Motors Ltd to operate Walsall Aero Club and manage Walsall aerodrome for the Walsall Corporation). A Hawk with 'B' mark **U11** (it is speculated this may have meant U2 as opposed to U11) was first flown by Miles on 18.3.34 and again (7 flights) between 11-13.4.34 and it is possible that U11 became G-ACRB, which was first flown, by Bob Milne (dual flight with Bayley) 21.4.34. CofA No.4263 issued 21.4.34 and dd same day. Regd 30.4.35 (CofR 5844) to Peter Bryan Charles King, Bromley, Kent (base given as Reading). Sold and dd 8.36 (although possibly earlier after CofA renewal 16.6.36) (but not regd) to AG Ortega, London NW1, (a Brooklands Flying Club private owner, later based at Reading); flown to Brooklands and return to Woodley by George Miles 1.8.36. Regn cld 1.37 as sold (probably abroad). Fate unknown (but not, as sometimes reported, sold to W Fairweather in India - this was c/n 22).

28 M.2 Hawk; painted silver. Possibly first flown under 'B' mark **U1** by Bob Milne between 25.4.34/3.5.34; not flown by Miles. Regd **G-ACSD** 11.5.34 (CofR 5033) to Mwldwen James Grover, Newmarket Road, Cambridge. CofA No.4287 issued (to Marshall Flying Services Ltd, Cambridge) 12.5.34; dd 17.5.34. Crashed in a severe snowstorm Kelshall, nr Royston, Herts 4.4.35, Mwldwen Grover killed. Regn cld 3.35 (sic).

29 M.2 Hawk; painted cream. Possibly first flown under 'B' mark **U1** by Bob Milne between 25.4.34/3.5.34; not flown by Miles. Regd **G-ACSL** 5.34 to Hardip Singh Uberoi, (a staff member of Phillips & Powis on an engineers course in 1933 who was also learning to fly) Woodley. CofA No.4288 issued 12.5.34; dd 22.5.34. Regd 20.1.36 (CofR 6650) to Phillips & Powis Aircraft Ltd for use by the Phillips & Powis School of Flying (but evidence suggests that it was in use by the School and Club from about 8.35; and had been repainted red at some time). On 26.2.36, Miles noted a 10-minute flight in G-ACSL with 'hydraulic flaps' but no further record of this modification (if this was indeed what was implied by his comment) can be found and it is by no means certain that these were in fact fitted to an otherwise standard Hawk. During 1936, G-ACSL was being flown by a member of the Reading Aero Club, when, blinded by the sun, he flew into a tree, totally destroying the aeroplane but remarkably not injuring himself! (Note, the official record does not show a CofA renewal beyond 11.5.35). Regn cld 12.36.

30 M.2D Hawk Three-seater; painted green. Regd **G-ACPD** 26.3.34 (CofR 4932) to Thomas Conyers Place, Chipperfield, Herts (based Woodley). CofA No.4238 issued 7.4.34; dd 8.4.34. Not flown by Miles; *(CLN states it was then dd to Marshalls Ltd but gave no date and he possibly confused it with c/n 28; then to Firm (Phillips & Powis) but again with no date)*. Operated in Sky Devils Air Circus (British Hospitals Air Pageants) and collided on the ground with Avro 504N G-ACLV at Hill Farm, Roding Lane, Claybury, Woodford, Essex 30.9.34. Believed not repaired and regn cld 4.37.

31 M.2 Hawk; painted red. Regd **G-ACRT** 2.5.34 (CofR 5020) to Mervyn Horatio Herbert, The Rt Hon Viscount Clive, Hanworth. First flown, by Miles 12.5.34. CofA No.4275 issued 12.5.34. CofA lapsed 17.6.36 (reason not known) and not renewed until 2.9.38. Reported at Heston on Air Ministry survey 31.8.39. Regn cld at census 1.12.46. Noted stored Kidlington by 1946; an advertisement appeared in *The Aeroplane* for 28.3.47 to the effect that a Miles 'Aries' Hawk was for sale - best offer over £60, replies to a Box No and it is thought to refer to this aircraft, since a similar advertisement in *Flight* for 3.4.47 referred to a Miles Cirrus Hawk. G-ACRT was scrapped at Kidlington in 1956 but some parts were still present in 1960 - and this was probably the last of the Cirrus Hawks to survive.

32 M.2 Hawk; painted in the orange and brown colours of the Herts & Essex Aero Club. Regd **G-ACTO** 5.34 (CofR 5114) to The Herts & Essex Aero Club Ltd, Broxbourne. CofA No.4313 issued 2.6.34. Not flown by Miles; dd 2.6.34. A report in *The Star* for 10.9.35 stated that *'The Frogleys own, or should I say owned, five Moths and a Miles Hawk, but someone hit the mast of the air stocking (windsock) with the Hawk on Sunday afternoon (the 8th) and 'wrote it off'*. Miraculously, nobody was hurt'. However, it could not have been too badly damaged as the accident was not reported to the Air Ministry and its annual CofA was renewed 5.5.36. Sold 4.37 and regd 4.5.37 (CofR 7859) to Humphry R Dimock (an ex-member of the Reading Aero Club), t/a Ely Aero Club, Ely (club renamed Cambridge Flying Services later in 1937). Accident on landing during Coronation Air Displays event at Stoke Park, Guildford 7.6.37 when wing struck and fatally injured 80-year old Charles Puttock; pilot Engert Brand & aircraft undamaged. Ely Aero Club incorporated as Cambridgeshire Flying Services Ltd 20.5.38 but aircraft not re-regd. Regd 20.2.39 to Airsales & Service Ltd, t/a The Kent Flying Club, Bekesbourne. Regn cld 26.6.40 as sold and impressed as **AW152**. Delivered to 46 MU Lossiemouth 5.7.40. To GI airframe **2626M** 26.6.41 and allocated to No.1005 (Radcliffe) ATC Squadron, Manchester Wing. SOC by 46 MU 1.8.41. *Note: regardless of other published information, this Hawk has no connection with CH-380, which was c/n 34.*

33 M.2D Hawk Three-seater; painted grey. Regd **G-ACPW** 13.4.34 (CofR 4968) to Edward Devereux Spratt, (c/o Air Service

Training Ltd), Hamble. CofA No.4253 issued to Phillips & Powis Aircraft (Reading) Ltd 11.5.34; dd 20.5.34. Flown by Miles 13.5.34. Regn cld 4.35 as sold abroad.

34 M.2 Hawk; painted red with silver wings and tailplane. First flown, by Miles 23.4.34. Regd **CH-380** 24.4.34 to Marc M Debrit, Geneva (P&P's Swiss agent). CofA No.4265 issued to Diffusion Industrielle SA, Geneva (of which Marc Debrit was part) 24.4.34; del 25.4.34. Flown out by Stephen Cliff (of P&P) for display by Debrit on the exhibition stand at the first International Aero Show, Geneva (organised by the Aero Club of Switzerland); not flown during the show, (although *Flight* 19.4.34 reported that FG Miles will *probably be demonstrating a Miles Hawk at the Geneva Show, which would start on 27.4.34*). Regd **HB-OXU** 1.35 to same owner. Regd 11.2.36 to Heinrich Muller, Rheinfelden (based at Basle/Birsfelden). Regd 5.37 to Heinrich Spillman, Basle/Birsfelden. Regd 3.10.38 to Air Club Yverdon, Yverdon. Met with an accident during 1938 (piloted by Dr E Savary, described as the owner and possibly connected with Air Club; in 6.39 Dr Savary c/o Aerodrome de Lausanne was advertising for a left wing; aircraft thus possibly repaired). Some time fitted with 105hp DH Gipsy II engine. Regd (.47) to Aero Club de Suisse, Section Vaudoise, Blecherette. Regn cld .47.

35 M.2D Hawk Three-seater; painted blue. Regd **G-ACSC** 5.34 to Humphry R Dimock, Ely (based Denny Abbey, Newmarket Road, Cambridge). CofA No.4297 issued 18.5.34. First flight unknown

In January 1935, CH-380 was re-registered HB-OXU. The aircraft is seen at the Geneve-Cointrin old airfield in 1934. [Via S Blandin]

Blossom christens G-ACOP. Note the hammer firmly grasped and the look of determination on her face! Jimmy Jeffs, on the left, seems quite amused by it all! [Via B Clarke]

(flown by Miles 19.5.34 - and CLN states dd 18.5.34). Regd 1.8.35 (CofR 6143) to Aircraft Distributors Ltd, Hanworth (based Skegness on joyriding/charter). CofA lapsed 17.7.36 and fate unknown. Regn cld at census 1.12.46.

37 M.2 Hawk; painted cream. Regd **G-ACTI** 5.34 (CofR 5088) to Phillip T Petley, Woodley. *An interesting entry in Miles' log-book for 24.4.34 has 'G-AC' firmly written with 'TI' added later in a lighter hand, but since G-ACTI was not regd until late 5.34 and the first recorded flight as G-ACTI was made, by Miles, on 14.6.34, the 'TI' could possibly have been a 'back entry' for record purposes and it could thus have been first flown with only the 'G-AC' part of the reg painted on, in preparation for the full regn when received.* CofA No.4307 issued 9.6.34 and dd same day; CofA lapsed 8.6.35 (reasons unknown); renewed 27.7.36. Regd 15.8.36 (CofR 7265) to Stephen Edgar Edwards, Folkestone (based Lympne) and re-painted silver. Regd 2.12.36 to Cyril Geoffrey Marmaduke Alington, Binbrook, Lincoln (based Hatfield). Entered by its owner for the 'International Circuit of the Oasis' meeting, held 22-26 February 1937 at Cairo, but while en route, Geoffrey was forced down by low cloud and made a forced landing on a patch of level sand near the railway line (and which turned out to be the top of an underground ammunition dump (!) in the Mareth Line) nr Gabes, Tunisia; the port wing was damaged when it struck a sand dune and the front spar root-end fitting to the centre section was torn out. G-ACTI was taken to Gabes on a lorry, where it was believed to have become an insurance write-off, and thus the property of the insurance company. It was returned to England, being seen at Hatfield in 1939, but was not repaired and was probably used for spares. Regn cld in 5.11.45 census.

38 M.2 Hawk; painted red. Regd **G-ACUD** 11.6.34 (CofR 5137) to Phillips & Powis Aircraft (Reading) Ltd. First flown, by Miles 14.8.34, with 'flaps' noted against the flight. CofA No.4331 issued 17.8.34 and dd to Phillips & Powis School of Flying same day. *(The Aeroplane for 22.8.34 showed a photograph of G-ACUD with 'the new Miles split flaps, which can be seen in the depressed position across the fuselage and inner sections of the wings').* Crashed West Woodhay 5.6.36. Regn cld 12.36.

39 M.2 Hawk; painted blue. Regd **G-ACVN** 3.7.34 (CofR 5201) to William Foster, Woodley. Not flown by Miles. CofA No.4405 issued 19.7.34; dd 9.8.34. Regn cld 7.35. Shipped by its owner Capt Dean to Brazil where it was assembled and based at Fazenda Estrella, Ligacao, Matto Grosso. Probably regd locally in Brazil. *(Note - the usually meticulous Bob Milne recorded a flight in 'G-ACVN' on 17.7.34 but noted that it was fitted with a 130hp Gipsy Major and this test flight almost certainly referred to the first flight of Hawk Major G-ACVM)*

40 M.2 Hawk; painted cream. First flown, by Miles 25.7.34, who noted 'climbed to 11,000ft in 23 min'. CofA No.4404 issued to Greek newspaper Ellhnikon Mellon 'Greek Future' 25.7.34. Dd 27.7.34 by Stephen Cliff (reportedly its joint owner) via Le Bourget, Lyon, Pisa, Rome, Brindisi and Ionnina to Athens/Tatoi in 17½ hrs, arriving on 28.7.34. Regd **SX-AAB** 7.9.34 (CofR 7) to Nikolaos P. Efstratiou, Athens and reportedly used by the newspaper throughout Greece under the name 'The Hawk of Acroplis' (*Acroplis* being a monthly news magazine). Regd 12.37 to Athens Municipality, Athens. Regd 14.2.40 to Aeroleschi Thessalonikis [Aero Club of Thessaloniki]. Presumed destroyed by Luftwaffe in late 4.41.

41 M.2 Hawk; painted red. Regd **G-ACOP** 28.5.34 (CofR 4904) to Phillips & Powis Aircraft (Reading) Ltd (advantage taken of the 'in-sequence' registration to have personal marks for C O Powis). First

This Miles M.2 Hawk was registered G-ACOP to Phillips & Powis Aircraft (Reading) Ltd, Woodley, on 28th May 1934 in honour of C O Phillips – a very early example of a personalised registration. [Via B Clarke]

flown, by Miles, 3.6.34; dd to Phillips & Powis School of Flying 3.6.34. CofA No.4312 issued 4.6.34. A contemporary report of the 'At Home' held by Reading Aero Club at Woodley on 9 June 1934 states that *'after tea, the new Miles Hawk of the Phillips & Powis School was baptised by Mrs Miles. The name chosen, 'Ruddy Duck' - an authentic ornithological designation which is vouched for by the club, - suggests that if the machine is never called anything worse by harassed pupils, its future will be as rosy as its exterior. G-ACOP - a very smart red Hawk with silver registration letters, etc - was the first Hawk to be supplied to the club'*. This was probably the Hawk flown with flaps by Miles for the first time on 16.7.34, and G-ACOP was confirmed fitted as such on 5.8.34. Crashed Woodley 4.8.35 when, on entering the second loop on the last flight of the day, the front stick (which had been taken 'in and out' all day) came out and in the ensuing accident both the instructor, F/O John F Lawn and his pupil, Emil M Boysen, were killed. Regn cld 12.35.

42 M.2D Hawk Three-seater; painted blue. Fitted with 90hp Cirrus IIIA *(although CLN states that it had a 110hp Hermes II)*. Regd **G-ACSX** 17.5.34 (CofR 5074) to Harlow Mill Ltd, Harlow, Essex (a company managed by Arthur Ashby and Henry Mayes and operated from North Weald by their newly-formed associated company, Harlow Flying Fields Ltd). CofA No.4296 was issued 18.5.34. Flown by Miles 19.5.34 and again on 22.5.34. *(CLN states dd 15.5.34 – but this seems unlikely)*. Crashed into high ground in heavy sea mist at Bilsdale, nr Stokesley, in the Cleveland Hills, Yorks 5.6.34; pilot Henry Mayes and passengers, Miss L Debham and Mr HS Morton, both from South Shields, were all slightly injured. Regn cld 7.34 (and replaced by G-ACVR).

101 Probably allocated to rebuild of G-ACIZ c/n 7 (which see)

103 M.2 Hawk; painted silver. Regd **G-ACTN** 6.34 (CofR 5113) to North Staffordshire Aero Club (which was owned by Stoke-on-Trent Corporation), Meir. CofA No.4314 issued 4.6.34 and dd same day. Regd 13.7.34 (CofR 5236) to Staffordshire Airplanes Ltd, which was formed to run the club 20.6.34). Hit tree in forced landing in field nr Meir Aerodrome in severe snowstorm 15.12.35; pilot Miss Dorothy Clive injured and her cousin Roger David Clive killed. Regn cld 12.35. *See comments under c/n 21.*

Miles M.2 Hawk G-ACTN at Bournemouth on 23rd June 1935.
[Via B Clarke]

Eric Brorup, as a 17-year-old, had a dual lesson in the Miles M.2 Hawk OY-DEI at Copenhagen in March 1935. The instructor was Lt Aage Hedahl, who later became a Senior Captain with SAS, and Eric also became a commercial pilot later, bush flying in Canada.
[E Brorup]

104 M.2 Hawk; painted silver. Regd **G-ACXZ** 10.9.34 (CofR 5341) to Capt Noel Alexis Blandford-Newson, Haddington, East Lothian (intended to be based at Macmerry for owner's 'embryonic' Edinburgh Flying Club). CofA No.4495 issued 12.9.34; dd 13.9.34 *(but CLN records that 'customer failed to pay' and this may have been the Hawk which was clandestinely recovered by Miles and his associates; it was certainly not used by the club which finally opened its doors on 21.1.35)*. Subsequent fate unknown; the initial CofA lapsed 11.9.35 and was not renewed. Regn cld 3.37 as sold (and if so, it may have been exported, perhaps in 1935 – and note that Blandford-Newson purchased Falcon Major G-ADBF in January 1935).

105 M.2 Hawk; painted silver. Regd **G-ACYA** 10.9.34 (CofR 5342) to Herbert Jacob Hardy, Manchester (based Barton). CofA No.4501 issued 1.10.34; (reportedly dd 22.9.34). Advertised for sale by Hardy 3.38 and regd 23.4.38 (CofR 8473) to Stanley Birley Wilmot, Steventon, Ludlow, Shropshire (and based at a field near his home 'The Woodlands'). By 31.8.39, it was reported in store with Martin Hearn Ltd, Hooton Park and was destroyed by fire there 8.7.40. Regn cld 7.10.40.

106 M.2 Hawk; painted blue. Regd **G-ACVO** 4.7.34 (CofR 5202) to Phillips & Powis Aircraft (Reading) Ltd. Regd 18.10.34 (CofR 5396) to Kenneth William Hole, Southgate, London N14 (based Hatfield); but Hole was probably simply a nominee for a Norwegian, Harold Vilen. CofA No.4366 issued (to Phillips & Powis Aircraft (Reading) Ltd) 20.10.34. *Flight* for 29.11.34 recorded that a Mr G Velen had taken delivery of a Cirrus III Hawk to fly it back to Norway with his pilot. In fact, it crashed at Flamstead, nr Harpenden, Herts 15.11.34 with the unidentified pilot escaping with a slight injury and the passenger (Harold Vilen, nominally of Southgate, London) being rendered unconscious. Regn cld 12.34 and no Norwegian regn allocated.

107 M.2 Hawk; painted silver. Regd **G-ACVP** 4.7.34 (CofR 5203) to Phillips & Powis Aircraft (Reading) Ltd. Regn cld 8.34 as sold abroad. CofA No.4367 issued (to Phillips & Powis Aircraft (Reading) Ltd) 28.11.34. Dd 1.12.34 and regd **OY-DEI** 4.1.35 (CofR 60) to Ljtn Flemming Lonborg, Hellerup, Denmark (based Kastrup, Copenhagen). (A contemporary report stated that *'Mr A Reedtz of Denmark had joined the Phillips & Powis Flying School and is taking delivery of a Hawk'*. *Flight* of 6.12.34 stated that Reedtz *'with another fellow-countryman, left Reading Aerodrome last Sunday in a Cirrus Hawk for Copenhagen'*). Regd 1.8.35 (CofR 76) to Aage Rasmussen, Charlottenlund (based Kastrup). Regd 16.4.36 to A/S Nordisk Lufttrafik, Copenhagen (based Kastrup). Crashed on landing at Kastrup 10.10.36 (or 11.10.36). Regn cld 29.12.37.

108 M.2D Hawk Three-seater; painted blue. Initially fitted with 90hp Cirrus IIIA (see below). Regd **G-ACVR** 4.7.34 to Harlow Flying Fields Ltd, North Weald. CofA No.4364 issued 18.7.34; dd 19.7.34 (but it possibly returned to Phillips & Powis shortly afterwards. Used by Miles to carry out trials of the 'Burble creators' designed by Italian Dr Mattioli, which consisted of slender rods carried in three wing fences wrapped around the leading edge of the wings. The fences each had a number of holes in them the same diameter as the rods, in order that the rods could be placed in various positions, either closer to or further back from the leading edge and closer to or raised above the surface of the wing. G-ACVR was flown by Miles on 13.9.34; and although he did not record the purpose of the flight, this could have been when he first tested the 'burble creators'. *(Miles flew G-ACNW twice on 15.10.34 on 'burble creator' tests which confirms that these were also tested on this machine)*. It is possible that G-ACVR was also used for early trials with flaps but this cannot be confirmed; Bob Milne flew G-ACVR on 7.6.35 and again in 8.35 (still with the Cirrus III fitted). Regd 2.8.35 (CofR 6148) to Aircraft Distributors Ltd, Hanworth (based at Skegness for joyriding/charter). Prior to 29.5.36, the Hawk was fitted with a 110hp Cirrus Hermes II (as it was a three-seater, this would have given more power for joyriding). CofA lapsed 15.8.36. Regn cld 22.2.39 as sold (to unknown party). Regd 29.8.39 to Airsales & Service Ltd, Bekesbourne. Stored there during war and noted as 'under repair' (by 1.10.40) and on 15.4.42 as 'not impressed as privately owned' (File AVIA 2/1617). When owner Robert Ramsay 'recovered' his airfield in April 1945 from the military, he found the four aircraft present had been reduced to 'junk'. Regn cld 5.11.45 in census.

117 M.2 Hawk; painted silver. Regd **G-ACYW** 24.10.34 (CofR 5392) to Mark Antony Lacayo, Old Windsor (based Heston) (owner was sales director of Aircraft Distributors Ltd). CofA No.4543 issued 10.11.34 *(Bob Milne made a 35-minute 'Test for CofA' flight in*

Miles M.2 Hawk SE-AFS with Knut K:son Mark at Göteborg in March 1937. *[Via B Clarke]*

133 M.2K Hawk; painted silver. Fitted with 110hp Cirrus Hermes II (to maintain compatibility with other club aircraft). Regd **VR-SAJ** 1.35 (CofR 10) to Royal Singapore Flying Club, Singapore. CofA No.4570 issued 10.1.35; (but dd 7.1.35). Stalled on approach to land and crashed Singapore Airport in 1935.

136 M.2 Hawk; painted silver. Regd **G-ACZW** 19.11.34 (CofR 5452) to Philip Clarence Kendall, London SW7 (based Heston). CofA No.4574 issued 16.1.35; dd 23.1.35. Officially, its regn was cld 12.34 as sold (but date likely to have been a clerical error). Fate unknown but presumed sold abroad.

142 M.2 Hawk; painted silver. Originally reserved for Intreprinderile Technica Romane (Rumania) but ntu. Regd **G-ADDM** 5.3.35 (CofR 5659) to Stanley John Hawley, Longton, Stoke-on-Trent (based Meir). CofA No.4636 issued 4.3.35; dd 7.3.35; operated by North Staffordshire Aero Club. Whilst paying a visit to the Leicestershire Flying Pou Club's rally at Sandy Lane, Melton Mowbray 2.5.36, a wing touched the ground on take-off and the Hawk turned over and caught fire; pilot Mr Peach and his son escaped with minor injuries. Regn cld 12.36.

146 M.2 Hawk; painted silver. Probably the first Hawk to be fitted with a long exhaust pipe from new. Regd **G-ADBK** 24.1.35 (CofR 5557) to Kent Flying Club, Bekesbourne. CofA No.4663 issued 26.1.35; dd 30.1.35. Stalled and crashed on turn on finals on landing Bekesbourne 9.8.37; Garnet Percival Lovatt killed. Regn cld 10.37.

170 M.2 Hawk; painted red/black. CofA test flight carried out by Bob Milne on 31.3.35 and CofA No.4738 issued to Aage Rasmussen, Charlottenlund, Denmark 2.4.35; dd 17.4.35. (Note that this is confirmed as being fitted with a Cirrus III and not a Gipsy Major). Regd **OY-DEK** 5.8 35 (CofR 77) to architect Aage Rasmussen, Charlottenlund (based at Kastrup) and Danish CofA 64 issued 12.8.35. Regd 16.4.36 to A/S Nordisk Lufttrafik, Copenhagen. Regd 17.5.38 to Anton Fehr and Holger Petersen, Odense. Crashed 1.9.38. Regn cld 9.38.

171 M.2 Hawk; painted silver. CofA No.4739 issued 9.5.35 to Interprinderile Technica Romane, Rumania. Regd **YR-ITR** but although the marks 'YR-' were painted on the fuselage, the regn was not taken up nor the aircraft delivered. Regd **G-ADVR** 10.35 to Phillips & Powis Aircraft Ltd; re-painted red and dd to Phillips & Powis School of Flying. Regd 7.36 to Staffordshire Airplanes Ltd, t/a North Staffordshire Aero Club, Meir (and repainted blue at same time). Regd 22.4.37 (CofR 7834) to Stanley John Hawley (the Secretary of North Staffordshire Aero Club), Meir Aerodrome, Stoke. Regd 2.5.38 (CofR 8477) to Airsales & Service Ltd, operated by Kent Flying Club, Bekesbourne. A photograph of it taken at Bekesbourne in 1938 shows it to have been fitted with a long exhaust pipe. Reportedly crashed Bekesbourne 30.8.39 (unconfirmed). Regn cld 30.8.39.

G-ACYW, with an 85hp Cirrus III engine on 13.11.34 (sic) so whilst reported elsewhere as a Hawk Major, this is incorrect). Regn cld 5.35 as sold. Dd 14.11.34 to Luis Moroder and flown to Spain under ferry regn **EC-W25**. Regd **EC-ZZA** (by 5.35) to Luis Moroder Gomez, Valencia. Regn cld 12.11.40 at census. Its history (as with so many Spanish aircraft) is obscure and confusing. One report suggests it was the 'private aeroplane' used by Moroder 'a notorious Fascist' to escape from Valencia to Mallorca and thereafter joining the Nationalist air force. The Nationalist serial **30-72** is commonly attributed to it, but this serial was not allocated until late 1938, whereas the Hawk would have been used in 1936. EC-ZZA was also reported as having been taken over by the Republicans in which case, it would have been given a serial commencing 'EN-' as a trainer/liaison aircraft. 30-72 was subsequently renumbered **L5-72** post-civil war and was regd **EC-AHZ** .52 to Aero Club de Sevilla before being written off in a crash in 1957 (but EC-AHZ was fitted with a Gipsy Major and it seems unlikely that a Cirrus-powered Hawk would have been rebuilt with a Gipsy Major, since those engines were in short supply in Spain in the 1940s/50s).

130 M.2 Hawk; painted silver. Regd **G-ACZD** 30.10.34 (CofR 5406) to The Eastern Counties Aeroplane Club Ltd, (also t/a Ipswich Aero Club), Ipswich. CofA No.4555 issued 13.12.34; dd 15.12.34. CofA lapsed 12.12.35 (reason unknown); renewed 12.8.36 (after club taken over by the Straight Corporation Ltd). Regd 9.2.37 (CofR 7651) to Phillips & Powis Aircraft Ltd, Woodley. Regn cld 12.37 as sold. Dd to Sweden in 3.37 and regd **SE-AFS** 11.5.37 to Knut K:son Mark, Göteborg. Crashed nr Styrso 18.7.37; owner and passenger Margareta Gustafsson killed. Regn cld 11.8.37.

Miles M.2 Hawk G-ADGI at Bekesbourne, Kent. Note the other two Hawks in the background. *[Via B Clarke]*

175 M.2 Hawk; painted orange and fitted with long exhaust pipe. Regd **G-ADGI** 30.4.35 (CofR 5788) to Kent Flying Club, Bekesbourne. CofA No.4823 issued 11.5.35; dd 18.5.35. Regd 12.4.39 to Airsales & Service Ltd, Bekesbourne (although this company had actually been formed 4.4.35 to acquire the Kent Flying Club). CofA lapsed 2.5.40. Regn cld as sold 26.6.40 and impressed 30.6.40 as **AW150**. Dd to 46 MU Lossiemouth 5.7.40 but not used and SOC there 1.8.41.

192 M.2 Hawk; painted blue. Regd **G-ADGR** 8.4.35 (CofR 5796) to Cyril Albert 'Bish' Nepean Bishop (Secretary of the Reading Aero Club), Woodley. CofA No.4781 issued 29.5.35; dd 2.6.35 to Reading Aero Club (as a gift from Mr & Mrs George Royle in memory of their son Gerald Royle, and with 'personalised' marks in his honour). Regd 2.12.36 (CofR 7504) to The Insurance Flying Club Ltd, Hanworth. Regd 24.6.37 (CofR 7959) to Julian Rowntree, Elvington (based York). On 18.7.37, returning from the International Meeting at Frankfurt, the aircraft stalled and crashed from about 100ft while landing at Evere Airport, Brussels; pilot Julian Rowntree sustained only slight facial injuries but his wife Mrs Beatrice Rowntree was killed. Regn cld 12.37.

200 M.2 Hawk. Recorded by CLN as a 'Standard Hawk' against Job No H3044 but with no further details. Probably allocated to a new fuselage for an unidentified Hawk which had crashed in April/May 1935.

212 M.2 Hawk; painted silver. CofA No.5393 issued 9.3.36 to HU Shepherdson, Australia; despatched by Dallas & Co 14.3.36. Regd **VH-UGQ** 28.5.36 (CofR 587) to Harold Urquart Shepherdson, c/o Methodist Mission, Port Darwin, Northern Territory. Crashed on take-off Groote Eylandt, NT 22.11.36. Regn cld 22.11.36. Remains stored but were destroyed in Qantas hangar fire at Archerfield 28.6.39.

214 M.2 Hawk Recorded by CLN as a 'Standard Hawk (Scrap)' against Job No H3986 but with no further details and assumed not completed.

Epilogue: An advertisement appeared in *The Aeroplane* for 28.9.45, from an anonymous vendor offering a Miles Hawk II (Cirrus Hermes) for sale; '*looked after and stored, airframe damaged but repairable. Quick sale £75*'. Which one it was, and what became of it, remains a mystery.

Production (only completed aircraft counted):

M.2 Standard Hawk	H.101, 3, 4, 5, 6, 7, 8, 9, (10), 11, 13, 15, 16, 17, 18, 21, 22, 23, 24, 25, 26, 27, 28, 29, 31, 32, 34, 37, 38, 39, 40, 41, 103, 104, 105, 106, 107, 117, 130, 136, 142, 146, 170, 171, 175, 192, 212	46
M.2A Cabin Hawk	14	1
M.2B Hawk Single-seater	12	1
M.2C Hawk Sports	19	1
M.2D Hawk Three-seater	20, 30, 33, 35, 42, 108	6
M.2K Hawk	133	1
Total M.2 Hawk variants built		56

Appendix 10:
Miles M.2E Hawk Gipsy VI, M.2L and M.2U Hawk Speed Six Production

43 M.2E Hawk Gipsy VI; painted cream and fitted with DH Gipsy VI engine. Regd **G-ACTE** (CofR 5084) 24.5.34 to Sir Charles Henry Rose, Whitchurch, Oxford (based Woodley and later Portsmouth). First flown, by Miles 28.6.34. CofA No.4362 issued 5.7.34. Although recorded as being dd 20.7.34, Sir Charles flew it in the King's Cup Air Race on 13.7.34. Sold 14.6.35 and regd (CofR 6241) 26.8.35 to William 'Bill' Humble, Worksop (based at Firbeck, later Sherburn-in-Elmet); entered and flown by Humble in the 1935 King's Cup Air Race at Hatfield on 6/7.9.35 (Race No.5), to gain 13th place at average speed of 177.79 mph. It is possible that c/n 243 was issued to cover the modification to open cockpit in late 1935/early 1936 and which may have been done during the CofA renewal in May 1936. Entered and flown by Humble in the 1936 King's Cup Air Race held at Hatfield on 10/11.7.36. Sold 12.36 and regd (CofR 7575) 4.1.37 to Leslie Charles Lewis (actually Leslie Charles Percy Stanynought), Croydon for onward (illegal) sale to the Spanish Republicans. G-ACTE departed Croydon for Spain 4.1.37. Regd cld 9.37 as sold.

160 M.2L Hawk Speed Six; painted black with the wings and empennage white and fitted with 240hp DH Gipsy VI. Regd **G-ADGP** (CofR 5795) 20.5.35 to Luis Goncelvis Fontes, Sonning-on-Thames, (based Woodley). First flown, for 10 min, by Miles 2.6.35 (who recorded it as a 'Hawk Gipsy VI'). The first entry in the airframe log book for G-ADGP was a 30-min test flight by Tommy Rose on 6.6.35 which has since been taken as its first flight, but Rose was not at the time employed as a test pilot to make first flights and his flight was a test flight for the CofA. Flown by Miles 22.6.35 and by Bob Milne 30.6.35 (neither of which flights are recorded in the airframe log book, though flights made on 21.6.35, 3.7.35 and 4.7.35 by anonymous pilots were recorded). CofA No.4893 issued 13.6.35; dd 29.6.35. Flown by Miles as the 'Hawk VI GP' 1.9.35 for 10 min. Entered and flown by Luis Fontes in the 1935 King's Cup Air Race at Hatfield 6/7.9.35 (Race No.6) but damaged in forced landed on 6.9.35 with engine trouble at Old Shotton, near Easington, Co Durham. (Fontes was jailed in 10.35 for manslaughter following a motor accident in which he was found to be drunk). Following repairs, flown by Bob Milne on a 30-min 'CofA test' 14.3.36. Aircraft used by Phillips & Powis until Fontes released in March 1938. Flown by Bob Milne in the 1936 King's Cup Air Race held at Hatfield 10/11.7.36, again unplaced. George Miles flew Hawk 'G-ACGP' from Woodley-Farnborough-Woodley on 1.9.36 (probably an error for G-ADGP). Flown by Tommy Rose in the 1937 King's Cup Air Race held at Hatfield 10/11.9.37 (Race No.8) but was unplaced. Flown by Luis Fontes in the 1938 King's Cup Air Race at Hatfield 2.7.38 (Race No.19) but was again unplaced. Flown by Fontes to 4th place in the London to Isle of Man Race on 27.5.39 and to

G-ACTE being flagged away at Brooklands on 8th August 1936.
['Flight' via B Clarke]

APPENDIX 10

G-ADGP seen at Woodley after being fitted with the larger 'blown' hood in 1947. [F G Miles Ltd via The Miles Aircraft Collection]

G-ADGP in 1967 having been modified and repainted by Tony Osborne for the British Historic Aircraft Museum, Southend. [Via P Amos]

G-ADGP at Stapleford Tawney in 1968 after having been modified to open cockpit. [B Woodcock via P H T Green Collection]

12th place in Manx Air Derby on 29.5.39 by Tommy Rose. Advertised for sale in May 1939 at £850 (115 airframe hours) but not sold. Extensively modified at Woodley and reflown 11.8.39 (CofA renewed 12.8.39) in time for the 1939 King's Cup Air Race at Elmdon on 2.9.39 but the race was cancelled (total time to 25.8.39 recorded as 115 hrs 35 min). Last flown by Fontes 25.8.39 and inspected by Air Ministry survey at Woodley 31.8.39 but not impressed. Aircraft dismantled and taken to Luis Fontes' London mews home at 3 Wimpole Street in 4.40 where it was stored in his garage *(Fontes then joined the ATA but was killed 12.10.40 nr Llandow while ferrying Wellington R1156)*. Advertised for sale at £750 by High Speed Motors Ltd (of 59 Lancaster Mews W2) in *The Aeroplane* 3.8.45 and purchased by Miles Aircraft Ltd, to whom it was regd 25.9.45. Reconditioned, it made its first post-war flight on 23.8.46; CofA renewed 24.8.46. Sometime, probably either between 10.5.47 and 24.5.47 or between 20.7.47 and 8.8.47, a larger 'bubble' type canopy was fitted (a note in the aircraft's log-book reads '20-min, new type hood fitted', but with no date recorded). Purchased 6.2.48 by Ron Paine from Miles Aircraft's Receiver & Manager; regd 2.3.48 to Ronald Royal Paine, Wolverhampton (based Burnaston, later Wolverhampton). Sold 5.5.65 and regd 15.10.65 to William Henry Todd, Wolverhampton. Delivered 23.3.67 and regd 22.6.67 to Mrs Lesley Ann Osborne, London, E11 (nominee for husband Anthony J 'Tony' Osborne who was founder of the British Historic Aircraft Museum, Southend). Put up for sale 10.67 by Osborne's company, Target Towing Aircraft Co Ltd, Wanstead for £2,000 (with new CofA); to Stapleford Tawney in 1968 for repainting for the BHAM, Southend (painted white overall but now with an open cockpit and a three-piece windscreen). Dismantled and moved to Blackbushe during 3.69 (but still being offered for sale by Osborne 8.69). On 6.2.70 (with the assured shown as AJ Osborne), the propeller was damaged, no further details. Sold 2.3.70 and regd 7.5.70 to London Sports Car Centre Ltd, Edgware (based Shoreham). CofA lapsed 7.8.70. Sold 15.12.70 and regd 12.1.71 to David Angus Hood, Bledington, Oxford (based Booker); to Old Warden in 1972 (CofA renewed 28.2.72) and flown by Ron Paine to 2nd place at 195.5 mph in 50th Anniversary King's Cup Air Race 15.7.72 (Race No.96). To Castle Donnington (by 1974). CofA expired 3.7.74 and it returned to Old Warden between 1975 and 1979. Overhauled and Permit to Fly issued 22.6.79; regd 13.7.79 to Andrew Martin Stow, Peppard Common, Oxford (remained based Old Warden). Sold 13.8.82 and regd 17.9.82 to Roger Howard Reeves, Macclesfield (but remained at Old Warden). Sold 2.85 to Desmond McCarthy, Florida and shipped to Miami (and UK permit lapsed 15.3.86). Sold .86 to Californian collector Tom Buffaloe (although not regd to either US owner and never regd in the USA). Returned to UK 12.86 and to Ron Souch t/a the Antique Aeroplane Company, Sarisbury Green, nr Southampton for restoration. Regd 8.12.88 to Ron Ivor Souch, Sarisbury Green; first flown after restoration by Martin Barraclough 23.5.89 from Lee-on-Solent. Permit to Fly renewed 29.6.89 and then loaned to the Shuttleworth Collection at Old Warden. In the late 90s, Tom Buffaloe sought a buyer for G-ADGP but as no sale had been concluded, he proposed to have it shipped to San Francisco. Ron Souch duly dismantled G-ADGP and it was taken to Cardiff docks for shipment via Rotterdam to the USA. Regn cld 8.7.99 as sold abroad in the USA. The ship had reached Rotterdam, when Capt. Roger Mills, a British Airways Concorde pilot sitting in his hotel room in New York, finally made up his mind. Although Buffaloe had not accepted his earlier offer, Roger remembered, as a small boy, being enthralled at the sight of G-ADGP being flown by Ron Paine at Wolverhampton and decided that he had to have it. Telephoning Buffaloe, he negotiated a deal and it was arranged to off-load G-ADGP at Rotterdam and return it on to the next ship back to England! G-ADGP was regd 20.7.99 to Roger Arthur Mills, Wokingham (originally based at Fairoaks but later White Waltham). Fitted with a tailwheel (in deference to operating from the Fairoaks runway) it then underwent a major inspection during which a number of minor irregularities were found and rectified. G-ADGP is maintained in airworthy condition and it is understood that legal arrangements have been put in place whereby it will remain forever in this country.

195 M.2U Hawk Speed Six; painted cream. Fitted with a mass balanced rudder. Regd **G-ADOD** (CofR 6191) 24.8.35 to Miss Ruth Fontes (sister of Luis Fontes), Sonning (based Woodley). First flown, by Miles 17.8.35. CofA No.5027 issued 26.8.35; dd 5.9.35. Entered and flown by Ruth Fontes under the pseudonym 'Miss R Slow' in the 1935 King's Cup Air Race at Hatfield 6/7.9.35 (Race No.7) but was unplaced. Miss Fontes became Mrs Norman Howard-Jones by 6.36 and in the 1936 King's Cup Air Race held at Hatfield on 10/11.7.36, G-ADOD was flown to 2nd place at 185 mph by Tommy Rose (having been officially entered by Viscountess Wakefield). Regd (CofR 7259) 13.8.36 to Flt Lt Arthur Edmond Clouston (of RAE Farnborough) & Frederick Edward Tasker (of East Barnet, Herts) and nominally based at Woodley. Purchased for entry in the Schlesinger Air Race (Race No.8). Flown to Farnborough 11.9.36 for CofA tests, returning to Woodley 13.9.36. Clouston departed Portsmouth 30.9.36 for Johannesburg but crashed at Felixburgh, nr Gwelo, Southern Rhodesia 1.10.36. Regn cld 8.38. Clouston had the instruments and the engine shipped back, returning the borrowed Sperry Bank indicator and P.4 Compass to the RAE (with the rest given to Tasker) and the engine - *the only recognisable piece of the aeroplane* - to Gravesend where Jack Cross looked after the latter until it was later acquired by the Shuttleworth Trust at Old Warden. In 1959 they swapped it with Mr Shelley, a motor engineer in Billericay, for a Hispano-Suiza engine wanted for their SE.5A and Mr Shelley's son Michael has recently restored the Gipsy Six engine to working condition (but he says it will not be run).

243 M.2? Hawk Speed Six. CLN records this as a fuselage only under Job No H5573 but with no further details; the date of the allocation places this at about the end of 1935 or early 1936. It is speculated that this c/n was used for the modification of G-ACTE (c/n 43) to open cockpit, which took place by May 1936.

Ruth ('Miss Slow') Fontes seated in her Miles M.2U Hawk Speed Six, G-ADOD, in preparation for the 1936 King's Cup Air Race, held at Hatfield on 6th September 1935.
[Keystone via P Jarrett]

Appendix 11: Miles M.2F, M.2G, M.2H, M.2J, M.2M, M.2N, M.2P, M.2S and M.2T Hawk Major Production

36 M.2F Hawk Major prototype; painted cream. Fitted with 110hp Gipsy III and apparently also with manually operated flaps (see later). Regd **G-ACTD** (CofR 5083) 24.5.34 to Capt Geoffrey Reginald Devereux Shaw, Rugby (based at Sywell). First flown, by Miles, as 'Hawk Gipsy III' 23.6.34. CofA No.4361 issued 5.7.34; dd 7.7.34; named 'Susan'. Modified with the front cockpit faired over, it was entered in the sixth heat of the 1934 King's Cup Air Race, held on Friday 13.7.34 and flown by Flt Lt Thomas Rose (Race No.42) to finish in second place at a speed of 147.78 mph. Regd (CofR 6030) 2.7.35 to Kenneth Crawford, Murrayfield, Edinburgh (based Macmerry). Sold 27.8.36 (but not regd) to 2nd Lt Robert 'Rex' King-Clark, of 1st Manchester Regiment, then at Strensall Camp, York. On Monday 31.8.36, King-Clark took-off from York Aerodrome at about 4.15 pm to fly to Doncaster and return with passenger 2nd Lt Henry NE Frisby (also of the Manchester Regt). While manoeuvring to land at Doncaster, at about 5.25 pm, the pilot turned sharply right and proceeded to make a left-hand circuit of the aerodrome. After one circuit and at a height of about 100 ft, he throttled back and lowered the flaps and then attempted a left-hand gliding turn into wind; the aircraft immediately falling into an incipient left-hand spin into the ground. Both occupants were injured, the passenger seriously (and he died on 25.9.36). The opinion of the Inspector of Accidents was that the accident was due to errors of judgement on the part of the pilot, errors that were probably accentuated by his limited experience in handling an aircraft equipped with landing flaps. Regn cld 4.38 but some parts of G-ACTD survived to be used in a fire-fighting demonstration at York 16.6.51.

109 M.2F Hawk Major; painted cream. First production aircraft; fitted with 130hp Gipsy Major I. Regd **G-ACVM** 3.7.34 to Phillips & Powis Aircraft (Reading) Ltd, Woodley. Almost certainly first test flown by Bob Milne on 17.7.34 (he recorded the regn 'G-ACVN' in his flying log-book but noted that it was fitted with a 130hp Gipsy Major). CofA No.4412 issued 27.7.34. Dd 15.8.34 and regd (CofR 5306) 16.8.34 to Sir John Valentine Carden Bt, Camberley (based Woodley). Following a flight made by Sir John from Heston to Nice soon after he bought it, he wrote a very complimentary letter to on 19 August 1934: *The Hawk Major is marvellous - Heston to Nice 5¾ hours flying time - at 2,000 to 2,050 RPM - taking things very easily and no short cuts, Rhine Valley way. Petrol used 40 gallons probably under seven to the hour if you allow for running up, taxying about etc. Capt Baker flew this machine and tested it very thoroughly before I left Heston and he is loud in its praise. Landed at Nice cross wind as landing ground is very narrow - no trouble. I think there is no doubt this is the ideal private owner's machine, and wonderful value at £750.* (Sir John Carden was tragically killed in the crash of SABENA Savoia S.73 OO-AGN at Tatsfield, Kent on 10.12.35). Sold 5.36 by his executors to George Lissant Beardmore, London W2. At about 5.10 pm on 2.6.36, Beardmore, occupying the front cockpit, took-off from Woodley on his first flight in the Hawk with the intention of flying for about half an hour. The weather was fine with good visibility and about 10 minutes later, the aeroplane was seen flying westwards near the village of Hurst, some 1½ mls east of Woodley aerodrome at a height between 1,000 and 1,500 ft. Eyewitnesses stated the engine sounded that it had been throttled down and it was observed making a 360° right-hand gliding turn, with very little bank. It then turned through a further 180°, with increasing bank, stalled and went into a spin. It came out of the spin at about 100 to 150 ft but, heading towards some high trees, it then dived at a steep angle into the ground. The pilot was killed instantly on impact. Regn cld 3.37. The Accidents Investigation Branch report stated that from August 1935 until the day of the accident the aircraft had only been flown once, that flight being made on 8.2.36 and its total time-in-air amounted to about 85 hrs. Prior to the accident, most of Beardmore's experience was in his single-engined Junkers F.13 G-EBZV fitted with a 480hp Jupiter engine. The Hawk Major was certified airworthy only for normal flying (not aerobatics) owing to the fact that it was fitted with an auxiliary fuel tank immediately aft of the rear cockpit so as to use the aeroplane for long-distance flights. As loaded for the flight in question (19 gallons of fuel in the auxiliary tank) the calculated total weight of the aircraft was about 30 lbs less than the maximum weight authorised in the CofA but about 120 lbs more than would be authorised in a CofA in the aerobatic category. The position of the centre of gravity was about 23 inches aft of the leading edge of the mainplane - i.e. about 1 inch forward of the furthest aft position allowable for this type of aircraft.

The first Miles M.2F Hawk Major in flight over the Berkshire countryside. ['Flight' via B Clarke]

It had been found on test that, when this type of aircraft was loaded up to the maximum for aerobatics and the centre of gravity was in its furthest aft position, recovery from a spin should normally be effected after about 1¼ turns. The opinion of the Inspector of Accidents was that the accident was due to errors of airmanship on the part of the pilot which resulted in the aircraft stalling when in a gliding turn and then spinning and that the amount and distribution of the load was such as to render recovery from the spin somewhat slower and less easy than usual and this, combined with the pilot's inexperience of the 'feel' of this particular type of aircraft, probably accounted for his failure to check the spin at a height sufficient to enable him to level-up from the ensuing dive.

110 M.2F Hawk Major; painted cream. Flown as **U1** by Bob Milne 9.8.34 (a 5-minute test which may have been its first flight). Regd **G-ACWV** (CofR 5260) 7.8.34 to Phillips & Powis Aircraft (Reading) Ltd. CofA No.4446 issued 23.8.34; dd 1.9.34 to firm for use as demonstrator. Almost certainly the Hawk Major used by Captain Neville Stack for his record breaking flight to Copenhagen, Denmark on Saturday 8.9.34, when he flew from Woodley via Heston to Copenhagen, resting for just 45 minutes before returning to Woodley via Heston the same day. Regd (CofR 6089) 18.7.35 to The British Instrument Co Ltd, (a company owned by Major Savage), Hendon and modified to single-seat with Savage smoke producing equipment for sky-writing, in which guise it was flown by Sydney St. Barbe; re-painted red. The CofA lapsed 9.11.38 and it is believed the Hawk Major was stored by Savage at Hendon, where it was noted in the Air Ministry survey on 1.9.39. On 11.3.40, a permit was issued for three sky-writing demonstration flights at Hendon. On 11.4.40, it was flown Hendon-Farnborough-Hendon by Sqdn Ldr Combe for aircraft recognition flights. It was not impressed and probably scrapped during the war. Regn cld 1.12.46 at census.

G-ACWV showing smoke-producing equipment modification for skywriting in 1936. Note the original style undercarriage fairings.
[Via B Clarke]

111 M.2F Hawk Major; painted maroon. Fitted with split trailing-edge flaps (a Phillips & Powis advertisement proclaimed that the selling price was £750 + £50 extra for flaps). Regd **G-ACWW** (CofR 5261) 7.8.34 to Phillips & Powis Aircraft (Reading) Ltd. Regd (CofR 5347) 10.9.34 to Mrs Helen Marcelle Barnes, Portslade (based Shoreham). First flown, by Miles 6.10.34 (no mention of flaps being fitted). CofA No.4447 issued 6.10.34; dd 6.10.34. Regd (CofR 7361) 19.9.36 to Giles Alexander Maysey Vandaleur of the Irish Guards (based Heston); re-painted blue and entered by owner in the *International Circuit of the Oasis* meeting at Cairo, held 22-26 February 1937, but he failed to start due to sickness. CofA lapsed 8.9.37 (reason unknown). Sold 6.2.39 and regd 20.2.39 to Leslie Robert Hiscock, Guildford (based Brooklands) and CofA renewed 10.2.39. Regd 18.4.40 to WS Shackleton Ltd, London, W1. Regn cld 19.4.40 on sale in Malaya. Regd **VR-RAV** 6.40 to Kuala Lumpur Flying Club, Kuala Lumpur. Impressed into Malayan Volunteer Air Force as serial "**31**" 12.41. Crashed at Pakanbaroe, Sumatra 1.2.42.

112 M.2F Hawk Major; painted maroon. Fitted with split trailing-edge flaps. Regd **G-ACWX** (CofR 5262) 7.8.34 to Phillips & Powis Aircraft (Reading) Ltd. To RAE for tests of flaps. Regd (CofR 5313) 20.8.34 to Dr Oliver Francis Haynes Atkey, London W1 (based Heston). Flown by Bob Milne with Mrs Dulcibella Atkey 25.8.34; flown by Milne 'testing flaps' 14.9.34 and flown by Miles four times on 21.9.34 with 'flaps' noted. CofA No.4448 issued 4.10.34 but handed over 2.10.34 and flown by both Atkeys 3.10.34. CofA lapsed 3.10.35 and reported as written-off in a crash on unrecorded date. Dr Atkey was the Director of the Sudan Medical Service in 1935 and it is therefore possible the aircraft was written off there (it was replaced by Hawk Major G-AEGE in April 1936).   Regn cld 6.36 as wfu.

113 M.2F Hawk Major; painted cream. Regd **G-ACWY** (CofR 5263) 7.8.34 to Phillips & Powis Aircraft (Reading) Ltd. CofA No.4449 issued 30.8.34. Regd (CofR 5337) 4.9.34 to Lady Blanche Scott Douglas, Manor Farm, Sherston, Wiltshire (kept at Whitchurch and at owners home); dd 5.9.34. Lady Blanche learned to fly at Bristol & Wessex Aeroplane Club in 1934 and, after she had only been flying some 6 to 8 weeks, she bought the Hawk Major to visit her friend the Maharajah of Cooch Behar in Bengal, India 'for her own amusement'! She invited her instructor F/O Cecil Victor 'Micky' Ogden to accompany her as co-pilot and they departed 20.11.34, arriving the following month after being marooned in the Arabian Desert following a forced landing with engine trouble and a subsequent landing accident at Bandar Abbas, when the Hawk ran into a hole and tipped on to its nose. They returned to Whitchurch 16.2.35. Stored at Sherston Farm on outbreak of war (CofA lapsed 2.10.39). Regd 27.3.42 to Frederick William Griffith, Waddon, Surrey (still stored at Sherston Farm). Impressed 18.2.43 as **NF748** and taken to 50 MU Salvage Centre, Oxford for assembly but instead SOC 21.3.43 (as Cat E1) for spares use. Regn cld 6.2.46 at census.

114 M.2F Hawk Major; painted cream. Regd **G-ACXL** (CofR 5307) 21.8.34 to Aircraft Sales Ltd, Heston; (company renamed Aircraft Distributors Ltd 6.9.34). First flown, by Miles 16.9.34 and handed over same day. CofA No.4471 issued 17.9.34. The company obtained an agency for the type and moved to Hanworth 6.35 and but also undertook joyriding/charter work at Skegness from 8.35. While en route from Skegness to Doncaster, it hit a hedge and crashed on take-off from a field at Armthorpe, nr Doncaster 21.2.36; Mr & Mrs M Robson injured. Rebuilt by Phillips & Powis with new fuselage c/n 219 under Job No.H3807. CofA lapsed 22.4.38 and fate unknown. Regn cld 1.12.46 at census.

115 M.2F Hawk Major; painted maroon. Regd **G-ACXM** (CofR 5308) 21.8.34 to Noshir Manesk Gazdar, Bombay (and who at time was on a training course at Air Service Training, Hamble). CofA No.4472 issued 2.10.34; dd 13.10.34 after owner received familiarisation flying training at Woodley. Owner flew the Hawk Major back to India in 2.35, accompanied by Tehmurusp H Dastur. Regn cld 11.35 as sold. Regd **VT-AGX** (CofR 265) 30.10.35 to Noshir M. Gazdar, Muryabanabad, Andheri (Bombay). Regn cld 26.8.41 (and recorded by CLN as 'crashed and written off' but no details).

116 M.2F Hawk Major; painted red. Regd **G-ACXN** (CofR 5309) 21.8.34 to Phillips & Powis Aircraft (Reading) Ltd. Regn cld 9.34 as sold. CofA No.4473 issued 19.10.34. Handed over to owner Brig Genl Arthur Corrie Lewin CB, CMG, DSO the same day painted as VP-KBL. On 23.10.34, both Miles and Bob Milne gave Lewin dual instruction, followed by more dual with Milne; again test flown by Bob Milne on 17.11.34 before its new owner carried out cross-country practice. Flown to Hatfield by Milne on Sunday 18.11.34 in preparation for Lewin's departure to Kenya and he departed Heston 25.11.34. Forced landed 20 mls from Kisumu 12.12.34. Finally formally regd **VP-KBL** (CofR 38) 4.2.35 to Brig-Genl Lewin and based Njoro. Flown back to England by Lewin 8.35 and crashed following engine failure on take-off at Tilesford Aerodrome, Pershore 19.8.35; Lewin and wife slightly injured. Not repaired and Kenyan regn cld 10.1.36.

118 M.2F Hawk Major; painted green. *(Recorded by CLN as the first M.2F fitted with split trailing-edge flaps but he probably meant first with flaps fitted with the Schrenk modification).* Regd **G-ACXT** (CofR 5326) 10.9.34 to Miss Ruth Fontes, Sonning (based Woodley). First flown, by Miles, 26.10.34. CofA No.4562 issued 3.11.34. Flown Woodley-Farnborough 22.11.34 by Miss Fontes for test of Schrenk flaps; collected by her 3.12.34. *(Miss Fontes became Mrs Norman Howard-Jones by June 1936; but he should not be confused with Norman Herbert Jones, later of The Tiger Club).* CofA lapsed 1.3.38. Advertised for sale in the *Miles Magazine* for May 1939 with 'About 200 hrs; 12 months CofA; Sperry Artificial Horizon and Directional Gyro mounted on Sperry Anti-Vibration Panel in rear cockpit; ASI and Cross Level in front cockpit; Aerobatic Harness'. Sold 18.5.39 and regd 31.5.39 to Staffordshire Airplanes Ltd, t/a North Staffordshire Aero Club, City Airport, Meir, Stoke-on-Trent; CofA renewed 25.5.39. Regn cld 7.1.41 as sold. Impressed 7.1.41 and dd to 24 MU Ternhill 5.2.41 as **DG577**. Fitted with new Gipsy Major 10.4.42 prior to being transferred to 5 MU Kemble 18.4.42. Relegated to instructional airframe **4020M** 6.8.43.

119 M.2F Hawk Major; painted silver. Regd **G-ACXU** (CofR 5328) 31.8.34 to Manawatu Aero Club, New Zealand (for MacRobertson England-Australia Air Race). British regn ntu and cld 9.34 as sold. Regd **ZK-ADJ** 21.9.34 to Manawatu Aero Club, Palmerston North. First flown, as ZK-ADJ, by Miles, 7.10.34. CofA No.4504 issued 12.10.34. Named *Spirit of Manawatu*, it was flown to Mildenhall for the start on 14.10.34 and given Race No.2. Originally to have been flown by GAC Cowper, in the event it was flown by Sqn Ldr Malcolm C MacGregor (the club's chief instructor) and Henry C Walker. They took off at 6.42 am on Saturday 20.10.34 and made their first stop at Rome at 2.52 pm. Departing at 3.22 pm; they arrived Athens at 8.35 am, staying the night. The next landing was at Baghdad at 2.00 pm on Sunday 21.10.34; after resting they took-off at 10.47 pm, arriving Jask at 6.45 am on Monday. Leaving at 7.20 am for Karachi (arrived 11.40 am, departed 12.30 pm), then Jodhpur (departing 3.45 pm) and arriving Calcutta 6.41 am on Tuesday 23.10.34. Taking off at 7.15 am for Rangoon (arrived 12.30 pm) and Alor Star (arrived 4.34 pm) before reaching Darwin at 9.55 pm on Thursday 25.10.34. Cloncurry was reached the next day and Narromine at 6.00 am on Saturday 27.10.34 before finally reaching their destination, Melbourne, at 9.50 pm the same day. They were credited with the fastest time for a single-engined aircraft, having beaten all other light aeroplanes in the race by reaching Australia in 5 days 15 hrs 13 mins, with a total flying time of 118 hrs 5 mins and 46 secs. They came 5th overall in the handicap section (the 'adjusted' flying time according to the Handicap Formula being 102 hrs 7 min 48 sec), with an average speed of 105 mph. The Hawk Major was then shipped to New Zealand, arriving 12.11.34. Reported to have been regd to Middle Districts Aero Club, Palmerston North (which was probably the correct name of the Manawatu club). Regd 20.1.36 to Wellington Aero Club, Wellington. Ran out of fuel en route Palmerston North to New Plymouth and badly damaged in forced landing Stratford, nr Maxwell 21.3.36; pilot HC Walker & passenger EB Firth unhurt. Returned to Wellington for repairs but regn cld 10.36; replaced by M.2Y Hawk Trainer ZK-AEQ and used for spares for that aircraft.

120 M.2G Hawk Major Three-seater; painted blue with light coloured wings & tail. Fitted with 130hp Gipsy Major; M.2A type undercarriage and initially with cabin over rear two seats (later removed). Regd **G-ACYB** (CofR 5343) 10.9.34 to Phillips & Powis Aircraft (Reading) Ltd. Regn not used and cld 9.34 on sale in Switzerland. Regd **HB-OAS** (CofR 404) 2.10.34 to JR Pierroz [of the Club George Chavez], Sion, Switzerland. CofA No.4533 issued 12.10.34 to Club George Chavez; dd 14.10.34. Regd 4.36 to Aero Club Suisse, Section Valais, Sion. Regn cld 7.47.

121 M.2F Hawk Major; painted blue. Fitted with early trousers to u/c, and had flaps and one-piece windscreens to both cockpits, raked back similar to the M.5. Regd **G-ACYO** (CofR 5370) 24.9.34 to Dr John Myles Bickerton, Denham, Bucks (initially based at Hanworth but soon based at his own Hawksridge Aerodrome, later re-named Denham). First flown, by Miles, 6.11.34. CofA No.4527 issued 9.11.34. Charles Hughesdon failed to finish in King's Cup Air Race 10.7.36 due to faulty oil pressure gauge. Test flown by George Miles from Woodley 16.7.36 (probably after rectification of fault). CofA lapsed 2.7.40 and stored initially at Denham. Impressed 29.9.43 and conveyed to Phillips & Powis Aircraft Ltd, Woodley for repair after storage and given serial **NF752**. To 43 Group Comm Flt, Hendon 23.10.43. To RAF Whitchurch 3.2.44; at RAF Roborough from 27.3.44 and to RAF St. Davids 13.5.44. To 5 MU Kemble 11.10.44 for storage and disposal. Sold to Miles Aircraft Ltd for £150 at the first sale of impressed light aircraft held at 5 MU Kemble 12.45, for 'business or sale'. Regd **G-ACYO** 9.9.46 to Miles Aircraft Ltd, Woodley (a post war photograph of G-ACYO taken at Woodley shows this now to have the later M.2H style u/c trousers and normal three-piece windscreens). Flown by Hugh Kendall (who entered it in his log-book as NF757 in error) 10/11/13.6.47. Regd 11.6.47 to The Reading Aero Club Ltd, Woodley. CofA renewed 27.6.47. Cld 11.2.48 and regd 5.3.48 to BC Barton & Son Ltd, Elmdon; raced at Wolverhampton 12.6.48 (Race No.7). Cld 13.11.50 and regd 15.11.50 to Harold Charles Blumenthal, Edgbaston, Birmingham (based Elmdon, possibly to Wolverhampton by 12.51). Cld 5.8.53 and regd 10.8.53 to Miss Freydis Mary Leaf, London W11 (based White Waltham/Elstree); flown by Miss Leaf to win the Air League Challenge Cup in 1954 at 138 mph. Cld 14.8.54 and regd 28.8.54 to Frederick Howard Stirling, London SW7 (based Elstree). Crashed on landing in a gale at Elstree Aerodrome at 3.20 pm on 28.11.54 following a local flight piloted by owner; passenger Jack Bosman killed. Regn cld 17.2.59 as destroyed.

122 M.2H Hawk Major; painted silver. First M.2H to be fitted with split trailing edge flaps as standard (early M.2H's retained the same u/c trousers as the M.2F before they were re-designed). Regd **G-ACZJ** 2.11.34 (probably) to Baron Herbert von Schinkel, Vasteras, Sweden to whom CofA No.4528 was issued 15.11.34. Regd (CofR 5449) 16.11.34 to Aircraft Distributors Ltd, Heston (but it is thought they were simply acting as nominee owner for the Baron) Flown to Sweden by von Schinkel, departing Woodley 17.11.34, but not regd locally. Possibly damaged in a forced landing in bad weather in Germany c.1.35. Probably involved in a (later?) accident (as CLN records that 'fuselage c/n 199 was used 'to rebuild G-ACZJ' and this would have been about 9.36 and thus possibly prior to CofA renewal 30.9.36). Regn cld 6.37 on sale in India. Regd **VT-AIR** (CofR 311) 20.5.37 to The Aeronautical Training Centre of India Ltd, New Delhi. Regn cld 18.2.43. Impressed into RAF in 1942 as **LV768** and to 223 Group Comm Flt, Peshawar. Swung on landing at Chaklala aerodrome 16.3.43 and port u/c collapsed. SOC 30.11.43 at the ACSEA census.

123 M.2H Hawk Major; painted cream. Regd **G-ACYZ** (CofR 5395) 16.10.34 to Sir Alfred Lane Beit Bt. MP (an Irish gold magnate), London SW1 (based Heston); purchased to replace crashed Hawk G-ACJD. First flown, by Miles, 20.12.34. CofA No.4541 issued 20.12.34; dd 26.12.34. CofA lapsed 17.1.38, having been replaced by Whitney Straight G-AEUY and presumed stored pending sale. CofA renewed 30.9.38 and regn cld 10.38 on sale in Australia. Regd **VH-ACC** (CofR 725) 9.12.38 to Royal Aero Club of South Australia, Parafield (as an M.2F). Sold but not regd to GH Michell, Ballarat. Impressed into RAAF 19.11.40 as **A37-4** (but also noted as given by owner to RAAF 10.2.41). To 1 EFTS Parafield after re-marking 11.2.41. To 1 Comm Flt 28.3.42 and subject to complete overhaul there 8.2.43. To Ansetts 19.2.43. To 1 Communications Unit, Laverton 24.4.44. Issued to Point Cook 23.7.45 and dd 26.7.45 for disposal through Commonwealth Disposals Commission. *(A letter was sent to GH Michell 29.7.45 offering the aircraft back to him as it was no longer required by the RAAF - he had gifted it to the RAAF).* On 8.2.46, noted that aircraft to be issued to JH Schutt, Fisher Parade, Ascot Vale, as agent for GH Michell, Hindmarsh, SA; issued on voucher 788 30.9.46. Regd **VH-ACC** 1.2.47 (with 904 hours) to GH Michell, Hindmarsh, Adelaide, SA. Regd to OL Lansell, Moulamein, NSW. Regd 7.6.51 to Noel Keith Green, Fairfield, Melbourne, Vic. Regd 15.6.53 to Brian James Garvan Hurley, Kew, Melbourne, Victoria. Regd 17.3.54 to Leslie Fuller, South Carlton, Melbourne, Victoria. Regd 24.4.54 to Leslie Elliot, Moorabbin, Victoria. Regd 20.10.54 to Brian Joseph Trethown, Melton, Victoria. Crashed in 1955; rebuilt in 1956 by Air Operators Pty Ltd, Moorabbin. Regd 14.3.61 to Dr Ralph Henry Capponi, Apollo Bay, Victoria. Regd 19.11.64 to L Hornsby, Donald, Victoria. Regd 8.3.68 to JC Lane & JE Pike, Morwell, Victoria. Regd 26.8.68 to Skyservice Aviation Pty Ltd, Camden, NSW. Regn cld 6.12.68 on sale in Canada. Shipped to Canada .69 by E Fleming and sold to Bob Diemert. Regd **CF-AUV** 17.4.70 to Robert Diemert, Carman, Manitoba (CofA issued 19.2.70). Sold late 71 and regd 24.1.72 to Clifford W Glenister, Weston, Ontario. CofA lapsed 13.7.74. Rebuilt 1974-1983 by Glenister, incorporating wings and undercarriage from Falcon VH-ABT, which had also been shipped to Canada; to Brampton, Ontario (by 8.79), now painted red with white trim as **C-FAUV**. CofA renewed 10.5.83 and officially re-regd C-FAUV 12.12.83. CofA lapsed and sold to International Vintage Aircraft Inc (Allan 'Al' Rubin), Markham, Ontario (but initially based at Hamilton Airport); regd to them 8.8.86. Under restoration (89/90) in a small workshop at Hamilton under a team led by Ken Elliot but reportedly held up awaiting new tyres, spark plugs and a propeller. The work was never completed and the aircraft was removed (after the 1991 Canadian Warplane Heritage fire there) and put into store in an articulated lorry trailer at Markham Airfield, NE of Toronto. Nominal regn change 3.4.95 to IVA International Vintage Inc. Still stored Markham (2007).

124 M.2H Hawk Major; colour not recorded. Regd **VT-AFR** (CofR 234) 5.9.34 to Flt Lt Gordon Valentine Carey, Aero Club of India & Burma Ltd, New Delhi. Flown by Bob Milne on CofA test flight 1.11.34. CofA No.4519 issued 6.11.34; although reportedly handed over 2.11.34. Flown by Carey from Bombay to Heston in 6.35 (to take up a post as an instructor for de Havilland School of Flying). Regn cld 4.36. Regd **G-AEFS** (CofR 6882) 8.4.36 to Flt Lt Carey, Ruislip (based Northolt); CofA renewed 10.9.36. Regn cld 12.36 as sold. Regd in Australia as **VH-AAH** (CofR 630) 22.3.37 to FW Hewson, Augathella, Qld (based Highfields); flown to 1st place by Hewson in the Sydney to Moree Air Race on 11.3.38. Regd 24.3.38 to AK Bates, Cooparoo, Qld. Crashed at Leura Station, Qld

8.6.38. Regn cld 19.10.38. Some parts of VH-AAH still survive with Graham Orphan in New Zealand who intends to build a Hawk Major 'around them' one day.

125	M.2J? Hawk Major 2/3 seat cabin?; painted blue. First flown, as **U1**, by Bob Milne 1.11.34 on CofA test. *This was possibly the mysterious 'Cabin Hawk' pictured on page 72 of Don Brown's Miles Aircraft since 1925.* CofA No.4508 issued 7.11.34 to Kuala Lumpur Flying Club, Malay States; dd 2.11.34. Regd **VR-RAH** (CofR 8) 20.12.34 to Kuala Lumpur Flying Club, Kuala Lumpur. Possibly crashed prior to 11.35 (see also c/n 227). Regn cld 9.3.38.

Miles M.2F Hawk Major HK863 shown while being used by Lydda Communications Flight in 1942. [Lord Apsley via J Havers]

126	M.2F Hawk Major; painted silver. CofA No.4520 issued 17.11.34 to Prince Omar Halim, Egypt; dd 25.11.34. Regd **SU-AAP** 12.34 to Prince Omar Halim; entered by its owner for the International Circuit of the Oasis meeting of 22-26.2.37 at Cairo (Competition No.27). While taking off from Hurghada 24.2.37 the undercarriage was damaged and the aeroplane overturned on touching down short of Luxor Airfield, Prince Halim and passenger Fuad Bey Sabet were slightly hurt. Presumed repaired prior to purchase 4.39 by Lord Apsley (a senior officer in the Arab Legion in Transjordan); used by him until 11.41. Impressed into RAF as **HK863** 17.2.42 for use by Lydda Communications Flight. (A Hawk, which must have been SU-AAP, arrived at 206 Maintenance Group, Middle East in 2.42; was purchased by RAF 16.6.42 and allotted to Comm Flight Levant for use by the Arab Legion). Forced landed Ziza 14.2.43 while en route Amman to Aqaba with choked petrol filters, undamaged. Forced landed Ben Shamen, 5 mls from Lydda 15.2.43 while en route Ziza-Ammanand to Lydda, due to choked carburettor jet, Cat.1 minor damage, pilot uninjured. Test flown by W/O Cuthbert 28.3.43 from Lydda and found satisfactory. Believed damaged at some time by flooding in Palestine. Ferried Lydda to Kolundia (Jerusalem) on delivery to W/C Macdonald, on what was probably its last flight, 9.4.43, for major inspection by the Engineering Officer from local R&SU (who may have been W/C Macdonald); grounded due to glue failure 30.4.43, probably as a result of the previous flood damage. SOC 1.8.43.

127	M.2F Hawk Major; painted black. CofA No.4566 issued 28.11.34 to John Michael Litsas (or Libas), Athens, Greece. Flown (possibly first flight), by Miles, as 'Greek Hawk' 5.12.34 and flown to RAE Farnborough same day by Tommy Rose; dd 13.12.34. Regd **SX-AAD** (CofR 8) 26.1.35 to L Kyriacou, Athens; owned and operated by APQIA (Morning) newspaper and carried its titles in white, in Greek, on the fuselage under the cockpits and 'No.1', also in white, behind the engine cowling (these titles being applied at Woodley). Also recorded as regd to Stephen Pesmazoglou, Athens. Spun in and crashed Athens/Dekelia 25.2.36 when the controls jammed while on a training flight; pilot Steve Tsicaliotes and pupil George Soutsos both killed. Regn cld 18.3.36 as destroyed.

Miles M.2F Hawk Major SX-AAD at Woodley prior to delivery to L. Kyriacou, Athens, Greece, on 13th December 1934.
[The A J Jackson Collection via P Jarrett]

128	M.2M Hawk Major Three-seat; painted red fuselage and polished cowling and silver wings and tailplane, with the name 'FEFOR' painted on the fuselage around the contour of the wing leading edge. Fitted with a coupe top over the rear two seats. CofA No.4593 issued 1.12.34 to CF Walther, Norway. Regd **LN-BAH** 1.12.34 to Christian F Walther, Fefor, Oslo. Departed Woodley on Sunday 2.12.34 en route to Norway via Holland, Germany, Denmark (with a few days stop and where Walther left to complete the journey by sea) and Sweden. Crashed on arrival Kjeller, nr Oslo at the completion of flight 10.12.34; occupants E Gran-Herriksen and Thor Bernhoft both killed. (Note: the engine was probably sold for $350 to the Toronto Flying Club in response to their advert in *The Aeroplane* for a Gipsy Major for their Puss Moth CF-CDN by 'a man in Lillestrom, Norway who claimed it was from a Hawk which crashed on delivery with only 16 hrs').

Miles M.2H Hawk Major G-ACYX at Hatfield during the 1938 King's Cup Air Race. Note the lowered windscreens.
['Flight' via B Clarke]

129	M.2H Hawk Major; painted green with light coloured (silver?) wings and tailplane and a broad strip down the side of the fuselage from the rear of the engine cowling. *(Although recorded as an M.2H, this machine had the early trousered u/c of the M.2F, probably an early M.2H with flaps fitted as standard).* Regd **G-ACYX** (CofR 5393) 16.10.34 to John Arthur Hampden Parker, Godalming (based Brooklands and at his private airstrip at his home, Featherstone, Hambledon). CofA No.4544 issued 18.12.34; dd 19.12.34. Operated by The Old Etonian Flying Club (of which Parker was founder/secretary), Heston. Regd (CofR 7468) 18.11.36 to The Old Etonian Flying Club Ltd, Heston. Hired by Leslie Harold Talbot Cliff (for £25) to fly in the King's Cup Air Race in July 1938 (Race No.2); modified for the race by Ron Paine at Brooklands with lower one-piece windscreen, new cowlings and fairings, a top overhaul for the engine and fitting a fine-pitch propeller to increase the revs. Cliff came 3rd and used the £450 prize money and £100 from tyre and spark-plug sponsors to purchase the aircraft with Ron Paine. Advertised for sale with 230 F/Hours 12.38 for £375 and sold to M Colledge (who was taking his 'B' Licence at Shoreham). Regd 1.2.39 to his mother, Mrs Margaret W Colledge, the wife of surgeon Lionel Colledge. Ron Paine continued to look after the engineering side and laid up the aircraft in June 1939 to be prepared for the next King's Cup to be held in 9.39. He fitted new cowlings, deleted and faired over the flaps and altered the decking to accept smaller windscreens/doors. He also faired in the tail section and the gaps between the ailerons and trailing edges of the wings. Entered into King's Cup by Miss Cecilia Colledge and with Leslie Cliff as pilot; but the race was abandoned due to the outbreak of war. The Hawk was taken back to John Parker's hangar at Hambledon on 23.10.39 where it was stored throughout the war. Taken to Broxbourne to be refurbished by Herts & Essex Aviation Ltd and CofA renewed 26.7.46. Sold to Yves Pougnet of Paris and painted as F-BCEX (Race No.29), it took part in the Siddeley Trophy Air Race, the first post war air race, held at Lympne over the weekend of 31.8.46/1.9.46. Regn cld 6.9.46 as sold, it was delivered from Heston in 9.46. Regd **F-BCEX** 13.1.47 to Yves Pougnet, Toussus-le-Noble (officially based Casablanca). It had a short French life, being written off almost immediately after regn (although its wings were still at St. Cyr 8.50). Regn cld 22.9.71 as destroyed.

132	M.2H Hawk Major; painted green. Regd **G-ACZI** (CofR 5413) 2.11.34 to Edward Hamilton Fleetwood Fuller, Hanworth (later based Heston). First flown, by Miles 21.12.34. CofA No.4569 issued 21.12.34; dd 1.1.35. Regd (CofR 6370) 4.10.35 to George Archibald McPhee, Stockton-on-Tees (based Woolsington). Regn cld 11.36 as sold. Regd in South Africa as **ZS-AFM** 3.11.36. Impressed into SAAF 3.40 as **1576**; later to GI airframe **IS117**.

APPENDIX 11

134 M.2F Hawk Major; painted green. Regd **G-ADAC** (CofR 5473) 29.11.34 to RK Dundas Ltd, Portsmouth/Heston. CofA No.4571 issued 2.1.35. Regn cld 9.35 on sale in India. Regd **VT-AGT** (CofR 261) 15.8.35 to RK Dundas Ltd, New Delhi. Regd 24.12.35 to The Aeronautical Training Centre of India Ltd, New Delhi. Flown by Flt Lt Nevill Vintcent in the Viceroy's Challenge Trophy Air Race, over the Madras-Hakimpet-Bombay-Ahmedabad-Jodhpur-Delhi course, on 14-15 February 1936, to gain 5th place at a speed of 148.3 mph. Regn cld 3.5.36 as crashed and written-off.

135 M.2N Hawk Major de luxe; painted blue. Originally regd as M.2H but fitted with a high comp ratio 140hp Gipsy Major I srs II; the airframe was 'cleaned up' and M.5 type windscreens fitted, with a headrest to the rear cockpit. Regd **G-ADGE** (CofR 5784) 30.4.35 to Arthur Harry Cook, Bletchley (based Hatfield). First flown, by Bob Milne as the 'Cook's Special' on a CofA test on 15.5.35 and by Miles as 'Hawk (Cook's King's Cup)' 17.5.35. CofA No.4916 issued 24.6.35; dd 9.7.35. Entered and flown by Cook in the King's Cup Air Race at Hatfield on 6/7.9.35, (Race No.22) and flown as single-seater. During the first day, a long course around Britain, Cook was forced to land near Girvan in Ayrshire with a faulty exhaust system; repaired with a section from an old lorry exhaust by a local garage, Cook continued in the race but finished just too late to qualify for the final. G-ADGE was used as a two-seater after the race with an identical windscreen fitted. Regd (CofR 7121) 11.6.36 to Edward Fraser 'Ted' Walter, London W6 (based Brooklands). Finished second in the 1936 London to Newcastle race at 155.25 mph behind Hawk Major G-ACYO, but both were disqualified following a dispute over cornering at Yeadon. Regn cld 12.37 as sold. Regd in Kenya as **VP-KCM** (CofR 64) 21.1.38 to WC Mitchell, Nairobi. When being flown by WC Mitchell and Major CA Hooper from Durban to Lourenco Marques 12.8.39, a propeller blade broke off causing the engine to be torn out; aircraft crashed between Lake Sibayi and Mselini, Maputaland; occupants unhurt.

G-ADAB, the Miles M.2H Hawk Major, with pilot Mrs. Elsie Battye and navigator Mrs. MacDonald, seen planning a route at Woodley in July 1935. [Via B Clarke]

137 M.2H Hawk Major; painted black. Regd **G-ADAB** (CofR 5469) 29.11.34 to Mrs Beatrice Shedden MacDonald, Hawkhurst, Kent (based Lympne). CofA No.4584 issued 14.1.35. Fitted with Schrenk flaps and flown to RAE Farnborough 5.12.34 by Flt Lt SR Ubee for tests (loaned per Contract No.354809/34). Flown by Ubee 14 times during January and three times during early February before returning to Woodley 4.2.35. Piloted by Mrs Elise Battye, and with Mrs MacDonald as navigator, they gained 3rd place in the Helene Boucher Cup, Paris to Cannes Air Race in 8.35. Owner moved to Eynsham, Oxford and aircraft based at Witney from 2.38. Regd 9.2.39 to John Maurice Houlder, Esher (based Brooklands). CofA lapsed 25.5.40. Regd 19.8.42 to Aubrey Jappe Cripps, West Molesey (and thought to have moved to his home). Regd 6.5.46 to Charles Newton Cooper (of Cooper's Garage, (Surbiton) Ltd), Surbiton, Surrey. Believed to have been taken to East Preston when owner moved his garage there in the 1950's. Sold to racing driver Dennis Poore and probably taken by him to Walsall where it was reported 12.47 as being 'derelict and wrapped in a tarpaulin' at Walsall Airport. Regn cld 17.8.48 as lapsed. Recorded still at Walsall in 1952 in dismantled state but thought scrapped soon after.

138 M.2H Hawk Major; painted blue. Regd **G-ADAS** (CofR 5505) 19.12.34 to Ernest Long Maddox (an insurance assessor), London W1 (based Heston). CofA No.4614 issued 29.1.35; dd 30.1.35. Operated by Maddox Airways Ltd at Brooklands for general

Miles M.2H Hawk Major CX-AAW seen at Montevideo, Uruguay, sometime prior to 1941. [Via B Clarke]

charter and air taxi service. Regd (CofR 6520) 25.11.35 to Malcolm & Farquharson Ltd, Heston. Sold 9.8.36 to Spanish Republicans via Union Founders Trust Ltd and departed Woodley same day for Bilbao, Spain, via France, where it arrived on 14.8.36. Entered service with the Basque Air Force (who fought for Republican Spain) and flown by Bert Acosta 11.12.36. Regn cld 1.37 as sold. Later shipped to South America (possibly as part of a 'deal'), where it was believed to have been sold first in Argentina. Regd in Uruguay as **CX-AAW** 13.9.37 to Juan M Acuna, Montevideo. During 1938, the Polish pilot Esteban Xavier Krajewski and a passenger flew from Uruguay to Brazil in this Hawk Major, which now wore the inscription *Pilsudski* (who was a Polish national hero, Marshall and dictator during the interwar period), Probably suffered an accident in Imbituba, Brazil and the damaged airframe returned to Uruguay by boat. Repaired and repainted white. Regd **CX-ACT** 26.6.41 (as it was decreed that the letter 'W' was no longer to be used). Regd .41 to JR Ricardi, Salto. Regd (12.47, 8.50) to Hector Salaverry, Montevideo. It is believed to have had four or five other owners in Uruguay over the years. Painted with black fuselage and silver wings at some time (possibly when still CX-AAW) and it was in these colours when it was acquired in .77 by Ariel Fabius of Montevideo, with a view to restoring it to airworthy condition. In a letter dated 7.1.78, he stated that the present state was rather sorry as one wing spar appears to be cracked (the right wing is not attached to its stub), the rear top fuselage decking is missing as are the landing gear fairings, instruments, control column and several ancillary accessories. In 1.79 Fabius decided to sell it in order to get married! Its last known owner was Elio Riveron. Regn cld 13.1.86. When inspected in the early 1990's, it still had its original engine and propeller fitted, but the u/c trousers were missing and its condition could only be described 'somewhat ropey' with the plywood covering stripped from both wings which had subsequently warped quite badly. In 1998, it was purchased by the Associacao Brasileira de Aeronaves (The Wings of a Dream aviation museum), Antigas e Classicas, Jundial, N of Sao Paulo, Brazil. Recent plans are to build two new flying Hawk Majors from scratch, using the original aircraft as a pattern, and the metal parts from it for one of them. The original aeroplane would then be rebuilt to static display condition for their new Museum at Americana, NE of Sao Paulo. Their workshop is in the city of São Carlos, São Paulo and the first 'replica' has been completed and finished in a maroon and cream colour scheme as G-ADAS. *Note: some sources have quoted the previous identity of CX-ACT as c/n 128, 134, or even 307, but these are incorrect.*

On 26th June 1941, CX-AAW was re-registered CX-ACT, and it is seen here less undercarriage fairings. [Via P Butler]

139 M.2H Hawk Major; painted blue/red. Regd **G-ADAW** (CofR 5511) 28.12.34 to Phillips & Powis Aircraft (Reading) Ltd. CofA No.4635 issued 5.3.35; dd 6.3.35, for use of P&P School of Flying. Flown to RAE Farnborough 25.4.35 for 'acrobatic' (aerobatic in modern parlance) handling trials before returning to Woodley 2.5.35 *(the RAE Report is in the National Archives under AVIA 6 4625)*. Stalled and crashed Sandford Mill, nr Woodley 3.5.36; Victor Neale, flying 'figure of 8' patterns for his 'A' licence test, was killed. Regn cld 5.36.

143 M.2H Hawk Major; painted cream. Regd **G-ADBG** (CofR 5539) 9.1.35 to British & Overseas Aircraft Ltd, London (a company formed by Thomas Claud Worth to act as agents for Miles and Airspeed aircraft in Buenos Aires, Argentina and also to sell and service yachts and boats; also reported to be based in El Palomar). First flown, as **U1**, by Bob Milne 19.1.35 on CofA test. CofA No.4637 issued 21.1.35; shipped to El Palomar, where erected and flown to Buenos Aires. Regn cld 12.36. Regd in Argentina as **R 276** 24.4.35 to Thomas Claud Worth, Buenos Aires. Regd 3.3.37 to Heriberto Van Deurs, Buenos Aires. Re-regd **LV-FCA** 2.7.38 to same owner (still regd 12.47). Regd 1.9.49 to Ricardo Brema. Regd 13.2.51 to Enrique Ortega. Regd 29.11.51 to Vito Politti. Regd 4.2.54 to Roberto Juarez. Regd 30.4.58 to Andres Cabrera, San Fernandoo. Regn cld 18.7.64.

Miles M.2H Hawk Major R 276 in Argentina in 1935. [*Via P Jarrett*]

144 M.2H Hawk Major; painted blue. CofA No.4604 issued 2.1.35 to H Mehta & Co, Bombay, India. Regd **VT-AGH** (CofR 249) 22.2.35 to Metro Motors, Bombay (believed to be owned by H Mehta & Co). Regd 25.1.36 to Jal H Mehta, Bombay. Flown by Noshir Manesk Gazdar (a Bombay Flying Club instructor) in the Viceroy's Challenge Trophy Air Race, over the Madras-Hakimpet-Bombay-Ahmedabad-Jodhpur-Delhi course but forced landed 195 miles from Madras on 14.2.36 and broke undercarriage (the aircraft was uninsured); repaired. Regn cld 24.4.37 as sold. Purchased by John M Gamble of Lambton Quay, Wellington who was the agent for Phillips & Powis Aircraft in New Zealand; arrived 7.6.37. Regd **ZK-AFK** (CofR 85/192) 19.5.37 to Wellington Aero Club, Rongotai. Impressed into RNZAF as **NZ578** 12.9.39. To 2 EFTS New Plymouth. Written off in forced landing after engine failure Whale Bay, nr Ohakea 5.5.41; en route Hobsonville to Ohakea, Flt Lt Johnstone & P/O James unhurt (some sources say aircraft was with 3 FTS).

145 M.2H Hawk Major; painted green with silver wings. Regd **G-ADBT** (CofR 5569) 31.1.35 to Miss Rosemary Rees, London SW1 (based Heston). CofA No.4670 issued 5.2.35; dd 26.2.35. Miss Rees 'flew it around Europe' and once to 'a field in nr Poznan in Poland' and photographs show it with two seats and also with the front cockpit faired over. Regd (CofR 8472) 23.4.38 to Stanley John Hawley, Draycott-in-the-Moors, Staffs (based Meir). Confirmed at Meir on outbreak of war; CofA lapsed 5.6.40. Regn cld at the 1.12.46 census. The remains of fuselage seen dumped at Blackbushe from 1950 to at least 15.7.53.

147 M.2F Hawk Major; painted red. Regd **G-ADCI** (CofR 5590) 8.2.35 to Turab Ali Khan Aga (an Indian or Persian student at Air Service Training, Hamble), Brooklands (later Hatfield). CofA No.4682 issued 9.2.35; dd 9.2.35. Regn cld 10.37 as sold. Regd in New Zealand as **ZK-AFM** (CofR 83/194) 28.5.37 to Western Federated Flying Club, New Plymouth. Name of club changed .38 to Wanganui Aero Club, Wanganui. Impressed into RNZAF as **NZ587** 10.10.39. To 2 EFTS New Plymouth. Engine failure and struck fence in forced landing at Kent Road Farm, 10 ml from New Plymouth 18.11.39 (still wearing its civil regn); pilot P/O CR Parker slightly injured. SOC 20.11.39.

Mlle Smaranda Braesco with her new Miles M.2H Hawk Major Coupé YR-ADB at Woodley on 2nd June 1935.

['*Flight*' *via B Clarke*]

Shell Mex & BP Ltd's green and silver Miles M.2H Hawk Major G-ADCF at Heston in September 1935.

[*J Edelsten via The Croydon Airport Society*]

148 Blank in the records kept by CLN and may have been reserved for a Hawk Major variant which was not taken up or built.

150 M.2H Hawk Major Coupe; painted red/blue. Long range single-seat version with canopy and long exhaust pipe. First flown, by Miles, 31.5.35 as a 'Hawk' (Mlle Braescu's) and on 1.6.35 he flew it as **YR-ADB**; handed over to customer 2.6.35. CofA No.4735 issued 26.7.35 to Mlle Smaranda Braescu, Bucharest, Romania; dd 27.7.35. Made a forced landing in Germany due to poor visibility in 10.35 and later the same month carried out another forced landing, this time due to engine trouble. Officially regd YR-ADB 10.3.36. Regn cld 23.5.40 and fate unknown. *(Note: this aircraft closely resembled the later c/n 194, which was designated M.2S and should therefore perhaps have had a new designation allotted)*.

152 M.2H Hawk Major; painted silver. Regd **G-ADCW** (CofR 5616) 19.2.35 to Arthur Sebag-Montefiore, London W1 (based Heston). CofA No.4683 issued 28.2.35. Spun in and crashed on turn to finals at Manston 28.4.35, killing both Sebag-Montefiore and passenger George Manchester Steavenson. Regn cld 12.35.

153 M.2H Hawk Major; painted red/green. Regd **G-ADCF** (CofR 5581) 6.2.35 to Shell Mex & BP Ltd, Croydon ("fixed" marks for Shell's aviation manager, Cecil Field). CofA No.4765 issued 2.4.35; dd 3.4.35 for use as a flying laboratory and flown by Richard 'Dick' Bentley. Sold 8.3.39 and regd 31.3.39 to Charles Desmond Brown, Skelmorlie, Ayrshire (based Carlisle & Brooklands). It was at Carlisle at the outbreak of war; CofA lapsed 28.4.40 and presumed stored. Regn cld 5.11.45 at census. Noted in a dismantled state at Squires Gate in 7.49 and burned in Guy Fawkes celebration 5.11.49.

154 M.2H Hawk Major; painted green. Regd **G-ADCJ** (CofR 5591) 13.2.35 to Albert Austen Voorsanger of Audenshaw (based Woodford). CofA No.4674 issued 16.2.35 (the ledger shows to Edward GH Forsyth, but this would seem to be a clerical error since he had G-ADCU). The true original ownership of this Hawk was the then well-known industrialist and MP for Mossley, Austin Hopkinson. Bob Milne noted 'Hopkinson' on a dual instruction flight in G-ADCJ on 23.2.35 and CLN says delivered to A Hopkinson 26.2.35. However, G-ADCJ was flown in to the official opening of Walsall aerodrome on 6.7.35 by Austen Voorsanger and it is possible that Hopkinson used this name as a pseudonym to avoid recognition. On the other hand, it is possible there was a 'real' Voorsanger and he may have simply been Hopkinson's pilot. Regd (CofR 6758) 21.2.36

to Austin Hopkinson MP, Audenshaw (based Woodford and Barton; to Heston 7.9.37). Overturned in forced landing in fog Fiddlers Ferry, Penketh, nr Warrington 10.3.36, wings and propeller damaged; Hopkinson slightly injured. Confirmed at Heston on outbreak of war; CofA lapsed 12.3.40. Fate unknown and regn cld 3.11.45 at census.

155   M.2H Hawk Major; painted green. CofA No.4732 issued to Aero Club of South Australia 5.3.35; dd 7.3.35. Regd **VH-UAI** (CofR 530) 18.6.35 to Aero Club of South Australia Ltd, Parafield. Owner's name changed to Royal Aero Club of South Australia Ltd. Crashed Port Lincoln, SA 1.5.36. Regn cld 18.6.36. Regd 25.11.36 to same owner (following repairs). Flown by H Plumridge in the Brisbane to Adelaide Air Race 16.12.36. (Some time named *Queen Adelaide*). Sold 19.11.40 to GH Michell, Hindmarsh, SA and regn cld same day. Handed over by owner to RAAF at 1 EFTS Parafield 10.2.41 (together with VH-ACC) and given serial **A37-5**. Struck fence on take-off 4.6.41 causing major damage to starboard mainplane and u/c leg; repaired. To 1 CF 28.3.42. To 1 Communications Unit, Laverton 22.5.44. Issued to Point Cook 26.7.45 for disposal by Commonwealth Disposals Commission. A letter was sent to GH Michell offering return of aircraft as no longer required by RAAF and aircraft transferred 5.8.45 to Central Flying School, Point Cook. Transferred back from CFS to Commonwealth Disposals Commission 8.2.46 for issue to W Schutt, Fisher Pt, Ascot Vale, Melbourne acting for GH Michell, 19.8.46. Regd **VH-UAI** 14.10.46 to GH Michell, Hindmarsh, SA. Regd 16.1.47 to HP Davis, t/a Hamilton Downs Pastoral Co, Hamilton Downs Station, Alice Springs, NT. Regn cld 19.12.47 at census. Regd 28.1.48 to Kurt Gerhardt Johannsen, Alice Springs. Stalled and crashed on landing at Mount Eba, SA at 6.55 pm on 10.3.48; pilot Kurt Johannsen slightly hurt and wife Kathleen Johannsen seriously injured. Regn cld 7.5.48.

Miles M.2H Hawk Major VH-UAI at Alice Springs 'Aerodrome' (as written on the back of the original photograph). [Via P Jarrett]

156   M.2M Hawk Major Three-seater; painted white overall. Fitted with cabin over both rear seats. Regd **G-ADCV** (CofR 5615) 20.2.35 to John Eric Duncan Holder, Croxley Green, Herts (based Heston). CofA No.4691 issued 6.4.35; dd 12.4.35. Crashed Weston Zoyland 2.6.35 (ex Hanworth); returned to Woodley for repairs. Flown by Miles 6.7.35 as 'Hawk CV - (Long distance); loaded 95 gals petrol'. Later based by owner at Denham, where located at outbreak of war. CofA lapsed 20.3.40. Later stored in a garage opposite the de Havilland factory at Hatfield where it was discovered by Ron Paine, to whom it was sold 13.6.44 and he took it to Wolverhampton by road (possibly his home at The Corner House, Murchall Road, Penn). Regd 23.6.44 to Ronald Royal Paine, Wolverhampton. Ron then set about rebuilding it, fitting new windscreens and rear canopy and repairing the ply covering which had been damaged while in store. Repainted black with white wings and trim, the CofA was renewed 16.3.46. Flown by Paine in the Siddeley Trophy Air Race (Race No.23), in the first post-war air races held at Lympne over the weekend of 31.8.46/1.9.46, and gained 2nd place at 129.5 mph (Paine's first post-war race). Then fitted with 2 x 16 gallon fuel tanks in the wing roots (which gave it a 4½ hr endurance) enabling Paine to participate in some of the earliest post-war rallies to the Continent. Sold early .47 and regd 3.4.47 to Laurence Aelred Harris, London NW1, a retired RAF pilot who flew it to Bombay, India. After his arrival in India, he advertised it for sale in a Bombay newspaper and it was purchased by Lt Col Gordon H Wotton, an officer serving in India (who had joined the Karachi Flying Club to learn to fly for his amusement) who had decided that it would be better to fly back home to Southampton rather than wait for a troopship. Wotton flew the Hawk Major to Poona where the Indian Air Force allowed him the use of a hangar, whilst he learned to fly it properly. Wotton left Poona on 17.11.47 and flew via Rajkot, Karachi, Jiwani, Sharjah, Bahrein, Basra, Baghdad, LGH 3, Lydda, Nicosia, Rhodes, Calato, Athens, Araxos, Brindisi, Rome, Nice, Lyon, Toussus-le-Noble, Paris, arriving

Miles M.2M Hawk Major G-ADCV at Croydon after the hangar wall collapsed on it on 4th February 1950. [C Holland via B Clarke]

at Southampton on 14.12.47, where he cleared customs and refuelled with just enough light for him to fly to Hamble, his destination. He had covered 5,510 miles in 55 hrs 30 min flying time, consumed 390 gallons of petrol at a total cost of about £150. Cld 3.1.48 and regd 16.1.48 to Lt Col Gordon Henry Wotton, Netley Abbey, Southampton. Sold by Wotton when due for CofA renewal (30.3.48) to an unidentified purchaser intending to use it for charter flying; flown to Croydon to have the necessary work done. For sale by WS Shackleton Ltd 9.48 for £125; price reduced to £95 by 12.48. Cld 13.1.49 and regd 18.1.49 to Basil Henry St Andrew Hurle Hurle-Hobbs, Littlehampton (based Croydon). On 4.2.50, before re-certification, gales blew down the wall of a hangar at Croydon Airport and although not badly damaged, it was not repaired. Regn cld 11.11.59 by Secretary of State.

158   M.2H Hawk Major; painted orange. Regd **G-ADCY** (CofR 5626) 20.2.35 to Robert Christian Ramsay, Canterbury (based Bekesbourne). CofA No.4694 issued 13.3.35; dd 15.3.35. Operated by Ramsay's company, Airsales & Service Ltd. Regn cld as sold 26.6.40. Impressed 30.6.40 as **BD141** but not taken on RAF charge. Recorded 15.4.42 as 'Not impressed as privately owned' (File AVIA 2/1617). Probably one of airframes recovered by Ramsay at Bekesbourne post-war but found to be beyond repair.

159   M.2H Hawk Major; painted silver. Regd in Switzerland as **HB-ERI** (CofR 413) 4.5.35 to Rudolf Herzig, Altenrhein. CofA No.4733 issued 13.5.35; dd 21.5.35. On 19.5.35 Miles flew a 'Hawk' which he recorded as HB-IRA (a Douglas DC-3 registration) and which must have been this machine. Regd .37 to Dr Fritz Schindler, *(not the famous Schindler of Schindler's List)* Kennelbach, Austria; post-war Locarno (based Altenrhein & Locarno). New CofR 1021 issued .46. Modified at some time to have a large, extended, headrest. Wore name 'Flying Baby' (in 1950s). Regn cld 5.7.56 as sold in Germany.

Miles M.2H Hawk Major HB-ERI, seen at Magadino, near Locarno, in the late 1940s. Note the large headrest. [Via B Clarke]

161   M.2H Hawk Major; painted silver. Regd **G-ADEN** (CofR 5692) 22.3.35 to Phillips & Powis Aircraft (Reading) Ltd. CofA No.4736 issued 26.3.35. Regn cld 5.35 as sold. Sold to Jose Adriano Pequito Rebello, Gaviao, Lisbon, Portugal; handed over 29.3.35 and departed same day flown by Rebello and John Lawn (a Reading Aero Club instructor), on delivery to the owner's private airstrip NE of Lisbon; arrived 30.3.35. Regd **CS-AAL** .35. In 8.36, Rebello flew it to Tablada, Seville and placed himself and the Hawk

Anglo-Portuguese help. The Miles M.2H Hawk Major CS-AAL of José Rebello in Spanish Nationalist markings. The photograph was probably taken in 1938.
['The Aeroplane' via The A J Jackson Collection]

Major at the disposal of the Spanish Nationalists. He was still flying in Northern Spain in October 1938, when CG Grey, the editor of The Aeroplane reported him operating on communications and liaison duties. He noted that Rebello carried a large map of Spain taped to the top surface of the port wing where it could be easily read from the cockpit! Fate unknown.

162 M.2H Hawk Major; painted blue. Regd **G-ADCU** (CofR 5614) 18.2.35 to Edward Gordon Houstoun Forsyth (jointly purchased with his wife Mrs Ivy Forsyth), London SW7 (based Heston). CofA No.4689 issued 9.3.35; dd 12.3.35. Regd (CofR 7254) 11.8.36 to Major Henry Philip Lawton Higmann, St. Austell, Cornwall (based Hooton Park). (Higmann also had an address in Barcelona and the aircraft was officially recorded as based there in July 1936, just prior to the civil war but probably then returned to UK) Confirmed at Hooton Park on outbreak of Second World War (with CofA expiring 10.8.40) but not impressed and fate unknown (possibly burnt in the fire at Hooton on 8.7.40, but not officially recorded as such). Regn cld at 1.12.46 census.

164 M.2H Hawk Major; painted black. Regd **G-ADDC** (CofR 5631) 12.3.35 to Capt John Carne Hargreaves (of 2nd Battalion, Grenadier Guards), London SW3 (based Heston). CofA No.4734 issued 16.4.35; dd 16.4.35. Purchased 7.10.36 and regd (CofR 7398) 16.10.36 to "Leslie Charles Lewis" (but whose real name was Leslie Charles Stanynought), ostensibly for Lady Heath's flying

Miles M.2H Hawk Major G-ADGL, which later became K8626 for RAE Farnborough. Note the front cockpit faired over.
[Via P Jarrett]

Miles M.2H Hawk Major K8626 (previously G-ADGL) at Woodley prior to delivery to the RAE Farnborough on 20th February 1936.
[Via P Jarrett]

school in Dublin and ferried Croydon-Brooklands 7.10.36. In fact Stanynought had been retained to purchase aircraft by the Spanish Republicans and G-ADDC actually departed Brooklands the same day (with G-ADDU) on delivery to Spain. Regn cld 9.37 as sold abroad.

165 M.2H Hawk Major; painted green. Regd **G-ADGD** (CofR 5783) 18.4.35 to Willoughby Rollo Norman, London EC3 (based Heston). CofA No.4791 issued 23.4.35; dd 23.4.35. Crashed out of fuel into the King George V Reservoir, Ponders End at about midnight 21/22.6.35; pilot Anthony Norman and passenger Miss Jacqueline Paravicini (returning from dinner in Dorset) swam ashore unhurt. Regn cld 12.35.

166 M.2H Hawk Major; painted red top to fuselage and down sides of cowling, fin, rudder and trousers, with the wings and tailplane a light colour. A photograph taken on delivery at Woodley shows the front cockpit faired over. Regd **G-ADGL** (CofR 5791) 24.5.35 to John Patrick Wakelyn Topham, Folkestone (based Lympne). CofA No.4824 issued 24.5.35; dd 24.5.35. Regn cld at census 16.11.45. Actually bought back by Phillips & Powis Aircraft (Reading) Ltd and supplied to the Air Ministry for trials at RAE Farnborough under Contract No.492695/36 dated 27.1.36, for the sum of £662 10s; ITP dated 18.2.36. Painted silver and given serial **K8626**, the cockpit sides were deepened and a lamp was fitted behind rear cockpit. Flown by Bob Milne 18.2.36; dd by him to the RAE 20.2.36 and TOC by DTD there 2.3.36. Principally used wef 27.2.36 on 'wire barrage tests' by F/O Arthur Clouston and others to fly into cables to investigate the feasibility of using barrage balloons and their steel cables as a defence against modern metal aircraft. Also used for flap research and testing trailing edge static tubes. Following a final flight on 13.5.38, K8626 was 'sufficiently deteriorated for it to be grounded' and was transferred to ground instruction use as **1081M** 20.5.38. Fate unknown *(and the reports that, as G-ADGL, it was destroyed in an air-raid at Lympne in June 1940 are incorrect).*

167 M.2H Hawk Major; painted red/silver. Regd **G-ADFC** [CofR 5714) 22.3.35 to Randle Kynaston Lloyd Salusbury 'Lloyd' Mainwaring, St Asaph (based at Brooklands and at strip at home). CofA No.4737 issued 16.4.35; dd 16.4.35. Advertised in The Aeroplane 7.7.37 for £575 with 100 F/Hrs and probably sold to aircraft dealers WS Shackleton & Co Ltd. Regd (CofR 8190) 5.11.37 to William Fairweather, Muzaffarpur, Bihar, India (initially based at Heston). Regn cld 6.38 as sold. Fairweather shipped it to India, completely erected, as personal baggage in the hold of the ship he travelled in. On arrival in Bombay, he flew it 1,200 miles to his home in just over 9 hours. Regd **VT-AKG** (CofR 352) 31.5.38 to William Fairweather, Muzaffarpur. An agricultural and motor engineer, Fairweather flew the Hawk Major widely throughout India on business up until early 1942. Declining an offer from the RAF to purchase the Hawk, he took a Squadron Leader commission (at the age of 48) continuing to fly as a civilian on duties "siting" airfields around Calcutta and throughout South India. Aircraft slightly damaged 15.5.45 when its Service hangar at Bangalore collapsed during a cyclone but repaired locally in two weeks. Flown back to Muzaffarpur on termination of his services and repainted silver with red trim and named "Robin". Fairweather returned to England after Indian independence in 1947 and the fate of Hawk is unknown, although it was still fully airworthy at the time. Regn cld 26.1.48.

168 M.2H Hawk Major; painted green. Regd **G-ADIT** (CofR 5943) 12.6.35 to 2nd Lt Vincent Alexander Prideaux Budge (of the Grenadier Guards), Parkstone, Dorset, (based Heston). CofA No.4897 issued 28.6.35. Flown by Miles 30.6.35; dd 30.6.35; flown by owner on his posting to Alexandria, Egypt. Entered by owner for the International Circuit of the Oasis meeting, held 22-26 February 1937 at Cairo but was unplaced. Returned to Heston (by 9.37); CofA renewed 12.8.37 (and put up for sale). Still being advertised for sale 9.38 'fitted Pioneer blind-flying instruments, navigation lights, extra tanks, range 800 miles, Schwartz propeller, two compasses, and many other extras; total hours 340; nil since complete overhaul of engine and airframe'. Finally sold 25.5.39 and regd 21.6.39 to The Cotswold Aero Club Ltd, Staverton; CofA renewed 14.6.39. Regn cld 16.1.40 as sold. Impressed as **X5125** at 10 MU Hullavington 19.12.39. Released to the ATA White Waltham 19.7.41. Relegated to instructional airframe **3017M** and dd 4.4.42 to 1851 Squadron ATC (Suffolk Wing), Norton, Suffolk.

169 M.2F Hawk Major; painted red. Regd **G-ADGA** (CofR 5780) 12.4.35 to The British Instrument Co Ltd (Major Jack Savage), Hendon. CofA No.4777 issued 3.5.35; dd 18.5.35. Fitted with

Miles M.2H Hawk Major PK-SAR (with front cockpit faired over), seen at Surabaya, Dutch East Indies, in the late 1930s.
[Via R Brown]

Mrs. Battye and Mrs. MacDonald in the Miles M.2H Hawk Major G-ADLA at Hatfield for the 1935 King's Cup Air Race.
[Via B Clarke]

special Savage smoke-producing equipment for skywriting and operated by Savage Skywriting. Shipped to Juhu, Bombay and flown for skywriting by Sydney St. Barbe from 2.38 for some months from Rawalpindi in North of India to Ceylon. CofA lapsed 22.12.38 and noted at Juhu on Air Ministry survey 1.9.39. Regn cld 15.5.41 as sold, and possibly impressed locally.

172 M.2H Hawk Major; painted blue. *Recorded (by CLN) as allocated G-ADFN but this is unconfirmed and seems unlikely.* CofA No.4837 issued 18.5.35 to Aero Club de Valencia; dd 18.5.35 under ferry marks **EC-W44**. Regd **EC-DDB** .35 to Jose Albinana Ferre; operated by the Aero Club de Valencia *(Note - a drawing of c/n 172 appears on an Italian internet site in a red overall colour scheme with the regn EC-DDB and a note 'Frente de Aragon' August 1936)* Used by the Spanish Republicans in the civil war before being taken over by the Nationalists with serial **30-145**. Returned to Jose Albinana Ferre (Aero Club de Valencia) and regd **EC-CAS** 21.5.43. Regd **EC-ABI** .47 (still with same owner). Wfu Valencia .70 and reported stored by Ferre in a private hangar near Valencia and available for £600, with full log books and 1,000 hours on the engine. Moved to Liria (inland from Valencia) where the wings were removed and the fuselage left outside to gradually rot away until it broke in half behind the rear seat, tipping onto its Gipsy Major engine with the rear of what remained of the fuselage then being supported by a spar from the tailplane. Shortly afterwards, the remnants of the airframe were acquired by José Chicarro Villar, Gerona but before he could move it, only the wings, flaps, controls, engine bearers, oil tank, one cockpit lever and fuel selector valve had survived. These components were collected from Spain by Tim Cox .03 with the intention of restoring the aircraft to airworthy condition with a new fuselage. Regd **G-CCMH** 20.10.03 to (José's friend) John Aubrey Pothecary, Salisbury to cover its flight testing when finally rebuilt.
Note: Displayed in the Spanish Air Force Museum at Cuatro Vientos is the Hawk Trainer Mk III HB-EEB, masquerading as 'Hawk Major 30-145'! The rear windscreen has been modified to three-piece as on the front cockpit but the undercarriage remains 'single-leg' as in the Magister. Because EC-ABI took part in the Spanish Civil War on both sides, this 'hybrid' machine is now painted in Republican colours on the port side and Nationalist colours on the starboard side!

173 M.2H Hawk Major; painted blue. Regd **G-ADHF** (CofR 5867) 11.5.35 to Capt Henry J 'Jerry' Shaw, London W1 (based Woodley). CofA No.4826 issued 11.5.35. Regn cld 9.35 as sold. Jerry Shaw was aviation manager of Asiatic Petroleum Co (a Royal Dutch Shell company) at the time and probably ordered the Hawk Major on behalf of its true owner. Handed over to Jhr HL Krayenhoff 12.7.35 and shipped to the Dutch East Indies, where regd **PK-SAR** 9.35 to Jhr H L Krayenhoff, Surabaya. The front cockpit was temporarily faired over at some time. Regd .39 to Tj de Cock Buning. Requisitioned by ML/KNIL in 1941 and scrapped Tasik, Malaja in 3.42. A known photograph shows PK-SAR in a light-colour with the name '...sudski' (the first part is unreadable) on the cowling and another of the same machine claims that the photograph was supplied by 'Victor Warner, son of the owner' and that it was taken at Tjililitan airstrip, Batavia in 1939. The caption also states that the aircraft was destroyed in 1942 to stop it falling into the hands of the Japanese.

174 M.2H Hawk Major; painted blue. Per CLN allocated to Interprinderile Technica Romane but apparently ntu and stored. CofA No.4884 issued 19.5.36 to Capt Spiridon Varnav, Bucharest, Roumania; dd as **YR-BOB** 27.5.36 (which regn may have been allocated 6.35) and officially regd 23.9.36. No further details.

176 M.2H Hawk Major; painted orange. Regd **G-ADLA** (CofR 6025) 8.7.35 to Mrs Elise Battye, Bracknell (based Woodley). CofA No.4883 issued 20.7.35; dd 22.7.35 but actually handed over by Charles Powis at a ceremony at Woodley on 31.7.35. Named *Aliboo* after her pet dachshund. Entered and flown by Mrs Battye in the King's Cup Air Race at Hatfield 6/7.9.35 (Race No.29), but unplaced. Regd (CofR 7127) 15.6.36 to Capt. James Knox Mathew (of the Irish Guards) (based Heston). Flown by Mathew in the 1936 King's Cup Air Race (Race No.23) although formally entered by Mrs Battye (presumably before she sold it); undershot on landing at Shoreham, during the elimination round 10.7.36, causing the u/c to be bent back; repaired. Entered for the International Circuit of the Oasis meeting, held Cairo 22 - 26 February 1937, but unplaced. Advertised for sale 5.37. Regd (CofR 8789) 4.10.38 to James M Rout, Roodepoort, Transvaal (based Brooklands). Regn cld 1.2.39 on sale in South Africa. Regd **ZS-APX** 6.2.39. Impressed into SAAF as **2008** in 3.40. Relegated to ground instruction as **IS118** at 73 Air School, Wonderboom.

177 M.2H Hawk Major; painted green. Regd **G-ADMW** (CofR 6124) 30.7.35 to Willoughby Rollo Norman, London EC3 (based Heston). First flown, by Charles Powis, 31.7.35. CofA No.4982 issued 1.8.35; dd 2.8.35. Probably jointly owned with brother, Antony Charles Lynyard Norman, who entered and flew it in the King's Cup Air Race at Hatfield 6/7.9.35 (Race No.28), to gain 20th place at an average speed of 139.24 mph. Regd (CofR 7958) 24.6.37 to RK Dundas Ltd, Portsmouth (possibly operated by Portsmouth Aero Club). Regd (CofR 8434) 5.4.38 to Portsmouth Aero Club Ltd, Portsmouth. Its last flight prior to the outbreak of war was made on 30.8.39, following which it was stored at the National Garage, Twyford Ave, Portsmouth (by 6.11.39). Moved to EMA Garage, Grove Rd, Southsea by 10.6.40 (but made two short test flights on 4th and 5th July 1940). Requisitioned for RAF service 27.1.41 and to 5 MU Kemble from Woodley 14.2.41. Regn cld as sold 15.2.41. Formally impressed 15.2.41 as **DG590**; painted in full service markings 21.3.41 and air tests at 5 MU 10/12.4.41. Flown to Herts & Essex Aviation Ltd, Broxbourne for major inspection 22.8.41; air test after overhaul 6.12.41; departed Broxbourne 10.12.41; to 46 MU Lossiemouth 2.1.42. To 5 MU Kemble 8.10.42. To HQ No.2 Group, Wyton Stn Flt 24.1.43. To RAF Swanton Morley Stn Flt 17.5.43. Damaged in taxiing accident Northolt when tailwheel caught in some hidden netting 12.7.43; repaired and flown from Northolt to 2 Group HQ Comm Flt, Swanton Morley 22.8.43. At 5 MU Kemble 4.12.43. SOC for spares use (Cat E1) 27.5.44 but brought back on charge 6.2.46 (probably error for 6.2.45). *(An undated document (ref: AVIA 57/28), titled 'Surplus Aircraft held by MASC at 5 MU Kemble' records that DG590 was 'u/s - for spares only')*. Sold to Miles Aircraft Ltd for £100 in 12.45 at the first sale of impressed light aircraft held at 5 MU Kemble. Regd **G-ADMW** 17.9.46 to Miles Aircraft Ltd, Woodley. 30-min test flight following overhaul 15.4.47; CofA renewed 25.4.47. Regd 12.6.47 to Reading Aero Club Ltd, Woodley. Cld 6.2.48 and regd 8.3.48 to T Shipside Ltd, Tollerton and also flown as a single-seater. Last flight for nearly the next 3 years was made on 12.6.48; CofA lapsed 24.3.49 and stored. Cld 14.5.51 and regd 18.5.51 to Arthur Edward Henry Coltman, Braunstone, Leicester; flown by Coltman from Leicester East from 19.5.51; CofA renewed 26.5.51. Two photographs of it taken in 1951 (both with Race No.44), show it in normal configuration and then with the front cockpit faired over and with a headrest fitted. Regn cld 6.5.52 as sold abroad to USA and to White Waltham on that date (possibly with unknown American owner – but not put on US register). CofA lapsed

25.5.52. Regd 2.9.52 to John Paul Gunner, Winchester, Hants (based Denham and White Waltham). Three-year CofA renewed 20.5.53. Damaged when taxied 'out of control' (by Gunner) into motor car at Croydon Airport at 4.58 pm on 21.5.53, sustaining damage to port wing tip and propeller. Repaired by WA Rollason Ltd, Croydon; test flown after repairs on 19.8.53. Gunner flew G-ADMW regularly, visiting Redhill on 25.8.53 and 15.11.53 (and where it had an engine run 30.7.54 implying that it had been there for a while). Further trips made to Le Touquet, Stuttgart and Kassel in 7.56. On 26.6.57, damaged following in-flight fire, due to a defective exhaust, off the East Frisian Islands; repaired. On 3.7.60, Gunner flew it to Sleap, where it was based until 4.64 (owner is believed to have been flying Vampires for Marshalls of Cambridge on contract to the School of Air Traffic Controllers at Shawbury). Returned to Denham 4.64 and thence to Westland's hangar at White Waltham - emerging on 16.5.64 to fly to France. To RAF Ternhill (by 21.6.64) (then the home of the Helicopter element of the CFS) and grounded for the next year. In June/July 65, it made 3 flights from Ternhill prior to CofA expiring on 30.7.65. One last flight, of 1 hr 25 min, was made on 1.8.65 before John Gunner presented it (seen in retrospect as a very bad move) to the Air Historical Branch 2.8.65. In August/September 1965 it was repainted in 'wartime RAF markings' (early 1940 RAF colour scheme) as DG590 at RAF Ternhill, making a 50-minute evening flight there 16.9.65, to be displayed at their Battle of Britain Open Day on 18.9.65, when it was flown by the station's Chief Flying Instructor. Flown by de Havilland test pilot Des Penrose at Ternhill 30.9.65 and then flown to RAF Henlow by W/C (later AVM) GC Cairns in 11.65. Stored at Henlow, in a leaking hangar with other aircraft destined for the RAF Museum and allotted instructional airframe serial **8379M** on 5.9.73 (implying that by then it had officially come on to RAF charge). Regn cld 16.9.86 by CAA. Remained at RAF Henlow until loaned 10.11.88 to the Museum of Army Flying, Middle Wallop (to represent a Miles Magister, showing the training phase of pilots for the Glider Pilot Regiment). Taken by road 8.91 to the RAF Museum Reserve Collection and Restoration Centre, RAF Cardington. Moved by road to Rotary Farm in 4.93 for a survey to be carried out by Tim Moore of Skysport Engineering Ltd (following a letter sent by him to Dr Michael Fopp, Director of the RAF Museum, offering a free survey with a view to returning the aircraft to airworthy condition for use by Dr Fopp). Nothing came of this proposal and aircraft returned to RAF Cardington 10.93 for further storage. Formally donated 3.8.98 by the Ministry of Defence to the RAF Museum. Following the closure of RAF Cardington, taken by road on 26.1.00 to the temporary RAF Museum Reserve Collection and Restoration Centre at RAF Wyton. Confirmed by the RAF Museum that, due to the Centre's many other commitments, it is unlikely that G-ADMW will be restored to static display condition in the foreseeable future. No conservation or restoration work of any description has been carried out since it was handed over by John Gunner in August 1965 and the RAF Museum's formal position remains; *'there are a considerable number of more interesting truly military aircraft, much more worthy of restoration'*. Subsequently moved to the RAF Museum's 'deep storage facility' at RAF Stafford for yet further storage, although it has been announced that RAF Stafford is earmarked for closure in the not too distant future..

178 Blank in CLN records and nothing known; possibly a replacement fuselage for use in a rebuild.

The sad remains of G-ADMW in deepest storage at the RAFM's Reserve Collection, Stafford, in 2007. [D Pearce]

179 M.2H Hawk Major; painted silver. Built in mid 1935 but not completed until later. Regd **G-AEEZ** (CofR 6856) 30.3.36 to Phillips & Powis Aircraft Ltd. CofA No.5419 issued 14.4.36; dd to Phillips & Powis School of Flying 28.4.36. Regd (CofR 8315) 24.1.38 to Reading Aero Club Ltd, Woodley. Regd (CofR 8526) 28.5.38 to Staffordshire Airplanes Ltd, t/a North Staffordshire Aero Club, Meir. Wrecked in crash on landing in gale Barton 8.3.39; Percy William Fell seriously injured. Regn cld 1.12.46 at census.

180 M.2H Hawk Major; painted red. Regd **G-ADDU** (CofR 5671) 15.11.35 to Ernest John Lambert, Byfleet (based Brooklands); probably as nominee for Dutchman Herman Theodor Van Marken. CofA No.5215 issued 15.11.35; dd 24.11.35. Regd (CofR 7574) 4.1.37 to "Leslie Charles Lewis" (actually Leslie Charles Stanynought, acting on behalf of the Spanish Republicans). Actually departed Brooklands 7.10.36 on delivery to the Spanish Republicans (although ostensibly for delivery to Lady Heath's flying school in Dublin). On 6.8.37, Stanynought handed the logbook over to the Croydon police saying 'I no longer need this'! Regn cld 9.37 as sold abroad.

182 M.2H Hawk Major; painted black. CofA No.4983 issued 13.8.35 to Interprinderila Technica Romana, Roumania; dd 29.8.35 as **YR-ITB**. Regd **YR-FIL** 23.1.36 to Teodor Filip. Later regd to Ion Cociasu. Believed flown to Bombay in 'raid' (fly-in competition) in 1939. Fate unknown.

183 M.2H Hawk Major; painted blue. Regd **G-AEFA** (CofR 6857) 30.3.36 to Phillips & Powis Aircraft Ltd. CofA No.5420 issued 28.4.36. Regn cld 4.36 as sold; dd 2.5.36. Regd **ZS-AHH** 3.7.36 to African Air Transport Ltd, Potchefstroom. Overturned on landing and wrecked Potchefstroom 23.10.39; pilot J McArthur and passenger injured.

Miles M.2H Hawk Major ZS-AHH, probably at Potchefstroom in the late 1930s. [Via P Jarrett]

184 M.2H Hawk Major; painted silver. Regd **G-ADZU** (CofR 6590) 19.12.35 to Lt Richard Hatt Noble Graham, High Wycombe (based Heston). CofA No.5286 issued 11.3.36. Regd (CofR 6826) 14.3.36 to Phillips & Powis Aircraft Ltd; dd 22.3.36 to Phillips & Powis Air Hire, Hanworth. Crashed at Linz, Vienna 15.5.37, while flown by Austrian pilot A Lasnavsky giving an evening pleasure flight to J Nadir-Tata. After performing 3 circles at low altitude over his parent's home nearby, he lost control and spun into the garden of a neighbouring house, killing both occupants. Regn cld 12.37. (Note: CLN has c/n 207 as 'G-ADZU Fuselage No. for Rebuild' under Job No. H3768; since c/n 207 is dated to early 1936, this might imply an earlier rebuild and may account for why the original owner sold it back to P&P).

185 M.2H Hawk Major; painted blue. Regd **G-AEKJ** (CofR 7066) 23.6.36 to William Foster, Fazenda Estrella, Ligacao, Matto Grosso, Brazil (based at Fazenda Estrella). CofA No.5574 issued 4.7.36; dd 14.7.36. Fate and local registration (if any) unknown. Regn cld 27.12.45 at census.

186 M.2H Hawk Major; painted silver. Regd **G-AEGP** (CofR 6933) 24.4.36 to Phillips & Powis Aircraft Ltd. CofA No.5456 issued 28.5.36; dd to Phillips & Powis Flying School 10.6.36. Hit hedge on take-off from field nr Tonbridge, Kent 1.11.36 en route Woodley to Sevenoaks; pilot Frederick George Malin slightly injured; repaired. Regd (CofR 8316) 24.1.38 to Reading Aero Club Ltd, Woodley. CofA lapsed 27.7.39; stored at Woodley. Regn cld 14.3.41 as sold. Impressed into RAF 10.3.41 and dd to 5 MU Kemble the same day.

Released to the ATA White Waltham with the serial **DP851** (which duplicated that of the Spitfire XV prototype). Relegated to instructional airframe **3016M** and dd to 1889 Squadron ATC (Berkshire Wing), Farringdon, Berks 4.4.42.

188 M.2H Hawk Major; painted silver. Regd **G-AEGR** (CofR 6934) 24.4.36 to Phillips & Powis Aircraft Ltd. CofA No.5457 issued 17.6.36; dd to Phillips & Powis School of Flying 26.6.36. *(George Miles recorded flights in 'Falcon' G-AEGR on 27.8.36 from Woodley-Heston-Woodley-Heston-Woodley and again on 14.10.36 from Woodley - Croydon - Woodley - Hanworth - Woodley - Hanworth - Woodley; it is possible that the Falcon he flew had the incorrect regn applied, or was he actually flying a Hawk Major!).* Crashed Bucklebury Place, Upper Woolhampton, nr Reading 27.4.37; pilot Allan Anderson injured and passenger Peter J Burra killed. Regn cld 12.37.

190 M.2P Hawk Major de luxe; painted red. Regd **G-ADDK** (CofR 5657) 21.5.35 to Stanley Kenneth Davies, Cardiff (based Pengam Moor, Cardiff). First flown, by Miles, 1.6.35. CofA No.4896 issued 12.6.35; dd 22.6.35. Sold 8.36 and regd (CofR 7403) 20.10.36 to The Cardiff Aeroplane Club Ltd, Pengam Moor. At Airwork Ltd, Whitchurch for CofA renewal (on Air Ministry survey 29.8.39); CofA renewed 13.11.39 and returned to Cardiff on Permit JP86 dated 9.11.39. Regn cld as sold 5.8.40. Impressed into RAF as **BD180** 5.8.40; to 38 MU Llandow 6.8.40. Released to the ATA White Waltham 8.7.41. To 5 MU Kemble 12.4.42. To 282 Squadron, Castletown 23.3.43 and remained at Castletown when the squadron was merged into 281 Squadron in 11.43. To RAF Davidstowe Moor 2.44. To 275 Squadron, Valley 10.3.44 and moved with the squadron to RAF Warmwell 4.44; RAF Bolt Head 8.44, RAF Exeter 10.44 and RAF Harrowbeer 1.45. The squadron disbanded 14.2.45 and the Hawk Major was SOC for spares use (Cat E1) 24.2.45.

194 M.2S Hawk Major Long-range single-seat cabin version; painted yellow. As M.2H with short exhaust stubs and different shaped u/c trousers. Regd **G-ADLH** (CofR 6038) 10.7.35) to John Henry 'Jack' Van (a 20 year old Canadian), Broxbourne. First flown, by Miles, as a 'Hawk (Atlantic Flight)' 17.8.35. CofA No.5020 issued 21.8.35; dd Broxbourne the same day. Intended by Van (who was a pilot for Shell Co of West Africa, based in Lagos) to fly the North Atlantic, but this was refused by the authorities and the aircraft returned to store at Woodley. CofA lapsed 20.8.36. Regd (CofR 8069) 23.8.37 to Govind Paramaswaran Nair, London NW3 (based Woodley). CofA renewed 26.8.37. Nair left Croydon 28.10.37 for an extended tour to include parts of America and at least two ocean crossings but got no further than Rouen in France, where he stalled on the approach to a field near Forges-les-Eaux and was killed in the ensuing crash. Regn cld 1.38.

198 M.2H Hawk Major; painted silver. Regd **G-AENS** (CofR 7347) 18.9.36 to Phillips & Powis Aircraft Ltd. CofA No.5673 issued 28.9.36; dd 2.10.36 to Phillips & Powis School of Flying. Regd (CofR 8317) 24.1.38 to Reading Aero Club Ltd, Woodley. CofA lapsed 27.2.40. Regn cld 15.3.41. Impressed into RAF 10.3.41 as **DP848**. To 5 MU Kemble 23.3.41. Released to the TFPP, ATA White Waltham 25.5.41. Crashed on a hillside 1½ miles west of Priddy, nr Wells, Somerset at 13.50 hrs on 20.11.41 and destroyed by fire, pilot S/O JR Baker killed; cause not established but possibly fuel starvation and/or persisting in flying in bad weather.

199 M.2H Hawk Major fuselage used in rebuild of G-ACZJ, c/n 122.

203 M.2T Hawk Major Single-seat, long range; painted red/cream. (P&P works records state 'Subs to Type M.2H G-ACYZ, but with Cirrus Major 'R' engine; engine supplied free by Blackburn for tests under racing conditions' - no details can be found for this engine, which was probably a modified 150 hp Cirrus I) Regd **G-ADNJ** (CofR 6160) 12.8.35 to Alexander Adolphus Dumfries Henshaw, Trusthorpe, Mablethorpe, Lincs (based Mablethorpe). CofA No.5045 issued 30.8.35; dd 2.9.35. Entered and flown by Alex Henshaw in the 1935 King's Cup Air Race held at Hatfield 6/7.9.35 (Race No.24) but, after leaving Ireland on 6.9.35, crankshaft failure caused the propeller to part company from the engine and he ditched into the Irish Sea, nr Donaghadee, 6 miles off Malin Head. Alex was rescued and the wreck was salvaged to Androssan, Glasgow, and thence to Renfrew (where presumed scrapped). Regn cld 1.38.

204 M.2P Hawk Major de luxe; painted blue. (P&P works records state 'Subs to G-ADDK', and probably laid down as M.2H). Regd **G-ADIG** (CofR 5923) 31.5.35 to Major Roland Hobhouse Thornton (a director of shipping companies Alfred Holt & Co and Glen Line Steamships), Liverpool (based Speke). CofA No.4901 issued 28.6.35; dd 1.7.35. Whilst flying from Budapest to Hamburg on 8.6.36, the propeller fractured and the subsequent out of balance force whipped the engine clean out of the airframe! The aircraft went into a flat spin and crashed into a pine copse on a hillside nr the Bohemian village of Jedlova, Czechoslavakia; the Major and his wife receivied only minor injuries. Regn cld 12.36.

205 M.2H Hawk Major; painted cream. (P&P works records state 'conforms to G-AEGR with the exception of two 15 gal aluminium fuel tanks (approved mod) and aluminium leading edge oil tank (approved mod) acrobatic category (approved mod)'). Regd **G-AEOX** (CofR 7428) 5.11.36 to William Leslie Gordon, Southampton (based Eastleigh). CofA No.5702 issued 7.11.36; dd 12.11.36. Regn cld as sold in 1.37. Regd in France as **F-APOY** (CofR 5069) 18.3.37 to Marie Nicole & Jeanne Gilbert, Paris. Regd 4.38 to Frederic Andre Engel, Paris. (CLN records to Mme Hotelais *a'Hotelans* but with no date).

207 M.2H Hawk Major – CLN recorded as fuselage used in rebuild of G-ADZU (c/n 184) under Job No.H3768 but actual use unconfirmed; (the page relating to c/n 207 in P&P works records is missing).

208 M.2H Hawk Major CLN recorded as Hawk Major with no further details. Possibly allocated to new aircraft which was not built, or for a replacement fuselage for the rebuild of a crashed aeroplane.

218 M.2H Hawk Major CLN recorded as fuselage sent to 'Aircraft Trade Exhibition, Liverpool' with Job No.H3895. No further details but would have been built in early 1936; dd in 4.36.

219 M.2F Hawk Major CLN recorded as fuselage used to rebuild G-ACXL (c/n 114) under Job No.H3897.

Miles M.2P Hawk Major de luxe G-ADLO outside the new Assembly Building in July 1935. ['Flight' via B Clarke]

220 M.2P Hawk Major de luxe; painted blue. (P&P works records state 'Subs to G-ADDK'; fitted with hydraulically operated split flaps). Regd **G-ADLO** (CofR 6058) 15.7.35 to Airwork Ltd, Heston Air Park. CofA No.4953 issued 23.7.35; dd 26.7.35. Regn cld 5.37 as sold. Regd **ZK-AFL** (CofR 86/188) 5.7.37 to Marlborough Aero Club, Blenheim (based Omaka). Damaged nr Port Waikato 29.6.39; returned to Blenheim by road and ship for repair. CofA reportedly renewed 4.7.39 (implying repaired) but taken over by RNZAF 14.10.39 in damaged condition and impressed as **NZ588** 5.10.39. Not repaired and SOC Rongotai 6.40 as reduced to produce.

221 M.2H Hawk Major Regd **G-AFKL** (CofR 8776) 5.10.38 to Reading Aero Club Ltd, Woodley. CofA No.6406 issued 12.11.38. (Aircraft actually built in 1935 and stored). Whilst being flown solo on 29.3.39 by Paul Uphill (a Police Constable from Wallingford, who had obtained his 'A' licence on 1.3.39) he lost control while attempting an unauthorised aerobatic manoeuvre and dived inverted at high speed into the ground at Woodley, with fatal results. Regn cld 28.8.39.

222 M.2T Hawk Major de luxe; painted cream. (P&P works records state M.2T Hawk Trainer 'Subs to G-ADNJ'). Regd **G-ADNK** (CofR 6161) 12.8.35 to Francis Delaforce Bradbrooke, London W14 (based Heston). CofA No.5046 issued 30.8.35; dd 5.9.35. Flown by Bradbrooke (but entered by F/O SA Sadler) in the 1935 King's Cup

Air Race at Hatfield on 6/7.9.35 (Race No.23); forced landed nr Blackpool 6.9.35 after engine failed over Irish Sea due to broken valve spring. Regn cld 10.35 as sold. Returned to Woodley and Works records then state 'c/n 222 OE-DKA M.2W Cirrus Major engine replaced by Gipsy Major engine; inverted flying system fitted. Subs to G-ADVF, with exception that larger area horn balanced rudder and elevator trim in front cockpit is not fitted. Acrobatic weight 1,650lb' *(although it was referred to as an M.2W, it appears to have more closely resembled an M.2Z Hawk Trainer Mk.II)*. Regd in Austria as **OE-DKA** to Smoliner & Kratky, Vienna. Amended CofA issued 7.3.36 and collected by Capt. von Brismowski 3.36. Advertised for sale at Wien-Aspern 1.3.38. Following the occupation of Austria by Germany in 1938, it was regd **D-EBNF** 2.39 with the owner shown as Leopold Schneider, Vienna.

223 M.2H (?) Hawk Major. CLN recorded as Hawk Major 'at Service with Job No. H3987' and probably to Phillips & Powis Repair & Service Dept for a repair or rebuild.

225 M.2H (?) Hawk Major. CLN recorded as Hawk Major with Job No.H3988; possibly allotted for a new aircraft which was not built or for a replacement fuselage for the rebuild of a crashed aeroplane.

227 M.2H (?) Hawk Major; unpainted. (P&P works records state that allotted to Kuala Lumpur Flying Club and supplied in component parts; dd 1.11.35). A contemporary report of the Kuala Lumpur Flying Club for 10.35 stated that a new Miles Hawk Major (minus engine) and a new DH Hornet Moth *(which was VR-RAI)* have been ordered from England, and it is hoped to have these aircraft in the air by the end of the year. Since no regn is known (or possible), this could have been used to rebuild M.2J (?) Hawk Major **VR-RAH** c/n 125 (which may have been crashed). Regn cld 9.3.38.

251 M.2P Hawk Major de luxe; painted red. (P&P works records state 'Subs to G-ADDK'). First flown as **U2,** by Bob Milne, 4.10.35 on CofA test. CofA No.5073 issued 7.10.35 to AC Lewin. Regd **VP-KBT** (CofR 45) 24.11.35 to Brig Genl Arthur Corrie Lewin, Njoro, Kenya; flown by owner to Njoro late 10.35/early 11.35. Believed returned to England in 1936/37; CofA renewed 9.6.37, prior to shipment to New Zealand by sea 7/8.37. Regn cld as sold New Zealand 4th qtr .37. Regd **ZK-AFJ** (CofR 91/193) .37 to Brig Genl Arthur Lewin; erected and reflown Wigram 21.8.37. Regd 21.8.38 to Canterbury Aero Club, Christchurch. Impressed into RNZAF as **NZ589** 13.10.39; to 2 EFTS New Plymouth. To 3 FTS Ohakea. To 1 (B) OTU Ohakea (or 1 FTS). Crashed Ohakea in 1943. SOC as reduced to parts Rongotai 8.7.43.

267 M.2P Hawk Major de luxe; painted blue. (P&P works records state 'Conforms to G-ADDK with the exception of two 22 gal aluminium fuel tanks'). Regd **G-AEGE** (CofR 6914) 17.4.36 to Dr Oliver Francis Haynes Atkey, London W6 (based Heston). CofA No.5463 issued 27.4.36; dd 29.4.36. Flown by Dr Oliver & Mrs (Dulcibella) Atkey to Greece in late 5.36. Based at Woodley at time of Air Ministry survey 12.9.39. Regn cld as sold 15.9.41. Impressed into RAF as **HL538** 14.10.41; to Station HQ Hunsdon. Damaged in crash 21.11.41; to Herts & Essex Aviation, Broxbourne for repair; to 29 MU High Ercall 10.2.42. Issued to ATA and undercarriage slightly damaged at Kemble on 1.8.42 at 16.15 hrs when swung violently on landing due to excessive friction in port axle bearing; ferry pilot T/O JH Stubbs; repaired on site. Flown by Flt Lt Hugh Kennedy from Woodley on a 'service test' 3.9.43; dd to RAF Watchfield 5.9.43 to become the personal mount of Wing Cdr HF Jenkins, the CO of 1 Beam Approach School. To School of Flying Control, Watchfield in 1946; painted silver and coded 'FD-YZ'. To 5 MU Kemble 27.8.47 for disposal. Sold to Aeronautical Educational Trust Ltd in 1948 and dd to Chelsea College of Aeronautical and Automobile Engineering, Wimbledon 15.4.48 for instructional purposes. Moved to Redhill Aerodrome when the college re-located there in 11.49. Acquired by John Gunner for use as spares for G-ADMW in 1954.

An M.2H Hawk Major was reserved as **VH-ACD** about 11.38 but the regn was NTU and it remains unidentified. Reservation cld 30.6.43.

Summary of Hawk Major production:

M.2F Hawk Major	c/n 36, 109, 110, 111, 112, 113, 114, 115, 116, 118, 119, 121, 126, 127, 134, 147, 169	17
M.2G Hawk Major Cabin	c/n 120	1
M.2H Hawk Major	c/n 122, 123, 124, 129, 132, 137, 138, 139, 143, 144, 145, 150, 152, 153, 154, 155, 158, 159, 161, 162, 164, 165, 166, 167, 168, 172, 173, 174, 176, 177, 179, 180, 182, 183, 184, 185, 186, 188, 198, 205, 218, 221, 227	43
C/n's allocated for rebuilds of M.2F and M.2H		
c/n	199 for rebuild of c/n 122	
c/n	207 for rebuild of c/n 184	
c/n	219 for rebuild of c/n 114	
M.2J Hawk Major Cabin?	c/n 125? (possibly)	1
M.2M Hawk Major Cabin	c/n 128, 156	2
M.2N Hawk Major de luxe	c/n 135	1
M.2P Hawk Major de luxe	c/n 190, 204, 220, 251, 267	5
M.2S Hawk Major	c/n 194	1
M.2T Hawk Major	c/n 203, 222	2
Total Hawk Major variants built		73

Miles M.2P Hawk Major de luxe ZK-AFJ in New Zealand in the late 1930s. [M E Kirkus Collection]

Appendix 12:
Miles M.2R Hawk Trainer I and M.2W, M.2X, M.2Y and M.2Z Hawk Trainer II Production

210   M.2R Hawk Trainer Mk.I single-seat; painted cream. Initially intended to be a modified M.2H Hawk Major de luxe by which it was also generally known (P&P works records state 'Subsequent to G-ADLN'). Regd **G-ADLB** (CofR 6026) 9.7.35 to Major George William Graham Allen (a director of Phillips & Powis Aircraft Ltd) of The Elms, Iffley, Clifton Hampden (based at Woodley and at private field at Iffley). Almost certainly first flown, by Miles, 31.7.35 (who just recorded 'Hawk Trainer' in his log book). CofA No.4940 issued 20.8.35; dd 30.8.35 (operated as company demonstrator) Entered by Major Allen and flown by Owen Cathcart Jones in the 1935 King's Cup Air Race held at Hatfield on 6/7.9.35 (Race No.32), to gain 3rd place at a speed of 157.52 mph. For sale by P&P 10.35 for £700. Regn cld 7.36 as sold. To Argentina and regd **R302** 21.2.36 to Eduardo Torney. Re-regd **LV-YCA** 1.7.38 to same owner. Regd 6.7.43 to Jorge Iturrospe. Regd .43 to Desiderio Jose Echeverry (or Echeverz), Buenos Aires. Re-regd **LV-FAO** 25.7.49 or 29.5.53 to Hector P Aguirre, Rio Negro. CofR lapsed 6.8.55. Regn cld 20.11.96.

G-ADLB, the Miles M.2R Hawk Trainer Mk.I.   *['Flight' via B Clarke]*

211   M.2R Hawk Trainer Mk.I; painted cream. (P&P works records state 'conforms to Type M.2P G-ADDK but with minor modifications'). Regd **G-ADLN** (CofR 6057) 13.7.35 to Reginald Cornwall & William Stratfield Verralls, London, SW1 & NW8 (but no base known). It is not known when, or by whom, this was first flown. CofA No.4951 issued 16.8.35; dd 4.9.35 (CLN recorded dd to 'Firm, Demonstration'). Entered by Cornwall in the 1935 King's Cup Air Race held at Hatfield on 6/7.9.35 (Race No.33) and flown (as a single-seater) by F/O HRA Edwards, to gain 2nd place at a speed of 157.48 mph. After the race, Edwards sent a very complimentary letter to Charles Powis: *This is just to tell you how much I enjoyed flying the Hawk Trainer in the King's Cup Air Race. It gave me great pleasure to handle such a nice aeroplane and the fact that its average speed on the first day, including four landings and refuelling, was over 153 mph speaks volumes for its efficiency, manoeuvrability and ease of handling on the ground. As an instructor, I think it is an ideal training machine. It is easy to fly, the controls are beautifully harmonised, it performs all evolutions easily and cleanly and it has all the characteristics of an ultra modern aeroplane. May I congratulate you on producing such an aeroplane?* George Miles recorded a flight from Woodley in a Hawk Trainer with a Gipsy Major engine on 25.9.35; almost certainly this aircraft. P&P works records show an undated 'application for approval of fitting Menasco B4 engine' *(actually a C4 - the B4 only gave 95hp)*. First flown with the Menasco C4 Pirate engine, by Miles, 21.3.36; presumed to A&AEE Martlesham Heath for tests since flown by Miles from Martlesham Heath to Woodley 8.4.36. Flown by George Miles, with Don Brown, on 22.5.36 and then to Shoreham from where he commenced endurance flying at full throttle for the next three months. These tests were done under manufacturer's certification, since the CofA was allowed to lapse 15.8.36 and was not renewed until 8.12.37. At some stage during these trials, G-ADLN was fitted with a mass-balance weight to the rudder. Regd (CofR 7922) 3.6.37 to Phillips & Powis Aircraft Ltd. While being flown on 9.6.37 by George Miles with the Menasco engine, he recorded: 'Propeller burst in air, landed in field nr White Waltham airfield, airframe and engine undamaged'; flown back to Woodley later the same day. Regd (CofR 8865) 3.11.38 to Reading Aero Club Ltd, Woodley (CLN records 5.9.38 - probably the hand-over date). Still fitted with the Menasco C4 engine on the Air Ministry survey at Woodley 31.8.39. CofA lapsed 9.1.40 and regn cld 6.2.41 as sold. Impressed 18.2.41 as **DG664** and to 5 MU Kemble 13.3.41. Not used and returned to Phillips & Powis for major inspection 6.6.41. Soc for spares use 24.7.41. (A report that it might have been refitted with a DH Gipsy Major engine can be dismissed; the Menasco engine itself was given Air Ministry serial A257362 on impressment of G-ADLN).

215   M.2W Hawk Trainer Mk.II; painted red. (P&P works records state 'Subs to G-ADVF'). Regd **G-ADWT** (CofR 6456) 6.11.35 to Phillips & Powis Aircraft Ltd. It was the second M.2W to fly and was first flown, by Bob Milne, on a 'CofA test' on 22.11.35. CofA No.4954 issued 23.11.35. Flown by Miles with Blossom Miles on instruction 24.11.35. The P&P operated 8 ERFTS (known locally simply as the Reserve Training School) was opened at Woodley on 25.11.35 and G-ADWT was dd the same day; later fleet no.1 was painted on the rudder. Fitted with horn balanced rudder in early 1936, it is possible that it was modified to M.2Y standard later (since CofA was renewed 'prematurely' on 24.7.36). Flown to A&AEE Martlesham Heath 6.37 to complete its programme of 28 spins thereby paving the way for the similar Magister to enter service. CofA lapsed 1.10.40 but not impressed into RAF service. Acquired (probably in 1942) by Cyril Geoffrey Marmaduke Alington and regd to him 12.1.43, although CofA renewed 26.10.42; Alington based the Hawk Trainer at the Austin factory airfield at Longbridge, Birmingham, where he had become chief test pilot. Regn cld 21.4.43 by Secretary of State. Reconditioned by P&P at Woodley, it was impressed into the RAF as **NF750** 12.1.43 and issued to 26 OTU, Wing, Bucks. To 5 MU Kemble 8.9.43 for storage. SOC 27.5.44 as Cat.E1 for spares use. It was amongst the impressed aircraft sold as surplus at 5 MU in December 1945 and listed as 'U/S for spares only'. Sold to Miles Aircraft Ltd for £150 and regd **G-ADWT** 9.9.46 to them. Hugh Kendall test flew a Hawk with serial 'NF730' on 26.11.46 and this must have been the first flight of NF750 after overhaul. CofA renewed 10.12.46. Cld 9.6.47 and regd 12.6.47 to The Reading Aero Club Ltd, Woodley. Cld 6.2.48 and regd 24.3.48 to T Shipside Ltd, Tollerton. CofA lapsed 24.3.49 and presumed stored until CofA renewed 15.1.52 (also reported as for sale with new CofA at Tollerton 11.50). Cld 4.3.52 and regd 21.3.52 to Michael Joseph Conry,

Pupils under instruction on Miles M.2W Hawk Trainer Mk.II G-ADWT of the Reserve Training School at Woodley in 1936.
[Phillips & Powis via The Miles Aircraft Collection]

Croydon. Cld 1.9.52 and regd 19.9.52 to Thomas Edward Parkinson, Knockholt, Kent (based Croydon). Cld 1.1.55 and regd 18.1.55 to Sqdn Ldr David Pitcairn Boulnois, t/a The 47 Squadron Flying Club, RAF Abingdon. Cld 17.5.55 and regd 23.5.55 to Cartwright Hamilton Aviation Ltd, London SW10 (based Croydon). Cld 1.9.56 and regd 23.10.56 to John Donaldson-Palmer, William John Searle & Charles Long, t/a Ruston Flying Club, Lincoln (based Bardney). Cld 26.2.58 and regd 3.3.58 to WS Shackleton Ltd, London, W1 (based Kidlington). Cld 6.5.58 to Edward Albert Clack & Bernard John Clack, Elm Park, Essex. Cld 15.5.58 and regd 19.5.58 (back) to WS Shackleton Ltd, London W1 (sale presumed cancelled). Regd 30.9.58 to Leonard Denman Blyth, Sawbridgeworth; operated by Panda Flying Group, Panshanger. Regd 18.8.60 to Albert Apthorpe Drew & Benjamin Herman Gaston, Watlington, Oxon (based Kidlington). Sold 5.9.61 by RK Dundas Ltd, for a reduced price of £650 and regd 31.10.61 to Flt Lt John MacGillivray (of Antiguniah, Nova Scotia), a Roman Catholic chaplain in the RCAF based at Baden-Sollingen (and later Zweibrucken) in West Germany. Nominally sold 1.10.62 and regd 4.3.63 to Terence David Peters of No.3 Wing, RCAF Zweibrucken (this 'token sale' was probably for legal reasons connected with its proposed importation into Canada). Seen at Denham in 4.64 and on 10.6.64 MacGillivray notified the Canadian DoT that it was to leave England the following day on *MT Caxton* for Quebec City. Regn **CF-NXT** allotted 25.2.64 to John MacGillivray, Chatham, NS. UK regn cld 25.5.64 as sold in Canada and UK Export CofA E.2519 issued 11.6.64. On 25.6.64 the Canadian DoT gave MacGillivray one month's authority to ferry it from Quebec to Charlo, Nova Scotia (although it seems he had already flown it there 24.6.64!). Application made 30.6.64 for renewal of CofA showing total time 2,380.55 hrs and Gipsy Major 1c engine No.80294 total time 1,101.35 hrs. A formal Bill of Sale dated 1.10.64 for $1.00 from Peters, of No.1 Wing, Merville, Meuse, France, to MacGillivray completed the 'legals'. On 5.10.64, authority issued for ferry Charlo to Moncton (valid till 16.10.64); CofR applied for 9.10.64; inspected Chatham and finally regd 14.10.64 (CofR No.29995) to MacGillivray (& with CofA No.5739 valid until 11.6.65). MacGillivray and the Hawk moved a number of times; to Winnipeg about 7.65; Saskatoon 1.9.65 (for winter hangarage and where it was repainted by Jack Statham); 'back' at Shearwater (by end 1968); to Smiths Falls, Ontario 6.8.73; to CFB Pettawa, Ontario (by 8.82 and probably much earlier). Final CofA issued 1.8.82 and on 25.6.84, MacGillivray signed a Deed of Gift of the Hawk Trainer to the EAA Aviation Foundation and the aircraft was moved to Oshkosh, USA, for display in their museum. Regn cld 19.11.84 (and John MacGillivray died about this time). Remained on static display until successful negotiations for its purchase by Brian Morris and Richard Earl (members of *The Miles Aircraft Collection*), commenced in 1999. The Hawk Trainer was packed into a container and shipped to England, arriving at The Aeroplane Company's workshop at Newbury, Berks on 29.3.00 for restoration to airworthy condition. Regd **G-ADWT** 14.4.00 to Brian Morris and Richard Earl (with total hrs of 2,865) and repainted in the original authentic red and silver colours in early 2004. Des Penrose made the first post-restoration flight from a strip southwest of Abingdon on 7.5.04 and its first public appearance was at the G-VFWE at Abingdon on Saturday 10.5.04. Permit to fly issued 28.5.04 and now based at Landmead Farm, Garford.

217 M.2W Hawk Trainer Mk.II; painted red. The first M.2W (P&P works records state 'Subs to G-ADDK but top decking to allow easy entry and exit with parachute, windscreens altered, dual control to flap operating pump (hydraulic), dual control to tail trim gear, parachute seats at back and front, mass balance on rudder and ailerons, acrobatic all up weight 1,650lb, control column shortened, dual control brake system fitted, dual control flap operating gear'). Regd **G-ADVF** (CofR 6386) 10.10.35 to Phillips & Powis Aircraft Ltd. *Miles recorded a number of cross-country flights in anonymous 'Hawk Trainers' and 'Trainers' in late 1935 and it is possible that the flight he made on 1.10.35 to 'Andover and local' which took 30 mins, could have been its first flight, as it only normally took 15 min to get to Andover from Woodley.* CofA No.4955 issued 10.10.35; dd 9.10.35 as firm's demonstrator (actually handed over to Service Dept on that date). Flown to RAE Farnborough (by King) 28.10.35 for type tests; returned to Woodley (by Smith) 8.11.35. Flown by Miles 29.11.35 and he also gave Blossom instruction on this 1.12.35. Formally designated 'type machine' 23.12.35 and handed over to the P&P Reserve Training School on 4.1.36. Fitted with horn balanced rudder in early 1936 and possibly modified to M.2Y standard later. CofA lapsed 13.10.39 and reduced to spares. Regn cld 31.3.41 as wfu.

224 M.2W Hawk Trainer Mk.II; painted red. (P&P works records state 'Subs to G-ADVF'); Regd **G-ADWU** (CofR 6457) 6.11.35 to Phillips & Powis Aircraft Ltd. CofA No.5202 issued 23.11.35. Flown by Miles, with Blossom on instruction 24.11.35. Handed over to the P&P Reserve Training School on 25.11.35. Fitted with horn balanced rudder early 1936; possibly modified to M.2Y standard later. Probably suffered major accident late 1936 since CLN stated that c/n 484 was fuselage number for 'rebuild of Hawk Trainer G-ADWU'; probably completed when CofA renewed 30.4.37. CofA lapsed 3.8.40 and reduced to spares. Regn cld 31.3.41 as wfu.

228 M.2W Hawk Trainer Mk II; painted red. (P&P works records state 'Subs to G-ADVF'). Regd **G-ADWV** (CofR 6458) 6.11.35 to Phillips & Powis Aircraft Ltd. CofA No.5203 issued 23.11.35. Flown by Miles, with Blossom on instruction 24.11.35. Handed over to the P&P Reserve Training School on 25.11.35. Fitted with horn balanced rudder early 1936, possibly modified to M.2Y standard later. CofA lapsed 23.1.40 and reduced to spares. Regn cld 31.3.41 as wfu.

235 M.2X Hawk Trainer Mk.II; painted red. (P&P works records state 'Subs to Type M.2W G-ADVF but fitted with Redwing undercarriage struts and Theed flap vacuum-operating gear'). Regd **G-ADYZ** (CofR 6561) 5.12.35 to Phillips & Powis Aircraft Ltd. (An article in *The Aeroplane* for 12.2.36 described the new Theed vacuum system for operating the flaps, developed by Miles from the Theed vacuum-operated brake system fitted to commercial vehicles. It had been fitted to a 'Miles aircraft' for experimental purposes but the article omitted to confirm the type to which it was fitted, but likely to have been this one). First flight not known. On 22.4.36, an unidentified 'Hawk' with the 'B' marks **U8**, (and almost certainly c/n 235) was flown to Farnborough (by Smith), where it carried out four 'flap tests' before returning to Woodley 30.4.36. (While CLN says that c/n 235 was del to the Reserve Training School 1.5.36, this is not recorded in the P&P works records and seems premature). CofA No.5572 issued 2.7.36 (delay probably due to the trials of both the Theed flap system and the Redwing undercarriage struts). Handed over to the P&P Reserve Training School on an unrecorded date (but possibly 9.7.36). The P&P works records also show 'M.2X Subs to G-ADZD' and since G-ADZD was an M.2Y this amendment implies that c/n 235 was modified to M.2Y standard, probably prior to issue of CofA. Flown to Farnborough (by Roxburgh) 19.10.36, returning to Woodley (by Moore) 22.10.36. CofA lapsed 18.8.40 and reduced to spares. Regn cld 31.3.41 as wfu.

237 M.2Z Hawk Trainer Mk.II; painted blue. (Laid down as an M.2W but completed as an M.2Z; P&P works records state 'Subs to G-AEHP'). Regd **G-AEHR** (CofR 6970) 6.5.36 to Phillips & Powis Aircraft Ltd. CofA No.5532 issued 12.6.36 and dd 16.6.36 to order from Roumanian Government as **No.2** for Roumanian Air Force. Regn cld 12.36 as sold abroad. Almost certainly regd **YR-AOJ** 10.8.44 to Clinceni Airport (which was regd with c/n '2').

Miles M.2Z Hawk Trainer Mk.IIs preparing to depart from Woodley on delivery to Roumania. ['Flight' via B Clarke]

241 M.2W Hawk Trainer Mk.II; painted red. (P&P works records state 'Subs to G-ADVF'). Regd **G-ADZA** (CofR 6561A) 5.12.35 to Phillips & Powis Aircraft Ltd. CofA No.5291 issued 31.12.35; dd to P&P Reserve Training School 10.1.36 and later given fleet No.4. Modified to M.2Y and CofA sent to Air Ministry 13.2.36 for endorsement for Acrobatics and auw increased to 1,750lb from 1,650lb. Regd 17.8.39 to Reading Aero Club Ltd, Woodley. CofA lapsed 14.2.40. Regn cld as sold 6.2.41. Impressed into RAF as **DG665** 18.2.41; to 5 MU Kemble 11.3.41. To 1 FPP White Waltham 15.6.41. Relegated to ground instructional airframe **3015M** and issued to No.1887 Squadron ATC, (County of London Wing) Lewisham, London 4.4.42.

242 M.2W Hawk Trainer Mk.II; painted red. (P&P works records state 'Subs to G-ADVF'). Regd **G-ADZB** (CofR 6562) 6.12.35 to Phillips & Powis Aircraft Ltd. CofA No.5306 issued 22.1.36; handed over to P&P Reserve Training School 23.1.36; later given fleet No.7. Probably modified to M.2Y standard later. Probably suffered major accident after 9.36 and CLN records that c/n 481 was 'Fuselage No for rebuild of G-ADZB'. CofA renewed 5.3.37. CofA lapsed 4.7.40 and reduced to spares. Regn cld 31.3.41 as wfu.

245 M.2Z Hawk Trainer Mk.II; painted blue. (P&P works records state 'Subs to G-AEHP'). Regd **G-AEHS** (CofR 6971) 6.5.36 to Phillips & Powis Aircraft Ltd. CofA No.5533 issued 12.6.36; dd 16.6.36 to order from Roumanian Government as **No.3** for Roumanian Air Force. Regn cld 12.36 as sold abroad.

246 M.2Y Hawk Trainer Mk.II; painted red. (P&P works records state 'Subs to G-ADZD'). Regd **G-AEAW** (CofR 6659) 29.1.36 to Phillips & Powis Aircraft Ltd. CofA No.5359 issued 14.4.36; aircraft handed over 27.4.36 to P&P Reserve Training School. Collided with Hawk Trainer G-ADZE over Woodley 26.8.36 and forced landed in allotment alongside Reading Road; pilot Roland Fearon injured. Regn cld 12.36.

G-ADZC, a Miles M.2W Hawk Trainer Mk.II of the Reserve Training School, Woodley, in 1936. ['The Aeroplane' *19th August 1936*]

249 M.2W Hawk Trainer Mk.II; painted red. (P&P works records state 'Subs to G-ADVF'). Regd **G-ADZC** (CofR 6563) 6.12.35 to Phillips & Powis Aircraft Ltd. CofA No.5300 issued 13.1.36; aircraft handed over to P&P Reserve Training School 25.1.36; later given fleet No.6. Probably modified to M.2Y standard prior to new CofA issued 18.9.36. Regd 12.5.39 to Reading Aero Club Ltd, Woodley (to service Civil Air Guard needs). CofA lapsed 17.11.39 and scrapped in 1940. Regn cld in census 5.11.45.

253 M.2Y Hawk Trainer Mk.II; painted red. (P&P works records state 'Conforms to G-ADVF but acrobatic category weight increased to 1,750lb' to become the first M.2Y and the first Miles aeroplane to have a horn balanced rudder). Regd **G-ADZD** (CofR 6564) 6.12.35 to Phillips & Powis Aircraft Ltd. First flown, as **U2**, by Bob Milne 1.2.36 on 'C of A test'. CofA No.5307 issued 4.2.36; (and per CLN) dd same day to P&P Reserve Training School. It is recorded in P&P works records that this aeroplane was the 'Roumanian demonstrator', although they did not record any modification to the fuel system to allow the aircraft to fly inverted nor that the throttle was modified to work in the opposite sense to conform with Roumanian practice. *The Aeroplane* 25.3.36 recorded that in the week to March 21, Captain Godwin von Brumowski came from Austria to fetch a Hawk Trainer for inverted flying. It is likely that this was the machine taken by him to Roumania for demonstration purposes; in the same week, it was also noted that Wing Cdr Freddie Stent (who helped Miles with test flying at Reading in the 1930's) returned it from Roumania. Assuming G-ADZD was the Roumanian demonstration aircraft, it was later returned to standard configuration for the Reserve School. CofA lapsed 4.4.40 and reduced to spares. Regn cld 31.3.41 as wfu.

254 M.2W Hawk Trainer Mk.II; painted red. (P&P works records state 'Subs to G-ADVF'). Regd **G-ADZE** (CofR 6565) 6.12.35 to Phillips & Powis Aircraft Ltd. CofA No.5308 issued 20.2.36; handed over 2.3.36 to P&P Reserve Training School. Probably later modified to M.2Y standard. Collided with Hawk Trainer G-AEAW over Woodley 26.8.36; tail severed and spun in; pilot RJ Shelard successfully parachuted. Regn cld 12.36.

257 M.2R Hawk Trainer Mk.I; painted silver. Long range version (P&P works records state 'Subs to G-ADLN'). Regd in Chile as **CC-FBB 319** to Franco Bianco (of Basillou 56, Magallanes, Punta Arenas). CofA No.5255 issued 10.12.35; aircraft handed to shippers Parkes 11.12.35. Named *Saturno* it was often flown by Bianco with the front cockpit faired over. Bianco made first non-stop Trans-Andean flight on 7.6.36, setting out from Bahia Catalina Aerodrome (Punta Arenas) before dawn on a direct flight routing over the Canal Region and crossing the Andes at 12,000ft to land at the Air Base of the Chamiza in Puerto Montt - a flight of 9 hr 55 min – a distance of 1,850 km (1,150 miles). The return journey to Magallanes began at Los Cerrillos on 24.6.36 and this time he crossed the Andes via Buenos Aires in Argentina but was delayed by adverse weather to arrive in Bahia Catalina at 3.45 pm on 7.7.36. The first air-mail flight Punta Arenas-Santiago-Punta Arenas was also reportedly made in 1936 by this machine. New CofR 0053 issued 2.7.43. Re-regd **CC-PFB-0053** 31.12.45 to same owner. Regn cld 16.8.49; it later became a feature in a children's playground, before being rescued and put on display in an engineless condition (but painted as CC-FBB) in the Museo Nacional Aeronautico y del Espacio, Los Cerrillos, Santiago, Chile. More recently, it has been completely refurbished to original condition by the museum's staff and is on static display in the museum.

258 M.2Z Hawk Trainer Mk.II; painted blue and first of export order for ten M.2Z for Roumanian Government. (P&P works records state 'Conforms to G-ADZD Type M.2Y with the exception of two 15-gal aluminium fuel tanks, inverted flying system incorporating auxiliary aluminium fuel tank and reversed throttle and mixture controls') Regd **G-AEHP** (CofR 6969) 6.5.36 to Phillips & Powis Aircraft Ltd. First flown, under 'B' marks **U3**, by Flt Lt JF Moir 9.6.36 and again on 12.6.36. CofA No.5531 issued 12.6.36; dd 16.6.36 to order from Roumanian Government as **No.1** for Roumanian Air Force. Regn cld 12.36 as sold abroad.

Ten Miles M.2Z Hawk Trainer Mk.IIs were delivered to the Roumanian government in 1936.
['The Aeroplane' *16th September 1936*]

Four of the Miles M.2Z Hawk Trainer Mk.IIs on their ferry flight to Roumania, seen just after they left Woodley. ['Flight' *via B Clarke*]

260 M.2Y Hawk Trainer Mk.II; painted red. (P&P works records state 'Subs to G-ADZD'). Regd **G-AEAX** (CofR 6660) 29.1.36 to Phillips & Powis Aircraft Ltd. Flown by Bob Milne on 24.2.36 on CofA test. CofA No No.5358 issued 25.2.36; aircraft handed over to P&P Reserve Training School, Woodley (CLN states dd 14.2.36); later given fleet No.10. Reportedly repainted yellow at or about time of CofA renewal 5.3.37. Flown to Farnborough by Miles 22.6.38 for handling trials; flown back to Woodley by Wilson 24.6.38. Regd 12.5.39 to Reading Aero Club Ltd, Woodley (for Civil Air Guard training). CofA lapsed 4.5.40. Regn cld 6.2.41 as sold. Impressed into RAF as **DG666** 18.2.41; to 5 MU Kemble 14.3.41. To ATA White Waltham 4.6.41. To Southern Aircraft (Gatwick) Ltd, Gatwick 30.12.41; returned to ATA White Waltham 22.2.42. To 29 MU High Ercall 28.3.42. To 5 MU Kemble 1.8.42. To 7 FPP ATA. While stationary at junction of two taxying tracks at Hawarden at 1425 hrs 11.2.43 on a ferry flight, it was struck by a taxying Proctor (pilot S/O HD Steel of 7 FPP); Cat.Ac and repaired locally. To 5 MU Kemble 30.7.43. SOC Cat E1 for spares use 11.12.43.

261 M.2Y Hawk Trainer Mk.II; painted red. (P&P works records state 'Subs to G-ADZD'). Regd **G-AEAY** (CofR 6661) 29.1.36 to Phillips & Powis Aircraft Ltd. CofA No.5328 issued 18.2.36; handed over to P&P Reserve Training School 13.2.36 (sic). Probably repainted yellow at time of CofA renewal 27.7.37. CofA lapsed 20.9.39 and reduced to spares. Regn cld 31.3.41.

265 M.2Z Hawk Trainer Mk.II; painted blue. (P&P works records state 'Conforms to G-AEHP'). Regd **G-AEHT** (CofR 6972) 6.5.36 to Phillips & Powis Aircraft Ltd. CofA No.5534 issued 12.6.36. Airscrew broke while carrying out test flights. Dd 15.7.36 to order from Roumanian Government as **No.5** for Roumanian Air Force. Regn cld 12.36 as sold abroad.

270 M.2Y Hawk Trainer Mk.II; painted red. (P&P works records state 'Subs to G-ADZD'). Regd **G-AEAZ** (CofR 6662) 29.1.36 to Phillips & Powis Aircraft Ltd. CofA No.5360 issued 13.3.36; handed over P&P Reserve Training School, Woodley 20.3.36; later given fleet No.8. Spun in and crashed Haines Hill, Hurst 5.6.37 during forced landing practice; pilot Sgt John Edmund Roe, RAFVR killed. Regn cld 12.37.

271 M.2Y Hawk Trainer Mk.II; painted red. (P&P works records state 'Subs to G-ADZD'). Regd **G-AEEL** (CofR 6824) 14.3.36 to Phillips & Powis Aircraft Ltd. CofA No.5399 issued 26.3.36; handed over to P&P Reserve Training School, Woodley 26.3.36; fleet No.13. Badly damaged in crash at Loddon Bridge Road, Reading 30.11.36; JPC Roll unhurt. Rebuilt and given fleet No.9. CofA lapsed 25.5.40 and reduced to spares. Regn cld 31.3.41.

277 M.2R Hawk Trainer Mk.I. (P&P works records state 'Subs to G-ADLN'; acrobatic wt 1,650lb). Supplied to RK Dundas Ltd (P&Ps Indian agents) in component parts to be assembled and built by The Aeronautical Training Centre of India; dd 1.11.35. This machine, which was possibly nearer to the M.2W spec than the M.2R, was regd **VT-AHF** (CofR 273) 24.12.35 to The Aeronautical Training Centre of India Ltd, New Delhi (with c/n ATC.9/1). (It is of interest to note that the regn date to ATCI is same as that for their M.2F VT-AGT). A photograph of VT-AHF taken at the Delhi aerodrome on 27.1.38 was probably the occasion of its first flight, and was published in the *Miles Magazine* for March 1938 stating it was successfully flight tested by Capt Pierce, Chief Instructor of the Centre. Fate unknown and regn cld 4.4.41.

Possibly the Miles M.2R Hawk Trainer Mk.I which was assembled in India as VT-AHF. [Via B Clarke]

292 M.2Z Hawk Trainer Mk.II; painted blue. (P&P works records state 'Conforms to G-AEHP'). Regd **G-AEHU** (CofR 6973) 6.5.36 to Phillips & Powis Aircraft Ltd. CofA No.5559 issued 10.7.36 and dd 15.7.36 to order from Roumanian Government as **No.7** for Roumanian Air Force. Regn cld 12.36 as sold abroad. A photograph of this aircraft in Roumanian Air Force colours and markings shows it to have the serial '7' in black on the fuselage and on the central white stripe of the rudder so presumably the other M.2Z's were painted in similar colours; the caption to the photograph read: *Les Hawk (No.7) ont êtê utilisês tant pour l'entrainement, que pour les missions de liaison pendant la guerre (Antoniu).* Probably regd **YR-AOL** 10.8.44 to Clinceni Airport (which was regd with c/n '7').

293 M.2Z Hawk Trainer Mk.II; painted blue. (P&P works records state 'Subs to G-AEHP'). Regd **G-AEHV** (CofR 6974) 6.5.36 to Phillips & Powis Aircraft Ltd. CofA No.5537 issued 15.6.36; dd 16.6.36 to order from Roumanian Government as **No.4** for Roumanian Air Force. Regn cld 12.36 as sold abroad.

No.7 Miles M.2Z Hawk Trainer Mk.II of the Roumanian Air Force (FARR). [H-H Stapfer via P Butler]

This advertisement shows G-AEEL, a Miles M.2Y Hawk Trainer Mk.II of the Reserve Training School, Woodley.
['Flight' 19th November 1936]

294 M.2Z Hawk Trainer Mk.II; painted blue. (P&P works records state 'Subs to G-AEHP'). Regd **G-AEHW** (CofR 6975) 6.5.36 to Phillips & Powis Aircraft Ltd. CofA No.5560 issued 2.7.36; dd 15.7.36 to order from Roumanian Government as **No.8** for Roumanian Air Force. Regn cld 12.36 as sold abroad.

295 M.2Z Hawk Trainer Mk.II; painted blue. (P&P works records state 'Subs to G-AEHP'). Regd **G-AEHX** (CofR 6976) 6.5.36 to Phillips & Powis Aircraft Ltd. CofA No.5552 issued 26.6.36; dd 15.7.36 to order from Roumanian Government as **No.10** for Roumanian Air Force. Regn cld 12.36 as sold abroad.

296 M.2Z Hawk Trainer Mk.II; painted blue. (P&P works records state 'Subs to G-AEHP'). Regd **G-AEHY** (CofR 6977) 6.536 to Phillips & Powis Aircraft Ltd. CofA No.5561 issued 10.7.36; dd 15.7.36 to order from Roumanian Government as **No.6** for Roumanian Air Force. Regn cld 12.36 as sold abroad. Probably regd **YR-AOK** 10.8.44 to Clinceni Airport (which was regd with c/n '6').

297 M.2Z Hawk Trainer Mk.II; painted blue. (P&P works records state 'Subs to G-AEHP'). Regd **G-AEHZ** (CofR 6978) 6.5.36 to Phillips & Powis Aircraft Ltd. CofA No.5562 issued 10.7.36; dd 15.7.36 to order from Roumanian Government as **No.9** for Roumanian Air Force. Regn cld 12.36 as sold abroad.

302 M.2Y Hawk Trainer Mk.II; painted yellow. (P&P works records state 'Subs to G-ADZD with exception of two 15-gal aluminium fuel tanks (has been cleared on other aircraft)'). Flown by Flt Lt JF Moir 21.8.36 (possibly first flight). CofA No.5661 issued 17.9.36 to Wellington Aero Club (Mr Gamble), New Zealand; dd 19.9.36 and arrived in NZ 10.36 (ordered as replacement for Hawk Major ZK-ADJ). Regd **ZK-AEQ** (CofR 91) .36 to Wellington Aero Club, Wellington. First flight after erection 3.12.36; NZ CofA No.177 issued with expiry 16.9.37. Later painted black and red. Crashed on Mount Egmont, nr Eltham 2.6.37 at 0955 hrs; passenger Phillip Joseph Nathan killed, pilot CH Dunford badly injured. Regn cld 6.38. Some parts survive with Graham Orphan in NZ.

328 M.2Y Hawk Trainer Mk.II; painted red with silver wings and tailplane. (Initial application for CofA was as M.2H Hawk Trainer, which has caused some confusion, compounded by the A&AEE referring to it as an M.2H Hawk Major when they had it for spinning trials and quoting a wingspan of 33ft 0in as used on the M.2H, whereas the Hawk Trainer variant had a 34ft 0in wingspan and increased wing area of 176 sq ft; however it was indeed an M.2Y Hawk Trainer Mk.II with a headrest to rear cockpit and a horn balanced rudder). Regd **G-AENT** (CofR 7348) 18.9.36 to Phillips & Powis Aircraft Ltd. CofA No.5684 issued 3.10.36; dd 7.10.36 to P&P Reserve Training School. Dd 15.12.36 to the A&AEE Martlesham Heath for spinning trials (ERFTS). Failed to recover from a right-hand spin on 19.12.36 and crashed into a school playground near Woodbridge, Suffolk; pilot Sgt Sammy Wroath baled out successfully. Regn cld 12.36.

480 M.2Z? Hawk Trainer Mk.II. First flown, with 'B' marks **U1**, by Flt Lt JF Moir, on u/c tests on 7.1.37 and 15.1.37. CofA No.5784 issued 5.2.37 to Arcos Ltd, the UK agent for the USSR; dd 4.2.37. *(The precise variant of this machine has never been established, but it was similar to the machines supplied to Roumania).*

481 Recorded by CLN as fuselage used in rebuild of G-ADZB (c/n 242)

484 Recorded by CLN as fuselage used in rebuild of G-ADWU (c/n 224).

Breakdown of production		
M.2R Hawk Trainer Mk.I	c/n 210, 211, 257, 277	4
M.2W Hawk Trainer Mk.II	c/n 215, 217, 224, 228, 241, 242, 249, 254	8
M.2X Hawk Trainer Mk.II	c/n 235	1
M.2Y Hawk Trainer Mk.II	c/n 246, 253, 260, 261, 270, 271, 302, 328, (c/n 241 was modified to M.2Y later so not counted)	8
M.2Z Hawk Trainer Mk.II	c/n 237, 245, 258, 265, 292, 293, 294, 295, 296, 297, 480 (type not confirmed but included here), (c/n 222 was converted from a M.2T later so is not counted here)	11
Total Hawk Trainer Mk.I variants built		4
Total Hawk Trainer Mk.II variants built		28

Appendix 13:
Miles M.3 Falcon and M.3A Falcon Major Production

102 M.3 Falcon prototype; painted silver with red trim. Fitted with 130hp Gipsy Major engine salvaged from Harold Brook's Puss Moth G-ABXY. Regd **G-ACTM** (CofR 5112) 11.6.34 to Harold Leslie Brook, Harrogate (based Sherburn-in-Elmet). First flown, by Miles 23.9.34 as **U3**. Flown by Miles to A&AEE Martlesham Heath 4.10.34. CofA No.4536 issued (with c/n F.1) 9.10.34; dd 12.10.34. Flown by Brook in MacRobertson Mildenhall to Melbourne air race (Race No.31 and named *Ovaltine* on the cowling, presumably in deference to his sponsor) with passenger Miss E Lay as passenger, departing 20.10.34 and finally arriving at Darwin on 20.11.34, having taken 26 days 20 hours. Before Brook set out on his return flight, he fitted extra fuel tanks to give a 2,000 mile range and, flying solo, left Darwin on 23.3.35 to arrive at Lympne on 31.3.35, to establish a record time of 7 days, 19 hrs, 50 mins. G-ACTM was later flown to Woodley where it was flown by Miles on 13.4.35. In July 1935, Brook left England in an attempt to establish a new England to Capetown record but on 18.7.35 he crashed on landing in the dark at Mersa Matruh, 260 miles ENE of Cairo, Egypt; G-ACTM was returned to England and repaired. Brook competed in the London - Newcastle Air Race on 8.8.36 and kept G-ACTM at Barton for 3 months during 1936. The CofA lapsed 14.11.36 but fate unknown. Regn cld 11.36 as wfu.

131 M.3A Falcon Major; painted red. First production (P&P works records - Subs & mods to G-ACTM with the AUW increased to 2,200lb; max speed 145 mph). Regd **G-ADBF** (CofR 5535) 9.1.35 to Capt Noel Alexis Blandford-Newson, Haddington, East Lothian (based Macmerry). First flown, by Miles 16.1.35 (possibly as **U5** but unconfirmed). An official flight trial of 1 hour duration was made on 15.2.35. CofA No.4704 issued 26.2.35; dd by Miles to Heston for customer 27.2.35. *(CLN recorded regd ZS-ATN to Capt Blandford-Newson but this seems unlikely as that regn was not allocated until 1941).* Blandford-Newson may have been posted to the Middle East whilst still owning this aircraft. Regn cld at census 1.1.39. Regd **HB-USU** (CofR 504) 16.3.39 to Heinrich Spillmann, Basel/ Birsfelden, Switzerland. Regd to Marius & Emile Felley, based at Sion. Regn cld 28.1.48 but still at Sion in 1949. *Note: A report that this Falcon became I-ZENA in 1935 is almost certainly an error - see c/n 163.*

140 M.3A Falcon Major; painted red. (P&P works records – Subs & mods to G-ADBF) Regd **G-ADBI** (CofR 5545) 1.35 to Phillips & Powis Aircraft Ltd. CofA No.4710 issued 29.4.35. Regd (CofR 5843) 30.4.35 to Air Hire Ltd, Heston; dd 5.5.35 and named *Ariel*. Flown 9.5.35 from Shoreham to Woodley and return by George Miles with Tommy Rose. Regn cld 3.37 as sold and with new CofA issued 23.3.37. Regd in Southern Rhodesia as **VP-YBN** 8.37 to MH Kassim, Salisbury. Crashed Bulawayo in 1938.

149 M.3A Falcon Major; painted cream. A Falcon was flown with 'B' marks **U9** by Miles to Farnborough and back on 25.3.35 and also

VR-HCV Miles M.3A Falcon Major at Hong Kong in June 1935.
[Via P Jarrett]

on 27.3.35 and 31.3.35 (twelve flights in total) and although it cannot be positively identified with this particular machine, no other candidate seems likely (and a Flight photo dated 1.4.35 shows U9 in formation with Hawk Major G-ADAW & Merlin U8). (P&P works records – Subs & mods to G-ADBF) CofA No.4711 issued 12.4.35 to Far East Aviation Co Ltd, Hong Kong (CLN records dd 4.4.35 but this probably the date of handover or sale). Regd **VR-HCV** (CofR 74) 27.5.35 to Far East Aviation Co Ltd, Hong Kong. Regn cld in 4th quarter of 1937 as sold to Federated Malay States. Regd **VR-RAP** (CofR 20) 17.11.37 to MT Stanley, c/o The Great Eastern Life Assurance Co Ltd, Kuala Lumpur. Regd 1.3.39 to RFC Markham, Seramban. Regd 3.1.40 (back) to MT Stanley, Kuala Lumpur. Impressed into Malayan Volunteer Air Force 12.41 as **32**. Flown by KG Hannett (former CFI of Kuala Lumpur Flying Club), with a Danish navigator, from Pekanbaroe via Djambi and Lahat to Batavia in advance of Japanese invasion and probably destroyed Batavia, Java in early 1942.

157 M.3A Falcon Major; painted blue. (P&P works records – Subs & mods to G-ADBF) Regd **G-ADER** (CofR 5699) 16.4.35 to Maddox Airways Ltd, Brooklands. CofA No.4780 issued 18.4.35. Flown by Miles on 24, 25 and 27.6.35 but with no reason noted. Regd (CofR 6519) 25.11.35 to Malcolm & Farquharson Ltd, Heston. Regd (CofR 6894) 8.4.36 to Staffordshire Airplanes Ltd, t/a North Staffordshire Aero Club, Meir. Regd (CofR 7473) 23.11.36 to Miss Barbara Maude Wenman, Cobham, Kent (based Gravesend). Regn cld 2.37 as sold. Regd in France as **F-AQER** (CofR 5024) 13.2.37 to Henry Leroy Beaulieu, Paris; entered by its owner for the 'International Circuit of the Oasis' meeting, held at Cairo 22-26 February 1937, but failed to start. Regd 7.37 to Robert Bonnefon, Arles. Regd 4.38 to Jean Bajollet, Nancy. Regd 3.39 to Henri Paland, Basse-Indre (based Nantes). Stored during the war and restored postwar (CofR 10938) 12.7.46. Regd 2.47 to Societe Vve Farbos et Huileries des Landes Reunies, Mont-de-Marsan. Last CofA inspection date was 28.6.55 and regn cld 1958/59.

163 M.3A Falcon Major; painted red. Initially allocated to Nikolae Kulcha but ntu. (P&P works records – Subs & mods to G-ADBF) Regd **G-ADHC** (CofR 5807) 22.5.35 to Galbraith Pembroke & Co Ltd, London EC3 (based Woodley). CofA No.4779 was issued 21.5.35; dd 23.5.35 (owners were probably the agents for the Italian owner). Almost certainly the Falcon flown by Sr Giorgio Parodi in late May 1935 when he won the Raduno Sahariano in Tripoli, Libya. Also flown by Sr Parodi in the 1st Raduno Aereo del Littorio 24-30.8.35. Regn cld 3.36 as sold. Regd in Italy as **I-ZENA** (CofR 1777) 23.9.35 to Aero Club 'L Olivari', Genova. Regd to RUNA Sede Provinciale di Genova. Regn cld 5.36 as destroyed in accident.

181 M.3A Falcon Major; painted blue. (P&P works records – Subs & mods to G-ADBF) Regd **G-ADHH** (CofR 5872A) 16.5.35 to Major Douglas William Gumbley, the Director of Civil Aviation, Palestine (based Lydda). First flown, by Miles 26.5.35, to Hanworth and return via Brooklands later same day, at 12,000 ft alt! CofA No.4832 issued 28.5.35; dd to Heston 30.5.35 and flown to Lydda Airport by Gumbley's pilot. On 16.5.35, this unidentified pilot wrote to Phillips & Powis: *Just a short line to tell you that I had a most successful trip to Palestine in Gumbley's Miles Falcon. Both the aircraft and engine behaved splendidly and I never touched either the whole way out. My flying time to Ramleh, which was 3,750 miles the way I came, was 37 hours, an average ground speed of 100 mph, which was good considering I had a head wind all along the North African coast. The petrol consumption worked out at 7 gallons per hour, running the engine just under 2000 rpm. I agree with you, she is a peculiar little aeroplane to fly first, but by the time I got to Ramleh I liked her very much indeed. The Italians admired her appearance very much, all the way out.* Regn cld 14.3.40 as sold locally (to Aviron Company). Regd **VQ-PAO** 21.3.40 to Palestine Aviation Ltd, Lydda. Skidded off runway at Ashdot-Yaakov in 1941 and hit a fence; repaired. Regd in Egypt as **SU-ADA** .45 to HE Alexander Kirk (the US Air Attaché in Egypt), Almaza, Cairo. Crashed at Marsa-Aalem (a remote gold mine on the Red Sea coast) and written off 12.2.48.

Miles M.3A Falcon Major G-ADHH at Lydda, Palestine, in about 1937.
[Via P Amos]

Miles M.3A Falcon Major SU-ADA belonging to the US Air Attaché in Egypt, at Almaza in 1946. *[LAC Elliott via N D Welch and J Havers]*

189 M.3A Falcon Major; painted blue. (P&P works records – Subs & mods to G-ADBF) Regd **G-ADHI** (CofR 5873) 14.5.35 to Continental Carrosseries Ltd, London W.10 (based Heston). CofA No.4841 issued 29.5.35. (CLN recorded dd to 'Carrosserie' Darrin 2.6.35). Advertised for sale by Darrin as 'new' prior to delivery; Darrin was probably Howard Darrin. Flown by Frenchman Frederick Blow (but who was actually a US pilot based in Paris and business partner with Darrin) in the 1st Raduno Aereo del Littorio in Tripoli, Libya 24-30.8.35. Regd (CofR 7135) 18.6.36 to Robert Christian Ramsay, Bekesbourne, and operated by his Kent Flying Club). Regd (CofR 7305) 28.8.36 to (Ramsay's company) Airsales & Service Ltd (remained in operation by Kent Flying Club). Sold 16.8.39 and regd 28.8.39 to Portsmouth Southsea & Isle of Wight Aviation Ltd, Portsmouth. Allocated to NAC following the Air Ministry survey at 31.8.39 and regn cld 14.4.40 as sold. Impressed as **X9300** 14.3.40 and to 20 MU Aston Down same day. To 81 (F) Group Sealand 23.12.40. SOC as Cat.E1(scrap) 31.3.41.

193 M.3A Falcon Major; painted red and green. (P&P works records – Subs & mods to G-ADBF) Regd **G-ADHG** (CofR 5872) 5.6.35 to Aircraft Distributors Ltd, Hanworth (as demonstrator). CofA No.4842 issued 13.6.35. Regd cld 5.37 as sold. Regd in Australia as **VH-AAT** (CofR 649) 7.7.37 to Royal Queensland Aero Club, Archerfield. Impressed into RAAF as **A37-3** 2.1.41. Dd ex Archerfield to 2 Communications Flt 7.4.41. Airframe to De Havilland, Australia for overhaul or repair 23.8.41; returned to 2 Comm Flt 9.1.42. Issued to 3 Comm Flt 3.8.42; dd 10.8.42. On 5.3.43 an accident was recorded but it was flying again by 17.5.43. On 19.2.44 it suffered a landing accident which damaged the starboard wing and u/c; repaired by 20.3.44. Ground looped on landing at Evans Head 9.4.44, sustaining minor damage to the starboard side of the airframe and both u/c legs, 'remainder of aircraft u/s'! To 3 CRD 11.4.44 for 'survey report with view to conversion'; on 14.7.44 the record stated 'recommend this aircraft be offered to Civil Aviation for disposal; available for disposal to civvie operators'. Sold 9.3.45 to SW Hecker; dd to him 16.3.45. Regd

VH-AAT 12.6.46 to Samuel William Hecker, Maryborough. Qld. Regd 29.12.49 to Schutt Aircraft Sales & Service, Essendon, Vic. Hit fence on landing Echuca, Victoria 9.2.51 damaging wing (no injuries); repaired. Regd 21.9.55 to Schutt Aircraft Pty Ltd, Cheltenham, Vic. Later based at Waikerie Airfield, South Australia. Regn cld 22.3.56. Regd 20.8.59 to Alan Lauder Hume, c/o Aero Club of Southern Tasmania, Hobart. Regd 20.8.65 to Harry Thomas Burgess, Huon, Tasmania. Regd 5.7.72 to Leo Smothers, Somerton Park, SA. Ground-looped Birdsville, Qld 1.9.72; minor damage repaired. Regd 20.6.73 to David John Russell Barker, Caloundra, Qld. Regd 20.10.75 Col Leslie Keith Hatfield, Point Cook, Victoria and dd to Berwick, Victoria for total refurbishment, mainly by David Squirrel. Regn cld 14.2.78 as wfu. Reflown 3.90 from Melbourne and regd 29.5.90 to same owner. In 4.91, painted to represent Harold Brook's Falcon G-ACTM during the filming of the Australian TV mini series, *Half a World Away*. By 1993, based at Casey Airfield, Berwick, Victoria where plans were formulated for David Squirrel and Doug Bowles to make a flight from Australia to England as a tribute to the pioneers of the England to Australia air route and to commemorate Harold Brook's MacRobertson Air Race flight of 1935; abandoned through lack of sponsorship. VH-AAT repainted cream and green and maintained in airworthy condition with the Drage Air Museum at Wangaratta, Northern Victoria. Regd 2.6.06 to Leslie Hatfield, Orchards, Victoria (believed to be the Colonel's son). In 7.06, Peter Holloway was offered 'first refusal' with a view to the Falcon returning to England but despite making an offer, nothing more was heard and the plan was abandoned.

196 M.3A Falcon Major; painted silver. (P&P works records – Subs & mods to G-ADBF) Regd **G-ADFH** (CofR 5730) 21.5.35 to Edward Devereux 'Bunny' Spratt, Hanworth/Heston. First flown, by Miles 28.6.35, on 'flap tests'. CofA No.4863 issued 28.6.35; dd 18.7.35 to Spratt at Air Service Training Ltd, Hamble where he was obtaining his 'B' Licence. Initially used by Bunny Spratt to tour round Europe and Middle East with his wife Isla. Entered by Spratt in the 'International Circuit of the Oasis' meeting 22-26.2.37 at Cairo, where he gained 10th place in the handicap event. Spratt met fellow entrant Geoffrey Alington in Cairo and this led to the formation of Air Touring in April 1938, an air taxi business which they based in the circular terminal building *The Beehive* at Gatwick. CofA lapsed 22.6.40 and regn cld 31.12.40 by Secretary of State. Impressed into the RAF as **HM496** 15.12.41; taken by road from Hamble to Woodley; test flight made by Flt Lt Hugh Kennedy 1.1.42. To RAF Turnhouse 6.1.42. To Stn Flt Northolt 30.5.43. To 5 MU Kemble 12.9.43; to Miles Aircraft Ltd for repairs 27.11.43; returned to 5 MU 21.7.44. *(Reported as having been flown by Lettice Curtis from White Waltham to Bourn, Cambridge 25.9.44, but the official record states that it remained in storage at 5 MU Kemble until 1946).* Sold in 1946 to The Champion Plug Co. Regd **G-ADFH** 31.8.46 to Hubert Granville Starley, Twickenham. Regd 1.10.46 to Thomas Charles Sparrow, Bournemouth. CofA renewed 13.6.47 but lapsed 12.6.48. Cld 16.3.50 & regd 24.4.50 Douglas Edward Bianchi, Staines (based at White Waltham) and CofA renewed 12.6.50. Flown by David Ogilvy in the Daily Express International Air Race (Race No.42), from Hurn to Herne Bay, Kent, on 16.9.50. Cld 16.1.51 & regd 19.1.51 to Raymond Allen Drean, Twickenham and hangared and 'cared for' by the College of Aeronautical Engineering at Redhill Aerodrome. CofA lapsed 11.6.51 and, in 1952, 'rotten timbers' were found in the cabin area and it was stripped for repairs, but it was found that similar problems prevailed elsewhere and the repairs were abandoned. The airframe remained in store until the closure of Redhill Aerodrome (in 4.54) and was later taken to Croydon Airport by road where it was still extant in 2.57. Regn cld 13.10.58 by Secretary of State.

197 M.3A Falcon Major; painted blue and silver. (P&P works records – Subs & mods to G-ADBF) CofA No.4882 issued 22.6.35 to Rafael de Mazarredo Trener, Madrid, Spain; regd **EC-W48** for the ferry flight 23.6.35. Regd **EC-BDD** 16.6.35 (also reported as 16.7.35) to Rafael de Mazarredo, the manager of the school at the Valencia Aero Club. Taken over by the Spanish Republicans as **EN-001** in July 1936 and served at Aragón at the Los Alcázares School during the first month of the Spanish Civil War. Reported as being 'knocked down in Teruel a month after the beginning of the fight' and then captured by Nationalists and repaired to be flown as **30-168**. Re-serialled post-civil war as **L5-168**. Spanish civil regn cld at census 12.11.40. As with so many Spanish aircraft, its history is uncertain and confused. If it did become L5-168, this Falcon was understood to have been damaged in 1947 in an accident at Jerez de la Frontera. It was then acquired without engine by the Huesca Aero Club in 1952 and regn **EC-AGY** reserved. Not repaired and scrapped in 1961. *In the alternative, there have been various statements that c/n 197 was modified with a Gipsy Six engine and is the Falcon which ultimately became EC-ACB and which still survives. The balance of probability however is that, despite being quoted as c/n 197, EC-ACB is actually c/n 231 – and this is where we record EC-ACB's known or reported history.*

201 M.3A Falcon Major; painted blue. (P&P works records – Subs & mods to G-ADBF) CofA No.4886 issued 21.6.35 to Aero Club de Andalucia, Seville, Spain; regd **EC-W45** for the ferry flight 23.6.35. Regd **EC-DBB** 10.10.35 to Aero Club de Andalucia, Tablada. Reported that it was 'taken to Seville by the well-known Fernando Flores Solis'. Taken over by the Nationalist Air Force at Seville in 10.36 as **30-55**. Spanish civil regn cld at census 12.11.40. Regd **EC-BAY** 27.3.41 to the Aero Club de Andalucia, Seville. Re-regd **EC-ABZ** 4.53 (noted at Cuatro Vientos, Madrid in 7.64). Regd to Emanuel Lopez Cases & Partner, Seville (still active in 1967). Regn cld in 1967 and fate unknown.

202 M.3A Falcon Major; painted blue. (P&P works records – Subs & mods to G-ADBF) Regd **G-ADIU** (CofR 5944) 13.6.35 to The Leicestershire Aero Club Ltd, Braunstone. CofA No.4899 issued 11.7.35; dd 11.7.35. Flown 7.1.37 by Flt Lt RL Bateman to Rollason Aircraft Ltd, Hanworth following sale. Regd cld 1.37 as sold. Regd in Australia as **VH-ACE** (CofR 627) 17.3.37 to FC Higginson & Co, Archerfield; repainted red and white. Regd 13.10.38 to Mrs Ethel Blanche Jones, t/a Airwork Co, Archerfield. Regd 8.5.40 to Thomas Horatio McDonald, Cairns, Qld. Regn cld 27.5.42 and impressed into RAAF as **A37-6**. To 33 Squadron, ex SHQ Townsville 11.6.42. Caught in a severe gust of wind on take-off 15.10.42 which caused it to swing and the u/c collapse, damaging the port aileron and bending the airscrew tip; to 12 RSU 16.10.42. Despite reportedly being under repair for return to 33 Squadron 21.10.42; on 25.10.42 it was recorded as 'being brought overland from Augustus Downs' and on 3.12.42 it was 'being repaired Augustus Downs - awaiting paint'. On 11.12.42 the allotment back to 33 Squadron was cancelled and it was allotted to 5 Comm Flt, but on 28.12.42 it was 'still under repair at Augustus Downs'. On 20.7.43 it was allotted to 13 ARD for completion of repair and was issued to them 28.7.43 but seemingly not received until 14.10.43. On 11.10.43 authority was requested 'to convert aircraft to components' as it was then considered that it was 'damaged beyond economical repair' (strange that they should only have been awaiting paint on 24.11.42!). Approval to reduce to spares was given on 18.12.43; no further details.

206 M.3A Falcon Major; painted blue. (P&P works records – Subs & mods to G-ADBF) Regd **G-ADLI** (CofR 6040) 10.7.35 to Kennings Ltd, Clay Cross (Kennings were motor dealers run by George Kenning who kept the Falcon at Skegness where it was flown by his personal pilot John Addinsell). CofA No.4943 issued 12.7.35; dd 13.7.35. Regd (CofR 8609) 5.7.38 to Kenneth James Dear, Liverpool (but based at Redhill). CofA lapsed 29.6.39. Cld 3.8.39 and regd 23.8.39 to Charles Brian Field, Kingswood, Surrey. Surveyed late 8.39 by Air Ministry and allocated to NAC.1 1.9.39. Not impressed and CofA renewed 31.10.40 to enable Field to operate it in relation to his test flying work for Airspeed Ltd at Portsmouth and Christchurch. Regd 17.11.41 to Marshall's Flying Schools Ltd, Cambridge and CofA renewed 11.12.41 (and renewed throughout war). Postwar CofA renewed 5.6.46 and regd 12.6.46 to Charles Philip Lloyd Godsal, London SW.1 (based Heston). Cld 25.5.48 & regd 27.5.48 to Leslie Thomas Edward Bradley, (t/a Falcon Airways), Birmingham. Cld 10.6.50 & regd 21.6.50 to James William Haggas, Hatfield. Cld 29.9.51 & regd 29.10.51 to James William Haggas & Timothy Robin Hilton, Hatfield. Sold 9.8.52 to unidentified party and not reregd. Hit tree 600 yds from take-off from

A post-war photograph of G-ADLI. [A J Jackson via B Clarke]

Elstree Aerodrome at 10.00 hrs 10.9.52 and completely destroyed; pilot Robert Pilkington and passengers Mrs Jenny Gilbert, Sir James Scott Douglas and Mr V Snell suffered minor injuries.

209 M.3A Falcon Major; painted blue. (P&P works records – Subs & mods to G-ADBF) CofA No.5071 issued 25.9.35 possibly to Phillips & Powis Aircraft (Reading) Ltd for unrecorded sale. Regd **G-ADZR** (CofR 6586) 13.12.35 to Miss Helen Marcelle Harrison, Yapton. (CLN records dd 31.10.35 but the first reference to Miss Harrison was when Bob Milne made 2 dual flights with her at Woodley in Hawk Major G-ADAW on 18.12.35). The works records show Miss Harrison as c/o Mrs Saunders, Lepp Street, Johannesburg, South Africa (and thus it is possible that the Falcon was initially intended for export to her in South Africa). Dd to Yapton 1.36 and based with the Yapton Aero Club, which Miss Harrison had joined to obtain her 'B' licence and instructor's certificate. Regn cld 6.36 as sold. Regd in Australia as **VH-AAS** (CofR 598) 24.7.36 to Arthur Henry Schutt, Essendon; named *The Rebel*. Regd 23.2.40 to RND Miller, Melbourne and dd 2.3.40 to Connellan Airways Ltd, Alice Springs. Crashed into tree on take-off on medical evacuation flight Hatches Creek, Northern Territories 23.7.40. Pilot, EJ Connellan and passengers unhurt but Falcon destroyed in bush fire before it could be recovered. Regn cld 21.8.45.

216 M.3A Falcon Major; painted silver. (P&P works records - conforms to G-ADBF but fitted with one pilot's seat and modified rudder bar). Regd **G-AEEG** (CofR 6813) 14.3.36 to Phillips & Powis Aircraft Ltd for their air taxi department at Hanworth. CofA No.5413 issued 25.3.36. (A Falcon Major with the 'B' marks U20 was recorded by AJ Jackson at Woodley 15.4.36 but, despite his assumption, it was NOT G-AEEG. Miles had flown G-AEEG to Martlesham Heath and back 27.3.36 and on the day AJJ visited, Miles flew G-AEEG on a 25-minute flight from Woodley). One of G-AEEG's early hirers was Geoffrey Last, chief pilot for British Air Transport Ltd, Redhill, who hired it to take Mr JM and Mrs Waller to Chateauroux in France on 4th June, returning on the 6th. On 7.8.36, Graham Head flew it to Newtownards in Northern Ireland via Heston and Hooton Park, but ran out of fuel over the Irish Sea, managing to glide in from 12,000 ft!. Flown by George Miles between 18.9.36 & 1.10.36. Regn cld 10.36 as sold. Regd in Sweden as **SE-AFN** (CofR 203) 26.11.36 to Karl E Sandberg, Stockholm, Sweden. Sandberg had learned to fly at Reading Aero Club and flew home to Norrköping in SE-AFN. The Falcon, by now red and cream, was repainted green and cream for the Swedish operator to match their other aircraft. Operated by Nordisk Aerotjänst, Norrköping, named *Kolmården*. Mishaps were reported on 9.1.37, 21.7.38 and in 8.39. lin addition a fatal incident occurred on 12.3.38, when flights were being made on skis from a frozen creek of the River Dalälven, just outside the skiing resort of Sälen. Rolling out following a landing, the propeller struck a 60-year old woman who was crossing the frozen river on foot on her way home; she died instantly. Sold 11.8.37 but not regd until 14.8.39 to Capt CA Nilsson, Norrköping. Another mishap occurred in 2.40 but no details known. Regn cld 24.5.40 as sold to the Flygvapnet (Swedish Air Force) (designation Tp7 - light transport); painted dark green. Dd to F1 Wing, Västerås/Hasslö as **Fv 913** 24.5.40 and used by the Bomber Group's Liaison Flight, with code 1-71. (A report that the serial was later changed to Fv.7001 is incorrect). To F8 Wing, Barkaby 12.5.41 for use by the Flygvapnet Central Staff Liaison Squadron, with code 8-58; following overhaul it was repainted silver and re-coded 8-80. On 30.11.43, the Falcon was put up for sale due to lack of spares and was SOC in 1944. Sold 29.1.44 (for SwKr 7,500) and regd **SE-AFN** 21.3.44 to Svensk Flygtjänst AB (Swedish Air Service), Stockholm; based as an air taxi at Bromma and Malmö/Bulltofta. On 21.9.44 it was involved in a collision on the ground with the Fw.200A Condor OY-DEM at Bulltofta, sustaining minor damage. Sold 16.9.46 and regd 27.11.46 to ATC Gosta Bergelin, Såtenäs. Sold 29.9.49 & regd 1.12.49 to Göte Rosen, Såtenäs, operated by Auto-Flyg Reklam of Linköping and Sundyberg. Sold 14.4.52 and regd 29.5.52 to Bengt A Eriksson, Bromma and Lennart Zandin, Dreylken. Regd 4.7.55 to Bengt A Eriksson, Bromma (sole). Sold 30.11.55 and regd 9.2.56 to Åke Laurell, Uppsala. Following CofA expiry 28.6.61, the Swedish authorities declined its renewal due to concerns as to the state of the glue. A 3-day permit was obtained to fly it to England 'in the hope that someone would give it a good home'; Laurell took off from Uppsala/Sundbro on 21.7.61 en route to FG Miles Engineering Ltd, Shoreham via Malmö/Bulltofta - Eslöv - Bussbafen - Groningen (calling in at an airshow at Siljänsnäs in Balärnas, Sweden en route), to arrive at Lympne 23.7.61 and flying on to Shoreham Airport the same day. Retained at Shoreham pending sale; it was taken by road to White Waltham 5.7.62. Laurell finalised a sale to Doug Bianchi of Personal Plane Services Ltd, White Waltham 10.10.62, reportedly for 'a tenner'! Swedish regn cld 12.10.62 as sold in UK. Regd **G-AEEG** (CofR R7760) 26.11.62 to Personal Plane Services Ltd, White Waltham and restoration commenced on the rear fuselage with some of the ply skinning being renewed and the bottom of the fin kingpost replaced and modified to accept an Auster Autocrat tailwheel assembly in place of the tailskid. Not completed and exchanged by Bianchi for Edward Eves' Alfa-Romeo 6C/1750 twin-cam Castagna coupe! Eves collected G-AEEG from White Waltham on a Dixon Bate trailer and delivered it to Frank Golding (of Brooklands Aviation Ltd) at Sywell on 12.3.64, for completion of rebuild. Cld 6.3.64 & regd 2.4.64 to Edward Eves Ltd, Leamington Spa. Rebuild completed by Brooklands Aviation, including 'Top and bottom longerons repaired at front of fuselage port and starboard, new ply panels fitted as required, fireproof bulkhead repaired'. It was also fitted with wheels and brakes from a Proctor (though this not recorded in logbook) but references to a wholesale replacement of forward fuselage due to 'the original being beyond redemption due to oil soakage' is not corroborated by the record in the 'Log Book Certificate' issued by Brooklands at the time (but this may have been unrecorded work done by PPS). Since it had been dd from Sweden 'trouserless', new 'trouser' fairings were made, modelled on those salvaged from the wreck of Geoffrey Marler's Falcon Six G-ADTD. The first post-restoration flight was made at Sywell on 15.4.65 by former Miles test pilot Hugh Kendall, (now Chief Test Pilot for the ARB) and G-AEEG was dd to Eves at Baginton 23.4.65. CofA renewed 27.4.65. In May (or July) 1966, the Gipsy Major engine was replaced by Gipsy Major 10 Mk.2 (No.10812), taken from Ernie Crabtree's racing Gemini G-ALUG. Flown in the King's Cup Air Race on 28th June 1970, but with the handicap of 158 mph probably based on the Falcon Six, Eves put up a very creditable average

SE-AFN in service with Nordisk Aerotjänst, Norköping, prior to 1939. [Via P Amos]

The Falcon Major (SE-AFN) in service with the Swedish Air Force, with F8 Wing at Barkaby during 1943. [Via P Holloway]

SE-AFN in service with Svensk Flygtjänst AB, Stockholm, after the war. [Via P Amos]

131.5 mph to come in 33rd. Stored at Baginton from June 1970 to November 1972. Cld 1.11.72 & regd 4.12.72 to Philip Ashley Mann, Greenwich, having been dd 8.11.72 to Personal Plane Services Ltd, Booker, for major re-skinning work. CofA renewed 10.7.73 and then based Biggin Hill. Regd 8.2.78 to Mann's company Shipping and Airlines Ltd, Biggin Hill. On 9.9.79, G-AEEG was flown to victory by Dr Ian Dalziel in the King's Cup Air Race, held at Jurby in the Isle of Man. Entered in Christies' auction held at Duxford in 1983, but failed to reach its reserve price, with a maximum bid of just £7,500. Regd 29.8.84 to Vintage Aircraft Magazine Ltd, Denham, (under the care of EMK Aeroplanes) and flown by Roger Hinchcliffe. Sold in 1989 to Anthony Pritchard of Aston Publications Ltd, Bourne End, Bucks, but on 14.4.89 he sent the Log Books and associated documents to Henry Labouchere of Holt, Norfolk, with a view to its sale. Probably sold to Roger Reeves and in June 1989 sold to Tim Moore of Skysport Engineering Ltd, Rotary Farm, Sandy, Beds. Completely refurbished by Skysport and repainted in its current attractive cream and blue colour scheme. Swung on aborted take-off with Hatch, Beds 18.9.97 damaging propeller, engine, left wing & starboard u/c; pilot Dr Michael Fopp (of RAF Museum) unhurt and aircraft repaired. Sold 10.98 & regd 24.11.98 to Gordon EJ Spooner, Colchester (based at Earls Colne). Sold 9.9.00 & regd 12.7.01 to Peter Holloway, initially based at Turweston but later at Old Warden, where it is maintained in airworthy condition (total 2,175 hours at 31.12.03).

226 M.3A Miles Falcon; painted silver. (P&P works records – Subs & mods to G-ADBF) Regd **G-AETN** (CofR 7654) 15.2.37 to Phillips & Powis Aircraft Ltd. CofA No.5874 issued 22.4.37; dd for use by their air taxi department 5.37. G-AETN had the shortest life of all Falcon Majors as it crashed at Duckaller, nr Starcross, Dawlish, Devon 17.5.37, the pilot William Bruce was killed and the two passengers injured. Regn cld 12.37.

229 M.3A Falcon Major; painted silver. (P&P works record - Conforms to G-AEEG with the exception of 17.5 gal aluminium fuel tank and M.3B Falcon Six type control box). Regd **G-AEFB** (CofR 6858) 30.3.36 to Phillips & Powis Aircraft Ltd. CofA No.5528 issued 4.6.36; dd for use by Air Hire department 14.6.36. Flown by Graham Head, from Woodley, on 17.6.36, where he commented 'Overshot twice!' and went on to fly it with passengers on many occasions between 19.6.36 and 27.8.36; George Miles also flew it many times between 30.10.36 and 1.3.37. Flown by Flt Lt Arthur Clouston in the King's Cup Air Race at Hatfield 10-11.7.36; finished 12th at 137.04 mph. Sold by Airwork Ltd, Heston for £800 and regd (CofR 8050) 12.8.37 to Denys Felton Peel, London W1 (based Heston); Peel worked in Egypt and was on leave in England and decided to fly back to Egypt 'with an unsuspecting Army friend returning to Cairo from home leave'. Leaving Heston in 8.37, they flew via Paris, Lyons, Cannes, Rome, Naples, Catania, Sicily and along the North African coast, with refuelling stops at Tripoli, Sirte, Benghazi and Amseat. Shortly after take-off on the last leg of the journey to Alexandria and whilst crossing the frontier into Egypt, it was noticed that a metal band which sealed the 2" gap between the centre section and the outer wing was loose and flapping. A hurried forced landing on what looked like a level and firm piece of desert near Solum proved to be a rough area and they hit a bump, at which the metal strip ripped off and the wing dropped. Peel managed to avoid a complete disaster but damage was sustained to the port undercarriage fairing and the tyre was punctured; the aircraft was successfully recovered several weeks later and repaired (a full account appeared in *Pilot* magazine for January 1969). Cld 1.8.39 and regd 15.8.39 to Portsmouth Southsea & Isle of Wight Aviation Ltd, Portsmouth. Regn cld 14.4.40 as sold. Impressed as **X9301** 14.3.40 and dd to 20 MU Aston Down the same day. To Phillips and Powis Aircraft Ltd, Woodley by road for major inspection in 1.41 but SOC (Cat.E1) spares use 10.2.41.

234 M.3A Falcon Major; painted silver. (P&P works records - Conforms to G-AEFB, with the exception of modified tail trim gear as fitted to M.3B Falcon Six aircraft, an approved modification). Regd **G-AENG** (CofR 7319) 10.9.36 to Phillips & Powis Aircraft Ltd. CofA No.5672 issued 28.9.36; dd for use by Phillips & Powis 'Firm Air Hire' 30.9.36. Flown by Wing Cdr Edward Hilton with Wing Cdr Percy Sherren in the King's Cup Air Race from Hatfield on 10.9.37 and crashed during a turn on the course near Scarborough Castle; both were killed. The cause of the accident was determined to be the abnormal squally weather conditions which resulted in one of the occupants being thrown through the cockpit canopy. The Falcon had been among the last of the 27 entrants to leave Hatfield in the 1,442 mile, two-day race around England and Ireland; Hilton was previously CO of the Performance Testing Squadron, A&AEE, Martlesham Heath and Sherren was the first CO of 15 Squadron at A&AEE. Regn cld 12.37.

236 M.3A Falcon Major. CLN recorded Falcon Major airframe and parts dd 9.2.37 to the Bristol Training Centre for Unemployed.

Note: Whilst we discount the possibility that U20 seen by A J Jackson on a Falcon Major at Woodley on 15.4.36 was c/n 216 (G-AEEG), the actual identity of the Falcon wearing these marks is not known. It is possibly the slotted Falcon Six c/n 252 (K5924) which Miles first flew on 12.2.36 but that must be considered purely speculative.

Production summary

M.3 Falcon	c/n 102	**1** prototype
M.3A Falcon Major	c/n 131, 140, 149, 157, 163, 181, 189, 193, 196, 197, 201, 202, 206, 209, 216, 226, 229, 234, 236*	**19** * c/n 236 a/f and parts
Total		**20**

Appendix 14: Miles M.3B, M.3C, M.3D, M.3E and M.3F Falcon Six Production

213 M.3B Falcon Six; painted cream. (P&P works records 'conforms to M.3A G-ADBF but fitted with modified controls, seating reduced from 4 to 3; fitted with 200hp DH Gipsy Six engine'). Regd **G-ADLC** (CofR 6027) 10.7.35 to Charles Owen Powis, Woodley. Built in the new experimental shop and first flown, by Miles 20.7.35. CofA No.4976 issued 27.7.35; handed over the same day. Application made 20.8.35 for fitting extra 10-gallon fuel tank in locker. Entered by Viscountess Wakefield of Hythe in the King's Cup Air Race at Hatfield 6/7.9.35 (Race No.16) and flown by Tommy Rose to 1st place at an average speed of 176.28 mph. AUW increased to 2,350lb on unrecorded date (some confusion exists here as this was the normal AUW, so should probably have read increased to 2,500lb). Modified flap gear and Harley leading edge type landing light fitted (no date recorded). Application made 2.12.35 for Special category CofA for Capetown flight, out and return (also for fitting a 75-gallon fuel tank) Tommy Rose left Lympne 7.1.36 on South Africa record attempt but was forced down by icing at Abbeville, France; aircraft damaged and Rose slightly injured. The second attempt by Tommy Rose on record breaking flight to South Africa was successful; he departed Lympne 6.2.36 and arrived Capetown 9.2.36; return flight departed 3.3.36 and arrived Croydon 9.3.36. G-ADLC was then transferred to Phillips & Powis Aircraft Hire Dept, Hanworth. Modified with elevator trimming tab operated by Teleflex control on CofA renewal 30.10.36. Regd (CofR 7444) 9.11.36 to Clive Branson, London SW.11 (based Woodley). Regd (CofR 7605) 18.1.37 to Germ Lubricants Ltd, Heston, named *Germ 3* (and operated by their aviation manager James 'Wattie' Watson). Regd (CofR 8607) 5.7.38 to Edward Devereux Spratt, t/a Air Touring, Gatwick. Hit hedge in forced landing due to engine failure Painsford Farm, Harbertonford, nr Totnes 28.7.38; pilot Norman Waugh and passengers Peter Morey & The Hon Victor Harvey all slightly injured. Regn cld 31.12.38.

G-ADLS, a Miles M.3C Falcon Six, at Hatfield for the 1935 King's Cup Air Race. [The A J Jackson Collection]

The much-modified Miles M.3C Falcon Six EC-ACB, photographed in the Aero Club hangar at Zaragosa, Spain, on 15th May 1976. [A Camarasa via J Havers]

231 M.3C Falcon Six; painted blue. (P&P works records state 'conforms to G-ADLC but with dual control, side-by-side seating and two passenger seats'). Regd **G-ADLS** (CofR 6062) 15.7.35 to Samuel Harris, Caterham, Surrey and operated by his company Croydon Airways, Croydon. CofA No.5039 issued 28.8.35; dd 5.9.35. Entered by Harris and flown by him and Laurence Lipton in the King's Cup Air Race at Hatfield 6/7.9.35 (Race No.15) to gain 5th place at an average speed of 166.88 mph. Regd (CofR 7123) 11.6.36 to Andrew Douglas Farquhar, Hermiston, Bridge of Weir (based Abbotsinch). Regn cld 1.37 as sold. In fact it was sold 9.8.36 to Union Founders Trust Ltd (on behalf of the Euskadi Government, allies of the Spanish Republicans) and dd same day via France, arriving Bilbao on 14.8.36 to join the Krone Circus Squadron. One unconfirmed report states it was delivered with the false regn I-BELA and as with many foreign aircraft involved in the civil war, its subsequent history is unconfirmed. Jaime Velarde Silio stated that it was the aircraft usually flown by Walter Scott Coates, an English pilot who, in September 1936, was involved in a 'melée' among various Republican aeroplanes and three Luftwaffe Condor Legion Heinkel He51's who took the Falcon for 'a captured monoplane of new design'. Reported as having Spanish military serial **30-117**. In late 1936/early 1937, 'G-ADLS' took part in the Basque offensive in Villarreal, with its crew even throwing four bombs over the Álava base of the Condor Legion. Also reported as being the Falcon flown by Elias Hernandez Carnison (with three passengers) which landed at Biarritz 20.10.37 while escaping from the Nationalist advance; ordered to leave France, it was wrecked in a forced landing at Grimaud, nr St. Tropez 7.11.37. After this, its history becomes more uncertain, since it becomes intertwined with that of M.3A Falcon Major EC-BDD (c/n 197). However, while unconfirmed, it is probably the Falcon Six regd **EC-CAO** 21.2.42 to Maria Mercedes Iriarte. Regd 5.5.42 to Compania Espanola de Trabajos Fotogrametricos Aereos (CETFA). Reregd **EC-ACB** (CofR 52) .47 to same owner (but also quoted as regd 13.11.50 to Ramon Munoz Soler with date shown in BV as 4.9.51, albeit with no owner). Fitted at some time with a very poorly made but more 'conventional-looking' windscreen and enlarged fin and rudder (and by now officially regd with c/n 197). Wfu by CETFA in 1959. Purchased in about 1969 by and regd 23.2.73 to General Manuel Buitron Fernandez, Zaragoza (reportedly flown by him until at least 1971). Stored in a hangar at Zaragoza Airport (where it was inspected in 10.86 by potential purchaser, Michael Abbott who reported that it was in generally good condition being painted cream with red internal trim and sporting the title *Virgen de Loreto*). Later acquired by a group of individuals who became the Fundacion Infante de Orleans (FIO), Madrid. Shipped 10.93 to Personal Plane Services Ltd, Booker for rebuild, which ultimately included a new centre section built by Tim Moore of Skysport at Rotary Farm, Beds (replacing the original oil-soaked one which was donated to The Miles Aircraft Collection). Assembled at Booker and reflown by Jonathon Whaley (chief pilot of Jet Heritage Foundation), on 15.5.97. Dd back to the FIO at Cuatro Vientos, Madrid 10.6.97; it is maintained in airworthy condition and flown on the Museum's flying days.

232 M.3B Falcon Six; colour not recorded. (P&P works records state 'conforms to G-ADLC but with low pressure u/c'). CofA No.4988 issued 26.10.35. Regd **OE-DVH** to Adrianus J van Hengel, Vienna, Austria; collected by Captain Godwin von Brumowski and dd 14.11.35. Flown (by von Brumowski) in the 1936 Verzeichnis der Teilnehmer am Pfingstflug from 28.5.36 to 4.6.36 (Competition No.26), which was organised by the Österreichischen Aero Club. Stalled on landing and crashed at Amsterdam 3.6.36; owner van Hengel and Brumowski both killed.

233 M.3B Falcon Six; colour not recorded. (P&P works records state 'subs to G-ADLC') CofA No.4989 issued 26.10.35. Regd **OE-DBB** 31.10.35 to Leopold Bloch-Bauer, Vienna. Flown by owner (with passenger Fr A Bloch-Bauer) in the 1936 Verzeichnis der Teilnehmer am Pfingstflug, from 28.5.36 to 4.6.36 (Competition No.33) which was organised by the Österreichischen Aero Club. Regn cld on sale in England. Regd **G-AFAY** (CofR 8032) 29.7.37 to Airwork Ltd, Heston. CofA renewed 24.8.37. Regd (CofR 8348) 8.2.38 to Brian Allen Aviation Ltd, Croydon. Sold 8.38 to Edward Devereux Spratt, t/a Air Touring, Gatwick (as replacement for written-off Falcon Six G-ADLC) and regd (CofR 8900) 24.11.38 to (his partner in Air Touring) Cyril Geoffrey Marmaduke Alington, Lincoln, named *Angela II*. Left Gatwick 18.1.39, flown by Alington (with his mother, Angela and the former Gatwick Control Officer JV Bowring as passengers) to Nairobi, Kenya, and on to Kitale and Nyeri before returning to England on a 14,000 mile round trip. CofA lapsed 17.4.40. CofA renewed 12.6.41 and regd 14.8.41 to Hawker Aircraft Ltd, Brooklands and flown on communications duties throughout war. MAP Permit No.13 lapsed 8.12.45. Regn cld 1.10.46 as sold; presumed used for spares.

Miles M.3B Falcon Six G-AFAY. [Via B Clarke]

247 M.3B Falcon Six; painted red fuselage and silver wings and tail unit (although contemporary reports quoted orange as the main colour). (P&P works records state 'subs to OE-DVH, less wireless'). CofA No.5072 issued 29.10.35 to Union Airways, New Zealand; dd to shippers 3.11.35. First flown after assembly in Wellington 18.12.35. Regd **ZK-AEI** 12.35 (CofR 46/CofA No.157) to Union Airways of NZ Ltd, Wellington (for air taxi and charter work). Struck anemometer mast or windsock in turbulent conditions and crashed on landing in bad weather at Rongotai (Wellington) 19.2.36 after aborted flight to Palmerston North; pilot Sqn Ldr Malcolm C McGregor (who had flown the Miles Hawk in the MacRobertson Race) died of his injuries and passenger, sheep farmer CLF Hamilton was injured but survived.

248 M.3D Falcon Six; painted yellow. (P&P works records state 'conforms to G-AEAO, with the exception of two 22-gall aluminium fuel tanks and RTE Homing Set not fitted'). Regd **G-AEKK** (CofR 7067) 23.6.36 to The Dunlop Rubber Co Ltd, Castle Bromwich. CofA No.5609 issued 27.7.36; dd 12.8.36. Regn cld 6.1.40 as sold. Impressed into Royal Navy as **W9373**; dd to 781 Squadron, RNAS Lee-on-Solent 2.3.40-(8.40). Possibly SOC following an accident on 16.10.41 while being flown by Lt Cdr AM Pilling (although this accident also recorded as involving Vega Gull W9375 of RNAS Evanton).

252 M.3B Falcon Six Special; painted silver. (P&P works records state 'subs to ZK-AEI but with modified flaps and leading edge slats' - but they give no details of the subsequent changes to configuration; a mass balanced rudder was initially fitted). Ordered (as one aeroplane plus one extra wing) by the Air Ministry to Contract No.414667/35 at a projected cost of £1,560 plus £411, with an instruction to proceed (ITP) dated 31.7.35 as **K5924** for RAE Farnborough for research into wings with leading-edge slats and modified flaps. Contract No.435096/35 issued to Phillips & Powis Aircraft Ltd for research flights using this aircraft at a projected cost of £300; ITP dated 1.8.35. First flown, by Miles as the 'Falcon VI (slotted)' 12.2.36 (some sources say that it carried 'B' mark U20 but this is not confirmed); dd to Farnborough 9.4.36 (although CLN records 27.3.36). Made 23 flights from Farnborough in April, 17 flights in May and 9 in June before being returning to Woodley 20.6.36. With effect from 4.7.36, it was then fitted, in turn, with three different sets of high-taper outer wings. Miles recorded a 10-min flight on 22.8.36 in a 'Falcon with taper wings' (presumed to be this aircraft); dd to RAE 1.9.36 'for flap research'. Contract No.523530/36 issued to Phillips & Powis Aircraft Ltd (date unknown) for the manufacture of one 'Falcon wing' at a projected cost of £320, presumably to cover part of this work. Returned to Woodley 30.7.38 where a special 'double flap' was fitted; flown back to RAE (by Pebody) 9.8.38 'for stalling tests' (not known whether the taper wings were still fitted at this time). Returned to Phillips & Powis (pilot Burke) 1.11.38 for 'fitting wing'; back to RAE 10.1.39. Returned to Phillips & Powis (pilot Wilson) 2.5.39 for 'fitting tapering wing'; back to RAE 26.5.39 by Bill Skinner for handling tests and communications. At some stage (and while still painted silver) it was fitted with a horn-balanced rudder. Returned to Phillips & Powis for fitting 'high speed' wings (which were the 3 sets of alternative wings of Piercy laminar flow section which Miles had built for the RAE for comparative tests to investigate the drag in differing thickness ratios); these wings were fitted in late 1939/early 1940. On 27/28.3.40 Hugh Kennedy carried out two 'test high speed wings' flights, followed on 28/29.3.40 by two 'test flight - loading to L13/15', after which he delivered the Falcon to Farnborough on 30.3.40. Flown by Hugh Kennedy 22.4.41 on a handling test from Woodley and dd by him 24.4.41 to Farnborough (camouflaged by this time and with a cylindrical device with a flat top installed on the top fuselage decking aft of the cabin). Presumed reverted to standard wings prior to dd to 4 Grp Comm Flight, Linton-on-Ouse (by 31.12.42). To 5 MU Kemble 17.12.43. To Miles Aircraft Ltd 29.3.44. SOC Cat.E 27.4.44. Still at Miles Aircraft Ltd 30.8.45.

255 M.3B Falcon Six; painted blue. (P&P works records state 'subs to G-ADLC') Regd **G-ADTD** (CofR 6306) 20.9.35 to Maddox Airways Ltd, Brooklands; not dd as company had probably ceased trading. Regd (CofR 6698) 1.2.36 to Vickers (Aviation) Ltd, Brooklands. CofA No.5195 issued 4.2.36. On 5.3.36, Vickers test pilot Jeffrey Quill flew chief test pilot Capt 'Mutt' Summers in G-ADTD from A&AEE Martlesham Heath to the Vickers-Supermarine works at Eastleigh for Summers to make the first flight of the prototype Spitfire K5054. Regn lapsed (accidentally) at census 1.1.39; regd 19.5.39 to Vickers-Armstrongs Ltd, Eastleigh and used as a hack by them throughout the war. On 6.3.41, G-ADTD forced landed in a field at Wisley while flown by Mutt Summers (a consequence of which led to a decision to build an aerodrome there for test flying). CofA lapsed 22.11.43. Regd 26.7.44 to Miles Aircraft Ltd and fitted with a one-piece moulded Perspex windscreen; MAP Permit No.27 issued from 22.12.44. CofA renewed 19.2.46. Cld 17.6.47 & regd 30.6.47 to Royce Sidney Turner, London, W8 (based at Woodley). CofA lapsed 5.5.49. Cld 17.8.50 & regd 14.9.50 to FG Miles Ltd, Redhill; test flown by Miles at Redhill 10.12.50 and CofA renewed 15.12.50. Regd 9.2.51 to The Wiltshire School of Flying Ltd, Thruxton. Flown by Geoffrey Marler (a Wiltshire School of Flying director) to 4th place in the 1955 King's Cup Air Race held at Baginton 20.8.55 and to 1st place in the SBAC Challenge Cup at Baginton 21.7.56 (Race No.92) at an average speed of 169 mph. Regd 2.4.59 to Geoffrey Cecil Marler, London W1 (based at Thruxton). Sustained minor damage when struck by another aircraft at Thruxton Aerodrome 6.3.60 (repaired and CofA renewed 3.3.61). Crashed into the sea about 150 yards off Angmering, Sussex 21.9.62, Geoffrey Marler killed. The aircraft was on a local flight from Shoreham and Marler had undertaken a low aerobatic loop over his Angmering home at 3.57 pm when the engine cut and he struck a 30 ft tree in his garden, slicing off part of one wing. Regn cld 6.11.62 as destroyed.

256 M.3B Falcon Six; painted blue. (P&P works records state 'subs to ZK-AEI'). CofA No.5193 issued 9.12.35. Supplied via Airwork Ltd, Heston and regd **D-EGYV** 1.36 to Karl Theodor Roechling, Krei Alsdorf, Aachen, Germany; handed over 15.12.35. Regn cld 8.37 as sold in England. Regd **G-AFBF** (CofR 8060) 14.10.37 to Airwork Ltd, Heston. CofA renewed 3.2.38. Sold 1.9.39 and regd 31.10.39 to Birkett Air Service Ltd, Speke. Regn cld 26.4.40 as sold. Impressed into RAF 26.4.40 as **AV973** and to 48 MU Hawarden 9.5.40. To ATA White Waltham 13.10.40 but possibly on delivery to Vickers Armstrongs (Aircraft) Ltd, Brooklands for communications duties. Cat.B; to Atlantic Coast Air Lines Ltd, Barnstaple for repairs 15.2.42; to RAF Marston Moor 21.8.42. To 5 MU Kemble 9.11.43; to Miles Aircraft Ltd, Woodley for Cat.B repairs 27.11.43. To CRD at Westland Aircraft Ltd, Yeovil 30.4.44. To 5 MU Kemble 11.45 for disposal. Sold to Holmwood Motors Ltd, Southsea, Hants in 1946. Regd **G-AFBF** 31.5.46 to John Dixon Habin, t/a Southampton Air Services, Eastleigh. CofA renewed 19.6.46 (possibly 9.6.46). Cld 25.6.47 & regd 30.6.47 to Cunliffe-Owen Aircraft Ltd, Eastleigh. Crashed in France shortly prior to 15.7.48; possibly in a forced landing in an orchard at Tours, while being flown by AJ Payne. The aircraft was acquired by The British Aviation Insurance Co Ltd, London EC.3 and regd to them 9.12.50 (but probably remained in France). Regn cld 20.12.53 as sold in France. Reportedly rebuilt by Aero Club de Touraine for owner M. Gregoire (by early 1954). Not regd in France and used for spares for F-BBCN c/n 269 at Lognes-Emerainvilles. Remains still present in 1972, in silver finish without markings but with its RAF serial AV973 still clearly visible underneath.

Miles M.3B Falcon Six G-AFBF seen at Eastleigh Airport in 1946.
[Via B Clarke]

259 M.3D Falcon Six; painted blue. (Laid down as M.3B but completed as an M.3D; P&P works records show originally 'conforms to ZK-AEI but fitted with RTE Transmitting and Receiving wireless set'; later amended to 'conforms to G-AEAO but fitted with RTE Receiving and Transmitting wireless set'). Regd **G-AEDL** (CofR 6770) 3.3.36 to Edward Gordon Houstoun Forsyth, London SW.7 (based Heston, later Croydon, possibly when he became a director of Personal Airways); type formally amended 15.4.36. CofA No.5362 issued 26.5.36; dd 3.6.36. Operated by Personal Airways Ltd, Croydon (by 10.37); still operated 4.39. Regn cld 3.9.39 but circumstances unknown.

G-ADTD, a Miles M.3B Falcon Six, seen in wartime colours.
[Via B Clarke]

Miles M.3D Falcon Six G-AEDL at Croydon Airport in 1936.
[Mr Lintott via N D Owen]

262 M.3D Falcon Six; painted red. (Laid down as M.3B but completed as M.3D; P&P works records state originally 'conforms to OE-DVH but fitted with Dunlop brake gear, retractable landing lights (Grimes) and RTE Homing set'; later amended to 'conforms to G-AEAO but fitted with Dunlop brake gear and Grimes retractable landing lights'). Regd **G-ADZL** (CofR 6576) 12.12.35 to Fairey Aviation Co Ltd, Great West Aerodrome, Hayes. CofA No.5485 issued 8.5.36 (after application amended 20.4.36 because Onions u/c legs to be fitted and AUW increased to 2,500lb); dd 9.5.36. CofA lapsed 29.10.37 but operated under 'dispensation' permit until CofA renewed 5.10.42. Operated under MAP Permit No.18 14.12.43-13.12.44. Regn cld in postwar census 22.11.45; fate unknown.

266 M.3D Falcon Six; painted blue. (P&P works records state 'conforms to ZK-AEI but with modified u/c and strengthened tailplane (Dowty) (believed error for tailwheel – Ed), AUW increased to 2,500 lb'). Regd **G-AEAG** (CofR 6632) 13.1.36 to Henry Deterding, Newham Grounds, Daventry (based at private field there and also at Sywell). CofA No.5410 issued 21.3.36; dd 22.3.36. Regd (CofR 7944) 17.6.37 to Arthur Wilfred Alexander Whitehead, London W1 (based at Woodley). Whitehead departed England 13.9.37 for Australia, but the aircraft was shipped ex Sourabaya to Darwin on *SS Merkur*, arriving Darwin 13.11.37; assembled & departed Darwin 19.11.37 and arrived Mascot Aerodrome, Sydney 21.11.37. Regn cld 31.12.38 as sold. Regd **VH-ABT** (CofR 701) 9.6.38 to RH Hamblin, Ganmain, NSW. Impressed into RAAF as **A37-1** 13.11.40; received from Hamblin and allotted to 2 Comm Flt 12.11.40. Port u/c collapsed whilst landing 12.5.41 damaging the port side of the centre section spar; to Qantas, Archerfield for repair 31.5.41. Progress was slow due to shortage of materials and it was finally dd back to 2 Comm Flt, Abascot 24.1.42. To 3 Comm Flt 10.8.42. On 3.11.44 a 'request allotment aircraft due for complete overhaul, hours flown since new 1,863.45 hrs since last complete overhaul' was recorded; to 2 CRD 6.12.44 for survey report. On 6.3.45, disposal offer accepted from W Stillard; collected by him 8.4.45. Regd **VH-ABT** 5.8.46 to Wallace Stillard, Marbor, Vic. Regd to EN Sutton, Sydney. Regd to Sutton's Motor Pty Ltd, Sydney. Regd 5.4.49 to Guinea Air Traders Ltd, Lae, NG. Regd 6.6.50 to Robert Smith & Charles Kenneth Rohr, Tooraweenah, NSW. Swung on take-off Coonabarabran, NSW 6.12.50 and went through fence, damaging port wing & u/c; repaired. Regd 17.5.51 to Howard Kennedy Morris, Dover Heights, NSW. Regd 16.7.51 to John Malcolm Bonney, Cobar, NSW. Regd 29.9.56 to WJ Robinson, Rowville, Vic. Regn cld 22.8.59 as wfu. The wings and u/c were shipped to Canada by Bob Diemart in about 1970 for use in the rebuild of the Hawk Major CF-AUV (c/n 123); the fuselage having 'long gone'. In May 1994, Col L Keith Hatfield of Moorabbin Airport, Victoria, confirmed that he still had the Airframe Log Book, the Gipsy Six engine (still 'in time' when taken out of service but without log book) complete with feet and mounts back to the fireproof bulkhead, the Fairey-Reed propeller complete with spinner cap, the tailplane, rudder, stern-post and tailwheel. These items were subsequently sold but to person/s unknown.

Miles M.3D Falcon Six VH-ABT, seen after the war.
[E D Coates Collection]

268 M.3B Falcon Six; colour not recorded. (P&P works records state 'conforms to OE-DVH but with Grimes retractable landing lights'). Flown twice by Bob Milne on CofA tests 17.2.36. CofA No.5376 issued 27.2.36 to African Air Transport Ltd; handed over to shippers 4.3.36. Regd **ZS-AFW** 4.36 to African Air Transport Ltd, Johannesburg, South Africa. Impressed into the SAAF as **1425** 3.40; used by 69 Air School, Germiston for ground instructional purposes from some time after 11.11.40. Allocated GI number **IS116** 11.42. To Zwartkop Air Station 5.7.43; probably sold for scrap by the end of the war.

269 M.3D Falcon Six; colour not recorded. (Laid down as M.3B but completed as M.3D; P&P works records originally state 'conforms to OE-DVH'; later amended after fitted with RTE Homing set 'conforms to G-AEAG'). After CofA received, Onions u/c struts were fitted and the max AUW increased to 2,500lb). Regd **G-AEAO** (CofR 6644) 18.1.36 to Flt Lt Thomas 'Tommy' Rose, Woodley; not dd. CofA No.5228 issued 18.1.36 as **PH-KGH** and actually handed over to owner K van de Heuvel 26.1.36 painted as PH-KGH but these marks were not available as they were still in use on owner's previous De Schelde S.12. UK regn cld 3.36 as sold. Regd **PH-EAO** (CofR 196) 20.3.36 to K van de Heuvel, Baam, Holland (based Schiphol). Sold 6.39 to Phillips & Powis Aircraft Ltd and Dutch regn cld 1.8.39. Not regd in UK and sold to Air Ministry under Contract No.8425/39 dated 12.8.39 and given serial **R4071**. Fitted with spoilers to become the M.3F Falcon Six and dd to AMDP at RAE Farnborough 30.11.39 for spoiler tests. Flown by Hugh Kennedy from Woodley on 16/17.4.40 and 6.5.40 and on 25.5.40 'test flight - ailerons'. To RAE Farnborough 31.5.40. To Flight Refuelling Ltd 30.10.40 (probably based at Perdiswell or Staverton). Returned to RAE 18.4.41. To Phillips & Powis 8.7.41; returned to RAE Farnborough (by census 31.12.41). To Phillips & Powis, Woodley for repairs 17.11.42; returned to RAE 25.2.43. To 5 MU Kemble for storage 3.6.44. Sold 12.45 to Holmwood Motors for £350 at the first sale of impressed light aircraft. Regd **G-AGZX** (CofR 9874) 22.1.46 to Leslie Ernest Gisborne, Holmwood Motors, Southsea (based Portsmouth). Regd 6.5.46 to John Ernest Coxon, Southern Aircraft (Gatwick) Ltd, Gatwick. Noted awaiting restoration at Gatwick 9.9.46 but no UK CofA issue known. Regn cld 23.12.46 as sold abroad. Regd in Belgium as **OO-FLY** (CofR 630) 28.1.47 to Octave Dierckx. Regd 31.12.48 to A Jaune, Anderlecht, Brussels. Regn cld 21.1.52 on sale in France. Regd **F-BBCN** (CofR 18318) 12.7.54 to Lucien Lecoq, Vincennes. Regd 10.9.54 to Aero Club de St Maur (Lucien Lecoq), Lognes-Emerainville. Regd 15.3.58 to Lucien Lecoq, Vincennes. Regd 27.3.58 to Association Club Aerien Les Gerfauts, Vincennes (based Lognes). Sometime fitted with a crude standard-type windscreen and sliding cabin. CofA lapsed 28.2.64. It languished in a corner of a hangar at Lognes in apparently good condition for a while thereafter, and in the summer of 1969 was photographed in a field on the southern boundary of the airfield. Regn cld 22.9.71 as destroyed. There were reports of it still being extant until about 1972 and rumours that it was at a private museum in Toulouse by 1999, but despite much sleuthing, no trace of it has been found.

The modified Miles M.3D Falcon Six F-BBCN, seen 'out to grass' at Lognes-Emerainville on 6th June 1969. [A B de Groot via N Braas]

279 M.3D Falcon Six 'Queen Bee' ordered by the Air Ministry to be fitted with radio control for experimental purposes at RAE Farnborough. (Originally laid down as M.3B; P&P works records state 'Subs to G-AEAG but fitted with Onions (Ribbesford Co) u/c, land type, also to be suitable for seaplane – floats). CLN records that it was a Queen Wasp mock-up only; (see also M.10 Queen Wasp c/n's 433 and 482) and believed that construction was probably started but for some reason it was not completed to flying standard. Its fate is unknown.

280 M.3B Falcon Six; painted red. (Originally laid down as M.3D but completed as M.3B) (P&P works records state 'Conforms to ZK-AEI but fitted with 2½ mm ply centre section fuselage sides; two 22-gal aluminium fuel tanks'). Regd **G-AECC** (CofR 6718) 10.2.36 to Arthur Niall Talbot Rankin, London NW.8 (based Heston, later RAF Stranraer); owned and operated jointly with his wife, Lady Jean Rankin. CofA No.5468 issued 27.4.36 (as an M.3D, so presumably the ARB had not caught up with the change of type); dd 16.5.36. Operated on charter work by Personal Airways Ltd, Croydon (1.37)

A Miles M.3B Falcon Six at Woodley in April 1936. Possibly G-AECC.
[Phillips & Powis via B Clarke]

while owner was in the USA. CofA lapsed 18.7.40 and regn cld 1.41 by Secretary of State. Impressed into RAF as **DG576** 1.1.41; to RAF Andover same day. To 51 MU Lichfield 27.9.41. To RAF Andover 30.5.42. To RAF Heston 2.6.42. To Phillips & Powis, Woodley for major inspection 30.6.42; ret to Heston for use by Col Turner's camouflage dept, Air Ministry; seen flying over North Lancashire in 9.42. To ADGB Comm Sqn, Northolt 5.44. To 5 MU Kemble 11.45 from where sold 12.45 to Miles Aircraft Ltd for £350, at the first sale of impressed light aircraft held at Kemble. Regd **G-AECC** 13.4.46 to Miles Aircraft Ltd, Woodley; fitted with a one-piece moulded Perspex windscreen. Regd 18.9.46 to Miles Aircraft (Northern Ireland) Ltd, Newtownards. CofA renewed 30.9.46. CofA lapsed 29.9.47. Cld 1.12.47 & regd 22.1.48 to Ulster Aviation Ltd, Newtownards. Cld 27.6.49 & regd 1.7.49 to Sqn Ldr James Rush AFC, Newcastle; CofA renewed 27.7.49 and painted pale blue. Rush gained 11th place in the 1949 King's Cup Air Race (Race No.32) held at Elmdon 29.7/1.8.49. Painted high gloss black with gold trim, probably in 1950; Rush won the Norton Griffiths Challenge Trophy at Woolsington on 29.7.50 at a speed of 165.50 mph. He also participated (Race No.64) in the Daily Express International Air Race on 16.9.50 from Hurn to Herne Bay, and although starting 58th out of 67 starters, gained 4th place at an average speed of 180.00 mph. Test flown by Miles 10.12.50 prior to CofA renewal by FG Miles Ltd, Redhill 2.1.51; still with FG Miles Ltd 22.4.51. Cld 30.3.51 & regd 23.5.51 to James Rush & Co (Northern) Ltd, Newcastle. Test flown by Miles at Redhill 15.2.52 prior to CofA renewal by FG Miles Ltd 29.2.52. Port leg fairing damaged through contact with rough ground while taking off from Coventry 24.8.52; repaired. Forced landed Tempsford Aerodrome, nr Biggleswade 18.6.54 with slight damage during the qualifying round for the King's Cup Air Race, flown by James Rush, after engine cowling flew off in air. Rush took part in each subsequent King's Cup Air Race (albeit unplaced) until 1957 when he gained 2nd place at Baginton 14.7.57 at 172.75 mph. Cld 17.12.57 & regd 1.1.58 to Crop Culture (Aerial) Ltd, Bembridge. Presumed ditched into the sea off the Dorset coast on a flight from Bembridge to Exeter 8.5.59; pilot A Cook killed; cause unknown. Regn cld as destroyed 8.5.59.

289 M.3E Falcon Six. Probably built early 1936 but not completed. Regd **G-AFCP** (CofR 8180) 17.11.37 to Phillips & Powis Aircraft Ltd; built and fitted with a 200hp Menasco Buccaneer engine but CofA application made 13.11.37 was not followed through and no CofA issued. Regn cld 4.38. In the February 1938 issue of the *Miles Magazine* was the following 'for sale' advertisement (which could only have referred to G-AFCP): *Falcon IV fitted with a 200hp super-charged Menasco Buccaneer engine. Exceptionally full equipment includes navigation and instrument lights, turn and bank indicator, hydraulic flaps, booster coil etc. A virtually new aeroplane at second-hand value; price on application. Stamped across the advertisement was the word 'sold'.* Actually sold to the Air Ministry under Contract No.734772/38 for test purposes. The ITP with a projected cost of £2,500 was dated 13.4.38, as was ITP accompanying Contract No.734771/38 for '4 sets of wings with a projected cost of £3,000'. Fitted with a 205hp DH Gipsy Six II engine and dd to the RAF as **L9705** 13.6.38 (per CLN although this date was crossed out). Ferried to Farnborough by RAE pilot Wilson 21.4.38. To Phillips & Powis 19.8.38 for 'new wing'; returned to RAE 1.9.38. Phillips & Powis built three sets of low-taper wings for fitment to L9705 for comparative testing with K5924 which was tested with three sets of high-taper wings. To Phillips & Powis 8.9.38 'new wing'; returned to RAE 9.11.38. To Phillips & Powis 10.2.39 'new wing'; back to RAE 23.3.39. These tests were probably completed in 8.39. To Philiips & Powis 31.12.40. Hugh Kennedy flew it on 28.4.41 on a 'handling and diving' flight from Woodley. Flown to RAE 2.5.41 and then back to Woodley by 6.5.41, when Hugh Kennedy made a 'climb & levels ADM.293' and, on 7.5.41, a 'performance ADM.293' before delivering it back to Farnborough on 9.5.41. To Phillips & Powis 14.8.41; back to RAE 24.8.41. Nothing further known until early 1944 when it was returned to Woodley to be modified for use to flight test the biconvex section wing (at low speeds) designed for the Miles M.52 High Speed Research aircraft. A 3-view general arrangement drawing dated 1.2.44 entitled 'Falcon Six A/c fitted with Biconvex wings' shows it to have the wings of similar section but with the originally proposed straight leading and trailing edges and a biconvex tailplane with elevators. The wing was manufactured in wood and had split flaps and a narrow tracked u/c directly attached to the fuselage in order to keep the wing free of all excrescences. The razor sharp leading and trailing edges of the wing led to L9705 being known in the factory as the "Gillette" Falcon and it was first flown in this guise, but with a standard Falcon tailplane, by Hugh Kennedy on 11.8.44. It was later fitted with a variable incidence, 'all-moving' M.52 tailplane but retaining the original elevators, in order for it to be more representative. Later fitted with an 'all-moving', variable-incidence tailplane with no elevators (believed to have been the first aeroplane in the world to fly with a variable-incidence tailplane). Kennedy last flew the 'M.52 Falcon Six' on 23.4.45. Hugh Kendall took over the testing, making his first flight on 1.10.45 and his last flight in L9705 on 4.1.46 when he flew it from Woodley to Farnborough. In 3.46, the M.52 project was suddenly and unexpectedly cancelled but L9705 stayed at Farnborough until it was flown back to Woodley by Capt Eric Brown RN of the Aero Flight, RAE on 20.6.46. L9705 made a public appearance at Woodley at the Miles 'At Home' day on 8.9.46 and was statically displayed at the last Miles 'At Home' day on 20.7.47. It was still in good condition when last seen at Woodley on 10.9.48, but was broken up later that year.

291 M.3B Falcon Six. CLN records as 'Falcon Six Service' with Job No.12030. 'Service' referred to the Repair & Service Dept, Woodley, so it is possible that this was a spare fuselage for use in the repair or rebuilding of a crashed aircraft.

Production summary:

M.3B Falcon Six	c/n 213, 232, 233, 247, 252 Special, 255, 256, 268, 280*	9
* 1 laid down as M.3D, not incl c/n 291 'Service'		
M.3C Falcon Six	c/n 231	1
M.3D Falcon Six	c/n 248, 259*, 262*, 266, 269*	5
* 3 laid down as M.3B, c/n 269 mod to M.3F, later mod to 'M.52 Gillette Falcon'		
M.3E Falcon Six	c/n 289	1
M.3F Falcon Six	c/n 269	(1)
Mod from M.3D		
Total		16

Known Job Nos:			
		H5639	c/n 268
		H5677	c/n 266
H3949	c/n 247	H5720	c/n 279
H5579	c/n 248	H5734	c/n 280
H5584	c/n 259	H3769 (5769?)	c/n 289
H5587	c/n 262	12030	c/n 291

Appendix 15: Miles M.4 and M.4A Merlin Production

141 M.4A Merlin (1st production); painted red. *Curiously, the c/n of the first production aircraft was allocated before that of the prototype!* Regd **VT-AGP** (CofR 256) 27.4.35 to Tata Sons Ltd, Bombay, India. First flown, by Miles, 29.5.35 as 'Merlin (Tata's)' with no regn noted. On 1.6.35 Miles made the first recorded flight as VT-AGP. While flying it on a demonstration on 4.6.35, Miles noted that 'the lid blew off'! This curtailed the flight to 15 minutes and immediately after landing back at Woodley, Miles took-off in the Hawk G-ACUD 'to look for the lid'! CofA No.4878 issued 6.6.35. Miles last flew VT-AGP on 12.6.35 and it was handed over to Tata the same day. Shipped to Bombay and erected by Tata Airlines Ltd at Yeravda Aerodrome, Poona by 10.35; operated on the Bombay-Goa-Cannanore-Trivandrum Mail route wef 26.11.35. Possibly operated later by Indian Transcontinental Airways Ltd. Regn cld 1.4.40; fate unknown.

151 M.4 Merlin prototype; painted blue with Birkett Air Services Ltd (sic) on the fuselage. First flown (without marks), by Miles 24.3.35 and flown as **U8** by Miles to Shoreham 31.3.35. Regd **G-ADFE** (CofR 5727) 5.4.35 to Birkett Air Service Ltd, Heston Airport. Flown by George Miles with FG Miles 4.5.35. On 10.5.35, Flt Lt JF Moir, who was a test pilot with the A&AEE at Martlesham Heath, flew U8 (with Miles Merlin written on starboard door) on a 10-min handling flight there. CofA No.4843 issued 15.5.35. First recorded flight as G-ADFE made by Miles 25.5.35. Flown to Farnborough and return by Tommy Rose 27.5.35; dd to Birkett 11.6.35. G-ADFE made many flights around Europe and several to Abyssinia before the war. Allocated to NAC.1 on outbreak of war whilst still with Birkett Air Service Ltd. Its fate is not recorded; its CofA lapsed 5.4.40 and it is possible that it was lost during the Nazi invasion of France and the Low Countries. Regn cld in census 1.12.46.

272 M.4A Merlin; painted blue. (P&P works records state 'Subsequent to G-ADFE but with 2.5 mm ply sides to centre section of the fuselage around the cabin. Modified flap gear and RTE homing set'). First flight not recorded and not flown by either Miles or Bob Milne. CofA No.5407 issued 16.3.36 to Capt Eric Chaseling (of Roberts Airways) as '**VH-UBN**'; dd as such same date. Erected at Essendon and test-flown 19.6.36. Regd in Australia as **VH-UXN** (CofR 590/CofA 547) 22.6.36 to Roberts Airways Pty Ltd, Melbourne (but retained VH-UBN in service until corrected in stages; a photo shows it with VH-UXN on the fuselage with VH-UBN still on the wings!). Named *Wilgul* and operated on services from Melbourne to Hay via Deniliquin. Regd 25.9.36 on change of name to Victorian & Interstate Airways Ltd. Impressed into the RAAF as **A37-2** 23.11.40 at 1 AD and issued to 1 Conversion Flt 1.12.40. To 1 AD 30.7.41 and to HQ Pearce 16.8.41. Converted to components 12.12.41.

274 M.4A Merlin; painted red. (P&P works records state 'Subsequent to VH-UBN but fitted with 5-gallon leading edge auxiliary fuel tanks. RTE Homing Set not fitted. AUW increased to 3,150lb'). Regd **VT-AHC** (CofR 269) 6.12.35 to Tata Sons Ltd, Bombay. First flown by Bob Milne on CofA test 4.3.36; appln made 11.3.36 for approval of 5-gallon leading edge auxiliary fuel tanks. CofA No.5416 issued 24.3.36; handed over to Tata Sons Ltd 23.3.36.
Appln made 20.5.36 for increased AUW to 3,150lb, approval received 19.8.36. Shipped to India and erected by Tata Airlines Ltd at Yeravda Aerodrome, Poona; operated on the Bombay-Goa-Cannanore-Trivandrum Mail route. Regn cld 16.5.41; fate unknown.

278 M.4A Merlin. P&P works records sheet is missing but CLN recorded 'Merlin' in pencil against c/n 278 with Job No.H5684, so possibly a reservation for a customer which was not proceeded with.

Production summary

M.4 Merlin	c/n 151	1
M.4A Merlin	c/n 141, 272, 274	3
M.4A Merlin	c/n 278	-
Total production		4

Appendix 16: Miles M.5 and M.5A Sparrowhawk Production

239 M.5 Sparrowhawk prototype; painted cream. Fitted with 140hp DH Gipsy Major HC engine. (P&P works records state conforms to type M.2H G-ACYZ but with reduced centre section span. Tanks in outer wings. Shortened H/P undercarriage. Front cockpit faired in). Regd **G-ADNL** (CofR 6162) 12.8.35 to Frederick George Miles, Woodley. First flown, by Miles, 19.8.35. CofA No.5018 issued 21.8.35. Entered by GA Hebden but flown by Miles in the 1935 King's Cup Air Race held at Hatfield on 6/7.9.35 (Race No.9) to gain 11th place at an average speed of 172.38 mph. Flown by George Miles 12.10.35. Regd (CofR 6638) 15.1.36 to Phillips & Powis Aircraft Ltd. Flown by Col Charles Lindbergh 5.5.36. Entered and flown by Pat Maxwell in Kings's Cup Air Race held at Hatfield 10/11.7.36 (Race No.9); finished 9th at average speed of 165.74 mph. Flown by George Miles 1.8.36. Flown to RAE Farnborough (by Dalrymple) 7.8.36 for CofA tests and by George Miles 12.8.36, returning to Woodley 13.8.36. Flown by Flt Lt James Moir to Rotterdam via Lympne on 14.8.36 for three 15-min demonstrations on 14/15/16.8.36 before returning to Woodley via Lympne on 16.8.36. Flown by George Miles from Shoreham to Woodley 27.11.36; flown by George Miles 8.7.37. Advertised for sale in the *Miles Magazine* for February 1938 for £675. Regd (CofR 8349) 8.2.38 to Miss Joan Mary Parsons, Leamington Spa (based at Bishop's Tachbrook, where she ran an 'informal' air taxi service, although believed also based some of the time at Woodley). Miss Parsons flew G-ADNL to South Africa in 5.38 and despite at least two forced landings (one on 5.7.38 200 mls north of Mbeya, Tanganyika), she arrived back in England safely in early 8.38. Sold 17.4.39 and regd 12.5.39 to Cyril Geoffrey Marmaduke Alington (of Air Touring), Gatwick; named *Angela III*. G-ADNL was entered for the 1939 King's Cup Air Race (2.9.39) by Mrs RH Dent, to be flown by James Rush, but the race was abandoned. Nothing is known about Mrs Dent or her connection with the owner, Geoffrey Alington. James Rush also demonstrated 'his' Sparrowhawk (reg not quoted) at the last Garden Party at Hanworth Park on 6.8.39. G-ADNL was not impressed on outbreak of war and was initially stored at Gatwick until Alington became chief test pilot at Austin Motors Ltd, when he obtained permission to use it as a personal 'hack'. CofA renewed 23.5.41 and based at Longbridge and later Elmdon and flown throughout war. CofA lapsed 27.5.46 but renewed 7.5.47. Thought to have been the 'Miles Hawk (Gipsy Major) with expired CofA, property of a famous test pilot at £150 - bargain for enthusiastic ground engineer' advertised in *Flight* 2.10.47 by WS Shackleton Ltd (although G-ADNL's CofA was current). Flown by Alington, as a member of the Hereford Aero Club, winning the Isle of Wight Air Race at Bembridge 5.6.49 at an average speed of 159.5 mph. Flown by Alington in the 1949 King's Cup Air Race, held at Elmdon 29.7-1.8.49, but achieved only 9th place at 163.50 mph. Gained 2nd place in the International Air Races at Sherburn-in-Elmet on 22.7.50 at a speed of 170 mph. Alington competed in G-ADNL in many post war air races (wore Race No.23 for at least one), at speeds averaging between 165-170 mph. Cld 9.10.50 and regd 5.1.51 to Oldham Tyre Cord Co Ltd (Fred Dunkerley), Barton and flown to Redhill Aerodrome shortly after

Miles M.5 Sparrowhawk G-ADNL at Austin Motors, Elmdon Aerodrome, while being used by test pilot Geoffrey Alington during the war. The photograph was taken in 1944. [Via B Clarke]

where he had arranged for FG Miles Ltd to convert it to take two Turbomeca Palas gas turbine engines of 330 lb static thrust. Flown by Miles at Redhill 29.1.51, a 'local' of 1 hr 15 min duration, on probably its last flight as the M.5 Sparrowhawk. The aircraft was then completely stripped down and work on its conversion commenced, albeit slowly. In 1951/52, Miles moved his company to Shoreham and G-ADNL was eventually completed there as the Miles M.77 Sparrowjet, with c/n FGM.77/1006. It was first flown, by George Miles as **G-35-2** on 14.12.53. CofA issued, as **G-ADNL**, 17.6.54. The Sparrowjet won the SBAC Challenge Trophy at Yeadon on 21.5.56 at an average speed of 197.50 mph and the King's Cup Air Race at Baginton on 13.7.57, at an average speed of 228 mph (the King's Cup being the race the Sparrowhawk had been designed for 22 years earlier!). CofA lapsed 13.5.58 and subsequently presented (by 10.60) to the RAeS as a museum piece. Owing to the lack of suitable display facilities it was stored, first at Heathrow, then later at RAF Upavon, where it was destroyed in a hangar fire in 7.64. Regn cld 14.3.73 as destroyed. On 8.11.91, Kathleen D Dunkerley of Bury, Lancs 'restored' G-ADNL to the register as an M.5 Sparrowhawk. On 24.9.97 it was registered to Alan G Dunkerley who then contracted Tim Cox to build a flying replica of the original Sparrowhawk based on the few surviving metal components and a pair of Magister wings. Regd 11.2.04 to Angela Patrice Pearson (Alan's No.1 daughter), Ramsbottom, Lancs. The airframe is complete and was awaiting larger premises for assembly when it was put on hold in 2008..

264 M.5 Sparrowhawk; 'standard' colour scheme of sky blue fuselage with silver empennage (although this was never a standard colour scheme – PA). (P&P works records state it was originally to have been an M.5A Sparrowhawk and 'Conforms to G-ADNL but with two seats and two 11 gal fuel tanks. Mass balance will be fitted to rudder and aileron (Standard Gipsy Major)' but this entry was later crossed out and replaced with 'Conforms to G-ADNL Type M.5 but fitted with 3 gal leading edge type aluminium oil tank. 11 gal tinned steel petrol tank in front cockpit compartment and mass balance weights to rudder and ailerons'). Fitted with 130/140hp DH Gipsy Major HC engine (No. 8662). Regd **G-ADWW** 6.11.35 to an unidentified customer (who allegedly did not pay for it, and completion probably deferred). Regd (CofR 7101) 27.5.36 to James Henry Gordon McArthur, Bournemouth, Hants (based initially at Yatesbury). Build completed 6.36 (and works records state - fitted with a 125/130hp Gipsy Major HC engine, crew 1, span 28ft 0in, length 23ft 6in, height 5ft 6in, max total weight 1,700lb, max speed 243 mph, flaps must not be used at speeds in excess of 75 mph). First flown 10.6.36; this machine was not flown by Miles. CofA No.5549 issued 23.6.36; handed to Inspection Dept 25.6.36 and to McArthur 28.6.36. Flown by McArthur in the 1936 King's Cup Air Race (Race No.10) at Hatfield on 10/11.7.36 but was disqualified. Flown 15.8.36 by George Miles from Woodley-Heston-Woodley-Shoreham-Wilmington-Shoreham-Woodley. Sold 14.9.36 by Phillips & Powis Aircraft Ltd to James Hopkins Smith Jnr (a pilot with Pan American Airways), New York (per invoice dated 15.9.36 as 'secondhand' for £730 plus shipping charges including insurance £73). Despatched in two crates by R&J Park Ltd aboard the SS Georgic for New York; arrived Long Island 27.9.36 and assembled by Aero Trades Inc, Roosevelt Field, Mineola, New York. UK regn cld 11.36 as sold (a letter from Air Ministry dated 18 November 1936 stated that: *Re: NC191M This is to certify that Miles Sparrowhawk aircraft, Constructor's No.264, formerly owned by John Henry Gordon McArthur, and bearing the nationality and registration marks G-ADWW, has upon sale to a foreign national been removed from the Register of Aircraft in Great Britain and Northern Ireland and the marks G-ADWW cancelled*). Regd **NC191M** 10.36 to James Hopkins Smith Jnr, New York. (The US Dept of Commerce registration application stated fitted with Fairey metal 2-bladed propeller but retained same engine, although noted as rated at 125/130 hp at 2,100 rpm). The US licence was revoked 22.12.36 pending further details from Great Britain, but operated on a temporary licence until an affidavit from Phillips & Powis Aircraft Ltd dated 23.3.37 confirmed that it met Aeronautics Bulletin 7A. A telegram was sent 24.4.37 to Smith confirming temporary licence extended to 15.5.37 and full licence reinstated 21.5.37. Placed in store for 8 months; temporary authority issued 1.6.38 for flight from Bridgehampton, New York to Roosevelt Field for inspection (actually flown 4.6.38); flying time to date 202 hours, and licensed to 1.7.39. Sold 1.5.39 and regd 8.5.39 to Perry Boswell Jnr, 401 Colesville Road, University Park, Maryland (later Hyattsville, Maryland). Repainted to white with red trim on the u/c trousers, fitted with a steerable tailwheel and named 'Electron'. On 27.6.39, NC191M was licensed to 1.7.40, with 256.50 hours to date. Flown by Boswell extensively in eastern states, also to California and New Mexico and he competed in the 1939 Langley Day Air Races at College Park, Maryland, where it won its class. On 16.1.40, Aero Trades wrote to the CAA (predecessor of FAA) requesting authority to convert NC191M into a two-seater and install a sliding pyralin hatch over the two cockpits but by reply 1.2.40, this was refused as no approval could be given without information from the manufacturers as to the structural integrity of the aeroplane with the proposed alterations etc. Flown to second place in its class at the 1940 Langley Day Air Races at Hybla Valley Airport, Washington, DC. Damaged in an accident (reportedly 3.6.40); presumed repaired. Boswell joined the RCAF in 12.40 and was posted as O/C Test Flight, 9 Repair Depot, St Johns, Quebec and obtained permission to operate NC191M under US marks. On 30.6.41, L Scholfield of Hyattsville reported to the CAA that he had completed repairs to nose ribs, right landing gear struts, right wing and spar. Flown 1.42 from Burlington to Canada; later fitted with cockpit canopy and raised rear decking (to cope with cold weather). License lapsed 1.7.42 and remained stored at 9 RD, St John (after Boswell joined the USAAC 5.42). Later moved to de Havilland Canada factory at Toronto. Returned 1.45 to Buffalo, New York. Sold 25.9.46 and regd 22.11.46 to Carl L Conrad, Maryland. Re-licensed 15.11.46 at Burlington (which entailed removal of canopy and rear decking as 'unapproved mods'). On 18.7.47, Conrad applied for licence, approved 8.8.47 for ferry flight from Burlington Airport to Cumberland Airport for inspection. On 2.8.48, the owner applied for licence, approved 21.9.48. License lapsed and aircraft stored in West Virginia. Purchased back by Perry Boswell, now resident at Delray Beach, Florida 6.8.54 and regd to him (as **N191M**) 12.8.54. Ferried by Boswell to his ranch at Pompano Beach, nr Miami, dismantled and trucked to Delray Beach for rebuild. Application for Airworthiness Certificate eventually made by Boswell on 4.12.56, noting engine No.86939 and 555.50 hours flying time (which meant that the engine had been changed at some time), and the CofA issued same day. Sold 12.56 to George Roberts and based at Lantana Airport, West Palm Beach, Florida. Sold 27.10.57 and regd 30.12.57 to Sunrise Sales Co, Palm Beach (possibly as nominee for George Roberts). On 5.4.59, Roberts allowed a friend, USAF Sgt Hamlin, to carry-out a 'fly-by' at Lantana air show but he instead undertook three steep pull-ups to a stall, on the third on which, he spun in from a low altitude and hit the runway almost vertically. The Sparrowhawk disintegrated and Hamlin was killed instantly; a few days later the owner burned all the remaining wreckage. Regn cld 15.5.59 as destroyed in accident.

273 M.5A Sparrowhawk; colour not recorded. Laid down in late 1935 but not completed. Regd **G-AFGA** (CofR 8397)16.3.38 to Lionel Edward Richard Bellairs, Shoreham. Order probably cancelled (Bellairs purchased the Whitney Straight G-AECT instead). Regd (CofR 8451) 12.4.38 to William 'Bill' Humble, Doncaster. Fitted with 140hp DH Gipsy Major engine and CofA No.6265 issued 20.5.38. Humble flew G-AFGA (Race No.3) in the Hatfield to Ronaldsway Race on 4.6.38 but only managed 11th place. Bill gained 7th place in the 1938 King's Cup Air Race (Race No.17) held at Hatfield on 2.7.38 at 169.50 mph. Regn cld 1.1.39 at census as having been sold. Actually dismantled at Woodley in 1938 and later modified to test large-chord trailing edge flaps and drooping ailerons on a redesigned wing of reduced span for trials at the RAE as a projected shipborne fighter. Painted silver and given 'B' marks **U5**; it had a three-piece windscreen but no headrest to the cockpit and was reflown prior to the outbreak of war. Flown at Brough 19.12.39 by Sqdn Ldr Wilson (probably from RAE Farnborough) – 30 min 'handling research'. This was probably related to the 'special Miles Sparrowhawk built to a order from Blackburns to test slip-wings and high-lift devices'. An anonymous (and possibly this) Sparrowhawk was flown by Hugh Kennedy on 12.7.40 and he also recorded a 'local' in Sparrowhawk U5 on 16.7.40, by which time it would have been camouflaged. Des Armour recalled seeing a camouflaged Sparrowhawk 'with double-flaps' being broken up in the Repair & Service hangar at Woodley in late 1941/early 1942 by an enthusiastic labourer, who was 'smashing it to pieces with a tailskid'. *Given the conversion state of U5, it would have required a completely new wing to convert it back to standard, which must be considered unlikely. While there was a report that the regn G-AGBI was reserved in mid-1940 for a Miles Sparrowhawk and which could have been for U5, the allocation has never been confirmed and was given to an ex-KLM Douglas DC-3 on 1.8.40.*

275 M.5A Sparrowhawk; painted blue. Fitted with 140hp DH Gipsy Major HC engine. (P&P works records state 'Conforms to G-ADWW, Type M.5 with the exception of fitting single-piece spars in wings, long range aluminium fuel tanks, Grimes retractable landing lights. Navigation lights. AUW increased to 2,200lb') (and which was decreased to 2,175lb later). Regd **G-AELT** (CofR 7216) 28.7.36 to Victor C Smith, Capetown, South Africa. CofA No.5656 issued 16.9.36; dd 18.9.36. Not flown by Miles. Entered and flown by Victor Smith in the Schlesinger African Air Race from Portsmouth to Johannesburg, departing 29.9.36. Although he arrived 4th at Belgrade, he had to forced land at Skopje in Yugoslavia soon after, due to a choked return valve in the oil tank which practically filled the engine with oil. This made him late in arriving at Cairo but he flew on to Khartoum, where he retired from the race. He later completed the journey at his own pace. Regn cld 15.8.45 in post-war census. Rebuilt by Victor Smith into a two-seater cabin machine and regd **ZS-ANO** 3.38 to him and operated on charter work. Sold .39 to Carl Erasmus, Durban. Impressed into the SAAF as **1427** 3.40 and to the SFTS at Bloemfontein. Engine failure and crashed on take-off Bloemfontein 16.7.43 and not repaired. *Reports that it was sold to Ben Dupreez in 1937 are incorrect; he was however a frequent client of Smith's charter service. It is also highly unlikely that the aircraft was restored post-war and survived until the 1960s as has also been reported.*

276 M.5A Sparrowhawk. Laid down in 1935 but not completed. Completed by Experimental Department with double-slotted high-lift flaps for RAE trials in 2.40 and originally fitted with 140hp DH Gipsy Major Srs I engine, but later replaced by a Gipsy Major Srs II HC. First flown, by Hugh Kennedy, as **U3** on 24.2.40; originally had no headrest (added later), a wrap-round windscreen and a horn balanced rudder (later deleted). Although Kennedy only recorded 'U3' on this first flight, it is probable that subsequent flights until 14.5.40 were also in this aircraft. Flown on 1.3.40 by Walter Kaepelli (became Capley) Woodley-Farnborough-Woodley (also on 8.6.40 & 10.6.40). Flown by Tommy Rose 13.3.40 (also on 22.6.40 & 26.6.40). Hugh Kennedy flew an anonymous (but probably this) Sparrowhawk on 26.4.40 and recorded 'Test flight - experimental flaps'. Kennedy's next recorded (anonymous) Sparrowhawk flight was to Northolt and return on 20.7.40 and 'local' anonymous Sparrowhawk flights were also made on 20th, 21st and 22nd and on 23.7.40 he made a return flight to Boscombe Down. Kennedy recorded no more 'experimental' Sparrowhawk flights after this; U3 was returned to Phillips & Powis in 1941 where it was converted back to standard configuration and took on the new B marks of **U-0223**. Kennedy flew to South Marston and return in U-0223 on 18.10.41. Regd **G-AGDL** (CofR 9345) 22.10.41 to Phillips & Powis Aircraft Ltd. CofA No.6916 issued 3.12.41. Kennedy carried out 'airscrew tests' in G-AGDL on 23.12.41 and used it for communications and for taxi flights to South Marston to test fly newly built Masters. Regd 5.10.43 to Miles Aircraft Ltd (and MAP Permit No.16 issued 15.12.43-14.12.45). G-AGDL made its first post war appearance on 1.6.46, in the Miles house colours of cream with red trim, at the Garden Party organised by the Reading Branch of the RAeS at Woodley, where it was shown off to its full extent by the Chief Test Pilot Ken Waller. It took part in the first post-war air races, held at Lympne over the weekend of 31.8.46/1.9.46 (Race No.31) although its postwar CofA was only renewed 23.12.46. Flown by Grp Capt AF 'Bush' Banditt on 26.5.47 in the Tynwald Air Race of the Manx Air Derby, at Ronaldsway for the Olley Challenge Trophy (Race No.17). Advertised for sale in *The Aeroplane* 5.9.47 for £800. Flown by Hugh Kendall on 25.2.48 and 1.3.48 for CofA renewal. Sold by the Receiver of Miles Aircraft Ltd and regd 11.3.48 to T Shipside Ltd, Tollerton and operated by the Nottingham Flying Club. Took-off with the petrol turned off, stalled and crashed Tollerton at 11.27 am on 19.6.48; pilot F Kirk seriously injured. Regn cld as crashed 19.6.48.

281 M.5A Sparrowhawk. CLN records 'Sparrowhawk' (in pencil) with no further details and no P&P works record sheet exists; possibly reserved for an aircraft which was not built.

Production summary

M.5 Sparrowhawk	c/n 239, 264	2
M.5A Sparrowhawk	c/n 273, 275, 276	3
'Sparrowhawk'	c/n 281	-
Total built		5

Miles M.5A Sparrowhawk G-AGDL, seen at Woodley after the war in Miles' 'house' colours of cream with crimson trim.
[G Swanborough Collection]

Appendix 17:
Miles M.7 and M.7A Nighthawk Production

263 M.7 Nighthawk; painted silver. First prototype; (P&P works records state 'Conforms to G-ADLC (M.3B c/n 213 – PA), but with 4 in wider fuselage, thrust line lowered 2 in, seating reduced to two, side by side dual control, fitted with two 'Harley 8' retractable landing lamps. Special double set of instruments for blind flying and two 16½ gallon fuel tanks – Spinning requirements). The first prototype had a mass balanced rudder with a straightish trailing edge and the tailwheel was attached to the rear edge of the rudder stern-post necessitating an angular cut-out to the bottom edge of the rudder. First flown, by Miles, 26.10.35 as 'Nighthawk'. Blossom made two flights in the first prototype Nighthawk with Miles on 4.12.35 and fortunately confirmed, by recording U1 in her log book, that it had made its early flights with the 'B' mark **U1** (not U5 as commonly asserted; which was in fact the 'B' mark used on the second prototype). No photographs showing the first prototype as U1 have been found. Regd **G-ADXA** (CofR 6468) 7.11.35 to Phillips & Powis Aircraft Ltd. CofA No.5243 issued 7.12.35. CLN records that it was dd on 8.12.35 to 'Firm, Demonstration' and on the same day, Miles carried out 'Night landings' and then flew it to Hendon and return. On 9.12.35 Miles, accompanied by 'M Miles' (Blossom), carried out a 'Lighting test', noting 'XA', for the first time, then, following a demonstration, he carried out two 'Night flying' trips with M Miles the same day. On 10.12.35, Tommy Rose flew it to RAF Upavon and return and on 11.12.35 Miles carried out further flights with M Miles, Christopher Clarkson (of Anglo-American Oil Co Ltd) and Alan Muntz (of Airwork) at Heston, before returning to Woodley via Hanworth Park. On 12.12.35 Miles flew it to RAF Upavon, returning on 13.12.35. (Although it is recorded that Miles demonstrated the Nighthawk at Brooklands about this time, he failed to record any flights to Brooklands in his log book). Miles flew the Nighthawk with MM on 5.1.36 and the works records then state 'Application for approval of fitting increased area horn-balanced rudder and strengthened tailplane 7.1.36'. Miles next flew it on 20.1.36 so the new horn-balanced rudder (but still with its straightish trailing edge) must have been fitted between these dates. On 22.1.36, two weeks before it was due to give an official demonstration to the Air Ministry, G-ADXA crashed near the lake in Stanlake Park, Twyford, just to the east of Woodley, while on a spin test with very aft cg. The pilot, Wing Cdr Freddie Stent, was forced to parachute to safety. CofR form returned to CA.2 (after crash) 6.2.36. Regn cld 3.36.

282 M.7 Nighthawk; painted silver. Second prototype; (P&P works records state 'Conforms to G-ADXA, with exception of two 17½ gall. aluminium fuel tanks. All up weight increased to 2,500lbs (afterwards 2,350lbs); Ribbesford Co Onions shock absorber u/c legs fitted). The tailwheel was moved forward to accommodate a vertical stern-post for the horn-balanced rudder of larger chord and with a curved trailing edge and a larger fin with a curved leading edge. Regd **G-AEBP** (CofR 6685) 3.2.36 to Phillips & Powis Aircraft Ltd. First flown, by Miles, 5.2.36 but he failed to record any regn, just noting 'Spins' but it is fairly certain that the early flights were carried out under 'B' mark **U5**. CLN records dd 9.2.36 (probably the date to A&AEE Martlesham Heath for CofA tests and suitability trials for service use) and would still have carried U5 at that time. CofA No.5412 issued 19.3.36. AJ Jackson saw it at Woodley on 15.4.36, still as U5. Soon after this the 'B' mark was changed to **U6** and it was flown as such by Charles Lindbergh at Woodley on 6.5.36 and from Woodley to Farnborough 20.6.36, where it was flown twice by RAE pilots, before being flown to Hendon for the RAF Display on 24.6.36. Flown by RAE pilot Flt Lt Burke at the Hendon Display, wearing the new type number '10' in black in large numerals on the cowling, RAF roundels on the fuselage (but possibly not on the wings) and with U6 on the rudder; returning to Woodley 27.6.36. Demonstrated, still as U6 and '10', at the fifth SBAC Show, held at Hatfield on 29.6.36 *(while some sources claim that it was shown at the SBAC show as L6846, this was not allotted until about 3.37)*. Miles flew it (without noting marks) on 6/7.7.36, 13.7.36 and 20.7.36 with the note 'Lavaud prop', this referred to the Sensand de Lavaud variable pitch propeller which had been fitted for trials (he also flew it 14.7.36 making no comment). Next flown by Miles on 18.7.36 Reading-Dover, Dover-Shoreham and Shoreham-Reading again with no comment). On 20.7.36 he flew it (noting Lavaud) and on 8.8.36, he flew it to Speke, returning to Reading via Abingdon and his last recorded flight in it was on 9.8.36, but these last three flights were made with no comment as to its identity. George Miles recorded flights in G-AEBP from Woodley-Shoreham and return to Woodley on 16.9.36 (which would suggest that the regn was then carried) and George flew it again on 28.10.36 and 5.11.36. Contract No. 559859/36 dated 12.3.37 was issued to Phillips & Powis Aircraft Ltd for the purchase of one 'second-hand' Nighthawk. CofA renewed 30.4.37 and regn cld 5.37. Dd to CFS Upavon as **L6846** 13.5.37. To Grantham 2.6.37. To 24 Sqdn, Hendon 2.7.37. To A&AEE Martlesham Heath 17.6.38. To 24 Sqdn, Hendon 25.6.38. Relegated to GI airframe at 6 SofTT Hednesford as **1587M** 14.6.39.

283 M.7A Nighthawk; painted cream. (P&P works records state 'Conforms to G-AEBP, with the exception of reversed throttle and mixture controls. Seating capacity raised from 2 to 4, all up weight increased to 2,650lbs). Regd **G-AEHN** (CofR 6967) 6.5.36 to Phillips and Powis Aircraft Ltd, Woodley. CofA No.5619 issued 14.8.36. First of two for the Roumanian Government; dd 26.8.36. Regn cld 12.36 as sold. Regd **YR-AUK** 20.8.37 to Ministry of Air & Navy/Marine (it is unclear if it served under military marks).

286 M.7A Nighthawk; painted in camouflage. This was a hybrid version of the M.7 Nighthawk, utilising the wings from the uncompleted M.12 Mohawk c/n 301; it had a moulded one-piece windscreen, a horn balanced rudder with the straightish trailing edge and the same shape fin as the modified first prototype, smaller rear cabin windows and larger windows in the cabin doors. First flown by Hugh Kennedy 11.3.40 on a 20-min local flight but with no identity recorded (but probably with 'B' mark U5). Used as a company hack, its last recorded flight as U5 by Don Brown was on 2.10.40. To **U-0225** and flown as such by Walter Capley 27.11.40 from Woodley to Farnborough and return. Permit to fly renewed in 10.42 and 3.43. Regd **G-AGWT** (CofR 9796) 28.12.45 to Miles Aircraft Ltd. CofA No.7436 issued 15.3.46. Regd 21.5.46 to The Wokingham Motor Co Ltd, Wokingham (based Woodley). Regd 24.7.46 to Hugh Vincent Kennedy, Wargrave (based Woodley); who had in fact been flying it regularly since 10.5.46. Cld 1.1.47 and regd 23.1.47 to Raceways Ltd, Wargrave (Hugh Kennedy's company based at Woodley) and used as a taxi for taking jockeys around the country. Hugh also gave 10/- joy-rides from Woodley every Saturday and Sunday from 14.00 hrs to dusk, weather permitting. Sold, probably shortly after its CofA renewal on 26.3.47, but regn only cld 16.8.47. Regd 5.9.47 to Edwin Williams, Teddington (based Fairoaks); actually dd 25.6.47. (A letter

Miles M.7 Nighthawk L6846 shortly after it was purchased by the Air Ministry in March 1937. Note the No. 54 Squadron Gladiators lined up behind. [Via P Jarrett]

The second Miles M.7 Nighthawk. This photograph confirms that it actually carried the registration G-AEBP. [Via P Jarrett]

G-AGWT, the Miles M.7A Nighthawk Hybrid, was based at Redhill from late 1948 to about the end of 1952. It is seen here outside the Redhill Flying Club's hangar. [Via P Amos]

G-AGWT was last seen in a scrapyard at Lignane, Southern France. [Via M Hooks]

Two M.7A Nighthawks were delivered to the Roumanian government in August 1936, G-AEHN and G-AEHO. The latter was registered YR-GIM in August 1946. [Via P Jarrett]

No.93), winning the 'hideous' Norton-Griffiths Trophy for the fastest aircraft not over 2,205lbs - at 158.5 mph, beating all the Proctors by a good 20 mph! He then flew her out to East Africa by the old Imperial route (still carrying Race No.93). Regn cld 8.6.54 as sold in East Africa. Regd **VP-KMM** 3.6.54 to Major Lovatt-Campbell. Regd 9.54 to Steel Brothers (Tanganyika Forests) Ltd, Lindi. Rebuilt in 1958/59 by Campling Bros & Vanderwal Ltd and fitted with a Gipsy Queen II engine and with a Proctor u/c and spats following the condemning of the original u/c due to sand abrasion of the legs - it had been regularly landed on beaches! Probably repainted red and black with gold regn at this time. Badly damaged on landing at private airstrip at Rondo, Tanganyika 5.2.60 in severe turbulence, causing it to drift off runway and then stall from 30 ft; pilot unhurt & aircraft repaired. Regd in Tanganyika as **VR-TCM** 9.6.61 to Steel Bros (Tanganyika Forests) Ltd. Flown back by Lovatt-Campbell to England from Tanganyika, arriving at Redhill on 12.8.61 and then flown to Biggin Hill 9.61 for overhaul. Regd **G-AGWT** 8.3.62 to Steel Brothers & Co Ltd, London EC3. CofA renewed 31.5.62. Cld 24.7.62 and regd 27.7.62 to Geoffrey Cecil Marler, London W1 (at Baginton 18.8.62 wearing Race No.94). Cld 14.12.62 and regd 27.12.62 to Whirlyfoto Ltd, Stoke, Coventry (based Baginton). On 15.3.63, G-AGWT, flown by Brian McAllister (believed of Whirlyfoto Ltd – since it is reported that he purchased it in 1962) departed Southend for Le Touquet en route Kenya (although also recorded as being en route to Singapore) but due to engine problems the aircraft got no further than Marseilles (Marignane) Airport in Southern France, where it was seized, by whom and for what reasons remain unknown. By 1965, G-AGWT had been moved to Lignane, 5 miles from Aix-en-Provence, Southern France, where it was photographed on 29.6.65 in what appears to be a scrapyard on what looked like the edge of an aerodrome. While the Nighthawk appeared to be still in reasonably good condition, it was reported at the time to have been sold for scrap. Its final demise is not known (although an unconfirmed report has it ending up in a children's playground) - a sad end for such an interesting aircraft. Regn cld 14.12.68 as pwfu.

287 Nighthawk/Mentor. CLN records this was used for mock-up only 1936/1937.

288 M.7A Nighthawk; painted cream. (P&P works records state 'Subsequent to G-AEHN'). Regd **G-AEHO** (CofR 6968) 6.5.36 to Phillips & Powis Aircraft Ltd. CofA No.5620 issued 14.8.36. Second of two for the Roumanian Government; dd 26.8.36. Regn cld 12.36 as sold. Regd **YR-GIM** 13.8.46 (probably an error for 13.8.36) to George Ionescu (local records also claim the c/n as 284 and it is unclear if it ever served under military marks).

dated 15.6.48 to Handley Page, Woodley from British Air Transport Ltd, Redhill asked for a copy of the Service Manual for Nighthawk as they were looking after the aircraft for its owner). Damaged its airscrew at Lyons, France 24.9.48; repaired. Cld 28.10.48 and regd 8.11.48 to Reginald Crewdson, Horley (based Redhill). Air tests flown by Miles at Redhill 10/15/26.3.50. Entered by owner and flown by Miss Jean Lennox Bird in the Air League Challenge Cup (Race No.51) held at Sherburn-in-Elmet on 22.7.50; also at Woolsington on 29.7.50, when flown into 2nd place in the Norton Griffiths Challenge Trophy by Jean Bird, the only lady participant, at a speed of 157.00 mph. Entered by owner in the Daily Express International Air Race on 16.9.50 from Hurn to Herne Bay, flown by Ian Forbes into 2nd place, at a speed of 175.00 mph. Cld 9.5.52 and regd 5.6.52 to Grp Capt Clair Mansell Maybury Grece DFC, Redhill. Grece had the aircraft repainted and then won the Concours d'Elegance at the 1954 Jersey Air Rally. Cld 18.5.54 and regd 3.6.54 to Major Percival Verney Lovatt-Campbell. MD of Steel Bros (Tanganyika Forests) Ltd, Lindi, Tanganyika, who was on home leave. Lovatt-Campbell flew her in the National Air Races at Baginton on 18/19.6.54 (Race

299 CLN records Nighthawk with no further details (and no details in the P&P works records).

Production summary		
M.7 Nighthawk	c/n 263, 282	2
M.7A Nighthawk	c/n 283, 288	2
M.7A Nighthawk Hybrid	c/n 286	1
Nighthawk/Mentor Mock-up	c/n 287	1
Nighthawk	c/n 299	-
Total built		6

Known Job Nos.		H5747/13668	c/n 286
H5740	c/n 283	H3771	c/n 288

Major P V Lovatt-Campbell standing by the Miles M.7A Nighthawk on the beach at Lindi, Tanganyika, in 1954. [M Barraclough]

Appendix 18:
Miles M.9 Rolls-Royce PV Trainer – Brochures

The M.9 Rolls-Royce Private Venture Trainer marked a turning point in the fortunes of the firm and fortunately a number of brochures on this aircraft have survived. The specification detail varies slightly between the undated brochures but such is the amount of hitherto unpublished information contained within them, that it worth quoting from what was possibly the latest one. This brochure, entitled **'Miles Kestrel high-speed Trainer'** was published (in English) by Phillips & Powis Aircraft Ltd, but it made no reference to the firm, either on the cover or within its contents. Instead the grey cover of the brochure carried the inscription 'Technische Handelmij. Hollinda NV (translates as 'Technical firm in The Netherlands'), Lange V, Ij Verberg 13, Den Haag' at the bottom. The cover had a logo at the top consisting of a stylised bird with the word 'Hollinda' within its outline and this brochure also differed from the others in having metric comparisons for the dimensions, weights and performance (some of which differed from their English counterpart).

It can only be assumed that Phillips & Powis were either trying to sell the concept of the multi-role aircraft to the Dutch, but for some reason they chose not to put their name on it, or they were told to try to sell it to the Dutch by the Air Ministry. Although the content of this brochure differed from the other brochures in many respects, the relevant sections of this and some details from the others make for very interesting reading:

The Miles Kestrel High Speed Trainer

INTRODUCTION

The aircraft described herein is designed to give a high performance for dual instruction and advanced training to the original specification T.6/36. It may also be considered as a two-seater fighter or a general purpose aircraft with a speed range of 5½ to 1 *(5 - 1 in the other brochures - PA)*.

After careful consideration of the various operational requirements the two-seater tandem arrangement was decided upon. This, it was considered, would give the best performance, whilst the somewhat restricted view of the instructor in the rear seat would be offset by his familiarity with the aircraft *(the other brochures state that 'The construction gives a particularly good view from the rear seat, as can be seen in the photographs! - PA)*. It was also felt that flying conditions for the pupil would more closely resemble those of the aircraft he would subsequently be flying.

The cooling duct for the engine has been developed to give minimum drag and throughout the design, aerodynamic cleanliness has been most carefully studied to produce an aircraft of exceptional performance without the extravagant use of Power.

Rolls-Royce Limited have developed the 'Kestrel' Series engines for many years and in its present form it is probably the most reliable power unit that exists today.

GENERAL DESCRIPTION

The aircraft is a high speed, low wing monoplane of wooden construction, powered with a Rolls-Royce Kestrel XVI engine, fitted with a three-bladed variable pitch airscrew *(or a two-bladed wooden airscrew in the other brochures - PA)*.

The design is such that provision has been made for the following operational duties:-
  (a) Training for dual instruction and blind flying.
  (b) Training in wireless operation with T.1083 and R.1082, TR.9 or TR.11 with D/F loop and photography.
  (c) Training in bombing and photography.
  (d) Training for rear gunner.

The aircraft may also be considered as a single or two-seater fighter.

Armament equipment consists of the following:-
(1) Browning Gun fitted in the starboard wing outside the propeller disc with 300 rounds of ammunition.
(2) Cinema Gun type GX 42, fitted in the port wing.
(3) The bomb load consists of sixteen practice smoke bombs, supported on electro-magnetic release units stowed in the wing root.
(4) A retracting gun turret hydraulically operated may be installed in the rear cockpit. This is equipped with two Browning guns with 300 rounds of ammunition for each gun. A mechanically compensated sight is fitted.

Careful attention has been given during the design stages to facilitate the rapid removal and installation of the various operational equipment.

CONSTRUCTION

The fuselage is of semi-monocoque construction and for the prototype, has been made with wood. The lines of the fuselage have been developed with a view to rapid production either in metal or wood.

Doors for the use of the Mark IX Course Setting Bomb Sight, hydraulically operated, are positioned in the flooring. The cockpit width is such as to provide for the easy accommodation of the equipment and the comfort of the crew. At the rear end, the lines have been arranged to give the maximum base for the attachment of the fin and tailplane. The engine mounting attachment plates are of high tensile steel to specification S.4. Between the cockpits is a strong metal former built out to the window fairing line for the protection of the crew should the aircraft nose over on landing. The cockpits are adequately warmed by a controlled flow of warm air from the engine cooling duct. The front windows are made to slide upwards on guide channels on the port side with fixed hinge on the starboard side, which assures easy entry into the front cockpit. Windows are fitted with emergency catches on both sides for use should the aircraft overturn on landing. The windscreen is of Perspex shaped vertically and flat in the region of the reflector sight to give an undistorted view forward on to the target. This screen has a small hinged panel fitted for use when flying in bad weather. At the rear end is a fixed window which allows the pilot an all-round view aft before taking off. Eye line makes contact with the ground approximately the length of the machine aft of the rudder in the tail down position.

WINGS

The wings are to NACA Section No 230 and of such depth at the root to allow the top boom to continue across the fuselage without interference to the pilot's use of the rudder bar. This depth accommodates in the centre section the fuel tanks, bomb cradles and the chassis in the retracted position. The thickness chord ratio is 23.8% with 7' 9" chord at the root and 6% with 50" developed chord at the tip. The aspect ratio is 7.25. The wings are fitted with flaps hydraulically operated. The outer trailing portion is depressed to an angle of approximately 23 degrees for take-off and 90 degrees for landing. At this position the centre balance flap is also depressed 90 degrees but is not brought into operation for take-off. The wing and centre section is completely braced with ply covering. All electrical leads and air lines are housed in a duct at the under surface and forward of the front spar and are readily inspected.

TAIL UNITS

Care and attention has been given to the design to provide for the maximum rigidity of the surface together with a clean entry. The tail plane is fully cantilever.

FLYING CONTROLS

To facilitate assembly each cockpit has a control unit which incorporates the rudder bar with its adjustment and control column with brake and gun firing attachment. The front unit has also a single lever locking device which locks all the flying controls in the central position. This is a top dead centre lever, spring loaded, which may be instantly released. It is connected to the throttle control or engine switch thus making it impossible to take-off with the controls locked. All three systems are operated by push-pull tubes adequately stabilised by ball races covered with rubber tyres. Attention has been paid to the rapid removal of the rear control unit together with its interconnection to the main control system. The elevator and rudder are fitted with damped trimming tabs with the irreversible units positioned as close to the tabs as possible. These tabs may be operated from either cockpit.

ENGINE INSTALLATION

The Rolls-Royce Kestrel engine is resiliently mounted to a steel tubular engine mounting structure which is designed to facilitate engine removal. The forward portion which remains with the engine is detached by unscrewing four nuts. The rear portion remains with the fuselage. A low drag cooling duct for Glycol and Oil Systems is positioned in the cowling under the engine sump with a variable opening outlet controlled from the front cockpit. Two 35 gallon fuel tanks of De Bergue construction are slung in the centre section near the root. These tanks are readily removable through the door in the under surface. Additional tanks maybe housed in the leading edge. All tanks are interconnected with a large diameter fuel pipe across the fuselage. Emergency hand pump is supplied by the Korect Depth Gauge Co. Ltd. and is capable of delivering 40 gallons per hour. Smith's electrical fuel capacity gauges are fitted. The oil system is of normal design and the tank is resiliently mounted on Silent-bloc attachments. The tank has a partial circulator incorporated and the filter may be withdrawn through the top of the tank. For draining a pipe is led to the side of the fuselage via a three-way cock in the tank sump. The Glycol header tank is above the engine gear box and the radiators are in the cooling duct. The engine has attachments for:-

(a) Frazer Nash Hydraulic pump for rear gun turret operation.
(b) Lockheed hydraulic pump for chassis, flap and bomb door operation.
(c) B.T.H. Air Compressor for firing Browning guns and brake operation.
(d) Pump for variable pitch airscrew.

CHASSIS & TAIL WHEEL

The chassis comprises two independent units made to retract into the centre-section operated by the Lockheed hydraulic circuit. Emergency pipe lines and hand pump are fitted in case of failure of the main circuit. The chassis position indicator is interconnected with the engine ignition switch and the throttle control. Wheels and tyres are Palmer 580 x 180 *(another brochure says 680 x 220 - PA)* and the brakes are pneumatically operated. The tail wheel will be made retractable and has Palmer 270 x 150.

OPERATIONAL EQUIPMENT

Dual & Blind Flying Instruction

The pupil will occupy the front cockpit which is provided with the blind flying hood. The pilot is able to throw out the pupils use of the brakes, throttle and engine switch. W/T sets, TR.9 or TR.11, may be fitted aft of the rear cockpit. Browning gun, cinema gun type GX 42 are fitted in the wings.

Wireless & Photography (Reconnaissance)

The wireless operator faces aft with the sets complete with power tray mounted in a quickly removable one unit crate. The Type F.24 camera is positioned aft of the sets and is within easy reach of the operator for drift correction and operation. The bomb sight doors are opened for photography but the W/T operator is protected from draught by an extra flooring integral with his seat.
The D/F loop is fitted above and aft of the sets and navigational equipment is also included. Browning gun and cinema gun, Type GX 42 are fitted in the wings.

Bombing & Photography

The Mark IX course setting bomb sight is attached to a movable mounting and may be lowered when the sighting doors have been opened. The dimensions of the door are such that a good view of the target may be obtained before training it along the ground bar of the sight. The Type F.24 camera is mounted forward of the bomb sight and photographs through the bomb sight slot. Power for opening the sighting doors is derived from the Lockheed Hydraulic Circuit. Navigational Equipment is fitted but not wireless. Browning gun and cinema gun Type GX 42 are fitted in the wings.

Rear Gun Turret

Case I. One or two Browning guns are mounted in the fuselage aft of the trailing edge, to give maximum view of the gunner in a downward and forward direction. Rotation, elevation and firing is controlled by the Frazer Nash type of control handle. The gunner is completely enclosed in a transparent hood which may be released for emergency exit. The turret and hood may be retracted hydraulically into the fuselage and the sliding windows wound forward which restores the aerodynamic efficiency of the aircraft. TR.9 or TR. 11 sets may be fitted. Browning gun and cinema gun, Type GX 42 are fitted in the wings.

Case II. The Browning guns and ammunition box in the turret may be removed and wireless crate housing R.1082 and T.1083 sets complete with power tray installed, with navigational equipment and D/F loop. Browning gun and cinema gun, Type GX 42 are fitted in the wings.

ARMAMENT. (Various)

Fixed gun

The Browning gun is mounted between the end ribs of the centre section and outer wing, on the starboard side of the aircraft. The ammunition box for 300 rounds integral with the link and empty case reservoir is in the outer wing. The sealing strip between the ribs is quickly detachable for the rapid removal of the gun and ammunition box, whilst the sealing strip on the top surface may be removed for servicing the gun from above.

Cinema Gun

The GX 42 Cinema gun is mounted between the end ribs of the centre section and the outer wing forward of the front spar, on the port side of the aircraft.

Bombs

The sixteen practice smoke bombs are on electro-magnetic release units, stowed inside the wing fillet and centre section. The racks are accessible through doors on the bottom surface. Dive bombing clearances have been designed for.

Rear gun and turret

One or two Browning guns are mounted aft of the centre section trailing edge to give maximum view for the gunner in a downward and forward direction. Rotation, elevation and firing is controlled by the Frazer Nash type of control handle.

(What appears to be an earlier brochure also mentions a 'Multi-Gun Single-Seater Fighter', where: 'The guns would be mounted in the wings outside the airscrew disc' - PA).

WEIGHT ESTIMATES OF AIRCRAFT WITH VARIOUS EQUIPMENT

Single-Seater Aircraft

Weight of aircraft structure	3,225lb
Pilot and parachute	200lb
Fuel (70 gallons)	540lb
Oil (8 gallons)	80lb
Basic weight	4,045lb
Equipment (full)	220lb
Armament	61lb
Ballast	121lb
AUW	4,447lb

Dual Control

Basic weight	4,045lb
Dual Equipment (full)	343lb
Pupil and parachute	200lb
Armament and wireless	111lb
AUW	4,699lb

APPENDICES 18 / 19

Bombing and Photography

Basic weight	4,045lb
Equipment (full) Fixed	220lb
Bomber and parachute	200lb
Armament	61lb
Bombs, cradles and photography	274lb
AUW	4,800lb

To each of the above configurations may be added extra fuel and tanks (45 gal) 370lb.

OVERALL DIMENSIONS etc.

Span 39ft 0in; length 30ft 0in; minimum height (tail down) 10ft 0in; wing area (nett) 210 sq ft;

PERFORMANCE ESTIMATES

Full throttle speed at different height

Sea level 230 mph; 5,000ft 251 mph; 10,000ft 270 mph and 16,500ft 295 mph
Cruising speed at 2,600 rpm (normal) at rated boost and 16,500ft 253 mph

Appendix 19:
Miles M.11 Whitney Straight Special and M.11A, M.11B and M.11C Whitney Straight Production

290 M.11 'Whitney Straight Special'; painted grey. Fitted with 130hp DH Gipsy Major Srs I engine. Regd **G-AECT** (CofR 6742) 18.2.36 to Phillips & Powis Aircraft Ltd. First flown, by Miles, as a 'Whitney Straight' 3.5.36. CofA No.5494 issued 9.5.36 (the application was made as an 'M.9' Straight Special, probably in error). Retained by makers; flown by George Miles from Woodley to Heston and return 28.5.36 (but with no regn recorded). Demonstrated at the fifth SBAC Show at Hatfield on 29.6.36. Flown by Charles Lindbergh in 5.36 and later loaned to him for his personal use while the Mohawk was being constructed; probably based at Penshurst (as he was living nearby). In July 1936, used for boundary layer suction experiments. Lindbergh also used it for some 5-6 weeks in July and August 1936 (and possibly later); The Aeroplane for 12 August 1936 reported that Lindbergh had flown it on his first visit to Germany on business, landing at Staaken Airport, Berlin. On 27.8.36, an application was made for 'modified aluminium oil tank, oval bulkhead type and modified Onions legs'. CofA renewed (early) 1.2.37, implying major modification. On 12.2.37, George Miles carried out an 'Altitude test - reached 20,000 ft' between Woodley and Shoreham, during a 1 hr 30 min flight. On 23.2.37 George Miles flew it to Martlesham Heath (it was recorded as being flown at RAF Wittering on 27.2.37 by Flt Lt's Edwardes-Jones & RDG Wright); George flew it back to Woodley from Martlesham on 1.3.37. On 9.3.37 George Miles carried out 'Take-off tests with new flaps' (these comprised two external flaps in tandem, known by Miles as a 'Venetian Blind' flap, mounted aft of the wing trailing edge), for an international competition for army co-operation aircraft to be held at Odiham; later that day he made a 'Check with standard flaps' in the same aircraft. Fitted with a 125hp Menasco C.4S Pirate engine; CLN recorded the date dd as 1.4.37, however on 20.3.37 George Miles had carried out a 1 hr 45 min 'Test with C4S engine of 150hp' with a considerable number of further flights commencing on 22.3.37, including one with Miles on 31.3.37. George carried on flying G-AECT until 24.4.37, when he 'Took U.B. (possibly operating under 'B' marks as U3 (?) and badly written in logbook – see below) up to 15,000 ft and allowed engine to oil plugs on glide. Had f/l in field nr Farnborough'!, he flew it out later that day. G-AECT visited Farnborough 20.6.37 (but not flown by George Miles on that day); on 2.7.37 George carried out a 'Test with cg forward extended' and on 6.7.37 he flew G-AECT (still fitted with the Menasco engine) to Odiham for a 'demonstration of slotted flaps'. Flown by George Miles 10.7.37, 'Test' 12.7.37 and 15.7.37; again visited Farnborough on 20.8.37. CofA was renewed 16.3.38 probably with a 130hp Gipsy Major engine re-fitted (but recorded by CLN as fitted in 4.38). Regd (CofR 8452) 12.4.38 Lionel Edward Richard Bellairs, Burwash, Sussex (based there & at Shoreham). CofA lapsed 17.3.40. Regn cld 27.10.40. Impressed into Royal Navy 21.10.40 (reportedly under Contract No.B.50801/39) as **BS755**. To 804 Sqdn, Skeabrae 12.40; later with unit to Skitten and Yeovilton. To 787 Sqdn, Yeovilton 3.41. Op by Fighter School, Yeovilton (10.41–5.42). Flown Yeovilton-Woodley-Yeovilton on 1.5.42 by Lt Alan Peter Goodfellow. To 792 Sqdn, St Merryn. Suffered heavy landing, damaging undercarriage 25.7.42; pilot Lt MRF Lemon. Later recorded at Yeovilton (23.11.42); RNARY Fleetlands; test flown at Gosport (4.6.43) and Lee-on-Solent (21.6.43). Fate unknown.

303 M.11A Whitney Straight; painted Whitney Straight blue and silver. First production (P&P works records state 'Conforms to G-AECT with the exception of two 15-gall aluminium fuel tanks & various production modifications including redesigned u/c fairings'). Regd **G-AENH** (CofR 7320) 10.9.36 to Whitney Straight Ltd, London, W1 (based Heston). First flown, by George Miles, 9.1.37. CofA No.5755 issued 16.1.37; dd 22.1.37 and demonstrated at Heston the same day. Flown by Miles with George Miles 31.3.37. Flown by George Miles 16.6.37. Regn cld 11.37 as sold. Regd **VH-ABN** (CofR 683) 21.12.37 to Falkiner William Hewson, Augathella, Qld (in exchange for his M.2H Hawk Major VH-AAH). Regd 31.12.42 to Royal Victorian Aero Club, Essendon. Regn cld 6.11.47. Regd 20.2.48 to same owner. Exported to New Zealand 12.2.51 & regn cld 22.3.51. Regd **ZK-AXN** 1.12.50 to Auckland Aero Club Inc, Mangere. NZ CofA issued 30.7.51. Regd 17.3.58 to George William Henry, Matamata. Regd 25.9.61 to Kea Syndicate. Regd 17.9.62 to Campbell Robert Feather, Otematata. Wfu on CofA expiry 20.10.62 and regn cld 12.9.63. Aircraft destroyed by fire in hangar at Otematata in 11.63 and remains scrapped in 5.64.

304 M.11A Whitney Straight; painted with grey fuselage with red flashing. Regd **G-AERS** (CofR 7559) 12.36 to Airwork Ltd, Heston. First flown, by George Miles, 9.1.37. CofA No.5760 issued 20.1.37; dd 22.1.37 to Heston by George Miles. Regd (CofR 7721) 2.3.37 to Second Lt Robert 'Rex' King-Clark, of 1st Manchester Regiment, Strensall Camp, York (based at Clifton Aerodrome, York); dd 1.3.37 and the tail bore the Fleur de Lys crest of his regiment. Flown by owner to Ismailia, Egypt 4.3.37-14.3.37 on his overseas posting; returned to UK 9.37. Departed to Palestine 14.1.38 and to Almaza, Cairo, where remained until flown by owner to Singapore, departing 8.10.38 and arriving 21.10.38. (CofA lapsed 3.1.39 and the official record does not show renewal). Departed Singapore for UK 9.3.39, shipped from Bombay to Port Said, Egypt and flown Almaza to Heston, departing 17.4.39 and arriving 12.5.39. Sold 18.8.39 and regd 28.9.39 to WS Shackleton Ltd, Denham. Regd 20.4.40 to Miss Eveline MC Townshend, Bodiam Manor, Bodiam, Sussex (jointly with WS Shackleton Ltd). CofA renewed 29.4.40 and flown to Leixlip, Ireland to continue flying training operations but these ceased in 6.40. For sale by WS Shackleton Ltd in Flight 9.1.41 for £675. Regn cld 9.6.41. Impressed into RAF 6.5.41 as **ES922** and to Stn Flt Northolt. To De Havillands at Witney for overhaul 29.8.42; returned to Stn Flt Northolt. Taken to Phillips & Powis, Woodley 19.4.43 by road for repairs but SOC (Cat.E1) for spares use 31.5.43.

305 M.11A Whitney Straight; painted blue and silver. Regd **G-AERC** (CofR 7486) 11.36 to unknown party (but possibly Phillips & Powis Aircraft Ltd). Demonstrated at Heston 22.1.37 and flown back to Woodley by George Miles later the same day. Fitted with the 135hp Amherst Villiers Maya engine 3.2.37 for flight tests to become the sole M.11B. CofA No.5785 issued 5.2.37. Regd (CofR 7785) 24.3.37 to Charles Amherst Villiers and Mrs Maya Villiers (of Villiers Hay Development Ltd), London W1 (based Heston). Gipsy Major engine re-fitted 1.38 and reverted to M.11A. Regd (CofR 8314) 24.1.38 to Thanet Aero Club Ltd, Ramsgate (part of The Straight Corporation); CofA renewed 29.1.38. Regd (CofR 8506) 19.5.38 to (associated company) Ipswich Aero Club Ltd, Ipswich. Permit to Fly No.74 issued 1.2.40 for Straight Corporation to fly in UK. Regn cld 16.4.40 as sold. Impressed into RAF 26.4.40 as **AV971** although was still being flown as G-AERC when issued to Stn Flt Hucknall 4.7.40 (and at least until 28.8.40). To Percival Aircraft Ltd, Luton (by road) (MI) 27.1.41; to 51 MU Lichfield 3.5.41. To RAF Northolt 6.5.41. Cat B and to Phillips & Powis by road 16.5.43 for repairs but SOC (Cat.E1) for spares use 9.6.43.

306 M.11A Whitney Straight; painted orange and silver. CofA No.5778 issued 3.2.37 to Camille Gutt, Belgium; dd 6.2.37. Regd **OO-UMK** (CofR 394) 12.2.37 to Camille Gutt (a director of Union Minière du Haut Katanga), Jette (based Brussels). Regn cld 27.7.38 as sold (Gutt traded it to Airwork Ltd in part-exchange for an M.17 Monarch, also regd OO-UMK). Regd **G-AFJJ** (CofR 8679) 19.8.38 to Airwork Ltd, Heston; CofA renewed 29.9.38. CofA lapsed 28.9.39. Permit CO1/84 issued 23.4.40 for flights at Staverton. Regn cld 23.8.40 as sold. Impressed into RAF 14.8.40 as **BD168**. To 271 Sqdn, Doncaster 14.8.40 (but probably retained regn until 2.41). To 60 MU York 30.9.43 (possibly after an accident) and taken to Phillips & Powis, Woodley for repairs 3.10.43. To 5 MU Kemble 16.4.44. To Maintenance Command Comm Sqdn, Andover 12.5.44. Damaged in forced landing nr Andover 1.8.44 following engine failure on take-off and SOC (Cat.E1).

Miles M.11A Whitney Straight OO-UMK of M Camille Gutt in 1937.
[Via P Amos]

307 M.11A Whitney Straight; painted silver. Regd **G-AERV** (CofR 7566) 30.12.36 to Harry William Horace Moore, London NW8 (based Heston). CofA No.5761 issued 30.1.37; dd 6.2.37. At Old Warden in 9.39 on Air Ministry survey. CofA lapsed 23.1.40. Regn cld 7.6.41 by Secretary of State. Impressed into RAF 7.6.41 as **EM999**. To Halton 21.6.41. To Andover 18.7.41 but to 39 MU Colerne 20.7.41 and issued instead to Abingdon 24.7.41. To 86 MU Sundridge, Kent by road 28.5.45 (possibly after accident) and to 13 MU Henlow for repairs. To 5 MU Kemble 22.7.46 for disposal. Sold (back) to pre-war owner, Harry William Henry Moore, London, W2 and regd **G-AERV** to him 3.6.47. CofA renewed 5.7.47; now named 'Anky Sim' and based Denham. Sustained damage to port wing, airscrew and undercarriage in forced landing at Poix, France 2.7.48; repaired. Stalled on approach while landing in a field at Manor Farm, Studland, near Swanage 24.5.53; port wing, propeller and undercarriage damaged, no casualties and repaired. Cld 17.7.56 & regd 20.7.56 to WS Shackleton Ltd, London, W1. Cld 29.8.56 & regd 7.9.56 to Robin Reginald Harrington, Edgware. Cld 7.6.57 & regd 29.6.57 to James Matthew Banks, London W1. Cld 12.5.58 & regd 16.5.58 to Charles Mortimer Tollemache Smith-Ryland, Warwick. Cld 20.5.59 & regd 26.6.59 to James William Arthur Hainge & Geoffrey Raymond Nixey, Kidlington. Cld 6.4.60 & regd 11.4.60 to Jack Braithwaite, Kidlington. Cld 3.4.62 & regd 12.4.62 to Robert Thompson Boyes, Newtownards. Forced landed following engine failure by owner/pilot on flat hilltop at Pond crossroads, nr Seskinore, Co Omagh 22.5.63; only minor damage to airframe but much damage to the undercarriage. After jury-rigging an undercarriage, it was towed to Newtownards on 24/25.5.63 for repairs. In early 1966, (with a current CofA) G-AERV was loaned by owner to the Belfast Transport Museum, Witham Street, Newtownards and on 30.1.66, its wings were removed and it was towed to the museum store. It went to the Folk & Transport Museum at Caltre, Co Down but was later towed back to Newtownards (by 1977). Regn cld 7.9.81 by the CAA. It was acquired by a Mr Sherry, the owner of a chicken farm at Upper Ballinderry, nr Crumlin, Co Antrim, who stored it in a dilapidated shed with a leaking roof, where its condition deteriorated. The farm was sold in early 2002 and G-AERV was purchased by Richard Allen Seeley, London SW13 and regd to him 10.5.02. Collected by Ron Souch of Aero Antiques, Durley, nr Southampton 4.02, it is being rebuilt by Ron Souch to airworthy condition, with all new wooden structure and a completely new undercarriage. To Airtime Paint, Hurn for respray 1.08.

308 M.11A Whitney Straight; colour not recorded. Sold by John M Gamble of Lambton Quay, Wellington the NZ agent for Phillips & Powis Aircraft Ltd. CofA No.5770 issued 5.2.37 to Canterbury Aero Club, Christchurch; handed over 4.2.37. Regd **ZK-AEO** (CofR 79/CofA 182) 14.4.37 to Canterbury Aero Club, Christchurch. Taken over by RNZAF 'for duration of war' 13.10.39; impressed into RNZAF 1.11.39 as **NZ576**. Regd **ZK-AJF** 28.3.46 to Canterbury Aero Club, Christchurch. CofA No.286 renewed 12.4.46. Converted to 3-seater by Christchurch Aero Club. Lost control and dived into ground under full power at North Loburn, Canterbury 26.6.50 at 11.00 hrs; pilot James Russell Kinggan and two passengers Gordon Brown and Henry Mead were killed. The aircraft had been on a sightseeing flight from Harewood and was obviously overloaded; it appears that the pilot had lost control in cloud at 1,000 ft when flying VFR. Weather conditions were described as good and there was no evidence of structural damage prior to the accident, although the door flew off on the way down.

ZK-AEO Miles M.11A Whitney Straight, seen in New Zealand before the war. *[Via P Amos]*

Impressed Miles M.11A Whitney Straights NZ576, NZ577 and NZ571, together with an assortment of other training aircraft of No.42 Squadron, at RNZAF Station Roncotai, in 1944.
[Reproduced with kind permission of the RNZAF Museum]

309 M.11A Whitney Straight; painted light blue and silver. Regd **G-AERY** 1.37 to British Air Transport Ltd, Redhill. CofA No.5781 issued 13.2.37; dd 13.2.37. Regd (CofR 7710) 26.2.37 to HB Legge & Sons Ltd, London EC4 (based Hamsey Green and Redhill); regularly flown to Sweden, Finland and Germany by Geoffrey Legge on business. Regd cld 4.38 as sold abroad. Regd in India as **VT-AKF** (CofR 351) 11.6.38 to Maharajkumar DC Bhanj Deo, Calcutta. Regn cld 6.1.43. Impressed into RAF locally as **MA944** 23.8.42. Flown from Dum Dum-Jessore-Khulna-Dum Dum on 16.1.43 in search for a Tiger Moth that had forced landed due to Japanese bombers (pilot Flt Lt JF Parry with Sgt R Barnby). To Bengal Comm Flt. Overturned in forced landing following engine failure Narayanpur, Bengal 11.3.43. SOC 2.7.43.

310 M.11A Whitney Straight; painted blue and silver. Regd **G-AETB** (CofR 7628) 4.2.37 to The Hon Thomas Sharon Fermor-Hesketh, London W1 (based Brooklands). CofA No.5792 issued 12.2.37; dd 13.2.37. While on a flight from Cannes to Croydon, via

Miles M.11A Whitney Straight G-AETB, seen prior to its crash on 21st June 1937. [Via B Clarke]

Le Bourget on 21.6.37, the aircraft entered a very violent thunderstorm and emerged from the clouds in a steep inverted dive. While some control was recovered near the ground it crashed at Cires-les-Mello, south east of Beauvais, France; the owner/pilot Lt Fermor-Hesketh and his passenger 2nd Lt BA Ludford-Astley were killed. Regn cld 12.37.

311 M.11A Whitney Straight; painted silver. Sold in New Zealand by John M Gamble of Wellington, the local agent for Phillips & Powis Aircraft Ltd. First flown, by George Miles, (recorded as '311') 3.4.37 and flown again by him on 5.4.37, 6.4.37 and 7.4.37; handed over 8.4.37. CofA No.5872 issued 14.4.37 to RNZAF and possibly initially operated under serial **311** on seeding and spraying trials. Although the regn **ZK-AFH** was reserved in early 1937, it was not regd (CofR 222) until 19.5.38 to the Public Works Department, Civil Aviation Branch of the New Zealand Government, Rongotai. Alan Pritchard, Chief Pilot for the Public Works Dept carried out the first lupin seed sowing by air near Ninety Mile Beach on 10.3.41; further seeding trials carried out in 7.46. Used in 8.47 to solve copper deficiency in swampy soils by spraying bluestone; hopper then fitted and used to spread cobalt solutions on hill country nr Taumaranui. Regd 22.7.49 to Civil Aviation Administration; last used for spraying in 1949. Regd 15.11.57 to G Hufton, New Plymouth. Regd 15.5.59 to NB Hoey, Hawera. Regd 10.4.61 to William Decimus Fleming, Auckland. Dd 24.2.62 & regd 26.7.62 to Des S Rowles, Omakau. Stored at Lauder in 1962 while owner was absent for a year in Antarctica but water got into the wings and froze; wfu due to frost damage and aircraft dilapidated beyond repair. Scrapped 65/66. Regn cld 29.8.66.

312 M.11A Whitney Straight; colour not recorded. Regd **G-AETS** (CofR 7662) 16.2.37 to Edward Gordon Houston Forsyth, London SW1 (based Hanworth, later Croydon). First flown, by George Miles, 23.3.37 (recorded as 'No.312') and flown again by him on 24.3.37 on a 'Radio Test'. CofA No.5854 issued 25.3.37; dd same day. To Renfrew 12.39; flown by ATA Training Pool from about 4.41 as G-AETS, then to the ATA Pool at Brockworth by 17.6.41 where used as a taxi to collect pilots ferrying newly built Hurricanes from the Gloster Aircraft factory. Regn cld 31.5.41 by the Secretary of State. Impressed into RAF 20.5.41 as **DR611** at Brockworth but still flown as G-AETS until at least 22.6.41. The serial was misapplied as **DR617** and it retained this for the remainder of its service career. To Shrager Bros, Old Warden (MI) 23.6.42. To 5 MU Kemble 5.8.42. Issued to Stn Flt Wyton 14.8.42. To 5 MU Kemble for disposal 10.5.45 and SOC (Cat.E2) the same day (an undated document (ref: AVIA 57/28) records that DR617 was 'U/S for spares only'). In fact, it was sold at the first sale of impressed aircraft held at 5 MU in 12.45 to Straight Corporation Ltd for £50, (for 'airline, charter and club work' although the price obviously reflected little more than its scrap value). Regd **G-AITM** (CofR 10957) 11.11.46 to Whitney Straight Ltd, Weston-super-Mare. CofA 8709 allocated but never issued. Not overhauled and used for spares at Weston-super-Mare. Regn cld 3.3.48 as reduced to parts.

313 M.11A Whitney Straight; colour not recorded. Regd **G-AEUJ** (CofR 7687) 19.2.37 to Major Charles Lamond Yarborough Parker, Crowborough, Sussex (based at Gatwick). CofA No.5797 issued (to Phillips & Powis Aircraft Ltd) 24.2.37; dd 25.2.37 (CLN records its delivery to Air Travel Ltd, Gatwick but there is no evidence that it was ever operated by them). George Miles made return flights to Shoreham in G-AEUJ on 20.7.37 and 22.7.37 (and CLN also records dd to Major CLY Parker, Sussex 15.3.38). Damaged in ground collision with DH.60X Moth G-EBUS Portsmouth 24.3.39; repaired. At Shoreham for Air Ministry survey in 9.39. CofA renewed 28.11.40. Regd 29.1.41 to Hawker Aircraft Ltd, Kingston-on-Thames (reportedly based at Gatwick). CofA lapsed 27.11.41; renewed 23.7.43. Issued with MAP Permit No.12 11.11.43-14.11.46. CofA renewed 8.11.46. Cld 15.7.57 and regd 17.7.57 to Clarence Louis John Reed, Mill Hill, London NW7 & Sydney Mark Aarons, Borehamwood, t/a The Mill Hill Flying Group (based at Elstree). Blown off the runway and ground-looped while landing at Elstree Aerodrome 18.7.57; the flap, propeller and undercarriage were damaged. Cld 1.8.57 and regd 12.8.57 (solely) to Clarence Louis John Reed, Mill Hill, NW.7. Cld 28.3.61 and regd 7.4.61 to Edward Eves Ltd, Leamington Spa (based at Baginton). Cld 10.9.62 and regd 19.9.62 to John Tullett, Warlingham, Suffolk (based at Swanton Morley). Cld 21.10.66 and regd 8.11.66 to Peter Hillwood, Southampton and Charles Herbert Parker, Christchurch (based at Hurn). Cld 30.11.67 and regd 18.3.68 (solely) to Charles Herbert Parker, Ferndown, Dorset (based Hurn). CofA lapsed 4.6.70. Noted 12.71, still stored under tarpaulins at rear of flying club hangar at Hurn. Taken by road 8.74 to Castle Donington (East Midlands Airport) for rebuild by Parker Aviation (reportedly for DA Hood). Nominal regn change 13.12.76 to Julia Anne Mary Parker, Ferndown. Cld 8.12.78 and regd 28.12.78 to Robert Evan (Bob) Mitchell, Sutton Coldfield (based at East Midlands). To Speedwell Sailplanes, Marple, nr Stockport in 1979 for considerable structural work and taken by road to a Warwickshire Farm 13.2.82 where it was 'hoped that it would be flying by the end of the year', but this did not happen and its completion appears to have been abandoned. In store near Sutton Coldfield (6.92); but moved to RAF Cosford during 1992, where it is still stored by Mitchell. In early 2008, it was moved to Sleap, where it is hoped that its rebuild will eventually be completed.

Miles M.11A Whitney Straight G-AEUJ in the house colours of dark blue and gold of Hawker Aircraft Ltd, seen at Baginton, on 22nd August 1953. [Via B Clarke]

314 M.11A Whitney Straight; colour not recorded. Regd **G-AEUX** (CofR 7702) 23.2.37 to John Bayly, Yelverton, Plymouth (based Roborough). First flown, by George Miles, 27.2.37 and flown again by him on 2.3.37. CofA No.5798 issued 2.3.37; dd 3.3.37. At Portsmouth for Air Ministry survey in 9.39; to Porthcawl 1.40; to Heston 2.40; to Denham. Regn cld as sold 12.1.41. Impressed into RAF as **DJ713** 12.1.41; to 1 FPP, ATA White Waltham retaining the last 2 letters 'UX' on the fuselage. To Herts & Essex Aviation Ltd, Broxbourne for overhaul 28.8.41; returned to White Waltham 2.11.41. Nosed over taxying in gusty weather St Athan 6.11.41 due to badly adjusted brakes; ATA pilot FR Tullett unhurt; minor damage only. Collided with an Anson after swung on landing at Staverton 30.11.41 due to port brake locking; pilot FR Tullett (again); minor damage. On 6.1.42 after being collected following repairs, it overshot on landing Brockworth and ran into obstruction; repaired by Gloster Aircraft Ltd. To Stn Flt Andover 20.5.42. To Phillips & Powis

Miles M.11A Whitney Straight G-AETS at Croydon Airport before the war. [Mr Lintott via N D Owen]

Miles M.11A Whitney Straight G-AEUX. [Via B Clarke]

for overhaul 3.10.42; returned to Andover 7.12.42. Flown by Hugh Kennedy at Woodley on 'service test' flights on 4.1.43, 8.1.43 and 15.1.43. To 5 MU Kemble for storage and disposal 4.10.44, from where it was sold to Air Cdre H Probyn for £511 at the first sale of impressed light aircraft held in 12.45. Regd **G-AEUX** 1.5.46 to Air Cdre Harold Melsome Probyn, Studham, Beds. CofA renewed 9.9.46. Regn cld 4.1.50 as sold in Kenya. Flown by Probyn to Nyeri, Kenya and regd **VP-KHO** 8.12.49 to him. Regd 23.10.50 to Mrs EG Jennings, Mombasa. Regd 25.10.51 to PG Nicholas, Nairobi West. Ran away when hand-started by pilot M Bearcroft without chocks or brakes at Nairobi West 2.3.52 and badly damaged in collision with Messenger VP-KHG and Macchi MB320 VP-KIT. Remains still present at Nairobi West 4.53.

315   M.11A Whitney Straight; colour not recorded. Regd **G-AEUY** (CofR 7703) 24.2.37 to Richard Exton Gardner Jnr, Hamsey Green. First recorded flight was by George Miles (recorded as 'No.315') 4.3.37. CofA No.5799 issued 3.3.37; reportedly dd the same day. Regd (CofR 8134) 12.10.37 to Sir Alfred Lane Beit, Woodley (or Heston). Regn cld 5.3.40 as sold. Impressed into RAF as **W7422** 4.3.40; to 24 Sqdn, Hendon and used by the British Air Attaché, Paris. Abandoned at Le Bourget during the evacuation from France in 5.40. SOC (Cat.Em - missing on sortie) 22.6.40.

316   M.11A Whitney Straight; colour not recorded. Regd **G-AEUZ** (CofR 7705) 5.3.37 to Rolls-Royce Ltd, Hucknall. First flown, by George Miles (recorded as 'No.316') 22.3.37 and again by him, on 23.3.37. CofA No.5810 was issued 24.3.37; dd 24.3.37. Remained in service with Rolls-Royce during the war; MAP Permit No.24 issued 31.7.44-30.7.46. Post-war CofA renewed 19.10.46. Cld 11.2.48 and regd 25.2.48 to Arthur Frank Eayrs, Oakham, Rutland. Regd 15.8.49 to H Tempest Ltd, Tollerton. Cld 11.6.51 and regd 14.6.51 to Henry Stockley Brown, Reading. Regn cld 10.10.52 as sold in Kenya (and noted at Old Warden as VP-KKF 4.11.52). Regd **VP-KKF** 20.10.52 to PG Nicholas, Nairobi West (bought as a replacement for VP-KHO). Spun in from 200 ft on landing at Mweija 7.3.53 on Police Air Wing duties; repaired on site. Regn cld prior to 1963.

Miles M.11A Whitney Straight G-AEUZ, seen at a race meeting after the war. [Via B Clarke]

317   M.11A Whitney Straight; colour not recorded. Regd **G-AEVF** (CofR 7716) 2.3.37 to Luke Theodore Lillingston, Elvaston Castle, Derby (based Braunstone). CofA No.5805 issued 10.3.37; dd to Derby 9.3.37. Regn cld 27.10.40. Impressed into RAF as **BS814** 21.9.40. To 24 Sqdn, Hendon. To Stn Flt Northolt 23.4.42. SOC (Cat.E1) for spares use (MR) 22.3.45 as 'beyond repair'.

318   M.11A Whitney Straight; colour not recorded. Regd **G-AEVA** (CofR 7706) 5.3.37 to Gerald Cohen, London W.7 (based Heston). CofA No.5811 issued 9.3.37; dd 9.3.37. Regd (CofR 7882) 20.5.37 to Airwork Ltd, Heston. Regd (CofR 8094) 28.8.37 to David de Crespigny Smiley, Sunningdale (based Heston). Cld 28.12.38 & regd 10.1.39 to Edward Alan Strouts, Biddenden, Kent (based Bekesbourne). Stored at Milsted's Garage, Tenterden in 9.39. CofA lapsed 13.2.40. Regn cld 20.5.41 as sold. Impressed into RAF as **DR612** 20.5.41. To 6 AONS, Staverton. Ferried by ATA from Ratcliffe to Bircotes 19.10.43; to 1 Grp Comm Flt, Bawtry 20.10.43. To 5 MU Kemble for storage and disposal 6.11.44, from where sold to Rice Caravans Ltd for £500 at the first sale of impressed light aircraft in 12.45. Regd **G-AEVA** 29.1.46 to John Cecil Rice, Cosby, Leicester (based at Walsall, later at Ratcliffe & Leicester East). Damaged on landing at Bruntingthorpe 22.4.50; repaired. CofA lapsed 31.7.51; renewed 16.9.52. Hit a mountain near Moutier (Switzerland) and burned at 5.30 pm 2.7.54; both occupants, Mr and Mrs Rice were killed. Regn cld 19.7.54 as destroyed.

319   M.11A Whitney Straight; colour not recorded. Regd **G-AEVG** (CofR 7718) 2.3.37 to Willoughby Rollo Norman and Antony Charles Lynyard Norman, Heston. First flown, by George Miles (recorded as 'No.319') 15.3.37. CofA No.5812 issued 15.3.37; supposedly dd 14.3.37. Regn cld at census 1.1.39. Regd 26.4.39 (solely) to Antony Norman, Heston. CofA lapsed 5.4.40. Regn cld as sold 18.3.41. Impressed into RAF as **DP854** 18.3.41 (but this serial had already been allotted to Puss Moth G-ABMS so it was changed to DP845; however, this belonged to the F.4/40 Spitfire III prototype, so it retained the original serial DP854!). To Stn Flt Northolt. To Stn Flt Andover 1.8.42. To 8 MU Little Rissington 13.1.44 for storage. To Fighter Command Comm Sqdn, Northolt (undated). Damaged in collision with hangar door at Hamble 2.7.46; abandoned there and later acquired by Air Service Training Ltd. Regd **G-AEVG** 25.3.47 to Air Service Training Ltd, Hamble. CofA renewed 17.6.47. Damaged on take-off after landing on downs two miles North of Southwick in heavy fog at 4.10 pm on 13.5.48; being flown by student pilot Jehuda Boupko, on cross-country navigation exercise; starboard wheel broken and leading edge of starboard wing damaged; repaired. Cld 5.12.52 and regd 9.12.52 to Hants & Sussex Aviation Ltd, Portsmouth. Cld 27.1.53 and regd 11.3.53 to Harry William Horace Moore, London W2. Cld 27.2.54 and regd 4.3.54 to William Allan Strauss, East St Kilda, Victoria, Australia. Named *Victor George* and fitted with long range fuel tanks, it was flown to Australia by owner, arriving at Moorabbin 16.10.54. Regn cld 27.5.55 as sold in Australia. Regd **VH-EVG** (CofR 2825) 27.5.55 to William Allan Strauss, East St Kilda, Victoria. Regd 3.7.56 to Desmond Terril, Ballarat, Victoria. Ground-looped on landing at Rutherglen, Vic 26.1.57 to avoid hitting fence; u/c broken off; owner/pilot D Terril unhurt. Not repaired and regn cld 13.9.58.

320   M.11A Whitney Straight; colour not recorded. Regd **G-AEWA** (CofR 7555) 9.3.37 to Lawrence Alfred Mervyn Dundas, Earl of Ronaldshay, Richmond, Yorks (based at Portsmouth). First flown, by George Miles, (recorded as 'No.320'), 11.3.37, with further flights as 'No.320' on 15.3.37 and 18.3.37. CofA No.5830 issued 19.3.37; dd 19.3.37. Flown as G-AEWA by George Miles 12.4.37. Cld 3.2.39 and regd 1.3.39 to Trevor Bertram Birkett, Portsmouth. CofA lapsed 22.2.40. Regn cld 13.3.41 as sold. Impressed into RAF 12.1.41 as **DJ714**. To ATA TFPP, White Waltham, where it was used for taxi work, initially as G-AEWA (serial applied by 7.6.41). To Shrager Bros, Old Warden for overhaul 23.6.42; to 10 OTU, Abingdon 26.7.42 (also used by Station Flt). To Stn Flt Northolt 7.3.43. To 5 MU Kemble 8.8.46. Sold .47 to Minchin Bros. Regd **G-AEWA** 27.5.47 to Harold Richard Minchin, Bracknell. CofA renewed 3.9.47. The next CofA renewal was undertaken by FG Miles Ltd, Redhill (the firm's first CofA renewal); it was flown by Miles on 8.4.49 and the CofA renewed the following day. Cld 12.4.49 and regd 4.5.49 to Stewart John Burt, Weybridge. CofA lapsed 24.8.59. Cld 13.6.60 and regd 23.6.60 to Aubrey William Offen & Kenneth Grundy, Walton-on-Thames (and CofA renewed 29.6.60). Crashed after hitting a hedge taking off from a field 5km from Neufchatel 3.3.61, following a forced landing due to bad weather en route Lympne to Toussus-le-Noble; pilot DJ Barrett & one passenger unhurt. Wreck collected by Doug Bianchi of Personal Plane Services Ltd, White Waltham but not repaired. Regn cld 15.11.61 as pwfu. A contemporary report claims that some parts used in rebuild of Falcon G-AEEG.

321   M.11A Whitney Straight: colour not recorded. Regd **G-AEVH** (CofR 7719) 2.3.37 to Capt Arthur Vere Harvey, London SW1 (based Heston). First flown, by George Miles, (recorded as 'No.321') 19.3.37. CofA No.5813 issued 24.3.37; dd 24.3.37. Flown by owner (with Race No.26) and came 4th in the 1937 King's Cup Air Race, held at Hatfield 10/11.9.37. Dd 30.4.38 and regd (CofR 8485) 4.5.38 to Reading Aero Club Ltd, Woodley. Crashed nr Ashmansworth, Newbury, Berks 28.1.39 when control lost in bad visibility; GR Hobbs and C Morris both killed. Regn cld 29.8.39.

322 M.11A Whitney Straight; colour not recorded. Regd **G-AEVL** (CofR 7728) 9.3.37 to Major Roland Hobhouse Thornton, Liverpool (based Speke). CofA No.5821 issued 7.4.37; dd 7.4.37. Flown Woodley-Speke-Woodley-Cardiff-Woodley by George Miles 14.5.37. Cld 11.3.39 and regd 30.3.39 to Major Herbert Musker, Thetford, Norfolk. At Bembridge, IoW at Air Ministry survey in 9.39. CofA lapsed 18.11.39. Regd cld 10.4.41. Impressed into RAF as **DP855** 18.3.41. To ATA White Waltham. To Shrager Bros, Old Warden for overhaul 23.6.42. To 5 MU Kemble 5.8.42. To Rolls-Royce Ltd, Hucknall 13.8.42 for communications duties. To 5 MU Kemble 18.2.43 where belatedly the serial was found to duplicate that of a Barracuda, so new serial **NF751** was issued. To Stn Flt Andover 14.3.43. To 48 MU Hawarden 13.1.44 for storage. To RAF Polebrook 11.10.45. To 5 MU Kemble. Sold to Capt GW Harben for £225 at the first sale of impressed light aircraft at 5 MU in 12.45. Regd **G-AEVL** 26.1.46 to Capt Guy Wilfred Harben, Bourne End (based Polebrook-on-Oundle). CofA renewed 1.6.46. Cld 18.10.46 and regd 18.11.46 to James Matthew Banks, London SW1. CofA lapsed 31.5.47; renewed 20.11.48. Cld 22.11.49 and regd 18.1.50 to Lady Margaret Frances Anne Vane-Tempest-Stewart, London EC4. Cld 24.8.51 and regd 1.9.51 to RK Dundas Ltd, London SW1. Regn cld 14.10.51 as sold in New Zealand. Regd **ZK-AZX** (CofR 845) 10.12.51 to PT Collins, Otane, Hawkes Bay. NZ CofA No.747 not issued as the aircraft was, rather stupidly, shipped out as deck cargo and on arrival in Wellington was found to be heavily waterlogged. It was taken to Hastings where it was 'scrapped for spare parts'. Regn cld 9.9.52.

323 M.11A Whitney Straight; colour not recorded. Sold in New Zealand by John M Gamble of Wellington, the local agent for Phillips & Powis Aircraft Ltd. CofA No.5859 issued 1.4.37 to Harold Edwards; reported as dd 25.3.37. Regd **ZK-AFG** (CofR 84/CofA 217) 28.5.37 to Harold Edwards, Wellington; operated by Wellington Aero Club, Rongotai. Taken over by RNZAF as **NZ571** 12.9.39. Returned to former owner and regd **ZK-AJZ** 20.5.46 to him. Regd 17.4.47 to Airwork NZ Ltd, Harewood Airport, Christchurch. Regd 22.2.50 to Hawkes Bay & East Coast Aero Club, Hastings. Crashed on landing Hastings 23.1.59; not repaired. Regn cld 8.10.59. Remains burnt at Ardmore in 1965.

324 M.11A Whitney Straight; colour not recorded. Regd **G-AEVM** (CofR 7729) 10.3.37 to John Arthur Hampden Parker, Godalming, Surrey (and based at home strip at Feathercombe, Hambledon). First flown, by George Miles, (recorded as '324') 13.4.37. CofA No.5831 issued 14.4.37. Regn cld 23.8.40 as sold. Impressed into RAF as **BS815** 21.9.40 and allotted to Stn Flt Northolt 28.9.40. Swung into hedge on forced landing in poor visibility at High Cross, nr Butterworth, Leics 18.11.40, port wing damaged; pilot P/O AH Rawnsley unhurt. Still marked as G-AEVM at RAF Kenley 30.3.41. Starboard wing and prop damaged and engine shock-loaded when swung into adjacent aircraft at Heston 10.6.43, while being hand-swung without chocks by Sqdn Ldr Stubbs; repaired. SOC as scrap (Cat.E2) Northolt 13.4.45.

325 M.11A Whitney Straight; colour not recorded. Regd **G-AEWK** (CofR 7771) 24.3.37 to Ernest Joseph Jobling-Purser, Sunderland (based Woolsington). First flown, by George Miles, 22.4.37. CofA No.5855 issued 22.4.37; dd same day. Entered in the 1937 King's Cup Air Race (with Race No.27) held at Hatfield 10/11.9.37. Regn cld as sold 19.4.40. Impressed into RAF 19.4.40 as **AV970**; to 13 Group Comm Flt, Woolsington. To 39 MU Colerne 20.2.41. To 3 Grp Comm Flt, Mildenhall 24.3.41, later detached to Newmarket from 8.41. Taken to Lundy & Atlantic Coast Airlines, Barnstaple by road for (Cat. B) repairs 16.11.41. To 51 Grp Comm Flt, Yeadon 3.12.41. To Fighter Command at RAF Longmans 18.12.41. Still marked as G-AEWK 5.42. To Stn Flt Northolt 8.9.43. Involved in a flying accident and crashed 24.11.44. SOC (Cat.E2) 1.12.44.

326 M.11A Whitney Straight; colour not recorded. Regd **G-AEWT** (CofR 7784) 30.3.37 to Whitney Willard Straight, London W2/Totnes (probably based Heston). First flown, by George Miles (recorded as '326') 30.4.37; flown again by him on 1.5.37, 3.5.37 and 4.5.37. CofA No.5858 issued 5.5.37; dd 7.5.37. Regn cld 10.37 as sold. Regd in France as **F-APPZ** (CofR 5439) 27.10.37 to Societe Pierre Genin et Cie, Lyon-Bron. Flown by Mme Genin to Indo-China and back in 1937. Stored in France during the second world war, it was restored as F-APPZ in the late 1940's. A regular visitor to England in the 50s, flown in the *Daily Express* South Coast Air Race at Shoreham 22.9.52 by Mme Pierre Genin, finished 49th. Regd (by early .54) to Aero Club du Rhone et du Sud-Est, Lyon. Crashed St. Didier on 7.7.54 and burnt out; pilot Bernard killed.

327 M.11A Whitney Straight; colour not recorded. First flown, by George Miles (recorded as '327') 15.5.37; flown again, by George (as YR-MZM), 20.5.37 and again on 21.5.37, 28.5.37, 7.6.37 and 14.6.37. CofA No.5897 issued 26.5.37. Regd **YR-MZM** 2.6.37 to Mihail Marinescu, Bucharest; dd 30.6.37. Still regd in 1942 but fate unknown.

341 M.11C Whitney Straight; colour not recorded. First flown in 5.37; fitted with 140hp DH Gipsy Major srs II and vp propeller. Regd **G-AEYI** (CofR 7903) 31.5.37 to Phillips & Powis Aircraft Ltd. Flown Woodley to Farnborough by George Miles 6.9.37 as G-AEYI and tested by Clouston 8.9.37. CofA No.6036 issued 8.9.37. Also recorded in Farnborough logs as G-AEYI 8.11.37 (George Miles), 11.11.37 (Miles), 12.11.37 (Bill Skinner), 25.3.38 and 27.5.38 (Kaepelli) – the latter flight being 'to and from Farnborough' (recorded as 'M.11.3' in movements log). But recorded as test flown as **U4** by Don Brown (with George Miles) at Woodley on 5.4.38 and at Shoreham (also with George Miles) on 19.6.38. While being flown from Woodley to Martlesham Heath (apparently in a full gale) by Wing Cdr Frederick 'Freddie' W Stent on 28.6.38, an eye witness nr Uxbridge stated he heard the engine stop and then saw 'the nose go down slightly', the aeroplane then suddenly started to loose height (possibly caught in a violent downdraught?) and before Freddie could recover the situation it hit the side of a gravel pit at Harefield, nr Hounslow, killing the pilot. Regn cld 7.38.

342 M.11A Whitney Straight; colour not recorded. Regd **G-AEYA** (CofR 7880) 20.5.37 to William Headlam, Whitby, Yorks (based at York (Clifton) Aerodrome). First flown, by George Miles (recorded as '342') 10.7.37 and flown again by him (as '342') 'Test' on 12.7.37. CofA No.5963 issued 13.7.37; recorded by CLN as dd 7.7.37. Flown by George Miles as G-AEYA 14.7.37. CofA lapsed 11.7.40. Regn cld 17.4.41 as sold. Impressed into RAF as **DP237** 17.4.41. To DH Hatfield for overhaul. To 24 Sqdn, Hendon 29.8.41 for personal use of HRH Prince Bernhard of the Netherlands but replaced later by the Beech Traveller PB-1. To Stn Flt Northolt 23.4.42. To Fighter Command Comm Sqdn, Northolt. Dbr in crash following engine failure on take-off from Biggin Hill 10.2.45.

343 M.11A Whitney Straight; colour not recorded. Regd **G-AEYJ** (CofR 7904) 31.5.37 to James Brockesby Turnbull (Airwork's sales manager), Heston. First flown, by George Miles (recorded as '343') 'Test' 13.7.37 and again by him (as '343') 'Test' 14.7.37. Take-off tests at 2,000lb on 15.7.37. CofA No.5938 issued 14.7.37. Regd (CofR 8760) 19.9.38 to Airwork Ltd, Heston. CofA renewed 9.12.38 and regn cld 12.38 as sold. Regd in Belgium as **OO-ZUT** (CofR 460) 16.1.39 to Octave Dierckx, Brussels. Regn cld 22.2.46 at post war census of register and annotated, somewhat cryptically, as 'Destroyed at ZO 10.5.40' ('ZO' was probably 'zone occupée' an abbreviation for occupied territories).

344 M.11A Whitney Straight; colour not recorded. First flown, by George Miles (recorded as '344') 'Test', 20.7.37. CofA No.5986 issued 23.7.37 to Dr Rene Clavel. Regd **HB-URO** (CofR 462) 26.7.37 to Dr Rene Clavel, Bale-Birsfelden. New CofR 1036 issued .46. Regd .47 to Robert Meyer, Geneva (based Cointrin). Brakes adjusted and battery changed by Airwork at Gatwick 23.8.50, left same day; also visited in 9.53. Regd .55 to Edmond Pernet, La Côte. Regd .56 to Aero Club de Suisse, Sektion Fricktal, Sisseln. Ran out of fuel and landed in sea 50 metres offshore, Beirut, Lebanon 17.10.57 en route Rhodes to Beirut; wings broke off and fuselage sank while being salvaged; pilot M Emerling but no casualties. Regn cld 5.11.57.

345 M.11A Whitney Straight; colour not recorded. CofA No.5982 issued 29.7.37 to Capt Roques; dd 6.8.37. Regd **F-AQMA** (CofR 5351) 13.8.37 to Jacques Rafael Roques, Paris (based Buc). Failed to survive the war.

346 M.11A Whitney Straight; colour not recorded. Regd **G-AFAB** (CofR 8000) 20.7.37 to Peregrine Philip Saillard Pratt, London W1 (based at Brooklands). CofA No.5983 issued 18.8.37; dd 17.8.37. Cld 13.2.39 and regd 27.2.39 to Venner Time Switches Ltd, New Malden (based Brooklands). To Sywell in 10.39. Regn cld 24.7.40 as sold. Impressed into RAF as **BD145** 24.7.40. To 24 Sqdn,

Hendon 25.7.40. While en route Hendon-Turnhouse on 10.12.41, the engine failed and it spun into the recreation ground in Park Street village, 2 ml N of Radlett, Herts and caught fire on impact. The remains were taken to Phillips & Powis, Woodley where it was SOC (Cat.E1) 1.42.

347 M.11A Whitney Straight; colour not recorded. Regd **G-AEZO** (CofR 7979) 12.7.37 to Brig Genl Arthur Corrie Lewin, Upton-upon-Severn (based Worcester). CofA No.5980 issued 21.8.37; dd 20.8.37. Flown by owner (with Race No.25) to 2nd place in the 1937 King's Cup Air Race, held at Hatfield 10/11.9.37. Flown by owner and wife from UK for Nairobi, Kenya 10.37 but crashed and overturned in forced landing 150 miles south of Malakal, Sudan 10.10.37; fortunately neither the Brig nor his wife were hurt and were eventually rescued on 20.10.37. Regn cld 12.37.

348 M.11A Whitney Straight; colour not recorded. CofA No.6023 issued 6.9.37 to Madame de Bohomoletz, France; dd 4.9.37. Regd **F-AQIK** (CofR 5462) 26.11.37 to Mme Marie C de Bohomoletz, Paris. Regd 22.1.38 to Paul Bernard Levy, Nancy. Regd 8.38 to Alain Leroux, Orchies (based Roubaix). Failed to survive the war.

349 M.11A Whitney Straight; colour not recorded. CofA No.6024 issued 27.8.37 to Dr Kurt Tschudi; dd 28.8.37. Regd **HB-EPI** (CofR 476) 31.8.37 to Dr Kurt Tschudi. Dr Tschudi, an explorer, departed Bergamo on 4.2.38 for a lengthy tour of North Africa which took him to Tunis, Tripoli, Benghazi, Mersa Matruh, Luxor (Cairo), Wadi Halfa, Abu Hamed, Khartoum, Kassala, Asmara, Kassala, Khartoum, El Obeid, El Fasher, Geneina, Fort Lamy-Kano, Niamey, Gao, Bidon 5 ('in the middle of this land of thirst'), Reggan, Colomb-Bechar, Laghouat, Tunis and back to Bergamo (the only things Dr Tschudi marked in his log-book were 'four times cleaned petrol-filters and changed oil, replaced one exhaust pipe washer'; quite an achievement considering the hostile territory over which he had flown). Regd .38 to Miss J Trumpy, Mitlodi (based Zurich/Dubendorf); still regd to her 12.47 (new CofR 1038 issued .46). Regd .49 to M Stocklin & M de Terra, Zollikon (based Kloten). Regd .56 to Dr Emanuele Bianda, Ascona. CofA lapsed 23.1.62. Regd 24.11.65 to Antonietti-Favre AG (based Basel). On 18.5.73, it struck Piper Cherokee G-BASF while taxying at Le Touquet Airport, France; presumed repaired. Regn cld 1.2.84. Remains extant at Basel in 1988 but believed burnt shortly afterwards.

350 M.11A Whitney Straight; colour not recorded. Purchased by the Australian Department of Civil Aviation under Contract No 653357/37, with an ITP dated 30.6.37 at a projected cost of £1,078 16s 0d. CofA No. 6064 issued 23.9.37 to the Royal Victorian Aero Club. (CLN records dd 28.8.37 to Victorian & Interstate Airways/Royal Victorian Aero Club 'for Australian Government'; the date would seem to be wrong). Regd **VH-UZA** (CofR 678/CofA 635) 19.11.37 to Royal Victorian Aero Club, Essendon. Crashed 1½ mls north of Essendon 12.9.42. Regn cld 2.12.42.

496 M.11A Whitney Straight; colour not recorded. CofA No.6056 issued 24.9.37 to PLE Salomon, France; probably dd 25.10.37. Regd **F-AQCZ** (CofR 5437) 27.10.37 to Pierre Louis E Salomon, Ruoms, Ardeche. Regd 18.12.46 to M Michel, Marseille. Regd 31.4.51 (sic) to Groupement Inter Régional des Pilotes d'Avions et de l'Association de Propagande Aéronautique de la Région de Pons, Pons-Avy. Regd (by 6.54) to Association Co-operative de Pilotage, ACS de Provence, Aix-Les-Milles. Regd 6.5.59 to Philippe le Pivain, Brest. CofA lapsed 5.11.60. Regn cld .63.

Miles M.11A Whitney Straight F-AQCZ of the Aero-Club de Paris, seen at Léognan (Gironde) in 1953. [S Blandin]

497 M.11A Whitney Straight; colour not recorded. Regd **G-AFBV** (CofR 8081) 2.10.37 to Albert Batchelor, Broadstairs (based Ramsgate). CofA No.6073 issued 4.10.37. Cld 20.1.39 and regd 28.1.39 to Ipswich Aero Club Ltd, Ipswich. Regd 9.2.39 to (associate) Thanet Aero Club Ltd, Ramsgate (but probably continued in operation by Ipswich Aero Club). While being flown from Ipswich on 15.6.39, the pilot Charles F Almond with trainee pilot passenger Henry Angwin Spray engaged in 'shooting-up' the aerodrome buildings, during the course of which the aircraft hit the windsock mast, crashed and caught fire; both occupants killed. Regn cld 7.7.39.

498 M.11A Whitney Straight; colour not recorded. CofA No.6076 issued 9.10.37 to Karl Roechling, Germany; dd 9.10.37. Regd **D-EKTR** 1.38 to Karl Theodor Roechling, Voelklingen, Saar. Believed taken over by the Luftwaffe (a record exists in the German military archives of it being repaired in an aircraft factory after the outbreak of war).

499 M.11A Whitney Straight; colour not recorded. Regd **G-AFCC** (CofR 8139) 14.10.37 to Tomas Saunders, Magallenes, Chile. CofA No.6084 issued 26.10.37. Regn cld 16.5.46 as sold abroad (but wef 20.3.46). Regd **CC-FBL** 12.37 to Tomas Saunders, Magallenes. (reported as re-regd CC-PBL but this was cld and re-allocated to M.2R CC-PFB). The September 1938 issue of the Miles Magazine, published a letter and photo from Mr Saunders *'taken in Southern Chile after a forced landing on 20.7.38 due to bad weather. I had to transport the Whitney Straight 33 kms to the rail head; this took four days in one steady downpour of rain. The natives in these parts, descendants of Spaniards mixed with Indians, thought that the Devil had arrived and it was sometime before I could gain their confidence to help me transport the machine to the river where I rafted it down to the village of Nacimiento. Apart from the undercarriage and propeller, very little damage was done.* Fate unknown.

500 M.11A Whitney Straight; colour not recorded. Regd **G-AEYB** (CofR 7881) 20.5.37 to Whitney Straight Ltd, London W.1 (based Heston). First flown, by George Miles (recorded as '500') 23.6.37; flown again by him, as G-AEYB 'Test' 26.6.37. George then flew it to Hatfield for the SBAC Display on 28.6.37, from where he made two return flights to Woodley on 29.8.37 before returning to Woodley the same day. On 30.6.37 George made a 'Test of 973 airscrew' and on 1.7.37 he made a return flight to Heston. CofA No.5946 issued 2.7.37; flown by George Miles on 6.7.37 and to Hatfield on 20.7.37. Recorded (erroneously) by CLN as dd 15.6.37. Regn cld 12.37 as sold. Regd in Italy as **I-BONA** (CofR 2174) 9.2.38 to Gaspare Bona, Carignano (based Turin). Regd 9.41 to Frederica Cassinas. Fate unknown.

501 M.11A Whitney Straight; colour not recorded. Regd **G-AEXJ** (CofR 7826) 20.4.37 to Air Service Training Ltd, Hamble (initially regd with c/n 341 but later amended). First flown, by George Miles (recorded as '501') 1.7.37 and flown again by him on two 'Test' flights on 2.7.37 and 3.7.37. CofA No.5880 issued 3.7.37; recorded erroneously by CLN as dd 23.6.37. CofA renewed during war on 2.9.40. Regn cld 29.10.40 as sold. Impressed into RAF as **BS818** 29.10.40. To 12 SFTS Grantham; renamed 12 (P)AFU 3.42. To Phillips & Powis Aircraft Ltd, Woodley for repairs 4.6.43. To 5 MU Kemble 28.11.43. To Stn Flt Heston 8.4.44. Engine failed and crashed during forced landing Lowfield Heath Road, Gatwick 12.5.44. SOC Cat.E2.

502 M.11A Whitney Straight; colour not recorded. Regd **G-AFCN** (CofR 8178) 8.11.37 to Whitney Willard Straight, London W2 (based Heston). CofA No.6120 issued 20.11.37; dd 26.11.37. Regd (CofR 8844) 26.10.38 to Henry Maynard Mitchell, Birmingham (based Castle Bromwich). Cld 21.3.39 & regd 25.3.39 to Southern Aircraft (Gatwick) Ltd, Gatwick. Regd 30.3.39 to Yacht Cruises Ltd, London EC2 (based Gatwick). Regd 25.8.39 to John Osmotherly Baines, Baghdad, Iraq (probably not dd). Regn cld 2.3.40 as sold. Impressed into RAF as **V4739** 12.10.39 and used by Stn Flt Ternhill. To AOC 24 Grp, Halton 5.2.40. To 51 Grp Comm Flt, Yeadon 14.11.40. To Belfast Comm Flt, Sydenham 28.7.41. To Torpedo Development Unit, Gosport 28.2.43. SOC (Cat.E1) 30.9.43 after 449.20 total flying hours.

503 M.11A Whitney Straight; colour not recorded. Sold in New Zealand by John M Gamble of Lambton Quay, Wellington, the local agent for Phillips & Powis Aircraft Ltd. CofA No.6175 issued to RNZAF 22.12.37; handed over to the RNZAF 4.1.38; arrived Wellington 10.2.38. Regd **ZK-AGB** (CofR 122) 3.3.38 to Wairarapa

Impressed Miles M.11A Whitney Straight NZ577 at RNZAF Station Ohakea on 12th November 1941.
[Reproduced with kind permission of the RNZAF Museum]

& Ruahine Aero Club, Masterton. Impressed into RNZAF as **NZ577** 9.10.39; to RNZAF Comm Flight. Regd **ZK-ALE** (CofR 288) 6.12.46 to JM Gould, Paraparaumu Beach, Wellington. CofA No.340 renewed 5.12.46. Regd 29.4.47 to Auckland Aero Club, Mangere. Crashed into the mist-covered Mamaku Hills, nr Fitzgerald Glade 7.9.53 while on a flight from Mangere to Rotorua on a supply dropping operation, killing the pilot L Brooker. Regn cld 7.4.54.

504 M.11A Whitney Straight; colour not recorded. CofA No.6204 issued 27.1.38 (as "SA-SBB") to the Royal Singapore Flying Club; dd 1.2.38. Regd **VR-SBB** (CofR 31) 18.3.38 to Royal Singapore Flying Club, Kallang. *The Miles Magazine* September 1938 reported that: *The longest 'out-and-home' flight by an aircraft of the Royal Singapore Flying Club was completed on June 22nd, when a fast new machine returned from a trip to Bangkok. The Whitney Straight was chartered and piloted by Mr EC Whiteley who had to visit Siam on business in connection with Metropolitan Vickers Electrical Co Ltd, for whom he is the local representative.* To 'A' Flight MVAF 1.12.41 with serial **12**. Badly damaged in heavy landing Andir, Bandoeng 20.2.42 and abandoned there.

505 M.11A Whitney Straight; colour not recorded. CofA No.6215 issued 16.2.38 to Société d'Enterprises Electro-Techniques, France; dd 17.2.38. Regd **F-AQLX** (CofR 5612) 16.3.38 to Societe d'Enterprises Electro-Techniques, Paris (based Buc). Failed to survive the war.

506 M.11A Whitney Straight; colour not recorded. Possibly the Whitney Straight with 'B' mark **U3** which was flown by Flt Lt JF Moir to A&AEE Martlesham Heath 9.5.39; U3 also made a return flight to RAE Farnborough on 10.6.39 (flown by Kaepelli). Regd **G-AFZY** (CofR 9257) 9.11.39 to Phillips & Powis Aircraft Ltd. CofA No.6790 issued 20.11.39. Retained by firm and allotted 'B' mark **U-0227** 10.40 (prior to CofA lapse 19.11.40). Reverted to **G-AFZY** on renewal of CofA 22.1.42. Hugh Kennedy recorded a CofA test flight in G-AFZY on 26.1.42. Regn cld 20.11.42 by the Secretary of State. Impressed into RAF as **NF747** 20.11.42. To 51 Grp Comm Flt, Yeadon 12.42. To RAF Sealand 6.7.44. To 51 Grp Comm Flt, Yeadon 6.1.45. To 50 Grp Comm Flt, Woodley 17.7.45. To 5 MU Kemble 7.3.46. Sold .46 to Southern Aircraft (Gatwick) Ltd. Regd **G-AFZY** 28.5.46 to Southern Aircraft (Gatwick) Ltd, Gatwick. CofA renewed 21.9.46. Regd 21.9.46 to Miss Beatricia McCormack, Worthing and named *Vagabond*. Damaged in gale Shoreham overnight 7/8.8.48 (when owned by JE Wettler). Regn cld 18.9.48 on 'Miss McCormack's marriage to a foreign national' (probably Mr Wettler). Regd 28.6.49 to Robert Charles Cox, Chobham. CofA renewed 1.9.49. Cld 21.1.50 & regd 31.1.50 to Lord Peter Waldon Somerset Gough-Calthorpe (Lord Calthorpe), Woking (based Fairoaks). Regn cld 20.10.50 on sale in New Zealand. Regd **ZK-AXD** (CofR 754) 28.9.50 to Hawkes Bay & East Coast Aero Club Inc, Hastings. CofA No.630 issued 2.2.51; lapsed 1.2.52. CofA renewed 21.12.56. CofA lapsed 3.4.58. Regd 25.9.64 to Auckland Flying School Ltd, Ardmore. Regd 13.2.65 to Peter Sheridan Lockley, Auckland. Wfu and stored in a garage at Drury, Papakura; later displayed on a rooftop at Mangatawhiri, where it steadily deteriorated, until it was scrapped in 1969. Regn cld 27.11.80 as wfu.

507 M.11A Whitney Straight; colour not recorded. Regd **G-AFJX** (CofR 8726) 30.8.38 to Brig Genl Arthur Corrie Lewin, Njoro, Kenya. CofA No.6383 issued 14.8.38; dd 13.9.38. Offered for sale by WS Shackleton Ltd as 'unused' 2.39 & 6.39 and in *Miles Magazine* for March 1939 with *only 30 minutes flying since delivered to owner who was suddenly called abroad*. (Lewin bought Puss Moth G-ABOC and flew this out to Kenya at the time). Cld 1.7.39 and regd 12.7.39 to John Alwyn Tweedale, Rochdale (based Woodford). Regd cld 18.8.40 as sold. Impressed into RAF as **BD183** 18.8.40. To Comm Flt, Skeabrae. To RAF Grimsetter 27.8.45. To 5 MU Kemble 24.5.46. Sold 12.46 to Warden Aviation & Engineering, Old Warden. Regd **G-AFJX** 3.3.47 to Mrs Dorothy Clotilda Shuttleworth & Allan Henry Wheeler, t/a Warden Aviation & Engineering Co, Old Warden. CofA renewed 5.9.47 but lapsed 4.9.48. Regn cld 21.3.50 as sold in New Zealand; CofA renewed 29.3.50. Sold 20.4.50 to Harry Wigley (later Sir Henry Wigley) of Mount Cook & Southern Lakes Tourist Co. Regd **ZK-AUK** 8.5.50 to Mount Cook Air Services Ltd, Timaru. Regd 25.11.55 to DH Withers, Christchurch. Regd 13.6.56 to AM Ferguson, Christchurch. Regd 17.8.59 to AV Martyn, Timaru. Regd 17.5.60 to Donald George Walther, Maungatua. Regd 28.3.63 to Ivan Adrian Theodore Strathern, Dunedin. Ground-looped on landing Christchurch 26.6.66, pulling the mainspar in the centre section away from the tank skin; aircraft deemed to be beyond economical repair. Stored at Forest Field, Christchurch and acquired 14.7.66 by Valentine Brian Murray, Timaru, but then remained at a farm (same farm?) until purchased by a Mr Macdonald and friends for NZ$200. Regn cld 29.7.68 as 'aircraft in a permanently unserviceable condition'. Some parts are believed to have gone to a Museum in Christchurch at unknown date. Remains then acquired by Greg W Macdonald (brother of the previous purchaser and an aircraft engineer with Mount Cook Airlines) and taken to Forest Field, Christchurch in about 1985. By 1988, he had restored all the metal parts, with the exception of the u/c, which was "basically two blobs of corroded steel with wheel forks attached to them!" The outer wings, folding sections and rear fuselage remained in good condition and although much of the remainder of the woodwork will have to be replaced, Greg intends to utilise as much of the original as possible in a restoration to airworthy condition. Still on slow rebuild in 2008.

508 M.11A Whitney Straight; colour not recorded. CofA No.6353 issued 26.7.38 to Aero Club de Syrie et du Liban; dd 10.8.38. Regd **F-AREQ** (CofR 6043) 9.9.38 to Aero Club de Syrie et du Liban, Damascus (operated by Section Maurice Nogues). *(According to CLN, this Whitney Straight was originally allocated to M Pirard (or NMJMH Pirara) as OO-ATC; but no trace of such can be found in Belgian records, nor an owner by such name).* A photograph of a Whitney Straight in Free French Air Force markings with a black Cross of Lorraine in a circle on the fin, in what appears to be a grey finish but with no serial, was taken by Howard Levy at El Kabrit, Egypt on 1.12.42 (see *Aeroplane Monthly* July 1983) and it is highly likely that this was F-AREQ. Further, whilst serial HK853 was officially allotted in 1.42 to an ex Royal Hellenic AF Avro Tutor which had escaped to the Middle East for 206 Grp, the British Airways Repair Unit (ME) referred to the use of **HK853** on a Miles M.11A Whitney Straight in 6.43 and may be the same aircraft.

509 M.11A Whitney Straight; colour not recorded. An M.11A with

G-AFZY, a late production Miles M.11A Whitney Straight, was used by Phillips & Powis Aircraft Ltd during the early part of the war with 'B' mark U-0227.
[B Wood via MAP]

G-AFZY was sold in New Zealand in late 1950 and registered ZK-AXD to the Hawkes Bay & East Coast Aero Club, Hastings.
[Via B Stainer]

dark coloured fuselage and silver wings and empennage was flown under 'B' mark **U1** and an undated photograph, probably taken at Heston, shows it with a number of new Stinson aircraft in the background. It is suspected that this was c/n 509. Regd **G-AFGK** (CofR 8425) 6.4.38 to Miss Rosemary Rees, London SW1 (based Heston). CofA No.6238 issued 11.4.38; dd 12.4.38. Reportedly flown by Charles Lindbergh 1938/39 while his Mohawk was receiving attention at Woodley. At Combermere Abbey, Whitchurch, Shropshire on Air Ministry survey in 9.39. Permit issued 4.1.40 for Miss Rees to fly from Chester to Heston to have a Maclaren Drift Landing Undercarriage fitted (which was installed by Alan Muntz & Co Ltd at Heston). Permit CO1/106 issued 2.5.40 to Airwork Ltd for Maclaren u/c tests (also Permit CO1/131 issued 19.7.40). Regd 29.8.40 to Airwork Ltd, Heston but 'owned' by Owen Finlay Maclaren (the designer of the Maclaren Drift Undercarriage) who was working for Airwork Ltd. Used for development work and flown frequently to and from RAE Farnborough (between 9.5.40 & 15.4.41) and demonstrated by Maclaren to Hawker Aircraft Ltd at Langley. Probably granted permit to fly 19.5.42 – since quoted as 'permit expires 18.5.43'. CofA renewed 19.2.43 and maintained throughout war and subsequently. Cld 14.8.48 & regd 6.10.48 to Edward Henry Thierry, Beverley. Cld 23.4.51 & regd 30.4.51 to Francis Henry Wilson, Speke. CofA lapsed 14.8.52. CofA renewed 1.12.54 and regd 29.12.54 to Edmund Gwyn Hordern, Chichester. Cld 24.3.56 & regd 26.3.56 to Harry Judd & Victor Andreanoff, t/a Aldenham Private Flying Group, Elstree. CofA lapsed 3.9.62. Cld 14.11.63 & regd 4.12.63 to Edmund Wilkinson, Pinner (later Burnham-on-Crouch) and CofA renewed 6.12.63; (based Denham by 12.65). CofA lapsed 8.7.70. Cld 3.1.72 & regd 14.1.72 to Walter Ian Scott-Hill & Cyril Alfred Herring, Kew, Surrey. Moved from Old Warden to Booker by 5.72 for a complete rebuild with its new owners, 'to be rebuilt to its pre-war glory'; fitted with a DH Gipsy Major 10 Mk.2 engine. CofA renewed 30.11.72; now based Panshanger. Regd 15.6.77 to Harold Best-Devereux, Welwyn Garden City. Regn cld 13.9.77 on sale in USA. Regd **N72511** 9.77 to John B Lawrence, Lahaina, Hawaii. Regd 12.85 to Daniel D Nelson, Torrance, California. Sold 26.9.88 to Stan Reynolds Museum, Canada. Regd **CF-FGK** 17.10.88 to Byron Reynolds of Stan Reynolds Museum, Wetaskiwin, Alberta. Sold 4.4.89 & regd 6.4.89 to Stan Reynolds Sales Ltd, Wetaskiwin. Last flown in 1992, it is now painted in a dreadful pseudo-wartime desert camouflage colour scheme with yellow undersurfaces and RAF roundels. By July 2008 it was stored with wings removed and the windscreen broken (storm damage) in the renamed Reynolds-Alberta Museum and will in due course be restored to airworthy condition but since it is provincial government owned, will not be allowed to fly.

510 M.11A Whitney Straight CLN recorded 'Fuselage only' but with no date or further details so presumed never completed.

Note: An as yet unidentified Whitney Straight with 'B' mark **U3** is shown in a photograph of Lympne Aerodrome published in *Flight* for 24.3.38. Although these marks have been linked with G-AECT, possibly because of the change of engine back to a Gipsy Major, this is considered to be highly unlikely. Flt Lt John Moir flew a Whitney Straight (recorded as 'V3' in his log book but almost certainly meant to be U3) to A&AEE Martlesham Heath on 9.5.39, returning the same day. U3, probably the same Whitney Straight, was also flown by Walter Kaepelli to and from Farnborough 10.6.39. This latter aircraft is probably c/n 506.

Summary of production		
M.11	c/n 290	1
M.11A	c/n 303-327, 342-350, 496-510	49
M.11B	c/n 305, later reverted to M.11A	(1)
M.11C	c/n 341	1
Total		51

Job Numbers:	
H5747	c/n 290
5001	c/n 303-327 – c/n 303 also had Job No 15153 and c/n's 303-312 had the individual numbers 23923-23932 prefixed to 5001.
5018	c/n 341
5010	c/n 342-350 & 496-510

Left: 'U1', an unidentified M.11A Whitney Straight, seen at Heston, which probably became G-AFGK, owned by Miss Rosemary Rees at Heston from April 1938.
[The A J Jackson Collection]

Below: Miles M.11A Whitney Straight G-AFGK at a post-war air race meeting. [Via B Clarke]

Appendix 20:
Miles M.14 Magister (and variants) and M.14 Hawk Trainer Mk.II and Mk.III Production to 3rd September 1939

Miles M.14 Hawk Trainer Mk.III NZ586 (ZK-AEY) after impressment into the RNZAF.
[Photo reproduced with kind permission of the RNZAF Museum]

329 Miles Magister - used for mock-up purposes only.

331 M.14 Hawk Trainer Mk.III. Regd **G-AETJ** (CofR 7636) 15.2.37 to Phillips & Powis Aircraft Ltd. (CLN recorded: Miles Magister First of Type; Customer 'High Commissoner for New Zealand'). First flown (after being christened by Blossom Miles before an assembled crowd of workers) by Miles, on 20.3.37. Painted in red primer with RAF roundels (the latter presumably for publicity purposes) with 'B' mark **U2**. Flown by George Miles on a 'Temperature Test' 22.3.37 and flown again by George on 5.4.37, 6.4.37 and 7.4.37. Regn cld 4.37 as sold. Sold via John M Gamble of Lambton Quay, Wellington, the agent for P&P in New Zealand; dd 10.6.37. Assembled locally; a Contractor's Flight Test Report for ZK-AEX was dated 21.7.37. CofA No.5987 issued 23.7.37 to RNZAF. Regd **ZK-AEX** (CofR 196) 2.9.37 (or 7.9.37) to Auckland Aero Club, Mangere. Crashed after engine failure on take-off from Great Barrier Island 17.3.39 and aircraft destroyed; pilot EM Walker injured, passenger WH Clavis killed. Regn cld 3.39.

332 M.14 Hawk Trainer Mk.III. Regd **G-AETL** (CofR 7647) 15.2.37 to Phillips & Powis Aircraft Ltd. (CLN recorded: Miles Magister; Customer 'High Commissioner for New Zealand'). Regn cld 4.37 as sold. Sold via John M Gamble of Lambton Quay, Wellington, the P&P agent in New Zealand; dd 10.6.37. CofA No.5988 issued 23.7.37 to RNZAF. Regd **ZK-AEY** (CofR 197/CofA 36) 2.10.37 (also quoted as 7.9.37 or 13.9.37) to Otago Aero Club, Dunedin. First flight in New Zealand 2.10.37. Impressed into RNZAF 26.9.39 as **NZ586**; probably TOC 24.10.39. To De Havilland, Rongotai 12.39 to 9.41. To 2 EFTS New Plymouth/Ashburton 10.42. SOC 19.9.46. Regd **ZK-ALO** (CofR 260/CofA 342) 29.10.46 to Otago Aero Club Inc, Dunedin. Regd 13.6.50 to FA Patterson, Waipawa. Regd 30.7.50 to Piako Aero Club, c/o JN Johnson, Waharoa. Crashed on landing Waharoa, Matamata 16.12.56; total time 3,115 hrs. Regn cld 3.3.58. Used as spares by Piako Aero Club. Parts later to Museum of Transport & Technology, Western Springs, Auckland. These parts, comprising one wing and parts of the tail unit, were sold to Stan Smith who eventually intends to rebuild it.

333 M.14 **L5912**; painted yellow with natural metal cowling. Dd 14.6.37 to A&AEE, Martlesham Heath. Abandoned by pilot in spin off Felixstowe, Suffolk 22.7.37 and crashed; pilot Flt Lt Eric W Simonds drowned.

334 M.14 **L5913**; painted red/white for the CFS Aerobatic Display Team and modified to participate in a formation of three in the Hendon Display on 26.6.37. This was to include inverted flying but the Magister was found to be too 'clean' aerodynamically to hold formation inverted and the CFS team reverted to Avro Tutors. Dd CFS Upavon 12.6.37 and TOC 14.6.37. To 1 FTS Leuchars 16.7.37. To Debden 6.9.37. To 504 Sqdn, Debden 12.10.39. Engine cut on take-off from proposed landing ground, hit wall and crashed Buldoo, 7 ml W of Thurso, Caithness 23.7.40. SOC Cat.E 2.8.40.

335 M.14 **L5914**; painted red/white for the CFS Aerobatic Display Team and modified for inverted flying (see L5913). Dd CFS Upavon 12.6.37 and TOC 14.6.37. To 1 FTS Leuchars 16.7.37. To Cranfield 6.9.37. To P&P (Cat.B) 4.6.40. To 5 MU Kemble 24.10.40. To Detling 23.11.40. To 9 MU Cosford 24.6.41. To 51 OTU Cranfield 16.8.41. To 32 MU St Athan 13.10.41. To P&P 22.10.41. To 46 MU Lossiemouth 24.1.42. To 22 MU Silloth 31.1.42. To 46 MU Lossiemouth 16.2.42. To 5 MU Kemble 12.12.45. Sold 26.2.48 to Gloster Flying Club. Regd **G-AKML** (CofR 12049) 27.11.47 to Laurence Dudley Trappitt, Whyteleafe, Surrey. Cld 22.7.48 & regd 27.10.48 to Robert Alan Short, Derek John Hayles & Martha Marodeen, Croydon/Portsmouth. Cld 5.11.48 & regd 31.1.49 to Robert Alan Short, Croydon. CofA No.10407 issued 18.3.49. Regn cld 2.9.49 as sold abroad - to **Egyptian Air Force**.

L5912, the first Miles M.14 Magister for the RAF. Note the following differences between this and U2, the first 'supposed' Magister: 1) the spats are not so long; 2) no pitot under wing; 3) the intake on the rear lower cowling on the starboard side is deeper; 4) the headrest is smaller; 5) blind flying hood installed. [Via B Clarke]

336 M.14 **L5915**; painted red/white for the CFS Aerobatic Display Team and modified for inverted flying (see L5913). Dd CFS Upavon 12.6.37 and TOC 16.6.37. To 1 FTS Leuchars 16.7.37. To Dishforth 3.9.37. To 10 Sqdn, Dishforth 30.11.39. To P&P (SAS) 24.8.41. To 37 MU Burtonwood 9.10.41. To 71 Sqdn, North Weald 26.11.41. To 64 Sqdn, Hornchurch 15.12.41. To 307 Sqdn, Exeter 17.1.43. To Herts & Essex, Broxbourne for repairs 26.4.43. To 51 MU Lichfield 5.7.43. Sold 20.5.46 to Miles Aircraft Ltd. To H Hennequin & Cia, Argentina with CofA No.8119 issued 18.7.46; regd **LV-XQH** 30.6.47 to Secretaria Estado Aeronautica, Buenos Aires. To Circ. Av. Rosario. Accidents 26.8.50; repaired. Accident 15.6.52. Regn cld 8.52 as wfu. *Note, LV-XRX also regd with c/n 0336 but is believed to be c/n 1822.*

337 M.14 **L5916**; painted red/white for the CFS Aerobatic Display Team and modified for inverted flying (see L5913). Dd CFS Upavon 6.6.37 (probably 16.6.37) and TOC 18.6.37. Demonstrated at the sixth SBAC Show at Hatfield on 28/29.6.37. To 1 FTS Leuchars 19.7.37. To Hemswell 14.9.37. To 61 Sqdn, Hemswell 30.11.39. To 27 MU Shawbury 21.7.41. To 317 Sqdn, Exeter 17.10.41. To 308 Sqdn, Exeter 29.4.42. To 245 Sqdn, Middle Wallop 25.5.42. To Atlantic Coast, Barnstaple for repairs 31.1.43. To 45 MU Kinloss 5.4.43. To HQ ATA White Waltham 22.1.44. To 51 MU Lichfield 7.3.45. To 4 SofTT, St Athan for GI as **5362M** 11.7.45. SOC 1.8.46.

338 M.14 **L5917**; painted yellow overall with natural metal cowling, as were all subsequent aircraft up to the outbreak of war. Dd 1 FTS Leuchars 24.6.37. To Abingdon 24.9.37. To 21 AD Service Flt, Henlow Camp; but at Bouguenais, France 30.11.39. Presumed lost in France in 6.40. SOC 26.3.42.

339 M.14 **L5918** Dd 1 FTS Leuchars 24.6.37. To Abingdon 20.8.37. To 21 AD Service Flt, Henlow Camp; but at Bouguenais, France 30.11.39. Presumed lost in France in 6.40. SOC 26.3.42.

340 M.14 **L5919** Dd 1 FTS Leuchars 25.6.37. To Finningley 27.8.37. To 76 Sqdn, Upper Heyford 30.11.39. To Finningley 16.1.40. To 15 EFTS Carlisle 4.7.41. Engine cut, hit hedge in forced landed Flash Park, Dumfriesshire 11.9.41. To P&P (15?)5.9.41 & SOC 19.9.41.

351 M.14 **L5920** Dd 1 FTS Leuchars 30.6.37. To Church Fenton 23.8.37. To 54 OTU Church Fenton 16.5.41. SOC (MR) 12.9.44.

352 M.14 **L5921** Dd 1 FTS Leuchars 3.7.37. To Kenley 19.8.37. To P&P 29.11.39. To 37 MU Burtonwood 2.9.40. To 15 EFTS Carlisle 21.6.41. To 29 EFTS Clyffe Pypard 24.5.42. To 51 MU Lichfield 4.4.43. Sold 10.3.47 to British Air Transport Ltd. Regd **G-AJRU** (CofR 11533) 30.4.47 to British Air Transport Ltd, Redhill. CofA No.9315 allocated but not issued. Used for spares for G-AJDR c10.53 and remains burnt Redhill 21.5.54. Regn cld 27.2.59 as destroyed.

353 M.14 **L5922** Dd 1 FTS Leuchars 8.7.37. To Bicester 29.8.37. To 21 AD Henlow Camp; but at Bouguenais, France 22.12.39. To 1 Salvage Section/21 AD 15.2.40. Abandoned in France in 6.40. SOC 26.3.42.

354 M.14 **L5923** Dd 1 FTS Leuchars 9.7.37. To Scampton 24.8.37. Crashed out of control Wetton, Staffs 27.2.40 while each pilot believed the other was flying the aircraft. SOC 7.3.40.

355 M.14 **L5924** Dd 1 FTS Leuchars 9.7.37. To Driffield 8.9.37. SOC 22.8.40.

356 M.14 **L5925** Dd 1 FTS Leuchars 15.7.37. To Leconfield 9.8.37. To Abingdon 20.11.39. To 15 EFTS Carlisle 1.7.41. To 21 EFTS Booker 12.6.42. To 22 EFTS Cambridge 5.9.43. To 3 (P)AFU South Cerney 13.9.43; post-war coded 'FBQ-K'. To 51 MU Lichfield. Sold 4.12.47 to Wolverhampton Flying School Ltd, Wolverhampton. Regd **G-AKPG** (CofR 12119) 31.1.48 to John Brooke Hall, Wolverhampton. CofA No.9961 allocated but not issued. Cld 15.12.56 & regd 17.12.56 to Derby Aviation Ltd, Burnaston. CofA A279 issued 28.6.57. Cld 21.5.59 & regd 28.5.59 to Ronald Herbert Richards, t/a The Propellers Flying Group, Hatfield, Herts. Cld 6.1.64 & regd 16.1.64 to Robert Bass Newton, Alan Wright Mowat & Roger Craig, t/a Bristol Mercury Flying Group, Bristol. Crashed Cranfield 12.11.64; JR Craig & Grant Parry-Jones killed. Regn cld 8.3.65 as destroyed wef 12.11.64.

357 M.14 **L5926** Dd 1 FTS Leuchars 21.7.37. To Feltwell 18.8.37. To 37 Sqdn, Feltwell. Flew into HT cables nr Brandon, Norfolk 1.8.39 and DBR.

358 M.14 **L5927** Dd 1 FTS Leuchars 21.7.37. To Harwell 16.8.37. To 105 Sqdn, Villeneuve-les-Vertus 30.11.39. Presumed abandoned in France in 5.40.

359 M.14 **L6001** Dd 12 FTS Grantham 24.2.38. To 27 MU Shawbury 25.10.40. To Martlesham Heath 19.11.40. To 9 MU Cosford 3.6.41. To 16 EFTS Burnaston 30.5.41. To 27 MU Shawbury 2.9.42. To 47 MU Sealand 10.2.43. Arrived in ME 12.6.43 (on SS Clan Lamont). To 107 MU Kasfareet. To 247 Wing 19.9.43. SOC 29.6.44.

360 M.14 **L5928** Dd Stn Flt Hucknall 21.7.37. To 1 CRU Cowley for repairs 19.5.41. To 46 MU Lossiemouth 29.7.41. To 15 EFTS Carlisle 11.8.41. Hit tree recovering from dive, Rockcliffe, Cumberland 28.1.42; Sgt B Astley killed & LAC JMM Waite injured.

361 M.14 **L5929** Dd St Flt Hucknall 26.7.37. To 72 Sqdn, Acklington. Crashed 10.4.41. To 1 CRU Cowley 18.4.41 and SOC 21.4.41.

362 M.14 **L5930** Dd Stn Flt Hucknall 27.7.37. Hit ditch landing in field near Uttoxeter, Staffs 9.1.40 at site of a Hawker Audax. To P&P for repairs 20.1.40. To 37 MU Burtonwood 20.8.40. To **Turkey** 20.7.41 as replacement for 1 of 6 lost at sea en route from original order.

363 M.14 **L5931** Dd 24 (Comm) Sqdn, Hendon 30.7.37. To 5 MU Kemble 20.12.39. To 5 OTU Chivenor 6.9.40. To 55 OTU Usworth 13.4.41. To 32 MU St Athan 23.7.41. Engine cut, ditched in Bristol Channel 25.9.41. SOC 13.10.41.

364 M.14 **L5932** Dd Stn Flt Northolt 30.7.37. To 3 Grp Comm Flt Mildenhall 10.10.39. To Stradishall 27.7.40. To 5 MU Kemble 11.6.41. To Shrager Bros, Old Warden (SAS) 8.7.41. To 27 MU Shawbury 2.5.42. To 316 Sqdn, Hutton Cranswick 14.8.42. To 32 MU St Athan 19.9.42; returned to Hutton Cranswick 1.10.42. To Herts & Essex, Broxbourne (MI) 25.10.42. To 51 MU Lichfield 23.1.43. Sold 27.5.48 to Aircraft & Engineering Services Ltd, Croydon. Regd **G-AKXN** (CofR 12299) 12.4.48 to Surrey Financial Trust Ltd, Old Coulsdon (based Croydon) (regd as N5932). Cld 19.1.49 & regd 20.1.49 to Robert Alan Short, Croydon. CofA No.10392 issued 21.1.49. Regn cld 2.9.49 as sold abroad - to **Egyptian AF**.

Miles M.14A Magister L5932 at an unknown airfield, after modification to M.14A, but still retaining the original cowling – note the leather strap for added security! [Via P Amos]

365 M.14 **L5933** Retained by P&P (originally allocated for dd to Stn Flt Northolt 31.7.37 but allotment cancelled). To A&AEE Martlesham Heath 25.8.37. Modified 10.37 to raise tailplane by 6", flatten the top to rear fuselage and fit strakes to top rear fuselage as a cure for spin recovery problems; which modifications amended the designation to M.14A. Following makers trials it was delivered 10.37 to A&AEE Martlesham Heath for spinning trials. Returned to P&P 11.4.38. Contract No.976444/38, ITP dated 1.12.38, called for 'Installation of mods' but no further details known. To 2 SofTT Cosford for GI as **1199M** (allotted 13.12.38). To 11 SofTT Hednesford 15.8.47.

366 M.14 **L5934** Dd Stn Flt Northolt 9.8.37. To 3 Grp Comm Flt, Mildenhall 9.10.39. To 33 MU Lyneham 14.10.40. To 603 Sqdn, Hornchurch 26.8.41. Hit obstacle on take-off from forced landing, Tonbridge, Kent 13.9.41. SOC 21.9.41.

APPENDIX 20

367 M.14 **L5935** Dd Stn Flt Northolt 9.8.37. To 24 Sqdn, Hendon 29.3.41. To 15 EFTS Carlisle 25.6.41. Engine cut, hit hedge in forced landing, 2 ml W of Kirkbride 9.1.42; pilot LAC H Burkiewicz. SOC 16.1.42.

368 M.14 **L5936** Dd Stn Flt Northolt 17.8.37. Damaged when taxied into Rapide G-AEXP at Northolt 25.1.39. To 141 Sqdn, Grangemouth 18.6.40. SOC 30.12.40. To P&P 3.1.41.

369 M.14 **L5937** Dd Stn Flt Northolt 19.8.37. To 145 Sqdn, Croydon 10.2.40. To Tangmere 21.6.40. To 4 MU (SAS) at unknown location 21.8.40. To 37 MU Burtonwood 15.10.40. To 15 EFTS Carlisle 5.6.41. To 21 EFTS Booker 17.6.42. To 22 EFTS Cambridge 7.9.43. To 3 (P)AFU South Cerney 13.9.43. To 4 SofTT St Athan for GI as **6172M** 4.11.46.

370 M.14 **L5938**; fitted with 'new type tailwheel'. Dd Stn Flt Northolt 27.8.37. Engine cut, hit ditch in forced landing and undercarriage collapsed near Hunsdon 18.7.38. SOC 7.10.38 (142:00 F/H).

371 M.14 **L5939** Dd Stn Flt Ternhill 31.8.37. To 9 AOS Penrhos 25.10.39. To 16 EFTS Burnaston 2.9.40. FA 18.10.40; to P&P for repairs 25.10.40. To 24 EFTS Luton 4.5.41. To 2 EFTS Staverton 31.10.41. To 6 FIS Staverton 1.11.41. To 4 FIS Cambridge 9.5.42. To 27 MU Shawbury 7.6.42. To 118 Sqdn, Zeals 7.9.42. To 32 MU St Athan 9.9.42. To 118 Sqdn, Zeals 17.9.42. To 537 Sqdn, Hibaldstow 8.10.42. To 51 MU Lichfield 8.2.43. SOC 2.4.46.

372 M.14 **L5940** Dd HQ Practice Flt, Digby 2.9.37. To 4 MU (SAS) at unknown location 25.8.39. To 24 MU Ternhill 15.3.40. To 5 EFTS Hanworth 28.5.40. Spun into ground during spinning practice, Nott Hill Farm, N of Uttoxeter, Staffs 11.8.41. SOC 18.8.41.

373 M.14 **L5941** Dd HQ Practice Flt, Tangmere 13.9.37. To 605 Sqdn, Tangmere 30.11.39. To 49 MU Faygate 25.8.40, by road after accident. To 1 CRU Cowley (Cat.B) for repairs. To 48 MU Hawarden 24.10.40. SOC 15.11.40 at 48 MU Hawarden, believed destroyed in an enemy air attack.

374 M.14 **L5942** Dd HQ Practice Flt, Boscombe Down 13.9.37. To 2/3 Servicing Flt (part of First Echelon, Advanced Striking Force) Reims-Champagne; to Trepany by 1.6.40; Dreux 8.6.40; Bretteville 10.6.40; Mathieu 11.6.40; Dinard 15.6.40. Presumed lost in France in 6.40. SOC 26.3.42.

375 M.14 **L5943** Dd DTD at HQ Duxford 14.9.37. Used by P&P until 1.38. To RNAS Ford 10.1.38 (or 21.12.37) & possibly op by 17 Grp Comm Flt Gosport. To 780 Sqdn, Eastleigh 10.39-12.40. To RNAS Ford. At RNAS Lee-on-Solent for servicing, test flown 18.5.41. To RNAS Ford. SOC 30.4.43. *Reported as operated by 218 Sqn in France 6.40; coded 'HA-C'.*

376 M.14 **L5944** Dd HQ Practice Flt, Upper Heyford 14.9.37. To 7 Sqdn, Upper Heyford. To 16 OTU Upper Heyford 18.4.40. Engine cut on take-off from forced landing, hit trees near Amersham, Bucks 26.11.40; F/O PG Sooby slightly hurt. To P&P (SAS) 1.12.40 but SOC 16.12.40.

377 M.14 **L5945** Dd HQ Practice Flt, Honington 23.9.37. FA 18.8.40; to P&P (SAS) 27.8.40. To 45 MU Kinloss 17.12.40. To Ayr 22.7.41. To 46 MU Lossiemouth 21.4.42. To 488 RNZAF Sqdn, Colerne 11.8.44. SOC (MR) 20.2.45.

378 M.14 **L5946** Dd HQ Practice Flt, Marham 23.9.37. To 26 MU Cardington 16.6.38 pending transfer to OTC (Officer's Training Corps?) as instructional aircraft for a school but ntu. To 1 SofTT Halton for GI as **1074M** 28.10.38.

379 M.14 **L5947** Dd HQ Practice Flt, Waddington 6.10.37. 'Anti-spinning mod incorporated 6.4.38'. Crashed into house Bracebridge Heath, 2 ml south of Lincoln 2.7.40 while being flown between Finningley and Waddington; the control column became jammed during aerobatics by passenger's parachute; 1 killed, 1 seriously injured. SOC 13.7.40.

380 M.14 **L5948** Dd HQ Practice Flt, Wyton 27.9.37. Dived into ground Godmanchester, Huntingdon 10.5.38; P/O Ivor J Fawdry & P/O Colin J Lee of 139 Sqdn killed, cause not known (19:10 F/H).

381 M.14 **L5949** Dd HQ Practice Flt, Catterick 29.9.37. To 27 MU Shawbury 17.6.41. To 153 Sqdn, Ballyhalbert (31.12.41 census). Cat.B 2.5.42; to Short Bros & Harland for repairs 5.5.42. To 27 MU Shawbury 7.6.42. To 421 RCAF Sqdn, Exeter 11.7.42. To ATA 27.4.43. Engine cut, successfully forced landed Waddesdon, Bucks at 15.55 hrs 28.5.43; Flt Sgt BF MacMurty unhurt. Damage assessed as Cat A (repair on site) but fate unrecorded.

382 M.14 **L5950** Dd HQ Practice Flt, Grantham 30.9.37. To 12 FTS Grantham (census). To 27 MU Shawbury 25.10.40. To 16 EFTS Burnaston 19.12.40; coded 'A'. Hit tree while forced landing in thunderstorm Draycott-in-the-Moors, Staffs 12.7.41. To P&P (SAS) 29.7.41 but SOC 28.8.41.

383 M.14 **L5961** Dd DTD at HQ Eastleigh (800 Sqdn) 6.12.37. To RNAS Lee-on-Solent. To RNAS Worthy Down 24.5.39. To 780 Sqdn, Eastleigh 1.40-1.41. To RNAS Worthy Down 9.43. To 757 Sqdn 1.44. Fate unknown.

384 M.14 **L5952** Dd Stn Flt Worthy Down 4.11.37. To Admiralty 24.5.39 (at Worthy Down). To 780 Sqdn, Eastleigh (9.39-7.40). Fate unknown.

385 M.14 **L5953** Dd Stn Flt Duxford 9.11.37. To 222 Sqdn, Duxford. Control lost in loop, dived into ground, Fowlmere 28.3.40. SOC 4.4.40.

386 M.14 **L5954** Dd HQ Practice Flt, Thornaby 28.10.37. To 15 Grp Comm Flt, Roborough 1.2.40. To 48 MU Hawarden 22.3.42. To 3 Sqdn, Hunsdon 6.8.42. Blown off course on take-off and wing hit fence, Great Sampford 17.1.43. SOC 17.1.43. To 71 MU (Salvage Centre), Premier Garage, Bath Road, Slough 20.1.43.

387 M.14 **L5951** Dd Upwood 9.11.37. To 17 OTU Upwood. Hit HT cables low flying and crashed near Sawtry, Hunts 25.2.41. SOC 7.3.41.

388 M.14 **L5955** Dd HQ Practice Flt, Manston 4.11.37. SOC 3.10.40 (believed to have been DBR in air raid).

389 M.14 **L5956** Dd Stn Flt Mildenhall 13.11.37. Engine cut, hit fence in forced landing, Worlington, Suffolk 4.8.39 & DBR.

390 M.14 **L5957** Dd Camp Catfoss (1 ATC) 9.11.37. To 3 B&GS Aldergrove (census). To 37 MU Burtonwood 2.6.40. To 21 OTU Moreton-in-Marsh 3.5.41. To 51 MU Lichfield 1.1.44. Sold 19.2.46 to Miles Aircraft Ltd. To H Hennequin & Cia, Argentina with CofA No.7894 issued 13.6.46; regd **LV-XOW** 30.6.47 to Secretaria Estado Aeronautica, Buenos Aires. To Aero Club San Juan. Accident San Agustin 8.6.48 (with 40% damage); probably repaired. Regn cld 7.54 as wfu.

391 M.14 **L5958** Dd HQ Practice Flt, Turnhouse 10.11.37. To Bassingbourn 24.5.38. To Bicester 30.11.39. To 10 OTU Abingdon. Control lost on approach, wing hit ground and cartwheeled, Abingdon 28.11.40. SOC 13.1.41.

392 M.14 **L5959** Dd HQ Practice Flt, Lympne 9.11.37. To 24 MU Ternhill 21.10.38. To 27 E&RFTS Tollerton 12.4.39. To 32 MU St Athan 14.9.39. To 6 MU Brize Norton 6.5.40. To 16 EFTS Burnaston 29.6.40. Hit tree on approach Burnaston 14.10.41. SOC 25.10.41.

393 M.14 **L5960** Dd Stn Flt Mildenhall 17.11.37. SOC 27.10.40 (believed to have been DBR in air raid). (Also shown as SOC 14.1.41).

394 M.14 **L5962** Dd HQ Practice Flt, Hawkinge 17.12.37. To 91 Sqdn, Hawkinge 25.5.41. Wingtip hit ground on landing in gust and flew into trees Lympne 24.4.42. SOC 1.5.42.

395 M.14 **L5963** Dd Stn Flt Hucknall 13.12.37. To 48 MU Hawarden 21.2.41. To 61 OTU Heston 23.4.41. To Rednal 15.4.42. To 51 OTU Cranfield 9.6.43. To 51 MU Lichfield 18.4.45. Sold 26.2.48 to Gloster Flying Club. Regd **G-AKMO** (CofR 12052) 27.11.47 to Laurence Dudley Trappitt, Whyteleafe, Surrey. Cld 22.7.48 & regd 14.10.48 to Robert Alan Short, Derek John Hayles & Martha Marodeen, Croydon/Portsmouth. Cld 5.11.48 & regd 31.1.49 to Robert Alan Short, Croydon. CofA No.10442 issued 6.4.49. Regn cld 2.9.49 as sold abroad - to **Egyptian AF**.

396 M.14 **L5964** Dd 2 ASU Cardington 11.12.37. To 25 E&RFTS Waltham 30.8.38. To 24 EFTS Sydenham 15.10.39, to Luton 22.7.40. Complete overhaul by 1 CRU Cowley completed

1.41. To 27 MU Shawbury 13.2.41. To 16 EFTS Burnaston 25.3.41. Had done 445 hrs flying with 16 EFTS (TT 1,445 hrs), when structural failure of starboard wing during aerobatics caused it to spin into ground Walton-on-Trent, near Burton-on-Trent, Derbyshire 8.12.41; P/O Prior killed, LAC Norris slightly injured on baling out.

397   M.14 **L5965**   Dd 2 ASU Cardington 11.12.37.   To RAE Farnborough 16.6.38 (sideslip tests in 1939). To 5 MU Kemble 14.3.46.   Cat.B (MR) 5.4.46.   Sold 27.8.47 to Laygold & Co, Blackbushe. Unconverted remains extant Blackbushe in 7.53.

398   M.14 **L5966**   Dd 2 ASU Cardington 11.12.37. To Henlow 22.3.39. To 13 MU Henlow 1.3.40. To Atlantic Coast, Barnstaple (MI) 12.1.42. To 37 MU Burtonwood 28.2.42. To Pocklington 30.4.42. To 46 MU Lossiemouth 15.7.42. To 51 MU Lichfield 13.5.44. Sold 20.2.46 to Miles Aircraft Ltd. Test flown 8.5.46 by Hugh Kendall as L5966. To H Hennequin & Cia, Argentina with CofA No.7804 issued 24.5.46; regd **LV-XOE** 30.6.47 to Secretaria Estado Aeronautica, Buenos Aires. To Aero Club Formosa. Regn cld 10.50 as wfu.

399   M.14 **L5967**   Dd 2 ASU Cardington 30.12.37.   To Bircham Newton 26.3.38. To Gosport 28.9.40. To 5 MU Kemble 1.7.41. To 32 MU St Athan 28.9.41. To 15 EFTS Carlisle 10.10.41. To 21 EFTS Booker 29.6.42. To 7 (P)AFU Peterborough 10.6.43. To 21 (P)AFU Wheaton Aston 3.8.43. SOC 27.4.47 as 'deteriorated'.

400   M.14 **L5968**   Dd 2 ASU Cardington 30.12.37. To 6 MU Brize Norton 2.2.39. To 16 OTU Upper Heyford 23.12.40. To 27 MU Shawbury 15.7.41. To 313 Sqdn, Leconfield 25.7.41. To 154 Sqdn, Hornchurch 8.6.42. To 51 MU Lichfield 29.11.42. To 4 SofTT St Athan for GI as **5363M** 14.7.45. SOC 1.8.46.

401   M.14 **L5969**   Dd 2 ASU Cardington 30.12.37. To 6 MU Brize Norton 2.2.39. To 24 EFTS Luton 10.1.41. DBR in heavy landing Luton 12.7.41. SOC 23.7.41.

402   M.14 **L5970**   Dd 1 ASU Waddington 3.1.38. To 5 MU Kemble 31.1.39. To 2 Sqdn, Bekesbourne 31.5.40. To 9 MU Cosford 22.5.41. To 15 EFTS Carlisle 16.8.41. To 21 EFTS Booker 10.6.42. SOC 25.5.44.

403   M.14 **L5971**   Dd 1 ASU Waddington 18.1.38. To 3 E&RFTS Hamble 9.3.38. To E&WS Cranwell 22.6.38. To 9 MU Cosford 30.8.40. To 16 EFTS Burnaston 9.9.40. To 32 MU St Athan 10.2.42 (unconfirmed, still at Burnaston 31.12.42 census). To FTC Comm Flt, Woodley 16.6.43. To Herts & Essex for repairs 22.7.43. To 51 MU Lichfield 6.11.43. Sold 11.11.46 to Miles Aircraft Ltd. CofA No.9484 issued 27.6.47 to Eelde Aviation. Regd in Belgium as **OO-MIC** (CofR 706) 29.7.47 to Ecole de Pilotage de Coninck SA, Ghent. Regd 15.12.52 to O Georis, J de Paepe & R Declercq, Deurne-Antwerp. Regd 23.9.53 to J de Paepe, Deurne-Antwerp. Crashed Delbruck, West Germany 4.9.55. Regn cld 13.7.56.

404   M.14 **L5972**   Dd 1 ASU Waddington 3.1.38. To 3 E&RFTS Hamble 9.3.38. Spun into ground Locks Heath, near Southampton 7.5.38; Sgt Stephen HD Leech, RAFVR killed. SOC 13.10.38.

405   M.14 **L5973**   Dd 1 ASU Waddington 6.1.38. To 5 MU Kemble 31.1.39. To 145 Sqdn, Croydon 10.2.40. To 12 OTU Benson 8.10.40. To 5 MU Kemble 24.12.40. To 24 EFTS Luton 5.1.41. To P&P (SAS) 20.7.41. To 20 MU Aston Down 29.8.41. To 6 EFTS Sywell 6.11.41. To 46 MU Lossiemouth 9.7.42. To St Davids 26.9.43. To Perranporth 11.5.44. To Dallachy 20.9.44. To Thornaby (no date). To 51 MU Lichfield 7.1.46. Sold 8.6.48 to BOAC. Regd **G-AKKT** (CofR 12006) 23.6.48 to Speedbird Flying Clubs Ltd (a BOAC subsidiary), Denham.   Regd 4.49 (on name change) to Airways Aero Associations Ltd, Denham.   Not converted – extant Denham (9.51). Regn cld 27.3.52 as wfu.

406   M.14 **L5974**   Dd 1 ASU Waddington 6.1.38. To 5 MU Kemble 31.1.39. To Franco-Belgian Air Training School, Odiham 10.1.41. Hit ground low flying, Upton Grey, Hants 17.1.41.

407   M.14 **L5975**   Dd 1 ASU Waddington 6.1.38. To 5 MU Kemble 1.2.39. To 24 EFTS Luton 17.10.40. To 1 CRU Cowley (SAS) 15.11.40. To 27 MU Shawbury 13.2.41. To 1 AACU Farnborough 30.3.41. To 48 MU Hawarden 28.5.41. To 15 EFTS Carlisle 25.6.41. To 7 FIS Upavon 31.7.42. Dived into ground during low aerobatics near Newbury, Berks 20.12.42 (864:55 F/H).

408   M.14 **L5976**   Dd 1 ASU Waddington 18.1.38. To 5 MU Kemble 23.1.39. To 24 EFTS Luton 17.10.40. To Shrager Bros, Old Warden (SAS) 8.10.41. To 37 MU Burtonwood 18.12.41. Ferried Ossington-Upavon 18.3.42 and forced landed en route, 20 mls S of Ossington with engine failure; undamaged. To CFS Upavon 29.3.42. To 7 FIS Upavon 1.4.42. To ECFS Hullavington 14.5.42. Cat.Ac 28.2.44. SOC Cat.E 23.11.44. Sold (undated) to Miles Aircraft Ltd. Test flown 21.2.46 by Hugh Kendall as L5976. To H Hennequin & Cia, Argentina with CofA No.7448 issued 4.3.46; regd **LV-XMP** 30.6.47 to Secretaria Estado Aeronautica, Buenos Aires. To Aero Club Corrientes. Crashed Corrientes 26.9.46; 2 killed.

409   M.14 **L5977**   Dd 1 ASU Waddington 18.1.38. To 5 MU Kemble 31.1.39. To 12 OTU Benson 8.10.40. To 5 MU Kemble 21.12.40. To 24 EFTS Luton 5.1.41. To 45 MU Kinloss 23.2.42. To Atlantic Coast, Barnstaple (MR) 15.3.42. To 16 EFTS Burnaston 12.4.42. To 46 MU Lossiemouth 11.7.42. To ATA 15.9.44. To 51 MU Lichfield 14.12.45. Sold 15.4.46 to Miles Aircraft Ltd. To H Hennequin & Cia, Argentina with CofA No.8046 issued 8.7.46; regd **LV-XSJ** 30.6.47 to Secretaria Estado Aeronautica, Buenos Aires. To Aero Club La Plata. Accident 15.8.48; repaired. Crashed Ojeda 7.9.55; 2 injured. Repaired? Regd (5.67) to Direccion General de Aeronautica Civil, Buenos Aires.

410   M.14 **L5978**   Dd 1 ASU Waddington 18.1.39. To 5 MU Kemble 26.1.39. To 24 EFTS Luton 17.10.40. To 1 CRU Cowley for repairs 27.5.41. To 46 MU Lossiemouth 26.7.41. To 15 EFTS Carlisle 5.11.41. To 29 EFTS Clyffe Pypard 26.6.42. To 45 MU Kinloss 15.4.43. To 51 MU Lichfield 11.2.44. Sold 28.2.46 to Miles Aircraft Ltd. Test flown 30.4.46 by Hugh Kendall as L5978. To H Hennequin & Cia, Argentina with CofA No.7767 issued 20.5.46; regd **LV-XNN** 30.6.47 to Secretaria Estado Aeronautica, Buenos Aires. To Aero Club Esquina. Crashed 1.3.52; repaired? Regn cld 10.53 as wfu.

411   M.14 **L5979**   Dd 115 Sqdn, Marham 19.1.38. Bounced on landing, stalled and flew into ground Newmarket 19.11.39 and DBR. SOC 20.11.39.

412   M.14 **L5980**   Dd 1 ASU Waddington 18.1.38. To 5 MU Kemble 24.1.39. To Andover 17.7.40. To 5 MU Kemble 4.2.41. To 24 EFTS Luton 21.5.41. To 2 FIS Montrose 22.3.42. Flying accident 6.5.43; to Herts & Essex, Broxbourne (Cat B) 13.7.43. To 51 MU Lichfield 22.10.43. Sold 29.1.46 to Miles Aircraft Ltd. To Portuguese AF **190** with CofA No.7548 issued 18.3.46; later re-serialled **1201**.

413   M.14 **L5981**   Dd 1 ASU Waddington 24.1.38. To 5 MU Kemble 16.1.39. To 24 EFTS Luton 17.10.40. Crashed 5.3.41; to 1 CRU Cowley (SAS) 10.3.41. To 5 EFTS Meir 15.4.41. To 71 Group Farnborough? 17.4.41. To 9 MU Cosford 16.6.41. To ATA Training Unit, White Waltham 9.8.41. To Atlantic Coast, Barnstaple (MR) 22.9.43. To 51 MU Lichfield 12.11.43. Sold by MAP to Miles for export to **Chile** in early 1945; flight tested as **U-0263**.

414   M.14 **L5982**   Dd 1 ASU Waddington 24.1.38. To 5 MU Kemble 26.1.39. To Andover 16.7.40. To 5 MU Kemble 1.2.41. To 7 AACU Castle Bromwich 6.3.41. To 27 MU Shawbury 5.9.42. To 51 MU Lichfield 21.3.44. Sold 5.2.46 to Miles Aircraft Ltd. Test flown 30.4.46 by Hugh Kendall as L5982.   To H Hennequin & Cia, Argentina with CofA No.7762 issued 15.5.46; regd **LV-XNU** 30.6.47 to Secretaria Estado Aeronautica, Buenos Aires. To Aero Club Salta. Accident 7.4.48; repaired. Regn cld 9.53 as wfu.

415   M.14 **L5983**   Dd 1 ASU Waddington 24.1.38. To 5 MU Kemble 26.1.39.   To 24 EFTS Luton 17.10.40.   To 46 MU Lossiemouth 3.42.   To Lundy & Atlantic Coast, Barnstaple (MI) 23.3.42. To 20 MU Aston Down 17.4.42. To 121 Sqdn, Southend 29.8.42. To 335 Sqdn, USAAF 1.10.42. To 515 Sqdn, Heston 13.1.43. To 51 MU Lichfield 22.7.45. Sold 28.8.46 to Miles Aircraft Ltd. CofA No.8598 issued 11.10.46 to Middle East Airlines. Regd in Lebanon as **LR-AAL** to Middle East Airlines, Beirut.   Re-regd **OD-AAL** 7.7.49. Regn cld .53 as wfu and seen on the dump at Beirut 8.5.56.

416   M.14 **L5984**   Dd 1 ASU Waddington 24.1.38. To 5 MU Kemble 26.1.39. To 24 EFTS Luton 17.10.40. Crashed and dbf following collision with Tiger Moth T7113 7.7.41; Flt Lt AC Dier & HR Hearne killed. SOC 11.7.41.

417   M.14 **L5985**   Dd 1 ASU Waddington 24.1.38. To 5 MU Kemble 26.1.39. To 616 Sqdn, Coltishall 6.9.40. To 65 Sqdn, Kirton-in-Lindsey 26.2.41. To 616 Sqdn, Westhampnett 28.7.41.

Stalled on take-off in strong up-current from cliff Perranporth 6.8.41. SOC 19.8.41.

418 M.14 **L5986** Dd 1 ASU Waddington 31.1.38. To 5 MU Kemble 26.1.39. To 24 EFTS Luton 17.10.40. To 16 EFTS Burnaston 11.2.42. To P&P for repairs 2.9.42. To 27 MU Shawbury 21.10.42. To 51 MU Lichfield 21.11.42. To 47 MU Sealand for packing 15.2.43. Shipped to Middle East on SS Clan Leonard 19.4.43; arrived 12.6.43. To 41 SAAF Shandur? Engine cut, stalled and hit trees recovering from dive, El Bassa, Palestine 13.8.44.

419 M.14 **L5987** Dd 1 ASU Waddington 31.1.38. To 5 MU Kemble 26.1.39. To 24 EFTS Luton 17.10.40. To 2 FIS Montrose 15.2.42. To Herts & Essex, Broxbourne (MI) 22.8.42. To 27 MU Shawbury 9.10.42. To 76 MU Wroughton for packing 18.5.43. To **Turkey** 2.6.43; arrived in ME 27.6.43.

420 M.14 **L5988** Dd 1 ASU Waddington 2.2.38. To 5 MU Kemble 26.1.39. To 24 EFTS Luton 17.10.40. Hit HT wires Haynes 21.3.41; Flt Lt P Cannam unhurt; to 1 CRU Cowley for repairs 22.3.41. To 15 EFTS Carlisle 28.5.41. To P&P 14.12.41. To 37 MU Burtonwood 26.1.42. To ECFS Hullavington 22.5.42. To 51 MU Lichfield 8.1.45. Sold 9.2.46 to Miles Aircraft Ltd. Test flown 10.4.46 by Hugh Kendall as L5988. To H Hennequin & Cia, Argentina with CofA No.7690 issued 3.5.46; regd **LV-XOR** 30.6.47 to Secretaria Estado Aeronautica, Buenos Aires. To Aero Club Santa Fe. Accidents 23.10.46 & 29.8.51; repaired. To Aero Club San Julian 21.4.52. Regn cld 7.53 as wfu.

421 M.14 **L5989** Dd 1 ASU Waddington 2.2.38. To 5 MU Kemble 26.1.39. To 8 EFTS Woodley 25.9.40. To 15 EFTS Carlisle 16.2.42. To 29 EFTS Clyffe Pypard 30.5.42. To 46 MU Lossiemouth 14.4.43. To 51 MU Lichfield 19.11.45. Sold 23.5.46 to Miles Aircraft Ltd. To H Hennequin & Cia, Argentina with CofA No.8188 issued 30.7.46; regd **LV-XRQ** 30.6.47 to Secretaria Estado Aeronautica, Buenos Aires. To Circ. Universitario de Rosario. Regd (5.67) to Direccion General de Aeronautica Civil, Buenos Aires.

422 M.14 **L5990** Dd 1 ASU Waddington 3.2.38. To 5 MU Kemble 26.1.39. To Franco-Belgian Air Training School, Odiham 10.1.41. To 32 MU St Athan 14.6.41. To Short Bros 15.7.41; allocated to their MAP Overseer 2.12.42. To 51 MU Lichfield 5.2.43. Sold 18.6.46 to Miles Aircraft Ltd. Test flown 25.7.46 by Hugh Kendall as L5990. To H Hennequin & Cia, Argentina with CofA No.8210 issued 7.8.46; regd **LV-XSM** 30.6.47 to Secretaria Estado Aeronautica, Buenos Aires. To Aero Club Pampeano. Regn cld 8.49 as wfu.

423 M.14 **L5991** Dd 1 ASU Waddington 3.2.38. To 5 MU Kemble 26.1.39. To 24 EFTS Luton 17.10.40. To 2 FIS Montrose 20.2.42. To 11 (P)AFU Shawbury 24.9.43. To 51 MU Lichfield 14.6.45. Sold 3.12.46 to The Wiltshire School of Flying Ltd. Regd **G-AITU** (CofR 10963) 11.11.46 to Anna Valley Motors (Salisbury) Ltd, Salisbury (based Thruxton). CofA No.10360 issued 3.5.49. Cld 8.6.49 as sold (probably to RA Short, Croydon). Regn cld 2.9.49 as sold abroad - to **Egyptian AF**.

424 M.14 **L5992** Dd 1 ASU Waddington 3.2.38. To 5 MU Kemble 30.1.39. To 12 OTU Benson 8.10.40. To Shrager Bros, Old Warden (SAS) 18.6.41. To 27 MU Shawbury 21.8.41. To ATA (31.12.41 census); op by ATA Training Unit (2.42), coded '25'. To 51 MU Lichfield 13.2.45. Sold 26.2.48 to Gloster Flying Club, Staverton. Regd **G-AKMN** (CofR 12051) 27.11.47 to Laurence Dudley Trappitt, Whyteleafe, Surrey (still at Staverton 7.48). Sold 20.11.50 & regd 12.12.50 to Wolverhampton Aviation Ltd, Wolverhampton. CofA No. A276 issued 4.5.51. Modified to single-seater for racing, with the rear cockpit faired over (Race No.57). Crashed Wheaton Aston, Staffs during practice forced landings 26.2.53; Geoffrey H Lloyd killed. Regn cld 13.10.53 as destroyed.

425 M.14 **L5993** Dd 1 ASU Waddington 10.2.38. To 5 MU Kemble 30.1.39. To 304 and 305 Sqdns, Bramcote 23.9.40. Flying accident 18.1.42; to P&P for repairs 27.1.42. To 9 (P)AFU Hullavington 1.4.42. To ECFS Hullavington (undated). To Atlantic Coast, Barnstaple (SAS) 22.11.42. To 51 MU Lichfield 1.1.43. Sold 22.4.49 to RA Short. Regd **G-ALNZ** (CofR 12708) 25.4.49 to Robert Alan Short & James Henry Tattersall, Croydon/Clitheroe, Lancs. Not converted; extant Elstree 7.51. Regn cld by MCA 12.9.52. Regd 21.1.53 to Robert Alan Short, Croydon. Regn cld 18.5.59 as pwfu.

426 M.14 **L5994** Dd Gosport 7.2.38. Damaged & to 49 MU Faygate (Cat.B) 3.9.40. To 1 CRU Cowley (SAS) 21.9.40. To 5 MU Kemble 7.11.40. To Pilotless Aircraft Unit, St Athan 9.1.41. To 145 Sqdn, Catterick 19.8.41. Hit fence on take-off from forced landing Lemsford, Herts 14.10.41. SOC 20.10.41.

427 M.14 **L5995** Dd 1 ASU Waddington 10.2.38. To 5 MU Kemble 30.1.39. To 8 EFTS Woodley 25.9.40. To 10 FIS Woodley 21.7.42. To 7 FIS Upavon 26.7.42. To 51 MU Lichfield 8.11.44. Sold 4.2.46 to Miles Aircraft Ltd. To H Hennequin & Cia, Argentina with CofA No.7635 issued 9.4.46; regd **LV-XMS** 30.6.47 to Secretaria Estado Aeronautica, Buenos Aires. To Aero Club Neuquen. Regn cld 10.56 as wfu.

428 M.14 **L5996** Dd 1 ASU Waddington 10.2.38. To 5 MU Kemble 30.1.39. To 305 Sqdn, Bramcote 1.10.40. To 18 OTU Bramcote (undated). Destroyed in air raid Bramcote 13.3.41. SOC 21.3.41 as 'burnt out'.

429 M.14 **L5997** Dd 1 ASU Waddington 10.2.38. To 5 MU Kemble 30.1.40. To 8 EFTS Woodley 25.9.40. To 32 MU St Athan 25.1.42. To 8 FTS Montrose 7.2.42. To 29 EFTS Clyffe Pypard 27.7.42. To 5 GTS Stoke Orchard 22.3.43. To 51 MU Lichfield 10.4.44. Sold 15.11.46 to Miles Aircraft Ltd. CofA No.9841 issued 1.1.48 to **Thai AF** (and shown as c/n 479).

430 M.14 **L5998** Dd 1 ASU Waddington 14.2.38. To 5 MU Kemble 30.1.39. To 24 EFTS Luton 6.10.40. Crashed 7.7.41; to 1 CRU Cowley (SAS) 10.7.41. To 27 MU Shawbury 12.9.41. To 32 MU St Athan 2.10.41. To 16 EFTS Burnaston 22.10.41. To 5 EFTS Meir 14.12.41. To 32 MU St Athan 14.4.42. To 51 MU Lichfield 7.12.42. Sold 5.2.46 to Miles Aircraft Ltd. To H Hennequin & Cia, Argentina with CofA No.7787 issued 20.5.46; regd **LV-XNW** 30.6.47 to Secretaria Estado Aeronautica, Buenos Aires. To Aero Club Rafaela. Accident 11.7.47; repaired. Accident 8.11.52; repaired? Regn cld 12.53 as wfu.

431 M.14 **L5999** Dd 1 ASU Waddington 14.2.38. To 5 MU Kemble 30.1.39. To 24 EFTS Luton 7.10.40. To 2 FIS Montrose 15.2.42. Ferried to Kirkbride 2.7.43; possibly en route to DH (CRO) (undated). To 51 MU Lichfield 30.11.43. Sold 22.9.46 to Miles Aircraft Ltd. CofA No.9183 issued 31.3.47 to Aeromnium SA, Geneva. Regd in Switzerland as **HB-EEB** (CofR 1172) 26.4.47 to NM Weber, Cointrin. Regd .47 to Aero Club de Suisse, Section Lugano (based Agno). Regd .49 to SALGA SA, Lugano (based Agno). Regn cld. 55. Regd .56 to Dr Hans Isler (and operated by Harold B Williams), Munchen; to Poitiers .64. Regn cld 20.12.66. To Amicale J-B Salis at La Ferte Alais, France (& later at Etampes). Sold to the Spanish Air Force Museum and 'modified' with dummy 'trousered' u/c fairings and three-piece rear windscreen to represent the Hawk Majors used in the Spanish Civil War; painted in Republican colours as 'EN-002' on the civil side and Nationalist colours as '30-145' on the starboard side; on display in the Spanish Air Force Museum at Cuatro Vientos, Madrid.

432 M.14 **L6000** Dd Stradishall 19.2.38 (CLN recorded dd to CC Gosport 19.2.38?). To Morris, Cowley for repairs 27.3.40. To 46 MU Lossiemouth 25.6.40. To 241 Sqdn, Inverness 23.10.40. To 45 MU Kinloss 5.1.41. To 15 EFTS Carlisle 20.2.41. To 21 EFTS Booker 30.6.42. SOC 25.5.44.

479 M.14 Hawk Trainer III CofA No.5985 issued 21.7.37 to Royal Singapore Flying Club (originally made for c/n 365 which was L5933 and amended to c/n 479 on 5.5.37). A Contractor's Flt Test Report for VR-SAY was dated 13.7.37; dd to Royal Singapore Flying Club 17.7.37. Regd **VR-SAY** 1.10.37 to Royal Singapore Flying Club. To MVAF with serial **39** in 12.41. Destroyed Bukit Timah 5.2.42.

Note: c/n 479 given as identity to a Thai AF Magister – see c/n 429

Miles M.14A Hawk Trainer Mk.III of the Royal Singapore Flying Club in 1937.

486 M.14A Hawk Trainer Mk.III CofA No.6060 issued 30.9.37 to the Royal New Zealand Air Force. Sold in New Zealand via John M Gamble, Wellington (P&P local agent); reported to have arrived in NZ 11.5.37 but actually dd 21.9.37. Regd **ZK-AEZ** (CofR 201/CofA 100) 30.9.37 to Wellington Aero Club, Wellington. CofA validated in NZ 20.12.37. Impressed into RNZAF as **NZ585** 12.9.39; to 2 EFTS New Plymouth. To 4 EFTS Whenuapai. To 1 (GR) Sqdn, Whenuapai. Destroyed when Hudson NZ2077 crashed into No.1 hangar at Whenuapai 17.12.42.

Miles M.14A Hawk Trainer Mk.III ZK-AFA after its arrival in New Zealand in 1937. [Via P Amos]

487 M.14A Hawk Trainer Mk.III CofA No.6057 issued 30.9.37 to the Royal New Zealand Air Force. Sold in New Zealand via John M Gamble, Wellington, (P&P local agent); reported to have arrived in NZ 16.8.37 but actually dd 2.10.37; assembled at Wigram. Regd **ZK-AFA** (CofR unkn/CofA 102) 30.9.37 to Canterbury Aero Club, Christchurch. Crashed during aerobatics near Prebbleton, Christchurch 27.2.38, pilot Clifford A Burminster killed. Total hours approx 25.

488 CLN recorded (in pencil) Magister with customer P&P. ARB recorded M.14 Hawk Trainer to New Zealand but no CofA issued and not regd in New Zealand. Presumed not built.

489 CLN recorded (in pencil) Magister with customer P&P. ARB recorded M.14 Hawk Trainer to New Zealand but no CofA issued and not regd in New Zealand. Presumed not built.

490 M.14A Hawk Trainer Mk.III CofA No. 6058 issued 30.9.37 to Aircraft Industries (Pty) Ltd; dd 2.10.37. Regd in South Africa as **ZS-AMR** 11.37 to Aircraft Industries (Pty) Ltd, Johannesburg. (A report in the South African magazine *The Fly Paper* for August 1937: *'We learn that the four new Miles Trainer two-seaters on order for the Witwatersrand Technical College Flying School (c/n 490, 491, 492 and 493, - PA) will arrive in Johannesburg at the end of September'*. Impressed into SAAF as **1476** 3.40. Collided with Hawk Trainer 1479 during formation practice 20.12.40.

Miles M.14A Hawk Trainer Mk.III ZS-AMR in 1937, less undercarriage fairings. [The A J Jackson Collection]

491 M.14A Hawk Trainer Mk.III CofA No. 6090 issued 14.10.37 to Aircraft Industries (Pty) Ltd; dd 9.10.37. Regd in South Africa as **ZS-AMS** 11.37 to Aircraft Industries (Pty) Ltd, Johannesburg. Impressed into SAAF as **1479** 3.40. Collided with Magister 1476 in formation practice 20.12.40.

492 M.14A Hawk Trainer Mk.III CofA No. 6091 issued 14.10.37 to Aircraft Industries (Pty) Ltd; dd 9.10.37. Regd in South Africa as **ZS-AMT** 11.37 to Aircraft Industries (Pty) Ltd, Johannesburg. Impressed into SAAF as **1505** 3.40.

493 M.14A Hawk Trainer Mk.III CofA No. 6092 issued 14.10.37 to Aircraft Industries (Pty) Ltd; dd 9.10.37. Regd in South Africa as **ZS-AMU** 11.37 to Aircraft Industries (Pty) Ltd, Johannesburg. Impressed into SAAF as **1448** 3.40. Undercarriage collapsed on landing Knysna 22.4.41 (or 22.4.42) and written off.

494 M.14B Hawk Trainer Mk.II Regd **G-AEZP** (CofR 7980) 17.7.37 to Blackburn Aircraft Ltd, Brough. A Contractor's Flt Test Report for G-AEZP dated 5.9.37 stated 'Gipsy Major II' presumably in error for Blackburn Cirrus Major engine. CofA No.6035 issued 6.9.37; dd 8.9.37. Flown by Sqdn Ldr ECT Edwards in 1937 King's Cup Air Race at Hatfield 10/11.9.37; retired at Macmerry with piston seizure. Regn cld 15.10.38 as sold. Transferred to AM Contract No.602395/37 and transferred 6.39 to 4 E&RFTS Brough as **P2150**. To 5 EFTS Hanworth. Spun into ground during aerobatics at South Godstone, Surrey 3.5.40. Wreck to 49 MU Faygate 14.5.40 and SOC 12.6.40.

495 M.14A Hawk Trainer Mk.III Regd **G-AEZR** (CofR 7981) 7.37 to Yorkshire Aviation Services Country Club Ltd, York. CofA No.6000 issued 31.7.37; dd 6.8.37. Regd (CofR 8089) 25.8.37 to John Morgan Barwick, York. Regn cld 10.38 as sold, but not re-regd despite CofA being renewed 24.3.39. Fate unknown.

Preparations for the King's Cup Air Race at Hatfield over 10th/11th September 1937. Miles M.14A Hawk Trainer Mk.III G-AEZR was flown by Capt Barwick. [Via B Clarke]

511 M.14A Hawk Trainer Mk.III CofA No. 6107 issued 27.10.37 to Aircraft Industries (Pty) Ltd; dd 18.10.37. Regd in South Africa as **ZS-AMV** 12.37 to Aircraft Industries (Pty) Ltd, Johannesburg (although CLN recorded it as ZS-AMW for Rand Flying Club, Rand Airport). Crashed while pilot was circling father's house at Randfontein 17.2.40; Walter Mathias injured.

512 M.14A Hawk Trainer Mk.III CofA No. 6108 issued 27.10.37 to Aircraft Industries (Pty) Ltd; dd 31.10.37. Regd in South Africa as **ZS-AMW** 12.37 to Aircraft Industries (Pty) Ltd, Johannesburg (although CLN recorded it as ZS-AMY for Rand Flying Club). Impressed into SAAF as **1483** 3.40. To 73 Air School for GI use as **IS119**.

513 M.14 Hawk Trainer Mk.III CofA No. 6115 issued 2.11.37 to Aircraft Industries (Pty) Ltd; dd 2.11.37. Regd in South Africa as **ZS-AMY** 12.37 to Aircraft Industries (Pty) Ltd, Johannesburg (although CLN recorded it as ZS-AMV for Technical College, Rand Airport). Impressed into SAAF as **1485** 3.40.

514 M.14A Hawk Trainer Mk.III Contractor's Flt Test Report for ZS-ALT was dated 2.11.37. CofA No.6132 issued 19.11.37 to Aircraft Industries (Pty) Ltd; dd 17.11.37. Regd **ZS-ALT** 1.38 probably to South African Airways. Impressed into SAAF as **263** 3.40.

515 M.14 **L6894** Dd 3 E&RFTS Hamble 10.11.37. Dived into ground Bursledon, near Hamble 13.6.38; pilot Gerald IF Colmore bailed out too low and was killed. SOC 13.10.38.

516 M.14 **L6895** Dd 3 E&RFTS Hamble 23.11.37. Abandoned in spin Sarisbury Green, near Southampton 25.1.38; pilot Desmond A Beauclaire bailed out unhurt. SOC 13.10.38.

517 M.14 **L6896** Dd 3 E&RFTS Hamble 13.11.37. To 1 ASU Waddington 13.6.38. To 5 MU Kemble 16.1.39. To 16 Sqdn, Weston Zoyland 24.10.40. To 9 MU Cosford 6.5.41. To 15 EFTS Carlisle 16.8.41. To 29 EFTS Clyffe Pypard 26.5.42. To 46 MU Lossiemouth 14.4.43. To ATA 29.11.43. To 39 MU Colerne 5.1.45. Sold 31.12.46 to J Neasham. Regd **G-AIYL** (CofR 11080) 9.12.46 to John Neasham, Darlington. CofA No.8976 issued 2.4.47. CofA lapsed 28.3.51. Cld 20.4.53 & regd 29.4.53 to Darlington & District Aero Club Ltd, Croft. CofA renewed 29.5.53. Regn cld 9.6.53 as sold to French Morocco. Regd **F-DADV** (CofR 52828) 30.12.54 to Air Police Ifrane, French Morocco. CofA suspended 18.8.60. Regn cld 22.9.71 as destroyed.

518 M.14 **L6897** Dd 3 E&RFTS Hamble 23.11.37. To 27 MU Shawbury 29.12.39. To 48 MU Hawarden 14.6.40. To Aldergrove 27.11.40. Engine cut on take-off, crash landed in water Crumlin, Co. Antrim 24.2.43. SOC 9.3.43.

519 M.14 **L6898** Dd 3 E&RFTS Hamble 26.11.37. To 9 E&RFTS Ansty (undated). To Andover 22.6.38. To 26 Sqdn, West Malling 8.7.40. To 9 MU Cosford 27.5.41. To 16 EFTS Burnaston 24.8.41. To 10 FIS Woodley 15.5.43. To 27 MU Shawbury 17.9.43. To 51 MU Lichfield 17.2.44. Sold 1.4.46 to Miles Aircraft Ltd. Test flown 7.6.46 by Hugh Kendall as L6898. To H Hennequin & Cia, Argentina with CofA No.7973 issued 2.7.46; regd **LV-XQL** 30.6.47 to Secretaria Estado Aeronautica, Buenos Aires. To IV Censo Gral de la Nacion. Accident 1.3.47. Regn cld 6.47 as wfu.

520 M.14 **L6899** Dd 3 E&RFTS Hamble 26.11.37. To 9 E&RFTS Ansty (undated). To Andover 23.6.38. To Cranfield 2.9.38. To Andover 30.11.39. To 4 Sqdn, Linton-on-Ouse 6.7.40. To P&P (SAS) 8.10.40. To 74 Sqdn, Biggin Hill 17.1.41. Stalled recovering from loop and crashed 1 ml W of Manston 18.3.41. SOC 31.3.41.

521 M.14 **L6900** Dd 3 E&RFTS Hamble 29.11.37. To Henlow 22.6.38. To Gosport 25.11.38. Damaged in ground collision with Swordfish K8426 at Gosport 28.11.38. To 6 MU Brize Norton 27.7.39. To 16 EFTS Burnaston 29.6.40. To 27 MU Shawbury 2.9.42. To ATA 8.6.43. To Atlantic Coast, Barnstaple for repairs 21.9.43. To 51 MU Lichfield 25.11.43. Sold 4.2.46 to Miles Aircraft Ltd. To H Hennequin & Cia, Argentina with CofA No.7629 issued 9.4.46; regd **LV-XMY** 30.6.47 to Secretaria Estado Aeronautica, Buenos Aires.To Aero Club S. del Estero. Accident 25.4.47 (70% damaged); 2 injured.

522 M.14 **L6901** Dd 3 E&RFTS Hamble 29.11.37. Dived into sea recovering from dive off Calshot 24.1.38; Sgt David L Middleton killed. SOC 13.10.38.

523 M.14 **L6902** Dd 3 E&RFTS Hamble 14.12.37. To 9 E&RFTS Ansty. Stalled avoiding a cart in forced landing in poor visibility, hit ground Hathern, near Loughborough, Leics 9.2.38; Archibald D Maclaren injured. SOC 13.10.38.

524 M.14 **L6903** Dd 9 E&RFTS Ansty 3.12.37. To 3 E&RFTS Hamble (undated). To RAF Staff College Flt, Andover 22.6.38; renamed College Stn Flt 10.9.38. To 47 MU Sealand 3.6.40. To Irish Air Corps as **73** 7.6.40. Scrapped in 3.46.

525 M.14 **L6904** Dd 9 E&RFTS Ansty 14.12.37. To 10 MU Hullavington 4.5.39. To Franco-Belgian Air Training School, Odiham 3.1.41. To 27 MU Shawbury 16.6.41. To 24 EFTS Luton 11.9.41. To 15 EFTS Carlisle 10.10.41. To 21 EFTS Booker 29.6.42. Collided with Tiger Moth T6686 on approach Cliffe Pypard 3.8.42; pilot Cpl CD James unhurt. SOC 11.8.42.

526 M.14A **L6905** First fully-modified Magister with tailplane raised 6 inches, flattened top to rear fuselage and fillets (anti-spin strakes) fitted. Dd 3 E&RFTS Hamble 18.12.37. To 27 MU Shawbury 3.11.38. To 15 OTU Harwell 27.5.40. To 27 MU Shawbury 2.10.41. To 6 FIS Staverton (census 31.12.41 - unconfirmed). To 27 MU Shawbury (undated). To 4 SS Madley 1.6.42. To 4 RS Madley 24.12.42. To St Athan for GI as **5360M** 14.7.45.

527 M.14A **L6906** Dd 3 E&RFTS Hamble 22.11.37. Abandoned after control lost in cloud, crashed Bursledon, near Hamble 16.3.38; K Baldry bailed out unhurt. SOC 13.10.38.

528 M.14A **L6907** Dd 3 E&RFTS Hamble 24.11.37. To 9 E&RFTS Ansty (undated). To HAD Henlow 12.7.38. To 13 MU Henlow 1.3.40. To Atlantic Coast, Barnstaple (MI) 12.1.42. To 130 Sqdn, Perranporth 19.3.42. To 27 MU Shawbury 2.11.42. To ATA 19.9.44. To 21 EFTS Booker 29.3.45. To RAF Staff College (Gerrard's Cross) 19.4.45. To TTC Comm Flt White Waltham 14.8.45. To RAF Staff College Flt 18.9.45; coded 'TBR-K'. To Bracknell/Reserve Cmd Comm Sqdn, White Waltham 16.10.46. To 5 MU Kemble 29.11.46. Sold 5.12.47 to WS Shackleton Ltd. Regd **G-AKJW** (CofR 11984) 17.12.47 to Short Bros & Harland Ltd, Rochester. Cld 25.6.48 & regd 6.7.48 to Harry John George Turner, London SE15. CofA No.9916 issued 24.6.48. Wing struck a tree at Hodore Farm, Hartfield, Sussex and crashed 20.7.48 on take-off after forced landing in bad weather; pilot Harry Turner injured. Regn cld 16.8.48 as crashed.

529 M.14A **L6908** Dd 9 E&RFTS Ansty 2.2.38. To 3 E&RFTS Hamble 10.2.38. To 24 Sqdn, Hendon 23.6.38. To 4 MU (SAS) at unknown location 30.10.39. To 6 MU Brize Norton 18.3.40. To 16 EFTS Burnaston 24.6.40. Hit tree in forced landing while lost Chapel-en-le-Frith, Derbyshire 27.8.41. SOC 5.9.41.

530 M.14B **L6909**, fitted with Cirrus Major engine. Dd 4 E&RFTS Brough 12.2.38. To 74 Sqdn, Hornchurch 24.6.38. To 27 MU Shawbury 19.8.38. To 4 E&RFTS Brough 6.3.39. To 5 EFTS Hanworth 15.10.39. Hit obstacle on take-off from forced landing Calton, Staffs 24.5.41 & DBR; F/O Lofthouse & LAC Horshall survived. SOC 17.6.41.

531 M.14B **L6910**, fitted with Cirrus Major engine. Dd 4 E&RFTS Brough 12.2.38. Stalled on approach to Brough and spun into River Humber 15.6.38; GAM Baird drowned. SOC 26.9.38 (48:20 F/H).

532 M.14B **L6911**, fitted with Cirrus Major engine. Dd 4 E&RFTS Brough 24.2.38. To 5 EFTS Hanworth 15.10.39. To 1 CRU Cowley 1.4.41. To 20 MU Aston Down 22.4.42. To 27 MU Shawbury 1.9.42. To ATA 27.7.43. SOC 7.9.44.

533 M.14B **L6912**, fitted with Cirrus Major engine. Dd 4 E&RFTS Brough 26.2.38. To Wyton 22.6.38. To 27 MU Shawbury 7.1.39. To 4 E&RFTS Brough 6.3.39. To repairs (unknown). To 5 EFTS Hanworth 4.11.39. To 1 SS Cranwell North 9.1.42. To 5 MU Kemble 4.3.43. To Atlantic Coast, Barnstaple (MI) 20.10.43. To 51 MU Lichfield 13.1.44. Sold by MAP to Miles Aircraft Ltd for **Chile** in early 1945; refurbished by Marshall's Flying School, Cambridge and test flown as **U-0264**. *(This is an oddity, since the Chilean contract called for Gipsy Major engines and this was one of the few surviving with a Cirrus Major engine).*

534 M.14B **L6913**, fitted with Cirrus Major engine. Dd 4 E&RFTS Brough 26.2.38. To HAD Henlow 23.6.38. To 27 MU Shawbury 10.9.38. To 4 E&RFTS Brough 26.10.38. To 5 EFTS Hanworth 15.10.39. To 37 MU Burtonwood 8.12.41. To ECFS Hullavington. To 15 MU Wroughton 4.12.42. To 51 MU Lichfield (1942 census). Sold 26.2.48 to Herts & Essex Aero Club. Regd **G-AKNA** (CofR 12063) 29.11.47 to Herts & Essex Aero Club (1946) Ltd, Broxbourne. Refurbished and fully painted in the club colours of silver with green trim, for the club's co-director Roger Frogley, it was test flown and issued with CofA No.10029 on 19.3.48. However, the CofA was then refused or withdrawn by ARB: *due to an M.14B not being recognised as either a service or civilian type* (they had obviously forgotten the civilian and service M.14B's of pre-war days). Used for spares at Broxbourne. Sold (.53) to Liverpool Flying Club, Speke and seen dismantled at Speke 25.7.53 being used for spares. Regn cld 20.4.54 as sold. Later taken to Liverpool Fire Brigade, Speke Station where it was extant (3.58) but subsequently scrapped.

535 M.14B **L6914** Allotted to a M.14B for delivery to Blackburn Aircraft (4 E&RFTS) but the individual aircraft record notes 'Not built'. (Recorded by CLN but with no engine or delivery details). Replaced on contract by N4557 (c/n 640) 8.6.38.

536 M.14A **L6915** Dd to 8 E&RFTS Woodley 3.1.38; renamed 8 EFTS 3.9.39. Crashed after control lost while low flying near Dene Camp, Camberley, Surrey 22.6.41; Sgt M Traisnel badly injured. SOC 27.6.41.

537 M.14A **L6916** Dd 8 E&RFTS, Woodley 15.1.38. Dived into ground Sonning Common, near Reading, Berks (also reported as Kidmore End, Oxford) 28.3.38; pilot LAC Herbert E Griffiths killed. SOC 2.9.38.

538 M.14A Hawk Trainer Mk.III Regd **G-AEZS** (CofR 7982) 12.7.37 to Phillips & Powis Aircraft Ltd. CofA No.5984 issued 21.7.37. Although CLN recorded dd to 8 E&RFTS Reading 21.7.37,

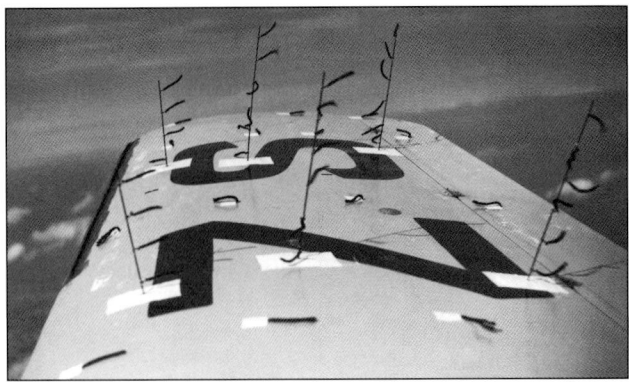

Wool tufts on the starboard wing of Miles M.14A Hawk Trainer Mk.III G-AEZS. [Phillips & Powis via The Miles Aircraft Collection]

'Diruptors' fitted to the leading edge of wing of G-AEZS. [Phillips & Powis via The Miles Aircraft Collection]

Deepened underside of fuselage of G-AEZS. [Phillips & Powis via The Miles Aircraft Collection]

Photograph showing what appears to be an instrumentation probe under the tailplane of G-AEZS. Note deep underside running out under tailplane. [Phillips & Powis via The Miles Aircraft Collection]

a later scribbled note read 'now Stock Plane'. Retained by P&P for trials. Three separate sets of photos show (i) with wool tufts attached to the upper surfaces of the wing, (ii) with 'Diruptors' attached to the leading edge of the wings and (iii) fitted with a very deep fairing to the underside of the fuselage centre section, tapering away to almost join the normal lines just forward of the tailwheel; the purpose for the fairing has not been established.

Following the loss of L5912 during spinning trials on 22.7.37, G-AEZS was flown to A&AEE Martlesham Heath to continue the trials. Loaded as per L5912, it was flown by a P&P pilot who, on 27.7.37, also failed to recover from a spin after 30 turns (but the aircraft was saved by using its tail parachute). CofA renewed 20.1.39. Fitted with an M.18 wing in 8.39 for trials with 'B' mark **U-6**; it was at A&AEE Boscombe Down in August/September 1939 before reverting to M.14 standard later. Flown at CFS Upavon 7.12.39 and 13.12.39 by Henry Woodhouse. Dd (as U6) 5.3.40 by Hugh Kennedy back to CFS Upavon and again flown by Woodhouse on 5, 7, 8 & 27.3.40. To RAF Upavon 22.4.40. Fitted with light bomb racks in 5.40 and, after a local flight (at Woodley) on 20.5.40, Hugh Kennedy carried out bomb drop tests on 21.5.40 (recording U6 in logbook) and again on 25.5.40 (now recording it as G-AEZS). It is believed that 'B' mark **U-0229** was allotted at some time for this aircraft but it is not known if it was carried. Regd .44 to Miles Aircraft Ltd, its fate is unknown. Regn cld 1.12.46 in census.

539 M.14A Hawk Trainer Mk.III Regd **G-AFBS** (CofR 8078) 17.9.37 to Phillips & Powis Aircraft Ltd. Contractor's Flt Test Report was dated 12.10.37. CofA No.6080 issued 15.10.37; dd 9.10.37 to 8 E&RFTS Woodley; re-named 8 EFTS 3.9.39. Regn cld 14.11.40 as sold. Impressed as **BB661** under AM Contract No. A113018/40 dated 17.9.40; coded 'A' and remained with 8 EFTS. To 32 MU St Athan 20.11.41. To 10 FIS Woodley 21.7.42 (postwar coded 'FDT-A'). To 51 MU Lichfield 28.5.46. Sold 8.6.48 to BOAC. Regd **G-AKKU** (CofR 12007) 23.6.48 to Speedbird Flying Clubs Ltd, Denham; reverted to **G-AFBS** 1.11.49 (backdated to 23.6.48) when true identity subsequently discovered. Regd 4.49 (on change of name) to Airways Aero Associations Ltd, Denham. CofA renewed 7.6.50. Cld 4.10.52 & regd 9.10.52 to Denham Flying Club Ltd, Denham. Damaged when forced landed out of fuel Lymster's Farm, West Hyde, Rickmansworth 5.9.54; pilot JD Scarlett. Fuselage and tail damaged Denham 2.10.55 when struck by ground-looping G-AIDF. Heavy landing at Denham 6.6.56 and u/c collapsed. Badly damaged in heavy landing out of fuel Blackbushe 14.6.59; pilot BT Baker. Struck drain taxying Blackbushe 13.9.62; u/c collapsed; possibly not repaired prior to CofA lapse 25.2.63. Cld 21.7.63 & regd 31.7.63 to Terrance Patrick Hurworth Jones, Aldershot (and being rebuilt in a garage nr Blackbushe; probably moved to garage at Aldershot by 7.64). Sold .64 to Bristol Mercury Flying Group, Lulsgate as spares for G-AKPG (but not used following w/off of that aircraft on 12.11.64). Cld 1.8.64 (notified 11.3.65) & regd 15.3.65 to Gilbert Denis Durbridge-Freeman, Blackbushe. Sold 4.65 to Skyfame Ltd (for £20); restored to static display condition by Graham Johnson at Kingswood, Bristol and taken to the Skyfame Aircraft Museum, Staverton by road 19.10.65. Transferred to the Imperial War Museum, Duxford in 1977. Regn cld by the CAA 22.12.95. Refurbished to static display condition by volunteers 2001/04.

540 M.14A Hawk Trainer Mk.III CofA No.6087 issued 13.10.37 to Egyptian Army Air Force. Dd as **L202** to Almaza, Cairo 9.10.37 (although CLN recorded c/n 540 as L201).

541 M.14A Hawk Trainer Mk.III CofA No.6085 issued 13.10.37 to Egyptian Army Air Force. Dd as **L201** to Almaza, Cairo 9.10.37 (although CLN recorded c/n 541 as L202).

A Miles M.14A Magister of the Royal Egyptian Air Force. [Via P Amos]

APPENDIX 20

542 M.14A Hawk Trainer Mk.III; fitted with special Dowty u/c legs. Regd **G-AFDB** (CofR 8214) 27.11.37 to Phillips & Powis Aircraft Ltd. CofA No.6140 issued 2.12.37; dd to 8 E&RFTS Woodley 30.11.37; re-named 8 EFTS 3.9.39. Regn cld 14.11.40 as sold. Impressed as **BB662** under AM Contract No A113018/40 dated 17.9.40 and remained with 8 EFTS. To 10 FIS Woodley 21.7.42. To 45 MU Kinloss (by road) 15.9.43. To 3 SofTT Squires Gate for GI as **4557M** 21.2.44. SOC 25.5.45. To 75 MU Wilmslow by road as scrap.

543 M.14A Hawk Trainer Mk.III CofA No.6131 issued 19.11.37 to Egyptian Army Air Force. Dd as **L203** to Almaza, Cairo 16.11.37.

544 M.14A Hawk Trainer Mk.III CofA No.6137 issued 24.11.37 to Egyptian Army Air Force. Dd as **L204** to Almaza, Cairo 22.11.37.

545 M.14A Hawk Trainer Mk.III CofA No.6138 issued 24.11.37 to Egyptian Army Air Force. Dd as **L205** to Almaza, Cairo 22.11.37.

546 M.14A Hawk Trainer Mk.III CofA No.6139 issued 24.11.37 to Egyptian Army Air Force. Dd as **L206** to Almaza, Cairo 22.11.37.

547 M.14A Magister To Royal Australian Air Force as **A15-1** under Contract 687916/37 for the supply of one Miles Magister to the Australian Air Ministry. Despatched from Woodley 12.1.38 and received 18.2.38 (the engine arrived later on SS Narkunda as the contractor had removed it due to a misunderstanding). Acceptance testing carried out at 1 Air Depot 3.38 by Flt Lt GE Douglas; then issued to the Training Depot (where any HQ or other officer could fly it). Flight Testing carried out at 1 FTS Point Cook and a comprehensive report on the suitability of the type for ab initio requirements prepared. Loaned to Commonwealth Aircraft Corporation 7.38; returned to 1 FTS 23.8.38. On 19.9.38, the starboard u/c oleo was damaged in a heavy landing at Point Cook by F/O GC Hartnell. Not recorded again in the unit diary until 10.39 or after 1.40. SOC 8.12.39 and converted to **Instructional Airframe Miles No.1**. Handed over 4.40 to the RAAF Engineering School, Ascot Vale, Vic.

556 M.14A Hawk Trainer Mk.III Regd **G-AFET** (CofR 8331) 3.2.38 to Ipswich Aero Club Ltd, Ipswich. CofA No.6205 issued 4.2.38. Contractor's Flt Test Report dated 22.2.38; dd 27.2.38. Regd (CofR 8953) 22.12.38 to Thanet Aero Club Ltd, Ramsgate. At Weston-super-Mare 9.39; permit issued 19.2.40 for flight to Ipswich for storage. Regn cld 1.5.40 as sold. Impressed as **AV978** 1.5.40. To 38 MU Llandow 4.5.40. By road to Herts & Essex, Broxbourne for overhaul 28.7.41; transferred to Percival Aircraft, Luton 14.8.41. To 46 MU Lossiemouth (for use as a station hack) 17.10.41. To 5 MU Kemble 11.9.42. To 8 Grp Stn Flt, Wyton 9.1.43. To 275 Sqdn, Eglinton 12.6.43; unit to Warmwell 14.4.44, to Bolt Head 7.8.44 and to Exeter 18.10.44. Undercarriage collapsed in forced landing on Salcombe Golf Course 18.11.44. SOC.

557 M.14A Hawk Trainer Mk.III Regd **G-AFEU** (CofR 8332) 5.2.38 to Thanet Aero Club Ltd, Ramsgate. CofA No.6209 issued 17.2.38; dd 27.2.38. Spun in after loop and crashed into sea off Cliftonville 17.7.38, pilot Edmund Betts and passenger Miss Marjorie Walk both killed. Regn cld 17.7.38.

558 M.14A Hawk Trainer Mk.III Regd **G-AFEV** (CofR 8333) 5.2.38 to Exeter Aero Club Ltd, Exeter. CofA No.6210 issued 23.2.38; dd 27.2.38. Regd (CofR 8874) 9.11.38 to Weston Aero Club Ltd, Weston-super-Mare. Cld 3.2.39 & regd 11.2.39 to Thanet Aero Club Ltd, Ramsgate. Inspected by AM at Exeter 29.8.39. Crashed near Lyme Regis 30.8.39. Regn cld at census 1.12.46.

559 M.14A Hawk Trainer Mk.III Regd **G-AFEW** (CofR 8334) 5.2.38 to Plymouth & District Aero Club Ltd, Roborough. CofA No.6211 issued 23.2.38; dd 27.2.38. Believed written off in unknown circumstances. Regn cld 31.12.38. *A note in the National Archives states 'repaired by Airwork General Trading Co'; no further details.*

560 M.14A **L8051** Dd 12 FTS Grantham 14.3.38. To 32 MU St Athan 1.12.41. To 11 EFTS Perth 11.12.41. To 12 MU Kirkbride 17.8.42. To 5 MU Kemble 15.11.44. To Power Jets Ltd, Bitteswell 12.8.45. To National Gas Turbine Establishment, Farnborough 30.11.46. Sold 5.12.47 to WS Shackleton Ltd. Regd **G-AKJX** (CofR 11985) 17.12.47 to Short Bros & Harland Ltd, Rochester. Cld 18.3.48 & regd 12.4.48 to Norman West, Gravesend. CofA No.9917 issued 6.5.48. Cld 8.3.49 & regd 28.3.49 to Hugh Charles Kennard; operated by Air Kruise (Kent) Ltd, Lympne. CofA lapsed 9.6.50. Scrapped Lympne 7.54. Regn cld 5.9.54 as w/off by owner.

561 M.14A **L8052** Dd Gosport 11.3.38; op 'C' Flt. To 3 (C)OTU Chivenor 28.11.40. To 1 CRU Cowley (SAS) 19.2.41. To 5 (C)OTU Chivenor 19.8.41. To Herts & Essex, Broxbourne (SAS) 18.9.41. To 5 MU Kemble 27.11.41. To 4 FIS Cambridge 18.1.42. To 32 MU St Athan 23.3.42. To 6 FIS Worcester 9.5.42. To 7 FIS Upavon 1.8.42. Stalled off turn and flew into ground Etchilhampton, Wilts 4.6.43. SOC 15.6.43 (950:25 F/H).

562 M.14A **L8053** Dd 17 Grp Comm Flt, Lee-on-Solent 14.3.38. To Gosport (31.12.39 census). To 6 MU Brize Norton 22.3.40. To 16 EFTS Burnaston 29.6.40. To Shrager Bros, Old Warden (MI) 1.9.42. To 51 MU Lichfield 4.11.42. Sold 6.1.46 to Miles Aircraft Ltd. Test flown 8.5.46 by Hugh Kendall as L8053. To H Hennequin & Cia, Argentina with CofA No.7801 issued 24.5.46; regd **LV-XNS** 30.6.47 to Secretaria Estado Aeronautica, Buenos Aires. To Aero Club Tucuman. Accident 8.9.46; repaired. Regn cld 5.52 as wfu.

563 M.14A **L8054** Dd 2 Grp Comm Flt, Cottesmore 15.3.38. To 1 CRU Cowley (SAS) 21.5.41. To 27 MU Shawbury 1.9.41. To 457 RAAF Sqdn, Jurby 13.9.41. To 452 RAAF Sqdn (undated). To 93 Sqdn, Andreas 4.6.42. To 538 Sqdn, Hibaldstow 27.11.42. To Herts & Essex, Broxbourne for repairs 13.1.43. To 45 MU Kinloss 14.3.43. To ATA 3.2.44. To 51 MU Lichfield 20.1.45. Sold 20.6.46 to Miles Aircraft Ltd. To H Hennequin & Cia, **Argentina** with CofA No.8285 issued 28.8.46; probably regd **LV-XRX** 30.6.47 to Secretaria Estado Aeronautica, Buenos Aires. Accident 29.3.48; repaired. Accident 26.11.50; repaired. Accident 10.2.52. Regn cld 7.54 as wfu. *(Note LV-XRX is shown in Argentine register with c/n 1822 deleted, this became LV-XSO; then regd with c/n 0336 but this was LV-XQH – and LV-XRX is only unidentified Magister which fits)*

564 M.14A **L8055** Dd 2 (AC) Sqdn, Hawkinge 14.3.38. To Stn Flt Digby 21.11.39. To 611 Sqdn, Digby (30.11.39 census). Cat.B 5.6.40; to 1 CRU Cowley (SAS). To 37 MU Burtonwood 24.7.40. To 15 EFTS Carlisle 5.6.41. To 29 EFTS Clyffe Pypard 24.5.42. To 3 GTS Stoke Orchard 18.3.43. To 22 EFTS Cambridge 19.3.44. To 51 MU Lichfield 6.5.44. Sold 15.3.46 to Miles Aircraft Ltd. To H Hennequin & Cia, Argentina with CofA No.7785 issued 20.5.46; regd **LV-XNY** 30.6.47 to Secretaria Estado Aeronautica, Buenos Aires. To Aero Club Villa Maria. Accident 30.3.47; repaired. Accident 2.7.50. Regn cld 5.52 as wfu.

565 M.14A **L8056** Dd 4 (AC) Sqdn, Odiham 14.3.38. To 613 Sqdn, Odiham (30.11.39 census). To Stn Flt Middleton St George 16.6.41. To P&P (MI) 1.7.42. To 27 MU Shawbury 27.8.42. To Stn Flt Northolt 24.9.42. To 51 MU Lichfield 20.6.44. Sold 9.2.46 to Miles Aircraft Ltd. To H Hennequin & Cia, Argentina with CofA No.7708 issued 7.5.46; regd **LV-XNG** 30.6.47 to Secretaria Estado Aeronautica, Buenos Aires. To Aero Club Dolores, Bolivar. Still regd 7.8.62.

566 M.14A **L8057** Dd 13 (AC) Sqdn, Odiham 14.3.38. To RAF Wittering 9.12.38. Engine cut, crashed in forced landing while avoiding cables Finchdale, Co Durham 22.7.39 & DBR.

567 M.14A **L8058** Dd 16 (AC) Sqdn, Old Sarum 15.3.38. To Stn Flt Hucknall 16.5.39. To 45 MU Kinloss 30.11.40. To 614 Sqdn, Macmerry 13.3.41. Flying accident (Cat.B) 13.4.41; to P&P (SAS). To 54 OTU Church Fenton 26.6.41. To 32 MU St Athan 25.7.41. To 15 EFTS Carlisle 27.9.41. Flying accident (Cat.B) 22.10.43; to Atlantic Coast, Barnstaple for repairs. To 51 MU Lichfield 24.1.44. To RAE Farnborough 18.4.45. Declared surplus to requirements; to 51 MU Lichfield 11.3.48. Sold 22.4.49 to RA Short. Regd **G-ALOA** (CofR 12709) 25.4.49 to Robert Alan Short and James Henry Tattersall, Croydon/Clitheroe, Lancs. CofA No.10729 issued 27.10.49. CofA lapsed 26.10.50. Extant Elstree (7.51). Regn cld 12.9.52 by MCA. Regn cld 18.5.59 as pwfu.

568 M.14A **L8059** Dd 16 (AC) Sqdn, Old Sarum 15.3.38. To Stn Flt Hucknall 21.8.39. To 1 CRU Cowley (SAS) 7.5.41. To 54 OTU Church Fenton 25.6.41. To 135 Sqdn, Baginton 26.8.41. To 1456 Flt, High Ercall 6.6.42. To 535 Sqdn, High Ercall (from formation of squadron on 2.9.43). To 51 MU Lichfield 8.2.43. Sold 27.5.46 to Miles Aircraft Ltd. Test flown 15.7.46 by Hugh Kendall (recorded as 'P8059'). To H Hennequin & Cia, Argentina with CofA No.8192 issued 5.8.46; regd **LV-XRR** 30.6.47 to Secretaria Estado Aeronautica, Buenos Aires. Accident 17.4.47; repaired. Accident 23.5.47 (18% damaged); repaired. Regn cld 6.53 as wfu.

569 M.14A **L8060** Dd 26 (AC) Sqdn, Catterick 21.3.38. Engine lost power, stalled in forced landing and spun into ground nr Richmond, Yorks 16.8.38; F/O Edward J Boyle and Dr William D Walker (a former Australian flying doctor) killed. SOC 13.11.38 (82:40 F/H).

570 M.14A **L8061** Dd 59 (AC) Sqdn, Old Sarum 15.3.38. To Stn Flt Andover 11.1.40. To 5 MU Kemble 1.2.41. To HQ ACC Farnborough 17.2.41. To 110 Wing, Filton 20.2.41. To 6 AACU Ringway 25.4.41. Flying accident 9.7.41. To 27 MU Shawbury 7.9.41. Forced landed nr Ringway on ferry flight by TFPP ATA at 0955 hrs 29.9.41 with engine failure; Cat.B 15.10.41. To 46 MU Lossiemouth (31.12.41 census). To 25 Sqdn, Church Fenton 26.1.43. To AFDU Wittering 19.5.43. To 51 MU Lichfield 12.4.44. Sold 30.5.46 to Miles Aircraft Ltd. Test flown 24.7.46 by Hugh Kendall as L8061. To H Hennequin & Cia, Argentina with CofA No.8214 issued 7.8.46; regd **LV-XSF** 30.6.47 to Secretaria Estado Aeronautica, Buenos Aires. Regd 17.6.60 to Aero Club Carhue, Carhue. Still regd 5.67.

571 M.14A **L8062** Dd 1 ASU Waddington 15.3.38. To 5 MU Kemble 14.2.39. To 25 Sqdn, Wittering 23.4.41. To 32 MU St Athan 12.10.41. To Stn Flt Wittering 20.10.41. To 25 Sqdn, Wittering (31.12.41 census). To 51 MU Lichfield 10.5.45. Sold 2.9.46 to Miles Aircraft Ltd. CofA No.8597 issued 11.10.46 to Middle East Airlines. Regd in Lebanon as **LR-AAK** .46 to Middle East Airlines. Swung on landing Beirut 17.9.48 damaging starboard wing and u/c. Written off when swung on landing Beirut 26.5.49 damaging wings, fuselage, u/c and propeller. Regd **OD-AAK** 7.7.49 but ntu. Regn cld .50.

572 M.14A **L8063** Dd 1 ASU Waddington 21.3.38. To 5 MU Kemble 8.2.39. To 22 EFTS Cambridge 26.8.40. To 24 EFTS Luton 30.8.41. To 2 FIS Montrose 2.2.42. Cat.B 15.5.42; to P&P for repairs. To 46 MU Lossiemouth 23.7.42. To ATA 19.5.43. SOC 7.9.44 after Cat.E1 flying accident. To Miles Aircraft Ltd. To H Hennequin & Cia, Argentina with CofA No.7444 issued 11.3.46; regd **LV-XMI** 30.6.47 to Secretaria Estado Aeronautica, Buenos Aires. Accidents on 28.9.47 & 21.7.50; repaired. To Aero Club Rosario. Crashed nr Est San Pedro 24.4.56; pilot killed. Regd 8.8.62 to Aero Club Paso de los Libres, Paso de los Libres, Corrientes.

573 M.14A **L8064** Dd 1 ASU Waddington 21.3.38. To 5 MU Kemble 14.2.39. To 22 EFTS Cambridge 26.8.40. (Recorded at 3 FIS (31.12.41 census), but unit did not then exist). To 32 MU St Athan 1.4.42. To 4 FIS Cambridge 10.4.42. To 22 EFTS Cambridge 13.5.43. SOC (MR) 18.3.44.

574 M.14A **L8065** Dd 1 ASU Waddington 21.3.38. To 8 EFTS Woodley 25.9.40. To 32 MU St Athan 13.2.42. To 10 FIS Woodley 21.7.42. To 3 GTS Stoke Orchard 8.9.43. To 51 MU Lichfield 23.4.44. Sold 10.3.47 to Miles Aircraft Ltd. CofA No.9832 issued 22.1.48 to **Thai AF**.

575 M.14A **L8066** Dd 1 ASU Waddington 23.3.38. To 27 E&RFTS Tollerton 16.8.38. Engine cut, hit hedge in forced landing near Hucknall 6.7.39 and DBR. SOC 31.7.39 by RAF Hucknall.

576 M.14A **L8067** Dd 1 ASU Waddington 23.3.38. To 5 MU Kemble 20.2.39. To Stn Flt Benson 3.7.40. To Stn Flt Hucknall 2.9.40. To 1 CRU Cowley (SAS) 19.1.41. To 24 EFTS Luton 31.3.41. To Herts & Essex, Broxbourne (B) 21.10.41. To 48 MU Hawarden 10.12.41. To 19 Sqdn, Perranporth 18.8.42. To 27 MU Shawbury 6.10.42. To 47 MU Sealand 3.2.43. Despatched to ME 4.2.43 on SS Blommersdyk; arr 27.4.43. To 216 Grp Comm Flt, Heliopolis. SOC 6.9.45.

577 M.14A **L8068** Dd 1 ASU Waddington 23.3.38. To 5 MU Kemble 1.2.39. To 24 EFTS Luton 17.10.40. To 1 CRU Cowley (SAS) 20.3.41. To 9 MU Cosford 21.5.41. To ATA Training Unit, White Waltham 9.8.41. Cat.B, to Herts & Essex, Broxbourne 24.6.43. To 45 MU Kinloss 25.9.43. To 51 MU Lichfield 1.2.44. Allotted to Miles by MAP for Chile in early 1945, allotment cancelled ("Marshalls 14.4.45"). Sold 16.12.46 to The Wiltshire School of Flying. Regd **G-AITV** (CofR 10964) 11.11.46 to Anna Valley Motors (Salisbury) Ltd, Salisbury (based Thruxton). CofA No.10280 issued 6.9.48. Regn cld 20.9.48 as sold abroad. Regd in French Morocco as **F-OAAQ** (CofR 246081) 1.2.49 to Pierre Macheras, Oujda. Regd (by .54) to Societe Maroc Air Service, Rabat. Regn cld 12.8.59 as sold abroad. *Note: Magister '8068' was test flown by Hugh Kendall 6.1.48; but was possibly L8065.*

578 M.14A **L8069** Dd 1 ASU Waddington 23.3.38. To 5 MU Kemble 1.2.39. To 1 AACU Farnborough 26.3.41. To 20 MU Aston Down 19.11.41. To 6 FIS Staverton 22.2.42. To 7 FIS Upavon 1.8.42. Control lost, flew into ground 1m NW of Collingbourne Kingston, Wilts 15.7.43. SOC 23.7.43.

579 M.14A **L8070** Dd 1 ASU Waddington 23.3.38. To 5 MU Kemble 1.2.39. To 24 EFTS Luton 17.10.40. Collided with Tiger Moth BB738 over St Albans, Herts 10.12.40; Sgt BP Buchegger & Acting LA GRD Tait killed. SOC 19.12.40.

580 M.14A **L8071** Dd 1 ASU Waddington 23.3.38. To 5 MU Kemble 1.2.39. To 234 Sqdn, St. Eval 7.11.40. Flying accident 2.4.41. SOC 9.4.41.

581 M.14A **L8072** Dd 27 MU Shawbury 28.3.38. To Stn Flt Halton 5.4.40. To 46 MU Lossiemouth 21.7.40. To 5 EFTS Meir 1.5.41. To CFS Upavon 11.1.42. To ECFS Hullavington 14.5.42. Cat.B 5.7.42, to P&P for repairs. To 27 MU Shawbury 28.8.42. To 234 Sqdn, Portreath 19.9.42. To 39 MU Colerne 27.10.42. To 51 MU Lichfield 2.11.42. Sold 6.1.46 to Miles Aircraft Ltd. To H Hennequin & Cia, Argentina with CofA No.7581 issued 27.3.46; regd **LV-XMW** 30.6.47 to Secretaria Estado Aeronautica, Buenos Aires. Collided with LV-XRE 1.6.47 (20% damaged); repaired. Regd 8.8.61 to Aero Club Junin, Buenos Aires. Still regd 5.67.

582 M.14A **L8073** Dd 27 MU Shawbury 28.3.38. To 15 EFTS Carlisle 11.8.40. Stalled in forced landing and crashed Lowther Castle, Cumberland 17.9.41. SOC 21.9.41.

583 M.14A **L8074** Dd 27 MU Shawbury 28.3.38. To 48 MU Hawarden 30.6.40. To Franco-Belgian Air Training School, Odiham 21.3.41. To 27 MU Shawbury 16.6.41. To 15 EFTS Carlisle 24.6.41. To 29 EFTS Clyffe Pypard 26.5.42. To 10 FIS Woodley 26.3.43. To 51 MU Lichfield 17.2.44. Sold 10.3.47 to Miles Aircraft Ltd. CofA No.9835 issued 2.2.48 to **Thai AF**.

584 M.14A **L8075** Dd 27 MU Shawbury 31.3.38. To 256 Sqdn 20.4.40 (unconfirmed – Sqdn only reformed 23.11.40 – at Catterick). To Stn Flt Pembrey 4.11.41. To 256 Sqdn, Woodvale (by 3.1.43). Cat B, to Atlantic Coast, Barnstaple (MR) 6.10.43. To 51 MU Lichfield 3.12.43. Sold 22.4.49 to RA Short. Regd **G-ALNY** (CofR 12707) 21.4.49 to Robert Alan Short and James Henry Tattersall, Croydon/Clitheroe, Lancs. Not converted. Regn cld 12.9.52 by MCA. Regd 21.1.53 to Robert Alan Short, Croydon. Burnt Elstree 5.11.53. Regn cld 18.5.59 as pwfu.

585 M.14A **L8076** Dd 27 MU Shawbury 31.3.38. To 15 EFTS Carlisle 11.8.40. Damaged when bounced on landing Kirkpatrick 6.8.41; to P&P (SAS) 6.8.41. To 5 MU Kemble 17.11.41. To 234 Sqdn, Ibsley 28.11.41. To 418 RCAF Sqdn, Debden 7.12.41. Flying accident 11.8.42; to Herts & Essex, Broxbourne for repairs. To 27 MU Shawbury 9.10.42. To 47 MU Sealand 31.1.43. To 76 MU Wroughton for packing 12.5.43. To **Turkey** 17.5.43; arr in ME 14.6.43.

586 M.14A **L8077** Dd 27 MU Shawbury 31.3.38. To 16 EFTS Burnaston 17.9.40. To 5 GTS Shobdon 1.5.43. To 51 MU Lichfield 10.4.44. Sold 22.11.46 to Miles Aircraft Ltd. CofA No.9398 issued 28.5.47 to Raoul Duval. Regd in France as **F-BDPJ** (CofR 20004) 30.8.47 to Aeronautique Havraise 'Jean Maridor', Le Havre. CofA suspended 7.1.58. Regd 21.5.58 to Aero Club de l'Aube, Troyes, but not re-certified. Regn cld 22.9.71 as destroyed.

587 M.14A **L8078** Dd 27 MU Shawbury 30.3.38. To 1 SofTT Halton 9.4.40. To 46 MU Lossiemouth 21.7.40. To 15 EFTS Carlisle 14.2.41. Accident 4.9.41; to P&P 11.9.41 and SOC 22.9.41.

588 M.14A **L8079** Dd 27 MU Shawbury 30.3.38. To 16 EFTS Burnaston 22.11.40. Bounced on landing, stalled and hit ground Burnaston 23.10.41 & DBR. SOC 3.11.41.

589 M.14A **L8080** Dd 27 MU Shawbury 4.4.38. To 48 MU Hawarden 30.6.40. To 15 EFTS Carlisle 15.11.40. To 29 EFTS Clyffe Pypard 24.5.42. To 21 EFTS Booker 20.3.43. To 51 MU Lichfield 1.2.44. Sold 28.2.47 to Southern Aircraft (Gatwick) Ltd. Regd **G-AJGM** (CofR 11277) 5.2.47 to Southern Aircraft (Gatwick) Ltd, Gatwick. CofA No.9186 issued 3.7.47. Cld 1.7.47 & regd 14.7.47 to Herts & Essex Aero Club (1946) Ltd, Broxbourne. Cld 26.1.48 & regd 3.2.48 to Denham Air Services Ltd, Denham. Overshot on landing Elstree 21.9.48 and struck fence damaging fuselage and u/c, possibly not repaired. Regd 25.1.49 to Speedbird Flying Clubs Ltd, Denham. Regd 4.49 (on name change) to Airways Aero Associations Ltd, Denham. (Unconfirmed report as stored at Heston 4.53). Regn cld 21.9.55 as reduced to produce. Remains at Croydon (.56) and to Burnaston for spares .56. Wings reported at Croydon 8.60 and may have been the Magister wing at the Timber Research & Development Association Laboratory, High Wycombe in 3.64.

APPENDIX 20

590 M.14A **L8081** Dd 27 MU Shawbury 4.4.38. To 11 OTU Bassingbourne 26.3.41. To 605 Sqdn, Honiley 29.9.41. To 257 Sqdn, Honiley 30.11.41. Engine cut, hit obstacles in forced landing Westwood Heath, near Kenilworth 12.2.42. SOC 1.3.42.

591 M.14A **L8082** Dd 27 MU Shawbury 4.4.38. To 48 MU Hawarden 30.6.40. To Franco-Belgian Air Training School, Odiham 24.11.40. Flying accident 21.5.41, to 1 CRU Cowley (SAS) 2.6.41. To 46 MU Lossiemouth 26.7.41. To 2 FIS Montrose 10.1.42. To 51 MU Lichfield 19.11.43. Sold 1.1.48 to Adie Aviation Ltd. Regd **G-AKPM** (CofR 12125) 12.1.48 to Adie Aviation Ltd, Croydon. Cld 29.9.48 & regd 19.10.48 to Hugh Charles Kennard, operated by Air Kruise (Kent) Ltd, Lympne. CofA No.10044 issued 4.4.49. Struck car on landing and crashed Lympne 22.10.50; pilot unhurt but 4 in car slightly injured. Regn cld 19.3.59 as pwfu.

592 M.14A **L8083** Dd 27 MU Shawbury 4.4.38. To 48 MU Hawarden 17.7.40. To 64 Sqdn, Hornchurch 9.4.41. To 32 MU St Athan 13.10.41. To 134 Sqdn, Catterick 19.12.41. To 133 Sqdn, Eglinton 28.12.41; unit to Kirton-in Lindsey 2.1.42, to Biggin Hill 3.5.42. To 332 Sqdn, North Weald 24.7.42. To 27 MU Shawbury 9.10.42. To 307 Sqdn, Fairwood Common 29.5.43. To 51 OTU Cranfield 3.7.43. To 39 MU Colerne 8.4.45. To 5 MU Kemble 4.4.46. Sold 22.11.46 to Southern Aircraft (Gatwick) Ltd. Regd **G-AIZL** (CofR 11105) 30.12.46 to Southern Aircraft (Gatwick) Ltd, Gatwick. CofA No.8873 issued 7.7.48. Cld 6.7.48 & regd 14.7.48 to John Victor Green, Driffield. CofA lapsed 6.7.49. Cld 1.6.50 & regd 3.10.50 to Wright Aviation Ltd, t/a Liverpool Flying Club, Speke. CofA renewed 26.10.50. Damaged propeller, u/c and wing in heavy landing Squires Gate 2.9.51; pilot Sidebottom. Returned to Speke and stored; overhaul commenced (9.53) using wings from G-AKRM but abandoned. Cld 12.2.54 & regd 16.2.54 to Dragon Airways Ltd, Speke. Regn cld 9.9.55 as wfu. Remains extant Speke 3.58.

593 M.14A **L8084** Dd 27 MU Shawbury 5.4.38. To 15 EFTS Carlisle 11.8.40. Damaged in heavy landing 22.12.40, repaired with new port wing. Damaged in heavy landing 24.2.41 which entailed more repairs to port wing. Damaged in heavy landing 25.3.41, repairs to port centre section. Abandoned after structural failure of port wing centre section on pull-up into climb and crashed, Park Broom, 3 ml NE of Carlisle 25.6.41; pilot Cpl Habela. SOC 2.7.41.

594 M.14A **L8085** Dd 27 MU Shawbury 5.4.38. To 5 EFTS Meir 23.9.40. Control lost, spun into ground, Wood Farm, Leigh, Shropshire 9.4.41; LAC LH Smith killed. SOC 29.4.41.

595 M.14A **L8086** Dd 27 MU Shawbury 8.4.38. To 5 EFTS Meir 25.9.40. To Atlantic Coast, Barnstaple (MI) 20.12.41. To 20 MU Aston Down 13.2.42. To 41 OTU Old Sarum 9.3.42. To Atlantic Coast, Barnstaple (MR) 6.1.44; re-assessed Cat.E and SOC 7.2.44.

596 M.14A **L8087** Dd 27 MU Shawbury 8.4.38. To 5 EFTS Meir 11.1.41. To Atlantic Coast, Barnstaple (MI) 20.12.41. To 37 MU Burtonwood 18.2.42. To 46 MU Lossiemouth 17.7.42. To AFDU Wittering 19.5.43. Hit hedge in forced landing after engine cut Kempston, Beds 12.3.44. SOC 23.3.44.

597 M.14A **L8088** Dd 27 MU Shawbury 8.4.38. To Stn Flt Yeadon 11.6.40. Cat.B, to Shrager Bros, Old Warden for repairs 10.12.41. To 46 MU Lossiemouth 30.1.42. To Stn Flt Dyce 4.7.42. To Stn Flt Vaagar (31.12.42 census). Flown by W/C Stewart from Vaagar 14.6.43 on a 'Recco'. To Stn Flt Leuchars 30.6.44. To Stn Flt Beccles 2.9.44. To Stn Flt Swingfield 13.9.44. SOC Cat.E 7.12.44. To Miles Aircraft Ltd. To Portuguese AF **198** with CofA No.7649 issued 9.4.46; later **1209**. Regd **CS-AFM**, wfu 12.62.

598 M.14A **L8089** Dd 27 MU Shawbury 12.4.38. To 601 Sqdn, Manston 8.5.41. To Stn Flt Manston 30.6.41. To 616 Sqdn, Westhampnett 25.8.41. Flying accident (Cat.B) 14.3.42; repaired. To 32 MU St Athan 23.5.42. To Stn Flt Wittering 27.5.42. Cat B (MI) 1.11.42; to Atlantic Coast, Barnstaple 8.12.44. To 51 MU Lichfield 6.2.43. Sold by MAP 13.4.45 to Miles Aircraft Ltd for **Chile**. To Marshall's Flying School, Cambridge 14.4.45 for refurbishment, test flown as **U-0265**.

599 M.14A **L8090** Dd 27 MU Shawbury 12.4.38. To 48 MU Hawarden 30.6.40. To 15 EFTS Carlisle 24.11.40. To 29 EFTS Clyffe Pypard 30.5.42. To 5 GTS Shobdon 22.3.43. Hit HT cables low flying Ludlow 20.4.43. SOC 3.5.43.

600 M.14A **L8091** Dd 27 MU Shawbury 12.4.38. To 48 MU Hawarden 11.6.40. To Franco-Belgian Air Training School, Odiham 8.11.40. Flying accident 21.3.41, to P&P (SAS) 12.6.41. To 46 MU Lossiemouth 19.7.41. To 15 EFTS Carlisle 14.11.41. To 21 EFTS Booker 8.6.42. To 18 (P)AFU Church Lawford 13.12.44. To 51 MU Lichfield 19.5.45. Sold 3.12.46 to Miles Aircraft Ltd. CofA No.9404 issued 28.5.47. To France & regd **F-BDPP** 27.6.47 to Octave Delasorne, Neuilly-sur-Seine. Regd 25.11.48 to Aero Club Langrois, Langres. Regd 20.1.49 to Service de l'Aviation Legere et Sportive (SALS), Nivert. Regn cld 16.4.51 as scrapped.

601 M.14A **L8092** Dd 27 MU Shawbury 12.4.38. To 5 EFTS Meir 25.9.40. To CFS Upavon 11.12.41. To 27 MU Shawbury 29.6.42. To 611 Sqdn, Redhill 19.9.42. To 530 Sqdn, Hunsdon 12.10.42. To 51 MU Lichfield 10.2.43. Sold 26.2.48 to Gloster Flying Club. Regd **G-AKMP** (CofR 12053) 27.11.47 to Laurence Dudley Trappitt, Whyteleafe, Surrey. Cld 22.7.48 & regd 14.10.48 to Robert Alan Short, Derek John Hayles & Martha Marobeen, Croydon/Portsmouth. Cld 5.11.48 & regd 31.1.49 to Robert Alan Short, Croydon. CofA No.10408 issued 4.2.49. Regn cld 2.9.49 as sold abroad - to **Egyptian AF**.

602 M.14A **L8093** Dd 27 MU Shawbury 20.4.38. To 16 EFTS Burnaston 3.7.40. To P&P (SAS) 29.10.40. To Stn Flt Odiham 19.1.41 for Franco-Belgian Air Training School. To 24 EFTS Luton 6.7.41. To Atlantic Coast, Barnstaple (MI) 20.12.41. To 27 MU Shawbury 16.4.42. To 32 MU St Athan 28.4.42. To 219 Sqdn, Tangmere 28.4.42. To 604 Sqdn, Church Fenton 28.4.43. Cat.B (MR) 18.5.44 and SOC 30.5.44.

603 M.14A **L8094** Dd 27 MU Shawbury 20.4.38. To Stn Flt Dyce 30.6.40. Cat B (MI) 30.7.42, to Shrager Bros. Old Warden 7.8.42. To 27 MU Shawbury 9.10.42. To 47 MU Sealand 27.2.43. Despatched to ME on SS City of Agra 4.5.43; arr 5.6.43. SOC 31.5.45.

604 M.14A **L8095** Dd 27 MU Shawbury 21.4.38. To 5 EFTS Meir 21.9.40. To CFS Upavon 10.2.42. Cat.B, to P&P for repairs 7.4.42. To 27 MU Shawbury 1.7.42. To 243 Sqdn, Ouston 20.7.42. To 51 MU Lichfield 13.11.44. Sold by MAP 13.4.45 to Miles Aircraft Ltd for Chile. To Marshall's Flying School, Cambridge 14.4.45 for refurbishment, test flown as **U-0266**. Regd in Chile as **CC-PAE-0180** 26.9.45 to Agustin R Edwards Budge. Regd 28.10.46 to John Hurleston Leche. Regd **CC-PPA** 24.11.48 to Patricio Arrigorriaga. Regd **CC-PTA 0097** 11.8.49 to Carlos Jose Grangila Litras. Regn cld 20.6.67.

605 M.14A **L8127** Dd 27 MU Shawbury 21.4.38. To 257 Sqdn, Martlesham Heath 26.11.40. To P&P 29.11.41. To 32 MU St Athan 23.3.42. To 16 EFTS Burnaston 7.4.42. To 32 Sqdn, West Malling 6.5.42. To 51 MU Lichfield 15.12.42. Sold 4.2.46 to Miles Aircraft Ltd. Test flown 8.4.46 by Hugh Kendall as L8127. To H Hennequin & Cia, Argentina with CofA No.7662 issued 24.4.46; regd **LV-XOQ** 30.6.47 to Secretaria Estado Aeronautica, Buenos Aires. To Aero Club Pigue. Accident 21.10.46 (40% damaged). Regn cld 11.50 as wfu.

606 M.14A **L8128** Dd 27 MU Shawbury 25.4.38. To 8 EFTS Woodley 10.2.41. To 10 FIS Woodley 21.7.42. To 3 GTS Stoke Orchard 5.9.43. To Stn Flt Dumfries 6.9.43. To 29 Grp Comm Flt, Dumfries (31.12.43 census). To 1 (O)AFU Wigtown 30.7.45. To 51 MU Lichfield 3.10.45. Sold 18.11.46 to The Wiltshire School of Flying Ltd. Regd **G-AIUD** (CofR 10972) 11.11.46 to The Wiltshire School of Flying Ltd, Thruxton. CofA No.9487 issued 3.10.47. Sold 22.1.49 to Robert Alan Short (but not regd). Regn cld 2.9.49 as sold abroad – to **Egyptian AF**.

607 M.14A **L8129** Dd 27 MU Shawbury 22.4.38. To 5 EFTS Meir 24.4.41. Hit tree on take-off and crashed Meir 28.7.41. SOC 6.8.41.

608 M.14A **L8130** Dd 27 MU Shawbury 4.5.38. To Stn Flt Debden 4.6.41. To 611 Sqdn, Drem 17.5.42, to Kenley with unit 3.6.42. Stalled during low aerobatics and crashed Woldingham, Surrey 5.7.42. SOC 11.7.42.

609 M.14A **L8131** Dd 27 MU Shawbury 25.4.38. To 5 EFTS Meir 24.3.41. Hit tree on approach and crashed Abbots Bromley 15.10.41; pilot LAC R Dunne. SOC 31.10.41.

610 M.14A **L8132** Dd 27 MU Shawbury 25.4.38. To 79 Sqdn, Pembrey 23.4.41. To 132 Sqdn, Skeabrae 14.3.42. To 3 Del Flt, High Ercall 7.5.43. To Stn Flt Acklington 30.11.43. To 1 SofTT Halton 25.2.44 for GI as **4526M** (allotted 8.2.44). SOC 26.7.44.

611 M.14A **L8133** Dd 27 MU Shawbury 10.5.38. To 16 EFTS Burnaston 20.11.40. Hit trees on overshoot Derby 23.6.42. SOC 2.7.42.

612 M.14A **L8134** Dd 27 MU Shawbury 26.4.38. To 24 Sqdn, Hendon 27.6.38. To Stn Flt Yeadon 18.10.39. To 607 Sqdn, Drem 13.3.41. Spun into ground after stalling on recovering from loop Skitten 1.5.41.

613 M.14A **L8135** Dd 27 MU Shawbury 26.4.38. To 24 Sqdn, Hendon 27.6.38. To 5 MU Kemble 10.12.39. To 8 EFTS Woodley 25.9.40. To 10 FIS Woodley 21.7.42. To 27 MU Shawbury 12.9.43. To 3 (P)AFU South Cerney 23.5.44. To 5 MU Kemble 14.2.46. Sold 19.11.46 to The Wiltshire School of Flying Ltd. Regd **G-AIUB** (CofR 10970) 11.11.46 to Anna Valley Motors (Salisbury) Ltd, Salisbury (based Thruxton). CofA No.8767 issued 27.3.47. Cld 19.4.47 & regd 24.7.47 to Universal Flying Services Ltd, Fairoaks. Crashed on take-off Cowes, Isle of Wight 18.7.50; AP Cooke killed. Regn cld 18.7.50 as destroyed.

614 M.14A **L8136** Dd Stn Flt Wittering 25.4.38. Control lost, spiralled into ground Shillington, Beds 11.11.38. To 13 MU Henlow after crash (possibly as instructional airframe?). SOC 26.1.41.

615 M.14A **L8137** Dd Stn Flt Thorney Island 2.5.38. Destroyed in air raid Thorney Island 16.8.40. SOC 3.9.40 as 'burnt out'.

616 M.14A **L8138** (CLN recorded dd 27 MU Shawbury 26.4.38 but unlikely). Dd to 1 AACU Farnborough 5.5.38. To W Flt, 1 AACU Kidsdale (5.41). To Stn Flt Langham (undated). To RAF Lasham 1.6.43. To 613 Sqdn, Lasham 8.6.44. To Stn Flt Lasham 4.1.45. To 51 MU Lichfield 23.3.45. Sold 11.12.46 to The Wiltshire School of Flying Ltd. Regd **G-AITY** (CofR 10967) 11.11.46 to Anna Valley Motors (Salisbury) Ltd, Salisbury (based Thruxton). CofA No.8881 issued 24.8.48. Cld 24.8.48 & regd 25.8.48 to John Haptist Barold Modica, London N10 (based Thruxton). Regn cld 23.6.49 as sold abroad. Regd in Italy as **I-AITY**. Crashed Valle Brembana, nr Bergamo 28.7.54.

617 M.14A **L8139** Dd 27 MU Shawbury 30.4.38 (although CLN recorded to Thorney Island same date). To 5 EFTS Meir 21.11.40. Engine cut during forced landing while lost, hit tree nr Leek, Staffs 20.2.41. SOC 30.4.41.

618 M.14A **L8140** Dd 27 MU Shawbury 4.5.38. To 16 EFTS Burnaston 25.3.41. To 32 MU St Athan 24.1.42 (unconfirmed and probably retained at 16 EFTS). To 27 MU Shawbury 2.9.42. SOC 25.2.44 Cat.E (MR).

619 M.14A **L8141** Dd 7 ATS Acklington 9.5.38. To Stn Flt Hucknall 10.5.38. Engine cut on ferry flight, overshot forced landing Hasland, near Chesterfield, Derbyshire 2.5.38 (note discrepancy with delivery date). To 1 SofTT Halton for GI as **1099M** 13.7.38. SOC 20.10.40 as RTP at Halton.

620 M.14A **L8142** Dd 27 MU Shawbury 4.5.38. To 16 EFTS Burnaston 28.5.41. Hit trees in forced landing while lost Rochester, Kent 25.8.41. To P&P (SAS) but SOC 6.9.41.

621 M.14A **L8143** Dd 27 MU Shawbury 7.5.38. To 16 EFTS Burnaston 13.8.40. Flying accident 21.8.41, to P&P (SAS) 25.8.41. To 46 MU Lossiemouth 2.11.41. To 2 FIS Montrose 10.4.42. To Shrager Bros. Old Warden (MI) 29.8.42. To 51 MU Lichfield 12.10.42. Sold 26.2.46 to Miles Aircraft Ltd. To H Hennequin & Cia, Argentina with CofA No.7661 issued 24.4.46; regd **LV-XOP** 30.6.47 to Secretaria Estado Aeronautica, Buenos Aires. To Aero Club Mendoza. Accident 11.2.47 (15% damaged); repaired. Regn cld 7.54 as wfu.

622 M.14A **L8144** Dd 27 MU Shawbury 10.5.38. To 5 EFTS Meir 25.9.40. To 16 EFTS Burnaston 26.1.42. To 45 MU Kinloss 27.5.43. To 51 MU Lichfield 24.1.44. Sold 29.11.46 to The Wiltshire School of Flying Ltd. Regd **G-AITR** (CofR 10960) 11.11.46 to Anna Valley Motors (Salisbury) Ltd, Salisbury (based Thruxton) (regd as ex R8144). CofA No.10362 issued 19.3.49. Sold 8.6.49 to Robert Alan Short, Croydon (but not regd). Regn cld 2.9.49 as sold abroad - to **Egyptian AF**.

623 M.14A **L8145** Dd 27 MU Shawbury 10.5.38. To 8 EFTS Woodley 10.2.41. To 10 FIS Woodley 21.7.42. To 9 (P)AFU Errol 15.10.43. To 5 (P)AFU Ternhill 1.7.45. To 21 EFTS Booker 13.7.45. To Staff College White Waltham 15.7.45. To TTC Comm Flt, White Waltham 14.8.45. To Staff College White Waltham 13.9.45. To 51 MU Lichfield 18.1.46. Sold 1.1.48 to WA Rollason Ltd. Regd **G-AKRI** (CofR 12145) 20.1.48 to WA Rollason Ltd, Croydon. CofA No.10060 issued 28.4.48. Regn cld 14.5.48 as sold abroad. Regd in Ireland as **EI-ADU** to Weston Ltd, Leixlip (regn applied for 17.4.48 but NTU – for owner Collistown Flying Club); dd ex Liverpool 17.5.48. Crashed nr Weston airfield, Leixlip, Co Kildare 23.5.48 prior to formal regn; pilot Arthur Cowan killed.

624 M.14A **L8146** Dd LRDU Upper Heyford 10.5.38. To 12 Grp Comm Flt, Old Sarum 17.10.38. To Westland Aircraft Ltd, Yeovil for MAP Overseer 2.5.39. SOC 25.5.44 Cat.E (MR).

625 M.14A **L8147** Dd 27 MU Shawbury 10.5.38. To 16 EFTS Burnaston 13.8.40. To 1 CRU Cowley for repairs 26.8.40. To 48 MU Hawarden 29.9.40. To 5 EFTS Meir 4.2.41. Crashed on landing Meir 3.9.41; Sgt Gadd & LAC Tomlin slightly injured. To P&P (SAS) 7.9.41 but SOC 29.9.41.

626 M.14A **L8148** Dd 27 MU Shawbury 18.5.38. To 48 MU Hawarden 14.6.40. To Franco-Belgian Air Training School, Odiham 14.11.40. To P&P (SAS) 1.1.41. To 60 OTU Leconfield 18.5.41. To 51 MU Lichfield 8.8.41. To 60 OTU East Fortune 29.8.41. To 32 MU St Athan 4.6.42. To 56 OTU/1 TEU Tealing 14.11.43. SOC 6.8.44 Cat. E1.

627 M.14A **L8149** Dd 27 MU Shawbury 19.5.38. To 5 EFTS Meir 21.9.40. To CFS Upavon 28.1.42. To 27 MU Shawbury 23.5.42. To 16 EFTS Burnaston 16.9.42. To 45 MU Kinloss 27.5.43. To 51 MU Lichfield 1.2.44. Sold 9.2.46 to Miles Aircraft Ltd. To H Hennequin & Cia, Argentina with CofA No.7663 issued 24.4.46; regd **LV-XON** 30.6.47 to Secretaria Estado Aeronautica, Buenos Aires. To Centro Universitario de Aviacion. Accident 10.3.50. Regn cld 3.52 as wfu.

628 M.14A **N2259** (replacement for unbuilt L6917). Dd 8 E&RFTS Woodley 23.5.38; renamed 8 EFTS 3.9.39; coded "13". To 1 CRU Cowley (SAS) 7.11.40. To 27 MU Shawbury 20.2.41. To Franco-Belgian Air Training School, Odiham 25.3.41. To Herts & Essex, Broxbourne for repairs 13.6.41. To 9 MU Cosford. To 15 EFTS Carlisle 16.8.41. To 29 EFTS Clyffe Pypard 26.5.42. To 3 GTS Stoke Orchard 18.3.43. To 51 MU Lichfield 23.4.44. Sold 10.3.47 to FK McIntyre. Regd **G-AJHB** (CofR 11291) 11.2.47 to Frank Keiller McIntyre, Darlington. CofA No.9054 issued 10.6.47. Cld 22.7.47 & regd 30.7.47 to Darlington & District Aero Club Ltd, Croft. CofA lapsed 9.6.48 and wfu Croft. Scrapped Sherburn-in-Elmet in 1957. Regn cld 24.3.59 as pwfu.
Note: the CofA ledger shows c/n 628 as going to Argentina with CofA No.7736 issued 9.5.46 but this should be c/n 811.

629 M.14A **L8151** Dd 27 MU Shawbury 12.5.38. To 616 Sqdn, Leconfield 10.8.40. To 65 Sqdn, Kirton-in-Lindsey 12.3.41. To 616 Sqdn, Kirton-in-Lindsey (undated). Flew into ground Thoresby Bridge, Lincs 15.11.41, both pilots believed that the other was in control of the aircraft. SOC 21.11.41.

630 M.14A **L6917** While CLN recorded dd 8 E&RFTS Reading (with no delivery date) the record states 'Not built'. Replaced by c/n 628 N2259.

631 M.14A **L6918** Dd 8 E&RFTS Woodley 20.12.37; renamed 8 EFTS 3.9.39. To 15 EFTS Carlisle 17.8.40. Collided with P2470 and crashed Newbie, Dumfries 5.5.41; LAC Potter killed. SOC 13.5.41.

632 M.14A **L6919** Dd 8 E&RFTS Woodley 29.12.37; renamed 8 EFTS 3.9.39. Damaged; to P&P for repairs 12.6.40. To 8 MU Little Rissington 9.10.40. To 24 EFTS Luton 8.11.40. To P&P (SAS) 2.1.41. To 403 RCAF Sqdn, Baginton 25.5.41. To Atcham 11.10.41. To 46 MU Lossiemouth 14.7.42. To ATA 8.9.44. To 51 MU Lichfield 17.10.44. Sold 13.3.46 to Miles Aircraft Ltd (on overhaul erroneously referred to as c/n 623). To H Hennequin & Cia, Argentina with CofA No.7870 issued 11.6.46; regd **LV-XQO** 30.6.47 to Secretaria Estado Aeronautica, Buenos Aires. To Aero Club Tres Arroyos. Regd 7.8.62 to Aero Club Mendoza, Lujan de Cuyo. Still regd 5.67.

633 M.14A Hawk Trainer Mk.III (Magister) CofA No.6157 issued 13.12.37 to Egyptian Army Air Force. Dd to Almaza, Cairo 11.12.37; became **L207**. (No serial recorded by CLN)

634 M.14A Hawk Trainer Mk.III (Magister) CofA No.6173 issued 21.12.37 to Egyptian Army Air Force. Dd to Almaza, Cairo 11.12.37; became **L208**. (CLN recorded as L207).

APPENDIX 20

635   M.14A Hawk Trainer Mk.III (Magister)   CofA No.6174 issued 21.12.37 to Egyptian Army Air Force.   Dd to Almaza, Cairo 18.12.37; became **L209**.   (CLN recorded as L208).

636   M.14A Hawk Trainer Mk.III (Magister)   CLN recorded as L209 to Egyptian Army Air Force (see c/n 635) but with no delivery date.   Possibly the otherwise unidentified Magister sold to Latvia with serial **181** (see also c/n 796) or possibly not built.

637   M.14A Hawk Trainer Mk.III (Magister)   CofA No.6292 issued 25.5.38 to Egyptian Army Air Force.   Dd as **L210** to Almaza, Cairo 7.5.38.

639   M.14A Magister.   A letter from L Hackett, Sales Dept, P&P, to The Chief of the Technical Section of the Ministry of War, Tallinn, Estonia, dated 9.5.38, stated that the Miles Magister being supplied to their Order No.1677, was being dd earlier than expected. It was being shipped on SS *Baltavia* from London on 19th May and due at Tallinn on 23rd.   A Flight Test Report for a Magister with s/n **159** was dated 9.5.38 and CofA No.6281 17.5.38 issued to the Estonian Air Force. (Note: it was not regd ES-AIN, as this was an OGL-2 monoplane of Estonian design).   To the Estonian Air Defence, 3rd Division.   A photograph of Magister 159 with two Estonian designed and manufactured PTO-4s, was taken at Riga/Spilve, Latvia, during a visit to the Latvian Aviation Regiment, in 1939.   In storage at Jagala airfield in summer 1941 when the Russians arrived, but was later destroyed by them.   Two Magisters, probably both coded letters 'SB' of the Luftwaffe Staffel Buschman, were recorded on strength from 6.42 to 9.42; the other was probably the ex Latvian 181.

640   M.14A **N4557** (replacement for L6914).   Dd 3 E&RFTS Hamble 8.6.38.   To 9 E&RFTS Ansty 20.6.38.   To Henlow 23.6.38. To Herts & Essex Broxbourne for repairs 9.7.43.   To 51 MU Lichfield 12.10.43.   Sold 16.2.49 to RA Short (error for Short Bros?).   Regd **G-ALIM** (CofR 12572) 21.2.49 to Short Bros & Harland Ltd, Rochester.   CofA No.10489 issued 30.5.49.   CofA lapsed 14.10.53. CofA renewed 23.12.54.   Cld 29.12.54 & regd 7.2.55 to The Scottish Flying Club Ltd, Renfrew.   Cld 1.5.56 & regd 13.8.56 to Scottish Aero Club Ltd, Perth.   Crashed Kirriemuir, near Perth 14.9.56 wrecking undercarriage; pilot John Ewart unhurt.   Regn cld 1.11.56 as reduced to produce.   Remains extant Perth 8.60.

642   M.14A **L8152**   Dd 27 MU Shawbury 19.5.38.   To 48 MU Hawarden 14.6.40.   To Franco-Belgian Air Training School, Odiham 8.11.40.   To 5 EFTS Meir 4.6.41.   Hit telegraph wire low flying in mist 1 ml N of Hilderstone, Staffs 16.9.41; Sgt Clarke injured.   SOC 21.9.41.

643   M.14A **L8153**   Dd 27 MU Shawbury 19.5.38.   To 1 SofTT Halton 9.4.40.   To 46 MU Lossiemouth 21.7.40.   To Stn Flt Ayr 21.3.41.   To 45 MU Kinloss 21.6.41.   To 5 FIS Perth 6.6.41.   To 32 MU St Athan 16.12.41.   To 11 EFTS/5 FIS Perth 5.1.42.   To 2 AGS Dalcross 27.12.42.   To 22 EFTS Cambridge 18.5.43.   Collided with Master Mk.II DL236 on approach Cambridge and crashed Fen Road, Chesterton 29.3.44; Sgt C Hausley & Sgt FE Lewer killed.

644   M.14A **L8154**   Dd 27 MU Shawbury 19.5.38.   To 5 EFTS Meir 25.9.40.   Stalled in forced landing while lost and hit tree Gisburn, near Settle, Yorks 3.11.40; pilot LAC Watson.   To P&P (SAS) 7.11.40 but SOC 1.12.40.

645   M.14A **L8155**   Dd 27 MU Shawbury 21.5.38.   To Stn Flt Kirton-in-Lindsey 10.7.41.   To 3 Del Flt, High Ercall 5.5.43.   To 307 Sqdn, Fairwood Common 3.7.43.   Engine cut on ferry flight; ditched 5 miles off South Devon coast 11.8.43.   SOC 13.8.43.

646   M.14A **L8156**   Dd 27 MU Shawbury 4.5.38.   To 5 EFTS Meir 25.9.40.   Hit tree low flying Abbots Bromley, Staffs 6.11.41; Sgt SA Holland & LAC Stock killed.   SOC 17.11.41.

647   M.14A **L8157**   Dd 27 MU Shawbury 25.5.38.   To 5 EFTS Meir 25.9.40.   To 1 CRU Cowley (SAS) 2.1.41.   To FPP 21.2.41.   To 5 EFTS Meir 11.6.41.   Engine cut, dived into ground on approach to forced landing 1½ ml S of Stone Road, Staffs 17.10.41.   SOC 26.10.41.

648   M.14A **L8158**   Dd 27 MU Shawbury 1.6.38.   To 5 EFTS Meir 25.9.40.   Spun into ground Groundslow Farm, near Tittensor, Staffs 1.5.41; AC Tuffin killed.

649   M.14A **L8159**   Dd 27 MU Shawbury 1.6.38.   To 5 EFTS Meir 21.9.40.   To CFS Upavon 11.12.41.   To 7 FIS Upavon (undated).   To ECFS Hullavington 19.8.42.   SOC 1.8.44 Cat.E (MR).

650   M.14A **L8160**   Dd Stn Flt Wyton 2.6.38.   To Stn Flt Horsham St Faith 29.6.40.   Flying accident 23.7.41; to 1 CRU Cowley (SAS) 29.8.41.   To 24 EFTS Luton 12.10.41.   To 2 FIS Montrose 22.4.42. To Stn Flt Dalcross 5.10.43.   To 51 MU Lichfield 21.11.45.   Sold 10.3.47 to FK McIntyre.   Regd **G-AJHD** (CofR 11293) 11.2.47 to Frank Keiler McIntyre, Darlington.   CofA No.9056 issued 14.8.47. Cld 12.5.48 & regd 20.5.48 to John Neasham, Croft.   CofA lapsed 7.4.50.   CofA renewed 12.4.57.   Cld 15.4.57 & regd 11.7.57 to John Colin Parry, Rhyl.   Cld 8.1.58 & regd 15.1.58 to Ronald Percy Mayes, Worthing.   Cld 13.1.59 & regd 4.2.59 to Elwyn McAully, t/a Fakenham Flying Group, Fakenham.   CofA lapsed 11.4.60.   Cld 28.6.60 & regd 8.7.60 to John Frederic Mercer, Southampton. Broken up Woolsington in 11.60.   Sold 23.8.61 & not regd.   Regn cld 21.10.61 as Rts.

651   M.14A **L8161**   Dd Practice Flt Abbotsinch 2.6.38.   To Stn Flt Limavady 26.9.42.   To 51 MU Lichfield 20.1.44.   Sold 11.3.46 to Miles Aircraft Ltd.   To H Hennequin & Cia, Argentina with CofA No.7825 issued 3.6.46; regd **LV-XPR** 30.6.47 to Secretaria Estado Aeronautica, Buenos Aires.   To Aero Club Mendoza.   Accident 2.4.50.   Regn cld 10.50 as wfu.

652   M.14A **L8162**   Dd Practice Flt Lee-on-Solent 31.5.38. Possibly to 17 Grp Comm Flt, Lee-on-Solent.   To Stn Flt Pocklington 16.7.41.   Dived into ground during aerobatics 1 ml W of Pocklington 4.4.42.

653   M.14A **L8163**   Dd Practice Flt Donibristle 8.6.38.   To Stn Flt Turnhouse 24.5.39.   To Stn Flt Drem 21.12.39.   To 19 Grp Comm Flt, Plymouth 15.5.41.   To Atlantic Coast, Barnstaple (MI) 17.1.42.   To 16 EFTS Burnaston 12.4.42.   To Comm Flt Woodley 1.5.43.   SOC 25.6.44 Cat.E (MR).

654   M.14A **L8164**   Dd Practice Flt North Weald 3.6.38.   Stalled while low flying and spun into ground Stumps Cross, near Duxford 11.12.41.   SOC 17.12.41.

655   M.14A **L8165**   Dd School of Army Co-operation, Old Sarum 2.6.38.   (CLN recorded dd to Wyton).   Banked on overshoot to avoid trees and hit hedge Old Sarum 21.1.40.   SOC 8.2.40, to 50 MU Salvage Centre Oxford 10.2.40.

656   M.14A **L8166**   Dd Practice Flt Odiham 9.6.38.   To 614 Sqdn, Odiham (30.11.39 census).   To School of Army Co-operation, Old Sarum 23.1.40.   To 5 MU Kemble 18.6.41.   To Herts & Essex, Broxbourne (SAS) 14.9.41.   To Stn Flt Marston Moor 8.11.41.   To Stn Flt Pocklington 11.4.42.   To 27 MU Shawbury 30.6.42.   To 76 MU Wroughton (undated).   Despatched 2.6.43 to ME on SS Fort Chipyewan for **Turkey**; arr Alexandria 27.6.43.

657   M.14A **L8167**   Dd Usworth 8.6.38.   To 55 OTU Aston Down 27.5.41.   To Llanbedr 30.5.44.   SOC 4.7.44 Cat.E (MR).

658   M.14A **L8168**   Fitted with the higher aspect ratio, taller, rudder to become the M.14A Magister Mk.I.   Dd to AMRD/A&AEE Martlesham Heath 13.6.38.   To RAE Farnborough 23.5.39.   To A&AEE Martlesham Heath 14.6.39.   To P&P 28.6.39.   To A&AEE Martlesham Heath 30.6.39 for tests of blistered wings.   Tail parachute installed to Contract No.10718/39, ITP dated 17.7.39. Remained with 'D' Flt Martlesham Heath.   To CFS Upavon 4.8.39. To A&AEE Boscombe Down 12.10.39.   To RAE Farnborough.   To 2 SofTT Cosford for GI as **5279M** 26.6.45.   To 11 SofTT Hereford; reformed at Hednesford 10.45.   To 6 SofTT Hednesford 15.8.47.

659   M.14A **L8169**   Dd 27 MU Shawbury 9.6.38.   To 10 Grp Comm Flt Colerne 3.5.41.   To Stn Flt Exeter.   Damaged 14.5.41, to 1 CRU Cowley (SAS) 28.5.41.   To 27 MU Shawbury 21.7.41.   To ATA 9.8.41; op ATA Training Unit (2.42); coded '20'.   To 27 MU Shawbury 22.8.43.   To 51 MU Lichfield 2.4.44.   Sold 20.2.46 to Miles Aircraft Ltd.   To H Hennequin & Cia, Argentina with CofA No.7717 issued 7.5.46; regd **LV-XNV** 30.6.47 to Secretaria Estado Aeronautica, Buenos Aires.   To Aero Club Pergamino.   Accident 19.4.47; repaired. Regn cld 3.53 as wfu.

660   M.14A **L8170**   Dd 27 MU Shawbury 9.6.38.   To 19 E&RFTS Gatwick 8.8.39.   To 30 EFTS Burnaston 15.10.39.   To 16 EFTS Burnaston 10.4.40.   To 3 EFTS Shellingford 14.6.42.   To 45 MU Kinloss 14.3.43.   To Herts & Essex, Broxbourne for repairs 21.4.43. To 51 MU Lichfield 28.6.43.   Sold 9.2.46 to Miles Aircraft Ltd.   To H Hennequin & Cia, Argentina with CofA No.7645 issued 10.4.46; regd **LV-XOS** 30.6.47 to Secretaria Estado Aeronautica, Buenos Aires.   To DA Deportiva.   Accident 15.12.47; repaired.   Regn cld 7.54 as wfu.

661 M.14A **L8171** Dd 27 MU Shawbury 9.6.38. To 19 E&RFTS Gatwick 11.8.39. To 30 EFTS Burnaston 15.10.39. To 16 EFTS Burnaston 10.4.40. To 27 MU Shawbury 2.9.42. To ATA 12.6.43. Cat.B (MR) 24.10.43. To 21 EFTS Booker 10.1.44. To 51 MU Lichfield 15.1.44. Sold 8.2.46 to Miles Aircraft Ltd. To H Hennequin & Cia, Argentina with CofA No.7664 issued 24.4.46; regd **LV-XOO** 30.6.47 to Secretaria Estado Aeronautica, Buenos Aires. To Aero Club Rio Cuarto. Accident 9.11.47; repaired. Crashed Cordoba 5.10.48; 2 killed. Regn cld 23.10.50.

662 M.14A **L8172** Dd 27 MU Shawbury 9.6.38. To 19 E&RFTS Gatwick 11.8.39. To 30 EFTS Burnaston 15.10.39. To 16 EFTS Burnaston 10.4.40. To RAF College Cranwell 15.3.43. To 1 (O)AFU Wigtown 29.3.44. Engine cut during aerobatics, hit wall in forced landing Kirkland of Longcastle, Wigtown 21.10.44. SOC 29.10.44.

663 M.14A **L8173** Dd 27 MU Shawbury 9.6.38. To 27 E&RFTS Tollerton 30.6.38. To 8 EFTS Woodley 15.10.39. While being flown by LAC LA Plummer on 10.1.41, L8173 overshot forced landing at Abingdon, hit a hedge and was written off, pilot uninjured. SOC 14.1.41.

664 M.14A **L8174** First flight test 4.6.38. Dd 27 MU Shawbury 9.6.38. To 27 E&RFTS Tollerton 30.6.38. To 8 EFTS Woodley 15.10.39. To 15 EFTS Carlisle 17.8.40. Crashed 26.10.41 (TT 1,480.25 hrs), to P&P (SAS) 3.11.41; repairs to port wing carried out by P&P at Willesden. To 2 FIS Montrose 21.2.42. On 12.4.42 (some 87.15 F/Hrs later), suffered structural failure of port wing in dive and crashed Craig Farm, Garvock, Kincardineshire, Angus 12.4.42, 2 killed (TT 1,575.30 hrs). SOC 20.4.42.

665 M.14A **L8175** Dd 27 MU Shawbury 9.6.38. To 27 E&RFTS Tollerton 30.6.38. Stalled recovering from spin and dived into ground Old Rectory Field, Holme Pierrepont, Notts 27.11.38; Sgt Edward Watkin-Thomas killed; (360:30 F/H). (Record card shows to Stn Flt Hucknall 8.3.39 in error)

666 M.14A **L8176** Dd 26 E&RFTS Kidlington 22.6.38. To 5 EFTS Hanworth 15.10.39. To 22 EFTS Cambridge 15.2.40. To 4 FIS Cambridge 21.8.40. To 22 EFTS Cambridge 13.5.43. To 51 MU Lichfield 6.5.44. Sold 15.1.46 to Miles Aircraft Ltd. To Portuguese AF **194** with CofA No.7569 issued 25.3.46, later **1205**.

667 M.14A **L8200** Dd 26 E&RFTS Kidlington 22.6.38. To 5 EFTS Hanworth 15.10.39. Soc 6.4.40.

668 M.14A **L8201** Dd 26 E&RFTS Kidlington 22.6.38. To 5 EFTS Hanworth 15.10.39; to Meir 17.6.40. To 48 MU Hawarden 8.12.41. Cat.B, to Lundy & Atlantic Coast, Barnstaple for repairs 16.12.41. To Stn Flt Valley 26.1.42. To 306 (Polish) Sqdn, Churchstanton (undated). Engine cut, hit tree in forced landing 3 ml E of Cullompton, Devon 1.5.42.

669 M.14A **L8202** Dd 25 E&RFTS Waltham 20.6.38. To 24 EFTS Sydenham 15.10.39; to Luton 22.7.40. To 1 CRU Cowley (SAS) 2.1.41. To 1 (Polish) FTS, Hucknall 22.2.41. To 24 EFTS Luton 27.7.41. To 16 EFTS Burnaston 22.10.41. To 27 MU Shawbury 11.9.42. SOC 9.5.44 Cat.E1.

670 M.14A **L8203** Dd 25 E&RFTS Waltham 20.6.38. To 24 EFTS Sydenham 15.10.39; to Luton 22.7.40. Stalled on landing and wing hit ground Luton 1.8.40. SOC 8.8.40.

671 M.14A **L8204** Dd 25 E&RFTS Waltham 21.6.38. To 32 MU St Athan 13.9.39. To repairs (unknown location) 4.11.39. To 8 MU Little Rissington 16.6.40. To RNAS Eastleigh 9.7.40. To 780 Sqdn, Eastleigh/Lee-on-Solent (6.40-5.41). To RNAS Machrihanish (10.41). To Stn Flt Fearn (3.43-5.43). To RNAS Turnhouse (6.44-7.44). SOC 16.1.45.

672 M.14A **L8205** Dd 25 E&RFTS Waltham 21.6.38. Engine cut on take-off, crash landed Waltham 23.7.38. To 1 SofTT Halton for GI as **1124M** 5.9.38. Reduced to produce 22.10.40.

673 M.14A **L8206** Dd 25 E&RFTS Waltham 20.6.38. To 24 EFTS Sydenham 15.10.39; to Luton 22.7.40. To P&P (SAS) 15.11.40. To 9 MU Cosford 20.4.41. To 60 OTU Leconfield 8.5.41. Flying accident (Cat.B) 22.7.41; to P&P for repairs 30.7.41. To Stn Flt Old Sarum 10.10.41. To 45 MU Kinloss 17.9.42. To 51 MU Lichfield 11.2.44. Sold 2.46 to Miles Aircraft Ltd. Test flown 12.4.46 by Hugh Kendall as L8206. To H Hennequin & Cia, Argentina with CofA No.7689 issued 3.5.46; regd **LV-XNF** 30.6.47 to Secretaria Estado Aeronautica, Buenos Aires. To Aero Club Pampeano. Accident 23.5.53; repaired. Regd 7.8.62 to Aero Club Comodoro Rivadavia. Still regd 5.67.

674 M.14A **L8207** Dd 25 E&RFTS Waltham 20.6.38. To 32 MU St Athan 14.9.39. To 8 MU Little Rissington 11.6.40. To 3 FPP ATA Hawarden 11.8.40. To Herts & Essex, Broxbourne (SAS) 16.7.41. To 27 MU Shawbury 26.8.41. To 15 EFTS Carlisle 7.9.41. To 46 MU Lossiemouth 9.8.42. To ECFS Hullavington 17.9.42. SOC 3.8.44 Cat.E1 (MR).

675 M.14A **L8208** Dd 604 Sqdn, Hendon 25.6.38. To 24 Sqdn, Hendon 8.9.38. To 13 EFTS White Waltham (30.11.39 census). To Stn Flt Halton 17.1.41. To 5 MU Kemble 21.2.41. To (S)FPP Kemble? 5.3.41. To P&P (SAS) 28.7.41. To Baginton 31.8.41. To 257 Sqdn, Honiley 15.5.42. To 153 Sqdn, Ballyhalbert 6.6.42. To 45 MU Kinloss 25.4.43. To RAF Blakehill Farm 17.3.44. To 51 MU Lichfield 15.5.44. Sold 28.2.46 to Miles Aircraft Ltd. To H Hennequin & Cia, Argentina with CofA No.7763 issued 15.5.46; regd **LV-XNR** 30.6.47 to Secretaria Estado Aeronautica, Buenos Aires. To Aero Club San Fransisco. Regn cld 7.54 as wfu.

676 M.14A **L8209** Dd 24 Sqdn, Hendon 25.6.38. To 2 EFTS Filton 15.10.39. To 8 MU Little Rissington 12.3.40. To Stn Flt Pembrey 21.7.40. To P&P (SAS) 12.12.40. To Stn Flt Turnhouse 28.5.41. To 123 Sqdn, Turnhouse 17.7.41. To Stn Flt Castletown (undated). To 165 Sqdn, Ayr 20.5.42. To P&P (MI) 6.7.42. To 27 MU Shawbury 20.8.42. To 47 MU Sealand 26.3.43. Despatched to ME 16.5.43 on SS City of Khios; arr 14.6.43. SOC 30.11.44.

677 M.14A **L8210** Dd CFS Upavon 21.6.38. To 9 ATS Stormy Down 12.7.39; became 7 AOS 3.9.39; renamed 7 B&GS 1.12.39; renamed 7 AGS 9.6.41. To 5 EFTS Meir 5.8.41. To Atlantic Coast, Barnstaple (MI) 2.12.41. To 20 MU Aston Down 19.2.42. To 32 MU St Athan 8.3.42. To 401 (RCAF) Sqdn, Biggin Hill 7.3.42. Flying accident (Cat.B); to Herts & Essex, Broxbourne for repairs 1.10.42. To 51 MU Lichfield 2.1.43. Sold 30.5.46 to Weston-super-Mare Flying Club. Regd **G-AHUL** (CofR 10357) 6.6.46 to Weston Aero Club Ltd, Weston-super-Mare & Home Counties Aero Club Ltd, Radlett. Not refurbished (although CofA No.7200 allocated). Scrapped at Weston-super-Mare in 2.47. Regn cld 2.5.47 as wfu.

678 M.14A **L8211** Dd 1 AACU Farnborough 28.6.38. To 'B' Flt 1 AACU, Carew Cheriton 1.6.39. To Parnall Aircraft for repairs 12.6.41. To A&AEE Boscombe Down 6.12.41. To Atlantic Coast, Barnstaple (MR) 6.2.42. To 20 MU Aston Down 22.4.42. To 10 FIS Woodley 8.2.43. To Atlantic Coast, Barnstaple for repairs 28.2.43. To Air Fighting Development Unit, Wittering 18.5.43. To CFE Wittering (absorbed AFDU 16.10.44). To 51 MU Lichfield 1.1.46. Sold 22.4.49 to RA Short. Regd **G-ALNX** (CofR 12706) 21.4.49 to Robert Alan Short and James Henry Tattersall, Croydon/Clitheroe, Lancs. Not refurbished; extant Elstree 7.51. Cld 12.9.52 by MCA. Regd 21.1.53 to Robert Alan Short, Croydon. Regn cld 18.5.59 as pwfu.

679 M.14A **L8212** Dd 24 Sqdn, Hendon 27.6.38. Destroyed in air raid Hendon 8.10.40. SOC 17.10.40.

680 M.14A **L8213** Dd 24 Sqdn, Hendon 27.6.38. Damaged 5.4.40; repaired at unknown location. To 45 MU Kinloss 10.7.40. To 15 EFTS Carlisle 8.11.40. To Hucknall 8.11.40 for 1 (P)FTS. To Shrager Bros, Old Warden (SAS) 23.7.41. To 37 MU Burtonwood 7.12.41. To Stn Flt Leconfield 5.1.42. To 2 (O)AFU Millom 27.3.42. To Staff Pilot Training Unit 27.4.42. To 51 MU Lichfield 30.7.44. Sold 7.3.46 to Miles Aircraft Ltd. Test flown 8.5.46 by Hugh Kendall as L8213. To H Hennequin & Cia, Argentina with CofA No.7802 issued 24.5.46; regd **LV-XNX** 30.6.47 to Secretaria Estado Aeronautica, Buenos Aires. To Aero Club Santa Fe. Accident 24.11.46. Regn cld 4.50 as wfu.

681 M.14A **L8214** Dd 27 MU Shawbury 28.6.38. To 19 E&RFTS Gatwick 8.8.39. Spun into ground during spinning practice near Horley, Surrey 30.8.39; Flt Sgt Thomas RW Owens & Roger J Gillingham (of CAG) killed.

682 M.14A **L8215** Dd 27 MU Shawbury 28.6.38. To 27 E&RFTS Tollerton 31.8.39. To 24 EFTS Sydenham 15.10.39; to Luton 22.7.40; coded '25'. To 45 MU Kinloss 9.2.42. To 51 MU Lichfield 28.2.44. Sold 6.2.46 to Miles Aircraft Ltd. To H Hennequin & Cia, Argentina with CofA No.7707 issued 7.5.46; regd **LV-XNH** 30.6.47 to Secretaria Estado Aeronautica, Buenos Aires. To Aero Club Canada de Gomez. Accident 29.10.46; repaired. Accident 29.5.47; repaired. Accident 12.6.55; repaired. Regd 7.8.62 to Aero Club Bahia Blanca. Still regd 5.67.

APPENDIX 20

683 M.14A **L8216** Dd 27 MU Shawbury 30.6.38. To 19 E&RFTS Gatwick 8.8.39. To 30 EFTS Burnaston 15.10.39. Hit by Battle L5283 while parked Cranfield 3.10.39. SOC 12.10.39.

684 M.14A **L8217** Dd 27 MU Shawbury 30.6.38. To 5 EFTS Hanworth 4.9.39. To CFS Upavon 11.12.41. To 27 MU Shawbury 25.5.42. To Atlantic Coast, Barnstaple (MI) 4.6.42. To 20 MU Aston Down 19.6.42. To 27 MU Shawbury 14.9.42. To 47 MU Sealand 7.3.43. Despatched to ME 5.3.43 on SS City of Worcester; arr 13.5.43. To Malta 1.8.43. To ASR & Comm Flt, Ta Kali. To Malta Comm Flt 1.3.44 (still op 11.45). SOC 28.3.46.

685 M.14A **L8218** Dd 27 MU Shawbury 2.7.38. To 19 E&RFTS Gatwick 11.8.39. To 30 EFTS Burnaston 15.10.39. To 16 EFTS Burnaston 21.6.40. Overshot landing and hit tree Derby 22.9.40. SOC 1.10.40.

686 M.14A **L8219** Dd Stn Flt Ternhill 2.7.38. To 8 EFTS Woodley 8.3.40. To 4 FIS Cambridge 19.6.41. To 22 EFTS Cambridge 15.4.43. To 51 MU Lichfield 15.5.44. Sold 6.1.46 to Miles Aircraft Ltd. To H Hennequin & Cia, Argentina with CofA No.7693 issued 3.5.46; regd **LV-XOM** 30.6.47 to Secretaria Estado Aeronautica, Buenos Aires. To Aero Club La Plata. Accident 26.5.48; repaired. Regd (5.67) to Direccion General de Aeronautica Civil, Buenos Aires.

687 M.14A **L8220** Dd 27 MU Shawbury 4.7.38. To 40 E&RFTS Mousehold Heath 31.8.39. To 30 EFTS Burnaston 15.10.39. To 16 EFTS Burnaston 10.4.40. To P&P (SAS) 26.8.40. To 5 MU Kemble 5.1.41. To 71 Grp Comm Flt, Hanworth 4.2.41. Damaged by fire in air raid at Hanworth 8.2.41. To 5 EFTS Meir 15.9.41. To Atlantic Coast, Barnstaple for repairs 12.12.41. To 20 MU Aston Down 18.1.42. To ATA Barton-in-the Clay 11.2.43. Engine lost power during practice forced landing; crash landed Shillington, Beds 19.6.43. Flying accident (Cat.Ac) 9.7.44; reportedly repaired on site by Southern Aircraft (Gatwick) Ltd 18.7.44. SOC 21.11.44 Cat.E.

688 M.14A **L8221** Dd 27 MU Shawbury 4.7.38. To Stn Flt Marham 13.12.39. Stalled in circuit and dived into ground Marham 25.4.40. SOC 28.4.40.

689 M.14A **L8222** Dd 27 MU Shawbury 4.7.38. To 5 EFTS Hanworth 4.9.39. SOC 6.4.40 Cat. E.

690 M.14A **L8223** Dd 27 MU Shawbury 7.7.38. To 5 EFTS Hanworth 4.9.39; to Meir 17.6.40. To 16 EFTS Burnaston 6.2.42. To P&P (MI) 8.7.42. To 27 MU Shawbury 3.10.42. To ATA Barton-in-the Clay 16.6.43. DBR in heavy landing Barton-in-the-Clay 16.8.44 after pupil froze on throttle; F/O JC Addams & T/O JS Chivas unhurt.

691 M.14A **L8224** Dd Shawbury 7.7.38. To 5 EFTS Hanworth 4.9.39; to Meir 17.6.40. Abandoned take-off from RLG at Stone, Staffs and hit trees 15.10.40. SOC 6.3.42.

692 M.14A **L8225** Dd E&WS Cranwell 12.7.38 (CLN recorded dd to RAF College Cranwell); renamed 1 E&WS 1.11.38. To Stn Flt (4 SofTT) St Athan 31.8.40. To 38 MU Llandow 1.7.41. To 247 Sqdn, Predannack 2.8.41. MR (Cat.B) 1.5.42; to P&P for repairs 23.5.42. To 46 MU Lossiemouth 1.8.42. To ATA 2.9.43. To 51 MU Lichfield 7.3.45. Sold 15.4.46 to Miles Aircraft Ltd. Test flown 6.6.46 by Hugh Kendall as L8225. To H Hennequin & Cia, Argentina with CofA No.7967 issued 1.7.46; regd **LV-XQN** 30.6.47 to Secretaria Estado Aeronautica, Buenos Aires. To Aero Club Cnel. Pringels. Crasehed nr Cnel Pringles 13.9.47 (95% damaged); passenger killed.

693 M.14A **L8226** Dd E&WS Cranwell 12.7.38; renamed 1 E&WS 1.11.38. To 9 MU Cosford 30.8.40. To 16 EFTS Burnaston 20.9.40. Flying accident; to P&P (SAS) 12.12.40. To 54 OTU Church Fenton 30.6.41. To 136 Sqdn, Kirton-in-Lindsey 24.8.41. To 253 Sqdn, Hibaldstow 14.12.41 (entry probably for next day 15.12.41). Stalled on approach to forced landing in fog 1 ml NE of Kirton-in-Lindsey, Lincs 29.8.42. SOC 5.9.42.

694 M.14A **L8227** Dd Farnborough 13.7.38. To Stn Flt Farnborough & SofP Farnborough 29.8.40. To 5 MU Kemble 4.5.41. To 16 EFTS Burnaston 25.7.41. Stalled during practice forced landing and hit ground on Long Lane, Dalbury Lees, Derbyshire 21.6.42; P/O Godfrey H Grantham killed & Cpl JP Ward badly injured. SOC 29.6.42.

695 M.14A **L8228** Dd HAD Henlow 14.7.38. To 13 MU Henlow 1.3.40. To Atlantic Coast, Barnstaple (MI) 12.1.42. To 20 MU Aston Down 13.2.42. To 51 OTU Cranfield 6.3.42. To 247 Sqdn, Exeter 10.7.42. Hit hedge on take-off from field 2 ml SE of Burton Bradstock, Devon 6.9.42 (678:20 F/H). SOC 7.9.42.

696 M.14A **L8229** Dd HAD Henlow 14.7.38. To 13 MU Henlow 1.3.40. To 27 MU Shawbury 3.8.43. To 51 MU Lichfield 20.6.44. Sold 19.2.46 to Miles Aircraft Ltd. To H Hennequin & Cia, Argentina with CofA No.7895 issued 13.6.46; regd **LV-XQQ** 30.6.47 to Secretaria Estado Aeronautica, Buenos Aires. To Aero Club Deportiva. Accident 28.9.47; repaired. Regd 8.8.62 to Aero Club Saenz Pena. Still regd 5.67.

697 M.14A **L8230** Dd 7 ATS Acklington 18.7.38. To 10 B&GS Warmwell (formed 1.1.40) (census). To 3 AONS Bobbington 22.3.41. Damaged 7.5.41, to P&P for repairs 14.5.41. To 5 MU Kemble 6.7.41. To 16 EFTS Burnaston 31.7.41. To 27 MU Shawbury 7.9.42. To 60 OTU High Ercall 5.8.43. To 1 SofTT Halton for GI as **4497M**, allotted 21.7.44.

698 M.14A **L8231** Dd 1 ASU Waddington 12.7.38. To 5 MU Kemble 14.7.39. To 238 Sqdn, Chilbolton 24.10.40. To Stn Flt Middle Wallop 16.5.41. To Stn Flt Ibsley 27.11.43. To Stn Flt Colerne 23.3.44. To Stn Flt Exeter 20.9.44. To Stn Flt Harrowbeer. SOC 29.1.45 Cat. E2.

699 M.14A **L8232** Dd Stn Flt Andover 16.7.38. To Stn Flt Church Fenton 15.6.41. Stalled during low aerobatics and spun into ground Beaumont Ave, Sandpit Lane, St Albans, Herts 16.8.41. SOC 24.8.41.

700 M.14A **L8233** Dd Stn Flt Andover 16.7.38. To 5 MU Kemble 1.2.41. To (S)FPP Kemble 5.3.41. To 15 MU Wroughton 2.1.42. To ATA Barton-in-the-Clay 16.6.42. Spun into ground and dbf on cross country 1½ ml SSW Baldock, Great North Road, near Letchworth, Herts 6.12.42 probably due to engine failure after fuel exhausted in starboard tank; pilot ATA Cadet FG Bowles killed. SOC 7.12.42.

701 M.14A **L8234** Dd 1 ASU Waddington 19.7.38. To 5 MU Kemble 20.2.39. To 24 EFTS Luton 17.10.40. To 2 FIS Montrose 19.2.42. Cat.B (MR) 16.11.43. To 51 MU Lichfield 10.1.44. Sold 19.2.46 to Miles Aircraft Ltd. Test flown 12.4.46 by Hugh Kendall as L8234. To H Hennequin & Cia, Argentina with CofA No.7691 issued 3.5.46; regd **LV-XNI** 30.6.47 to Secretaria Estado Aeronautica, Buenos Aires. To Aero Club Suarez. Accident 24.8.46; repaired. Crashed San Martin 22.6.51; pilot killed.

702 M.14A **L8235** Dd 1 ASU Waddington 19.7.38. To 5 MU Kemble 14.2.39. To Stn Flt Hucknall 4.7.40. To 1 CRU Cowley (SAS) 9.1.41. To 2 SofAC Old Sarum 21.3.41. To 414 RCAF Sqdn, Croydon 11.12.41. To Atlantic Coast, Barnstaple for repairs 29.1.42. To 27 MU Shawbury 14.2.42. To ECFS Hullavington 6.4.42. To 29 EFTS Clyffe Pypard 26.9.42. To 46 MU Lossiemouth 14.4.43. To ATA Initial Flying Training School, Thame 8.6.43. Cat.B 8.12.43. To 51 MU Lichfield 31.1.44. To RAE Farnborough 26.3.45. To 51 MU Lichfield 9.11.45. Sold 30.5.46 to Miles Aircraft Ltd. To H Hennequin & Cia, Argentina with CofA No.8212 issued 7.8.46; regd **LV-XRU** 30.6.47 to Secretaria Estado Aeronautica, Buenos Aires. To Aero Club Pigue. Accident 18.5.54.

703 M.14A **L8236** Dd 1 ASU Waddington 19.7.38. To 5 MU Kemble 14.2.39. To AMDP 20.8.40. To 249 Sqdn, Boscombe Down 20.8.40. To Stn Flt North Weald 10.5.41. To 121 Sqdn, North Weald 21.4.42. To 335 USAAF Sqdn 1.10.42. To 27 MU Shawbury 6.10.42. To 47 MU Sealand 21.2.43. Despatched to ME 5.5.43 on SS Tactician; arr Port Suez 26.6.43. SOC 29.5.47.

704 M.14A **L8237** Dd 1 ASU Waddington 19.7.38. To 5 MU Kemble 20.2.39. To 249 Sqdn, North Weald 8.9.40. SOC 15.11.40.

705 M.14A **L8249** Dd 1 ASU Waddington 21.7.38. To 5 MU Kemble 31.1.39. To Stn Flt Benson 3.7.40 (departed); to Stn Flt Hucknall 2.7.40 (arrived!). To 24 EFTS Luton 27.7.41. Undershot landing and hit tree Halton 24.1.42; Acting LA J Hayward unhurt. SOC 30.1.42.

706 M.14A **L8250** Dd 1 ASU Waddington 21.7.38. To Stn Flt Driffield 27.7.38. To Blackburn Aircraft 2.5.39 for MAP Overseer. To 92 Grp Comm Flt, Little Horwood 5.10.44. To 51 MU Lichfield 27.7.45. Sold 6.6.46 to Miles Aircraft Ltd. To H Hennequin & Cia, Argentina with CofA No.8215 issued 7.8.46; regd **LV-XRT** 30.6.47 to

707 M.14A **L8251** Dd 28 E&RFTS Meir 28.7.38. Hit hedge on take-off from forced landing Bradley, Derbyshire 17.11.38; F/O KC Richardson & Sgt RG Sawyer unhurt. To 2 SofTT Cosford for GI as **1254M** 3.2.39. SOC 13.3.39, reduced to parts.

708 M.14A **L8252** Dd 28 E&RFTS Meir 29.7.38. To 30 EFTS Burnaston 20.9.39. To 16 EFTS Burnaston 10.4.40. Stalled on approach and spun into ground Abbots Bromley 5.6.42; Sgt GHT Evans & Cpl SJ Graham killed. SOC 17.6.42.

709 M.14A **L8253** Dd 28 E&RFTS Meir 29.7.38. To 30 EFTS Burnaston 20.9.39. To 16 EFTS Burnaston 10.4.40. To 32 MU St Athan 20.4.42. To 27 MU Shawbury 29.4.42. To 16 EFTS (undated). To 46 MU Lossiemouth 11.7.42. To Stn Flt Dalcross (31.12.42 census). To 46 MU Lossiemouth 17.12.44. To A&AEE Boscombe Down 18.12.44. To 4 SofTT St Athan for GI as **5947M** 1.6.46.

710 M.14A **L8254** Dd 28 E&RFTS Meir 29.7.38. To 30 EFTS Burnaston 20.9.38. To 16 EFTS Burnaston 10.4.40. Converted to light bomber. To 1 CRU Cowley (SAS) 18.4.41. To 5 MU Kemble 6.6.41. To 85 Sqdn, Hunsdon 22.7.41. To 51 MU Lichfield 21.8.42. To 34 MU Monkmoor 18.2.44. SOC 25.2.44.

711 M.14A **L8255** Dd 28 E&RFTS Meir 29.7.38. To 30 EFTS Burnaston 20.9.38. To 16 EFTS Burnaston 10.4.40. Caught fire in hangar during inspection 28.10.42 (Cat.B) (but also recorded as flying accident). SOC as re-cat.E '26'.10.42(?) (TT:1769:25 F/H).

712 M.14A **L8256** Dd 28 E&RFTS Meir 29.7.38. To 30 EFTS Burnaston 20.9.39. To 16 EFTS Burnaston 10.4.40. To 5 GTS Shobdon 18.5.43. To 51 MU Lichfield 10.4.44. Sold 11.11.46 to Miles Aircraft Ltd. CofA No.9399 issued 28.5.47. Regd in France as **F-BDPK** 17.6.47 to Suzanne Goute, Roquebrune-sur-Argens. Regd 29.10.48 to Aero Club d'Aix en Provence, Aix-en-Provence. Regn cld 3.11.65 as scrapped.

713 M.14A **L8257** Dd 29 E&RFTS Luton 29.7.38. To 30 EFTS Burnaston 15.10.39. To 16 EFTS Burnaston 10.4.40. Cat B; to P&P for repairs 27.9.40. To 8 EFTS Woodley 24.2.41. To 10 FIS Woodley 21.7.42. Hit tree in practice forced landing Waltham St Lawrence LG 16.3.43. SOC 18.3.43 (2052:15 F/H).

714 M.14A **L8258** Dd 29 E&RFTS Luton 29.7.38. To 5 EFTS Hanworth 5.10.39; to Meir 18.6.40. Engine cut, hit wires in forced landing Cheddleton, Staffs 4.9.40. SOC 20.9.40.

715 M.14A **L8259** Dd 29 E&RFTS Luton 16.8.38. Failed to recover from spin and abandoned near Luton 3.7.39; Sgt Allan Bold injured. Wreck to HAD Henlow 5.7.39.

716 M.14A **L8260** Dd 29 E&RFTS Luton 16.8.38. To 5 EFTS Hanworth 15.10.39. To 1 CRU Cowley (Cat.B) 18.3.40. To 45 MU Kinloss 2.7.40. To 614 Sqdn, Grangemouth 24.8.40. To 1 CRU Cowley for repairs 14.11.40. To 46 MU Lossiemouth 7.8.41. To 51 OTU Cranfield 14.8.41. To 32 MU St. Athan 8.10.41. To 60 OTU East Fortune (31.12.41 census). To 51 OTU Cranfield 8.3.44. To 51 MU Lichfield 26.3.45. Sold 29.1.46 to Miles Aircraft Ltd. To H Hennequin & Cia, Argentina with CofA No.7634 issued 9.4.46; regd **LV-XMV** 30.6.47 to Secretaria Estado Aeronautica, Buenos Aires. To Aero Club Pigue. Regn cld 5.53 as wfu.

717 M.14A **L8261** Dd 29 E&RFTS Luton 17.8.38. To 8 EFTS Woodley 5.10.39. To 16 EFTS Burnaston 11.9.40. Cat.B 12.7.41; to P&P (SAS) 29 7.41. To 16 EFTS Burnaston 31.8.41. To 3 EFTS Shellingford 14.6.42. To 45 MU Kinloss 14.3.43. To Atlantic Coast, Barnstaple for repairs 20.4.43. To 27 MU Shawbury 10.6.43. To ATA 30.6.43. To 51 MU Lichfield 31.1.45. Sold 18.11.46 to Miles Aircraft Ltd. CofA No.9367 issued 19.5.47. Regd in France as **F-BDPE** (CofR 19999) 13.6.47 to Xavier Beau, Paris. Regd 1.3.48 to Roger Hutz, Le Havre. Regd 29.3.57 to Aeronautique Havraise Jean Maridor, Le Havre. CofA suspended 25.7.63. Regn cld 15.11.67 as destroyed.

718 M.14 **L8262** Dd 29 E&RFTS Luton 18.8.38. To 8 EFTS Woodley 5.10.39. To 5 EFTS Meir 12.8.40. To P&P (SAS) 28.12.40. To Coltishall 11.5.41. To 27 MU Shawbury 8.9.41. To 6 FIS Staverton 24.11.41. To 2 FIS Montrose 27.5.42. To RAF College Cranwell 15.7.43. To 17 SFTS Cranwell; coded 'FCC-Z'. To CFS Upavon 14.11.46; coded 'FDK-H'. To 21 (P)AFU Moreton-in-Marsh 19.12.46. To 51 MU Lichfield 15.7.47. Sold 29.4.48 to LW Farrer (possibly for spares use for G-AKPF). Regd **G-ANWO** (CofR R6518) 31.12.58 to Ronald Royal Paine, Anthony Insley Topps & John Samuel Everatt t/a The Derby Airways Flying Group, Burnaston, CofA issued 19.4.60. Cld 25.2.62 & regd 26.2.62 to John Thomas Hayes & David John Walters, t/a Lincoln Aero Club, Kirton-in-Lindsey. Cld 16.4.62 & regd 24.4.62 to Lincoln Aero Club Ltd, Kirton-in-Lindsey. Crashed Kirton-in-Lindsey 21.4.62 damaging u/c, flaps and propeller; not repaired. Regn cld 9.6.64 as pwfu. Used for spares (for G-AKAT) at Kirton-in-Lindsey; still present 1.67. Reportedly acquired by Cambridge Aircraft Preservation Society .71. Later to Sandy Topen at Cranfield and fuselage reduced to a collection of small pieces of wooden structural members in a tea chest. Remains together with wings were acquired by Adrian Brook and moved to his home in West Sussex 12.5.87 and were used in other restoration projects.

719 M.14A **L8263** Dd 213 Sqdn, Wittering 18.8.38. Collided with Gauntlet K7840 and abandoned near Stamford, Lincs 15.9.38; F/O JEJ Sing & AC R Humphreys parachuted safely (as did pilot of Gauntlet) but Mary Russell killed by falling wreckage.

720 M.14A **L8264** Dd 1 AACU Farnborough 19.8.38; to 'E' Flt West Freugh (by 9.38). Stalled and dived into ground during aerobatics Glenwilly, Galloway 2.3.40. SOC 9.3.40.

721 M.14A **L8265** Dd 1 AACU Farnborough 19.8.38. To Henlow 15.1.39. To 13 MU Henlow 1.3.40. To 37 MU Burtonwood 26.5.42. To ATA 21.6.42. To Herts & Essex, Broxbourne for repairs 17.5.43. To 51 MU Lichfield 22.7.43. Sold 7.12.46 to Miles Aircraft Ltd. CofA No.9403 issued 28.5.47. Regd in France as **F-BDPO** (CofR 20009) 18.7.47 to Roger Goldet, Paris. Regd 20.10.48 to Jean Louis Chavy, Neuilly-sur-Seine, Roger Cuchet, Nancy & Louis Garnier, Le Puy-en-Velat. Regd 9.5.52 to Association Cooperative de Pilotage de l'Aero Club du Puy, Le Puy-en-Velat. Regd 5.11.53 to Aero Club d'El Oued, El Oued. Regd 14.11.58 to Association Comite d'Aviation de Tourisme de Biskra, Biskra. Regn cld 26.5.71 as scrapped.

722 M.14A **L8266** Dd 1 AACU Farnborough 19.8.38. SOC Cat.E 20.7.40.

723 M.14A **L8267** Dd Stn Flt Farnborough 17.8.38. Flying accident 14.7.40 (Cat.B); to 1 CRU Cowley for repairs 20.7.40. To 37 MU Burtonwood 26.8.40. To 317 Sqdn, Ouston 10.5.41. Flying accident (Cat.B) (10.6.41?); to P&P (SAS) 26.6.41. To 16 EFTS Burnaston 30.8.41. To 27 MU Shawbury 11.9.42. To 51 MU Lichfield 21.9.44. Sold 8.1.46 to Miles Aircraft Ltd. To H Hennequin & Cia, Argentina with CofA No.7786 issued 20.5.46; regd **LV-XNP** 30.6.47 to Secretaria Estado Aeronautica, Buenos Aires. To Aero Club Parana. Accident 17.8.46; repaired. Accident 8.5.48; probably repaired. Reported as donated .58 but to whom/where not known.

724 M.14A **L8268** Dd Stn Flt Farnborough 18.8.38. To 13 Sqdn, Speke 4.7.40. To 24 EFTS Luton 15.4.41. While on delivery to 15 EFTS Carlisle, stalled off a turn and hit trees near Annan at 16.00 hrs 10.10.41; ATA pilot S/O TJ Corsellis killed. SOC Cat.E 'Burnt' 16.10.41.

725 M.14A **L8269** Dd 2 Grp Comm Flt, Wyton 18.8.38. Stalled on approach and hit ground Wyton 13.9.38 (22:35 F/H).

726 M.14A **L8270** Dd 74 Sqdn, Hornchurch 25.8.38. To Stn Flt Hornchurch 22.7.40. To 9 MU Cosford 26.9.41. To 411 RCAF Sqdn, Digby 25.7.41. To 32 MU St Athan 3.4.42. To 411 Sqdn, Digby 30.6.42. To Herts & Essex, Broxbourne (MI) 2.10.42. To 51 MU Lichfield 28.11.42. To 76 MU Wroughton for packing. Despatched to ME 12.5.43 on SS Katuma. To **Turkey** 13.6.43.

727 M.14A **L8271** Dd 1 AACU Farnborough 22.8.38; to 'Z' Flt at Watchet (by 9.38). To 'A' Flt, 1 AACU, Weston Zoyland (3.39-8.40). To 1 CRU Cowley (SAS) 20.8.40. To 24 EFTS Luton 6.2.41. To 2 FIS Montrose 19.2.42. To 51 MU Lichfield 12.10.43. Sold 8.2.46 to Miles Aircraft Ltd. To H Hennequin & Cia, Argentina with CofA No.7644 issued 10.4.46; regd **LV-XOT** 30.6.47 to Secretaria Estado Aeronautica, Buenos Aires. To Aero Club Argentino. Accident 28.11.4x; repaired. Regn cld 7.54 as wfu.

728 M.14A **L8272** Dd Ternhill 23.8.38. To 2 SofTT Cosford 16.6.39. To 1 AAS Manby 10.12.39. To 44 Grp Comm Flt, Staverton 12.10.41. To 8 FTS Montrose 20.10.41. To 2 FIS Montrose 31.5.42. To Atlantic Coast, Barnstaple (MR) 2.10.43. To 51 MU Lichfield

APPENDIX 20

5.12.43. Sold 8.4.46 to Miles Aircraft Ltd. Test flown 24.5.46 by Hugh Kendall as L8272. To H Hennequin & Cia, Argentina with CofA No.7872 issued 11.6.46; regd **LV-XOY** 30.6.47 to Secretaria Estado Aeronautica, Buenos Aires. To Aero Club Argentino. Destroyed in hangar fire San Justo 5.5.47.

729 M.14A **L8273** Dd Ternhill 23.8.38. To 2 SofTT Cosford 16.6.39. To 9 MU Cosford 17.9.40. To 15 EFTS Carlisle 24.9.40. Engine cut, stalled on approach to practice forced landing and hit hedge 1 ml S of Gretna, Dumfries 9.5.41; LAC Chaplin injured. To P&P (SAS) 12.5.41 but SOC.

730 M.14A **L8274** Dd 46 Sqdn, Digby 23.8.38. To 229 Sqdn, Wittering 6.8.40. To 1 CRU Cowley (SAS) 27.1.41. To 24 EFTS Luton 15.4.41. To 16 EFTS Burnaston 21.10.41. To 4 FIS Cambridge 24.8.42. To 22 EFTS Cambridge 13.5.43. To 51 MU Lichfield 6.5.44. Sold 26.2.48 to Gloster Flying Club. Regd **G-AKMT** (CofR 12056) 27.11.47 to Laurence Dudley Trappitt, Whyteleafe, Surrey. Cld 22.7.48 & regd 14.10.48 to Robert Alan Short, Derek John Hayles & Martha Marodeen, Croydon/Portsmouth. Cld 5.11.48 & regd 31.1.49 to Robert Alan Short, Croydon. CofA No.10391 issued 26.3.49. Regn cld 2.9.49 as sold abroad - to Egyptian AF.

731 M.14A **L8275** Dd 41 Sqdn, Catterick 24.8.38. To 32 MU St Athan 25.12.41. To 11 EFTS Perth 5.1.42. To 41 Sqdn, Westhampnett or Merston (undated). To Herts & Essex, Broxbourne (MR) 14.6.42. To 46 MU Lossiemouth 27.7.42. To 41 OTU Old Sarum or Hawarden (undated). To 63 OTU Honiley 21.9.43. To 51 MU Lichfield 11.5.44. Sold 13.2.46 to Miles Aircraft Ltd. Test flown 23.4.46 by Hugh Kendall as L8275. To H Hennequin & Cia, Argentina with CofA No.7737 issued 9.5.46; regd **LV-XNE** 30.6.47 to Secretaria Estado Aeronautica, Buenos Aires. To Circ Universitario de Aviation. Accident 24.9.47; repaired. Regn cld .54 as wfu.

732 M.14A **L8276** Dd 54 Sqdn, Hornchurch 24.8.38. To 48 MU Hawarden 16.7.40. To Franco-Belgian Air Training School, Odiham 19.11.40. To P&P for repairs 1.1.41. To 5 MU Kemble 15.3.41. To DGRD at Bristol Aeroplane Co 12.5.41 for use of MAP Overseer. To Gloster Aircraft Co, Moreton Valence 11.1.47 for use of MoS Overseer. To 51 MU Lichfield 9.5.47; dd 15.5.47. Sold 29.4.48 to WC Jemmett. Regd **G-ALOG** (CofR 12714) 24.2.50 to Derek Malcolm Brown, Merstham, Surrey. CofA No.10825 issued 9.6.50. Cld 25.8.50 & regd 7.9.50 to Wright Aviation Ltd, t/a Liverpool Flying Club, Speke. Cld 12.2.54 & regd 16.2.54 to Dragon Airways Ltd; op by Liverpool Flying Club, Speke. Damaged beyond repair in heavy landing Speke Airport 29.7.55. Regn cld 9.9.55 as wfu.

733 M.14A **L8277** Dd 56 Sqdn, North Weald 23.8.38. To RAF North Weald 21.6.40. To 25 Sqdn, North Weald 18.9.40. To 417 RCAF Sqdn, Charmy Down 8.1.42; to Tain 28.3.42. To 242 Sqdn, Drem 21.6.42. Flying accident (Cat.B) 26.9.42; to P&P. To 27 MU Shawbury 16.11.42. To 47 MU Sealand 16.3.43. Despatched to ME 5.4.43 on SS City of Agra; arr 5.6.43. To Ein Shemer (census 31.12.47). To 160 MU Aqir. To Levant Comm Sqn 27.5.48. To RAF Amman 29.7.48. *A postwar photograph taken of L8277 shows it in silver finish with a Group Captain's pennant painted on the side of the fuselage near the front cockpit and with a Spitfire FR Mk.XVIII of 32 Sqdn in the background.* SOC 14.10.48 and possibly sold locally in Amman.

734 M.14A **L8278** Dd 32 Sqdn, Biggin Hill 23.8.38. To 501 Sqdn, Gravesend 19.8.40. To 32 Sqdn, Acklington 2.10.40. To 1 CRU Cowley (SAS) 19.10.40. To 19 MU St Athan 11.12.40. To ATA (Training) Ferry Pool, White Waltham 2.3.41; coded '12'. To Blackburn Aircraft for repairs 4.6.41. Returned to ATA TFPP, White Waltham; undamaged in forced landing nr Sywell 20.8.41 due to engine failure. To (S)FPP Kemble (31.12.41 census). To Herts & Essex, Broxbourne for repairs 21.5.43. To 51 MU Lichfield 24.7.43. Sold 12.1.46 to Miles Aircraft Ltd. To H Hennequin & Cia, Argentina with CofA No.7582 issued 27.3.46; regd **LV-XMZ** 30.6.47 to Secretaria Estado Aeronautica, Buenos Aires. To Aero Club Mar del Plata.

735 M.14A **L8279** Dd 239 Sqdn, Debden 27.8.38. To 51 MU Lichfield 9.11.42. SOC Cat.E (MR) 25.2.44.

736 M.14A **L8280** Dd 111 Sqdn, Northolt 30.8.38. To 92 Sqdn, Croydon 10.3.40. To 1 CRU Cowley (SAS) 6.1.41. To 45 MU Kinloss 31.3.41. To 263 Sqdn, Filton 15.4.41. To Atlantic Coast, Barnstaple (MR Cat.B) 20.10.43. To 51 MU Lichfield 28.12.43. Sold 23.1.46 to Miles Aircraft Ltd. To H Hennequin & Cia, Argentina with CofA No.7643 issued 10.4.46; regd **LV-XOU** 30.6.47 to Secretaria Estado Aeronautica, Buenos Aires. To Aero Club Argentino. Crashed Estancia La Concepcion 17.2.47; 2 killed.

737 M.14A **L8281** Dd 3 Sqdn, Kenley 30.8.38. To 615 Sqdn, Old Sarum 14.6.39. To 27 MU Shawbury 30.11.39. To 615 Sqdn, Kenley 1.6.40. Destroyed in air raid Kenley 14.8.40. SOC 20.8.40.

738 M.14A **L8282** Dd 1 Sqdn, Tangmere 27.8.38. Overshot forced landing while lost and hit hedge Little Hallingbury, Essex 10.11.38. SOC from North Weald 3.1.39. Sold to British National Films Ltd and presumed used as film prop.

739 M.14A **L8283** Dd 19 Sqdn, Duxford 31.8.38. Cat.B at 54 MU Cambridge 18.5.40; to 1 CRU Cowley (SAS) 24.5.40. To 48 MU Hawarden 18.7.40. To Franco-Belgian Air Training School, Odiham 21.3.41. To 27 MU Shawbury 6.6.41. To 5 EFTS Meir 27.6.41. To 32 MU St Athan 2.10.41. To 8 EFTS Woodley 7.10.41. To 15 EFTS Carlisle 24.10.41. To 46 MU Lossiemouth 4.8.42. To 18 TSP B.56 Brussels/Evère 19.9.44. To 51 MU Lichfield (undated). Sold 4.2.46 to Miles Aircraft Ltd. To H Hennequin & Cia, Argentina with CofA No.7665 issued 24.4.46; regd **LV-XOV** 30.6.47 to Secretaria Estado Aeronautica, Buenos Aires. To DA Deportiva. Accident 21.12.51. Regn cld 4.52 as wfu.

740 M.14A **L8284** Dd 64 Sqdn, Church Fenton 1.9.38. Flying accident 25.2.41; to P&P for repairs 28.2.41. To 45 MU Kinloss 8.6.41. To 16 EFTS Burnaston 31.7.41. To 32 MU St Athan 26.3.42 (unconfirmed). To 3 EFTS Shellingford 17.8.42. To 45 MU Kinloss 14.5.43. To CRO 2.9.43. SOC Cat.E (MR) 22.6.44.

741 M.14A **L8285** Dd Stn Flt Northolt 31.8.38. To 141 Sqdn, Grangemouth 26.4.40. To 219 Sqdn, Catterick 7.6.40. To 600 Sqdn, Catterick 14.10.40. To 57 OTU Hawarden 29.4.41. Cat.B (MR) 28.12.43; repaired. To 51 MU Lichfield 18.2.44. Sold 3.5.49 to WA Rollason Ltd. Regd **G-AMMD** (CofR R3402) 15.9.51 to WA Rollason Ltd, Croydon. CofA issued 15.11.51. Regn cld 16.11.51 as sold abroad to New Zealand. Regd **ZK-AWX** (CofR 862) 11.2.52 to Piako Aero Club Inc, Matamata Airport; first flight in NZ 15.2.52. CofA No.730 issued 18.2.52. Regd 13.10.56 to Kawerau Aero Club. Crashed Kawerau 11.2.57. Regn cld 26.3.58.

742 M.14A **L8286** Dd 26 Sqdn, Catterick 1.9.38. To 607 Sqdn, Croydon 3.6.40. To RAF Kenley 15.7.40. Destroyed in air raid Kenley 20.8.40.

743 M.14A **L8287** Dd 1 ASU Cardington 31.8.38. To RAF Henlow 27.3.39. To 13 MU Henlow 1.3.40. Crashed avoiding building while low flying Letchworth Gasworks, Herts 20.7.40. To 54 MU Cambridge 28.7.40 and SOC 1.8.40.

744 M.14A **L8288** Dd 26 MU Cardington 5.9.38. To 27 E&RFTS Tollerton 31.8.39. To 24 EFTS Sydenham 15.10.39; coded '10'. Flying accident 21.6.40; to P&P for repairs 26.6.40. To 403 RCAF Sqdn, North Weald (31.12.41 census). To 32 MU St Athan 28.3.42. To 403 RCAF Sqdn, North Weald 8.4.42. To 539 Sqdn, Acklington 30.9.42. To Atlantic Coast, Barnstaple (MR) 2.2.43. To 51 MU Lichfield 30.4.43. Sold 10.3.47 to British Air Transport Ltd. Regd **G-AJRT** (CofR 11532) 30.4.47 to British Air Transport Ltd, Redhill. CofA No.9314 reserved but not issued. CofA No.A296 issued 3.1.52. Cld 14.6.54 & regd 16.6.54 to WS Shackleton Ltd, London W1. Cld 22.2.55 & regd 10.3.55 to John Robertson Johnston, Chobham. Fitted with canopy. Flown by owner to 3rd place (at 140 mph) in the 1955 King's Cup Air Race (Race No.43) and to 2nd place in the Kemsley Challenge Trophy Air Race at Baginton 20.8.55. Later fitted with two separate canopies. Cld 9.8.57 & regd 26.8.57 to Reginald Joseph Horace Parkes, Birmingham. Cld 14.5.59 & regd 22.5.59 to James Richard Walgate, Phyllis Mary Bird & Denise Imison t/a Grimsby Flying Club, Grimsby. Crashed Waltham, Grimsby 15.8.59. Sold to Experimental Flying Group, Croydon for spares; to Biggin Hill 1.60; extant 7.60. Regn cld 3.9.62 as pwfu.

745 M.14A **L8289** On static display at British Empire Exhibition, Glasgow prior to delivery. Dd 27 MU Shawbury 29.11.38. To Yeadon 10.6.40. To 24 EFTS Luton 13.9.41. To 2 EFTS Filton/Worcester 31.10.41. To 6 FIS Staverton 1.11.41. To 2 EFTS Worcester 22.7.42. To 46 MU Lossiemouth 11.8.42. To ECFS Hullavington 13.9.42. To Herts & Essex, Broxbourne (MR) 8.1.43. To 51 MU Lichfield 18.5.43. Sold 28.2.46 to Miles Aircraft Ltd. To H Hennequin & Cia, Argentina with CofA No.7784 issued 20.5.46; regd **LV-XOI** 30.6.47 to Secretaria Estado Aeronautica, Buenos Aires. Accident 26.9.46; repaired. Regd 7.8.62 to Aero Club Carhue. Still regd 5.67.

746 M.14A **L8290** Dd 26 MU Cardington 5.9.38. To 19 E&RFTS Gatwick 6.10.38. To 30 EFTS Burnaston 18.8.39. To 16 EFTS Burnaston 10.4.40. To 45 MU Kinloss 21.4.43. SOC 1.3.44.

747 M.14A **L8291** Dd 26 MU Cardington 7.9.38. To 19 E&RFTS Gatwick 3.10.38; coded 'F'. To 30 EFTS Burnaston 15.10.39. Damaged 8.12.39; repaired. To 9 MU Cosford 25.5.40. To 222 Sqdn, Kirton-in-Lindsey 28.8.40. To Hornchurch 29.8.40. To P&P (SAS) 9.12.40. To 15 EFTS Carlisle 13.6.41. To P&P for repairs 27.2.42 following incident previous day when aircraft entered steep dive after LAC WB Jordon's parachute slipped and pushed control column forward; pilot LAC WT Jenkins. To 16 EFTS Burnaston 22.4.42. To HQ FTC Woodley 21.4.43. To Herts & Essex, Broxbourne for repairs 22.7.43. To 51 MU Lichfield 18.10.43. Sold 25.11.46 to Miles Aircraft Ltd. CofA No.9840 issued 1.1.48 to **Thai AF**.

748 M.14A **L8292** Dd 26 MU Cardington 7.9.38. To 19 E&RFTS Gatwick 3.10.38; coded 'H'. To 30 EFTS Burnaston 15.10.39. To 16 EFTS Burnaston 10.4.40. Crashed 13.3.41. SOC 25.3.41.

749 M.14A **L8293** Dd Kenley 12.9.38. To Vickers-Armstrongs, Weybridge 2.5.39. To Armstrong Whitworth Aircraft, Whitley .42. To 51 MU Lichfield 14.5.43. Sold 15.2.46 to Miles Aircraft Ltd. To H Hennequin & Cia, Argentina with CofA No.7709 issued 7.5.46; regd **LV-XND** 30.6.47 to Secretaria Estado Aeronautica, Buenos Aires. To Circ. Universitario de Aviation. Accident 15.6.47; repaired. Accident 15.11.47.

750 M.14A **L8294** Dd Wittering 7.9.38. To Armstrong Whitworth Aircraft, Whitley 2.5.39. SOC Cat.E 5.4.44.

751 M.14A **L8295** Dd 23 Grp, Stn Flt Ternhill 13.9.38. To Boulton Paul Aircraft, Wolverhampton 24.4.39. To A&AEE Boscombe Down 5.11.41. To 45 MU Kinloss 13.10.43. To 51 MU Lichfield 28.2.44. Sold 9.2.46 to Miles Aircraft Ltd. To H Hennequin & Cia, Argentina with CofA No.7692 issued 3.5.46; regd **LV-XNJ** 30.6.47 to Secretaria Estado Aeronautica, Buenos Aires. To Aero Club Mar del Plata. Accident 22.9.46; repaired. Accident 5.4.47; repaired. Destroyed by fire 10.49.

752 M.14A **L8326** Dd 24 Sqdn, Hendon 13.9.38. To Handley Page Ltd, Radlett 22.11.38. To RAE Farnborough 14.3.41. To P&P for modifications (to take the Malinowski towed wing) 15.3.41. To A&AEE, Boscombe Down, where the first flight took place on 12.5.41 *(although the record shows that it was not delivered to Boscombe Down until 23.5.41!)* To RAE Farnborough 25.5.41. To P&P Woodley 29.5.41 to have the rudder and rudder horn balance extended. To RAE Farnborough 14.6.41. Accident at Farnborough 24.6.41 (also reported incorrectly as at Woodley) when swung into obstruction at side of runway on take-off; repaired. To P&P for fitment of Blackburn Cirrus Major engine. Trials ended due to further accident 22.8.41 (or 19.8.41). Ferried from RAE to P&P Woodley on 7.11.41 for re-fitment of Gipsy Major engine and return to standard configuration. Returned to RAE 21.2.42. To Atlantic Coast, Barnstaple (MI) 3.6.43. To 51 MU Lichfield 19.7.43. To RAE Farnborough for TCTF (flying practice) 27.3.45; again from 51 MU for flying practice on 18.4.45. Sold per CS(A) Return 8.5.50; dd Fairoaks Aero Club 12.5.50. Regd **G-AMBN** (CofR 15021) 10.5.50 to Universal Flying Services Ltd, op by Fairoaks Aero Club, Fairoaks. CofA No.10903 issued 26.7.50. Cld 10.3.54 & regd 17.3.54 to Wolverhampton Aviation Ltd, Wolverhampton. Crashed Wightwick, near Wolverhampton 25.4.54; pilot LE Mason killed & passenger Frederick Simmonds injured. Regn cld 8.5.54 as wfu.

753 M.14A **L8327** Dd Gosport 13.9.38. To Southampton 7.10.38. To Vickers-Armstrongs, Southampton 28.1.41. To P&P for repairs 8.8.41. To 5 EFTS Meir 22.9.41. To CFS Upavon 11.12.41. To 27 MU Shawbury 24.5.42. To 7 FIS Upavon. To 2 FIS Montrose 29.5.42. Accident 15.7.42; to P&P (Cat.B) 24.7.42. To 27 MU Shawbury 8.9.42. To 47 MU Sealand 26.2.43. Despatched to ME on SS Clan Lamont 19.4.43; arr 16.6.43. SOC 31.5.45.
*Note: Erroneously quoted becoming **G-AJGP** which was actually c/n 1979 ex T9692 (but regd with c/n 80327, which may have been its engine number).*

754 M.14A **L8328** Dd 30 E&RFTS Burnaston 21.9.38; renamed 30 EFTS 3.9.39. To 1 CRU Cowley (SAS) 3.4.40. To 45 MU Kinloss 24.6.40. To 15 EFTS Carlisle 8.11.40. Spun into wood during low aerobatics 1 ml NW of Southwaite Station, near Carlisle 18.4.41; pilot LAC Kibart. To P&P 24.4.41 and SOC 'total wreck'.

755 M.14A **L8329** Dd 30 E&RFTS Burnaston 22.9.38. Crashed into Rolls-Royce factory, Derby 3.1.39; pilot Sgt George A 'Tony' Bell killed and 5 girls hurt by flying debris and 5 would-be rescuers injured when the aircraft burst into flames (probably as a result of one of the rescuers smoking) (77:30 F/H). Bell had previously worked at the factory and had made his will the previous evening but no evidence was produced at the inquest to suggest suicide.

756 M.14A **L8330** Dd 30 E&RFTS Burnaston 22.9.38; renamed 30 EFTS 3.9.39. To 16 EFTS Burnaston 10.4.40. Stalled at low altitude and crashed near Stone, Staffs 28.10.41 and burnt out; P/O Lewis rescued from burning aircraft. SOC 12.11.41.

757 M.14A **L8331** Dd 30 E&RFTS Burnaston 22.9.38; renamed 30 EFTS 3.9.39. To 16 EFTS Burnaston 10.4.40. Spun into ground during stalling practice Etwall, 4 ml W of Burnaston 15.4.40; Sgt Miskelly killed, Sgt Bailey seriously injured. To 58 MU Newark 17.4.40 for salvage.

758 M.14A **L8332** Dd 30 E&RFTS Burnaston 22.9.38; renamed 30 EFTS 3.9.39. To 16 EFTS Burnaston 10.4.40. Damaged 8.7.41; to 1 CRU Cowley 11.7.41. To P&P 23.8.41. To 92 Sqdn, Gravesend 26.9.41. Engine cut, stalled and hit wires and crashed 1 ml SW of Wellingore 19.11.41 (783:50 F/H).

759 M.14A **L8333** Dd 30 E&RFTS Burnaston 27.8.38; renamed 30 EFTS 3.9.39. To 16 EFTS Burnaston 10.4.40. To 32 MU St Athan 16.2.42 but returned to 16 EFTS 24.2.42. Hit Magister N3849 on landing Burnaston 5.4.42 and DBR. SOC Cat.E1 (from 21 EFTS Booker (?) 12.8.42 (1586:55 F/H).

760 M.14A **L8334** Dd 26 MU Cardington 13.9.38. To 19 E&RFTS Gatwick 3.10.38. To 30 EFTS Burnaston 15.10.39. To 16 EFTS Burnaston 10.4.40. To 46 MU Lossiemouth 11.7.42. To ATA 30.11.43. Cat.B (MR) 7.1.44; re-Cat.E and SOC 7.2.44.

761 M.14A **L8335** Dd 24 MU Ternhill 19.9.38. To 15 E&RFTS Redhill 10.2.39. To 5 EFTS Hanworth 15.12.39. To 1 CRU Cowley (Cat.B) 24.7.40. To 37 MU Burtonwood 21.9.40. To 15 EFTS Carlisle 21.6.41. Stalled during practice forced landing and side-slipped into ground Burnfoot LG 2.9.41. SOC 6.9.41.

762 M.14A **L8336** Dd 24 MU Ternhill 19.9.38. To 15 E&RFTS Redhill 7.2.39; renamed 15 EFTS 3.9.39; to Carlisle 2.6.40. Stalled on landing and hit ground Kingstown, Carlisle 26.7.40. To P&P 31.7.40 and SOC 23.9.40.

763 M.14A **L8337** Dd RAF Staff College Andover 19.9.38. To 5 MU Kemble 1.2.41. To ATA (TFPP) White Waltham 5.2.41. To Herts & Essex, Broxbourne (SAS) 16.7.41. To 27 MU Shawbury 22.8.41. To 15 EFTS Carlisle 17.9.41. To 7 FIS Upavon 31.7.42. To 2 SofTT Cosford for GI as **4770M** 3.5.44 (allotted 22.4.44). To Belgian TTS, RAF Snailwell 1.1.45.

764 M.14A **L8338** Dd 3 ATS Sutton Bridge 21.9.38. To Stn Flt Sutton Bridge (30.11.39 census). To 264 Sqdn, Martlesham Heath 25.1.40. To Stn Flt Sutton Bridge 21.6.40. To 27 MU Shawbury 17.7.41. To ATA 6.10.41. To 5 MU Kemble 17.11.41. To 418 RCAF Sqdn, Debden 25.11.41. To 234 Sqdn, Ibsley 17.12.41, to Predannack 24.12.41, to Ibsley 31.12.41, to Warmwell 23.3.42. Cat.B 14.4.42; to P&P for repairs 21.4.42. To 45 MU Kinloss 9.6.42. To ECFS Hullavington 29.12.42. To 51 MU Lichfield 7.2.45. Sold 21.1.46 (or 29.1.46) to Miles Aircraft Ltd. To H Hennequin & Cia, Argentina with CofA No.7583 issued 27.3.46; regd **LV-XMQ** 30.6.47 to Secretaria Estado Aeronautica, Buenos Aires. Damaged by storm Jujuy 9.12.47; repaired. Regn cld 7.55 as wfu. Reported as regd 8.8.62 to Aero Club Trelew.

765 M.14A **L8339** Dd 4 SofTT St Athan 29.9.38. To Stn Flt Northolt 18.8.39. Collided on ground with Q.6 P5638 at Northolt 18.8.39. To 1 CRU Cowley for repairs 10.11.39. To 24 MU Ternhill 21.3.40. To 8 EFTS Woodley 20.5.40. To 5 EFTS Meir 12.8.40. Damaged in collision with L8149 21.6.41; to 1 CRU Cowley (SAS) 27.6.41. To 5 EFTS Meir 1.10.41. To 16 EFTS Burnaston 16.1.42. To 46 MU Lossiemouth 11.7.42. To 19 TSP, Ypres, France 19.9.44. To 51 MU Lichfield. Sold 18.3.46 to Miles Aircraft Ltd. Test flown 10.5.46 by Hugh Kendall as L8339. To H Hennequin & Cia, Argentina with CofA No.7827 issued 30.5.46; regd **LV-XOG** 30.6.47 to Secretaria Estado Aeronautica, Buenos Aires. To Aero Club Corrientes. Regn cld 7.54 as wfu.

766 M.14A **L8340** Dd 4 SofTT St Athan 28.9.38. Cat.B 5.10.41; repaired. To 6 FIS Staverton (31.12.41 census). To 154 Sqdn, Fowlmere 8.2.42. To 313 Sqdn, Churchstanton 8.6.42. To 256 Sqdn, Woodvale 7.10.42. To 51 MU Lichfield 5.2.43. To **3641M** 26.3.43 (allotted 23.3.44 for 6 SofTT Hednesford for GI). SOC Cat.E 25.10.44.

767 M.14A **L8341** Dd 4 SofTT St Athan 30.9.38. To 9 B&GS Penrhos 15.4.40. To 4 SofTT St Athan 3.10.40. To 41 OTU Old Sarum 2.10.41. To 45 MU Kinloss 18.9.42. To ATA 26.5.44. SOC Cat.E (MR) 20.7.44.

768 M.14A **L8342** Dd 24 MU Ternhill 20.9.38. To 15 E&RFTS Redhill 7.2.39; renamed 15 EFTS 3.9.39; coded '20'; to Carlisle 2.6.40. To 32 MU St Athan 1.10.41 (unconfirmed). To 15 EFTS Carlisle 11.10.41. To 21 EFTS Booker 16.6.42. Cat.B 3.7.42; to P&P for repairs 8.7.42. To 27 MU Shawbury 27.8.42. To ATA 11.4.43. SOC Cat.E1 4.10.44. To Miles Aircraft Ltd. To Irish Air Corps as **131** with CofA No.7455 issued 14.2.46; dd 9.3.46. Scrapped 6.52.

769 M.14A **L8343** Dd 24 MU Ternhill 20.9.38. To 15 E&RFTS Redhill 10.2.39; renamed 15 EFTS 3.9.39; to Carlisle 2.6.40. SOC Cat.E 23.7.40.

770 M.14A **L8344** Dd 24 Sqdn, Hendon 27.9.38. To 51 Grp Comm Flt, Yeadon 16.10.39. Stalled avoiding trees in forced landing in bad visibility near Yeadon 7.11.39. SOC 5.1.40.

771 M.14A **L8345** Dd 24 Sqdn, Hendon 27.9.38. Destroyed in air raid Hendon 8.10.40. SOC 17.10.40.

772 M.14A **L8346** Dd 24 Sqdn, Hendon 27.9.38. To Yeadon 16.10.39. To Digby 5.12.40. Cat.B 7.5.41; to 1 CRU Cowley (SAS). To 24 EFTS Luton 1.7.41. Lost height while low flying and hit ground 1/2 ml NE of Higham Gobion, Beds 4.8.41; Sgt GE Chugg & LAC A Bell killed. To 1 MPRD Cowley 6.8.41 and SOC 9.8.41.

773 M.14A **L8347** Dd 24 Sqdn, Hendon 27.9.38. To 73 Sqdn, Le Havre. To Octeville 17.9.39. Undercarriage collapsed in forced landing near Amiens, France 24.9.39.

774 M.14A **L8348** Dd 24 MU Ternhill 26.9.38. To 16 Grp Comm Flt, Rochester 14.2.39. Hit hedge on undershoot, stalled and hit ground Liton, Chatham 4.8.40. SOC 12.8.40.

775 M.14A **L8349** Dd 24 MU Ternhill 26.9.38. To 29 E&RFTS Luton 7.3.39. To 30 EFTS Burnaston 15.10.39. To 16 EFTS Burnaston 10.4.40. To 27 MU Shawbury 28.8.42. Cat.Ac (minor) accident with 27 MU 28.3.44. To ATA 3.5.44. To 51 MU Lichfield 4.6.44. Sold 26.2.48 to Gloster Flying Club, Staverton. Regd **G-AKMU** (CofR 12057) 27.11.47 to Laurence Dudley Trappitt, Whyteleafe, Surrey. Cld 22.7.48 & regd 14.10.48 to Robert Alan Short, Derek John Hayles & Martha Marodeen, Croydon/Portsmouth. Cld 2.6.49 & regd 7.6.49 to Christopher John de Vere, Ealing, London W5. CofA No.10627 issued 2.7.49. DBR in night landing Weston-super-Mare 4.11.49. Wings used on G-AJDR (10.50). Stored Belle Vue Garage, Bagshot. To Redhill for spares and scrapped 5.2.52. Regn cld 7.9.59 as pwfu.

776 M.14A **L8350** Dd 24 MU Ternhill 26.9.38. To 19 E&RFTS Gatwick 6.10.38. To 9 MU Cosford 12.10.39. To 5 EFTS Meir 29.8.40. To Atlantic Coast, Barnstaple (MI) 20.12.41. To 20 MU Aston Down 13.2.42. To ECFS Hullavington 24.5.42. To 46 MU Lossiemouth 4.7.42. To 51 MU Lichfield 27.12.43. Sold 20.2.46 to Miles Aircraft Ltd. Test flown 8.5.46 by Hugh Kendall as L8350. To H Hennequin & Cia, Argentina with CofA No.7803 issued 24.5.46; regd **LV-XOJ** 30.6.47 to Secretaria Estado Aeronautica, Buenos Aires. To Aero Club Posadas. Regn cld 7.54 as wfu.

777 M.14A **L8351** Dd 24 MU Ternhill 26.9.38. To 19 E&RFTS Gatwick 6.10.38; coded 'N'. To 30 EFTS Burnaston 15.10.39. Cat.B 8.12.39; to P&P (SAS) 5.1.40. To 46 MU Lossiemouth 21.6.40. To 241 Sqdn, Inverness 23.10.40. To 27 MU Shawbury 18.7.41. To ATA 9.8.41. To 15 MU Wroughton 26.12.41. To Herts & Essex, Broxbourne 5.7.43. To 51 MU Lichfield 20.3.46. Sold 4.12.47 to JE Tomlinson. Regd **G-AKMJ** (CofR 12047) 26.11.47 to John Edward Tomlinson, Burton-on-Trent. CofA No.9908 issued 22.1.48. Regn cld 8.12.48 as sold in South Africa. Flown to South Africa by owner, regd **ZS-DBF** 18.11.48 to JWJ Claessens, Durban. Regn cld 6.50 as sold in Kenya. Regd **VP-KIK** (3.52) to J Wright & B Hurle-Hobbs, Nairobi. Regd 12.53 to Miss J Wright, Kabete. Regn cld as sold in Belgian Congo. Regd **OO-CRU** (CofR 1087/C.263) 25.7.56 to Aéro Club de Ruanda-Urundi, Usumbura. Regn cld 23.2.57 or 6.3.57.

778 M.14A **L8352** Dd 24 MU Ternhill 27.9.38. To 15 EFTS Redhill 6.4.40; to Carlisle 2.6.40. To Harrowbeer 19.10.41. SOC Cat.E 5.9.44. To Miles Aircraft Ltd. To Irish Air Corps as **137** with CofA No.7504 issued 24.2.46; dd 9.3.46. Scrapped 11.52.

779 M.14A **L8353** Dd 24 MU Ternhill 3.10.38. To 27 E&RFTS Tollerton 12.4.39. To 8 EFTS Woodley 15.10.40. To 10 FIS Woodley 21.7.42. To 29 EFTS Clyffe Pypard 18.1.43. *(Reported as coded 'FDT-E' of 10 FIS in 1946).* To 8 EFTS Woodley (undated). To 51 MU Lichfield 28.5.46. Sold 3.5.49 to WA Rollason Ltd. Regd **G-AMMC** (CofR R3401) 15.9.51 to WA Rollason Ltd, Croydon. CofA issued 1.10.53. Regn cld 1.10.53 as sold New Zealand. Regd **ZK-AYW** (CofR 1015) 18.11.53 to Waitomo Aero Club Inc, Te Kuiti. CofA No.1013 validated 3.12.53. Regd 14.2.61 to RC Perry, Te Kuiti. Regd 23.7.61 to Piako Aero Club Inc, Matamata. Wfu after last flight made on 2.9.62. To Museum of Transport & Technology, Auckland 20.11.67 and on static display (TT 4,302 hrs). Regn cld 19.12.80 as wfu.

780 M.14A **L8354** Dd 24 MU Ternhill 3.10.38. To 16 EFTS Burnaston 13.7.40. To 27 MU Shawbury 29.8.42. To IFTS ATA Barton-in-the-Clay 10.6.43. Engine failure and hit HT cable during forced landing and crashed Burton, near Aylesbury, Bucks at 11.35 on 24.8.43; Cadet S Fong slightly injured.

781 M.14A **L8355** Dd 24 MU Ternhill 3.10.38. To 15 EFTS Redhill 6.4.40; to Carlisle 2.6.40. Wing touched the surface while low flying over Lake Ullswater and aircraft crashed into the lake 3 ml SW of Pooley Bridge, Cumberland on 31.12.40; LAC J Senior unhurt. SOC 8.1.41.

782 M.14A **L8356** Dd 24 MU Ternhill 6.10.38. To 19 E&RFTS Gatwick 18.10.38. To 30 EFTS Burnaston 15.10.39. To 16 EFTS Burnaston 10.4.40. Spun into ground Draycott-in-the-Clay, Staffs 22.5.42; Cpl JH Allan killed. SOC 31.5.42.

783 M.14A **L8357** Dd 24 MU Ternhill 6.10.38. To 19 E&RFTS Gatwick 18.10.38. To 30 EFTS Burnaston 15.10.39. To 16 EFTS Burnaston 10.4.40. Cat.B; to 58 MU Newark 31.5.40; to P&P 2.6.40. To 27 MU Shawbury 30.9.40. To 65 Sqdn, Turnhouse 25.11.40; to Westhampnett 7.10.41. Cat. B 10.10.41; repaired. To 27 MU Shawbury 21.12.41. To 6 FIS Staverton (31.12.41 census). To 22 EFTS/4 FIS Cambridge 25.1.42. To 22 EFTS Cambridge 13.5.43. MR (Cat.B) 23.11.43; repaired. To 51 MU Lichfield 4.1.44. Sold 29.10.46 to Miles Aircraft Ltd. CofA No.9037 issued 20.2.47 to John M Gamble, Wellington, NZ. Regd **ZK-ANJ** (CofR 361) 21.7.47 to Wellington Aero Club Inc, Wellington. NZ CofA validated 17.7.47. Badly damaged when struck leeward boundary of aerodrome in forced landing competition Rongotai 19.8.50; pilot CE Pasducci & instructor passenger JF Wright; not repaired. Regn cld 1.3.57.

784 M.14A **L8358** Dd 24 MU Ternhill 6.10.38. To 16 EFTS Burnaston 5.7.40. To 1 CRU Cowley for repairs 28.9.40. To 27 MU Shabury 26.11.40. To 8 EFTS Woodley 10.2.41. To 10 FIS Woodley 21.7.42. Cat.B; to Atlantic Coast, Barnstaple for repairs 14.9.43. To 51 MU Lichfield 18.10.43. Sold 13.11.46 to Miles Aircraft Ltd. CofA No.9400 issued 28.5.47. Regd in France as **F-BDPL** (CofR 20006) 13.6.47 to Michel Vernier, Paris. Regd 18.6.48 to Societe Compagnie Marocaine d'Aviation, Casablanca. Regd 28.6.50 to Aero Club de Casablanca les Ailes Cherifiennes, Casablanca. Regd 5.11.52 to Societe Maroc Air Service, Rabat. Regd 18.2.54 to Aero Club de Casablanca les Ailes Cherifiennes, Casablanca. Regd 8.6.55 to Societe Piperavia Maroc, Casablanca. Regn cld 22.9.71 as destroyed.

785 M.14A **L8359** Dd 24 MU Ternhill 6.10.38. To 74 Sqdn, Hornchurch 8.5.40. To 302 Polish Sqdn, Leconfield 19.7.40. To 8 MU Little Rissington 2.10.40. To P&P (SAS) 9.10.40. To 5 MU Kemble 20.4.41. To 4 FIS Cambridge 10.6.41. To 22 EFTS Cambridge 13.5.43. To 4 AGS Morpeth 24.4.44. To Wymeswold 13.12.44. To 108 (T)OTU Wymeswold 21.12.44; became 1382 TSTU 10.8.45. To 1333 TSCU Syerston 10.5.46. To 51 MU Lichfield 17.1.47. Sold 1.1.48 to WA Rollason Ltd. Regd **G-AKRH** (CofR 12144) 20.1.48 to WA Rollason Ltd, Croydon. CofA No.10130 issued 27.5.48. Cld 27.5.48 & regd 29.6.48 to Stanley James Bartlam, Birmingham. Cld 4.9.50 & regd 25.9.50 to WS Shackleton Ltd, London W1. Cld 25.4.51 & regd 27.4.51 to Michael Rupert Thomas Chandler, RAF Cottesmore. Cld 20.11.51 & regd 24.11.51

to Loxham's Flying Services Ltd, Lancaster. Cld 30.3.53 & regd 1.4.53 to Blackpool & Fylde Aero Club Ltd, Squires Gate. CofA lapsed 25.6.53. CofA renewed 11.11.55. CofA lapsed 10.11.56. Stored (5.61) (by owner Russell L Whyham) in a garage in Smithy Lane, Blackpool. Scrapped Squires Gate in 1963, parts reportedly to G-AKAT. Regn cld 20.2.63 as pwfu.

796 M.14A Hawk Trainer Mk.III? CLN records nothing against c/n 796 but no c/n, previous regn, or CofA, has ever been linked to the Latvian Magister s/n **181**. This machine was purchased in 1938 on behalf of the Latvian VEF factory by Phillip de W. Avery, their British representative, however, see also c/n 636. According to one source, it was flown to Latvia by Phillip de W. Avery, although other sources say that it was shipped out in the summer of 1939 and assembled by VEF. Evaluated first by the VEF factory pilots Rudzitis and Mikelsons, it was later handed over to the Latvian Aviation Regiment, where it was given s/n **181**. It was acquired only as a temporary measure as it was intended that the indigenous VEF I-17 would fulfill the monoplane trainer function, it being claimed by VEF that the I-17 would be a similar and better trainer. S/n **181** was flown for familiarisation by Capt Augusts Graudins in 9.39, and it apparently remained hidden in the VEF factory after the Russian occupation of Latvia in 6.40. However, it was found and evaluated at Riga in Luftwaffe marks coded AW+12 by the Ausbildungswesen unit (the letters underlined being used in the code). On 21.5.42, it was flown by Luftwaffe (former Estonian AD) pilot Habel from Riga/Kalnciems, Latvia, to Riga/Spilve and on to Tallinn, Estonia. There it was coded SB+AF of the Estonian Staffel Buschman (Gerhard Buschman was a pre-war Baltic German sports pilot who was by then in the Luftwaffe). SB was not strictly a part of the Luftwaffe but a group of former Estonian Air Defence pilots, commanded by Buschman, and was used on coastal patrols in the Gulf of Finland and loosely under the auspices of the German police. SAGr.127, which listed aircraft in service with the Luftwaffe, showed two Magisters (probably both with the unit code letters 'SB') as being on strength from 6.42 to 9.42; the other would have been the ex Estonian s/n 159. However, **181**, along with a DH Dragon Rapide and 4 Stampe SV-5s, were ferried by SB personnel to Ülemiste in 5-6.42. At 12.17 on 11.11.42, the Estonian pilot Eduard Lepp, with Nomm, left Tallinn on a recce flight but, at 14.47, due to a dead engine, he had to carry out a forced landing in a field at Harku, 20 km from Tallinn. Near the end of the landing run, the left undercarriage leg hit a small stone and broke, but the aircraft was otherwise undamaged and the crew were uninjured. The engine failure was caused by 'a bit of ice in the fuel tank which stopped the carburettor feeding pipe'; the undercarriage was never repaired and nothing further is known.

797 M.14A Hawk Trainer Mk.III; fitted with tall rudder. CofA No.6420 issued 21.10.38 to HM King Ghazi I of Iraq. Regd **YI-GFH** (CofR 15) 15.8.38 to HM King Ghazi I; dd with the name 'Shatt Al Arab' painted on the top cowling.

798 M.14A Hawk Trainer Mk III/Magister. Contractor's Flt Test Report was dated 1.12.38. CofA No.6453 issued 1.12.38 to Egyptian Army Air Force; dd as **L220** to Almaza, Cairo in 1938.

811 M.14A **L8150** Dd 27 MU Shawbury 1.6.38. To 48 MU Hawarden 4.7.40. To 15 EFTS Carlisle 1.6.41. To 29 EFTS Clyffe Pypard 26.3.42. To 45 MU Kinloss 15.4.43. To 51 MU Lichfield 28.2.44. Sold 19.2.46 to Miles Aircraft Ltd (and erroneously assumed by them to be c/n 628 when refurbished). Test flown 23.4.46 by Hugh Kendall as L8150. To H Hennequin & Cia, Argentina with CofA No.7736 issued 9.5.46; regd **LV-XOL** 30.6.47 to Secretaria Estado Aeronautica, Buenos Aires. To Aero Club Rosario. Accidents 10.2.52; repaired. Accident 27.9.56; repaired. Regd 19.10.60 to Aero Club San Luis. Regd 1.9.70 to Lloveras Quenon & Marco Nicanor, Rio Cuarto.

812 M.14A Hawk Trainer Mk.III/Magister. CofA No.6347 issued 12.7.38 to Egyptian Army Air Force; dd as **L211** 20.6.38 to Almaza, Cairo.

813 M.14A Hawk Trainer Mk.III/Magister. CofA No.6349 issued 12.7.38 to Egyptian Army Air Force; dd as **L212** 29.6.38 to Almaza, Cairo.

814 M.14A Hawk Trainer Mk.III/Magister. CofA No.6357 issued 20.7.38 to Egyptian Army Air Force; dd as **L213** 6.7.38 to Almaza, Cairo.

815 M.14A Hawk Trainer Mk.III/Magister. CofA No.6380 issued 19.8.38 to Egyptian Army Air Force; dd as **L214** 13.7.38 to Almaza, Cairo.

816 M.14A Hawk Trainer Mk.III/Magister. CofA No.6374 issued 19.8.38 to Egyptian Army Air Force; dd as **L215** 20.7.38 to Almaza, Cairo.

817 M.14A Hawk Trainer Mk.III/Magister. CofA No.6389 issued 31.8.38 to Egyptian Army Air Force; dd as **L216** 27.7.38 to Almaza, Cairo

818 M.14A Hawk Trainer Mk.III/Magister. CofA No.6392 issued 9.9.38 to Egyptian Army Air Force; dd as **L217** 19.9.38 to Almaza, Cairo.

819 M.14A Hawk Trainer Mk.III/Magister. CofA No.6405 issued 28.9.38 to Egyptian Army Air Force; dd as **L218** 20.9.38 to Almaza, Cairo.

820 M.14A Hawk Trainer Mk.III/Magister. CofA No.6413 issued 14.10.38 to Egyptian Army Air Force; dd as **L219** 1.10.38 to Almaza, Cairo.

821 M.14A **N3773** Dd 24 MU Ternhill 12.10.38. To 19 E&RFTS Gatwick 18.10.39. To 30 EFTS Burnaston 15.10.39. To 16 EFTS Burnaston 10.4.40. Stalled at low altitude and spun into ground near Longford, Derby 3.4.40. SOC 10.4.40.

822 M.14A **N3774** Dd 24 MU Ternhill 12.10.38. To 16 EFTS Burnaston 13.7.40; coded "60". To 3 EFTS Shellingford 14.8.42. To 16 (P)SFTS Newton 20.2.43. Engine cut, overshot forced landing into ditch 1/2 ml N of Bingham, Notts 26.9.43. SOC 6.10.43.

823 M.14A **N3775** Dd 24 MU Ternhill 12.10.38. To 30 EFTS Burnaston 30.9.39. To 16 EFTS Burnaston 10.4.40. To 27 MU Shawbury 11.9.42. To ATA 27.4.43. To Atlantic Coast, Barnstaple (MI) 18.10.43. To 51 MU Lichfield 28.12.43. Sold 4.12.47 to DC Jemmett. Regd **G-AKPE** (CofR 12117) 27.1.48 to Douglas Charles Jemmett, Bridgnorth. CofA No.9959 issued 26.2.48. Cld 7.7.55 & regd 12.7.55 to Derby Aviation Ltd, Burnaston. Cld 7.5.56 & regd 15.5.56 to Henry Ellis Riley, Mansfield, Notts. Cld 1.2.58 & regd 15.4.58 to Dennis Stanley Ascott & Abraham Cooper, Derby. Cld 7.3.60 & regd 14.3.60 to David & John Adlington, Chesterfield, Derby. Cld 16.3.62 & regd 6.4.62 to John William Stobart, Brampton & George Lionel Spencer Lightfoot, Carlisle. Wfu Crosby-on-Eden with current CofA. Regn cld 12.12.63 as pwfu. Donated to local school at Austin Friars, Carlisle. Reportedly scrapped in 1966 after gale damage.

824 M.14A **N3776** Dd 'H' MU Waddington 5.11.38. To 5 MU Kemble 13.2.39. To 5 EFTS Hanworth 3.4.40. Spun into ground off turn Meir, Staffs 2.2.41; P/O GC Blunt & LAC JA Ashley killed. SOC 10.2.41.

825 M.14A **N3777** Dd 24 MU Ternhill 12.10.38. To 8 EFTS Woodley 5.3.40. To 24 EFTS Sydenham 9.6.40; to Luton 22.7.40. Flying accident (Cat.B) 9.7.41; to 1 CRU Cowley (SAS) 13.7.41. To 27 MU Shawbury 12.8.41. To 74 Sqdn, Acklington 29.9.41. To 232 Sqdn, Atcham 8.4.42. To Atlantic Coast, Barnstaple (Cat.B) 17.12.42. To 51 MU Lichfield 6.2.43. Sold 12.2.47 to FK McIntyre. Regd **G-AJHE** (CofR 11294) 11.2.47 to Frank Keiller McIntyre, Darlington. CofA No.9057 issued 14.4.47. CofA lapsed 13.4.48. Cld 12.5.48 & regd 20.5.48 to John Neasham, Croft. Cld 29.11.49 & regd 10.12.49 to Darlington & District Aero Club Ltd, Croft. CofA renewed 8.12.49. Regn cld 21.12.49 as sold abroad. Regd in French Morocco as **F-OAFU** 24.3.50 to Aero Club de Meknes, Meknes. Regn cld 19.12.56 as destroyed.

826 M.14A **N3778** Dd 24 MU Ternhill 12.10.38. To 30 EFTS Burnaston 2.11.39. To 16 EFTS Burnaston 10.4.40. To 3 EFTS Shellingford 14.8.42. To 16 (P)SFTS Hucknall 20.2.43. To 51 MU Lichfield 20.11.43. Sold 3.5.49 to WA Rollason Ltd, Croydon but not registered.

827 M.14A **N3779** Dd 24 MU Ternhill 12.10.38. To 15 EFTS Redhill 2.4.40; to Carlisle 2.6.40. To 29 EFTS Clyffe Pypard 26.5.42. To 21 EFTS Booker 22.3.43. Engine cut on take-off, hit hedge Booker 27.9.43. To Atlantic Coast, Barnstaple and SOC 9.11.43.

828 M.14A **N3780** Dd 24 MU Ternhill 12.10.38. To 15 EFTS Redhill 5.4.40; to Carlisle 2.6.40; coded '49'. To 21 EFTS Booker 8.6.42. SOC 25.5.44.

829 M.14A **N3781** Dd 24 MU Ternhill 12.10.38. To 16 EFTS 16.4.40. To 46 MU Lossiemouth 9.7.42. To 63 OTU Honiley 4.9.43. To 60/63 OTU Combined Gunnery Flt, Chedworth 12.43-1.44. To 51 MU Lichfield 12.4.44. Sold 13.3.46 to Miles Aircraft Ltd. Test flown 25.4.46

APPENDIX 20

by Hugh Kendall as N3781. To H Hennequin & Cia, Argentina with CofA No.7733 issued 9.5.46; regd **LV-XNQ** 30.6.47 to Secretaria Estado Aeronautica, Buenos Aires. Accident 2.7.46; repaired. Crashed San Justo 2.2.47 (30% damage); repaired. Regn cld 8.52 as wfu.

830   M.14A **N3782**   Dd 27 MU Shawbury 18.10.38. To 15 E&RFTS Redhill 21.11.38. To 5 EFTS Hanworth 15.10.39. To 37 MU Burtonwood 9.12.41. To 2 FIS Montrose 13.1.42. To Atlantic Coast, Barnstaple (Cat.B) 1.6.42. To 46 MU Lossiemouth 9.8.42. To A&AEE Boscombe Down 22.12.44. To 5 MU Kemble 9.10.46. SOC 10.12.47 as scrap.

831   M.14A **N3783**   Dd 27 MU Shawbury 18.10.38. To 15 E&RFTS Redhill 21.11.38. To 5 EFTS Hanworth 15.10.39. To 2 FIS Montrose 9.1.42. To 12 MU Kirkbride 11.8.42. SOC 5.10.45.

832   M.14A **N3784**   Dd 27 MU Shawbury 18.10.38. To 19 E&RFTS Gatwick 24.11.38. Engine cut, hit hedge in forced landing Sidlow Bridge, Surrey 18.5.39 (TT 70:40 F/H).

833   M.14A **N3785** Dd 27 MU Shawbury 18.10.38. To 28 E&RFTS Meir 30.12.38. To 30 EFTS Burnaston 20.9.39. To 16 EFTS Burnaston 10.4.40. To 17 (P)AFU Calveley 1.5.43. To 51 MU Lichfield 23.2.44. Sold 7.6.46 to Miles Aircraft Ltd. To H Hennequin & Cia, Argentina with CofA No.8259 issued 20.8.46; regd **LV-XSP** 30.6.47 to Secretaria Estado Aeronautica, Buenos Aires. To Censo General de la Nacion. Crashed Goya 5.8.47 (95% damage); pilot injured.

834   M.14A **N3786**   Dd 27 MU Shawbury 18.10.38. To 19 E&RFTS Gatwick 24.11.38. To 30 E&RFTS Burnaston (undated); renamed 30 EFTS 3.9.39. Hit trees recovering from dive and crashed Culland Hall, Hollington, Derby 21.9.39. SOC 6.12.39.

835   M.14A **N3787**   Dd 27 MU Shawbury 18.10.38. To 19 E&RFTS Gatwick 24.11.38. To 30 EFTS Burnaston 15.10.39. To 16 EFTS Burnaston 10.4.40. Spun into ground on approach Burnaston 21.5.41. To P&P 29.5.41 and SOC 3.6.41.

836   M.14A **N3788**   Dd 27 MU Shawbury 18.10.38. To 27 E&RFTS Tollerton 28.12.38. To 8 EFTS Woodley 15.8.39. To 5 EFTS Meir 2.8.40. To 37 MU Burtonwood 8.12.41. To 2 FTS Brize Norton 13.1.42 To 2 FIS Montrose 12.2.42. To P&P 1.7.42. To 46 MU Lossiemouth 12.8.42. To 169 Sqdn, Little Snoring 13.5.44. To 51 MU Lichfield 27.8.45. Sold 27.8.47 to Laygold & Co, Blackbushe. Regd **G-ANLT** (CofR R4384) 7.1.54 to Wolverhampton Aviation Ltd, Wolverhampton. Not refurbished but fuselage, centre section, tailplane, fin and rudder, flaps and u/c to G-AKPF at Burnaston in 5.55, what remained was burnt there in 1957. Regn cld 21.7.59 as pwfu.
Note: G-AKPF was repainted as N3788 in late 2004 to reflect its 'true' ancestry.

837   M.14A **N3789**   Dd 27 MU Shawbury 18.10.38. To 29 E&RFTS Luton 10.2.39. To 30 EFTS Burnaston 15.10.39. To 16 EFTS Burnaston 10.4.40. Stalled on approach Tatenhill and hit ground 24.4.41. SOC 7.5.41.

838   M.14A **N3790**   Dd 27 MU Shawbury 18.10.38. To 25 E&RFTS Waltham 4.5.39. To 24 EFTS Sydenham 15.10.39; to Luton 22.7.40. To 1 CRU Cowley for repairs 4.11.40. To Stn Flt Syerston 10.1.41. To P&P for repairs 18.3.41. To 27 MU Shawbury 2.10.41. To 131 Sqdn, Atcham 13.10.41. To 51 MU Lichfield 21.10.42. SOC 25.2.44.

839   M.14A **N3791**   Dd 27 MU Shawbury 18.10.38. To 29 E&RFTS Luton 10.2.39. Hit trees in low cloud and crashed Pepperstock, Slip End, 2 ml S of Luton, Beds 21.2.39; Flt Lt Donaldson & F/Sgt Basson injured. To 3 MU Milton, Didcot for reduction to spares.

840   M.14A **N3792** Allocation to 27 MU Shawbury cancelled and dd direct to 15 E&RFTS Redhill 20.10.38. Overshot landing, finished up in bushes Redhill 22.5.39. SOC for 'weight of metal'.

841   M.14A **N3793** Allocation to 27 MU Shawbury cancelled and dd direct to 15 E&RFTS Redhill 20.10.38; renamed 15 EFTS 3.9.39; to Carlisle 2.6.40. To 29 EFTS Clyffe Pypard 26.5.42. To 3 GTS Stoke Orchard 18.3.43. To Herts & Essex, Broxbourne (MR) 25.8.43. To 51 MU Lichfield 10.1.44. Sold 22.9.47 to Miles Aircraft Ltd. CofA No.9877 issued 22.3.48 to **Thai Navy**.

842   M.14A **N3794**   Dd 27 MU Shawbury 26.10.38. To 30 E&RFTS Burnaston 10.3.39; renamed 30 EFTS 3.9.39. Stalled during forced landing while lost near Kneesall, Notts 16.3.40. Wreck to 58 MU Newark 16.3.40 and SOC 27.3.40.

843   M.14A **N3795**   Dd 27 MU Shawbury 26.10.38. To 25 E&RFTS Waltham 23.3.39. To 24 EFTS Sydenham 15.10.39; to Luton 22.7.40. To 2 EFTS Filton 31.10.41. To 6 FIS Filton 1.11.41. Reverted to 2 EFTS Worcester 22.7.42. To 46 MU Lossiemouth 11.8.42. To 104 SP Landing Ground B.6 Coulombs 19.9.44. To 51 MU Lichfield 25.11.44. Sold 3.12.46 to The Wiltshire School of Flying Ltd. Regd **G-AIUC** (CofR 10971) 11.11.46 to The Wiltshire School of Flying Ltd, Thruxton. CofA No.9486 issued 24.9.47. Cld 7.2.50 & regd 1.3.50 to WS Shackleton Ltd, London W1. Regn cld 24.3.50 as sold abroad. Regd in French Morocco as **F-OAGQ** (CofR 246237) 19.12.50 to Aero Club de Tanger, Tangier/Agadir. Regd to Henri Carton, Casablanca. Regd to Aero Club d'Agadir, Agadir. Regd 31.10.62 to Michel Elias, Casablanca/Tit Mellil. Regn cld 24.12.63 as sold abroad. Regd **CN-TZE** (CofR 13499) 4.2.64 to Michel Elias, Casablanca/Tit Mellil. CofA lapsed 31.12.64. Regn cld .65. Reported extant at Marrakesh c73.

844   M.14A **N3796**   Dd 27 MU Shawbury 26.10.38. To 26 E&RFTS Kidlington 3.1.39. To 5 EFTS Hanworth 15.10.39; to Meir 17.6.40. To 2 FIS Montrose 9.1.42. To P&P 26.4.42 and SOC 23.5.42.

845   M.14A **N3797**   Dd 27 MU Shawbury 26.10.38. To 27 E&RFTS Tollerton 28.12.38. To 32 MU St Athan 14.9.39. To 6 MU Brize Norton 25.5.40. To 16 EFTS Burnaston 29.6.40. Undercarriage collapsed in heavy landing Burnaston 11.2.42. SOC 12.2.42.

846   M.14A **N3798**   Dd 27 MU Shawbury 26.10.38. To 27 E&RFTS Tollerton 28.12.38. To 8 EFTS Woodley 15.10.39. Damaged in accident at Theale RLG 19.7.40. To 1 CRU Cowley (Cat.B) 26.7.40. To 37 MU Burtonwood 26.9.40. To 15 EFTS Carlisle 5.6.41. To 29 EFTS Clyffe Pypard 24.5.42. To 46 MU Lossiemouth 14.4.43. To 1 (O)AFU Wigtown 30.10.44. To 22 SFTS Calveley 7.11.45; coded "FCJ-G", moved to Ouston 22.5.46. To 51 MU Lichfield 13.7.47. Sold 27.5.48 to Short Bros & Harland Ltd, Rochester but not registered. Noted dismantled at Rochester 7.51.

847   M.14A **N3799**   Dd 27 MU Shawbury 26.10.38. To 25 E&RFTS Waltham 20.3.39. To 24 EFTS Sydenham 15.10.39; to Luton 22.7.40. Hit wires low flying and crashed Sandridge, Herts 4.4.41; Acting LA V Morley killed. SOC 14.4.41.

848   M.14A **N3800**   Dd 27 MU Shawbury 26.10.38. To 25 E&RFTS Waltham 20.3.39. To 32 MU St Athan 14.9.39. To 9 MU Cosford 10.6.40. To 8 EFTS Woodley 16.2.41. To 10 FIS Woodley 21.7.42. To 7 FIS/CFS Upavon 6.5.46; coded "FDK-D". To 51 MU Lichfield 15.7.47. Sold 27.5.48 to Short Bros & Harland Ltd, Rochester but not registered. Noted dismantled at Rochester 7.51. The wings were noted in storage at Biggin Hill 12.4.59 (along with four other sets of wings) having been brought from Croydon.

849   M.14A **N3801**   Dd 27 MU Shawbury 26.10.38. To 26 E&RFTS Kidlington 28.3.39. To 5 EFTS Hanworth 15.10.39. To P&P 6.4.40. To 46 MU Lossiemouth 11.7.40. To 614 Sqdn, Grangemouth 26.10.40. To 46 MU Lossiemouth 30.5.41. To Stn Flt Peterhead 13.7.41. To 416 RCAF Sqdn, Peterhead (detached Dyce) 5.5.42. To 602 Sqdn, Peterhead 15.7.42. To 164 Sqdn, Peterhead 6.9.42.   To 51 MU Lichfield 27.10.42. SOC 25.2.44 Cat.E (MR). *Reportedly seen at Croydon with Willis Hole Aviation Ltd in 1950 by author but probably misrecording of N3805.*

850   M.14A **N3802** Dd 27 MU Shawbury 26.10.38. To 3 Grp Comm Flt/Stn Flt Mildenhall 21.11.39. To 3 Grp Comm Flt, Newmarket 1.8.41. To Prestwick 5.8.42. To 46 MU Lossiemouth 6.8.42. To ATA 15.10.43. Cat.B (MR) 11.11.43. To 51 MU Lichfield 28.12.43. To RAE Farnborough 26.6.46. To Reserve Command Comm Sqdn/Staff College Flight, White Waltham 16.10.46. To 51 MU Lichfield 9.8.49. Sold (by gift) 8.12.49 to Lucien Lecoq, Toussus-le-Noble, France. Lecoq (thought at time to be an agent working for Germany) had flown his Comper Swift F-AOTP to Marchwood, nr Southampton 12.42 (or 9.42) to offer his services to the Allies; this was an attempt to recompense him for the loss of his Swift! Reportedly issued with a permit to fly on 26.1.50 for ferrying from Lympne to France but CofA No.10886 actually issued 4.7.50 to Pierre Dundal. Regd **F-BCDU** (CofR 19027) 18.1.52 to Lucien Lecoq, Vincennes. Regd 18.1.52 to Aero Club Paris Nord, Persan. Regd 1.12.55 to Armand Maze, Taverny (based Persan-Beaumont). CofA suspended 9.6.60. Regn cld 22.9.71 as destroyed.

851 M.14A **N3803** Dd 27 MU Shawbury 31.10.38. To 15 E&RFTS Redhill 10.11.38; renamed 15 EFTS Redhill 3.9.39; to Carlisle 2.6.40. SOC 23.7.40.

852 M.14A **N3804** Dd 27 MU Shawbury 28.10.38. To 15 E&RFTS Redhill 10.11.38. To 5 EFTS Hanworth 15.10.39, later Meir 18.6.40. To P&P (Cat.B) 11.7.40. To 27 MU Shawbury 22.8.40. To Abingdon 19.11.40. To 610 Sqdn/Stn Flt Acklington 25.11.40. To Herts & Essex, Broxbourne for repairs 28.7.42. To 27 MU Shawbury 8.9.42. To 76 MU Wroughton 13.5.43 for packing. To **Turkey** 20.5.43 on SS Clan McIver; arr ME 14.6.43.

853 M.14A **N3805** Dd 27 MU Shawbury 31.10.38. To 15 E&RFTS Redhill 16.11.38. To 5 EFTS Hanworth 15.10.39. To 2 FIS Montrose 9.1.42. MR 6.7.42; to Shrager Bros, Old Warden 18.7.42. To 46 MU Lossiemouth 28.8.42. To 57 OTU Eshott 30.9.43. To 51 MU Lichfield 23.7.45. To Staff College White Waltham (by 21.3.46 census). To Reserve Command Comm Sqdn, White Waltham 16.10.46. To Maintenance Command Comm Sqdn, Andover (for Staff College, Brackley) 2.3.48; coded "TBR". Cat.B 17.5.49. To 5 MU Kemble 26.8.49. Sold 2.1.50 to RA Short, Croydon but not registered.

854 M.14A **N3806** Dd Stn Flt Donibristle 12.11.38. To Stn Flt Turnhouse 24.5.39. Dived into ground 3 ml SW of Turnhouse, 24.12.39, cause not known. SOC 6.6.40.

855 M.14A **N3807** Dd Stn Flt Donibristle 8.11.38. To Stn Flt Turnhouse 24.5.39. To Stn Flt Leuchars (same date). To Stn Flt Port Ellen 4.5.41. To 1 CRU Cowley 31.5.41. To Stn Flt Islay 14.8.41. To Stn Flt Skeabrae 17.12.41. To Stn Flt Peterhead 7.4.43. SOC 15.5.45. To Miles Aircraft Ltd. To Portuguese AF **191** with CofA No.7549 issued 18.3.46; later **1202**.

856 M.14A **N3808** Dd Stn Flt Donibristle 8.11.38. Control lost in cloud, dived into ground near Laurencekirk, Kincardine 16.11.38; F/O Peter M Hamilton-Hall killed (F/H 6.40).
Record card shows allocated from RAF Ouston to 8 FTS Montrose 5.12.38 and then marked as crashed 27.2.39 (sic); also reported as being on charge of RAF Lee-on-Solent at time of 16.11.38 accident.

857 M.14A **N3809** Dd Stn Flt Donibristle 8.11.38. To Stn Flt Leuchars 19.4.39. To Stn Flt Thorney Island 24.9.40. Damaged 13.5.41; to 1 CRU Cowley 23.5.41. To 5 MU Kemble 9.7.41. To 16 EFTS Burnaston 31.7.41. To 46 MU Lossiemouth 9.7.42. To IFTS ATA Barton-in-the-Clay 11.11.43. Failed to recover from spin at 1,000 ft on training flight and crashed Toddington, Beds at 11.45 on 6.9.44, pilot T/O FW Christie died from injuries. SOC 25.9.44.

858 M.14A **N3810** Dd 27 MU Shawbury 31.10.38. To 19 E&RFTS Gatwick 26.11.38. To 30 EFTS Burnaston 15.10.39. To 16 EFTS Burnaston 10.4.40. Damaged 27.6.40; to 1 CRU Cowley (SAS) 2.7.40; test flown by Hugh Kennedy (assistant test pilot P&P) at Cowley 30.7.40. To 27 MU Shawbury 2.8.40. To English Electric Co 13.12.40; to AID at English Electric to assist in ferrying 18.12.40. To 27 MU Shawbury 17.6.41. To 122 Sqdn, Ouston 17.7.41. Engine cut, crash landed in field Elsenham, Essex 18.4.42. SOC 24.4.42.

859 M.14A **N3811** Dd 27 MU Shawbury 1.11.38. To 19 E&RFTS Gatwick 24.11.38. To 30 EFTS Burnaston 15.10.39. To 16 EFTS Burnaston 10.4.40. Hit wall in forced landing while lost Darley Dale, Matlock, Derbyshire 31.7.40. To P&P 9.8.40 but SOC 7.8.40 (sic).

860 M.14A **N3812** Dd 27 MU Shawbury 1.11.38. To 15 E&RFTS Redhill 10.11.38. To 5 EFTS Hanworth 15.10.39. Converted to light bomber. Stalled and dived into ground, Wombourne, Staffs 6.7.41; Sgt HDB Jones & LAC Golder killed. SOC 13.7.41.

861 M.14A **N3813** Dd 27 MU Shawbury 1.11.38. To 19 E&RFTS Gatwick 26.11.38. To 30 EFTS Burnaston 15.10.39. To 16 EFTS Burnaston 10.4.40. Ran out of fuel, stalled and crashed 1 ml S of Roston, Derby 18.8.40; LAC Brownlie badly injured. SOC 27.8.40

862 M.14A **N3814** Dd 27 MU Shawbury 18.11.38. To 19 E&RFTS Gatwick 3.12.38. To 30 EFTS 15.10.39. Converted to light bomber. To 16 EFTS Burnaston 10.4.40. Flying accident (Cat.B) 23.8.41; to P&P (SAS) 28.8.41. To 27 MU Shawbury 11.10.41. To 2 EFTS Staverton (flying from Worcester) 4.11.41. To 6 FIS Staverton (31.12.41 census). To ECFS Hullavington 27.5.42. To 46 MU Lossiemouth 15.7.42. To ATA 18.9.44. To 51 MU Lichfield 12.12.45. Sold 26.2.46 to Miles Aircraft Ltd. To H Hennequin & Cia, Argentina with CofA No.7832 issued 30.5.46; regd **LV-XNT** 30.6.47 to Secretaria Estado Aeronautica, Buenos Aires. To Aero Club Concepcion del Uruguay. Accident 12.11.52; repaired. Accident 11.7.54.

863 M.14A **N3815** Dd 'H' MU Waddington 5.11.38. To 5 MU Kemble 16.2.39. To 15 E&RFTS Redhill 27.6.39; renamed 15 EFTS 3.9.39; to Carlisle 2.6.40. Converted to light bomber. Stalled in sideslip on landing and spun into ground Kirkpatrick LG 18.3.41; pilot A/Cpl Semmerling. To 1 CRU Cowley (SAS) 27.3.41 and SOC 8.4.41.

864 M.14A **N3816** Dd 'H' MU Waddington 7.11.38. To 5 MU Kemble 31.1.39. To 15 E&RFTS Redhill 27.6.39; renamed 15 EFTS 3.9.39; to Carlisle 2.6.40. To 29 EFTS Clyffe Pypard 28.5.42. To 3 GTS Stoke Orchard 18.3.43. To 51 MU Lichfield 23..4.44. Sold 27.5.48 to Aircraft & Engineering Services Ltd, Croydon. Regd **G-AKXM** (CofR 12298) 12.4.48 to Surrey Financial Trust Ltd, Coulsdon (based Croydon). CofA No.10358 issued 17.1.49. Cld as sold 18.1.49 (but not regd) to Robert Alan Short, Croydon. Sold 2.9.49 to **Egyptian AF**.

865 M.14A **N3817** Dd 'H' MU Waddington 7.11.38. To 5 MU Kemble 31.1.39. To 15 E&RFTS Redhill 27.6.39; renamed 15 EFTS 3.9.39; to Carlisle 2.6.40. Converted to light bomber. Abandoned in spin near Kirkpatrick LG 10.7.41; pilot Cpl WH Szajdi. To 1 MPRD Cowley and SOC 1.10.41.

866 M.14A **N3820** Dd 'H' MU Waddington 7.11.38. To 5 MU Kemble 31.1.39. To 19 E&RFTS Gatwick 29.6.39. To 30 EFTS Burnaston 15.10.39. To 16 EFTS Burnaston 10.4.40. Dived into ground during aerobatics Milton Sawmills, near Repton, Derbyshire 5.11.41; LAC JAMacD Teasher baled out but parachute failed to open. SOC 20.11.41.

867 M.14A **N3821** Dd 'H' MU Waddington 8.11.38. To 5 MU Kemble 14.2.39. To 19 E&RFTS Gatwick 29.6.39. To 30 EFTS Burnaston 15.10.39. To 16 EFTS Burnaston 10.4.40. Flying accident (Cat.B) 22.7.41; to P&P (SAS) 27.7.41. To 32 MU St Athan 12.10.41. To 8 EFTS Woodley 17.10.41. To 16 EFTS Burnaston 25.10.41. To 5 FIS Perth 7.11.41. To 11 (P)AFU Shawbury 12.4.42. To 32 MU St Athan 1.5.42. To 6 FIS Worcester 8.5.42. To 32 MU St Athan 1.6.42. To 51 MU Lichfield 7.12.42. Sold 3.9.46 to Miles Aircraft Ltd. CofA No. 8630 issued 22.10.46 to Lars Oskarsson. Regd in Iceland as **TF-BLU** 26.11.46. Crashed into control tower at Reykjavik Airport 23.10.58 during attempted start with no-one in the cockpit! Scrapped in 1962. Some parts survived (not the u/c legs as at one time thought) and it was planned to build a 'new' Magister around these parts but nothing has come of it.

868 M.14A **N3822** Dd 'H' MU Waddington 8.11.38. To 5 MU Kemble 31.1.39. To 39 E&RFTS Weston-Super-Mare 8.8.39. To 5 MU Kemble 4.3.40. To 8 EFTS Woodey 23.7.40. To 5 EFTS Meir 12.8.40. Flying accident (Cat.B) 24.6.41; to P&P (SAS) 5.7.41. To 5 EFTS Meir 24.8.41. Flying accident (Cat.B) 2.9.41. To 2 EFTS Staverton (flying from Worcester); redesignated 6 FIS 1.11.41. To 16 EFTS Burnaston 26.1.42. To 46 MU Lossiemouth 9.7.42. To 18 (P)AFU Church Lawford 17.10.44. To 51 MU Lichfield 10.5.45. Sold 13.5.46 to Fairey Flying Club. Regd **G-AHYK** (CofR 10456) 12.7.46 to The Fairey Aviation Co Ltd, op by The Fairey Flying Club, White Waltham. CofA No.8207 issued 26.3.47. Crashed in fog and rain at Uppark, Harting, West Sussex 7.9.47; pilot Grant and passenger killed. Regn cld 15.3.48 as crashed.

869 M.14A **N3823** Dd 'H' MU Waddington 8.11.38. To 5 MU Kemble 20.2.39. To 39 E&RFTS Weston-Super-Mare 8.8.39. To 24 EFTS Sydenham 15.10.39; to Luton 22.7.40. To Herts & Essex, Broxbourne (SAS) 9.10.41. To 48 MU Hawarden 24.12.41. To 32 MU St Athan 26.3.42. To 310 Czech Sqdn, Perranporth 22.4.42. To 130 Sqdn, Perranporth 26.4.42. To 234 Sqdn, Portreath 26.5.42. Dived into ground near Portreath, Cornwall 17.6.42, cause not known. SOC 5.7.42.

870 M.14A **N3824** Dd 'H' MU Waddington 12.11.38. To 5 MU Kemble 24.1.39. To 5 EFTS Hanworth 4.4.40. Undercarriage collapsed in forced landing while lost in bad weather Wiley, 6 ml N of Rugby 5.12.40. To P&P for repairs 11.12.40 but SOC 22.2.41.

871 M.14A **N3825** Dd 'H' MU Waddington 12.11.38. To 5 MU Kemble 13.2.39. To 19 E&RFTS Gatwick 29.6.39. To 30 EFTS Burnaston 15.10.39. To 16 EFTS Burnaston 10.4.40. Flying

APPENDIX 20

accident (Cat.B) 16.2.41; to unknown CRU (SAS) 19.2.41. To 24 EFTS Luton 14.6.41. To 37 MU Burtonwood 16.2.42. To 46 MU Lossiemouth 30.6.42. To 122 Sqdn, Hornchurch (undated). Reported as flying accident (Cat.E1) 25.8.42 but either erroneous or repaired. To 51 MU Lichfield 10.5.45. Sold 1.1.48 to Adie Aviation Ltd. Regd **G-AKPL** (CofR 12124) 12.1.48 to Adie Aviation Ltd, Croydon. CofA No.10043 issued 11.9.48. Cld 11.9.48 & regd 19.10.48 to Hugh Charles Kennard, operated by Air Kruise Ltd, Lympne. Cld 18.6.50 & regd 28.6.50 to Eagle Aviation Ltd, London W1. CofA lapsed 16.2.51. Cld 20.3.52 & regd 23.7.52 to Robert Alan Short, Croydon. CofA renewed 6.11.52. CofA lapsed 5.11.53. Cld 20.3.58 & regd 26.3.58 to Denham Flying Club Ltd, Denham. CofA renewed 11.7.58. Cld 12.7.58 & regd 24.7.58 to Frank Thomas Darton Moore, Pinner, Middlesex. Cld 30.8.59 & regd 25.9.59 to Kenneth Jones, Loudwater, Herts. Cld 9.9.61 & regd 21.9.61 to Herbert Henry, Plymouth. Cld 25.6.63 & regd 26.7.63 to Edward Vincent Gullery & Joseph Trainor, Belfast. Crashed in a ploughed field at Sion Mills, Strabane, Ireland 26.1.64; pilot V Adams and passenger injured. Regn cld 26.1.64 as destroyed.

872   M.14A **N3826**   Dd 26 MU Cardington 14.11.38. To 26 E&RFTS Kidlington 19.11.38. To 5 EFTS Hanworth 15.10.39. To 1 CRU Cowley (SAS) 18.6.40. To 37 MU Burtonwood 1.9.40. To Valley 3.5.41. To P&P for repairs 10.1.42. To 2 FIS Montrose 1.4.42. To Herts & Essex, Broxbourne (MI) 25.5.42. To 27 MU Shawbury 3.10.42. To IFTS ATA, Thame/Barton-in-the-Clay 10.6.43. SOC Cat.E1 15.10.44. Sold to Miles Aircraft Ltd. To H Hennequin & Cia, Argentina with CofA No.7450 issued 11.3.46; regd **LV-XMO** 30.6.47 to Secretaria Estado Aeronautica, Buenos Aires. To Aero Club Cordoba. Reported as wfu 2.55. Regd (5.67) to Direccion General de Aeronautica Civil, Buenos Aires.

873   M.14A **N3827**   Dd 'H' MU Waddington 18.11.38. To 5 MU Kemble 31.1.39. To 26 E&RFTS Kidlington 10.7.39. To 5 EFTS Hanworth 15.10.39; to Meir 17.6.40. To 2 FIS Montrose 9.1.42. To 12 MU Kirkbride 28.8.42. To 125 Sqdn, Middle Wallop 28.8.44 (appeared in the Battle of Britain Display at Church Fenton 15.9.45). To 51 MU Lichfield 12.12.45. Sold 12.6.46 to Miles Aircraft Ltd. To H Hennequin & Cia, Argentina with CofA No.8287 issued 28.8.46; regd **LV-XSG** 30.6.47 to Secretaria Estado Aeronautica, Buenos Aires. To Aero Club Villa Maria. Regd 7.8.62 to Aero Club San Martin. Slight accident .82; repaired. Maintained in airworthy condition with the Aero Club San Martin, Mendoza, it has recently been painted in 1940 style RAF camouflage as N3827; another photograph shows it with the regn **LV-X246**. Presently for sale.

874   M.14A **N3828**   Dd 'H' MU Waddington 18.11.38. To 5 MU Kemble 31.1.39. To 29 E&RFTS Luton 15.7.39. To 30 EFTS Burnaston 15.10.39. To 16 EFTS Burnaston 10.4.40. Stalled and hit trees on take-off Burnaston 3.9.41. To P&P (SAS) 9.9.41 and SOC 11.9.41.

875   M.14A **N3829**   Dd 'H' MU Waddington 18.11.38. To 5 MU Kemble 24.1.39. To 23 E&RFTS Rochester 18.8.39. To 24 EFTS Sydenham 15.10.39; to Luton 22.7.40. To 2 EFTS Staverton (flying from Worcester) 31.10.41. To 6 FIS Staverton 31.12.41. Dived into ground Combe Court, Worcester 14.2.42. SOC 24.2.42.

876   M.14A **N3830**   Dd 'H' MU Waddington 18.11.38. To 5 MU Kemble 31.1.39. To 23 E&RFTS Rochester 21.8.39. To 24 EFTS Sydenham 15.10.39, to Luton 22.7.40. To 1 CRU Cowley (SAS) 20.3.41. To 9 MU Cosford 5.5.41. To 15 EFTS Carlisle 16.8.41. To 29 EFTS Clyffe Pypard 26.5.42. To Herts & Essex, Broxbourne (MI) 11.4.43. To 51 MU Lichfield 7.7.43. Sold 10.3.47 to FK McIntyre. Regd **G-AJHC** (CofR 11292) 11.2.47 to Frank Keiller McIntyre, Darlington. CofA No.9055 issued 14.8.47. Cld 12.5.48 & regd 20.5.48 to John Neasham, Croft. CofA lapsed 13.8.48. Scrapped Croft in 1952. Regn cld 9.4.52 as reduced to spares.

877   M.14A **N3831**   Dd 'H' MU Waddington 28.11.38. To 5 MU Kemble 31.1.39. To 23 E&RFTS Rochester 19.8.39. To 24 EFTS Sydenham 15.10.39; to Luton 22.7.40. To 1 CRU Cowley (SAS) 16.1.41. To 2 SofAC Andover 31.3.41. To 8 AACU Pengam Moors 25.11.41. Engine cut, undershot on forced landing Bishops Lydeard, Somerset 26.6.42. SOC 2.7.42.

878   M.14A **N3832**   Dd 'H' MU Waddington 22.11.38. To 5 MU Kemble 31.1.39. To 23 E&RFTS Rochester 19.8.39. To 24 EFTS Sydenham 15.10.39; to Luton 22.7.40. To 48 MU Hawarden 28.2.42. To EFTS ATA, Barton-in-the-Clay 14.5.42. To De Havilland Aircraft Co (undated). To Hendon 13.1.43. To Atlantic Coast, Barnstaple (MR) 22.9.43. To 51 MU Lichfield 12.11.43. Sold 2.4.46 to Miles Aircraft Ltd. Test flown 8.5.46 by Hugh Kendall as N3832. To H Hennequin & Cia, Argentina with CofA No.7800 issued 24.5.46; regd **LV-XNM** 30.6.47 to Secretaria Estado Aeronautica, Buenos Aires. To Aero Club Jujuy. Accident 6.2.48; repaired. Accident 14.9.53; repaired. Regd 25.3.59 to Aero Club Cruz Alta. Still regd 5.67.

879   M.14A **N3833**   Dd 'H' MU Waddington 22.11.38. To 5 MU Kemble 16.2.39. To 23 E&RFTS Rochester 21.8.39. To 24 EFTS Sydenham 15.10.39; to Luton 22.7.40. Hit bird and lost fabric, hit hedge in forced landing 1 ml SE of Barton Hill Farm 6.8.41. SOC 14.8.41.

880   M.14A **N3834**   Dd 'H' MU Waddington 22.11.38. To 5 MU Kemble 14.2.39. To 23 E&RFTS Rochester 21.8.39. To 24 EFTS Sydenham 15.10.39; to Luton 22.7.40. Collided with Magister R1818 on landing Luton (Cat.B) 10.9.41; LAC Rodaway unhurt. To P&P (SAS) for repairs 15.9.41 but SOC 22.9.41.

881   M.14A **N3835**   Dd 'H' MU Waddington 22.11.38. To 5 MU Kemble 16.2.39. To Stn Flt Speke 10.10.40. To RAF Woodvale 13.2.42. To 32 MU St Athan 28.2.42. To 122 Sqdn, Hornchurch 25.4.42; to Fairlop 8.6.42; to Martlesham Heath 29.6.42; to Fairlop 6.7.42. Forced landed while lost on navex, hit obstruction on take-off Gadding Moor, near Barnsley, Yorks 25.8.42 & SOC.

882   M.14A **N3836**   Dd 'H' MU Waddington 22.11.38. To 15 E&RFTS Redhill 24.1.39; renamed 15 EFTS 3.9.39; to Carlisle 2.6.40. Engine cut on take-off, stalled and spun into ground Kingstown, Carlisle 19.3.41; LAC Milne badly injured. SOC 6.6.41.

883   M.14A **N3837**   Dd 'H' MU Waddington 22.11.38. To 15 E&RFTS Redhill 30.8.39; renamed 15 EFTS 3.9.39; to Carlisle 2.6.40. Flying accident Tarporley (Cat.B) 7.4.41; to 1 CRU Cowley (SAS) 10.4.41. To 5 MU Kemble 24.5.41. To 32 MU St Athan 11.8.41. To 15 EFTS Carlisle 27.9.41. To 29 EFTS Clyffe Pypard 24.5.42. To 21 EFTS Booker 22.3.43. SOC 25.5.44.

884   M.14A **N3838**   Dd 'H' MU Waddington 22.11.38. To 15 E&RFTS Redhill 24.1.39. To 5 EFTS Hanworth 15.10.39. To P&P (undated) & converted to light bomber. To 5 EFTS Meir 20.6.40. To 48 MU Hawarden 8.12.41. To ECFS Hullavington; coded "3"; transferred to (9)AFU Hullavington 28.3.42. To Atlantic Coast, Barnstaple for repairs 30.6.42. To 45 MU Kinloss 10.8.43 (coded '3' in 1943?). To Fighter Leaders School, Milfield 1.2.44. To 51 MU Lichfield 5.11.44. Sold 18.3.46 to Miles Aircraft Ltd. To H Hennequin & Cia, Argentina with CofA No.7831 issued 3.6.46; regd **LV-XPL** 30.6.47 to Secretaria Estado Aeronautica, Buenos Aires. To Aero Club Bahia Blanca. Accident 5.4.47. Regn cld .48 as wfu.

885   M.14A **N3839**   Dd 'H' MU Waddington 22.11.38. To 15 E&RFTS Redhill 30.1.39. To 5 EFTS Hanworth 15.10.39. Hit trees low flying in turbulence and crashed Horton, near Windsor 27.2.40; P/O J McClintock & Sgt FCH Taylor injured. SOC 7.3.40.

886   M.14A **N3840**   Dd 'H' MU Waddington 22.11.38. To 15 E&RFTS Redhill 30.1.39; coded "14"; renamed 15 EFTS 3.9.39; to Carlisle 2.6.40. To 46 MU Lossiemouth 9.8.42. SOC (MR) 19.10.45. BBOC 8.3.46 but SOC (again) 5.4.46.

887   M.14 **N3841**   Dd 'H' MU Waddington 24.11.38. To 15 E&RFTS Redhill 30.1.39; renamed 15 EFTS 3.9.39; to Carlisle 2.6.40; coded '54'. To 21 EFTS Booker 16.6.42. To Atlantic Coast, Barnstaple for repairs 29.7.43. To 51 MU Lichfield 4.10.43. Sold 7.3.46 to Miles Aircraft Ltd. Test flown 15.5.46 by Hugh Kendall (who recorded it as N3441). To H Hennequin & Cia, Argentina with CofA No.7826 issued 3.6.46; regd **LV-XPI** 30.6.47 to Secretaria Estado Aeronautica, Buenos Aires. To Aero Club Parana. Regn cld 7.53 as wfu.

888   M.14A **N3842**   Dd 'H' MU Waddington 24.11.38. To 15 E&RFTS Redhill 30.1.39. To 5 EFTS Hanworh 15.10.39. Stalled on approach and spun into ground, Hanworth 16.4.40; Sgt A McPherson killed. SOC 4.40.

889   M.14A **N3843**   Dd 'H' MU Waddington 24.11.38. To 15 E&RFTS Redhill 24.1.39; renamed 15 EFTS 3.9.39. Crashed during low level aerobatics Brockham, near Dorking, Surrey 27.9.39; 2 killed.

890 M.14A **N3844** Dd 36 MU Sealand 9.1.39 (delivery out of sequence 'defective on flight trials'). To docks for shipping to Middle East 8.3.39. To 208 Sqdn, probably at Mersa Matruh. To Comm Flt Heliopolis (2.40). Flying accident & SOC 12.6.42.

891 M.14A **N3845** Dd 36 MU Sealand 29.11.38. To Middle East. To 208 Sqdn, probably at Mersa Matruh. Engine cut, stalled in forced landing and hit ground & dbf Koubbeh, near Heliopolis 27.4.39; F/O David R Hopper & F/O Derek LR Hutchinson killed.

892 M.14A **N3846** Dd Abingdon 1.12.38. To AASF France 13.10.39. Presumed lost in France in 5.40. SOC 26.3.42.

893 M.14A **N3847** Dd Stn Flt Linton-on-Ouse 1.12.38. Ran short of fuel, forced landed and hit tree in rainstorm near Hay-on-Wye, Powys 9.12.38; en route Linton-on-Ouse to Cardiff; pilot F/O Lawrence G Belcham injured. Collected by 4 SofTT St Athan 17.12.38, but no 'M' serial known. SOC at census.

894 M.14A **N3848** Dd 24 Grp Comm Flt, Grantham 6.12.38. To 3 FTS South Cerney 10.10.39. To 24 EFTS Luton 3.6.41; coded '32'. To 2 FIS Montrose 14.2.42. To 12 MU Kirkbride 11.8.42. To CFS Upavon 8.12.42. To 27 MU Shawbury 14.1.43. To AFDU Wittering 31.5.43. To 51 MU Lichfield by 12.12.45. Sold 16.9.46 to Miles Aircraft Ltd. CofA No.9368 issued 19.5.47. To France and regd **F-BDPF** 19.6.47 to Jacques Habert, Paris. Regd 24.8.48 to Stephane Grandjean, Perpignan. Regn cld 5.2.52 as destroyed.

895 M.14A **N3849** Dd Andover 29.11.38. To 1 CRU Cowley (SAS) 26.11.40. To 27 MU Shawbury 20.2.41. To 5 EFTS Meir 6.4.41. To 16 EFTS Burnaston 16.1.42. Hit by Magister L8333 while awaiting take-off Burnaston 5.4.42. SOC 9.4.42.

896 M.14A **N3850** Dd Northolt 29.11.38. To 141 Sqdn, Grangemouth 25.4.40. To 219 Sqdn, Catterick 7.6.40. To 600 Sqdn, Catterick 16.10.40. To 59 OTU Crosby-on-Eden (undated). To 58 OTU Grangemouth 12-13.5.41. Flying accident (Cat.B) 9.12.42; to Herts & Essex, Broxbourne for repairs 3.1.43. To 51 MU Lichfield 8.8.43. Sold 10.3.47 to Southern Aircraft (Gatwick) Ltd. Regd **G-AJGN** (CofR 11278) 5.2.47 to Southern Aircraft (Gatwick) Ltd, Gatwick. Cld 2.11.48 & regd 18.1.49 to Robert Alan Short, Croydon. CofA No. 9187 issued 16.2.49. Regn cld 2.9.49 as sold abroad - to **Egypt AF**.

897 M.14A **N3851** Dd 'H' MU Waddington 29.11.38. To 5 MU Kemble 16.2.39. To 24 EFTS Luton 17.10.40. To 1 CRU Cowley (SAS) 12.11.40. To 48 MU Hawarden 21.2.41. To 51 (M) Wing, Broughton 17.4.41. To 48 MU Hawarden (31.12.41 census). To 51 MU Lichfield 28.11.43. Sold 26.2.48 to Gloster Flying Club. Regd **G-AKMK** (CofR 12048) 27.11.47 to Laurence Dudley Trappitt, Whyteleafe, Surrey (probably at Staverton). CofA No.10125 issued 26.5.49. Regn cld 21.6.49 as sold abroad. Regd in New Zealand as **ZK-ATD** 14.9.49 to Canterbury Aero Club, Christchurch. NZ CofA issued 15.10.49. At some time the aircraft was fitted with a DH.82C Tiger Moth style canopy with the rear headrest and top fuselage decking also lowered. Regd 25.6.58 to WJ Telford, Balclutha. Regd 14.4.60 to Robert Adams, Pukekoma, Queenstown. Badly damaged in heavy landing Pukekoma 19.5.60 and not repaired. Broken up by Piako Aero Club at Matemata in 1961. Regn cld 7.12.62.

898 M.14A **N3852** Dd 'H' MU Waddington 29.11.38. To 5 MU Kemble 16.2.39. To 5 EFTS Hanworth 3.4.40. Hit HT cables low flying and spun into ground Trentham Park, Staffs 3.7.40; P/O RC Sankey & Sgt DA Green killed. SOC 7.7.40.

899 M.14A **N3853** Dd 'H' MU Waddington 29.11.38. To 5 MU Kemble 20.2.39. To 5 EFTS Hanworth 4.4.40. Crashed 26.10.40; to P&P (SAS) 14.11.40. SOC 17.3.41.

900 M.14A **N3854** Dd 'H' MU Waddington 29.11.38. To 5 MU Kemble 20.2.39. To 5 EFTS Hanworth 4.4.40; to Meir 17.6.40. To 37 MU Burtonwood 8.12.41. To 32 MU St Athan 25.3.42. To 68 Sqdn, Coltishall 12.4.42. To 306 (Polish) Sqdn, Churchstanton 24.4.42; to Kirton-in-Lindsey 3.5.42; to Northolt 16.6.42. To P&P for repairs 6.7.42. To 27 MU Shawbury 3.9.42. To 51 MU Lichfield 29.1.44. Sold by MAP to Miles Aircraft Ltd early 1945 for Chile; to Marshalls Flying School, Cambridge for refurbishment; test flown as **U-0267**. Regd **CC-KZB-0184**. Crashed when right wing struck ground turning back following engine failure on take-off Los Angeles airfield 6.4.52.

901 M.14A **N3855** Dd 'H' MU Waddington 1.12.38. To 5 MU Kemble 13.2.39. To 5 EFTS Hanworth 5.4.40; to Meir 17.6.40. Engine cut, hit pole and building in forced landing Meir 16.12.40; pilot LAC Potter. SOC 1.1.41 (305:55 F/H).

902 M.14A **N3856** Dd 26 MU Cardington 3.12.38. To 6 MU Brize Norton 16.2.39. To 15 EFTS Redhill (30.11.39 census); to Carlisle 2.6.40. To 32 MU St Athan 6.1.42. To 46 MU Lossiemouth 4.8.42. To Stn Flt Llanbedr 8.7.44. SOC Cat.E1 4.10.44. To Miles Aircraft Ltd. Allocated 'B' marks **U-0252** in 1945 but use not confirmed. Regd **G-AGZR** (CofR 9868) 26.2.46 to Miles Aircraft Ltd. CofA No.7524 issued 12.4.46. Regn cld 12.11.47 as sold abroad. CofA renewed 2.2.48 and to **Thai AF** in 3.48 (also reported as to Thai Navy, but unlikely).

903 M.14A **N3857** Dd 9 MU Cosford 5.12.38. To 264 Sqdn, Sutton Bridge 8.11.39. Flying accident (Cat.B) 28.9.40; to 1 CRU Cowley 30.9.40. To 27 MU Shawbury 5.12.40. To 24 EFTS Luton 16.4.41. To 48 MU Hawarden 28.2.42. To ECFS Hullavington (undated). To 9 (P)AFU Hullavington 29.3.42. To 46 MU Lossiemouth 15.7.42. To ATA 1.9.44. SOC Cat.E1 5.10.44. To Miles Aircraft Ltd. To H Hennequin & Cia, Argentina with CofA No.7445 issued 5.3.46; regd **LV-XMF** 30.6.47 to Secretaria Estado Aeronautica, Buenos Aires. To Aero Club Argentino. Destroyed in hangar fire at San Justo 5.5.47.

904 M.14A **N3858** Dd 9 MU Cosford 5.12.38. To 15 EFTS Redhill 10.10.39. To 234 Sqdn, Leconfield 3.11.39. Lost 19.9.40, circumstances not recorded.

905 M.14A **N3859** Dd 9 MU Cosford 5.12.38. To 234 Sqdn, Leconfield 3.11.39. SOC 9.3.40. *Reported as to P&P for repairs 26.7.42 - clerical error.*

906 M.14A **N3860** Dd 9 MU Cosford 5.12.38. To 234 Sqdn, Leconfield 9.11.39. Spun into ground during aerobatics 10 ml W of Leconfield 15.12.39. SOC 28.12.39. Wreck to 60 MU York 7.1.40 for salvage.

907 M.14A **N3861** Dd 9 MU Cosford 5.12.38. To 245 Sqdn, Leconfield 6.11.39 (card marked as 254 Sqdn). Blown over on landing and tipped up Church Fenton 6.12.41. SOC Cat.E 30.12.41.

908 M.14A **N3862** Diverted to Royal Egyptian Air Force, Almaza, Cairo and dd as **L221** 14.12.38.

909 M.14A **N3863** Diverted to Royal Egyptian Air Force, Almaza, Cairo and dd as **L222** 14.12.38.

910 M.14A **N3864** Diverted to Royal Egyptian Air Force, Almaza, Cairo and dd as **L223** 24.12.38.

911 M.14A **N3865** Diverted to Royal Egyptian Air Force, Almaza, Cairo and dd as **L224** 14.12.38.

912 M.14A **N3866** Diverted to Royal Egyptian Air Force, Almaza, Cairo and dd as **L225** 24.12.38.

913 M.14A **N3867** Dd 9 MU Cosford 8.12.38. To 264 Sqdn, Sutton Bridge 8.11.39. To 5 OTU Aston Down 9.4.40. Flying accident (Cat.B) 12.6.41; to P&P (SAS) 24.6.41. To 27 MU Shawbury 17.8.41. To 16 EFTS Burnaston 22.8.41; coded "34". To 32 MU St Athan 10.2.42; unconfirmed as still at 16 EFTS at 31.12.42 census. To 5 GTS Shobdon 1.5.43. To 51 MU Lichfield 10.4.44. Sold 1.4.46 to Miles Aircraft Ltd. To H Hennequin & Cia, Argentina with CofA No.8045 issued 8.7.46; regd **LV-XRF** 30.6.47 to Secretaria Estado Aeronautica, Buenos Aires. To Aero Club Villaguay. Accident 4.9.51; repaired. Accident 5.5.53. Regn cld 9.53 as wfu after accident.

914 M.14A **N3868** Dd 9 MU Cosford 9.12.38. To 264 Sqdn, Sutton Bridge 8.11.39. To P&P 26.9.40. To 5 MU Kemble 18.2.41. To HQ SFP Kemble 1.3.41. To 5 FPP Hatfield (5.41). To TFPP ATA White Waltham. Forced landed in field nr Enfield 31.1.42 (Cat.B); pilot ATA Cadet Miss Margot Duhalde (Chilean) unhurt. To P&P for repairs 5.2.42. To 16 EFTS Burnaston 1.4.42. To 46 MU Lossiemouth 9.7.42. To IFTS ATA, Thame/Barton-in-the-Clay 26.8.43. SOC Cat.E1 14.8.44.

915 M.14A **N3869** Dd 9 MU Cosford 9.12.38. To 245 Sqdn, Leconfield 6.11.39 (card marked as 254 Sqdn). Flying accident

APPENDIX 20

(Cat.B) 13.5.42; to P&P for repairs 17.5.42. To 46 MU Lossiemouth 14.7.42. To ECFS Hullavington 13.9.42. To Comm Flt Woodley 24.7.44. SOC Cat.E 7.11.44. To Miles Aircraft Ltd. To Irish Air Corps as **130** with CofA No.7460 issued 14.2.46; dd 17.2.46. Scrapped 1.53.

916 M.14A **N3875** Diverted to Royal Egyptian Air Force, Almaza, Cairo and dd as **L226** 14.12.38.

917 M.14A **N3876** Diverted to Royal Egyptian Air Force, Almaza, Cairo and dd as **L227** 24.12.38.

918 M.14A **N3877** Diverted to Royal Egyptian Air Force, Almaza, Cairo and dd as **L228** 24.12.38.

919 M.14A **N3878** Diverted to Royal Egyptian Air Force, Almaza, Cairo and dd as **L229** 24.12.38.

920 M.14A **N3879** Diverted to Royal Egyptian Air Force, Almaza, Cairo and dd as **L230** 24.12.38.

921 M.14A **N3880** Dd RAF College, Cranwell 9.12.38. To 6 AO&NS Staverton 22.9.39. To 1 CRU Cowley (SAS) 20.7.40. To 37 MU Burtonwood 8.9.40. To 258 Sqdn, Kenley 3.5.41. To 157 Sqdn, Castle Camps 24.12.41. To 29 MU High Ercall 21.3.42. To 157 Sqdn, Castle Camps 25.3.42. MI 11.6.42; to P&P for repairs 14.6.42. To 46 MU Lossiemouth 10.8.42. To Andover 3.8.44. To 51 MU Lichfield 11.10.44. Sold 5.2.46 to Miles Aircraft Ltd. Test flown 10.4.46 by Hugh Kendall as N3880. To H Hennequin & Cia, Argentina with CofA No.7688 issued 3.5.46; regd **LV-XOD** 30.6.47 to Secretaria Estado Aeronautica, Buenos Aires. Accident 31.7.46 (15% damage); presumed repaired. Regn cld 10.49 as wfu.

922 M.14A **N3881** Dd 1 SofTT Halton 5.12.38. To 5 MU Kemble 9.11.40. To 24 EFTS Luton 27.11.40. Bounced on landing, lost height on attempted overshoot and hit ground Luton 27.6.41. SOC 5.7.41.

923 M.14A **N3882** Dd 6 MU Brize Norton 9.2.39. To Stn Flt Northolt 3.6.40. To 1 CRU Cowley (SAS) 11.9.40. To 8 MU Little Rissington 19.10.40. To 257 Sqdn, Martlesham Heath 7.12.40. To P&P (SAS) 21.2.41. To 24 EFTS Luton 11.8.41. To 6 FIS Staverton 16.2.42. To ECFS Hullavington 27.5.42. To 46 MU Lossiemouth 15.7.42. To EAAS Manby 26.9.44; coded 'FGC-N'. To RAF Staff College, Bracknell (probably based Woodley) 22.7.46. To Reserve Command Comm Flt, White Waltham 16.10.46. To 5 MU Kemble 29.11.46. Sold 9.12.47 to SRC Partridge. Regd **G-AKOL** (CofR 12099) 4.12.47 to Stanley Robert Cross Partridge, Fraserburgh. CofA No.9938 issued 27.5.48. Dbf Dyce 12.5.49. Regn cld 20.5.49.

924 M.14A **N3883** Dd 1 SofTT Halton 3.1.39. To 5 MU Kemble 9.11.40. To 24 EFTS Luton 27.11.40. Flying accident (Cat.B) 16.2.41; to 1 CRU Cowley 20.3.41. To 5 EFTS Meir 26.4.41. To Atlantic Coast, Barnstaple (MI) 12.12.41. Flying accident (Cat.B) 12.1.42; to P&P for repairs 19.1.42. To 9 (P)AFU Hullavington 1.4.42. To 7 FIS Upavon 7.11.42. To 51 MU Lichfield 27.5.44. Sold 26.2.46 to Miles Aircraft Ltd. Test flown 15.5.46 by Hugh Kendall as N3883. To H Hennequin & Cia, Argentina with CofA No.7829 issued 30.5.46; regd **LV-XPU** 30.6.47 to Secretaria Estado Aeronautica, Buenos Aires. Accident 6.11.4x; repaired. Crashed Bella Vista 20.11.46 (60% damaged); presumed repaired. Regn cld 7.54 as wfu.

925 M.14A **N3884** Dd 9 MU Cosford 31.12.38. To 24 EFTS Luton 27.9.40. To 2 FIS Montrose 26.3.42. Dived into ground near Laurencekirk, Kincardine 17.8.43.

926 M.14A **N3885** Diverted to Royal Egyptian Air Force, Almaza, Cairo and dd as **L239** 4.1.39.

927 M.14A **N3886** Diverted to Royal Egyptian Air Force, Almaza, Cairo and dd as **L240** 24.12.38.

928 M.14A **N3887** Diverted to Royal Egyptian Air Force, Almaza, Cairo and dd as **L241** 4.1.39.

929 M.14A **N3888** Diverted to Royal Egyptian Air Force, Almaza, Cairo and dd as **L242** 4.1.39.

930 M.14A **N3889** Diverted to Royal Egyptian Air Force, Almaza, Cairo and dd as **L243** 24.12.38.

931 M.14A **N3890** Dd 1 SofTT Halton 5.1.39. To 5 MU Kemble 8.11.40. To 1 AACU Farnborough 11.12.40. To 48 MU Hawarden 28.5.41. To 15 EFTS Carlisle 25.6.41. To 21 EFTS Booker 12.6.42. To 50 Grp Comm Flt, Woodley 25.7.43. To 7 FIS Upavon 29.9.44. To FTC Comm Sqdn, Woodley 29.10.44; coded 'FKN-N'. Probably fitted with enclosed cabin over both cockpits by P&P/Miles Aircraft soon after. To 51 MU Lichfield 21.5.47. Sold 8.1.48 to Short Bros & Harland Ltd. Regd **G-AKRW** (CofR 12158) 24.1.48 to Short Bros & Harland Ltd, Rochester. CofA No.10041 issued 6.5.48. Damaged when u/c collapsed on crosswind landing at Sherburn-in-Elmet 22.7.50 after the International Air Races there; repaired. CofA lapsed 19.6.52. CofA renewed 12.6.53. Engine failed and badly damaged in forced landing Hawkhurst, Kent 11.7.53; pilot Macey. Regn cld 26.8.53 as destroyed. Remains burned at Rochester 11.53.

932 M.14A **N3891** Dd AMDP RTO P&P Reading 26.1.39. Modified to take BLS tailwheel to Contract No.11314/39 (ITP dated 20.7.39); retained by makers. To 32 MU St Athan 23.7.41. To 8 EFTS Woodley 1.8.41. To CRD at P&P 29.9.41; flown by Hugh Kennedy on 24.4.42 on 'Syn. Landing Tests'. To RAE Farnborough 14.5.42. To Herts & Essex, Broxbourne for repairs 27.1.43. To 51 MU Lichfield 22.3.43. To 76 MU Wroughton for packing; desp to ME for **Turkey** 12.5.43 on SS 'Katuma'; arr 13.6.43.

933 M.14A **N3892** Dd 24 Grp HQ, Cardington 31.12.38. To 2 E&WS Yatesbury (30.11.39 census). To 16 EFTS Burnaston 19.6.41. Hit tree low flying near Burnaston Hall in practice attack on Home Guard unit 28.6.42; P/O Leslie Stanynought and P/O EJ Hurst both killed. (Stanynought had been responsible, under the pseudonym LC Lewis, for the export of Hawk Majors G-ADDC & G-ADDU and Hawk Speed 6 G-ACTE to the Spanish Republicans). SOC 9.7.42.

934 M.14A **N3893** Dd 24 Grp HQ Cardington 31.12.38. To 2 E&WS Yatesbury (30.11.39 census). To 5 MU Kemble 11.6.41. To TFPP ATA HQ, White Waltham 31.7.41; coded "3". While approaching to land on a training flight at Little Rissington at 1605 hrs on 3.8.43, a Wellington cut in front, causing the Magister to stall and dive in to the ground; cadet pilot Mrs J Mullineaux killed. SOC Cat.E 11.8.43.

935 M.14A **N3894** Dd 24 Grp HQ Cardington 31.12.38. To 20 MU Aston Down 31.8.39. To P&P 6.6.40. To 16 EFTS Burnaston 11.6.40. To 8 EFTS Woodley 11.9.40. To 10 FIS Woodley 21.7.42. To 5 GTS Shobdon 5.9.43. To 51 MU Lichfield 10.4.44. Sold 25.10.46 to Miles Aircraft Ltd. CofA No.9038 issued 20.2.47 to John Gamble. Regd in New Zealand as **ZK-ANK** (CofR 367) 31.7.47 to John M Gamble, Wellington. NZ CofA No.428 validated 31.7.47. Regd 11.8.48 to Wellington Aero Club (Inc), Wellington. Ditched in sea in hazy conditions off Kaikoura coast 6.11.48; pilot Mrs N Broads and passenger Mr A Foster both unhurt and rescued by dinghy. Regn cld 13.1.49.

936 M.14A **N3895** Diverted to Royal Egyptian Air Force, Almaza, Cairo and dd as **L231** 4.1.39.

937 M.14A **N3896** Diverted to Royal Egyptian Air Force, Almaza, Cairo and dd as **L232** 4.1.39.

938 M.14A **N3897** Diverted to Royal Egyptian Air Force, Almaza, Cairo and dd as **L233** 4.1.39.

939 M.14A **N3898** Diverted to Royal Egyptian Air Force, Almaza, Cairo and dd as **L234** 14.1.39.

940 M.14A **N3899** Diverted to Royal Egyptian Air Force, Almaza, Cairo and dd as **L235** 14.1.39.

941 M.14A **N3900** Dd 24 Group HQ Cardington 3.1.39. To 2 E&WS Yatesbury (30.11.39 census). To 5 MU Kemble 15.10.42. To IFTS ATA Barton-in-the-Clay 4.3.44. Struck rising ground at Stone, nr Aylesbury, Bucks at 1450 hrs on 12.8.44; pilot T/O AD Cuningham badly injured. SOC 1.9.44.

942 M.14A **N3901** Dd Cardington 3.1.39. To 20 MU Aston Down 14.8.39. To 47 MU Sealand 3.6.40. To Irish Air Corps, Baldonnel as **75** 7.6.40. Scrapped 8.46.

943 M.14A **N3902** Dd Cardington 3.1.39. To 2 E&WS Yatesbury (30.11.39 census). To 5 MU Kemble 30.6.41. To 66 Sqdn, Perranporth 2.7.41; to Portreath 14.12.41; Ibsley 27.4.42; Zeals 24.8.42. To 51 MU Lichfield 28.11.42. To 4 SofTT St Athan for GI as **5364M** 14.7.45 (allotted 11.7.45).

944 M.14A **N3903** Dd 5 SofTT Locking 5.1.39. To 16 EFTS Burnaston 22.9.41. Undershot and hit tree on approach Burnaston 4.4.42. SOC 13.4.42.

945 M.14A **N3904** Dd 5 SofTT Locking 12.1.39. To 16 EFTS Burnaston 22.9.41. To 2 FIS Montrose 5.7.42. To 19 (P)AFU Dalcross 30.9.43. To 21 (P)AFU Tatenhill 19.2.44. To 7 SofTT Innsworth for GI as **4675M** 16.3.44. To 75 MU Wilmslow 24.5.45 and SOC 25.5.45.

946 M.14A **N3905** Dd 5 SofTT Locking 13.1.39. To 5 MU Kemble 27.3.41. To 8 MU Little Rissington 15.4.41. To 20 MU Aston Down 6.5.41. To 49 Sqdn, Scampton 7.5.41. To 8 MU Little Rissington 23.10.41. To 51 MU Lichfield 31.12.42. Sold 11.3.46 to Miles Aircraft Ltd. To H Hennequin & Cia, Argentina with CofA No.7871 issued 11.6.46; regd **LV-XPW** 30.6.47 to Secretaria Estado Aeronautica, Buenos Aires. To Aero Club Deportiva. Accident 6.10.56. Regd 9.8.60 to Aero Club Pampeno, Santa Rosa.

947 M.14A **N3906** Dd 5 SofTT Locking 24.1.39. To 5 MU Kemble 27.3.41. To HQ SFP Kemble 16.4.41. To Shrager Bros, Old Warden (SAS) 16.7.41. To 37 MU Burtonwood 4.11.41. To 615 Sqdn, Angle 13.12.41. To 486 Sqdn, Kirton-in-Lindsey 4.4.42. To 51 MU Lichfield 9.1.44. SOC Cat.E (MR) 25.2.44.

948 M.14A **N3907** Dd 1 Sqdn, Tangmere 4.1.39. Dived into ground out of cloud Walberton, 3 ml E of Tangmere 14.1.39; F/O Alexander C Douglas & Sgt John J Cooper killed (8:00 F/H).

949 M.14A **N3908** Dd 5 MU Kemble 9.1.39. To Andover 4.9.39. To Church Fenton 12.6.41. Hit marker on night take-off Macmerry 29.8.41. SOC 15.9.41 by 60 OTU East Fortune.

950 M.14A **N3909** Dd 5 MU Kemble 9.1.39. To 24 EFTS Luton 17.10.40. Stalled at low altitude and hit ground Palgrave Farm, Sporle, Norfolk 22 6.41; F/O WE Cater killed & LAC MH Jones badly injured. SOC 30.6.41.

951 M.14A **N3910** Dd 5 MU Kemble 10.1.39. To Andover 4.9.39. To 2 SofAC Old Sarum. Stalled and spun into ground in bumpy conditions Thruxton, 3 ml W of Andover 9.6.40; Sgt Worth killed & Sgt Drake badly injured. SOC 17.6.40.

952 M.14A **N3911** Dd 5 MU Kemble 9.1.39. To Andover 4.9.39. Flew into lighthouse at Detling 30.8.40; P/O Anderson & Sgt Wyse unhurt; to 1 CRU Cowley for repairs 10.9.40. To 5 MU Kemble 1.11.40. To 56 Sqdn, Boscombe Down 14.11.40. Flew into ground while neither pilot was in control near Collingbourne, Wilts 16.11.40. SOC 26.1.41 (138:45 F/H).

953 M.14A **N3912** Diverted to Royal Egyptian Air Force, Almaza, Cairo and dd as **L236** 14.1.39.

954 M.14A **N3913** Diverted to Royal Egyptian Air Force, Almaza, Cairo and dd as **L237** 14.1.39.

955 M.14A **N3914** Diverted to Royal Egyptian Air Force, Almaza, Cairo and dd as **L238** 14.1.39.

956 M.14A **N3918** Dd 5 MU Kemble 10.1.39. To 50 Sqdn, Lindholme 30.9.40. To 27 MU Shawbury 8.9.41. To 603 Sqdn, Hornchurch 1.10.41. To 164 Sqdn, Peterhead 10.4.42. To 602 Sqdn, Skeabrae 10.9.42. To 51 MU Lichfield 21.10.42. Sold 18.4.46 to Miles Aircraft Ltd. To H Hennequin & Cia, Argentina with CofA No.8047 issued 8.7.46; regd **LV-XRO** 30.6.47 to Secretaria Estado Aeronautica, Buenos Aires. To Circ Universitario de Aviation. Accident 3.8.48; repaired. Accident 8.12.52; repaired. Regd 7.8.62 to Aero Club Bahia Blanca.

957 M.14A **N3919** Dd 5 MU Kemble 10.1.39. To 30 E&RFTS Burnaston 23.5.39; renamed 30 EFTS 3.9.39. To 16 EFTS Burnaston 10.4.40. To 27 MU Shawbury 19.2.42. To ATA 2.2.44. SOC Cat.E 7.11.44. To Miles Aircraft Ltd. Test flown 7.3.46 by Hugh Kendall as N3919. To H Hennequin & Cia, Argentina with CofA No.7449 issued 14.3.46; regd **LV-XMH** 30.6.47 to Secretaria Estado Aeronautica, Buenos Aires. To Aero Club La Plata. Crashed 28.7.48; probably repaired. Regn cld 7.53 as wfu.

958 M.14A **N3920** Dd 5 MU Kemble 10.1.39. To 24 EFTS Sydenham 14.10.39; to Luton 22.7.40. Collided (on ground?) with Magister L5981 5.3.41; to 1 CRU Cowley for repairs 10.3.41. To 54 OTU Church Fenton 14.6.41. To 60 OTU East Fortune (31.12.41 census). To 2 FIS Montrose 10.5.42. To 17 (P)AFU Calveley 5.10.43. To 51 MU Lichfield 23.2.44. Sold 18.4.46 to Miles Aircraft Ltd. To H Hennequin & Cia, Argentina with CofA No.7923 issued 19.6.46; regd **LV-XPG** 30.6.47 to Secretaria Estado Aeronautica, Buenos Aires. To Aero Club Cordoba. Accident 7.7.51; repaired. Regn cld 8.56 as wfu.

959 M.14A **N3921** Dd 5 MU Kemble 12.1.39. To 24 EFTS Luton 17.10.40; coded "11". Crashed following collision 15.7.41; LAC FF Grebby badly injured. SOC 23.7.41.

960 M.14A **N3922** Dd 5 MU Kemble 23.1.39. To 15 E&RFTS Redhill 27.6.39; renamed 15 EFTS 3.9.39; coded "27". To P&P (SAS) 17.6.40. To 5 MU Kemble 10.12.40. To 601 Sqdn, Northolt 21.12.40. Engine cut, overshot forced landed on football field and hit fence Northwood Hills, Middx 27.3.41. SOC 9.4.41.

961 M.14A **N3923** Dd 5 MU Kemble 23.1.39. To 24 EFTS Sydenham 14.9.39; to Luton 22.7.40. Bounced on landing, stalled and crashed Luton 13.8.41; Sgt KF Dacre injured & Acting LA KA Remmington unhurt. SOC 21.8.41.

962 M.14A **N3924** Dd 5 MU Kemble 23.1.39. To 24 EFTS Sydenham 14.9.39; to Luton 22.7.40. To 1 CRU Cowley (SAS) 17.10.40. To 27 MU Shawbury 22.11.40. To 5 EFTS Meir 3.1.41. To CFS Upavon 11.12.41. To 16 EFTS Burnaston 19.4.42. To 27 MU Shawbury 11.9.42. To Down Ampney 22.3.44. To 51 MU Lichfield 30.5.44. Sold 18.6.46 to Miles Aircraft Ltd. Test flown 30.7.46 by Hugh Kendall as N3924. To H Hennequin & Cia, Argentina with CofA No.8254 issued 20.8.46; regd **LV-XSS** 30.6.47 to Secretaria Estado Aeronautica, Buenos Aires. To Circ. Universitario de Rosario. Regn cld 4.52 as wfu.

963 M.14A **N3925** Dd 5 MU Kemble 23.1.39. To 30 E&RFTS Burnaston 23.5.39; renamed 30 EFTS 3.9.39. To 16 EFTS Burnaston 10.4.40. Landed upon by P6370 (Cat.B) 28.3.42; pilot Cpl Malcolm slightly injured; repaired. To 46 MU Lossiemouth (31.12.42 census). To ATA 20.11.43. To 51 MU Lichfield 7.3.45. Sold 18.11.46 to Miles Aircraft Ltd. CofA No.9365 issued 22.5.47. Regd in France as **F-BDPC** 5.6.47 to Camille Balabaud, La Baule-les-Pins. Regd 31.5.48 to Aero Club du Haut Rhin, Mulhouse/Habsheim. Regn cld 2.4.54 as wfu.

964 M.14A **N3926** Dd 5 MU Kemble 23.1.39. To 24 E&RFTS Sydenham 14.9.39; renamed 24 EFTS 3.9.39; to Luton 22.7.40. To 1 CRU Cowley (SAS) 20.3.41. To Duxford 3.5.41. MI 12.8.42; to Herts & Essex, Broxbourne 27.8.42. To 51 MU Lichfield 6.11.42. Sold 22.4.49 to RA Short. Regd **G-ALOE** (CofR 12712) 21.2.49 to Robert Alan Short & James Henry Tattersall, Croydon/Clitheroe (regd as ex N3936). CofA No.10732 issued 7.12.49. CofA lapsed 6.12.50. Regn cld 12.9.52 by MCA. (Seen at Southend as OO-ACH 27.11.53; possibly on delivery). Regd in Belgium as **OO-ACH** (CofR 998) 23.3.54 to G de Coster, Brussels and J Manquoy, Louvain/Keerbergen (based at Keerbergen) (now regd as "3926"). Fitted with canopy over both cockpits but when and by whom is not known. CofA lapsed 6.3.56. Regn cld 29.12.67.

965 M.14A **N3927** Dd 5 MU Kemble 23.1.39. To 12 MU Kirkbride 19.9.39. To 610 Sqdn, Acklington 29.11.40. To 615 Sqdn, Kenley 10.2.41 (but still recorded at 610 Sqdn at 31.12.41 census). To 51 MU Lichfield (1942 census). To ATA 19.10.43. To 51 MU Lichfield 2.12.43. To 4 SofTT St Athan for GI as **5366M** 14.7.45. SOC 24.4.47 by 34 MU Montford Bridge.

966 M.14A **N3928** Dd 5 MU Kemble 23.1.39. To Binbrook 30.9.40. To 46 MU Lossiemouth (31.12.42 census). To ATA 15.12.43. To 51 MU Lichfield 16.2.44. Sold 13.3.46 to Miles Aircraft Ltd. Test flown 21.5.46 by Hugh Kendall as N3928. To H Hennequin & Cia, Argentina with CofA No.7869 issued 11.6.46; regd **LV-XQP** 30.6.47 to Secretaria Estado Aeronautica, Buenos Aires. To Aero Club Rio Gallegos. Regd 8.8.62 to Aero Club Salliquello.

967 M.14A **N3929** Dd 5 MU Kemble 23.1.39. To 609 Sqdn, Middle Wallop 1.8.40. To High Ercall 22.10.41. To 133 Sqdn, Biggin Hill 16.7.42. Engine cut, hit tree in forced landed near Guildford, Surrey 26.7.42 & dbf. SOC 27.7.42.

968 M.14A **N3930** Dd 5 MU Kemble 23.1.39. To 10 Grp Comm Flt, Colerne 1.8.40. To 5 MU Kemble 19.2.41. To HQ SFP Kemble 1.3.41. Flying accident (Cat.B) 19.7.41; to P&P (SAS) 29.7.41. To 27 MU Shawbury 7.9.41. To 600 Sqdn, Colerne 13.9.41. To 51 MU Lichfield 11.1.43. Sold 15.4.46 to Miles Aircraft Ltd. Test flown

APPENDIX 20

7.6.46 by Hugh Kendall as N3930. To H Hennequin & Cia, Argentina with CofA No.7966 issued 1.7.46; regd **LV-XRD** 30.6.47 to Secretaria Estado Aeronautica, Buenos Aires. To Aero Club Corrientes. Accident 12.12.46; repaired. Accident 20.9.47; repaired. Accident 30.10.47; repaired. Regn cld 7.54 as wfu.

969 M.14A **N3931** Dd 5 MU Kemble 23.1.39. To 5 EFTS Hanworth 4.4.40; to Meir 17.6.40. Stalled on overshoot and hit tree Stone, Staffs 22.5.41; LAC Heard badly injured. SOC 3.6.41.

970 M.14A **N3932** Dd 5 MU Kemble 23.1.39. To HQ SFP Kemble 18.2.41. To 1 CRU Cowley (SAS) 19.4.41. To 5 EFTS Meir 4.6.41. To CFS Upavon 30.1.42. To 16 EFTS Burnaston 19.4.42. To 46 MU Lossiemouth 9.7.42. To 51 MU Lichfield 13.5.44. Sold 26.2.48 to Brookmoor Foundry Co but not regd. Seen at Wolverhampton Aviation awaiting overhaul in 12.51 and still present 5.53. Ultimate fate not known.

971 M.14A **N3933** Dd 5 MU Kemble 23.1.39. To 253 Sqdn, Turnhouse 19.8.40. To 1 CRU Cowley (SAS) 24.12.40. To 24 EFTS Luton 20.2.41. To 2 FIS Montrose 15.2.42. To 7 or 17 SFTS 15.7.43 (writing on card is unclear and neither seem likely). To CFS Upavon 14.11.46. To 21 (P)AFU Moreton-in-Marsh 19.12.46. To 51 MU Lichfield 30.6.47. Sold 15.2.49 to RA Short. Regd **G-ALHB** (CofR 12536) 17.2.49 to Sqdn Ldr Kenneth James Nalson, Croydon. CofA No.10716 prepared but not issued. Extant Croydon 10.50 but later scrapped. Regn cld 11.12.59 by MCA.

972 M.14A **N3934** Dd 5 MU Kemble 23.1.39. To 603 Sqdn, Turnhouse 15.8.40. Cat.B 13.3.41; to P&P (SAS) 16.3.41. To 54 OTU Church Fenton 29.6.41. To 32 MU St Athan 25.7.41. To 402 RCAF Sqdn, Southend (undated). Dived into ground after take-off Friston 18.10.41, presumed due to turbulence. SOC 28.10.41.

973 M.14A **N3935** Dd 5 MU Kemble 23.1.39. To 5 EFTS Hanworth 5.4.40; to Meir 17.6.40. To P&P 26.9.40. To 615 Sqdn, Valley 26.5.41. Stalled at low altitude and hit ground Rawburgh, near Horsham St Faith 5.6.41, en route Valley-Coltishall; 2 killed. SOC 14.6.41.

974 M.14A **N3936** Dd 5 MU Kemble 27.1.39. To 24 EFTS Luton 17.10.40. To 15 EFTS Carlisle 22.10.41. Flying accident (Cat.B) 9.1.42; repaired. To 2 FIS Montrose 1.4.42. To 12 MU Kirkbride 11.8.42. To ATA 19.9.44. To 51 MU Lichfield 12.12.45. Sold 18.4.46 to Miles Aircraft Ltd. Test flown 6.6.46 by Hugh Kendall as N3936. To H Hennequin & Cia, Argentina with CofA No.7965 issued 1.7.46; regd **LV-XPV** 30.6.47 to Secretaria Estado Aeronautica, Buenos Aires. To Aero Club Santa Fe. Crashed Campos Santos 9.9.53; pilot injured.

975 M.14A **N3937** Dd 5 MU Kemble 27.1.39. To 248 Sqdn, Hendon 29.12.39. To 500 Sqdn, Detling 28.3.40. To 3 General Recommaissance Unit, Thorney Island 1.6.40. SOC Cat.E 29.8.40. To 3 SofTT Blackpool 12.9.40 for GI as **2212M** (allocated 28.8.40).

976 M.14A **N3938** Dd 5 MU Kemble 27.1.39. To 1 SofTT Halton 27.9.40. To 27 MU Shawbury 27.2.43. To ATA 10.6.43. SOC 4.12.44.

977 M.14A **N3939** Dd 5 MU Kemble 27.1.39. To 5 EFTS Hanworth 9.4.40. To P&P for repairs 29.6.40. To 5 MU Kemble 30.11.40. To 23 Sqdn, Ford 12.12.40. Flying accident (Cat.B) 26.5.41; to P&P (SAS) 31.5.41. To 46 MU Lossiemouth 19.7.41. To 51 OTU Debden 14.8.41; to Cranfield 17.8.41. To P&P for repairs 16.2.42. To CRD at Heston Aircraft Ltd 13.4.42 and fitted with Maclaren Drift Undercarriage Installation by Airwork Ltd, Denham. To A&AEE Boscombe Down 1.9.42 for brief check tests which were completed by 7.9.42. To 613 Sqdn, Ouston for service trials 10.9.42. To 4 Sqdn, Clifton for service trials. To 26 Sqdn, Gatwick 21.12.42. To 268 Sqdn, Weston Zoyland 15.2.43. Ran short of fuel and undercarriage collapsed in forced landing 1 ml NE of Northchapel, Sussex 8.4.44. SOC Cat.E 20.4.44.

978 M.14A **N3940** Dd 5 MU Kemble 27.1.39. To 24 EFTS Luton 17.10.40. To 15 EFTS Carlisle 10.10.41. To 29 EFTS Clyffe Pypard 26.5.42. To 10 FIS Woodley 26.3.43. To 6 (P)AFU Little Rissington 28.9.43. To 51 MU Lichfield 23.4.44. Sold 25.11.46 to Miles Aircraft Ltd. CofA No.9401 issued 28.5.47. Regd in France as **F-BDPM** (CofR 20007) 17.6.47 to Albert Japy, Paris. Regd 2.12.47 to Louis de la Rochette, Paris. Regd 26.2.48 to Air Tourist sarl, Paris. Regd 19.6.50 to Heynrickx, Lille. Regd 14.4.51 to Societe Anselme Dewavrin Fils et Cie, Tourcoing. Regd 7.3.57 to Association Union Aerienne Lille, Roubaix, Tourcoing, Lille/Marcq. Regd 10.2.58 to Marc Jouret, Roubaix. Regd 10.2.59 to Association Interclubs de Parachutisme de la Region Nord, Lille. Regd 18.3.63 to Jean Dieu & Hubert Brundt, Lille. Wfu .65. Regn cld 22.9.71 as destroyed.

979 M.14A **N3941** Dd 5 MU Kemble 30.1.39. To 5 EFTS Hanworth 4.4.40; to Meir 17.6.40. To 48 MU Hawarden 8.12.41. To ECFS Hullavington (undated). To 9 (P)AFU Hullavington 29.3.42. To ECFS Hullavington 16.5.42. To P&P for repairs 24.6.42. To 27 MU Shawbury 12.8.42. To 408 RCAF Sqdn, Balderton 7.9.42. To 32 MU St Athan 9.9.42. To 418 RCAF Sqdn, Bradwell 21.9.42. Overshot forced landing while short of fuel Ilkeston, Notts 29.7.44. SOC 16.8.44.

980 M.14A **N3942** Dd 5 MU Kemble 30.1.39. To 8 EFTS Woodley 25.9.40. Hit tree during low aerobatics and crashed Warren Row, near Maidenhead, Berks 20.11.41; P/O JN Sellers killed & Cpl J Wood badly injured. SOC 28.11.41.

981 M.14A **N3943** Dd 5 MU Kemble 30.1.39. To Andover 14.9.39. To Church Fenton 15.6.41. To 402 RCAF Sqdn, Southend 5.10.41. To Hunsdon 22.10.41. To 85 Sqdn, Hunsdon 1.11.41. To 63 OTU Honiley 27.8.43. SOC 2.4.44.

982 M.14A **N3944** Dd 5 MU Kemble 30.1.39. To 5 EFTS Hanworth 4.4.40; to Meir 17.6.40. To 1 CRU Cowley (Cat.B) 11.7.40. To 27 MU Shawbury 13.8.40. To Swanton Morley 19.11.40. To 9 MU Cosford 23.5.41. To 51 OTU Debden 16.8.41; to Cranfield 17.8.41. Blown aside on landing and undercarriage collapsed Cranfield 2.4.42; repaired locally. To Herts & Essex, Broxbourne for repairs 16.3.43. To 51 MU Lichfield 31.5.43. Sold by MAP to Miles Aircraft Ltd in early 1945 for **Chile**. To Marshalls Flying School, Cambridge 14.4.45 for refurbishment; test flown as **U-0268**. Believed later sold to Argentina (either this or c/n 2188 (V1015/U-0269) is thought to have been regd LV-RYD 7.12.48 and crashed 9.11.49).

983 M.14A **N3945** Dd 5 MU Kemble 30.1.39. To 5 EFTS Hanworth 4.4.40; to Meir 17.6.40. To P&P 2.10.40. To 5 MU Kemble 14.3.41. To HQ SFP Kemble 6.4.41. To TFPP ATA, White Waltham (5.41); coded '22'. To 1 SofTT Halton for GI as **4527M** 14.2.44 (allotted 8.2.44).

984 M.14A **N3951** Dd 5 MU Kemble 30.1.39. To 1 Camouflage Unit, Cosford 11.10.40. To Stn Flt Baginton. Engine cut, hit wall landing on road Bolton, Lancs 10.2.41. SOC 18.2.41.

985 M.14A **N3952** Dd Kemble 31.1.39. To 53 Sqdn, Detling 19.8.40. Abandoned in fog near Ormstead Hall 19.11.40. SOC 18.12.40.

986 M.14A **N3953** Dd Grantham 31.1.39. To 3 (S)FTS South Cerney 10.10.39. To 16 EFTS Burnaston 14.6.41. Lost and hit wires in attempted forced landing at Gobowen, NE of Oswestry, Shropshire 13.4.42; pilot Cpl Sunter. Not repaired and SOC Cat.E.

987 M.14A **N3954** Dd Eastchurch 31.1.39. To Brize Norton 16.9.39. To Ternhill 1.3.40. To 5 FTS Meir 10.12.40. To 8 EFTS Woodley 1.9.41. To 10 FIS Woodley 21.7.42. To 8 EFTS Woodley 7.5.46; coded 'FDT-G'. To 51 MU Lichfield 28.5.46. Sold 8.6.48 to BOAC. Regd **G-AKKV** (CofR 12008) 23.6.48 to Speedbird Flying Clubs Ltd (a BOAC subsidiary), Denham. CofA No.10426 issued 15.2.49. Regd 4.49 on change of name to Airways Aero Associations Ltd, Denham. Cld 1.3.53 & regd 10.6.53 to Wolverhampton Aviation Ltd, Wolverhampton. Cld 26.6.56 & regd 29.11.56 to Beaver Flying Club Ltd, RCAF Langar. CofA lapsed 31.3.59. Cld 8.12.59 & regd 7.3.60 to Robert Brian Flitton, Brian Richard Luesley & Ronald Mullens, t/a Hemswell Flying Group, RAF Hemswell. CofA renewed 27.4.60. Struck fence on landing Hemswell 13.6.60; pilot R Kolburn & passenger unhurt. Regn cld 8.1.61 as pwfu.

988 M.14A **N3955** Dd Eastchurch 6.2.39. To Brize Norton 16.9.39. To Ternhill 1.3.40. To 5 FTS Meir 10.12.40. To 8 EFTS Woodley 21.8.41. To 10 FIS Woodley 21.7.42. To 7 FIS Upavon 28.7.42. To 5 (P)AFU Ternhill 14.6.44. To 7 SFTS Kirton-in-Lindsey 16.4.46. Damaged 21.6.46 and to 5 MU Kemble 28.6.46. Sold 15.11.46 to TC Sparrow. Regd **G-AIUG** (CofR 10975) 11.11.46 to Thomas Charles Sparrow, Bournemouth. Regd 28.11.46 to Benjamin Gerald Heron, Bournemouth. CofA No.8953 issued 10.6.47. Cld 23.6.48 & regd 12.7.48 to Paul Rene Lamer, London SW1. Cld 15.11.48 & regd 9.2.49 to Ronald George Sturges,

London W11. Cld 11.6.49. CofA lapsed 3.12.49. Regd 2.50 to RG Forbes-Bassett. Remains to WA Rollason Ltd, Croydon 9.51 and used for spares. Wings burned at Hamsey Green 5.11.56.

989 M.14A **N3956** Dd 3 Grp Comm Flt, Mildenhall 2.2.39. To Herts & Essex, Broxbourne for repairs 4.1.43. To 51 MU Lichfield 14.3.43. Allotted by MAP to Miles Aircraft Ltd for **Chile** in early 1945 but allotment cancelled. Sold 22.9.47 to Miles Aircraft Ltd. CofA No.9878 issued 20.2.48 to **Thai AF**.

990 M.14A **N3957** Dd 1 E&WS, Cranwell 6.2.39. To P&P (SAS) 29.11.40. To 9 MU Cosford 20.4.41. To 60 OTU Leconfield 8.5.41; to East Fortune 4.6.41. To 132 (C)OTU East Fortune 24.11.42. To 9 (C)OTU Crosby-on-Eden 5.1.43. To RAF Crosby 27.4.44. To 51 MU Lichfield 19.2.46. Sold 26.5.48 to Short Bros & Harland Ltd, Rochester but not regd. Noted dismantled at Rochester 7.51.

991 M.14A **N3958** Dd 1 E&WS, Cranwell 6.2.39. Damaged 10.10.40; to 1 CRU Cowley (SAS) 11.11.40. To 9 MU Cosford 2.1.41. To 315 (Polish) Sqdn, Acklington 22.2.41. To 303 (Polish) Sqdn, Speke 16.7.41. Flying accident (Cat.B) 3.10.41; repaired. To 48 MU Hawarden 24.2.42. To 32 MU St Athan 25.3.42. To 402 RCAF Sqdn, Fairwood Common 8.4.42; to Kenley 14.5.42. To 27 MU Shawbury 6.10.42. To 47 MU Sealand 21.1.43. Despatched to Middle East 4.2.43 on SS Blommersdyk; arr Suez 27.4.43. SOC 31.5.45.

992 M.14A **N3959** Dd 1 E&WS, Cranwell 6.2.39. To Atlantic Coast, Barnstaple (MI) 9.3.42. To 2 FIS Montrose 13.4.42. Flying accident (Cat.B) 20.7.42; to P&P for repairs 30.7.42. To 27 MU Shawbury 27.9.42. To 47 MU Sealand 7.3.43. Despatched to Middle East 4.5.43 on SS City of Agra; arr 5.6.43. SOC 31.5.45.

993 M.14A **N3960** Dd 1 E&WS, Cranwell 6.2.39. To 15 EFTS Carlisle 16.7.41. To 32 MU St Athan 23.12.41. To 11 EFTS Perth 5.1.42. To 7 FIS Upavon 31.7.42. SOC 25.3.44.

994 M.14A **N3961** Dd Biggin Hill 9.2.39. To 16 Grp Comm Flt, Rochester (30.11.39 census). To Short Bros & Harland Ltd for MAP Overseer 15.3.40. To 5 MU Kemble 24.6.41. To 32 MU St Athan 8.7.41. To 16 Grp Comm Flt, Detling 14.8.41. To 27 MU Shawbury 26.8.41. To 15 EFTS Carlisle 7.2.42. Hit trees on attempted overshoot (or swung on obstacle on landing) Burnfoot RLG 25.2.42; pilot LAC T Pliszka. SOC 1.3.42.

995 M.14A **N3962** Dd 6 MU Brize Norton 6.2.39. To 44 E&RFTS Elmdon 3.5.39. To 5 EFTS Hanworth 15.10.39. Converted to light bomber by P&P. To 5 EFTS Meir 18.6.40. To 16 EFTS Burnaston 10.1.42. To 46 MU Lossiemouth 11.7.42. To 51 MU Lichfield 6.11.42 (unconfirmed). To 46 MU Lossiemouth (31.12.42 census). To 10 FIS Woodley 12.12.44. To 8 EFTS Woodley 7.5.46; coded 'FDT-H'. To 5 MU Kemble 28.3.46 (sic). Sold 15.11.46 to TC Sparrow. Regd **G-AIUE** (CofR 10973) 11.11.46 to Thomas Charles Sparrow, Bournemouth. CofA No.8975 issued 14.7.47. CofA lapsed 13.7.48. Cld 16.3.50 & regd 24.4.50 to Douglas Edward Bianchi, Ashford, Middlesex. CofA renewed 2.5.50. Front cockpit faired over for King's Cup Air Race; tested by ARB at Blackbushe 13.6.50 by GL Howitt. Flown in King's Cup 17.6.50 by entrant Hugh Scrope (Race No.13) but finished 26th at 126 mph. Cld 27.3.51 & regd 31.3.51 to James Thomas Barnett, Hayes. Regd 27.10.51 to Henry Turner Armstrong, Kew. CofA lapsed 20.6.52. Regn cld 20.10.52 by Secretary of State. Regd 10.52 to ACT Carey, Wooburn Green, Bucks. CofA renewed 25.6.53. Regd 3.54 to Fenhurst Properties Ltd, Hayes. Regd 7.55 to VB Nightscale, t/a Nightscale Aircraft Services, Denham. Regd 9.57 to RA Baines of Baines Car Sales Ltd, Grantham. Regd 6.58 to A Turley & partners, t/a Lincoln Aero Club. Regd 5.6.62 to A&D Turley Ltd, t/a Magister Aero Club, Ipswich. Crashed on take-off Seething 26.8.62; Flt Lt A Turley unhurt. Regn cld 11.10.62 as destroyed.

996 M.14A **N3963** Dd 6 MU Brize Norton 6.2.39. To 44 E&RFTS Elmdon 4.5.39. To 5 EFTS Hanworth 15.10.39. Converted to light bomber by P&P. To 5 EFTS Meir 20.6.40. To 16 EFTS Burnaston 16.1.42. MI 10.9.42; to Herts & Essex, Broxbourne 11.9.42. To 51 MU Lichfield 24.10.42. To 76 MU Wroughton 12.5.43. Despatched to Middle East for **Turkey** 17.5.43; arr 13.6.43.

997 M.14A **N3964** Dd 6 MU Brize Norton 6.2.39. To 44 E&RFTS Elmdon 3.5.39. To 32 MU St Athan 14.9.39. To Morris Cowley for repairs (30.11.39 census). To 9 MU Cosford 17.4.40. To 16 EFTS Burnaston 13.9.40. Engine cut on night approach and hit chimney of house in Burton Road, Burnaston 30.3.42; pilot Lt Col J Rock slightly hurt. SOC 9.4.42.

998 M.14A **N3965** Dd 6 MU Brize Norton 9.2.39. To 44 E&RFTS Elmdon 4.5.39. To 5 EFTS Hanworth 15.10.39. SOC 6.4.40.

999 M.14A **N3966** Dd 6 MU Brize Norton 9.2.39. To 25 E&RFTS Waltham 11.5.39. To 24 EFTS Sydenham 15.10.39; to Luton 22.7.40. Hit HT wires low flying and crashed into trees Cheddington, Beds 8.9.40 & dbf. SOC 21.9.40.

1000 M.14A **N3967** Dd 6 MU Brize Norton 9.2.39. To 28 E&RFTS Meir 23.5.39. To 30 EFTS Burnaston 20.9.39. To 16 EFTS Burnaston 10.4.40. MI 25.8.42; to P&P 2.9.42. To 27 MU Shawbury 15.10.42. To ATA White Waltham 29.7.43. On 12.9.44, it struck Anson EG374 while taxying at White Waltham; T/O G Herman unhurt and 50 mins later was itself struck by Magister N5438; repaired on site. To 51 MU Lichfield (21.3.46 census). Sold 10.3.47 to J Neasham. Regd **G-AJHH** (CofR 11297) 11.2.47 to John Neasham, Croft. CofA No.9060 issued 8.5.47. Cld 16.7.47 & regd 30.7.47 to Darlington & District Aero Club Ltd, Croft. Regn cld 29.3.49 as sold abroad. Regd in French Morocco as **F-OAFV** 21.4.50 to Aero Club de Meknes, Meknes. CofA suspended in 6.54. Regn cld 17.9.71 as wfu.

1001 M.14A **N3968** Dd 6 MU Brize Norton 9.2.39. To 28 E&RFTS Meir 23.5.39. To 30 EFTS Burnaston 20.9.39. To 16 EFTS Burnaston 10.4.40. To 1 CRU Cowley (SAS) 22.12.40. To 24 EFTS Luton 21.3.41; coded '23'. Flying accident (Cat.B) 22.9.41; to P&P for repairs. To 37 MU Burtonwood 17.12.41. To 16 EFTS Burnaston 10.1.42. To 2 FIS Montrose 30.1.42. Structural failure, wing broke off and crashed Mill of Barns, near Laurencekirk, Kincardineshire 25.6.42, 2 killed. SOC 5.7.42.

1002 M.14A **N3969** Dd 6 MU Brize Norton 10.2.39. To 39 E&RFTS Weston-Super-Mare 29.6.39. To 8 EFTS Woodley 15.10.39. To 15 EFTS Carlisle 11.2.42. To 29 EFTS Clyffe Pypard 30.5.42. To 5 GTS Shobdon 22.3.43. To Herts & Essex, Broxbourne (Cat.B) 13.5.43. To 27 MU Shawbury 25.8.43. To ATA 9.11.43. To 51 MU Lichfield 7.3.45. Sold 28.11.46 to Miles Aircraft Ltd. CofA No.9364 issued 19.5.47. Regd in France as **F-BDPB** (CofR 19996) 20.6.47 to Roland Baudon, Paris. Regd 3.12.47 to Roger Maire, Nancy. Regd 15.3.54 to Aero Club de l'Est, Nancy. Regd 12.6.56 to Jean de Lescazes, Verdun. Regd 1.8.60 to Aero Club Robert Thierry, Verdun. Wfu .63 and regn cld 22.9.71 as destroyed.

1003 M.14A **N3970** Dd 6 MU Brize Norton 10.2.39. To 39 E&RFTS Weston-Super-Mare 29.6.39. To 8 EFTS Woodley 15.10.39. To 5 EFTS Meir 12.8.40. Stalled on take-off with flaps down and spun into ground Meir 5.3.41. SOC 18.3.41.

1004 M.14A **N3971** Dd 6 MU Brize Norton 10.2.39. To 39 E&RFTS Weston-Super-Mare 29.6.39. To 8 EFTS Woodley 15.10.39. To 15 EFTS Carlisle 8.2.42. To 21 EFTS Booker 16.6.42. SOC 25.2.44.

1005 M.14A **N3972** Dd 6 MU Brize Norton 11.2.39. To 39 E&RFTS Weston-Super-Mare 29.6.42. To 30 EFTS Burnaston 15.10.39. To 16 EFTS Burnaston 10.4.40. Damaged 30.6.41; to P&P for repairs 2.7.41. To 24 EFTS Luton 24.8.41. To 46 MU Lossiemouth 28.2.42. To ATA 8.6.43. Cat.B (MR) 8.12.43. To 51 MU Lichfield 12.2.44. Sold 11.3.46 to Miles Aircraft Ltd. To H Hennequin & Cia, Argentina with CofA No.7830 issued 30.5.46; regd **LV-XPN** 30.6.47 to Secretaria Estado Aeronautica, Buenos Aires. To Centro Universitario de Aviacion. Accident 30.10.47; repaired. Regn cld .66 as wfu.

1006 M.14A **N3973** Dd 6 MU Brize Norton 11.2.39. To 23 E&RFTS Rochester 8.8.39. To 24 EFTS Sydenham 15.10.39; to Luton 22.7.40. To 15 EFTS Carlisle 9.10.41. To 2 FIS Montrose 11.7.42. To 12 MU Kirkbride 13.8.42. SOC 29.9.45.

1007 M.14A **N3974** Dd 6 MU Brize Norton 11.2.39. To 23 E&RFTS Rochester 9.8.39. To 24 EFTS Sydenham 15.10.39; to Luton 22.7.40. To 37 MU Burtonwood 16.2.42. To 9 (P)AFU Hullavington 26.3.42. To ECFS Hullavington 23.5.42. To 7 FIS Upavon 7.11.42. To Comm Flt Woodley 10.5.44. SOC 20.1.45.

1008 M.14A **N3975** Dd 6 MU Brize Norton 20.2.39. To 26 E&RFTS Kidlington 23.5.39. To 5 EFTS Hanworth 15.10.39. Converted to light bomber by P&P. To 5 EFTS Meir 20.6.40. Flying accident (Cat.B) 4.11.41; to P&P 10.11.41. To 20 MU Aston Down 23.3.42. To ECFS Hullavington 6.8.42. To 46 MU Lossiemouth 13.8.42. No fate recorded.

APPENDIX 20   395

1009 M.14A **N3976** Dd Gosport 13.2.39. To 6 MU Brize Norton 22.5.39. To Stn Flt Northolt 3.6.40. Starboard wing broke off during aerobatics & crashed near Sunningdale, Berks 25.11.40, en route Tangmere-Northolt; 2 killed. SOC 12.1.41.

1010 M.14A **N3977** Dd Gosport 13.2.39. To 6 MU Brize Norton 22.5.39. To Northolt 3.6.40. To 24 Sqdn, Hendon 26.3.41. Flying accident (Cat.B) 22.5.41; to P&P (SAS) 31.5.41. To 45 MU Kinloss 28.7.41. To 60 OTU East Fortune 28.8.41. Damaged 21.9.41; to P&P (SAS) 29.9.41. To 27 MU Shawbury 7.12.41. To 6 FIS Staverton (31.12.41 census). To 2 FIS Montrose 26.4.42. MI 6.9.42; to Lundy & Atlantic, Barnstaple 5.10.42. To 51 MU Lichfield 8.11.42. Sold 28.2.46 to Miles Aircraft Ltd. Test flown 25.4.46 by Hugh Kendall as N3977. To H Hennequin & Cia, Argentina with CofA No.7764 issued 15.5.46; regd **LV-XOH** 30.6.47 to Secretaria Estado Aeronautica, Buenos Aires. To Circ. De Aviacion Rosario. Regn cld 4.50 as wfu.

1011 M.14A **N3978** Dd Lee-on-Solent 13.2.39. To Gosport 3.8.39. To 6 MU Brize Norton 22.2.40. To Bramcote 28.6.40. To 304 (Polish) Sqdn, Bramcote 11.10.40; to Syerston 2.12.40; to Lindholme 19.7.41. To Holme-on-Spalding Moor 16.4.42. To 46 MU Lossiemouth 12.7.42. To CRD at MAEE Helensburgh by 30.12.44. To 51 MU Lichfield 24.8.45. Sold 3.6.46 to Miles Aircraft Ltd. Test flown 30.7.46 by Hugh Kendall as N3978. To H Hennequin & Cia, Argentina with CofA No.8258 issued 20.8.46; regd **LV-XSR** 30.6.47 to Secretaria Estado Aeronautica, Buenos Aires. To Aero Club Argentino. Accident 24.9.47; repaired. Accident 10.12.50. Regn cld 3.52 as wfu.

1012 M.14A **N3979** Dd 501 Sqdn, Filton 13.2.39. To CRD at Bristol Aeroplane Co Ltd 2.5.39. SOC 29.8.39.

1013 M.14A **N3980** Dd 6 MU Brize Norton 13.2.39. To 32 MU St Athan 14.9.39. To Morris Motors/1 CRU, Cowley (30.11.39 census). To 5 MU Kemble 6.5.40. To Hucknall 6.7.40. To 48 MU Hawarden 21.2.41. To 12 MU Kirkbride 2.6.41. To 302 (Polish) Sqdn, Churchstanton 17.8.41. To 32 MU St Athan 25.2.42. Crashed near Wareham, Dorset while low flying 6.5.42 and dbf. SOC 16.5.42.

1014 M.14A **N3981** Dd 6 MU Brize Norton 13.2.39. To 23 E&RFTS Rochester 8.8.39. Dived into ground out of cloud Cassington, near Oxford 2.9.39, en route Rochester-Sydenham; 2 killed.

1015 M.14A **N3982** Dd 6 MU Brize Norton 13.2.39. To 23 E&RFTS Rochester 9.8.39. To 24 EFTS Sydenham 15.10.39. Hit hedge on take-off from forced landing Ballyclare, Co Antrim 6.1.40. To 47 MU Sealand 8.3.40. SOC 29.7.40.

1016 M.14A **N3983** Dd 10 MU Hullavington 17.2.39. To 24 EFTS Luton 2.10.40. To 8 EFTS Woodley 22.10.41. To 10 FIS Woodley 21.7.42. To 7 FIS Upavon 28.7.42. SOC 25.3.44.

1017 M.14A **N3984** Dd 10 MU Hullavington 17.2.39. To 254 Sqdn, Stradishall 6.11.39. To 245 Sqdn, Leconfield (30.11.39 census). To 5 MU Kemble 13.6.41. To 15 EFTS Carlisle 19.8.41. Flying accident (Cat.B) 15.10.41; repaired. To 27 MU Shawbury 16.12.41. To 6 FIS Staverton (31.12..41 census). To 16 EFTS Burnaston 13.3.42. To 27 MU Shawbury 11.9.42. To 76 MU Wroughton 18.5.43. Despatched to Middle East for **Turkey** 2.6.43 on SS Chipewyan; arr Alexandria 27.6.43.

1018 M.14A **N3985** Dd 10 MU Hullavington 17.2.39. To 24 EFTS Luton 2.10.40. To 46 MU Lossiemouth 28.2.42. To ATA 17.5.43. Cat.B 8.12.43; repaired at CRO. To 51 MU Lichfield 24.1.44. Sold 7.3.46 to Miles Aircraft Ltd. To H Hennequin & Cia, Argentina with CofA No.7828 issued 30.5.46; regd **LV-XPS** 30.6.47 to Secretaria Estado Aeronautica, Buenos Aires. To Aero Club Neuquen. Crashed Neuquen 3.4.47 (90% damaged); 2 killed.

1019 M.14A **N3986** Dd 10 MU Hullavington 18.2.39. To 40 E&RFTS Mousehold 31.7.39. To 30 EFTS Burnaston 15.10.39. To 16 EFTS Burnaston 10.4.40. Stalled and hit ground 1 ml N of Sutton-on-the-Hill, Derbyshire 9.9.41. SOC 18.9.41.

1020 M.14A **N3987** Dd 10 MU Hullavington 18.2.39. To 30 EFTS Burnaston 15.10.39. To 16 EFTS Burnaston 10.4.40. Damaged 9.6.40; to Surrey Flying Services for repairs 14.6.40. To 37 MU Burtonwood 24.7.40. To 15 EFTS Carlisle 5.6.41. To 21 EFTS Booker 8.6.41. To 6 SofTT Hednesford for GI as **4551M** 16.2.44.

1021 M.14A **N3988** Dd 10 MU Hullavington 18.2.39. To 40 E&RFTS Mousehold 28.7.39. To 30 EFTS Burnaston 15.10.39. To 16 EFTS Burnaston 10.4.40. To 1 CRU Cowley (SAS) 6.12.40. To 8 EFTS Woodley 28.2.41. To 10 FIS Woodley 21.7.42. Flying accident (Cat.B) 13.8.42; to P&P for repairs 13.8.42. To 27 MU Shawbury 24.9.42. To Air Navigation & Bombing School, Jurby 24.5.44. To 5 ANS Jurby 31.5.45. To 51 MU Lichfield 3.9.47. Sold 1.1.48 to The Fairey Flying Club. Regd **G-AKUA** (CofR 12212) 2.3.48 to Fairey Aviation Co Ltd, op by The Fairey Flying Club, White Waltham. CofA No.10013 issued 6.5.48. CofA lapsed 17.11.50. CofA renewed 27.3.52. Cld 27.4.53 & regd 8.5.53 to Douglas Edward Bianchi, Ashford, Middlesex. Cld 7.5.54 & regd 21.5.54 to Wolvehampton Aviation Ltd, Wolverhampton; operated by Air Schools Ltd/Derby Aero Club Ltd, Burnaston. Crashed during balloon-bursting display Burnaston 21.7.57; pilot Norman Green killed. Regn cld 21.8.57 as destroyed.

1022 M.14A **N3989** Dd 10 MU Hullavington 18.2.39. To 40 E&RFTS Mousehold 4.8.39. To 30 EFTS Burnaston 15.10.39. To 16 EFTS Burnaston 10.4.40. Flying accident; to P&P (SAS) 31.3.41 and SOC 26.4.41 (651:35 F/H).

1023 M.14A **N3990** Dd 10 MU Hullavington 20.2.39. To 26 E&RFTS Kidlington 22.7.39. To 5 EFTS Hanworth 15.10.39; to Meir 17.6.40. Hit tree on take-off with flaps down and crashed in road Meir 19.7.40. SOC 26.7.40.

1024 M.14A **N3991** Dd 10 MU Hullavington 20.2.39. To 26 E&RFTS Kidlington 22.7.39. To 5 EFTS Hanworth 15.10.39; to Meir 17.6.40. To 16 EFTS Burnaston 10.1.42. To RAF College Cranwell 3.3.43. To 4 (O)AFU West Freugh 11.3.44. To 51 MU Lichfield 7.8.45. Sold 7.6.46 to Miles Aircraft Ltd. Test flown 30.7.46 by Hugh Kendall as N3991. To H Hennequin & Cia, Argentina with CofA No.8257 issued 20.8.46; regd **LV-XQY** 30.6.47 to Secretaria Estado Aeronautica, Buenos Aires. Regn cld 4.50 as wfu.

1025 M.14A **N5389** Diverted to Irish Air Corps, Baldonnel as **31** 22.2.39. DBR 2.3.46.

1026 M.14A **N5390** Diverted to Irish Air Corps, Baldonnel as **32** 22.2.39. Scrapped 1.46.

1027 M.14A **N5391** Diverted to Irish Air Corps, Baldonnel as **33** 22.2.39. Written off 1.7.42.

1028 M.14A **N5392** Diverted to Irish Air Corps, Baldonnel as **34** 22.2.39. Instructional airframe from 11.3.52. Preserved as a static exhibit Baldonnel (with terrible spats!).

1029 M.14A **N5393** Diverted to Irish Air Corps, Baldonnel as **35** 22.2.39. Written off 6.9.42.

1030 M.14A **N5394** Dd Kenley 27.2.39. To Hawker Aircraft Ltd for MAP Overseer 5.5.39. To 51 MU Lichfield 10.7.43. To ATA 1.2.44. To 51 MU Lichfield 6.6.44. Sold 20.6.46 to Miles Aircraft Ltd. To H Hennequin & Cia, Argentina with CofA No.8286 issued 28.8.46; regd **LV-XQU** 30.6.47 to Secretaria Estado Aeronautica, Buenos Aires. Regn cld 3.52 as wfu.

1031 M.14A **N5395** Dd 6 MU Brize Norton 27.2.39. To Gloster Aircraft Ltd for MAP Overseer 2.5.39. To Vickers-Armstrong Aircraft Ltd 30.8.45. To 4 SofTT, St Athan for GI as **5965M** 1.7.46.

1032 M.14A **N5396** Dd 10 MU Hullavington 27.2.39. To 40 E&RFTS Mousehold 1.8.39. To 30 EFTS Burnaston 15.10.39. To 16 EFTS Burnaston 10.4.40. Flying accident (Cat.B) 5.9.40; to 1 CRU Cowley 18.9.40. To 5 MU Kemble 8.12.40. To HQ 71 Group Farnborough 1.2.41. To 8 AACU Pengam Moors 25.4.41. To P&P (SAS) 21.5.41. To 46 MU Lossiemouth 28.7.41. To 15 EFTS Carlisle 10.8.41. To 29 EFTS Clyffe Pypard 24.5.42. To 45 MU Kinloss 15.5.43. SOC 1.3.44.

1033 M.14A **N5397** Dd 10 MU Hullavington 27.2.39. To 40 E&RFTS Mousehold 24.7.39. To 1 CRU Cowley 17.11.39. To 24 MU Ternhill 15.3.40. To 9 EFTS Ansty (undated). To 16 EFTS Burnaston 7.7.40. To 1 CRU Cowley 23.9.40. To 5 MU Kemble 2.11.40. To 600 Sqdn, Catterick 3.1.41. Damaged 5.9.41; to P&P (SAS) 9.9.41. To 46 MU Lossiemouth 2.11.41. To 15 EFTS Carlisle 9.3.42. To 7 FIS Upavon 31.7.42. Lost power during aerobatics, hit ground and cart-wheeled near Littlecote, Wilts 7.3.44.

1034 M.14A **N5398** Dd 10 MU Hullavington 27.2.39. To 27 E&RFTS Tollerton 31.7.39. To 8 EFTS Woodley 15.10.39. To P&P (SAS) 31.1.40. To 9 MU Cosford 8.8.40. To 24 EFTS Luton

26.9.40. Hit trees on overshooting forced landing ground 1 ml N of Luton 30.10.40; Sgt Kiley killed & Acting LA Eddie seriously injured. SOC 15.11.40.

1035 M.14A **N5399** Dd 10 MU Hullavington 1.3.39. To St Athan 3.10.39. To P&P 15.9.40. To 222 Sqdn, Coltishall 13.1.41. Cat.B 26.8.42; to Shrager Bros, Old Warden for repairs 1.9.42. To 51 MU Lichfield 15.11.42. To 76 MU Wroughton for packing; despatched to Middle East for **Turkey** 12.5.43 on SS Katuma; arr 13.6.43.

1036 M.14A **N5400** Diverted to Irish Air Corps, Baldonnel as **36** 8.3.39. Scrapped in 10.46.

1037 M.14A **N5401** Diverted to Irish Air Corps, Baldonnel as **37** 8.3.39. Written off 15.5.44.

1038 M.14A **N5402** Diverted to Irish Air Corps, Baldonnel as **38** 8.3.39. Written off 28.2.44.

1039 M.14A **N5403** Diverted to Irish Air Corps, Baldonnel as **39** 8.3.39. Scrapped in 8.46.

1040 M.14A **N5404** Diverted to Irish Air Corps, Baldonnel as **40** 8.3.39. Scrapped in 2.46.

1041 M.14A **N5405** Dd 19 MU St Athan 3.3.39. To 1 SofTT Halton 13.6.39. To Shrager Bros, Old Warden (SAS) 1.6.41. To 27 MU Shawbury 21.8.41. To 2 EFTS Staverton (flying from Worcester) 4.11.41. To 6 FIS Staverton (31.12.41 census). Engine cut, stalled during forced landing and crashed on river bank Severn Stoke, Worcester 30.3.42.

1042 M.14A **N5406** Dd 19 MU St Athan 3.3.39. To 15 E&RFTS Redhill 30.5.39; renamed 15 EFTS 3.9.39; to Carlisle 2.6.40. To 1 CRU Cowley (SAS) 18.9.40. To 27 MU Shawbury 5.12.40. To 16 EFTS Burnaston 12.3.41. To 46 MU Lossiemouth 9.7.42. To IFTS ATA, Barton-in-the-Clay 26.8.43. To 51 MU Lichfield 31.1.45. Sold 26.2.46 to Miles Aircraft Ltd. To H Hennequin & Cia, Argentina with CofA No.7714 issued 9.5.46; regd **LV-XNO** 30.6.47 to Secretaria Estado Aeronautica, Buenos Aires. To Aero Club San Rafael. Accident 21.5.54.

1043 M.14A **N5407** Dd 19 MU St Athan 3.3.39. To 15 E&RFTS Redhill 30.5.39; renamed 15 EFTS 3.9.39; coded '23'; to Carlisle 2.6.40. To 29 EFTS Clyffe Pypard 30.5.42. Structural failure of wing in cloud and abandoned over Swindon, Wilts 26.1.43. SOC 30.1.43 as 'missing'.

1044 M.14A **N5408** Dd 19 MU St Athan 3.3.39. To 15 E&RFTS Redhill 30.5.39; renamed 15 EFTS 3.9.39; to Carlisle 2.6.40. To 29 EFTS Clyffe Pypard 24.5.42. To 46 MU Lossiemouth 14.4.43. To 7 Grp Comm Flt, Bottesford 15.11.44. To 51 MU Lichfield 27.12.45. Sold 22.4.49 to RA Short. Regd **G-ALOB** (CofR 12710) 25.4.49 to Robert Alan Short and James Henry Tattersall, Croydon/Clitheroe. Refurbished & CofA No.10731 prepared but not issued. Regn cld 30.4.52 by MCA. Reportedly stored at Elstree 6.52. Regd 21.1.53 to Robert Alan Short, Croydon. Regn cld 18.5.59 as pwfu.

1045 M.14A **N5409** Dd 19 MU St Athan 3.3.39. To 15 E&RFTS Redhill 30.5.39; renamed 15 EFTS 3.9.39; to Carlisle 2.6.40. To 29 EFTS Clyffe Pypard 24.5.42. To 5 GTS Shobdon 22.3.43. To Herts & Essex, Broxbourne (MR) 5.5.43. To 46 MU Lossiemouth 5.7.43. To 51 MU Lichfield 13.6.45. Sold 22.9.47 to Miles Aircraft Ltd. CofA No.9879 issued 13.2.48 to **Thai Navy**.

1046 M.14A **N5410** Dd 19 MU St Athan 3.3.39. To 15 E&RFTS Redhill 7.6.39; renamed 15 EFTS 3.9.39; to Carlisle 2.6.40. To 1 CRU Cowley 18.9.40. To 45 MU Kinloss 25.11.40. To 603 Sqdn, Drem 18.1.41. Cat.B 22.5.41; to P&P for repairs 4.6.41. To 5 MU Kemble 24.12.41. To 4 FIS Cambridge 18.1.42. To 6 FIS Worcester 10.5.42. To 2 EFTS Staverton (flying from Worcester) (undated). To 46 MU Lossiemouth 11.8.42. To ECFS Hullavington 13.9.42. SOC Cat.E1 (MR) 8.8.44.

1047 M.14A **N5411** Dd 19 MU St Athan 6.3.39. To 19 E&RFTS Gatwick 6.6.39. To 30 EFTS Burnaston 15.10.39. To 16 EFTS Burnaston 10.4.40. To 21 EFTS Booker 17.4.43. To 6 SofTT Hednesford for GI as **4550M** 16.2.44.

1048 M.14A **N5412** Dd 19 MU St. Athan 6.3.39. To 45 E&RFTS Ipswich 25.6.39. To 5 EFTS Hanworth 15.10.39; to Meir 17.6.40. Crashed 10.7.41. SOC 31.8.41 (996:15 F/H).

1049 M.14A **N5413** Dd 19 MU St Athan 1.3.39. To 45 E&RFTS Ipswich 28.6.39. To 5 EFTS Hanworth 15.10.39; to Meir 17.6.40. To 37 MU Burtonwood 8.12.41. To 2 FIS Montrose 5.2.42. To 12 MU Kirkbride 11.8.42. To 4 Signals School, Madley 22.11.42; renamed 4 Radio School 1.1.43. To 1 Radio School, Cranwell 19.1.45. To 4 Radio School, Madley 6.6.45. To 4 SofTT St Athan for GI as **5359M** 14.7.45.

1050 M.14A **N5414** Dd 19 MU St Athan 6.3.39. To 45 E&RFTS Ipswich 28.6.39. To 32 MU St Athan (undated). To 1 CRU Cowley (30.11.39 census). To 6 MU Brize Norton 11.6.40. To Watchfield 13.9.40. To 6 MU Brize Norton 17.10.40. To Middle Wallop 29.4.41. To Farnborough 2.5.41. To Middle Wallop 15.5.41. To West Freugh 11.10.41. To TFU Christchurch (31.12.41 census). To Defford 4.5.42. To 51 MU Lichfield 7.11.42. SOC Cat.E (MR) 25.2.44.

1051 M.14A **N5415** Dd 19 MU St Athan 7.3.39. To 45 E&RFTS Ipswich 28.6.39. To 5 EFTS Hanworth 15.10.39; to Meir 17.6.40. To 2 FIS Montrose 28.1.42. MI 3.8.42; to P&P for repairs 18.8.42. To 27 MU Shawbury 7.10.42. To ATA 30.6.43. SOC Cat.E 30.11.44. To Miles Aircraft Ltd. To H Hennequin & Cia, Argentina with CofA No.7447 issued 14.3.46; regd **LV-XMM** 30.6.47 to Secretaria Estado Aeronautica, Buenos Aires. To Aero Club Bahia Blanca. Accident 7.10.47; repaired. Regd 21.10.60 to Aero Club Puerto Madryn, Chubut.

1052 M.14A **N5416** Dd 19 MU St Athan 8.3.39. To 29 E&RFTS Luton 9.6.39. To 24 EFTS Sydenham 15.10.39; to Luton 22.7.40. To 1 CRU Cowley (SAS) 20.3.41. To 39 MU Colerne 23.4.41. To 10 Grp Comm Flt, Colerne (31.12.41 census). Ran out of fuel while lost and hit HT cables in forced landing Longbridge 3.1.42. To 1282 Sqdn ATC, Leadenhall Street, London 17.2.42 but no M number traced.

1053 M.14A **N5417** Dd 19 MU St Athan 8.3.39. To 29 E&RFTS Luton 9.6.39. To 30 EFTS Burnaston 15.10.39. To 16 EFTS Burnaston 10.4.40. Engine cut, stalled while forced landing and spun into ground near Repton, Derbyshire 12.10.41; LAC EW Pierce killed. SOC 19.10.41.

1054 M.14A **N5418** Dd 19 MU St Athan 14.3.39. To 1 SofTT Halton 13.6.39. To 5 MU Kemble 9.11.40. To 145 Sqdn, Tangmere 25.11.40. To 1 CRU Cowley (SAS) 21.12.40. To P&P (SAS) 16.8.41. To HQ SFP 14.9.41. To ATA HQ (Class 1 Training), White Waltham (2.42); coded '43'. To Atlantic Coast, Barnstaple (Cat.B) 11.9.43. To 51 MU Lichfield 8.11.43. Allotted by MAP to Miles for Chile in early 1945 but cld. Sold 22.4.49 to RA Short. Regd **G-ALOC** (CofR 12711) 25.4.49 to Robert Alan Short and James Henry Tattersall, Croydon/Clitheroe. CofA No.10711 issued 27.10.49. Regn cld 24.2.50 as sold abroad (suggested to Egyptian AF but cancellation details not consistent with other sales).

1055 M.14A **N5419** Dd 19 MU St Athan 10.3.39. To 1 SofTT Halton 13.6.39. To 1 CRU Cowley (SAS) 9.1.41. To 5 MU Kemble 6.2.41. To 32 Sqdn, Ibsley 25.2.41. Hit hedge on take-off from forced landing Bow Street, Aberystwyth 4.12.41. SOC 9.12.41.

1056 M.14A **N5420** Dd 19 MU St Athan 13.3.39. To 1 SofTT Halton 13.6.39. To Shrager Bros, Old Warden (SAS) 11.6.41. To 27 MU Shawbury 29.8.41. To 2 EFTS Staverton (flying from Worcester) 4.11.41. To 6 FIS Staverton (31.12.41 census). To 4 FIS Cambridge 9.5.42. To 27 MU Shawbury 7.6.42. To Herts & Essex, Broxbourne for repairs 12.8.42. To 27 MU Shawbury 10.9.42. To 47 MU Sealand 18.2.43. Despatched to Middle East 19.4.43 on SS Clan Lamont; arr 12.6.43. SOC 31.5.45.

1057 M.14A **N5421** Dd 19 MU St Athan 13.3.39. To 1 SofTT Halton 13.6.39. To 9 MU Cosford 12.6.41. To 412 RCAF Sqdn, Digby 27.7.41. To 133 Sqdn, Coltishall 2.8.41. To 336 Sqdn, USAAF 1.10.42. To 51 MU Lichfield 17.10.42. To ATA 18.9.43. Cat.B MR 12.1.44 and SOC Cat.E 7.2.44.

1058 M.14A **N5422** Dd 19 MU St Athan 13.3.39. To 52 MU Cardiff 9.4.40. Despatched to Middle East 13.5.40. To 254 Wing Comm Flt, Khartoum; to Erkowett 7.6.40. To 223 Sqdn. To 103 MU Aboukir (41/42). To 201 Grp Comm Flt, Mariut (formed 1.6.43). To 135 MU Burg-el-Arab 12.6.43 (still there 1.44). To Eastern Med Comm Flt, Mariut 1.2.44. To ME Comm Sqdn, Heliopolis (reformed from Stn Flight Heliopolis 31.5.45). SOC 31.12.46.

1059 M.14A **N5423** Dd 19 MU St Athan 13.3.39. To 1 SofTT Halton 13.6.39. To Stn Flt Halton. To 51 MU Lichfield 4.12.43. Sold

APPENDIX 20

1060 M.14A **N5424** Dd 19 MU St Athan 14.3.39. To 52 MU Cardiff 9.4.40. Despatched to Middle East 13.5.40. To Western Desert Comm Flt. Destroyed by enemy action. SOC 6.5.41.

15.5.46 to Miles Aircraft Ltd. To H Hennequin & Cia, Argentina with CofA No.8150 issued 24.7.46; regd **LV-XSE** 30.6.47 to Secretaria Estado Aeronautica, Buenos Aires. Accident 17.12.52; repaired. Crashed La Paz 18.8.54; pilot killed.

1061 M.14A **N5425** CLN recorded 'Dd (36 MU) Sealand 16.3.39 but allotment cancelled'. Dd 10 FTS Ternhill 16.3.39 and hit fence during forced landing while lost in bad weather 1/2 ml W of Ternhill 16.3.39 while on delivery by 2 FPP. To **1529M** 5.6.39 to 5 Plumber Street, Nottingham (probably an ATC or similar training unit). Reduced to 'group assemblies' 26.9.40.

1062 M.14A **N5426** Dd 36 MU Sealand 16.3.39. To 3 AACU Kalafrana, Malta 5.39. Unit disbanded 19.9.40. To 202 Sqdn, Gibraltar. To 830 Sqdn. Destroyed in air raid 13.5.41 (103:30 F/H).

1063 M.14A **N5427** Dd 36 MU Sealand 20.3.39. To 3 AACU Kalafrana, Malta 8.5.39. Unit disbanded 19.9.40. To 202 Sqdn, Gibraltar. To 830 Sqdn. SOC 5.9.41.

1064 M.14A **N5428** Dd 36 MU Sealand 20.3.39. To 3 AACU Kalafrana, Malta 8.5.39. To 261 Sqdn, Takali, Malta. Destroyed in air raid Takali, Malta 19.3.42.

1065 M.14A **N5429** Dd 19 MU St Athan 18.3.39. To 1 SofTT Halton 13.6.39. To 27 MU Shawbury 27.2.43. Despatched to Middle East for **Turkey** 12.5.43 on SS Katuna; arr 13.6.43.

1066 M.14A **N5430** Dd 19 MU St Athan 18.3.39. To 1 SofTT Halton 13.6.39. To 1 CRU Cowley (SAS) 21.12.40. To 3 SofGR Squires Gate 21.2.41. To Herts & Essex, Broxbourne (SAS) 26.6.41. To 27 MU Shawbury 27.7.41. To 24 EFTS Luton 2.10.41. To 2 FIS Montrose 15.2.42. MR (Cat.B); repaired at unknown CRO 26.7.43. To 51 MU Lichfield 6.11.43. Sold 22.4.49 to RA Short. Regd **G-ALOF** (CofR 12713) 25.4.49 to Robert Alan Short and James Henry Tattersall, Croydon/Clitheroe. Regn cld 12.9.52 by MCA. Regd 21.1.53 to Robert Alan Short, Croydon. Not refurbished and remains burnt Elstree 5.11.53. Regn cld 18.5.59 as pwfu.

1067 M.14A **N5431** Dd 19 MU St Athan 18.3.39. To 1 SofTT Halton 13.6.39. To HQ 24 Group Halton. Dived into ground Weston Turville, Bucks 10.8.40, believed to have stalled in bumpy conditions, 1 killed. SOC 26.8.40.

1068 M.14A **N5432** Dd 19 MU St Athan 18.3.39. To 1 SofTT Halton 13.6.39. To 37 MU Burtonwood 30.7.40. To 15 EFTS Carlisle 21.6.41. To P&P (Cat.B) 23.10.41. To 46 MU Lossiemouth 28.1.42. To ATA 20.11.43. To 51 MU Lichfield 7.3.45. Sold 15.5.46 to Miles Aircraft Ltd. To H Hennequin & Cia, Argentina with CofA No.8117 issued 18.7.46; regd **LV-XQE** 30.6.47 to Secretaria Estado Aeronautica, Buenos Aires. To Aero Club Posadas. Crashed Estancia Corin 23.3.47; pilot killed.

1069 M.14A **N5433** Dd 19 MU St Athan 18.3.39. To 23 E&RFTS Rochester 2.8.39. To 24 EFTS Sydenham 15.10.39; to Luton 22.7.40. To 1 CRU Cowley (SAS) 22.1.41. To 2 SofAC Andover 21.3.41. To 42 OTU Andover 18.7.41. To 263 Sqdn, Warmwell 29.11.43. Engine cut and crashed in forced landing Addington, near Liskeard, Cornwall 5.4.44 or 5.6.44? SOC 18.6.44.

1070 M.14A **N5434** Dd 19 MU St Athan 20.3.39. To 23 E&RFTS Rochester 1.8.39. To 24 EFTS Sydenham 15.10.39; to Luton 22.7.40. To Shrager Bros, Old Warden (SAS) 8.10.41. To 37 MU Burtonwood 28.12.41. To 4 FPP ATA Prestwick. Crashed when hit signal mast on dummy deck while taking-off on wrong runway Arbroath at 1323 hrs on 30.1.42; ATA pilot S/O H Kindberg badly injured. SOC (Cat.E) 6.2.42.

1071 M.14A **N5435** Dd 19 MU St Athan 20.3.39. To 23 E&RFTS Rochester 18.8.39. To 24 EFTS Sydenham 15.10.39. To 47 MU Sealand 8.3.40. To 'G' Temporary MU Mousehold 5.5.40. SOC 30.4.43.

1072 M.14A **N5436** Dd 15 Grp Comm Flt, Lee-on-Solent 20.3.39; to Mount Batten (Plymouth) 7.6.39. To 37 MU Burtonwood 30.7.40. To 15 Grp Comm Flt, Hooton Park 17.2.41. To 272 Sqdn, Aldergrove 18.3.41. To 233 Sqdn, Aldergrove 10.4.41. To 27 MU Shawbury 27.2.42. To 122 Sqdn, Hornchurch 3.9.42. To 27 MU Shawbury 3.10.42. To 76 MU Wroughton 13.5.43 for packing. Despatched to Middle East for **Turkey** 2.6.43 on SS Chipewyan; arr 3.7.43.

1073 M.14A **N5437** Dd 15 Grp Comm Flt, Lee-on-Solent 20.3.39; to Mount Batten (Plymouth) 7.6.39. To 19 Grp Comm Flt, Roborough 15.4.41. Flying accident (Cat.Ac) 1.7.41; repaired. To 20 MU Aston Down 20.3.42. To 1 AACU Farnborough 13.4.42. To 8 AACU Pengam Moors 11.12.42. To 1607 (AACU) Flt, Carew Cheriton. To Stn Flt Farnborough 18.3.43. To 70 Grp Comm Flt Farnborough (undated). SOC 13.3.45.

1074 M.14A **N5438** Dd 19 MU St Athan 20.3.39. To 23 E&RFTS Rochester 2.8.39. To 24 EFTS Sydenham 15.10.39; to Luton 22.7.40. To 9 (P)AFU Hullavington (undated). MR (Cat.B) 26.5.42; to P&P for repairs 11.7.42. To 46 MU Lossiemouth 23.7.42. To IFTS ATA 20.11.43. To 51 MU Lichfield 4.3.45. Sold 11.12.46 to The Wiltshire School of Flying Ltd. Regd **G-AITZ** (CofR 10968) 11.11.46 to Anna Valley Motors (Salisbury) Ltd, Salisbury (based Thruxton). CofA No.8882 issued 5.9.47. Cld 5.5.48 & regd 11.5.48 to Hyman Victor Behar, Portsmouth. CofA lapsed 19.9.49. Regn cld 30.10.51 as pwfu. Under overhaul by Hants & Sussex Aviation Ltd, Portsmouth 11.51 but abandoned.

1078 M.14A Hawk Trainer Mk.III. Regd **G-AFTR** (CofR 9103) 4.5.39 to Phillips & Powis Aircraft Ltd. CofA No 6575 issued 10.5.39; dd to 8 E&RFTS Woodley; renamed 8 EFTS 3.9.39. Crashed Theale Forced Landing Ground 8.9.39; pupil pilot Cpl R Murgatroyd uninjured; aircraft repaired. Regn cld 14.11.40 as sold. Impressed under Contract A.113018/40 dated 17.9.40 as **BB663**, retained by 8 EFTS. (Recorded as to 32 MU St Athan 25.12.41 and 11 EFTS Perth 5.1.42 but still on charge of 8 EFTS at time of final accident). Crashed into tree Poyle Park, Tongham, Surrey 25.4.42 during attempted overshoot on precautionary landing after pilot became lost on cross-country flight; LAC RW Hibbs injured. SOC Cat.E.

1079 M.14A Hawk Trainer Mk.III. Regd **G-AFTS** (CofR 9104) 4.5.39 to Philips & Powis Aircraft Ltd. CofA No 6576 issued 18.5.39; dd to 8 E&RFTS Woodley; renamed 8 EFTS 3.9.39. Regn cld 14.11.40 as sold. Impressed under Contract A.113018/40 dated 19.9.40 as **BB664**, retained by 8 EFTS. U/c hit HT cables on demonstration forced landing and crashed Farley Hill, near Woodley 15.8.41; F/O RM Dryden & LAC FC Collins uninjured. Wreckage recovered to Woodley and SOC Cat.E 15.9.41.

1080 M.14A Hawk Trainer Mk.III. Regd **G-AFWY** (CofR 9182) 27.7.39 to Phillips & Powis Aircraft Ltd. CofA No 6716 issued 4.8.39; dd to 8 E&RFTS Woodley; renamed 8 EFTS 3.9.39. Regn cld 14.11.40 as sold. Impressed under Contract A.113018/40 dated 19.9.40 as **BB665**, retained by 8 EFTS. Suffered three minor accidents during the next 12 months including forced landings at Faringdon, Berks 24.9.40 and Sindlesham, Berks 1.11.40, but repaired in each case. To 37 MU Burtonwood 19.10.41. To Stn Flt Honiley 30.12.41. To 54 OTU Charter Hall 16.3.43. SOC Cat.E1 2.11.44.

1081 M.14A Hawk Trainer Mk.III. Regd **G-AFXA** (CofR 9184) 11.8.39 to Phillips & Powis Aircraft Ltd. CofA No.6735 issued 11.8.39; dd to 8 E&RFTS Woodley; renamed 8 EFTS 3.9.39. Regn cld 25.11.40 as sold. Impressed under Contract A.113018/40 dated 17.9.40 as **BB666**, retained by 8 EFTS. To 10 FIS Woodley 7.42. To 51 MU Lichfield 23.7.46. To A&AEE Boscombe Down 13.8.46. To 11 RFS Perth 26.4.47. To 51 MU Lichfield 6.11.47. Sold 22.4.49 to RA Short. Regd **G-ALOG** (CofR 12714) 25.4.49 to Robert Alan Short & James Henry Tattersall, Croydon/Clitheroe. Regn NTU and regd **G-AFXA** wef 25.4.49 to same owners. Still marked as G-ALOG at Croydon 12.9.49 but as G-AFXA on cowling at Croydon 20.10.50. Regn cld by MCA 30.4.52. Regd 21.1.53 to Robert Alan Short, Croydon. Not refurbished and scrapped Croydon 3.56. Regn cld 18.5.59 as pwfu.

1082 M.14A Hawk Trainer Mk.III. Regd **G-AFXB** (CofR 9185) 11.8.39 to Phillips & Powis Aircraft Ltd. CofA No.6736 issued 19.8.39; dd to 8 E&RFTS Woodley; renamed 8 EFTS 3.9.39. Regn cld 14.11.40 as sold. Impressed under Contract A.113018/40 dated 19.9.40 as **BB667**, retained by 8 EFTS. To 10 FIS Woodley 7.42. To A&AEE Boscombe Down 4.6.46. To 51 MU Lichfield 3.7.46. To 5 MU Kemble 4.2.47. SOC Cat.E2 10.12.47. Wings reportedly at Croydon 10.58.

1083 M.14A Hawk Trainer Mk.III. Regd **G-AFYV** (CofR 9229) 22.8.39 to Phillips & Powis Aircraft Ltd. Not built and regn cld 19.10.39 as pwfu.

1084 M.14A Hawk Trainer Mk.III. Regd **G-AFYW** (CofR 9230) 22.8.39 to Phillips & Powis Aircraft Ltd. Not built and regn cld 19.10.39 as pwfu.

1085 M.14A Hawk Trainer Mk.III. Regd **G-AFYX** (CofR 9231) 22.8.39 to Phillips & Powis Aircraft Ltd. Not built and regn cld 19.10.39 as pwfu.

1086 M.14A Hawk Trainer Mk.III. Regd **G-AFYY** (CofR 9232) 22.8.39 to Phillips & Powis Aircraft Ltd. Not built and regn cld 19.10.39 as pwfu.
Note: Contrary to previous speculation, serials BB668 to BB671 were never allotted to the above unbuilt aircraft.

1611 M.14A **P2374** Dd 19 MU St Athan 24.3.39. To 23 E&RFTS Rochester 1.8.39. To 24 EFTS Sydenham 15.10.39; to Luton 22.7.40. To 37 MU Burtonwood 16.2.42. To Boulton Paul Aircraft Ltd for MAP Overseer 20.4.42. SOC 15.3.45.

1612 M.14A **P2375** Dd 19 MU St Athan 24.3.39. To 219 Sqdn, Catterick 6.11.39. Spun into ground Bolton-on-Swale 13.3.40 (Cat.B). To 60 MU York 19.3.40 & SOC Cat.E 24.3.40.

1613 M.14A **P2376** Dd 19 MU St Athan 24.3.39. To 152 Sqdn, Acklington 16.3.40. Crashed in forced landing near Knaresborough, Yorks 16.3.40. SOC 2.4.40.

1614 M.14A **P2377** Dd 19 MU St Athan 29.3.39. To 52 MU Cardiff 9.4.40. Despatched to Middle East 6.6.40. To 267 Sqdn, probably Heliopolis. To 41 SAAF Sqdn. SOC 26.4.45 as scrap.

1615 M.14A **P2378** Dd 19 MU St Athan 29.3.39. To 52 MU Cardiff 9.4.40. Despatched to Middle East 6.6.40. To 38 Sqdn. To 12 SoFP ?. To 238 Sqdn, Gamil (after 12.1.43). *A photograph of P2378 taken by Howard Levy at El Kabrit, Egypt on 18.6.43 shows it to be in good condition, with spats, and still with the large pre-war style roundels under the wings.* SOC 1.12.43.

1616 M.14A **P2379** Dd 19 MU St Athan 29.3.39. To 52 MU Cardiff 9.4.40. Despatched to Middle East 13.5.40. To Comm Flt Western Desert. Destroyed by enemy action 6.5.41.

1617 M.14A **P2380** Dd 4 Grp Finningley 30.3.39 and allocated to AV Roe & Co Ltd 2.5.39. To AV Roe & Co Ltd (28.1.41). To General Aircraft Ltd, Hanworth 5.6.44. SOC 31.10.44.

1618 M.14A **P2381** Dd AMDP Reading 3.4.39 for noise-reducing exhaust system. To AAEE Martlesham Heath 26.7.39. To 6 MU Brize Norton 4.6.40. To 16 EFTS Burnaston 29.6.40. To 27 MU Shawbury 2.9.42. To 47 MU Sealand 9.2.43. Despatched to Middle East on SS Turkmenstan 14.4.43; arr 29.6.43. SOC 31.12.46.

1619 M.14A **P2382** Dd 6 Grp Comm Flt, Mousehold 30.3.39. To HQ Flt AASF France 16.10.39. To 1 Salvage Unit Reims/Champagne 24.10.39. To 21 AD Nantes/Bouguenais 12.2.40. Presumed abandoned in France. SOC 26.3.42.

1620 M.14A **P2383** Dd 6 Grp Comm Flt, Mousehold 30.3.39. To HQ Flt AASF France 19.10.39. To 218 Sqdn, Suippes, France 31.10.39. Destroyed in air raid 23.5.40. SOC 'enemy action' 28.5.40.

1621 M.14A **P2384** Dd 6 Grp Comm Flt, Mousehold 30.3.39. To Stn Flt Northolt 25.10.39. To 141 Sqdn, Grangemouth 23.4.40. To 52 OTU Debden 2.4.41. Cat.B 15.8.41; to P&P (SAS) 2.9.41. To 46 MU Lossiemouth 8.11.41. To 2 FIS Montrose 10.1.42. To 14 (P)AFU Ossington 25.10.42. To 1 SofTT Halton for GI as **4513M** 31.1.44 (allotted ex HQ ATA White Waltham).

1622 M.14A **P2385** Dd 19 MU St Athan 4.4.39. To 219 Sqdn, Catterick 6.11.39. Hit HT cable during AA co-operation and flew into ground Piercebridge, near Darlington 6.8.40. SOC 15.8.40.

1623 M.14A **P2386** Dd 36 MU Sealand 6.4.39. Despatched to Malta 8.5.39. Destroyed in air raid in 5.42. SOC 1.5.42 'enemy action' (162:20 F/H).

1624 M.14A **P2387** Dd Stn Flt Wyton 4.4.39. To RAF Coningsby 24.2.41. To P&P (SAS) 19.6.41. To Shrager Bros, Old Warden (SAS) 8.7.41. To 37 MU Burtonwood 18.10.41. To ATA IFTS, Thame/Barton-in-the-Clay 24.4.42. Still in service 12.43 and ultimate fate unknown.

1625 M.14A **P2388** Dd Stn Flt Wyton 3.4.39. To 27 MU Shawbury 16.7.41. To ATA 9.8.41; op by ATA Training Unit (2.42); coded '32'. MI (Cat.B) 21.11.43 at unknown CRO. To 51 MU Lichfield 24.1.44. To RAE Farnborough 20.1.48 (as communication aircraft). Damaged Cat.2 13.10.49; repaired. Sold per CS(A) Return 8.5.50 to Fairoaks Aero Club (and dd 12.5.50). Regd **G-AMBM** (CofR 15020) 10.5.50 to Universal Flying Services Ltd, Fairoaks. CofA No.A41 issued 27.3.51. Cld 28.3.51 & regd 29.3.51 to All-Power Transformers Ltd, Byfleet. Cld 18.10.52 & regd 10.11.52 to Jean Lennox Bird, Alton, Hants. Regd 7.8.53 to Experimental Flying Group Ltd, Redhill. Flown by Rex Nicholls and the author, Peter Amos, on the 'Grand Tour' of France and Switzerland in the summer of 1953. Crashed at Deols, France 14.9.54, wrecking u/c and damaging propeller and rudder. Regn cld 12.1.55 as destroyed. Later taken to St. Cyr where it was still to be seen 7.65.

1626 M.14A **P2389** Dd Stn Flt Andover 3.4.39. To 1 CRU Cowley (SAS) 29.5.41. To 46 MU Lossiemouth 24.8.41. To 331 Sqdn, Skeabrae 14.10.41. To 27 MU Shawbury 6.10.42. To 47 MU Sealand 24.1.43. Despatched to Middle East 4.2.43 on SS Blommersdyk; arr Suez 27.4.43. To 201 Grp Comm Flt Mariut (formed 1.6.43). SOC 26.4.45.

1627 M.14A **P2390** Dd 19 MU St Athan 4.4.39. To 52 MU Cardiff 9.4.40. Despatched to Middle East 13.5.40. To 71 OTU Ismailia (formed 1.6.41). SOC 1.3.44.

1628 M.14A **P2391** Dd 19 MU St Athan 4.4.39. To 52 MU Cardiff 9.4.40. Despatched to Middle East 6.5.40. To Stn Flt Ismailia. To 71 OTU Ismailia (formed 1.6.41); present 8.41. SOC 1.9.43.

1629 M.14A **P2392** Dd 19 MU St Athan 4.4.39. To 219 Sqdn, Catterick 6.11.39. To 607 Sqdn, Usworth 3.6.40. Crashed on take-off Sherburn-in-Elmet 9.10.40. SOC 18.10.40.

1630 M.14A **P2393** Dd 19 MU St Athan 12.4.39. To 52 MU Cardiff 9.4.40. Despatched to Middle East 6.6.40. To Ismailia (.41). To 13 Sqdn. To Comm Flt Western Desert. To 204 Group Comm Flt, Heliopolis. To Ma'aten Bagush. Cat.B flying accident 16.12.42; to Adv. Salvage Unit and repaired. Crashed near Benghazi 7.12.43. SOC 1.1.47.

1631 M.14A **P2394** Dd 103 Sqdn, Benson 11.4.39; unit to Challerange, France 2.9.39. Presumably lost in France in 5.40.

1632 M.14A **P2395** Dd 19 MU St Athan 12.4.39. To 219 Sqdn, Catterick 6.11.39; unit to Redhill 12.10.40 and to Tangmere 10.12.40. Hit by incendiary bomb and burnt out Tangmere 11.5.41. To 1 CRU Cowley (SAS) 16.5.41 and SOC 19.5.41.

1633 M.14A **P2396** Dd 19 MU St Athan 12.4.39. To 52 MU Cardiff 9.4.40. Despatched to Middle East 13.5.40. To 33 Sqdn. To 112 Sqdn. At 103 MU Aboukir .41. Burnt on evacuation El Adem in 4.41 (possibly 22.4.41).

1634 M.14A **P2397** Dd 19 MU St Athan 12.4.39. To 52 MU Cardiff 9.4.40. Despatched to Middle East 13.5.40. To 33 Sqdn. At 103 MU Aboukir (.41). To 204 Group Comm Flt, Heliopolis. To Ma'aten Bagush. To 213 Sqdn. To 132 MU Gaza (12.42-6.43). To 206 CU (probably 206 Group). To 119 MU Shaibah (3.44). SOC 31.5.45.

1635 M.14A **P2398** Dd 19 MU St Athan 14.4.39. To 52 MU Cardiff 9.4.40. Despatched to Middle East 13.5.40. To 33 Sqdn. To TU&RP Ismailia (redesignated 70 OTU 10.12.40). To 208 Sqdn. To Comm Flt Western Desert. To 70 OTU Ismailia. SOC 17.4.41.

1636 M.14A **P2399** Dd 19 MU St Athan 14.4.39. To 52 MU Cardiff 9.4.40. Despatched to Middle East 6.6.40. To 108 Sqdn. SOC 1.9.43.

1637 M.14A **P2400** Dd 19 MU St Athan 14.4.39. To 52 MU Cardiff 10.4.40. Despatched to Middle East 6.5.40. To 274 Sqdn. To 252 Wing Improvised HQ at Seagull Camp, Mex (just outside Alexandria) from 3.6.40. Loaned to Pool Sqdn, RNAS Dekheila and forced-landed 20.1.41 (pilot Sub Lt DW Phillips). Hit wires and crashed near Cairo 6.10.41. SOC 9.10.41. (Also recorded as SOC Flying Accident 27.8.41).

1638 M.14A **P2401** Dd 19 MU St Athan 14.4.39. To 52 MU Cardiff 9.4.40. Despatched to Middle East 13.5.40. To 39 Sqdn, Heliopolis. Flew into ground in bad visibility 10 ml NE of Cairo 15.2.41. SOC 20.2.41.

APPENDIX 20

1639 M.14A **P2402** Dd Stn Flt Aldergrove 25.4.39. To 3 B&GS Aldergrove (30.11.39 census). To Comm Flt Woodley 3.7.40. Cat.B; to P&P (SAS) 10.10.40. To 5 MU Kemble 20.3.41. To ATA 16.4.41. Crashed 28.6.41. SOC 4.7.41.

1640 M.14A **P2403** Dd 4 AOS West Freugh 18.4.39; renamed 4 B&GS 1.11.39. To 15 EFTS Carlisle 3.6.41. To 2 FIS Montrose 11.7.42 To SFTS/RAF College, Cranwell 15.7.43. To 7 (O)AFU Bishops Court 17.3.44. To 7 EFTS Desford 26.10.45. To 5 MU Kemble 7.10.46. SOC 10.12.47.

1641 M.14A **P2404** Dd 5 ATS Penrhos 19.4.39. To 9 B&GS Penrhos 1.11.39. To P&P 17.11.39. To 46 MU Lossiemouth 26.6.40. To 241 Sqdn, Inverness 23.10.40. To 45 MU Kinloss 5.1.41. To 15 EFTS Carlisle 9.3.41. To 32 MU St Athan 30.9.41. To 8 EFTS Woodley 10.41. To 10 FIS Woodley 21.7.42. To 51 MU Lichfield 15.9.43. Sold 15.11.46 to Miles Aircraft Ltd. Regd **G-AJZH** (CofR 11721) 4.7.47 to Cecil Robert Jackson, Wanstead, London E11. CofA No.9555 issued 8.7.47. Cld 9.12.49 & regd 18.1.50 to Patrick Joseph McNamara, James Charles Milli & Bruno Peter Pini, Denham; named 'Milli's Maggie'. Cld 30.4.51 & regd 2.5.51 to Frank Walter Seymour, James Charles Milli & Bruno Peter Pini; op by Denham Flying Club, Denham. Cld 31.8.53 & regd 3.9.53 to Denham Flying Club Ltd, Denham. Crashed Nuthampstead, Herts 22.8.57. Wreck still stored Denham (4.58); one wing probably used in rebuild of G-AJDR (.59). Regn cld 2.8.63 as destroyed 1.58.

1642 M.14A **P2405** Dd 6 ATS Warmwell 18.4.39; became 10 AOS 3.9.39; became 10 B&GS 1.1.40; to Dumfries 10.7.40; reformed as 10 AOS 13.9.41; reformed as 10 (O)AFU 1.5.42. To 7 SofTT Insworth for GI as **4111M** 31.8.43.

1643 M.14A **P2406** Dd 8 ATS Evanton 18.4.39; became 8 AOS 3.9.39; became 8 B&GS 1.11.39. To 5 EFTS Meir 10.8.41. To 16 EFTS Burnaston 10.1.42. To 3 EFTS Shellingford 14.6.42. To HGCU Brize Norton 20.2.43. To 3 EFTS Shellingford 25.2.43. To 45 MU Kinloss 20.3.43. To Herts & Essex, Broxbourne (MI) 21.4.43. To 27 MU Shawbury 19.6.43. To ATA 3.7.43. To 51 MU Lichfield 31.1.45. Sold 11.3.46 to Miles Aircraft Ltd. Test flown 23.4.46 by Hugh Kendall as P2406. To H Hennequin & Cia, Argentina with CofA No.7734 issued 9.5.46; regd **LV-XOF** 30.6.47 to Secretaria Estado Aeronautica, Buenos Aires. Probably to Aero Club Mendoza. Crashed following engine failure 17.8.46 (40% damage); possibly not repaired.

1644 M.14A **P2407** Dd 1 AACU Farnborough 18.4.39. Damaged 14.10.39; to 1 CRU Cowley (SAS) 15.11.39. To 8 MU Little Rissington 11.6.40. To CRD Rootes Group, Speke 27.6.40. To Cosford (probably 2 SofTT) for GI as **5355M** 10.7.45.

1645 M.14A **P2408** Dd 1 AACU Farnborough 19.4.39. To 9 MU Cosford 23.6.41. To ATA 9.8.41; op by ATA Training Unit (2.42); coded '40'. Cat.B (MR); to Atlantic Coast, Barnstaple 30.7.43. To 51 MU Lichfield 18.10.43. To 39 MU Colerne 4.8.45. To 5 MU Kemble 3.4.46. Sold 28.11.46 to Southern Aircraft (Gatwick) Ltd. Regd **G-AIZJ** (CofR 11103) 30.12.46 to Southern Aircraft (Gatwick) Ltd, Gatwick. CofA No.8872 prepared but not issued. Cld 2.11.48 & regd 18.1.49 to Robert Alan Short, Croydon. Scrapped Croydon in 1949. Regn cld 18.5.59 as pwfu.

1646 M.14A **P2409** Dd 1 AACU Farnborough 19.4.39. To 48 MU Hawarden 6.6.41. To 15 EFTS Carlisle 26.6.41. To 21 EFTS Booker 10.6.42. To 439 RCAF Sqdn, Wellingore 29.3.43. Flying accident 28.3.43; to Herts & Essex, Broxbourne for repairs. To 27 MU Shawbury 28.6.43. To ATA 13.8.43. To 51 MU Lichfield 1.3.45. Sold 20.5.46 to Miles Aircraft Ltd. Test flown 28.6.46 by Hugh Kendall as P2409. To H Hennequin & Cia, Argentina with CofA No.8115 issued 16.7.46; regd **LV-XQI** 30.6.47 to Secretaria Estado Aeronautica, Buenos Aires. To Aero Club Alvear. Accident 21.2.47; repaired. Regn cld 8.52 as wfu.

1647 M.14A **P2410** Dd 1 AACU Farnborough 24.4.39. DBR 20.11.41. SOC 27.11.41.

1648 M.14A **P2426** Dd 8 MU Little Rissington 24.4.39. To 30 EFTS Burnaston 16.1.40. To 16 EFTS Burnaston 10.4.40. Reported at ATA Training Unit (6.41); coded '27'. Damaged 2.6.41; to P&P (SAS) 28.8.41. To 37 MU Burtonwood 24.10.41. To Bottesford 21.12.41. To 46 MU Lossiemouth 13.7.42. To ATA 15.12.43. To 51 MU Lichfield 7.3.45. Sold 18.3.46 to Miles Aircraft Ltd. Test flown 17.4.46 by Hugh Kendall as P2426. To H Hennequin & Cia, Argentina with CofA No.7735 issued 9.5.46; regd **LV-XNZ** 30.6.47 to Secretaria Estado Aeronautica, Buenos Aires. To Association Aerea Azul. Accident 20.6.47; repaired. Accident 19.3.48; repaired. Regn cld 4.52 as wfu.

1649 M.14A **P2427** Dd 8 MU Little Rissington 24.4.39. To 8 EFTS Woodley 15.10.39. To 5 EFTS Meir 13.8.40. Damaged u/c in forced landing at Leek while lost 26.11.40; to P&P (SAS) 13.12.40. To 5 MU Kemble 30.3.41. To HQ SFP Kemble 16.4.41; op by ATA Training Unit (6.41); coded '23'. To 1 CRU Cowley (SAS) 14.6.41. To 24 EFTS Luton 22.8.41. To 37 MU Burtonwood 16.2.42. To ATA 21.6.42. To Atlantic Coast, Barnstaple (Cat.B MR) 24.6.43. To 27 MU Shawbury 31.8.43. To ATA Flying School, Barton-in-the-Clay & Thame (8.43-12.43). To 51 MU Lichfield 16.2.44. Sold 3.5.49 to WA Rollason Ltd. Regd **G-ALUX** (CofR 14855) 10.8.49 to WA Rollason Ltd, Croydon (regd as ex T2427). CofA No.10666 issued 5.10.49. Cld 25.1.50 & regd 13.2.50 to William Arthur Rollason, Croydon. CofA lapsed 4.10.50; renewed 29.6.51. Cld 29.6.51 & regd 30.6.51 to Universal Flying Services Ltd, Fairoaks. Cld 20.10.53 & regd 22.10.53 to William Edwin Thomas Way, Wimbledon. Cld 13.4.59 & regd 15.4.59 to Henry Justin Pelham, London WC2. CofA lapsed 7.4.63 and cld 14.8.63 as sold (but not regd). Wfu Fairoaks. Regn cld 6.10.64 as pwfu. Burnt at Fairoaks as film prop in late 1967.

1650 M.14A **P2428** Dd 8 MU Little Rissington 24.4.39. To 24 EFTS Sydenham 15.10.39; to Luton 22.7.40. To 8 EFTS Woodley 9.10.41. To 10 FIS Woodley 21.7.42. To 5 GTS Shobdon 5.9.43. To 51 MU Lichfield 10.4.44. Sold 26.2.48 to Gloster Flying Club. Regd **G-AKMS** (CofR 12055) 27.11.47 to Laurence Dudley Trappitt, Whyteleafe, Surrey. Cld 22.7.48 & regd 14.10.48 to Robert Alan Short, Derek John Hayles & Martha Marodeen, Croydon/Portsmouth. CofA No.10388 issued 21.1.49. Regn cld 2.9.49 as sold - to **Egypt AF**.

1651 M.14A **P2429** Dd 8 MU Little Rissington 24.4.39. To 30 EFTS Burnaston 15.10.39. To 16 EFTS Burnaston 10.4.40. To 27 MU Shawbury 2.7.42. To 51 MU Lichfield 24.6.44. Sold 13.5.46 to Miles Aircraft Ltd. To H Hennequin & Cia, Argentina with CofA No.8120 issued 16.7.46; regd **LV-XRZ** 30.6.47 to Secretaria Estado Aeronautica, Buenos Aires. Accident 4.5.47; repaired. Regn cld 1.53 as wfu.

1652 M.14A **P2430** Dd 8 MU Little Rissington 24.4.39. To 30 EFTS Burnaston 15.10.39. To 16 EFTS Burnaston 10.4.40. Hit tree in circuit Ashbourne 15.10.40. SOC 25.10.40.

1653 M.14A **P2431** Dd 8 MU Little Rissington 24.4.39. To Stn Flt Hucknall 2.7.40. To 1 CRU Cowley (SAS) 19.1.41. To 2 SofAC Andover 5.4.41; reformed as 42 OTU 18.7.41. To RAF Llanbedr 29.12.43. Ditched after engine failure on take-off Llanbedr 22.5.44.

1654 M.14A **P2432** Dd 8 MU Little Rissington 24.4.39. To 8 EFTS Woodley 27.3.40. To 15 EFTS Carlisle 17.8.40. Crashed in forced landing 2 ml NNW of Annan, Dumfries 15.3.41 (524:50 F/H). *Accident possibly 28.2.41 and SOC 15.3.41?*

1655 M.14A **P2433** Dd 8 MU Little Rissington 25.4.39. To 6 EFTS Sywell 18.4.40. To 16 EFTS Burnaston. Crashed in forced landing Galley Common, Nuneaton, Warwicks 5.3.41. To P&P (SAS) 7.3.41 but SOC Cat.E 17.3.41.

1656 M.14A **P2434** Dd 8 MU Little Rissington 25.4.39. To 1 Sqdn, Vassincourt, France 26.3.40. Presumed lost in France in 5.40.

1657 M.14A **P2435** Dd 8 MU Little Rissington 25.4.39. To AASF Blois, France 12.6.40; moved to Chaeau de la Cressonniere Muides 13.6.40. Lost in France in 6.40. SOC 26.3.42.

1658 M.14A **P2436** Dd 8 MU Little Rissington 2.5.39. To 43 Sqdn, Northolt 28.8.40. To 1 CRU Cowley (SAS) 9.10.40; to P&P 12.10.40. To 5 MU Kemble 30.3.41. To HQ SFP Kemble 16.4.41; op by ATA Training Unit (2.42); coded '27'. To 46 MU Lossiemouth 14.7.42. To ATA (31.12.42 census). To Herts & Essex, Broxbourne (Cat.B MR) 4.6.43. To 27 MU Shawbury 19.9.43. To 155 (GR) Wing, Manston 10.6.44. To 51 MU Lichfield 5.9.45. Sold 21.11.46 to The Wiltshire School of Flying Ltd. Regd **G-AITT** (CofR 10962) 11.11.46 to Anna Valley Motors (Salisbury) Ltd, Thruxton. CofA No.10359 issued 22.12.48. Cld 25.2.49 as sold - to Robert Alan Short, Croydon (but not regd). To **Egypt AF** 2.9.49.

1659 M.14A **P2437** Dd 8 MU Little Rissington 2.5.39. To Stoke Orchard 13.10.40. To 8 MU Little Rissington 19.10.40. To 268 Sqdn, Bury St Edmunds 1.11.40. To 1 CRU Cowley (SAS) 17.5.41. To

24 EFTS Luton 1.7.41. To 16 EFTS Burnaston 12.2.42. Spun and dived into ground Dolbury Lees, near Derby 25.3.42; Cpl E Chadaway killed.

1660 M.14A **P2438** Dd 8 MU Little Rissington 2.5.39. To AASF Blois, France 12.6.40; moved to Chaeau de la Cressonniere Muides 13.6.40. Lost in France in 6.40. SOC 26.3.42.

1661 M.14A **P2439** Dd 8 MU Little Rissington 2.5.39. To 247 Sqdn, Roborough 1.2.41. Crashed in forced landing near Roborough 21.5.41. To P&P (SAS) 28.5.41 but SOC 31.5.41.

1662 M.14A **P2440** Dd 8 MU Little Rissington 2.5.39. To Stoke Orchard 13.10.40. To 8 MU Little Rissington 22.10.40. To 268 Sqdn, Bury St Edmunds 1.11.40. To 5 MU Kemble 20.4.41. To 24 EFTS Luton 27.4.41. To Atlantic Coast, Barnstaple (MI) 20.12.41. To 27 MU Shawbury 10.4.42. To 111 Sqdn, Debden 20.8.42. To 51 MU Lichfield 9.12.42. SOC Cat.E (MR) 25.2.44.

1663 M.14A **P2441** Dd 10 MU Hullavington 27.7.39. To 242 Sqdn, Church Fenton 15.11.39. To 219 Sqdn, Catterick 27.3.40; moved to Redhill 12.10.40 and to Tangmere 10.12.40. Crashed in forced landing Barnham Court, Sussex 10.10.41. SOC 22.10.41.

1664 M.14A **P2442** Dd 1 AOS North Coates 8.5.39. To 5 AOS Jurby 10.10.39. To 5 EFTS Meir 20.8.41. To CFS Upavon 30.1.42. To 16 EFTS Burnaston 19.4.42. To 5 GTS Shobdon 1.5.43. SOC Cat.E (MR) 27.5.44.

1665 M.14A **P2443** Dd 1 AACU Farnborough 8.5.39. To 1 CRU Cowley (SAS) 31.5.41. To 132 Sqdn, Peterhead 7.8.41. To P&P (Cat.B) 2.3.42. To 20 MU Aston Down 5.5.42. To 27 MU Shawbury 31.8.42. To 47 MU Sealand 22.2.43. Despatched to Middle East 26.4.43; arr 26.5.43. SOC 29.6.44.

1666 M.14A **P2444** Dd 1 AACU Farnborough 9.5.39. To 9 MU Cosford 30.6.41. To 32 MU St Athan 26.11.41. To 11 EFTS Perth 5.1.42. To 157 Sqdn, Castle Camps 18.2.42. SOC Cat.E (MR) 13.6.44.

1667 M.14A **P2445** Dd 36 MU Sealand 8.5.39. Allotted to Middle East 7.7.39 for Stn Flt Aboukir. To 253 Wing ME. To 202 Grp Comm Flt, Cairo. To 3 Sqdn, RAAF Middle East. To 267 Sqdn. Presumed SOC 1.1.47.

1668 M.14A **P2446** Dd Wattisham 9.5.39 for 6 Grp Comm Flt, Mousehold. To 107 Sqdn, Wattisham (30.11.39 census). To 110 Sqdn, Horsham St Faith 14.3.41. To Herts & Essex, Broxbourne (SAS) 17.8.41. To 16 EFTS Burnaston 9.9.41. To ECFS Hullavington 20.5.43. SOC Cat.E1 3.8.44.

1669 M.14A **P2447** Dd Hucknall 9.5.39. To 1 CRU Cowley 15.11.39. To 8 MU Little Rissington 25.5.40. To Stoke Orchard 13.10.40. To 8 MU Little Rissington 19.10.40. Spun into ground near Bury St Edmunds 2.11.40.

1670 M.14A **P2448** Dd 2 FPP Filton 8.5.39. Stalled on take-off Kemble 7.2.41. SOC 13.2.41.

1671 M.14A **P2449** Dd 36 MU Sealand 8.5.39. Allotted to Middle East for Stn Flt Aboukir 7.7.39. To 253 Wing. To 202 Grp Comm Flt, Cairo. To Ismailia (.41). To 450 RAAF Sqdn. To 263 Wing, Beirut. Crashed into aircraft pen on landing Dekheila 14.2.42. Salvaged 1.7.42.

1672 M.14A **P2450** Dd 36 MU Sealand 11.5.39. Allotted to Middle East for Stn Flt Aboukir 7.7.39. To Comm Flt Heliopolis. To Stn Flt Aboukir. To Comm Flt Heliopolis. To 267 Sqdn. Flying accident and SOC 15.6.41.

1673 M.14A **P2451** Dd 36 MU Sealand 11.5.39. Allotted to Middle East for Stn Flt Aboukir 7.7.39. To Comm Flt Heliopolis. Crashed 1.12.39.

1674 M.14A **P2452** Dd 36 MU Sealand 13.5.39. Allotted to Middle East for Stn Flt Aboukir 7.7.39. To Comm Flt Heliopolis. To 71 OTU Ismailia (7.41-8.41). DBR by enemy action 5.7.41. SOC 11.9.41 (281:30 F/H).

1675 M.14A **P2453** Dd 36 MU Sealand 13.5.39. Allotted to Middle East for Aboukir Depot 7.7.39. To Comm Flt Heliopolis (9.39-2.40). To 267 Sqdn. To 127 Sqdn. To 103 MU Aboukir (41-42). SOC 31.5.45.

1676 M.14A **P2454** Dd Stn Flt Eastchurch 19.5.39. To Brize Norton 24.5.39. To 24 MU Ternhill 25.1.40. To 10 FTS Ternhill 13.2.40. To 5 FTS Ternhill 10.12.40. To 8 EFTS Woodley 16.5.41. Hit tree on approach Waltham St Lawrence 1.10.41; LAC F Birchwood killed. SOC 8.10.41.

1677 M.14A **P2455** Dd Stn Flt Eastchurch 19.5.39. To 10 FTS Ternhill 1.3.40. To 5 FTS Ternhill 10.12.40. To 8 EFTS Woodley 16.8.41. To 10 FIS Woodley 21.7.42. To 51 Grp Comm Flt Yeadon 7.7.44. To 51 MU Lichfield 3.8.45. Sold as scrap 29.4.49 to Air Kruise Ltd, Lympne; not regd.

1678 M.14A **P2456** Dd Comm Flt Andover 15.5.39. Flew into ground 4 ml S of Andover 29.11.39 (but also recorded as damaged 15.11.39). SOC 21.12.39.

1679 M.14A **P2457** Dd Andover 15.5.39 for SofAC Old Sarum. Forced landed Weyhill 15.7.40; to 1 CRU Cowley (Cat.B) 22.7.40. To 9 MU Cosford 22.8.40. To 24 EFTS Luton 26.9.40. Damaged 5.4.41; to P&P (SAS) 13.5.41. To 9 MU Cosford 11.7.41. To ATA 9.8.41. To Atlantic Coast, Barnstaple (Cat.B MR) 11.9.43. To 51 MU Lichfield 19.11.43. Sold 27.5.48 to Short Bros & Harland Ltd, Rochester but not regd. Noted dismantled at Rochester 7.51.

1680 M.14A **P2458** Dd HAD Henlow 18.5.39. To 13 MU Henlow 1.3.40. To 1 CRU Cowley (SAS) 12.1.41. To 2 SofAC Andover 4.4.41; reformed as 42 OTU 5.41. To 1526 (BAT) Flt, Thruxton 11.11.41. To 20 MU Aston Down 18.8.42. To 7 FIS Upavon 24.10.42. SOC 25.3.44.

1681 M.14A **P2459** Dd West Raynham 24.5.39. To Wyton 4.12.39. To 27 MU Shawbury 17.7.41. To TFPP ATA, White Waltham 9.8.41. Damaged when struck hedge on take-off after forced landing Alford 20.9.41; ATA pilot S/O DG Bach unhurt. To Cat.B 23.9.41 and repaired. To 5 MU Kemble 25.11.41. To 15 MU Wroughton 5.1.42. To 32 MU St Athan 7.1.42. To 71 Sqdn, Martlesham Heath 19.2.42. To 334 Sqdn, USAAF 1.10.42. To 51 MU Lichfield 15.10.42. To ATA 15.12.43. To 51 MU Lichfield 5.1.45. Sold 6.5.46 to Miles Aircraft Ltd. To H Hennequin & Cia, Argentina with CofA No.8053 issued 8.7.46; regd **LV-XRE** 30.6.47 to Secretaria Estado Aeronautica, Buenos Aires. To Aero Club S Del Estero. Accident 25.3.47; repaired. Accident 28.5.47; repaired. Collided with LV-XMW 1.6.47 (15% damaged); repaired. Regn cld 1.54 as wfu.

1682 M.14A **P2460** Dd 2 SofTT Cosford 22.5.39. To 5 FTS Sealand 15.6.40. DBR in gale Sealand 5.12.40. To P&P (SAS) 9.2.41. SOC 8.10.41 by 5 FTS.

1683 M.14A **P2461** Dd 2 SofTT Cosford 23.5.39. To 9 B&GS Penrhos 6.12.39. To 24 EFTS Luton 11.7.41. To Atlantic Coast, Barnstaple (MI - Cat.B) 20.12.41. To 37 MU Burtonwood 23.3.42. To CRD at Heston Aircraft Ltd 23.5.42. To 27 MU Shawbury 13.6.42. To 453 RAAF Sqdn, Drem 25.7.42. To 488 RNZAF Sqdn, Ayr 18.2.43. SOC Cat.E (MR) 22.7.44.

1684 M.14A **P2462** Dd 2 SofTT Cosford 24.5.39. To 9 MU Cosford 17.9.40. To 8 EFTS Woodley 22.2.41. To 10 FIS Woodley 21.7.42. To HQ FTC Woodley 18.8.42. Cat.B; to Atlantic Coast, Barnstaple for repairs 13.4.43. To 27 MU Shawbury 14.6.43. To ATA 24.11.43. To 51 MU Lichfield 9.2.45. Sold 22.9.47 to Miles Aircraft Ltd. CofA No.9880 issued 22.3.48 to **Thai AF**.

1685 M.14A **P2463** Dd 2 SofTT Cosford 24.5.39. To 12 FTS Spittlegate 7.12.39. To 11 EFTS Perth 19.7.41. To 12 MU Kirkbride 13.8.42. SOC Cat.E (MR) 7.2.44.

1686 M.14A **P2464** Dd 2 SofTT Cosford 25.5.39. To 5 FTS Sealand 29.6.40. Flying accident (Cat.B) 21.7.41; to P&P (SAS) 14.8.41. To 456 RAAF Sqdn, Valley 15.9.41. Spun into ground Winterslow, Wilts 23.5.43. SOC 31.5.43.

1687 M.14A **P2465** Dd 32 Sqdn, Biggin Hill 22.5.39. To 15 EFTS Redhill (undated). To Middle Wallop 29.6.40. To 51 OTU Cranfield 16.6.42. To 72 Sqdn, Biggin Hill 7.7.42. To 51 MU Lichfield 13.11.42. Sold 18.3.46 to Miles Aircraft Ltd. To H Hennequin & Cia, Argentina with CofA No.7890 issued 14.6.46; regd **LV-XPO** 30.6.47 to Secretaria Estado Aeronautica, Buenos Aires. To Aero Club Salta. Accident 7.12.47; repaired. Regn cld 12.51 as wfu.

1688 M.14A **P2466** Dd 32 Sqdn, Biggin Hill 22.5.39. To 1 CRU Cowley (Cat.B) 10.9.40. To 8 MU Little Rissington 29.10.40. To

141 Sqdn, Gravesend 15.11.40. Flying cccident (Cat.B) 28.7.41; to P&P (SAS) 1.8.41. To 27 MU Shawbury 12.10.41. To 2 EFTS Staverton (flying from Worcester) 4.11.41; became 6 FIS 1.11.41. Flying accident (Cat.E) 28.1.42. SOC 5.2.42.

1689 M.14A **P2467** Dd 29 Sqdn, Debden 23.5.39. To 85 Sqdn, Debden 26.6.40; unit to Croydon 19.8.40. Destroyed in air raid Croydon 31.8.40. SOC 17.9.40. To 54 MU Cambridge for scrap 18.9.40.

1690 M.14A **P2468** Dd 29 Sqdn, Debden 25.5.39. To P&P 19.4.40. To 37 MU Burtonwood 25.8.40. To 15 EFTS Carlisle 22.6.41. Flew into ground in bad weather Raffles near Carlisle 13.1.42 on ferry flight from Sealand to Carlisle; ATA pilot S/O BL Pelham badly injured. SOC (Cat.E) 17.1.42.

1691 M.14A **P2469** Dd 25 Sqdn, Hawkinge 25.5.39. To 602 Sqdn, Drem 29.11.39. To 406 RCAF Sqdn, Acklington 1.6.41. Flying accident (Cat.B) 7.9.41; to P&P (SAS) 10.9.41 but SOC 22.9.41.

1692 M.14A **P2470** Dd 25 Sqdn, Hawkinge 26.5.39. To P&P 12.1.40. To 48 MU Hawarden 12.6.40. To 15 EFTS Carlisle 16.11.40. Collided with L6918 and crashed Newbie, Dumfries 5.5.41. To P&P (SAS) 8.5.41 but SOC 15.5.41.

1693 M.14A **P2493** Dd 65 Sqdn, Hornchurch 26.5.39. To 46 Sqdn, Digby 28.7.40. To 1 CRU Cowley 13.8.40. To 37 MU Burtonwood 26.9.40. To 15 EFTS Carlisle 5.6.41. To 21 EFTS Booker 17.6.42. To 24 EFTS Sealand 14.8.43; to Rochester 1.3.46; coded 'FJG-W'. To 11 EFTS Perth 17.3.47; redesignated 11 RFS 18.3.47. To 51 MU Lichfield 6.11.47. Sold 3.5.49 to WA Rollason Ltd. Regd **G-ALUW** (CofR 14854) 10.8.49 to WA Rollason Ltd, Croydon (regd as T2493). CofA No.10671 issued unknown date but mid.51. Regn cld 13.3.52 as sold in Belgium (although at Croydon as OO-AJT 3.6.51). Regd **OO-AJT** (CofR 889) 17.4.52 to R. Heuvelmans, St Lambrechts-Woluwe (based Grimbergen). Regd 5.5.53 to C de Paepe, Deurne-Antwerp. Accident in 1954. Regn cld 17.10.56.

1694 M.14A **P2494** Dd 65 Sqdn, Hornchurch. Flying accident 31.8.40; to P&P 14.9.40. To 5 MU Kemble 7.2.41. To 91 Sqdn, Hawkinge 19.3.41. To 253 Sqdn, Skeabrae 26.8.41. Flying accident (Cat.B) 7.11.41; to P&P 13.11.41. To 48 MU Hawarden 25.1.42. To 32 MU St. Athan 19.5.42. To 257 Sqdn, Honiley 30.5.42. To FIU Ford 7.10.43. SOC 31.1.45.

1695 M.14A **P2495** Dd 151 Sqdn, North Weald 26.5.39. Hit house while low flying Goring-on-Thames, Oxon 4.8.39; P/O Peter Phillips killed & P/O John F Pettigrew badly injured.

1696 M.14A **P2496** Dd 151 Sqdn, North Weald 31.5.39. To 1 CRU Cowley 23.2.40. To 48 MU Hawarden 16.6.40. To 18 OTU Bramcote 30.3.41. To P&P (SAS) 10.6.41. To 92 Sqdn, Biggin Hill 20.8.41. To 615 Sqdn, Manston 30.10.41. To 32 Sqdn, Manston 26.11.41. Blown over on take-off Manston 28.4.42. SOC 30.4.42.

1697 M.14A **P2497** Dd 1 Sqdn, Tangmere 31.5.39. To 87 Sqdn, Lille/Seclin, France 17.11.39. Crashed during aerobatics 15.3.40. SOC 30.4.43.

1698 M.14A **P2498** Dd 1 Sqdn, Tangmere 31.5.39. To Vassincourt, France 9.10.39. Damaged in flying accident 3.40; to 1 Salvage Unit Reims-Champagne 19.3.40. To 21 AD Nantes/ Bouguenais 29.3.40. Abandoned at Chateau Bougon 6.40. SOC 26.3.42.

1699 M.14A **P2499** Dd 64 Sqdn, Church Fenton 5.6.39. To 72 Sqdn, Church Fenton 10.6.39. To 74 Sqdn, Acklington 9.7.41. Flying accident (Cat.B) 8.9.41; to P&P (SAS) 14.9.41. To 46 MU Lossiemouth 2.11.41. To 2 FIS Montrose 10.4.42. To 12 MU Kirkbride 28.8.42. To 105 (T)OTU Bramcote 31.1.45; unit became 1381 (T)CU 10.8.45; to Desborough 19.11.45. To 51 MU Lichfield 20.4.46. To RAE Farnborough for training duties (and to replace R1818) 13.11.46; flown to Farnborough 21.11.46. SOC as reduced to spares 5.3.49. Sold 28.6.50 and collected by 49 MU Colerne 30.6.50.

1700 M.14A **P2500** Dd 64 Sqdn, Church Fenton 5.6.39. Flying accident 2.7.40; collected by 49 MU Faygate 29.8.40 and to 1 CRU Cowley 4.9.40. To 48 MU Hawarden 13.10.40. To 68 Sqdn, Catterick 14.3.41; to High Ercall 17.4.41. Damaged 4.9.41; to P&P (SAS) 9.9.41. To 46 MU Lossiemouth 2.11.41. To 2 FIS Montrose 10.1.42. To 14 (P)AFU Banff 1.10.43. To 2 FIS Montrose (31.12.43 census). To 17 SFTS Spitalgate 15.6.45. To 51 MU Lichfield 10.1.46. Sold 16.2.49 to RA Short, Croydon. Regd **G-ALIO** (CofR 12574) 21.2.49 to Short Bros & Harland Ltd, Rochester. CofA No.10488 issued 1.6.49. Cld 3.6.49 & regd 27.6.49 to Major-Genl JEC McCandlish & Major Ian Hallam Lyall-Grant, t/a The Royal Engineers Flying Club, Rochester. Cld 3.7.51 & regd 26.7.51 to Major Lionel Gordon Sherriff Thomas & Major Ian Hallam Lyall-Grant, t/a The Royal Engineers Flying Club, Rochester. Cld 22.2.53 & regd 25.2.53 to Charles William Francis Buck, Salfords; dd to Redhill 22.2.53 for use by the Experimental Flying Group. Named 'Pongo' by Rex Nicholls 28.3.53 in memory of his time as a signalman in the army! Regd 7.8.53 to Experimental Flying Group Ltd, Redhill, later Croydon, then Biggin Hill. Crashed in poor visibility Brook Place Farm, Ide Hill, near Sevenoaks, Kent 6.7.58; pilot James Gainsford & passenger unhurt. Regn cld 30.7.61 as reduced to spares.

1701 M.14A **P2501** Dd 73 Sqdn, Digby 5.6.39. To 611 Sqdn, Digby 27.10.39. Hit tree on take-off from field Woodend Farm, Fife 18.4.42. SOC 23.4.42.

1702 M.14A **P2502** Dd 73 Sqdn, Digby 5.6.39. To 24 Sqdn, Hendon 17.9.39. To Stn Flt Yeadon 18.10.39. To 1 CRU Cowley 12.5.40. To 48 MU Hawarden 16.7.40. To 15 EFTS Carlisle 16.11.40. DBR in forced landing on navex 5.6.41; pilot LAC M Lopato. SOC 12.6.41.

1703 M.14A **P2503** Dd 66 Sqdn, Duxford 3.6.39. To 242 Sqdn, Martlesham Heath 27.3.41. To P&P 16.10.41. To 27 MU Shawbury 16.12.41. To 6 FIS Staverton (31.12.41 census). Flown from 22 MU Silloth to 4 FIS Cambridge 25.1.42. To 6 FIS Worcester 9.5.42; unit became 2 EFTS 22.7.42. To 46 MU Lossiemouth 11.8.42. To 51 MU Lichfield 14.12.43. Sold 18.3.46 to Miles Aircraft Ltd. To H Hennequin & Cia, Argentina with CofA No.7833 issued 30.5.46; regd **LV-XNL** 30.6.47 to Secretaria Estado Aeronautica, Buenos Aires. To Aero Club Rio Cuarto. Regn cld 5.52 as wfu.

1704 M.14A **P2504** Dd 66.Sqdn, Duxford 5.6.39. Damaged 16.6.41; to P&P (SAS) 26.6.41. To 20 MU Aston Down 29.8.41. To 27 MU Shawbury 4.9.41. To 607 Sqdn, Martlesham Heath 12.9.41. To P&P 13.11.41. To 46 MU Lossiemouth 25.1.42. Flying accident 9.2.42. To CRD at Scottish Aviation Ltd 13.10.42 (allotted 20.5.42). Flying accident (Cat.B) 29.11.43; to unidentified CRO and repairs completed 15.1.44. To 51 MU Lichfield 31.1.44. Sold 27.5.46 to Miles Aircraft Ltd. To H Hennequin & Cia, Argentina with CofA No.8185 issued 30.7.46; regd **LV-XQV** 30.6.47 to Secretaria Estado Aeronautica, Buenos Aires. To Aero Club Deportiva. Accident 31.7.48.

1705 M.14A **P2505** Dd 23 Sqdn, Wittering 5.6.39. Crashed in forced landing Hill Farm, Thorney, Hunts 8.5.40. SOC 11.5.40.

1706 M.14A **P2506** Dd 23 Sqdn, Wittering 5.6.39. Damaged taxying Toussus-le-Noble, France 11.6.40; to 1 CRU Cowley for repairs 8.7.40. To 27 MU Shawbury 13.8.40. To 16 EFTS Burnaston 25.3.41. To 3 EFTS Shellingford 14.6.42. To 45 MU Kinloss 14.3.43. To 51 MU Lichfield 11.2.44. Sold 6.12.46 to Southern Aircraft (Gatwick) Ltd. Regd **G-AIZK** (CofR 11104) 30.12.46 to Southern Aircraft (Gatwick) Ltd, Gatwick. CofA No.8874 issued 14.5.48. Cld 14.5.48 & regd 18.5.48 to Aero Publicity Ltd, London EC2. Used by ARB at Gatwick 20.8.48 to test smoke equipment fittings. Cld 18.12.48 & regd 20.12.48 to Alistair Pringle Frazer, Southall, Middlesex. Cld 14.2.49 & regd 28.2.49 to Francis Alexander Frazer, Kensington. CofA lapsed 13.5.49. Cld 19.6.49 & regd 17.10.49 to John Paul Gunner, London W1. CofA renewed 20.3.50. Cld 7.2.51 & regd 3.3.51 to FG Miles Ltd, Redhill, first noted at Shoreham 29.6.52. CofA lapsed 14.11.56. Cld 8.58 as sold to Mercer (but not regd). Reportedly to Foulsham for spares; stored Little Snoring (1.60). Regn cld 12.10.61 as reduced to spares. Donated c.60 to a primary school in Little Snoring *on which the children can play and learn the use of some of the instruments from their headmaster, a member of the local aero-club.* Still in use 2.64.

1707 M.14A **P2507** Dd 79 Sqdn, Biggin Hill 9.6.39. To 10 MU Hullavington 28.6.40. To Warmwell 8.1.41. To 5 MU Kemble 4.6.41. To Herts & Essex, Broxbourne (SAS) 8.7.41. To 27 MU Shawbury 31.7.41. To ATA 10.8.41; op ATA Training Unit, Barton-in-the-Clay (2.42); coded '39'. Damaged on overshoot in forced landing Shrob Lodge Farm, Stoney Stratford 21.5.42; Cadet SW Stroud unhurt; to P&P (Cat.B) 29.5.42. To 46 MU Lossiemouth 9.7.42. To 51 MU

Lichfield 14.12.43. Sold by MAP to Miles Aircraft Ltd for **Chile** in early 1945, test flown as **U-0272**. Regd in Argentina as **LV-RYE** 7.12.48. Regn cld 10.56 as broken up.

1708   M.14A **P2508** Dd 79 Sqdn, Biggin Hill 9.6.39. Dived into ground during aerobatics near Margate, Kent 2.1.40, 2 killed. SOC 17.1.40.

1709   M.14A **P2509**   Dd 111 Sqdn, Hornchurch 10.6.39.   To 141 Sqdn, Gatwick 26.9.40. To Ayr 2.5.41. Hit wires and crashed near Ayr 11.12.41. SOC 19.12.41.

1710   M.14A **P2510** Dd 85 Sqdn, Debden 10.6.39; to Rouen/Boos, France 9.39. Lost in France in 5.40. SOC as 'No trace'.

1711   M.14A **P6343** Dd 85 Sqdn, Debden 10.6.39; to Rouen/Boos, France 9.39. To 607 Sqdn, Vitry-en-Artois 22.12.39. Lost in France (probably at Norrent Fontes) in 5.40. SOC 'No trace'.

1712   M.14A **P6344**   Dd 74 Sqdn, Hornchurch 13.6.39.   Flying accident 5.8.40; to 1 CRU Cowley 16.8.40. To 48 MU Hawarden 21.9.40. To 54 Sqdn, Catterick 22.1.41. To 124 Sqdn, Biggin Hill 18.11.41. To 51 MU Lichfield 12.10.42. Sold 26.2.48 to Gloster Flying Club, ferried from 51 MU Lichfield to Staverton by Jack Meaden.   Regd **G-AKMM** (CofR 12050) 27.11.47 to Laurence Dudley Trappitt, Whyteleafe, Surrey. Cld 22.7.48 & regd 27.10.48 to Robert Alan Short, Derek John Hayles & Martha Marodeen, Croydon/Portsmouth. Cld 5.11.48 & regd 31.1.49 to Robert Alan Short, Croydon. CofA No.10389 issued 24.6.49. Regn cld 2.9.49 as sold abroad - to **Egypt AF**.

1713   M.14A **P6345** Dd 74 Sqdn, Hornchurch 13.6.39. Hit HT wires and crashed Rayleigh, Essex 7.11.39.   To Marshalls Cambridge 21.11.39 for repairs? To 51 MU Lichfield. To 4 SofTT St Athan for GI as **5361M** 11.7.45.

1714   M.14A **P6346** Dd 46 Sqdn, Digby 12.6.39. To Hucknall 19.6.39. To P&P 4.6.40. To 5 MU Kemble 30.10.40. To 24 EFTS Luton 22.11.40. Flying accident (Cat.B) 8.6.41; to P&P (SAS) 11.6.41. To 24 EFTS Luton 19.7.41. Damaged u/c and starboard wing in forced landing nr Halton 22.9.41; LAC BC Gray unhurt; to P&P (Cat.B) 25.10.41. To 2 FIS Montrose 21.3.42. MI 3.8.42; to P&P (possibly at their Montrose repair facility) 16.8.42. To 27 MU Shawbury 2.10.42. To 76 MU Wroughton undated. Despatched to Middle East on SS Katuna, for **Turkey** 12.5.43; arr 13.6.43.

1715   M.14A **P6347** Dd 46 Sqdn, Digby 12.6.39. To 229 Sqdn, Digby 4.5.40. To 46 Sqdn, Digby 25.6.40. To 1 CRU Cowley (Cat.B) 29.7.40. To 48 MU Hawarden 26.9.40. To Pembrey 29.9.40. Hit balloon cable and crashed Langley, Bucks 8.10.40. SOC 17.10.40.

1716   M.14A **P6348** Dd 213 Sqdn, Wittering 15.6.39. Hit trees in forced landing Stoneyburn, West Lothian 6.5.41. To P&P (SAS) 9.5.41 but SOC 15.5.41 (by 218 Sqdn?).

1717   M.14A **P6349** Dd 213 Sqdn, Wittering 15.6.39. To 23 Sqdn, Collyweston 12.7.40. To 605 Sqdn, Ford 5.8.42. Crashed 19.9.44 and SOC Cat.E 21.9.44.

1718   M.14A **P6350** Dd 41 Sqdn, Catterick 25.6.39.   Flew into ground Tunstall, Catterick 12.11.39. SOC by 60 MU York 15.12.39.

1719   M.14A **P6351** Dd 87 Sqdn, Debden 16.6.39. To 73 Sqdn, Rouvres 10.11.39. Probably lost in France (although record card shows to Middle East). PSOC 1.1.47.

1720   M.14A **P6352**   Dd 17 Sqdn, North Weald 19.6.39; to Martlesham Heath 26.2.41; to Croydon 28.2.41. Flying accident 17.7.41; repaired. To 20 OTU Lossiemouth (31.12.41 census). MR 2.9.42; to P&P but to Cat.E 26.9.42. SOC 13.10.42.

1721   M.14A **P6353** Dd 17 Sqdn, North Weald 19.6.39. To 5 OTU Aston Down 17.4.40. To 118 Sqdn, Filton 2.3.41. To 1 CRU Cowley (SAS) 9.5.41. To 51 MU Lichfield 8.7.41. To 5 EFTS Meir 16.8.41. To CFS Upavon 31.1.42. To 2 FIS Montrose 9.5.42.   Stalled on approach near Laurencekirk, Kincardine 8.12.42.

1722   M.14A **P6354** Dd 56 Sqdn, North Weald 19.6.39. To RAF North Weald 21.6.40. Damaged in air raid and SOC 10.10.40.

1723   M.14A **P6355**   Dd 72 Sqdn, Church Fenton 19.6.39. To 406 RCAF Sqdn, Acklington 31.5.41. To 32 MU St Athan 21.9.41. To Abingdon 2.10.41. To 406 RCAF Sqdn, Acklington 3.11.41. To Atlantic Coast, Barnstaple for repairs 15.12.42. To 51 MU Lichfield 6.2.43. Sold by MAP in early 1945 to Miles Aircraft Ltd for Chile. To Marshalls Flying School, Cambridge [4.45] and test flown as **U-0262**. Diverted from Chilean contract and dd 1.46 to Fuerza Aerea Argentina, El Palomar Air Base.   Regd in Argentina as **LV-RUX** 16.6.48 to H Hennequin & Cia. Crashed Villa Madero 12.12.48; pilot Cayetano & passenger injured.

1724   M.14A **P6356** Dd 19 Sqdn, Duxford 19.6.39. To Shrager Bros, Old Warden (SAS) 13.7.41. To 37 MU Burtonwood 15.2.42. To ATA 24.4.42. MR (Cat.B) at unidentified unit 30.7.43. To 46 MU Lossiemouth 11.10.43.   To ATA 30.11.43.   To 51 MU Lichfield 17.3.45. Sold 25.3.46 to Miles Aircraft Ltd. To H Hennequin & Cia, Argentina with CofA No.7867 issued 11.6.46; regd **LV-XPH** 30.6.47 to Secretaria Estado Aeronautica, Buenos Aires. To Aero Club San Juan. Accident 29.11.46; repaired. Crashed Gral Acha 4.11.51; pilot killed.

1725   M.14A **P6357**   Dd 3 Sqdn, Biggin Hill 24.6.39; to Kenley 29.5.40. Destroyed in air raid 20.8.40.

1726   M.14A **P6358** Dd 54 Sqdn, Hornchurch 26.6.39. Hit tree while low flying Upminster, Essex 15.3.40. SOC 21.3.40.

1727   M.14A **P6359** Dd 43 Sqdn, Tangmere 26.6.39; unit to Acklington 18.11.39; to Wick 26.2.40; returning to Tangmere 31.5.40. Crashed after take-off Inverness 9.7.40. SOC 17.7.40.

1728   M.14A **P6360** Dd 43 Sqdn, Tangmere 26.6.39; to Acklington 18.11.39; to Wick 26.2.40. To 3 Sqdn, Wick 1.6.40; to Castletown 2.9.40; to Martlesham Heath 3.4.41; to Stapleford 23.6.41; to Hunsdon 9.8.41. Cat.B 17.7.42; to P&P for repairs 23.7.42. To 27 MU Shawbury 24.9.42. To 47 MU Sealand 9.3.43. Despatched to Middle East on SS City of Agra 4.5.43; arr 5.6.43. SOC 31.12.46.

1729   M.14A **P6361**   Dd Fairey Aviation Ltd, Hayes 26.6.39; probably based Heston. To 27 MU Shawbury 4.1.43. To 47 MU Sealand 26.2.43. Despatched to Middle East on SS Tactician 3.5.43; arr Port Suez 22.6.43. SOC 31.12.46.

1730   M.14A **P6362** Dd 32 MU St Athan 26.6.39. To HQ Practice Flt, St Athan (30.11.39 census). Hit cliffs north side of Emmetts Hill, Worth Matravers LG, Dorset 14.9.40. SOC 27.9.40.

1731   M.14A **P6363** Dd 32 MU St Athan 26.6.39. To Practice Flt, St Athan (30.11.39 census).   To 38 MU Llandow 1.7.41.   To 130 Sqdn, Portreath 27.7.41. To P&P for repairs 3.2.42. To Bottesford 8.4.42. Flying accident (Cat.B); to Atlantic Coast, Barnstaple for repairs 5.3.43. To 46 MU Lossiemouth 11.10.43. To ATA 20.11.43. DBR 25.6.44 and SOC Cat E 16.7.44.

1732   M.14A **P6364** Dd 32 MU St Athan 26.6.39. To Practice Flt, St Athan (30.11.39 census). Hit HT cable in bad visibility 4 ml E of St Athan 23.4.40. SOC 30.4.40.

1733   M.14A **P6365** Dd HQ Practice Flt, St Athan 26.6.39. To 32 MU St Athan 16.6.41. To 303 Polish Sqdn, Northolt 24.10.41. To Herts & Essex, Broxbourne for repairs 15.9.42. To 51 MU Lichfield 26.11.42. No fate recorded and SOC 21.6.47 at census.

1734   M.14A **P6366**   Dd 20 MU Aston Down 30.6.39. To P&P 4.6.40. To 16 EFTS Burnaston 11.6.40. To 8 EFTS Woodley 11.9.40. Crashed at Theale 10.2.41, causing extensive damage to aircraft but pilot Capt DPD Oldman was uninjured; accident due to shortage of petrol in port tank and pilot failed to switch to full starboard tank. Repaired (by P&P?). To 5 MU Kemble 11.6.41. To 4 FIS Cambridge 25.6.41. To 6 FIS Worcester 10.5.42; unit became 2 EFTS 22.7.42. To 46 MU Lossiemouth 11.8.42. To 3 EFTS Shellingford 20.6.44. To 51 MU Lichfield 19.10.45. Sold 26.7.46 to Bournemouth Flying Club. Regd **G-AICD** (CofR 10549) 13.8.46 to Bournemouth Flying Club Ltd, Christchurch.   CofA No.8368 prepared but not issued. Stored in prewar clubhouse at Christchurch and wrecked 22.6.52 where an Army lorry ran into the building. Regn cld undated as scrapped.

1735   M.14A **P6367**   Dd 20 MU Aston Down 30.6.39.   To 29 E&RFTS Luton 7.7.39.   To 30 EFTS Burnaston 15.10.39. To 16 EFTS Burnaston 10.4.40. To 3 EFTS Shellingford 14.6.42. To 10 FIS Woodley 27.2.43. To 51 MU Lichfield 15.9.43. Sold 4.11.46 to Miles Aircraft Ltd. CofA No.9370 issued 20.5.47. Regd in France as **F-BDPH** (CofR 20002) 3.7.47 to Louis Kahn & Michel Bernard,

Paris. Regd 7.2.50 to Association Touring Club de France, Paris. Regd 2.4.57 to Aero Club de Touraine, Tours. CofA suspended 5.4.58. Regn cld 22.9.71 as destroyed. *F-BDPH and LV-XMN are both quoted as c/n 1735 but latter probably c/n 1739; also note that Hugh Kendall test flew 'P6377' 31.12.46.*

1736 M.14A **P6368** Dd 20 MU Aston Down 30.6.39. To 29 E&RFTS Luton 7.7.39. To 30 EFTS Burnaston 15.10.39. To 16 EFTS Burnaston 10.4.40. To 2 FIS Montrose 20.3.43. Dived into ground near Gourdon, Kincardine 8.8.43.

1737 M.14A **P6369** Dd 20 MU Aston Down 30.6.39. To P&P 6.6.40 and converted to light bomber. To 8 EFTS Woodley 17.6.40. To 10 FIS Woodley 21.7.42. To 27 MU Shawbury 17.9.43. To 51 MU Lichfield 18.7.44. Sold 10.3.47 to Miles Aircraft Ltd. CofA No.9834 issued 13.2.48 to **Thai AF**.

1738 M.14A **P6370** Dd 20 MU Aston Down 30.6.39. To 15 EFTS Redhill 23.3.40; to Carlisle 2.6.40. To Polish FTS Hucknall 8.12.40. To Shrager Bros, Old Warden (SAS) 23.7.41. To 37 MU Burtonwood 30.10.41. To 16 EFTS Burnaston 8.12.41. To 27 MU Shawbury 2.9.42. To 45 MU Kinloss 15.3.43. To RAF Halton 23.3.43. To 7 SofTT, Innsworth for GI as **4047M** 15.8.43. SOC 15.1.45.

1739 M.14A **P6371** Dd 6 MU Brize Norton 3.7.39. To 15 EFTS Redhill 23.3.40; to Carlisle 2.6.40. MI 30.7.42; to Herts & Essex, Broxbourne 23.8.42. To 27 MU Shawbury 3.10.42. To IFTS ATA, Thame/Barton-in-the-Clay 27.7.43. SOC Cat.E 7.12.44 and sold to Miles Aircraft Ltd. To H Hennequin & Cia, Argentina with CofA No.7446 issued 5.3.46; believed regd **LV-XMN** 30.6.47 to Secretaria Estado Aeronautica, Buenos Aires (but with c/n 1735). To Aero Club Commodora Rivadivia. Accident 18.9.46; repaired. Accident 14.10.53. Regn cld 7.54 as wfu.
Sometimes reported as becoming LV-XME, this is thought to be erroneous and the identity of this is unconfirmed.

1740 M.14A **P6372** Dd 6 MU Brize Norton 3.7.39. To 5 EFTS Hanworth 20.3.40; to Meir 17.6.40. To 16 EFTS Burnaston 26.1.42. To 46 MU Lossiemouth 11.7.42. To IFTS ATA, Thame 2.9.44. On take-off from Thame 14.2.45, it struck propeller of taxying Spitfire R7261 and crashed; ATA pilot T/O JC Finlayson slightly injured. SOC (Cat.E) 17.3.45.

1741 M.14A **P6373** Dd 6 MU Brize Norton 4.7.39 and loaned for display at the Brussels Exhibition 3.7.39-26.7.39; painted yellow overall including the cowling. To 16 EFTS Burnaston 22.4.40. To 3 EFTS Shellingford 14.6.42. To 45 MU Lossiemouth 14.3.43. To Herts & Essex, Broxbourne for repairs 21.4.43. To 27 MU Shawbury 28.6.43. To IFTS ATA, Thame/Barton-in-the-Clay 29.7.43. SOC Cat.E 27.11.44. To Miles Aircraft Ltd. To Portuguese AF **192** with CofA No.7550 issued 18.3.46; later **1203**.

Miles M.14A Magister I P6373 at the Brussels Exhibition in July 1939. Note the yellow painted cowling. [Via MJ Hooks]

1742 M.14A **P6374** Dd 6 MU Brize Norton 6.7.39. To P&P 5.6.40. To 8 EFTS Woodley 17.6.40. To 21 EFTS Booker 26.7.42. To 6 (P)AFU Little Rissington 8.11.43. To 6 SFTS Little Rissington (21.3.46 census); coded 'FBJ-Z'. To 51 MU Lichfield 24.6.47. Sold 27.5.48 to Short Bros & Harland Ltd. Regd **G-ALGK** (CofR 12520) 15.1.49 to Short Bros & Harland Ltd, Rochester. CofA No.10424 issued 16.3.49 and fitted with coupe. Flown by Flt Lt Raymond into 3rd place in the Daily Express Air Race on 16.9.50 from Hurn to Herne Bay, at a speed of 143 mph. Forced-landed in water on a cross-country exercise shortly after noon at Burnham-on-Crouch 21.1.51; pupil pilot Mamsukhani was rescued and the aircraft moored to a barge pending salvage but was abandoned on site 10.2.51. Regn cld 27.1.51 as destroyed.

1743 M.14A **P6375** Dd 6 MU Brize Norton 6.7.39. To P&P 5.6.40. To 15 EFTS Carlisle 12.6.40. SOC 23.7.40.

1744 M.14A **P6376** Dd 6 MU Brize Norton 6.7.39. To P&P 5.6.40. To 8 EFTS Woodley 17.6.40. To 10 FIS Woodley 21.7.42. To 51 MU Lichfield 15.9.43. Sold 25.3.46 to Miles Aircraft Ltd. To H Hennequin & Cia, Argentina with CofA No.8044 issued 8.7.46; regd **LV-XSI** 30.6.47 to Secretaria Estado Aeronautica, Buenos Aires. To Aero Club Gral Alvear. Regn cld 1.53 as wfu. Regd (5.67) to Direccion General de Aeronautica Civil, Buenos Aires.

1745 M.14A **P6377** Dd 24 Sqdn, Hendon 6.7.39. Loaned to Air Attaché Paris. To 601 Sqdn, Tangmere 2.5.40; to Exeter 7.9.40. Blown over on take-off from forced landing Sidford, Devon 6.12.40. SOC 25.12.40.

1746 M.14A **P6378** Dd Bristol Aeroplane Co Ltd 7.7.39. To P&P (SAS) 28.2.41. To 15 EFTS Carlisle 21.6.41. To 7 FIS Upavon 31.7.42. Allotted to 2 SofTT Cosford for GI as **4774M** 8.5.44 but not progressed. To 10 SofTT Kirkham for GI as **4801M** 25.5.44. To spares 25.1.45.

1747 M.14A **P6379** Dd 36 MU Sealand 7.7.39. To Middle East 30.8.39 for 208 Sqdn. To Ismailia (.41). To 103 MU Aboukir (41/42). To 267 Sqdn (5.42). To 74 OTU. SOC 1.1.47.

1748 M.14A **P6380** Dd 9 MU Cosford 10.7.39. To Hucknall 4.6.40. To Rolls-Royce Ltd, Hucknall 25.1.41. To Rotol 3.2.42. To P&P (31.12.43 census). To 51 MU Lichfield 15.4.44. Sold by MAP in early 1945 to Miles Aircraft Ltd for Chile; but not progressed. Sold later in 1945 to Miles Aircraft Ltd, allotted **U-0259** (but possibly not used). Regd **G-AGVW** (CofR 9774) 15.11.45 to Miles Aircraft Ltd. CofA No.7265 issued 19.11.45. Dd to Denmark in 11.45 for demonstrations. On Easter Day 21.4.46 it was being 'aero-batted' at Kalveboderne, near Copenhagen for air-to-air photography for the picture paper 'Billad-Bladot'; it came from below and struck the starboard wing of photo-ship Auster Autocrat OY-DGA which lost its fin and rudder. The Hawk Trainer dived in from 3,000ft, killing Capt. Jorgen Edsberg (the owner of OY-DGA). *The Auster entered a spin but recovered sufficiently to land with 80% damage, its two passengers being slightly injured.* Regn cld 24.5.46 as wfu.

1749 M.14A **P6381** Dd 9 MU Cosford 10.7.39. To 16 EFTS Burnaston 22.9.40. To 3 EFTS Shellingford 14.6.42. To HGCU Brize Norton 25.2.43. Engine cut, crash landed Hanney Road, Berks 28.12.44. SOC 2.1.45.

1750 M.14A **P6382** Dd 9 MU Cosford 10.7.39. To 16 EFTS Burnaston 22.9.40. To 3 EFTS Shellingford 14.6.42. To 45 MU Kinloss 14.3.43. To 51 MU Lichfield 24.2.44. Sold 10.3.47 to British Air Transport Ltd. Regd **G-AJRS** (CofR 11531) 30.4.47 to British Air Transport Ltd, Redhill. CofA No.9313 prepared but not issued. Regn cld 27.2.59 as destroyed. In fact, the airframe was refurbished at Redhill late 53/early 54 but adopted the identity of G-AJDR (T9976); CofA 'renewed' 21.1.54. (For history as G-AJDR up until its CofA lapsed 14.9.64 – see under c/n 2169). Rebuilt by Shoreham Aviation Services with CofA renewed 4.9.69 (still as G-AJDR). Acquired by the Shuttleworth Trust in May 1970 and repainted in an all-yellow c/s as P6382 to reflect its true identity; dd Old Warden 1.8.70. G-AJDR's regn was cld 1.3.71 as "transferred to military marks" and it was thereafter flown without formal registration under the special Shuttleworth permit. When this matter was required to be regularised, the Hawk Trainer was regd with its true identity G-AJRS 29.10.81 to The Shuttleworth Collection, Old Warden. Regd 8.4.93 to Richard Shuttleworth Trustees, Old Warden. Total 2,371 hrs at 31.12.03. Maintained in airworthy condition at Old Warden, retaining military c/s as P6382.

1751 M.14A **P6396** Dd 9 MU Cosford 10.7.39. To 24 EFTS Luton 26.9.40. To 15 EFTS Carlisle 12.2.42. To 21 EFTS Booker 17.6.42. SOC Cat.E1 24.8.44. To Miles Aircraft Ltd and to Portuguese AF **199** with CofA No.7648 issued 9.4.46; later **1210**.

1752 M.14A **P6397** Dd 9 MU Cosford 13.7.39. To Northolt 14.8.39. To 71 Sqdn, Church Fenton 8.11.40; to Kirton-in-Lindsey 23.11.40. To Martlesham Heath 9.4.41; to North Weald 23.6.41. Crashed during low level aerobatics North Weald 28.8.41. SOC 5.9.41.

1753 M.14A **P6398** Dd 9 MU Cosford 13.7.39. To Northolt 14.8.39. To 46 MU Lossiemouth 21.7.40. To 15 EFTS Carlisle 16.2.41. To 29 EFTS Clyffe Pypard 24.5.42. To 5 GTS Shobdon 22.3.43. SOC Cat.E (MR) 27.5.44.

1754 M.14A **P6399** Dd 612 Sqdn, Dyce 14.7.39. To 12 MU Kirkbride 17.1.40. To 1 CRU Cowley 8.2.40. To 48 MU Hawarden 16.6.40. To Ouston (undated). To 37 MU Burtonwood 22.6.41. To 'Sundry Units Servicing Aircraft Section (SAS)' 4.8.41 and shipped to **Turkey** as replacement for 1 of 6 lost at sea en route from the original order.

1755 M.14A **P6400** Dd 612 Sqdn, Dyce 14.7.39. To 12 MU Kirkbride 17.1.40. To Marham 3.5.40. To 99 Sqdn, Waterbeach 26.4.41. To 37 MU Burtonwood 5.6.41. To Shrager Bros, Old Warden 12.6.41. To Colerne 25.8.41. To 2 FIS Montrose 28.2.42. Crashed in practice forced landing Laurencekirk, Kincardine 11.4.43. SOC 26.4.43 (643:35 F/H).

1756 M.14A **P6401** Dd 612 Sqdn, Dyce 14.7.39. To 12 MU Kirkbride 19.1.40. To Leconfield 19.7.40. To 485 RNZAF Sqdn, Leconfield 14.5.41. To 129 Sqdn, Leconfield 29.6.41. To Skeabrae 1.10.42. To 27 MU Shawbury 15.12.42. SOC 28.6.45 'deteriorated beyond repair'.

1757 M.14A **P6402** Dd 612 Sqdn, Dyce 14.7.39. To 12 MU Kirkbride 17.1.40. To Woolsington 20.6.40. Cat.B 5.3.41; to P&P (SAS) 2.4.41. To 13 Grp Comm Flt, Ouston 19.4.41. Cat.B; to Atlantic Coast, Barnstaple (MR) 24.9.43. To 51 MU Lichfield 13.11.43. To BLEU Martlesham Heath 6.1.48 for flying training of civilians 12.1.48, probably dd 7.1.48. To RAE Farnborough for communication/training duties 17.1.49. Sold per CS(A) Return 8.5.50 (RAE record card states sold to Fielding Industries Ltd, Elstead 23.5.50). Regd **G-AMBP** (CofR 15023) 10.5.50 to William Joseph Nobbs, Fairoaks; dd Farnborough to Fairoaks 12.5.50. CofA No.10904 issued 13.7.50. Cld 2.9.50 & regd 9.9.50 to Universal Flying Services Ltd, Fairoaks. Stalled on landing and crashed Fairoaks 2.6.51; pilot slightly hurt. Regn cld 18.6.51 as destroyed.

1758 M.14A **P6403** Dd Northolt 17.7.39. To North Weald 15.8.39. Crashed North Weald 15.8.39 and w/off (4:00 F/H).

1759 M.14A **P6404** Dd Northolt 17.7.39. Probably to Brussels Exhibition 17.7.39-19.7.39. To Biggin Hill 8.5.41. To 24 EFTS Luton 22.7.41. To 2 FIS Montrose 12.2.42. Crashed in forced landing Stracathro 10.5.42. SOC Cat.E 28.5.42.

1760 M.14A **P6405** Dd Northolt 17.7.39; op Stn Flt (8.40-8.41). To 51 MU Lichfield 12.3.43. Sold 6.5.46 to Miles Aircraft Ltd. To H Hennequin & Cia, Argentina with CofA No.8121 issued 16.7.46; regd **LV-XRG** 30.6.47 to Secretaria Estado Aeronautica, Buenos Aires. To Aero Club Esquina. Crashed San Justo 11.7.47; pilot killed. Also reported as crashed nr Campillo 22.10.54; pilot injured and passenger killed.

1761 M.14A **P6406** Dd 20 MU Aston Down 22.7.39. To P&P (SAS) 6.6.40 and converted to light bomber. To 8 EFTS Woodley 17.6.40. To 10 FIS Woodley 21.7.42. To 27 MU Shawbury 12.9.43. SOC Cat.E (MR) 25.2.44.

1762 M.14A **P6407** Dd 20 MU Aston Down 22.7.39. To 23 E&RFTS Rochester 8.8.39. To 24 EFTS Sydenham 15.10.39; to Luton 22.7.40. To 15 EFTS Carlisle 11.10.41. To 21 EFTS Booker 10.6.42. To 8 (O)AFU Mona 5.12.43. To FTC Comm Flt Woodley 3.7.45 (coded 'FKN-N') and fitted with sliding canopy over both cockpits by Miles Aircraft Ltd. To 5 MU Kemble 22.1.47. Sold 5.12.47 to WS Shackleton Ltd. Regd **G-AKJV** (CofR 11983) 17.12.47 to Short Bros & Harland Ltd, Rochester. CofA No.9915 issued 6.4.48 (canopy retained). CofA lapsed 5.4.49; renewed 15.5.50. Cld 4.3.52 & regd 17.3.52 to Redhill Flying Club Ltd, Redhill (by which time the sliding canopy had been removed). CofA lapsed 5.2.54 (possibly after accident). Broken up and burnt Redhill 11.45 am 24.5.54. Regn cld 7.4.59 as destroyed.

1763 M.14A **P6408** Dd 20 MU Aston Down 22.7.39. To 23 E&RFTS Rochester 8.8.39. To 24 EFTS Sydenham 15.10.39; to Luton 22.7.40. To 45 MU Kinloss 22.2.42. Cat.B; to Atlantic Coast, Barnstaple (MI) 4.3.42. To 20 MU Aston Down 2.4.42. To 416 RCAF Sqdn, Martlesham Heath 5.8.42. To 27 MU Shawbury 31.10.42. To 7 FIS Upavon 23.10.43. Hit HT wires and crashed near Collingbourne Kingston, Wilts 7.1.44.

1764 M.14A **P6409** Dd 20 MU Aston Down 22.7.39. To 23 E&RFTS Rochester 8.8.39. To 24 EFTS Sydenham 15.10.39; to Luton 22.7.40. To 45 MU Kinloss 2.2.42. To Atlantic Coast, Barnstaple (MI) 19.3.42. To 16 EFTS Burnaston 12.4.42. To 46 MU Lossiemouth 11.7.42. To 105 (M)SP Broadwell 19.9.44. SOC Cat.E 21.11.44 and to Miles Aircraft Ltd. To Portuguese AF **193** with CofA No.7551 issued 18.3.46; later **1204**.

1765 M.14A **P6410** Dd 20 MU Aston Down 22.7.39. To 23 E&RFTS Rochester 8.8.39. To 24 EFTS Sydenham 15.10.39; to Luton 22.7.40. Damaged 19.4.41; to 1 CRU Cowley 12.5.41. To 60 OTU East Fortune 20.6.41. To 485 RNZAF Sqdn, Kenley 11.4.42. MI 5.6.42; to Atlantic Coast, Barnstaple 7.6.42. To 27 MU Shawbury 5.7.42. To 46 MU Lossiemouth 30.7.42. To 7 FIS Upavon 6.6.43. To Staff Pilot Training Unit, Cark 22.6.44. To ECFS Hullavington 5.12.45; coded 'FCV-A'. To 5 MU Kemble 12.4.46. Front cockpit faired over and rear cockpit fitted with a teardrop canopy with a sliding hood late in the war at RAF Barrow. Sold 14.11.46 to Bournemouth Flying Club Ltd. Regd **G-AKGS** (CofR 11906) 8.9.47 to Bournemouth Flying Club Ltd, Christchurch. CofA No.9799 issued 13.5.48; mod to standard with open cockpits. CofA lapsed 12.5.49. Cld 1.1.53 & regd 12.1.53 to Thomas Hutton Marshall, Christchurch. CofA renewed 18.9.53. Cld 11.3.54 & regd 16.3.54 to Southern Flying Schools Ltd, Portsmouth. Cld 20.3.54 & regd 27.3.54 to Ronald Vivian Courtnay Jones, Oxford. Crashed near Granada, Spain 1.4.54; pilot Seeley & passenger Mullin. Regn cld 26.5.54 as wfu.

1766 M.14A **P6411** Dd 20 MU Aston Down 22.7.39. To 23 E&RFTS Rochester 9.8.39. To 24 EFTS Sydenham 15.10.39; to Luton 22.7.40. Damaged when hit hedge on landing Luton 29.3.41; Acting LA D Brooker unhurt; to 1 CRU Cowley for repairs 1.4.41. To 68 Sqdn, High Ercall 21.5.41. To 32 MU St. Athan 6.11.41. To High Ercall 13.12.41. Cat.B; to P&P 16.2.42. To 68 Sqdn, Coltishall 12.4.42. To RAF Church Fenton for their Battle of Britain Display on 15.9.45, To 51 MU Lichfield 12.12.45. Sold 23.8.46 to Royal Aero Club (on behalf of Portsmouth Aero Club). Regd **G-AIDF** (CofR 10576) 21.8.46 to Portsmouth Aero Club, Portsmouth. CofA No.8459 issued 28.11.46. Cld 1.11.46 & regd 15.1.47 to Hunting Flying Clubs Ltd, Luton. CofA lapsed 27.11.47. Cld 25.7.50 & regd 1.8.50 to William James Twitchell, St Albans. CofA renewed 18.8.50. Cld 18.4.51 & regd 20.4.51 to Arthur Ernest Tuck Allen, Peterborough (Southsea wef 8.9.53). Cld 16.10.53 & regd 21.10.53 to Southern Flying Schools Ltd, Portsmouth. Cld 10.2.54 & regd 19.2.54 to John Patrick Desmond Hayward, Hayes, Middlesex. Cld 6.4.54 & regd 13.4.54 to John Henry Sauvage, Camberley t/a Blackbushe Flying Club, Blackbushe. Cld 29.1.55 & regd 8.2.55 to Denham Flying Club Ltd, Denham. CofA lapsed 2.5.59. Cld 16.5.60 & regd 24.5.60 to William George Boylett, Plymouth. CofA renewed 9.5.60. Regn cld 2.3.61 as sold (notified 27.7.62) and regd 29.8.62 to Herbert Henry, Plymouth. Cld 24.6.63 & regd 12.7.63 to William Henry Wanstall Lucas, Plymouth. Cld 25.4.64 & regd 12.5.64 to Herbert Henry & Samuel Richman, Plymouth. CofA lapsed 27.1.66. Acquired .67 by the British Historic Aircraft Museum at Southend, where, with its restoration as P6411 almost complete, it was wrecked by a gale on 5.9.67. Regn cld 6.7.73 as pwfu.

1767 M.14A **P6412** Dd 20 MU Aston Down 24.7.39. To P&P (SAS) 5.6.40; converted to light bomber. To 15 EFTS Carlisle 12.6.40. To P&P (Cat.B) 30.7.40. To 27 MU Shawbury 1.9.40. To 5 EFTS Meir 21.11.40. To 1 CRU Cowley (SAS) 13.1.41. To 24 EFTS Luton 14.3.41. To 2 FIS Montrose 18.2.42. To 9 (P)AFU Hullavington 30.9.43. To 51 MU Lichfield 4.5.44. Sold 18.6.46 to Miles Aircraft Ltd. To H Hennequin & Cia, Argentina with CofA No.8283 issued 28.8.46; regd **LV-XSV** 30.6.47 to Secretaria Estado Aeronautica, Buenos Aires. To Aero Club Apostoles. Regn cld 8.50 as wfu.

1768 M.14A **P6413** Dd 20 MU Aston Down 24.7.39. To P&P (SAS) 3.6.40; converted to light bomber. To 15 EFTS Carlisle 12.6.40. To 8 EFTS Woodley 19.8.40; (carried code 'B-27' - possibly while here). Cat.B 18.5.41; to P&P (SAS) 31.5.41. To 46 MU Lossiemouth 24.7.41. To 15 EFTS Carlisle 10.8.41. To 7 FIS Upavon 31.7.42. Dived into ground low flying near Chinton, Wilts 19.10.42. SOC 29.10.42.

1769 M.14A **P6414** Dd 20 MU Aston Down 25.7.39. To 47 MU Sealand 3.6.40. Dd to Irish Air Corps Baldonnel 7.6.40 as **76**. Scrapped 8.46.

1770 M.14A **P6415** Dd 20 MU Aston Down 27.7.39. To P&P (SAS) 6.6.40; converted to light bomber. To 16 EFTS Burnaston 11.6.40. To P&P (Cat.B) 9.8.40. To 48 MU Hawarden 5.10.40. To 310 Czech Sqdn, Duxford 10.10.40. Flying accident 5.12.40; to P&P

APPENDIX 20

(SAS) 13.12.40. To 5 MU Kemble 12.5.41. To 4 (S)FIS Cambridge 8.6.41. Flew into high ground in low cloud Therfield, near Royston, Cambs 16.11.41. SOC 24.11.41.

1771   M.14A **P6416**   Dd Stn Flt Linton-on-Ouse 27.7.39. To 4 Grp Comm Flt, York 7.10.40. MI at Atlantic Coast, Barnstaple 18.6.42. To 27 MU Shawbury 6.7.42. To 266 Sqdn, Duxford 11.9.42. To 51 MU Lichfield 3.1.43. Sold 6.5.46 to Miles Aircraft Ltd. To H Hennequin & Cia, Argentina with CofA No.8123 issued 16.7.46; regd **LV-XRI** 30.6.47 to Secretaria Estado Aeronautica, Buenos Aires. To Aero Club Trelew. Destroyed in hangar fire Trelew 29.12.47.

1772   M.14A **P6417**   Dd 20 MU Aston Down 1.8.39. To 16 EFTS Burnaston 21.6.40. To 27 MU Shawbury 11.9.42. To 51 MU Lichfield 28.3.44. Sold 11.3.46 to Miles Aircraft Ltd. To H Hennequin & Cia, Argentina with CofA No.7824 issued 3.6.46; regd **LV-XPM** 30.6.47 to Secretaria Estado Aeronautica, Buenos Aires. To Aero Club Rosario. Crashed Santo Tome 1.7.47; pilot injured. Regn cld 10.62 as wfu.

1773   M.14A **P6418**   Dd 36 MU Sealand 28.7.39. To Royal Egyptian AF Almaza; probably as **L244**.

1774   M.14A **P6419**   Dd 20 Aston Down 1.8.39. To 16 EFTS Burnaston 21.6.40; coded '26'. Cat.B; to Atlantic Coast, Barnstaple (MR) 9.6.43. To 27 MU Shawbury 31.8.43. To 51 MU Lichfield 29.1.44. Sold 10.3.47 to RC Hobbs. Regd **G-AJRV** (CofR 11534) 30.4.47 to Reginald Carne Hobbs, Maidenhead. Cld 17.9.47 & regd 17.10.47 to Maurice Dumont, Stockton-on-Tees. CofA No.9865 issued 3.4.48. Cld 13.9.48 & regd 9.10.48 to Darlington & District Aero Club Ltd, Croft. CofA lapsed 2.4.49. Cld 17.6.52 & regd 20.6.52 to The Scottish Flying Club Ltd, Renfrew. CofA renewed 21.8.52. One u/c leg collapsed on landing Renfrew 17.6.53. CofA lapsed 20.8.53; renewed 22.10.54. Crashed 4 ml E of Alexandria, Dumbartonshire 29.9.55; FJ Osborne killed. Regn cld 11.1.56 as destroyed.

1775   M.14A **P6420**   Dd 20 MU Aston Down 1.8.39. To 8 EFTS Woodley 27.4.40. To FTC Comm Flt Woodley 5.8.41. To 8 EFTS Woodley (31.12.41 census). To MAP Overseer at Miles Aircraft Ltd for communications duties 13.10.42. To Fairey Avn Ltd 15.11.43. To 51 MU Lichfield 9.5.46. To Syerston 14.6.46. To 51 MU Lichfield 11.7.47. Sold 22.12.47 to Herts & Essex Aero Club. Regd **G-AKMZ** (CofR 12082) 29.11.47 to Herts & Essex Aero Club (1946) Ltd, Broxbourne. Not refurbished. To Squires Gate and burnt there 1956. Regn cld 20.4.59 as pwfu.

1776   M.14A **P6421**   Dd 20 MU Aston Down 1.8.39. To Pembrey 22.1.41. To 5 MU Kemble 1.6.41. To 16 EFTS Burnaston 25.7.41. To SFTS Cranwell 1.5.43; reformed as 17 SFTS 20.3.44; coded 'FCC-Y'. To 21 (P)AFU Moreton-in-Marsh 13.1.47. To 51 MU Lichfield 24.11.47. Sold 8.6.48 to BOAC. Regd **G-AKKW** (CofR 12009) 23.6.48 to Speedbird Flying Clubs Ltd, Denham. CofA No.10429 issued 23.3.49. Regd 4.49 (on name change) to Airways Aero Associations Ltd, Denham. CofA lapsed 22.3.50 & wfu Denham. Regn cld 27.3.52 as wfu.

1777   M.14A **P6422**   Dd 20 MU Aston Down 1.8.39. To 47 MU Sealand 3.6.40. Delivered to Irish Air Corps at Baldonnel 7.6.40 as **77**. Scrapped 8.46.

1778   M.14A **P6423**   Dd 20 MU Aston Down 1.8.39. To P&P (SAS) 5.6.40; converted to light bomber. To 5 EFTS Hanworth 11.6.40; to Meir 17.6.40. To 8 EFTS Woodley 12.8.40. To 10 FIS Woodley 21.7.42. To 27 MU Shawbury 12.9.43. To 141 Sqdn, West Raynham 14.6.44. To 51 MU Lichfield 27.9.45. Sold 26.11.46 to Miles Aircraft Ltd. CofA No.9366 issued 19.5.47. Regd in France as **F-BDPD** (CofR 19998) 16.7.47 to Aero Club d'Arras, Arras/Roclincourt. Regd 13.8.51 to Bertrand Lionel du Bouexic, St Malo de Phily. Regd 14.6.55 to Aero Club de Cherbourg et de la Manche, Caen. Regd 13.3.58 to Aero Club Les Ailes Basques, Biarritz. Regn cld 22.9.71 as destroyed.

1779   M.14A **P6424**   Dd 20 MU Aston Down 1.8.39. To P&P (SAS) 3.6.40; converted to light bomber. To 15 EFTS Carlisle 12.6.40. To 8 EFTS Woodley 19.8.40. To 7 FIS Upavon (undated). To 2 FIS Montrose 27.4.42. Flying accident (Cat.A) 5.8.42; to P&P for repairs 12.8.42. To 27 MU Shawbury 26.9.42. To 604 Sqdn 13.7.44. SOC Cat.E1 24.8.44. To Miles Aircraft Ltd. To Irish Air Corps, Baldonnel as **127** with CofA No.7458 issued 14.2.46; dd 17.2.46. Scrapped May 1952.

1780   M.14A **P6436**   Dd 20 MU Aston Down 8.8.39. To P&P (SAS) 3.6.40; converted to light bomber. To 15 EFTS Carlisle 12.6.40. To 8 EFTS Woodley 31.8.40. Crashed on overshoot Woodley 14.9.40; LAC WA Bonney slightly hurt. To P&P (SAS) 16.9.40 but SOC 21.9.40. *Note – accident date also given as 14.9.41.*

1781   M.14A **P6437**   Dd 20 MU Aston Down 8.8.39. To P&P (SAS) 3.6.40; converted to light bomber. To 15 EFTS Carlisle 12.6.40. Crashed while low flying Hollee, Dumfries 23.7.40. SOC 31.7.40.

1782   M.14A **P6438**   Dd 20 MU Aston Down 8.8.39. To P&P 3.6.40; converted to light bomber. To 15 EFTS Carlisle 11.6.40. To 8 EFTS Woodley 21.8.40. To 10 FIS Woodley 21.7.42. To 5 MU Kemble 16.6.46. Sold 14.11.46 to TC Sparrow. Regd **G-AIUF** (CofR 10974) 11.11.46 to Thomas Charles Sparrow, Bournemouth. Regn cld 3.1.48 as sold & regn lapsed. Sold .51 to Bournemouth Flying Club Ltd, Christchurch. Regd 3.12.52 to Albert Edward Hawes, Emsworth, Hants. CofA A3882 issued 21.4.53. Cld 27.7.53 & regd 7.8.53 to Short Bros & Harland Ltd, Rochester. Cld 31.3.55 & regd 5.4.55 to Joan Esme Dickinson, Cheltenham. CofA lapsed 3.4.56. Sold .56 to US Army Aviation Club, Germany (based Illesheim 1.58). Regn cld 16.11.59 as transferred to US Army Aviation Club. No US registration known and probably wfu in West Germany.

1783   M.14A **P6439**   Dd Watton 15.8.39. To Horsham St. Faith 3.5.41. To 1 CRU Cowley for repairs 29.5.41. To 5 EFTS Meir 3.9.41. To CFS Upavon 10.2.42. To ECFS Hullavington 14.5.42. To 29 EFTS Clyffe Pypard 23.9.42. To 51 MU Lichfield 14.4.43. Sold 1.11.46 to Miles Aircraft Ltd. CofA No.9369 issued 19.5.47. Regd in France as **F-BDPG** (CofR 20001) 16.10.47 to Aero Club du Rhone et de Sud Est, Lyon-Bron. Regd 14.11.50 to Association Cooperative de Pilotage de l'Ouest (Coopavia), La Rochelle. Regd 30.9.52 to Aero Club de la Rochelle et de la Charente Maritime (ACLR), La Rochelle. Regd 16.5.56 to Aero Club Yonnais, La Roche sur Yon. Regd 20.6.56 to Jacques Jaunet, La Roche sur Yon. Regd 16.5.62 to Armand Maze, Persan-Beaumont. Regd 12.6.64 to Michel Paris, Lavau. Regn cld 22.9.71 as destroyed.

1784   M.14A **P6440**   Dd 20 MU Aston Down 8.8.39. To 47 MU Sealand 3.6.40. Dd to Irish Air Corps Baldonnel 7.6.40 as **74**. Scrapped 9.45.

1785   M.14A **P6441**   Dd 9 MU Cosford 8.8.39. To 8 EFTS Woodley 14.2.41. To 10 FIS Woodley 21.7.42. MR at Atlantic Coast, Barnstaple 13.9.43. To 51 MU Lichfield 18.10.43. Sold 20.11.46 to Miles Aircraft Ltd. CofA No.9363 issued 9.5.47. Regd in France as **F-BDPA** .47 to Aero Club Roland Garros. Fate unknown.

1786   M.14A **P6442**   Dd 9 MU Cosford 14.8.39. To 16 EFTS Burnaston 20.9.40. MI 7.7.42; to P&P 29.7.42. To 27 MU Shawbury 16.9.42. To 47 MU Sealand 10.2.43. Despatched to Middle East on SS City of Worcester 5.3.43; arr 13.5.43. To Comm Flt Amman. SOC 10.1.46.

1787   M.14A **P6443**   Dd Feltwell 12.8.39. To 9 MU Cosford 5.6.41. To ATA Training Unit 11.8.41; coded '41'. Cat.B, to Atlantic Coast, Barnstaple 14.11.43. To 51 MU Lichfield 31.1.44. Sold 10.3.47 to The Wiltshire School of Flying Ltd, Thruxton. Not regd but used in the rebuild of G-AITO (R1841) in about 1947/48.

1788   M.14A **P6444**   Dd 18 Sqdn, Upper Heyford 12.8.39. To 6 MU Brize Norton 15.9.39. To 83 Sqdn, Scampton 9.3.40. To 37 MU Burtonwood 24.6.41. To 16 EFTS Burnaston 22.8.41. Badly damaged 17.3.42 and SOC 1.5.42.

1789   M.14A **P6445**   Dd 57 Sqdn, Upper Heyford 10.8.39. To 6 MU Brize Norton 27.9.39. To 16 EFTS Burnaston 11.6.40. To 1 CRU Cowley 5.8.40. To 27 MU Shawbury 8.9.40. To 8 EFTS Woodley 10.2.41. To 21 EFTS Booker 26.7.42. To 7 FIS Upavon 27.7.43. SOC 7.7.44.

1790   M.14A **P6446**   Dd 9 MU Cosford 14.8.39. To 605 Sqdn, Croydon 29.6.40. To 22 MU Silloth 11.10.40. To 605 Sqdn, Martlesham Heath 26.2.41. Flying accident (Cat.B) 5.6.41; to P&P 2.11.41. To 2 FIS Montrose 1.4.42. To 18 (P)AFU Church Lawford 1.10.43. To FTC Comm Flt Woodley 5.5.45. To 10 FIS Woodley 18.7.45; coded 'FDT-J'. To 7 FIS Upavon 6.5.46. To CFS Upavon 6.5.46; coded 'FDK-B'. To 51 MU Lichfield 25.4.47. Sold 8.6.48 to BOAC. Regd **G-AKKS** (CofR 12005) 23.6.48 to Speedbird Flying Clubs Ltd, Denham (initially regd as ex P6466). CofA No.10425 issued 23.3.49. Regd 4.49 (on name change) to Airways Aero

Associations Ltd, Denham. Cld 10.10.52 & regd 31.12.52 to The Scottish Flying Club Ltd, Renfrew. CofA lapsed 17.10.52 and used for spares at Renfrew. Regn cld 11.9.54 as wfu.

1791 M.14A **P6447** Dd 9 MU Cosford 14.8.39. To 16 EFTS Burnaston 20.9.40. Collided with Magister P6373 and crashed Hilton, nr Repton, Derby 8.5.42; Cpl WA Scudamore bailed out too low and was killed. SOC 15.5.42.

1792 M.14A **P6448** Dd 9 MU Cosford 14.8.39. To 8 EFTS Woodley 15.2.41. To 21 EFTS Booker 14.7.42. To 51 MU Lichfield 18.11.44. Sold 25.3.46 to Miles Aircraft Ltd. To H Hennequin & Cia, Argentina with CofA No.7855 issued 6.6.46; regd **LV-XPP** 30.6.47 to Secretaria Estado Aeronautica, Buenos Aires. Crashed Paganini 12.11.46 (20% damaged); repaired. Regn cld 2.51 as wfu.

1793 M.14A **P6449** Dd 9 MU Cosford 14.8.39. To 16 EFTS Burnaston 29.9.40. To 5 MU Kemble 13.4.41. To 234 Sqdn, Warmwell 22.4.41; to Ibsley 26.1.42. To 118 Sqdn, Ibsley 12.3.42. To Herts & Essex, Broxbourne for repairs 12.8.42. To 27 MU Shawbury 18.9.42. To 47 MU Sealand 26.2.43. Despatched to Middle East on SS Clan Lamont 29.4.43; arr 12.6.43. To 104 MU Aleppo area. SOC 20.9.45.

1794 M.14A **P6450** Dd 9 MU Cosford 14.8.39. To 16 EFTS Burnaston 20.9.40. Crashed in forced landing 1 ml S of Banbury, Oxon 8.12.40. SOC 25.12.40.

1795 M.14A **P6451** Dd 9 MU Cosford 18.8.39. To 16 EFTS Burnaston 20.9.40. To 4 FIS Cambridge 24.8.42. To 22 EFTS Cambridge 13.5.43. To 51 MU Lichfield 6.5.44. Sold 8.4.46 to Miles Aircraft Ltd. To H Hennequin & Cia, Argentina with CofA No.7889 issued 14.6.46; regd **LV-XPX** 30.6.47 to Secretaria Estado Aeronautica, Buenos Aires. To Aero Club Concepcion del Uruguay. Accident 27.3.48; repaired. Regd 10.8.62 to Aero Club Concepcion del Uruguay.

1796 M.14A **P6452** Dd 9 MU Cosford 18.8.39. To 24 EFTS Sydenham 30.3.40; to Luton 22.7.40; coded '47'. Cat.B; to Atlantic Coast, Barnstaple (MI) 20.12.41. To 15 EFTS Carlisle 10.4.42. To 21 EFTS Booker 17.6.42. MR (Cat.B) 24.9.42; to P&P 27.9.42. To 51 MU Lichfield 6.11.42. To 76 MU Wroughton 14.5.43. Despatched to Middle East for **Turkey** on SS Clan McIver. 20.5.43; arr 14.6.43.

1797 M.14A **P6453** Dd 9 MU Cosford 18.8.39. To 24 EFTS Sydenham 30.3.40. To 8 EFTS Woodley 9.4.40 (unconfirmed). To 24 EFTS Luton 24.6.40. Spun into ground near Luton 6.3.41; P/O Gittins killed & Acting LA Watson badly injured. To P&P (SAS) 10.3.41 but SOC 19.3.41.

1798 M.14A **P6454** Dd 9 MU Cosford 18.8.39. To 16 EFTS Burnaston 20.9.40. Flying accident 13.11.40; to 1 CRU Cowley (SAS) 25.11.40. To 27 MU Shawbury 20.2.41. To Franco-Belgian Air Training School, Odiham 28.3.41. To 5 EFTS Meir 6.6.41. To CFS Upavon 10.2.42. To 2 FIS Montrose 26.4.42. To 12 MU Kirkbride 4.8.42. SOC Cat.E (MR) 7.2.44.

1799 M.14A **P6455** Dd Wyton 18.8.39. To 27 MU Shawbury 16.7.41. To ATA Training Unit, Barton-in-the-Clay 25.9.41; coded '5'. Damaged when stalled on landing Barton-in-the-Clay 25.7.43 (Cat.B); to Atlantic Coast, Barnstaple for repairs 11.8.43. To 51 MU Lichfield 18.10.43. To Reserve Command Comm Sqdn, White Waltham 11.7.47. To 5 MU Kemble 27.9.49. Sold 2.1.50 to RA Short but not regd.

1800 M.14A **P6456** Dd Wyton 21.8.39. To 1 CRU Cowley (30.11.39 census). To 5 MU Kemble 26.4.40. To 8 EFTS Woodley 19.6.40. To 5 EFTS Meir 12.8.40. To 37 MU Burtonwood 8.12.41. To CRD Denham 12.4.42; fitted with Maclaren Drift Undercarriage Installation by Airwork Ltd, Denham. To A&AEE Boscombe Down for tests which were completed by 7.9.42 '13.9.42'. To 4 Sqdn, Clifton for service trials. To 613 Sqdn, Ouston for service trials. To 778 Sqdn, FAA 30.9.42. To Recorded to 21 EFTS Booker in 1942 (but unconfirmed). To CRD Airwork Ltd, Denham 2.2.43. Flown by Flt Lt Hugh Kennedy at Woodley 30.3.43. To 43 OTU Old Sarum 12.5.43. Possibly returned to Airwork (no date recorded). To ECFS Hullavington 17.9.43. Sold 11.7.46 to OF Maclaren Ltd. Regd **G-AIAI** (CofR 10504) 24.7.46 to OF MacLaren Ltd, Heston (regd with c/n 'PP38384'). CofA No.8172 issued 13.11.46. CofA lapsed 12.11.47. Regd 28.1.49 to Speedbird Flying Clubs Ltd, Denham. Regd 4.49 on name change to Airways Aero Associations Ltd, Denham. Remains at Northolt (6.51). Regn cld 27.3.52 as wfu.

1801 M.14A **P6457** Dd Mildenhall 21.8.39. To Newmarket 1.8.41. To 3 Grp Comm Flt Newmarket 3.6.42. To Prestwick 5.8.42. To 46 MU Lossiemouth 6.8.42. To 51 MU Lichfield 19.1.45. Sold 27.5.46 to Miles Aircraft Ltd. Test flown 15.7.46 by Hugh Kendall as P6457. To H Hennequin & Cia, Argentina with CofA No.8194 issued 5.8.46; regd **LV-XRP** 30.6.47 to Secretaria Estado Aeronautica, Buenos Aires. To Aero Club San Rafael. Accident 23.11.47; repaired. Accident 25.3.56; repaired. Regd 15.7.59 to Aero Club Comodoro Rivadavia. Still regd 5.67.

1802 M.14A **P6458** Dd Abingdon 21.8.39. To 'Service Flight' 30.11.39. To 81 Sqdn, Amiens/Montjois. Lost in France in 5.40. SOC 26.3.42.

1803 M.14A **P6459** Dd 3 Sqdn 31.8.39. To 257 Sqdn, Hendon 18.5.40. Hit anti-glider pole in forced landing Brentwood, Essex 18.10.40. To P&P (SAS) 4.11.40. SOC 27.3.41.

1804 M.14A **P6460** Dd 111 Sqdn, Northolt 1.9.39. Crashed in forced landing Exted Farm, Kent 2.7.42. SOC 10.7.42.

1805 M.14A **P6461** Dd Filton 1.9.39. Damaged 23.10.39; 1 CRU Cowley 22.11.39. To 9 MU Cosford 17.4.40. To 16 EFTS Burnaston 29.9.40. Crashed 6.2.41. To P&P (SAS) 17.2.41. SOC 22.9.41.

1806 M.14A **P6462** Dd 149 Sqdn, Mildenhall 3.9.39. Blown over on take-off Cleave 8.11.39. SOC 17.1.40.

1807 M.14A **P6463** Dd 213 Sqdn, Wittering 3.9.39. To Tangmere 13.5.41. To 1 Sqdn, Redhill 23.5.41; to Tangmere 1.7.41. Crashed near Arundel, Sussex 1.11.41. To P&P (SAS) 9.11.41 but SOC 18.11.41.

1808 M.14A **P6464** Dd 9 MU Cosford 3.9.39. To Wyton 8.10.39. To 82 Sqdn, Watton 25.6.40. To Lossiemouth 27.6.40. To Herts & Essex, Broxbourne (SAS) 17.8.41. To ATA Training Unit 16.10.41; coded '25'. Cat.B; to Atlantic Coast, Barnstaple (MR) 5.6.43. To 27 MU Shawbury 31.7.43. To Halton 4.9.43. To 51 MU Lichfield 4.12.43. Sold 11.3.46 to Miles Aircraft Ltd. Test flown 21.5.46 by Hugh Kendall as P6464. To H Hennequin & Cia, Argentina with CofA No.7868 issued 11.6.46; regd **LV-XPT** 30.6.47 to Secretaria Estado Aeronautica, Buenos Aires. To Aero Club Esquel. Accident 13.9.50. Regd 10.8.60 to Aero Club Trenque Lauquen. Regd (5.67) to Aero Club Pampeano, Santa Rosa, La Pampa.

1809 M.14A **P6465** Dd 9 MU Cosford 3.9.39. To Stn Flt Gosport 26.11.39. To 16 Grp Comm Flt Rochester 11.4.41. To 32 MU St Athan 19.6.41. To 125 Sqdn, Colerne 20.7.41. SOC Cat.E (MR) 19.8.44. To 51 MU for salvage.

1810 M.14A **P6466** Dd 10 MU Hullavington 4.9.39. To 24 EFTS Luton 17.9.40. To 1 CRU Cowley (SAS) 5.3.41. To 5 MU Kemble 8.4.41. To 25 Sqdn, Wittering 23.4.41. To 604 Sqdn, Middle Wallop 27.11.41. To 174 Sqdn, Manston 5.3.42. To 264 Sqdn, Colerne 4.1.43. To 142 Wing, B.17 Caen/Carpiquet 21.9.44. To 264 Sqdn, Odiham 21.12.44. To 85 M&T Group Blankenese/Hamburg 8.2.45. SOC 20.2.45.

Known Job Numbers:	
5003	c/n 333-340, 351-432
5009	c/n 479, 486, 487, 490-495, 511-529, 538, 539
5022	c/n 530-535
5038	c/n 556-559
5040	c/n 630
5043	c/n 547
5072	c/n 560-584
5073	c/n 585-609
5074	c/n 610-629, 642-646, 811
5075	c/n 647-671
5076	c/n 672-696
5077	c/n 697-721
5078	c/n 722-748
5079	c/n 749-771
5080	c/n 772-785
5348	c/n 821-1074
5676	c/n 797
5822	c/n 1611-1701

Appendix 21:
Miles M.16 Mentor Production

45 aircraft manufactured under Contract No. 534848/36 were delivered between March 1938 and March 1939 as follows:-.

434 L4392 Prototype. First flown, by Bill Skinner 5.1.38. To A&AEE Martlesham Heath 28.3.38; formally TOC 29.3.38. To AMRD at A&AEE 4.4.38 for trials; returned to Phillips & Powis, Woodley 11.7.38. Flown to RAE Farnborough by Miles 15.7.38. To AMRD at Woodley 20.7.38; to A&AEE 22.8.38; AMRD at A&AEE 26.8.38. To AMRD at Phillips & Powis, Woodley 19.11.38, probably dd 29.11.38. Flown by Hugh Kennedy on a 'local' from Woodley on 29.4.40. (A date of 20.8.40 is recorded on the record card with no comment). To Phillips & Powis, Woodley 25.12.40. Damaged 13.1.41; to Phillips & Powis, Woodley (SAS) 16.1.41. To 18 MU Dumfries 14.9.41. To Stn Flt Wyton 26.9.41. To 3 Grp Comm Flt, Newmarket 16.11.43. Cat.Ac 13.6.44 (following accident?); re-Cat.E 27.6.44 and SOC 27.6.44.

435 L4393 To A&AEE Martlesham Heath ('A' Performance Sqdn) 12.12.38. Fitted experimentally with a two-pitch propeller in mid-1939, which marginally improved its handling and take-off performance. To 19 MU St Athan 10.5.40. To A&AEE Boscombe Down 18.8.41. Operated by ETPS, Boscombe Down, where the main spar was damaged in a heavy landing 26.6.44. SOC 7.7.44.

436 L4394 To CFS Upavon 19.10.38. To 24 Sqdn, Hendon 4.11.38. To P&P 15.11.39. To 18 MU Dumfries 26.3.41. To 27 MU Shawbury 11.4.41. To Stn Flt St Eval 16.14.41. Cat B, to Lundy & Atlantic Coast, Barnstaple 12.11.41. To Stn Flt Halton 29.12.41. To 18 MU Dumfries 11.11.43. SOC at 18 MU 30.9.44.

437 L4395 To 24 Sqdn, Hendon 19.10.38. Dived into ground out of cloud Burbage Wood, nr Hinckley, Leics 21.3.40; Sgt Edwards & Mr Tanqueray killed. SOC 29.3.40.

438 L4396 To 24 Sqdn, Hendon 20.10.38. DBR in air raid Hendon 8.10.40. SOC 14.10.40.

Miles M.16 Mentor Mk.I L4396 was in service with No.24 Squadron, Hendon, from 20th October 1938 until it was damaged beyond repair in an air raid on 8th October 1940.   ['The Aeroplane' via B Clarke]

439 L4397 To 24 Sqdn, Hendon 28.10.38. To Special Duties Flt, St Athan 25.3.40. To 18 MU Dumfries 18.10.40. To 19 Grp Comm Flt, Roborough 29.12.41. Wing hit ground in low cloud nr Roborough 15.3.42 and SOC same date.

440 L4398 To 24 Sqdn, Hendon 3.11.38. Allocated 43 Group following damage 10.7.41; probably an accident. SOC 29.8.41. To GI Airframe **2681M** on 15.9.41 and to No.1373 Sqdn ATC at Barking, Essex.

441 L4399 To 24 Sqdn, Hendon 10.11.38. To 11 Group Pool, St Athan 5.11.39; reformed as 6 OTU Sutton Bridge 6.3.40. To Stn Flt Leuchars 7.9.40. To Lundy & Atlantic Coast, Barnstaple for repairs 15.7.43. To 18 Grp Comm Flt, Leuchars 26.7.43. SOC at 18 MU Dumfries 30.9.44.

442 L4400 To 24 Sqdn, Hendon 18.11.38. To 11 Group Pool, St Athan 5.11.39; reformed as 6 OTU Sutton Bridge 6.3.40. To 24 Sqdn, Hendon 24.6.40. DBR in air raid Hendon 8.10.40. SOC 16.10.40.

443 L4401 To 59 Sqdn, Old Sarum 19.1.39. To Stn Flt Andover 30.11.39. To 18 MU Dumfries 24.10.40. To Stn Flt Andover 3.5.41. SOC Cat. E 25.6.43.

444 L4402 To 24 Sqdn, Hendon 1.12.38 (actually dd 2.12.38). Engine cut after take-off for test flight and hit fence Hendon 24.6.39 and DBR (F/H 100:20).

445 L4403 To 24 Sqdn, Hendon 10.1.39. To 11 Group Pool, St Athan 5.11.39; reformed as 6 OTU Sutton Bridge 6.3.40. To 56 OTU Sutton Bridge 1.11.40. To Lundy & Atlantic Coast, Barnstaple for repairs 22.5.42. To 18 MU Dumfries 18.9.42. To Defford 11.12.42. To 3 (P)AFU South Cerney 10.6.43. SOC Cat.E 2.6.44.

446 L4404 To 24 Sqdn, Hendon 1.12.38. To 11 Group Pool, St Athan 5.11.39; reformed as 6 OTU Sutton Bridge 6.3.40. To 420 Flt, Middle Wallop 6.11.40; re-desig 93 Sqdn 7.12.40. To Stn Flt Abingdon 21.5.41. To 32 MU St Athan 29.6.41. To Abingdon 2.10.41. To 18 MU Dumfries 12.10.41. To Stn Flt Tiree 11.2.42. To Stn Flt Benbecula (no date). To Stn Flt Stornoway (31.12.42 census; but also still Stn Flt Benbecula at 31.12.42 census!). To Stn Flt Tiree 26.11.43. Cat.Ac 6.4.44 and SOC Cat.E 20.5.44. Brought back on charge Reading 15.2.45 - cancelled.

447 L4405 To 22 (AC) Grp Fighter Command for 4 (AC) Sqdn, Odiham 29.12.38. To 614 Sqdn, Odiham (30.11.39 census). To 59 Sqdn, Thorney Island 3.7.40. SOC 22.9.41 (by 59 Sqdn); Cat.E2. (SAS Cat. B P&P 22.4.41?).

448 L4406 To 22 (AC) Grp Fighter Command for 4 (AC) Sqdn, Odiham 13.12.38; actually dd to 13 Sqdn, Odiham (possibly to France with BEF). To 5 OTU Aston Down 19.6.40. To 9 Grp Comm Flt, Samlesbury, 21.12.40. To 71 Wing (Calibration Flt), Dyce/Turnhouse (31.12.41 census). To 26 EFTS Theale 27.10.42. To 18 MU Dumfries 24.10.43. SOC at 18 MU 30.9.44.

449 L4407 To 22 (AC) Grp Fighter Command for 4 (AC) Sqdn, Odiham 14.12.38; actually to 16 Sqdn, Old Sarum. To Stn Flt Old Sarum 13.2.40. To Fighter Interception Unit, Tangmere 1.5.40. To P&P 6.9.40 for repairs; presumed returned to FIU. SOC 27.7.41 by FIU Tangmere.

450 L4408 To 22 (AC) Grp Fighter Command for 4 (AC) Sqdn, Odiham 20.12.38, actually to 16 Sqdn, Old Sarum. To Stn Flt Old Sarum 13.2.40. To Stn Flt Andover 30.5.40. To P&P 10.5.41. To 18 MU Dumfries 12.10.41. To 3 Grp Comm Flight, Newmarket 18.11.41. To 52 Wing, Cheltenham (no record & possibly erroneous). To Newmarket 12.3.42. To Stn Flt Northolt 22.4.42. To 24 Sqdn, Hendon 25.4.42. To 15 MU Wroughton 9.7.42 (for overseas shipment); to 76 MU Wroughton 26.7.42 (but allocation for overseas shipment cancelled). To 18 MU Dumfries 1.8.43. SOC at 18 MU 30.8.44.

451 L4409 To 22 (AC) Grp Fighter Command for 2 (AC) Sqdn, Hawkinge 10.1.39. To Stn Flt Hawkinge (no date). To P&P 4.10.40. To 18 MU Dumfries 4.3.42. To Stn Flt Newmarket 5.6.42. Flew into hill in bad weather Martinsell Hill, Clinch (a stately home nr Newmarket, Suffolk) on a ferry flight from Haldon to 3 Grp HQ, RAF Exning, nr Newmarket 27.12.42.

452 L4410 To 22 (AC) Grp Fighter Command for 26 (AC) Sqdn, Catterick 5.1.39. At P&P 29.12.39. To 18 MU Dumfries 20.6.41. To Stn Flt Northolt 20.7.41. To 9 Grp Comm Flt, Samlesbury 8.10.42. To 18 MU Dumfries 5.3.44. SOC at 18 MU 30.9.44.

453 L4411 To Stn Flt Ternhill 10.1.39. To 8 EFTS Woodley 8.3.40. To Training Command Comm Flt, White Waltham 10.4.40. To Comm Flt, Reading 30.6.41. To 82 Grp Comm Flt, Ballyherbert 25.8.41. To Stn Flt Newtownards 6.1.42. To Atlantic Coast Air Lines (MI) 11.12.42. To 15 MU Wroughton 13.4.43. SOC Cat E due to major repairs 27.6.44.

G-AHKM was the only Miles M.16 Mentor to grace the Civil Register after the war. [G Swanborough Collection]

454 L4412 To Stn Flt/23 Grp Comm Flt Ternhill 10.1.39. To 1 AAS Manby 23.11.39. To Stn Flt Andover 7.3.40. Crashed (Cat.Ac) 24.4.41; repaired. To DGRD 9.9.41, possibly AFEE or another MAP administered unit. To TFU Christchurch (31.12.41 census). Cat.B 19.1.44; repaired by 21.3.44. To 18 MU Dumfries 30.3.44. SOC at 18 MU 30.9.44.

455 L4413 To Stn Flt/25 Grp Comm Flt Eastchurch 16.1.39. To Stn Flt Brize Norton 24.8.39. To Stn Flt Ternhill 29.2.40. To 18 MU Dumfries 16.10.40. To 71 Grp Comm Flt, Hanworth 13.1.41 (recorded as at ACC Comm Flt at 31.12.41 census – possibly same unit). To 101 (G)OTU Kidlington 19.4.42. To 1 PTS Ringway 26.5.42. To unidentified CRO (Cat.B) 7.4.44. SOC Cat.E 19.5.44.

456 L4414 To Stn Flt Eastchurch 10.1.39. Hit tree in bad visibility Tangham Forest, Butley, 8 ml ENE Martlesham Heath 17.1.39; pilot F/O Geoffrey Beavis killed & AVM Henry M Cave-Brown-Cave badly injured. SOC with just 6 hr 10 min flying time!

457 L4415 To Stn Flt Northolt 19.1.39. Flying accident (Cat.B); to P&P for repairs 3.7.41. To Atlantic Coast, Barnstaple for repairs 29.7.41. To Stn Flt Wyton 11.9.41. To 12 MU Kirkbride 29.3.41. To 15 MU Wroughton 19.7.42. SOC 6.11.42.

458 L4416 To Stn Flt Northolt 20.1.39. To 24 Sqdn, Hendon 8.8.39. To 11 Group Pool, St Athan 14.11.39; reformed as 6 OTU Sutton Bridge 6.3.40. To 24 Sqdn Hendon 24.6.40. To Stn Flt Leuchars 15.9.41. To 18 MU Dumfries 17.8.43. SOC at 18 MU 30.8.44.

459 L4417 To Stn Flt Linton-on-Ouse (4 Bomber Group) 18.1.39. To 1 PRU Benson 26.2.41. To Stn Flt Tiree 22.11.41. To Stn Flt Abbotsinch (no date). To 18 MU Dumfries 26.11.42. To Stn Flt Sealand 23.5.43. Undershot landing and hit bank, Sealand 13.7.43, Cat.B but recat.E 9.8.43. SOC 14.8.43.

460 L4418 To Stn Flt Grantham (23 Grp Comm Flt) 16.2.39. To Stn Flt Andover 26.10.39. SOC 6.1.43 'P&P Cat.E major repair'.

461 L4419 To Stn Flt Grantham (23 Grp Comm Flt or 5 Grp Comm Flt) 27.1.39. To 12 FTS Grantham (31.12.39 census). To Stn Flt Andover 3.3.40. SOC Cat.E (MR) 7.3.44.

462 L4420 To Stn Flt Hucknall (2 Grp Comm Flt) 30.1.39. To 24 Sqdn Hendon 30.11.39. To 6 OTU, Sutton Bridge 18.3.40. To Stn Flt Bircham Newton 2.9.40. To 5 MU Kemble 17.7.45. Sold 3.5.46. Regd **G-AHKM** (CofR 10135) 18.4.46 to Laurence Dudley Trappitt, South Kensington, London. Cld and regd 5.8.46 to Peter Wellburn Bayliss, Brierley Hill, Staffs. Believed to have been 'civilianised' by the Miles Aircraft Ltd Repair & Service Department, Woodley and first flown, by Hugh Kendall, on a 'Handling test', at Woodley on 18.3.47. After another 'Handling test', on 20.3.47, Hugh delivered it to White Waltham the same day. CofA No.9338 issued 8.5.47. Competed for the Siddeley Trophy in 1948. Minor damage at Toussus le Noble c26.5.48. Advertised for sale in *The Aeroplane* 18.6.48 as: 'Miles Mentor, 4-seater cabin Gipsy Queen, low hours, dual control, full instruments, 125-130 mph cruising, 12 months CofA'. Cld 26.11.48 and regd 7.12.48 to John Charles Turnill, Old Hill, Staffs (based Wolverhampton). Flown to 8th place at a speed of 150.5 mph by Turnill in the 1949 King's Cup Air Race at Elmdon (Race No.31). Crashed in bad visibility at Clayhidon, Devon 1.4.50 at about 18.00 hrs while en route Wolverhampton to Exeter; pilot and one passenger killed. Regn cld 14.12.50 as destroyed.

463 L4421 To Stn Flt Abingdon (1 Grp Comm Flight) 31.1.39. To Service Flt (part of 21 AD?) 30.11.39. SOC 7.40 as abandoned in France in 6.40.

464 L4422 To Stn Flt Mildenhall (3 Grp Comm Flight) 7.2.39, later at Exning, nr Newmarket. MI 8.11.42; to Atlantic Coast Air Lines, Barnstaple 14.11.42. To Stn Flt Andover 9.1.43. To RAE Farnborough 26.1.43. To 18 MU Dumfries 20.2.44. SOC at 18 MU 30.9.44.

465 L4423 To Comm Flt Lee-on-Solent 9.2.39. To Stn Flt Northolt 8.8.39. To Stn Flt Bircham Newton 22.1.43. To 18 MU Dumfries 25.2.43. To Stn Flt Andover 3.7.43. SOC Cat.E (MR) 6.10.43.

466 L4424 To Stn Flt Farnborough (22 Grp Comm Flt) 11.2.39. To 231 Sqdn Aldergrove 4.7.40. To BSP Belfast (no date). To 18 MU Dumfries 26.3.43. To CRD at Martin Hearn Ltd, Hooton Park 11.9.43. SOC 30.8.44.

467 L4425 To unidentified unit 3.3.39. To 16 Grp Comm Flt, Rochester (formed 3.9.39). Opened up to avoid dip in landing area but hit ground, Ford 3.6.40. SOC 26.7.40. To 8 SofTT Weeton as GI airframe **2145M**, allotted 29.7.40.

468 L4426 To Stn Flt Gosport 25.3.39. To Stn Flt Turnhouse 19.2.42. To 13 Grp Comm Flt, Woolsington 14.10.42 (but still recorded at Stn Flt Turnhouse at 31.12.42 census). To 18 MU Dumfries 9.5.43. SOC at 18 MU 30.9.44.

469 L4427 To Stn Flt Donibristle 4.3.39 (also reported as 17 Grp Comm Flt, Gosport). To Stn Flt Turnhouse 24.5.39. To ACC Comm Flt/Stn Flt Farnborough 17.1.41. To 18 MU Dumfries 26.8.43. SOC 30.8.44.

470 L4428 To 1 EWS Cranwell 7.3.39. (Allotted Usworth 31.8.39 but not delivered). To 1 Sqdn, Tangmere (census 31.12.39). To 43 Sqdn, Acklington (undated & unconfirmed). SOC by RAF Tangmere Cat.E 19.6.42.

471 L4429 To Stn Flt Eastchurch 15.3.39. To Stn Flt Brize Norton 24.8.39. To Stn Flt Ternhill 29.2.40. To 18 MU Dumfries 16.10.40. To 19 Grp Comm Flt, Roborough 15.1.41. To 18 Grp Comm Flt, Leuchars 18.5.41. To Stn Flt Leuchars 29.11.41. To 19 Grp Comm Flt, Roborough 12.4.42. Cat.B (MR) 1.9.43; to de Havilland Aircraft Co Ltd; repairs completed 16.11.43. To 18 MU Dumfries 18.11.43. SOC 30.9.44.

472 L4430 To 19 MU St Athan 17.3.39. To Stn Flt Andover 9.6.39. To Atlantic Coast Air Lines, Barnstaple (MR) 2.9.42. To 5 MU

Kemble 19.10.42 (confirmed there 20.7.43). Flown by Don Brown on a 'local' from Woodley 22.9.44 (circumstances unknown). SOC 28.9.44.

473 L4431 To Lee-on-Solent 23.3.39. To 16 Grp Comm Flt, Rochester 3.6.39. To 18 MU Dumfries 13.10.40. To 12 Grp Comm Flt/Stn Flt Hucknall 9.12.41. To 13 Grp Comm Flt, Woolsington 5.2.43. To Stn Flt Andover 21.10.43. To Cat.B 28.4.44 and SOC as Cat.E 19.5.44.

474 L4432 To 22 Grp Comm Flt, Andover 9.6.39 (possibly via 19 MU but no date entered). Hit hedge on take-off and collided with truck at Andover 16.8.39; (F/H 54:40).

475 L4433 To 19 MU St Athan 9.5.39. To Stn Flt Manston 7.6.39. To P&P (Cat. B) 10.10.40; upgraded to Cat. E 18.10.40.

476 L4434 To 19 MU St Athan 4.4.39. To Stn Flt Andover 12.6.39. To 18 MU Dumfries 24.10.40. To Stn Flt Halton 3.5.41. DBR when engine lost power on take-off from Halton, lost height in downdraught and hit trees 29.8.41.

477 L4435 To Stn Flt Wyton 12.4.39. To 18 MU Dumfries 25.10.40. To Stn Flt Andover 29.3.41. Op by 6 FPP, ATA 29.10.41 when suffered minor (Cat A) taxying accident in strong winds at Ratcliffe; pilot F/O Whitnell. SOC Cat.E1 by Miles Aircraft 15.10.44.

478 L4436 Dd Stn Flt Andover/1 Grp Comm Flt 14.4.39. SOC 16.8.41.

Summary of production		
M.16	c/n 434-478	45
Total built:		45

Job Numbers:	
5002	L4392
5021	L4393 – L4395
5023	L4396 – L4436

Serials L4437 to L4440 have been quoted for this batch but this is believed to be in error.

A further 40 M.16 Mentors were ordered under Contract No 706918/37 with serials L9068-9091; L9126-L9141 but the contract was cancelled before any were built.

Appendix 22:
Miles M.17 Monarch Production

638 M.17 Monarch prototype built in the Experimental Department; painted cream with red trim. Regd **G-AFCR** (CofR 8181) 17.11.37 to Phillips & Powis Aircraft Ltd. First flown by Bill Skinner 21.2.38, in primer finish and with 'B' mark **U1** *(at least the ninth known use of U1)*. CofA No.6295 issued 1.6.38; (delayed probably due to prolonged testing of the new Glide Control Flaps). Retained by firm as demonstrator and entered in King's Cup Air Race at Hatfield 2.7.38 (Race No.5) to be flown by Wing Cdr Freddie Stent but withdrawn due to his death 28.6.38. Flown to the Paris Air Show (held 25.11.38-11.12.38) by Ranald Porteus and Bill Skinner. Offered for sale in *Miles Magazine* for May 1939 for £1,150 with 12 months CofA - *Specially equipped with Pioneer turn & bank indicator, Air Log, navigation and instrument lights and Harley Landing Light in leading edge of wing; Special finish.* It was not sold and was still at Woodley 1.9.39 for Air Ministry survey. Regn cld 23.9.39 as sold. Impressed 30.11.39 as **W6461**; although earmarked for Air Member for Supply and Organisation (AMSO) it was dd 30.11.39 to 13 EFTS White Waltham. To Halton Comm Flt 17.1.41. Skidded on crossing tarmac on landing Linton-on-Ouse 4.6.41 and u/c collapsed; taken to Phillips & Powis, Woodley by road for repairs. To FTC Comm Flt Woodley 21.1.42. To 10 Grp Comm Flt, Colerne 18.8.43. To CRD at Vickers-Armstrong Ltd 10.43. Cat Ac, taken to Miles Aircraft Ltd, Woodley by road for repairs 23.3.44. To 5 MU Kemble 24.6.44. To CRD at Vickers-Armstrong Ltd 4.3.45. To 5 MU Kemble for disposal 1.8.45. Sold to Air Schools Ltd, for £1000, at the first sale of impressed light aircraft held at 5 MU in 12.45. Regd **G-AFCR** 26.1.46 to Norman Roy Harben of Air Schools Ltd, Burnaston. CofA renewed 5.9.46. Regd 10.9.46 to Air Schools Ltd, Burnaston. Cld 20.3.52 & regd 3.4.52 to Herbert Hodson Mould, Wolverhampton. The undercarriage was severely damaged and the tips of the propeller were bent in a cross-wind landing by owner Mould at Toussus-le-Noble, near Paris 24.5.52; repaired. Badly damaged in forced landing in darkness near Este, Padua, Venice on the evening of 2.7.57; pilot Mould and passenger uninjured. Regn cld 10.5.58 as destroyed.

The next five aircraft, c/n 786 to 790, were built in the Final Assembly of the main factory.

786 M.17 Monarch. Regd **G-AFGL** (CofR 8426) 12.4.38 to Airwork Ltd, Heston and used as a demonstrator. CofA No.6336 issued 30.6.38. Probably the aircraft fitted with an R Sensand de Lavaud automatically variable pitch propeller (as it was Airwork who carried out the trials at Heston). Regn cld 7.3.39 as sold. Regd in France as **F-ARPE** (CofR 6302) 25.3.39 to Georges Lilloz, Paris. Fate unknown.

787 M.17 Monarch. CofA No.6378 issued 18.8.38 to Camille Gutt (Belgian Minister of Finance from 11.34 to 3.35 and from 2.39). Originally allotted **OO-SCM** but actually regd **OO-UMK** (CofR 449) 1.9.38 *(2nd use - see M.11A c/n 306)* to Camille Gutt, Jette (based at Brussels). Probably flown to England by owner (who remained Minister of Finance in the Government in exile in London) just prior to 10.5.40 (when the Belgian Government evacuated to France by road). Loaned to Phillips & Powis Aircraft Ltd and operated with 'B' mark **U-0226** (first recorded use by Hugh Kennedy 16.11.40). CofA renewed 6.5.41. Fitted with the first single-control automatic pilot; Hugh Kennedy records a flight in this machine on 27.9.41 noting, 'Test - autopilot' in his log book. CofA renewed 22.6.42. Regd **G-AGFW** (CofR 9404) 29.12.42 to Phillips & Powis Aircraft Ltd. New CofA No.6956 issued 28.1.43. Regd to Miles Aircraft Ltd, subsequent to name change 5.10.43. MAP Permit to Fly No.15 issued 15.12.43. Regn cld 7.10.44 as sold to Belgium. Formally transferred to military marks as **TP819** on 30.5.44 on its return to use by Camille Gutt; Gutt had a minor accident at Exeter 6.6.44, aircraft ferried back to Woodley by Don Brown. Continued flying as G-AGFW on attachment to the Allied Flight of the Metropolitan Comm Sqdn, Hendon (and serviced by Belgian AF mechanics); first recorded as TP819 in MCS ORB 25.2.45. SOC 1.9.46 on return to the Belgian Government, although retained by Camille Gutt. Restored as **OO-UMK** (CofR 639) 7.2.47 to Joannes JMJ 'John'

U-0226, the Miles M.17 Monarch once OO-UMK, at Woodley while in service with Phillips & Powis as a communications aircraft.
[Phillips & Powis via The Miles Aircraft Collection]

Mahieu, Haren Airport, Brussels. Severely damaged landing Het Zoute 20.3.47 (port wing torn away). Quoted in Lloyds as a total loss but it was repaired. Operated by John Mahieu Aviation scrl following its formation on 1.3.48. Regd 12.11.52 to Leopold Vlieghe, Lier (based Antwerp Deurne). Regd 19.1.57 to Mrs S Van den Broecke (jointly with husband RE Declercq), Oostende. Reported 7.57 as at Grimbergen 'now serviceable again after ten years'. Regd 15.9.59 to Publiciel sprl, Grimbergen. Regd 17.11.60 to Jan Calmeyn, Ghent. Crashed and dbf nr St. Denis-Westrem, Ghent 21.11.60, killing the owner and passenger. Regn cld 7.12.60.

788 M.17 Monarch CofA No.6372 issued 17.8.38 to Dr Hans Christian Hagedorn. Regd **OY-DIO** (CofA 92) 31.8.38 to Dr Hans Christian Hagedorn, Gentofte (based at Kastrup). CofA deposited 9.9.39, due to wartime restrictions. A German report dated 19.12.43 stated that OY-DIO was to be dd to the Luftwaffe for local use; also reported that it was seized by Wehrmacht sometime after 19.10.43 and stored at Vælose in 12.43. Fate unknown.

789 M.17 Monarch. Regd **G-AFJU** (CofR 8706) 25.8.38 to Sir Victor Alexander George Anthony Warrender (MP for Grantham and Financial Secretary to the War Office), Heston. CofA No.6373 issued 2.9.38. Regn cld 1.4.40 as sold. Impressed into RAF 31.3.40 as **X9306**; to 20 MU Aston Down 1.4.40. To HQ, SFPP Kemble 24.2.41. Reported on charge Ringway in 8.41. At FTU Honeybourne (census 31.12.41). To Comm Sqdn, Lyneham .42 and involved in accident there 23.11.42. SOC Cat.E 26.11.42. To 5 MU Kemble and repaired. To Vickers-Armstrong Ltd, Hawarden in 1943. To 5 MU Kemble 11.45. Sold 8.2.46 to Lt Cdr H Kidston, Tolpuddle, Dorset for £700, at the sale of impressed light aircraft at 5 MU. Regd **G-AFJU** 2.3.46 to Lt Cdr Home Ronald Archibald Kidston (based at Woodley). CofA renewed 18.4.46. Regd 2.9.46 to Miss Betty John, Llanelli (based at Hanworth). Regd 17.6.48 to Arthur Robert Pilgrim, Totteridge (based Elstree). Cld 4.9.51 & regd 6.9.51 to Furzehill Laboratories Ltd, Borehamwood (based Elstree). Cld 5.2.53 & regd 7.3.53 to Frederick Roger Milsom, Staverton. Cld 7.7.54 & regd 13.7.54 to S Smith & Sons (England) Ltd, Cricklewood (based Staverton). Cld 16.3.55 & regd 22.3.55 to The Cotswold Aero Club Ltd, Staverton. Cld 6.8.57 & regd 9.8.57 to Donald Jackson, operated by Cambridge Private Flying Group, Cambridge. Damaged in heavy landing at Cambridge 19.8.62; probably not repaired. CofA lapsed 18.5.64 (application for renewal made 18.6.64 but cld 21.3.68). Cld 20.7.64 & regd 30.7.64 to Kenneth Arthur Hudson, Staverton. Cld 1.1.66 & regd 10.2.66 to Arthur Horton Luscombe, Swindon (still stored Staverton 9.66). Fuselage moved to Netheravon (by 8.67) and to Lasham (by 6.70). Reported 5.71 as being restored by Bill Townsend at Lasham but with many components missing. Cld 19.6.71 as sold. Regn cld 18.11.74 as pwfu. Sold in 1976 to Sir William Roberts, Strathallan for use as spares for his Monarch G-AFLW. The 'sadly dilapidated remains' of G-AFJU, less engine and propeller, were offered as Lot 63 in the Strathallan auction by Christies on 14.7.81. Sold to R Menage, Providence Hill, Husthwaite, York for just £50. Changing circumstances forced the new owner to abandon his plans and he offered it to The Aircraft Preservation Society of Scotland, who completed its purchase, with financial assistance from the Government's Local Museums Purchase Fund, for £200. Moved to East Fortune Aerodrome 14.11.83; it was fully restored to static display condition over a period of years by members of the APSS (with technical assistance from The Miles Aircraft Collection) for display in The Museum of Flight at East Fortune. Following a change in policy in 2006 with their recently acquired Concorde, the museum made the Monarch available for disposal and Peter Bishop heard of its plight. Following inspection by Rex Coates and the local CAA representative, it was found to be in a reasonable enough state for rebuild to flying condition and Bishop therefore purchased it. Regd 18.7.06 to Peter William Bishop, Woodley, Berks. The fuselage and centre section were loaded on to a flat trailer and attached to a truck containing the wings and smaller parts and left East Fortune on 10.9.06 en route to Oaksey Park Airfield, near Malmesbury, Wiltshire, where John Eagles of Air Stratus Ltd had been contracted to undertake the work necessary to restore it to airworthy condition. Work was started and a new engine had been found but in late 2007, John gave up restoration work to concentrate on his new responsibility for the Miles Type Approval from the CAA. The Monarch will now be restored to airworthy condition by Aero Classique in France, to where it was moved by road on 14.1.08.

790 M.17 Monarch. Regd **G-AFJZ** (CofR 8736) 5.9.38 to Edward Ormonde Liebert, Blackpool (based Stanley Park). CofA No.6391 issued 15.9.38. Cld 13.6.39 & regd 16.6.39 to Gerard Evan Wallace, RAF Manston. Flown by Hugh Kennedy to High Post and return on 23.8.39. At Woodley for Air Ministry survey 1.9.39. Regn cld 9.39 as sold. Impressed into RAF as **W6462** 1.11.39 and to 91 Grp Comm Flt, Abingdon 1.11.39. Unit renamed 21 Grp Comm Flt 1.5.47. To 60 MU Salvage Centre, York 16.12.48 and SOC as Cat.E2 scrap.

791 M.17 Monarch (construction probably commenced in Final Assembly in the main factory but was almost certainly completed in the Repair & Service Department). CofA No.6433 issued 9.11.38 to Aero Services (Pty) Ltd, South Africa. Regd **ZS-AOY** 31.12.38 to Aero Services (Pty) Ltd. Sold to Mrs Ethel Louisa Phillips, Mossel Bay. Stalled and spun in Brooklyn, Cape Town 19.2.40 when canopy came open after take-off; Mrs Phillips killed.

792 M.17 Monarch (construction commenced in Final Assembly in the main factory, but a photograph taken in 1938 shows this machine, less outer wings and cabin, in primer finish, with '792' in chalk on the fin, being pushed along a track near the Reserve Training School on its way to the Repair & Service Department for completion). Regd **G-AFLW** (CofR 8861) 2.11.38 to Phillips & Powis Aircraft Ltd. CofA No.6435 issued 18.11.38. Probably the Monarch statically displayed on the Phillips & Powis stand at the Paris Aeronautical Exhibition in 1939. Cld 9.3.39 and regd 20.3.39 to Rolls-Royce Ltd, Hucknall; dd 27.2.39. Retained and operated by Rolls-Royce during war. On 27.6.40, it was flown by AVM Mallory from Hucknall to Debden and Coltishall. Issued with MAP Permit No.25 5.9.44. Postwar CofA renewed 27.3.47. CofA lapsed 20.9.57. Cld 25.3.58 & regd 27.3.58 to Derby Aviation Ltd, Burnaston; CofA renewed 23.7.59. Regd 26.8.59 to Blackpool & Fylde Aero Club Ltd, Squires Gate. Undercarriage, propeller and port wing damaged when ground-looped on landing Squires Gate 15.4.60; no injuries and repaired. Cld 3.1.61 & regd 6.1.61 to Dennis Bircher, Wellington, Salop. Cld 23.9.62 & regd 28.9.62 to Alan Frederick Jarman, Oadby, Leics. Cld 22.2.63 & regd 5.3.63 to Maurice Fortescue Kirk, Leicester. CofA lapsed 20.1.67. Regd 19.6.68 to Rex Edward Coates, Blackbushe; restored and CofA renewed 27.6.69. Cld 29.9.71 & regd 2.11.71 to John Edwin Randall, Crowthorne, Berks (based at Blackbushe); John Randall tragically died at the controls while awaiting take-off clearance from a private strip in Sussex. Regd 15.7.76 to Sir William James Denby Roberts, t/a The Strathallan Aircraft Collection, Strathallan. CofA permit lapsed 9.7.79. Sold 14.7.81 & regd 8.10.81 to Neil Campbell Jensen, Oxted (based Biggin Hill); inspected on site by Neil Jensen and Cobby Moore and flown to Biggin Hill on a ferry permit for reconditioning. Cld 15.7.82 as sold. CofA renewed 3.11.82 but regn cld 4.11.82 by Secretary of State. Regd 22.3.83 to Dr Norman Ian Dalziel, Heston, Middlesex (based Biggin Hill). Flown to 3rd place by Dalziel in the 1983 King's Cup and to 5th place in the 1984 King's Cup; based at Roundwood Farm (by 6.88); based White Waltham from 1992. Total hours 2,819 at 31.12.94. CofA lapsed 30.7.98; but placed into storage at White Waltham prior to this. Regn cld 3.5.01 by CAA. Remains in storage at White Waltham.

A post-war photograph of the Miles M.17 Monarch G-AFJU at an unknown location. [Via P Amos]

Miles M.17 Monarch G-AFLW at the Air-Britain Fly-In at Wroughton on 28th June 1987. [P Amos]

A fine post-war photograph of Miles M.17 Monarch G-AIDE at Shoreham. [G Swanborough Collection]

793 M.17 Monarch (construction commenced in Final Assembly in the main factory but a photograph taken in 1938 shows this machine, less outer wings and cabin, in primer finish, with '793' (indistinct) in chalk on the fin, being pushed along a track near the Reserve Training School on its way to the Repair & Service Department for completion). Regd **G-AFRZ** (CofR 9061) 24.3.39 to Lord Malcolm Avendale Douglas-Hamilton, Shaftesbury, Dorset (based at Heston). CofA No.6525 issued 22.4.39. Regn cld 1.3.40 as sold. Impressed into RAF as **W6463** 30.11.39 and to 13 EFTS White Waltham. To Halton Comm Flt 17.1.41. To FTC Comm Flt, Woodley 20.2.41. To 10 Grp Comm Flt, Colerne 16.6.43. To 5 MU Kemble 16.7.43. To CRD at Vickers-Armstrong Ltd, Winchester 19.3.44. To 5 MU Kemble 18.3.46. Sold to BG Heron in 1946. Regd **G-AIDE** (CofR 10575) 23.8.46 to Benjamin Gerald Heron, Christchurch. CofA No.8439 issued 4.1.47. Crash-landed at Christchurch Aerodrome 4.10.51; damaging u/c and breaking its propeller; repaired. Cld 30.3.56 and regd 16.4.56 to Walter Purkis Bowles, Pinner (based at Elstree). Cld 21.5.59 and regd 26.5.59 to Charles Mortimer Tollemache Smith-Ryland, Warwick (based Baginton). Cld 17.4.62 and regd 12.6.62 to John Fricker, t/a Aviation Advisory Services, Stapleford. Cld 18.8.64 and regd 25.9.64 to James Hay Stevens, Charing, Kent (based on his strip). Cld 22.7.68 and regd 24.7.68 to Christopher David Cyster, Northiam, Rye (remained based at the Charing strip; later to Church Fenton, moving to Cambridge on 12.2.69; later with owner to RAF Valley). CofA lapsed 29.6.70. A contemporary report stated that (by 7.71) G-AIDE was owned by a Lightning pilot at Coltishall, but he abandoned it following failure of the glue joints. Regd **G-AFRZ** 14.10.82 to Robert Evan Mitchell, Sutton Coldfield and stored, dismantled and in very poor condition, at RAF Cosford for many years. Moved (with Mitchell's collection) to Sleap early 2008.

The final two Monarchs were built in the Repair & Service Department at Woodley:

794 M.17 Monarch. CofA No.6526 issued 18.4.39 to NV Nationale Luchtvaart School, Netherlands. Regd **PH-ATP** (CofR 342) 27.4.39 to NV NLS; dd 3.6.39. Destroyed on invasion of Holland 10.5.40. Regn cld 4.6.40.

795 M.17 Monarch. Regd **G-AFTX** (CofR 9109) 11.5.39 to William Henry Whitbread, London EC1 (based Heston). Offered for sale (but marked as sold) in the *Miles Magazine* for May 1939: *Only 6 hours demonstration flying; virtually a new aeroplane. Special finish. Colour - Ivory fuselage and wings with red lines. Standard equipment.* CofA No.6577 issued 31.5.39; dd 1.6.39. Cld 27.7.39 & regd 12.9.39 to Airwork Ltd, Heston. Regn cld 30.11.39 as sold. Regd in France as **F-ARRL** to M. Berthie.

799-810 Recorded by CLN as reserved for the next batch of twelve M.17 Monarchs but due to the outbreak of war in 9.39 these were not built.

Production summary		
Miles M.17 Monarch	c/n 638	**1** prototype
	c/n 786-795	**10** production
Total		**11**

Appendix 23:
Miles M.18 Production

1075 M.18 Mk.I prototype; fitted with 130hp DH Gipsy Major I engine. First flown as **U2** by Miles 4.12.38. To Martlesham Heath for trials 3.39. Regd **G-AFRO** (CofR 9051) 11.3.39 to Phillips & Powis Aircraft Ltd (but marks never used). No CofA issued. Fitted with wrap-round single-piece windscreens. A return flight was made by Walter Kaepelli (Capley) to RAE Farnborough 23.5.39. Later camouflaged with yellow undersides. Twice test flown by Hugh Kennedy as U2 on 28.2.40 with 'P & P airscrew'. Became **U-0222** after 17.10.40. Converted to a glider at some time but for what reason and for how long is not known. Modified to tricycle u/c in 1941 (and given the RAF Experimental Aircraft No.173 for identification purposes); flown by Hugh Kennedy on an 'u/c test' 11.8.41. Flown by Don Brown on 9.10.41 who recorded 'Tri u/c'. Many other modifications were carried out on this aircraft as detailed in the text. Don Brown last flew U-0222, noting it as the 'M.18T' on 13.9.42. Hugh Kennedy first flew the 'M.18 Tricycle' on 17.9.42 (with no 'B' mark noted) and on his next flight in the 'M.18 Tricycle' on 29.10.42, he recorded it as **U-0239** for the first time. Fitted with a shortened tricycle u/c and much reduced wingspan at about this time and this may have been why the 'B' mark was changed; a Permit to Fly for U-0239 was renewed (or issued) 10.42. By 16.11.42 Hugh was recording 'T.18/Tricycle' when he flew it on 't/o tests' and made his last flight in U-0239 on 8.12.42. The Permit to Fly for **U-0222** was renewed 1.43 and it was flown to RAE Farnborough 30.1.43. Hugh Kennedy commenced a series of test flights in 'T.18' U-0222 from 25.2.43 until 2.3.43. Given the RAF Experimental Aircraft No.204 for identification purposes at some stage (but it is not known for what purpose). *(It should be noted that sales demonstration pilot Sqdn Ldr James Nelson recorded flying M.18 U-0239 on a considerable number of occasions between 14.4.45 and 26.9.45 but it is suspected that this was an error for the M.18 Mk.III U-0238).* Regn G-AFRO cld 1.12.46 at the census. U-0222 was scrapped Woodley in 12.47

4426 M.18 Mk.II; fitted with 150hp Blackburn Cirrus Major III engine. First flown by unrecorded pilot in 11.39 as **U8** and then flown by Don Brown 18.11.39. Although the windscreens were originally made in three pieces, the rear one was later changed to a single-piece wrap-round screen; a headrest was fitted to the rear cockpit, as was a blind flying hood; the u/c was fitted with spats with detachable sides and these improvements increased the max speed from 130 to 135 mph. Flown by Hugh Kennedy on 'Spinning trials' 19.5.40. The last recorded use of U8 was on 29.10.40; shortly thereafter it became **U-0224** and was dd to 759 (Fleet Requirements) Sqdn, Yeovilton 1.41 on 'short loan, possibly for Service Trials'. To CFS 2.41; flown by Flt Lt HdeCA Woodhouse 5.2.41. Fitted with a 145hp DH Gipsy Major III (and/or 130hp DH Gipsy Major) engine before being dd to A&AEE Boscombe Down 5.41. Returned to Woodley where the Blackburn Cirrus Major III engine was refitted before being dd 13.10.41 to 'A' Performance Sqdn, A&AEE Boscombe Down. Flown by Hugh Kennedy on 'Flame damper tests at night' from Upper Culham (Henley) 23.3.42. A blind flying hood was then fitted to the rear cockpit and it was allotted on 10.11.42 to ECFS Hullavington for service assessment. Allocated serial **HM545** on 11.12.42. Probably returned to Miles Aircraft Ltd, Woodley by 8.5.43 and reverted to **U-0224**. Probably sold late .45

to Miles Aircraft Ltd. However, the 41 Group Return detailing RAF aircraft held in storage at 6.12.45 records one M.18 in long-term storage and is presumed to refer to this aircraft. Hugh Kendall flew an M.18, which he recorded as 'U-0227' on an air-to-ground photo sortie on 1.11.45 and on an air-to-air photo sortie on 7.11.45; however, the cine film of the air-to-air photo sortie still survives and clearly shows the aircraft painted yellow overall, in full RAF markings and as U-0224. Regd **G-AHKY** (CofR 10146) 26.4.46 to Miles Aircraft Ltd. CofA No.8379 issued 29.8.46. Acquired by Ron Paine from the Receiver and Manager in early 1948 (but not regd to him). Cld 17.2.48 & regd 9.3.48 to Flt Lt Henry Brian Iles, Burnaston (Iles was an old pupil of Ranald Porteous and who became air-racing champion with G-AHKY). Raced by Brian Iles (Race No.35), to victory in the 1961 King's Cup Air Race at Baginton 15.7.61 at a speed of 142 mph. CofA lapsed 28.5.64 and kept in gliding hangar at 71 MU Bicester from about 5.65 until moved to Redhill by 5.68 (probably after Iles moved to Walton-on-Thames and later Claygate, Surrey). Cld 25.10.69 & regd 14.11.69 to Rex Edward Coates, Camberley; stored/overhauled by Rex at the British Airways Engineering Base at Heathrow. Cld 23.7.74 & regd 2.8.74 to Etienne Marie Rene Deniel & Patricia Yveline Annick Etiennette Coates, Isleworth. Ferried mid 2.75 to White Waltham and CofA finally renewed 2.5.75. Regd 17.12.76 to Sir William James Denby Roberts t/a The Strathallan Aircraft Collection, Strathallan. CofA lapsed 10.6.77. Sold at auction by Christies at Strathallan on 14.7.81 to The Scottish Aircraft Collection Trust Ltd, Haddington, East Lothian (and moved to Perth airfield). Cld as sold 8.10.81 & regn cld 9.11.81 as wfu. Restored to airworthy condition over the next 2 years, which required 'much painstaking work' and a £3,000 engine refit. Regd 23.3.83 to The Scottish Aircraft Collection Trust Ltd, Haddington; its first post-restoration flight was made on 26.5.83 by Wing Cdr Colin Bider (to whom the Trust's address was amended 30.4.84). Permit to Fly issued 31.5.83. Flown regularly until Permit to Fly lapsed 20.9.89, when it was grounded and stored at Perth

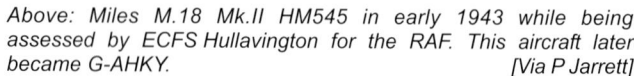

Above: Miles M.18 Mk.II HM545 in early 1943 while being assessed by ECFS Hullavington for the RAF. This aircraft later became G-AHKY. [Via P Jarrett]

Miles M.18 Mk.III U-3, which was registered G-AHOA to Miles Aircraft Ltd on 13th May 1946.
[Miles Aircraft via The Miles Aircraft Collection]

The Miles M.18 Mk.III G-AHOA, complete with spats, before its unfortunate demise on 25th May 1950. [A Madden via D Sykes]

This Miles M.18 Mk.II was registered G-AHKY to Miles Aircraft Ltd on 26th April 1946. [G Swanborough Collection]

Miles M.18 Mk.IV H.L. U11 at Woodley in 1946.
[G Swanborough Collection]

(TT at 31.12.88 1,357 hrs). The Scottish Aircraft Collection Trust Ltd collapsed in 1991 and G-AHKY was then, apparently 'given' to The Museum of Flight at East Fortune (reportedly by a lady who told the museum that she wished it to stay for all time grounded in the museum). Regn cld 19.3.92 as pwfu.

4432 M.18 Mk.III; fitted with 150hp Blackburn Cirrus Major III engine and with a sliding canopy over both cockpits. First flown, by Don Brown (accompanied by George Miles) as **U-0238** on 17.10.42. *(Don noted the engine as a 'ditto' from the previous aircraft flown, Monarch U-0226 with a 'Gipsy Major 130hp' but this was probably in error; the next flight he made in U-0238, on 15.12.42, he recorded 'Cirrus Major 150hp').* Given the RAF Experimental Aircraft No.194 for identification purposes (and noted as being fitted with a 'Gipsy Major III' engine but believed always to have had a Blackburn Cirrus Major III engine). Used as a general hack by the firm during the war. (Its last 'recorded' flight as U-0238 was made by Hugh Kendall on 12.4.46 but he may have recorded the marks in error as it had become U-3 earlier in the year). First flown as **U-3** by sales and demonstration pilot Sqdn Ldr James C Nelson on 9.2.46. Hugh Kendall carried out tests in 'U-3 M.18 (V.P.)' with a manually-operated, variable pitch airscrew on 22/23.7.46. Regd **G-AHOA** (CofR 10223) 13.5.46 to Miles Aircraft Ltd (as an M.18 Series II) but still being flown as U-3 until at least 9.1.47. Ron Paine (of Wolverhampton Aviation Ltd) acquired G-AHOA in early 1948 along with six unfinished sets of Gemini components, as part of the deal, from the Receiver and Manager of Miles Aircraft Ltd. Cld 17.2.48 & regd 9.3.48 to Miss Elaine Menzies Denman, Derby (based Burnaston) (Miss Denman shortly after became Mrs Elaine Porteous on marrying Ranald Porteous). Ranald Porteous recalled that: *Its arrival at Burnaston caused quote a stir, for the wrong reasons. I had ferried it up from Woodley and, full of enthusiasm, took Elaine up for her initial flight in it. Halfway round our first circuit the engine (a Cirrus Major III of 155 hp) slowly died, leaving us with a windmilling propeller and no 'feel' to the throttle, whose linkage had in fact become disconnected. Luckily, although downwind, we were within gliding distance of the airfield, on which I landed crosswind with no difficulty, trying hard to convey that such trifling inconveniences were part of the routine lives of competent, masterful flying instructors.* ARB handling and climb tests at Woodley by Hugh Kendall 12.3.48. CofA No.10062 issued 3.4.48. Cld 10.5.50 & regd 24.5.50 to Thomas William 'Tom' Hayhow, Bagshot. Crashed into a cloud-covered Pennine hillside at Littondale, Yorkshire 25.5.50 with Tom escaping with little more than a broken ankle. Regn cld 31.5.50 as wfu.

unknown M.18 Mk.IV High Lift; fitted with a 150hp Blackburn Cirrus Major III engine. First flown, by Tommy Rose as **U-0236** in 10.42. Allocated serial **JN703** and flown by Hugh Kennedy, between 3/13.1.43 and dd to RAE Farnborough later in 1.43. Hugh Kennedy flew JN703 again on 20.7.43. At the completion of the tests, it was returned to Woodley and given 'B' mark **U11** on or after 29.1.46. Only one flight with these marks is recorded, this being by Hugh Kendall, a 40-min 'local' on 17.12.46. Broken up at Woodley in 1948.

Miles M.18 Trainer (replacement for the Miles M.14A Magister)
Contracts/Acft/1272, undated, for 200 single-engined Trainer (M.18) aircraft. Although serials were allotted, the Contracts Ledger entry was all crossed out (in red ink!), with the note; 'The Contract was not placed'. The serials allocated during 7.41 were: DW515-DW562; DW578-DW614; DW635-DW684; DW712-DW756; DW777-DW796. None were built.

Appendix 24:
Work Carried Out by the Drawing Office and the Experimental Department at Woodley in 1937

Two documents, probably written by Miles for the P&P Board, gave details of the large amount of work carried out by the Drawing Office and Experimental Department at Woodley during 1937.

The first document was entitled: DRAWING OFFICE - 1937.

The drawing office now consists of 71 draughtsmen. This must be considered low for the amount of work and number of types undertaken. The design of the Kestrel was an outstanding performance when it is considered that the number of men employed on this job did not exceed 8 and the machine was flown within a year. For instance the Westland Lysander was started in January 1936 and took 15 men 1¼ years to design. The Spitfire took 24 senior men 2¼ years and it was over 2½ years from the beginning of design before it was flown. The approximate difference in cost for the design would be £9,400 for the Spitfire, £6,800 for the Lysander and £2,500 for the Kestrel. These figures are up to the date of flight and do not include the preparation of the machine for production.

The 196 hours shown on the Magister represents the cleaning up we had to do in 1937 after the aircraft was supposed to be completed, a good deal was due to lack of co-ordination from the works. The Twin is now complete in the Drawing Office and is, as you know, a most elaborate job.

The re-organisation of the Drawing Office is proceeding but I do not wish to hurry it too much as men in this capacity are both rare and touchy. Nevertheless, I have started off the new system without undue resistance and hope it will be running smoothly in a month or so.

Different work undertaken through the year is listed below :-

Aircraft	Man Weeks	
Miles Whitney Straight	48	
M.11C	29	
Kestrel	325	
Production schemes for Kestrel	58	
Magister (military)	196	
Magister (civil)	8	
Magister (inverted flying mods) (for CFS aerobatic team - Ed)	12	
Magister (Cirrus Hermes) (probably error for Cirrus Major? – Ed)	4	
Queen Wasp	120	
Mentor	630	
M.17	116	
Research Twin	370	Senior men
	240	Junior men
Miscellaneous work	441	

In addition to the above, my own Drawing and Design section consisting of one senior lad and three to five juniors have prepared about eight specifications, carried out much of the research work listed under that heading, completed the preliminary design for the T.1/37 (after winning the tender to the Air Ministry) and have very much advanced the design work on the "X".

The second document was entitled: EXPERIMENTAL DEPARTMENT - 1937.

Most of the work listed under the Research and Drawing Office was carried out in the Experimental Department. Types built or complete in 1937 include the Kestrel Trainer, the Mentor I, Queen Wasp (part), Research Twin, M.17.

Research or modification work was carried out on the following :-

a) Hawcon (new wings)
b) Falcon (written down on last balance sheet)
c) Hobby (experimental work on wings)
d) Magister (work on improvements to meet complaints, such as draughty cockpits)
e) Mohawk (various modifications)
f) T.1/37 (mock-up)

g) M.11c (modification and rectification)
h) Menasco Hawk (experimental work on deck landing gear)

In this department the metal working section is particularly good. Much work has been done on castings and a pattern making section developed. Various models and test rigs were built. All test work (including all techniques) were developed in this department.

(The 'experimental work on deck landing gear' for the Menasco Hawk is of interest but, unfortunately, nothing further is known of this project – PA).

The above accounts give a very good indication of the considerable amount of work which was being undertaken at Woodley - just four years since the first M.2 Hawk had flown.

Appendix 25:
Under 'B' Conditions – 'B' Marks Used Between 1934 and 1940

Prior to the introduction of the 'B' conditions marks, aircraft manufacturers were able to test fly their aircraft without markings of any description and Phillips & Powis were no exception to this practice. Witness, the first flights of the Miles M.2 Hawk, which were all made without any form of identifying marks at all!

While this book takes the story of Miles Aircraft up to the outbreak of the Second World War, for completeness, details are included of all the aircraft known to have used the 'first series' of 'B' marks prior to the change to 'U-0xxx' on 17 October 1940. The Phillips & Powis (Reading) Ltd Approval to carry a special identifying letter when test flying their aircraft 'under 'B' conditions', was contained in a letter to the firm dated 5 February 1934 (and is reproduced below):

5th February, 1934
281527/33/DDCA

Gentlemen,

With reference to your letter of the 17th January (FGM/ES), I am commanded by the Air Council to state that Phillips & Powis Aircraft (Reading) Ltd have been specially approved under conditions (2) of the B conditions contained in paragraph 60 of the Air Navigation Directions 1932 (AND.11)

2. For the purpose of identifying your aircraft when being flown under these Conditions and not bearing the prescribed registration marks, a special identification letter followed by a digit must be employed.

*3. The letter **U** has been allotted to you in this connexion. The special markings '-1, '2 etc. must be painted on both sides of the fuselage and conform with the requirements of the schedule I to the Air Navigation (Consolidation) Order 1923 in regard to size and colour in relation to the background on which they are painted.*

4. I am to invite your attention in particular to the condition that no part of such flights shall take place over any populous area.

5. It will be observed that as the flight of aircraft for the purpose of demonstration or delivery does not satisfy the B conditions, the approval conveyed to you in the first paragraph of this letter does not extend to such flights.

I am, Gentlemen, Your obedient Servant,
(sgd.) CR BRIGSTOCKE

The idea of using identifying markings for aircraft flying under 'B Conditions' arose from discussions between the Society of British Aircraft Constructors (SBAC) and the Civil Aviation Department of the Air Ministry. The object being to provide a means of 'identifying' (not 'registering') prototype aircraft prior to them receiving a Certificate of Airworthiness, or other aircraft undergoing tests beyond their existing Type CofA Approval. The system was agreed in 1929, to come into force on 1st January 1930. The identities comprised a letter indicating the design firm, followed by an identifying number. The list issued in December 1929 allotted identification letters to the firms who were SBAC members at that time. The letter 'U' for Phillips & Powis was added to the list on 5th February 1934 and indicated that the Civil Aviation Department considered that the company was competent to conduct its own design developments and trials without direct Air Ministry supervision.

The later additions to the list included firms who were not necessarily members of the SBAC at the time of their approval and this included Phillips & Powis Aircraft (Reading) Ltd. The 'B Conditions' referred to in the Air Navigation Order set out the rules under which the approved design firms worked and, amongst other things, set limits as to who could fly these aircraft and where they could be flown.

The first numerical sequence remained in use until 17th October 1940. However, the same 'B' numbers could be (and often were) used more than once – for example, there are 14 recorded different uses of the mark U1 by Phillips & Powis!

The summary which follows records all known First Series 'B' marks used by Phillips & Powis, in numerical order. Regrettably, the actual identity of many Miles aircraft with 'B' marks remains unknown.

U1 M.2D Hawk Three-Seater (c/n 20) First use. First flown, by Miles 4.2.34, one day before the 'B' marks came into official use. Also flown by Bob Milne 25.2.34 and 1.3.34 and by Miles 18.3.34.

U1 M.2 Hawk; second use. First flown, by Bob Milne 25.4.34 and probably either M.2D (c/n 28) or M.2 (c/n 29). Probably the same U1 flown again by Bob Milne 1.5.34 and 3.5.34.

U1 M.2 Hawk; possibly the third use. Flown by Miles 27.5.34.

U1 M.2F Hawk Major (c/n 110); possibly the fourth use. Flown by Bob Milne 9.8.34.

U1 M.2 Hawk; possibly the fifth use. Flown by Miles 21.8.34 (instruction), 22.8.34 and 23.8.34.

U1 M.2 Hawk; possibly the sixth use. Flown by Miles on 30.8.34 (five times) and on 31.8.34 with 'flaps' noted against all flights.

U1 M.2 Hawk; possibly the seventh use. Flown by Miles three times on 19.9.34.

U1 M.2H Hawk Major (c/n 125); possibly the eighth use. Flown by Bob Milne on CofA test 1.11.34.

U1 M.2H Hawk Major (c/n 143); possibly the ninth use. Flown by Bob Milne on CofA test 19.1.35.

U1 M.7 Nighthawk prototype (c/n 263); possibly the tenth use. Flown by Miles 26.10.35 and with Blossom as passenger 4.12.35.

U1 M.2Z? Hawk Trainer II; possibly the eleventh use. A 'U1' was flown by JF Moir on four 'u/c test' flights on 7.1.37 and 15.1.37 and was possibly c/n 480, later sold to Russia.

U1 M.11A Whitney Straight; possibly the twelfth use. Photographed as such and possibly c/n 509 (although the reason for issuing a 'B' mark to a production M.11A is unknown), regd G-AFGK 6.4.38.

U1 M.17 Monarch (c/n 638); possibly the thirteenth use. There is no confirmation that this mark was used on the prototype M.17 Monarch; first flown by Bill Skinner 21.2.38 (it was regd G-AFCR 17.11.37).

APPENDIX 25

U1 M.15 Trainer (c/n 641); possibly the fourteenth use. Flown by Bill Skinner in 1938 and by JF Moir on a 10 min 'test' 22.9.38.

'U11' M.2 Hawk A Hawk, recorded as 'U11', was flown by Miles on 14.3.34 and 18.3.34. As 'U1' was still flying on this date, the significance of 'U11' is not known, unless it was his way of entering U2 (ie UII) in his flying log book. While the first entry on 14.3.34 could have been an error in entering the mark in his log book Miles then made eleven flights in 'U11' between 11.4.34 and 13.4.34.

U2 M.2P Hawk Major de luxe (c/n 251); first known use. Flown by Bob Milne on CofA test 4.10.35.

U2 M.2Y Hawk Trainer II (c/n 253) second known use. Flown by Bob Milne on CofA test 1.2.36.

U2 M.13 Hobby (c/n IY); third known use. Flown first without marks, by Miles 4.9.37.

U2 M.14 Magister (c/n 331?); fourth known use. The prototype 'M.14 Magister' was flown in what is believed to be an overall red primer finish with RAF roundels and U2, by Miles at the Christening Ceremony of the new aircraft held at Woodley on 20.3.37. It has not been possible to confirm if this machine was actually the first production M.14 Magister (c/n 333) or (as suspected) the first M.14 Hawk Trainer Mk.III (c/n 331) 'dressed up' for the occasion as the first Magister for the RAF.

U2 M.18 Mk.I (c/n 1075); fifth known use. Flown, by Miles 4.12.38 and by JF Moir later that day; by JF Moir on 'test' flights on 5.12.38 and 7.2.39; flown by Hugh Kennedy on 2.12.39 and again by him, on 28.2.40; on 31.5.40 Hugh Kennedy flew it to Middle Wallop, returning to Woodley 1.6.40 and on 14.6.40, Hugh Kennedy made a return flight to Cowley to test fly Magisters. Became U-0222 after 17.10.40.

U3 M.3 Falcon (c/n 102); first use of U3. Flown by Miles 23.9.34.

U3 M.2Z Hawk Trainer II (c/n 258?); second use. Flown by JF Moir 9.6.36 (almost certainly c/n 258, which was the first M.2Z for the Roumanian Government).

U3 M.11A Whitney Straight; third use. A photograph taken at Lympne Aerodrome, published in *Flight* for 24.3.38, shows a Whitney Straight with U3 but which remains unidentified.

U3 M.11A Whitney Straight (c/n 506?); fourth use. JF Moir recorded a return flight to the A&AEE Martlesham Heath on 9.5.39 in a Whitney Straight (which he noted as 'V3'). It has not been possible to positively identify this particular aircraft, but it could possibly have been G-AFZY.

U3 M.5A Sparrowhawk (c/n 276); fifth use. Fitted with double-slotted high-lift flaps for RAE trials and was flown by Hugh Kennedy on 24.2.40. U3 was camouflaged and was probably the first of two Sparrowhawks to be modified for trials. U3 also apparently visited RAE Farnborough on 8.6.40 (but not flown by Kennedy). Returned to Phillips & Powis in 1941 and converted back to standard as U-0223 by 27.9.41. (U3 has also been quoted as being used on M.5 G-ADNL (c/n 239) but as Miles made the first flight of this on 19.8.35 and noted 'Sparrowhawk "NL"' in his log-book, this is highly unlikely).

U4 M.11C Whitney Straight (c/n 341); only known use of U4. First flown in 5.37.

U5 M.3A Falcon Major; first use. Flown by Bob Milne 29.1.35 (and thus unlikely to have been G-ADBF, because Miles flew 'Falcon BF' five times in 1.35 with the last entry on 22.1.35). An otherwise unidentified Falcon was also flown twice by Miles 29.1.35 and against the second flight he noted 'tapes'.

U5 M.2H Hawk Major; second use. Flown by Miles 28.2.35 to Sealand and Eaton Hall, returning to Woodley 2.3.35; probably c/n 155 (later VH-UAI dd 7.3.35).

U5 M.7 Nighthawk (c/n 282); third use. First flown, by Miles 5.2.36 and later U6.

U5 M.9 Rolls-Royce PV Trainer (c/n 330); fourth use. Flown, by Miles 3.6.37 without the 'B' mark, but which was applied by 23.6.37. Later modified to become the M.9A Master prototype, retaining the 'B' mark U5. Allocated N3300 5.38, it was dd to A&AEE Martlesham Heath for service trials 10.38.

U5 M.5A Sparrowhawk (c/n 273 & regd G-AFGA 12.4.38); fifth use. Photographs show that it was painted silver overall before September 1939 and fitted with a redesigned wing of reduced span with wide chord flaps but with no wing tips or trousers to the u/c. Last recorded flight was made by Hugh Kennedy 16.7.40 (so it must have been camouflaged by then) a 30-min local flight. On 20.7.40, he made a return flight to Northolt in an unidentified Sparrowhawk, probably U5 and local flights followed on 20th, 21st and 22nd. The identity of U5 has never been confirmed categorically, but can only have been c/n 273. Broken up in the Repair & Service Dept in late 1941/early 1942.

U5 M.7A Nighthawk (c/n 286); sixth use. A 'hybrid' aircraft, first flown, on a 20 min 'local' flight, but with no marks recorded, by Hugh Kennedy 11.3.40. Last recorded flight as U5, by Don Brown 2.10.40; became U-0225 after 17.10.40.

U6 M.7 Nighthawk (c/n 282); first use (and previously U5). First flown, by Miles 5.2.36 (marks unconfirmed). U6 was flown from Woodley to Farnborough 20.6.36, returning to Woodley 24.6.36; also flown by RAE pilot Flt Lt Burke at the Hendon Display between 25/27.6.36, with the new type no '10' on the cowling, 'U6' on the rudder and RAF roundels on the fuselage. Purchased by Air Ministry 13.5.37 and became L6846.

U6 M.14A Hawk Trainer III (c/n 538 – G-AEZS); second use. U6, a 'Miles Trainer' was flown by JF Moir on a 'Handling' flight 16.8.39. Believed to have been fitted with an M.18 wing at this time; with 'A' Performance Sqdn at A&AEE Boscombe Down in 9.39, being one of the first aircraft tested there after the A&AEE moved in; departed A&AEE Boscombe Down between 1-8.12.39. U6 dd by Hugh Kennedy to an unrecorded destination 30-min flying time from Woodley 5.3.40. On 20.5.40 Kennedy made a local flight in a 'Magister' G-AEZS; on 21.5.40 he flew U6 on a bomb drop test and on 25.5.40 he flew G-AEZS on a 5 min bomb drop test. This could imply either that U6 was not G-AEZS (although there is no other likely candidate) or simply that Hugh Kennedy was aware of both marks and used them indiscriminately in his log book. It has been claimed that c/n 538 later used U-0229 but this is not confirmed.

U7 No record of this 'B' mark having been used.

U8 M.4 Merlin (c/n 151); first use. First flown without marks by Miles 24.3.35; first recorded use was when Miles flew U8 to Shoreham 31.3.35. U8 was flown by JF Moir to Martlesham Heath 10.5.35 and by Miles on 15.5.35.

U8 M.2X Hawk Trainer Mk.II; second use. A Hawk with 'B' mark U8 arrived at Farnborough from Woodley 22.4.36, made 4 'flap tests' and returned to Woodley 30.4.36. Since c/n 235 (G-ADYZ) was the one-off M.2X which was fitted with Redwing u/c struts and Theed flap vacuum operating gear, it is possible that U8 could have been this aircraft. However, the Theed system had been incorporated in a Miles aeroplane by 12.2.36 so it is strange that this Hawk should not have been tested by the RAE Farnborough before 4.36.

U8 M.12 Mohawk (c/n 298); probably third use. Flown by Miles without marks 22.8.36. At some time U8 was applied to the fuselage in chalk and later in masking tape.

U8 M.18 Mk.II (c/n 4426) probably fourth use. First flown in 11.39; flown by Don Brown 18.11.39; flown on spinning tests/trials by Hugh Kennedy on 19.5.40, 28.6.40 and twice on 30.6.40; on 6.7.40, it was flown to Cowley and return by Hugh Kennedy to make two test flights on Magisters. On 28.7.40 Hugh Kennedy made a 30 min 'Demonstration to Canadians' flight and on 31.7.40 he was 'trailing static lead'; take-off tests followed later the same day and on 13.8.40 he made a 20 min flight 'during an air-raid'! Flown by Hugh Kennedy, still as U8, on 29.10.40, before becoming U-0224 shortly after.

U9 M.3A Falcon Major (c/n 149); first use. Flown by Miles between 27.3.35 and 31.3.35 this was almost certainly c/n 149 (later VR-HCV). On 28.3.35 Miles put up a record time in U9 between Redhill and Brooklands; a *Flight* photograph of a loose formation of three Miles aircraft (the prototype Miles M.4 Merlin U8, Miles M.3A Falcon Major U9 and Miles M.2H Hawk Major G-ADAW), was dated 1.4.35.

U9 M.8 Peregrine I (c/n 300); second use. First flown, by Charles Powis 12.9.36. Last known flight made by JF Moir on a 'Test' flight on 27.5.37 with a Sqdn Ldr Carter. Dismantled Woodley in 12.37.

U9 M.20/2; third use. First flown, by Tommy Rose 15.9.40. First flown as AX834 by Hugh Kennedy 27.10.40.

U10 No record of this 'B' mark being used.

U11 No record of this 'B' mark being used

U12 M.5 Sparrowhawk (?). A recently discovered photograph shows a Sparrowhawk (or possibly a Hawk Speed Six type) with a low wrap round windscreen and with U12 in white on the fuselage, standing outside the Repair & Service hangar. Unfortunately the photograph is undated and it has not yet been possible to establish its identity.

U13 - U19 No record of these marks having been used.

U20 M.3A Falcon Major/M.3B Falcon Six? The 'B' mark U20 was recorded on a Falcon at Woodley during a visit by AJ Jackson on 15.4.36. AJJ later wrote that this became G-AEEG but G-AEEG had been dd 20.3.36 to Phillips & Powis Air Hire Company and was flown as such by Miles on 27.3.36 when he made a 1 hr 40 min return flight to Martlesham Heath. It is possible that U20 was used on the M.3B Falcon Six (slotted) c/n 252 (K5924), which Miles first flew as 'Falcon VI (slotted)' on 12.2.36. A contemporary, but undated, photograph of the assembly shop at Woodley shows a Miles Falcon of indeterminate type about to be pushed out of the factory, with the U of the 'B' mark just visible by a worker's right hand which, unfortunately, obscures the number - could this have been the mysterious U20?

Appendix 26:
Miles M.14 Magister in Egypt

Magisters were supplied to the Egyptian Air Force from 1937 to 1949, latterly from RAF stocks and by RA Short. For clarity, the full story is set out in this book on Miles pre-war.

Towards the end of 1937, the Egyptian Army Air Force formed a Flying Training School at Almaza, Cairo to cater for both elementary and advanced training, based on RAF methods. Its initial strength was 9 Magisters with 4 RAF qualified flying instructors on loan. Flying training commenced during January 1938 and by the end of that year the number of Magisters on strength had increased to 20 and the service had been renamed the Royal Egyptian Air Force.

The first order for 9 M.14 'Hawk Trainer Mk.IIIs' were delivered to O/C EAAF Almaza, Cairo between October and December 1937. There are some discrepancies in the information as recorded by CLN and that shown on the Air Ministry CofA records (see Appendix 20) but it seems the aircraft concerned were c/n's 540, 541, 543-546, 633-635 and became L.201 to L.209.

A second order for 11 aircraft soon followed and were delivered to Almaza between May and December 1938. The aircraft concerned were c/n's 637, 812-820 and 798 and they became L.210 to L.220. In the information now recorded, there remain unexplained anomalies because the official CofA dates, normally issued following flight-testing at Woodley, often post-date the delivery dates recorded by CLN (and in the later examples by some considerable time) but which could simply be due to paperwork 'catching up' with the aircraft. It is not known if the aircraft were flown or shipped to Egypt.

The much-published photograph of a row of brand new Magisters awaiting delivery outside the factory at Woodley (e.g. in *The Aeroplane* for 12th October 1938) shows a consignment of 10 Magisters for the RAF (L8334-L8343 - all delivered during September 1938) and one last machine on the extreme end of the line for the Royal Egyptian Air Force. The serial is not visible but should be L.217, L.218 or L.219.

It was soon found that 20 aircraft were not sufficient to fulfil the needs of the expanding Air Force and authority was given to Phillips & Powis Aircraft Ltd in late 1938 to divert 23 Magisters from RAF Contract No.778435/38. *The Aeroplane* for 28th December 1938 reported that: *Twenty-three Miles Magister two-seat initial training aeroplanes (130hp Gipsy Major) have been ordered from Phillips & Powis Aircraft Ltd by the Royal Egyptian Army Air Force. Delivery will start soon.* A local newspaper carried a photograph, the caption of which read: *Christmas present for Egypt. Aircraft of the batch of 23 being loaded at Reading for shipment.*

The *Miles Magazine* for March 1939 also reported that *The REAF have recently taken delivery of a further 23 Magisters, making a total at Almaza of 42.* This last figure should have read 43, unless one had already been written off by then.

This third order (which were not given UK CofAs) comprised N3862-3866 (c/n 908-912); N3875-3879 (c/n 916-920); N3895-3899 (c/n 936-940); N3912-3914 (c/n 953-955) & N3885-3889 (c/n 926-930) which became L.221 to L.243 and they were delivered between December 1938 to January 1939. In August 1939, a final example was shipped out – possibly in replacement for one lost and was P6418 (c/n 1773); it may have become L.244 although this has never been confirmed.

Two REAF Magisters were sold to Airwork Ltd (for Misr-Airwork SAE) in 1943 as part of an £E50,000 expenditure for 3 Ansons, 2 DH.89A Dragon Rapides and 1 DH.82A Tiger Moth from RAF and UK sources (Ref AVIA 2/2305). These were registered SU-ACV and SU-ACW by September 1943 and were used by the Misr-Airwork Flying School, Almaza, but their previous identities were not recorded. SU-ACV collided with Tiger Moth SU-ACZ in about January 1946 and was seen at Almaza on 15 December 1946 by Conway Longworth-Dames, who also saw SU-ACW there on 22 January 1946. One of these machines was extensively damaged around the tailplane when the Master DL979 (whose pilot was to blame) collided with it on 6th November 1945 and it is not known if it was repaired.

Some sources state that the Magister strength by March 1940 stood at 43 aircraft (which tends to confirm that one had been written off); by April 1941 there were 36 on strength and on 30th April 1943 these had reduced to 34. However, it is possible that further Magisters were supplied by the RAF from stocks in the Middle East as surviving records are incomplete and a considerable number of RAF Magisters were SOC there with no further details as to their fate. It should be noted that figures on strength and the numbers of aircraft actually serviceable on those dates differ considerably!

An REAF Magister Order of Battle (document dated 3rd June 1950) taken from Air Attaché reports shows the figures reproduced in the table at the top of the next page, and is likely to be the most accurate available data. *Note: The number in the last column is that of the report.*

A report in *Air-Britain Digest* for May 1966, accompanying a photograph of L234 stated that about 120 Magisters went into service with the Royal Egyptian Air Force during the Second World War. This figure must be treated with some caution and may simply have comprised all Magisters dispatched to the Middle East including those serving with the RAF and others en route Turkey.

The section on Egypt in the *World Sales Review*, (an internal Miles report by Lionel A Hackett, the Chief Sales Executive of Miles Aircraft Ltd) on 12th April 1945 stated that:
We sold about 50 Magisters (actually 43 – PA) to the Royal Egyptian Air Force in 1938 and 1939 as an ordinary commercial transaction in which the British Government were not concerned. The contract was a profitable one for the Company and we were

30.4.38	9		AIR 20/243, AIR 23/741	3
25.5.38			9 more ordered. AIR 23/741	
31.10.38	FTS: 19 (1 crashed - low aerobatics, pilot killed)		23 more ordered. AIR 23/741	4
2.39			25 more delivered = 42 at Almaza	
30.4.39	46		AIR 23/741	5
30.4.41	36 (1 spun-in)		AIR 23/741	9
30.4.43	34			
1.5.43 to 31.10.43	ETS; 30. 3 Squadron: 1		1 crashed Khanka - low flying, pilot killed. AIR 23/1158	14
1.5.44 to 31.10.44	21 + 9 u/s as at 1.2.45		1 spun-in 19.6.44 and was written off, pilot OK. AIR 23/1158	16
1.11.44 to 30.4.45			L205, L217, L224, L241 and L243 slightly damaged. AIR 23/1158	17
1.5.45 to 31.10.45	1 authorised for write off		21.7.45 FTS: L237 forced landed. 23.10.45 FTS: L224 slightly damaged in undershoot. AIR 23/8346	18
1.11.45 to 30.4.46	At 20.6.46 FTS: 28. King's Flight: 1		27.11.45 L237 written off landing Khanka. 24.1.46 L202 slightly damaged. AIR 23/8346	19
1.5.46 to 31.10.46	29 (not including King's Flight and Misc.)		7.7.46 FTS: L205 slightly damaged Khanka LG. AIR 23/8346	20
1.1.47	EFTS: 28. 3 Squadron (including King's Flt): 1		AIR 23/8346	
1.1.50	FTS: 10 + 10 in poor condition and due for scrapping		AIR 23/8346	

repaid for the services we rendered by sending an engineer on two visits to Cairo to assist with the assembly and maintenance of the aeroplanes. About 25 of the original aeroplanes are still being operated by the REAF and they have given a completely satisfactory service.

Soon after the outbreak of war the Air Ministry took over the supply of spares with the result that for the first time the REAF received unsatisfactory service. This will continue until after the war, when it is only right that the British Government should cease their competition with us and we can resume business with our own customers. This is a question of policy which I have raised with the Air Ministry and which is still under consideration by them. We have never had an agent in Egypt and all our negotiations with the Egyptian Government have been carried out directly with them or through the Egyptian Offices in London. However, post war I consider it will be necessary for us either to have an agent in Egypt or for us to have a representative permanently resident out there.

With the formation of the African Company (Miles Aircraft of South Africa Pty Ltd – PA) *Egypt was handed over to that company as coming within the African continent. This policy, over which I was not consulted, will in my opinion lead to the loss of possible trade with Egypt unless it is quickly revised. It is quite illogical that Egypt, because it happens not to be separated by water from South Africa, should be looked after by a branch company of ours whose offices are in Johannesburg, and I am quite satisfied that the Egyptian Government would resent having to, if not refuse to, deal with the African Company, as opposed to the parent company in England.*

The great point is that the African Company is doing nothing about our business, or its own business, in Egypt and while this extraordinary policy exists we are virtually prohibited from protecting our own interests in that country. This is a question which should be settled without delay. In the meantime I have instructed an Iraqi, Sqdn Ldr Ibrahim, at present in the Royal Air Force who is going on a month's visit to the Middle East to carry out a survey in Egypt as well as in Palestine, Persia, Saudi Arabia, Iraq, Syria and the Lebanon.

A further 18 refurbished ex RAF (but British civil registered) Magisters were supplied by Robert Alan Short to the Egyptian Government for the Egyptian Air Force in 1949 and a report on this sale appeared in *Aerosphere* for May 1949: *Most of the 18 ex RAF Magisters going to Port Said in Egypt have now been dispatched* (only 14 were listed, marked thus * - Ed), *they were all overhauled by Aircraft & Engineering Services Ltd.*

The majority are shown in the ARB register as having been cancelled 2.9.49 as 'sold abroad' but this followed long after the actual event. A better guide to delivery is the CofA date:

January 1949:   G-AKXN/L5932*; G-AIUD/L8128*;
                G-AKXM/N3816*; G-AKMS/P2428*;
                G-AITX/R1842*
February 1949: G-AKMP/L8092*; G-AJGN/N3850*;
                G-AITT/P2436*; G-AITW/R1898*;
                G-AJGL/R1962*; G-AJGK/R1970*
March 1949:   G-AKML/L5914*; G-AITR/L8144; G-AKMT/L8274*
April 1949:   G-AKMO/L5963*;
May 1949:   G-AITU/L5991;
June 1949:   G-AKMM/P6344

This list shows only 17 Magisters; the eighteenth (if delivered) cannot be identified but is not G-AITV/L8068 as sometimes quoted – this was not an RA Short machine and had been sold to Morocco as F-OAAQ in September 1948.

No Egyptian serials have been traced for the RA Short aircraft and he is also believed to have hoped to sell a further 12 ex RAF Magisters to Egypt in 1949/50. These are listed below, but in the event most were not refurbished and ended up being scrapped at Elstree in 1953 after a period of storage:

G-AFXA/BB666; G-ALNX/L8211; G-ALNY/L8075; G-ALNZ/L5993; G-ALOA/L8058; G-ALOB/N5408; G-ALOC/N5418; G-ALOE/N3926; G-ALOF/N5430; G-ALOH/T9758; N3805 (unregd); P6455 (unregd).

G-ALOC is an anomaly, since it was issued with a CofA in October 1949 and then sold abroad to an unrecorded destination in February 1950; it could therefore conceivably have been the 18th to Egypt.

Nothing further is known of the Egyptian Magisters. It was reported that Magister '908' (possibly the c/n for L221?) was seen on the dump at Embaba on 5th December 1963. This was the last recorded sighting of a Magister in Egypt.

Appendix 27:
Miles M.14A Magister Mk.Is Supplied to Eire

Details of all the Magisters supplied to Eire are included, even though it extends the scope of the story to 1946. In 1938, the Department of Defence of the Irish Air Corps ordered ten M.14A Magisters to replace its ageing Avro Cadets as elementary trainers.

In order to fulfill this order, Phillips & Powis were instructed by the Air Ministry to divert ten Magisters from Contract No.778435/38 and these were delivered to the O/C Irish Air Corps at Baldonnel, Co Dublin from Woodley in February and March 1939, as follows:-

IAC serial	c/n	RAF serial	dd	Comments
31	1025	N5389	22.2.39	damaged beyond repair 2.3.46 and written off
32	1026	N5390	22.2.39	withdrawn from service in 1.46 and scrapped
33	1027	N5391	22.2.39	crashed 1.7.42 and written off
34	1028	N5392	22.2.39	withdrawn from service 11.3.52; to Technical Training Squadron as inst a/f until 1968. Preserved (see below).
35	1029	N5393	22.2.39	crashed 6.9.42 and written off
36	1036	N5400	8.3.39	withdrawn from service in 9.46 and scrapped
37	1037	N5401	8.3.39	crashed 15.5.44 and written off, pilot killed in crash
38	1038	N5402	8.3.39	crashed 28.2.44 and written off
39	1039	N5403	8.3.39	withdrawn from service in 8.46 and scrapped
40	1040	N5404	8.3.39	withdrawn from service in 2.46 and scrapped

A further five M.14A Magisters were ordered by the Department of Defence and were supplied in June 1940 from RAF stocks held in store by 36 MU Sealand, as follows:-

IAC serial	c/n	RAF serial	dd	Comments
73	524	L6903	7.6.40	withdrawn from service in 3.46 and scrapped
74	1784	P6440	7.6.40	withdrawn from service in 9.45 and scrapped
75	942	N3901	7.6.40	withdrawn from service in 8.46 and scrapped
76	1769	P6414	7.6.40	destroyed in crash 7.10.41
77	1777	P6422	7.6.40	withdrawn from service in 8.46 and scrapped

An Irish Air Corps Return of Strength as at 1st January 1945 gave a total of 13 Magisters (incl one recently written off), of which 10 were serviceable and 3 were unserviceable. In fact, the u/s ones had already been written off.

Twelve more Magisters were ordered by the Department of Defence in 1945 as replacements for attrition and were drawn from aircraft surplus to RAF requirements.

A photograph showing the first four completed Magisters (with serials 127-130) at Woodley awaiting delivery appeared in *The Aeroplane Spotter* for 21st February 1946. However, there were problems with this batch of aircraft as, when the Irish Air Corps Officers went to Woodley to look at them, they found that there were no log-books available for inspection. This was hardly surprising seeing as they had all been SOC by the RAF for various reasons in 1944! All that appeared to be known about them was that they had each flown a minimum of 1,000 hours but it was not known if any of them had been involved in any flying accidents etc. Miles Aircraft Ltd could give no assurances as to their condition prior to refurbishment as there was no record of what work had been carried out by the RAF. They had been completely reconditioned by the Repair and Service Department in one of the two Bellman Hangars at Woodley, on an assembly line basis, with the fuselages in a row down one side, supplied with any wings from racks as they became available. All the reconditioned Magisters were therefore issued with CofA's prior to delivery and CofA Application No.'s 8260-8271 were received on 15th February 1946 from Miles Aircraft Ltd for the batch.

IAC serial	c/n	RAF serial	CofA No	CofA issued	dd	Comments
127	1779	P6424	7458	14.2.46	17.2.46	SOC 24.8.44; test flown, as P6424, by Hugh Kendall 26.1.46 & 7.2.46; withdrawn from service in 5.52
128	1827	R1826	7459	14.2.46	17.2.46	SOC 27.5.44; test flown, as R1826, by Hugh Kendall 26.1.46; withdrawn from service 11.51
129	2000	T9733	7461	14.2.46	17.2.46	SOC 21.11.44; no record of test flight; withdrawn from service in 12.52
130	915	N3869	7460	14.2.46	17.2.46	SOC 7.11.44; test flown, as N3869, by Hugh Kendall 5.2.46; withdrawn from service in 1.53
131	768	L8342	7455	14.2.46	9.3.46	SOC 4.10.44; test flown, as L8342, by Hugh Kendall 4.2.46 & as 131, again by Hugh, on 21.2.46; withdrawn from service in 6.52
132	2044	T9807	7456	14.2.46	20.2.46	SOC 21.11.44; test flown, as T9807, by Hugh Kendall 7.2.46; withdrawn from service in 9.52
133	2242	V1089	7457	14.2.46	20.2.46	SOC 29.8.44; test flown, as V1089, by Hugh Kendall 11.2.46; crashed 7.2.47 and damaged beyond repair
134	2189	V1016	7454	14.2.46	20.2.46	SOC 30.11.44; test flown, as V1016, by Hugh Kendall 12.2.46; withdrawn from service in 1.53
135	2040	T9803	7503	24.2.46	9.3.46	SOC 4.12.44; no record of test flight; withdrawn from service in 8.51
136	2247	V1094	7505	24.2.46	9.3.46	SOC 11.10.44; test flown, as V1094, by Hugh Kendall 21.2.46; withdrawn from service in 11.52
137	778	L8352	7504	24.2.46	9.3.46	SOC 5.9.44; no record of test flight; withdrawn from service in 11.52
138	1835	R1834	7502	24.2.46	9.3.46	SOC 7.11.44; test flown, as R1834, by Hugh Kendall 15.2.46; withdrawn from service in 6.52

An ex Irish Air Corps Magister was reported as being seen at Borrisoleigh in 5.61, but no identity is known.

Just one of the Irish Air Corps Magisters, serial 34, survives to this day and perhaps oddly, this was one from the first batch of 10 aircraft delivered in 1939. In 1969, its restoration to static display condition was commenced by Paul Duffey, Dublin, joined in August 1970 by volunteer apprentices of the Irish Air Corps at Baldonnel. It was completed in June 1981 and then presented to the Irish Aviation Museum. Later, it was completely refurbished to static display condition by the Irish Air Corps, and completed just in time for the 75th Anniversary of the IAC celebrations, held at Baldonnel in 1997. A set of spats had to be made from scratch and these appear, unfortunately, to have been manufactured without the benefit of drawings or even photographs. In consequence, they bear little relationship to the originals and rather tend to spoil the appearance of an otherwise immaculate job.

Appendix 28:
The EMK Miles M.2 Hawk Proposal

In 1986, John F Evetts, Managing Director of EMK Aeroplane, prepared a proposal, which is of sufficient interest to be reproduced here in full. It must be emphasised that this brochure is reproduced 'as written' and as such, should not be taken as factual when it comes to the number of variants and designations. These should not, therefore, be confused with the actual figures and designations used in the main text.

INTRODUCTION

1986 is the tenth anniversary of the death of F G Miles, founder of Miles Aircraft. He had been involved in aviation since taking his first joyride in 1925, but the firm's first real success began in 1933, with the appearance of the Miles M.2 Hawk monoplane, which he designed jointly with his wife Blossom, who died in 1984. The proposal outlined in this brochure is put forward as a tribute to them both.

We are most grateful to George Miles, younger brother of F G, a member of the Miles team from its earliest years and later himself Chief Designer, for kindly writing our Foreword.

John F Evetts
Managing Director
EMK Aeroplane, White Hall, Watton-At-Stone,
Hertfordshire SG14 3RP

FOREWORD

Having been asked to write a foreword to EMK's brochure on their proposed M.2 Hawk replica, I first re-read the foreword to *The Book of Miles Aircraft* written by that most renowned of aviation journalists, CG Grey. This contains a far better remembrance of the spirit and atmosphere of early days at Shoreham and Woodley than I could aspire to write in such short space so I will content myself with a brief comment on the EMK venture and similar reconstructions.

The faithful replication and demonstration of past aeroplanes serves a useful current purpose in addition to satisfying the nostalgia of those once closely associated with them and the curiosity of subsequent generations. I discovered this for myself when making and flying the Bristol Boxkite replica for the film *Those Magnificent Men*. I found, slightly to my surprise, that there was much to be learned from the design features and handling characteristics of aircraft that were in production before I was born and which could be adapted to improve current aircraft.

I hope the EMK Hawk replica may also help to stimulate designers of the 1990s to realise that some aspects of progress can be illusory. For example, manufacturers are now learning the hard way that the application of fatigue theory is no substitute for adequate provision for the regular inspection of structural components. Similarly, operators must be realising that their schedules and financial forecasts should take account of ample downtime and labour content to maintain and improve safety standards on scheduled services. They would do well to study the editorials of the above-mentioned C G Grey written during the 1930s.

Meantime, I wish EMK's personnel many happy hours in building and flying the project described in this pamphlet.

George Miles
7th February 1986

THE MILES M.2 HAWK

The pre-war *Golden Age* of British light aviation really began in 1925, with the appearance of the de Havilland Moth, the first practical aeroplane for the private owner. For several years the Moth and its rivals dominated the scene, but by the early 1930s these biplanes, with their cumbersome struts and rigging wires, were beginning to look decidedly old-fashioned, and progressive designers were planning clean cantilever low-wing monoplanes.

One such forward-looking team was the combination of F G Miles and his wife Blossom, who together had already designed a successful single-seat aerobatic biplane, the Miles M.1 Satyr, in 1931. For their next project they joined Phillips & Powis Aircraft Ltd, at Woodley aerodrome near Reading, where they produced an inexpensive yet elegant and robust monoplane, the M.2 Hawk. By adopting all-wooden construction with two open cockpits in tandem, and by acquiring a batch of cheap Cirrus engines from a bankrupt order, they were able to offer the new machine at the incredible price of £395 when it first came out in 1933. This made the Hawk a most attractive proposition, for it was 25% faster than the £650 Moth, while the new cabin monoplanes such as the 3-seat Percival Gull and the high-wing de Havilland Puss Moth, were priced at around £1,000.

The Hawk was an immediate and unqualified success. Within a week of its first flight on March 29th 1933, the prototype had been flown by more than 50 pilots, including the Geoffrey de Havillands, father and son, and orders began to come in (allowing a price increase to £450 to include a small profit). The basic Hawk design proved very adaptable; as well as the 'standard' M.2 with upright Cirrus engine, there was a series of 24 variants, from M.2A to M.2Y incorporating different seating arrangements, powerplants or other modifications. The single-seat Sparrowhawk racer of 1935 was a modified Hawk, whose ultimate derivative was the M.14A Magister, the RAF's first monoplane basic trainer, of which well over a 1,000 were produced.

The Hawk's success was such that, within two years of its first flight, the Miles organisation had grown from a handful of enthusiasts to become second only to de Havilland as a producer of light aeroplanes. Indeed, 1935 was a vintage year for Miles: no fewer than 13 of their aircraft were entered in the King's Cup air-race that September, in which they gained 1st, 2nd, 3rd and 5th places, a feat as yet unrivalled. The Hawk and other Miles designs, all with a distinct family resemblance, continued to play a prominent part in British sporting, record-breaking and ordinary private aviation until the 1939 war put an untimely stop all such activities.

There is no doubt that the original M.2 Cirrus Hawk, from which all this began, deserves a place of honour in any representative flypast of aeroplanes of the Golden Age. But time has not been kind. The only Hawk known to have survived the war was last reported in 1960, lying derelict at Kidlington aerodrome near Oxford, where it had been since 1946. Recent enquires have produced no indication of its ultimate fate, so it seems that not a single example remains today.

That might well have been the end of the story. In the difficult conditions of the post-war period, and despite producing a number of successful aircraft, both large and small, Miles Aircraft Ltd suffered a financial crisis in Autumn 1947, when for a variety of reasons, all overdraft facilities were withdrawn and the business was reorganised during a period of receivership. The company's aviation activities were taken over by Handley Page (Reading) Ltd, but this too closed down in 1963. However, the manufacture of certain other products had been retained by the Miles organisation, under the name Western Manufacturing Estate Ltd. Over the course of time this has become Adwest Group plc, the present owners of the Woodley factory and former aerodrome. During a research project on the history of the site, undertaken by Julian Temple, sponsored by Adwest, several sealed parcels of Miles documents have come to light, which had lain undisturbed in the vaults of a Reading bank since being deposited there in 1935. These turned out to include a virtually complete set of drawings for the M.2 'standard' Hawk, as well as some for the Falcon and Merlin. There are some 1,350 sheets in all, ranging from the smaller components to complete assemblies and general arrangements.

This discovery offers an unexpected and unique opportunity to recreate an important missing link in British aviation history.

EMK AEROPLANE

EMK Aeroplane is a small company specialising in the building, restoration and preservation of vintage and sporting aircraft. The workshops and offices are at Watton-at-Stone, Hertfordshire, and the company also operates a maintenance facility at Denham Airfield, Buckinghamshire.

Originally formed as a partnership in 1980, EMK was from the outset involved in replica-building as well as restoration. One of the first projects was a pair of Sopwith Pup replicas; another was the part-restoration, part-rebuilding of a Bleriot monoplane for a French customer. More recent work has included the rebuild of a de Havilland Queen Bee radio-controlled target aircraft and the conversion of a replica World War 1 DH.2 to resemble a Vickers Gunbus for a film. During the early part of 1985 the firm carried out a thorough check on the wing-spar moisture content of the world's oldest DH.60 Cirrus Moth of 1925, saving it from becoming a non-flying museum exhibit and enabling it to take an active part in the celebrations to mark the 60th anniversary of the Moth's first flight. A most interesting restoration during 1984-85 has been the repair of a 1937 Percival Gull Six, whose designer, Edgar Percival, was one of Miles' chief competitors in the 1930s. Following a take-off accident, both wings and ailerons have been rebuilt, taking the opportunity to correct a twist, which has been responsible for awkward flying characteristics. After the first subsequent flight, the test pilot reported a 'transformed' aeroplane, which can now be flown virtually 'hands-off'.

Over the last few years, EMK has developed a special interest in Miles aeroplanes. Three years ago a Gemini was purchased on behalf of Mr A C Pritchard, Publisher of *Vintage Aircraft* magazine, and this machine is undergoing long-term restoration in our White Hall workshops. Through our connection with the Russavia Collection at Duxford we were able in 1984 to acquire the entire remaining stock of spare parts for Cirrus Minor engines as used in the Gemini, which should enable us to keep it flying, besides offering a spares service to others. The 1936 Falcon, operated by the same magazine, is in our care at Denham. In 1985 we began restoring a World War II Magister trainer, already mentioned as a direct descendant of the Hawk, whose basic structure is very similar.

Even before we at EMK had heard about the discovery of the Miles drawings, we had considered the possibility of constructing a Hawk replica, as it was such an important aeroplane in its time. However, we had reluctantly concluded that there was probably insufficient information available. As soon as we learnt of the find at Reading, we wrote at once for further details, and our Managing Director John Evetts, together with Ian Harwood our Technical Librarian, were kindly invited by Julian Temple of Adwest to examine the drawings at first hand. It soon became clear that here was everything necessary for a reconstruction, and we made a formal request for permission to make use of the material. This request was most generously granted, on the understanding that the drawings must first be catalogued and microfilmed, and that the originals must not leave the security of Woodley.

The cataloguing has now been completed, by a former student at the Miles Technical School, Geoffrey Beckett. We are most grateful to him for allowing us to use his work.

Appendix 29:
Agreement Between Phillips & Powis Aircraft Ltd and Rolls-Royce Ltd in April 1936

Dave Birch of the Rolls-Royce Heritage Trust has provided a copy of the important Agreement made between Phillips & Powis Aircraft Ltd, Charles Owen Powis, Frederick George Miles, George William Graham Allen and Rolls-Royce Limited. The full text of this document is reproduced below:

THIS AGREEMENT is made the 6th day of April One Thousand nine hundred and thirty-six
BETWEEN
PHILLIPS AND POWIS AIRCRAFT LIMITED whose registered office is situate at Reading Aerodrome, Woodley, Reading in the County of Berks (hereinafter called "Phillips and Powis") of the first part; and
CHARLES OWEN POWIS of Westwinds, Woodley, Reading in the County of Berks, Aeronautical Engineer; and
FREDERICK GEORGE MILES of Lands End House, Sonning, Reading in the County of Berks, Aircraft Engineer; and
GEORGE WILLIAM GRAHAM ALLEN of The Elms, Iffley, Oxon, Director of Companies (hereinafter called "the Directors") of the second part; and
ROLLS-ROYCE LIMITED whose registered office is situate at Nightingale Road, Derby in the County of Derby (hereinafter called "Rolls-Royce") of the third part.

WHEREAS :-
(a) Phillips and Powis was incorporated on the Eleventh day of March One thousand nine hundred and thirty-five with a nominal share capital of One hundred and twenty-five thousand Pounds divided into Five hundred thousand Ordinary shares of five Shillings each all of which have been issued and are fully paid up.
(b) The Directors are desirous of raising further capital for Phillips and Powis and have approached Rolls-Royce for this purpose.
(c) Rolls-Royce have agreed to subscribe further capital in manner hereafter set out upon the terms and subject to the conditions hereinafter appearing.

NOW IT IS HEREBY AGREED as follows:-
1. As soon as practicable after the date hereof and in any event by the Thirty-first May One thousand nine hundred and thirty-six an Extraordinary General Meeting of Phillips and Powis shall be held for the purpose of considering and if thought fit passing the following resolutions as Special Resolutions :-

(i) That the capital of the Company be increased to Two hundred and fifty thousand Pounds by the creation of Five hundred thousand Three and one-half per cent Cumulative Convertible Preference Shares of Five Shillings each carrying such rights and privileges as are provided by the Articles of Association of the Company as altered by the next following Resolution.
(ii) That the Articles of Association of the Company be altered by deleting Article 8 and substituting the following Article therefor:
  "8. (a) The capital of the Company is £250,000 divided into 500,000 3½% Cumulative Convertible Preference Shares of 5/- each and 500,000 Ordinary Shares of 5/- each.
  (b) The said 3½% Cumulative Convertible Preference shares shall confer the right to a fixed cumulative preferential dividend at the rate of 3½% per annum on the capital for the time being paid up or credited as paid up thereon payable half-yearly on every 30th day of June and 31st day of December and in a winding-up to payment off of such capital together with all arrears of the said fixed dividend accrued up to the date of commencement of the winding up (whether earned or declared or not) in priority to the Ordinary Shares but shall not confer any further right to participate in profits or assets.
  (c) The Company shall not create or issue any shares ranking in priority to or pari passu with the said 3½% Cumulative Convertible Preference Shares except with the sanction of an Extraordinary Resolution passed at a separate meeting of the holders of the said 3½% Cumulative Convertible Preference Shares in accordance with Article 10 of the Articles of Association of the Company.

(d) Any holder of the said 3½% Cumulative Convertible Preference Shares may at any time during the period ending on the 30th April 1939 give written notice to the Company of his desire to convert the Preference Shares held by him or any part thereof into Ordinary Shares and upon receipt by the Company of any such notice at its registered office together with a payment of 3/- for every share in respect whereof such notice shall be given all shares included in such notice shall ipso facto be deemed to be converted into Ordinary Shares accordingly and shall thenceforth confer the same rights and privileges as the other Ordinary Shares in the Company's Capital and shall for the purposes of dividend count as Ordinary Shares for the whole of the financial year in which such conversion takes place"

2. A copy of Phillips and Powis' Balance Sheet shall accompany the notices calling the Meetings to pass the said Resolutions and the Balance Sheet and notices shall be in such a form as shall previously have been approved in writing by Rolls-Royce.

3. The Directors shall do everything in their power to ensure that the said Resolutions shall be duly passed by the date aforesaid and shall use all the votes conferred by shares held by them in favour of the said Resolutions.

4. Within seven days after the passing of the said Resolutions Rolls-Royce shall subscribe the said Five hundred thousand Three and one-half per cent Cumulative Convertible Preference Shares at par and shall pay for the same in full and Phillips and Powis shall thereupon issue and allot the same to Rolls-Royce or as it shall direct.

5. As soon as Rolls-Royce has complied with the provisions of Clause 4 hereof the Directors will procure that two persons nominated by Rolls-Royce shall be appointed as Directors of Phillips and Powis.

6. The Directors shall at all times before the Meeting to pass the said Resolutions keep this Agreement secret from all other persons (including shareholders and employees of Phillips and Powis) save and except as my be expressly authorised in writing by Rolls-Royce.

7. This Agreement shall be interpreted as a joint and several covenant by Phillips and Powis and each of the Directors individually with Rolls-Royce.

8. If the said Resolutions shall not be passed by the Thirty-first May One thousand nine hundred and thirty six or by such later date as the parties shall mutually agree this Agreement and all the provisions hereof (save and except Clauses 6 and 7 hereof) shall become null and void and no party shall (save as aforesaid) have any claim against any other under or by virtue of this Agreement or anything herein contained.

IN WITNESS whereof these presents have been entered into the day and year first before written.

THE COMMON SEAL of Rolls-Royce Limited was hereto affixed in the presence of:-

Signed.....A F Sidgreaves   Director

Signed.....John DeLooze   Secretary

Appendix 30: No.8 Elementary & Reserve Flying Training School, Woodley

The Phillips & Powis run Reserve Training School was opened on 25th November 1935 and in the period to 31st August 1939 completed 19 Ab Initio Courses. There were 2 fatal accidents and a few minor accidents over this period of approximately 4 years during which time 45,000 flying hours were completed. On the outbreak of war, the unit was redesignated 8 Elementary Flying Training School and retained only the Magisters and Tiger Moths – the other types being flown away to other RAF units.

The strength and establishment of the unit as at 31.8.39 was as follows:-

16 Flying Instructors:
CFI - S/Ldr James F Moir (RAFO)
F/Lt G C O'Donnell DFC (Retired)   F/Lt A L Duke (RAFO)
F/Lt W E Hooper (RAFO)   F/Lt J D H Slade (RAFO)
F/Lt R V Bucknall (RAFO)   F/Lt E P P Gibbs (RAFO)
F/Lt M RD Trewby (RAFO)   F/Lt H V Kennedy (RAFO)
F/O H A Crommelin (RAFO)   F/O H A Roxburgh (RAFO)
F/O A L Collins (RAFO)   F/O A L Alington (RAFO)
P/O R L Porteous (RAFO)   P/O W H L Dudley (RAFO)
Sgt E F J L'Estrange (RAFVR)

4 Ground Instructors:
CGI - F/O J A Manning-Fox (Retired)
A G Broad   J M Arnott   F R Fane

Aircraft:-
25 Elementary -
6 Magisters; 6 Tiger Moths; 13 Hawk Trainers.
18 Service -
6 Hart Bombers; 3 Hart Trainers; 2 Audax; 2 Hinds; 3 Battles; 2 Ansons.
(Note – there are anomalies in reconciling these numbers to those set out below – PA!)

Pupils under Training:
20th Regular Course   27 Pupil Pilots (commenced 24.7.39); 5 Airmen Pilots.
RAFVR Pilots   112 Sergeant Pilots.
RAFVR Aircrew   46 Air Observers; 79 Air Gunners and W/T Operators.

Aircraft known to have been operated by 8 ERFTS prior to 3rd September 1939:

DH.82 Tiger Moth
G-ADJB   Toc 18.11.35. Crashed Hurst, nr Woodley 12.3.36
G-ADJC   Toc 18.11.35; later coded "1". To 8 EFTS 3.9.39 (later BB815)
G-ADJD   Toc 18.11.35; later coded "2". To 8 EFTS 3.9.39 (later BB816)

W/C J F Moir, CFI of the Reserve Training School, later No.8 E&RFTS and No.8 EFTS.   [Miles Aircraft via The Miles Aircraft Collection]

DH Tiger Moths and Miles Hawk Trainer Mk.IIs of the Reserve Training School outside the HQ RTS building in 1936. ['Flight' via B Clarke]

G-ADJE	Toc 19.11.35; later coded "3". Sold to Malling Aviation Ltd 7.39.
G-ADJF	Toc 19.11.35. Sold to Brooklands Aviation Ltd 1.36
G-ADJG	Toc 20.11.35; later coded "4". To 8 EFTS 3.9.39 (later BB817)
G-ADJH	Toc 19.11.35. Sold to Horton Kirby Flying Club Ltd 7.39.
G-ADJI	Toc 20.11.35; later coded "6". To 8 EFTS 3.9.39 (later BB818)
G-ADJJ	Toc 19.11.35; later coded "7". To 8 EFTS 3.9.39 (later BB819)

Miles M.2X/Y Hawk Trainer II

G-ADWT	Toc 25.11.35; later coded "1". CofA lapsed 1.10.40.
G-ADWU	Toc 25.11.35. CofA lapsed 3.8.40.
G-ADWV	Toc 25.11.35. CofA lapsed 23.1.40.
G-ADVF	Toc 4.1.36. CofA lapsed 7.10.39.
G-ADZA	Toc 10.1.36; later coded "4". CofA lapsed 14.2.40 (later DG665)
G-ADZB	Toc 23.1.36; later coded "7". CofA lapsed 4.7.40.
G-ADZC	Toc 25.1.36; later coded "6". CofA lapsed 17.11.39.
G-ADZD	Toc 4.2.36. CofA lapsed 4.4.40.
G-ADZE	Toc 2.3.36. Collided with G-AEAW over Woodley 26.8.36.
G-AEAX	Toc 25.2.36 (or 14.2.36); later coded "10". Sold to Reading Aero Club Ltd 5.39.
G-AEAY	Toc 13.2.36. CofA lapsed 20.9.39.
G-AEAZ	Toc 20.3.36; later coded "8". Spun in & crashed Haines Hill, Hurst 5.6.37.
G-AEEL	Toc 26.3.36; later coded "13". Crashed Lodden Bridge Rd, Woodley 30.11.36; rebuilt and coded "9". CofA lapsed 25.5.40
G-AEAW	Toc 27.4.36. Collided with G-ADZE over Woodley 26.8.36.
G-ADYZ	Toc 9.7.36. CofA lapsed 18.8.40.

Miles Magisters (Hawk Trainer III)

G-AFBS	Toc 9.10.37. To 8 EFTS 3.9.39 (later BB661).
G-AFDB	Toc 30.11.37. To 8 EFTS 3.9.39 (later BB662).
L6918	Toc 20.12.37. To 8 EFTS 3.9.39.
L6919	Toc 29.12.37. To 8 EFTS 3.9.39.
L6915	Toc 3.1.38. To 8 EFTS 3.9.39.
L6916	Toc 15.1.38. Dived into ground Sonning Common, nr Reading 28.3.38.

Part of the RTS Miles Hawk Trainer Mk.II fleet outside the HQ building in February 1936. ['Flight' via B Clarke]

The RTS building with pupils in February 1936. ['Flight' via B Clarke]

N2259	Toc 23.5.38; later coded "13". To 8 EFTS 3.9.39.	
G-AFTR	Toc 10.5.39. To 8 EFTS 3.9.39 (later BB663).	
G-AFTS	Toc 18.5.39. To 8 EFTS 3.9.39 (later BB664).	
G-AFWY	Toc 4.8.39. To 8 EFTS 3.9.39 (later BB665).	
G-AFXA	Toc 11.8.39. To 8 EFTS 3.9.39 (later BB666).	
G-AFXB	Toc 19.8.39. To 8 EFTS 3.9.39 (later BB667).	

Avro Anson I
- N5251 Toc 3.7.39. To 5 CANS 3.9.39
- N5252 Toc 3.7.39. To 5 CANS 3.9.39

Fairey Battle
- K7604 Toc 2.1.39. To 7 FTS 3.9.39.
- K7645 Toc 11.1.39. To 7 FTS 3.9.39.
- K7565 Toc 12.1.39. To 7 FTS 3.9.39

Hawker Audax
- K8316 Toc 28.2.38. To 11 FTS 3.9.39
- K7440 and/or K7446 were also reported as operated but this is not confirmed from RAF records.

Hawker Demon I
- Flt Lt HV Kennedy flew K3790 but this is not shown as allocated to 8 ERFTS.

Hawker Hart (Bomber)
- K3853 Toc 9.6.38. To General Aircraft Ltd, Hanworth 27.7.39.
- K3881 Toc 9.6.38. To General Aircraft Ltd, Hanworth 27.7.39
- K3847 Toc 28.6.38. To General Aircraft Ltd, Hanworth 27.7.39.
- K3866 Toc 28.6.38. To 2 FTS 3.9.39.
- K3879 Toc 28.6.38. To 2 FTS 3.9.39.
- K2981 Toc 30.6.38. To 2 FTS 3.9.39.

Hawker Hart Trainer
- K4769 Toc 31.8.37. To 2 FTS 3.9.39
- K6506 Toc 31.8.37. To 2 FTS 3.9.39.
- K6472 Toc 13.4.38. To 2 FTS 3.9.39.

Hawker Hind
- L7209 Toc 24.2.39. To 9 Sub MU 13.11.39.
- L7211 Toc 24.2.39. To 11 FTS 3.9.39
- K6720 Toc 28.6.39. To 12 FTS 3.9.39
- K5471 Toc 6.7.39. To 12 FTS 3.9.39.
- K6774 Toc 6.7.39. To 12 FTS 3.9.39.

Miles Hawk Trainer Mk.II, G-ADZA, in post-Munich 1938 part camouflage colours. [Via P H T Green Collection]

Appendix 31: Summary of Constructors Numbers (Pre-War)

Miles' first design from scratch was the M.1 Satyr, which was built by George Parnall & Co, Yate and initially given their c/n J7. This was later changed to c/n 1 by Miles, and from this simple act stemmed confusion regarding the c/ns given to the first few M.2 Hawks.

C L Nash joined Phillips & Powis Aircraft (Reading) Ltd in October 1933 kept basic details of aircraft constructed at Woodley in small hardback notebooks, which had been professionally printed for that express purpose. The columns were headed:

Machine No.	Regn.	Job No.	Type
Customer	Colour	Engine	Delivered

A number of aeroplanes had been completed before Nash joined the firm and it is apparent from study of his notebooks that he must have decided to fill in what he thought were the details for these machines from various sources and in consequence, some of this has been found to include inaccuracies.

After an initial 'hiccup', due to the first Miles M.2 Hawk being numbered H.101 by Miles, the Machine Nos. (conventionally known in the trade as c/ns) ran consecutively, with occasional gaps. In particular, there is a gap from c/n 43 to 101 and it is understood that Miles thought that it would look as if they had sold more aeroplanes if subsequent c/ns were numbered in the 100s. It should be noted that Phillips & Powis also gave c/ns to mock-ups.

Machine No (c/n)	Comments
H.101	M.2 Hawk prototype. This was equivalent to c/n 2, which was never used as such
3-43	M.2 Hawk variants.
44-100	Not used.
101	Fuselage No for Rebuild Hawk No.7
102	M.3 Falcon prototype.
103-147	Production of aircraft of various types.

148	Blank.	
149-177	Production aircraft of various types.	
178	Blank.	
179-190	Production of aircraft of various types, including c/n 187 'Hawcon Mock-up.'	
191	Job No.H3038 recorded with no other details.	
192-199	Production of aircraft of various types.	
200	'Standard Hawk' (in pencil), with Job No.H3044.	
201-207	Production of aircraft of various types.	
208	'Hawk Major' (in pencil).	
209-213	Production of aircraft of various types.	
214	'Standard Hawk (Scrap)' (in pencil), with Job No.H3986 (possibly too late for the period, but could have been in error for H3896).	
215-222	Production of aircraft of various types.	
223	'Hawk Major at Service' (in pencil), with Job No.H3987.	
224	M.2W Hawk Trainer Mk.II.	
225	'Hawk Major' (in pencil), with Job No.H3988.	
226-239	Production of aircraft of various types.	
240	Job No.H5571 recorded with no further details.	
241-242	M.2Y & W Hawk Trainer Mk.II's.	
243	'Hawk Speed Six Fuse only' (in pencil), with Job No.H5573.	
244	Job No.H5574 recorded with no further details.	
245-249	Production of aircraft of various types.	
250	Job No.H5580 recorded with no further details.	
251-277	Production of aircraft of various types.	
278	'Merlin' (in pencil), with Job No.H5684.	
279	'Miles Queen Wasp - Mock up only', with Job No.H5720.	
280	M.3B Falcon Six.	
281	'Sparrowhawk' - in pencil.	
282-283	M.7 Nighthawks.	
284	Job No.H5742 recorded with no further details.	
285	Blank.	
286	M.7A Nighthawk.	
287	'Nighthawk/Mentor used for mock-up only 1936/1937.'	
288-290	Production of aircraft of various types.	
291	'Miles Falcon Six Service', with Job No.12030.	
292-298	Production of aircraft of various types.	
299	'Nighthawk' - with no further details.	

A mock-up of the M.10 Queen Wasp was built but there is no record of its c/n although it could well have been one of the blank c/ns above.

Apart from the mysterious c/n (1-Y) used for the Miles M.13 Hobby, for which no logical explanation has yet been found, the c/ns of new machines as they were ordered/built/reserved continued as follows:-

300-328	Production of aircraft of various types.
329	'Miles Magister used for mock-up only.'
330-482	Production of aircraft and mock-ups of various types.
483	'Hawk Trainer Fuselage only.'
484-487	Production of aircraft and fuselage for rebuild of various types.
488-489	'Magister - P & P' (in pencil), with a left-handed 'tick' in the Regn. column.
490-547	Production of aircraft of various types.
548-555	Blank. Possibly reserved for M.14 Hawk Trainer Mk.IIIs - not built.
556-795	Production of aircraft of various types.
796	Blank. Possibly reserved for the M.15 mock-up.
797-798	M.14 Magisters.
799-810	Blank. Reserved for production of 12 M.17 Monarchs - not built.
811-1074	Production of M.14 Magisters.
1075	M.18 first prototype.
1076-1077	M.15, 2 prototypes to Spec T.1/37.
1078-1082	M.14 Hawk Trainer Mk.IIIs for 8 E&RFTS.
1083-1086	Allotted to 4 M.14 Hawk Trainer Mk.IIIs for 8 E&RFTS - not built.
1087-1110	Blank. Possibly reserved for production of 24 M.14 Hawk Trainer Mk.IIIs - not built.
1111-1610	Production of 500 M.9B Master Mk.Is.
1611-2255	Production of 645 M.14A Magisters, completed in 1.41.

Appendix 32:
Registration, Type and C/n Cross-Reference

Reg	Type	C/n	Reg	Type	C/n	Reg	Type	C/n	Reg	Type	C/n
G-AAII	Martlet	200	G-ACOC	M.2	25	G-ACXT	M.2F	118	G-ADDK	M.2P	190
'G-AAJW'	MMart	31/1	G-ACOP	M.2	41	G-ACXU	M.2F	119	G-ADDM	M.2	142
G-AAVD	Martlet	201	G-ACPC	M.2D	20	G-ACXZ	M.2	104	G-ADDU	M.2H	180
G-AAYX	Martlet	202	G-ACPD	M.2D	30	G-ACYA	M.2	105	G-ADEN	M.2H	161
G-AAYZ	Martlet	203	G-ACPW	M.2D	33	G-ACYB	M.2G	120	G-ADER	M.3A	157
G-ABBN	Martlet	204	G-ACRB	M.2	27	G-ACYO	M.2F	121	G-ADFC	M.2H	167
G-ABIF	Martlet	205	G-ACRT	M.2	31	G-ACYW	M.2	117	G-ADFE	M.4	151
G-ABJW	MMart	31/1	G-ACSC	M.2D	35	G-ACYX	M.2H	129	G-ADFH	M.3A	196
G-ABMM	MMart	31/2	G-ACSD	M.2	28	G-ACYZ	M.2H	123	G-ADGA	M.2F	169
G-ACGH	M.2	H101	G-ACSL	M.2	29	G-ACZD	M.2	130	G-ADGD	M.2H	165
G-ACHJ	M.2	3	G-ACSX	M.2D	42	G-ACZI	M.2H	132	G-ADGE	M.2N	135
G-ACHK	M.2	4	G-ACTD	M.2F	36	G-ACZJ	M.2H	122	G-ADGI	M.2	175
G-ACHL	M.2	5	G-ACTE	M.2E	43	G-ACZW	M.2	136	G-ADGL	M.2H	166
G-ACHZ	M.2	6	G-ACTI	M.2	37	G-ADAB	M.2H	137	G-ADGP	M.2L	160
G-ACIZ	M.2	7	G-ACTM	M.3	102	G-ADAC	M.2F	134	G-ADGR	M.2	192
G-ACJC	M.2	8	G-ACTN	M.2	103	G-ADAS	M.2H	138	G-ADHC	M.3A	163
G-ACJD	M.2	9	G-ACTO	M.2	32	G-ADAW	M.2H	139	G-ADHF	M.2H	173
G-ACJY	M.2	10	G-ACUD	M.2	38	G-ADBF	M.3A	131	G-ADHG	M.3A	193
G-ACKI	M.2	11	G-ACVM	M.2F	109	G-ADBG	M.2H	143	G-ADHH	M.3A	181
G-ACKW	M.2B	12	G-ACVN	M.2	39	G-ADBI	M.3A	140	G-ADHI	M.3A	189
G-ACKX	M.2	13	G-ACVO	M.2	106	G-ADBK	M.2	146	G-ADIG	M.2P	204
G-ACLA	M.2	15	G-ACVP	M.2	107	G-ADBT	M.2H	145	G-ADIT	M.2H	168
G-ACLB	M.2	16	G-ACVR	M.2D	108	G-ADCF	M.2H	153	G-ADIU	M.3A	202
G-ACLI	M.2A	14	G-ACWV	M.2F	110	G-ADCI	M.2F	147	G-ADLA	M.2H	176
G-ACMH	M.2	17	G-ACWW	M.2F	111	G-ADCJ	M.2H	154	G-ADLB	M.2R	210
G-ACMM	M.2	18	G-ACWX	M.2F	112	G-ADCU	M.2H	162	G-ADLC	M.3B	213
G-ACMX	M.2	23	G-ACWY	M.2F	113	G-ADCV	M.2M	156	G-ADLH	M.2S	194
G-ACNW	M.2	26	G-ACXL	M.2F	114	G-ADCW	M.2H	152	G-ADLI	M.3A	206
G-ACNX	M.2	24	G-ACXM	M.2F	115	G-ADCY	M.2H	158	G-ADLN	M.2R	211
G-ACOB	M.2C	19	G-ACXN	M.2F	116	G-ADDC	M.2H	164	G-ADLO	M.2P	220

Appendix 32

Reg	Type	No	Reg	Type	No	Reg	Type	No	Reg	Type	No
G-ADLS	M.3C	231	G-AEVH	M.11A	321	G-AIZK	M.14A	1706	U-0239	M.18/1	1075
G-ADMW	M.2H	177	G-AEVL	M.11A	322	G-AIZL	M.14A	592	U-0252	M.14A	902
G-ADNJ	M.2T	203	G-AEVM	M.11A	324	G-AJGM	M.14A	589	U-0259	M.14A	1748
G-ADNK	M.2T	222	G-AEVW	M.11A	320	G-AJGN	M.14A	896	U-0262	M.14A	1723
G-ADNL	M.5	239	G-AEWK	M.11A	325	G-AJHB	M.14A	628	U-0263	M.14A	413
G-ADOD	M.2U	195	G-AEWT	M.11A	326	G-AJHC	M.14A	876	U-0264	M.14B	533
G-ADTD	M.3B	255	G-AEXJ	M.11A	501	G-AJHD	M.14A	650	U-0265	M.14A	598
G-ADVF	M.2W	217	G-AEYA	M.11A	342	G-AJHE	M.14A	825	U-0266	M.14A	604
G-ADVR	M.2	171	G-AEYB	M.11A	500	G-AJHH	M.14A	1000	U-0267	M.14A	900
G-ADWT	M.2W	215	G-AEYI	M.11C	341	G-AJRS	M.14A	1750	U-0268	M.14A	982
G-ADWU	M.2W	224	G-AEYJ	M.11A	343	G-AJRT	M.14A	744	U-0272	M.14A	1707
G-ADWV	M.2W	228	G-AEZO	M.11A	347	G-AJRU	M.14A	352	G-35-2	M.77	239
G-ADWW	M.5	264	G-AEZP	M.14B	494	G-AJRV	M.14A	1774			
G-ADXA	M.7	263	G-AEZR	M.14A	495	G-AJZH	M.14A	1641	USSR	M.2Z	480
G-ADYZ	M.2X	235	G-AEZS	M.14A	538	G-AKGS	M.14A	1765			
G-ADZA	M.2W	241	G-AFAB	M.11A	346	G-AKJV	M.14A	1762	CC-FBB	M.2R	257
G-ADZB	M.2W	242	G-AFAW	M.13	1-Y	G-AKJW	M.14A	528	CC-FBL	M.11A	499
G-ADZC	M.2W	249	G-AFAY	M.3B	233	G-AKJX	M.14A	560	CC-FBN	M.2	Note
G-ADZD	M.2Y	253	G-AFBF	M.3B	256	G-AKKS	M.14A	1790	CC-PBL	M.2	Note
G-ADZE	M.2W	254	G-AFBS	M.14A	539	G-AKKT	M.14A	405	CC-PFB	M.2R	257
G-ADZL	M.3D	262	G-AFBV	M.11A	497	G-AKKU	M.14A	539	CC-PTA	M.14A	604
G-ADZR	M.3A	209	G-AFCC	M.11A	499	G-AKKV	M.14A	987			
G-ADZU	M.2H	184	G-AFCN	M.11A	502	G-AKKW	M.14A	1776	CF-AUV	M.2H	123
G-AEAG	M.3D	266	G-AFCP	M.3E	289	G-AKMJ	M.14A	777	CF-FGK	M.11A	509
G-AEAO	M.3D	269	G-AFCR	M.17	638	G-AKMK	M.14A	897	CF-NXT	M.2W	215
G-AEAW	M.2Y	246	G-AFDB	M.14A	542	G-AKML	M.14A	335			
G-AEAX	M.2Y	260	G-AFET	M.14A	556	G-AKMM	M.14A	1712	CH-380	M.2	34
G-AEAY	M.2Y	261	G-AFEU	M.14A	557	G-AKMN	M.14A	424			
G-AEAZ	M.2Y	270	G-AFEV	M.14A	558	G-AKMO	M.14A	395	CN-TZE	M.14A	843
G-AEBP	M.7	282	G-AFEW	M.14A	559	G-AKMP	M.14A	601			
G-AECC	M.3B	280	G-AFGA	M.5A	273	G-AKMS	M.14A	1650	CS-AAL	M.2H	161
G-AECT	M.11	290	G-AFGK	M.11A	509	G-AKMT	M.14A	730	CS-AFM	M.14A	597
G-AEDE	M.8	300	G-AFGL	M.17	786	G-AKMU	M.14A	775			
G-AEDL	M.3D	259	G-AFJJ	M.11A	306	G-AKMZ	M.14A	1775	CX-AAW	M.2H	138
G-AEEG	M.3A	216	G-AFJU	M.17	789	G-AKNA	M.14A	534	CX-ACT	M.2H	138
G-AEEL	M.2Y	271	G-AFJX	M.11A	507	G-AKOL	M.14A	923			
G-AEEZ	M.2H	179	G-AFJZ	M.17	790	G-AKPE	M.14A	823	D-EBNF	M.2W	222
G-AEFA	M.2H	183	G-AFKL	M.2H	221	G-AKPG	M.14A	356	D-EGYV	M.3B	256
G-AEFB	M.3A	229	G-AFLW	M.17	792	G-AKPL	M.14A	871	D-EKTR	M.11A	498
G-AEFS	M.2H	124	G-AFRO	M.18/1	1075	G-AKPM	M.14A	591			
G-AEGE	M.2P	267	G-AFRZ	M.17	793	G-AKRH	M.14A	785	EC-ABI	M.2H	172
G-AEGP	M.2H	186	G-AFTR	M.14A	1078	G-AKRI	M.14A	623	EC-ABZ	M.3A	201
G-AEGR	M.2H	188	G-AFTS	M.14A	1079	G-AKRW	M.14A	931	EC-ACB	M.3C	231
G-AEHN	M.7A	283	G-AFTX	M.17	795	G-AKUA	M.14A	1021	EC-AGY	M.3A	197
G-AEHO	M.7A	288	G-AFWY	M.14A	1080	G-AKXM	M.14A	864	EC-AHZ	M.2	117
G-AEHP	M.2Z	258	G-AFXA	M.14A	1081	G-AKXN	M.14A	364	EC-BAY	M.3A	201
G-AEHR	M.2Z	237	G-AFXB	M.14A	1082	G-ALGK	M.14A	1742	EC-BDD	M.3A	197
G-AEHS	M.2Z	245	G-AFYV	M.14A	1083	G-ALHB	M.14A	971	EC-CAO	M.3C	231
G-AEHT	M.2Z	265	G-AFYW	M.14A	1084	G-ALIM	M.14A	640	EC-CAS	M.2H	172
G-AEHU	M.2Z	292	G-AFYX	M.14A	1085	G-ALIO	M.14A	1700	EC-DBB	M.3A	201
G-AEHV	M.2Z	293	G-AFYY	M.14A	1086	G-ALNX	M.14A	678	EC-DDB	M.2H	172
G-AEHW	M.2Z	294	G-AFZY	M.11A	506	G-ALNY	M.14A	584	EC-W25	M.2	117
G-AEHX	M.2Z	295	G-AGDL	M.5A	276	G-ALNZ	M.14A	425	EC-W44	M.2H	172
G-AEHY	M.2Z	296	G-AGFW	M.17	787	G-ALOA	M.14A	567	EC-W45	M.3A	201
G-AEHZ	M.2Z	297	G-AGVW	M.14A	1748	G-ALOB	M.14A	1044	EC-W48	M.3A	197
G-AEKJ	M.2H	185	G-AGWT	M.7A	286	G-ALOC	M.14A	1054	EC-ZZA	M.2	117
G-AEKK	M.3D	248	G-AGZR	M.14A	902	G-ALOE	M.14A	964			
G-AEKW	M.12	298	G-AGZX	M.3D	269	G-ALOF	M.14A	1066	EI-AAV	M.2	23
G-AEKX	M.12	301	G-AHKM	M.16	462	G-ALOG'	M.14A	1081	EI-AAX	M.2	24
G-AELT	M.5A	275	G-AHKY	M.18/2	4426	G-ALOG"	M.14A	732	EI-ABG	Martlet	200
G-AENG	M.3A	234	G-AHOA	M.18/3	4432	G-ALUW	M.14A	1693	EI-ABQ	M.2	23
G-AENH	M.11A	303	G-AHUL	M.14A	677	G-ALUX	M.14A	1649	EI-ADU	M.14A	623
G-AENS	M.2H	198	G-AHYK	M.14A	868	G-AMBM	M.14A	1625			
G-AENT	M.2Y	328	G-AIAI	M.14A	1800	G-AMBN	M.14A	752	F-AMZW	M.2C	19
G-AEOC	M.9	330	G-AICD	M.14A	1734	G-AMBP	M.14A	1757	F-APOY	M.2H	205
G-AEOX	M.2H	205	G-AIDE	M.17	793	G-AMMC	M.14A	779	F-APPZ	M.11A	326
G-AERC	M.11A	305	G-AIDF	M.14A	1766	G-AMMD	M.14A	741	F-AQCZ	M.11A	496
G-AERS	M.11A	304	G-AITM	M.11A	312	G-ANLT	M.14A	836	F-AQER	M.3A	157
G-AERV	M.11A	307	G-AITR	M.14A	622	G-ANWO	M.14A	718	F-AQIK	M.11A	348
G-AERY	M.11A	309	G-AITT	M.14A	1658	G-CCMH	M.2H	172	F-AQLX	M.11A	505
G-AETB	M.11A	310	G-AITU	M.14A	423				F-AQMA	M.11A	345
G-AETJ	M.14	331	G-AITV	M.14A	577	U-1 etc	see App 25		F-AREQ	M.11A	508
G-AETL	M.14	332	G-AITY	M.14A	616	U-0222	M.18/1	1075	F-ARPE	M.17	786
G-AETN	M.3A	226	G-AITZ	M.14A	1074	U-0223	M.5A	276	F-ARRL	M.17	795
G-AETS	M.11A	312	G-AIUB	M.14A	613	U-0224	M.18/2	4426	F-BBCN	M.3D	269
G-AEUJ	M.11A	313	G-AIUC	M.14A	843	U-0225	M.7A	286	F-BCDU	M.14A	850
G-AEUX	M.11A	314	G-AIUD	M.14A	606	U-0226	M.17	787	F-BCEX	M.2H	129
G-AEUY	M.11A	315	G-AIUE	M.14A	995	U-0227	M.11A	506	F-BDPA	M.14A	1785
G-AEUZ	M.11A	316	G-AIUF	M.14A	1782	U-0229	M.14A	538	F-BDPB	M.14A	1002
G-AEVA	M.11A	318	G-AIUG	M.14A	988	U-0234	M.15	1077	F-BDPC	M.14A	963
G-AEVF	M.11A	317	G-AIYL	M.14A	517	U-0236	M.18/4	unkn	F-BDPD	M.14A	1778
G-AEVG	M.11A	319	G-AIZJ	M.14A	1645	U-0238	M.18/3	4432	F-BDPE	M.14A	717

Reg	Type	No	Reg	Type	No	Reg	Type	No	Reg	Type	No
F-BDPF	M.14A	894	LV-XOP	M.14A	621	OY-DEK	M.2	170	ZK-ADJ	M.2F	119
F-BDPG	M.14A	1783	LV-XOQ	M.14A	605	OY-DIO	M.17	788	ZK-AEI	M.3B	247
F-BDPH	M.14A	1735	LV-XOR	M.14A	420				ZK-AEO	M.11A	308
F-BDPJ	M.14A	586	LV-XOS	M.14A	660	PH-ATP	M.17	794	ZK-AEQ	M.2Y	302
F-BDPK	M.14A	712	LV-XOT	M.14A	727	PH-KGH	M.3D	269	ZK-AEX	M.14	331
F-BDPL	M.14A	784	LV-XOU	M.14A	736				ZK-AEY	M.14	332
F-BDPM	M.14A	978	LV-XOV	M.14A	739	PK-SAL	M.2	13	ZK-AEZ	M.14A	486
F-BDPO	M.14A	721	LV-XOW	M.14A	390	PK-SAR	M.2H	173	ZK-AFA	M.14A	487
F-BDPP	M.14A	600	LV-XOY	M.14A	728				ZK-AFG	M.11A	323
F-DADV	M.14A	517	LV-XPG	M.14A	958	R276	M.2H	143	ZK-AFH	M.11A	311
F-OAAQ	M.14A	577	LV-XPH	M.14A	1724	R302	M.2R	210	ZK-AFJ	M.2P	251
F-OAFU	M.14A	825	LV-XPI	M.14A	887				ZK-AFK	M.2H	144
F-OAFV	M.14A	1000	LV-XPL	M.14A	884	SE-AFN	M.3A	216	ZK-AFL	M.2P	220
F-OAGQ	M.14A	843	LV-XPM	M.14A	1772	SE-AFS	M.2	130	ZK-AFM	M.2F	147
			LV-XPN	M.14A	1005				ZK-AGB	M.11A	503
HB-EEB	M.14A	431	LV-XPO	M.14A	1687	SU-AAP	M.2F	126	ZK-AJF	M.11A	308
HB-EPI	M.11A	349	LV-XPP	M.14A	1792	SU-ADA	M.3A	181	ZK-AJZ	M.11A	323
HB-ERI	M.2H	159	LV-XPR	M.14A	651				ZK-ALE	M.11A	503
HB-OAS	M.2G	120	LV-XPS	M.14A	1018	SX-AAB	M.2	40	ZK-ALO	M.14	332
HB-OXU	M.2	34	LV-XPT	M.14A	1808	SX-AAD	M.2F	127	ZK-ANJ	M.14A	783
HB-URO	M.11A	344	LV-XPU	M.14A	924				ZK-ANK	M.14A	935
HB-USU	M.3A	131	LV-XPV	M.14A	974	TF-BLU	M.14A	867	ZK-ATD	M.14A	897
			LV-XPW	M.14A	946				ZK-AUK	M.11A	507
I-AITY	M.14A	616	LV-XPX	M.14A	1795	VH-AAC	M.2H	123	ZK-AWX	M.14A	741
I-BONA	M.11A	500	LV-XQE	M.14A	1068	VH-AAH	M.2H	124	ZK-AXD	M.11A	506
I-ZENA	M.3A	163	LV-XQH	M.14A	336	VH-AAS	M.3A	209	ZK-AXN	M.11A	303
			LV-XQI	M.14A	1646	VH-AAT	M.3A	193	ZK-AYW	M.14A	779
LN-BAH	M.2M	128	LV-XQL	M.14A	519	VH-ABN	M.11A	303	ZK-AZX	M.11A	322
			LV-XQN	M.14A	692	VH-ABT	M.3D	266			
LR-AAK	M.14A	571	LV-XQO	M.14A	632	VH-ACE	M.3A	202	ZS-AFM	M.2H	132
LR-AAL	M.14A	415	LV-XQP	M.14A	966	VH-EVG	M.11A	319	ZS-AFW	M.3B	268
			LV-XQQ	M.14A	696	VH-UAI	M.2H	155	ZS-AHH	M.2H	183
LV-FAO	M.2R	210	LV-XQU	M.14A	1030	'VH-UBN'	M.4A	272	ZS-ALT	M.14A	514
LV-FCA	M.2H	143	LV-XQV	M.14A	1704	VH-UGQ	M.2	212	ZS-AMR	M.14A	490
LV-RUX	M.14A	1723	LV-XQY	M.14A	1024	VH-UXN	M.4A	272	ZS-AMS	M.14A	491
LV-XME	M.14A	1739	LV-XRD	M.14A	968	VH-UZA	M.11A	350	ZS-AMT	M.14A	492
LV-XMF	M.14A	903	LV-XRE	M.14A	1681				ZS-AMU	M.14A	493
LV-XMH	M.14A	957	LV-XRF	M.14A	913	VP-KBL	M.2F	116	ZS-AMV	M.14A	511
LV-XMI	M.14A	572	LV-XRG	M.14A	1760	VP-KBT	M.2P	251	ZS-AMW	M.14A	512
LV-XML	M.14A	1051	LV-XRI	M.14A	1771	VP-KCM	M.2N	135	ZS-AMY	M.14A	513
LV-XMN	M.14A	1733	LV-XRO	M.14A	956	VP-KHO	M.11A	314	ZS-ANO	M.5A	275
LV-XMO	M.14A	872	LV-XRP	M.14A	1801	VP-KIK	M.14A	777	ZS-AOY	M.17	791
LV-XMP	M.14A	408	LV-XRQ	M.14A	421	VP-KKF	M.11A	316	ZS-APX	M.2H	176
LV-XMQ	M.14A	764	LV-XRR	M.14A	568	VP-KMM	M.7A	286	ZS-DBF	M.14A	777
LV-XMS	M.14A	427	LV-XRT	M.14A	706	VP-RAC	M.2	6/11			
LV-XMV	M.14A	716	LV-XRU	M.14A	702	VP-YBN	M.3A	140	**MILITARY SERIALS**		
LV-XMW	M.14A	581	LV-XRZ	M.14A	1651						
LV-XMY	M.14A	521	LV-XSE	M.14A	1059	VQ-PAO	M.3A	181	**United Kingdom**		
LV-XMZ	M.14A	734	LV-XSF	M.14A	570				K5924	M.3B	252
LV-XND	M.14A	749	LV-XSG	M.14A	873	VR-HCV	M.3A	149	K5925	M.6	238
LV-XNE	M.14A	731	LV-XSI	M.14A	1744	VR-RAH	M.2J	125	K8626	M.2H	166
LV-XNF	M.14A	673	LV-XSJ	M.14A	409	VR-RAP	M.3A	149	K8889	M.10	433?
LV-XNG	M.14A	565	LV-XSM	M.14A	422	VR-RAV	M.2F	111	K8890	M.10	482?
LV-XNH	M.14A	682	LV-XSP	M.14A	833	VR-SAJ	M.2K	133	L4392	M.16	434
LV-XNI	M.14A	701	LV-XSR	M.14A	1011	VR-SAY	M.14	479	to		
LV-XNJ	M.14A	751	LV-XSS	M.14A	962	VR-SBB	M.11A	504	L4436	M.16	478
LV-XNL	M.14A	1703	LV-XSV	M.14A	1767	VR-TCM	M.7A	286	L5912	M.14	333
LV-XNM	M.14A	878	LV-X246	M.14A	873				to		
LV-XNN	M.14A	410	LV-YCA	M.2R	210	VT-AES	M.2B	12	L5927	M.14	358
LV-XNO	M.14A	1042				VT-AFD	M.2	22	L5928	M.14	360
LV-XNP	M.14A	723	NC191M	M.5	264	VT-AFR	M.2H	124	to		
LV-XNQ	M.14A	829	N72511	M.11A	509	VT-AGH	M.2H	144	L5950	M.14	382
LV-XNR	M.14A	675				VT-AGP	M.4A	141	L5951	M.14	387
LV-XNS	M.14A	562	OD-AAK	M.14A	571	VT-AGT	M.2F	134	L5952	M.14	384
LV-XNT	M.14A	862	OD-AAL	M.14A	415	VT-AGX	M.2F	115	L5953	M.14	385
LV-XNU	M.14A	414				VT-AHC	M.4A	274	L5954	M.14	386
LV-XNV	M.14A	659	OE-DBB	M.3B	233	VT-AHF	M.2R	277	L5955	M.14	388
LV-XNW	M.14A	430	OE-DKA	M.2W	222	VT-AIR	M.2H	122	to		
LV-XNX	M.14A	680	OE-DVH	M.3B	232	VT-AKF	M.11A	309	L5960	M.14	393
LV-XNY	M.14A	564				VT-AKG	M.2H	167	L5961	M.14	383
LV-XNZ	M.14A	1648	OO-ACH	M.14A	964				L5962	M.14	394
LV-XOD	M.14A	921	OO-AJT	M.14A	1693	YI-GFH	M.14A	797	to		
LV-XOE	M.14A	398	OO-CRU	M.14A	777				L6000	M.14	432
LV-XOF	M.14A	1643	OO-FLY	M.3D	269	YR-ADB	M.2H	150	L6001	M.14	359
LV-XOG	M.14A	765	OO-MIC	M.14A	403	YR-AOJ	M.2Z	237	L6346	M.8A	485
LV-XOH	M.14A	1010	OO-SCM	M.17	787	YR-AUK	M.7A	283	L6846	M.7	282
LV-XOI	M.14A	745	OO-UMK'	M.11A	306	YR-BOB	M.2H	174	L6894	M.14	515
LV-XOJ	M.14A	776	OO-UMK"	M.17	787	YR-FIL	M.2H	182	to		
LV-XOL	M.14A	811	OO-ZUT	M.11A	343	YR-GIM	M.7A	288	L6908	M.14A	529
LV-XOM	M.14A	686				YR-ITB	M.2H	182	L6909	M.14B	530
LV-XON	M.14A	627	OY-DAK	M.2	8	YR-ITR	M.2	171	to		
LV-XOO	M.14A	661	OY-DEI	M.2	107	YR-MZM	M.11A	327	L6914	M.14B	535

Appendix 32

Serial	Type	c/n
L6915	M.14A	536
L6916	M.14A	537
L6917	M.14A	630
L6918	M.14A	631
L6919	M.14A	632
L7714	M.15	641
L7717	M.15	1076
L8051 to L8095	M.14A	560 604
L8127 to L8149	M.14A	605 627
L8150	M.14A	811
L8151	M.14A	629
L8152 to L8176	M.14A	642 666
L8200 to L8237	M.14A	667 704
L8249 to L8295	M.14A	705 751
L8326 to L8359	M.14A	752 785
L9705	M.3E	289
L9706	M.13	1-Y
N2259	M.14A	628
N3300	M.9A	330
N3773 to N3817	M.14A	821 865
N3820 to N3869	M.14A	866 915
N3875 to N3914	M.14A	916 955
N3918 to N3945	M.14A	956 983
N3951 to N3991	M.14A	984 1024
N4557	M.14A	640
N5389 to N5438	M.14A	1025 1074
P2150	M.14B	494
P2374 to P2410	M.14A	1611 1647
P2426 to P2470	M.14A	1648 1692
P2493 to P2510	M.14A	1693 1710
P6326	M.15	1077
P6343 to P6382	M.14A	1711 1750
P6396 to P6424	M.14A	1751 1779
P6436 to P6466	M.14A	1780 1810
R4071	M.3D	269
V4739	M.11A	502
W6461	M.17	638
W6462	M.17	790
W6463	M.17	793
W7422	M.11A	315
W9373	M.3D	248
X5125	M.2H	168
X9300	M.3A	189
X9301	M.3A	229
X9306	M.17	789
AV970	M.11A	325
AV971	M.11A	305
AV973	M.3B	256
AV978	M.14A	556
AW150	M.2	175
AW152	M.2	32
BB661	M.14A	539
BB662	M.14A	542
BB663	M.14A	1078
BB664	M.14A	1079
BB665	M.14A	1080
BB666	M,14A	1081
BB667	M.14A	1082
BD141	M.2H	158
BD145	M.11A	346
BD168	M.11A	306
BD180	M.2P	190
BD183	M.11A	507
BS755	M.11	290
BS814	M.11A	317
BS815	M.11A	324
BS818	M.11A	501
DG576	M.3B	280
DG577	M.2F	118
DG578	M.2	24
DG590	M.2H	177
DG664	M.2R	211
DG665	M.2Y	241
DG666	M.2Y	260
DJ713	M.11A	314
DJ714	M.11A	320
DP237	M.11A	342
DP848	M.2H	198
DP851	M.2H	186
DP854	M.11A	319
DP855	M.11A	322
DR611	M.11A	312
DR612	M.11A	318
'DR617'	M.11A	312
EM999	M.11A	307
ES922	M.11A	304
HK853	M.11A	508
HK863	M.2F	126
HL538	M.2P	267
HM496	M.3A	196
HM503	M.12	298
HM545	M.18/2	4426
JN703	M.18/4	unkn
LV768	M.2H	122
MA944	M.11A	309
NF747	M.11A	506
NF748	M.2F	113
NF750	M.2W	215
NF751	M.11A	322
NF752	M.2F	121
TP819	M.17	787
1074M	M.14A	378
1081M	M.2H	166
1099M	M.14A	619
1124M	M.14A	672
1199M	M.14A	365
1254M	M.14A	707
1529M	M.14A	1061
1587M	M.7	282
2145M	M.16	467
2212M	M.14A	975
2617M	M.2	24
2626M	M.2	32
2681M	M.16	440
3015M	M.2Y	241
3016M	M.2H	186
3017M	M.2H	168
3641M	M.14A	766
4020M	M.2F	118
4047M	M.14A	1738
4111M	M.14A	1642
4497M	M.14A	697
4513M	M.14A	1621
4526M	M.14A	610
4527M	M.14A	983
4550M	M.14A	1047
4551M	M.14A	1020
4557M	M.14A	542
4675M	M.14A	945
4770M	M.14A	763
4774M	M.14A	1746
4801M	M.14A	1746
5279M	M.14A	658
5355M	M.14A	1644
5359M	M.14A	1049
5360M	M.14A	526
5361M	M.14A	1713
5362M	M.14A	337
5363M	M.14A	400
5364M	M.14A	943
5366M	M.14A	965
5947M	M.14A	709
5965M	M.14A	1031
6172M	M.14A	369
8379M	M.2H	177

Australia

Serial	Type	c/n
A15-1	M.14A	547
A37-1	M.3D	266
A37-2	M.4A	272
A37-3	M.3A	193
A37-4	M.2H	123
A37-5	M.2H	155
A37-6	M.3A	202
Inst 1	M.14A	547

Dutch East Indies

Serial	Type	c/n
M1	M.2	13

Egypt

Serial	Type	c/n
L201	M.14A	541
L202	M.14A	540
L203	M.14A	543
L204	M.14A	544
L205	M.14A	545
L206	M.14A	546
L207	M.14A	633
L208	M.14A	634
L209	M.14A	635
L210	M.14A	637
L211 to L219	M.14A	812 820
L220	M.14A	798
L221 to L225	M.14A	908 912
L226	M.14A	916
L230	M.14A	920
L231 to L235	M.14A	936 940
L236	M.14A	953
L237	M.14A	954
L238	M.14A	955
L239 to L243	M.14A	926 930
L244	M.14A	1773

Estonia

Serial	Type	c/n
159	M.14A	639

Ireland

Serial	Type	c/n
31 to 35	M.14A	1025 1029
36 to 40	M.14A	1036 1040
73	M.14A	524
74	M.14A	1784
75	M.14A	942
76	M.14A	1769
77	M.14A	1777
127	M.14A	1779
130	M.14A	915
131	M.14A	768
137	M.14A	778

Latvia

Serial	Type	c/n
181	M.14A	636

Malaya

Serial	Type	c/n
12	M.11A	504
31	M.2F	111
32	M.3A	149
39	M.14	479

New Zealand

Serial	Type	c/n
311	M.11A	311
NZ571	M.11A	323
NZ576	M.11A	308
NZ577	M.11A	503
NZ578	M.2H	144
NZ585	M.14A	486
NZ586	M.14	332
NZ587	M.2F	147
NZ588	M.2P	220
NZ589	M.2P	251

Portugal

Serial	Type	c/n
190/1201	M.14A	412
191/1202	M.14A	855
192/1203	M.14A	1741
193/1204	M.14A	1764
194/1205	M.14A	666
198/1209	M.14A	597
199/1210	M.14A	1751

Roumania

Serial	Type	c/n
1	M.2Z	258
2	M.2Z	237
3	M.2Z	245
4	M.2Z	293
5	M.2Z	265
6	M.2Z	296
7	M.2Z	292
8	M.2Z	294
9	M.2Z	297
10	M.2Z	295

South Africa

Serial	Type	c/n
263	M.14A	514
1425	M.3B	268
1427	M.5A	275
1448	M.14A	493
1476	M.14A	490
1479	M.14A	491
1483	M.14A	512
1485	M.14A	513
1505	M.14A	492
1576	M.2H	132
2008	M.2H	176
IS116	M.3B	268
IS117	M.2H	132
IS118	M.2H	176
IS119	M.14A	512

Spain

Serial	Type	c/n
30-55	M.3A	201
30-72	M.2	117
30-117	M.3C	231
30-145	M.2H	172
30-168	M.3A	197
EN-001	M.3A	197
L5-72	M.2	117
L5-168	M.3A	197

Sweden

Serial	Type	c/n
913	M.3A	216

Unidentified Aircraft:
CC-FBN, an M.2 Hawk, which later became CC-PBL-0054 and may have been c/n 16. VH-ACD reserved for an unidentified M.2H.

A Selected Bibliography

A Sound in the Sky	Geoffrey Alington (R K Hudson 1994)
Aviation in Doncaster 1909-1992	Geoffrey Oakes (G H Oakes 1995)
Black Country Transport: Aviation in Old Photographs	Alec Brew (Alan Sutton 1994)
Blossom – Biography of Mrs FG Miles	Jean M Fostekew (Cirrus Associates 1998)
British Aircraft (vols 1 & 2)	R A Saville-Sneath (Penguin 1944)
British Aircraft Specifications File	Ken Meekcoms and Eric Morgan (Air-Britain 1994)
British Flight Testing - Martlesham Heath 1920-1939	Tim Mason (Putnam 1993)
British Racing and Record Breaking Aircraft	Peter Lewis (Putnam 1971)
Gypsies to Jets	Keki R Gazder (K R Gazder 2001)
Hanworth Air Park 1916-1949	Feltham Arts Association (1998)
History of Britain's Military Training Aircraft	Ray Sturtivant (Haynes 1987)
Lindbergh: Flight's Enigmatic Hero	Von Hardesty (Harcourt Brace 2002)
Miles Aircraft since 1925	Don L Brown (Putnam 1970)
Milestones (3 volumes)	Don L Brown (Miles Aircraft 1944-46)
More Tails of the Fifties	Peter G Campbell (Cirrus Associates 1998)
Parnall Aircraft since 1914	Kenneth E Wixey (Putnam 1990)
RAF Flying Training and Support Units since 1912	Ray Sturtivant with John Hamlin (Air-Britain 2007)
Shoreham Airport, Sussex	Tim M A Webb and D L Bird (Cirrus Associates 1996)
Somewhere in the West Country – Whitchurch Airport	Ken Wakefield (Crecy 1997)
Squadrons of the Royal Air Force & Commonwealth 1918-1988	James J Halley (Air-Britain 1988)
Sussex Aviation 1920-1939	Roy Brooks
Tail Ends of the Fifties	Peter G Campbell (Cirrus Associates 1999)
Tails of the Fifties	Peter G Campbell (Cirrus Associates 1997)
The Book of Miles Aircraft	A H Lukins (Harborough 1946)
The Fifties Revisited - An Aerobiography	Peter G Campbell (P G Campbell 1994)
The Menasco Story 1926 through 1991	Ralph J Schmidt (Aerofax 1944)
The Miles Magister	Graham H R Johnson (Newark Museum 1975)
Walsall Aviation	Edwin Shipley and John W T Jeffries (E Shipley 1990)
Wings over Nazeing – Broxbourne Aerodrome 1929-45	Leslie A Kimm (L A Kimm 2005)
Wings over Woodley	Julian C Temple (Aston Publications 1987)

Ranald Porteous and Tommy Rose with a representative selection of Miles aircraft outside the Falcon Hotel, Woodley, in about 1938.
[Phillips & Powis via The Miles Aircraft Collection]

Index

References in this index apply only to the text in Chapters 1 to 31 and not to Appendices. However all illustrations throughout the book are included.
Page numbers in *italics* refer to illustrations.

AERO magazine 142
Aero Trades Inc 169, 170-171
Aeropilot 135, 264-265
Aeroplane, The
    1931   45-46, 53-54
    1932   67
    1933   98-99, 100, 101, *102*
    1934   71, 74, 121-122, 123, *124*, 141
    1935   128, 153, 183-184, 186
    1936   41-42, 81 85, *86*, 135, *152*, 153-154, 155,
                165-166, 186, 190, 192, 221, 223
    1937   85, *86*, 115, 128, 131, *220*, 231, 264
    1938   93, 173-174, 206, 235-237, 248, 288, 289
    1939   212, 215
    1944   174
    1945   116
    1947   116
    1948   297
Aeroplane Monthly 115, 249-251, 258
Aeroplane Spotter, The 215, 249
Aeroplastics Ltd 293
Air Ministry and Master Mk.I 207, 211, 212
Air Ministry and trainer aircraft 87, 88, 90, 91, 194, 200, 201, 205, 206, 207, 263-264
Air Pictorial 55
Air Touring 147, 178
Airwork Ltd 288, 289
Alfriston (Drusilla's) aerodrome 32-33, 39, 40
Alington, Geoffrey 147, 175, *176*, 178
Allen, Maj George William Graham 74
Allen, R E H 67
Ames, Capt David 43
Anderson, Andy *229*
Anderson, Byron 244, 248
Angell, Jack 211
Arscott, P/O Gordon Owen 288
Ashdown (AID Inspector) 19
Atcherley, Flt Lt Richard Llewellyn Roger 'Batchy' 38
Australia, MacRobertson Air Race from England to 73, 74, *121*, *124*, 141-142
Australian Air Board 271
Aviation Historical Society of New Zealand Journal 154-155
Avro (A V Roe & Co Ltd) 27, 29, 30, 35, 49, 50
    504K 16, *21*, *26*, 33-34, 41-42, 45
        G-EAAY 25-26
        G-EAJU 27, *27*, 35
        G-EATU 17, 19, 24-25, *26*, 33
        G-EBJE *18*, 25, 29, 31, *31*, 32, *38*, 47
        G-EBVL 40, 49
        G-EBWO 61, 62
        G-EBYB 33, 41, 42
        G-AACW 33, 35
        G-AACX 35
        G-AAGG 35, 62
    504N G-EBHD *63*
    534 Baby G-EAUM 29-30, *30*, *31*, 49
    547A Triplane G-EAUJ 30-31, 32
    548A G-EBKN 32, *33*, 39
    552A G-ABGO 27
    594 Avian 64
    594 Avian IIIA G-EBVA 33
    594 Avian IV G-AADF 33, 55, 56
    616 Avian IVM G-AAHJ 47
                  G-AAWH 81, 223
                  G-ABIE *High Test* 74, 153
    621 Tutor 267

Balbo, AM Italo 248
Bancroft, Denis 174
Bandidt, Grp Capt 'Bush' 174
Barraclough, Martin 120
Barrett, A G 106
BAT Crow 139
Battye, Mrs 67, *331*
Beamont, Wing Cdr Roland 32
Bebb, C W H 42, 45
Bellairs, Lionel Edward Richard *26*, 33, 35, 38, 45, 47, 49, 50, 53, 173
Bergerie, Gaston 147
Berks, Bucks and Oxon Flying Club 57, 59, 61, 62, 67, 70, 71
Berkshire Chronicle, The 71
Bianco, Franco 133-134, *134*
Birkett, Flt Lt George 74, 161, 162, 163
Birkett, Nancy B 33, 41, *43*, 44, 45, 47
Birkett Air Service Ltd 161, 162, 163
Bishop, Cyril Albert Nepean 'Bish' 72, 108
Blackburn Aeroplane & Motor Co Ltd 71
Blackburn Aircraft Ltd 271
Bona, Gaspare 233
Book of Miles Aircraft, The 244
Boswell, Perry 170, 171-172
Botting, Tommy 87, *98*, 257, 258
Boucher, Cecil 25
Boucicault, Dion 44
boundary layer control research 92, 194, 195, *197*, 198, 223, *224*, 225
Bournon, Ray 190
Bowles, Walter 290
Boyes, Robert T 233-234
Bradbrooke, F D 45-46, 53-54, 100, 131
Brancker, AVM Sir John William 21, 23-24, 41
Bremridge, P A R 67
Brighton, Annual Gala air display (1930) 41
Brighton-Shoreham Aero Club 17
Bristol 142 *Britain First* 199
       142M Blenheim 199
       Blenheim I 87, 199
       Bulldog 199
       Fighter 24
Bristol, Whitchurch Airport 63
Bristol Air Race (1951) 118
Bristol Air Taxis 63
Bristol Evening Post 124
Bristow, Mr 60
British Aircraft, Volume I 127-128
British Hospitals Air Pageants 59
Brook, Harold Leslie 74, 141-142, *142*, *143*, *144*
Brooke-Smith, Tom 89
Brooklands Aviation Ltd 48, 81
Brown, Cdr Eric 297-298
Brown, Don 15, 16, *26*, 32, 35, 67, 75, 85, 127, 174
    The Aeroplane article 41-42
    and Avro 504Ks 19, 25, 27, 29, 40
    and Avro 534 Baby 30
    career 74

    and Desoutter 44
    and Hawk 68, 83, 100
    and Hawk Major 73
    and Hobby 257-258, 262
    and Hornet Baby 50
    and joy-riding 33, 40
    learns to fly 33, 45, 47
    and M.9 Trainer 201
    and M.18 Trainer 291, 297
    and Nighthawk 186, 188
    and Peregrine 190
    and pusher aircraft 139, 140
    and Queen Wasp 219
    and Satyr 60
    and Whitney Straight aircraft 223, 227, 233
Brown, Vernon 267-269
Browning, Neville 251
Bruce, the Hon Mrs Victor 59, 60, 72
Buffaloe, Tom 119, 120
Burke, Flt Lt 261, 262
Burnett, Victor *85*, 209
Bystander, The 101

Cape Town 47
    record flights to and from (1936) 81, 153-154
Capley (formerly Kaepelli), Walter Gustav 92, *93*, 291
Cardigan, Rt Hon Earl of 67
Carson, Samuel A 234
Casey, Louis S 251, 253
Cathcart-Jones, Lt Owen 75, *78*, 122-123, 133
Centaur IV
    G-EABI 21
    G-EALL 19, 21
Central Aircraft Company 21
Chapman, G 102
Chichester, Summersdale Copse 32
Chichester flying display (1930) 42
Chichester Observer 42
Cierva C.8V Autogiro G-EBTX 27
Cinque Ports Flying Club 81
Cinque Ports International Meeting, Wakefield Cup race (1934) 112
Cirrus-Hermes Engineering Co Ltd 71, 97
Civil Aviation Authority (US) 170-171
Clarke, Bert 60, 67
Clarke, F M 155
Clarke Cheetah G-AAJK 42
Cliff, Stephen 63, 108
Clouston, Flt Lt Arthur 115, 181
Cobb, John 25
Cobham, Sir Alan 45
Coles, Jack 61
Commonwealth Aircraft Corporation Wackett 271
Comper
    Streak 101
    Swift 100
        G-ABRE 172-173
        G-ABUS 118
Conrad, Carl L 171
Contact magazine 227
Cook, Monty 212
Cowley, Sqn Ldr A T 54
Cox, Tim 178
Cranfield, College of Aeronautics 271
Croxford, Lt Cdr C W 'Stiffey' 63, *64*

Dack, Ron 298
Daily Express International Air Races (1950/52) 118
Daily Herald 100

Daily Mail 199
Daily Mirror *69*, 95, 100
Daily Telegraph 125, 206, 227, 288-289
Dalziel, Dr Ian 290
Darby, Lt Col Maurice Ormonde 81, 90
de Havilland
    DH.60 Moth *61*, 64, 100
        G-EBOT 61-62
    DH.60G Moth
        G-AAHI 67
        G-AAJJ 74
        G-AAKN *315*
    DH.60M Moth
        G-AATB 83
        G-ABWM 56
    DH.60X Moth
        G-EBVC *316*
        G-EBVD *315*
        G-EBZG *Jemimah* 33, *44*, 45, 47
        G-EBZI or L *316*
        G-EBZL 65
    DH.80A Puss Moth
        G-ABKD 74
        G-ABNZ 83, 221
        Heart's Content 141, 142
    DH.82A Tiger Moth 81, 89, 264, 271
        G-ADJB to G-ADJJ 78, 81
        G-ADJI 78, *80*, 81
    DH.86 biplane airliner 255
    DH.88 Comet G-ACSS 115
    DH.93 Don 88, 201, 202
    DH.100 Vampire 198
de Havilland, Geoffrey, senior and junior 68
de Valera, Eamon 247
Dermot, Adelaide 43-44
Dermot, Thomas 43, 44
Desoutter I G-AATF and G-AAWT 44, 45
Desoutter II G-AAVO 63
Dinkas 231
Dismore, Captain 21
Dominion Aircraft Ltd 35
Don, John Colquhoun 35
Douglas and Clydesdale, Marquess of 38
drag, forms of 181, *181*
drag research 261-262
Dunkerley, Alan 178
Dunkerley, Fred 178, 290

Earle, Wg Cdr 251
East Fortune, Museum of Flight 298
Eastern Evening News 106
Edgar (Norman) & Co 63
Edwards, Connie 251
Edwards, Flt Lt Hugh R A 75, *78*, 193-194
Edwards, 'Tich' *213*
Elliott, Gertrude 44
Elliott, Maxine 44
Emmet, A M 67
Evetts, John 108

Fairey Battle 87, 199
Fairey Fox G-ACAS *60*
Fairweather, William 127, *128*
Falla, N S 155
Farman Pusher biplane *17*
Farnborough, Royal Aircraft Establishment 75, 92, 155, 157, ...
    179, 180, 182, 261-262, *262*
Ferguson, Hon Mr 54
Findlay, M H 67

flaps, advances in 73, *107*, 108, 123, *124*, 126-127, 135, 174, *175*, *177*, 229, *230*, 231
Flight (film) 42
Flight International (1981) 271
Flight magazine 233
    1933 70, 99, 100-101, 111
    1934 48, 72, 73, 74, *105*, 111, 123, 125-127
    1935 142-145, 162-163
    1936 186, 193, 245
    1937 *265*
    1939 95, 195, 198, 223
    1960 38-39
Flying 97-98
Flying to the Limit 32
Fontes, Luis 112, 115, 116
Fontes, Ruth 112, *326*
Forbes-Robertson, Diana 43
Forbes-Robertson, Maxine 44
Freeman-Thomas, the Hon Inigo Brassey (later Lord Ratendone) *37*, 41, 42, 43, *43*, 45, 48
Freeman-Thomas, the Hon Maxine Mary 'Blossom' *37*, 38, 41, 42-43, *43*, 45 *see also* Miles, Maxine Mary 'Blossom'
    autobiographical details 43-44
    divorced 48
    falls in love with 'Miles' 42, 47
    marries 'Miles' 48, 57
    and Metal Martlet 53, 54

Gage, Viscount 45
Gamble, John 265
Garage and Motor Agent, The 72-73
Gates, Charles 16
Gates, Grahame 262, 280
Gazder, Keki R 166
Gear, Mr 17
General Strike (1926) 21
Georgic, SS 169
Giddy, S W 'Pat' 39
Gloster Gauntlet 199
Gnat *15*, 16, *17*
Gnat Aero & Motor Company 19, 21
Gnat Aero Co Ltd 25, 31 35 *see also* Southern Aircraft Ltd
Godfrey, K P 258
Goodfellow, Lt Alan Peter, RNVR 159-160
Gordon, Mr *43*
Gower, Miss Pauline 63, *64*
Grahame-White GW.15 Boxkite *18*, 19, *20*, 21, 26
Grahame-White GWE.6 Bantam G-EAFL 19, *20*, 21
Grey, C G 24, 42, 81, 84, 85, 123, 128, 155, 173-174, 206, 248
Guest, Sqn Ldr the Hon Frederick Edward 38, 47
Gutt, Camille 289
Gypsies to Jets 166

Hackett, Leonard 81, *189*
Haizelden, Constable 32
Halford, Maj Frank 112
Hall-Warren, Norman 55
Hamilton, C L F 155
Hammond, Darren 253
Handley Page 198
Hannett, K G 148
Hanworth Air Park 50
Hart, Bert 25
Hart, Ruben *15*, 16
Hart-Still, S 293
Hawes, A E 'Ted', OBE 27
Hawes, A H 'Jimmy' 19, 25, 27
    wife 27
Hawker
    Demon 199
    Fury 199
    Hart Trainer 200
    Hurricane 81, 87, 88, 199, 200, 210, 215
    Hurricane Mk.I 93, 95
    Nimrod 59
Hawker Siddeley 125: 198
Hayter, Mr 67
Head, (Anthony) Graham 32, 33, *33*, 39-40, 41, 47
Head, Don 135
Heinkel He.70 G-ADZF 204
Henderson, Basil Balfour 42, 53, 221
Henderson, Col G L P 25
Henderson Falcon IV 42
Henderson School of Flying, Brooklands 61
Hendon RAF Display (1937) 201-202, *203*, 267, *268*
Hendy Heck 83, 154
    G-ACTC 221
Henshaw, Alex 131
Herbert Engineering Ltd 61
Heston Aircraft 275, 280
Hillson Bi-Mono 174
Hollinda NV 204
Holloway, Peter 148
Hood, David 119
Hornet Baby 26, *26*, 35, 49-50, *49*, *50 see also* Southern Martlet
    G-AAII 35, *36*, 50, *51*
Howbrook, Cpl 231
Hull, Harry 25, *26*, 35, 67, 98, 100, *101*, 186
Humble, William 'Bill' 112, 115, 173-174, *174*

India, Aeronautical Training Centre of 134
Indian Aviation 134
Innes, Veronica 89-90

Jameson, J L, Ltd 295
Jeffs, J J 'Jimmy' 29, 45, 47, 67
Johnson, Amy 63
Johnson, 'Johnny' *213*
joy-riding 27, 31, 32, 33, *38*, 39-40, 42

Kaepelli (later Capley), Walter Gustav 92, *93*, 291
Keep, Capt Stuart 235
Kendall, Hugh 116, 117
Kennedy, Flt Lt Hugh V *96*, *96*, 159-160, 188, 210, 218, 262, 280, 293, 294
Kent, Reg 288
King-Clark, 2nd Lt Rex 231
King's Cup Air Races 118, 222
    1930 38
    1934 73, 111, *111*, 112, 121, *122*
    1935 75, *78*, 112, 131, *133*, 149, *149*, *150*, 151, *151*, 153, 154, 167, 168, *168*
    1936 115, 169, *169*
    1937 115, 169, 231, *232*, 257
    1938 115, *116*, 174
    1949 118
    1950 118, 178
    1952 118
    1957 178, 290
    1958 290
    1972 119
    1983/84 290
Kirk, F 175
Koolhoven, Frederik 'Frits' 139

Lacayo, Mr 288
Langley Day Air Races (1940) 171
Lawn, F/O John F 'Laddie' 57, 59, 63, *64*, 65, 67, 72

leading edge slats experiment 210, *210*
Leamington & District Flying Club 178
Leech, Flt Lt Halliburton H 'Girlie' 30, *30*, 38, 39, 47, 50-51, 53, 54, 100, 101
Lewin, Brig Gen and Mrs Arthur 231, *232*
Lewis, Leslie 26
Light Plane, The 251
Lilley, Bert 250
Lindbergh, Anne 247-248, 250
Lindbergh, Col Charles 85, 223, 243-244, *243*, 245, *246*, 247-249, *247*, 250
Lister, Sir Phillip Cunliffe *151*
Llewellyn, David 154
Lukins, A H 244
Luxury Air Tours Ltd *57*, 59

M.1 Satyr G-ABVG 47, 48, 56, 57, *57*, *58*, 59-60, *59*, *60*, 63-64, 67
    specification and performance data 60
M.2 'Colonial' Hawk 106, *106*, 108
M.2 Hawk 59, *66*, 67-68, *68*, *69*, 70, 71, 73, 75, *80*, 97-102, *97*, *101*
    advertisement *102*, *105*
    brochure details 102, 106
    dispute with purchaser 74
    flaps, split *107*, 108
    manufacture *103*, *104*
    production 102, 108
    running costs 106
    specification 110
    wing fold *98*
    CH-380 *321*
    G-ACGH (G-ACJY) 101, 106, *317*
    G-ACHJ 102, *104*
    G-ACHL 102, *104*
    G-ACMX *319*
    G-ACNX *320*
    G-ACOP *Ruddy Duck* 72, *321*
    G-ACTN *322*
    G-ACUD *107*, 108
    G-ADGI *323*
    G-ADGR 108, *108*
    OY-DEI *322*
    PK-SAL *318*
    SE-AFS *323*
M.2A Cabin Hawk G-ACLI 108, *109, 318*
M.2B Hawk Single-seater VT-AES 108, *109*
M.2C Hawk Sports Two-seater 108
M.2D Hawk Three-seater 108, *110*
    flaps 123
    G-ACPC and G-ACPD *72*
    U1 *319*
M.2E Hawk (Gipsy VI) G-ACTE 70, 111-112, *111*, *112*, 113, 115, *324*
    specification and performance 120
M.2F Hawk Major 73, 108, 121-124, 138
    advertisement 123, *124*
    specification 131-132
    G-ACTD 73, 121, *122*, *327*
    G-ACWV 124, *125, 328*
    G-ACWY 124, 125
    HK863 *330*
    SX-AAD *330*
    ZK-ADJ *Spirit of Manawatu* 73, *121*, 124
M.2G Hawk Major G-ACYB/HB-OAS *125*, *126*
M.2H Hawk Major 73, 108, 125-127, *125*
    flaps 126-127
    RAF service 127
    CS-AAL *334*
    CX-AAW
    CX-ACT *331*
    DG590 *336*
    G-ACYX *126*, *330*
    G-ADAB *331*
    G-ADCF *332*
    G-ADFC/VT-AKG 127, *128*
    G-ADGL *334*
    G-ADLA *335*
    G-ADMW 127, *129*
    HB-ERI *333*
    K8626 *334*
    PK-SAR *335*
    R 276 *332*
    VH-UAI *333*
    YR-ADB *332*
    ZS-AHH *336*
M.2J Hawk Major 127, *129*
M.2K Hawk 108
M.2L Hawk Speed Six G-ADGP 111, 112, *114*, 115-120, *116*, *117*, *119*, *120*, *325*
    specification and performance 115, 120
M.2M Hawk Major Three-seater 127
    specification 132
    G-ADCV *333*
    LN-BAH *129*
M.2N Hawk Major de luxe G-ADGE 127-128, *130*
M.2P Hawk Major de luxe 128, *130*, 135, 138
    specification 132
    G-ADLO *337*
    ZK-AFJ *338*
M.2R Hawk Trainer Mk.I 128, 133-134, *133*
    CC-FBB 133-134, *134*
    G-ADLB 75, *80*, 133, *339*
    G-ADLN 75, *79*, 83, 133, 134, 223
    VT-AHF 134, *342*
M.2S Hawk Major 128, 131
M.2T Hawk Major 131
M.2U Hawk Speed Six G-ADOD 111, 112, *114*, 115, *173, 326*
    specification and performance 120
M.2W Hawk Trainer Mk.II 75, 78, 83, *88*, 89, 135, 138, 265
    G-ADWT 78, 127, 138, *339*
    G-ADWU 78
    G-ADWV 78
    G-ADZA *423*
    G-ADZC *341*
    G-ADZD *84*
M.2X Hawk Trainer Mk.II G-ADYZ 85, 135, 265
M.2Y Hawk Trainer Mk.II 135, 137-138, 265-266
    G-ADVD 135
    G-ADVF 135, 168
    G-AEEL *342*
    G-AENT 135, *136*, 137-138
M.2Z Hawk Trainer Mk.II 135, 210
    G-AEHP & others *341*
    G-AEHR *340*
    Romanian AF '7' *342*
M.3 Falcon 73-74, 75, *80*, *88*, 141-144, *152*
    G-ACTM 141-142, *141*, *142*, *143*
    G-ADLS 75
M.3A Falcon Major 74, 144-145, 147-148
    cabin structural details *145*
    specification and performance data 148
    G-ADBF 144
    G-ADFH 147, *147*
    G-ADHC//I-ZENA 145, *147*
    G-ADHH *344*
    G-ADLI *345*
    G-AEEG 144, *146*, 148

INDEX 433

SE-AFN/8-80 *346*
SU-ADA *344*
VH-AAT 148, *148*
VR-HCV *344*
VR-RAP 148
M.3B Falcon Six 74, 149, *149*, 151, 153-155, 159-160
    cabin assembly *83*
    general arrangement *157*, *159*
    specification and performance data 160
    wing experiments 155, 157, *157*, 159, *159*
        G-ADLC *Preggers* 75, *78*, 81, 149, *149*, *150*, 151, 153-154, 168
        G-ADTD *349*
        G-AECC 290, *351*
        G-AFAY *348*
        G-AFBF/AV973 159-160, *349*
        K5924 155, *156*
        ZK-AEI 154-155
M.3C Falcon Six EC-ACB *348*
        G-ADLS 151, 154, *348*
M.3D Falcon Six 154
        F-BBCN *350*
        G-AEDL *219, 349*
        VH-ABT *350*
M.3E Falcon Six L9705 ('Gillette' Falcon) 155, 157, *158*, 159
M.3F Falcon Six R4071 159
M.4 Merlin G-ADFE 74, 161-163, *161*, *162*, *163*
    specification and performance data 166
M.4A Merlin 165-166, *165*
        VH-UBN/VH-UXN *164*, 165
        VT-AGP 165-166, *165*
M.5 Sparrowhawk 167-172
    specification and performance 178
        G-ADNL 75, 81, 167, *167*, 168-169, *168*, *169*, 175, *176*, 178, 244, *353 see also* M.77 Sparrowjet
        G-ADWW/NC191M 169-172, *170*, *171*
M.5A Sparrowhawk 172-175, 178
        flap experiments 174, *175*, *177*
        G-AELT 172, *172*, 173, *173*
        G-AFGA/U5 173, 174, *174*, *175*
        G-AGDL 174-175, *178, 354*
        U3 174, *177*
M.6 Hawcon K5925 75, 179-182, *179*, *180*, *182*
    specification and performance data 182
M.7 Nighthawk 81, 159, 183, 184, *184*, 186, 188, 223
    cabin *184*
    specification and performance data 188
        G-ADXA *183*, 186
        G-AEBP/U5 186, *187*, 188, *355*
        L6846 *355*
M.7A Nighthawk 188
        U6/10 *84*, 244
        VP-KMM *356*
        YR-GIM *356*
M.7A Nighthawk Hybrid G-AGWT 188, *188, 356*
M.8 Peregrine Mk.I 85, 92, 189-190, *190*, 192-194, 198, 264
    cabin *190*, *191*
    specification and performance data 198
        G-AEDE/U9 189-190, *189*, *191*, 193-194, *193*, 198
M.8A Peregrine Mk.II L6346 91, 194-195, *194*, *195*, *196*, *197*, 198
    specification and performance data 198
M.9 Kestrel 88, 90, 92, *199*, 202-203, 204, *204*, 208, 259, 261
    Second Phase 205-206, *205*, 207, 208-209
    specification and performance data 210
M.9 Master 87, 90, 93, 95
    N7408 93
M.9 Rolls-Royce Private Venture Trainer 81, 87, 88, *89*, *199*, 200, *200*, 201-203, *202*, *203*, 204, 205, 206

    specification and performance data 210
    G-AEOC 201
M.9A Master prototype N3300 207-208, *207*, *208*, *209*, 210
    leading edge slats 210, *210*
    specification and performance data 210
M.9B Master Mk.I 206-207, 211-212, *212*, *213*, 215, *217*, 218
    cockpit *215*
    instructor's position *216*
    specification and performance data 218
    throttle and control box *217*
    N7408 *211*, 212, *213*, 215, *218*
    N7410 *214*, 215
    N7422 218
M.9C Master Mk.IA *217*
M.10 Queen Wasp r/c target 159, 200, 219
    specification and performance data 219
M.11 Whitney Straight 194, 200, 285
    specification and performance data 234
M.11 Whitney Straight Special G-AECT 84, *84*, *203*, 221-224, *221*, *222*, 227, 229, *229*
    boundary layer control experiments 223, *224*, 225
    flap experiments 229, *230*, 231
M.11A Whitney Straight 84, 89, 90, 91, 135, *220*, 224, *226*, 227, *227, 228*, 231, 233-234
    in RAF service 233
    F-AQCZ *364*
    G-AENH *226*, 227
    G-AERS 227, 231
    G-AERV 233-234
    G-AETB *361*
    G-AETS *361*
    G-AEUJ *361*
    G-AEUX *362*
    G-AEUZ *362*
    G-AEVH 231
    G-AEZO 231, *232*
    G-AFGK 233, *366*
    G-AFZY *365*
    HB-URO 233
    NZ576 & others *360*
    NZ577 *365*
    OO-UMK *360*
    U1 *366*
    VR-SBB 233, *234*
    ZK-AEO *360*
    ZK-AXD *365*
M.11B Whitney Straight G-AERC 227, 229, *229*
M.11C Whitney Straight G-AEYI 93, 233
M.12 Majestic? 244, *253*
M.12 Mohawk 135, 243-251, *243*, *246*, *247*, *249*, *250*, *252*, 253-254
    general arrangement *244*
    specification and performance data 254
    G-AEKW/U8 83, 85, 245, 247, *247*, 248-251, *249*, *250*, *252*, 253-254, *254*
    G-AEKX 254
M.13 Hobby G-AFAW/L9706 *256*, 257-259, *257*, *258*, *259*, *260*, 261-262, *261*, *262*
    specification and performance data 262
M.14 Hawk Trainer Mk.III 264-265, *264*, 273
    specification and performance data 138, 274
    G-AETJ 264
    G-AEZR *372*
    G-AEZS 266, *274, 374*
    G-AHNU 117
    G-AKRV 118
M.14 Magister 85, 87-89, *88*, 90, 91, *95*, 135, 194, 200, 263-265, *263*, *264*, *265*, 266, 271, 273, 291
    modifications 269, 271

RAF contracts 264, 273-274
  sideslip problem 269
  specification and performance data 274
  spinning accidents 265-266, 267-269, 271
    L5912 265-266, *266*, 267, *367*
    L5913 267
    L5916 *268*
    L5932 *368*
    L8168 269
    NZ586 *367*
M.14A Hawk Trainer Mk.III 271, 273
  specification and performance data 274
  Egypt *374*
  VR-SAY *371*
  ZK-AFA *372*
  ZS-AMR *372*
M.14A Magister Mk.I 264, 267, *270*, 271, 273
  c/n 159: *272*
  No.29 E&RTFS Luton *273*
  specification and performance data 274
    A15-1 271, *272*
    L5933 266, *266*
    L5990 to L5996 *272*
    L8334 to L8343 *273*
    L8338 *270*, *273*
    N3806 *290*
    P6373 *403*
M.14B Hawk Trainer Mk II 271, 273
  specification and performance data 274
M.15 Trainer Mk.I 275, *275*, *276*, 277, *278*, 279-280, *280*
  design requirements 277, 279
  operational requirements 275
  specification and performance data 280
M.15 Trainer Mk.II 280
  specification and performance data 280
M.16 Mentor 186, 281, 284
  general arrangement *283*
  specification and performance data 284
    L4392 281, *282*, 284, *292*
    L4393 *281*, 284
    L4396 *407*
    L4420/G-AHKM *284, 408*
M.17 Monarch 93, 135, 285-290, *286*
  c/n 793 *288*
  'Glide Control' 285-286, *286*, 287
  interior *287*
  RAF aircraft 290
  specification and performance data 290
    G-AFCR/W6461 90, 93, *285*, 286, *286*, 290
    G-AFGL *288, 289*
    G-AFJU 289, 290, *410*
    G-AFLW/c/n 792 *288*, 290, *290, 410*
    G-AIDE/G-AFRZ 290, *411*
    OO-UMK/G-AGFW 289
    U-0226 *409*
M.18 Trainer Mk.I 92-93, 291-295, *292*
  dimensions 294
  specification and performance data 298
  wing *292*
  U2/U-0222 291, *291*, 293, *293*, 294, *294*, 295
M.18 Trainer Mk.II 295-296, *295*, *296*, 298
  HM545/G-AHKY *296, 412*
M.18 Trainer Mk.III 297, *298*
  U3/G-AHOA *412*
M.18 Trainer Mk.IV HL 297-298, *298*
  U11 *412*
M.19 Master Mk.II 95, 212, 218
M.20/1 'Munich Fighter' 93, *94*, 95
M.28 231

M.38 Messenger 294
M.38A Mariner 294
M.47, M.47A and M.49 pilotless r/c target aircraft projects 219
M.52 high-speed research aircraft 157, 159
M.60 Marathon 255
M.65 Gemini G-AISM 117
M.65 Gemini 3 G-AKDC 118
M.77 Sparrowjet 178, 290 *see also* M.5 Sparrowhawk G-ADNL
McArthur, James 'Gordon' 169, 244, 254
McCarthy, Desmond 119
MacGregor, Sqn Ldr Malcolm 124, 154, 155
Maclaren, Owen 233
McMullen, Mr 67
MacRobertson England to Australia Air Race 73, 74, *121*, 124, 141-142
McWade, Chief Engineer *166*
Malayan Volunteer Air Force 148
Manx Air Derby, Tynwald Air Race (1947) 116, 117
Martinsyde F.4 G-EBMI 38-39, *40*
Martlesham Heath, A&AEE 135, 137-138, 188, 205, 206, 208, 265, 266, 281, 284, 291-293, 295
Mason, H H 41
Maxine 43
Maxwell, Pat 169, *169*
Maxwell, W R 'Roy' 54
Menasco, Al 81, 85, 90, 169, 243, 244
Menasco aero engines 81, 83
Menasco Story, The 243
Menezes, Peter *166*
Mermagen, Air Cdre A W 267
Milan Aeronautical Exhibition (1937) 233
Miles, Dennis 15, *15*, 16
Miles, Esther 15
Miles, Frederick Gaston 15, 21, 25, 48
Miles, Frederick George ('Miles') *4, 9*, 15, *15, 18, 26, 43*, 88, *94, 144, 185*
  and advanced trainer 199, 200, 206
  agreement with Charles Powis 65-67
  and air display (1926) 21, 24
  applies for Air Ministry approval as manufacturer 74
  autobiographical details 15-16
  becomes managing director of Phillips & Powis Aircraft Ltd 90
  and Bill Skinner 95
  boundary layer control research 92, 194, 195, *197*, 198, 223, 225
  Brighton air display (1930) 41
  and caravan 47, 57
  and Charles Lindbergh 85
  cinema jobs 15-16
  Clarke Cheetah flight 42
  and Desoutter monoplane 44
  as director of Phillips & Powis Aircraft Ltd 74
  and Dominion Aircraft Ltd 35
  emigrates to and returns from Cape Town 47, 54
  and Falcon 73-74, 141, 142, 144, 149
  falls in love with 'Blossom', Lady Ratendone 42, 47
  flap experiments 73, 108, 123, 229
  and flying instruction 183-184, 186
  forms Gnat Aero & Motor Company with Cecil Pashley 17, 19
  grandparents 15
  and Hawcon 180, 181
  and Hawk 67-68, *69*, 70, 72, 97-98, 99, 100, 101, *101*, 108
  and Hawk Major 121
  and Hawk Speed Six 111, 112
  and Hobby 257, 258, *259*
  Hornet Baby (Southern Martlet) conversion 35, 50, 51

joy-riding 40
and Kestrel 88
and King's Cup Air Race (1935) 75
learns to fly 19, 33
leases Easter's Field 19
licences obtained 27, 29, 33
and M.15 Trainer 275, 280
and Magister/Hawk Trainer 263, 264, *265*
and Marathon transport project (1936) 255
marries Blossom 48, 57
and Martinsyde F.4 38-39
and Master 212, *217*
meeting with Charles Powis 64-65, 97
and Merlin 74, 162
modifies Avro 534 Baby 29-30, *30*
and Mohawk *243*, 244, 245
and Monarch 285, 289
night flying 47
and Nighthawk 184, 186, *187*
and Peregrine 189-190, 192, 193, 194, *196*
purchases Avro 504Ks 25-26
purchases Grahame-White and Centaur aircraft 19, 21
purchases spares from A V Roe 27, 29
and pusher aircraft 139-140
and RAE Farnborough 75
at Reading 62, 63-64, 67
and Rolls-Royce Trainer 81, 87, 88, 200, 201, 202, *202*, 205, 208, 209
and Satyr 57, *58*, 63-64
Shoreham flying display (1934) 48
Shoreham luncheon and flying display (1931) 45, *46*, 47, 53, 54
and Sparrowhawk 167, 168, *168*, *169*
summons for low flying 32, 40
test flying with Parnall 55-56, 57
visit from AVM Brancker 21, 23, 24
visit to USA 85
and Whitney Straight 83, 84
and Whitney Straight aircraft 222, 223, *226*, 227
and X.2 project 235-237
Miles, George Herbert 15, 19, 33, 67, *83*, *85*, 90, 298
and Bill Skinner 210
and Hawk 101
and Hawk Speed Six 115
and Hawk Trainer 134-135, 138
joins brother at Woodley 81, 83
learns to fly 33
and M.9 Trainer 201
and Magister 273
management of Southern Aero Club 47, 48
and Master Mk.II 95
and Merlin 163
and Mohawk 244, 245, *246*, *247*, 248
and Monarch 93, 285, 286-287, 288
obtains 'A' Licence 47
and Peregrine 190, 192, 194, 195
and pusher aircraft 139, 140
at Shoreham *26*, 39, 41
and Sparrowjet 178
and Whitney Straight aircraft *84*, 227, *229*, 231
Miles, Horace 50, 53
Miles, Jeremy 16, 47, 192
Miles, Maxine Mary 'Blossom' 67, *69*, 72, *85*, *94*, 99, *101*, 288
see also Freeman-Thomas, the Hon Maxine Mary 'Blossom'
and Hobby 257
and King's Cup Air Race (1935) 75
and Magister/Hawk Trainer 88, *88*, *263*, 264
marries 'Miles' 48, 57
and Master *217*

and Mohawk *243*
and Nighthawk 186
and Sparrowhawk 167, *168*, 172
Miles, Reginald 15
Miles Aeroplane Company 67, 98
Miles aircraft *see also* Gnat; M.1 etc. *entries*; Southern Martlet; Southern Metal Martlet; X2 entries; 'X' Minor
deliveries, weekly, in 1938 *92*
'Kestrel high speed trainer' proposal 203-204
Marathon transport project (1936) 255, *255*
production, 1933-39 92
Pusher 139-140, *139*
Miles Aircraft since 1925 127, 251
Miles Magazine 19, 289
1938 38-39, 93, *96*, 134, 165, 233
1939 115
1942 138
1946 127, 148
1947 117
Milk, E S 50
Miller, Jay 251
Mills, Capt Roger 120
Milne, Bob 71, 72
Mitford, Unity Valkyrie 147
Moir, Flt Lt James F 78, 81, 168-169, 186, 247, 280, *421*
More Tails of the Fifties 89-90, 174-175

Nair, Govind 128, 131
Nash, C L 92, 219, 227, 233, 244, 259, 261, 289
National Air Races, Baginton 290
National Flying Services 70-71
Nevada State Journal 248
New Zealand Government 194
Newsletter, The 234
Nine Lives Plus 59
Norfolk and Norwich Aero Club 106
North American Harvard Mk.I 207, 208
Northamptonshire Aero Club 81
Northesk, Rt Hon Earl of 62, 63

Ogden, F/O Cecil Victor 'Micky' 124-125
Ogilvie, David 271
Old Warden *see* Shuttleworth Trust
Ontario Provincial Air Service 54
Osborne, Tony 119
Overton, Dolph 251

Paine, Ron 117, 118, 119, *119*, 120
Painter, John A 249-251
Parnall (George) & Co 48, 55, 56, 57
Elf G-AAFH 55, 56
Elf II G-AAIN *55*, 56, 57
Elf II G-AAIO 56, 57
Imp G-EBTE 55, *56*
J1 275, 280
Parodi, Giorgio 145, 147, *147*
Pashley, Cecil Lawrence 15, 16-17, *17*, *18*, 21, *43*, 47
after dinner speech (1939) 21, 23
air display (1926) 24
air display (1934) 48
autobiographical details 17
crash in Avro 504K 24-25, *26*
employed by Brooklands Aviation Ltd 81
forms Gnat Aero & Motor Company with 'Miles' 19
joyriding flights 27, 32, 39-40
Pashley, Eric Clowes 16
Pashley, Vera 27
Patterson, Mrs 72
Payne, Grp Capt L G S 288-289

Payne, Stanley 40
Pepper, Mr 67
Percival, Edgar 168
Percival Gull 64-65, 271
Percival Mew Gull
       G-ACND 168
       ZS-AHM *173*
Phillips, E 'Jack' 61, 66
Phillips & Powis Aircraft Ltd 75, 168 *see also* M.1 etc. *entries*;
Miles aircraft
       advertisements *78*
       Experimental Department 74, 91, *194*, *261*, *282*
       fuselage production shop *83*
       Head Office *96*
       licence agreement to build Menasco aero engines 81, 83
       'Miles' becomes managing director after Powis's
          resignation 90
       1937 company report 90-91
       Phillips & Powis Aircraft (Reading) Ltd changed to 74
       The Reading Standard supplement 74-75
       Repair & Service Dept. *290*
       and Reserve Training School 75, 78, *80*, 81, *82*, 90, *96*,
          136, 138
       and Rolls-Royce relationship 81, 200, 203
Phillips & Powis Aircraft (Reading) Ltd 35, 57, 61, *61*, 63,
       65-67, 70, 71-72, 73 *see also* M.1 etc. *entries*;
       Miles aircraft
       changes to Phillips & Powis Aircraft Ltd 74
       and Hawk 97, 98, 99, 101, 102, 106
Phillips & Powis Ltd 61, 71, 72
Phillips & Powis School of Flying 61-62, 63, *64-65*, 67, *70*,
       71, 72, 73
       advertisement *70*
Pini, Bruno 251
Plant, Mr *43*
"Pontius" 121-122
Porteous, Ranald 89-90, 174-175, *428*
Portslade-by-Sea, Star Model Laundry 15, *15*, 19
Portsmouth to Johannesburg Air Race, Schlesinger *114*, 115, 173,
       173, 193-194
Powis, Charles Owen 9, 57, 61, 62, *62*, *64*, 65, 69, 74, *101*, 144
       agreement with 'Miles' 65-66, 97
       demonstration tour abroad 81
       flies Peregrine 85. 190
       interview with *The Garage and Motor Agent* 72-73
       meeting with 'Miles' 64-65, 97
       negotiates to build Menasco aero engines 81
       resignation 90, 202
       and Rolls-Royce Trainer 200
Powis, Oliver 90
Powis, Pauline 61, 62, *62*, 90
Price, Robert B 249
Probyn, Wg Cdr H 72, *104*
Pugh, Flt Lt John *57*, 59

Raceways Ltd 188
RAF Hendon, Pageant (1937) 201-202, *203*, *267*, *268*
RAF Museum 83, 253, 254
RAF Reserve Training School (later 8 EFTS), Woodley 75, 78,
       80, 81, *82*, 90, *96*, *136*, 138, *288*, *422*
Rawson, Sir Cooper 41, 45
Reading 56
Reading Aero Club 57, 59, 62-63, 67, 71, 72, 108
       clubhouse *58*, *62*, 63, *63*, 64, *66*, 85
Reading aerodrome *see* Woodley aerodrome, Reading
Reading Air Pageant (1928) 61
Reading Chronicle, The 62-63
Reading Motor Exchange, The 61
Reading Review 151, 153, 154

Reading Standard, The 70, 74-75
Reeves, Roger 119
Relf, E F 223
Roberts, George 172
Roberts Airways *164*
Robertson, F H 297
Rodd, Lt Cdr 67
Roedean School 32, *38*, 39
Rolls-Royce Ltd 81, 87, 90, 200, 201, 203, 204, 211-212
Rose, Sir Charles 70, 111, *111*, 112
Rose, Flt Lt Thomas 'Tommy' 84, 89, *94*, 117, 153-154,
       159-160, 280, *428*
       appointed chief test pilot 74, 95
       and Hawk Speed Six 112, 115, 116
       and King's Cup Air Races 73, 75, *78*, 115, 121, *122*,
          149, *149*, *150*, 151, *151*, 168
       overseas trips 81
       as Phillips & Powis sales manager 74
Rothermere, Lord 199
Royal Aeronautical Society Garden Party, Heathrow (1936) 84
Royle, Gerald 108, *108*
Rumanian Air Force 135
Rush, Sqn Ldr James 290

Sainsbury, F G 63
Sale, Jack 48
Saville-Sneath, R A 127-128
SBAC Displays *84*, 202, 223
SBAC Trophy Race, Bristol (1934) 112
Schlesinger, Isidor 173
Schlesinger Portsmouth to Johannesburg Air Race *114*, 115, 173,
       173, 193-194
Schmidt, Ralph J 243
Schofield, Flt Lt and Mrs H M 47
Scott, Bateman 193, 194
Scott Douglas, Lady Blanche 124, 125
Seaford, East Blatchington aerodrome, licence for 27, *28*, *29*
Seager Evans & Co Ltd 84
Self, A H 206
Selway, AM Sir Anthony 50-52
Sempill, Col The Master of 67
Sephton, Andy 52
Sevenoaks, drawing office 47, 48, 56
Shackleton, W S, Ltd 289
Shackleton-Murray SM.1 G-ACBP 139, 140
Shaw, Capt Geoffrey 121
'Shell-Mex' *85*
Shelmerdine, Lt Col Francis Claude 24, 63
Shelmerdine, Mrs 63
Shepherd, Capt Ronald 204
Sherry, Mr 234
Shoreham, Easter's Field *18*, 19, *20*, 21, *21*
Shoreham Aero Club 17
Shoreham Aerodrome 21, *22*, *23*, 24, 32, *43*, 44-45
       air display (1926) 21, 24, *25*
       air display (1934) 48
       conditions for engineers 27
       flarepath 27
       home-made hangar *26*
       luncheon and flying display (1931) 45, *46*, 47, 53-54
       Miles Tower 21, 23, *23*, *43*, 48
       Southern Aircraft site *23*, 24, *43*, 48
       UK Air Pilot (1929) entry 35, *36*
Shoreham Airport 21, 45, 112
Shoreham Herald 32
Short S.32 242
Shuttleworth Trust 52, 56, 83, 119, 148, 271
Sidgreaves, A F 81, 200
Simmonds, Oliver Edwin 61

Simmonds Spartan 61
    G-AAGY 32, 48, 56, *56*
    G-AAHA 47
Simonds, Flt Lt E W 266
Singh, Man Mohan 108, *109*
Skinner, Harold William Chetwynd 'Bill' 93, *94*, 95, *95*, 209-210
    and Kestrel *204*, 205, 207, 208
    and M.15 280
    and M.16 Trainer 293
    and Magister 273
    and Master *217*
    and Mentor 281
    and Monarch 74, 286, *286*
    and Peregrine 194, *196*
Skysport Engineering Ltd 253
Smith, James Hopkins 169-170
Smith, Victor 172-173, *172*
Sonning, White Hart pub 160
Souch, Ron 115, 119, 234
Sound in the Sky, A 147, 175, 178
South African Rugby Team 47
South Coast Air Trophy Race (1936) 112, 115
South Coast Flying Club 48, 81
Southend Cup Air Race (1947) 117
Southern Aero Club 21, 23-24, *23*, 41, 45, 81
    changes name to South Coast Flying Club after takeover 48
    Flying Meeting (1928) *34*, *35*
    luncheon and flying display (1931) 45, *46*, 47, 53-54
Southern Aero Club Ltd 45
Southern Aircraft (Gatwick) Ltd *250*, 251
Southern Aircraft Ltd 42, 44, 48, 81 *see also* Gnat Aero Co Ltd
    Gnat Aero Co Ltd name changed to 35
    Shoreham site *23*, *24*, *43*, 48
Southern Martlet 35, *37*, 38, 47, 50-52, *52*, 53, 54, 71
    see also Hornet Baby
    specification and performance data 52
    G-AAVD 35, 38, 52, *52*, 54, *309*
    G-AAYX 38, 45, 52, *310*, *311*
    G-AAYZ 38, *46*, 52, *311*
    G-ABBN 38, 52, *311, 312*
    G-ABIF 38, *46*, 52, *312*
Southern Metal Martlet 47, 53, *53*, 54, *54*
    specification and performance data 54
    G-ABJW (G-AAJW) *46*, 53, 54, *54*
    G-ABMM 54
Spartan Three-Seater G-ABJS 56
spins, flat 186, 188
Spooner, Winifred 38, 63
Spratt, Edward 'Bunny' 147
Stack, Capt Neville 124, *125*
Stainforth, Flt Lt and Mrs G H 47
Starling, Capt Eric 38
Stent, Wg Cdr Frederick W 'Freddie' 93, *93*, 169, 186, 233
Stewart, Capt E W 78
Stewart, Maj Oliver 72
Stieger, J Helmut 174
Stisted, Henry C G 67
Stow, Mike 119
Straight, Whitney Willard 83, 84, 221, *226*, 227
Straight Corporation Ltd, The 83-84, 221, 271
Straightaway Review 221
Strange, Lt Col L A 47
Sullivan, Jack 257
Sunday Express 248
Supermarine
    S.12-40 Seagull 297
    Spitfire 81, 87, 199, 200, 210, 215
    Spitfire I 199
    Spitfire Mk.IA P9444 93
Sussex Daily News 48
Swann, Brian 257
Swinton, Lord 88, 202

Tamplins 21
Tata Sons Ltd/Tata Airlines 165-166, *165*
Tatler, The 72
Taylor, H A 125-127, 233
Theale Relief Landing Ground (Sheffield Farm) 78
Theed, Mr 135
Thorn, F/O Sydney Albert 'Bill' 45, 47, 54
Thynne, Sqn Ldr Brian 193-194
Todd, William 118-119
Turner, Maj C C 227
Twyford, Goody's scrapyard 54
Tyrone Constitution, The 233-234
Tysoe, Lionel *84*

Underwood, John 90, 172
Union Airways of New Zealand Ltd 154-155, 194

Villiers, Charles Amherst 227
Vintage Aircraft 140, 251, 253
Volk, Magnus H 41, 45

Wackett, Laurence 271
Walker, Henry C 124
Waller, Ken 116, 117, 159-160
Wallis, Frederick 'Wally' *15*, 16
Wannock Glen, nr Eastbourne 27
Ward, H 67
Warrender, Sir Victor 289
Watts, Dudley 41
Westhead, Walter Retlaw 54
Westland Dreadnought J6986 235
Westland Widgeon G-EBRO 81
White Waltham 120
Wikner, Geoffrey Neville 139-140
Williams, L 102
Willingdon, Lord *37*, 41, 42
Wilson, Terry 269
wing experiments
    M.3B Falcon Six 155, 157, *157*, 159, *159*
    M.6 Hawcon 179-180, *179*, *180*, 181-182, *182*
Wingfield, G Arthur 45
Wood, Sir Kingsley 93, *94*, 206, 212, 289
Woodley aerodrome, Reading 61, 63, 67, 68, 71-72, *71*, *72*, 75, *88*
    assembly building 74, 75, *79*, *80*, *140*
    'At Home' days (1932) 57, *58*, 67
    details 78
    factory extensions 93, 212, *212*, *213*
    Falcon hotel 63, 85, *85*, *86*, *188*, *259*, *285*
    fires in factory 95
    fuselage production shop *83*
    hangar *61*, *63*, *66*, *103*, *104*
    'Hawkhurst' (adjacent house) 78
    in 1930 *62*
    in 1935 *79*
    in late 1930s *91*
    layout of buildings *77*
    location *77*
    Museum of Berkshire Aviation 60
    Phillips & Powis Aircraft Ltd Head Office *96*
    plan (1937) *87*
    RAF Reserve Training School (later 8 EFTS) 75, 78, *80*, 81, *82*, 90, *96*, *136*, 138, *288*
    Reading Aero Club clubhouse *58*, *62*, 63, *63*, 64, *66*, 85

Repair & Service Department Hangar 96
'Westwinds' (adjacent house) 78
Wormald, A 81
Worthing, Pier Pavilion 32
Woyevodsky (Russian inventor) 235
Wroath, Sgt 'Sammy' 135, 137-138, 265-266
Wyndham, Jill 154

X.2 Bomber project 235, *236*
X.2 Transport Aeroplane project 235-242, *235*
   cabin layout options 238-239, *239*
   design summary 237
   dimensions and aerodynamic data 240
   engines 239
   equipment 240
   performances, estimated 241-242
   structure 237-238
'X' Minor 237, 242, *242*

Yate airfield, Glos 55, 56, 57

Left: A view of the Repair & Service Department, circa 1937.
[P H T Green Collection]

Below: The factory undergoes expansion. G-AEZS (centre left, bottom of picture) was granted its CofA on 21st July 1937, so this photograph must have been taken in late 1937. Note the original hangar centre left and the old clubhouse centre right.
[Phillips & Powis via R W Simpson]

The beautifully-restored Martlet G-AAYX in its pre-war colour scheme, seen performing at the Shuttleworth Collection's display in June 2005.
[John Morris]

More pure nostalgia at Old Warden in June 2005, with the M.2W G-ADWT acting as spectator while Gemini G-AKKH performs in the background.
[John Morris]

Original Miles Aircraft colour film images.....

Thanks to Fred Waters (of the instrument department of Miles Aircraft Ltd and later Handley Page (Reading) Ltd, who 'acquired' a can of 16mm cine film after Miles Aircraft Ltd went into Receivership in late 1947, and to Bob Screen, an apprentice with Handley Page (Reading) Ltd who saved them from being thrown out, I am pleased to be able to show this selection of air-to-air colour images taken by the Miles Aircraft Film Unit in the early post-war period. [Stills captured by Richard Parry & Steve Partington]

This unmarked Southern Martlet (G-AAYX) was first flown, after refurbishment by Miles Aircraft in 1947, by Don Brown on 4th June 1947. Immediately afterwards it was flown by F G Miles following a trim adjustment; then again by Don Brown and finally by George Miles. On 10th and 13th June it was flown by Hugh Kendall and it received its CofA on the 20th June 1947. This film would have been taken during one of those flights.

Two views of the M.7A Nighthawk Hybrid in RAF camouflage airborne as U-0225.

M.5A Sparrowhawk G-AGDL in wartime camouflage wearing civil registration and compulsory red, white and blue horizontal civilian aircraft recognition stripes.

Colour Images

An atmospheric view of Woodley Aerodrome with the camouflaged M.3B Falcon Six G-ADTD in the foreground. Note the modification to a one-piece moulded windscreen.

The M.3E 'Gilette' Falcon Six L9705 showing the bi-convex wings fitted to explore the low speed handling characteristics of this aerofoil section which was to be fitted to the M.52 High Speed Research Aircraft.

Three views of the Miles M.18 Mk.III U-0238 which later became U3 and then G-AHOA.

No colour photographs which may have been taken of Miles aircraft prior to September 1939 have been found, so a representative selection of some of the colour schemes which adorned the surviving pre-war aircraft into the post-war era are reproduced on the following pages.

Left: Previously registered G-ACYZ and VH-ACC, this M.2H Hawk Major was probably photographed at Brampton, Ontario in about 1984 after becoming C-FAUV. [The Miles Aircraft Collection]

Right: M.2H Hawk Major G-ADMW at Denham in the early 1950s while owned by John Gunner. John presented it to the Air Historical Branch at Henlow in 1965 but it is currently stored in very poor condition with the RAFM Reserve Collection at RAF Stafford. [via Bert Clarke]

HAWKS

Two views of M.2L Hawk Speed Six G-ADGP c/n 160. Above: In familiar post-war racing colours during the 1950s when owned by Ron Paine and fitted with the enclosed 'blown' canopy. [The Miles Aircraft Collection]

Left: G-ADGP was rebuilt by Ron Souch with a canopy and colour scheme that resembled its pre-war appearance, as seen here in 1990. [Peter Amos]

Right: The second M.2W Hawk Trainer Mk.II, G-ADWT, shown in its 1950's colour scheme. The standard three-piece windscreens had been replaced by large one-piece curved ones by the time this photograph was taken (see their original shape on page 339).
[via Bert Clarke]

Below: Sold to a Canadian owner in Europe in 1961, G-ADWT later moved to Canada and became CF-NXT in 1964. It was given to the EAA Aviation Foundation at Oshkosh, WI, where it was on show in 1999 prior to returning to the UK. [Les Vowles]

Above: Back in the UK G-ADWT was refurbished and repainted in its pre-war colour scheme of the Reserve Training School at Woodley. It is seen here on 28.5.06 at its private strip in Oxfordshire.
[Dave Partington]

Right: M.2R Hawk Trainer Mk.I CC-FBB, having been sold in Chile, made the first non-stop trans-Andean flight in June 1936. This alone made it worthy of preservation in the Museo Nacional de Aeronautica de Chile at Los Cerillos, Santiago.
[via John Morris]

Left: Not what it seems! The M.3A Falcon Major VH-AAT was painted to represent Harold Brook's G-ACTM in April 1991 for the filming of an Australian TV mini-series "Half a World Away". [David Freeman]

Below: The real thing - VH-AAT in airworthy condition hangared at Drage Air Museum, Wangaratta, Vic, early in 2000. [Dave Welch]

Above: Cockpit interior of M.3A VH-AAT, May 1998. [Dick Green]

Right: The brightly-coloured Falcon EC-ACB of the Fundacion Infante de Orleans at Madrid is officially recorded as c/n 197 although it is actually believed to be M.3C c/n 231. [John Morris]

FALCONS

Right: M.3B Falcon Six G-ADTD, seen at Panshanger in 1961, had been fitted with a new rounded windscreen at some time, probably by Miles Aircraft. [via Bert Clarke]

Below: M.3A Falcon Major G-AEEG was sold in Sweden before the war and was used by the Royal Swedish Air Force during the war. It returned to England in 1961 and was restor4ed to the UK register in 1962. It is seen here at Sywell on 3.7.77. [via The Miles Aircraft Collection]

COLOUR IMAGES

Right: M.11A Whitney Straight G-AERV, date and place unknown. [via Bert Clarke]

Below: Photographed at Wisley in 1966, with G-AEUJ behind, M.11A G-AFGK had the name 'Whitney Straight' painted on the engine cowling.
[via The Miles Aircraft Collection]

WHITNEY STRAIGHTS

Above: M.11A G-AEUJ visiting Sandown, IoW, on 8.8.65. [via The Miles Aircraft Collection]

Right: CF-FGK (previously G-AFGK) was last flown in 1992 in a pseudo-wartime desert camouflage RAF colour scheme. It was later stored in the Reynolds-Alberta Museum with the wings removed but the windscreen had been broken by storm damage by July 2008.
[via The Miles Aircraft Collection]

Above: Miles M.14A Hawk Trainer Mk.III G-AFBS at The Imperial War Museum collection, Duxford on 3rd May 2008. [Dave Partington]

Left: The Miles M.14A Hawk Trainer Mk.III LV-X246 of the Aero Club San Martin, probably at Mendoza, Argentina, painted in a fairly authentic 1940 style RAF camouflage scheme when photographed in 2002. [C Osvaldo Viggiani]

Below: M.14A Magister '34' of the Irish Air Corps at Baldonnel on 2nd January 1998 following refurbishment to static display condition - a pity about the spats though! [P Dent]

M.14s

Left: Miles M.14A Magister P6382 was civilianised after the war as G-AJRS and has long been in the hands of the Shuttleworth Trust at Old Warden where it is still active, as seen in its RAF marks on 6th August 2006.
[Trevor Thornton]

COLOUR IMAGES

Above: A post war photograph of M.17 Monarch G-AFJU at a flying meeting, date and venue unknown.
[via Bert Clarke]

Right: Monarch G-AFJU being prepared for departure from East Fortune on a trailer, 10th September 2006.
[Greg Connell]

Above: This photograph of Miles M.17 Monarch G-AFLW wearing race number 108 may have been taken at a fly-in in the 1970s or early 1980s.
[via Bert Clarke]

Right: By the time that M.17 Monarch G-AFLW visited the Air-Britain Fly-in at Middle Wallop on 30th June 1985 it was wearing race number 54 and a striped cowling representing the multi-national colours of Scandinavian Airlines by whom a previous owner was employed.
[Bill Teasdale]

MONARCHS

Right: The sole surviving Miles M.18 Mk.II G-AHKY, which now languishes in the Museum of Flight at East Fortune.
[via The Miles Aircraft Collection]

Below: G-AHKY seen visiting the PFA Rally at Northampton-Sywell on 5th July 1975. This colour scheme has been retained to the present day, with the exception that the fin flashes are now reversed, with red on the leading edge.
[Bill Teasdale]

M.12 & M.18

Below: The refurbished Miles M.12 Mohawk G-AEKW now on static display at the RAF Museum at Hendon. [RAF Museum]

Left: A painting by, C T Howard, of the 'King's Cup Winner', the Miles M.3B Falcon Six G-ADLC, which appeared on a pre-war postcard printed and published by J Salmon, Sevenoaks. The card was posted from Liverpool at 9.30 PM (day not shown) in 1937 and the description on the back read:

"MILES FALCON" SIX Winner of King's Cup Race, 1935. "Gipsy Six" engine of 200 H.P. 3 Seats. Maximum Speed 180 m.p.h. Built by Phillips & Powis Aircraft Ltd., of Reading. All-wood construction.
[via Peter Amos]